Acta Numerica 2024

Acta Numerica

Volume 33 2024

CAMBRIDGE
UNIVERSITY PRESS

Published by the Press Syndicate of the University of Cambridge
The Pitt Building, Trumpington Street, Cambridge CB2 1RP
One Liberty Plaza, Floor 20, New York, NY 10006, USA
10 Stamford Road, Oakleigh, Melbourne 3166, Australia

First published 2024

Printed in Great Britain at the University Press, Cambridge

Library of Congress cataloguing in publication data available

A catalogue record for this book is available from the British Library

ISBN 978-1-009-576819
ISSN 0962-4929

Contents

Splitting methods for differential equations 1
Sergio Blanes, Fernando Casas and Ander Murua

Adaptive finite element methods 163
Andrea Bonito, Claudio Canuto, Ricardo H. Nochetto and Andreas Veeser

The geometry of monotone operator splitting methods 487
Patrick L. Combettes

Numerical analysis of physics-informed neural networks and related models in physics-informed machine learning 633
Tim De Ryck and Siddhartha Mishra

Optimal experimental design: Formulations and computations 715
Xun Huan, Jayanth Jagalur and Youssef Marzouk

The Moment-SOS hierarchy: Applications and related topics 841
Jean B. Lasserre

Acta Numerica (2024), pp. 1–161 Printed in the United Kingdom
doi:10.1017/S0962492923000077

Splitting methods for differential equations

Sergio Blanes
Instituto de Matemática Multidisciplinar,
Universitat Politècnica de València, 46022-Valencia, Spain
E-mail: serblaza@imm.upv.es

Fernando Casas
Departament de Matemàtiques and IMAC, Universitat Jaume I,
12071-Castellón, Spain
E-mail: Fernando.Casas@uji.es

Ander Murua
Konputazio Zientziak eta A.A. Saila, Informatika Fakultatea, EHU/UPV,
Donostia/San Sebastián, Spain
E-mail: Ander.Murua@ehu.es

This overview is devoted to splitting methods, a class of numerical integrators intended for differential equations that can be subdivided into different problems easier to solve than the original system. Closely connected with this class of integrators are composition methods, in which one or several low-order schemes are composed to construct higher-order numerical approximations to the exact solution. We analyse in detail the order conditions that have to be satisfied by these classes of methods to achieve a given order, and provide some insight about their qualitative properties in connection with geometric numerical integration and the treatment of highly oscillatory problems. Since splitting methods have received considerable attention in the realm of partial differential equations, we also cover this subject in the present survey, with special attention to parabolic equations and their problems. An exhaustive list of methods of different orders is collected and tested on simple examples. Finally, some applications of splitting methods in different areas, ranging from celestial mechanics to statistics, are also provided.

2020 Mathematics Subject Classification: Primary 65L05, 65L20, 65P10
Secondary 65M12, 65M70

CONTENTS

1	Introduction	2
2	High-order splitting and composition methods	29
3	Splitting methods for special problems	50
4	Qualitative properties of splitting methods	60
5	Highly oscillatory problems	72
6	Splitting methods for PDEs	86
7	Splitting methods with complex coefficients	106
8	A collection of splitting methods	114
9	Splitting everywhere: some relevant applications	123
Appendix: Testing splitting methods on matrices		143
References		149

1. Introduction

1.1. Lie–Trotter and Strang methods

If, as has sometimes been argued, there are only ten big ideas in numerical analysis and all the rest are merely variations on those themes, *splitting* is undoubtedly one of them (MacNamara and Strang 2016). In fact, we could say that, at least since Descartes stated his four rules of logic in the *Discours de la méthode*,[1] the notion of subdividing a complicated problem into its simpler constituent parts, solving each one of them separately and combining those separated solutions in a controlled way to get a solution to the original overall problem, constitutes a guiding principle in all areas of science and philosophy.

In the realm of numerical analysis of differential equations, this basic principle can be stated as follows. Suppose we have the abstract initial value problem

$$x' \equiv \frac{\mathrm{d}x}{\mathrm{d}t} = f(x), \quad x(0) = x_0 \tag{1.1}$$

associated to an ordinary differential equation (ODE) or a partial differential equation (PDE), in which case f is a certain spatial partial differential operator. Furthermore, suppose that f does not depend explicitly on time and can be decomposed as

$$f = f_1 + \cdots + f_m, \quad m \geq 2, \tag{1.2}$$

so that each initial value problem $x' = f_j(x)$, $x(0) = x_0$ is easier to solve than (1.1).

[1] 'Le second, de diviser chacune des difficultés que j'examinarais, en autant de parcelles qu'il se pourrait, et qu'il serait requis pour les mieux résoudre. Le troisième, de conduire par ordre mes pensées, en commençant par les objets les plus simples et les plus aisés à connaître, pour monter peu à peu, comme par degrés, jusques à la connaissance des plus composés; et supposant même de l'ordre entre ceux qui ne se précèdent point naturellement les uns les autres.' (René Descartes, *Discours de la méthode*, Seconde partie).

Most commonly, they can be integrated exactly in closed form. Then it is possible to take advantage of the decomposition (1.2) to get accurate approximations to the solution of (1.1) by means of *splitting methods*.

The best example to start with is perhaps the case of a linear differential equation defined in $\mathbb{R}^D$ and $m = 2$, namely

$$x' = f_1(x) + f_2(x) = F_1 x + F_2 x, \quad x(0) = x_0, \tag{1.3}$$

where F_1 and F_2 are $D \times D$ matrices and $x \in \mathbb{R}^D$. The solution reads

$$x(t) = \mathrm{e}^{t(F_1+F_2)} x_0,$$

so that by computing this matrix exponential directly we would have solved (1.3) without requiring splitting methods. Associated with (1.3), we have the matrix differential equation

$$\frac{\mathrm{d}X}{\mathrm{d}t} = (F_1 + F_2)X, \quad X(0) = I, \tag{1.4}$$

in the sense that

$$x(t) = X(t)\, x_0 \quad \text{and} \quad X(t) = \mathrm{e}^{t(F_1+F_2)}.$$

This is useful (mainly) for theoretical purposes, since we usually try to compute $\mathrm{e}^{t(F_1+F_2)} x_0$ directly instead of first computing the matrix exponential and then multiplying it by x_0.

It often happens that evaluating the action of $X(t)$ on x_0 is difficult or computationally expensive. If, however, this is not the case for each e^{tF_j} separately, then one may use the well-known Lie product formula (Reed and Simon 1980, p. 295)

$$\mathrm{e}^{t(F_1+F_2)} = \lim_{n\to\infty} \left(\mathrm{e}^{\frac{t}{n}F_2}\, \mathrm{e}^{\frac{t}{n}F_1}\right)^n. \tag{1.5}$$

To get an approximate solution of (1.3) at the final time $t = t_f$, we subdivide the interval $[0, t_f]$ into N steps of length h, with $Nh = t_f$, and compute the sequence

$$x_{n+1} = \mathrm{e}^{hF_2}\, \mathrm{e}^{hF_1}\, x_n, \quad n \geq 0, \tag{1.6}$$

so that $x_{n+1} \approx x(t_{n+1} = (n+1)h)$. This is the so-called *Lie–Trotter scheme*. When the matrices commute, the sequence produces the exact solution. To put it another way, if the commutator $[F_1, F_2] \equiv F_1F_2 - F_2F_1 = 0$, then $\exp(h(F_1 + F_2)) = \exp(hF_2)\exp(hF_1)$. Otherwise, a direct calculation shows that

$$\mathrm{e}^{h(F_1+F_2)} - \mathrm{e}^{hF_2}\mathrm{e}^{hF_1} = \frac{1}{2}h^2[F_1, F_2] + O(h^3)$$

as $h \to 0$, and hence the previous approximation is only of first order of accuracy. Another version of the method is possible, of course, by reversing the order of F_1 and F_2, namely

$$x_{n+1} = \mathrm{e}^{hF_1}\, \mathrm{e}^{hF_2}\, x_n, \quad n \geq 0 \tag{1.7}$$

has the same order of accuracy and properties as (1.6). Needless to say, for any

$m > 2$, it results in

$$x_{n+1} = \mathrm{e}^{hF_m} \cdots \mathrm{e}^{hF_2}\,\mathrm{e}^{hF_1}\, x_n$$

(or any other permutation of the matrices F_j).

A higher-order approximation can be achieved by considering a symmetrized version of (1.6),

$$\mathrm{e}^{h(F_1+F_2)} - \mathrm{e}^{\frac{1}{2}hF_1}\mathrm{e}^{hF_2}\mathrm{e}^{\frac{1}{2}hF_1} = Ch^3 + O(h^4), \qquad (1.8)$$

where the constant C can be obtained either by comparing Taylor series or by applying the Baker–Campbell–Hausdorff (BCH) formula as (Varadarajan 1984) $C = \frac{1}{24}([F_1,[F_1,F_2]] + 2[F_2,[F_1,F_2]])$. Therefore the sequence

$$x_{n+1} = \mathrm{e}^{\frac{1}{2}hF_1}\mathrm{e}^{hF_2}\mathrm{e}^{\frac{1}{2}hF_1}\, x_n, \quad n \geq 0 \qquad (1.9)$$

produces a second-order approximation for the solution of (1.3). This corresponds to the *Strang* (splitting) *scheme*. Again, if the role of F_1 and F_2 is interchanged, we have another version of the Strang splitting scheme.

Simple generalizations to the case $m > 2$ include in particular the product

$$\mathrm{e}^{\frac{1}{2}hF_1}\,\mathrm{e}^{\frac{1}{2}hF_2} \cdots \mathrm{e}^{hF_m} \cdots \mathrm{e}^{\frac{1}{2}hF_2}\,\mathrm{e}^{\frac{1}{2}hF_1}.$$

Higher-order splitting methods could in principle be constructed by including more exponentials with their corresponding coefficients in a time step, namely

$$\Psi(h) = \mathrm{e}^{a_{s+1}hF_1}\,\mathrm{e}^{b_s hF_2}\,\mathrm{e}^{a_s hF_1} \cdots \mathrm{e}^{b_1 hF_2}\,\mathrm{e}^{a_1 hF_1}. \qquad (1.10)$$

The number s as well as the coefficients a_j, b_j are chosen so that

$$\Psi(h) = \mathrm{e}^{h(F_1+F_2)} + O(h^{r+1})$$

as $h \longrightarrow 0$ for a given order r. In (1.10), the first and last exponentials correspond to F_1. This format is convenient for implementation, since the last exponential in one step can be concatenated with the first one at the next step, thus reducing the number of evaluations by one. This corresponds to the well-known FSAL (first same as last) property. The situation when we have an exponential of F_2 as the first and last term is recovered by taking $a_1 = a_{s+1} = 0$.

1.2. Flows and differential operators

The Lie–Trotter scheme can be easily generalized to any system (1.1)–(1.2) when the solution is no longer given by exponentials, as in the linear case. If $m = 2$, it is equivalent to the following.

Algorithm 1.1 (Lie–Trotter). Starting from $x_0 = x(0)$, for $n \geq 0$,

- solve $y_1' = f_1(y_1)$, $y_1(t_n) = x_n$, in $[t_n, t_{n+1}]$;
- set $y_{n+1/2} = y_1(t_{n+1})$;
- solve $y_2' = f_2(y_2)$, $y_2(t_n) = y_{n+1/2}$, in $[t_n, t_{n+1}]$;

- finally, set $x_{n+1} = y_2(t_{n+1})$.

Alternatively, if we denote the solution of equation (1.1) for each $t \in \mathbb{R}$ as

$$x(t) = \varphi_t^{[f]}(x_0),$$

then Algorithm 1.1 can be formally expressed as

$$x_{n+1} = \chi_h(x_n) \equiv \varphi_h^{[2]}\big(\varphi_h^{[1]}(x_n)\big) = \big(\varphi_h^{[2]} \circ \varphi_h^{[1]}\big)(x_n), \tag{1.11}$$

where we have used the simplified notation $\varphi_t^{[j]}(x_0)$ for the solutions $y(t) = \varphi_t^{[f_j]}(x_0)$ of the subproblems $y' = f_j(y)$, $y(0) = x_0$. Analogously, the Strang splitting (1.9) is generalized as follows.

Algorithm 1.2 (Strang). From $x_0 = x(0)$, for $n \geq 0$,

- solve $y_1' = f_1(y_1)$, $y_1(t_n) = x_n$, in $[t_n, t_{n+1/2}]$, with $t_{n+1/2} = (n + \frac{1}{2})h$;
- set $y_{n+1/2} = y_1(t_{n+1/2})$;
- solve $y_2' = f_2(y_2)$, $y_2(t_n) = y_{n+1/2}$, in $[t_n, t_{n+1}]$;
- set $\hat{y}_{n+1/2} = y_2(t_{n+1})$;
- solve $y_1' = f_1(y_1)$, $y_1(t_{n+1/2}) = \hat{y}_{n+1/2}$, in $[t_{n+1/2}, t_{n+1}]$;
- finally, set $x_{n+1} = y_1(t_{n+1})$

or, in short,

$$x_{n+1} = S_h^{[2]}(x_n) \equiv \big(\varphi_{h/2}^{[1]} \circ \varphi_h^{[2]} \circ \varphi_{h/2}^{[1]}\big)(x_n). \tag{1.12}$$

If equation (1.1) corresponds to a (nonlinear) ordinary differential equation evolving in $\mathbb{R}^D$,

$$x' = f(x), \quad x(0) = x_0 \in \mathbb{R}^D, \tag{1.13}$$

f is called the *vector field*. If (1.13) admits for each $x_0 \in \mathbb{R}^D$ a unique solution $x(t)$ defined for all $t \in \mathbb{R}$, the map

$$\begin{aligned} \varphi_t^{[f]}:\ \mathbb{R}^D &\longrightarrow \mathbb{R}^D \\ x_0 &\longmapsto x(t) = \varphi_t^{[f]}(x_0) \end{aligned}$$

is referred to as the *t-flow* (Arnold 1989). Thus, for each value of the real parameter t, $\varphi_t^{[f]}$ maps $\mathbb{R}^D$ in $\mathbb{R}^D$ in such a way that $\varphi_t^{[f]}(z)$ is the value at time t of the solution of the system with initial value z at time 0, whereas, for fixed x_0 and varying t, $\varphi_t^{[f]}(x_0)$ is the solution of the initial value problem (1.13).

It is worth mentioning that the solution $x(t)$ of (1.1) is in general defined for a *maximal time interval* $(t_{\min}(x_0), t_{\max}(x_0))$ (with $-\infty \leq t_{\min}(x_0) \leq 0 \leq t_{\max}(x_0) \leq +\infty$). Furthermore, the vector field f of many systems of ordinary differential equations is singular (or undefined) for some $x \in \mathbb{R}^D$. Thus, in general, f is defined for some open set $\mathcal{U} \subset \mathbb{R}^D$. Hence, for a given $t \in \mathbb{R}$, the t-flow $\varphi_t^{[f]}$ is a map from

$$\mathcal{D}_t = \{x_0 \in \mathcal{U} : t \in (t_{\min}(x_0), t_{\max}(x_0))\}$$

to $\mathcal{U}$. In general, one may have different domains of definition $\mathcal{U}$ (resp. $\mathcal{D}_t$) for each f_j (resp. $\varphi_t^{[f_j]}$). In this general situation the compositions in (1.11) and (1.12) are not well-defined for all x_0, h and n. In order to avoid these technicalities, we will assume in what follows that $\mathcal{U} = \mathbb{R}^D$, $t_{\min} = -\infty$ and $t_{\max} = +\infty$, for each vector field f_j.

Associated with the vector field f is the Lie derivative or *Lie operator* F (Arnold 1989), mapping smooth functions $g\colon \mathbb{R}^D \longrightarrow \mathbb{R}$ into the real-valued function $F g\colon \mathbb{R}^D \longrightarrow \mathbb{R}$ such that, for $x \in \mathbb{R}^D$,

$$(F g)(x) = \left.\frac{\mathrm{d}}{\mathrm{d}t}\right|_{t=0} g\bigl(\varphi_t^{[f]}(x)\bigr),$$

that is,

$$(F g)(x) = f(x) \cdot \nabla g(x).$$

Then the flow of (1.13) verifies (Sanz-Serna and Calvo 1994, Hairer, Lubich and Wanner 2006)

$$g\bigl(\varphi_t^{[f]}(x)\bigr) = \bigl(\mathrm{e}^{tF} g\bigr)(x) \equiv \left(\sum_{k\geq 0} \frac{t^k}{k!} F^k g\right)(x).$$

The operator $X(t) \equiv \mathrm{e}^{tF}$ is called *Lie transformation*, and can be seen as the formal solution of the operator equation

$$\frac{\mathrm{d}X}{\mathrm{d}t} = X\,F, \qquad X(0) = I. \tag{1.14}$$

This can be seen as follows: on the one hand,

$$\frac{\mathrm{d}}{\mathrm{d}t} g\bigl(\varphi_t^{[f]}(x)\bigr) = \frac{\mathrm{d}}{\mathrm{d}t}(X(t)g)(x) = \left(\frac{\mathrm{d}X(t)}{\mathrm{d}t} g\right)(x),$$

and on the other hand

$$\frac{\mathrm{d}}{\mathrm{d}t} g\bigl(\varphi_t^{[f]}(x)\bigr) = (Fg)\bigl(\varphi_t^{[f]}(x)\bigr) = X(t)(Fg)(x).$$

Lie operators satisfy some remarkable properties (Arnold 1989). In particular, although they do not commute, their commutator is nevertheless a first-order linear differential operator. Specifically, let F and G be the Lie operators associated with f and g, respectively, and $u\colon \mathbb{R}^D \longrightarrow \mathbb{R}$ a given smooth function. Then

$$[F, G]u = (FG - GF)\,u = \sum_{i,j=1}^{D} \left(f_j \frac{\partial g_i}{\partial x_j} - g_j \frac{\partial f_i}{\partial x_j} \right) \frac{\partial u}{\partial x_i}$$

and it is possible to associate a new vector field to this differential operator, $w = (f, g)$, with components

$$w_i = (f, g)_i = \sum_{j=1}^{D} \left(f_j \frac{\partial g_i}{\partial x_j} - g_j \frac{\partial f_i}{\partial x_j} \right).$$

It is called the *Lie–Poisson bracket* of f and g, and its Lie operator W satisfies $W = [F, G]$.

Now suppose that $f(x) = f_1(x) + f_2(x)$, so that each part $x' = f_j(x)$ is exactly solvable (or can be numerically solved up to round-off accuracy) with flow $x(t) = \varphi_t^{[j]}(x_0)$. Letting F_1 and F_2 denote the Lie operators associated with f_1 and f_2, respectively, it holds that

$$g\big(\varphi_t^{[1]}(x)\big) = \big(\mathrm{e}^{tF_1} g\big)(x), \quad g\big(\varphi_t^{[2]}(x)\big) = \big(\mathrm{e}^{tF_2} g\big)(x).$$

Then, for the first-order approximation $\chi_h = \varphi_h^{[2]} \circ \varphi_h^{[1]}$ furnished by Algorithm 1.1, we have $g(\chi_h(x)) = (\Psi(h)g)(x)$, where $\Psi(h)$ is a series of linear differential operators defined as

$$\Psi(h) = \mathrm{e}^{hF_1}\, \mathrm{e}^{hF_2}. \tag{1.15}$$

Notice that the exponentials of Lie derivatives in (1.15) appear in reverse order with respect to the maps in the integrator (Hairer *et al.* 2006, p. 88). Of course, the same procedure can be applied to the Strang splitting, resulting in the product

$$\Psi(h) = \mathrm{e}^{\frac{h}{2}F_1}\, \mathrm{e}^{hF_2}\, \mathrm{e}^{\frac{h}{2}F_1}. \tag{1.16}$$

These considerations show that: (i) splitting methods for the problem (1.1)–(1.2) defined in a certain function space (e.g. with partial differential equations) can be formulated in terms of the solution of each subproblem (either exact or approximate) by means of Algorithms 1.1 and 1.2, and (ii) splitting methods applied to nonlinear ODEs evolving in $\mathbb{R}^D$ can also be formally expressed as products of exponentials of differential operators, since it is possible to transform the original nonlinear problem into a linear one with the Lie formalism. This observation is very useful when analysing the order conditions for a method to be of a given order. In particular, we have the same order conditions for linear and nonlinear ODEs (see Section 2).

The previous integrators are sometimes called *multiplicative operator-splitting methods*, especially in the literature concerning the numerical treatment of partial differential equations. In that area, we still have to specify how to solve each initial value sub-problem in Algorithms 1.1–1.2 as well as the boundary conditions. Moreover, we should take into account that, for a given differential equation, different ways to carry out the splitting in fact lead to different integrators.

1.3. Adjoint method, conjugate method

The flow $\varphi_t^{[f]}$ of (1.13) verifies $\left(\varphi_{-t}^{[f]}\right)^{-1} = \varphi_t^{[f]}$, but this property is not shared by many numerical integrators, and in particular by the map χ_h corresponding to the Lie–Trotter scheme.

In general, if $\psi_h(x)$ represents a numerical method of order at least one, i.e. $\psi_h(x) = x + h\, f(x) + O(h^2)$, then $(\psi_{-h})^{-1}(x) = x + h\, f(x) + O(h^2)$, so that

$$\psi_h^* \equiv (\psi_{-h})^{-1}$$

is also a numerical method of order at least one. It is called the *adjoint method* of ψ_h (Sanz-Serna and Calvo 1994). In words, stepping forwards with the given method ψ_h is the same as stepping backwards with the inverse of its adjoint ψ_h^*. If $\psi_h \equiv \chi_h = \varphi_h^{[2]} \circ \varphi_h^{[1]}$, then clearly $\chi_h^* = \varphi_h^{[1]} \circ \varphi_h^{[2]}$. Additional examples are the explicit and implicit Euler methods

$$x_{n+1} = \psi_h^e(x_n) = x_n + hf(x_n), \quad x_{n+1} = \psi_h^i(x_n) = x_n + hf(x_{n+1}),$$

since $\psi_h^i = (\psi_h^e)^*$.

Whenever an integrator satisfies

$$\psi_h = \psi_h^* = (\psi_{-h})^{-1},$$

it is called a *time-symmetric* or *self-adjoint* method. Alternatively, $x_{n+1} = \psi_h(x_n)$ is time-symmetric if and only if, exchanging $h \leftrightarrow -h$ and $x_n \leftrightarrow x_{n+1}$, we get the same expression, i.e. $\psi_{-h}(x_{n+1}) = x_n$. The Strang scheme (1.12) is an example of a time-symmetric method.

It is in fact straightforward to construct time-symmetric methods using the adjoint: given an arbitrary method ψ_h of order $r \geq 1$, then the compositions

$$\psi_{h/2} \circ \psi_{h/2}^* \quad \text{and} \quad \psi_{h/2}^* \circ \psi_{h/2} \tag{1.17}$$

are time-symmetric methods of order $r \geq 2$ (Sanz-Serna and Calvo 1994). Further, symmetric methods are necessarily of even order, as we will show in Section 2. Notice that the Strang method (1.12) is simply

$$S_h^{[2]} = \chi_{h/2}^* \circ \chi_{h/2},$$

where χ_h is given by (1.11). Additional examples are the trapezoidal rule $\psi_h^t = \psi_{h/2}^i \circ \psi_{h/2}^e$ and the midpoint rule $\psi_h^m = \psi_{h/2}^e \circ \psi_{h/2}^i$.

The Strang scheme can also be expressed as

$$\begin{aligned} S_h^{[2]} &= \varphi_{-h/2}^{[1]} \circ \left(\varphi_h^{[1]} \circ \varphi_h^{[2]}\right) \circ \varphi_{h/2}^{[1]} \\ &= \pi_h^{-1} \circ \chi_h^* \circ \pi_h, \end{aligned} \tag{1.18}$$

with $\pi_h = \varphi_{h/2}^{[1]}$. In the terminology of dynamical systems, the Strang and Lie–Trotter schemes are said to be *conjugate* to each other by the ($O(h)$-near to identity) map $\pi_h = \varphi_{h/2}^{[1]}$, which can be considered as a change of coordinates. Furthermore, the result of n applications of the Strang scheme can be recovered from n applications of Lie–Trotter by carrying out just an initial transformation at the initial step and its *inverse* at the final step.

Since many dynamical properties are invariant under changes of coordinates, conjugate methods provide the same characterization of these properties. Other examples of conjugate methods are the trapezoidal and midpoint rules, the map π_h being in this case the implicit Euler method (Hairer *et al.* 2006). We will treat conjugate methods in detail in Sections 4 and 5.

1.4. The mathematical pendulum

The simple mathematical pendulum constitutes a standard example of a nonlinear Hamiltonian system. In appropriate units, the corresponding Hamiltonian function reads

$$H(q,p) = \frac{1}{2}p^2 + (1 - \cos q), \tag{1.19}$$

where q denotes the angle from the vertical suspension point and p is the associated momentum.

As is well known, the equations of motion of a generic Hamiltonian system with Hamiltonian $H(q,p)$, and $q, p \in \mathbb{R}^d$, are given by (Goldstein 1980)

$$\frac{\mathrm{d}q}{\mathrm{d}t} = \nabla_p H, \quad \frac{\mathrm{d}p}{\mathrm{d}t} = -\nabla_q H, \tag{1.20}$$

the function $H(q,p)$ remains constant along the evolution, and the corresponding t-flow, denoted $\varphi_t^{[H]}$, is a *symplectic* transformation (Arnold 1989): its Jacobian matrix $\varphi_t'^{[H]}$ verifies the identity

$$(\varphi_t'^{[H]})^\top J \varphi_t'^{[H]} = J \quad \text{for } t \geq 0,$$

where J is the basic canonical matrix

$$J = \begin{pmatrix} 0_d & I_d \\ -I_d & 0_d \end{pmatrix}. \tag{1.21}$$

In the particular case of (1.19), $d = 1$, and the equations of motion are

$$\frac{\mathrm{d}q}{\mathrm{d}t} = p, \quad \frac{\mathrm{d}p}{\mathrm{d}t} = -\sin q.$$

Given a Hamiltonian $H(q,p)$ that can be decomposed as

$$H(q,p) = H_1(q,p) + H_2(q,p), \tag{1.22}$$

it makes sense to split the equations of motion (1.20) as

$$\frac{\mathrm{d}}{\mathrm{d}t}\begin{pmatrix} q \\ p \end{pmatrix} = \begin{pmatrix} \nabla_p H_1(q,p) \\ -\nabla_q H_1(q,p) \end{pmatrix} + \begin{pmatrix} \nabla_p H_2(q,p) \\ -\nabla_q H_2(q,p) \end{pmatrix},$$

so that each subsystem is itself Hamiltonian. In that case, we can then apply

Algorithm 1.1 by composing both maps and form the first-order scheme

$$x_{n+1} = \chi_h(x_n) \equiv \big(\varphi_h^{[H_2]} \circ \varphi_h^{[H_1]}\big)(x_n), \quad n = 0, 1, 2, \ldots, \tag{1.23}$$

where $x_n = (q_n, p_n)^\top$. Similarly, Algorithm 1.2 gives the second-order scheme

$$x_{n+1} = S_h^{[2]}(x_n) \equiv \big(\varphi_{h/2}^{[H_1]} \circ \varphi_h^{[H_2]} \circ \varphi_{h/2}^{[H_1]}\big)(x_n), \quad n = 0, 1, 2, \ldots. \tag{1.24}$$

Notice that, since both χ_h and $S_h^{[2]}$ are defined as compositions of flows of Hamiltonian systems and the composition of symplectic maps is also symplectic (Arnold 1989), then both (1.23) and (1.24) are *symplectic integrators* (Sanz-Serna and Calvo 1994).

The fact that schemes (1.23) and (1.24) share the symplectic property with the exact flow has remarkable consequences in practice concerning the preservation of properties and the error propagation for long-time integrations, as we will shortly illustrate.

For the Hamiltonian (1.19) describing the pendulum (and in fact for many other mechanical systems), one can separate the contributions of the kinetic energy $T(p) = \frac{1}{2}p^2$ and the potential energy $V(q) = 1 - \cos q$, so that a natural splitting of the form (1.22) is then

$$H(q, p) = T(p) + V(q). \tag{1.25}$$

This corresponds to splitting the equations of motion (1.20) into the subsystems

$$\begin{pmatrix} q' \\ p' \end{pmatrix} = \begin{pmatrix} \nabla_p T(p) \\ 0 \end{pmatrix} \quad \text{and} \quad \begin{pmatrix} q' \\ p' \end{pmatrix} = \begin{pmatrix} 0 \\ -\nabla_q V(q) \end{pmatrix}, \tag{1.26}$$

which in turn implies that

$$\varphi_t^{[T]}: \begin{pmatrix} q_0 \\ p_0 \end{pmatrix} \longmapsto \begin{pmatrix} q_0 + t\,\nabla T_p(p_0) \\ p_0 \end{pmatrix} \tag{1.27}$$

and

$$\varphi_t^{[V]}: \begin{pmatrix} q_0 \\ p_0 \end{pmatrix} \longmapsto \begin{pmatrix} q_0 \\ p_0 - t\,\nabla V_q(q_0) \end{pmatrix}. \tag{1.28}$$

Then the first-order scheme (1.23) reduces to

$$q_{n+1} = q_n + h\,\nabla_p T(p_n), \quad p_{n+1} = p_n - h\,\nabla_q V(q_{n+1}), \quad n = 0, 1, 2 \ldots \tag{1.29}$$

Compared to the explicit Euler method

$$q_{n+1} = q_n + h\,\nabla_p T(p_n), \quad p_{n+1} = p_n - h\,\nabla_q V(q_n),$$

it only differs in that $\nabla_q V$ is evaluated at the updated value q_{n+1} instead of q_n. It makes sense, then, to call scheme (1.29) the symplectic Euler–VT method: we first compute the gradient of the kinetic energy T and then compute the gradient of the potential energy V.

In accordance with our treatment in Section 1.3, the adjoint of (1.29) corresponds to composing the maps $\varphi_t^{[T]}$ and $\varphi_t^{[V]}$ in reverse order,

$$p_{n+1} = p_n - h\,\nabla_q V(q_n), \qquad q_{n+1} = q_n + h\,\nabla T_p(p_{n+1}), \tag{1.30}$$

so we call it the symplectic Euler–TV method.

Obviously, our discussion of schemes (1.29) and (1.30) above applies for any Hamiltonian system whose Hamiltonian function can be written in the (so-called separable) form (1.25). That is, the Lie–Trotter scheme leads to the two variants of the symplectic Euler method when it is applied to separable Hamiltonian systems.

As for the Strang splitting scheme, described in Algorithm 1.2 in general, and in (1.24) for Hamiltonian problems, when $H(q,p) = T(p) + V(q)$ it reduces to the much celebrated *Störmer–Verlet method* (Hairer, Lubich and Wanner 2003). Specifically, depending on the order in which the parts are evaluated, we have the following two variants.

Algorithm 1.3 (Störmer–Verlet–VTV). From $(q_0, p_0) = (q(0), p(0))$, for $n \geq 0$,

- $p_{n+1/2} = p_n - \frac{h}{2}\nabla_q V(q_n)$;
- $q_{n+1} = q_n + h\nabla_p T(p_{n+1/2})$;
- $p_{n+1} = p_{n+1/2} - \frac{h}{2}\nabla_q V(q_{n+1})$.

Algorithm 1.4 (Störmer–Verlet–TVT). From $(q_0, p_0) = (q(0), p(0))$, for $n \geq 0$,

- $q_{n+1/2} = q_n + \frac{h}{2}\nabla_p T(p_n)$;
- $p_{n+1} = p_n - h\nabla_q V(q_{n+1/2})$;
- $q_{n+1} = q_{n+1/2} + \frac{h}{2}\nabla_p T(p_{n+1})$.

Clearly, Algorithms 1.3 and 1.4 correspond to *time-symmetric* methods and can be obtained by composing the Euler–TV method and its adjoint. Specifically, if χ_h corresponds to method (1.30), then

$$S_h^{[2]} \equiv \chi_{h/2}^* \circ \chi_{h/2} = \varphi_{h/2}^{[V]} \circ \varphi_h^{[T]} \circ \varphi_{h/2}^{[V]} \tag{1.31}$$

recovers Störmer–Verlet–VTV, whereas the TVT version corresponds to $\varphi_{h/2}^{[T]} \circ \varphi_h^{[V]} \circ \varphi_{h/2}^{[T]}$.

Figure 1.1(a) shows trajectories of the pendulum (1.19), starting from three different initial conditions $(q_0, p_0) = (-5, \frac{5}{2}), (1, 1), (\frac{1}{10}, 0)$, corresponding to different regions of the phase space. For $(q_0, p_0) = (-5, \frac{5}{2})$ it holds that $T(p(t)) > V(q(t))$, whereas for $(q_0, p_0) = (1, 1)$ we have $T(p(t)) \simeq V(q(t))$ on average. Finally, for $(q_0, p_0) = (\frac{1}{10}, 0)$, the system can be seen as a slightly perturbed harmonic oscillator. In this case we could consider the following decomposition:

$$H = \frac{1}{2}(p^2 + q^2) + \left(1 - \frac{1}{2}q^2 - \cos q\right) \equiv H_1(q,p) + H_2(q), \tag{1.32}$$

where $H_1(q,p) = \frac{1}{2}(p^2 + q^2)$ corresponds to the harmonic oscillator, whose exact solution is known (a rotation in phase space). Moreover, $|H_2(q)| = \varepsilon |H_1(q,p)|$, with $\varepsilon \approx 10^{-3}$ along the orbit originated in $(\frac{1}{10}, 0)$. With splitting (1.32), the map $\varphi_h^{[H_1]} \circ \varphi_h^{[H_2]}$ reads

$$\begin{pmatrix} q_{n+1} \\ p_{n+1} \end{pmatrix} = R(h) \begin{pmatrix} q_n \\ p_n + h(q_n - \sin q_n) \end{pmatrix}, \quad \text{with } R(h) = \begin{pmatrix} \cos h & \sin h \\ -\sin h & \cos h \end{pmatrix}, \tag{1.33}$$

whereas the second-order scheme $\varphi_{h/2}^{[H_1]} \circ \varphi_h^{[H_2]} \circ \varphi_{h/2}^{[H_1]}$ can be formulated as

$$\begin{aligned} \begin{pmatrix} q_{n+1/2} \\ p_{n+1/2} \end{pmatrix} &= R(h/2) \begin{pmatrix} q_n \\ p_n \end{pmatrix}, \\ p^*_{n+1/2} &= p_{n+1/2} + h(q_{n+1/2} - \sin q_{n+1/2}), \\ \begin{pmatrix} q_{n+1} \\ p_{n+1} \end{pmatrix} &= R(h/2) \begin{pmatrix} q_{n+1/2} \\ p^*_{n+1/2} \end{pmatrix}. \end{aligned} \tag{1.34}$$

To illustrate the different splittings, we take $(\frac{1}{10}, 0)$ as the initial condition, integrate until the final time $t_f = 500$ and measure the relative error in energy, $|H(q_n, p_n) - H(q_0, p_0)|/|H(q_0, p_0)|$, along the trajectory. The step size is taken so that all the methods tested require the same number of evaluations of the potential (and thus essentially involve the same computational cost): 1200 evaluations (Figure 1.1(b)) and 2400 evaluations (Figure 1.1(c)). The schemes we test are as follows: on the one hand, the Störmer–Verlet algorithm (1.31) (denoted by S_2 in the graphs) and the fourth-order Runge–Kutta–Nyström splitting method proposed in Blanes and Moan (2002) ($\text{RKN}_6 4$), as representatives of the $T + V$ splitting; on the other hand, the specially adapted schemes (1.34), denoted $(2, 2)$, and the $(10, 6, 4)$ integrator presented in Blanes *et al.* (2013*b*), both for the case where $H = H_1 + H_2$, with H_2 small compared with H_1. The notation $(10, 6, 4)$ refers to the fact that the local error of the method is of order $O(\varepsilon h^{11} + \varepsilon^2 h^7 + \varepsilon^3 h^5)$ if H_2 is ε times smaller than H_1. For analogy, we label the Strang method applied to the perturbed harmonic oscillator (1.32) as $(2, 2)$. Notice that the splitting (1.32) is more advantageous for this initial condition (the simple method (1.34) behaves better than the fourth-order scheme), and that the improvement with respect to the $T + V$ splitting is approximately of the size of ε. We therefore see that for this type of problem it is possible to construct integrators, providing much more accurate results with the same computational effort.

Since all the schemes are symplectic integrators and the evolution is taking place in a compact domain, the error in energy is bounded (in contrast to standard non-symplectic methods, whose error in energy usually grows linearly with t), whereas the error in phase space (q, p) grows linearly when applied to near-integrable Hamiltonian systems (Hairer *et al.* 2006). Notice, however, that if the scheme is conjugate to another more accurate integrator, then the global error in phase space will remain bounded for some time interval before it starts growing linearly; see

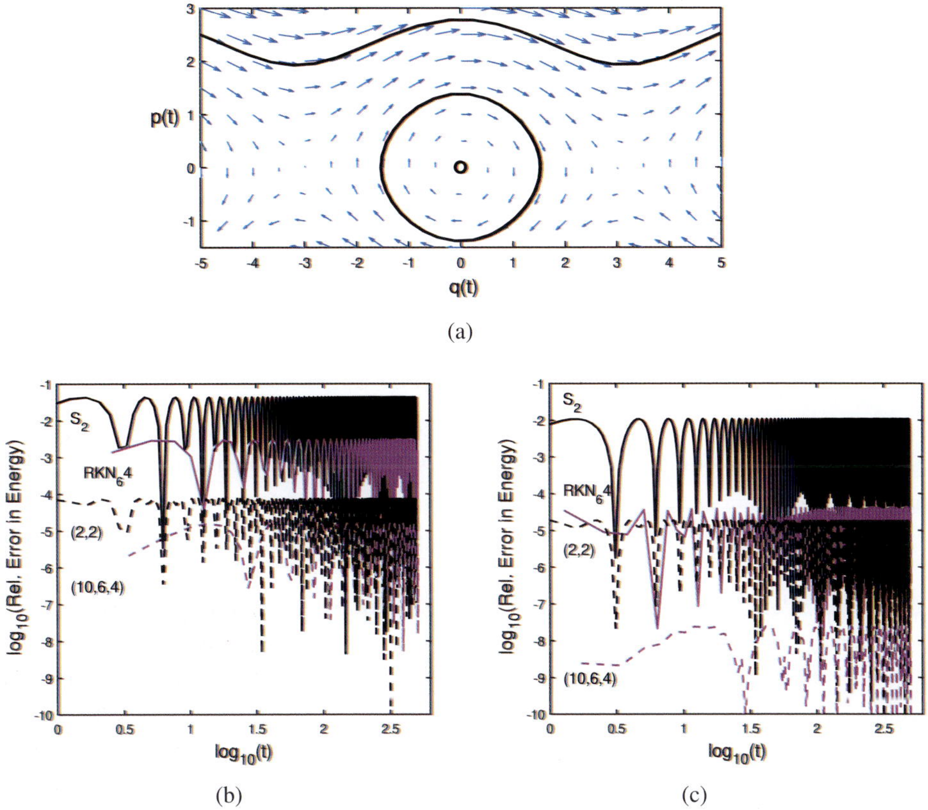

Figure 1.1. Simple pendulum. (a) Phase space and three trajectories with initial conditions $(q_0, p_0) = (-5, \frac{5}{2}), (1, 1), (\frac{1}{10}, 0)$. (b,c) Relative error in energy committed by different splitting methods along the solution with initial condition $(q_0, p_0) = (\frac{1}{10}, 0)$ in the interval $t \in [0, 500]$ with (b) 1200 evaluations and (c) 2400 evaluations of the potential.

Section 4.5 for more details. This feature is illustrated in Figure 1.2: we integrate the system starting with the same initial condition and final time $t_f = 500$ with $h = \frac{5}{12}$ and compute the relative error

$$\|(q(t_n), p(t_n)) - (q_n, p_n)\| / \|(q(t_n), p(t_n))\|,$$

with different schemes. The reference solution $(q(t_n), p(t_n))$ is computed numerically with very high accuracy. Specifically, we test the following integrators: the Lie–Trotter method, equation (1.23) (denoted LT in the graph) and the Störmer–Verlet (S_2) method for the splitting $H = T + V$, and scheme (2, 2) and (1.33) ($\mathrm{LT_{pert}}$), which corresponds to the Lie–Trotter method applied to the perturbed harmonic

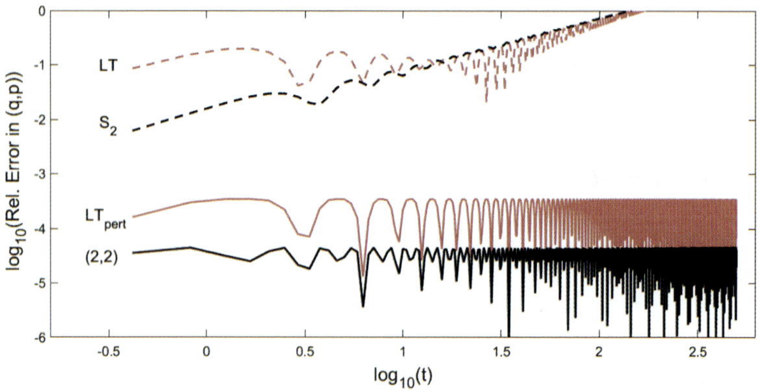

Figure 1.2. Pendulum. Relative error in phase space for different splitting methods along the solution with initial condition $(q_0, p_0) = (\frac{1}{10}, 0)$ in the interval $t \in [0, 500]$ and time step $h = \frac{5}{12}$.

oscillator (1.32). We observe that, since LT and S_2 are conjugate to each other (see (1.18)), their errors are quite similar after some time interval. On the other hand, $(2, 2)$ and LT_{pert} (which are also conjugate to each other) show a different behaviour: no linear growth is visible, and the error of LT_{pert} is larger than that of (2,2) by approximately the same factor for the whole time interval considered in Figure 1.2. This is related to the fact that $(2, 2)$ and LT_{pert} are conjugate to another method with a local error essentially $O(h^3\varepsilon^2)$; see Sections 4.5 and 5.7. More comments on these observations along with additional explanations will be given in Section 4.5.

1.5. The gravitational N-body problem

Another popular example to illustrate the behaviour and performance of splitting methods corresponds to the important problem in classical mechanics of a planetary system modelled as N bodies (a massive star and $N - 1$ planets) under mutual gravitational Newtonian interaction. This is also a Hamiltonian system with

$$H(q, p) = \sum_{i=0}^{N-1} \frac{1}{2m_i} p_i^\top p_i - G \sum_{i=1}^{N-1} \sum_{j=0}^{i-1} \frac{m_i m_j}{\|q_i - q_j\|}. \tag{1.35}$$

Here (q, p) denote the 'supervectors' composed by the positions $q_i \in \mathbb{R}^3$ and momenta $p_i \in \mathbb{R}^3$ of the 'Sun' $(i = 0)$ and the $N-1$ planets $(i = 1, \ldots, N-1)$ in some Cartesian coordinate system: $q = (q_0, q_1, \ldots, q_{N-1})^\top$, $p = (p_0, p_1, \ldots, p_{N-1})^\top$. In (1.35), m_i is the mass of the ith body, and G is the universal gravitational constant.

Now the equations of motion (1.20) read

$$\begin{aligned}\frac{\mathrm{d}q_i}{\mathrm{d}t} &= \nabla_{p_i} H = \frac{1}{m_i} p_i, \quad i = 0, \dots, N-1, \\ \frac{\mathrm{d}p_i}{\mathrm{d}t} &= -\nabla_{q_i} H = -G \sum_{j \neq i} \frac{m_i m_j}{\|q_i - q_j\|^3}(q_i - q_j), \quad i, j = 0, \dots, N-1.\end{aligned} \tag{1.36}$$

Since the kinetic energy $T(p)$ and the potential energy $V(q)$ are in this case

$$T(p) = \sum_{i=0}^{N-1} \frac{1}{2m_i} p_i^\top p_i, \quad V(q) = -G \sum_{i=1}^{N-1} \sum_{j=0}^{i-1} \frac{m_i m_j}{\|q_i - q_j\|}, \tag{1.37}$$

it also makes sense to separate the Hamiltonian (1.35) as $H(q, p) = T(p) + V(q)$, so that the symplectic Euler and the Störmer–Verlet schemes can be applied to arbitrary configurations of the bodies. However, this choice is suboptimal for planetary systems, where planets describe near-Keplerian orbits around the central star. An alternative procedure taking advantage of the hierarchical nature of the motion of the planets around the central massive body was first proposed in Wisdom and Holman (1991), and is known in the literature as the Wisdom–Holman integration map. It essentially consists in changing coordinates so that the transformed H can be written as an integrable part H_1 (corresponding to the Keplerian motion of the planets) and a small perturbation H_2 (that accounts for the gravitational interaction of the planets among themselves), and then applying the second-order scheme (1.24) to this new Hamiltonian.

Specifically, Wisdom and Holman (1991) consider a linear canonical change of variables to rewrite (1.35) in the so-called Jacobi coordinates $(\hat{q}_i, \hat{p}_i)$, $i = 0, 1, \dots, N-1$. Here $\hat{q}_0$ is the position of the centre of mass of the system, $\hat{q}_1$ is the relative position of the first planet with respect to the central star, and for $i = 2, \dots, N-1$, $\hat{q}_i$ is the position of the ith planet relative to the centre of mass of the central star and the planets with lower indices. That is,

$$\hat{q}_0 = \frac{1}{M_{N-1}} \sum_{j=0}^{N-1} m_j q_j, \quad \hat{q}_i = q_i - \frac{1}{M_i} \sum_{j=0}^{i-1} m_j q_j, \quad i = 1, \dots, N-1, \tag{1.38}$$

where $M_i = \sum_{j=0}^{i} m_j$ for $i = 0, \dots, N-1$. This can be written in a more compact way as $\hat{q} = Aq$, where A is an $N \times N$ invertible matrix with mass-dependent entries, $\hat{q} = (\hat{q}_0, \dots, \hat{q}_{N-1})^\top$ and $q = (q_0, \dots, q_{N-1})^\top$.

The conjugate momenta $\hat{p}_i$, $i = 0, 1, \dots, N-1$, are uniquely determined by the requirement that the change of variables be canonical (i.e. $p = A^\top \hat{p}$), so that the transformed Hamiltonian function is obtained by rewriting (1.35) in the new variables: $\hat{H}(\hat{q}, \hat{p}) \equiv H(A^{-1}\hat{q}, A^\top \hat{p})$.

It is straightforward to check that the kinetic energy, expressed in terms of $\hat{p}$, has the same diagonal structure as in p:

$$T = \sum_{i=0}^{N-1} \frac{1}{2m_i} p_i^\top p_i = \sum_{i=0}^{N-1} \frac{1}{2\hat{m}_i} \hat{p}_i^\top \hat{p}_i,$$

where

$$\hat{m}_0 = M_{N-1}, \quad \hat{m}_i = \frac{M_{i-1}}{M_i} m_i, \quad i = 1, \ldots, N-1.$$

The Hamiltonian in the new variables $\hat{H}(\hat{q}, \hat{p})$ can be split as $\hat{H}(\hat{q}, \hat{p}) = H_1(\hat{q}, \hat{p}) + H_2(\hat{q})$, where

$$\begin{aligned} H_1(\hat{q}, \hat{p}) &= \frac{1}{2\hat{m}_0} \hat{p}_0^\top \hat{p}_0 + \sum_{i=1}^{N-1} \left(\frac{1}{2\hat{m}_i} \hat{p}_i^\top \hat{p}_i - \frac{m_0 m_i}{\|\hat{q}_i\|} \right), \\ H_2(\hat{q}) &= V(q) + G \sum_{i=1}^{N-1} \frac{m_0 m_i}{\|\hat{q}_i\|} \\ &= G m_0 \sum_{i=1}^{N-1} m_i \left(\frac{1}{\|\hat{q}_i\|} - \frac{1}{\|q_i - q_0\|} \right) - G \sum_{i=2}^{N-1} \sum_{j=1}^{i-1} \frac{m_i m_j}{\|q_i - q_j\|}, \end{aligned} \tag{1.39}$$

and q has to be expressed in terms of $\hat{q}$ according to $q = A^{-1}\hat{q}$.

Observe that the potential energy does not depend on $\hat{q}_0$, so $\hat{p}_0$ is constant (in fact it is the linear momentum of the system) and therefore we can remove it from $\hat{H}$ if we assume that the centre of the mass is at rest.

Clearly, for fixed $\hat{q}$ and varying mass ratios,

$$\hat{q}_i = q_i - q_0 + O(\varepsilon), \quad H_2(\hat{q}) = O(\varepsilon) \quad \text{as } \varepsilon \equiv \frac{1}{m_0} \max_{1 \le i \le N-1} m_i \to 0.$$

Hamiltonian H_1 can then be considered as a collection of $N-1$ two-body problems, and H_2 as a perturbation. It turns out that the flow $\varphi_h^{[H_1]}$ can be computed with the algorithm proposed in Danby (1988, p. 165), for example, whilst H_2 only depends on $\hat{q}$, and thus its flow $\varphi_h^{[H_2]}$ can be explicitly evaluated in an efficient way. Notice that the number of terms in H_1 grows linearly with the number of bodies N, whereas the number of terms in H_2 grows quadratically.

As an illustrative example, we next consider the outer Solar System modelled as a six-body system with the inner Solar System ($i = 0$), the four giant planets ($i = 1, 2, 3, 4$) and Pluto ($i = 5$), all considered as point masses. The initial conditions for each planet are taken at Julian time (TDB) 2440400.5 (28 June 1969), obtained from the DE430 ephemerides (Folkner *et al.* 2014) and normalized so that the centre of mass of the system is at rest. A schematic diagram of the trajectories is shown in Figure 1.3(a) with the initial (circles) and final (stars) positions of each object after 200 000 days. For this problem we test the same methods as for the pendulum in Figure 1.1: on the one hand, Störmer–Verlet (1.31), S_2,

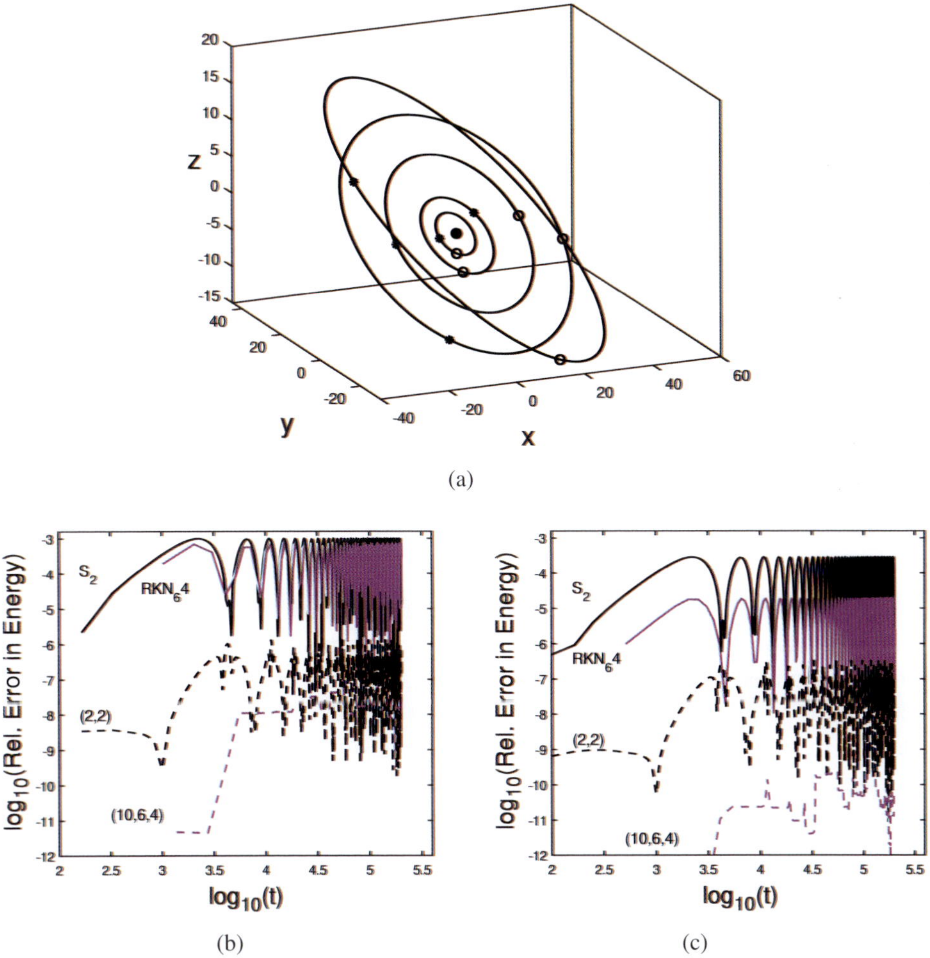

Figure 1.3. (a) Trajectories of the six-body system modelling the outer Solar System. (b,c) Relative error in energy as a function of time for an interval of 200 000 days obtained with different splitting methods with (b) 1200 evaluations and (c) 2400 evaluations of the force.

and the fourth-order Runge–Kutta–Nyström splitting method of Blanes and Moan (2002), $\text{RKN}_6 4$ when the Hamiltonian (1.35) is separated into kinetic and potential energy (1.37); on the other hand, the specially adapted schemes (1.34) (called in this setting the Wisdom–Holman integrator, and denoted (2,2) as before) and the $(10,6,4)$ integrator presented in Blanes *et al.* (2013*b*) when H is expressed as $H(\hat{q},\hat{p}) = H_1(\hat{q},\hat{p}) + H_2(\hat{q})$, with (1.39) in Jacobi coordinates. We integrate for a relatively short time interval, $t_f = 200\,000$ days (or approximately 46 periods of Jupiter and two periods of Pluto), and compute the relative error in the energy with each integrator with a step size so that all of them require the same number

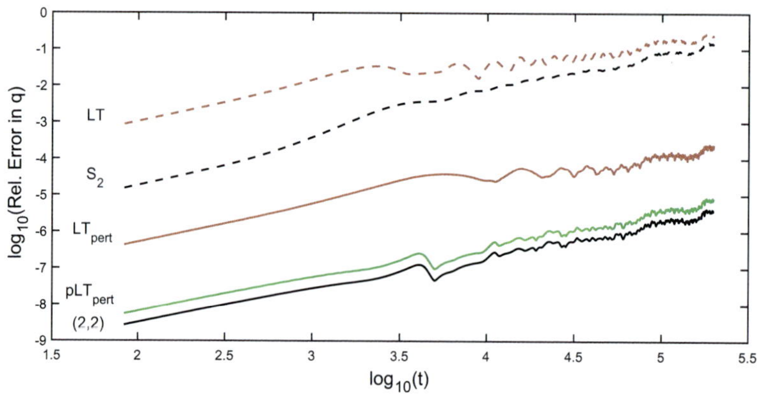

Figure 1.4. Outer Solar System. Relative error in position as a function of t for a time interval $[0, t_f = 200\,000]$ days with $h = t_f/1200$ obtained with different splitting methods.

of force evaluations. The results are displayed in Figure 1.3 with (b) 1200 and (c) 2400 evaluations. As in the previous example of the pendulum, the error in energy remains bounded in all cases, and scheme $(2,2)$ provides an error energy almost 1000 times smaller than S_2. This illustrates the fact that, by taking a splitting adapted to the structure of the problem and designing integrators taking these specific features into account, it is possible to greatly improve the efficiency.

In Figure 1.4 we display the relative errors in positions $q = (q_0, \ldots, q_5) \in \mathbb{R}^{18}$ along the time integration for the same methods as in Figure 1.2. As in the pendulum problem, the error for LT is larger than for S_2 at the beginning of the integration interval, but they become similar after some time. Here $(2,2)$ is also more accurate than S_2 by a factor significantly smaller than ε. Compared to the pendulum problem, the error grows linearly right from the beginning for schemes $\mathrm{LT_{pert}}$ and $(2,2)$ (although with a smaller slope than LT and S_2). The curve labelled by $\mathrm{pLT_{pert}}$ corresponds to the relative error obtained by applying $\mathrm{LT_{pert}}$ with initial conditions $(\bar{q}_0, \bar{p}_0) = \varphi_{h/2}^{[H_1]}(q_0, p_0)$. Notice that the phase errors of $\mathrm{pLT_{pert}}$ are very similar to those of $(2,2)$. All these observations will be accounted for in Section 4.5.

1.6. The time-dependent Schrödinger equation

The basic object to study the time evolution of a system in quantum mechanics is the (time-dependent) Schrödinger equation. In the case of one particle of unit mass in a potential $V(x)$, it reads (Messiah 1999)

$$i\hbar \frac{\partial}{\partial t}\psi(x,t) = -\frac{1}{2}\Delta\psi(x,t) + V(x)\psi(x,t), \qquad \psi(x,0) = \psi_0(x). \tag{1.40}$$

Here $\psi: \Omega \subset \mathbb{R}^3 \times \mathbb{R} \longrightarrow \mathbb{C}$ is the wave function, representing the state of the system, and $\hbar$ is the reduced Planck constant. The quantity $|\psi(x,t)|^2$ represents a probability density for the position of the particle, in the sense that the probability of the particle to be located in $S \subset \Omega$ at time t is $\int_S |\psi(x,t)|^2 \, \mathrm{d}x$. The equation is then defined in the Hilbert space $L^2(\Omega, \mathbb{C})$.

If we introduce the self-adjoint operators $\hat{T}$, $\hat{V}$ acting on $\psi \in L^2(\Omega, \mathbb{C})$ as

$$\hat{T}\psi = -\frac{1}{2}\Delta\psi, \qquad \hat{V}\psi = V(x)\psi,$$

then a straightforward calculation shows that $[\hat{V}, [\hat{T}, \hat{V}]]\psi = |\nabla V|^2\psi$, and therefore

$$[\hat{V}, [\hat{V}, [\hat{T}, \hat{V}]]]\psi = 0. \tag{1.41}$$

A standard approach for applying splitting methods in this setting consists in first discretizing the equation in space. If we consider the one-dimensional case for simplicity and if the wave function is negligible outside a space interval $[a, b]$ on the time interval of interest, then we can limit ourselves to the study of the equation on that finite interval with periodic boundary conditions (Lubich 2008). After rescaling, the periodic interval can always be restricted to $[-\pi, \pi]$. In this way, the original problem is transformed into ($\hbar = 1$)

$$\mathrm{i}\frac{\partial}{\partial t}\psi(x,t) = -\frac{1}{2}\frac{\partial^2\psi}{\partial x^2}(x,t) + V(x)\psi(x,t), \qquad x \in [-\pi, \pi], \tag{1.42}$$

with $\psi(-\pi, t) = \psi(\pi, t)$ for all t.

The wave function is then approximated by a trigonometric polynomial $u(x,t)$ whose coefficients are obtained by requiring that the approximation satisfies (1.42) in a grid of M equispaced points $x_j = -\pi + j \cdot 2\pi/(M-1)$ on the interval $[-\pi, \pi]$. The vector $u = (u_0, \ldots, u_{M-1})^\top \in \mathbb{C}^M$ formed by the grid values $u_j \approx \psi(x_j, t)$, $j = 0, 1, \ldots, M-1$, then verifies the M-dimensional linear ODE

$$\mathrm{i}\frac{\mathrm{d}}{\mathrm{d}t}u(t) = H\,u(t) = (T+V)u(t), \qquad u(0) = u_0 \in \mathbb{C}^M. \tag{1.43}$$

Here $V = \mathrm{diag}(V(x_j))$ and $T = -\frac{1}{2}D$, where D is the second-order periodic spectral differentiation matrix (Trefethen 2000). As is well known, $Tu = \mathcal{F}^{-1}D_T\mathcal{F}u$, where $\mathcal{F}$ and $\mathcal{F}^{-1}$ are the forward and backward discrete Fourier transform, and D_T is also diagonal. The transformations $\mathcal{F}$ and $\mathcal{F}^{-1}$ are computed with the fast Fourier transform (FFT) algorithm, requiring $O(M \log M)$ operations.

Notice that solving equations $\mathrm{i}u' = Tu$ and $\mathrm{i}u' = Vu$ is done trivially by using exponentials of diagonal matrices and FFTs, namely,

$$\left(\mathrm{e}^{\tau V}u\right)_j = \mathrm{e}^{\tau V(x_j)}u_j, \qquad \mathrm{e}^{\tau T}u = \mathcal{F}^{-1}\mathrm{e}^{\tau D_T}\mathcal{F}u,$$

for a time step h, with $\tau = -\mathrm{i}h$. Therefore, splitting methods constitute a valid alternative way to approximate the solution $u(t) = \mathrm{e}^{\tau H}u_0 = \mathrm{e}^{\tau(T+V)}u_0$, which may be prohibitively expensive to evaluate for large values of M. Thus the Lie–Trotter scheme reads

$$e^{\tau(T+V)} = e^{\tau T}\, e^{\tau V} + O(\tau^2), \tag{1.44}$$

whereas the second-order Strang splitting constructs the numerical approximation u_{n+1} at time $t_{n+1} = t_n + \Delta t$ by

$$u_{n+1} = e^{\tau/2V}\, e^{\tau T}\, e^{\tau/2V}\, u_n. \tag{1.45}$$

The resulting scheme is called the *split-step Fourier* method in the chemical literature (Feit, Fleck Jr and Steiger 1982) and has some remarkable properties. In particular, it is unitary and time-symmetric (Lubich 2008), as is the exact solution $e^{\tau H}$.

Relation (1.41) still holds for the matrices T and V if the number of points M in the space discretization is sufficiently large, and in fact $[V, [T, V]]$ is diagonal if the derivatives of the potential are computed first and then evaluated at the corresponding space grid.

We next illustrate the procedure with the one-dimensional double-well potential

$$V(x) = \frac{1}{80}(x^2 - 20)^2, \tag{1.46}$$

and the initial wave function $\psi(x, 0) = \psi_0(x) = \sigma \cos^2(x)\, e^{-\frac{1}{2}(x-1)^2}$, where σ is an appropriate normalizing constant. We take $M = 256$ discretization points on the interval $x \in [-13, 13]$ and integrate the resulting linear ODE (1.43) in the interval $0 \le t \le t_f = Nh = 10$. Figure 1.5(a) shows $|\psi_0(x)|^2$, $|\psi(x, t_f)|^2$ and the potential $V(x)$, whereas Figure 1.5(b) shows an efficiency diagram. Specifically, we display the error in energy measured at the final time, $|u_N^\top H u_N - u_0^\top H u_0|$, as a function of the number of FFT calls (and its inverse) as an estimate of the computational effort of each method. The lines correspond to the Strang splitting (1.45), S_2, with time steps $h = 10/2^k$, $k = 1, 2, \ldots, 12$, the fourth-order RKN splitting method RKN$_6$4 from Blanes and Moan (2002), already illustrated in Figures 1.1 and 1.3, and another fourth-order scheme including the double commutator $[V, [T, V]]$ into its formulation (denoted RKNm$_4$4). In this diagram the slope of each line for sufficiently small h indicates the order of the scheme. As in the previous examples, by taking into account the specific features of the problem at hand it is possible to construct more accurate and efficient numerical approximations.

The error in energy also remains bounded for these unitary integrators, as in the previous examples involving classical Hamiltonian problems, whereas the error in the wave function grows linearly with t, unless the scheme is conjugate to another more accurate one, in which case it is bounded for some time before linear growth takes place. To illustrate this feature, in Figure 1.6 we depict how the error in the solution $\|u(t_n) - u_n\|$ evolves with time for a longer integration interval $t \in [0, 1000]$, with step size $h = \frac{1}{20}$. As usual, the reference solution is computed numerically with sufficiently high accuracy, and the tested schemes are as follows: Lie–Trotter, equation (1.44) (LT), Strang, equation (1.45), and a variant of Strang

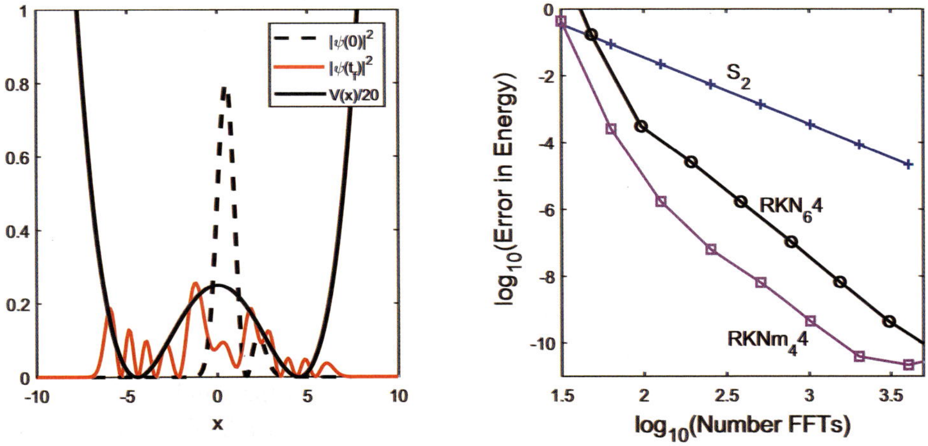

(a) (b)

Figure 1.5. Time-dependent Schrödinger equation with a double-well potential. (a) $V(x)$, initial and final wave function with $M = 256$ discretization points. (b) Relative error in energy at the final time vs. number of FFTs for different values of the time step obtained with the Strang method S_2, and two fourth-order splitting schemes: one involving six evaluations of V ($\mathrm{RKN}_6 4$) and another with four evaluations of V, and incorporating in addition the double commutator $[V, [T, V]]$ ($\mathrm{RKNm}_4 4$).

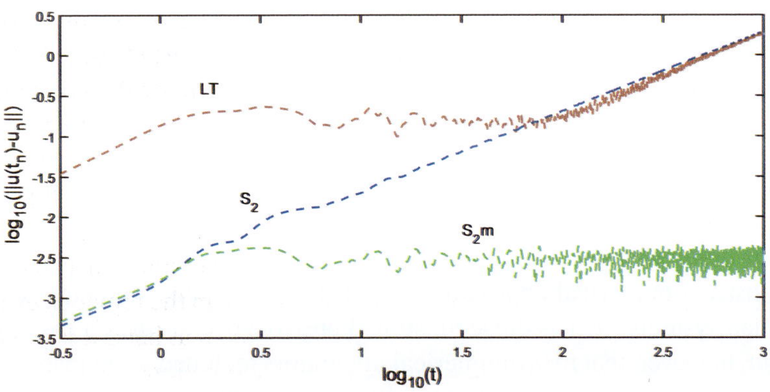

Figure 1.6. Time-dependent Schrödinger equation with a double-well potential: relative error in the wave function as a function of t with $h = 1/20$ for Strang (S_2), Lie–Trotter (LT) and scheme S_2m involving a double commutator and conjugate to a method of order four.

involving a double commutator, namely

$$S_2m: \quad u_{n+1} = \mathrm{e}^{\tau/2V+(\tau^3/48)[V,[T,V]]}\,\mathrm{e}^{\tau T}\,\mathrm{e}^{\tau/2V+(\tau^3/48)[V,[T,V]]}\,u_n. \tag{1.47}$$

This scheme does not require any additional FFTs, and in addition it is actually conjugate to a method of order four. We observe that LT approaches S_2 after a transition time, whereas the error for S_2m remains bounded for the whole interval. In fact, its linear error growth only appears when the time interval is doubled.

1.7. Splitting methods as geometric numerical integrators

As the examples of Sections 1.2–1.6 illustrate, splitting schemes, even of low order of consistency such as Lie–Trotter and Strang methods, preserve by construction structural properties of the exact solution, such as symplecticity (in classical Hamiltonian dynamics) and unitarity (in quantum evolution problems). This feature gives them qualitative superiority with respect to other standard integrators in practice, especially when long time intervals are concerned. In this sense, splitting methods constitute an important class of *geometric numerical integrators*.

Although the idea that numerical integrators applied to an ordinary differential equation should preserve as many properties of the system as possible has been implicitly assumed since the early days of numerical analysis, it is fair to say that in the classical consistency/stability approach the emphasis has been on other issues. In particular, the goal has mainly been to compute the solution of (1.1) at time $t_f = Nh$ with a global error $\|x_N - x(t_f)\|$ smaller than a prescribed tolerance and as efficiently as possible. To do that, we choose the class of method (one-step, multistep, extrapolation, etc.), the order (fixed or adaptive) and the time step (constant or variable). This approach has proved to be very fruitful, giving rise to highly tuned and thoroughly tested software packages generally available to solve a great variety of problems.

On the other hand, there are special types of problems arising in many fields of science and applied mathematics that possess an underlying geometric structure which influences the qualitative character of their solutions, so we naturally aim to construct numerical approximations that preserve this structure. Classical Hamiltonian systems such as those illustrated previously constitute a case in point. It turns out, however, that many numerical integrators included in standard software packages do not take into account these distinctive features of the equations to be solved, and the question is whether it is possible to design, analyse and apply new schemes providing approximate solutions that share one or several geometric properties with the exact solution. This is precisely the realm of geometric numerical integration, a terminology introduced in Sanz-Serna (1997).

According to McLachlan and Quispel (2006):

> 'Geometric integration' is the term used to describe numerical methods for computing the solution of differential equations, while preserving one or more physical/mathematical properties of the system *exactly* (i.e., up to round-off error).

Thus, rather than primarily taking into account prerequisites such as consistency and stability, the aim is to reproduce the qualitative features of the solution of the differential equation being discretized, in particular its geometric properties, such as the symplectic character (for Hamiltonian systems) and unitarity (quantum mechanics), but also the phase-space volume (for divergence-free vector fields), time-reversal symmetries, first integrals of motion (energy, linear and angular momentum), Casimirs, Lyapunov functions, etc. In these structure-preserving methods we try to incorporate as many of these properties as possible and, as a result, they exhibit improved qualitative behaviour. In addition, they typically allow for a significantly more accurate integration for long-time intervals than with general-purpose methods (Hairer *et al.* 2006, Blanes and Casas 2016).

Although splitting methods have a long history in numerical mathematics and have been applied, sometimes with different names, in many different contexts (e.g. partial differential equations, quantum statistical mechanics, chemical physics and molecular dynamics), it is fair to say that the interest in splitting has revived with the advent of geometric numerical integration, and new and very efficient schemes have been put to use to solve a wide variety of problems. The reason is clear: if the problem (1.3) has some property that is deemed to preserve (symplectic, unitary, volume-preserving, etc.) and each subproblem $x' = f_j(x)$, $x(0) = x_0$ can be integrated exactly or by means of a numerical method preserving these properties, then the splitting method constructed by composing the solution of the subproblems is also symplectic, unitary, volume-preserving, etc. In other words, splitting methods provide by construction approximations lying in the same group of diffeomorphisms as the system $x' = f(x)$ (McLachlan and Quispel 2002). Here we assume of course that each subproblem $x' = f_j(x)$ possesses the same characteristic feature as the total problem considered.

1.8. Relevance of splitting methods

Given a certain differential equation $x' = f(x)$, the application of splitting methods to solve the corresponding initial value problem involves three main steps (McLachlan and Quispel 2002).

(1) Choosing the set of terms f_j such that $f = f_1 + \cdots + f_m$. Different ways of decomposing f may give rise to integrators with different qualitative behaviour and efficiency, as we have seen in the previous examples.

(2) Solving each subproblem $x' = f_j(x)$ either exactly or approximately.

(3) Combining these solutions to get an approximation for the original overall problem.

Being such a simple idea, it is hardly surprising that the splitting principle can be used in so many different settings. In particular, one may

- split the differential equation into linear and nonlinear parts;

- in an ODE describing a Hamiltonian system with an additional small dissipation, separate the Hamiltonian part and the dissipation;
- decompose into parts describing different physical processes, for example diffusion and reaction in partial differential equations;
- get approximations to the original problem by solving one space direction at a time (dimensional splitting in PDEs).

In addition, splitting methods possess some advantages concerning their implementation, in particular the following.

- They are typically explicit.
- Their storage requirements are quite small. The algorithms are sequential and the solution at intermediate stages can be stored in the solution vectors.
- Programming higher-order schemes is no more difficult than Lie–Trotter and Strang splitting methods, at least in the context of ODEs. Usually, a few more lines of code is all that is required to deal with the additional stages.
- As stated earlier, they can preserve a wide variety of structures possessed by the differential equation.

They also present some disadvantages, of course. Among them, we can mention the following.

- Splitting schemes of order three or higher necessarily involve negative coefficients. In other words, they require substeps that go backwards in time, and this has severe repercussions when applying them to, for instance, reaction–diffusion equations (see Section 6).
- Although it is possible to construct accurate high-order splitting methods, stability can be an issue, in the sense that their stability interval might be reduced to render them useless in practice. This aspect has to be taken seriously when designing new methods.
- Ordinary splitting does not capture the correct steady-state solutions (where these solutions exist) (MacNamara and Strang 2016), in the sense that the numerical solutions obtained converge to limits that are not steady-state solutions but just approximations of them. This can lead to unacceptable errors, for instance in the simulation of combustion. In this context, balanced splitting techniques have been introduced to correct this flaw (Speth, Green, MacNamara and Strang 2013).

Splitting methods constitute an important tool in different areas of science where the evolution of systems is governed by differential equations. In addition to Hamiltonian systems, they can be successfully applied in the numerical study of Poisson systems, systems possessing integrals of motion (such as energy and angular momentum) and systems with (continuous, discrete and time-reversal) symmetries. In fact, splitting methods have been designed (often independently) and extensively used in fields as distant as molecular dynamics, simulation of storage rings in particle accelerators, celestial mechanics, astronomy, quantum (statistical) mechanics, plasma physics, hydrodynamics and Markov chain Monte Carlo methods.

Operator-splitting methods also appear outside the realm of differential equations, and in particular in optimization, in a variety of different special forms and different denominations (gradient-projection, proximal-gradient, alternating direction method of multipliers or ADMM, split Bregman, etc.). All of them are related to special types of splitting methods, such as Douglas–Rachford and Peaceman–Rachford schemes. More details can be found in several contributions collected in the comprehensive book by Glowinski, Osher and Yin (2016*a*).

1.9. Some historical remarks

There is ample consensus that the beginning of splitting is related to the product formula (1.5). What is not so clear is the origin of the formula itself. Thus Reed and Simon (1980, p. 295) establish it as 'the classical theorem of Lie', but give no exact source, whereas Chorin, Huges, Marsden and McCracken (1978) call it 'the 1875 formula of S. Lie', citing the classical treatise of Lie (1888), and, based on this reference, Glowinski *et al.* (2016*a*) even ascribe to Lie himself 'the first operator-splitting scheme recorded in history'. The problem is that the reference Lie (1888) is clearly not from 1875, and it is not evident (at least to us) that this formula appears there explicitly.

On the other hand, as pointed out in Cohen, Friedland, Kato and Kelly (1982), the result (1.5) can be found in several references published during the 1950s, namely Butler and Friedman (1955) and Golden (1957), whereas it was Trotter (1959) who generalized it to self-adjoint linear operators, without mentioning Lie or these previous references. Subsequently, formula (1.5), even in the matrix case, has been attributed to Trotter (Bellman 1970, p. 181). In view of the situation, we believe we are not committing an act of historical injustice by referring to the approximation (1.6) and Algorithm 1.1 as the *Lie–Trotter* scheme.

With respect to the splitting method (1.9), it first appeared in print in Strang (1968) as an alternative way to solve multidimensional problems with one-dimensional operators. We have already seen that, when applied to Hamiltonian systems of the form $H(q, p) = T(p) + V(q)$, it leads to Algorithm 1.4 when composing the flows associated to $T(p)$ and $V(q)$. It is called the *Störmer–Verlet method* since it was used by the astronomer Carl Störmer (1907) in his computations of the motion of ionized particles in the Earth's magnetic field, and by Loup Verlet (1967) in molecular dynamics. It is also referred to as the *leapfrog method* in the context of PDEs describing wave propagation, and as the *Wisdom–Holman method* when applied to the splitting (1.39) (Wisdom and Holman 1991). In fact, it can be found in several classical references, the oldest one being perhaps Newton's *Principia*.[2] For a detailed account the reader is referred to the enlightening review by Hairer *et al.* (2003).

As well as symplecticity when applied to Hamiltonian systems, the Störmer–Verlet method preserves many other geometric properties of the exact flow associated with an ordinary differential equation. This includes the preservation of all linear first integrals (such as the linear momentum), and quadratic first integrals of the form $I(q, p) = p^\top Cq$ for Hamiltonian systems, where C is a symmetric matrix. In other words, $I(q, p)$, computed along the numerical trajectory, is constant. A classical example is the angular momentum in N-body problems if the forces only depend on the distances of the particles.

All these favourable properties, in addition to its optimal stability property and reversibility, help us to understand why this method is probably the most widely used splitting scheme and geometric integrator, especially in molecular dynamics (Schlick 2010), condensed matter simulations (Ceperley 1995) and sampling with the hybrid Monte Carlo method (Neal 2011).

The convenience of designing numerical integration methods that, by construction, preserve the symplectic structure when applied to Hamiltonian systems was duly recognized during the 1950s in the field of accelerator physics. Thus, in the words of an early pioneer, 'if one wishes to examine solutions to differential equations, adoption of a "Hamiltonian" or "canonical" integration algorithm would be reassuring' (Laslett 1986). This was the point of view adopted in a pioneering paper by de Vogelaere (1956),[3] where he devoted himself to the task of designing 'a method of integration which, if there was no round-off error, would give a solution with the contact transformation property'. Here *contact transformation* has to be understood as *symplectic transformation*. The first-order schemes proposed by de Vogelaere (1956), although implicit in general, turn out to be explicit when $H(q, p) = T(p)+V(q)$, in which case they reproduce the symplectic Euler schemes (1.29) and (1.30).

It was another accelerator physicist, Ronald Ruth (1983), who presented what is probably the first splitting method of order three. This paper can be considered as the actual starting point in the systematic exploration of symplectic integrators along several parallel avenues: (i) the use of generating functions in the context of Hamiltonian mechanics to produce appropriate canonical transformations approximating the exact flow in each integration step (Feng and Qin 1987, Channell and Scovel 1990); (ii) the conditions that Runge–Kutta methods have to satisfy to be symplectic (Sanz-Serna 1988, Lasagni 1988, Suris 1988); (iii) the design of explicit symplectic methods of order four and higher for Hamiltonian systems that can be split into two pieces which can be solved exactly when considered as

[2] *Philosophiae Naturalis Principia Mathematica*, Book 1, Section 2, Proposition 1.

[3] See also Skeel and Cieśliński (2020) for the context of the work, and the preprint itself, typeset in LaTeX.

independent systems (Neri 1988, Forest 1992), with the help of the Lie formalism. This approach was further elaborated in Yoshida (1990). Working in the context of the hybrid Monte Carlo algorithm for dynamical fermions, a splitting method of order four was also independently proposed around the same time in Campostrini and Rossi (1990). In parallel developments, what is now called the Suzuki–Yoshida composition technique for increasing the order of numerical integrators appeared in Creutz and Gocksch (1989) for Monte Carlo simulations and in Suzuki (1990) and Yoshida (1990).

We should also mention the papers by de Raedt and de Raedt (1983) and Takahashi and Imada (1984), who pioneered the use of double commutators to get approximations of higher order than those obtained by the Störmer–Verlet method in path-integral Monte Carlo simulations: in the first case by constructing a fourth-order splitting scheme, and in the second, a method that it is also of order four by conjugation. In fact, scheme (1.47) is conjugate to the one proposed in Takahashi and Imada (1984). Ruth (1983) also presents a third-order method using double commutators.

That splitting and composition methods could be used to construct integrators for problems evolving in groups other than the symplectic group was emphasized in Forest and Ruth (1990) and further developed in Feng (1992), with the aim of constructing schemes able to preserve different structures.

The 1990s saw a dramatic increase in the interest and applications of splitting integrators in several fields, often with spectacular results. We should mention in particular those achieved in Wisdom and Holman (1991), revealing the existence of chaotic phenomena in the Solar System by numerically integrating the planetary equations of motion over very large time intervals.

The state of the art of splitting methods in the context of geometric numerical integration was masterfully summarized in a review paper by McLachlan and Quispel (2002), which has greatly influenced subsequent investigation in the field, as testified by its growing number of citations over the years, in many different areas.

Among the huge number of published works on splitting methods, the following surveys are worth highlighting.

- The monograph by Yanenko (1971) (an English translation of the Russian edition published in 1967) was perhaps the first to be devoted to the method of splitting (or *method of fractional steps*) for solving ‘complicated problems of mathematical physics in several variables’. Those include the numerical treatment of parabolic and hyperbolic equations, as well as boundary value problems for the Laplace and Poisson equations, and several applications in elasticity theory and hydrodynamics. It is based on the early contributions of Peaceman, Rachford, Douglas and several authors of the Soviet school (Dyakonov, Marchuk, Samarskii, Yanenko and others).

- The exhaustive review article by Marchuk (1990), included in Volume I of the *Handbook of Numerical Analysis* (Ciarlet and Lions 1990), can be seen as an update of the previous work, with a systematic study of operator splitting and alternating direction methods for solving linear and nonlinear partial differential equations. It includes convergence analyses and new applications to problems in hydrodynamics, meteorology and oceanography.
- The review paper by McLachlan and Quispel (2002) mainly focused on the application of splitting methods as geometric numerical integrators for various classes of ordinary differential equations. In that context, they carried out a classification of ODEs and their integration methods into different categories, and also examined the question of how to decompose a given vector field into much simpler vector fields, as well as the composition of these elementary flows.
- Books and monographs dealing with geometric numerical integration and structure-preserving algorithms contain plenty of material on splitting methods. Among them, we can cite the influential work of Sanz-Serna and Calvo (1994), the canonical reference by Hairer *et al.* (2006), as well as Leimkuhler and Reich (2004), Leimkuhler and Matthews (2015) and Blanes and Casas (2016).
- The multi-author book by Glowinski *et al.* (2016*a*) constitutes an excellent illustration of the ample scope and wide range of applications that today's operator-splitting methods are able to deal with. These include the numerical solution of problems modelled by linear and nonlinear partial differential equations and inequalities, problems in information sciences and image processing, and large-scale optimization problems, among others.

1.10. Plan of the paper

In this paper we will focus on splitting methods applied to evolutionary problems, mostly described by ordinary differential equations. These can directly model the problem we are interested in, or they can result from evolutionary PDEs previously discretized in space. Particular attention will be addressed to problems possessing special properties, very often from a geometric origin, that are worth preserving via the numerical methods. In so doing, we will follow a strategy similar to that in Blanes, Casas and Murua (2008*b*), trying to avoid any duplication of the material already gathered in the classic references cited above, and including new results, schemes and applications which have appeared in the literature during the last few years.

In particular, no general rule is provided here on how to split the defining operator in (1.1). As mentioned earlier, this issue is further analysed in McLachlan and Quispel (2002), and in fact some of the open problems listed there are related to it. We have already seen in the examples provided in this section that several splittings of the same problem are possible, often leading to methods with very different performances. Moreover, in certain cases, the original system has several geometric properties that are simultaneously preserved along the evolution, whereas different splittings preserve different properties, and it is generally difficult to find one splitting that preserves most of them.

With these considerations in mind, the rest of the paper is organized as follows. In Section 2 we first review the general composition technique and then provide a detailed analysis of the order conditions required by splitting and composition methods to achieve a given order of accuracy. There are some relevant problems, however, whose particular structure allows us to design specially adapted methods, and some of them are reviewed in Section 3, where we also show how to adapt existing splitting methods to non-autonomous problems.

In Section 4 we summarize some of the qualitative properties possessed by splitting methods in the context of geometric numerical integration of ordinary differential equations, with special attention to the idea of processing, whereas Section 5 is devoted to the treatment of highly oscillatory problems.

Splitting methods are particularly well adapted to deal with partial differential equations whose defining operator contains contributions coming from very different physical sources, so they have a long history in this area. Section 6 contains a brief survey, with special emphasis on Schrödinger equations and general parabolic evolution equations. The existence of negative coefficients in the methods, however, leads to an order barrier for parabolic equations, and Section 7 reviews splitting methods with complex coefficients as a possible way to overcome this order barrier.

In Section 8 we present an extended list of existing methods, classifying them into different families and giving the appropriate references. Their corresponding coefficients are also provided as supplementary material at the website www.gicas.uji.es/SplittingMethods.html. These methods are numerically tested on simple examples in the Appendix. Finally, some relevant applications of splitting methods in different fields are discussed in Section 9.

2. High-order splitting and composition methods

The Lie–Trotter and Strang splitting methods, despite their low order of accuracy, provide a fairly good description of the systems they are approximately solving. In fact, for many problems, including molecular dynamics applications and reaction–diffusion equations, Verlet splitting and Strang splitting are the most popular integrators, perhaps an illustration that, according to MacNamara and Strang (2016), 'it is a meta-theorem of numerical analysis that second-order methods often achieve the right balance between accuracy and complexity'. There are other areas, however, where a higher degree of precision is required, in addition to the preservation of qualitative properties. A classical example is the long-term numerical integration of the Solar System, both forwards (e.g. to analyse the existence of chaos; see Laskar 1989, Sussman and Wisdom 1992) and backwards in time (to study the insolation quantities of the Earth; see Laskar *et al.* 2004). Thus, in this section, after reviewing a general technique to get high-order integrators by composing low-order ones, we analyse from different perspectives the order conditions that have to be satisfied by a splitting method to achieve a given order. This analysis allows us to provide complementary information about the integrators: e.g. number of order conditions, explicit expressions, and remainders in the asymptotic expansions.

2.1. *Raising the order by composition*

2.1.1. *Composition of Strang maps*

Starting from the Strang splitting $S_h^{[2]} = \varphi_{h/2}^{[1]} \circ \varphi_h^{[2]} \circ \varphi_{h/2}^{[1]}$, the composition

$$\psi_h = S_{\gamma_s h}^{[2]} \circ S_{\gamma_{s-1} h}^{[2]} \circ \cdots \circ S_{\gamma_1 h}^{[2]} \tag{2.1}$$

is at least of order three if

$$\sum_{j=1}^{s} \gamma_j = 1 \quad \text{and} \quad \sum_{j=1}^{s} \gamma_j^3 = 0. \tag{2.2}$$

The smallest value of s for which equations (2.2) admit real solutions is $s = 3$. In that case, by imposing the symmetry $\gamma_1 = \gamma_3$, we indeed get a method of order four, sometimes called the *triple jump*:

$$S_h^{[4]} = S_{\gamma_3 h}^{[2]} \circ S_{\gamma_2 h}^{[2]} \circ S_{\gamma_1 h}^{[2]}, \quad \text{with} \quad \gamma_1 = \gamma_3 = \frac{1}{2 - 2^{1/3}}, \quad \gamma_2 = 1 - 2\gamma_1. \tag{2.3}$$

In general, the recursion

$$S_h^{[2k]} = S_{\gamma_1 h}^{[2k-2]} \circ S_{(1-2\gamma_1)h}^{[2k-2]} \circ S_{\gamma_1 h}^{[2k-2]}, \quad \text{with} \quad \gamma_1 = \frac{1}{2 - 2^{1/(2k-1)}} \tag{2.4}$$

can be used to get methods of arbitrarily high order $2k$ ($k \geq 2$) (Creutz and Gocksch 1989) starting from the Strang map $S_h^{[2]}$ (notice that such methods can be written in the form (2.1)). The price to be paid is the existence of large positive and negative coefficients γ_j and the great number of elementary flows in (2.4) for high orders. The alternative formed by the five maps composition (*quintuple jump*)

$$S_h^{[2k]} = S_{\gamma_1 h}^{[2k-2]} \circ S_{\gamma_1 h}^{[2k-2]} \circ S_{(1-4\gamma_1)h}^{[2k-2]} \circ S_{\gamma_1 h}^{[2k-2]} \circ S_{\gamma_1 h}^{[2k-2]}, \quad \gamma_1 = \frac{1}{4 - 4^{1/(2k-1)}} \tag{2.5}$$

also gives methods $S_h^{[2k]}$ of order $2k$ of the form (2.1) with relatively smaller coefficients γ_j but even larger numbers of elementary flows.

In general, other choices for the coefficients γ_j in (2.1) are more appropriate if we are interested in achieving orders ≥ 6 with a lower number of elementary flows and relatively small coefficients.

Condition (2.2), using the approach based on linear differential operators discussed in Section 1.2, can be derived as follows: the Lie transformation associated with the Strang map $S_h^{[2]}$, (1.16), can be written as $e^{h/2F_1}\, e^{hF_2}\, e^{h/2F_1} = e^{Y(h)}$, where

$$
\begin{aligned}
Y(h) &= \sum_{n=1}^{\infty} h^{2n-1} Y_{2n-1} \\
&:= \log(e^{h/2F_1}\, e^{hF_2}\, e^{h/2F_1}) \\
&= h(F_1+F_2) - \frac{h^3}{24}[F_1,[F_1,F_2]] - \frac{h^3}{12}[F_2,[F_1,F_2]] + \cdots,
\end{aligned}
$$

that is, $Y_1 = F_1 + F_2$, $Y_3 = -\frac{1}{24}[F_1,[F_1,F_2]] - \frac{1}{12}[F_2,[F_1,F_2]]$, and for each $n > 2$, Y_{2n-1} is a certain linear combination of $(2n-1)$-fold commutators of F_1 and F_2. In consequence, the Lie transformation $\Psi(h)$ of (2.1) formally satisfies

$$
\Psi(h) = e^{Y(\gamma_1 h)} \cdots e^{Y(\gamma_s h)}, \tag{2.6}
$$

so that $g(\psi_h(x)) = (\Psi(h)g)(x)$ for any $x \in \mathbb{R}^D$ and any smooth function $g\colon \mathbb{R}^D \longrightarrow \mathbb{R}$. It is straightforward to check that

$$
\begin{aligned}
\Psi(h) &= e^{\gamma_1 h Y_1 + \gamma_1^3 h^3 Y_3 + \cdots} \cdots e^{\gamma_s h Y_1 + \gamma_s^3 h^3 Y_3 + \cdots} \\
&= e^{h(\sum_{j=1}^{s}\gamma_j)(F_1+F_2)} + h^3\left(\sum_{j=1}^{s}\gamma_j^3\right) Y_3 + O(h^4).
\end{aligned} \tag{2.7}
$$

This shows that the composition (2.1) of Strang maps is of order at least three if condition (2.2) holds.

In fact, (2.7) is also true if in (2.1) the Strang map $S_h^{[2]}$ is replaced by any second-order time-symmetric map. Furthermore, the triple jump recursion (2.4) (resp. the quintuple jump recursion (2.5)) also gives rise to $2k$th-order maps $S_h^{[2k]}$ starting from an arbitrary time-symmetric second-order map $S_h^{[2]}$. Indeed, this is a consequence of the following four statements.

(1) Given an arbitrary near-identity map $\chi_h\colon \mathbb{R}^D \longrightarrow \mathbb{R}^D$ (i.e. $\chi_h(x) = x + O(h)$ as $h \longrightarrow 0$), there exists a series

$$
Y(h) = \sum_{n \geq 1} h^n Y_n
$$

of (first-order) differential operators acting on smooth functions such that, formally, $g(\chi_h(x)) = (e^{Y(h)} g)(x)$ for each $x \in \mathbb{R}^D$ and $g \in C^\infty(\mathbb{R}^D, \mathbb{R})$. Moreover, χ_h is an rth-order integrator for the ODE system $x' = f_1(x) + f_2(x)$ if and only if

$$
Y_1 = F_1 + F_2, \quad Y_n = 0 \quad \text{for } 2 \leq n \leq r. \tag{2.8}
$$

This statement can be proved as follows. Given a basic integrator $\chi_h\colon \mathbb{R}^D \longrightarrow \mathbb{R}^D$, consider the linear differential operators X_n ($n \geq 1$) acting on smooth functions $g \in C^\infty(\mathbb{R}^D, \mathbb{R})$ as

$$X_n g(y) = \frac{1}{n!} \frac{\mathrm{d}^n}{\mathrm{d}h^n}\bigg|_{h=0} g(\chi_h(y)), \quad y \in \mathbb{R}^D, \tag{2.9}$$

so that formally $g(\chi_h(x)) = (X(h)g)(x)$, where

$$X(h) = I + \sum_{n\geq 1} h^n X_n, \tag{2.10}$$

and I denotes the identity operator. Each X_n is an nth-order differential operator. Thus, the integrator χ_h is of order r if

$$X_n = \frac{1}{n!}(F_1 + F_2)^n, \quad 1 \leq n \leq r.$$

Now consider the series of differential operators

$$Y(h) = \sum_{n\geq 1} h^n Y_n := \log(X(h)) = \sum_{m\geq 1} \frac{(-1)^{m+1}}{m}(hX_1 + h^2 X_2 + \cdots)^m,$$

that is,

$$Y_n = \sum_{m\geq 1}^{n} \frac{(-1)^{m+1}}{m} \sum_{j_1+\cdots+j_m=n} X_{j_1}\cdots X_{j_m},$$

so that $X(h) = \exp(Y(h))$, and formally, $g(\chi_h(x)) = (\exp(Y(h))g(x)$. It can be shown that each Y_n is a first-order differential operator. Clearly, the basic integrator is of order r if (2.8) holds.

(2) The map χ_h is time-symmetric if and only if $Y(h) = hY_1 + h^3 Y_3 + \cdots$, which implies that time-symmetric methods *are necessarily of even order*. Indeed, for the adjoint integrator $\chi_h^* = \chi_{-h}^{-1}$, we obviously get $g(\chi_h^*(x)) = \mathrm{e}^{-Y(-h)} g(x)$. Hence χ_h is time-symmetric if and only if $-Y(-h) \equiv Y(h)$.

(3) If $S_h^{[2k-2]}$ is a time-symmetric integrator of order at least $2k-2$, then the composition

$$\psi_h = S_{\gamma_s h}^{[2k-2]} \circ S_{\gamma_{s-1} h}^{[2k-2]} \circ \cdots \circ S_{\gamma_1 h}^{[2k-2]} \tag{2.11}$$

is of order at least $2k-1$ if

$$\sum_{j=1}^{s} \gamma_j = 1, \quad \sum_{j=1}^{s} \gamma_j^{2k-1} = 0. \tag{2.12}$$

In fact, if $\mathrm{e}^{h(F_1+F_2)+h^{2k-1}Y_{2k-1}+\cdots}$ is the Lie transform of $S_h^{[2k-2]}$, then $g(\psi_h(x)) = (\Psi(h)g)(x)$ for any $x \in \mathbb{R}^D$ and any smooth function $g\colon \mathbb{R}^D \to \mathbb{R}$, where

$$\begin{aligned}\Psi(h) &= \mathrm{e}^{\gamma_1 h(F_1+F_2)+\gamma_1^{2k-1}h^{2k-1}Y_{2k-1}+O(h^{2k})} \cdots \mathrm{e}^{\gamma_s h(F_1+F_2)+\gamma_s^{2k-1}h^{2k-1}Y_{2k-1}+O(h^{2k})} \\ &= \mathrm{e}^{(\sum_{j=1}^{s}\gamma_j)(F_1+F_2)} + h^{2k-1}\left(\sum_{j=1}^{s}\gamma_j^{2k-1}\right)Y_{2k-1} + O(h^{2k}).\end{aligned}$$

(4) If $S_h^{[2k-2]}$ is a time-symmetric integrator and the sequence $(\gamma_1, \ldots, \gamma_s)$ is palindromic, in the sense that for all j

$$\gamma_{s-j+1} = \gamma_j,$$

then clearly the composition (2.11) is time-symmetric, and hence of even order.

2.1.2. *Composition of Lie–Trotter maps*

We could consider an analogous composition to (2.1), but this time with the Lie–Trotter scheme $\chi_h = \varphi_h^{[1]} \circ \varphi_h^{[2]}$ as the basic method. In that case, the Lie transform of χ_h is of the form $\mathrm{e}^{Y(h)}$ with

$$Y(h) = h(F_1 + F_2) + h^2 Y_2 + h^3 Y_3 + \cdots, \tag{2.13}$$

so that the Lie transform $\Psi(h)$ associated to the composition $\chi_{\gamma_s h} \circ \cdots \circ \chi_{\gamma_2 h} \circ \chi_{\gamma_1 h}$ is

$$\begin{aligned}\Psi(h) &= \mathrm{e}^{Y(\gamma_1 h)} \cdots \mathrm{e}^{Y(\gamma_s h)} \\ &= \mathrm{e}^{(\sum_{j=1}^{s} \gamma_j)(F_1+F_2)} + h^2 \left(\sum_{j=1}^{s} \gamma_j^2 \right) Y_2 + O(h^3).\end{aligned}$$

This shows that such a scheme is of order two if

$$\sum_{j=1}^{s} \gamma_j = 1, \quad \sum_{j=1}^{s} \gamma_j^2 = 0. \tag{2.14}$$

Obviously, such a system of equations does not admit real solutions.

The situation is different, however, if we compose χ_h with its adjoint $\chi_h^* = \varphi_h^{[2]} \circ \varphi_h^{[1]}$, that is,

$$\psi_h = \chi_{\alpha_{2s} h} \circ \chi^*_{\alpha_{2s-1} h} \circ \cdots \circ \chi_{\alpha_2 h} \circ \chi^*_{\alpha_1 h}. \tag{2.15}$$

If $\mathrm{e}^{Y(h)}$, with $Y(h)$ given by (2.13), is the Lie transform of χ_h, then the Lie transform of its adjoint χ_h^* is $\mathrm{e}^{-Y(-h)}$ and the Lie transform $\Psi(h)$ of (2.15) satisfies

$$\begin{aligned}\Psi(h) &= \mathrm{e}^{-Y(-\alpha_1 h)} \mathrm{e}^{Y(\alpha_2 h)} \cdots \mathrm{e}^{-Y(-\alpha_{2s-1} h)} \mathrm{e}^{Y(\alpha_{2s} h)} \\ &= \mathrm{e}^{(\sum_{j=1}^{2s} \alpha_j)(F_1+F_2)} + h^2 \left(\sum_{j=1}^{s} \left(\alpha_{2j}^2 - \alpha_{2j-1}^2 \right) \right) Y_2 + O(h^3),\end{aligned} \tag{2.16}$$

whence composition (2.15) is at least of order two if the coefficients satisfy

$$\sum_{j=1}^{2s} \alpha_j = 1, \quad \sum_{j=1}^{s} \left(\alpha_{2j}^2 - \alpha_{2j-1}^2 \right) = 0.$$

The argument above also holds when χ_h is a first-order integrator other than Lie–Trotter.

As mentioned earlier, the simplest situation corresponds to $s = 1$, in which case we recover the Strang splitting, $S_h^{[2]} = \chi_{h/2} \circ \chi^*_{h/2}$. In fact, the general scheme (2.15) can be rewritten as the splitting method

$$\psi_h = \varphi^{[1]}_{a_{s+1}h} \circ \varphi^{[2]}_{b_s h} \circ \varphi^{[1]}_{a_s h} \circ \cdots \circ \varphi^{[1]}_{a_2 h} \circ \varphi^{[2]}_{b_1 h} \circ \varphi^{[1]}_{a_1 h}, \tag{2.17}$$

where $a_1 = \alpha_1$, and for $j = 1, \ldots, s$,

$$a_{j+1} = \alpha_{2j} + \alpha_{2j+1}, \quad b_j = \alpha_{2j-1} + \alpha_{2j} \tag{2.18}$$

(with $\alpha_{2s+1} = 0$). Conversely, any integrator of the form (2.17) satisfying the condition $\sum_{j=1}^{s+1} a_j = \sum_{j=1}^{s} b_j$ can be expressed in the form (2.15), as shown in McLachlan (1995*b*).

Clearly, any splitting scheme (2.17) with a palindromic sequence of coefficients, that is, satisfying

$$a_{s-j+2} = a_j, \quad b_{s-j+1} = b_j \quad \text{for all } j,$$

is time-symmetric, and thus of even order. Written in the composition format (2.15), it is time-symmetric if

$$\alpha_{2s-j+1} = \alpha_j.$$

2.2. *Order conditions I: splitting schemes with the BCH formula*

In the analysis of the order conditions for splitting methods, and without loss of generality, we will consider the linear case (1.4), so that the treatment is essentially based on matrices, or more generally, on linear operators. Thus, the integrator (2.17) corresponds in this setting to the product of exponentials

$$\Psi(h) = \mathrm{e}^{a_{s+1}hF_1}\, \mathrm{e}^{b_s hF_2}\, \mathrm{e}^{a_s hF_1} \cdots \mathrm{e}^{a_2 hF_1}\, \mathrm{e}^{b_1 hF_2}\, \mathrm{e}^{a_1 hF_1}, \tag{2.19}$$

intended to approximate $\mathrm{e}^{h(F_1+F_2)}$.

In the case of systems of ODEs $x' = f_1(x) + f_2(x)$ with (not necessarily linear) vector fields f_1 and f_2, one may consider the Lie operators F_j of f_j, and compare the Lie transformation $\mathrm{e}^{h(F_1+F_2)}$ with the operator

$$\Psi(h) = \mathrm{e}^{a_1 hF_1}\, \mathrm{e}^{b_1 hF_2}\, \mathrm{e}^{a_2 hF_1} \cdots \mathrm{e}^{a_s hF_1}\, \mathrm{e}^{b_s hF_2}\, \mathrm{e}^{a_{s+1} hF_1} \tag{2.20}$$

associated with the map ψ_h (in the sense that $g(\psi_h(x)) = \Psi(h)g(x)$ for arbitrary smooth $g \in C^\infty(\mathbb{R}^D, \mathbb{R})$). Thus all the formulas derived below for the linear case can be applied to the Lie transformation (2.20) associated with the map (2.17) just by reversing the order of the sequence $(a_1, b_1, a_2, \ldots, a_s, b_s, a_{s+1})$ of coefficients of the splitting scheme. Alternatively, we may reverse the order of the products of the operators in all the expressions involved.

Equivalent results could also be obtained for the ODE case using the concept of *word series* and related techniques introduced in Murua and Sanz-Serna (2017). The Lie transform approach is more general, however, as it can be directly applied to the case of splitting methods for differential equations on manifolds, as discussed at the end of Section 4.1.

Generally speaking, the order conditions for a method of order r are systems of polynomial equations in the coefficients obtained by requiring that the Taylor expansions in the step size h of both the exact and numerical solution agree up to terms in h^r. A standard approach to obtain the order conditions of scheme (2.19) consists in formally using the Baker–Campbell–Hausdorff (BCH) formula to express $\Psi(h)$ as one exponential of a series of operators in powers of h, and finally to compare this series with $\mathrm{e}^{h(F_1+F_2)}$. In this way, we get

$$\begin{aligned}\log(\Psi(h)) &= h(w_1F_1 + w_2F_2) + h^2 w_{12}F_{12} + h^3(w_{122}F_{122} + w_{112}F_{112}) \\ &\quad + h^4(w_{1222}F_{1222} + w_{1122}F_{1122} + w_{1112}F_{1112}) + O(h^5), \qquad (2.21)\end{aligned}$$

where

$$\begin{aligned} F_{12} &= [F_1, F_2], & F_{122} &= [F_{12}, F_2], & F_{112} &= [F_1, F_{12}], \\ F_{1222} &= [F_{122}, F_2], & F_{1122} &= [F_1, F_{122}], & F_{1112} &= [F_1, F_{112}], \end{aligned}$$

and $w_1, w_2, w_{12}, w_{122}, \ldots$ are polynomials of homogeneous degree in the parameters a_j, b_j. In particular,

$$\begin{aligned} w_1 &= \sum_{i=1}^{s+1} a_i, \quad w_2 = \sum_{i=1}^{s} b_i, \quad w_{12} = \frac{1}{2}w_1w_2 - \sum_{1\le i<j\le s+1} b_i a_j, \\ w_{122} &= \frac{1}{12}w_1w_2^2 - \frac{1}{2}\sum_{1\le i\le j<k\le s} b_i a_j b_k, \\ w_{112} &= \frac{1}{12}w_1^2w_2 - \frac{1}{2}\sum_{1\le i<j\le k\le s+1} a_i b_j a_k, \end{aligned}$$

with $b_{s+1} = 0$. From (2.21), it is clear that a characterization of the order of the splitting scheme (2.17) is obtained by requiring the consistency conditions $w_1 = w_2 = 1$, that is,

$$\sum_{j=1}^{s+1} a_j = \sum_{j=1}^{s} b_j = 1 \qquad (2.22)$$

(ensuring that the scheme (2.17) is at least of order one) and $w_{12} = w_{122} = w_{112} = \cdots = 0$ up to the required order. The set of order conditions thus obtained will be independent in the general case if the operators F_1, F_2, F_{12}, F_{122}, $F_{112}, \ldots$ form a basis of the free Lie algebra in the alphabet $\{1, 2\}$. In (2.21) we considered the so-called *Lyndon basis*, associated to the set of Lyndon words in the alphabet $\{1, 2\}$:

$$\{1, 2, 12, 122, 112, 1222, 1122, 1112, \ldots\}. \qquad (2.23)$$

Table 2.1. Number of independent order conditions for general splitting methods, c_n, and for RKN-type splitting methods, d_n. The number m_n corresponds to the number of order conditions for compositions of a second-order time-symmetric method.

Order n	1	2	3	4	5	6	7	8	9	10	11
c_n	2	1	2	3	6	9	18	30	56	99	186
d_n	2	1	2	2	4	5	10	14	25	39	69
m_n	1	0	1	1	2	2	4	5	8	11	17

They are defined as follows (Reutenauer 1993): a word $\ell_1 \cdots \ell_m$ is a Lyndon word if $(\ell_1 \cdots \ell_k) < (\ell_{k+1} \cdots \ell_m)$ for each $1 \leq k < m$, where $<$ is the lexicographical order (i.e. the order used when ordering words in the dictionary) on the set of words in the alphabet $\{1, 2\}$. For instance, 112 is a Lyndon word, while neither 211 nor 121 are, as $2 \not< 11$ and as $12 \not< 1$, respectively.

The element of the basis associated to a Lyndon word $\ell_1 \cdots \ell_m$ with $m \geq 2$ is given as $F_{\ell_1 \cdots \ell_m} = [F_{\ell_1 \cdots \ell_n}, F_{\ell_{n+1} \cdots \ell_m}]$, where n is the smallest number such that both $\ell_1 \cdots \ell_n$ and $\ell_{n+1} \cdots \ell_m$ are themselves Lyndon words.

Casas and Murua (2009) have presented an efficient algorithm (based on the results in Murua 2006) for the BCH formula and related calculations in the Lyndon basis (and some other basis) that allows us to obtain (2.21) up to terms of arbitrarily high degree.

Of course, if another basis of the free Lie algebra in the alphabet $\{1, 2\}$ is used to expand $\log(\Psi(h))$ in (2.21), a different characterization of the order conditions will be obtained, with a different set of polynomial functions on $(a_1, b_1, \ldots, a_s, b_s, a_{s+1})$. In any case, the number of such independent conditions arising at each order n can be obtained just by determining the dimension of $\mathcal{L}_n(F_1, F_2)$, the linear span of all commutators containing n operators F_1, F_2. This number, denoted c_n, is given in Table 2.1; see Munthe-Kaas and Owren (1999) and McLachlan and Quispel (2002).

2.3. *Order conditions II: splitting schemes with Lyndon words*

Whereas the previous characterization of the order of the splitting scheme (2.19) allows us to easily get the number of order conditions, obtaining explicit expressions for the polynomials $w_{\ell_1 \cdots \ell_n}$ is much more difficult when the considered order increases. The following alternative characterization, based on the direct comparison of the power series expansions of $e^{h(F_1+F_2)}$ and (2.19), tries to ameliorate this difficulty.

2.3.1. Basic expansions and necessary order conditions

Both $e^{h(F_1+F_2)}$ and (2.19) admit an expansion in series indexed by the set

$$\mathcal{W} = \{1, 2, 11, 12, 21, 22, 111, 112, 121, 211, 122, \ldots\}$$

of words in the alphabet $\{1, 2\}$. More precisely, $e^{h(F_1+F_2)}$ can be expanded as

$$I + h\alpha_1 F_1 + h\alpha_2 F_2 + h^2\alpha_{11} F_1 F_1 + h^2\alpha_{12} F_1 F_2 + h^2\alpha_{21} F_2 F_1 + h^2\alpha_{22} F_2 F_2 + \cdots, \quad (2.24)$$

with $\alpha_{\ell_1\cdots\ell_n} = 1/n!$. As for (2.19), it can be expanded, for arbitrary s, as (2.24), where for each word $\ell_1 \cdots \ell_n \in \mathcal{W}$ with n letters, the corresponding coefficient

$$\alpha_{\ell_1\cdots\ell_n} = u_{\ell_1\cdots\ell_n}(a_1, b_1, \ldots, a_s, b_s, a_{s+1})$$

is a homogeneous polynomial of degree n in the variables $a_1, b_1, \ldots, a_s, b_s, a_{s+1}$.

It is straightforward to check that such polynomials satisfy the following relations, which allows us to compute them recursively.

- If $\ell_1 = \cdots = \ell_j = 1$ and $\ell_{j+1} \neq 1$ with $j \geq 1$,

$$u_{\ell_1\cdots\ell_n}(a_1, b_1, \ldots, a_s, b_s, a_{s+1}) = \sum_{k=0}^{j} u_{\ell_{k+1}\cdots\ell_n}(a_1, b_1, \ldots, a_s, b_s, 0)\,\frac{a_{s+1}^k}{k!},$$

$$u_{\ell_1\cdots\ell_n}(a_1, b_1, \ldots, a_s, b_s, 0) = u_{\ell_1\cdots\ell_n}(a_1, b_1, \ldots, a_s).$$

- If $\ell_1 = \cdots = \ell_j = 2$ and $\ell_{j+1} \neq 2$ with $j \geq 1$,

$$u_{\ell_1\cdots\ell_n}(a_1, b_1, \ldots, a_s, b_s, a_{s+1}) = \sum_{k=0}^{j} u_{\ell_{k+1}\cdots\ell_n}(a_1, b_1, \ldots, a_s)\,\frac{b_s^k}{k!}.$$

- If $\ell_n = \cdots = \ell_1 = 1$,

$$u_{\ell_1\cdots\ell_n}(a_1) = \frac{a_1^n}{n!}.$$

- If $\ell_1 \neq 1$,

$$u_{\ell_1\cdots\ell_n}(a_1) = 0.$$

In this way (2.19) is at least of order r if and only if the conditions

$$u_{\ell_1\cdots\ell_n}(a_1, b_1, \ldots, a_s, b_s, a_{s+1}) = \frac{1}{n!} \quad (2.25)$$

hold for each word $\ell_1 \cdots \ell_n$ with $n \leq r$ letters in the alphabet $\{1, 2\}$. For illustration, in Table 2.2 we explicitly give these conditions corresponding to words with up to two letters. Notice that (2.25) for the single-letter words 1 and 2 coincide with the consistency conditions (2.22).

However, such order conditions are not all independent. For instance, from Table 2.2, we can check that

$$u_1^2 = 2u_{11}, \quad u_2^2 = 2u_{22}, \quad u_1 u_2 = u_{12} + u_{21}.$$

Table 2.2. Conditions (2.25) corresponding to words with up to two indices.

Word	Condition	Word	Condition
1	$\sum_{j=1}^{s+1} a_j = 1$	2	$\sum_{j=1}^{s} b_j = 1$
11	$\frac{1}{2}\sum_{j=1}^{s+1}\frac{a_j^2}{2} + \sum_{1\le i<j\le s+1} a_i a_j = \frac{1}{2}$	12	$\sum_{1\le i<j\le s+1} b_i a_j = \frac{1}{2}$
21	$\sum_{1\le j\le i\le s} b_i a_j = \frac{1}{2}$	22	$\frac{1}{2}\sum_{j=1}^{s}\frac{b_j^2}{2} + \sum_{1\le j<i\le s} b_j b_i = \frac{1}{2}$

For a consistent method, $u_1 = u_2 = 1$, hence $u_{11} = \frac{1}{2}$, $u_{22} = \frac{1}{2}$, and

$$u_{12} - \frac{1}{2} = \frac{1}{2} - u_{21},$$

which implies that if $u_{12} = \frac{1}{2}$, then automatically $u_{21} = \frac{1}{2}$.

A complete characterization of the relations among the order conditions (2.25) will be obtained in Section 2.3.3 below. As a previous step, we obtain integral representations of both $e^{h(F_1+F_2)}$ and (2.19) , which in addition give useful expressions for the remainders of their truncated series expansions.

2.3.2. *Integral representation and remainders*

Consider the solution $Y(\tau)$ of the initial value problem

$$\frac{d}{d\tau}Y(\tau) = hA(\tau)Y(\tau), \quad Y(0) = I, \tag{2.26}$$

with $A(\tau) = d_1(\tau)F_1 + d_2(\tau)F_2$ and

$$(d_1(\tau), d_2(\tau)) = \begin{cases} (a_i, 0) & \text{if } \tau \in [2i-2, 2i-1], i \in \{1, \dots, s+1\}, \\ (0, b_i) & \text{if } \tau \in [2i-1, 2i], i \in \{1, \dots, s\}, \\ (0, 0) & \text{if } \tau > 2s+1. \end{cases} \tag{2.27}$$

It is straightforward to check that

$$Y(\tau) = \begin{cases} e^{(\tau-(2i-2))ha_{i+1}F_1}\, e^{b_i hF_2}\, e^{a_i hF_1} \cdots e^{a_2 hF_1}\, e^{b_1 hF_2}\, e^{a_1 hF_1}, \\ \qquad\qquad\qquad\qquad\qquad\qquad\qquad\qquad \tau \in [2i-2, 2i-1], \\ e^{(\tau-(2i-1))hb_i F_2}\, e^{a_i hF_1} \cdots e^{a_2 hF_1}\, e^{b_1 hF_2}\, e^{a_1 hF_1}, \quad \tau \in [2i-1, 2i]. \end{cases}$$

In particular, $\Psi(h) = Y(2s+1)$. The solution $Y(\tau)$ of (2.26) satisfies

$$Y(\tau) = I + h \int_0^{\tau} A(\tau_1) Y(\tau_1)\,\mathrm{d}\tau_1. \tag{2.28}$$

From that, we obtain

$$Y(\tau) = I + h \int_0^{\tau} A(\tau_1)\,\mathrm{d}\tau_1 + h^2 \int_0^{\tau}\int_0^{\tau_1} A(\tau_1)A(\tau_2)Y(\tau_2)\,\mathrm{d}\tau_2\,\mathrm{d}\tau_1,$$

and more generally

$$Y(\tau) = I + \sum_{n=1}^{m} h^n \int_0^{\tau}\int_0^{\tau_1}\cdots\int_0^{\tau_{n-1}} A(\tau_1)\cdots A(\tau_n)\,\mathrm{d}\tau_n\cdots\mathrm{d}\tau_1 + h^{m+1}\mathcal{R}_{m+1}(\tau,h), \tag{2.29}$$

where for each n the remainder $\mathcal{R}_n(\tau,h)$ satisfies

$$\mathcal{R}_n(\tau,h) = \int_0^{\tau}\int_0^{\tau_1}\cdots\int_0^{\tau_{n-1}} A(\tau_1)\cdots A(\tau_n)Y(\tau_n)\,\mathrm{d}\tau_n\cdots\mathrm{d}\tau_1. \tag{2.30}$$

By substituting $A(\tau_j) = d_1(\tau_j)F_1 + d_2(\tau_j)F_2$, then expanding all the products and taking constant linear operators out from integral signs, we obtain

$$Y(\tau) = I + \sum_{n=1}^{m} h^n \sum_{\ell_1,\dots,\ell_n\in\{1,2\}} \alpha_{\ell_1\cdots\ell_n}(\tau)\,F_{\ell_1}\cdots F_{\ell_n} + h^{m+1}\mathcal{R}_{m+1}(\tau,h),$$

where

$$\alpha_{\ell_1\cdots\ell_n}(\tau) = \int_0^{\tau}\int_0^{\tau_1}\cdots\int_0^{\tau_{n-1}} d_{\ell_1}(\tau_1)\cdots d_{\ell_n}(\tau_n)\,\mathrm{d}\tau_n\cdots\mathrm{d}\tau_1. \tag{2.31}$$

In particular, we have

$$\begin{aligned}\Psi(h) = I + \sum_{n=1}^{m} h^n &\sum_{\ell_1,\dots,\ell_n\in\{1,2\}} u_{\ell_1\cdots\ell_n}(a_1,b_1,\dots,a_s,b_s,a_{s+1})\,F_{\ell_1}\cdots F_{\ell_n}\\ &+ h^{m+1}\mathcal{R}_{m+1}(2s+1,h),\end{aligned}$$

where $u_{\ell_1\cdots\ell_n}(a_1,b_1,\dots,a_s,b_s,a_{s+1}) = \alpha_{\ell_1\cdots\ell_n}(2s+1)$, that is,

$$\begin{aligned}&u_{\ell_1\cdots\ell_n}(a_1,b_1,\dots,a_s,b_s,a_{s+1})\\ &\quad= \int_0^{2s+1}\int_0^{\tau_1}\cdots\int_0^{\tau_{n-1}} d_{\ell_1}(\tau_1)\cdots d_{\ell_n}(\tau_n)\,\mathrm{d}\tau_n\cdots\mathrm{d}\tau_1.\end{aligned}$$

We now consider (2.26) with $d_1(\tau) \equiv d_2(\tau) \equiv 1$. Clearly, in that case, $Y(\tau) = \mathrm{e}^{\tau h(F_1+F_2)}$, and

$$\begin{aligned}
&\mathrm{e}^{h(F_1+F_2)} \\
&= I + \sum_{n=1}^{m} h^n \int_0^1 \int_0^{\tau_1} \cdots \int_0^{\tau_{n-1}} (F_1+F_2)^n \, \mathrm{d}\tau_n \cdots \mathrm{d}\tau_1 + h^{m+1}\overline{\mathcal{R}}_{m+1}(1,h) \\
&= I + \sum_{n=1}^{m} h^n \sum_{\ell_1,\dots,\ell_n \in \{1,2\}} \frac{1}{n!} F_{\ell_1} \cdots F_{\ell_n} + h^{m+1}\overline{\mathcal{R}}_{m+1}(1,h),
\end{aligned}$$

where for each n, the remainder $\overline{\mathcal{R}}_n(\tau,h)$ is given by

$$\overline{\mathcal{R}}_n(\tau,h) = \int_0^{\tau} \int_0^{\tau_1} \cdots \int_0^{\tau_{n-1}} (F_1+F_2)^n \mathrm{e}^{\tau_n h(F_1+F_2)} \, \mathrm{d}\tau_n \cdots \mathrm{d}\tau_1 .$$

We finally arrive at the following expression for the local error of the splitting scheme (2.26):

$$\begin{aligned}
&\Psi(h) - \mathrm{e}^{h(F_1+F_2)} \\
&= \sum_{n=1}^{m} h^n \sum_{\ell_1,\dots,\ell_n \in \{1,2\}} \left(u_{\ell_1\cdots\ell_n}(a_1,b_1,\dots,a_s,b_s,a_{s+1}) - \frac{1}{n!} \right) F_{\ell_1} \cdots F_{\ell_n} \\
&\quad + h^{m+1}(\mathcal{R}_{m+1}(2s+1,h) - \overline{\mathcal{R}}_{m+1}(1,h)).
\end{aligned}$$

Hence, if the scheme is of order r (i.e. (2.25) holds for each word $(\ell_1,\dots,\ell_n)$ with $n \le r$ letters in the alphabet $\{1,2\}$), then

$$\Psi(h) - \mathrm{e}^{h(F_1+F_2)} = h^{n+1}(\mathcal{R}_{n+1}(2s+1,h) - \overline{\mathcal{R}}_n(1,h)).$$

2.3.3. *Iterated integrals and shuffle relations*

Iterated integrals of the form (2.31) were first considered and studied in Chen (1957). It is well known that the integration-by-parts formula gives (for an arbitrary integrable path $(d_1(\tau), d_2(\tau))$) the relations

$$\begin{aligned}
\alpha_{\ell_1}(\tau)\alpha_{\ell_2}(\tau) &= \alpha_{\ell_1\ell_2}(\tau) + \alpha_{\ell_2\ell_1}(\tau), \\
\alpha_{\ell_1}(\tau)\alpha_{\ell_2\ell_3}(\tau) &= \alpha_{\ell_1\ell_2\ell_3}(\tau) + \alpha_{\ell_2\ell_1\ell_3}(\tau) + \alpha_{\ell_2\ell_3\ell_1}(\tau),
\end{aligned}$$

and more generally,

$$\alpha_{\ell_1\cdots\ell_n}(\tau)\alpha_{\ell_{n+1}\cdots\ell_{n+m}}(\tau) = \sum_{\sigma \in \mathrm{Sh}(n,m)} \alpha_{\ell_{\sigma(1)}\cdots\ell_{\sigma(n+m)}}(\tau), \tag{2.32}$$

where $\mathrm{Sh}(n,m)$ is the set of the $(n+m)!/(n!m!)$ permutations σ of $(1,\dots,n+m)$ that are obtained by interleaving $(1,\dots,n)$ and $(n+1,\dots,n+m)$ while preserving their respective ordering.

It will be useful to interpret the relations (2.32) in terms of the so-called shuffle product of words: the shuffle product $\sqcup\!\sqcup$ of two words $\ell_1 \cdots \ell_n$ and $\ell_{n+1} \cdots \ell_{n+m}$ is defined as the following formal sum of words:

$$\ell_1 \cdots \ell_n ⧢ \ell_{n+1} \cdots \ell_{n+m} = \sum_{\sigma \in \mathrm{Sh}(n,m)} \ell_{\sigma(1)} \cdots \ell_{\sigma(n+m)}.$$

By extending $\alpha_w(\tau)$ linearly to the case where w is a linear combination of words, the relations (2.32) can be interpreted as

$$\alpha_w(\tau)\alpha_{w'}(\tau) = \alpha_{w ⧢ w'}$$

for arbitrary words $w = \ell_1 \cdots \ell_n$ and $w' = \ell_{n+1} \cdots \ell_{n+m}$ in the alphabet $\mathcal{A} = \{1, 2\}$.

The shuffle product ⧢ defines a commutative algebra (the so-called shuffle algebra) over the vector space of formal linear combinations of words in the alphabet $\mathcal{A}$. The shuffle algebra is freely generated by the set of *Lyndon words* (Reutenauer 1993).

The fact that the coefficients of the series expansions of both $\mathrm{e}^{h(F_1+F_2)}$ and $\Psi(h)$ satisfy the shuffle relations, together with the fact that the set of Lyndon words freely generate the shuffle algebra, implies that a set of independent conditions for a consistent splitting scheme to attain order r can be obtained by considering (2.25) for each Lyndon word $(\ell_1, \ldots, \ell_n)$ of length $n \leq r$.

2.4. *Order conditions III: splitting methods with Lyndon multi-indices*

Blanes *et al.* (2013*b*) have obtained yet another characterization of the order conditions in terms of explicitly given polynomial equations. We next describe this alternative formulation. To do that, we always assume that the consistency conditions (2.22) hold, so that the method is at least of first order. In this case, the polynomial equations are expressed in terms of the coefficients $b_1, \ldots, b_s$ and the coefficients $c_1, \ldots, c_s$ given by

$$c_i = \sum_{j=1}^{i} a_j, \quad i = 1, 2, \ldots, s. \tag{2.33}$$

We begin by rewriting (2.19) as

$$\begin{aligned}\Psi(h) &= \mathrm{e}^{a_{s+1}hF_1}\,\mathrm{e}^{b_shF_2}\,\mathrm{e}^{a_shF_1} \cdots \mathrm{e}^{b_1hF_2}\,\mathrm{e}^{a_1hF_1} \\ &= \mathrm{e}^{hF_1}\left(\prod_{j=s}^{1} \mathrm{e}^{-c_jhF_1}\,\mathrm{e}^{b_jhF_2}\,\mathrm{e}^{c_jhF_1}\right) \\ &= \mathrm{e}^{hF_1}\mathrm{e}^{b_shC(c_sh)} \cdots \mathrm{e}^{b_1hC(c_1h)},\end{aligned}$$

where

$$C(h) = \mathrm{e}^{-hF_1}\,F_2\,\mathrm{e}^{hF_1} = \sum_{n=1}^{\infty} h^{n-1}C_n = C_1 + hC_2 + h^2C_3 + h^3C_4 + \cdots, \tag{2.34}$$

with $C_1 = F_2$, and

$$C_k = \frac{1}{(k-1)}[C_{k-1}, F_1] \quad \text{for } k > 1.$$

Now the order of the scheme (2.17) is established by comparing the expansion in powers of h of $\mathrm{e}^{-hF_1}\Psi(h)$ with that of $\mathrm{e}^{-hF_1}\mathrm{e}^{h(F_1+F_2)}$.

Clearly, $Y(\tau) := \mathrm{e}^{-h\tau F_1}\mathrm{e}^{\tau h(F_1+F_2)}$ is the solution of (2.26) with $A(\tau) = C(\tau h)$. Since the solution $Y(\tau)$ of (2.26) admits the representation (2.29) with remainder (2.30), and $\mathrm{e}^{-hF_1}\mathrm{e}^{h(F_1+F_2)} = Y(1)$, we conclude that

$$\begin{aligned}\mathrm{e}^{-hF_1}\mathrm{e}^{h(F_1+F_2)} = I + \sum_{k=1}^{n} h^k \int_0^1 \int_0^{\tau_k} \cdots \int_0^{\tau_2} C(\tau_k h)\cdots C(\tau_1 h)\,\mathrm{d}\tau_1 \cdots \mathrm{d}\tau_k \\ + h^{n+1}\mathcal{R}_{n+1}(1,h),\end{aligned}$$

where for each k,

$$\mathcal{R}_k(1,h) = \int_0^1 \int_0^{\tau_k} \cdots \int_0^{\tau_2} C(\tau_k h)\cdots C(\tau_1 h) Y(\tau_1)\,\mathrm{d}\tau_1 \cdots \mathrm{d}\tau_k .$$

By substitution of (2.34), we obtain that for each k,

$$\begin{aligned}&h^k \int_0^1 \int_0^{\tau_k} \cdots \int_0^{\tau_2} C(\tau_k h)\cdots C(\tau_1 h)\,\mathrm{d}\tau_1 \cdots \mathrm{d}\tau_k \\ &\quad = \sum_{i_1,\ldots,i_k \geq 1} h^{i_1+\cdots+i_k}\left(\int_0^1 \int_0^{\tau_k} \cdots \int_0^{\tau_2} \tau_1^{i_1-1}\cdots\tau_k^{i_k-1}\,\mathrm{d}\tau_1 \cdots \mathrm{d}\tau_k\right) C_{i_k}\cdots C_{i_1} \\ &\quad = \sum_{i_1,\ldots,i_k \geq 1} \frac{h^{i_1+\cdots+i_k}}{(i_1+\cdots+i_k)\cdots(i_1+i_2)i_1} C_{i_k}\cdots C_{i_1}.\end{aligned}$$

As for $\mathrm{e}^{-hF_1}\,\Psi(h)$,

$$\begin{aligned}&\mathrm{e}^{-hF_1}\,\Psi(h) \\ &= \prod_{j=s}^{1} \mathrm{e}^{b_j h C(c_j h)} = \prod_{j=s}^{1}\left(I + \sum_{k\geq 1}\frac{h^k b_j^k}{k!} C(c_j h)^k\right) \\ &= I + \sum_{k\geq 1}\frac{h^k}{k!}\sum_{j=1}^{s} b_j^k C(c_j h)^k + \sum_{k_1,k_2\geq 1}\frac{h^{k_1+k_2}}{k_1!k_2!}\sum_{1\leq j_1<j_2\leq s} b_{j_1}^{k_1} b_{j_2}^{k_2} C(c_{j_2}h)^{k_2} C(c_{j_1}h)^{k_1} \\ &\quad + \sum_{k_1,k_2,k_3\geq 1}\frac{h^{k_1+k_2+k_3}}{k_1!k_2!k_3!}\sum_{1\leq j_1<j_2<j_3\leq s} b_{j_1}^{k_1} b_{j_2}^{k_2} b_{j_3}^{k_3} C(c_{j_3}h)^{k_3} C(c_{j_2}h)^{k_2} C(c_{j_1}h)^{k_1} + \cdots \\ &= I + \sum_{k\geq 1} h^k \sum_{1\leq j_1\leq\cdots\leq j_k\leq s}\frac{b_{j_1}\cdots b_{j_k}}{\sigma(j_1,\ldots j_k)} C(c_{j_k}h)\cdots C(c_{j_1}h),\end{aligned} \tag{2.35}$$

where

$$\begin{aligned} \sigma(j_1,\ldots,j_k) &= 1 && \text{if } j_1 < \cdots < j_k, \\ \sigma(j_1,\ldots,j_k) &= \ell!\,\sigma(j_{\ell+1},\ldots,j_k) && \text{if } j_1 = \cdots = j_\ell < j_{\ell+1} \leq \cdots \leq j_k. \end{aligned}$$

Since

$$h^k C(c_{j_k}h)\cdots C(c_{j_1}h) = \sum_{i_1,\ldots,i_k\geq 1} h^{i_1+\cdots+i_k} c_{j_1}^{i_1-1}\cdots c_{j_k}^{i_k-1} C_{i_k}\cdots C_{i_1},$$

we arrive at

$$\mathrm{e}^{-hF_1}\Psi(h) = I + \sum_{k\geq 1}\sum_{i_1,\ldots,i_k\geq 1} h^{i_1+\cdots+i_k} v_{i_1,\ldots,i_k}(b_1,c_1,\ldots,b_s,c_s)\,C_{i_k}\cdots C_{i_1}, \tag{2.36}$$

where

$$v_{i_1,\ldots,i_k}(b_1,c_1,\ldots,b_s,c_s) = \sum_{1\leq j_1\leq\cdots\leq j_k\leq s} \frac{b_{j_1}\cdots b_{j_k}}{\sigma(j_1,\ldots,j_k)} c_{j_1}^{i_1-1}\cdots c_{j_k}^{i_k-1}. \tag{2.37}$$

In this way, a consistent splitting method is at least of order r if and only if

$$\sum_{1\leq j_1\leq\cdots\leq j_k\leq s} \frac{b_{j_1}\cdots b_{j_k}}{\sigma(j_1,\ldots,j_k)} c_{j_1}^{i_1-1}\cdots c_{j_k}^{i_k-1} = \frac{1}{(i_1+\cdots+i_k)\cdots(i_1+i_2)i_1} \tag{2.38}$$

holds for each multi-index $(i_1,\ldots,i_k)$ such that $i_1+\cdots+i_k \leq r$. For illustration, in Table 2.3 we give explicit conditions (2.38) corresponding to multi-indices with up to three indices.

However, such order conditions are not independent. The situation is very similar to that of the previous subsection: instead of series indexed by the set of words in the alphabet $\{1,2\}$, now they are indexed by the set of words in the alphabet $\mathbb{N} = \{1,2,3,4,\ldots\}$. To distinguish the words of both sets, we will keep referring to the words in the alphabet $\mathbb{N}$ as multi-indices, and will write them as $(i_1,\ldots,i_k)$ instead of $i_1\cdots i_k$. Analogously to the previous subsection, the corresponding coefficients of the series expansions of both $\mathrm{e}^{-hF_1}\Psi(h)$ and $\mathrm{e}^{-hF_1}\mathrm{e}^{h(F_1+F_2)}$ satisfy the shuffle relations

$$v_{i_1,\ldots,i_n} v_{i_{n+1},\ldots,i_{n+m}} = \sum_{\sigma\in\mathrm{Sh}(n,m)} v_{i_{\sigma(1)},\ldots,i_{\sigma(n+m)}}. \tag{2.39}$$

This can be seen by showing that such coefficients can be written in both cases as iterated integrals. From the discussion above, we already know that, for each multi-index $(i_1,\ldots,i_k)$, the coefficients

$$\frac{1}{(i_1+\cdots+i_k)\cdots(i_1+i_2)i_1} = \int_0^1\int_0^{\tau_k}\cdots\int_0^{\tau_2} \tau_1^{i_1-1}\cdots\tau_k^{i_k-1}\,\mathrm{d}\tau_1\cdots\mathrm{d}\tau_k$$

of the series expansion of $\mathrm{e}^{-hF_1}\mathrm{e}^{h(F_1+F_2)}$ are indeed iterated integrals.

We next show that the coefficients (2.37) of the series expansion (2.36) can also be defined as iterated integrals.

Table 2.3. Conditions (2.38) for multi-indices with up to three indices.

Multi-index	Condition
(i)	$\sum_{j=1}^{s} b_j\, c_j^{i-1} = \frac{1}{i}$
(i_1, i_2)	$\frac{1}{2}\sum_{j=1}^{s} b_j^2 c_j^{i_1+i_2-2} + \sum_{1\le j_1<j_2\le s} b_{j_1} b_{j_2} c_{j_1}^{i_1-1} c_{j_2}^{i_2-1} = \frac{1}{(i_1+i_2)i_1}$
(i_1, i_2, i_3)	$\frac{1}{6}\sum_{j=1}^{s} b_j^3 c_j^{i_1+i_2+i_3-3} + \frac{1}{2}\sum_{1\le j_1<j_3\le s} b_{j_1}^2 b_{j_3} c_{j_1}^{i_1+i_2-2} c_{j_3}^{i_3-1} + \frac{1}{2}\sum_{1\le j_1<j_2\le s} b_{j_1} b_{j_2}^2 c_{j_1}^{i_1-1} c_{j_2}^{i_2+i_3-2} + \sum_{1\le j_1<j_2<j_3\le s} b_{j_1} b_{j_2} b_{j_3} c_{j_1}^{i_1-1} c_{j_2}^{i_2-1} c_{j_2}^{i_3-1} = \frac{1}{(i_1+i_2+i_3)(i_1+i_2)i_1}$

We begin by showing that the infinite series expansion (2.35) of $\mathrm{e}^{-hF_1}\,\Psi(h)$ can be represented as a truncated series plus a remainder. For that purpose, we consider (2.26) with $A(\tau)$ defined as follows:

$$A(\tau) = \begin{cases} b_i C(hc_i) & \text{if } \tau \in [i-1, i],\ i \in \{1, \dots, s\}, \\ 0 & \text{if } \tau > s. \end{cases} \tag{2.40}$$

In that case, if $\tau \in [i, i+1]$ with $i < s$,

$$Y(\tau) = \mathrm{e}^{(\tau-i)hb_{i+1}C(hc_{i+1})}\mathrm{e}^{hb_i C(hc_i)} \cdots \mathrm{e}^{hb_1 C(hc_1)},$$

and in particular $\mathrm{e}^{-hF_1}\,\Psi(h) = Y(s)$. Since the solution $Y(\tau)$ of (2.26) admits the representation (2.29) with remainder (2.30), then

$$\mathrm{e}^{-hF_1}\Psi(h) = I + \sum_{k=1}^{n} h^k \int_0^s \int_0^{\tau_k} \cdots \int_0^{\tau_2} A(\tau_k)A(\tau_{k-1})\cdots A(\tau_1)\,\mathrm{d}\tau_1 \cdots \mathrm{d}\tau_k$$
$$+ h^{n+1}\mathcal{R}_{n+1}(s, h),$$

where for each n,

$$\mathcal{R}_n(s, h) = \int_0^s \int_0^{\tau_n} \cdots \int_0^{\tau_2} A(\tau_n)A(\tau_{n-1})\cdots A(\tau_1)Y(\tau_1)\,\mathrm{d}\tau_1 \cdots \mathrm{d}\tau_n.$$

By comparison with (2.35), we conclude that

$$\sum_{1\le j_1\le\cdots\le j_k\le s} \frac{b_{j_1}\cdots b_{j_k}}{\sigma(j_1,\ldots,j_k)} C(c_{j_k}h)\cdots C(c_{j_1}h)$$
$$= \int_0^s \int_0^{\tau_k} \cdots \int_0^{\tau_2} A(\tau_k)A(\tau_{k-1})\cdots A(\tau_1)\, \mathrm{d}\tau_1 \cdots \mathrm{d}\tau_k.$$

Now, $A(\tau)$ can be written as

$$A(\tau) = d_1(\tau)C_1 + hd_2(\tau)C_2 + h^2 d_3(\tau)C_3 + \cdots$$

where $d_j(\tau) = b_i c_i^{j-1}$ if $\tau \in [i, i+1]$ with $i < s$. Proceeding as in the previous section, we obtain

$$v_{i_1,\ldots,i_k}(b_1,c_1,\ldots,b_s,c_s) = \int_0^s \int_0^{\tau_1} \cdots \int_0^{\tau_{k-1}} d_{i_1}(\tau_1)\cdots d_{i_k}(\tau_k)\, \mathrm{d}\tau_k \cdots \mathrm{d}\tau_1, \tag{2.41}$$

which implies that the shuffle relations (2.39) hold for the polynomials $v_{i_1,\ldots,i_k}$.

As in the previous subsection, a set of independent conditions that imply the order conditions (2.38) can be obtained by considering (2.38) for each Lyndon multi-index $(i_1,\ldots,i_k)$ such that $1 < i_1 + \cdots + i_k \le r$. Here, we exclude the multi-index (1) as in that case (2.38) coincides with the second equality in the consistency condition (2.22). For instance, the subset of Lyndon multi-indices $(i_1,\ldots,i_k)$ such that $1 < i_1 + \cdots + i_k \le 5$ is

$$\{(2),(3),(4),(5),(1,2),(1,3),(1,4),(2,3),(1,1,2),(1,1,3),(1,2,2),(1,1,1,2)\}.$$

For time-symmetric splitting methods, a set of independent order conditions will be obtained by considering (2.38) restricted to Lyndon multi-indices $(i_1,\ldots,i_k)$ with odd weight $i_1 + \cdots + i_k$. For instance, the subset of Lyndon multi-indices $(i_1,\ldots,i_k)$ such that $1 < i_1 + \cdots + i_k \le 5$ with odd weight $i_1 + \cdots + i_k$ is

$$\{(3),(5),(1,2),(1,4),(2,3),(1,1,3),(1,2,2),(1,1,1,2)\}.$$

Notice that the treatment carried out in this subsection may also be formally applied when F_1 is an unbounded operator. In that case, however, we have to get rigorous estimates of the remainders to prove stability and convergence of the corresponding schemes, as is done in Thalhammer (2008), for example.

2.5. *Order conditions IV: composition methods with Lyndon multi-indices*

We now turn our attention to compositions (2.1) of a basic second-order time-symmetric scheme $S_h^{[2]}$ with appropriate coefficients $\gamma_1,\ldots,\gamma_s$ chosen to achieve higher orders. Of course, a set of conditions that guarantee that the scheme (2.1) attains a given order can be obtained by rewriting that composition in terms of basic maps, such as (2.19), and using the characterization of the order of the splitting method (2.19) described in Section 2.4. The corresponding parameters a_j, b_j can be obtained in terms of $\gamma_1, \ldots, \gamma_s$ as follows: $a_1 = \gamma_1/2$, and

$$a_{j+1} = \frac{\gamma_j + \gamma_{j+1}}{2}, \qquad b_j = \gamma_j \quad \text{for } j = 1, \ldots, s,$$

with $\gamma_{s+1} = 0$. However, the resulting polynomial equations for any given order r, once written in terms of the coefficients $\gamma_1, \ldots, \gamma_s$, are no longer independent. An alternative formulation of the order of (2.1) in terms of explicit independent algebraic equations in the coefficients $\gamma_1, \ldots, \gamma_s$ will be presented in Section 2.5.2 below. This characterization is based on the treatment of the more general composition (2.15), which is treated next.

2.5.1. Order conditions of compositions of a basic method and its adjoint

The composition (2.15) is at least of order r if $\Psi(h) - \exp(h(F_1 + F_2)) = O(h^{r+1})$, where $\Psi(h)$ is the associated Lie transformation (2.16).

To get the series expansion of $\Psi(h)$, we first consider the expansion in powers of h of the Lie transformation $\mathcal{X}(h) = I + hX_1 + h^2X_2 + \cdots$ associated to the basic integrator χ_h. If χ_h is the Lie–Trotter scheme, $\chi_h = \varphi_h^{[1]} \circ \varphi_h^{[2]}$, then

$$\mathcal{X}(h) = \mathrm{e}^{hF_2}\mathrm{e}^{hF_1} = I + h(F_1 + F_2) + h^2\left(\frac{1}{2}F_1^2 + F_2F_1 + \frac{1}{2}F_2^2\right) + \cdots.$$

To deal with the most general problem, however, from now on we only assume that χ_h is a smooth consistent integrator, so that $X_1 = F_1 + F_2$, and each X_n can be defined so that for each smooth function g, $X_n g$ is a new smooth function given by (2.9).

Let us consider $\Psi_0(h) = I$, and for each $j \geq 1$,

$$\begin{aligned} \Psi_{2j-1}(h) &= \mathcal{X}(-\alpha_1 h)^{-1}\mathcal{X}(\alpha_2 h)\cdots\mathcal{X}(-\alpha_{2j-1}h)^{-1}, \\ \Psi_{2j}(h) &= \mathcal{X}(-\alpha_1 h)^{-1}\mathcal{X}(\alpha_2 h)\cdots\mathcal{X}(-\alpha_{2j-1}h)^{-1}\mathcal{X}(\alpha_{2j}h), \end{aligned} \tag{2.42}$$

so that in particular, $\Psi(h) = \Psi_{2s}(h)$. Notice that

$$\begin{aligned} \mathcal{X}(-h)^{-1} &= I + \sum_{k\geq 1}(-1)^k(-hX_1 + h^2X_2 - h^3X_3 + \cdots)^k \\ &= I + hX_1 + h^2(X_1^2 - X_2) + h^3(X_1^3 - X_1X_2 - X_2X_1 + X_3) + \cdots, \end{aligned}$$

which implies that for each $k \geq 1$ there exist polynomials $w_{i_1,\ldots,i_m}(\alpha_1, \ldots, \alpha_k)$ on the coefficients $\alpha_1, \ldots, \alpha_k$ such that

$$\Psi_k(h) = I + \sum_{n\geq 1} h^n \sum_{m\geq 1} \sum_{i_1+\cdots+i_m=n} w_{i_1,\ldots,i_m}(\alpha_1, \ldots, \alpha_k)\, X_{i_1} \cdots X_{i_m}. \tag{2.43}$$

We next determine the polynomial coefficients $w_{i_1,\ldots,i_m}(\alpha_1, \ldots, \alpha_k)$ recursively from the relations $\Psi_1(h) = \mathcal{X}(-\alpha_{2j-1}h)^{-1}$ and

$$\Psi_{2j-1}(h) = \Psi_{2j-2}(h)\mathcal{X}(-\alpha_{2j-1}h)^{-1}, \quad \Psi_{2j}(h) = \Psi_{2j-1}(h)\mathcal{X}(\alpha_{2j}h),$$

or equivalently, $\Psi_1(h)\mathcal{X}(-\alpha_1 h) = I$ and

$$\Psi_{2j-1}(h)\mathcal{X}(-\alpha_{2j-1}h) = \Psi_{2j-2}(h), \quad \Psi_{2j}(h) = \Psi_{2j-1}(h)\mathcal{X}(\alpha_{2j}h). \tag{2.44}$$

Specifically, for arbitrary coefficients $w_{i_1,\dots,i_m}$ and λ we have

$$\begin{aligned}\Bigg(I + \sum_{n\geq 1} h^n \sum_{m\geq 1} \sum_{i_1+\cdots+i_m=n} w_{i_1,\dots,i_m} X_{i_1}\cdots X_{i_m}\Bigg)\mathcal{X}(\lambda h) \\ = I + \sum_{i\geq 1} h^i (w_i + \lambda^i) X_n \\ + \sum_{n\geq 1} h^n \sum_{m\geq 2} \sum_{i_1+\cdots+i_m=n} (w_{i_1,\dots,i_m} + \lambda^{i_m} w_{i_1,\dots,i_{m-1}}) X_{i_1}\cdots X_{i_m},\end{aligned}$$

so that, taking this expression into account, (2.44) leads to the following identities:

$$\begin{aligned} w_i(\alpha_1) &= -(-\alpha_1)^i, \\ w_i(\alpha_1,\dots,\alpha_{2j-1}) &= w_i(\alpha_1,\dots,\alpha_{2j-2}) - (-\alpha_{2j-1})^i, \\ w_i(\alpha_1,\dots,\alpha_{2j}) &= w_i(\alpha_1,\dots,\alpha_{2j-1}) + \alpha_{2j}^i, \\ w_{i_1,\dots,i_m}(\alpha_1) &= -(-\alpha_1)^{i_m} w_{i_1,\dots,i_{m-1}}(\alpha_1), \\ w_{i_1,\dots,i_m}(\alpha_1,\dots,\alpha_{2j-1}) &= w_{i_1,\dots,i_m}(\alpha_1,\dots,\alpha_{2j-2}) \\ &\quad - (-\alpha_{2j-1})^{i_m} w_{i_1,\dots,i_{m-1}}(\alpha_1,\dots,\alpha_{2j-1}), \\ w_{i_1,\dots,i_m}(\alpha_1,\dots,\alpha_{2j}) &= w_{i_1,\dots,i_m}(\alpha_1,\dots,\alpha_{2j-1}) \\ &\quad + \alpha_{2j}^{i_m} w_{i_1,\dots,i_{m-1}}(\alpha_1,\dots,\alpha_{2j-1}). \end{aligned}$$

Clearly, (2.43) holds for the coefficients $w_{i_1,\dots,i_m}(\alpha_1,\dots,\alpha_k)$ determined by the relations above. Equivalently, the functions $w_{i_1,\dots,i_m}$ can be defined as

$$\begin{aligned} w_i(\alpha_1,\dots,\alpha_{2\ell}) &= \sum_{j=1}^{s} \left(\alpha_{2j}^i - (-\alpha_{2j-1})^i\right), \\ w_{i_1,\dots,i_m}(\alpha_1,\dots,\alpha_{2\ell}) &= \sum_{j=1}^{s} \left(\alpha_{2j}^{i_m} - (-\alpha_{2j-1})^{i_m}\right) w_{i_1,\dots,i_{m-1}}(\alpha_1,\dots,\alpha_{2j-1}). \end{aligned} \tag{2.45}$$

Notice that

$$w_{i_1,\dots,i_m}(\alpha_1,\dots,\alpha_k,0,\dots,0) = w_{i_1,\dots,i_m}(\alpha_1,\dots,\alpha_k),$$

as expected from (2.42), (2.43) and $\mathcal{X}(0) = I$.

Comparing the series expansion of $\Psi(h) = \Psi_{2s}(h)$ with

$$\exp(h\,(F_1 + F_2)) = \exp(h\,X_1) = I + \sum_{n\geq 1} \frac{h^n}{n!} X_1^n,$$

we finally conclude that the scheme (2.15) is of order at least r if and only if

$$w_{i_1,\ldots,i_m}(\alpha_1,\ldots,\alpha_{2s}) = \begin{cases} \dfrac{1}{m!} & \text{if } (i_1,\ldots,i_m) = (\overbrace{1,\ldots,1}^{m}), \\ 0 & \text{otherwise,} \end{cases} \tag{2.46}$$

for each multi-index $(i_1,\ldots,i_k)$ such that $1 \le i_1 + \cdots + i_k \le r$.

Furthermore, the order conditions (2.46) are not all independent. For instance, it is straightforward to check from (2.45) that $w_{i_1,i_2} + w_{i_2,i_1} + w_{i_1+i_2} = w_{i_1}w_{i_2}$ for arbitrary indices i_1, i_2. In particular, $2w_{1,1} + w_2 = w_1^2$, which implies that if the order conditions (2.46) for the multi-indices (1) and (2) are satisfied, then the condition for the multi-index $(1,1)$ is automatically fulfilled. Actually, such dependences are similar to the shuffle relations (2.32) that hold for the coefficients $v_{i_1,\ldots,i_m}$ considered in Section 2.4. Indeed,

$$w_{i_1,\ldots,i_n} w_{i_{n+1},\ldots,i_{n+m}} = \sum_{\sigma\in \mathrm{Sh}(n,m)} w_{i_{\sigma(1)},\ldots,i_{\sigma(n+m)}} + \cdots, \tag{2.47}$$

where $\cdots$ refers to sums of products of coefficients corresponding to multi-indices with $m-1$ or fewer indices. In fact, as shown in Chartier and Murua (2009), the dependences (2.47) are directly related to the quasi-shuffle product $*$ on the linear span of multi-indices introduced in Hoffman (2000), and due to such dependences, it is enough to consider (2.46) for Lyndon multi-indices $(i_1,\ldots,i_m)$ such that $i_1+\cdots+i_k \le r$. That is, the scheme (2.15) is of order at least r if $\alpha_1+\cdots+\alpha_{2s} = 1$ and for each Lyndon multi-index $(i_1,\ldots,i_k)$ such that $1 < i_1 + \cdots + i_k \le r$,

$$w_{i_1,\ldots,i_m}(\alpha_1,\ldots,\alpha_{2s}) = 0. \tag{2.48}$$

In particular, a method of order three must satisfy, besides consistency, the conditions $w_2 = w_3 = w_{1,2} = 0$.

This provides an alternative characterization of the order conditions of general splitting schemes (2.17) in terms of the coefficients α_j obtained from the method parameters a_j, b_j from (2.18). Most importantly, this also allows us to characterize the order of the scheme (2.1) obtained by composing Strang maps. This will be presented next in Section 2.5.2.

2.5.2. Explicit characterization of the order conditions of scheme (2.1)

Murua and Sanz-Serna (1999) have obtained such an explicit characterization in terms of a set of polynomials indexed by certain sets of rooted trees decorated by the set of odd positive integers. We now describe a related formulation in terms of polynomials indexed by the set of Lyndon multi-indices with odd indices based on the formalism developed in the previous subsection for the more general composition (2.15).

We begin by considering (2.1) as the particular case of (2.15) with

$$\alpha_{2j-1} = \alpha_{2j} = \gamma_j/2 \quad \text{for } j = 1,\ldots,s. \tag{2.49}$$

For each multi-index $(i_1, \ldots, i_k)$, we define the function $u_{i_1,\ldots,i_m}$ on the set of finite sequences $(\gamma_1, \ldots, \gamma_s)$ of real numbers as follows:

$$u_{i_1,\ldots,i_m}(\gamma_1, \ldots, \gamma_s) = 2^{i_1+\cdots+i_m-m} w_{i_1,\ldots,i_m}(\alpha_1, \ldots, \alpha_{2s}), \tag{2.50}$$

with $\alpha_{2j-1} = \alpha_{2j} = \gamma_j/2$. Clearly, scheme (2.1) is of order at least r if and only if $\gamma_1 + \cdots + \gamma_s = 1$ and

$$u_{i_1,\ldots,i_m}(\gamma_1, \ldots, \gamma_s) = 0 \tag{2.51}$$

for each Lyndon multi-index $(i_1, \ldots, i_k)$ such that $1 < i_1 + \cdots + i_k \leq r$. However, by definition, $u_{i_1,\ldots,i_m}(\gamma_1, \ldots, \gamma_s) \equiv 0$ if i_m is even. Moreover, for any multi-index $(i_1, \ldots, i_k)$ with some even index, (2.51) holds provided that it holds for every Lyndon multi-index with fewer indices.

Therefore, scheme (2.1) is of order at least r if and only if $\gamma_1 + \cdots + \gamma_s = 1$ and (2.51) for each Lyndon multi-index $(i_1, \ldots, i_k)$ with odd indices such that $1 < i_1 + \cdots + i_k \leq r$. For instance, the set of Lyndon multi-indices $(i_1, \ldots, i_k)$ of odd indices such that $1 < i_1 + \cdots + i_k \leq 7$ is

$$\{(3), (5), (7), (1,3), (1,5), (1,1,3), (1,1,5), (1,3,3), (1,1,1,3), (1,1,1,1,3)\}.$$

The resulting number of order conditions, denoted as m_n, is gathered in Table 2.1.

For multi-indices $(i_1, \ldots, i_m)$ with odd indices, the functions $u_{i_1,\ldots,i_m}$ can be written more explicitly as follows:

$$u_i(\gamma_1, \ldots, \gamma_s) = \sum_{j=1}^{s} \gamma_j^i, \tag{2.52}$$

$$u_{i_1,i_2}(\gamma_1, \ldots, \gamma_s) = \sum_{j_2=1}^{s} \gamma_{j_2}^{i_2} \sum_{j_1=1}^{j_2*} \gamma_{j_1}^{i_1}, \tag{2.53}$$

$$u_{i_1,i_2,i_3}(\gamma_1, \ldots, \gamma_s) = \sum_{j_3=1}^{s} \gamma_{j_3}^{i_3} \sum_{j_2=1}^{j_3*} \gamma_{j_2}^{i_2} \sum_{j_1=1}^{j_2*} \gamma_{j_1}^{i_1}, \tag{2.54}$$

$$u_{i_1,i_2,i_3,i_4}(\gamma_1, \ldots, \gamma_s) = \sum_{j_4=1}^{s} \gamma_{j_4}^{i_4} \sum_{j_3=1}^{j_4*} \gamma_{j_3}^{i_3} \sum_{j_2=1}^{j_3*} \gamma_{j_2}^{i_2} \sum_{j_1=1}^{j_2*} \gamma_{j_1}^{i_1}, \tag{2.55}$$

and so on. Here, as in Murua and Sanz-Serna (1999), we have used the notation

$$\sum_{j=1}^{k*} A_j = \frac{A_k}{2} + \sum_{j=1}^{k-1} A_j$$

for $A_1, \ldots, A_k \in \mathbb{R}$.

For time-symmetric integration schemes (2.1), a set of independent conditions for even order r will be obtained by considering (2.51) restricted to Lyndon multi-indices $(i_1, \ldots, i_k)$ of odd indices and odd weight $i_1 + \cdots + i_k < r$. For instance, for order eight, we only need to consider the following subset of Lyndon multi-indices:

$$\{(3), (5), (7), (1, 1, 3), (1, 1, 5), (1, 3, 3), (1, 1, 1, 1, 3)\}.$$

The above characterization of the order of the scheme (2.1) is also true in the more general case where the Strang map $S_h^{[2]}$ is replaced by an arbitrary time-symmetric integrator of order 2ℓ. In that case, only Lyndon multi-indices with indices from the set $\{1, 2\ell + 1, 2\ell + 3, 2\ell + 5, \dots\}$ have to be taken into account.

2.6. *Negative time steps*

Splitting and composition methods of order $r \geq 3$ necessarily involve some negative coefficients. This can already be observed in the simple triple jump scheme (2.3), and in fact has been established as a general theorem by Goldman and Kaper (1996), Sheng (1989) and Suzuki (1991). A simple proof can be obtained as follows (Blanes and Casas 2005): given the existing relationship between the splitting method (2.17) and the composition (2.15), it is clear that any splitting scheme of order $r \geq 3$ has to verify the condition $w_3 = 0$, where, by virtue of (2.45),

$$w_3(\alpha_1, \dots, \alpha_{2s}) = \sum_{j=1}^{s} \left(\alpha_{2j}^3 - (-\alpha_{2j-1})^3\right), \tag{2.56}$$

where the coefficients α_j are related to a_j, b_j via (2.18). Since, for all $x, y \in \mathbb{R}$, it is true that $x^3 + y^3 < 0$ implies $x + y < 0$, then there must exist some $j \in \{1, \dots, s\}$ in the sum of (2.56) such that

$$\alpha_{2j-1}^3 + \alpha_{2j}^3 < 0 \quad \text{and thus} \quad \alpha_{2j-1} + \alpha_{2j} = b_j < 0.$$

Obviously, we can also write (by taking $\alpha_0 = 0$, $\alpha_{2s+1} = 0$)

$$w_3(\alpha_1, \dots, \alpha_{2s}) = \sum_{i=0}^{s} \left(\alpha_{2j}^3 + \alpha_{2j+1}^3\right) = 0$$

just by grouping terms in a different way, and thus, by repeating the argument, there must exist some $j \in \{0, \dots, s\}$ such that

$$\alpha_{2j} + \alpha_{2j+1} = a_j < 0.$$

This proof clearly shows the origin of the existence of backward time steps: the equation $w_3 = 0$ can only be satisfied if at least one a_j *and* one b_j are negative.

3. Splitting methods for special problems

Whereas the analysis carried out in Section 2 is completely general, there are important problems arising in applications whose particular structure allows us to simplify the treatment and design schemes without taking into account all the order conditions. Some of them are reviewed in this section, where we also show how to adapt splitting methods to deal with explicitly time-dependent systems.

3.1. RKN splitting methods

Many differential equations of practical interest are of the form

$$y'' = g(y), \tag{3.1}$$

where $y \in \mathbb{R}^d$ and $g\colon \mathbb{R}^d \longrightarrow \mathbb{R}^d$. An example in point corresponds to Hamiltonian systems of the form $H(q,p) = T(p) + V(q)$, where the kinetic energy $T(p)$ is quadratic in the momenta p, i.e. $T(p) = \frac{1}{2}p^\top M^{-1}p$ for a constant invertible symmetric matrix M, and $V(q)$ is the potential. In that case, the corresponding Hamiltonian system can be written in the form (3.1) with $y = q$, $g(y) = -M^{-1}\nabla V(y)$.

By transforming (3.1) into a first-order ODE system (of dimension $D = 2d$) in the new variables $x = (y, v)^\top$, with $v = y'$, it is clear that the resulting equation

$$x' = \frac{\mathrm{d}}{\mathrm{d}t}\begin{pmatrix} y \\ v \end{pmatrix} = \begin{pmatrix} v \\ g(y) \end{pmatrix}$$

can be expressed as $x' = f_1(x) + f_2(x)$, with

$$f_1(y, v) = (0, g(y))^\top, \quad f_2(y, v) = (v, 0)^\top, \tag{3.2}$$

and splitting methods of the form (2.17) can be applied, with the exact h-flows $\varphi_h^{[1]}$ and $\varphi_h^{[2]}$ given by

$$\varphi_h^{[1]}\colon \begin{pmatrix} y_0 \\ v_0 \end{pmatrix} \longmapsto \begin{pmatrix} y_0 \\ v_0 + h\,g(y_0) \end{pmatrix}, \quad \varphi_h^{[2]}\colon \begin{pmatrix} y_0 \\ v_0 \end{pmatrix} \longmapsto \begin{pmatrix} y_0 + h\,v_0 \\ v_0 \end{pmatrix}. \tag{3.3}$$

Just as the class of Runge–Kutta methods can be conveniently adapted to (3.1) to get more efficient schemes (the so-called Runge–Kutta–Nyström or RKN methods; see Hairer, Nørsett and Wanner 1993), special splitting methods can also be designed to improve the accuracy whilst reducing the computational cost with respect to the general composition (2.17). For analogy, they are sometimes called splitting methods of RKN type. The key point here is that the differential operators F_1 and F_2 associated with (3.2) satisfy $[F_1, [F_1, [F_1, F_2]] = 0$ identically. In other words, $F_{1112} = 0$ in (2.21), which introduces linear dependences among higher-order terms in the expansion of $\log(\Psi(h))$ (McLachlan and Quispel 2002) and therefore contributes to a reduction in the number of order conditions. In the notation of Section 2.4, $6C_4 = -[F_1, [F_1, [F_1, F_2]]$, and thus we have that $C_k = 0$ for $k \geq 4$. Hence, the order conditions (2.38) for multi-indices $(i_1, \ldots, i_k)$ with some index $i_j \geq 4$ need not be considered in that case. In particular, a splitting scheme is at least of order five provided that (2.38) holds for the Lyndon multi-indices

$$\{(2), (3), (1,2), (1,3), (2,3), (1,1,2), (1,1,3), (1,2,2), (1,1,1,2)\}.$$

Here, we have excluded the Lyndon multi-indices $(4), (5), (1,4)$ from the set of Lyndon multi-indices $(i_1, \ldots, i_k)$ with $i_1 + \cdots + i_k \leq 5$. For order six, in addition, we have to consider (2.38) for the following Lyndon multi-indices:

$$(1,1,1,1,2), (1,1,1,3), (1,2,3), (1,1,2,2), (1,3,2).$$

They are obtained by excluding from the set of nine Lyndon multi-indices with $i_1 + \cdots + i_k = 6$, the Lyndon multi-indices $(2, 4), (1, 1, 4), (1, 5), (6)$. Similarly, for order seven, in addition, we have to consider (2.38) for the following Lyndon multi-indices:

$$(1, 1, 1, 1, 1, 2), (1, 1, 1, 1, 3), (1, 1, 2, 3), (1, 1, 1, 2, 2), (1, 1, 3, 2),$$
$$(2, 2, 3), (1, 2, 1, 3), (1, 1, 2, 3), (1, 3, 3), (1, 2, 2, 2).$$

For orders higher than seven, more reductions of the order conditions occur, in addition to those obtained by excluding Lyndon multi-indices $(i_1, \dots, i_k)$ having some index $i_j \geq 4$, due to additional dependences among nested commutators of C_1, C_2, C_3. For instance, it is straightforward to check that $[C_3, [C_2, C_3]]$ vanishes identically, which implies that for order eight we can also exclude the Lyndon multi-index $(2, 3, 3)$.

The class of problems for which the reduction in order conditions discussed above is in fact more general than (3.2). In particular, it includes the situation where f_1 depends only on y and $f_2(y, v)$ is linear in v. For Hamiltonian systems, this generalization corresponds to $H(q, p) = V(q) + T(q, p)$, where $T(q, p)$ is quadratic in p. In that case, the flow associated with $T(q, p)$ should be easily computed for the RKN splitting methods to be advantageous.

Actually, such a reduction of the order conditions also holds for certain PDEs, and, in particular, for the time-dependent Schrödinger equation considered in Section 1.6 (with $F_1 = \hat{V}$ and $F_2 = \hat{T}$), since the corresponding graded Lie algebra is isomorphic to the classes of problems discussed above (McLachlan and Murua 2019). In fact, McLachlan and Murua (2019) conjectured (and checked up to order 20) that the case (3.2) (where $f_2(y, v) = v$) and the more general case (where $f_2(y, v)$ is linear in v but may depend on y) give rise to the same reduction in the order conditions.

The actual number d_r of order conditions for orders $r \leq 11$ are given in Table 2.1. Since the order conditions up to order three are identical as in the general case, the results for negative time steps still apply.

This reduction in the number of order conditions allows us to design schemes involving a smaller number of elementary flows than in the general case, eventually leading to greater efficiency. We have already illustrated a popular fourth-order method within this class (scheme $\mathrm{RKN}_6 4$ in the examples of Section 1).

3.2. *Methods with commutators*

Another possible way to improve the efficiency of splitting methods consists in incorporating into the scheme not only the flows of F_1 and F_2 but also the flows of some of their commutators $[F_1, F_2]$, $[[F_1, F_2], F_1]$, etc., or convenient approximations of these flows. Of course, the strategy makes sense if the gain in accuracy, stability or any other favourable property compensates the extra computational cost due to the presence of these additional flows. A popular fourth-order method belonging to this class is

$$\psi_h = \varphi_{a_1 h}^{[1]} \circ \varphi_{b_1 h}^{[2]} \circ \varphi_{a_2 h}^{[1]} \circ \varphi_{d_2 h^3}^{[112]} \circ \varphi_{a_2 h}^{[1]} \circ \varphi_{b_1 h}^{[2]} \circ \varphi_{a_1 h}^{[1]}, \tag{3.4}$$

with

$$a_1 = \frac{1}{6}, \quad b_1 = \frac{1}{2}, \quad a_2 = \frac{1}{3}, \quad d_2 = -\frac{1}{72}, \tag{3.5}$$

first proposed by Koseleff (1993) and Chin (1997). Here $\varphi_h^{[112]}$ denotes the h-flow corresponding to $F_{112} = [F_1, [F_1, F_2]]$ (or $F_{112} = 2C_3$ in the notation of Section 2.4).

We next analyse the family of schemes (3.4). The Lie transformation $\Psi(h)$ is

$$\Psi(h) = \mathrm{e}^{ha_1 F_1}\, \mathrm{e}^{hb_1 F_2}\, \mathrm{e}^{ha_2 F_1}\, \mathrm{e}^{h^3 2 d_2 C_3}\, \mathrm{e}^{ha_2 F_1}\, \mathrm{e}^{hb_1 F_2}\, \mathrm{e}^{ha_1 F_1}.$$

Under the assumption that $a_1 + a_2 = \frac{1}{2}$, we have

$$\mathrm{e}^{-hF_1}\Psi(h) = \mathrm{e}^{hb_1 C(hc_3)}\, \mathrm{e}^{h^3 d_2 D(hc_2)}\, \mathrm{e}^{hb_1 C(hc_1)},$$

where $c_1 = a_1$, $c_2 = a_1 + a_2$, $c_3 = a_1 + 2a_2$, $C(h)$ is given by (2.34), and

$$D(h) = 2\mathrm{e}^{-hF_1}\, C_3\, \mathrm{e}^{hF_1} = \sum_{n=1}^{\infty} h^{n-1}(n+1)nC_{n+2},$$

so that

$$\mathrm{e}^{-hF_1}\Psi(h) = \mathrm{e}^{hb_1(C_1 + hc_3 C_2 + h^2 c_3^2 C_3)}\, \mathrm{e}^{2h^3 d_2 C_3}\, \mathrm{e}^{hb_1(C_1 + hc_1 C_2 + h^2 c_1^2 C_3)} + O(h^4).$$

Expanding the right-hand side in powers of h and comparing the coefficients multiplying hC_1, $h^2 C_2$, $h^3 C_1 C_2$ and $h^3 C_3$ respectively (corresponding to the Lyndon multi-indices $(1), (2), (1,2), (3)$) with those in the expansion

$$\mathrm{e}^{-hF_1}\mathrm{e}^{h(F_1+F_2)} = I + hC_1 + \frac{h^2}{2}C_2 + \frac{h^3}{3}C_3 + \frac{h^3}{6}C_1 C_2 + \frac{h^3}{3}C_2 C_1 + O(h^4),$$

we conclude that the time-symmetric scheme (3.4) is at least of order four if

$$b_1 = \frac{1}{2}, \quad b_1(c_1 + c_3) = \frac{1}{2}, \quad \frac{1}{2}\, b_1^2(c_1 + c_3) + b_1^2 c_1 = \frac{1}{6},$$
$$b_1\left(c_1^2 + c_3^2\right) + 2d_2 = \frac{1}{3}.$$

That system of polynomial equations has a unique solution, corresponding to the choice (3.5), as expected. The order conditions of more general products of scaled exponentials of hF_1, hF_2 and $h^3[F_1, [F_1, F_2]]$ can be derived similarly.

Recall that in the RKN case $C_4 = [C_3, F_1] = 0$, which implies that $\varphi_h^{[1]}$ and $\varphi_h^{[112]}$ commute. Hence, the three central terms in (3.4) can be merged into one, the h-flow of the differential operator $2a_2 F_1 + d_2 h^2 F_{112}$, which is of the form

$$\sum_j G_j(q, h) \frac{\partial}{\partial v^j}.$$

In the case of (3.2),

$$\varphi_{a_2 h}^{[1]} \circ \varphi_{d_2 h}^{[112]} \circ \varphi_{a_2 h}^{[1]}(y, v) = \left(y,\ v + 2ha_2\, g(y) + h^3 d_2 \frac{\partial g}{\partial y}(y) g(y) \right). \tag{3.6}$$

For Hamiltonian systems $H(p, q) = V(q) + \frac{1}{2} p^\top M^{-1} p$, F_{112} is the operator associated to the Hamiltonian function $(\nabla V)^\top M^{-1} \nabla V$, depending only on q. Thus (3.6) is the h-flow of the Hamiltonian

$$2a_2\, V(q) + d_2\, h^2 (\nabla V(q))^\top M^{-1} \nabla V(q), \tag{3.7}$$

which reduces to the potential $V(q)$ when $a_2 = 1/2$ and $d_2 = 0$. This explains the term 'splitting methods with modified potentials' frequently used in the literature (López-Marcos, Sanz-Serna and Skeel 1997, Rowlands 1991, Wisdom, Holman and Touma 1996). One such method has been illustrated in practice in Section 1.6 (Figure 1.5).

In addition to the reduction in the number of force evaluations, including flows associated with commutators has another advantage: since the coefficients a_i, b_i do not have to satisfy all the order conditions at order $r \geq 3$, the results for negative time steps do not apply here, and in fact methods of order greater than two do exist within this class. In addition to (3.4), other 'forward' fourth-order methods (i.e. with all $a_j > 0$ and $b_j > 0$) involving second derivatives of the potential have been published (Omelyan, Mryglod and Folk 2002, 2003) and applied to systems where the presence of negative coefficients leads to severe stability problems (Bader, Blanes and Casas 2013). Although it has been shown that achieving order six in general requires some negative coefficients (Chin 2005), it is indeed possible to construct a sixth-order processed method for cubic potentials with all a_j, b_j being positive (Blanes, Casas, González and Thalhammer 2023).

Additional flows corresponding to commutators involving more operators can in principle be incorporated into the scheme. Thus, for instance, the operator $[F_{12}, F_{112}] = -2[C_2, C_3]$ is also of the form

$$\sum_j g_j^{[5]}(q, h) \frac{\partial}{\partial v^j},$$

which allows us to compute its h-flow explicitly; see Blanes, Casas and Ros (2001*a*) and Blanes, Casas and Murua (2008*b*) for more details. Again, this procedure allows us to introduce additional free parameters into the scheme and construct more efficient integrators as long as the simultaneous evaluation of $g(y)$, $(\partial g/\partial y)(y)g(y)$, $g^{[5]}(y)$, etc., is not substantially more expensive than the evaluation of $g(y)$ itself.

3.3. *Near-integrable systems*

Very often in applications we have to deal with differential equations such as

$$x' = f_1(x) + \varepsilon f_2(x), \tag{3.8}$$

where $|\varepsilon| \ll 1$ and the h-flows $\varphi_h^{[1]}$, $\varphi_h^{[2]}$ corresponding to f_1 and εf_2, respectively, are readily available. In classical Hamiltonian mechanics, in particular, it is rather common to have a Hamiltonian function H which is a small perturbation of an exactly integrable Hamiltonian H_1, that is, $H = H_1 + \varepsilon H_2$, with $0 < \varepsilon \ll 1$ (Goldstein 1980, Pars 1979). A canonical example corresponds to the gravitational N-body problem in Jacobi coordinates, already considered in Section 1.5.

For this type of problem, splitting methods of the form (2.17),

$$\psi_h = \varphi_{a_{s+1}h}^{[1]} \circ \varphi_{b_s h}^{[2]} \circ \varphi_{a_s h}^{[1]} \circ \cdots \circ \varphi_{a_2 h}^{[1]} \circ \varphi_{b_1 h}^{[2]} \circ \varphi_{a_1 h}^{[1]},$$

are especially well adapted. On the one hand, the error is at most $O(\varepsilon)$ and vanishes with ε, since in that case the scheme reproduces the exact solution. On the other hand, typically, $|\varepsilon| \ll h$ (or at least $|\varepsilon| \approx h$), so that we are mainly interested in eliminating error terms with small powers of ε, and its number grows as a polynomial in the order r, rather than exponentially. Thus there is only one error term of order εh^k (namely, the term $h^k \varepsilon w_{11\cdots12} F_{11\cdots12}$ in the expansion (2.21)), $\lfloor \frac{1}{2}(k-1) \rfloor$ terms of order $O(\varepsilon^2 h^k)$ and $\lfloor \frac{1}{6}(k-1)(k-2) \rfloor$ terms of order $O(\varepsilon^3 h^k)$ (McLachlan 1995*a*).

In the treatment of splitting methods for near-integrable systems it is convenient to introduce the notion of *generalized order*, following McLachlan (1995*a*). Thus, we say that ψ_h is of generalized order $(r_1, r_2, \ldots, r_m)$, with $r_1 \geq r_2 \geq \cdots \geq r_m$, if

$$\psi_h(x) - \varphi_h(x) = O\left(\varepsilon h^{r_1+1} + \varepsilon^2 h^{r_2+1} + \cdots + \varepsilon^m h^{r_m+1}\right) \quad \text{as } (h, \varepsilon) \longrightarrow (0, 0).$$

With this notation, a method such that the local error is $O(\varepsilon h^{2n+1} + \varepsilon^2 h^3)$ is said to be of generalized order $(2n, 2)$. In this sense, the Strang scheme is of order $(2, 2)$, whereas the $(10, 6, 4)$ integrator ψ_h used in Sections 1.4 and 1.5 satisfies

$$\psi_h(x) - \varphi_h(x) = O\left(\varepsilon h^{11} + \varepsilon^2 h^7 + \varepsilon^3 h^5\right).$$

The general analysis of the order conditions carried out in Section 2.4 in terms of Lyndon multi-indices readily allows us to characterize the generalized order of a given splitting scheme. We simply need to observe that, by replacing F_2 with εF_2 in the expansions derived in Section 2.4, a series expansion of $\mathrm{e}^{-hF_1}(\Psi(h) - \mathrm{e}^{h(F_1+\varepsilon F_2)})$ is obtained, where the term associated to a multi-index $(i_1, \ldots, i_k)$ with k indices is affected by a kth power of ε. In particular, we get Table 3.1, which contains the Lyndon multi-indices involved in several consistent palindromic splitting methods of the given generalized order.

3.4. *Splitting methods for linear systems*

In the numerical integration of the Schrödinger equation implemented in Section 1.6, we separated the system into kinetic and potential energy and then applied several schemes based on this splitting. It has long been recognized, however, that there exist other possibilities for splitting such a system. Given the N-dimensional

Table 3.1. Lyndon multi-indices corresponding to consistent palindromic splitting methods of a given generalized order.

Generalized order	Lyndon multi-indices
$(2n, 2)$	$(3), (5), \ldots, (2n-1)$
$(8, 4)$	$(3), (5), (7), (1, 2)$
$(10, 4)$	$(3), (5), (7), (9), (1, 2)$
$(8, 6, 4)$	$(3), (5), (7), (1, 2), (1, 4), (2, 3)$
$(10, 6, 4)$	$(3), (5), (7), (9), (1, 2), (1, 4), (2, 3)$

linear ODE

$$\mathrm{i}\frac{\mathrm{d}}{\mathrm{d}t}u(t) = H\,u(t), \qquad u(0) = u_0 \in \mathbb{C}^N \tag{3.9}$$

resulting from the space discretization of equation (1.42), with H real and symmetric, Gray and Verosky (1994) and Gray and Manolopoulos (1996) separate u into its real and imaginary parts, $q = \mathrm{Re}\,(u)$, $p = \mathrm{Im}\,(u)$. Then, in terms of q, p, equation (3.9) leads to

$$\frac{\mathrm{d}}{\mathrm{d}t}q = H\,p, \qquad \frac{\mathrm{d}}{\mathrm{d}t}p = -H\,q, \tag{3.10}$$

so that they can be seen as the classical evolution equations corresponding to the Hamiltonian function

$$\hat{H}(q, p) = \frac{1}{2}p^{\mathsf{T}}Hp + \frac{1}{2}q^{\mathsf{T}}Hq \tag{3.11}$$

in terms of canonical variables q and p. The exact solution of (3.10) is given by

$$\begin{pmatrix} q(t) \\ p(t) \end{pmatrix} = O(tH)\begin{pmatrix} q_0 \\ p_0 \end{pmatrix}, \quad \text{where} \quad O(tH) = \begin{pmatrix} \cos(tH) & \sin(tH) \\ -\sin(tH) & \cos(tH) \end{pmatrix} \tag{3.12}$$

is an orthogonal and symplectic $2N \times 2N$ matrix. As with the formal solution $u(t) = \mathrm{e}^{-\mathrm{i}tH}u_0$ of (3.9), $O(tH)$ may be very expensive to compute, so that suitable approximations might be necessary, such as those provided by splitting methods applied to (3.11). In this respect, notice that if we introduce the nilpotent matrices A and B,

$$A \equiv \begin{pmatrix} 0 & H \\ 0 & 0 \end{pmatrix}, \quad B \equiv \begin{pmatrix} 0 & 0 \\ -H & 0 \end{pmatrix}, \tag{3.13}$$

then it is clear that the symplectic Euler–VT method of Section 1.4 is simply

$$\begin{pmatrix} q_{n+1} \\ p_{n+1} \end{pmatrix} = \mathrm{e}^{hB}\,\mathrm{e}^{hA}\begin{pmatrix} q_n \\ p_n \end{pmatrix}, \tag{3.14}$$

whereas the Störmer–Verlet Algorithm 1.3 corresponds to

$$x_{n+1} = \mathrm{e}^{\frac{h}{2}B}\,\mathrm{e}^{hA}\,\mathrm{e}^{\frac{h}{2}B}\,x_n, \quad x = (q,p)^\top \tag{3.15}$$

for a given time step h. Notice that all schemes based on this splitting are automatically symplectic.

At this point, nothing prevents us from using any of the RKN splitting methods treated in Section 3.1, even with nested commutators as in Section 3.2. It turns out, however, that the particularly simple algebraic structure of the system (3.11) makes it possible to design more efficient schemes. Specifically, there is only one independent condition to increase the order from $r = 2k - 1$ to $r = 2k$, and only two to increase the order from $r = 2k$ to $r = 2k + 1$ for a given k; see Blanes, Casas and Murua (2008*a*) for more details. As a result, splitting methods of the form

$$x_{n+1} = \mathrm{e}^{ha_{s+1}A}\,\mathrm{e}^{hb_sB}\,\cdots\,\mathrm{e}^{ha_2A}\,\mathrm{e}^{hb_1B}\,\mathrm{e}^{ha_1A}x_n \tag{3.16}$$

of order r for $r = 2, 4, 6, 8, 10$ and 12 can be obtained with $s = r$ exponentials e^{hb_jB} (Gray and Manolopoulos 1996). By contrast, at least 15 and 31 exponentials are needed in general to attain orders eight and ten, respectively.

A couple of comments are worth making. First, this class of symplectic methods do not preserve the orthogonal character of the exact solution given by $O(tH)$ (or alternatively the unitarity of (3.9)). Nevertheless, Blanes *et al.* (2008*a*) have shown that the average relative errors due to the lack of preservation of orthogonality or unitarity do not grow with time, since the schemes are conjugate to orthogonal or unitary methods for sufficiently small values of h. Second, although initially motivated by the time integration of the Schrödinger equation, methods (3.16) can be generalized in several ways. Thus they have been used to construct an algorithm to approximate $\mathrm{e}^{-\mathrm{i}tH}v$ for any real symmetric matrix H and any complex vector v by only carrying out matrix–vector products of the form Hv. As shown in Blanes, Casas and Murua (2015), the algorithm is more efficient than schemes based on Chebyshev polynomials for all tolerances and values of h. These methods can also be adapted for systems of the form

$$x' = My, \quad y' = -Nx,$$

with $x \in \mathbb{R}^{d_1}$, $y \in \mathbb{R}^{d_2}$, $M \in \mathbb{R}^{d_1 \times d_2}$ and $N \in \mathbb{R}^{d_2 \times d_1}$.

3.5. *Splitting methods for non-autonomous systems*

So far we have restricted our attention to splitting methods for autonomous differential equations. The question we analyse next is whether the same techniques can be applied when there is an explicit time dependence in the equation to integrate. The ideal situation would be that the methods designed for $x' = f(x)$ could also be used (maybe with only minor modifications) when we have $x' = f(t,x)$. In addition, we would like the schemes previously considered in this section for special problems to remain valid in the non-autonomous case.

Let us first consider the general situation, corresponding to a system of the form

$$x' = f(t,x) = f_1(t,x) + \cdots + f_m(t,x), \quad x(0) = x_0, \tag{3.17}$$

that is, when the explicit time dependence is present in each part. Then we can take t as a new coordinate and transform (3.17) into an equivalent autonomous equation to which standard splitting algorithms can subsequently be applied. More specifically, equation (3.17) is equivalent to the enlarged system

$$\frac{\mathrm{d}}{\mathrm{d}t}\begin{pmatrix} x \\ x_{d+1} \end{pmatrix} = \underbrace{\begin{pmatrix} 0 \\ 1 \end{pmatrix}}_{\hat{f}_0} + \underbrace{\begin{pmatrix} f_1(x_{d+1},x) \\ 0 \end{pmatrix}}_{\hat{f}_1} + \cdots + \underbrace{\begin{pmatrix} f_m(x_{d+1},x) \\ 0 \end{pmatrix}}_{\hat{f}_m}, \tag{3.18}$$

with $x_{d+1} \in \mathbb{R}$. If the resulting (autonomous) equations

$$y' = \hat{f}_i(y), \quad i = 0, 1, \ldots, m \quad \text{with} \quad y = (x, x_{d+1})$$

can be solved, then we may use any splitting method of the form (2.17), since x_{d+1} advances only with $\hat{f}_0$ and remains constant for the rest of the system.

It turns out that, for problems which are separable into just two parts, that is,

$$x' = f(t,x) = f_1(t,x) + f_2(t,x), \quad x(0) = x_0, \tag{3.19}$$

we can do better: if t is taken as a new coordinate *twice*, and we write

$$\frac{\mathrm{d}}{\mathrm{d}t}\begin{pmatrix} x \\ x_{d+1} \\ x_{d+2} \end{pmatrix} = \underbrace{\begin{pmatrix} f_1(x_{d+1},x) \\ 0 \\ 1 \end{pmatrix}}_{\hat{f}_1} + \underbrace{\begin{pmatrix} f_2(x_{d+2},x) \\ 1 \\ 0 \end{pmatrix}}_{\hat{f}_2}, \tag{3.20}$$

with $x_{d+1}, x_{d+2} \in \mathbb{R}$, then we can apply the same splitting schemes designed for autonomous systems separable into two pieces to

$$y' = \hat{f}_1(y), \quad y' = \hat{f}_2(y) \quad \text{with} \quad y = (x, x_{d+1}, x_{d+2}).$$

This is so because x_{d+1} is constant when integrating the first equation and x_{d+2} is constant when solving the second one. The procedure can be viewed as a generalization of the one proposed in Sanz-Serna and Portillo (1996) for Hamiltonian systems $H(t,q,p) = T(t,p) + V(t,q)$: by introducing two new coordinates q_{d+1}, q_{d+2} and their associated momenta p_{d+1}, p_{d+2}, we instead deal with the formally autonomous Hamiltonian

$$\tilde{H}(q_{d+1}, q_{d+2}, q, p_{d+1}, p_{d+2}, p) = (T(p_{d+2}, p) + p_{d+1}) + (V(q_{d+1}, q) - q_{d+2}).$$

In the special case of the non-autonomous second-order differential equation

$$y'' = g(t,y), \tag{3.21}$$

it is convenient to split the system in the extended phase space as

$$\frac{\mathrm{d}}{\mathrm{d}t}\begin{pmatrix} y \\ v \\ y_{d+1} \end{pmatrix} = \begin{pmatrix} v \\ 0 \\ 1 \end{pmatrix} + \begin{pmatrix} 0 \\ g(y_{d+1}, y) \\ 0 \end{pmatrix},$$

since then we have an autonomous system with the same algebraic structure as those considered in Section 3.1, so that RKN splitting methods (even including commutators) can also be used.

For Hamiltonian systems $H(t, q, p) = T(p) + V(t, q)$, this is equivalent to introducing a new coordinate $q_{d+1} = t$ and its associated momentum $p_{d+1} = -H$, and considering the extended (autonomous) Hamiltonian function

$$\tilde{H}(q_{d+1}, q, p_{d+1}, p) = (T(p) + p_{d+1}) + V(q_{d+1}, q),$$

which is still quadratic in momenta, so that symplectic RKN methods can be used. Notice that $\tilde{H}$ is only linear in p_{d+1}, and modified potentials only involve derivatives of the potential with respect to q but not with respect to q_{d+1}, i.e. they do not require time derivatives. In this case the evolution for p_{d+1} is irrelevant, so there is no need to compute it.

Finally, if we take t as two new coordinates in the non-autonomous near-integrable system

$$x' = f(t, x) = f_1(t, x) + \varepsilon f_2(t, x), \quad x(0) = x_0, \tag{3.22}$$

then the special structure of a near-integrable system is destroyed. A partial remedy consists in separating the system as

$$\frac{\mathrm{d}}{\mathrm{d}t}\begin{pmatrix} x \\ x_{d+1} \end{pmatrix} = \underbrace{\begin{pmatrix} f_1(x_{d+1}, x) \\ 1 \end{pmatrix}}_{\hat{f}_1} + \underbrace{\varepsilon \begin{pmatrix} f_2(x_{d+1}, x) \\ 0 \end{pmatrix}}_{\varepsilon \hat{f}_2}, \tag{3.23}$$

which requires us to numerically solve the non-autonomous unperturbed system.

We should bear in mind that taking time as an additional coordinate is of interest only if the time dependence in f_i, $i = 1, \ldots, m$ is cheap to compute. Otherwise the resulting algorithm may be computationally costly, since these functions have to be evaluated s times per time step. This drawback can be avoided by approximating the exact solution at each step by a composition of maps that in some sense incorporates average values of the vector fields with different weights on the subinterval $[t_n, t_{n+1}]$ (Blanes and Casas 2006). Specifically, the schemes read

$$\psi_h = \varphi_h^{[\hat{A}_{s+1}]} \circ \varphi_h^{[\hat{B}_s]} \circ \varphi_h^{[\hat{A}_s]} \circ \cdots \circ \varphi_h^{[\hat{A}_2]} \circ \varphi_h^{[\hat{B}_1]} \circ \varphi_h^{[\hat{A}_1]}, \tag{3.24}$$

where the maps $\varphi_h^{[\hat{A}_i]}$, $\varphi_h^{[\hat{B}_i]}$ are the exact 1-flows corresponding to the time-independent differential equations

$$x' = \hat{A}_i(x), \quad x' = \hat{B}_i(x), \quad i = 1, 2, \ldots \tag{3.25}$$

respectively, with

$$\hat{A}_i(x) \equiv h \sum_{j=1}^{k} \rho_{ij} f_1(\tau_j, x), \quad \hat{B}_i(x) \equiv h \sum_{j=1}^{k} \sigma_{ij} f_2(\tau_j, x). \tag{3.26}$$

Here $\tau_j = t_n + c_j h$ and the (real) constants c_j, ρ_{ij}, σ_{ij} are chosen in such a way that ψ_h provides an approximation of order r. These methods have the additional advantage that, when applied to (3.19) with the time frozen, they reproduce the standard splitting (2.17), since $\sum_j \rho_{ij} = a_i$ and $\sum_j \sigma_{ij} = b_i$. The same technique can be applied to the splitting methods analysed in Section 3.4 when the linear system (3.9) is explicitly time-dependent. In that case, the resulting scheme involves linear combinations of H evaluated at some intermediate times (Blanes, Casas and Murua 2017*a*).

4. Qualitative properties of splitting methods

4.1. Changes of variables and differential equations on smooth manifolds

Given a smooth autonomous differential equation in $\mathbb{R}^D$,

$$x' = f(x), \tag{4.1}$$

a smooth change of variables $x = \theta(\hat{x})$ transforms (4.1) into a new autonomous differential equation in $\mathbb{R}^D$,

$$\hat{x}' = \hat{f}(\hat{x}),$$

such that their t-flows are related as follows: for all $x_0 \in \mathbb{R}^D$,

$$\varphi_t^{[f]}(\theta(\hat{x}_0)) = \theta\big(\varphi_t^{[\hat{f}]}(\hat{x}_0)\big), \quad \text{with} \quad \hat{x}_0 = \theta^{-1}(x_0).$$

This implies that a similar property holds for the map $\psi_h \approx \varphi_h^{[f]}$ defined by the splitting method (2.17) applied to (4.1) when $f(x) = \sum_{j=1}^{2} f_j(x)$. That is, if the change of variables $x = \theta(\hat{x})$ transforms each equation $x' = f_j(x)$ into $\hat{x}' = \hat{f}_j(\hat{x})$, then the map $\hat{\psi}_h \approx \varphi_h^{[\hat{f}]}$ obtained by applying the splitting method with the same a_j, b_j coefficients in the new variables $\hat{x}$ is related to ψ_h by

$$\psi_h(\theta(\hat{x}_0)) = \theta(\hat{\psi}_h(\hat{x}_0)).$$

In words, the following two procedures give exactly the same numerical results: (i) applying the splitting method to the ODE corresponding to the new variables $\hat{x}$ with the initial condition $\hat{x}(0) = \hat{x}_0 \in \mathbb{R}^D$ and then transforming the result to the old variables x, and (ii) applying the splitting method to the ODE formulated in the old variables x with the initial condition $x(0) = x_0 := \theta(\hat{x}_0) \in \mathbb{R}^D$. This property does not hold in general for other integration schemes. For instance, in the case of Runge–Kutta methods it is true if the change of variables $\theta \colon \mathbb{R}^D \to \mathbb{R}^D$ is an affine map, but not in general.

The above property can be stated in a more abstract way by saying that splitting methods constitute a particular class of numerical integrators for differential equations defined on smooth manifolds: since the t-flow of a smooth vector field on a smooth manifold $\mathcal{M}$ is coordinate-independent, then splitting methods applied to a differential equation on $\mathcal{M}$ can also be naturally defined in a coordinate-independent way. More precisely, $\psi_h\colon \mathcal{M} \to \mathcal{M}$ defined by (2.17) gives a natural approximation of the h-flow $\varphi_h^{[f_1+f_2]}$ on the smooth manifold $\mathcal{M}$, described only in terms of objects related to the manifold itself. This is in contrast to other approaches for the numerical integration of differential equations on manifolds, which depend on particular choices of global or local charts, or particular embeddings of the manifold in a higher-dimensional Euclidean space (Hairer *et al.* 2006).

In fact, the treatment of the order conditions for splitting methods carried out in Section 2 is also valid in this more abstract setting. Suppose we have two smooth vector fields f_1 and f_2 on $\mathcal{M}$, and let F_i $(i = 1, 2)$ be the linear operators on $C^\infty(\mathcal{M}, \mathbb{R})$ defined as follows: for each $g \in C^\infty(\mathcal{M}, \mathbb{R})$, $F_i g \in C^\infty(\mathcal{M}, \mathbb{R})$ is the f_i-directional derivative of g. Then the t-flow $\varphi_t^{[f_i]}$ (or $\varphi_t^{[i]}$ for short) satisfies, for all $g \in C^\infty(\mathcal{M}, \mathbb{R})$ and $t \in \mathbb{R}$,

$$\frac{\mathrm{d}}{\mathrm{d}t} g(\varphi_t^{[i]}) = F_i g(\varphi_t^{[i]}).$$

This implies that the series in powers of t of $g(\varphi_t^{[i]})$ can be represented as $\mathrm{e}^{tF_i} g$, that is, we are in the same situation as in the case of differential equations on $\mathbb{R}^D$, discussed in Section 1.2. Consequently, the series in powers of h of $g(\psi_h(x))$ for the map $\psi_h\colon \mathcal{M} \to \mathcal{M}$ defined by (2.17) can be obtained by expanding $\Psi(h)g$, where $\Psi(h)$ is given by (2.20), so the analysis of the order conditions of Section 2 can be formally applied here as well.

4.2. *Stability*

An important characteristic of numerical integrators is their *stability*. Generally speaking, the numerical solution provided by a stable method does not tend to infinity when the exact solution is bounded. To analyse the (linear) stability of a given integrator, a model problem is typically chosen, so that both the numerical and exact solutions can be explicitly written out. In the case of a splitting method like (2.17), the model problem is the simple harmonic oscillator $y'' + \omega^2 y = 0$, $\omega > 0$ (López-Marcos, Sanz-Serna and Skeel 1996*b*, McLachlan and Gray 1997), with the standard $(x = (q, p) = (\omega y, y'))$ splitting

$$\begin{pmatrix} q' \\ p' \end{pmatrix} = \Bigg[\underbrace{\begin{pmatrix} 0 & \omega \\ 0 & 0 \end{pmatrix}}_{A} + \underbrace{\begin{pmatrix} 0 & 0 \\ -\omega & 0 \end{pmatrix}}_{B}\Bigg] \begin{pmatrix} q \\ p \end{pmatrix}, \tag{4.2}$$

and exact solution at time $t = h$

$$\begin{pmatrix} q(h) \\ p(h) \end{pmatrix} = M_z \begin{pmatrix} q(0) \\ p(0) \end{pmatrix}, \quad M_z = \begin{pmatrix} \cos z & \sin z \\ -\sin z & \cos z \end{pmatrix}, \quad z = h\omega. \tag{4.3}$$

There are at least two reasons for this choice of model. First, if a numerical method already provides unbounded numerical solutions for a given h on this system, we cannot expect good behaviour for more general problems. Second, there are physically relevant problems that, once formulated in appropriate coordinates, are expressed as a system of uncoupled harmonic oscillators. Thus, a precise characterization of the stability of a splitting method on (4.2) can be useful to build accurate and stable algorithms for their numerical treatment. The linear system (3.10) considered in Section 3.4 belongs to this category. This is also the case of the more general equation

$$q' = M^{-1}p, \quad p' = -Nq, \tag{4.4}$$

where M and N are $d \times d$ symmetric positive definite matrices. Writing $M = LL^\top$ and introducing new variables $\tilde{q} = L^\top q$, then $\tilde{q}'' = -L^{-1}NL^{-T}\tilde{q}$. Since N is symmetric positive definite, then $L^{-1}NL^{-T}$ is diagonalizable with positive eigenvalues. A new change of variables reduces the system to a set of d uncoupled scalar harmonic oscillators $y_i'' = -\omega_i^2 y_i$, with ω_i^2 the eigenvalues of $L^{-1}NL^{-T}$ (Bou-Rabee and Sanz-Serna 2018).

Application of the splitting method (3.16) to (4.2) results in the map

$$\begin{pmatrix} q_{n+1} \\ p_{n+1} \end{pmatrix} = \tilde{M}_z \begin{pmatrix} q_n \\ p_n \end{pmatrix}, \quad \tilde{M}_z = \begin{pmatrix} K_1(z) & K_2(z) \\ K_3(z) & K_4(z) \end{pmatrix}, \tag{4.5}$$

where $K_1(z)$ and $K_4(z)$ (resp. $K_2(z)$, $K_3(z)$) are even (resp. odd) polynomials in z,

$$\det \tilde{M}_z = 1, \quad K_1(0) = K_4(0) = 1,$$

and, if the scheme is time-symmetric, then $K_1(z) = K_4(z)$. An essential role in the analysis is played by the stability polynomial, defined as

$$p(z) = \frac{1}{2}\operatorname{tr} \tilde{M}_z = \frac{1}{2}(K_1(z) + K_4(z)).$$

The eigenvalues of $\tilde{M}_z$ are the zeros of $\lambda^2 - 2p(z)\lambda + 1$ and they determine the stability of the given method: if z is such that $|p(z)| < 1$, then $\tilde{M}_z$ has complex conjugate eigenvalues of modulus 1 and the powers $\tilde{M}_z^n$, $n \geq 0$, remain bounded, whereas if $|p(z)| > 1$, then $\tilde{M}_z^n$ grows exponentially with n. Linear instability occurs when $+1$ or -1 is an eigenvalue with multiplicity 2 and $\tilde{M}_z$ is not diagonalizable.

Notice that the stability polynomial of a consistent splitting method is an even polynomial satisfying

$$p(z) = 1 - \frac{z^2}{2} + O(z^4) \quad \text{as } z \to 0,$$

so that for sufficiently small $z = h\omega > 0$ the scheme will be stable. The stability interval is defined as the longest interval $(0, z_*)$ such that $\tilde{M}_z^n$ is bounded for all the iterations n.

In the particular case of the Störmer–Verlet method (3.15), we have

$$K_1(z) = K_4(z) = p(z) = 1 - \frac{z^2}{2}, \quad K_2(z) = z, \quad K_3(z) = -z + \frac{z^3}{4},$$

and thus $z_* = 2$. This is also true for Algorithm 1.4, or $x_{n+1} = \mathrm{e}^{\frac{h}{2}A}\,\mathrm{e}^{hB}\,\mathrm{e}^{\frac{h}{2}A}x_n$, since both have the same stability polynomial $p(z)$.

It is important to stress that for systems of the form (4.4) that can be reduced to a collection of d scalar harmonic oscillators with frequencies ω_i, $i = 1, \ldots, d$, the stability interval of Störmer–Verlet is restricted to $h < 2/\omega_{\max}$, where $\omega_{\max}$ is the largest frequency of the system.

Suppose now that, given an integer k, we concatenate k steps of length h/k of the Störmer–Verlet method to our model problem (4.2). The resulting scheme is stable for $0 < z/k < 2$, or alternatively, its stability interval is $(0, 2k)$. It is remarkable that this is in fact the longest stability interval we can achieve by considering any splitting method (2.17) with $s = k$ stages. An elementary proof of this statement is presented in Bou-Rabee and Sanz-Serna (2018). In consequence, the Störmer–Verlet method may be applied with longer scaled time steps z/k than any other splitting method with k stages. This makes it the method of choice in applications such as molecular dynamics, where high accuracy is not required and we are interested in using time steps as large as possible (Leimkuhler, Reich and Skeel 1996).

The problem of designing splitting methods of order, say, $2r$, with extended stability intervals can be addressed by first determining the coefficients c_j in

$$p(z) = \sum_{j=0}^{r} (-1)^j \frac{z^{2j}}{(2j)!} + \sum_{j=r+1}^{s} c_j z^{2j}, \tag{4.6}$$

so that $p(z)$ has the largest possible value of z_*. Thus a fourth-order integrator with maximal stability interval is presented in López-Marcos *et al.* (1996*b*), whereas in McLachlan and Gray (1997) the analysis is generalized to any order and number of stages. On the other hand, Blanes *et al.* (2008*a*) propose a different strategy to determine the coefficients c_j in (4.6), based on interpolatory conditions and minimization of the difference $(p(z) - \cos z)/z^{2r+2}$ in the stability interval. This results in high-order methods with a large number of stages whose stability and accuracy do not deteriorate for larger values of z. It also allows us to construct very efficient second-order methods for linear systems that outperform high-order methods for a wide range of values of the time step.

Problems of the form

$$q' = M^{-1}p, \quad p' = -Nq + f(q) \tag{4.7}$$

derived from the Hamiltonian function

$$H(q,p) = \frac{1}{2}p^\top M^{-1}p + \frac{1}{2}q^\top Nq + U(q), \tag{4.8}$$

with $f(q) = -\nabla_q U(q)$, also appear frequently in applications. The simple pendulum considered in Section 1.4 belongs to this class. Instead of the usual splitting into kinetic and potential energy, it may be advantageous to split H as $H(q,p) = H_1(q,p) + H_2(q)$, with

$$H_1(q,p) = \frac{1}{2}p^\top M^{-1}p + \frac{1}{2}q^\top Nq, \quad H_2(q) = U(q), \tag{4.9}$$

or alternatively to separate (4.7) as

$$\begin{aligned} q' &= M^{-1}p, \\ p' &= -Nq, \end{aligned} \qquad \text{and} \qquad \begin{aligned} q' &= 0, \\ p' &= f(q), \end{aligned}$$

and consider Strang integrators

$$S_h^{[RKR]} = \varphi_{h/2}^{[R]} \circ \varphi_h^{[K]} \circ \varphi_{h/2}^{[R]}, \quad S_h^{[KRK]} = \varphi_{h/2}^{[K]} \circ \varphi_h^{[R]} \circ \varphi_{h/2}^{[K]}, \tag{4.10}$$

based on the maps

$$\begin{aligned} \varphi_t^{[R]}&: \begin{pmatrix} q_0 \\ p_0 \end{pmatrix} \longmapsto \mathrm{e}^{tA}\begin{pmatrix} q_0 \\ p_0 \end{pmatrix}, \quad \text{with} \quad A = \begin{pmatrix} 0 & M^{-1} \\ -N & 0 \end{pmatrix}, \\ \varphi_t^{[K]}&: \begin{pmatrix} q_0 \\ p_0 \end{pmatrix} \longmapsto \begin{pmatrix} q_0 \\ p_0 + tf(q_0) \end{pmatrix}. \end{aligned} \tag{4.11}$$

Integrators (4.10) and other splitting methods based on sequences of *rotations* $\varphi_t^{[R]}$ and *kicks* $\varphi_t^{[K]}$ are specially suitable when $f(q)$ is a small perturbation of $-Nq$, since they provide the exact solution when the perturbation vanishes. This happens in particular in the Hamiltonian Monte Carlo method, when we deal with target densities that are perturbations of a Gaussian density (see Section 9.6). Section 5 is devoted to the analysis of splitting methods for this type of system.

In view of the applications, it is relevant to analyse the stability of compositions of $\varphi_t^{[R]}$ and $\varphi_t^{[K]}$. As in the previous case, a sequence of linear transformations render (4.7) into a more simplified form which is used as a model problem. Specifically, the one-dimensional oscillator

$$q'' = -q - \varepsilon q, \quad \varepsilon > -1$$

is the appropriate model here (Bou-Rabee 2017). It turns out that the Strang integrators (4.10) are also optimal concerning stability, in the following sense (Casas, Sanz-Serna and Shaw 2023). Let h_k be the smallest positive root of the equation

$$\frac{kh}{2}\sin\left(\frac{h}{k}\right) = \cos\left(\frac{\pi}{k}\right) - \cos\left(\frac{h}{k}\right).$$

Then, for each fixed $h^* < h_k$, $h^* \neq \pi, 2\pi, \ldots, (k-1)\pi$, the intersection of the stability region in the (ε, h) domain with the line $h = h^*$ is strictly larger for a sequence of k integrators (4.10) than for any other splitting method with k stages based on rotations and kicks. One could say, therefore, that for each given step size h, schemes (4.10) remain stable for larger perturbations than any other splitting method based on rotations and kicks. The value of ε where instabilities arise may be very small indeed, as shown in Casas *et al.* (2023) for the particular case of $k = 3$ stages.

4.3. Modified equations

The concept of *backward error analysis* (BEA), arising in several branches of numerical analysis, has also shown its effectiveness for explaining the good behaviour of splitting and composition methods in long-time integrations (Sanz-Serna 1992, Hairer 1994, Reich 1999). Generally speaking, given a problem $\mathcal{P}$ with true solution $\mathcal{S}$, when a suitable numerical solver is applied, we end up with an approximate solution $\tilde{\mathcal{S}}$. Backward error analysis thus consists in showing that $\tilde{\mathcal{S}}$ is indeed the *exact* solution of a problem $\tilde{\mathcal{P}}$ which is in some sense close to $\mathcal{P}$. This is in contrast to *forward error analysis*, where the aim consists in estimating an appropriate distance between $\tilde{\mathcal{S}}$ and $\mathcal{S}$.

In the domain of numerical analysis of differential equations, what lies at the heart of BEA is the idea of a *modified differential equation*: given the initial value problem $x' = f(x)$, $x(0) = x_0$ and a consistent numerical integrator ψ_h producing the sequence of approximations x_n at $t_n = nh$, $n = 0, 1, \ldots$, we look for another differential equation

$$\tilde{x}' = f_h(\tilde{x}) \tag{4.12}$$

whose vector field is defined as a formal series in powers of h,

$$f_h(\tilde{x}) \equiv f(\tilde{x}) + hf^{[2]}(\tilde{x}) + h^2 f^{[3]}(\tilde{x}) + \cdots \tag{4.13}$$

and such that $x_n = \tilde{x}(t_n)$ (Griffiths and Sanz-Serna 1986, Hairer *et al.* 2006). In this way, by analysing the difference of the vector fields $f(x)$ and $f_h(x)$, it is possible to extract useful information about the qualitative behaviour of the numerical solution and the global error $e_n = x_n - x(t_n) = \tilde{x}(nh) - x(t_n)$.

In the case of a splitting method, obtaining f_h is quite straightforward if we use the BCH formula to get the formal operator associated with the whole method, as was done in Section 2.2 for the order conditions. Thus, for the operator $\Psi(h) = \exp(hF(h))$ associated with a scheme (2.17) of order r, we have

$$F(h) = h(F_1 + F_2) + \sum_{i=1}^{\infty} h^{r+i} \sum_{j=1}^{c_{r+i}} w_{i,j} E_{r+i,j}, \tag{4.14}$$

where $E_{r+i,j}$ denotes the element j of the Lyndon basis of the subspace $\mathcal{L}_{r+i}(F_1, F_2)$ and $w_{i,j}$ are fixed real numbers determined by the actual coefficients of the method.

Since $F(h)$ lies in the same Lie algebra as F_1 and F_2, the numerical solution inherits the properties of the exact flow associated with this feature (e.g. Hamiltonian or volume-preserving). Now, from (4.14), we can easily determine the expression of each $f^{[j]}$ in (4.13).

A similar procedure can be applied to composition methods by determining the formal series $h^{-1} \log \Psi(h)$, with $\Psi(h)$ the operator associated to the scheme.

In the case of a Hamiltonian system of the form $H(q, p) = T(p) + V(q)$, the operators F_1 and F_2 are the Lie derivatives associated with the kinetic and potential energy, respectively, so that (4.14) is itself an operator associated with a modified Hamiltonian $\tilde{H}$. This is yet another reflection of the fact that splitting methods applied to a Hamiltonian system produce maps that are symplectic.

For linear problems, the series defining the modified Hamiltonian $\tilde{H}$ associated with the numerical solution is no longer formal, and $\tilde{H}$ can be explicitly determined in closed form. Thus the matrix $\tilde{M}_z$ in the map (4.5), obtained with a time-symmetric splitting method when $|K_1(z)| < 1$, can be expressed as

$$\tilde{M}_z = \begin{pmatrix} \cos\theta_z & \gamma_z \sin\theta_z \\ -\gamma_z^{-1} \sin\theta_z & \cos\theta_z \end{pmatrix},$$

where θ_z and γ_z are real functions such that $p(z) = \cos\theta_z$ and $K_2(z) = -\gamma_z^2 K_3(z)$ (see (4.5) and the subsequent discussion), and $\theta_{-z} = -\theta_z$, $\gamma_{-z} = -\gamma_z$. It is then straightforward to verify that the map $\tilde{M}_z$ is precisely the z-flow of the modified Hamiltonian (Blanes, Casas and Sanz-Serna 2014, Bou-Rabee and Sanz-Serna 2018)

$$\tilde{H}(q, p) = \frac{\theta_z}{2z} \left(\gamma_z p^2 + \frac{1}{\gamma_z} q^2 \right).$$

4.4. Modified equations and long-term behaviour

Convergence of the series (4.13) defining the modified equation, apart from the linear case, is the exception rather than the general rule. In consequence, an alternative strategy has to be pursued to get rigorous estimates concerning the long-time behaviour of the numerical solutions. Specifically, we first give bounds on the coefficient functions $f^{[j]}(x)$ of the modified equation, then determine an optimal truncation index, and finally estimate the difference between the numerical solution x_n and the exact solution $\tilde{x}(t_n)$ of the truncated modified equation. Here we summarize only the main results, and refer the reader to Hairer *et al.* (2006), Moan (2002) and references therein for a more comprehensive treatment.

Suppose $f(x)$, $f_1(x)$ and $f_2(x)$ are analytic in a complex neighbourhood of x_0 verifying $\|f(x)\| \leq K$ for all $x \in B_{2\rho}(x_0)$, where $B_\rho(x_0) = \{x \in \mathbb{C}^d : \|x - x_0\| \leq \rho\}$, and the same is true for the functions $f^{[j]}(x)$ of the modified equation on $B_{\rho/2}(x_0)$. If a suitable truncation index for the formal series (4.13) is selected, so that we have

$$\tilde{x}' = f(\tilde{x}) + h f^{[2]}(\tilde{x}) + h^2 f^{[3]}(\tilde{x}) + \cdots + h^{N-1} f^{[N]}(\tilde{x}), \tag{4.15}$$

with $\tilde{x}(0) = x_0$ and exact flow $\tilde{\varphi}_t^{[N]}$, then there exist constants h_0 with $h \le h_0/4$ and $\gamma > 0$ such that

$$\left\|\psi_h(x_0) - \tilde{\varphi}_h^{[N]}(x_0)\right\| \le h\gamma K \mathrm{e}^{-h_0/h}. \tag{4.16}$$

In other words, the difference between the numerical solution $\psi_h(x_0)$ and the exact solution $\tilde{\varphi}_h^{[N]}(x_0)$ of the truncated modified equation (4.15) is exponentially small.

Based on this result it is possible to get some insight into the long-time behaviour of the numerical scheme. Thus, for instance, suppose our splitting method of order r is applied to a Hamiltonian system with step size h. Then the modified equation can be derived from a (truncated) Hamiltonian

$$\tilde{H}(x) = H(x) + h^r H_{r+1}(x) + \cdots + h^{N-1} H_N(x),$$

where now $x = (q, p)$. Letting $\tilde{\varphi}_t^{[N]}$ denote the flow of the truncated modified equation as before, it is clear that $\tilde{H}(\tilde{\varphi}_t^{[N]}(x_0)) = \tilde{H}(x_0)$ for all t. Taking into account (4.16) and the bounds on the functions appearing in the modified equation (derivatives of the $\tilde{H}$ in this case), it follows that

$$\tilde{H}(x_{n+1}) - \tilde{H}\big(\tilde{\varphi}_h^{[N]}(x_n)\big) = O(h\mathrm{e}^{-h_0/h})$$

and

$$\tilde{H}(x_n) = \tilde{H}(x_0) + O(\mathrm{e}^{-h_0/2h}) \quad \text{for } nh \le \mathrm{e}^{h_0/2h}.$$

If we assume in addition that the numerical solution stays in a compact set $\mathcal{K}$, then $H_{r+1}(x) + \cdots + h^{N-r-1} H_N(x)$ is uniformly bounded on $\mathcal{K}$ independently of h and N (Hairer *et al.* 2006, p. 367) and finally

$$H(x_n) = H(x_0) + O(h^r).$$

Equivalently, the error in the energy corresponding to the numerical solution is of order r over exponentially long time intervals when a splitting method is applied with constant step size in a compact region of the phase space (Moan 2004).

With respect to the behaviour of the error in position, as shown in Calvo and Hairer (1995) and Hairer *et al.* (2006), if the Hamiltonian system is integrable and certain conditions on the frequencies at the initial point are satisfied, then

$$\|(q_n, p_n) - (q(t), p(t))\| \le C t h^r \quad \text{for } t = nh \le h^{-r},\ C = \text{const.},$$

that is, the global error grows at most linearly in time, whereas first integrals $I(q, p)$ that only depend on the action variables are well preserved on exponentially long-time intervals,

$$\|I(q_n, p_n) - I(q_0, p_0)\| \le C h^r \quad \text{for } t = nh \le h^{-r}.$$

In contrast, for a non-symplectic method (non-conjugate to a symplectic one) of order r we have

$$H(x_n) - H(x_0) = O(nh^{r+1}) = O(th^r),$$

that is, the error of the energy grows linearly, whereas the global error in the solution typically increases quadratically with time. We have seen illustrations of this feature in Section 1.

It is important to remark that the modified differential equation of a numerical scheme depends explicitly on the step size used, so that if h is changed, then we have a different modified equation. This fact helps to explain the poor long-time behaviour observed in practice when a symplectic scheme is implemented directly with a standard variable step size strategy; see e.g. Calvo and Sanz-Serna (1993*a*).

4.5. *Processing and long-term precision*

The concept of *conjugacy* plays a fundamental role in the study of the long term behaviour of both discrete and continuous dynamical systems. In the context of splitting methods (or more generally, numerical integration methods) for systems of ODEs, we replace the h-flow φ_h of the original system with a map ψ_h depending on the small parameter h (the step size) such that $\psi_h \approx \varphi_h$ for h small enough. The precision of the numerical approximations $x_n = \psi_h^n(x_0) \approx \varphi_h^n(x_0) = x(nh)$ can be analysed in one of two ways.

(i) We use standard techniques of numerical integration of ODEs to estimate the local error $\|\psi_h(x) - \varphi_h(x)\|$ and then study how this local error is propagated (Hairer *et al.* 1993),

(ii) As described in the previous subsection, we consider a truncated modified equation (4.15) of the numerical integration map ψ_h, and then study the effect of replacing the original system with the modified one, in addition to the propagation of the modified local error (4.16) (the local error between the map ψ_h and the h-flow $\varphi_t^{[N]}$ of the truncated modified equation).

In both cases, a better understanding of the long-term behaviour of the numerical error (of the application of a given integration scheme with constant step size) can be obtained by combining such techniques with the idea of *processing* a numerical integrator (López-Marcos, Sanz-Serna and Skeel 1996*a*). The main idea consists in analysing how close ψ_h and φ_h are to being conjugate to each other, and using that to estimate the long-term evolution of the errors. This is closely related to the concept of effective order (Butcher 1969, Butcher and Sanz-Serna 1996) and the idea of enhancing numerical integrators with correctors (Wisdom *et al.* 1996). Essentially, the procedure is as follows.

- Given ψ_h, we find a near-identity conjugacy map $\pi_h\colon \mathbb{R}^D \to \mathbb{R}^D$ (i.e. π_0 is the identity map) such that $\hat{\psi}_h := \pi_h^{-1} \circ \psi_h \circ \pi_h$ is as close as possible to the h-flow φ_h of the original system of ODEs.
- We estimate the propagated error $\|\psi_h^n(x_0) - \varphi_h^n(x_0)\|$ as

$$\|\psi_h^n(x_0) - \varphi_h^n(x_0)\| \le \|\psi_h^n(x_0) - \hat{\psi}_h^n(x_0)\| + \|\hat{\psi}_h^n(x_0) - \varphi_h^n(x_0)\|.$$

- We analyse the propagated error $\|\hat{\psi}_h^n(x_0)-\varphi_h^n(x_0)\|$ of the *processed integrator* $\hat{\psi}_h$, either by standard techniques or with the modified equation of $\hat{\psi}_h$.
- We estimate the difference between the original numerical solution $x_n = \psi_h^n(x_0)$ and the numerical approximation $\hat{x}_n = \hat{\psi}_h^n(x_0)$ that would be obtained if the processed integrator were used instead of the original integration map ψ_h. More precisely, using the notation

$$x_n = \psi_h^n(x_0), \quad \bar{x}_n = \psi_h^n(\pi_h(x_0)), \quad \hat{x}_n = \hat{\psi}_h^n(x_0) = \pi_h^{-1}(\bar{x}_n), \quad x(nh) = \varphi_h^n(x_0),$$

for $n \geq 0$, we have

$$\|\hat{x}_n - x_n\| \leq \|\pi_h^{-1}(\bar{x}_n) - \bar{x}_n\| + \|\bar{x}_n - x_n\|.$$

Summing up, the propagated error of the integrator ψ_h can be estimated as

$$\|x_n - x(nh)\| \leq \|\pi_h^{-1}(\bar{x}_n) - \bar{x}_n\| + \|\bar{x}_n - x_n\| + \|\hat{x}_n - x(nh)\|. \tag{4.17}$$

If the original method is of order r, that is, $\psi_h(x) - \varphi_h(x) = O(h^{r+1})$ as $h \to 0$, then it makes sense to choose a conjugacy map satisfying $\pi_h(x) = x + O(h^r)$. Hence, provided that the sequence $\{\bar{x}_n\}$ stays in a compact set, we can see that $\|\hat{x}_n - \bar{x}_n\| = \|\pi_h^{-1}(\bar{x}_n) - \bar{x}_n\| = O(h^r)$ with a constant independent of n. Therefore, for sufficiently large time intervals, the right hand side of (4.17) will be dominated either by $\|\bar{x}_n - x_n\| = \|\psi_h^n(\bar{x}_0) - \psi_h^n(x_0)\|$ (the propagation along successive iterations of ψ_h of a perturbation of size $O(h^r)$ in x_0) or by $\|\hat{x}_n - x(nh)\| = \|\hat{\psi}_h^n(x_0) - \varphi_h^n(x_0)\|$ (the sum of the propagated local errors of the processed method $\hat{\psi}_h$). Typically, the latter dominates over the former if the integration interval is sufficiently large. In that case, the precision of the numerical scheme ψ_h for sufficiently long-term integrations will depend on the size of the local errors $\|\pi_h^{-1} \circ \psi_h(x) \circ \pi_h - \varphi_h(x)\|$ of the processed method for an appropriately chosen conjugacy map (or processor map) π_h, rather than on the local errors $\|\psi_h(x) - \varphi_h(x)\|$ of the method itself.

This is illustrated by the evolution of the error in phase space of LT and S_2 displayed in Figures 1.2 and 1.4 for the pendulum problem and the six-body problem, respectively. Recall that the Lie–Trotter method is conjugate to the Strang splitting (see Section 1.3). In this case, $x_n = (q_n, p_n)^\top$ is the numerical solution provided by LT, whereas $\hat{x}_n = (\hat{q}_n, \hat{p}_n)^\top$ corresponds to S_2, with the processor map $\pi_h = \varphi_{h/2}^{[T]}$. In both examples, $\|\pi_h^{-1}(\bar{x}_n) - \bar{x}_n\| = O(h)$, which is bounded for all n provided that $\|\bar{p}_n\|$ remains bounded.

For the pendulum problem, the error $\|\bar{x}_n - x_n\| = \|\psi_h^n(\bar{x}_0) - \psi_h^n(x_0)\|$ due to the propagation of the initial difference $\|(\bar{q}_0 - q_0, \bar{p}_0 - p_0)\| = O(h)$ does not exhibit any significant increment, because (for the considered initial value) the pendulum behaves as a perturbed harmonic oscillator. On the other hand, the global error $\|\hat{x}_n - x(nh)\|$ of S_2 behaves as $O(th^2)$. Therefore the global error $\|x_n - x(nh)\|$ of LT is dominated at the beginning of the integration interval by $\|\hat{x}_n - x_n\| \approx \|\bar{x}_n - x_n\|$ (with no clear growth over time), until it is overcome by the linearly increasing global error of S_2, resulting in errors of similar size at the end of the interval for LT and S_2.

For the six-body problem, the propagation error $\|\bar{x}_n - x_n\| = \|\psi_h^n(\bar{x}_0) - \psi_h^n(x_0)\|$ increases linearly, because now H_1 corresponds to a collection of Keplerian problems, where perturbations in initial states are propagated linearly. However, the slope of that linear increase is smaller than that of the propagation of the global error $\|\hat{x}_n - x(nh)\|$ of S_2. Therefore, the global error for LT is dominated by $\|\hat{x}_n - x_n\|$ during most of the integration interval in Figure 1.4, and only at the end does it reach the global error of S_2. For longer times (not shown there), the global errors of LT and S_2 will be of similar size.

So far we have focused on studying the long-term performance of a given splitting method ψ_h with the help of a conjugacy map π_h. In practice, we may actually enhance the performance of a given splitting method (Rowlands 1991, Wisdom *et al.* 1996, McLachlan 1996, Laskar and Robutel 2001), effectively integrating the problem with the processed integration map $\hat{\psi}_h = \pi_h^{-1} \circ \psi_h(x) \circ \pi_h$. Indeed, if output is needed only every m steps, the computation of

$$\hat{x}_{n,m} = \pi_h^{-1} \circ \psi_h^m \circ \pi_h(\hat{x}_{(n-1),m}), \qquad n = 1, 2, 3, \dots \tag{4.18}$$

(with $\hat{x}_0 = x_0$), will not require substantially more CPU time than computing

$$x_{n,m} = \psi_h^m(x_{(n-1),m}), \qquad n = 1, 2, 3, \dots,$$

provided that the evaluation of $\pi_h(x)$ is computationally cheap compared to m evaluations of $\psi_h(x)$. Moreover, even if frequent output is required, we might approximate $x(nh)$ by

$$\bar{x}_n = \psi_h^m(\bar{x}_{n-1}), \qquad n = 1, 2, 3, \dots,$$

with $\bar{x}_0 = \pi_h(x_0)$. For sufficiently long integrations, this will cost essentially the same as applying the original integrator ψ_h in a standard way, and will be nearly as accurate as $\hat{x}_n$ (the full application of the processed integrator), since $\|\bar{x}_n - \hat{x}_n\| = \|\pi_h^{-1}(\bar{x}_n) - \bar{x}_n\| = O(h^r)$ will be negligible compared to $\|\bar{x}_n - x(nh)\|$ for n large enough.

This is again illustrated in Figure 1.4 for the six-body problem written in Jacobi coordinates as a perturbation of Keplerian problems. Indeed, if x_n is the numerical solution labelled by $\mathrm{LT_{pert}}$ and $\pi_h = \varphi_{h/2}^{[H_1]}$, then $\hat{x}_n$ corresponds to scheme $(2,2)$ while $\mathrm{pLT_{pert}}$ stands for $\bar{x}_n$. Notice that the position error of $\mathrm{pLT_{pert}}$ is very similar to that of $(2,2)$, as expected from the preceding discussion.

The enhancement of splitting integrators by processing is particularly effective for problems of the form $x' = f_1(x) + \varepsilon f_2(x)$ with $|\varepsilon| \ll 1$. Such enhancing was first considered in Wisdom *et al.* (1996) in the context of N-body problems modelling planetary systems. Several processors π_h were constructed for the Strang method, leading to processed (corrected) methods $\hat{\psi}_h$ of generalized order $(2k, 2)$ for several $k > 1$ (see also McLachlan 1996, Laskar and Robutel 2001), so that their local errors are $O(\varepsilon h^{2k+1} + \varepsilon^2 h^3)$ for a prescribed time integration.

In Section 5.7 we construct a different processor map that effectively reduces the local error to $O(\varepsilon^2 h^3)$ in the case where f_1 is derived from a Hamiltonian $H_1(q, p)$ of a harmonic oscillator (or a collection of harmonic oscillators whose frequencies are integer multiples of a basic frequency) and f_2 comes from a Hamiltonian $\varepsilon H_2(q)$ which is a polynomial in q. This means, going back to the results displayed in Figure 1.2 for the pendulum problem, that both methods $\mathrm{LT}_{\mathrm{pert}}$ and $(2, 2)$ are conjugate to a more accurate integrator $\hat{\psi}_h$ (with global error of order $O(t\varepsilon^2 h^2)$) obtained from them with an appropriate processor map π_h.

Assume that, in the previous notation, $(2, 2)$ (resp. $\mathrm{LT}_{\mathrm{pert}}$) corresponds to the numerical solution x_n, and $\hat{x}_n$ to the processed method. Then the error (4.17) is dominated by $\|\bar{x}_n - x_n\|$, which does not show secular growth in this case. Eventually, for long enough integration intervals, (4.17) will be dominated by the linearly increasing error $\|\hat{x}_n - x(nh)\|$ of the more accurate processed integrator.[4] The situation is very similar for the evolution of errors in position displayed for the six-body problem in Figure 1.4, the only difference being that the propagation $\|\bar{x}_n - x_n\|$ of differences in initial values now grows linearly. The error $\|\hat{x}_n - x(nh)\|$ will eventually dominate in (4.17) because $\|\bar{x}_n - x_n\|$ increases linearly with a smaller slope than $\|\hat{x}_n - x(nh)\|$.

Instead of enhancing a previously existing rth-order integrator by processing, we may also design a processed splitting method from scratch (López-Marcos *et al.* 1997, Blanes, Casas and Ros 1999*b*): determine the h-parametric maps ψ_h and π_h as compositions of basic flows $\varphi^{[1]}_{a_j h}$ and $\varphi^{[2]}_{b_j h}$ with different sequences of coefficients a_j and b_j, such that the processed map $\hat{\psi}_h = \pi_h^{-1} \circ \psi_h \circ \pi_h$ is a good approximation of φ_h for sufficiently small step sizes h. Typically, we require that $\pi_h^{-1} \circ \psi_h \circ \pi_h(x) - \varphi_h(x) = O(h^{r+1})$, so that the processed integrator is of order r. In that case, if we intend to compute the approximations (4.18) of $x(km)$ (for $k = 1, 2, 3 \ldots$), there is no need for the map ψ_h (referred to in this context as the kernel) to be an rth order approximation of φ_h.

In any of the situations considered above (either analysing the performance of a given splitting integrator with the help of a conjugacy map π_h, or enhancing an existing rth-order splitting integrator by processing, or designing a splitting processing integrator from scratch), we need to study the *effective order conditions* of ψ_h. These are the conditions on the parameters a_i, b_i that guarantee that there exists a processor map π_h such that $\pi_h^{-1} \circ \psi_h \circ \pi_h(x) - \varphi_h(x) = O(h^{r+1})$. A general treatment of the effective order conditions of several classes of numerical integrators including splitting methods and composition methods can be found in Blanes, Casas and Murua (2004, 2006*a*). That treatment is based on the series expansion (2.21) of the formal logarithm of the Lie transformation $\Psi(h)$, and it is shown that the conditions for effective order r can be written in terms of the coefficients w_1, w_2, w_{12}, $w_{122}, \ldots$ featuring in (2.21). In addition, the formal

[4] This can also be checked in Figure 5.1, where the same problem is integrated with scheme (2, 2) and processed Strang with a larger time step and a longer time interval.

logarithm of the Lie transformation $P(h)$ of the map π_h is determined as

$$\log(P(h)) = h(p_1F_1 + p_2F_2) + h^2 p_{12}F_{12} + h^3(p_{122}F_{122} + p_{112}F_{112}) + h^4(p_{1222}F_{1222} + p_{1122}F_{1122} + p_{1112}F_{1112}) + \cdots + O(h^{r+1}),$$

with the coefficients $p_{\ell_1\cdots\ell_k}$ given as polynomials of the coefficients in (2.21). Alternative ways of obtaining the effective order conditions of splitting methods can be derived following the different approaches considered in Section 2 for analysing the standard order conditions. However, this is beyond the scope of the present work. In any case, the analysis shows that many of the order conditions of the processed method $\hat{\psi}_h$ can be satisfied by π_h, so that ψ_h must fulfil a greatly reduced set of restrictions, also of lower complexity. As a result, it is possible to construct processed schemes as compositions of basic maps with a reduced computational cost.

5. Highly oscillatory problems

In this section we consider Hamiltonian systems of the form (4.8) with the splitting (4.9), that is,

$$H = H_1 + H_2, \quad H_1(q,p) = \frac{1}{2}p^\top M^{-1}p + \frac{1}{2}q^\top N q, \quad H_2(q,p) = U(q), \tag{5.1}$$

when M and N are real symmetric matrices, and $U\colon \mathbb{R}^d \to \mathbb{R}$ is a polynomial function. The corresponding equations of motion (4.7) can be rewritten as

$$x' = f_1(x) + f_2(x), \quad \text{with} \quad f_1(x) = Ax,\ f_2(x) = \begin{pmatrix} 0 \\ -\nabla U(q) \end{pmatrix} \tag{5.2}$$

in terms of $x = (q,p)^\top$ and the matrix A given in (4.11). Splitting methods are advantageous for system (5.1) provided $\mathrm{e}^{tA}x$ can be cheaply computed for each $x \in \mathbb{R}^D$, $D = 2d$.

For the time being, we assume that A is fully diagonalizable and the eigenvalues of A are integer multiples of $\omega\,\mathrm{i}$ (with i the imaginary unit). (The more general case where the eigenvalues of A lie on the imaginary axis will be considered in Section 5.8.) This implies that e^{tA} is $2\pi/\omega$-periodic in t. For system (5.1), the present assumption is equivalent to stating that the matrix $M^{-1}N$ is fully diagonalizable with all its eigenvalues of the form $-(\omega k)^2$, with $k \in \mathbb{Z}$. In other words, H_1 in (5.1) is just a collection of harmonic oscillators whose frequencies are integer multiples of a basic frequency ω.

Splitting methods applied to (5.2) can be analysed by considering series expansions in powers of h and using standard tools, in particular the material presented in the previous sections. Thus Section 3.3 is particularly relevant if the basic frequency ω is large compared to the size of the potential $U(q)$ (or more generally, the size of the components of $f_2(x)$). Indeed, rescaling time from t to $\tau = \omega t$, system

(5.2) is transformed into

$$\frac{\mathrm{d}}{\mathrm{d}\tau}x = \hat{f}_1(x) + \varepsilon f_2(x), \quad \hat{f}_1(x) = \hat{A}x,$$

where $\varepsilon = 1/\omega$ and $\hat{A} = \varepsilon A$, so that all the eigenvalues of $\hat{A}$ are integer multiples of the imaginary unit i. This approach nevertheless has an important limitation: it is meaningful only when $\omega h = h/\varepsilon$ is sufficiently small.

Different approaches have been adopted in the literature to overcome this limitation and obtain results that remain valid when ωh is large (provided that the step size h is small enough compared to the size of the perturbing potential $U(q)$ and its partial derivatives). Among them, we mention modulated Fourier expansions (see Hairer *et al.* 2006 and references therein) and extended word series (Murua and Sanz-Serna 2017, 2016). Extended word series were introduced in Murua and Sanz-Serna (2017) to analyse splitting methods for a class of problems that is equivalent to (5.2) under the more general assumption that all the eigenvalues of A lie on the imaginary axis. Such expansions were further used in Murua and Sanz-Serna (2016) to analyse normal forms and formal invariants of more general classes of problems. The formalism of extended word series allows us to work with asymptotic expansions valid for step sizes h that are sufficiently small independently of the frequencies of e^{tA}.

In the present section we provide an elementary derivation of second-order versions in h of such expansions, taking the Strang splitting as a case study. In particular, we provide a theoretical justification for the results presented in Section 1.4 for the simple pendulum and analyse the processing technique as a way to further improve those results, which are indeed valid for the general system (5.1). This is done by first constructing an asymptotic expansion of the exact solution of (5.2) and then comparing with the expansion corresponding to the numerical approximation obtained by a general splitting method. We also get the modified equation satisfied by the numerical scheme (exact up to terms in h^3), with explicit formulas for the coefficients, and the corresponding modified Hamiltonian. Furthermore, we explicitly construct a processor for the Strang splitting so that the resulting scheme leads to a better approximation to the solution of (5.2) (in particular, with a better preservation of the energy H for large time intervals). Finally, we indicate how the preceding results can be generalized to the more general case where the eigenvalues of A lie on the imaginary axis (so that, in general, e^{tA} is quasi-periodic in t).

For the analysis it is convenient to reformulate the problem (5.2) into the new variables $y(t)$ given through $x(t) = \mathrm{e}^{tA}y(t)$, so that now $y' = \mathrm{e}^{-tA}f_2(\mathrm{e}^{tA}y)$. Let us write

$$\mathrm{e}^{-tA}f_2(\mathrm{e}^{tA}x) = \sum_{k\in\mathbb{Z}} \mathrm{e}^{\mathrm{i}k\omega t}\, g_k(x), \tag{5.3}$$

that is, the right-hand side of (5.3) is the Fourier series expansion of $\mathrm{e}^{-tA}f_2(\mathrm{e}^{tA}x)$.

The map f_2 being real implies that $g_k : \mathbb{R}^D \to \mathbb{C}^D$ is such that each component of $g_{-k}(x)$ is the complex conjugate of the corresponding component of $g_k(x)$.

Since we have assumed that $U(q)$ is a polynomial, each component of $f_2(x)$ is also a polynomial in the variables x, which guarantees that there is a finite number of non-zero terms in (5.3). We will use the notation

$$\mathcal{I} = \{k \in \mathbb{Z} \colon g_k \neq 0\}. \tag{5.4}$$

Remarks.

- The assumption that $U(q)$ and each component of $f_2(x)$ are polynomials in the variables x might seem too restrictive. However, for more general assumptions (e.g. real-analyticity), we can always replace $U(q)$ and $f_2(x)$ with sufficiently accurate polynomial approximations. Furthermore, the material in the present section is formally valid for more general assumptions on $U(q)$ and $f_2(x)$ if we allow the set (5.4) to be infinite. In that case, (5.3) will be an infinite series, several of the formulae derived here will also involve infinite series, and appropriate assumptions should be made so as to guarantee convergence.
- The assumptions that A is a real matrix and that $f_2 \colon \mathbb{R}^D \to \mathbb{R}^D$ are not essential. We could consider a complex matrix A and $f_2 \colon \mathbb{C}^D \to \mathbb{C}^D$, with no changes in the formulae that follow.

By substitution of $t = 0$ into (5.3), we get $f_2(x) = \sum_{k\in\mathcal{I}} g_k(x)$. Thus, the solution of the initial value problem defined by (5.2) and $x(0) = x_0 \in \mathbb{R}^D$ can be expressed as $x(t) = \mathrm{e}^{tA} y(t)$, where $y(t)$ is the solution of

$$\frac{\mathrm{d}}{\mathrm{d}t} y = \sum_{k\in\mathcal{I}} \mathrm{e}^{\mathrm{i}k\omega t} g_k(y), \quad y(0) = x_0. \tag{5.5}$$

From the definition of the Fourier coefficients g_k in (5.3), we can prove that

$$\mathrm{e}^{-tA} g_k(\mathrm{e}^{tA} x) = \mathrm{e}^{\mathrm{i}k\omega t}\, g_k(x), \quad \text{for } k \in \mathcal{I}, \tag{5.6}$$

and, by applying the operator $(\mathrm{d}/\mathrm{d}t)|_{t=0}$ to both sides, this is equivalent to (Murua and Sanz-Serna 2016)

$$(f_1, g_k) = \mathrm{i}k\,\omega\, g_k, \quad \text{for } k \in \mathcal{I}. \tag{5.7}$$

Here, $(\cdot,\cdot)$ represents the usual Lie–Poisson bracket already defined in Section 1: $(f_1, g_k)(x) = g_k'(x) f_1(x) - f_1'(x) g_k(x)$.

5.1. *Expansion of the exact solution*

We next obtain an approximate representation of the flow $\varphi_h^{[f_1+f_2]}$ of (5.2) with initial condition $x_0 \in \mathbb{R}^D$ that is valid for sufficiently small values of $|h|$ independently of the basic frequency ω. This is done by getting an expansion of the solution $y(t)$ of (5.5) valid for $|t| \leq h$. To begin with, we apply the substitution $y(t) = x_0 + O(t)$ on the right-hand side of the integral form

$$y(t) = x_0 + \sum_{k \in \mathcal{I}} \int_0^t e^{ik\omega s} g_k(y(s))\,ds$$

of the initial value problem. This gives

$$y(t) = x_0 + \sum_{k \in \mathcal{I}} \int_0^t e^{ik\omega s} g_k(x_0)\,ds + O(t^2). \tag{5.8}$$

Furthermore, $y(h)$ admits the estimate

$$y(h) = x_0 + \sum_{\ell \in \mathcal{I}} \int_0^h e^{i\ell\omega t}(g_\ell(x_0) + g'_\ell(x_0)(y(t) - x_0))\,dt + O(h^3), \tag{5.9}$$

where $g'_\ell(x_0)$ is the value at $x = x_0$ of the Jacobian matrix of $g_\ell(x)$.

Substitution of (5.8) into the right-hand side of (5.9) finally leads to

$$\begin{aligned} y(h) = x_0 &+ \sum_{k \in \mathcal{I}} \left(\int_0^h e^{ik\omega t}\,dt \right) g_k(x_0) \\ &+ \sum_{k,\ell \in \mathcal{I}} \left(\int_0^h \int_0^t e^{i(\ell t + ks)\omega}\,ds\,dt \right) g'_\ell(x_0) g_k(x_0) + O(h^3). \end{aligned}$$

Equivalently,

$$y(h) = x_0 + h \sum_{k \in \mathcal{I}} \alpha_k(h) g_k(x_0) + h^2 \sum_{k,\ell \in \mathcal{I}} \alpha_{k\ell}(h) g'_\ell(x_0) g_k(x_0) + O(h^3), \tag{5.10}$$

where the coefficients are defined as follows.

- For $k \in \mathcal{I}$,

$$\alpha_k(h) = \int_0^1 e^{ik\omega h\tau}\,d\tau = \begin{cases} 1 & \text{if } k = 0, \\ \dfrac{e^{ik\omega h} - 1}{ik\omega h} & \text{otherwise.} \end{cases} \tag{5.11}$$

- For $k, \ell \in \mathcal{I}$,

$$\alpha_{k\ell}(h) = \int_0^1 \int_0^{\tau_2} e^{ih\omega(k\tau_1 + \ell\tau_2)}\,d\tau_1\,d\tau_2. \tag{5.12}$$

  Specifically, $\alpha_{00}(h) = 1/2$, and the following recursions hold:

$$\alpha_{0k}(h) = \frac{e^{ik\omega h} - \alpha_k(h)}{ik\omega h}, \quad \alpha_{k\ell}(h) = \frac{\alpha_{k+\ell}(h) - \alpha_\ell(h)}{ik\omega h} \quad \text{for } k, \ell \in \mathcal{I} \backslash \{0\}. \tag{5.13}$$

- The constant in the $O(h^3)$ term depends on upper bounds of the norm of g_k and its partial derivatives, but is independent of the basic frequency ω.

From this we conclude that

$$\varphi_h(x_0) = \mathrm{e}^{hA}\left(x_0 + h \sum_{k\in\mathcal{I}} \alpha_k(h) g_k(x_0) + h^2 \sum_{k,\ell\in\mathcal{I}} \alpha_{k\ell}(h) g'_\ell(x_0) g_k(x_0) + O(h^3)\right), \tag{5.14}$$

where we have dropped the upper index in φ_h for clarity. Observe that

$$|\alpha_k(h)| = |\mathrm{sinc}(kh\omega/2)| \le 1, \quad |\alpha_{k\ell}(h)| \le \frac{1}{2}.$$

5.2. *Expansion of the discrete solution given by Strang splitting*

We next proceed to construct an analogous expansion for the (discrete) solution furnished by the Strang splitting given in (4.10), namely

$$S_h^{[RKR]} = \varphi_{h/2}^{[R]} \circ \varphi_h^{[K]} \circ \varphi_{h/2}^{[R]},$$

based on the maps (4.11). In fact, it is straightforward to check that the approximation $\tilde{x}(h) = S_h^{[RKR]}(x_0)$ satisfies $\tilde{x}(h) = \mathrm{e}^{hA}\tilde{y}(h)$, where $\tilde{y}(t)$ is the solution of

$$\frac{\mathrm{d}}{\mathrm{d}t}\tilde{y} = \mathrm{e}^{-\frac{h}{2}A} g(\mathrm{e}^{\frac{h}{2}A}\tilde{y}(t)) = \sum_{k\in\mathcal{I}} \mathrm{e}^{\mathrm{i}k\omega h/2}\, g_k(\tilde{y}(t)), \quad \tilde{y}(0) = x_0.$$

In consequence, proceeding as in the previous subsection, we get

$$\tilde{y}(h) = x_0 + h \sum_{k\in\mathcal{I}} \tilde{\alpha}_k(h) g_k(x_0) + h^2 \sum_{k,\ell\in\mathcal{I}} \tilde{\alpha}_{k\ell}(h) g'_\ell(x_0) g_k(x_0) + O(h^3),$$

where for $k, \ell \in \mathcal{I}$,

$$\tilde{\alpha}_k(h) = \mathrm{e}^{\mathrm{i}k\omega h/2}, \quad \tilde{\alpha}_{k\ell}(h) = \frac{1}{2}\mathrm{e}^{\mathrm{i}(k+\ell)\omega h/2}, \tag{5.15}$$

and finally

$$S_h(x_0) = \mathrm{e}^{hA}\left(x_0 + h \sum_{k\in\mathcal{I}} \tilde{\alpha}_k(h) g_k(x_0) + h^2 \sum_{k,\ell\in\mathcal{I}} \tilde{\alpha}_{k\ell}(h) g'_\ell(x_0) g_k(x_0) + O(h^3)\right), \tag{5.16}$$

where we have also dropped the upper index in S_h. Notice that, from (5.15),

$$|\tilde{\alpha}_k(h)| \le 1, \quad |\tilde{\alpha}_{k\ell}(h)| \le \frac{1}{2}$$

and

$$\tilde{\alpha}_{k\ell}(h) + \tilde{\alpha}_{\ell k}(h) = \tilde{\alpha}_k(h)\tilde{\alpha}_\ell(h), \quad k, \ell \in \mathcal{I}. \tag{5.17}$$

5.3. *Composition formulae*

If we are interested in extending the previous analysis to more general splitting methods of the form (2.17), then a composition rule concatenating the expansions corresponding to different basic methods is clearly needed. This can be obtained as follows.

Suppose the map $\psi_h : \mathbb{R}^D \to \mathbb{R}^D$ can be expanded as

$$\psi_h(x) = x + h \sum_{k\in\mathcal{I}} \kappa_k\, g_k(x) + h^2 \sum_{k,\ell\in\mathcal{I}} \kappa_{k\ell}\, g'_\ell(x) g_k(x) + O(h^3), \tag{5.18}$$

for some coefficients $\kappa_k, \kappa_{k,\ell}$. Then, from (5.6), we have for all $s \in \mathbb{R}$ that

$$\begin{aligned}\psi_h(\mathrm{e}^{sA}x) = \mathrm{e}^{sA}\Bigg(& x + h \sum_{k\in\mathcal{I}} \mathrm{e}^{\mathrm{i}k\omega s} \kappa_k\, g_k(x) \\ & + h^2 \sum_{k,\ell\in\mathcal{I}} \mathrm{e}^{\mathrm{i}(k+\ell)\omega s} \kappa_{k\ell}\, g'_\ell(x) g_k(x) + O(h^3)\Bigg).\end{aligned} \tag{5.19}$$

If in addition the map $\hat\psi_h : \mathbb{R}^D \to \mathbb{R}^D$ can be expanded as

$$\hat\psi_h(x) = x + h \sum_{k\in\mathcal{I}} \hat\kappa_k\, g_k(x) + h^2 \sum_{k,\ell\in\mathcal{I}} \hat\kappa_{k\ell}\, g'_\ell(x) g_k(x) + O(h^3),$$

then we have the following expression for the composition $\psi_h \circ \hat\psi_h$:

$$\begin{aligned}\psi_h(\hat\psi_h(x)) = x &+ h \sum_{k\in\mathcal{I}} (\kappa_k + \hat\kappa_k)\, g_k(x) \\ &+ h^2 \sum_{k,\ell\in\mathcal{I}} (\kappa_{k\ell} + \hat\kappa_k\, \kappa_\ell + \hat\kappa_{k\ell})\, g'_\ell(x) g_k(x) + O(h^3).\end{aligned}$$

This is also valid if the coefficients κ_k, $\hat\kappa_k$, $\kappa_{k\ell}$ and $\hat\kappa_{k\ell}$ depend on h, although in that case the constant in the $O(h^3)$ term will also depend on the bounds of the coefficients.

By combining the previous results we finally arrive at the following composition rule:

$$\mathrm{e}^{sA}\psi_h\big(\mathrm{e}^{\hat sA}\hat\psi_h(x)\big) = \mathrm{e}^{(s+\hat s)A}\Bigg(x + h \sum_{k\in\mathcal{I}} \gamma_k\, g_k(x) + h^2 \sum_{k,\ell\in\mathcal{I}} \gamma_{k\ell}\, g'_\ell(x) g_k(x) + O(h^3)\Bigg), \tag{5.20}$$

where

$$\begin{aligned}\gamma_k &= \mathrm{e}^{\mathrm{i}\hat s\omega k}\, \kappa_k + \hat\kappa_k, \\ \gamma_{k\ell} &= \mathrm{e}^{\mathrm{i}\hat s\omega(k+\ell)}\, \kappa_{k\ell} + \mathrm{e}^{\mathrm{i}\hat s\omega\ell}\, \hat\kappa_k\, \kappa_\ell + \hat\kappa_{k\ell}.\end{aligned} \tag{5.21}$$

5.4. *Expansions for arbitrary splitting methods*

We now have all the required ingredients to extend the expansion (5.16) for Strang to a more general splitting method of the form (2.17) based on kicks and rotations, i.e. on the maps (4.11). Specifically, if ψ_h denotes such a splitting, then $\psi_h(x_0)$ can be expanded as

$$e^{h(a_1+\cdots+a_{s+1})A}\left(x_0 + h\sum_{k\in\mathcal{I}}\tilde{\alpha}_k(h)g_k(x_0) + h^2\sum_{k,\ell\in\mathcal{I}}\tilde{\alpha}_{k\ell}(h)g'_\ell(x_0)g_k(x_0) + O(h^3)\right) \tag{5.22}$$

with the following coefficients:

- for $k \in \mathcal{I}$,

$$\tilde{\alpha}_k(h) = \sum_{j=1}^{s} b_j\, e^{ikc_j\omega h}, \quad \text{with} \quad c_j = a_1 + \cdots + a_j, \quad \text{for } j = 1,\ldots,s; \tag{5.23}$$

- for $k,\ell \in \mathcal{I}$,

$$\tilde{\alpha}_{k\ell}(h) = \sum_{1\le j<m\le s} b_j b_m\, e^{i(c_jk+c_m\ell)\omega h} + \sum_{1\le j\le r}\frac{1}{2}b_j^2\, e^{ic_j(k+\ell)\omega h}, \tag{5.24}$$

so that they again satisfy the relation (5.17), since it is preserved under compositions (5.20)–(5.21).

We can now estimate the local error of a consistent splitting method by taking into account (5.22) and (5.14) as

$$\psi_h(x)-\varphi_h(x) = e^{hA}\left(x_0+h\sum_{k\in\mathcal{I}}\delta_k(h)g_k(x_0)+h^2\sum_{k,\ell\in\mathcal{I}}\delta_{k\ell}(h)g'_\ell(x_0)g_k(x_0)+O(h^3)\right),$$

where

$$\begin{aligned}\delta_k(h) &= \sum_{j=1}^{s} b_j\, e^{ikc_j\omega h} - \int_0^1 e^{ik\omega h\tau}\, d\tau,\\ \delta_{k\ell}(h) &= \sum_{1\le j<m\le s} b_j b_m\, e^{i(c_jk+c_m\ell)\omega h} + \sum_{1\le j\le r}\frac{1}{2}b_j^2\, e^{ic_j(k+\ell)\omega h}\\ &\quad - \int_0^1\int_0^{\tau_2} e^{ih\omega(k\tau_1+\ell\tau_2)}\, d\tau_1\, d\tau_2.\end{aligned} \tag{5.25}$$

It is worth remarking that if we expand the exponentials in (5.25) in series of powers of h up to a certain degree r, then we recover the order conditions obtained in Section 2.4 for multi-indices with one and two indices. In other words, using the terminology introduced in Section 3.3 to analyse perturbed problems of the form (3.8), we obtain the conditions for a splitting method to be of generalized order $(r_1, r_2, 1)$ (or $(r_1, r_2, 2)$ in the case of time-symmetric splitting methods). Notice, however, that replacing these exponentials with such truncated series expansions will only give useful information about the size of the local error coefficients $\delta_k(h)$ and $\delta_{k\ell}(h)$ provided that the scaled step sizes $\{|k|\omega h\colon\ k \in \mathcal{I}\}$ are sufficiently small.

5.5. *Modified ODE for the discrete flow furnished by a splitting method*

At this point it is useful to formulate a modified differential equation whose h-flow is closer in a certain sense to the map ψ_h corresponding to a general splitting method admitting an expansion of the form (5.22). More precisely, the idea is to construct a modified ODE whose h-flow is $O(h^3)$ close to ψ_h.

From the composition formula (5.20), the nth iterate of the map ψ_h admits an expansion of the form

$$\psi_h^n(x_0) = \mathrm{e}^{hnA}\Big(x_0 + h\sum_{k\in\mathcal{I}} \gamma_k(n,h) g_k(x_0) + h^2 \sum_{k,\ell\in\mathcal{I}} \gamma_{k\ell}(n,h) g_\ell'(x_0) g_k(x_0) + O(h^3)\Big),$$

where $\gamma_k(0,h) = 0$, $\gamma_{k\ell}(0,h) = 0$, $\gamma_k(1,h) = \tilde{\alpha}_k(h)$, $\gamma_{k\ell}(1,h) = \tilde{\alpha}_{k\ell}(h)$. Thus, to find a suitable modified ODE, it makes sense to assume that its corresponding t-flow $\tilde{\varphi}_t$ can be expanded for all $t \in \mathbb{R}$ as

$$\tilde{\varphi}_t(x_0) = \mathrm{e}^{tA}\Big(x_0 + h\sum_{k\in\mathcal{I}} \gamma_k(t/h,h) g_k(x_0) + h^2 \sum_{k,\ell\in\mathcal{I}} \gamma_{k\ell}(t/h,h) g_\ell'(x_0) g_k(x_0) + \cdots\Big). \tag{5.26}$$

Then the right-hand side of the corresponding ODE must be of the form

$$\frac{\mathrm{d}}{\mathrm{d}t}\tilde{\varphi}_t(x)\bigg|_{t=0} = Ax + \sum_{k\in\mathcal{I}} \beta_k(h) g_k(x) + h \sum_{k,\ell\in\mathcal{I}} \beta_{k\ell}(h) g_\ell'(x) g_k(x) + \cdots,$$

where

$$\beta_k(h) = \frac{\mathrm{d}}{\mathrm{d}\tau}\gamma_k(\tau,h)\bigg|_{\tau=0}, \quad \beta_{k\ell}(h) = \frac{\mathrm{d}}{\mathrm{d}\tau}\gamma_{k\ell}(\tau,h)\bigg|_{\tau=0}. \tag{5.27}$$

We are then bound to study the family of perturbed ODEs

$$\frac{\mathrm{d}}{\mathrm{d}t}x = Ax + \sum_{k\in\mathcal{I}} \beta_k(h)\, g_k(x) + h \sum_{k,\ell\in\mathcal{I}} \beta_{k\ell}(h)\, g_\ell'(x) g_k(x), \tag{5.28}$$

with initial condition $x(0) = x_0$. The coefficients $\beta_k(h)$ and $\beta_{k\ell}(h)$ are, for fixed h, arbitrary complex numbers. Notice that this equation generalizes the original ODE (5.2), which corresponds to the case where $\beta_k(h) = 1$ and $\beta_{k\ell}(h) = 0$.

Assuming that the t-flow $\tilde{\varphi}_t$ of (5.28) can be expanded as in (5.26), the group property of the flow (i.e. $\tilde{\varphi}_{t+s} = \tilde{\varphi}_t \circ \tilde{\varphi}_s$), together with the composition formula (5.20), lead to

$$\begin{aligned}\gamma_k(\tau+\sigma,h) &= \mathrm{e}^{\mathrm{i}\omega h\tau k}\,\gamma_k(\sigma,h) + \gamma_k(\tau,h),\\ \gamma_{k\ell}(\tau+\sigma,h) &= \mathrm{e}^{\mathrm{i}\omega h\tau(k+\ell)}\,\gamma_{k\ell}(\sigma,h) + \mathrm{e}^{\mathrm{i}\omega h\tau\ell}\,\gamma_k(\tau,h)\,\gamma_\ell(\sigma,h) + \gamma_{k\ell}(\tau,h).\end{aligned} \tag{5.29}$$

Now, applying the operator $(\mathrm{d}/\mathrm{d}\sigma)|_{\sigma=0}$ to both sides of (5.29) and using (5.27), we obtain the following system of ODEs for the coefficients $\gamma_k(\tau,h)$, $\gamma_{k\ell}(\tau,h)$:

$$\frac{\mathrm{d}}{\mathrm{d}\tau}\gamma_k(\tau,h) = \mathrm{e}^{\mathrm{i}\omega h\tau k}\,\beta_k(h),$$
$$\frac{\mathrm{d}}{\mathrm{d}\tau}\gamma_{k\ell}(\tau,h) = \mathrm{e}^{\mathrm{i}\omega h\tau(k+\ell)}\,\beta_{k\ell}(h) + \mathrm{e}^{\mathrm{i}\omega h\tau\ell}\,\beta_\ell(h)\,\gamma_k(\tau,h).$$

This, together with the initial conditions $\gamma_k(0,h) = 0$, $\gamma_{k\ell}(0,h) = 0$, allows us to express $\gamma_k(\tau,h)$, $\gamma_{k\ell}(\tau,h)$ in terms of the coefficients $\beta_k(h)$, $\beta_{k\ell}(h)$ of the perturbed ODE (5.28):

$$\gamma_k(\tau,h) = \beta_k(h)\int_0^\tau \mathrm{e}^{\mathrm{i}\omega h\sigma k}\,\mathrm{d}\sigma = \tau\,\beta_k(h)\,\alpha_k(\tau h),$$
$$\gamma_{k\ell}(\tau,h) = \tau\,\beta_{k\ell}(h)\,\alpha_{k+\ell}(\tau h) + \tau^2\,\beta_\ell(h)\,\alpha_{k\ell}(\tau h).$$

Specifying these equations to the case $\tau = 1$, and replacing $\gamma_k(1,h)$ and $\gamma_{k\ell}(1,h)$ with $\tilde{\alpha}_k(h)$ and $\tilde{\alpha}_{k\ell}(h)$ (the coefficients corresponding to ψ_h), respectively, we have

$$\begin{aligned}\tilde{\alpha}_k(h) &= \beta_k(h)\,\alpha_k(h),\\ \tilde{\alpha}_{k\ell}(h) &= \beta_{k\ell}(h)\,\alpha_{k+\ell}(h) + \beta_k(h)\,\beta_\ell(h)\,\alpha_{k\ell}(h),\end{aligned} \tag{5.30}$$

where the expression of $\alpha_k(h)$, for all $k \in \mathbb{Z}$, is given by (5.11). Hence $\alpha_0(h) = 1$ and, for any $k \neq 0$, $\alpha_k(h) \neq 0$ if and only if $kh \neq 2\pi j$ for all $j \in \mathbb{Z}$. Therefore, if we assume that $h \in \mathbb{R}$ is such that, for all $k, \ell \in \mathcal{I}\backslash\{0\}$,

$$\frac{k\omega h}{2\pi} \notin \mathbb{Z}\backslash\{0\} \quad \text{and} \quad \frac{(k+\ell)\omega h}{2\pi} \notin \mathbb{Z}\backslash\{0\} \tag{5.31}$$

(so that $\alpha_k(h) \neq 0$ and $\alpha_{k+\ell}(h) \neq 0$), then equations (5.30) can be solved in the $\beta_k(h)$ and $\beta_{k+\ell}(h)$ coefficients. Thus, for each $k, \ell \in \mathcal{I}$, we have

$$\beta_k(h) = \frac{\tilde{\alpha}_k(h)}{\alpha_k(h)}, \quad \beta_{k\ell}(h) = \frac{\tilde{\alpha}_{k\ell}(h) - \alpha_{k\ell}(h)\,\beta_k(h)\beta_\ell(h)}{\alpha_{k+\ell}(h)}, \tag{5.32}$$

and the h-flow of the corresponding modified equation (5.28) agrees up to terms of order $O(h^3)$ with the expansion (5.22) of the splitting method ψ_h. Moreover, the identity (5.17) satisfied by the coefficients in the expansions of both the splitting method ψ_h and the exact solution, equations (5.11)–(5.12), implies

$$\beta_{k\ell}(h) + \beta_{\ell k}(h) = 0, \quad k, \ell \in \mathcal{I}, \tag{5.33}$$

so that the modified equation (5.28) can also be expressed as

$$\begin{aligned}\frac{\mathrm{d}}{\mathrm{d}t}x &= Ax + \sum_{k\in\mathcal{I}}\beta_k(h)\,g_k(x) + \frac{h}{2}\sum_{k,\ell\in\mathcal{I}}\beta_{k\ell}(h)\,(g_\ell'(x)g_k(x) - g_k'(x)g_\ell(x))\\ &= Ax + \sum_{k\in\mathcal{I}}\beta_k(h)\,g_k(x) + \frac{h}{2}\sum_{k,\ell\in\mathcal{I}}\beta_{k\ell}(h)\,(g_k, g_\ell)(x).\end{aligned} \tag{5.34}$$

In fact, equation (5.34) is itself Hamiltonian, in the sense that there exists a Hamiltonian function $\tilde{H}(x;h)$ such that

$$\frac{\mathrm{d}}{\mathrm{d}t}x = J\,\nabla\tilde{H}(x;h), \tag{5.35}$$

where J is the canonical symplectic matrix (1.21). This can be seen as follows. First, H_1 in the Hamiltonian (5.1) can be written as

$$H_1(x) = \frac{1}{2}x^\top Qx, \quad \text{with} \quad Q = \begin{pmatrix} N & 0 \\ 0 & M^{-1} \end{pmatrix}, \tag{5.36}$$

whereas $H_2(\mathrm{e}^{tA}x)$ has a Fourier expansion of the form

$$H_2(\mathrm{e}^{tA}x) = \sum_{k\in\mathcal{I}} \mathrm{e}^{\mathrm{i}k\omega t} G_k(x), \quad \text{with} \quad g_k(x) = J\,\nabla G_k(x), \tag{5.37}$$

and furthermore, for all $k \in \mathcal{I}$,

$$G_k(\mathrm{e}^{t\,A}x) = \mathrm{e}^{\mathrm{i}k\,\omega t} G_k(x), \tag{5.38}$$

so that (5.7) is equivalent to

$$\{H_1, G_k\} = \mathrm{i}k\omega\, G_k, \quad \text{for } k \in \mathcal{I}. \tag{5.39}$$

Here $\{A, B\}$ stands for the Poisson bracket of $A, B \in C^1(\mathbb{R}^D)$ defined as follows: for each $x \in \mathbb{R}^D$,

$$\{A, B\}(x) = (\nabla A(x))^\top J\,\nabla B(x).$$

It is then clear that the modified ODE (5.34) can be written like (5.35) with

$$\tilde{H}(x;h) = \frac{1}{2}x^\top Qx + \sum_{k\in\mathcal{I}} \beta_k(h)\,G_k(x) + \frac{h}{2}\sum_{k,\ell\in\mathcal{I}} \beta_{k\ell}(h)\,\{G_k, G_\ell\}(x). \tag{5.40}$$

We should stress that both (5.28) and the modified Hamiltonian $\tilde{H}(x;h)$ are well-defined as long as the non-resonance assumptions (5.31) hold.

5.6. Splitting methods with processing

In the spirit of Section 4.5, we now consider a *processed* splitting integrator

$$\hat{\psi}_h = \pi_h^{-1} \circ \psi_h \circ \pi_h,$$

where ψ_h is a composition of type (2.17) based on kicks and rotations and $\pi_h\colon \mathbb{R}^D \to \mathbb{R}^D$ is a near-to-identity map with an expansion of the form

$$\pi_h(x) = x + h\sum_{k\in\mathcal{I}} \kappa_k(h)\,g_k(x) + h^2 \sum_{k,\ell\in\mathcal{I}} \kappa_{k\ell}(h)\,g'_\ell(x)g_k(x) + O(h^3). \tag{5.41}$$

In contrast to Section 4.5, where the processor map π_h is analysed by its power series expansion in h, the coefficients $\kappa_k(h)$ and $\kappa_{k\ell}(h)$ featuring in (5.41) will depend on h and the frequency ω in a non-polynomial way. This will allow us to find appropriate processors valid for step sizes h that are not necessarily small compared with ω.

Application of the composition rule (5.20)–(5.21) to both sides of the identity $\pi_h \circ \hat{\psi}_h = \psi_h \circ \pi_h$ implies that $\hat{\psi}_h$ can be expanded as

$$\hat{\psi}_h(x) = \mathrm{e}^{hA}\left(x + h\sum_{k\in\mathcal{I}} \hat{\alpha}_k(h) g_k(x) + h^2 \sum_{k,\ell\in\mathcal{I}} \hat{\alpha}_{k\ell}(h) g'_\ell(x) g_k(x) + O(h^3)\right), \quad (5.42)$$

where

$$\hat{\alpha}_k(h) = (1 - \mathrm{e}^{\mathrm{i}k\omega h})\,\kappa_k(h) + \tilde{\alpha}_k(h),$$
$$\hat{\alpha}_{k\ell}(h) = (1 - \mathrm{e}^{\mathrm{i}(k+\ell)\omega h})\,\kappa_{k\ell}(h) + \kappa_k(h)\,\tilde{\alpha}_\ell(h) - \mathrm{e}^{\mathrm{i}\ell\omega h}\kappa_\ell(h)\,\hat{\alpha}_k(h) + \tilde{\alpha}_{k\ell}(h). \quad (5.43)$$

The expansion of the processed scheme $\hat{\psi}_h$ coincides with that of the exact flow (5.14) if $\hat{\alpha}_k(h) = \alpha_k(h)$ for each $k \in \mathcal{I}\backslash\{0\}$, and this is possible only when the following non-resonance condition holds,

$$\frac{k\omega h}{2\pi} \notin \mathbb{Z}, \quad (5.44)$$

in which case

$$\kappa_k(h) = \frac{\tilde{\alpha}_k(h) - \alpha_k(h)}{\mathrm{e}^{\mathrm{i}k\omega h} - 1}. \quad (5.45)$$

Observe that, for $k = 0$, $\hat{\alpha}_k(h) = \alpha_k(h)$ regardless of the chosen value of $\kappa_0(h)$, since $\tilde{\alpha}_0(h) = \alpha_0(h) = 1$. For simplicity, it makes sense to choose the processor map in such a way that $\kappa_0(h) = 0$.

In consequence, if h satisfies the non-resonance condition (5.44) for all $k \in \mathcal{I}\backslash\{0\}$, and the coefficients $\kappa_k(h)$ for $k \neq 0$ are chosen as (5.45), then the local error of the processed scheme reads

$$\hat{\psi}_h(x) - \varphi_h(x) = \mathrm{e}^{hA}\left(h^2 \sum_{k,\ell\in\mathcal{I}} (\hat{\alpha}_{k\ell}(h) - \alpha_{k\ell}(h)) g'_\ell(x)\, g_k(x) + O(h^3)\right)$$

and the h-flow of $\hat{\psi}_h$ is also Hamiltonian, with the modified Hamiltonian function

$$\hat{H}(x;h) = \frac{1}{2}x^{\mathsf{T}}Qx + \sum_{k\in\mathcal{I}} \hat{\beta}_k(h)\, G_k(x) + \frac{h}{2}\sum_{k,\ell\in\mathcal{I}} \hat{\beta}_{k\ell}(h)\, \{G_k, G_\ell\}(x),$$

where

$$\hat{\beta}_k(h) = 1, \quad \hat{\beta}_{k\ell}(h) = \frac{\hat{\alpha}_{k\ell}(h) - \alpha_{k\ell}(h)}{\alpha_{k+\ell}(h)}. \quad (5.46)$$

Equivalently,

$$\hat{H}(x;h) = H(x) + \frac{h}{2}\sum_{k,\ell\in\mathcal{I}} \hat{\beta}_{k\ell}(h)\, \{G_k, G_\ell\}(x), \quad (5.47)$$

which is an $O(h)$ perturbation of the original Hamiltonian, and thus we may expect that the value of the $H(x)$ (typically the energy of the original system) will be well approximated for relatively large time intervals. A word of caution is in order here: by construction, the difference between the processed map $\hat{\psi}_h$ and the h-flow of the modified Hamiltonian (5.47) is formally of order $O(h^3)$, but the constant in $O(h^3)$ depends on the size of the modulus of the coefficients

$$\kappa_k(h), \quad \kappa_{k,\ell}(h), \quad \hat{\beta}_{k\ell}(h) \quad \text{for } k,\ell \in \mathcal{I}.$$

For near-resonant step sizes h, that is, for step sizes such that $\mathrm{e}^{\mathrm{i}k\omega h} - 1$ is small for some index k belonging to $\mathcal{I}$ or $\mathcal{I}+\mathcal{I}$, the size of some of these critical coefficients may become large. In such cases, we cannot expect that the processed map $\hat{\psi}_h$ will be close from the h-flow of the modified Hamiltonian (5.47).

5.7. *A processed Strang scheme*

In the particular case in which ψ_h is the Strang splitting $S_h^{[RKR]}$, the coefficients $\kappa_k(h)$ in the corresponding expansion (5.41) verifying (5.45) read

$$\kappa_k(h) = \frac{1}{\mathrm{i}k\omega h}(1 - \operatorname{sinc}(k\omega h/2)^{-1}), \quad k \in \mathcal{I}\backslash\{0\}, \tag{5.48}$$

and $\kappa_{-k}(h) = -\kappa_k(h)$ for all $k \neq 0$.

Assuming that the potential function $U(q)$ in (5.1) is a polynomial of degree m, so that

$$\mathcal{I} \subset \{-m,\dots,-1,0,1,\dots m\},$$

we next construct a fully explicit processor π_h for $S_h^{[RKR]}$. We define it as a composition of basic flows as follows:

$$\pi_h = \varphi_\alpha^{[R]} \circ \varphi_{b_{2m}(h)}^{[K]} \circ \varphi_\alpha^{[R]} \circ \varphi_{b_{2m-1}(h)}^{[K]} \circ \cdots \circ \varphi_{b_2(h)}^{[K]} \circ \varphi_\alpha^{[R]} \circ \varphi_{b_1(h)}^{[K]} \circ \varphi_\alpha^{[R]}, \tag{5.49}$$

where $\alpha = (2\pi)/(2m+1)$, and the coefficients $b_j(h)$, $j = 1,\dots,2m$, depend on h. Since $S_h^{[RKR]}$ is time-symmetric, then it makes sense to construct the processor π_h such that $\hat{\psi}_h = \pi_h^{-1} \circ \psi_h \circ \pi_h$ is also time-symmetric. This can be achieved by requiring that $\pi_{-h} = \pi_h$, or equivalently, by requiring that $b_j(-h) = b_j(h)$ for $j = 1,\dots,2m$. This condition, together with

$$\kappa_0(h) = 0, \quad \tilde{\alpha}_k(h) = \alpha_k(h) \quad \text{for all } k \in \{-m,\dots,-1,0,1,\dots m\},$$

uniquely determines the $b_j(h)$ coefficients as

$$b_j(h) = -b_{2m-j+1}(h) = \frac{2}{2m+1}\sum_{k=1}^{m}\frac{1}{k}(\operatorname{sinc}(k\omega h/2)^{-1} - 1)\sin\left(\frac{2kj\pi}{2m+1}\right), \tag{5.50}$$

for $j = 1,\dots,m$. This can be seen as follows: successive application of the composition rule (5.21) shows that (5.49) admits the expansion (5.41) with

$$h\,\kappa_k(h) = \sum_{j=1}^{2m} b_j(h)\,\mathrm{e}^{2\mathrm{i}kj\pi/(2m+1)}, \qquad k = -m, \dots, -1, 0, 1, \dots m, \tag{5.51}$$

and for $k, \ell \in \mathcal{I}$,

$$\begin{aligned} h^2 \kappa_{k\ell}(h) = & \sum_{1\le j<n\le 2m} b_j(h)\, b_n(h)\, \mathrm{e}^{2\pi(jk+n\ell)/(2m+1)} \\ & + \sum_{1\le j\le 2m} \frac{1}{2} b_j(h)^2\, \mathrm{e}^{\mathrm{i}2\pi j(k+\ell)/(2m+1)}. \end{aligned} \tag{5.52}$$

We can check that $\kappa_0(h) = 0$ and the symmetry condition $\kappa_{-k}(h) = -\kappa_k(h)$ $(k \neq 0)$ that holds for (5.48) imply that $b_j(h) = -b_{2m-j+1}(h)$ for $j = 1, \dots, 2m$. Now, (5.51) means that application of the inverse discrete Fourier transform to the vector $(0, b_1, b_2, \dots, b_m, -b_m, \dots, -b_1)$ gives the vector

$$\frac{h}{2m+1}(0, \kappa_1, \dots, \kappa_m, -\kappa_m, \dots, -\kappa_1).$$

Equivalently, the former is obtained by applying the discrete Fourier transform to the latter. Rearranging terms, we finally arrive at (5.50).

The modified Hamiltonian (5.47) can be obtained in the following way: first the coefficients (5.52) can be used to compute $\hat{\alpha}_{k\ell}(h)$ from (5.43), and then the coefficients $\hat{\beta}_{k\ell}(h)$ are determined from (5.46).

Example: simple pendulum. As an illustrative example we again consider the simple pendulum of Section 1.4, described by the Hamiltonian function (1.19). As we saw there, for initial conditions in a neighbourhood of the stable equilibrium (0, 0) it is advantageous to decompose H as in (1.32), so it constitutes a particular example of system (5.1):

$$H = H_1 + H_2, \quad H_1(q,p) = \frac{1}{2}p^2 + \frac{1}{2}q^2, \quad H_2(q,p) = U(q) = 1 - \frac{1}{2}q^2 - \cos q.$$

Although $U(q)$ is not a polynomial, and therefore the set $\mathcal{I}$ in (5.4) is infinite, we can truncate the Fourier expansion (5.37) and work instead with $\mathcal{I} = \{-m, \dots, -1, 0, 1, \dots, m\}$ for a given m to construct a processed Strang scheme based on kicks and rotations as proposed earlier. Specifically, we take $m = 4$, a step size $h = 5/6$, then determine the corresponding coefficients $b_j(h)$ and form the integrator $\hat{\psi}_h = \pi_h^{-1} \circ S_h^{[RKR]} \circ \pi_h$, with π_h given by (5.49) (with $m = 4$), whereas $S_h^{[RKR]}$ corresponds to the map (1.34).

Figure 5.1 shows the relative error in energy (a) and phase space (b) over the time interval $[0, 500]$ corresponding to the solution initiated at $(q_0, p_0) = (1/10, 0)$. As usual, S_2 denotes the Störmer–Verlet method applied to (1.19), (2,2) is the Strang splitting $S_h^{[RKR]}$, and P(2, 2) is the processed $S_h^{[RKR]}$ scheme with step size $h = 5/6$. Since the integrator (2, 2) is conjugate to P(2, 2), the error of the former is eventually

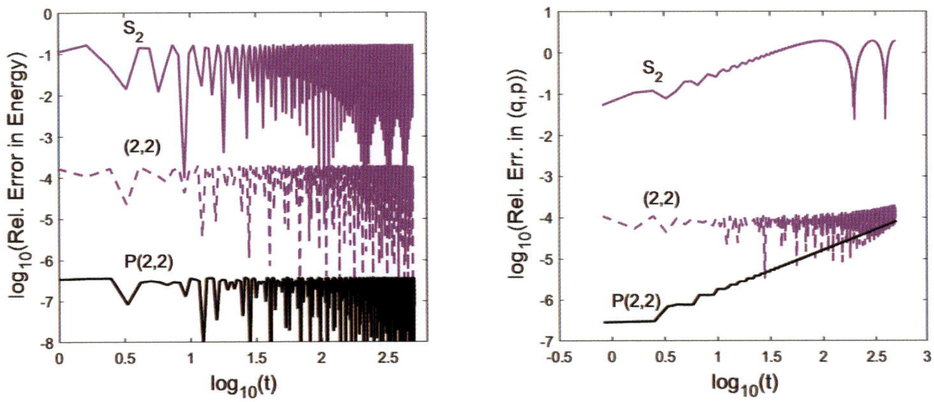

(a) (b)

Figure 5.1. Pendulum. Evolution of relative errors in energy (a) and in phase space (b) obtained with Störmer–Verlet, S_2, scheme $S_h^{[RKR]}$, denoted (2,2), and the processed (2,2), P$(2,2)$ with initial state $(q_0, p_0) = (1/10, 0)$ and step size $h = 5/6$.

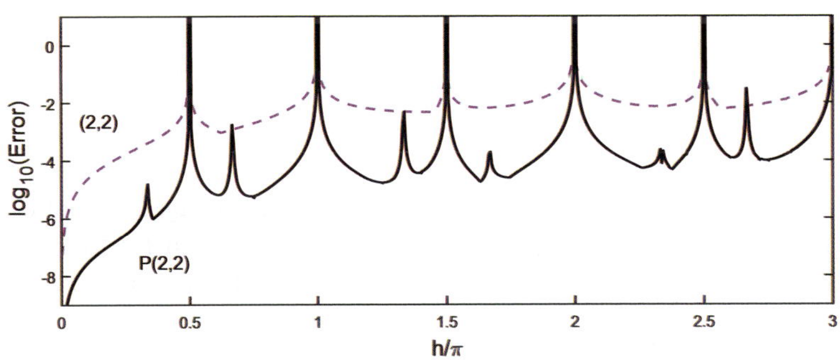

Figure 5.2. Pendulum. Maximum relative energy error in the time interval $[0, 500]$, initial state $(q_0, p_0) = (1/10, 0)$, and step sizes in the range $[0, 3\pi]$ committed by the Strang splitting $S_h^{[RKR]}$ and its processed version P$(2, 2)$.

dominated by the error of the latter, as expected from the discussion in Section 4.5. Observe that the step size used here is larger than those taken in Figure 1.1.

Next, we consider the Strang method $S_h^{[RKR]}$ and its processed version P$(2, 2)$ for the same initial condition and a time interval $[0, 500]$, and show the maximum energy error in this interval for step sizes in the range $[0, 3\pi]$ (Figure 5.2). The spikes in the curve of the error of Strang correspond to the step sizes violating the non-resonance condition (5.44) for $\omega = 1$ and $k = 2, 4$. Indeed, the potential $U(q)$ can be well approximated near the origin by $q^4/24$, which implies that $\mathcal{I} = \{-4, -2, 0, 2, 4\}$. The curve of the error of processed Strang also has spikes for such resonant step sizes, and additionally, for the step sizes h such that $3h/(2\pi) \in \mathbb{Z}$, which is due to the fact that the processor map π_h has been designed to work well for problems for which $\mathcal{I} = \{-4, -3, -2, -1, 0, 1, 2, 3, 4\}$. Away from such resonant step sizes, the energy error of processed Strang is considerably smaller than that of unprocessed Strang.

5.8. *More general assumptions*

The results given in the present section are also valid with small modifications in the more general setting of systems (5.2) such that the eigenvalues of A lie on the imaginary axis. In the most general case, where such eigenvalues are not integer multiples of $\omega\,\mathrm{i}$, the exponential e^{tA} is quasi-periodic with a finite number of basic frequencies $(\omega_1, \ldots, \omega_r) \in \mathbb{R}^r$. Under such conditions, $\mathrm{e}^{-tA} f_2(\mathrm{e}^{t\,A} x)$ admits a multi-variable Fourier expansion

$$\sum_{k \in \mathbb{Z}^r} \mathrm{e}^{\mathrm{i}\langle k, \omega\rangle t}\, g_k(x), \tag{5.53}$$

where k now denotes an r-tuple of integers $k = (k_1, \ldots, k_r) \in \mathbb{Z}^r$, $\omega = (\omega_1, \ldots, \omega_r)$ is the vector of basic frequencies, and $\langle k, \omega\rangle = k_1\omega_1 + \cdots + k_r\omega_r$. The set of indices $\mathcal{I}$ is then defined as

$$\mathcal{I} = \{k \in \mathbb{Z}^r : g_k \neq 0\}.$$

Under these more general assumptions and notation, all previous formulae are valid if each occurrence of $k\omega$ with $k \in \mathcal{I}$ is replaced by $\langle k, \omega\rangle$, the only exception being the explicit construction of the processing map π_h carried out in Section 5.7. It is worth remarking that, compared to the periodic case, in the quasi-periodic one there are typically more resonant step sizes, that is, step sizes h such that

$$\frac{\langle k, \omega\rangle h}{2\pi} \in \mathbb{Z}\backslash\{0\} \quad \text{or} \quad \frac{\langle (k+\ell), \omega\rangle h}{2\pi} \in \mathbb{Z}\backslash\{0\}$$

for some $k, \ell \in \mathcal{I}$.

6. Splitting methods for PDEs

6.1. *Splitting, LOD and ADI methods*

Splitting methods can also be applied to partial differential equations (PDEs), in which case equation (1.1) has to be viewed as the abstract system associated with the PDE initial value problem in autonomous form, and f as a spatial partial differential operator. For clarity, in this section we write

$$u_t(x, t) = f(x, u(x, t)), \qquad u(x, 0) = u_0(x) \tag{6.1}$$

to distinguish between the unknown $u(x, t)$ defined in a certain function space and the spatial variable $x \in \mathbb{R}^d$, but for notational purposes it is convenient to drop the dependence of u and f on x. To introduce the basic concepts and methods it is not necessary at this stage to specify the dimension d, the number of components of u, the relevant function space and the boundary conditions. These will be considered when analysing particular applications. Moreover, for simplicity in the presentation, only the autonomous case will be treated. If f depends explicitly on time, then we can take t as a new coordinate, as is done in Section 3.5.

Very often, the operator f contains contributions coming from very different physical sources, so we may decompose it into two (or more) parts, and use different schemes to solve each sub-problem approximately. For instance, in a reaction–diffusion system, $f(u) = \nabla \cdot (D\nabla u) + g(u)$, where D and g may also depend on x, it makes sense to split the diffusion from the reaction terms, that is,

$$f(u) = f_1(u) + f_2(u), \quad \text{with} \quad f_1(u) = \nabla \cdot (D\nabla u), \; f_2(u) = g(u).$$

Algorithm 1.1 can of course be used in this setting, but we still have to specify how to solve each initial value sub-problem in practice with appropriate boundary conditions. A simple possibility consists in applying the backward Euler scheme, thus resulting in the so-called *Marchuk–Yanenko* operator-splitting scheme

$$\begin{aligned} u_{n+1/2} &= u_n + hf_1(u_{n+1/2}), \\ u_{n+1} &= u_{n+1/2} + hf_2(u_{n+1}), \quad n = 0, 1, 2, \ldots, \end{aligned} \tag{6.2}$$

where, for consistency with the rest of the paper, we have denoted $h \equiv \Delta t$, the time step size. In spite of its low order of consistency (order one)[5] and the fact that the intermediate stage $u_{n+1/2}$ is not a consistent approximation to the exact solution, its simplicity and robustness make it a useful alternative way to deal with complicated problems and even non-smooth operators (Glowinski *et al.* 2016*a*). It is also appropriate for parabolic problems, since it incorporates the damping properties of the backward Euler method (Hundsdorfer and Verwer 2003).

If, instead of using the backward Euler scheme to integrate each sub-problem in Algorithm 1.1, we apply the second-order implicit trapezoidal rule, it results in Yanenko's Crank–Nicolson method (Hundsdorfer and Verwer 2003, Marchuk 1990):

$$\begin{aligned} u_{n+1/2} &= u_n + \frac{h}{2} f_1(u_n) + \frac{h}{2} f_1(u_{n+1/2}), \\ u_{n+1} &= u_{n+1/2} + \frac{h}{2} f_2(u_{n+1/2}) + \frac{h}{2} f_2(u_{n+1}), \quad n = 0, 1, 2, \ldots \end{aligned} \tag{6.3}$$

In the end, however, it is also of first order of consistency.

Another widely popular class of spitting methods in the domain of PDEs is the Peaceman–Rachford scheme and its variants. Although initially designed for the numerical solution of elliptic and parabolic equations (Peaceman and Rachford 1955, Douglas and Rachford 1956), they also apply to more general situations. The procedure goes as follows. Given an approximation u_n for the solution of (6.1)

[5] Here 'order' should be understood as the order of consistency with respect to the solution of the ODE problem on a fixed spatial grid, not with respect to the underlying PDE solution (Hundsdorfer and Verwer 2003).

at $t = t_n$, the approximation u_{n+1} is computed using the backward (resp. forward) Euler scheme with respect to f_1 (resp. f_2) on the sub-interval $[t_n, t_{n+1/2}]$. Then the roles of f_1 and f_2 are interchanged on the sub-interval $[t_{n+1/2}, t_{n+1}]$. In other words, the *Peaceman–Rachford* scheme corresponds to the sequence

$$
\begin{aligned}
u_{n+1/2} &= u_n + \frac{h}{2} f_1(u_{n+1/2}) + \frac{h}{2} f_2(u_n), \\
u_{n+1} &= u_{n+1/2} + \frac{h}{2} f_1(u_{n+1/2}) + \frac{h}{2} f_2(u_{n+1}), \quad n = 0, 1, 2, \ldots,
\end{aligned}
\tag{6.4}
$$

which is of second order of consistency. Notice that, in contrast to methods (6.2) and (6.3), both f_1 and f_2 appear in each of the two stages, and thus the intermediate value $u_{n+1/2}$ provides a consistent approximation at $t = t_{n+1/2}$. On the other hand, it does not possess a natural formulation where f is split into more than two operators. In general, f_1 and f_2 can be nonlinear, unbounded and even multi-valued. For a more detailed treatment, the reader is addressed to Glowinski, Pan and Tai (2016*b*) and references therein.

A classical alternative to scheme (6.4), of first order, is the *Douglas–Rachford* method (Douglas and Rachford 1956), which instead reads

$$
\begin{aligned}
\hat{u}_{n+1} &= u_n + h f_1(\hat{u}_{n+1}) + h f_2(u_n), \\
u_{n+1} &= u_n + h f_1(\hat{u}_{n+1}) + h f_2(u_{n+1}), \quad n = 0, 1, 2, \ldots,
\end{aligned}
\tag{6.5}
$$

and can be generalized to decompositions of f involving more than two operators. Notice that the roles of f_1 and f_2 in (6.5) are not symmetric, in contrast to the Peaceman–Rachford method. On the basis of many numerical experiments, Glowinski *et al.* (2016*b*) conclude that the scheme (6.5) is faster and more robust than (6.4) for problems where one of the operators is non-smooth, in particular when we are interested in approximating steady-state solutions.

Let us now analyse the particular case when f in (6.1) is a linear spatial differential operator. Assuming that an appropriate semidiscretization of (6.1) in the space variable x has been carried out, we end up with the system

$$
\frac{\mathrm{d}U}{\mathrm{d}t} = F_1 U + F_2 U, \tag{6.6}
$$

where $F_1, F_2 \in \mathbb{C}^{M\times M}$, $F_1 F_2 \neq F_2 F_1$ in general, and $U \in \mathbb{C}^M$ approximates u on the space grid points $x_1, \ldots, x_M$ (see below). Then a step of the Marchuk–Yanenko scheme (6.2) reads

$$
U_{n+1} = (I - hF_2)^{-1} (I - hF_1)^{-1} U_n, \tag{6.7}
$$

where $U_n \approx (u(x_1, t_n), \ldots, u(x_M, t_n))^\top$. Notice that (6.7) corresponds to applying the $[0/1]$ Padé approximant to the exponentials in the Lie–Trotter scheme $U_{n+1} = \mathrm{e}^{hF_2}\,\mathrm{e}^{hF_1} U_n$, whence the first order of the approximation is obtained at once. On the other hand, applying (6.3) results in the sequence

$$U_{n+1} = \left(I - \frac{h}{2}F_2\right)^{-1}\left(I + \frac{h}{2}F_2\right)\left(I - \frac{h}{2}F_1\right)^{-1}\left(I + \frac{h}{2}F_1\right)U_n. \tag{6.8}$$

In other words, it corresponds to the application of the [1/1] Padé approximant to the exponentials in the Lie–Trotter scheme. It is then clear that, although a second-order Crank–Nicolson method is carried out for each exponential, their combination (6.8) is of first order, since in the end we are using only a variant of the Lie–Trotter scheme.

By the same token, a straightforward computation shows that the Peaceman–Rachford scheme (6.4) applied to the linear equation (6.6) can be written as

$$U_{n+1} = \left(I - \frac{h}{2}F_2\right)^{-1}\left(I + \frac{h}{2}F_1\right)\left(I - \frac{h}{2}F_1\right)^{-1}\left(I + \frac{h}{2}F_2\right)U_n. \tag{6.9}$$

As a matter of fact, all these algorithms can be formulated by applying properly chosen compositions of the implicit and explicit Euler methods. This observation may eventually lead to the construction of methods of higher order or improved behaviour, but in the same family. Suppose f in (6.1) is of the form $f = f_1 + f_2$ and that the solution of each equation $u_t = f_k(u)$, $k = 1, 2$, is numerically approximated by the maps

$$\begin{aligned} u_{n+1} &= \psi_h^{ke}(u_n) \equiv u_n + hf_k(u_n), && \text{explicit Euler,} \\ u_{n+1} &= \psi_h^{ki}(u_n) \equiv u_n + hf_k(u_{n+1}), && \text{implicit Euler,} \end{aligned}$$

so that by combining all variants, we form the following first-order schemes (and their corresponding adjoints):

$$\begin{aligned} \phi_h^1 &= \psi_h^{1e} \circ \psi_h^{2e}, & \phi_h^{1*} &= \psi_h^{2i} \circ \psi_h^{1i}, \\ \phi_h^2 &= \psi_h^{1e} \circ \psi_h^{2i}, & \phi_h^{2*} &= \psi_h^{2e} \circ \psi_h^{1i}, \\ \phi_h^3 &= \psi_h^{1i} \circ \psi_h^{2e}, & \phi_h^{3*} &= \psi_h^{2i} \circ \psi_h^{1e}, \\ \phi_h^4 &= \psi_h^{1i} \circ \psi_h^{2i}, & \phi_h^{4*} &= \psi_h^{2e} \circ \psi_h^{1e}. \end{aligned} \tag{6.10}$$

Then we can conclude the following.

- The Marchuk–Yanenko operator-splitting scheme (6.2) can be expressed as $u_{n+1} = \phi_h^{1*}(u_n)$. Therefore compositions

  $$u_{n+1} = \phi_{h/2}^{1*} \circ \phi_{h/2}^{1}(u_n) \quad \text{and} \quad u_{n+1} = \phi_{h/2}^{1} \circ \phi_{h/2}^{1*}(u_n)$$

  yield time-symmetric second-order approximations.
- Yanenko's Crank–Nicolson method (6.3) corresponds to the composition

  $$u_{n+1} = \psi_{h/2}^{2i} \circ \psi_{h/2}^{2e} \circ \psi_{h/2}^{1i} \circ \psi_{h/2}^{1e}(u_n), \tag{6.11}$$

  which is not time-symmetric and therefore only of first order. Notice, however, that (6.11) can also be expressed as

$$u_{n+1} = \psi_{h/2}^{2i} \circ \left(\phi_{h/2}^{2*} \circ \phi_{h/2}^{2}\right) \circ \left(\psi_{h/2}^{2i}\right)^{-1}(u_n),$$

so it is conjugate to a time-symmetric second-order method. This feature could account for its observed good behaviour in practice.

- The Peaceman–Rachford scheme (6.4) is just the symmetric composition

$$u_{n+1} = \phi_{h/2}^{3*} \circ \phi_{h/2}^{3}(u_n),$$

and is therefore of second order. It is also worth mentioning that whereas all symmetric second-order compositions

$$\phi_{h/2}^{\ell*} \circ \phi_{h/2}^{\ell} \quad \text{and} \quad \phi_{h/2}^{\ell} \circ \phi_{h/2}^{\ell*}, \qquad \ell = 1, \dots, 4$$

provide consistent approximations at the midpoint, the steady-state solution is captured only when $\ell = 3$: if $f(w) = 0$, then

$$\phi_{h/2}^{3*} \circ \phi_{h/2}^{3}(w) = \phi_{h/2}^{3} \circ \phi_{h/2}^{3*}(w) = w.$$

The Douglas–Rachford method (6.5) can be alternatively formulated as a splitting method in an extended space as follows. Consider the enlarged system

$$w_t = \begin{pmatrix} \hat{u}_t \\ v_t \end{pmatrix} = \begin{pmatrix} f_1(\hat{u}) + f_2(v) \\ 0 \end{pmatrix} + \begin{pmatrix} 0 \\ f_1(\hat{u}) + f_2(v) \end{pmatrix} = g_1(w) + g_2(w), \qquad (6.12)$$

with initial condition $w(0) = (\hat{u}(0), v(0)) = (u_0, u_0)$ and solution $w(t) = (\hat{u}(t), v(t)) = (u(t), u(t))$. Take $w_n = (u_n, u_n)$ and form the composition

$$\hat{w}_{n+1} = \phi_h^{1*}(w_n) = \psi_h^{2i} \circ \psi_h^{1i}(w_n), \qquad n \geq 0,$$

providing $\hat{w}_{n+1} = (\hat{u}_{n+1}, v_{n+1})$. Then scheme (6.5) is recovered by taking $u_{n+1} = v_{n+1}$, and considering $w_{n+1} = (u_{n+1}, u_{n+1})$ as the starting point for the next iteration.

Example. The two-dimensional heat equation with source term and Dirichlet boundary conditions on the unit square may serve as an illustration of these methods (Hundsdorfer and Verwer 2003). Specifically, the system reads

$$\begin{aligned} u_t &= u_{xx} + u_{yy} + g(x, y, t) && \text{on } \Omega = (0,1) \times (0,1), \\ u(x, y, t) &= u_\Gamma(x, y, t) && \text{on } \Gamma = \partial\Omega, \\ u(x, y, 0) &= u_0(x, y) && \text{on } \Omega. \end{aligned} \qquad (6.13)$$

Suppose we take a Cartesian grid in Ω based on $M + 1$ equally spaced intervals in the x and y directions, so that $\Delta x = \Delta y = 1/(M+1)$ and apply finite differences to approximate the space derivatives. Then we end up with M^2 interior points and the aim is to get approximations of u at these points, i.e. to determine $u_{i,j}(t) \simeq u(x_i, y_j, t)$ for $i, j = 1, 2, \dots, M$. The problem can be conveniently formulated in terms of the 'supervector' $U = (u_1^\top, \dots, u_M^\top)^\top$, with $u_i = (u_{i,1}, \dots, u_{i,M})^\top$, and M^2 components

$$U_\ell(t) = u_{i,j}(t), \qquad \ell = j + M(i-1).$$

Using standard second-order finite differences for ∂_{xx} and ∂_{yy} we get a linear system of the form $U' = F\,U + s(t)$, where $s(t)$ contains sources and boundary data, and $F = F_1 + F_2$, with

$$F_1 = I_M \otimes B_M, \qquad F_2 = B_M \otimes I_M, \qquad B_M = \frac{1}{(\Delta x)^2}\begin{pmatrix} -2 & 1 & \cdots & 0 \\ 1 & -2 & \ddots & \vdots \\ \vdots & \ddots & \ddots & 1 \\ 0 & & 1 & -2 \end{pmatrix}.$$

Here I_M is the identity matrix, B_M the differentiation matrix and $\otimes$ denotes the tensor product. Notice that F_1 acts in the x-direction and F_2 in the y-direction, so both F_1 and F_2 are essentially one-dimensional. Since F_1 is a tridiagonal matrix, the linear system resulting from the application of the previous implicit schemes can be solved in an efficient way, whereas F_2 is equivalent to a tridiagonal matrix, so the same considerations apply (Iserles 1996).

First consider the homogeneous equation with zero Dirichlet boundary conditions. The solution for one time step is

$$U(t_n + h) = \mathrm{e}^{h(F_1+F_2)}\, U(t_n),$$

whereas the application of the Peaceman–Rachford scheme leads to the approximation (6.8):

$$U_{n+1} = (\phi^{3*}_{h/2} \circ \phi^{3}_{h/2})\, U_n = \mathrm{e}^{h(F_1+F_2)}\, U_n + O(h^3).$$

When a dimensional splitting is done, as in this case, applying schemes (6.7) and (6.8) corresponds essentially to carrying out computations in only one dimension. This is the reason why (6.2) and (6.3) are called *locally one-dimensional* (LOD) methods. Analogously, given the alternate use of F_1 and F_2 in this setting, the name *alternating direction implicit* (ADI) is usually attached to methods (6.4) and (6.5).

Regarding these ADI/LOD methods as composition schemes allows us to get approximations in the non-homogeneous case too, with only minor changes, while keeping any favourable properties (if no order reduction occurs due to the Dirichlet boundary conditions). To illustrate this point, we again apply the explicit and implicit Euler methods to the equation $U' = s(t)$ corresponding to the non-homogeneous term

$$U_{n+1} = \psi^{Se}_h(U_n) = U_n + h\, s(t_n), \qquad U_{n+1} = \psi^{Si}_h(U_n) = U_n + h\, s(t_{n+1}),$$

and consider the first-order scheme $\hat{\phi}^3_h = \psi^{1i}_{h/2} \circ \psi^{2e}_{h/2} \circ \psi^{Se}_h$. Then

$$\begin{aligned} U_{n+1} &= \left(\hat{\phi}^{3*}_{h/2} \circ \hat{\phi}^{3}_{h/2}\right) U_n = \left(\psi^{Si}_{h/2} \circ \psi^{2i}_{h/2} \circ \psi^{1e}_{h/2} \circ \psi^{1i}_{h/2} \circ \psi^{2e}_{h/2} \circ \psi^{Se}_{h/2}\right) U_n \\ &= \frac{h}{2} s(t_{n+1}) + \left(I - \frac{h}{2}F_2\right)^{-1} \left(I + \frac{h}{2}F_1\right) \left(I - \frac{h}{2}F_1\right)^{-1} \left(I + \frac{h}{2}F_2\right) \left(U_n + \frac{h}{2}s(t_n)\right) \end{aligned}$$

produces a symmetric second-order scheme.

It is not only reaction–diffusion problems that can be treated in this way. For instance, systems of hyperbolic conservation laws in three dimensions, such as

$$u_t + \nabla \cdot f(u) = 0, \qquad u(x, y, z, 0) = u_0(x, y, z),$$

can be numerically approximated with dimensional splitting by applying a specially tailored numerical scheme to each scalar conservation law $u_t + f(u)_x = 0$, etc. (Holden, Karlsen, Lie and Risebro 2010).

The one-dimensional convection–diffusion problem

$$u_t + f(u)_x = A(u)_{xx}, \qquad u(x, 0) = u_0(x)$$

with a scalar non-decreasing function $A(\cdot)$, $A(0) = 0$, possesses a rich set of phenomena depending on the interplay of the different nonlinearities. In this case we can split the problem into a convective and a diffusion part and apply Algorithm 1.1. This formally results in the so-called Godunov split (Holden *et al.* 2010). Specifically, letting $\mathcal{U}_h^{[1]}$ denote the solution operator corresponding to the scalar conservation law $u_t + f(u)_x = 0$, and letting $\mathcal{U}_h^{[2]}$ denote the solution operator corresponding to the (weak) solution of the nonlinear heat equation $u_t = A(u)_{xx}$, the scheme then reads

$$u(x, t_n + h) \approx u_{n+1} = \mathcal{U}_h^{[2]}\big(\mathcal{U}_h^{[1]}(u_n)\big).$$

Of course, we can also use the Strang splitting

$$u(x, t_n + h) \approx u_{n+1} = \mathcal{U}_{h/2}^{[2]}\big(\mathcal{U}_h^{[1]}\big(\mathcal{U}_{h/2}^{[2]}(u_n)\big)\big).$$

To get a numerical approximation, each of the two operators must be approximated. This can be done, for example, by a front-tracking method for $\mathcal{U}_h^{[1]}$ and by a standard implicit finite-difference method for the parabolic operator $\mathcal{U}_h^{[2]}$. A convergence analysis of such schemes has been carried out in Holden *et al.* (2010).

Although the analysis of splitting methods can be done by power series expansions and the formalism of Lie operators, there are fundamental differences with respect to the ODE case. Nonlinear PDEs in general possess solutions that exhibit complex behaviour in small regions of space and time, such as sharp transitions and discontinuities, and thus they lack the usual smoothness required for the analysis. Moreover, even if the exact solution of the original problem is smooth, it may well happen that the composition defining the splitting method provides non-smooth approximations. Therefore, it is necessary to develop an appropriate mathematical framework to analyse the convergence of the numerical solution to the correct solution of the original problem, and this has to be done very often on a case by case basis; see e.g. Holden *et al.* (2010) and references therein. Thus, in particular, the first- and second-order convergence of the Godunov and Strang splitting methods on the Korteweg–de Vries equation $u_t - uu_x + u_{xxx} = 0$ has been proved in Holden, Karlsen, Risebro and Tao (2011) if the initial data are sufficiently regular, whereas the result has been extended in Holden, Lubich and Risebro (2013) to equations of the form $u_t = Au + uu_x$, when A is a linear differential operator such that the full equation is well-posed. More recently, convergence results and error estimates have also been obtained for initial conditions with low regularity, namely $u_0 \in H^s$ with $0 < s \leq 3/2$, where H^s denotes the Sobolev space (Rousset and Schratz 2022).

Another source of difficulties related to the application of splitting methods to PDEs is the treatment of boundary conditions. In this respect, we should take into account that the boundary conditions are defined for the whole operator f in (6.1), and they do not necessarily hold for the subproblems defined by each part f_1 and f_2. Therefore, we cannot expect the numerical solution obtained by a splitting method to belong to the domain of f. This results in severe order reductions in reaction–diffusion problems when Dirichlet or Neumann boundary conditions are considered (Hansen and Ostermann 2009*b*, Hundsdorfer and Verwer 2003). In particular, the order reduction for the Strang splitting is one in the infinity norm. Similar order reductions for advection–reaction problems have also been reported (Hundsdorfer and Verwer 1995).

Several procedures have been considered in the literature to avoid this order reduction in the case of reaction–diffusion problems. One possibility, proposed in Einkemmer and Ostermann (2015, 2016), consists in introducing a smooth correction function in such a way that the new reaction flow is compatible with the prescribed boundary conditions. For time-invariant Dirichlet boundary conditions, this correction can be computed only once at the beginning of the simulation, but for time-dependent Dirichlet, Neumann or Robin boundary conditions, the correction is time-dependent and has to be computed at each time step. Various techniques to deal with this problem are explored in Einkemmer, Moccaldi and Ostermann (2018). An alternative approach requiring additional calculations with grid values on the boundaries, and not on grid values on the total domain, is proposed in Alonso-Mallo, Cano and Reguera (2018, 2019).

6.2. *IMEX methods*

IMEX schemes are suitable combinations of implicit and explicit schemes and constitute a popular technique for approximating the solution of PDEs that involve terms of a different nature (Ascher, Ruuth and Wetton 1995, Ascher, Ruuth and Spiteri 1997, Hundsdorfer and Verwer 2003). Thus, for convection–diffusion or reaction–diffusion problems where the convection or reaction terms are moderately stiff, it might be appropriate to use an explicit scheme for these parts and an implicit scheme for the diffusion term. We next analyse the connections of some popular IMEX methods with splitting and composition methods.

Suppose we have the semidiscrete system $u_t = f_1(u) + f_2(u)$, where f_1 is (still) a diffusion term and f_2 a nonlinear term suitable for explicit integration. Then the simple composition (see (6.10))

$$u_{n+1} = \phi_h^3(u_n) = \psi_h^{1i} \circ \psi_h^{2e}(u_n) = u_n + h\,(f_1(u_{n+1}) + f_2(u_n))$$

corresponds to the linear one-step IMEX scheme of Ascher *et al.* (1997).

Letting $\phi_h^{[2,k]}$, $k \geq 2$, denote an explicit kth-order (Runge–Kutta or multistep) method for the equation $u_t = f_2(u)$, it turns out that many IMEX methods from the literature have the structure

$$u_{n+1} = \psi_{h/2}^{1i} \circ \phi_h^{[2,k]} \circ \psi_{h/2}^{1e}(u_n), \tag{6.14}$$

thus yielding second-order approximations.

Let us consider, for instance, the popular Crank–Nicolson–leapfrog (IMEX-CNLF) method (Hundsdorfer and Verwer 2003)

$$u_{n+1} = u_{n-1} + 2hf_2(u_n) + h(f_1(u_{n+1}) + f_1(u_{n-1})), \tag{6.15}$$

to be initiated with, for example, $u_1 = u_0 + h(f_1(u_0) + f_2(u_0))$. It is equivalent to the one-step method

$$u_{n+1} = u_n + h^* f_2(u_{n+1/2}) + \frac{h^*}{2}(f_1(u_{n+1}) + f_1(u_n)), \tag{6.16}$$

with $h^* = 2h$, which requires us to compute the approximation at the midpoint and so advances every half time step. If we take the explicit second-order midpoint rule as $\phi_h^{[2,2]}$ in (6.14), then we get the following sequence of maps:

$$\begin{aligned}
\psi_{h/2}^{1e}: &\quad U = u_n + \frac{h}{2} f_1(u_n), \\
\phi_h^{[2,2]}: &\quad \left\{ \begin{aligned} V &= U + \frac{h}{2} f_2(U) \\ \hat{V} &= U + h f_2(V) \end{aligned} \right\}, \\
\psi_{h/2}^{1i}: &\quad u_{n+1} = \hat{V} + \frac{h}{2} f_1(u_{n+1}),
\end{aligned}$$

or equivalently

$$\begin{aligned}
U &= u_n + \frac{h}{2} f_1(u_n), \\
V &= U + \frac{h}{2} f_2(U), \\
u_{n+1} &= u_n + h f_2(V) + \frac{h}{2}(f_1(u_n) + f_1(u_{n+1})).
\end{aligned} \tag{6.17}$$

This scheme is in fact quite similar to (6.16) for $h = h^*$ since V is a first-order approximation to $u_{n+1/2}$, the solution at the midpoint. Notice that by replacing $f_2(u_{n+1/2})$ in (6.16) with $f_2(V)$, where V depends explicitly on u_n, it allows us to advance from u_{n+1} to u_{n+2} without evaluating the solutions at $u_{n+3/2}$ and therefore halving the computational cost to solve the implicit equations involved.

Although the method is not symmetric (due to the lack of symmetry of the explicit scheme $\phi_h^{[2,2]}$), by instead using an explicit method of order $k > 2$, the overall scheme will be time-symmetric up to this order k and therefore we can apply extrapolation to get a method of order k in an efficient way.

Another IMEX Runge-Kutta method that combines the implicit and explicit trapezoidal methods and shows a fairly good performance on examples is (Hundsdorfer and Verwer 2003)

$$\begin{aligned} U &= u_n + h(f_1(u_n) + f_2(u_n)), \\ u_{n+1} &= u_n + \frac{h}{2}(f_1(u_n) + f_2(u_n)) + \frac{h}{2}(f_1(u_{n+1}) + f_2(U)). \end{aligned} \tag{6.18}$$

If we take the system

$$w' = \begin{pmatrix} u \\ U \end{pmatrix}' = \begin{pmatrix} f_1(u) + f_2(U) \\ 0 \end{pmatrix} + \begin{pmatrix} 0 \\ f_1(u) + f_2(u) \end{pmatrix} = g_1(w) + g_2(w), \tag{6.19}$$

and form, as before, the composition

$$w_{n+1} = \psi^{1i}_{h/2} \circ \phi^{[2]}_h \circ \psi^{1e}_{h/2}(w_n), \tag{6.20}$$

where $\phi^{[2]}_h$ denotes the exact solution of $w' = g_2(w)$, and $w_n = (U_n, u_n) = (u_n, u_n)$, then we get

$$\begin{aligned} u_{1/2} &= u_n + \frac{h}{2}(f_1(u_n) + f_2(u_n)), \\ U_{n+1} &= u_n + h(f_1(u_{1/2}) + f_2(u_{1/2})), \\ u_{n+1} &= u_n + \frac{h}{2}(f_1(u_n) + f_2(u_n)) + \frac{h}{2}(f_1(u_{n+1}) + f_2(U_{n+1})). \end{aligned} \tag{6.21}$$

Scheme (6.18) is recovered by replacing $u_{1/2}$ with u_n in the expression of U_{n+1}. However, if in (6.21) we do not restart the value of U_n at each step, the method will be symmetric and therefore we can apply extrapolation to increase its order.

Higher-order IMEX methods have been built involving implicit multistep or Runge–Kutta methods. Although splitting methods are no longer appropriate to advance the diffusion term due to the presence of negative coefficients for orders higher than two, we can incorporate higher derivatives or complex coefficients and form new higher-order splitting IMEX methods (or consider extrapolation from a basic symmetric second-order method).

6.3. Schrödinger equations

T–V splitting. Until now, in the context of PDEs, we have dealt with splitting methods of orders one and two. There are, however, relevant problems where high-order splitting methods can be and have been safely used, and where a rigorous convergence analysis can be established. This is the case, in particular, for the time-dependent Schrödinger equation, already considered in Section 1.6. The numerical experiments presented there clearly indicate that the Strang splitting based on kinetic and potential energy in combination with a pseudo-spectral space discretization, i.e. method (1.45), provides approximations of order two in the time step, in accordance with

$$e^{\frac{\tau}{2}V}\, e^{\tau T}\, e^{\frac{\tau}{2}V} = e^{\tau(T+V)} + O(\tau^3(\|T\| + \|V\|)^3), \tag{6.22}$$

where $\tau = -ih$. Be aware, however, that this error estimate only makes sense for bounded T and V. In fact, when the norm of T or V is very large (as is usually the case when the number of space discretization points M is large), then (6.22) is of no practical use, and thus other estimates are necessary to explain the observed good behaviour.

Error bounds for the Strang splitting are actually derived in Jahnke and Lubich (2000). They clearly show that the method is indeed of order two when applied to pseudo-spectral discretizations of the time-dependent Schrödinger equation under some regularity conditions and periodic boundary conditions. Specifically, assume that the potential $V(x)$ is C^5-smooth, periodic and bounded, $\|V\psi\| \le \beta\|\psi\|$, $\beta > 0$. Then, if $u(x,t)$ denotes the trigonometric interpolation polynomial of the solution of the pseudo-spectral method and $u_n(x)$ is the corresponding trigonometric interpolation polynomial built from the numerical approximations obtained from the Strang splitting (1.45) at time $t = t_n = nh$, for the local and global errors we get the following bounds:

$$\begin{aligned} \|u_1 - u(\cdot,\tau)\|_{L^2} &\le C_1|\tau|^3\|u_0\|_{H^2}, \\ \|u_n - u(\cdot,t_n)\|_{L^2} &\le C_2|\tau|^2\|u_0\|_{H^2}, \end{aligned} \tag{6.23}$$

respectively. Here $\|\cdot\|_{H^2}$ denotes the usual Sobolev norm, and the constants C_1, C_2 are independent of the initial data u_0 and the discretization parameters M, n and τ, with $0 \le t_n \le t_f$ for some finite t_f. The case when V is time-dependent and bounded for any t has been recently treated in del Valle and Kropielnicka (2023), where new schemes are proposed and analysed.

The previous results for the Strang splitting have been extended in Thalhammer (2008) to splitting methods of the general form

$$u_{n+1} = \prod_{j=1}^{s} e^{b_j\tau V}\, e^{a_j\tau T}\, u_n = e^{b_s\tau V}\, e^{a_s\tau T} \cdots e^{b_1\tau V}\, e^{a_1\tau T}\, u_n, \tag{6.24}$$

whose coefficients a_j, b_j satisfy the order conditions up to order r. In that case

$$\|u_n - u(\cdot,t_n)\|_{L^2} \le C\|u(\cdot,0) - u_0\|_{L^2} + C|\tau|^r\|u(\cdot,0)\|_{H^r}, \quad 0 \le t_n \le t_f \tag{6.25}$$

is valid with some constant C depending on t_f, but not on n and h. This error bound implies, in particular, that the splitting methods of Section 8 retain their order of convergence when applied to the Schrödinger equation with periodic boundary conditions, provided that the data are sufficiently differentiable; see also Hansen and Ostermann (2009*a*). Otherwise, an order reduction may occur.

We can also take advantage of the property (1.41) and include the commutator $[V,[T,V]]$ in the composition (6.24), as in RKN splitting methods, so that we end up with

$$u_{n+1} = \prod_{j=1}^{s} \mathrm{e}^{b_j \tau V + c_j \tau^3 [V,[T,V]]} \, \mathrm{e}^{a_j \tau T} \, u_n. \tag{6.26}$$

Even in this case the resulting schemes retain their order of convergence if the solution is sufficiently regular, as shown in Kieri (2015). It is then also possible to apply the RKN schemes presented in Section 8.

Symplectic splitting. In Section 3.4 we reviewed the symplectic structure involved in the (semidiscretized) Schrödinger equation and illustrated how the Strang splitting (and, for that matter, all methods presented in Section 8) can be applied if H is a real and symmetric matrix. They are formulated as products of exponentials of the nilpotent matrices A and B given in (3.13),

$$\begin{pmatrix} q_{n+1} \\ p_{n+1} \end{pmatrix} = \prod_{j=1}^{s} \mathrm{e}^{b_j h B} \, \mathrm{e}^{a_j h A} \begin{pmatrix} q_n \\ p_n \end{pmatrix} = \exp\left[h \begin{pmatrix} 0 & H \\ -H & 0 \end{pmatrix} \right] \begin{pmatrix} q_n \\ p_n \end{pmatrix} + O(h^{r+1}), \tag{6.27}$$

and orders $r = 2, 4, 6, 8, 10$ and 12 have been achieved with only $s = r$ exponentials $\mathrm{e}^{a_j h A}$ and $\mathrm{e}^{b_j h B}$ (Gray and Manolopoulos 1996, Zhu, Zhao and Tang 1996, Liu, Ding, Hong and Wang 2005).

The processing technique has also been used to construct splitting schemes with two different goals in mind: to attain maximal stability and maximal accuracy. They have the general structure $P^{-1}(hH)K(hH)P(hH)$, where K (the kernel) is built as a composition (6.27) with a large number of stages s, and P (the processor) is taken as a polynomial. Although these methods are neither unitary nor unconditionally stable, they are symplectic and conjugate to unitary schemes. In consequence, neither the average error in energy nor the norm of the solution increases with time. Specifically, Blanes, Casas and Murua (2006*b*, 2008*a*) have proposed kernels with up to 19, 32 and 38 stages, either to construct methods of orders $r = 10$, 16 and 20, or to bring highly accurate *second*-order methods with an enlarged stability domain.

This approach to approximating $\mathrm{e}^{\tau H} u$ is closely related to other polynomial approximations of the form

$$\mathrm{e}^{\tau H} u \approx P_m(hH) u, \tag{6.28}$$

where $P_m(y)$ is a polynomial in y approximating the exponential $\mathrm{e}^{-\mathrm{i}y}$. Different choices for such $P_m(y)$ are available: truncated Taylor or Chebyshev series expansions of $\mathrm{e}^{-\mathrm{i}y}$ for an appropriate real interval of y, or a Lanczos approximation, where the polynomial is determined by a Galerkin approximation on the Krylov space spanned by $u, Hu, \ldots, H^{m-1}v$ (Lubich 2008).

Given a prescribed error tolerance, some appropriate estimates of the upper and lower bounds of the eigenvalues of the matrix H, $E_{\min}$ and $E_{\max}$, and a time integration interval, $[t_0, t_f]$, in the Chebyshev approach we choose, according to some known error bounds, the lowest-degree polynomial that provides the solution with such accuracy. The coefficients of the polynomial are determined for each case and the action of the polynomial on a vector is computed recursively using the Clenshaw algorithm (Lubich 2008). On the other hand, in the Taylor approach we have to adjust the maximum degree allowed with the time step h to reach the desired accuracy with the minimum number of matrix–vector products. As a result, in general, Chebyshev turns out to be between two and three times faster than Taylor, depending on the final time at which the output is desired. Chebyshev and Lanczos approximations have quite similar error bounds (see Lubich 2008) and their relative performance depends on the particular problem considered.

It is worth remarking that, whereas in the approach (6.28) the approximation of $\mathrm{e}^{\tau H}u$ is constructed by computing products of the form Hu, where $u \in \mathbb{C}^M$, with symplectic splitting methods of the form (6.27) we proceed by successively computing *real* matrix–vector products Hq and Hp with different weights. With splitting methods, the real and imaginary parts of $\mathrm{e}^{\tau H}u \equiv \mathrm{e}^{-\mathrm{i}hH}(q + \mathrm{i}p)$ are approximated in a different way, with a considerably reduced computational cost.

Blanes *et al.* (2015) construct several optimized symplectic splitting methods and present an algorithm that automatically selects the most efficient one for a prescribed error tolerance under the same conditions as when using the Chebyshev method. The resulting algorithm is between 1.4 and 2 times faster than the Chebyshev method for the same accuracy, with reduced energy and unitarity errors for large values of h. The computation of the coefficients of the schemes is largely based on the stability and error analysis of splitting methods carried out in Blanes, Casas and Murua (2008*a*, 2011).

In contrast to T–V splitting methods, which preserve unitarity by construction and are thus unconditionally stable, the previous polynomial approximations suffer from a step size restriction. Given h, Δx and m, the degree of the polynomial, these methods must satisfy the restriction

$$\frac{h}{m\Delta x^2} \leq C,$$

so the time interval that one can advance per matrix–vector product is proportional to Δx^2 or, equivalently, the number of matrix–vector products to reach the final time is inversely proportional to Δx^2.

Time-dependent potentials. We have assumed so far that the potential in the Schrödinger equation (1.40) does not depend explicitly on time or, if it does, it only varies slowly with time, so that in each sub-interval $[t_n, t_{n+1}]$ the corresponding matrix V is obtained from the average of $V(x, t)$ on this interval. In general, however, we have to deal with situations in which this approximation is no longer valid. In that case, and in contrast to other approaches based on Chebyshev or Lanczos approximations, splitting methods can still be applied with some appropriate modifications.

In a similar way to classical Hamiltonian problems, one may take t in the potential as an additional coordinate, $x_{d+1} = t$, introduce its canonical momentum, $p_{d+1} = -i\,\partial_{x_{d+1}}$, and deal with the system in the extended phase space, $\tilde{H} = (T + p_{d+1}) + V(x, x_{d+1})$. Since the action of the operator p_{d+1} just corresponds to a shift in the variable x_{d+1}, i.e. $\mathrm{e}^{ah\partial_t} V(t,x) = V(t+ah, x)$, and moreover

$$\mathrm{e}^{ah\partial_t}\,\mathrm{e}^{-\mathrm{i}bhV(t,x)}\,\mathrm{e}^{-ah\partial_t} = \mathrm{e}^{-\mathrm{i}bh\mathrm{e}^{ah\partial_t}V(t,x)} = \mathrm{e}^{-\mathrm{i}bhV(t+ah,x)},$$

composition (6.24) applied to the corresponding non-autonomous problem now simply reads (notice that T and p_{d+1} commute)

$$u_{n+1} = \prod_{j=1}^{s} \mathrm{e}^{b_j\tau V(t_n+c_jh)}\,\mathrm{e}^{a_j\tau T}\,u_n, \quad \text{with} \quad c_j = \sum_{k=1}^{j} a_k,\ j \in \{1,\dots,s\}, \tag{6.29}$$

and analogously if the scheme includes modified potentials; see also e.g. Chin and Chen (2002) and references therein.

For a generic time-dependent Hamiltonian $H(t)$, the Schrödinger equation $\mathrm{i}\partial_t\psi = H(t)\psi$ can be recast as a non-autonomous evolution equation of the form

$$u'(t) = A(t)u(t), \quad u(t_0) = u_0, \quad t \in [t_0, t_f], \tag{6.30}$$

defined by a family of time-dependent linear operators $(A(t))_{t\in[t_0,t_f]}$, which, assuming a spatial discretization has been carried out, are generally complex matrices of large dimension and large norm.

It turns out that standard rth-order splitting methods defined by coefficients $(a_j, b_j)_{j=1}^{s}$ can be applied in this setting simply by adding the trivial relation $(\mathrm{d}/\mathrm{d}t)t = 1$ to equation (6.30). This results in the scheme

$$u_{n+1} = \prod_{j=1}^{s} \mathrm{e}^{hb_jA(t_n+c_jh)}u_n, \quad \text{with} \quad c_j = \sum_{k=1}^{j} a_k,\ j \in \{1,\dots,s\} \tag{6.31}$$

of the same formal order r as the method originally designed for autonomous problems.

A different approach is based on the use of the Magnus expansion (Magnus 1954) to get a formal solution representation of (6.30) as the exponential of an infinite series:

$$u(t_n + h) = \mathrm{e}^{\Omega(h)}u_n, \quad \Omega(h) = \sum_{m=1}^{\infty} \Omega_m(h), \tag{6.32}$$

where each term Ω_m involves multiple integrals of nested matrix-commutators (Blanes, Casas, Oteo and Ros 2009). By appropriately truncating this series and approximating the integrals by quadratures, efficient integrator schemes can be constructed (Iserles, Munthe-Kaas, Nørsett and Zanna 2000). Thus, for instance, taking the two-stage Gauss–Legendre quadrature rule with nodes

$$c_{1,2} = \frac{1}{2} \mp \frac{\sqrt{3}}{6}$$

results in the scheme

$$\begin{aligned} \Omega^{[4]}(h) &= \frac{1}{2}h(A_1 + A_2) + \frac{\sqrt{3}}{12}h^2[A_2, A_1], \\ u_{n+1} &= \mathrm{e}^{\Omega^{[4]}(h)}u_n, \end{aligned} \tag{6.33}$$

with $A_j = A(t_n + c_j h)$, $j = 1, 2$. For the Schrödinger equation with $A(t) = -\mathrm{i}(T + V(t))$ and a smooth time-dependent potential $V(t)$, Hochbruck and Lubich (2003) showed that Magnus integrators retain their full order of convergence (without bounds on T in the error bound) for sufficiently regular solutions, uniformly with respect to the space discretization.

There are, however, several issues related to Magnus integrators due to the presence of iterated commutators. Thus, computing the action of iterated commutators on vectors can be very costly due to the number of matrix–vector products required. This is particularly relevant when considering problems in two and three space dimensions (Bader, Iserles, Kropielnicka and Singh 2016). In addition, the evolution equations defining high-order Magnus integrators in general involve differential operators of different nature from the original problem (Blanes, Casas, González and Thalhammer 2021*b*).

A different class of exponential integrators that circumvent these difficulties whilst still retaining the favourable properties of Magnus integrators is formed by the so-called commutator-free quasi-Magnus (CFQM) methods: the basic idea is to replace the single exponential in (6.32) with a composition of several exponentials involving linear combinations of the values of the operator A at certain nodes, c_k, of a quadrature rule:

$$\begin{aligned} u_{n+1} &= \mathrm{e}^{h\,B_{nJ}} \cdots \mathrm{e}^{h\,B_{n1}}\,u_n \;\approx\; u(t_{n+1}) = \mathrm{e}^{\Omega(h)}\,u(t_n), \\ c_k &\in [0,1], \quad A_{nk} = A(t_n + c_k h), \quad k \in \{1,\dots,K\}, \\ B_{nj} &= a_{j1}\,A_{n1} + \cdots + a_{jK}\,A_{nK}, \quad j \in \{1,\dots,J\}. \end{aligned}$$

Particular examples of CFQM exponential integrators are the exponential midpoint rule (order two)

$$\begin{aligned} J = K = 1, \quad c_1 = \tfrac{1}{2}, \quad a_{11} = b_1 = 1, \\ u_{n+1} = \mathrm{e}^{h\,A(t_n + \frac{1}{2}h)}\,u_n \end{aligned} \tag{6.34}$$

and the fourth-order scheme

$$\begin{gathered} J = K = 2, \quad \alpha = \tfrac{\sqrt{3}}{6}, \quad c_1 = \tfrac{1}{2} - \alpha, \quad c_2 = \tfrac{1}{2} + \alpha, \\ a_{11} = a_{22} = \tfrac{1}{4} + \alpha, \quad a_{12} = a_{21} = \tfrac{1}{4} - \alpha, \\ B_j(h) = a_{j1}\, A(t_n + c_1 h) + a_{j2}\, A(t_n + c_2 h), \quad j \in \{1, 2\}, \\ u_{n+1} = \mathrm{e}^{h\, B_2(h)}\, \mathrm{e}^{h\, B_1(h)}\, u_n. \end{gathered} \tag{6.35}$$

A detailed treatment and specific schemes up to order six can be found in Blanes, Casas and Thalhammer (2017*b*, 2018) and up to order eight in Alvermann and Fehske (2011), whereas Blanes *et al.* (2021*b*) have proved that CFQM methods applied to the Schrödinger equation with Hamiltonian $H(t) = -\frac{1}{2}\Delta + V(t)$ are unconditionally stable in the underlying Hilbert space and retain full order of convergence under low regularity requirements on the initial state.

Semiclassical regime. The so-called semiclassical Schrödinger equation

$$\mathrm{i}\varepsilon\partial_t\psi(x,t) = \left(-\frac{\varepsilon^2}{2}\Delta + V(x)\right)\psi(x,t) \tag{6.36}$$

(in atomic units), with a small parameter $\varepsilon \ll 1$, arises in particular when applying the time-dependent Born–Oppenheimer approximation for the motion of nuclei as driven by the potential energy surface of the electrons (Lubich 2008). In that case ε^2 represents the mass ratio of nuclei and electrons. Recall that (6.36) has highly oscillatory solutions with wavelength $\sim \varepsilon$, so grid-based numerical schemes require a resolution of this order in both space and time, which is computationally very expensive. One of the challenges, therefore, is to construct numerical methods that are robust in the limit $\varepsilon \to 0$.

Several options have been proposed and analysed in detail; see, for instance, the recent review by Lasser and Lubich (2020). Among others, we can recount the following.

- Split the equation into the usual kinetic and potential energy parts and apply the Strang splitting in time in combination with trigonometric spectral methods (Bao, Jin and Markowich 2002). Although the resulting scheme is unconditionally stable, time-reversible and preserves the position density, it requires very fine resolution, in both space and time, for small ε (Jin, Markowich and Sparber 2011).
- Use a Gaussian wave packet as an approximation for the wave function $\psi(x,t)$ depending on certain parameters and apply a variational splitting to get approximate solutions for the differential equations they satisfy (Faou and Lubich 2006). The resulting algorithm is symplectic, time-reversible and preserves the unit L^2-norm of the wave packets.
- Another variant of this approach instead consists in taking Hagedorn wave packets. They provide a spectral approximation in space with a time-dependent set of basis functions giving the exact solution of the Schrödinger

equation with the potential locally approximated by a quadratic function. The potential $V(x)$ is split into the quadratic term $U(q(t), x)$ in the Taylor expansion of V, around the time-dependent classical position $q(t)$ and the remainder (Faou, Gradinaru and Lubich 2009). The overall algorithm has a number of conservation and limit properties, as listed in Faou and Lubich (2006), for example. In addition, highly efficient splitting methods for perturbed problems can also be applied (Blanes and Gradinaru 2020).

- A different class of exponential splittings is proposed in Bader, Iserles, Kropielnicka and Singh (2014) for the one-dimensional case. Essentially, the formal solution of the space-discretized equation $u' = \mathrm{i}(\varepsilon T - \varepsilon^{-1} V)u$ is approximated as

$$\mathrm{e}^{\mathrm{i}h(\varepsilon T - \varepsilon^{-1} V)} \approx \mathrm{e}^{R_0} \mathrm{e}^{R_1} \cdots \mathrm{e}^{R_s} \mathrm{e}^{T_{s+1}} \mathrm{e}^{R_s} \cdots \mathrm{e}^{R_1} \mathrm{e}^{R_0}$$

  with error $O(\varepsilon^{2s+2})$. Here $R_0 = O(\varepsilon^0)$, $R_k = O(\varepsilon^{2k-2})$, $k \geq 1$, and $T_{s+1} = O(\varepsilon^{2s})$. In this approach the number of exponentials grows linearly with s and the exponentials can be computed efficiently, although the terms R_k and T_{s+1} contain nested commutators.

Nonlinear Schrödinger equations. Introducing nonlinear effects in the Schrödinger equation allows us to model some relevant physical phenomena taking place in nonlinear optics, quantum superfluids, plasmas, water waves, etc.; see e.g. Sulem and Sulem (1999) and references therein. Consider in particular a Bose–Einstein condensate (BEC), the ground state of a system of interacting bosons very close to zero temperature. It was first predicted by Einstein in 1925 and experimentally realized by Anderson *et al.* (1995). Mathematically, a BEC of an atomic species trapped in an external potential $V(x)$ is modelled by the (normalized) Gross–Pitaevskii equation (GPE)

$$\mathrm{i}\partial_t \psi(x,t) = \left(-\frac{1}{2}\Delta + V(x) + \sigma |\psi(x,t)|^2 \right) \psi(x,t), \tag{6.37}$$

with asymptotic boundary conditions $\psi(x,t) \to 0$ as $|x| \to \infty$. Here the parameter σ originates from the mean-field interaction between the particles: repulsive forces lead to $\sigma > 0$, whereas $\sigma < 0$ represents attractive forces. Equation (6.37) has been the subject of many different studies, including the existence of solutions and its numerical treatment. Concerning the first aspect, we refer to Cazenave (2003), Carles (2008) and references therein.

With respect to the numerical integration of the GPE equation, a combination of spectral discretization in space with splitting methods in time constitutes a natural option, and in fact has been explored in detail in the literature; see e.g. Bao, Jin and Markowich (2003*b*), Bao, Jaksch and Markowich (2003*a*, 2004) and Thalhammer, Caliari and Neuhauser (2009). If (6.37) is expressed as

$$\mathrm{i}\partial_t \psi(x,t) = (A + B(x,\psi))\psi(x,t), \quad \psi(x,0) = \psi_0(x), \tag{6.38}$$

with

$$A = -\frac{1}{2}\Delta, \quad B(x,\psi) = V(x) + \sigma|\psi|^2, \tag{6.39}$$

it is clear that the solution of the initial value problem $\mathrm{i}\psi_t = A\psi$, $\psi(x,0) = \psi_0(x)$ is

$$\psi(x,t) = \mathrm{e}^{-\mathrm{i}tA}\,\psi_0(x)$$

and (an approximation to) $\psi(x,t)$ is obtained by representing the initial value with respect to the (truncated) Fourier basis functions.

On the other hand, given a real function G, the solution of

$$\mathrm{i}\partial_t\psi(x,t) = G(x, |\psi(x,t)|)\,\psi(x,t) \tag{6.40}$$

leaves the norm invariant, i.e. $|\psi(x,t)| = |\psi(x,0)|$, and therefore

$$\psi(x,t) = \mathrm{e}^{-\mathrm{i}tG(x,|\psi(x,0)|)}\,\psi(x,0). \tag{6.41}$$

In consequence, the initial value problem $\mathrm{i}\partial_t\psi = B\psi$, $\psi(x,0) = \psi_0(x)$ is also solvable. Splitting methods can also be applied in the more general situation when the potential V is explicitly time-dependent, as explained earlier for the linear Schrödinger equation.

6.4. *Parabolic evolution equations*

Let us now consider the evolution equation

$$u'(t) = Lu(t) = Au(t) + Bu(t), \quad t \geq 0, \quad u(0) = u_0, \tag{6.42}$$

where the linear, possibly unbounded, operators A, B and L generate C_0 semigroups over an infinite-dimensional Banach space X. We recall that if L is the infinitesimal generator of the C_0 semigroup $T(t)$ on X and $u_0 \in \mathcal{D}(L)$, the domain of L (which is dense in X), then $u(t) = T(t)u_0$ is a classical solution of (6.42) (Partington 2004). Since in the special case of a bounded linear operator L the solution is given by the familiar expression $u(t) = \mathrm{e}^{tL}u_0$ (Engel and Nagel 2006), the semigroup $T(t)$ is also denoted by the symbol e^{tL}; see Engel and Nagel (2006), Pazy (1983) and Yosida (1971) for an introduction to the theory of C_0 semigroups.

A prototypical example is the linear heat equation with potential

$$u_t(t,x) = \frac{1}{2}\Delta u(t,x) - V(x)u(t,x). \tag{6.43}$$

Here $V(x) \geq 0$, $t \geq 0$ and $x \in \mathbb{R}^d$ (or $x \in \mathbb{T}^d$). In that case $(Au)(x) = \frac{1}{2}\Delta u(x)$, $(Bu)(x) = -V(x)u(x)$ and A generates only a C_0 semigroup. This can be seen by considering the equation $u_t = \frac{1}{2}\Delta u$ on $0 < x < 1$ with Dirichlet boundary conditions: the kth Fourier mode of the solution is $c_k\mathrm{e}^{-\frac{1}{2}(k\pi)^2 t}$, which is generally not well-defined for $t < 0$.

Hansen and Ostermann (2009*a*) have established the following result. Assuming that $\|\mathrm{e}^{tA}\| \leq \mathrm{e}^{\omega t}$, $\|\mathrm{e}^{tB}\| \leq \mathrm{e}^{\omega t}$ for the same value of $\omega \geq 0$ and all $t \geq 0$, and that for any operator E_{r+1} obtained as the product of exactly $r+1$ factors chosen amongst A and B, there is a constant $\tilde{C} > 0$ such that

$$\|E_{r+1}\mathrm{e}^{t(A+B)}u_0\| \leq \tilde{C}.$$

Then a splitting method $\Psi(h) = \prod_{j=1}^{s} \mathrm{e}^{b_j h B}\, \mathrm{e}^{a_j h A}$, with all $a_j \geq 0$ and $b_j \geq 0$ and of classical order r, retains its order when applied to (6.42):

$$\|(\Psi(h)^n - \mathrm{e}^{nhL})u_0\| \leq C h^r, \quad \text{for } n\, h \leq t_f, \tag{6.44}$$

where the constant C is independent of n and h on the bounded time interval $[0, t_f]$.

In practice, however, the positivity requirement on the coefficients restricts the splitting method to be of at most order two. If, in addition, $[B, [A, B]]$ is a bounded operator and $\|\mathrm{e}^{t[B,[A,B]]}\| \leq \mathrm{e}^{\omega t}$, then the same result (6.44) also holds for splitting methods involving double commutators,

$$\Psi(h) = \prod_{j=1}^{s} \mathrm{e}^{b_j h B + c_j h^3 [B,[A,B]]}\, \mathrm{e}^{a_j h A}$$

and positive a_j, b_j (Kieri 2015). Efficient schemes within this class up to order four specially tailored to the problem at hand have recently been proposed in Blanes *et al.* (2023).

Example: imaginary time propagation. The imaginary-time evolution method is a well-known approach to computing the ground state (and its corresponding eigenvalue) of a quantum system with Hamiltonian $H = -\frac{1}{2}\Delta + V$ (Auer, Krotscheck and Chin 2001, Lehtovaara, Toivanen and Eloranta 2007, Bader *et al.* 2013). Essentially, under the time transformation $t = -\mathrm{i}s$, the time-dependent Schrödinger equation (1.40) is transformed into ($\hbar = 1$)

$$\partial_s \psi(x, s) = \frac{1}{2}\Delta\psi(x, s) - V(x)\psi(x, s), \quad \psi(x, 0) = \psi_0(x), \tag{6.45}$$

that is, a linear heat equation of the form (6.43). If we denote the (real) eigenvalues of H as E_j, with $E_0 < E_1 < \cdots$, and the corresponding eigenfunctions as ϕ_j, $j = 0, 1, 2, \ldots$, the initial wave function $\psi_0(x)$ can be expanded in the orthonormal basis $\{\phi_j\}$,

$$\psi_0(x) = \sum_{j\geq 0} c_j\, \phi_j(x), \quad c_j = \langle \phi_j \,|\, \psi(\cdot, 0)\rangle,$$

where $\langle\cdot\,|\,\cdot\rangle$ is the usual L^2-scalar product. Then the solution of (6.45) can be written as

$$\psi(x, s) = \mathrm{e}^{-sH}\psi(x, 0) = \sum_{j\geq 0} \mathrm{e}^{-sE_j}\, c_j\, \phi_j(x). \tag{6.46}$$

Notice that for sufficiently large s we get $\psi(x, s) \longrightarrow \mathrm{e}^{-sE_0} c_0 \phi_0$, since the other exponentials decay more rapidly. In other words, any given wave function at $s = 0$ for which $c_0 \neq 0$, converges towards the ground state solution when $s \longrightarrow \infty$.

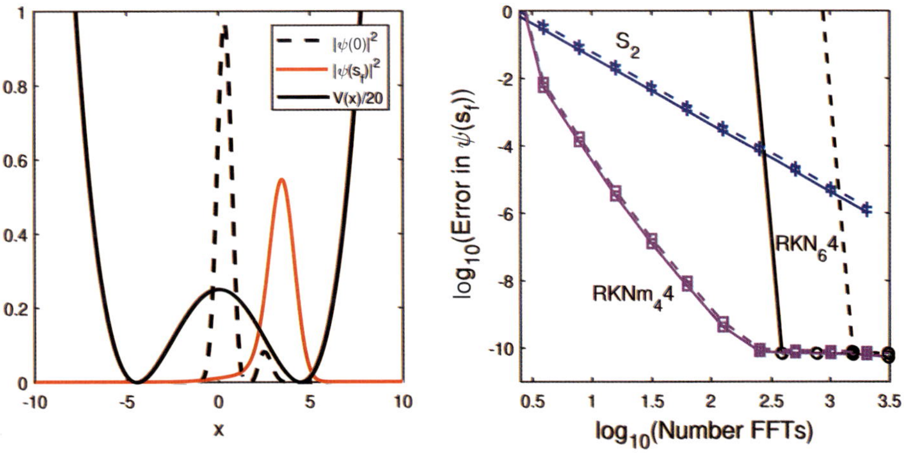

(a) (b)

Figure 6.1. Imaginary time propagation of the Schrödinger equation. (a) Double-well potential, initial and (normalized) final wave function obtained with a mesh with $N = 256$ and repeated with $N = 510$ (the two results overlap visually). (b) Discrete L_2-norm error in the normalized solution at the final time, $s_f = 1$, vs. number of FFTs for different values of the time step obtained with the same methods as in Figure 1.5: solid lines for $M = 256$ and dashed lines for $M = 512$ discretization points of the space interval.

Once an accurate approximation to ϕ_0 is obtained, the associated eigenvalue E_0 is easily obtained by computing $E_0 = \langle \phi_0 \,|\, H\phi_0 \rangle$. Other functions ϕ_j can also be approximated, for example by propagating different wave functions simultaneously in time (Aichinger and Krotscheck 2005).

To illustrate the technique, we next consider the same example as in Section 1.6, but now integrating in imaginary time up to $s_f = 1$. We take the same initial conditions (which is slightly closer to one of the minima of the potential) and normalize the solution at the final time. Figure 6.1(a) shows the potential, initial conditions and the normalized wave function at the final time when the spatial interval, $x \in [-13, 13]$, is divided into $M = 256$ and $M = 512$ parts, leading to the same visual results. In Figure 6.1(b) we show the L_2-norm error in the normalized wave function at the final time versus the number of FFTs required by the same methods as previously, both for $M = 256$ (solid lines) and $M = 512$ (dashed lines). We notice that one scheme suffers from step size restriction because it involves negative coefficients, and this is inversely proportional to Δx^2. However, the schemes with positive coefficients are insensitive to the spatial mesh.

One possible way to circumvent this order barrier for splitting methods when applied to problem (6.42) consists in considering schemes with *complex* coefficients a_i, b_i having *positive* real part. As shown in Castella, Chartier, Descombes and

Vilmart (2009) and Hansen and Ostermann (2009*b*), any splitting method

$$\Psi(h) = \prod_{j=1}^{s} \mathrm{e}^{b_j h B}\, \mathrm{e}^{a_j h A}$$

within this class still retains its classical order if the previous assumptions are conveniently modified. Specifically, we have to extend the notion of a C_0 semigroup $T(t)$ to the sector Σ_θ in the complex plane, for some angle $0 < \theta < \pi/2$,

$$\Sigma_\theta = \{t \in \mathbb{C} \colon |\arg(t)| < \theta\},$$

so that $T(t)$ is analytic in t for all $t \in \Sigma_\theta$ (Engel and Nagel 2006, Pazy 1983). Thus, if

- L, A and B generate analytic semigroups on X (now a complex Banach space) in the sector Σ_θ, $0 < \theta < \pi/2$, with

  $$\|\mathrm{e}^{tA}\| \le \mathrm{e}^{\omega|t|}, \quad \|\mathrm{e}^{tB}\| \le \mathrm{e}^{\omega|t|}$$

  for $\omega \ge 0$ and all $t \in \Sigma_\theta$,
- $\|E_{r+1}\mathrm{e}^{t(A+B)}u_0\| \le C$, where E_{r+1} is a composition of the operators A and B that consist of exactly $r + 1$ factors,

then the splitting method of classical order r with coefficients $a_i, b_i \in \Sigma_\theta \subset \mathbb{C}$ satisfies

$$\|(\Psi(h)^n - \mathrm{e}^{nhL})u_0\| \le Ch^r, \quad 0 \le nh < t_f. \tag{6.47}$$

Since $\Psi(h)u_0$ is complex-valued, this approach cannot be applied in principle when the operators A and B are real, i.e. for problems defined in a real Banach space X. The most straightforward remedy consists in projecting the numerical solution after each time step on the real axis, i.e. computing the approximations $u_n \approx u(t_n)$, as $u_n = \mathrm{Re}(\Psi(h)u_{n-1})$. In that case we still have the bound (6.47) for the resulting integration scheme (Hansen and Ostermann 2009*b*).

Although this is only valid in the linear case, similar results are observed in practice in the nonlinear heat equation $\partial_t u = \Delta u + F(u)$ with periodic boundary conditions, at least when F is analytic (Castella *et al.* 2009, Blanes, Casas, Chartier and Murua 2013*a*), so they constitute a strong motivation to study splitting methods with complex coefficients in general. This is precisely the subject of the following section.

7. Splitting methods with complex coefficients

It has been known for a long time that, besides real solutions, the order conditions arising from splitting and composition methods (see Section 2) also admit complex solutions (Bandrauk and Shen 1991, Suzuki 1990, 1991, 1995). In fact, some of the resulting methods were explored in the context of Hamiltonian (Chambers 2003) and quantum mechanics (Bandrauk, Dehghanian and Lu 2006, Prosen and Pizorn 2006) before being used more recently to overcome the order barrier in parabolic partial differential equations (Castella *et al.* 2009, Hansen and Ostermann 2009*b*).

The use of the complex plane to solve problems formulated in the real line has proved to be very fruitful in many branches of mathematics, as illustrated by Painlevé's famous dictum.[6] By applying the same logic in the particular case of splitting methods, it is reasonable to ask what benefits (if any) might result from carrying out the integration along paths in the complex plane.

We have already seen at the end of Section 6 that a strategy to develop effective methods of order higher than two for systems evolving in a semigroup, such as the heat equation,[7] consists precisely in applying splitting schemes with complex coefficients having positive real part. Moreover, the large number of complex solutions for the order conditions might offer more flexibility in the final choice of coefficients, and perhaps lead to schemes with smaller truncation errors and new symmetries.

On the other hand, when dealing with problems formulated in the real line, the use of complex arithmetic introduces an additional computational cost with respect to methods with purely real coefficients. From a theoretical point of view, the vector field appearing in the differential equation has to be analytic at least in a domain containing the path where the actual integration is carried out in the complex plane. Otherwise, order reductions are to be expected unless the implementation is not conveniently adapted.

In this section we summarize some of the issues involved in the construction and analysis of splitting methods with complex coefficients, and also review some of their most salient properties with regard to preservation of qualitative properties, in both classical Hamiltonian and quantum systems.

For brevity, henceforth we denote a complex number a with positive real part by writing $a \in \mathbb{C}_+$, so we are only interested in methods whose coefficients $a_j, b_j \in \mathbb{C}_+$.

7.1. *Compositions*

Most of the existing splitting methods with complex coefficients have been constructed by applying the composition technique of Section 2.1. Here, for completeness, we review the process, starting with the Lie–Trotter scheme $\chi_h = \varphi_h^{[2]} \circ \varphi_h^{[1]}$ as the basic method and composing it with different weights. The simplest situation corresponds of course to

$$\phi_h^{[2]} \equiv \chi_{\gamma_{1,2}h} \circ \chi_{\gamma_{1,1}h}. \tag{7.1}$$

[6] 'Il apparut que, entre deux vérités du domaine réel, le chemin le plus facile et le plus court passe bien souvent par le domaine complexe' (Painlevé 1900).

[7] This in fact constitutes Problem 10 in the list of open problems posed in McLachlan and Quispel (2002).

In accordance with Section 2.1.2, $\phi_h^{[2]}$ is of order two if the coefficients satisfy equation (2.14), that is,

$$\gamma_{1,1} = \frac{1}{2} \pm \frac{1}{2}\mathrm{i}, \quad \gamma_{1,2} = \overline{\gamma}_{1,1} = \frac{1}{2} \mp \frac{1}{2}\mathrm{i}.$$

If we try to construct a method of order three, a composition of at least four maps χ_h is needed, since in that case there are four order conditions. Although there are eight solutions (+cc), only four have positive real part. In fact, two of these solutions result from composing $\phi_h^{[2]}$, namely

$$\phi_h^{[3]} = \phi_{\gamma_{2,2}h}^{[2]} \circ \phi_{\gamma_{2,1}h}^{[2]}, \tag{7.2}$$

by requiring that $\gamma_{2,1} + \gamma_{2,2} = 1$, $\gamma_{2,1}^3 + \gamma_{2,2}^3 = 0$. Notice that in (7.2) $\gamma_{2,2} = \overline{\gamma}_{2,1}$. This procedure can in principle be repeated by considering the recurrence

$$\phi_h^{[r+1]} = \phi_{\gamma_{r,m_r}h}^{[r]} \circ \cdots \circ \phi_{\gamma_{r,1}h}^{[r]}, \quad r = 1, 2, \ldots \tag{7.3}$$

to construct higher-order methods with *any* first-order integrator $\phi_h^{[1]}$ (not necessarily the Lie–Trotter scheme). It turns out, however, that $\phi_h^{[4]}$ and higher-order schemes obtained from (7.3) have at least one coefficient with negative real part (Blanes *et al.* 2013*a*). Thus, if we are interested in schemes having only coefficients in $\mathbb{C}_+$, $r = 3$ indeed constitutes an order barrier for this type of composition (Hansen and Ostermann 2009*b*).

Let us now consider the Strang splitting as the basic method. Then composition (2.1) with $s = 2$ already provides a method of order three:

$$S_h^{[3]} = S_{\gamma_2 h}^{[2]} \circ S_{\gamma_1 h}^{[2]}, \quad \text{with} \quad \gamma_1 = \frac{1}{2} \pm \mathrm{i}\frac{\sqrt{3}}{6}, \ \gamma_2 = \overline{\gamma}_1. \tag{7.4}$$

Again, higher-order methods can be obtained by applying the recursive procedure

$$S_h^{[r+1]} = S_{\gamma_{r,2}h}^{[r]} \circ S_{\gamma_{r,1}h}^{[r]}, \quad r = 2, 3, \ldots, \tag{7.5}$$

with $\gamma_{r,1} + \gamma_{r,2} = 1$, $\gamma_{r,1}^{r+1} + \gamma_{r,2}^{r+1} = 0$ (see (2.52)). The solution with the smallest phase is given by

$$\gamma_{k,1} = \frac{1}{2} \pm \frac{\mathrm{i}}{2}\tan\left(\frac{\pi}{2(r+1)}\right), \quad \gamma_{r,2} = \overline{\gamma}_{r,1}, \tag{7.6}$$

but only schemes up to order six with coefficients in $\mathbb{C}_+$ are possible with this approach (Hansen and Ostermann 2009*b*).

As we have already seen, the triple jump (2.4) allows us to raise the order by two. Thus, starting from $S_h^{[2]}$, the composition

$$S_h^{[4]} = S_{\gamma_3 h}^{[2]} \circ S_{\gamma_2 h}^{[2]} \circ S_{\gamma_1 h}^{[2]} \tag{7.7}$$

has, apart from the real solution (2.3),

$$\gamma_1 = \gamma_3 = \frac{1}{2 - 2^{1/3} e^{2i\ell\pi/3}}, \quad \gamma_2 = 1 - 2\alpha_1, \quad \ell = 1, 2, \tag{7.8}$$

corresponding to a time-symmetric composition, and

$$\gamma_1 = \overline{\gamma}_3 = \frac{1}{4} \pm i \frac{1}{4} \sqrt{\frac{5}{3}}, \quad \gamma_2 = \frac{1}{2}, \tag{7.9}$$

possessing the same symmetry as (7.2) and (7.4). Palindromic methods up to order eight with coefficients having positive real part are possible by applying the triple jump composition, whereas if we instead consider the *quadruple jump*,

$$S^{[2k]}_{\gamma_{k,1}h} \circ S^{[2k]}_{\gamma_{k,2}h} \circ S^{[2k]}_{\gamma_{k,2}h} \circ S^{[2k]}_{\gamma_{k,1}h},$$

we can achieve order 14 with all coefficients in $\mathbb{C}_+$. These order barriers have been rigorously proved in Blanes *et al.* (2013*a*).

Compositions (7.4) and (7.7) can be reformulated as splitting methods when $S_h^{[2]}$ is the Strang splitting. Thus

$$S_h^{[3]} = \varphi^{[1]}_{\overline{a}_1 h} \circ \varphi^{[2]}_{\overline{b}_1 h} \circ \varphi^{[1]}_{a_2 h} \circ \varphi^{[2]}_{b_1 h} \circ \varphi^{[1]}_{a_1 h}, \tag{7.10}$$

with

$$a_1 = \frac{1}{4} \pm i \frac{\sqrt{3}}{12}, \quad b_1 = \frac{1}{2} \pm i \frac{\sqrt{3}}{2}, \quad a_2 = \frac{1}{2},$$

whereas the palindromic version of (7.7) reads

$$S^{[4]}_{h,P} = \varphi^{[1]}_{a_1 h} \circ \varphi^{[2]}_{b_1 h} \circ \varphi^{[1]}_{a_2 h} \circ \varphi^{[2]}_{b_2 h} \circ \varphi^{[1]}_{a_2 h} \circ \varphi^{[2]}_{b_1 h} \circ \varphi^{[1]}_{a_1 h}, \tag{7.11}$$

with

$$a_1 = \frac{\gamma_1}{2}, \quad b_1 = \gamma_1, \quad a_2 = \frac{1}{2} - a_1, \quad b_2 = 1 - 2b_1,$$

respectively. Finally, solution (7.9) leads to

$$S^{[4]}_{h,C} = \varphi^{[1]}_{\overline{a}_1 h} \circ \varphi^{[2]}_{\overline{b}_1 h} \circ \varphi^{[1]}_{\overline{a}_2 h} \circ \varphi^{[2]}_{b_2 h} \circ \varphi^{[1]}_{a_2 h} \circ \varphi^{[2]}_{b_1 h} \circ \varphi^{[1]}_{a_1 h}, \tag{7.12}$$

with

$$a_1 = \frac{b_1}{2}, \quad b_1 = \frac{1}{4} \pm i \frac{1}{4} \sqrt{\frac{5}{3}}, \quad a_2 = \frac{1}{4}(2b_1 + 1), \quad b_2 = \frac{1}{2}.$$

Before proceeding further, it is useful to illustrate how the different symmetries of these third- and fourth-order schemes manifest in practice on a simple example.

Example: two-level system. We consider the time evolution of a two-level quantum system, described by (Messiah 1999)

$$i \frac{dU}{dt} = HU = (\sigma_1 + \sigma_2) U, \quad \text{with} \quad \sigma_1 = \begin{pmatrix} 0 & 1 \\ 1 & 0 \end{pmatrix}, \ \sigma_2 = \begin{pmatrix} 0 & -i \\ i & 0 \end{pmatrix} \tag{7.13}$$

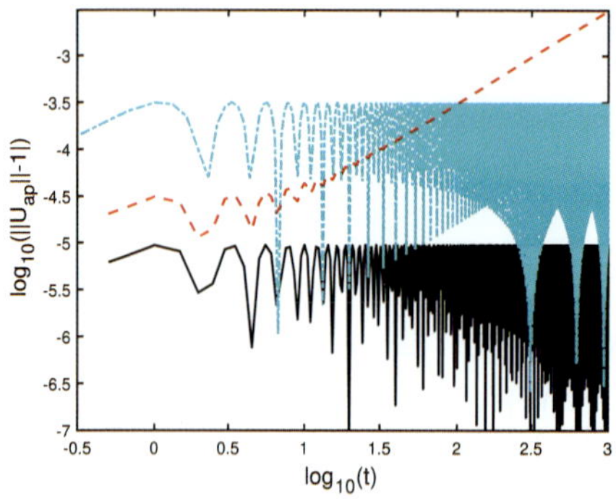

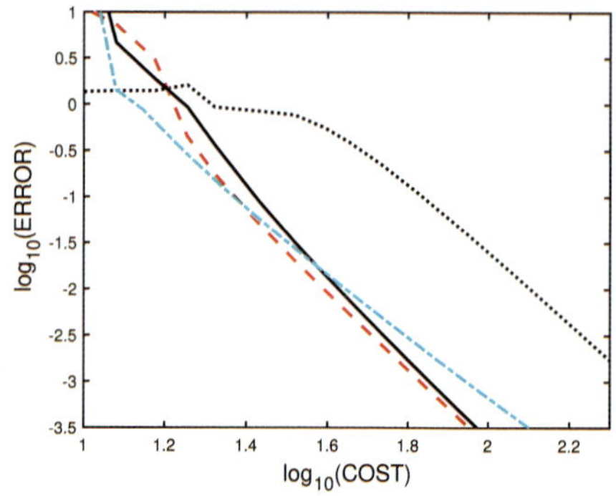

Figure 7.1. (a) Evolution of error in unitarity for system (7.13) integrated with third- and fourth-order splitting methods with complex coefficients. (b) Efficiency diagram computed for a final time $t_f = 10$. Different patterns in the coefficients lead to different qualitative behaviour.

and $U(0) = I$. Clearly, $U(t)$ is a 2×2 unitary matrix and $\det U(t) = 1$. We test both the preservation of unitarity in the numerical approximations and the computational efficiency of the previous third- and fourth-order schemes. To this end, we take $\varphi_h^{[j]} = \mathrm{e}^{\tau\sigma_j}$, $j = 1, 2$, with $\tau = -ih$, and compare with the exact solution $\mathrm{e}^{\tau(\sigma_1+\sigma_2)}$. Notice that for problems formulated in complex variables, the use of complex coefficients does not generally increase the overall cost of the algorithm.

We fix $t_f = 1000$ as the final time, and adjust h (and therefore the number of steps n) so that all the methods require the same computational effort. Specifically, $n = 6000$ ($h = 1/6$) for scheme (7.10) and $n = 4000$ ($h = 1/4$) for the fourth-order methods (7.11) and (7.12). Finally we compute $|\|U_{\mathrm{app}}(jh)\| - 1|$, $j = 1, 2, \ldots, n$, where $U_{\mathrm{app}}(jh)$ denotes the approximate solution after j steps, which is shown in Figure 7.1(a). The dashed (red) line corresponds to scheme (7.11), the solid (black) line is produced by (7.12), and finally the dot-dashed line is the result obtained with the third-order scheme (7.10). Notice the different behaviour of different classes of integrators with complex coefficients: whereas error in the preservation of unitarity does not grow with time for methods (7.12) and (7.10), it certainly does for the palindromic scheme (7.11).

Figure 7.1(b) shows an efficiency diagram of these integrators, together with the classical triple jump with real coefficients (dotted line). It is produced by integrating (7.13) with different values of h and computing the error of the approximation (in the 2-norm) at the final time $t_f = 10$ as a function of the computational cost (estimated as the number of exponentials involved in the whole integration). We observe the surprisingly good performance of scheme (7.10) and the superior efficiency of the fourth-order methods with complex coefficients in comparison to their real counterpart. This is related to the difference in size of the main error term in the expansion of the modified operator associated with each fourth-order method, which is typically associated with the coefficient multiplying $\mathcal{E}_5 = |\sum_{j=1}^{3} \gamma_j^5|$. Thus, $\mathcal{E}_5$ is about 200 times smaller for the schemes (7.8) and (7.9) than for the triple jump with real coefficients (Blanes, Casas and Escorihuela-Tomàs 2022*a*).

7.2. *Symmetric-conjugate methods and unitary problems*

The previous example shows that the time-symmetric character of a method with complex coefficients does not necessarily guarantee good preservation properties, whereas the situation is different for schemes (7.12) and (7.10). They belong to the general class of methods

$$\psi_h = \varphi_{a_{s+1}h}^{[1]} \circ \varphi_{b_s h}^{[2]} \circ \varphi_{a_s h}^{[1]} \circ \cdots \circ \varphi_{a_2 h}^{[1]} \circ \varphi_{b_1 h}^{[2]} \circ \varphi_{a_1 h}^{[1]}, \tag{7.14}$$

whose coefficients verify

$$a_{s+2-j} = \overline{a}_j, \quad b_{s+1-j} = \overline{b}_j, \quad j = 1, 2, \ldots \tag{7.15}$$

and can be properly called *symmetric-conjugate*. They possess the following distinctive feature: assume our differential equation $x' = f(x)$ is reversible with respect to complex conjugation, in the sense that $\overline{f(x)} = -f(\overline{x})$ for all $x \in \mathbb{C}$, and the same holds for each piece f_j in $f = f_1 + f_2$. Then scheme (7.14)–(7.15) verifies

$$\overline{\psi}_h = \psi_h^{-1}, \tag{7.16}$$

so that the map ψ_h is also reversible with respect to complex conjugation (Hairer *et al.* 2006, sect. V.1). Notice that (7.16) is not satisfied by palindromic compositions involving complex coefficients.

Property (7.16) has some major consequences. Suppose we are dealing with the linear problem

$$\mathrm{i}\frac{\mathrm{d}u}{\mathrm{d}t} = Hu, \quad u(0) = u_0, \quad u \in \mathbb{C}^N, \tag{7.17}$$

where H is an $N \times N$ real matrix of the form $H = H_1 + H_2$, with H_1, H_2 also real, so that the symmetric-conjugate method (7.14)–(7.15) reads

$$\psi_\tau = \mathrm{e}^{\overline{a}_1 \tau H_1} \mathrm{e}^{\overline{b}_1 \tau H_2} \mathrm{e}^{\overline{a}_2 \tau H_1} \cdots \mathrm{e}^{a_2 \tau H_1} \mathrm{e}^{b_1 \tau H_2} \mathrm{e}^{a_1 \tau H_1} \tag{7.18}$$

with $\tau = -\mathrm{i}h$. Bernier, Blanes, Casas and Escorihuela-Tomàs (2023) have proved that if all the eigenvalues of H are real and simple, then, for sufficiently small h, there exist real matrices D_τ (diagonal) and P_τ (invertible) such that $\psi_\tau^n = P_\tau \, \mathrm{e}^{n\tau D_\tau} P_\tau^{-1}$, all the eigenvalues of the linear map ψ_τ have modulus 1 and the norm $|u|^2$ and the energy $\overline{u}^\top H u$ are almost preserved for long times. In other words, any symmetric-conjugate splitting method ψ_τ applied to (7.17) is similar to a unitary method for sufficiently small values of the step size h.

We should recall that the time-dependent Schödinger equation, once discretized in space, leads to an equation of the form (7.17), with both H_1 and H_2 real and symmetric matrices: $H_1 = T$ is associated with the second-order periodic spectral differentiation matrix and $H_2 = V$ is the (diagonal) matrix corresponding to the potential evaluated at the grid points. It is then admissible to use symmetric-conjugate methods in this setting, since they also guarantee preservation of invariants for sufficiently small values of h. We should notice, however, that both positive and negative imaginary parts are present in the scheme if $a_j \in \mathbb{C}$ (since the method is consistent), and this may lead to severe instabilities due to the unboundedness of the Laplace operator (Castella *et al.* 2009, Hansen and Ostermann 2009*b*). It then seems appropriate to consider in this setting only symmetric-conjugate splitting methods with $0 < a_j < 1$ and $b_j \in \mathbb{C}_+$. Different schemes of orders three to six within this family have been proposed and tested in Bernier *et al.* (2023), also showing promising results in terms of efficiency.

If, on the other hand, the equation to integrate is of the form

$$\frac{\mathrm{d}u}{\mathrm{d}t} = Hu = (H_1 + H_2)u, \quad u(0) = u_0, \quad u \in \mathbb{R}^N \tag{7.19}$$

with H_1, H_2 real symmetric matrices, then a symmetric-conjugate splitting method ψ_h of order r satisfies

$$(\overline{\psi}_h)^\top = \psi_h,$$

for all values of h. In consequence, there exists a family of unitary matrices U_h and a family of real diagonal matrices D_h depending smoothly on h such that

$$(\psi_h)^n = U_h\, \mathrm{e}^{nhD_h}\, (\overline{U}_h)^\top,$$

where $\mathrm{Im}(U_h) = O(h^r)$ and D_h is a perturbation of order r of the matrix D_0 diagonalizing H. If $u_n = (\psi_h)^n u_0$, then the quantity $\|\mathrm{Im}(u_n)\|/\|u_n\|$ remains bounded along the numerical trajectory, since the error in the imaginary part is only due to the transformation U_h. Again, this is not true for palindromic schemes with complex coefficients.

We can then conclude that symmetric-conjugate splitting methods can be safely used to integrate equation (7.19) when H_1 and H_2 are real symmetric, as for palindromic (or time-symmetric) schemes with real coefficients, with one important difference: whereas in splitting methods with real coefficients at least one a_j and at least one b_j are negative when the order $r \geq 3$, symmetric-conjugate methods of order $r \geq 3$ do exist with coefficients having positive real part.

Several parabolic PDEs lead to equation (7.19) after space discretization. This is the case, in particular, for the time-dependent Schrödinger equation in imaginary time (6.45).

7.3. Projecting on the real axis

In the general case of an equation $x' = f(x) = f_1(x) + f_2(x)$ with f_1, f_2 real, the usual practice consists in projecting the numerical evolution after each time step to its real part, since we are interested in getting real approximations to the exact solution. Thus, if ψ_h is a splitting method of order r, by following this approach, we are actually applying the scheme

$$R_h = \frac{1}{2}(\psi_h + \overline{\psi}_h), \tag{7.20}$$

which is also at least of order r, in agreement with the comments at the end of Section 6 (Hansen and Ostermann 2009*b*). In fact, the order of R_h is actually $r+1$ if ψ_h is symmetric-conjugate and r is odd. This was already noticed by Chambers (2003) for the third-order method (7.10) and proved in general by Blanes, Casas, Chartier and Escorihuela-Tomàs (2022*c*).

This feature of symmetric-conjugate methods has some effects, especially when constructing high-order schemes to be used by projecting on the real axis. Although apparently a symmetric-conjugate composition requires us to solve the same number of order conditions to achieve order r as a general composition (and more than a palindromic one for orders higher than four), additional reductions take place for the projected method (7.20) (Blanes *et al.* 2022*c*), and in fact the resulting R_h requires fewer stages. Thus, in particular, it is possible to construct a composition of the form

$$\psi_h = S^{[2]}_{\overline{\alpha}_1 h} \circ S^{[2]}_{\overline{\alpha}_2 h} \circ \cdots \circ S^{[2]}_{\alpha_2 h} \circ S^{[2]}_{\alpha_1 h}, \tag{7.21}$$

with $S^{[2]}_h$ a time-symmetric second-order method, with only five stages so that its projected part R_h is of order six, whereas schemes based on palindromic compositions require at least seven stages. The reduction is more notable for higher orders: thus, a composition (7.21) of order five involving nine appropriately chosen stages leads to a projected method R_h of order eight. By contrast, 15 stages are required by palindromic compositions.

One of the salient features of splitting and composition methods is that they preserve by construction qualitative features of the exact solution, as seen in Section 4. These favourable properties are of course lost when the special linear combination (7.20) is considered, but nevertheless the resulting scheme R_h still preserves them up to an order much higher than the order of the method itself. More specifically, suppose $S^{[2]}_h$ is a time-symmetric second-order and symplectic method applied to a Hamiltonian system. Then we have the following result (Blanes *et al.* 2022*c*).

- If the symmetric-conjugate composition ψ_h given in (7.21) is of odd order $r = 2k-1$, $k \geq 2$, then the method R_h (of order $2k$) preserves time-symmetry up to order $4k-1$, that is,

  $$R_h \circ R_{-h} = \mathrm{id} + O(h^{4k}),$$

  and symplecticity up to the same order,

  $$(R'_h)^\top J \, R'_h = J + O(h^{4k}),$$

  where J denotes the canonical symplectic matrix and R'_h is the Jacobian.

- If ψ_h is symmetric-conjugate and of even order $r = 2k$, then R_h (which is also of order $2k$) preserves time-symmetry and symplecticity up to order $4k + 3$.
- If ψ_h is palindromic of order $r = 2k$, then the resulting R_h (also of order $2k$) preserves time-symmetry and symplecticity up to order $4k + 1$.

8. A collection of splitting methods

Once the order conditions for a splitting or composition method of a given order r have been explicitly obtained, the next step in the construction of particular schemes consists of course in solving these polynomial equations to determine the coefficients. To begin, we consider compositions with as many parameters as equations and try to use a computer algebra system to determine all real solutions. Nevertheless, solving the order conditions in this way is only possible for moderate values of r, and thus we have to turn to numerical techniques. Since the number of real solutions usually increases with the order, the problem is how to select the particular solution expected to give the best performance when the integrator is applied to practical problems. This is typically done by minimizing some objective function, depending on the particular class of schemes considered. Thus, in the case of splitting methods of the form (2.17), we try to minimize the leading error term in the asymptotic expansion of $\log(\Psi(h))$ in (2.21), namely $\sum_{j=1}^{c_{r+1}} \beta_j F_{r+1,j}$, where $F_{r+1,j}$ denote the elements of the Lyndon basis associated with Lyndon words with $r + 1$ letters. The objective function is then

$$E_{r+1} = \left(\sum_{j=1}^{c_{r+1}} |\beta_j|^2 \right)^{1/2}.$$

However, we have to take into account that E_{r+1} will change if another basis of Lie brackets is considered. We are also assuming that all these brackets contribute in a similar way, something that is not guaranteed to take place in general. It makes sense, then, to introduce other quantities as possible estimators of the error committed. In particular, it has been noticed that large coefficients a_j, b_j in the splitting method usually lead to large truncation errors, since higher-order terms in $\Psi(h)$ depend on increasingly higher powers of these coefficients. For this reason, it is also convenient to keep track of the quantities

$$\Delta \equiv \sum_j (|a_j| + |b_j|) \quad \text{and} \quad \delta \equiv \max_j (|a_j|, |b_j|) \tag{8.1}$$

and eventually discard solutions with large values of Δ or δ. In the case of compositions (2.15) of a basic first-order method and its adjoint, and compositions of Strang maps, (2.1), a frequently used criterion is to choose the solution that minimizes $\sum_{j=1}^{2s} |\alpha_j|$ and $\sum_{j=1}^{s} |\gamma_j|$, respectively. Other possibilities include minimizing only those terms in the truncation error that are not removable by a processor, in accordance with the analysis carried out in Sections 4 and 5.

Including additional maps (stages) in the composition provides additional parameters that may lead to smaller values of the chosen objective function. It is important at this point to remark that the efficiency of a method is measured by taking into account the computational cost required to achieve a given accuracy. Thus, if we have several methods of order r with different computational cost (usually measured as the number of stages or evaluations of the functions involved), the most efficient method does not necessarily correspond to the cheapest method: the extra cost of some methods can be compensated by an improvement in the accuracy obtained. In fact, this is what usually happens in practice, although solving the polynomial equations with additional stages (and free parameters) is by no means a trivial task. Continuation techniques have been shown to be very useful in this context; see e.g. Blanes *et al.* (2013*b*) and Alberdi, Antoñana, Makazaga and Murua (2019).

The use of the processing technique also allows us to construct methods of a given order typically requiring a reduced computational effort, so that the overall efficiency of the resulting schemes is enlarged if the output is not frequently computed.

With all these considerations in mind, our purpose in this section is to present a comprehensive overview of (most of) the existing methods with real coefficients by classifying them into different families and giving the appropriate references. The corresponding coefficients can be found at www.gicas.uji.es/SplittingMethods.html.

8.1. Symmetric compositions of time-symmetric second-order schemes

Perhaps the first method in this family corresponds to the fourth-order method obtained from the triple jump composition (2.4) when $k = 2$. It was known and internally used in the accelerator physics community during the 1980s (Forest and Ruth 1990) but was independently discovered afterwards in several settings: in Candy and Rozmus (1991), in Campostrini and Rossi (1990) as an algorithm for hybrid Monte Carlo simulations, and also in Creutz and Gocksch (1989) as the more general composition

$$S_h^{[2k+2]} = \left(S_{\gamma_1 h}^{[2k]}\right)^p \circ S_{(1-2p\gamma_1)h}^{[2k]} \circ \left(S_{\gamma_1 h}^{[2k]}\right)^p, \quad \text{with} \quad \gamma_1 = \frac{1}{2p - (2p)^{1/(2k+1)}}. \tag{8.2}$$

This recursion was also obtained in Suzuki (1990, 1991), whereas the case $p = 1$ became immensely popular following Yoshida (1990). It was in a certain sense generalized by McLachlan (2002), who provided a rule of thumb to select the optimal value of p for different orders, also valid for processed methods.

As stated in Section 2.1, other choices for the coefficients γ_j in the more general composition

$$\psi_h = S_{\gamma_s h}^{[2]} \circ S_{\gamma_{s-1} h}^{[2]} \circ \cdots \circ S_{\gamma_1 h}^{[2]} \tag{8.3}$$

lead to more efficient schemes when $r \geq 6$. It turns out that virtually all published methods of this form correspond to time-symmetric compositions, i.e. $\gamma_{s+1-j} = \gamma_j$,

Table 8.1. Symmetric compositions of second-order symmetric methods of order $r = 4, 6, 8, 10$ published in the literature. We indicate the number of stages **s** and the pertinent reference. Processed methods are preceded by **P**. The recommended methods, in accordance with the experiments in the Appendix, are $\boxed{\text{framed}}$.

$$\psi_h = S^{[2]}_{\gamma_s h} \circ S^{[2]}_{\gamma_{s-1} h} \circ \cdots \circ S^{[2]}_{\gamma_1 h}$$

$r = 4$	$r = 6$	$r = 8$	$r = 10$
3-CR90 $\boxed{\mathbf{5}}$-SUZ90,MCL95b	**7**-YOS90 **9**-MCL95b,KR97 **11**-$\boxed{\mathbf{13}}$-SS05	**15**-YOS90,SUZ93, MCL95b,KR97 **17**-MCL95b,KR97 **19**-$\boxed{\mathbf{21}}$-SS05 **24**-CSS93	**31**-SUZ93,KR97,SS05 **33**-KR97,TSI99, HLW02,SS05 $\boxed{\mathbf{35}}$-SS05
P:3-17-MCL02	**P:**$\boxed{\mathbf{11}}$-**13**-BCM06	**P:**$\boxed{\mathbf{13}}$-**19**-BCM06	**P:19**-$\boxed{\mathbf{23}}$-BCM06

and are therefore of even order. There are at least two reasons for that: (i) the task of constructing new schemes is simplified, since the number of order conditions to be solved is much reduced, and (ii) the resulting methods also require a smaller number of stages s, since the number of order conditions at even orders (automatically satisfied by a time-symmetric composition) is greater than the total number of order conditions divided by 2, for sufficiently large r. For instance, constructing a non-symmetric eighth-order scheme requires at least $s = 16$ (and therefore to solve a system of 16 polynomial equations), whereas $s = 15$ is the minimum number for a symmetric composition, so that only eight polynomial equations have to be solved.

In Table 8.1 we present the most relevant schemes of this type found in the literature. At each order, r, we label each method by the number of stages **s** and an acronym indicating the author(s) and year when it was first published. We also include processed methods, referred to as **P**:**s**, where **s** is the number of stages of the kernel. The following list provides additional information about the collected schemes.

CR90. Collective name given to the triple jump composition, (8.2) with $k = p = 1$.

SUZ90. Recursion (8.2) with $k = 1$, $p = 2$, proposed in Suzuki (1990).

YOS90. Yoshida (1990) gives three sixth-order compositions with $s = 7$ and five eighth-order methods with $s = 15$.

CSS93. Calvo and Sanz-Serna (1993*b*) obtain a seventh-order non-symmetric scheme which can be written as composition (8.3) with $s = 12$. After symmetrization, it provides a 24-stage eighth-order method.

SUZ93. Suzuki and Umeno (1993) provide one sixth-order method with $s = 14$, six eighth-order methods with $s = 15$ and four tenth-order schemes with $s = 31$.

MCL95b. McLachlan (1995*b*) gives several optimized schemes: one fourth-order method with $s = 5$, one sixth-order method with $s = 9$, and two eighth-order methods with $s = 15$ and $s = 17$.

KR97. Kahan and Li (1997) also present several optimized schemes: two sixth-order methods with $s = 9$, three eighth-order methods (one with $s = 15$ and two with $s = 17$), and five tenth-order methods (two with $s = 31$ and three with $s = 33$).

TSI99. Tsitouras (1999) constructs one optimized tenth-order method with $s = 33$.

MCL02. Construction of kernel with recurrence (8.2) for several values of p in McLachlan (2002).

HLW02. Hairer, Lubich and Wanner (2006, first edition published in 2002) provide one optimized tenth-order method with $s = 33$, the most efficient at the time.

SS05. Sofroniou and Spaletta (2005) carry out an exhaustive search of sixth-order schemes with $s = 11, 13$, eighth-order methods with $s = 19, 21$, and tenth-order methods with $s = 31, 33, 35$.

BCM06. Blanes, Casas and Murua (2006*a*) give several processed methods: two sixth-order kernels with $s = 11$ and $s = 13$, two eighth-order kernels with $s = 13$ and $s = 19$, and two tenth-order kernels with $s = 19$ and $s = 23$.

8.2. *Splitting into two parts / composition of a basic first-order method and its adjoint*

Although the order conditions for the two types of scheme,

$$\psi_h = \varphi^{[1]}_{a_{s+1}h} \circ \varphi^{[2]}_{b_s h} \circ \varphi^{[1]}_{a_s h} \circ \cdots \circ \varphi^{[1]}_{a_2 h} \circ \varphi^{[2]}_{b_1 h} \circ \varphi^{[1]}_{a_1 h} \tag{8.4}$$

and

$$\psi_h = \chi_{\alpha_{2s}h} \circ \chi^*_{\alpha_{2s-1}h} \circ \cdots \circ \chi_{\alpha_2 h} \circ \chi^*_{\alpha_1 h}, \tag{8.5}$$

are equivalent by virtue of the relationship (2.18), the optimization procedures to get the most efficient schemes may differ. In consequence, a particular method optimized for systems that are separable into two parts is not necessarily the best scheme when written as (8.5), although in practice their performances are closely related.

Since schemes of order $r \geq 6$ require more stages than taking the composition (8.3) with $S^{[2]}_h = \chi_{h/2} \circ \chi^*_{h/2}$ or $S^{[2]}_h = \varphi^{[1]}_{h/2} \circ \varphi^{[2]}_h \circ \varphi^{[1]}_{h/2}$, only methods with $r \leq 4$ seem promising. Note, however, that a sixth-order symmetric composition (8.3) with $s = 7$ has only three real solutions for the γ_j, whereas symmetric versions of (8.4) and (8.5) require at least $s = 9$ stages to solve the nine order conditions. This

Table 8.2. Symmetric composition schemes of the form (8.4) (appropriate when the ODE is split in two parts) and (8.5) (composition of a first-order method and its adjoint) of order $r = 2, 3, 4, 6$ with real coefficients. The notation is the same as in Table 8.1.

$$\psi_h = \chi_{\alpha_{2s}h} \circ \chi^*_{\alpha_{2s-1}h} \circ \cdots \circ \chi_{\alpha_2 h} \circ \chi^*_{\alpha_1 h},$$
$$\psi_h = \varphi^{[1]}_{a_{s+1}h} \circ \varphi^{[2]}_{b_s h} \circ \varphi^{[1]}_{a_s h} \circ \cdots \circ \varphi^{[1]}_{a_2 h} \circ \varphi^{[2]}_{b_1 h} \circ \varphi^{[1]}_{a_1 h}$$

$r = 2$	$r = 3$	$r = 4$	$r = 6$
$\boxed{2}$-MCL95b,OMF03,BCS14 **3**-BCS14	**3**-RUT83,SUZ92	**4**-**5**-MCL95b $\boxed{6}$-BM02	**9**-FOR92 $\boxed{10}$-BM02
		P:3,$\boxed{4}$-BCR99 **P:**$\boxed{6}$-BCM06	**P:5**-BCR99 **P:**$\boxed{9}$-BCM06

enlarged number of equations might then provide some solution leading to smaller error terms.

The schemes presented in Table 8.2 have been specifically designed to deal with these problems, and cannot be obtained as particular cases of the composition (8.3). In more detail, they correspond to the following.

RUT83. Ruth (1983) gives the first three-stage third-order non-symmetric method for systems separable into two parts.

SUZ92. Suzuki (1992) presents a family of three-stage third-order methods for the composition (8.5) with $s = 3$ and $\alpha_6 = 0$.

FOR92. Forest (1992) gives a sixth-order method for separable systems with $s = 9$.

MCL95b. McLachlan (1995*b*) gives an optimized two-stage second-order method and two optimized fourth-order schemes of the form (8.4) with $s = 4$ and $s = 5$.

BCR99. Blanes, Casas and Ros (1999*b*) present several processed methods for separable systems: fourth-order methods with $s = 3$ and $s = 4$, one fifth-order method with a non-symmetric kernel and $s = 4$, and finally one sixth-order scheme with $s = 5$.

BM02. Blanes and Moan (2002) obtain one fourth-order method with $s = 6$ and one sixth-order method with $s = 10$, both optimized for separable systems.

OMF03. Omelyan, Mryglod and Folk (2003) rediscover the optimized two-stage second-order scheme of McLachlan (1995*b*) as an efficient method with positive coefficients.

BCM06. Blanes, Casas and Murua (2006*a*) provide one fourth-order processed method with $s = 6$ and two sixth-order processed schemes with $s = 9$ and $s = 10$.

BCS14. Blanes, Casas and Sanz-Serna (2014) propose efficient symmetric second-order methods with $s = 2$ and $s = 3$, designed for hybrid Monte Carlo simulations.

8.3. *Runge–Kutta–Nyström methods*

Given the relevance of the problem $y'' = g(y)$, and the simplifications arising when splitting methods are applied to this system (see Section 3.1), many RKN splitting schemes have been obtained in the literature. Since F_1 and F_2 play different roles here, it is convenient to classify them according to the sequence of coefficients as

$$\begin{aligned} \text{AB}: \quad & \varphi_{a_s h}^{[A]} \circ \varphi_{b_s h}^{[B]} \circ \cdots \circ \varphi_{a_1 h}^{[A]} \circ \varphi_{b_1 h}^{[B]}, \\ \text{BA}: \quad & \varphi_{b_s h}^{[B]} \circ \varphi_{a_s h}^{[A]} \circ \cdots \circ \varphi_{b_1 h}^{[B]} \circ \varphi_{a_1 h}^{[A]}, \end{aligned} \tag{8.6}$$

although in fact the two types are conjugate to each other, thus providing similar efficiency. A word of caution is necessary: *in all the RKN splitting methods published in the literature the operator F_1 as defined in Section 3.1 is in fact associated with the map $\varphi_t^{[B]}$, and F_2 is associated with $\varphi_t^{[A]}$*, so that $[F_1, [F_1, [F_1, F_2]]] = 0$. Thus, in particular, in classical and quantum mechanics, A corresponds to the kinetic energy and B to the potential energy. This should be taken into account when implementing the schemes collected here.

It is also convenient to take profit of the FSAL (first same as last) property, and thus we may consider the non-equivalent compositions

$$\begin{aligned} \text{ABA}: \quad & \varphi_{a_{s+1} h}^{[A]} \circ \varphi_{b_s h}^{[B]} \circ \varphi_{a_s h}^{[A]} \circ \cdots \circ \varphi_{b_1 h}^{[B]} \circ \varphi_{a_1 h}^{[A]}, \\ \text{BAB}: \quad & \varphi_{b_{s+1} h}^{[B]} \circ \varphi_{a_s h}^{[A]} \circ \varphi_{b_s h}^{[B]} \circ \cdots \circ \varphi_{a_1 h}^{[A]} \circ \varphi_{b_1 h}^{[B]}, \end{aligned} \tag{8.7}$$

as well as their time-symmetric versions ($a_{s+2-i} = a_i, b_{s+1-i} = b_i$ for ABA and $b_{s+2-i} = b_i, a_{s+1-i} = a_i$ for BAB, respectively).

In Table 8.3 we present the most representative methods within this class. We add **S** or **N** to distinguish symmetric from non-symmetric schemes and the subscript AB, ABA and BAB to denote the different alternatives (8.6), (8.7). Processed methods have also been included. Explicitly, they correspond to the following.

FOR92. Forest (1992) gives a sixth-order method with $s = 7$.

MA92. McLachlan and Atela (1992) present non-symmetric fourth- and fifth-order methods with $s = 4$ and $s = 6$, respectively.

CSS93. Calvo and Sanz-Serna (1993*a*) obtain one fourth-order non-symmetric BAB method with $s = 4$.

OS94. Okunbor and Skeel (1994) obtain four fifth-order methods with $s = 5$ and sixteen sixth-order symmetric methods with $s = 7$.

Table 8.3. RKN splitting integrators of order $r = 4, 5, 6, 8$. Since the role of the flows $\varphi_h^{[A]}$ and $\varphi_h^{[B]}$ is not interchangeable here, we distinguish symmetric **S** and non-symmetric **N** compositions with a subscript AB, ABA, BAB. As usual, processed methods are preceded by **P** and the recommended schemes are [framed].

RKN splitting methods			
$r = 4$	$r = 5$	$r = 6$	$r = 8$
4N$_{\text{AB}}$-MA92 **4N**$_{\text{BAB}}$-CSS93 **4,5S**$_{\text{ABA}}$-MCL95b **[6]S**$_{\text{ABA,BAB}}$-BM02 **4,5S**$_{\text{ABA,BAB}}$-OMF03	**5N**$_{\text{ABA}}$-OS94 **[6]N**$_{\text{AB}}$-MA92 **6N**$_{\text{AB}}$-CHO00	**7S**$_{\text{ABA}}$-FOR92,OS94 **7S**$_{\text{BAB}}$-FOR92 **11S**$_{\text{BAB}}$-BM02 **[14]S**$_{\text{ABA}}$-BM02	**17S**$_{\text{ABA}}$-OL94 **17,18,[19]S**$_{\text{ABA, BAB}}$-BCE22
P:[2]N$_{\text{AB}}$-BCR99		**P:4-6S**$_{\text{ABA,BAB}}$-BCR01a **P:[7]S**$_{\text{BAB}}$-BCR01b	**P:9S**$_{\text{ABA}}$-BCR01a **P:[11]S**$_{\text{BAB}}$-BCR01b

OL94. Okunbor and Lu (1994) obtain one eighth-order ABA method with $s = 17$.

MCL95b. McLachlan (1995*b*) gives optimized RKN methods of order four with $s = 4$ and $s = 5$, and of order six with $s = 7$.

BCR99. Blanes, Casas and Ros (1999*b*) present a processed fourth-order method with a non-symmetric kernel and $s = 2$.

CHO00. Chou and Sharp (2000) get a non-symmetric scheme of order five with $s = 6$.

BCR01a. Blanes, Casas and Ros (2001*a*) present one processed sixth-order method with $s = 6$

BCR01b. Blanes, Casas and Ros (2001*b*) give one sixth- and one eighth-order processed method with $s = 7$ and $s = 11$, respectively.

BM02. Blanes and Moan (2002) obtain one BAB fourth-order method with $s = 6$ and two sixth-order methods: one BAB with $s = 11$ and one ABA with $s = 14$.

OMF03. Omelyan, Mryglod and Folk (2003) propose fourth-order methods of type ABA and BAB with $s = 4$ and $s = 5$.

BCE22. Blanes, Casas and Escorihuela-Tomàs (2022*b*) obtain optimized eighth-order methods with $s = 17, 18$ and 19 of type ABA and BAB.

As we have seen in Section 3.2, the particular structure of problem $y'' = g(y)$ allows us to include the flows corresponding to nested commutators in the previous compositions (8.6)–(8.7). This may on one hand lead to more efficient schemes, and on the other hand allow us to achieve order four with only positive coefficients.

Table 8.4. RKN splitting methods of order r = 4, 6, 8 with modified potentials. Schemes are coded as in Table 8.3.

RKN splitting methods with commutators		
$r = 4$	$r = 6$	$r = 8$
2S$_{\mathrm{ABA,BAB}}$-KOS93,CHI97 **4S**$_{\mathrm{ABA,BAB}}$-SUZ95 **3,$\boxed{4}$,5S**$_{\mathrm{ABA,BAB}}$-CHI06,OMF02	**4,$\boxed{5}$S**$_{\mathrm{ABA,BAB}}$-OMF02	**11S**$_{\mathrm{ABA,BAB}}$-OMF02
P:$\boxed{1}$S$_{\mathrm{BAB}}$-TI84,ROW91, WHT96,BCR99 **P:2S**$_{\mathrm{BAB}}$-LSS97	**P:$\boxed{3}$S**$_{\mathrm{ABA,BAB}}$-BCR99	**P:4S**$_{\mathrm{ABA}}$-BCR01a **P:5S**$_{\mathrm{BAB}}$-BCR01a,BCR01b

In Table 8.4 we present some relevant methods within this class, both processed and non-processed. They correspond to the following.

TI84. Takahashi and Imada (1984) obtain the kernel of a fourth-order processed method with $s = 1$.

ROW91. Rowlands (1991) rediscovers the kernel of the fourth-order processed method with $s = 1$.

KOS93. Koseleff (1993) obtains the first fourth-order BAB method with positive coefficients with $s = 2$ (but with a misprint in one of the coefficients).

SUZ95. Suzuki (1995) gives fourth-order ABA and BAB methods with $s = 3$.

WHT96. Wisdom, Holman and Touma (1996) obtain one fourth-order processed method with $s = 1$.

LSS97. López-Marcos, Sanz-Serna and Skeel (1997) obtain fourth-order processed methods with $s = 2$.

CHI97. Chin (1997) rediscovers the fourth-order method of Koseleff (1993) (but now with the correct coefficients) and shows its efficiency in practice.

BCR99. Blanes, Casas and Ros (1999*b*) obtain the following processed methods: one fourth-order method with $s = 1$, one non-symmetric fifth-order method with $s = 2$ and one sixth-order method with $s = 3$.

BCR01a. Blanes, Casas and Ros (2001*a*) present one sixth-order method with $s = 2$ and higher-order commutators and two sixth-order methods with $s = 3$.

BCR01b. Blanes, Casas and Ros (2001*b*) propose one optimized sixth-order method with $s = 3$ and one eighth-order method with $s = 5$.

OMF03. Omelyan, Mryglod and Folk (2003) present a detailed treatment of fourth-order methods with $2 \leq s \leq 5$ and sixth-order methods with $s = 4, 5$ both ABA and BAB and different number of modified potentials.

CHI06. Chin (2006) obtains families of fourth-order schemes with coefficients written analytically in terms of free parameters for $s > 2$.

8.4. *Methods for near-integrable systems*

As we have seen in Section 3.3, splitting methods are outstanding for near-integrable systems of the form $x' = f_1(x) + \varepsilon f_2(x)$: the presence of two parameters, h (the step size) and ε (the size of the perturbation), allows us to reduce the number of order conditions and still construct highly efficient schemes. There are problems where, in addition, we can use the same techniques as for RKN methods and also incorporate the flows of nested commutators into the algorithm. This is the case, in particular, for Hamiltonian systems of the form $H = H_1 + \varepsilon H_2$, when H_1 is the sum of two-body Kepler problems and H_2 depends only on coordinates. If we let F_1 denote the operator associated with H_1, and let F_2 denote the operator associated to the perturbation H_2, then $[F_2, [F_2, [F_2, F_1]]] = 0$, and the analysis done in Section 3.1 can be applied here (with the obvious interchange $F_1 \leftrightarrow F_2$).

In Table 8.5 we separate, as usual, non-processed from processed schemes (preceded by **P**). We also include methods when $[F_2, [F_2, [F_2, F_1]]] = 0$ (second row) and schemes with nested commutators (denoted by a label *) in the last two rows of processed methods.

WH91. Wisdom and Holman (1991) present the first $(2, 2)$ method.

MCL95a. McLachlan (1995*a*) gives families of symmetric $(2s, 2)$ schemes of type ABA and BAB with $s \leq 5$ and positive coefficients, one BAB (6,4) method with $s = 4$ and two (8,4) ABA and BAB schemes with $s = 5$.

WHT96. Wisdom, Holman and Touma (1996) present one processed $(k, 2)$ method with $k = 16$ and $s = 1$.

BCR00. Blanes, Casas and Ros (2000) give several processed methods: (i) one non-symmetric (6,4,3) method with $s = 2$, one (7,6,4) method with $s = 3$, one (7,6,5,4) method with $s = 4$, and (ii) several RKN methods, one (6,4) with $s = 2$ and one (7,6,5) with $s = 3$ both non-symmetric. In addition, with modified potentials, they give one (6,4) with $s = 1$ and a non-symmetric (7,6,5) with $s = 2$.

LR01. Laskar and Robutel (2001) carry out a systematic study of $(2s, 2)$ methods, and obtain new schemes up to $s = 10$ with positive coefficients. The order of these methods is increased to $(2s, 4)$ by including an appropriate modified potential.

BCF13. Blanes *et al.* (2013*b*) present several schemes of type ABA: one (10,4) with $s = 7$, one (8,6,4) with $s = 7$ and one (10,6,4) with $s = 8$.

Table 8.5. Splitting methods for near-integrable systems of the form $x' = f_1 + \varepsilon f_2$. For processed methods we also include methods applicable when $[F_2, [F_2, [F_2, F_1]]] = 0$ (the last two rows of processed methods) and schemes with modified flows (labelled *).

Splitting methods of generalized order		
$(n,2)$	$(n,4)$	$(n,6,4)$
$\boxed{\mathbf{1}}$(2,2)**S**-WH91 $\boxed{\mathbf{n}}$(2n,2)$\mathbf{S}_{\mathrm{ABA,BAB}}$-MCL95a,LR01	**4**(6,4)$\mathbf{S}_{\mathrm{BAB}}$-MCL95a $\boxed{\mathbf{5}}$(8,4)$\mathbf{S}_{\mathrm{ABA,BAB}}$-MCL95a **7**(10,4)$\mathbf{S}_{\mathrm{ABA}}$-BCF13	$\boxed{\mathbf{7}}$(8,6,4)$\mathbf{S}_{\mathrm{ABA}}$-BCF13 $\boxed{\mathbf{8}}$(10,6,4)$\mathbf{S}_{\mathrm{ABA}}$-BCF13
P:1(17,2)-WHT96		**P:**$\boxed{\mathbf{3}}$(7,6,4)$\mathbf{S}_{\mathrm{ABA}}$-BCR00
	P:$\boxed{\mathbf{2}}$(6,4)$\mathbf{S}_{\mathrm{AB}}$-BCR00	**P:**$\boxed{\mathbf{3}}$(7,6,5)$\mathbf{S}_{\mathrm{AB}}$-BCR00
	P:1(6,4)$\mathbf{S}^*_{\mathrm{ABA}}$-BCR00 **P:n**(n,4)$\mathbf{S}^*_{\mathrm{ABA}}$-LR01	**P:2**(7,6,5)$\mathbf{S}^*_{\mathrm{AB}}$-BCR00

9. Splitting everywhere: some relevant applications

9.1. Splitting and extrapolation

Extrapolation methods constitute a class of efficient high-order schemes for solving initial value problems when the local error of the basic integrator has an asymptotic expansion containing only even powers of the step size h. In this subsection we review how some of the previous techniques used in the construction and analysis of splitting and composition methods can also be applied to simplify the analysis of extrapolation.

9.1.1. The Gragg/GBS method

Given the initial value problem $x' = f(x)$, $x(t_0) = x_0$, Gragg proposed the following algorithm to produce the quantity $T_{h^*}(t)$. Denoting $t = t_0 + 2nh^*$, $t_i = t_0 + ih^*$, it reads

$$\begin{aligned} x_1 &= x_0 + h^* f(x_0), \\ x_{i+1} &= x_{i-1} + 2h^* f(x_i), \quad i = 1, 2, \ldots, 2n, \\ T_{h^*}(t) &= \frac{1}{4}(x_{2n-1} + 2x_{2n} + x_{2n+1}). \end{aligned} \tag{9.1}$$

Subsequently, he proved that $T_{h^*}(t)$ possesses an asymptotic expansion in even powers of h^* (Gragg 1965), so that it can then be used for Richardson extrapolation. The original proof was simplified by Stetter (1970) when he realized that (9.1) can be interpreted as a one-step algorithm. Later on, Hairer *et al.* (1993) showed that this scheme is consistent with the differential equation in the extended phase space

$$\begin{aligned} u' &= f(v), & u(t_0) &= u_0 = y_0, \\ v' &= f(u), & v(t_0) &= v_0 = y_0, \end{aligned} \tag{9.2}$$

whose exact solution is $u(t) = v(t) = x(t)$. In fact, to prove that $T_{h^*}(t)$ only contains even powers of h^* is trivial by noticing that system (9.2) can be expressed as

$$\frac{\mathrm{d}}{\mathrm{d}t}\begin{pmatrix} u \\ v \end{pmatrix} = \begin{pmatrix} 0 \\ f(u) \end{pmatrix} + \begin{pmatrix} f(v) \\ 0 \end{pmatrix} = f_1(u) + f_2(v), \qquad \begin{pmatrix} u(t_0) \\ v(t_0) \end{pmatrix} = \begin{pmatrix} x_0 \\ x_0 \end{pmatrix}, \tag{9.3}$$

and that the Strang splitting with step size $h = 2h^*$ gets

$$\begin{aligned} v_{1/2} &= v_{i-1} + \frac{h}{2} f(u_{i-1}), \\ u_i &= u_{i-1} + h f(v_{1/2}), \qquad i = 1, 2, \ldots, n, \\ v_i &= v_{1/2} + \frac{h}{2} f(u_i). \end{aligned}$$

Then, clearly, $T_{h^*} = \frac{1}{2}(u_n + v_n)$, and the result is obtained by noticing that the Strang splitting is time-symmetric.

On the other hand, application of the Gragg method (9.1) to the second-order equation

$$y'' = g(y), \quad y(t_0) = y_0, \quad y'(t_0) = y_0', \tag{9.4}$$

produces the equivalent formula (Hairer *et al.* 1993)

$$y_{i+1} - 2y_i + y_{i-1} = h^2 g(y_i), \quad i = 1, 2, 3, \ldots, \tag{9.5}$$

again with $h = 2h^*$. As is well known, by considering the equivalent first-order system

$$\frac{\mathrm{d}}{\mathrm{d}t}\begin{pmatrix} y \\ y' \end{pmatrix} = \begin{pmatrix} y' \\ 0 \end{pmatrix} + \begin{pmatrix} 0 \\ g(y) \end{pmatrix} = f_1(y') + f_2(y), \tag{9.6}$$

and then applying the Strang splitting $\varphi_{h/2}^{[2]} \circ \varphi_h^{[1]} \circ \varphi_{h/2}^{[2]}$,

$$\begin{aligned} y_{n+1} &= y_n + h y_n' + \frac{h^2}{2} g(y_n), \\ y_{n+1}' &= y_n' + \frac{h}{2}(g(y_n) + g(y_{n+1})), \end{aligned}$$

reproduces Störmer's rule (9.5) after simplification. In fact, one can also apply the other variant of the Strang splitting, i.e. $\varphi_{h/2}^{[1]} \circ \varphi_h^{[2]} \circ \varphi_{h/2}^{[1]}$, to (9.6), thus yielding

$$\begin{aligned} y_{n+1}' &= y_n' + hg\left(y_n + \frac{h}{2} y_n'\right), \\ y_{n+1} &= y_n + \frac{h}{2}(y_n' + y_{n+1}'). \end{aligned} \tag{9.7}$$

This (slightly cheaper) scheme could also be used as the basic method for extrapolation, as for the Störmer rule, but since $f_1(y')$ and $f_2(y)$ may have different qualitative properties, the resulting schemes would also be different.

9.1.2. *Multi-product expansions*

Linear combinations of splitting methods

$$\sum_k c_k \left(\varphi^{[2]}_{b_{k,i_k}h} \circ \varphi^{[1]}_{a_{k,i_k}h} \circ \cdots \circ \varphi^{[2]}_{b_{k,1}h} \circ \varphi^{[1]}_{a_{k,1}h}\right)$$

offer much freedom in the choice of parameters $\{c_k, a_{k,i_j}, b_{k,i_j}\}$ to approximate the flow of $x' = f_1(x) + f_2(x)$, although they no longer provide approximations with the same preservation properties as the original system. As shown in Sheng (1989), if all $\{a_{k,i_j}, b_{k,i_j}\}$ were to be positive, then any individual composition has to be at most of second order, and some coefficients c_k must be negative. The choice of the Strang splitting $S_h^{[2]}$ (or more generally a time-symmetric second-order integrator) as the basic scheme in the composition is particularly relevant due to the structure of the truncation error. Suppose $\{k_1, k_2, \ldots, k_m\}$ denotes a given set of m integer numbers, and form the linear combination

$$\phi_h \equiv \sum_{i=1}^{m} c_i \left(S^{[2]}_{h/k_i}\right)^{k_i}. \tag{9.8}$$

Then ϕ_h furnishes an approximation of order $2m$ if the coefficients satisfy the linear equations

$$G_0 = \sum_{i=1}^{m} c_i = 1, \quad G_{2+2j} = \sum_{i=1}^{m} \frac{c_i}{k_i^{2+2j}} = 0, \quad j = 0, 1, \ldots, m-2,$$

with solutions

$$c_i = \prod_{j=1(\neq i)}^{m} \frac{k_i^2}{k_i^2 - k_j^2}.$$

Schemes of the form (9.8) are referred to as multi-product expansions (Chin and Geiser 2011) and allow us to construct in an easy way efficient high-order integrators. In general, if we start with a basic time-symmetric geometric method of order $2n$, the corresponding linear combination (9.8) provides a scheme of order $2n + 2\ell$, $\ell = 1, \ldots, n$, which, quite remarkably, preserves the geometric properties of the system (e.g. symplecticity for Hamiltonian problems) up to order $4n + 1$ or higher (Blanes, Casas and Ros 1999*a*, Chan and Murua 2000). This can be shown by appropriately rewriting the differential operator associated with (9.8) as a product of exponentials of operators (Blanes and Casas 2016).

9.2. *Crouch–Grossman methods*

The nonlinear differential equation

$$Y' = A(Y)Y, \quad Y(0) = Y_0 \in \mathcal{G}, \tag{9.9}$$

where $A \in \mathbb{R}^{d\times d}$ and $\mathcal{G}$ is a matrix Lie group, appears in relevant physical fields such as rigid body mechanics, in the calculation of Lyapunov exponents ($\mathcal{G} \equiv \mathrm{SO}(d)$) and other problems arising in Hamiltonian dynamics ($\mathcal{G} \equiv \mathrm{Sp}(d)$). In fact, it can be shown that every differential equation evolving on a matrix Lie group $\mathcal{G}$ can be written in the form (9.9) (Iserles *et al.* 2000).

A class of methods providing by construction approximations in $\mathcal{G}$ was proposed in Crouch and Grossman (1993) by appropriately modifying Runge–Kutta schemes as follows; see also Hairer *et al.* (2006, sect. IV.8).

Let $b_i, a_{i,j}$, $(i, j = 1, \ldots, s)$ be real numbers. Then an explicit s-stage Crouch–Grossman method is given by

$$\begin{aligned}
Y^{(1)} &= Y_n, & K_1 &= A(Y^{(1)}), \\
&\vdots \\
Y^{(j)} &= \mathrm{e}^{ha_{j,j-1}K_{j-1}} \cdots \mathrm{e}^{ha_{j,1}K_1}\, Y_n, & K_j &= A(Y^{(j)}), \quad 2 \le j \le s, \\
Y_{n+1} &= \mathrm{e}^{hb_sK_s} \cdots \mathrm{e}^{hb_1K_1} Y_n.
\end{aligned} \tag{9.10}$$

As usual, the order conditions to be satisfied by the coefficients $b_i, a_{i,j}$, $(i, j = 1, \ldots, s)$ can be found by comparing the Taylor series expansions of the exact and numerical solutions. It has been shown that the order conditions for classical Runge–Kutta methods form a subset of those for the Crouch–Grossman methods (Owren and Marthinsen 1999). Notice that the s-stage scheme (9.10) involves a total of $s(s+1)/2$ matrix exponentials.

In Crouch and Grossman (1993) there are several three-stage methods of order three, whereas Owren and Marthinsen (1999) present a fourth-order scheme with $s = 5$ stages (or equivalently 15 exponentials). Achieving higher order within this approach is by no means simple, and the resulting methods require a large number of exponentials. It turns out, however, that splitting can also be used here to construct schemes of any order with a reduced number of matrix exponentials. To illustrate the technique, we consider the separable system in the extended phase space

$$\frac{\mathrm{d}}{\mathrm{d}t}\begin{pmatrix} U \\ V \end{pmatrix} = \begin{pmatrix} A(V)\,U \\ 0 \end{pmatrix} + \begin{pmatrix} 0 \\ A(U)\,V \end{pmatrix}, \tag{9.11}$$

to which we apply the Strang splitting. This leads to

$$\begin{aligned}
U_{1/2} &= \mathrm{e}^{\frac{h}{2}A(V_n)}\, U_n, \\
V_{n+1} &= \mathrm{e}^{hA(U_{1/2})}\, V_n, \\
U_{n+1} &= \mathrm{e}^{\frac{h}{2}A(V_{n+1})}\, U_{1/2}.
\end{aligned} \tag{9.12}$$

Notice that (9.12) is simply a particular example of (9.10) with $s = 3$, that is,

$$\begin{aligned} Y^{(1)} &= Y_n, & K_1 &= A(Y^{(1)}), \\ Y^{(2)} &= e^{ha_{2,1}K_1}\, Y_n, & K_2 &= A(Y^{(2)}), \\ Y^{(3)} &= e^{ha_{3,2}K_2}\, e^{ha_{3,1}K_1}\, Y_n, & K_3 &= A(Y^{(3)}), \\ Y_{n+1} &= e^{hb_3K_3}\, e^{hb_2K_2}\, e^{hb_1K_1}\, Y_n, \end{aligned} \tag{9.13}$$

with

$$a_{2,1} = \frac{1}{2}, \quad a_{3,2} = 1, \quad a_{3,1} = 0, \quad b_1 = \frac{1}{2}, \quad b_2 = 0, \quad b_3 = \frac{1}{2}.$$

Since $a_{3,1} = b_2 = 0$ and $a_{2,1} = b_1$, then the whole scheme can be computed with only three exponentials, instead of six by applying the recursion (9.10). Higher-order methods can be obtained in the same way by applying appropriate splitting schemes to the enlarged system (9.11). Thus, in particular, it is possible to get a Crouch–Grossman scheme of order four with seven exponentials.

9.2.1. Positivity-preserving splitting methods

Differential equations $x' = f(x)$ modelling population dynamics must satisfy positivity and frequently mass preservation, that is,

$$\begin{array}{llll} \text{(i)} & x_i(0) \geq 0 \quad \Rightarrow \quad x_i(t) \geq 0 \ \forall t,\ i = 1, \ldots, D & & \text{(positivity preservation)} \\ \text{(ii)} & \mathbf{1}^\top x(t) = \mathbf{1}^\top x(0), \quad \text{with} \quad \mathbf{1} = (1, \ldots, 1)^\top & & \text{(mass conservation)} \end{array}$$

Both of them are automatically fulfilled if the differential equation can be written as $x' = A(x)x$, where A is a matrix such that

- if $x_i \geq 0$, $i = 1, \ldots, D$, then $A(x)_{k,\ell} \geq 0$ when $k \neq \ell$, and $A_{k,k} \leq 0$, for $k, \ell = 1, \ldots, D$ (to guarantee positivity);
- $\sum_{k=1}^{D} A_{k,\ell} = 0$ for $\ell = 1, \ldots, D$ (to ensure mass conservation).

In such situations, it is of course expedient to use a numerical integrator that also satisfies these features at the discrete level when carrying out simulations. Unfortunately, there is also an order barrier here for two of the most used families of schemes, namely Runge–Kutta and multistep methods: if they unconditionally preserve positivity, then they cannot be better than first-order (Bolley and Crouzeix 1978).

It turns out, however, that splitting methods are able to overcome this order barrier. To see this point, again consider the problem in the extended phase space

$$\begin{aligned} y' &= A(z)y, & y(0) &= y_0 = x_0, \\ z' &= A(y)z, & z(0) &= z_0 = x_0, \end{aligned} \tag{9.14}$$

where $x(t) = y(t) = z(t)$. Now, as in the previous subsection, we apply the Strang splitting to the system (9.14), that is,

$$\begin{aligned} y_{1/2} &= \mathrm{e}^{\frac{1}{2}hA(z_0)} y_0, \\ z_1 &= \mathrm{e}^{hA(y_{1/2})} z_0, \\ y_1 &= \mathrm{e}^{\frac{1}{2}hA(z_1)} y_{1/2}. \end{aligned} \tag{9.15}$$

Then both y_1 and z_1 are second-order approximations to the exact solution $x(h)$ preserving mass and positivity. This can be easily checked by realizing that $\mathbf{1}^\top A = 0$ and so $\mathbf{1}^\top \mathrm{e}^A x = \mathbf{1}^\top x$. On the other hand, for $h > 0$, if $x_i \geq 0$ then $(\mathrm{e}^{hA(x)})_{jk} \geq 0$ for all i, j, k.

If $h < 0$, positivity is no longer preserved with this approach; see Blanes, Iserles and Macnamara (2022*d*) for a more detailed treatment of this topic.

9.3. *Exact splitting*

There are equations for which a clever decomposition of the vector field in fact provides the exact solution. The resulting factorization is called *exact splitting* and, in the case of partial differential equations, allows us to get very accurate approximations once each subsystem is solved by pseudo-spectral methods and pointwise multiplication. Early applications of this idea to the time integration of the Gross–Pitaevskii equation are contained in Chin and Krotscheck (2005) and Bader and Blanes (2011), and a convenient generalization to a large class of PDEs, called inhomogeneous quadratic differential equations, has been recently proposed in Bernier (2021). These PDEs are of the form

$$u_t(t,x) = -p^w u(t,x), \quad u(0,x) = u_0(x), \quad t \geq 0,\ x \in \mathbb{R}^d, \tag{9.16}$$

where $d \geq 1$, $u_0 \in L^2(\mathbb{R}^d)$ and p^w is the quadratic differential operator

$$p^w = (x, -\mathrm{i}\nabla_x) Q \begin{pmatrix} x \\ -\mathrm{i}\nabla_x \end{pmatrix} + Y^\top \begin{pmatrix} x \\ -\mathrm{i}\nabla_x \end{pmatrix} + c.$$

Here Q is a constant $2d \times 2d$ symmetric matrix with complex coefficients, $Y \in \mathbb{C}^{2d}$ and $c \in \mathbb{C}$ is constant. If $(-p^w)$ is such that the real part of the polynomial $p(X) = X^\top Q X + Y^\top X + c$ is bounded from below on $\mathbb{R}^{2d}$, then it generates a strongly continuous semigroup, denoted $\mathrm{e}^{-tp^w} u_0$. It was shown in Bernier (2021) that e^{-tp^w} can be factorized as a product of operators that can be exactly evaluated. Notice that some relevant equations are of the form (9.16), so this technique allows us to construct numerical methods that are spectral in space and exact in time using only a small number of fast Fourier transforms. This has been illustrated in Bernier, Crouseilles and Li (2021) in the case of kinetic and nonlinear Schrödinger equations.

As an illustrative example of this technique, let us consider the Schrödinger equation with a quadratic potential

$$\mathrm{i}u_t = \frac{1}{2}(-\partial_{xx} + x^2)\, u \equiv \frac{1}{2}(P^2 + X^2)\, u.$$

Then, for $|h| < \pi$, we have

$$u(h,x) = \mathrm{e}^{-\mathrm{i}h(P^2/2+X^2/2)}\, u(0,x) = \mathrm{e}^{-\mathrm{i}f(h)(X^2/2)}\, \mathrm{e}^{-\mathrm{i}g(h)(P^2/2)}\, \mathrm{e}^{-\mathrm{i}f(h)(X^2/2)}\, u_0(x),$$

with

$$f(h) = \frac{1-\cos(h)}{\sin(h)}, \qquad g(h) = \sin(h),$$

so the action of each exponential can be easily evaluated.

9.4. Gravitational N-body problem

The long-time numerical integrations of the whole Solar System carried out in Sussman and Wisdom (1992) and Wisdom and Holman (1991), suggesting its chaotic nature, represented a major boost for the development of new and efficient symplectic algorithms. The integrator used in these simulations, referred to in the dynamical astronomy literature as the Wisdom–Holman map, is simply the Strang splitting method applied to the Hamiltonian system

$$\hat{H}(\hat{q},\hat{p}) = H_1(\hat{q},\hat{p}) + H_2(\hat{q}), \tag{9.17}$$

where H_1 and H_2 are given by (1.39) in terms of Jacobi coordinates, as described in Section 1.5. Here H_1 represents the $N-1$ independent Keplerian motions of the planets (which can be efficiently solved) and H_2 accounts for the gravitational interactions among the planets. Moreover, as shown in Section 1.5, $H_2(\hat{q}) = O(\varepsilon)$ with $\varepsilon \equiv (1/m_0)\max_{1\le i\le N-1} m_i$. For the Solar System, $\varepsilon \approx 10^{-3}$. In consequence, splitting methods for near-integrable systems, such as those presented in Sections 3.3 and 8.4, are particularly appropriate. In fact, the Wisdom–Holman map, in the terminology of Section 3.3, corresponds to a method of generalized order $(2,2)$.

Very often, splitting methods are used when very long-time integrations are involved and getting high accuracy is not of particular concern, but rather when having good behaviour of the error over the whole integration period is the crucial point. For instance, in the situation analysed in Sussman and Wisdom (1992), the important issue was to decide whether the motion is regular or chaotic, and not to compute accurately the state of the Solar System after a long time.

The experiment we describe next illustrates that even when short-time integrations are involved and very high accuracy is required, splitting methods can outperform standard integrators once the particular structure of the problem is incorporated into their very formulation.

Specifically, we again deal with the simplified model of the outer Solar System addressed in Section 1.5, with the same initial conditions and an integration interval of 2×10^5 days, and compare splitting methods specially tailored for near-integrable systems with some popular standard integrators. Thus, Figure 9.1 shows efficiency diagrams for the error in energy (a) and in positions (b) at the final time vs. the number of force evaluations, obtained with three different splitting methods

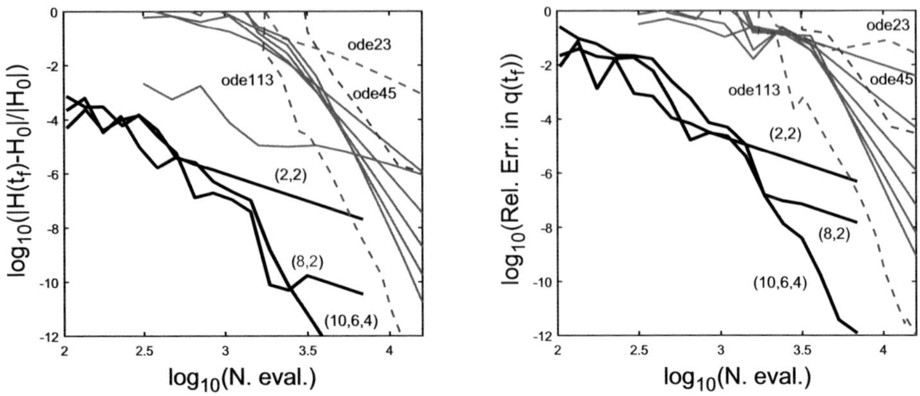

(a) (b)

Figure 9.1. Outer Solar System. Efficiency diagrams showing the relative error in energy (a) and in positions q (b) at the final time $t_f = 2 \times 10^5$ days vs. the number of force evaluations, obtained with different schemes: extrapolation of order $r = 2$, 4, 6, 8, 10, 12 (grey curves), Matlab routines `ode23`, `ode45` and `ode113`, and splitting methods of generalized order (2,2), (8,2) and (10,6,4).

of generalized orders (2,2), (8,2) and (10,6,4), in comparison with extrapolation methods of orders 2-12 (grey lines) and the Matlab routines `ode23`, `ode45` and `ode113`.

As stated before, extrapolation methods are particularly appropriate for second-order differential equations when high accuracy is required (Hairer *et al.* 1993). They take the symmetric second-order scheme (9.5) as the basic method in the linear combination (9.8) with the harmonic sequence $k_i = i$, $i = 1, 2, \ldots, m$, to get a method of order $r = 2m$ at the cost of $m(m+1)/2 + 1$ evaluations of the force,[8] On the other hand, `ode23` and `ode45` solve the ODE with variable time step using embedded Runge–Kutta methods of orders 2–3 and 4–5, respectively, whereas `ode113` does it with variable time step and variable order using multistep methods up to order 13. Due to the smoothness of the problem and the low cost of evaluating the highest order, the latter procedure provides better results than extrapolation methods of high order, as shown in the figure. Nevertheless, none of them can compete with the specially designed splitting methods for this particular problem at all accuracies. This is so even when a very short integration interval is considered. In fact, this relative superiority will only increase for longer integration intervals due to their different error propagation mechanism.

[8] One evaluation can be saved if the basic scheme (9.7) is taken instead. This is the actual implementation we use in our experiments here.

9.5. Molecular simulations

In the simulation of the dynamics of large molecules, two classes of algorithms are commonly used: (i) molecular dynamics, and (ii) those based on stochastic differential equations and the Monte Carlo method.

In classical molecular dynamics, the motion of the atoms in the molecule is determined by integrating Newton's second law,

$$Mq'' = -\nabla_q U(q),$$

where q is a vector containing all positions (in Cartesian coordinates), M is a diagonal matrix whose elements are the atomic masses and $U(q)$ is the (empirical) potential function modelling the inter-particle interactions. Popular models include the Lennard-Jones and Morse potentials, although many other choices are being used for different molecules; see e.g. Leimkuhler and Matthews (2015) for more details. In any case, the number of atoms may be exceedingly large (up to 100 000) and the resulting system is highly nonlinear, exhibiting sensitive dependence on perturbations. In addition, the initial velocities are typically assigned randomly, so there is no point in trying to obtain accurate trajectories. Thus, although in principle Runge–Kutta–Nyström splitting methods can be used in this setting, the method of choice in practice is the Störmer–Verlet scheme, given its excellent stability properties and low computational cost (Leimkuhler *et al.* 1996). Other variants include the use of multiple time-stepping (and in particular the mollified impulse method: García-Archilla, Sanz-Serna and Skeel 1999), and symplectic schemes specially designed for Hamiltonian systems with constraints, such as SHAKE (Ryckaert, Ciccotti and Berendsen 1977) and RATTLE (Andersen 1983).

To take into account the influence of a medium (say, a solvent or air) on a molecular system, typically as external random impacts, a common practice consists in extending the molecular dynamics approach by incorporating a stochastic component in the equations of motion. The resulting system is referred to as Langevin dynamics and is described by the stochastic differential equation

$$\begin{aligned} \mathrm{d}q &= M^{-1} p \,\mathrm{d}t, \\ \mathrm{d}p &= -\nabla_q U(q)\,\mathrm{d}t - \gamma p \,\mathrm{d}t + \sqrt{2\gamma k_B T} M^{1/2}\,\mathrm{d}W \end{aligned} \tag{9.18}$$

in terms of coordinates $q \in \mathbb{R}^d$ and momenta $p \in \mathbb{R}^d$. Here $\gamma > 0$ is the friction coefficient or collision frequency, k_B is the Boltzmann constant, T is the temperature and W is a Wiener process. We may regard (9.18) as modelling a system of particles immersed in a fluid bath consisting of many particles, giving rise to much weaker interactions than those modelled by the potential U (Leimkuhler and Matthews 2015).

As with deterministic equations, splitting methods can be constructed to deal with Langevin dynamics. One possible approach consists in decomposing (9.18) into three parts as follows:

$$\begin{pmatrix} \mathrm{d}q \\ \mathrm{d}p \end{pmatrix} = \underbrace{\begin{pmatrix} M^{-1}p \\ 0 \end{pmatrix} \mathrm{d}t}_{A} + \underbrace{\begin{pmatrix} 0 \\ -\nabla_q U(q) \end{pmatrix} \mathrm{d}t}_{B} + \underbrace{\begin{pmatrix} 0 \\ -\gamma p \, \mathrm{d}t + \sigma M^{1/2} \, \mathrm{d}W \end{pmatrix}}_{O}, \tag{9.19}$$

with $\sigma \equiv \sqrt{2\gamma k_B T}$. Each of the parts can be exactly solved: the solution of the first is a *drift* in position,

$$\varphi_h^{[A]}(q,p) = (q + hM^{-1}p, p),$$

the second corresponds to a *kick* in momentum,

$$\varphi_h^{[B]}(q,p) = (q, p - h\nabla_q U(q)),$$

and the third piece defines an Ornstein–Uhlenbeck process in p which can be exactly sampled,

$$\varphi_h^{[O]}(q,p) = (q, \mathrm{e}^{-\gamma h} p + \sqrt{k_B T(1-\mathrm{e}^{-2\gamma h})} M^{1/2} R),$$

where R is a vector of independent and identically distributed normal random numbers. In Leimkuhler and Matthews (2013*a*,*b*), several splitting methods constructed along these lines are designed and tested on numerical experiments. They are denoted by the acronym resulting from concatenating the previous symbols A, B and O. Thus, BAOAB corresponds to

$$\varphi_{h/2}^{[B]} \circ \varphi_{h/2}^{[A]} \circ \varphi_h^{[O]} \circ \varphi_{h/2}^{[A]} \circ \varphi_{h/2}^{[B]},$$

and so on. The analysis carried out in Leimkuhler and Matthews (2013*a*,*b*) and Leimkuhler, Matthews and Stolz (2016) for the large friction limit, and in Alamo and Sanz-Serna (2016) for the general case, shows that BAOAB offers an improved behaviour with respect to other members of this family. Other Langevin integrators are given in Leimkuhler and Matthews (2015).

Another approach to molecular simulations, rather than approximating Hamiltonian trajectories, consists in studying the paths originating from the collection of all initial conditions within a given set. This perspective allows us to apply statistical mechanics for calculating averages.

Specifically, if $H(q,p)$ denotes the Hamiltonian function of the system in thermal equilibrium at temperature T, then the probability measure μ in $\mathbb{R}^{2d}$ with density

$$Z^{-1}\mathrm{e}^{-\beta H(q,p)}, \quad \text{where} \quad Z = \int_{\mathbb{R}^{2d}} \mathrm{e}^{-\beta H(q,p)} \, \mathrm{d}q \, \mathrm{d}p < \infty, \tag{9.20}$$

is preserved by the flow of $H(q,p)$. Here $\beta = 1/(k_B T)$. The measure μ is called the Boltzmann–Gibbs distribution and Z is the partition function. Intuitively, μ provides the distribution of (q,p) over an ensemble of many copies of the given system when it is embedded in a much larger system acting as a 'heat bath' at constant temperature. In other words, $Z^{-1}\mathrm{e}^{-\beta H(q,p)} \, \mathrm{d}q \, \mathrm{d}p$ represents the fraction of copies with momenta between p and $p + \mathrm{d}p$, and configuration between q and $q + \mathrm{d}q$ (Bou-Rabee and Sanz-Serna 2018). Then the average energy is given by

$$\tilde{E} = Z^{-1} \int_{\mathbb{R}^{2d}} H(q,p)\, \mathrm{e}^{-\beta H(q,p)}\, \mathrm{d}q\, \mathrm{d}p.$$

In general, $\tilde{E}$ cannot be obtained analytically, and even the use of numerical quadratures is unfeasible if the dimension d is large. A common practice to approximate this class of integrals then consists in applying Monte Carlo methods, and more specifically, Markov chain Monte Carlo methods (Brooks, Gelman, Jones and Meng 2011).

9.6. *Hamiltonian Monte Carlo*

For Hamiltonian systems of the form

$$H(q,p) = \frac{1}{2} p^\top M^{-1} p + U(q), \tag{9.21}$$

we can factorize the (non-normalized) density (9.20) as

$$\mathrm{e}^{-\beta H(q,p)} = \mathrm{e}^{-\frac{1}{2}\beta p^\top M^{-1} p}\, \mathrm{e}^{-\beta U(q)},$$

so that q and p are stochastically independent (Bou-Rabee and Sanz-Serna 2018): the marginal distribution of the configuration variables q has probability density $\propto \mathrm{e}^{-\beta U(q)}$, whereas the momenta p obey a Gaussian distribution with zero mean and covariance matrix M. Therefore, samples from the p-marginal are easily obtained, and thus one may concentrate on generating samples with probability density

$$Z_q^{-1} \mathrm{e}^{-\beta U(q)}, \quad \text{where} \quad Z_q = \int_{\mathbb{R}^d} \mathrm{e}^{-\beta U(q)}\, \mathrm{d}q. \tag{9.22}$$

This can be carried out with the so-called Hamiltonian (or hybrid) Monte Carlo method (HMC). In fact, HMC was proposed in the landmark paper by Duane, Kennedy, Pendleton and Roweth (1987), not in the context of molecular simulation but in lattice quantum chromodynamics. Later on, it was used in data science and statistics (Liu 2008, Neal 2011). More generally, HMC can be used to sample from any continuous probability distribution on $\mathbb{R}^d$ for which the density function can be evaluated (perhaps up to an unknown normalizing constant): given a target distribution $\Pi(q)$, if $U(q)$ denotes the negative logarithm of the (not necessarily normalized) probability function of the target, then it is clear that $\Pi(q)$ is given by (9.22) (with $\beta = 1$). HMC then generates samples $(q_i, p_i) \in \mathbb{R}^{2d}$ from the Boltzmann–Gibbs distribution corresponding to H, that is,

$$P(q,p) = (2\pi)^{-d/2}\, |\det M|^{-1/2} \mathrm{e}^{-\frac{1}{2} p^\top M^{-1} p}\, Z_q^{-1}\, \mathrm{e}^{-U(q)}, \tag{9.23}$$

by means of a Markov chain so that $P(q,p)$ is an invariant of this chain. The corresponding marginal $q_i \in \mathbb{R}^d$ chain then leaves invariant the target distribution

Table 9.1. HMC algorithm. The function $H = (1/2)p^\top M^{-1}p + U(q)$ is the Hamiltonian. The algorithm generates a Markov chain $q^{(0)} \mapsto q^{(1)} \mapsto \cdots \mapsto q^{(m_{\max})}$ reversible with respect to the target probability distribution $\propto \exp(-U(q))$.

Given $q^{(0)} \in \mathbb{R}^d$, $m_{\max} \geq 1$, set $m = 0$.

1 (Momentum refreshment.) Draw $p^{(m)} \sim \mathcal{N}(0, M)$.

2 (Integration leg.) Compute (q^*, p^*) (q^* is the proposal) by integrating, by means of a reversible, volume-preserving integrator with step size h, the equations of motion derived from the Hamiltonian (9.21) over an interval $0 \leq t \leq Nh$. The initial condition is $(q^{(m)}, p^{(m)})$.

3 (Accept/reject.) Calculate

$$a^{(m)} = \min(1, \exp(H(q^{(m)}, p^{(m)}) - H(q^*, p^*)))$$

and draw $u^{(m)} \sim \mathcal{U}(0, 1)$. If $a^{(m)} > u^{(m)}$, set $q^{(m+1)} = q^*$ (acceptance); otherwise set $q^{(m+1)} = q^{(m)}$ (rejection).

4 Set $m = m + 1$. If $m = m_{\max}$ stop; otherwise go to step 1.

$\Pi(q)$, and this allows us to estimate the multidimensional integral of a certain function F with respect to Π by averaging F at the points (q_i) of the Markov chain.

Since the HMC algorithm has been thoroughly reviewed in the excellent survey by Bou-Rabee and Sanz-Serna (2018), we only summarize here those aspects most closely related to splitting methods. The basic HMC procedure is described in Table 9.1, although more elaborate possibilities exist. Notice that, when generating the proposal, the integrator to be used has to be both reversible and volume-preserving. Given the structure of H, it is clear that palindromic splitting methods constitute the natural choice for this task.

One of the most salient features of the algorithm is that it is able to generate proposal moves that, while being far away from the current state of the Markov chain, has a high probability of acceptance, thus reducing the correlation between samples.

As the main contribution to the computational cost of HMC resides in the numerical integration of the equations of motion, it is of paramount interest to use methods requiring as few evaluations of the force $-\nabla_q U(q)$ as possible, with small energy errors (to avoid rejections) and able to use large step sizes h. Taking these considerations into account, it is hardly surprising that the Störmer–Verlet algorithm is the method of choice, especially in low dimensions. If d increases, however, there are more favourable alternatives. In particular, the three-stage method

$$\varphi_{a_1 h}^{[T]} \circ \varphi_{b_1 h}^{[V]} \circ \varphi_{a_2 h}^{[T]} \circ \varphi_{b_2 h}^{[V]} \circ \varphi_{a_2 h}^{[T]} \circ \varphi_{b_1 h}^{[V]} \circ \varphi_{a_1 h}^{[T]},$$

with

$$a_1 = 0.11888010966548, \quad b_1 = 0.29619504261126, \quad a_2 = \frac{1}{2} - a_1, \quad b_2 = 1 - 2b_1,$$

provides better results at the same computational cost as the standard Störmer–Verlet method (Blanes *et al.* 2014). The rationale behind this and other multi-stage splitting methods presented in Blanes *et al.* (2014) is that the coefficients a_j, b_j are chosen to minimize the energy error in the proposal for relatively large values of h (and not, as in the usual numerical integration domain, to increase the accuracy in the limit $h \to 0$). An extension of this idea is the so-called adaptive integration approach (AIA) (Fernández-Pendás, Akhmatskaya and Sanz-Serna 2016, Akhmatskaya, Fernández-Pendás, Radivojević and Sanz-Serna 2017): here the user chooses the value of h to be used, and then the AIA algorithm itself finds the coefficients of the method within a given family of s-stage splitting integrators providing the best acceptance rate. The efficiency can be further enhanced by using a conveniently modified version of processing so that time-reversible kernels provide time-reversible integrations (Blanes, Calvo, Casas and Sanz-Serna 2021*a*).

In many situations of interest in statistics, the target density is a perturbation of a Gaussian density, so the corresponding Hamiltonian is given by (5.1), that is,

$$H(q,p) = \frac{1}{2}p^\top M^{-1} p + \frac{1}{2}q^\top N q + U(q),$$

where N is a constant symmetric, positive definitive $d \times d$ matrix whose spectral radius typically grows with the dimension d. In that case it is advantageous to precondition the dynamics by choosing the mass matrix (which is free in this setting) as $M = N$, since now all the d frequencies of the preconditioned system are 1, and the only restriction on the step size used by Störmer–Verlet is $h < 2$, independently of h (Bou-Rabee and Sanz-Serna 2018). Furthermore, the considerations on stability exposed in Section 4.2 indicate that the Strang integrators (4.10) based on the exact solution of the quadratic part of H and the perturbation $U(q)$ constitute the best option. In fact, the numerical experiments collected in Casas, Sanz-Serna and Shaw (2022) show that these methods, together with preconditioning, dramatically reduce the computational cost in all test problems and all observables considered, in comparison with the standard Störmer–Verlet scheme. It is worth remarking that all the theory developed in Section 5 is valid here, and in particular, the explicit processor built in Section 5.7 for the Strang splitting is expected to reduce the error in H, and thus it might contribute to increasing the acceptance probability.

9.7. *Quantum statistical mechanics*

The description of a quantum system in thermal equilibrium at temperature T with Hamiltonian H is based on the thermal density matrix

$$\rho = \mathrm{e}^{-\beta H},$$

where, again, $\beta = 1/(k_B T)$ (Feynman 1972). In fact, most of the properties of the system can be obtained from ρ, but now as an average on the different quantum states. Thus the equilibrium value of an operator $\hat{O}$ corresponding to a physical observable O for a system of N quantum particles in a volume V is given by

$$\langle \hat{O} \rangle = Z^{-1}\,\mathrm{Tr}(\mathrm{e}^{-\beta H}\hat{O}) = Z^{-1}\sum_n \langle n|\mathrm{e}^{-\beta H}\hat{O}|n\rangle, \tag{9.24}$$

where the partition function Z now reads

$$Z = \mathrm{Tr}\,(\mathrm{e}^{-\beta H}) = \sum_n \langle n|\mathrm{e}^{-\beta H}|n\rangle, \tag{9.25}$$

and the states $|n\rangle$ form a complete, orthonormal basis set (Ceperley 1995, Landau and Binder 2005). Since the eigenvalues of the Hamiltonian H are not generally known, we try to evaluate the traces in (9.24) and (9.25) without diagonalizing the Hamiltonian. This can be done with the Feynman path-integral approach. To proceed, we consider the position representation where the particle is labelled. Then the density matrix is given by $\rho(R, R'; \beta) \equiv \langle R|\mathrm{e}^{-\beta H}|R'\rangle$, where $R \equiv \{r_1, \ldots, r_N\}$, r_i is the position of the ith particle and the elements of $\rho(R, R'; \beta)$ are positive and can be interpreted as probabilities. The partition function is then

$$Z = \int \mathrm{d}R\,\langle R|\mathrm{e}^{-\beta H}|R\rangle = \int \mathrm{d}R\,\rho(R, R; \beta). \tag{9.26}$$

Since $\mathrm{e}^{-\beta H} = (\mathrm{e}^{-\varepsilon H})^M$, with $\varepsilon = \beta/M$ for any positive integer M, the density matrix can be expressed as

$$\begin{aligned}\rho(R_0, R_M; \beta) = \int \cdots \int \mathrm{d}R_1\,\mathrm{d}R_2 \cdots \mathrm{d}R_{M-1}\,\rho(R_0, R_1; \varepsilon) \\ \times\ \rho(R_1, R_2; \varepsilon) \cdots \rho(R_{M-1}, R_M; \varepsilon).\end{aligned} \tag{9.27}$$

The action for a given link k is defined as

$$S_k \equiv S(R_{k-1}, R_k; \varepsilon) = -\ln(\rho(R_{k-1}, R_k; \varepsilon)),$$

so (9.27) becomes

$$\rho(R_0, R_M; \beta) = \int \cdots \int \mathrm{d}R_1\,\mathrm{d}R_2 \cdots \mathrm{d}R_{M-1}\,\exp\Bigl(-\sum_{k=1}^{M} S_k\Bigr), \tag{9.28}$$

and the goal is then to construct a sufficiently accurate approximation to ρ while minimizing the number of integrals involved in (9.28) (the number of *beads* M).

On the other hand, the partition function (9.26) can be written as

$$\begin{aligned}Z = \int \cdots \int \mathrm{d}R_0\,\mathrm{d}R_1\,\mathrm{d}R_2 \cdots \mathrm{d}R_{M-1}\,\rho(R_0, R_1; \varepsilon) \\ \times\ \rho(R_1, R_2; \varepsilon) \cdots \rho(R_{M-1}, R_0; \varepsilon),\end{aligned} \tag{9.29}$$

where the first $|R_0\rangle$ and the last $|R_M\rangle$ elements are identified as required by the trace operations.

In practical applications, one must generally use approximations to ρ. To this end, notice that, typically

$$H = T + V = -\frac{1}{2}\sum_{i=1}^{N}\Delta_i + V. \tag{9.30}$$

It then makes sense to approximate $\mathrm{e}^{-\varepsilon H}$ by symmetric products of $\mathrm{e}^{-\varepsilon T}$ and $\mathrm{e}^{-\varepsilon V}$. The simplest approximation is given, of course, by the Lie–Trotter scheme, known in this setting as the *primitive action*, in which case

$$\rho(R_0, R_2; \varepsilon) \approx \int \mathrm{d}R_1 \langle R_0|\mathrm{e}^{-\varepsilon T}|R_1\rangle \langle R_1|\mathrm{e}^{-\varepsilon V}|R_2\rangle.$$

The operator V is diagonal in the position representation, whereas the kinetic matrix can be evaluated by using the eigenfunction expansion of T (Ceperley 1995). It is then possible to arrive at the discrete path-integral expression for the density matrix

$$\begin{aligned}\rho(R_0, R_M; \beta) = \int \cdots \int \mathrm{d}R_1\, \mathrm{d}R_2 \cdots \mathrm{d}R_{M-1} \left(\frac{1}{2\pi\varepsilon\hbar^2}\right)^{3NM/2} \\ \times \exp\left(-\sum_{j=1}^{M}\left(\frac{1}{2\varepsilon\hbar^2}\|R_{j-1} - R_j\|^2 + \varepsilon V(R_j)\right)\right),\end{aligned} \tag{9.31}$$

providing an approximation of order $O(\varepsilon^2)$, since the Lie–Trotter method is of effective order two. A significant improvement with respect to the primitive approximation is achieved by considering the so-called Takahashi–Imada action (Takahashi and Imada 1984), i.e. the approximation

$$\mathrm{e}^{-\varepsilon(T+V)} \approx \mathrm{e}^{-\frac{\varepsilon}{2}T}\, \mathrm{e}^{-\varepsilon V - \frac{\varepsilon^3}{24}[V,[T,V]]}\, \mathrm{e}^{-\frac{\varepsilon}{2}T}$$

of effective order four in ε.

A typical method to compute the multidimensional integrals appearing in (9.26) and (9.29) is to apply a Monte Carlo sampling according to the probability density $\pi \propto \exp(-\sum_{k=1}^{M} S_k)$, where Z normalizes π. In this respect, notice that for the Hamiltonian (9.30) we have $\langle R_{k-1}|\mathrm{e}^{-a_i\varepsilon T}|R_k\rangle \propto \exp(-\|R_{k-1} - R_k\|^2/(2a_i\varepsilon))$, so the coefficient a_j cannot be negative for a probabilistic based simulation. Here again the order barrier for splitting methods having positive coefficients is important: no splitting method of order higher than two can be used for doing quantum statistical calculations, unless nested commutators enter into their formulation. In this context, the fourth-order scheme (3.4) is widely used, as well as the more general two-parameter family of fourth-order methods

$$\mathrm{e}^{a_1\varepsilon T}\, \mathrm{e}^{\varepsilon W_{b_1,c_1}}\, \mathrm{e}^{a_2\varepsilon T}\, \mathrm{e}^{\varepsilon W_{b_2,c_2}}\, \mathrm{e}^{a_2\varepsilon T}\, \mathrm{e}^{\varepsilon W_{b_1,c_1}}\, \mathrm{e}^{a_1\varepsilon T}, \tag{9.32}$$

with modified potential

$$W_{b_i,c_i} = b_i V + c_i \varepsilon^2 [V,[T,V]]. \tag{9.33}$$

In this case, the presence of two free parameters makes it possible to minimize some of the sixth-order error terms and thus yield more efficient schemes. Numerical simulations carried out in Sakkos, Casulleras and Boronat (2009) show that the required number M in (9.28) to reproduce the exact energy of the system at low temperatures is much smaller with scheme (9.32) than with the Lie–Trotter and Takahashi–Imada methods; see Chin (2023) for a recent review.

9.8. Vlasov–Poisson equations

When a gas is brought to a very high temperature, electrons leave the atoms, thus leading to an overall mixture of electrically charged particles (ions and electrons) usually called *plasma*. There is a hierarchy of models to describe plasmas, ranging from those based on N particles evolving with the laws of classical relativistic mechanics and forces due to external and self-consistent electromagnetic fields, to kinetic equations and fluid models, more appropriate when the plasma is in thermodynamic equilibrium.

In kinetic models each species s in the plasma is characterized by a distribution function $f_s(x,v,t)$, so that $f_s \, \mathrm{d}x \, \mathrm{d}v$ is the average number of particles of species s with position and velocity in a box of volume $\mathrm{d}x \, \mathrm{d}v$ centred at (x,v). In the limit where the collective effects are dominant over collisions between particles, the kinetic equation describing the system (in the non-relativistic regime) is the Vlasov equation (Vlasov 1961)

$$\frac{\partial f_s}{\partial t} + v \cdot \nabla_x f_s + \frac{q_s}{m_s}(E + v \times B) \cdot \nabla_v f_s = 0, \tag{9.34}$$

where q_s and m_s denote the charge and mass of the particles of species s, and E and B stand for the electric and magnetic field, respectively. Equation (9.34) just expresses the fact that the distribution function f_s is conserved along the trajectories of the particles, and is typically coupled with the Maxwell equations to take into account the self-consistent electromagnetic field generated by the particles (Sinitsyn, Dulov and Vedenyapin 2011).

The so-called Vlasov–Poisson equation describes a plasma with only one atomic species (alternatively, electrons and positive ions) in the mean electric field derived from the potential ϕ created by the particles. Since $m_e \ll m_i$, then the effect of the ions can be treated as a uniform neutralizing background. Taking all constants equal to 1 and denoting the distribution by f, the relevant system of equations describing the electron dynamics reads

$$\begin{aligned} &\frac{\partial f}{\partial t} + v \cdot \nabla_x f - \nabla_x \phi(f) \cdot \nabla_v f = 0 \\ &\Delta_x \phi(f)(x) = -\left(\int_{\mathbb{R}^d} f(x,v) \, \mathrm{d}v - \frac{1}{(2\pi)^d} \int_{\mathbb{T}^d \times \mathbb{R}^d} f(x,v) \, \mathrm{d}x \, \mathrm{d}v \right) \end{aligned} \tag{9.35}$$

in the domain $(x, v) \in \mathbb{T}^d \times \mathbb{R}^d$, $d = 1, 2, 3$. The energy associated with equations (9.35) is

$$\begin{aligned} \mathcal{H}(f) &= \int_{\mathbb{T}^d\times\mathbb{R}^d} \frac{|v|^2}{2} f(x, v)\,\mathrm{d}x\,\mathrm{d}v + \int_{\mathbb{T}^d} \frac{1}{2}|E(f)(x)|^2\,\mathrm{d}x \\ &= \mathcal{T}(f) + \mathcal{U}(f) \end{aligned} \tag{9.36}$$

and is preserved along the solution. In fact the system has infinitely many invariants, but not all of them can be preserved by numerical schemes. Taking into account that the solution of the equations associated with $\mathcal{T}$ and $\mathcal{U}$ can be solved exactly (up to a phase space discretization), it is then natural to consider splitting methods for the time integration. Specifically, letting $\varphi_t^{\mathcal{T}}(f)$ denote the solution of $\partial_t f + v\cdot\nabla_x f = 0$, that is,

$$f(t, x, v) = f(0, x - tv, v),$$

and letting $\varphi_t^{\mathcal{U}}(f)$ be the solution of $\partial_t f - \nabla_x\phi(f)\cdot\nabla_v f = 0$, which reads

$$f(t, x, v) = f(0, x, v - tE(f(0))),$$

where $E(f(0))$ is the value of the electric field at time $t = 0$, the numerical integrators are of the form

$$\varphi^{\mathcal{U}}_{b_{s+1}h}(f) \circ \varphi^{\mathcal{T}}_{a_s h}(f) \circ \varphi^{\mathcal{U}}_{b_s h}(f) \circ \cdots \circ \varphi^{\mathcal{U}}_{b_2 h}(f) \circ \varphi^{\mathcal{T}}_{a_1 h}(f) \circ \varphi^{\mathcal{U}}_{b_1 h}(f).$$

The convergence of these schemes can be established by requiring the appropriate smoothness of f and using the Hamiltonian structure of the system. Moreover, the functionals $\mathcal{T}$ and $\mathcal{U}$ in the decomposition (9.36) satisfy the formal relation

$$[[[\mathcal{T},\mathcal{U}],\mathcal{U}],\mathcal{U}](f) = 0 \quad \text{for all } f,$$

where $[\cdot,\cdot]$ is the Poisson bracket associated with the infinite-dimensional Poisson structure of the system (see Casas, Crouseilles, Faou and Mehrenberger 2017 for a detailed treatment), so that Runge–Kutta–Nyström splitting methods can be safely used in this setting. In addition, when $d = 1$,

$$[[\mathcal{T},\mathcal{U}],\mathcal{U}](f) = 2m(f)\mathcal{U}(f), \quad \text{where} \quad m(f) = \frac{1}{2\pi}\int_{\mathbb{T}\times\mathbb{R}} f(x, v)\,\mathrm{d}x\,\mathrm{d}v.$$

Since $m(f)$ is a constant of the motion, this introduces additional simplifications. In particular, methods involving nested commutators only require the evaluation of $\mathcal{U}$ with appropriate coefficients, and thus methods up to order six can be designed involving a reduced number of maps that, when combined with semi-Lagrangian techniques in phase space, provides high efficiency (Casas *et al.* 2017)

9.9. *Quantum simulation of quantum systems*

As noted by Feynman (1982), simulating the full time evolution of arbitrary quantum systems on a classical computer requires exponential amounts of computational resources: the states of the system are wave functions that belong to a vector space whose dimension grows exponentially with the size of the system, so that merely to record the state of the system is already cumbersome. For this reason, he conjectured the possibility of using a specific quantum system (a quantum computer) to simulate the behaviour of arbitrary quantum systems whose dynamics are determined by local interactions. This conjecture was later shown to be correct (Lloyd 1996). Today, quantum computers can simulate a variety of systems arising in quantum chemistry, quantum field theory and many-body physics, etc.; see e.g. Berry *et al.* (2015), Childs and Su (2019) and the recent review by Miessen, Ollitrault, Tacchino and Tavernelli (2023).

Simulating the time evolution of a quantum system with Hamiltonian H requires approximating $\mathrm{e}^{-\mathrm{i}tH}$ ($\hbar = 1$). If the Hamiltonian is the sum over many local interactions, then one can also use splitting methods for this purpose. Specifically, suppose that for a system composed of n variables, H can be decomposed as

$$H = \sum_{j=1}^{\ell} H_j, \tag{9.37}$$

where each H_j acts on a space of dimension m_j encompassing at most k of the variables, and ℓ is a polynomial in n. For instance, the well-known Hubbard, Ising and Heisenberg models belong to this class (Nielsen and Chuang 2010). The important point is that, whereas $\mathrm{e}^{-\mathrm{i}tH}$ is difficult to compute, each $\mathrm{e}^{-\mathrm{i}tH_j}$ acts on a much smaller subsystem and can be straightforward to evaluate by quantum circuits. In fact, the explicit quantum simulation algorithm proposed in Lloyd (1996) is based on the Lie–Trotter method

$$\chi_t = \mathrm{e}^{-\mathrm{i}tH_\ell} \cdots \mathrm{e}^{-\mathrm{i}tH_2}\, \mathrm{e}^{-\mathrm{i}tH_1},$$

whereas subsequent proposals include the Strang splitting

$$S_t^{[2]} = \mathrm{e}^{-\mathrm{i}\frac{t}{2}H_1}\, \mathrm{e}^{-\mathrm{i}\frac{t}{2}H_2} \cdots \mathrm{e}^{-\mathrm{i}tH_\ell} \cdots \mathrm{e}^{-\mathrm{i}\frac{t}{2}H_2}\, \mathrm{e}^{-\mathrm{i}\frac{t}{2}H_1},$$

and especially the quintuple jump recursion (2.5)

$$S_t^{[2k+2]} = \left(S_{\gamma_{2k}t}^{[2k]}\right)^2 \circ S_{(1-4\gamma_{2k})t}^{[2k]} \circ \left(S_{\gamma_{2k}t}^{[2k]}\right)^2 \tag{9.38}$$

with $\gamma_{2k} = 1/(4 - 4^{1/(2k+1)})$. These schemes are known in this setting as *product formulas* (Childs and Su 2019, Chen *et al.* 2022) and the procedure is called *Trotterization*.

Product formulas provide approximations U_{app} to the exact evolution $\mathrm{e}^{-\mathrm{i}tH}$ and the goal is, given a time t and a maximum simulation error ε, to find an algorithm (a quantum circuit) U_{app} such that $\|U_{\mathrm{app}} - \mathrm{e}^{-\mathrm{i}tH}\| < \varepsilon$. In consequence, it is of paramount importance to analyse and eventually provide tight bounds for the error ε committed by product formulas / splitting methods when applied to Hamiltonian systems of the form (9.37). Thus, Childs *et al.* (2021) have shown that

$$\big\|S_t^{[2k]} - \mathrm{e}^{-\mathrm{i}tH}\big\| = O\big((B_H)^{2k+1}\big), \quad \text{with} \quad B_H \equiv \sum_{j=1}^{\ell} \|H_j\|\, t$$

if H_j are Hermitian. As usual, if t is large, then the whole interval is divided into N steps and (9.38) is applied within each step. In that case,

$$\big\|\big(S_{t/N}^{[2k]}\big)^N - \mathrm{e}^{-\mathrm{i}tH}\big\| = O(\varepsilon)$$

provided that

$$N = O\left(\frac{(B_H)^{1+1/(2k)}}{\varepsilon^{1/(2k)}}\right).$$

As we have already pointed out, whereas the recursion (9.38) (as well as the triple jump) constitutes a systematic way to achieve high-order approximations, it is not necessarily the most efficient, both in terms of errors and the number of exponentials involved. It makes sense, then, to consider some other methods in this setting, such as those collected in Section 8. This requires, in particular, a detailed analysis to achieve more stringent bounds for the corresponding errors than those obtained in Childs *et al.* (2021).

Hamiltonian systems of the form $H = -\frac{1}{2}\Delta + V(x)$ can also be simulated on quantum computers by using an appropriate representation of the states and the previous product formulas with the quantum Fourier transform (Nielsen and Chuang 2010).

9.10. *Other topics*

Space and time constraints prevent us from including in this review additional relevant applications where splitting methods have shown their merits, as well as other closely related important issues. Let us briefly mention some of them.

- Except for a brief incursion into Langevin dynamics, we have restricted our treatment to deterministic problems, although splitting methods have been widely applied to both ordinary and partial stochastic differential equations. Recent references in this area include Bréhier, Cohen and Jahnke (2023), Bréhier, Cohen and Giordano (2024) and Foster, dos Reis and Strange (2024).
- The evolution of a particle of mass m and charge q in a given electromagnetic field is modelled by the Lorentz equation $mx'' = q(E + x' \times B)$. By introducing the velocity $v = x'$ as a new variable, the resulting system of first-order ODEs can be split into three explicitly solvable parts, so that composition methods can be applied preserving volume in phase space (He, Sun, Liu and Qin 2015, 2016). The treatment can also be generalized to relativistic charged particles (Zhang *et al.* 2015).

- In addition to the Schrödinger equation, both linear and nonlinear, splitting methods have been applied to other relevant partial differential equations arising in quantum physics. This is the case, in particular, for the Dirac equation (Bao, Cai and Yin 2020) and the Klein–Gordon equation (Bao, Cai and Feng 2022).
- One important theoretical aspect not treated here refers to the convergence analysis of splitting methods applied to the nonlinear Schrödinger equation and other semi-linear Hamiltonian PDEs, as well as the use of the Birkhoff normal forms. This fascinating topic is the subject of much attention in the recent literature (see e.g. Faou 2012, Faou and Grébert 2011, Bernier and Grébert 2022) and probably deserves a review in its own right. It might be of interest to explore whether the formalism of extended word series, successfully applied in Murua and Sanz-Serna (2016) to construct formal invariants and normal forms of general classes of finite-dimensional Hamiltonian systems, could be adapted and applied for semi-linear Hamiltonian PDEs.
- Here we have only considered the application of splitting and composition methods with constant step size h. This is essential if we are interested in preserving qualitative properties of the system. As we have seen in Section 4, the modified equation corresponding to the numerical scheme depends explicitly on h, so that if h is changed then so is the modified equation, and the preserving properties that geometric integrators possess when h is constant are no longer guaranteed. In other problems, such as those defined by PDEs with no particular structure, this is not a matter of concern. There are systems which do possess a geometric structure that is advantageous to preserve by the numerical scheme and where the use of an adaptive step size is of the utmost importance to get efficient approximations. A relevant example is the gravitational N-body problem when there are close encounters between some of the bodies. In that case, one may apply splitting methods with variable step size by using some specifically designed transformations involving the time variable, in such a way that in the new variables the resulting time step is constant; see e.g. Mikkola (1997), Calvo, López-Marcos and Sanz-Serna (1998), Blanes and Budd (2005), Blanes and Iserles (2012) and Hairer *et al.* (2006, sect. VIII.2).

Forse altro canterà con miglior plectro.

Acknowledgements

The authors wish to thank J. Bernier, J. Boronat, A. Iserles, A. Pérez, J. M. Sanz-Serna, L. Shaw and M. Thalhammer for reading parts of this manuscript and for their comments and suggestions. They would also like to express their gratitude to their long-suffering families for their patience, understanding and wholehearted support. This work has been funded by Ministerio de Ciencia e Innovación (Spain) through projects PID2022-136585NB-C21 and PID2022-136585NB-C22,

MCIN/AEI/10.13039/501100011033/FEDER, UE, and also by Generalitat Valenciana (Spain) through project CIAICO/2021/180. AM has also received funding from the Department of Education of the Basque Government through the Consolidated Research Group MATHMODE (ITI456-22).

Appendix: Testing splitting methods on matrices

In Section 8 we collected more than 80 different splitting and composition methods that are available in the literature. Although they have been classified into different categories according to the structure of the system to be integrated (RKN splitting methods, near-integrable systems, etc.) and the type of scheme (composition of a basic second-order time-symmetric method, splitting for a system separated into two parts, etc.), it is not obvious in advance which particular method is the most suitable for carrying out a given integration of a specific problem. There are many factors involved in the final choice: type of problem, the qualitative properties we are interested in preserving, computational cost, desired accuracy, time integration interval and even initial condition.

For these reasons, it might be illustrative to fix a list of examples and apply to them the most representative numerical integrators gathered in Section 8, just to get some clues about their relative performances. To this end, we take the linear system

$$\frac{\mathrm{d}X}{\mathrm{d}t} = F\,X, \qquad X(0) = I, \tag{A.1}$$

where F is a constant real $d \times d$ matrix and I is the identity, so that the exact solution $X(t) = \mathrm{e}^{tF}$ can be easily computed with any desired accuracy. Obviously, none of the methods we test here should be used, in general, to approximate the exponential of a matrix or its action on a vector (unless additional information on the structure of the matrices is known which could make splitting an efficient technique). Our purpose is rather to compare the relative performance of the different schemes.

Let Φ_h^N, with $t_f = Nh$, denote the matrix obtained with a given method applied N times with step size h that approximates e^{tF} at $t = t_f$. We then compute the following relative errors:

$$\mathcal{E}_1 = \frac{\|\mathrm{e}^{tF} - \Phi_h^N\|}{\|\mathrm{e}^{tF}\|}, \qquad \mathcal{E}_2 = \frac{|\mathrm{tr}(\mathrm{e}^{tF}) - \mathrm{tr}(\Phi_h^N)|}{|\mathrm{tr}(\mathrm{e}^{tF})|}.$$

The first, $\mathcal{E}_1$, measures the accuracy of the method, whereas $\mathcal{E}_2$ provides an estimate of the accuracy of the scheme in the case where it is used as a kernel with an ideal processor. In all cases, $\|\cdot\|$ refers to the 2-norm of the matrices considered.

The numerical experiments reported here have been carried out with MATLAB, and the function `randn( )` has been used to construct the matrix elements randomly from a normal distribution, initiated with the seed `rng(1)` for reproducibility. Additional material related to these simulations can be found at our website accompanying this paper.

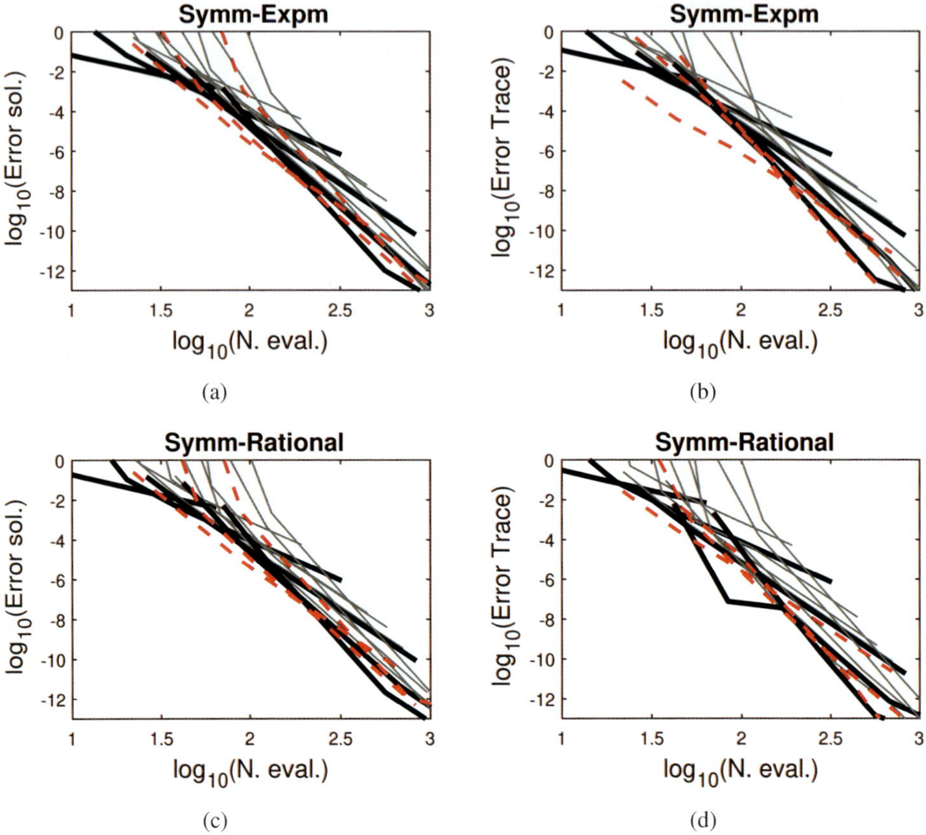

Figure A.1. Errors $\mathcal{E}_1$ (a,c) and $\mathcal{E}_2$ (b,d) vs. the number of evaluations of the basic scheme for $S_h^{[2,1]}$ (a,b) and $S_h^{[2,2]}$ (c,d). The non-processed schemes selected in Table 8.1 are shown in thick solid lines and the selected processed methods are drawn as thick red dashed lines.

Symmetric compositions of time-symmetric second-order schemes. We take $F = A + B + C$, with A, B, C matrices of dimension 50×50 constructed as previously indicated, that is,

```
d=50; rng(1);  A=randn(d);   A=A/norm(A); B=randn(d); ...
```

To illustrate the role that the basic scheme may play in the overall performance of the composition, we take the following two second-order time-symmetric methods:

$$S_h^{[2,1]} = \mathrm{e}^{hA/2}\mathrm{e}^{hB/2}\mathrm{e}^{hC}\mathrm{e}^{hB/2}\mathrm{e}^{hA/2} \tag{A.2}$$

and

$$S_h^{[2,2]} = \left(I - \frac{h}{2}A\right)^{-1}\left(I - \frac{h}{2}B\right)^{-1}\left(I - \frac{h}{2}C\right)^{-1}\left(I + \frac{h}{2}C\right)\left(I + \frac{h}{2}B\right)\left(I + \frac{h}{2}A\right). \tag{A.3}$$

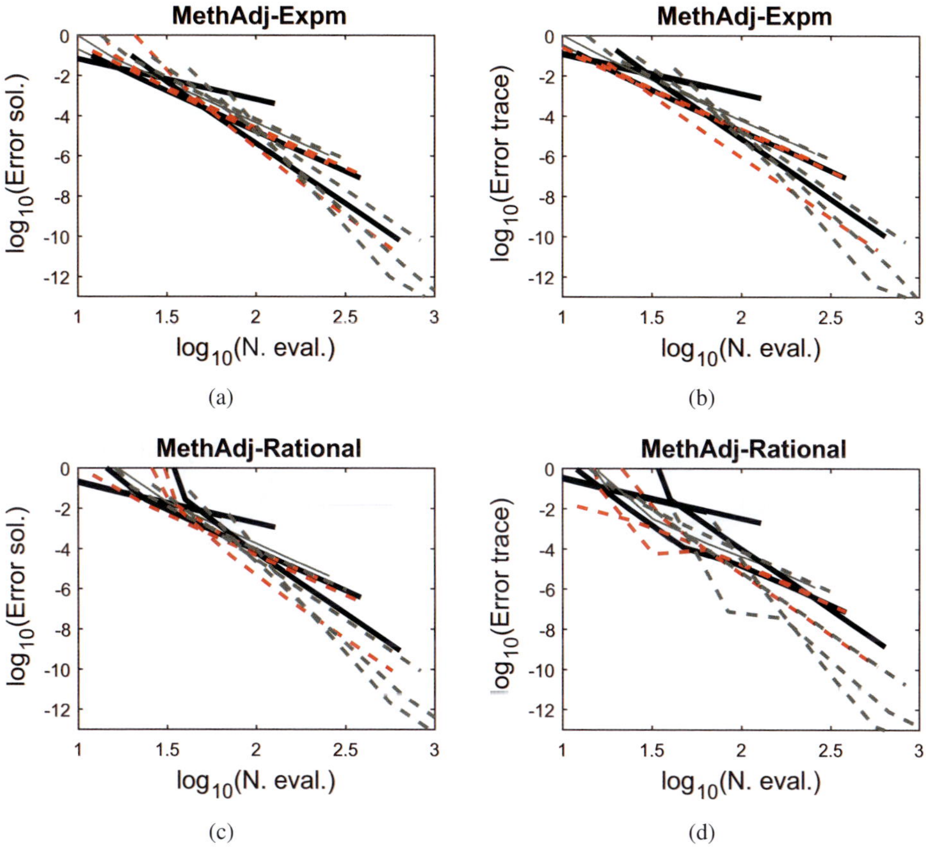

Figure A.2. Errors $\mathcal{E}_1$ (a,c) and $\mathcal{E}_2$ (b,d) vs. the number of evaluations of the basic scheme for $\chi_h^{[1,1]}$ (a,b) and $\chi_h^{[1,2]}$ (c,d). Now the schemes are compositions of $\chi_h^{[1,j]}$ and $\left(\chi_h^{[1,j]}\right)^*$.

The two differ in their computational cost and error terms. For the most relevant methods presented in Table 8.1, we plot the errors $\mathcal{E}_1$ and $\mathcal{E}_2$ as a function of the total number of evaluations of the basic scheme. Figure A.1 shows the results obtained for $t_f = 10$ when the basic scheme is $S_h^{[2,1]}$ (a,b) and $S_h^{[2,2]}$ (c,d). Compositions with $s = 5$, 13, 19 and 35 stages of order $r = 4$, 6, 8 and 10, respectively, which on average show good performance in most tested examples, correspond to the thick black lines, whereas the remaining non-processed methods are shown as thin lines. On the other hand, the recommended processed methods in Table 8.1 with kernels having $s = 13$, 19 and 23 stages of order 6, 8 and 10, respectively, are represented by thick red dashed lines.

Looking at these figures, it is hardly possible to recommend a particular scheme as the one leading to the best results for the example considered. We may only conclude that for some intervals of accuracy, there are certain methods exhibiting

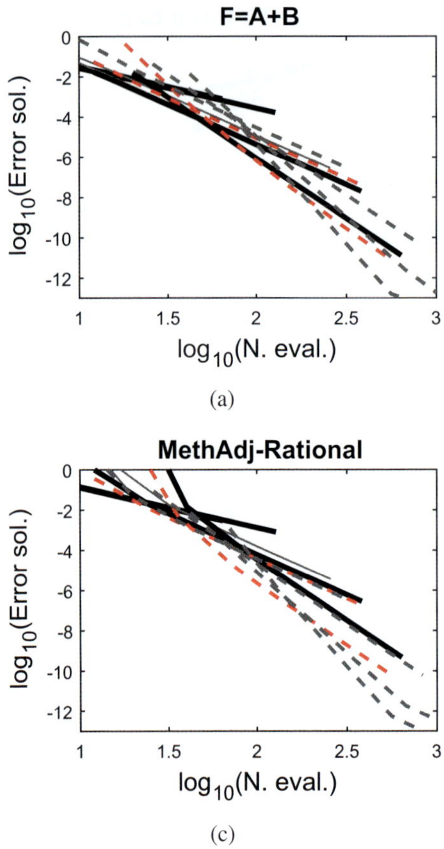

Figure A.3. The same as in Figure A.2 but now for the splitting $F = A + B$.

the best performance. Moreover, this performance also depends on the particular basic scheme chosen as $S_h^{[2]}$ in the compositions and the nature of the composition itself: whether it is intended to be used as the kernel of a processed method (with an optimal processor). In many situations, the relative performance displayed in Figure A.1(a,c) should be close to that of Figure A.1(b,d), when long-time integrations are considered and the contribution from the processable error terms can be neglected.

Splitting into two parts / composition of a basic method and its adjoint. We again consider $F = A + B + C$, with A, B, C the same matrices as in the previous example, but instead we take as basic methods the first-order compositions

$$\chi_h^{[1,1]} = \mathrm{e}^{hA}\mathrm{e}^{hB}\mathrm{e}^{hC}, \quad \left(\chi_h^{[1,1]}\right)^* = \mathrm{e}^{hC}\mathrm{e}^{hB}\mathrm{e}^{hA},$$

and

$$\chi_h^{[1,2]} = (I - hA)^{-1}(I - hB)^{-1}(I - hC)^{-1}, \quad \left(\chi_h^{[1,2]}\right)^* = (I + hC)(I + hB)(I + hA).$$

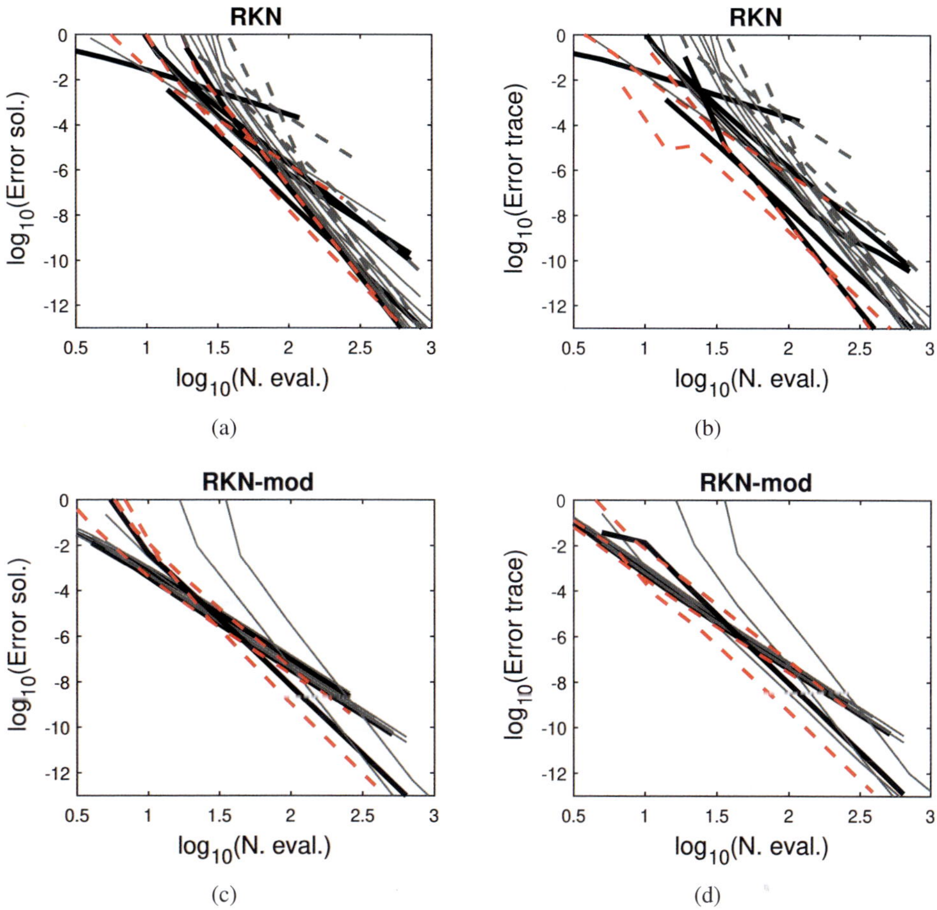

Figure A.4. RKN methods without (a,b) and with (c,d) modified potentials.

We plot the errors $\mathcal{E}_1$ and $\mathcal{E}_2$ versus the total number of evaluations of the basic scheme for most of the methods presented in Table 8.2. Figure A.2 shows the results obtained for $t_f = 10$. As before, the thick black solid lines correspond to the selected non-processed schemes and the thick red dashed lines correspond to the selected processed ones. The dashed grey lines correspond to the most efficient non-processed symmetric–symmetric schemes of orders 6, 8 and 10.

We repeat the same numerical experiments, but now taking $C = 0$, while keeping A, B the same matrices as previously. Hence, the problem has to be seen as separable into two parts. Notice that some methods have been optimized for this particular case. Figure A.3 shows the results obtained, where the superiority of the methods for this class of problems is clear for low-to-medium accuracy.

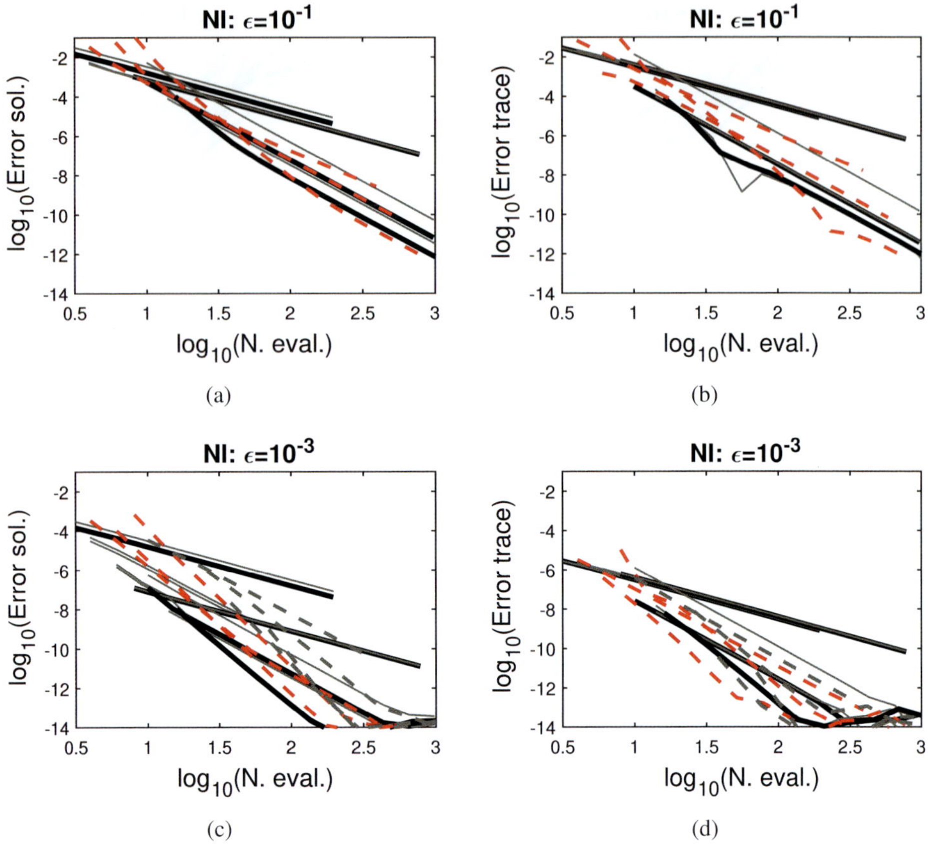

Figure A.5. Methods for near-integrable systems.

Runge–Kutta–Nyström methods. To test the RKN splitting methods gathered in Tables 8.3 and 8.4, we take $F = A + B$, with

$$A = \begin{pmatrix} O_d & O_d \\ A_1 & O_d \end{pmatrix}, \quad B = \begin{pmatrix} B_1 & B_2 \\ B_3 & B_4 \end{pmatrix},$$

where $A_1, B_i, i = 1, 2, 3, 4$ are matrices of dimension $d = 50$ with elements chosen as in the previous cases, with the same seed, and O_d is the null matrix. With this choice we have

$$[A, [A, B]] = \begin{pmatrix} O_d & O_d \\ 2A_1 B_2 A_1 & O_d \end{pmatrix}$$

and $[A, [A, [A, B]]] = 0$. For this particular choice, the computation of e^{hB} dominates the total cost of the method and the cost of evaluating e^{hA} and $\mathrm{e}^{hA+h^3[A,[A,B]]}$ can be neglected. Obviously, we can add an artificial cost to these exponentials as a test for different problems where this term can be more expensive to evaluate.

Figure A.4(a,b) shows the results for most of the methods from Table 8.3 for $t_f = 10$ where, as previously, the thick black lines correspond to the selected non-processed schemes, the thin lines correspond to the remaining non-processed schemes, the thick red dashed lines are the processed ones and the dashed grey lines correspond to the selected high-order symmetric–symmetric methods. Figure A.4(c,d) shows, for the same matrices, the results for the methods from Table 8.4 with modified potentials.

Methods for near-integrable systems. Let $F = A+\varepsilon B$ with A, B the same matrices as for the separable problem in two parts and ε a small parameter which corresponds to the relative norm of the matrices. We analyse the performance of methods tailored for perturbed problems for two choices of the small parameter: $\varepsilon = 10^{-1}$ and $\varepsilon = 10^{-3}$. Figure A.5 shows the results for the methods from Table 8.5 for $t_f = 10$. For small values of ε none of the previous splitting methods are competitive against the most efficient ones from this family.

References

M. Aichinger and E. Krotscheck (2005), A fast configuration space method for solving local Kohn–Sham equations, *Comput. Mater. Sci.* **34**, 188–212.

E. Akhmatskaya, M. Fernández Pendás, T. Radivojević and J. M. Sanz-Serna (2017), Adaptive splitting integrators for enhancing sampling efficiency of modified Hamiltonian Monte Carlo methods in molecular simulations, *Langmuir* **33**, 11530–11542.

A. Alamo and J. M. Sanz-Serna (2016), A technique for studying strong and weak local errors of splitting stochastic integrators, *SIAM J. Numer. Anal.* **54**, 3239–3257.

E. Alberdi, M. Antoñana, J. Makazaga and A. Murua (2019), An algorithm based on continuation techniques for minimization problems with highly non-linear equality constraints. Available at arXiv:1909.07263.

I. Alonso-Mallo, B. Cano and N. Reguera (2018), Avoiding order reduction when integrating linear initial boundary value problems with exponential splitting methods, *IMA J. Numer. Anal.* **38**, 1294–1323.

I. Alonso-Mallo, B. Cano and N. Reguera (2019), Avoiding order reduction when integrating reaction–diffusion boundary value problems with exponential splitting methods, *J. Comput. Appl. Math.* **357**, 228–250.

A. Alvermann and H. Fehske (2011), High-order commutator-free exponential time-propagation of driven quantum systems, *J. Comput. Phys.* **230**, 5930–5956.

H. C. Andersen (1983), Rattle: A 'velocity' version of the Shake algorithm for molecular dynamics calculations, *J. Comput. Phys.* **52**, 24–34.

M. H. Anderson, J. R. Ensher, M. R. Mathews, C. E. Wieman and E. A. Cornell (1995), Observations of Bose–Einstein condensation in a dilute atomic vapor, *Science* **269**, 198–201.

V. I. Arnold (1989), *Mathematical Methods of Classical Mechanics*, second edition, Springer.

U. Ascher, S. Ruuth and R. Spiteri (1997), Implicit–explicit Runge–Kutta methods for time-dependent partial differential equations, *Appl. Numer. Math.* **25**, 151–167.

U. Ascher, S. Ruuth and B. Wetton (1995), Implicit–explicit methods for time-dependent partial differential equations, *SIAM J. Numer. Anal.* **32**, 797–823.

J. Auer, E. Krotscheck and S. A. Chin (2001), A fourth-order real-space algorithm for solving local Schrödinger equations, *J. Chem. Phys.* **115**, 6841–6846.

P. Bader and S. Blanes (2011), Fourier methods for the perturbed harmonic oscillator in linear and nonlinear Schrödinger equations, *Phys. Rev. E* **83**, art. 046711.

P. Bader, S. Blanes and F. Casas (2013), Solving the Schrödinger eigenvalue problem by the imaginary time propagation technique using splitting methods with complex coefficients, *J. Chem. Phys.* **139**, art. 124117.

P. Bader, A. Iserles, K. Kropielnicka and P. Singh (2014), Effective approximation for the semiclassical Schrödinger equation, *Found. Comput. Math.* **14**, 689–720.

P. Bader, A. Iserles, K. Kropielnicka and P. Singh (2016), Efficient methods for linear Schrödinger equation in the semiclassical regime with time-dependent potential, *Proc. R. Soc. A* **472**, art. 20150733.

A. D. Bandrauk and H. Shen (1991), Improved exponential split operator method for solving the time-dependent Schrödinger equation, *Chem. Phys. Lett.* **176**, 428–432.

A. D. Bandrauk, E. Dehghanian and H. Lu (2006), Complex integration steps in decomposition of quantum exponential evolution operators, *Chem. Phys. Lett.* **419**, 346–350.

W. Bao, Y. Cai and Y. Feng (2022), Improved uniform error bounds on time-splitting methods for long-time dynamics of the nonlinear Klein–Gordon equation with weak nonlinearity, *SIAM J. Numer. Anal.* **60**, 1962–1984.

W. Bao, Y. Cai and J. Yin (2020), Super-resolution of time-splitting methods for the Dirac equation in the nonrelativistic regime, *Math. Comp.* **89**, 2141–2173.

W. Bao, D. Jaksch and P. A. Markowich (2003*a*), Numerical solution of the Gross–Pitaevskii equation for Bose–Einstein condensation, *J. Comput. Phys.* **187**, 318–342.

W. Bao, D. Jaksch and P. A. Markowich (2004), Three dimensional simulation of jet formation in collapsing condensates, *J. Phys. B* **37**, 329–343.

W. Bao, S. Jin and P. A. Markowich (2002), On time-splitting spectral approximations for the Schrödinger equation in the semiclassical regime, *J. Comput. Phys.* **175**, 487–524.

W. Bao, S. Jin and P. A. Markowich (2003*b*), Numerical study of time-splitting spectral discretizations of nonlinear Schrödinger equations in the semiclassical regimes, *SIAM J. Sci. Comput.* **25**, 27–64.

R. Bellman (1970), *Introduction to Matrix Analysis*, second edition, McGraw-Hill.

J. Bernier (2021), Exact splitting methods for semigroups generated by inhomogeneous quadratic differential operators, *Found. Comput. Math.* **21**, 1401–1439.

J. Bernier and B. Grébert (2022), Birkhoff normal forms for Hamiltonian PDEs in their energy space, *J. Éc. Polytech. Math.* **9**, 681–745.

J. Bernier, S. Blanes, F. Casas and A. Escorihuela-Tomàs (2023), Symmetric-conjugate splitting methods for linear unitary problems, *BIT* **63**, art. 58.

J. Bernier, N. Crouseilles and Y. Li (2021), Exact splitting methods for kinetic and Schrödinger equations, *J. Sci. Comput.* **86**, art. 10.

D. W. Berry, A. M. Childs, R. Cleve, R. Kothari and R. D. Somma (2015), Simulating Hamiltonian dynamics with a truncated Taylor series, *Phys. Rev. Lett.* **114**, art. 090502.

S. Blanes and C. J. Budd (2005), Adaptive geometric integrators for Hamiltonian problems with approximate scale invariance, *SIAM J. Sci. Comput.* **26**, 1089–1113.

S. Blanes and F. Casas (2005), On the necessity of negative coefficients for operator splitting schemes of order higher than two, *Appl. Numer. Math.* **54**, 23–37.

S. Blanes and F. Casas (2006), Splitting methods for non-autonomous separable dynamical systems, *J. Phys. A* **39**, 5405–5423.

S. Blanes and F. Casas (2016), *A Concise Introduction to Geometric Numerical Integration*, CRC Press.

S. Blanes and V. Gradinaru (2020), High order efficient splittings for the semiclassical time-dependent Schrödinger equation, *J. Comput. Phys.* **405**, art. 109157.

S. Blanes and A. Iserles (2012), Explicit adaptive symplectic integrators for solving Hamiltonian systems, *Celestial Mech. Dynam. Astronom.* **114**, 297–317.

S. Blanes and P. C. Moan (2002), Practical symplectic partitioned Runge–Kutta and Runge–Kutta–Nyström methods, *J. Comput. Appl. Math.* **142**, 313–330.

S. Blanes, M. P. Calvo, F. Casas and J. M. Sanz-Serna (2021*a*), Symmetrically processed splitting integrators for enhanced Hamiltonian Monte Carlo sampling, *SIAM J. Sci. Comput.* **43**, A3357–A3371.

S. Blanes, F. Casas and A. Escorihuela-Tomàs (2022*a*), Applying splitting methods with complex coefficients to the numerical integration of unitary problems, *J. Comput. Dyn.* **9**, 85–101.

S. Blanes, F. Casas and A. Escorihuela-Tomàs (2022*b*), Runge–Kutta–Nyström symplectic splitting methods of order 8, *Appl. Numer. Math.* **182**, 14–27.

S. Blanes, F. Casas and A. Murua (2004), On the numerical integration of ordinary differential equations by processed methods, *SIAM J. Numer. Anal.* **42**, 531–552.

S. Blanes, F. Casas and A. Murua (2006*a*), Composition methods for differential equations with processing, *SIAM J. Sci. Comput.* **27**, 1817–1843.

S. Blanes, F. Casas and A. Murua (2006*b*), Symplectic splitting operator methods tailored for the time-dependent Schrödinger equation, *J. Chem. Phys.* **124**, art. 234105.

S. Blanes, F. Casas and A. Murua (2008*a*), On the linear stability of splitting methods, *Found. Comput. Math.* **8**, 357–393.

S. Blanes, F. Casas and A. Murua (2008*b*), Splitting and composition methods in the numerical integration of differential equations, *Bol. Soc. Esp. Mat. Apl.* **45**, 89–145.

S. Blanes, F. Casas and A. Murua (2011), Error analysis of splitting methods for the time dependent Schrödinger equation, *SIAM J. Sci. Comput.* **33**, 1525–1548.

S. Blanes, F. Casas and A. Murua (2015), An efficient algorithm based on splitting for the time integration of the Schrödinger equation, *J. Comput. Phys.* **303**, 396–412.

S. Blanes, F. Casas and A. Murua (2017*a*), Symplectic time-average propagators for the Schrödinger equation with a time-dependent Hamiltonian, *J. Chem. Phys.* **146**, art. 114109.

S. Blanes, F. Casas and J. Ros (1999*a*), Extrapolation of symplectic integrators, *Celestial Mech. Dynam. Astronom.* **75**, 149–161.

S. Blanes, F. Casas and J. Ros (1999*b*), Symplectic integrators with processing: A general study, *SIAM J. Sci. Comput.* **21**, 711–727.

S. Blanes, F. Casas and J. Ros (2000), Processing symplectic methods for near-integrable Hamiltonian systems, *Celestial Mech. Dynam. Astronom.* **77**, 17–35.

S. Blanes, F. Casas and J. Ros (2001*a*), High-order Runge–Kutta–Nyström geometric methods with processing, *Appl. Numer. Math.* **39**, 245–259.

S. Blanes, F. Casas and J. Ros (2001*b*), New families of symplectic Runge–Kutta–Nyström integration methods, in *Numerical Analysis and its Applications*, Vol. 1988 of Lecture Notes in Computer Science, Springer, pp. 102–109.

S. Blanes, F. Casas and J. M. Sanz-Serna (2014), Numerical integrators for the Hybrid Monte Carlo method, *SIAM J. Sci. Comput.* **36**, A1556–A1580.

S. Blanes, F. Casas and M. Thalhammer (2017*b*), High-order commutator-free quasi-Magnus exponential integrators for non-autonomous linear evolution equations, *Comput. Phys. Commun.* **220**, 243–262.

S. Blanes, F. Casas and M. Thalhammer (2018), Convergence analysis of high-order commutator-free quasi-Magnus integrators for nonautonomous linear evolution equations, *IMA J. Numer. Anal.* **38**, 743–778.

S. Blanes, F. Casas, P. Chartier and A. Escorihuela-Tomàs (2022*c*), On symmetric-conjugate composition methods in the numerical integration of differential equations, *Math. Comp.* **91**, 1739–1761.

S. Blanes, F. Casas, P. Chartier and A. Murua (2013*a*), Optimized high-order splitting methods for some classes of parabolic equations, *Math. Comp.* **82**, 1559–1576.

S. Blanes, F. Casas, A. Farrés, J. Laskar, J. Makazaga and A. Murua (2013*b*), New families of symplectic splitting methods for numerical integration in dynamical astronomy, *Appl. Numer. Math.* **68**, 58–72.

S. Blanes, F. Casas, C. González and M. Thalhammer (2021*b*), Convergence analysis of high-order commutator-free quasi-Magnus integrators for nonautonomous linear Schrödinger equations, *IMA J. Numer. Anal.* **41**, 594–617.

S. Blanes, F. Casas, C. González and M. Thalhammer (2023), Efficient splitting methods based on modified potentials: Numerical integration of linear parabolic problems and imaginary time propagation of the Schrödinger equation, *Commun. Comput. Phys.* **33**, 937–961.

S. Blanes, F. Casas, J. A. Oteo and J. Ros (2009), The Magnus expansion and some of its applications, *Phys. Rep.* **470**, 151–238.

S. Blanes, A. Iserles and S. Macnamara (2022*d*), Positivity-preserving methods for ordinary differential equations, *ESAIM Math. Model. Numer. Anal.* **56**, 1843–1870.

C. Bolley and M. Crouzeix (1978), Conservation de la positivité lors de la discrétisation des problèmes d'évolution paraboliques, *RAIRO Anal. Numér.* **12**, 237–245.

N. Bou-Rabee (2017), Cayley splitting for second-order Langevin stochastic partial differential equations. Available at arXiv:1707.05603.

N. Bou-Rabee and J. M. Sanz-Serna (2018), Geometric integrators and the Hamiltonian Monte Carlo method, *Acta Numer.* **27**, 113–206.

C.-E. Bréhier, D. Cohen and G. Giordano (2024), Splitting schemes for Fitzhugh–Nagumo stochastic partial differential equations, *Discrete Contin. Dyn. Syst. Ser. B* **29**, 214–244.

C.-E. Bréhier, D. Cohen and T. Jahnke (2023), Splitting integrators for stochastic Lie–Poisson systems, *Math. Comp.* **92**, 2167–2216.

S. Brooks, A. Gelman, G. L. Jones and X.-L. Meng, eds (2011), *Handbook of Markov Chain Monte Carlo*, CRC Press.

J. C. Butcher (1969), The effective order of Runge–Kutta methods, in *Proceedings of the Conference on the Numerical Solution of Differential Equations* (J. L. Morris, ed.), Vol. 109 of Lecture Notes in Mathematics, Springer, pp. 133–139.

J. C. Butcher and J. M. Sanz-Serna (1996), The number of conditions for a Runge–Kutta method to have effective order p, *Appl. Numer. Math.* **22**, 103–111.

S. T. Butler and M. H. Friedman (1955), Partition function for a system of interacting Bose–Einstein particles, *Phys. Rev.* **98**, 287–293.

M. P. Calvo and E. Hairer (1995), Accurate long-term integration of dynamical systems, *Appl. Numer. Math.* **18**, 95–105.

M. P. Calvo and J. M. Sanz-Serna (1993*a*), The development of variable-step symplectic integrators, with application to the two-body problem, *SIAM J. Sci. Comput.* **14**, 936–952.

M. P. Calvo and J. M. Sanz-Serna (1993*b*), High-order symplectic Runge–Kutta–Nyström methods, *SIAM J. Sci. Comput.* **14**, 1237–1252.

M. P. Calvo, M. A. López-Marcos and J. M. Sanz-Serna (1998), Variable step implementation of geometric integrators, *Appl. Numer. Math.* **28**, 1–16.

M. Campostrini and P. Rossi (1990), A comparison of numerical algorithms for dynamical fermions, *Nuclear Phys. B* **329**, 753–764.

J. Candy and W. Rozmus (1991), A symplectic integration algorithm for separable Hamiltonian functions, *J. Comput. Phys.* **92**, 230–256.

R. Carles (2008), *Semi-Classical Analysis for Nonlinear Schrödinger Equations*, World Scientific.

F. Casas and A. Murua (2009), An efficient algorithm for computing the Baker–Campbell–Hausdorff series and some of its applications, *J. Math. Phys.* **50**, art. 033513.

F. Casas, N. Crouseilles, E. Faou and M. Mehrenberger (2017), High-order Hamiltonian splitting for the Vlasov–Poisson equations, *Numer. Math.* **135**, 769–801.

F. Casas, J. M. Sanz-Serna and L. Shaw (2022), Split Hamiltonian Monte Carlo revisited, *Statist. Comput.* **32**, art. 86.

F. Casas, J. M. Sanz-Serna and L. Shaw (2023), A new optimality property of Strang's splitting, *SIAM J. Numer. Anal.* **61**, 1369–1385.

F. Castella, P. Chartier, S. Descombes and G. Vilmart (2009), Splitting methods with complex times for parabolic equations, *BIT* **49**, 487–508.

T. Cazenave (2003), *Semilinear Schrödinger Equations*, American Mathematical Society.

D. M. Ceperley (1995), Path integrals in the theory of condensed helium, *Rev. Modern Phys.* **67**, 279–355.

J. E. Chambers (2003), Symplectic integrators with complex time steps, *Astron. J.* **126**, 1119–1126.

R. Chan and A. Murua (2000), Extrapolation of symplectic methods for Hamiltonian problems, *Appl. Numer. Math.* **34**, 189–205.

P. J. Channell and J. C. Scovel (1990), Symplectic integration of Hamiltonian systems, *Nonlinearity* **3**, 231–259.

P. Chartier and A. Murua (2009), An algebraic theory of order, *ESAIM Math. Model. Numer. Anal.* **43**, 607–630.

K.-T. Chen (1957), Integration of paths, geometric invariants and a generalized Baker–Hausdorff formula, *Ann. of Math.* **65**, 163–178.

Y.-A. Chen, A. M. Childs, M. Hafezi, Z. Jiang, H. Kim and Y. Xu (2022), Efficient product formulas for commutators and applications to quantum simulations, *Phys. Rev. Res.* **4**, art. 013191.

A. M. Childs and Y. Su (2019), Nearly optimal lattice simulation by product formulas, *Phys. Rev. Lett.* **123**, art. 050503.

A. M. Childs, Y. Su, M. C. Tran, N. Wiebe and S. Zhu (2021), Theory of Trotter error with commutator scaling, *Phys. Rev. X* **11**, art. 011020.

S. A. Chin (1997), Symplectic integrators from composite operator factorizations, *Phys. Lett. A* **226**, 344–348.

S. A. Chin (2005), Structure of positive decomposition of exponential operators, *Phys. Rev. E* **71**, art. 016703.

S. A. Chin (2006), Complete characterization of fourth-order symplectic integrators with extended-linear coefficients, *Phys. Rev. E* **73**, art. 026705.

S. A. Chin (2023), Anatomy of path integral Monte Carlo: Algebraic derivation of the harmonic oscillator's universal discrete imaginary-time propagator and its sequential optimization, *J. Chem. Phys.* **159**, art. 134109.

S. A. Chin and C. R. Chen (2002), Gradient symplectic algorithms for solving the Schrödinger equation with time-dependent potentials, *J. Chem. Phys.* **117**, 1409–1415.

S. A. Chin and J. Geiser (2011), Multi-product operator splitting as a general method of solving autonomous and nonautonomous equations, *IMA J. Numer. Anal.* **31**, 1552–1577.

S. A. Chin and E. Krotscheck (2005), Fourth-order algorithms for solving the imaginary-time Gross–Pitaevskii equation in a rotating anisotropic trap, *Phys. Rev. E* **72**, art. 036705.

A. Chorin, T. J. R. Huges, J. E. Marsden and M. McCracken (1978), Product formulas and numerical algorithms, *Commun. Pure Appl. Math.* **31**, 205–256.

L. Y. Chou and P. W. Sharp (2000), Order 5 symplectic explicit Runge–Kutta–Nyström methods, *J. Appl. Math. Decision Sci.* **4**, 143–150.

P. G. Ciarlet and J. L. Lions, eds (1990), *Handbook of Numerical Analysis*, Vol. I, North-Holland.

J. E. Cohen, S. Friedland, T. Kato and F. P. Kelly (1982), Eigenvalue inequalities for products of matrix exponentials, *Linear Algebra Appl.* **45**, 55–95.

M. Creutz and A. Gocksch (1989), Higher-order hybrid Monte Carlo algorithms, *Phys. Rev. Lett.* **63**, 9–12.

P. E. Crouch and R. Grossman (1993), Numerical integration of ordinary differential equations on manifolds, *J. Nonlinear Sci.* **3**, 1–33.

J. M. A. Danby (1988), *Fundamentals of Celestial Mechanics*, Willmann-Bell.

H. de Raedt and B. de Raedt (1983), Applications of the generalized Trotter formula, *Phys. Rev. A* **28**, 3575–3580.

R. de Vogelaere (1956), Methods of integration which preserve the contact transformation property of the Hamiltonian equations. Report no. 4, Department of Mathematics, University of Notre Dame, IN.

J. C. del Valle and K. Kropielnicka (2023), Family of Strang-type exponential splitting in the presence of unbounded and time dependent operators. Available at arXiv:2310.01556.

J. Douglas and H. H. Rachford (1956), On the numerical solution of heat conduction problems in two and three space variables, *Trans. Amer. Math. Soc.* **82**, 421–439.

S. Duane, A. D. Kennedy, B. J. Pendleton and D. Roweth (1987), Hybrid Monte-Carlo, *Phys. Lett. B* **195**, 216–222.

L. Einkemmer and A. Ostermann (2015), Overcoming order reduction in diffusion–reaction splitting, Part I: Dirichlet boundary conditions, *SIAM J. Sci. Comput.* **37**, A1577–A1592.

L. Einkemmer and A. Ostermann (2016), Overcoming order reduction in diffusion–reaction splitting, Part II: Oblique boundary conditions, *SIAM J. Sci. Comput.* **38**, A3741–A3757.

L. Einkemmer, M. Moccaldi and A. Ostermann (2018), Efficient boundary corrected Strang splitting, *Appl. Math. Comput.* **332**, 76–89.

K.-J. Engel and R. Nagel (2006), *A Short Course on Operator Semigroups*, Springer.

E. Faou (2012), *Geometric Numerical Integration and Schrödinger Equations*, European Mathematical Society.

E. Faou and B. Grébert (2011), Hamiltonian interpolation of splitting approximation for Hamiltonian PDEs, *Found. Comput. Math.* **11**, 381–415.

E. Faou and C. Lubich (2006), A Poisson integrator for Gaussian wavepacket dynamics, *Comput. Vis. Sci.* **9**, 45–55.

E. Faou, V. Gradinaru and C. Lubich (2009), Computing semi-classical quantum dynamics with Hagedorn wavepackets, *SIAM J. Sci. Comput.* **31**, 3027–3041.

M. D. Feit, J. A. Fleck Jr and A. Steiger (1982), Solution of the Schrödinger equation by a spectral method, *J. Comput. Phys.* **47**, 412–433.

K. Feng (1992), Formal power series and numerical algorithms for dynamical systems, in *Proc. Conf. Scientific Computation Hangzhou, 1991* (T. Chan and Z.-C. Shi, eds), World Scientific, pp. 28–35.

K. Feng and M. Z. Qin (1987), The symplectic methods for the computation of Hamiltonian equations, in *Numerical Methods for Partial Differential Equations* (Y. L. Zhu and B. Y. Gao, eds), Vol. 1297 of Lecture Notes in Mathematics, Springer, pp. 1–37.

M. Fernández-Pendás, E. Akhmatskaya and J. M. Sanz-Serna (2016), Adaptive multi-stage integrators for optimal energy conservation in molecular simulations, *J. Comput. Phys.* **327**, 434–449.

R. P. Feynman (1972), *Statistical Mechanics: A Set of Lectures*, Addison-Wesley.

R. P. Feynman (1982), Simulating physics with computers, *Internat. J. Theoret. Phys.* **21**, 467–488.

W. Folkner, J. Williams, D. Boggs, R. Park and P. Kuchynka (2014), The planetary and lunar ephemerides DE430 and DE431. IPN Progress Report 42-196.

E. Forest (1992), Sixth-order Lie group integrators, *J. Comput. Phys.* **99**, 209–213.

E. Forest and R. Ruth (1990), Fourth-order symplectic integration, *Phys. D* **43**, 105–117.

J. Foster, G. dos Reis and C. Strange (2024), High order splitting methods for SDEs satisfying a commutativity condition, *SIAM J. Numer. Anal.* **62**, 500–532.

B. García-Archilla, J. M. Sanz-Serna and R. D. Skeel (1999), Long-time-step methods for oscillatory differential equations, *SIAM J. Sci. Comput.* **20**, 930–963.

R. Glowinski, S. J. Osher and W. Yin, eds (2016*a*), *Splitting Methods in Communication, Imaging, Science, and Engineering*, Springer.

R. Glowinski, T.-W. Pan and X.-C. Tai (2016*b*), Some facts about operator-splitting and alternating direction methods, in *Splitting Methods in Communication, Imaging, Science, and Engineering* (R. Glowinski, S. J. Osher and W. Yin, eds), Springer, pp. 19–94.

S. Golden (1957), Statistical theory of many-electron systems: Discrete bases of representation, *Phys. Rev.* **107**, 1283–1290.

D. Goldman and T. J. Kaper (1996), *N*th-order operator splitting schemes and nonreversible systems, *SIAM J. Numer. Anal.* **33**, 349–367.

H. Goldstein (1980), *Classical Mechanics*, second edition, Addison-Wesley.

W. B. Gragg (1965), On extrapolation algorithms for ordinary differential equations, *SIAM J. Numer. Anal.* **2**, 384–403.

S. Gray and D. E. Manolopoulos (1996), Symplectic integrators tailored to the time-dependent Schrödinger equation, *J. Chem. Phys.* **104**, 7099–7112.

S. Gray and J. M. Verosky (1994), Classical Hamiltonian structures in wave packet dynamics, *J. Chem. Phys.* **100**, 5011–5022.

D. F. Griffiths and J. M. Sanz-Serna (1986), On the scope of the method of modified equations, *SIAM J. Sci. Statist. Comput.* **7**, 994–1008.

E. Hairer (1994), Backward analysis of numerical integrators and symplectic methods, *Ann. Numer. Math.* **1**, 107–132.

E. Hairer, C. Lubich and G. Wanner (2003), Geometric numerical integration illustrated by the Störmer–Verlet method, *Acta Numer.* **12**, 399–450.

E. Hairer, C. Lubich and G. Wanner (2006), *Geometric Numerical Integration: Structure-Preserving Algorithms for Ordinary Differential Equations*, second edition, Springer.

E. Hairer, S. P. Nørsett and G. Wanner (1993), *Solving Ordinary Differential Equations I: Nonstiff Problems*, second revised edition, Springer.

E. Hansen and A. Ostermann (2009*a*), Exponential splitting for unbounded operators, *Math. Comp.* **78**, 1485–1496.

E. Hansen and A. Ostermann (2009*b*), High order splitting methods for analytic semigroups exist, *BIT* **49**, 527–542.

Y. He, Y. Sun, J. Liu and H. Qin (2015), Volume-preserving algorithms for charged particle dynamics, *J. Comput. Phys.* **281**, 135–147.

Y. He, Y. Sun, J. Liu and H. Qin (2016), Higher order volume-preserving schemes for charged particle dynamics, *J. Comput. Phys.* **305**, 172–184.

M. Hochbruck and C. Lubich (2003), On Magnus integrators for time-dependent Schrödinger equations, *SIAM J. Numer. Anal.* **41**, 945–963.

M. Hoffman (2000), Quasi-shuffle products, *J. Algebraic Combin.* **11**, 49–68.

H. Holden, K. H. Karlsen, K.-A. Lie and N. H. Risebro (2010), *Splitting Methods for Partial Differential Equations with Rough Solutions*, European Mathematical Society.

H. Holden, K. H. Karlsen, N. E. Risebro and T. Tao (2011), Operator splitting for the KdV equation, *Math. Comp.* **80**, 821–846.

H. Holden, C. Lubich and N. E. Risebro (2013), Operator splitting for partial differential equations with Burgers nonlinearity, *Math. Comp.* **82**, 173–185.

W. Hundsdorfer and J. G. Verwer (1995), A note on splitting errors for advection–reaction equations, *Appl. Numer. Math.* **18**, 191–199.

W. Hundsdorfer and J. G. Verwer (2003), *Numerical Solution of Time-Dependent Advection–Diffusion–Reaction Equations*, Springer.

A. Iserles (1996), *A First Course in the Numerical Analysis of Differential Equations*, Cambridge University Press.

A. Iserles, H. Z. Munthe-Kaas, S. P. Nørsett and A. Zanna (2000), Lie-group methods, *Acta Numer.* **9**, 215–365.

T. Jahnke and C. Lubich (2000), Error bounds for exponential operator splittings, *BIT* **40**, 735–744.

S. Jin, P. Markowich and C. Sparber (2011), Mathematical and computational methods for semiclassical Schrödinger equations, *Acta Numer.* **20**, 121–209.

W. Kahan and R. C. Li (1997), Composition constants for raising the order of unconventional schemes for ordinary differential equations, *Math. Comp.* **66**, 1089–1099.

E. Kieri (2015), Stiff convergence of force-gradient operator splitting methods, *Appl. Numer. Math.* **94**, 33–45.

P.-V. Koseleff (1993), Calcul formel pour les méthodes de Lie en mécanique hamiltonienne. PhD thesis, École Polytechnique.

D. P. Landau and K. Binder, eds (2005), *A Guide to Monte Carlo Simulations in Statistical Physics*, Cambridge University Press.

F. M. Lasagni (1988), Canonical Runge–Kutta methods, *Z. Angew. Math. Phys.* **39**, 952–953.

J. Laskar (1989), A numerical experiment on the chaotic behaviour of the solar system, *Nature* **338**, 237–238.

J. Laskar and P. Robutel (2001), High order symplectic integrators for perturbed Hamiltonian systems, *Celestial Mech. Dynam. Astronom.* **80**, 39–62.

J. Laskar, P. Robutel, F. Joutel, M. Gastineau, A. C. M. Correia and B. Levrard (2004), A long-term numerical solution for the insolation quantities of the Earth, *Astron. Astrophys.* **428**, 261–285.

L. J. Laslett (1986), Nonlinear dynamics: A personal perspective, in *Nonlinear Dynamics Aspects of Particle Accelerators* (J. M. Jowett, M. Month and S. Turner, eds), Vol. 247 of Lecture Notes in Physics, Springer, pp. 519–580.

C. Lasser and C. Lubich (2020), Computing quantum dynamics in the semiclassical regime, *Acta Numer.* **29**, 229–401.

L. Lehtovaara, J. Toivanen and J. Eloranta (2007), Solution of time-independent Schrödinger equation by the imaginary time propagation method, *J. Comput. Phys.* **221**, 148–157.

B. Leimkuhler and C. Matthews (2013*a*), Rational construction of stochastic numerical methods for molecular sampling, *Appl. Math. Res. Express* **2013**, 34–56.

B. Leimkuhler and C. Matthews (2013*b*), Robust and efficient configurational molecular sampling via Langevin dynamics, *J. Chem. Phys.* **138**, art. 174102.

B. Leimkuhler and C. Matthews (2015), *Molecular Dynamics*, Springer.

B. Leimkuhler and S. Reich (2004), *Simulating Hamiltonian Dynamics*, Cambridge University Press.

B. Leimkuhler, C. Matthews and G. Stolz (2016), The computation of averages from equilibrium and nonequilibrium Langevin molecular dynamics, *IMA J. Numer. Anal.* **36**, 13–79.

B. Leimkuhler, S. Reich and R. D. Skeel (1996), Integration methods for molecular dynamics, in *Mathematical Approaches to Biomolecular Structure and Dynamics* (J. P. Mesirov, K. Schulten and D. W. Sumners, eds), Springer, pp. 161–185.

S. Lie (1888), *Theorie der Transformationsgruppen*, Druck un Verlag.

J. S. Liu (2008), *Monte Carlo Strategies for Scientific Computing*, second edition, Springer.

X. Liu, P. Ding, J. Hong and L. Wang (2005), Optimization of symplectic schemes for time-dependent Schrödinger equations, *Comput. Math. Appl.* **50**, 637–644.

S. Lloyd (1996), Universal quantum simulators, *Science* **273**, 1073–1078.

M. A. López-Marcos, J. M. Sanz-Serna and R. D. Skeel (1996*a*), Cheap enhancement of symplectic integrators, in *Proceedings of the Dundee Conference on Numerical Analysis* (D. F. Griffiths and G. A. Watson, eds), Longman, pp. 107–122.

M. A. López-Marcos, J. M. Sanz-Serna and R. D. Skeel (1996*b*), An explicit symplectic integrator with maximal stability interval, in *Numerical Analysis: A.R. Mitchell 75th Birthday Volume* (D. F. Griffiths and G. A. Watson, eds), World Scientific, pp. 163–176.

M. A. López-Marcos, J. M. Sanz-Serna and R. D. Skeel (1997), Explicit symplectic integrators using Hessian-vector products, *SIAM J. Sci. Comput.* **18**, 223–238.

C. Lubich (2008), *From Quantum to Classical Molecular Dynamics: Reduced Models and Numerical Analysis*, European Mathematical Society.

S. MacNamara and G. Strang (2016), Operator splitting, in *Splitting Methods in Communication, Imaging, Science, and Engineering* (R. Glowinski, S. J. Osher and W. Yin, eds), Springer, pp. 95–114.

W. Magnus (1954), On the exponential solution of differential equations for a linear operator, *Commun. Pure Appl. Math.* **VII**, 649–673.

G. I. Marchuk (1990), Splitting and alternating direction methods, in *Handbook of Numerical Analysis* (P. G. Ciarlet and J. L. Lions, eds), Vol. 1, North-Holland, pp. 197–462.

R. I. McLachlan (1995*a*), Composition methods in the presence of small parameters, *BIT* **35**, 258–268.

R. I. McLachlan (1995*b*), On the numerical integration of ODE's by symmetric composition methods, *SIAM J. Sci. Comput.* **16**, 151–168.

R. I. McLachlan (1996), More on symplectic correctors, in *Integration Algorithms and Classical Mechanics* (J. E. Marsden, G. W. Patrick and W. F. Shadwick, eds), Vol. 10 of Fields Institute Communications, American Mathematical Society, pp. 141–149.

R. I. McLachlan (2002), Families of high-order composition methods, *Numer. Algorithms* **31**, 233–246.

R. I. McLachlan and P. Atela (1992), The accuracy of symplectic integrators, *Nonlinearity* **5**, 541–562.

R. I. McLachlan and S. K. Gray (1997), Optimal stability polynomials for splitting methods, with applications to the time-dependent Schrödinger equation, *Appl. Numer. Math.* **25**, 275–286.

R. I. McLachlan and A. Murua (2019), The Lie algebra of classical mechanics, *J. Comput. Dyn.* **6**, 198–213.

R. I. McLachlan and R. Quispel (2002), Splitting methods, *Acta Numer.* **11**, 341–434.

R. I. McLachlan and R. Quispel (2006), Geometric integrators for ODEs, *J. Phys. A* **39**, 5251–5285.

A. Messiah (1999), *Quantum Mechanics*, Dover.

A. Miessen, P. J. Ollitrault, F. Tacchino and I. Tavernelli (2023), Quantum algorithms for quantum dynamics, *Nat. Comput. Sci.* **3**, 25–37.

S. Mikkola (1997), Practical symplectic methods with time transformation for the few-body problem, *Celestial Mech. Dynam. Astronom.* **67**, 145–165.

P. C. Moan (2002), On backward error analysis and Nekhoroshev stability in the numerical analysis of conservative systems of ODEs. PhD thesis, University of Cambridge.

P. C. Moan (2004), On the KAM and Nekhorosev theorems for symplectic integrators and implications for error growth, *Nonlinearity* **17**, 67–83.

H. Munthe-Kaas and B. Owren (1999), Computations in a free Lie algebra, *Philos. Trans. Roy. Soc. A* **357**, 957–981.

A. Murua (2006), The Hopf algebra of rooted trees, free Lie algebras, and Lie series, *Found. Comput. Math.* **6**, 387–426.

A. Murua and J. M. Sanz-Serna (1999), Order conditions for numerical integrators obtained by composing simpler integrators, *Philos. Trans. Roy. Soc. A* **357**, 1079–1100.

A. Murua and J. M. Sanz-Serna (2016), Computing normal forms and formal invariants of dynamical systems by means of word series, *Nonlinear Anal.* **138**, 326–345.

A. Murua and J. M. Sanz-Serna (2017), Word series for dynamical systems and their numerical integrators, *Found. Comput. Math.* **17**, 675–712.

R. M. Neal (2011), MCMC using Hamiltonian dynamics, in *Handbook of Markov Chain Monte Carlo* (S. Brooks *et al.*, eds), CRC Press, pp. 113–174.

F. Neri (1988), Lie algebras and canonical integration. Technical report, University of Maryland.

M. A. Nielsen and I. L. Chuang (2010), *Quantum Computation and Quantum Information*, 10th anniversary edition, Cambridge University Press.

D. I. Okunbor and E. J. Lu (1994), Eighth-order explicit symplectic Runge–Kutta–Nyström integrators. Technical report CSC 94-21, University of Missouri-Rolla.

D. I. Okunbor and R. D. Skeel (1994), Canonical Runge–Kutta–Nyström methods of orders five and six, *J. Comput. Appl. Math.* **51**, 375–382.

I. P. Omelyan, I. M. Mryglod and R. Folk (2002), On the construction of high order force gradient algorithms for integration of motion in classical and quantum systems, *Phys. Rev. E* **66**, art. 026701.

I. P. Omelyan, I. M. Mryglod and R. Folk (2003), Symplectic analytically integrable decomposition algorithms: Classification, derivation, and application to molecular dynamics, quantum and celestial mechanics simulations, *Comput. Phys. Commun.* **151**, 272–314.

B. Owren and A. Marthinsen (1999), Runge–Kutta methods adapted to manifolds and based on rigid frames, *BIT* **39**, 116–142.

P. Painlevé (1900), *Analyse des Travaux Scientifiques*, Gauthier-Villars.

L. A. Pars (1979), *A Treatise on Analytical Dynamics*, Ox Bow Press.

J. Partington (2004), *Linear Operators and Linear Systems: An Analytical Approach to Control Theory*, Cambridge University Press.

A. Pazy (1983), *Semigroups of Linear Operators and Applications to Partial Differential Equations*, Springer.

D. W. Peaceman and H. H. Rachford (1955), The numerical solution of parabolic and elliptic differential equations, *J. Soc. Indust. Appl. Math.* **3**, 28–41.

T. Prosen and I. Pizorn (2006), High order non-unitary split-step decomposition of unitary operators, *J. Phys. A* **39**, 5957–5964.

M. Reed and B. Simon (1980), *Methods of Modern Mathematical Physics I: Functional Analysis*, Academic Press.

S. Reich (1999), Backward error analysis for numerical integrators, *SIAM J. Numer. Anal.* **36**, 1549–1570.

C. Reutenauer (1993), *Free Lie Algebras*, Oxford University Press.

F. Rousset and K. Schratz (2022), Convergence error estimates at low regularity for time discretizations of KdV, *Pure Appl. Anal.* **4**, 127–152.

G. Rowlands (1991), A numerical algorithm for Hamiltonian systems, *J. Comput. Phys.* **97**, 235–239.

R. Ruth (1983), A canonical integration technique, *IEEE Trans. Nucl. Sci.* **30**, 26–69.

J.-P. Ryckaert, G. Ciccotti and H. J. C. Berendsen (1977), Numerical integration of the Cartesian equations of motion of a system with constraints: Molecular dynamics of *n*-alkanes, *J. Comput. Phys.* **23**, 327–341.

K. Sakkos, J. Casulleras and J. Boronat (2009), High order Chin actions in path integral Monte Carlo, *J. Chem. Phys.* **130**, art. 204109.

J. M. Sanz-Serna (1988), Runge–Kutta schemes for Hamiltonian systems, *BIT* **28**, 877–883.

J. M. Sanz-Serna (1992), Symplectic integrators for Hamiltonian problems: An overview, *Acta Numer.* **1**, 243–286.

J. M. Sanz-Serna (1997), Geometric integration, in *The State of the Art in Numerical Analysis* (I. S. Duff and G. A. Watson, eds), Vol. 63 of Institute of Mathematics and its Applications Conference Series: New Ser., Oxford University Press, pp. 121–143.

J. M. Sanz-Serna and M. P. Calvo (1994), *Numerical Hamiltonian Problems*, Chapman & Hall.

J. M. Sanz-Serna and A. Portillo (1996), Classical numerical integrators for wave-packet dynamics, *J. Chem. Phys.* **104**, 2349–2355.

T. Schlick (2010), *Molecular Modelling and Simulation: An Interdisciplinary Guide*, second edition, Springer.

Q. Sheng (1989), Solving linear partial differential equations by exponential splitting, *IMA J. Numer. Anal.* **9**, 199–212.

A. Sinitsyn, E. Dulov and V. Vedenyapin (2011), *Kinetic Boltzmann, Vlasov and Related Equations*, Elsevier.

R. D. Skeel and J. L. Cieśliński (2020), On the famous unpublished preprint 'methods of integration which preserve the contact transformation property of the Hamilton equations'. Available at arXiv:2003.12268.

M. Sofroniou and G. Spaletta (2005), Derivation of symmetric composition constants for symmetric integrators, *Optim. Methods Softw.* **20**, 597–613.

R. Speth, W. Green, S. MacNamara and G. Strang (2013), Balanced splitting and rebalanced splitting, *SIAM J. Numer. Anal.* **51**, 3084–3105.

H. J. Stetter (1970), Symmetric two-step algorithms for ordinary differential equations, *Computing* **5**, 267–280.

C. Störmer (1907), Sur les trajectoires des corpuscules électrisés, *Arch. Sci. Phys. Nat. Genève* **24**, 5–18, 113–158, 221–247.

G. Strang (1968), On the construction and comparison of difference schemes, *SIAM J. Numer. Anal.* **5**, 506–517.

C. Sulem and P.-L. Sulem (1999), *The Nonlinear Schrödinger Equation: Self-Focusing and Wave Collapse*, Springer.

Y. B. Suris (1988), Preservation of symplectic structure in the numerical solution of Hamiltonian systems, in *Numerical Solution of Differential Equations* (S. S. Filippov, ed.), pp. 148–160. In Russian.

G. Sussman and J. Wisdom (1992), Chaotic evolution of the Solar System, *Science* **257**, 56–62.

M. Suzuki (1990), Fractal decomposition of exponential operators with applications to many-body theories and Monte Carlo simulations, *Phys. Lett. A* **146**, 319–323.

M. Suzuki (1991), General theory of fractal path integrals with applications to many-body theories and statistical physics, *J. Math. Phys.* **32**, 400–407.

M. Suzuki (1992), General nonsymmetric higher-order decomposition of exponential operators and symplectic integrators, *J. Phys. Soc. Japan* **61**, 3015–3019.

M. Suzuki (1995), Hybrid exponential product formulas for unbounded operators with possible applications to Monte Carlo simulations, *Phys. Lett. A* **201**, 425–428.

M. Suzuki and K. Umeno (1993), High-order decomposition theory of exponential operators and its applications to QMC and nonlinear dynamics, in *Computer Simulation Studies in Condensed Matter Physics VI* (D. P. Landau, K. K. Mon and H. B. Schüttler, eds), Springer, pp. 74–86.

M. Takahashi and M. Imada (1984), Monte carlo calculation of quantum system II. Higher order correction, *J. Phys. Soc. Japan* **53**, 3765–3769.

M. Thalhammer (2008), High-order exponential operator splitting methods for time-dependent Schrödinger equations, *SIAM J. Numer. Anal.* **46**, 2022–2038.

M. Thalhammer, M. Caliari and C. Neuhauser (2009), High-order time-splitting Hermite and Fourier spectral methods, *J. Comput. Phys.* **228**, 822–832.

L. N. Trefethen (2000), *Spectral Methods in MATLAB*, SIAM.

H. F. Trotter (1959), On the product of semi-groups of operators, *Proc. Amer. Math. Soc.* **10**, 545–551.

C. Tsitouras (1999), A tenth order symplectic Runge–Kutta–Nyström method, *Celestial Mech. Dynam. Astronom.* **74**, 223–230.

V. S. Varadarajan (1984), *Lie Groups, Lie Algebras, and Their Representations*, Springer.

L. Verlet (1967), Computer 'experiments' on classical fluids I: Thermodynamical properties of Lennard-Jones molecules, *Phys. Rev.* **159**, 98–103.

A. A. Vlasov (1961), *Many-Particle Theory and its Application to Plasma*, Gordon & Breach.

J. Wisdom and M. Holman (1991), Symplectic maps for the n-body problem, *Astron. J.* **102**, 1528–1538.

J. Wisdom, M. Holman and J. Touma (1996), Symplectic correctors, in *Integration Algorithms and Classical Mechanics* (J. E. Marsden, G. W. Patrick and W. F. Shadwick, eds), Vol. 10 of Fields Institute Communications, American Mathematical Society, pp. 217–244.

N. N. Yanenko (1971), *The Method of Fractional Steps*, Springer.

H. Yoshida (1990), Construction of higher order symplectic integrators, *Phys. Lett. A* **150**, 262–268.

K. Yosida (1971), *Functional Analysis*, third edition, Springer.

R. Zhang, J. Liu, H. Qin, Y. Wang, Y. He and Y. Sun (2015), Volume-preserving algorithm for secular relativistic dynamics of charged particles, *Phys. Plasmas* **22**, art. 044501.

W. Zhu, X. Zhao and Y. Tang (1996), Numerical methods with a high order of accuracy applied in the quantum system, *J. Chem. Phys.* **104**, 2275–2286.

Acta Numerica (2024), pp. 163–485 Printed in the United Kingdom
doi:10.1017/S0962492924000011

Adaptive finite element methods

Andrea Bonito
Department of Mathematics, Texas A&M University,
College Station, TX 77843, USA
E-mail: bonito@tamu.edu

Claudio Canuto
Dipartimento di Scienze Matematiche, Politecnico di Torino,
Corso Duca degli Abruzzi 24, 10129 Torino, Italy
E-mail: claudio.canuto@polito.it

Ricardo H. Nochetto
Department of Mathematics and Institute for Physical Science and Technology,
University of Maryland, College Park, MD 20742, USA
E-mail: rhn@umd.edu

Andreas Veeser
Dipartimento di Matematica, Università degli Studi di Milano,
Via Saldini 50, 20133 Milano, Italy
E-mail: andreas.veeser@unimi.it

This is a survey of the theory of adaptive finite element methods (AFEMs), which are fundamental to modern computational science and engineering but whose mathematical assessment is a formidable challenge. We present a self-contained and up-to-date discussion of AFEMs for linear second-order elliptic PDEs and dimension $d > 1$, with emphasis on foundational issues. After a brief review of functional analysis and basic finite element theory, including piecewise polynomial approximation in graded meshes, we present the core material for coercive problems. We start with a novel *a posteriori* error analysis applicable to rough data, which delivers estimators fully equivalent to the solution error. They are used in the design and study of three AFEMs depending on the structure of data. We prove linear convergence of these algorithms and rate-optimality provided the solution and data belong to suitable approximation classes. We also address the relation between approximation and regularity classes. We finally extend this theory to discontinuous Galerkin methods as prototypes of non-conforming AFEMs, and beyond coercive problems to inf-sup stable AFEMs.

2020 Mathematics Subject Classification: Primary 65N30, 65N50
Secondary 65N35, 65N41

CONTENTS

1	Introduction: overview of AFEMs	164
2	Linear elliptic boundary value problems: basic theory	173
3	*A priori* approximation theory	185
4	*A posteriori* error analysis	217
5	Convergence of AFEM for coercive problems	286
6	Convergence rates of AFEM for coercive problems	325
7	Data approximation	355
8	Mesh refinement: the bisection method	381
9	Discontinuous Galerkin methods	397
10	AFEMs for inf-sup stable problems	429
Acknowledgements		475
Index		475
References		480

1. Introduction: overview of AFEMs

This is a survey of the theory of adaptive finite element methods (AFEMs), which are fundamental to modern computational science and engineering. We present a self-contained and up-to-date discussion of AFEMs for linear second-order elliptic PDEs in dimension $d > 1$, with emphasis on foundational issues rather than applications of AFEMs. This paper builds on and expands the older surveys by Nochetto, Siebert and Veeser (2009) and Nochetto and Veeser (2012). In fact, we decided to incorporate several new aspects into the theory described below.

The paper develops the theory of AFEMs gradually, and is meant to be accessible to advanced students and researchers interested in learning the fundamental aspects of adaptivity and why AFEMs outperform classical FEMs. We quantify the superior performance of AFEMs with precise mathematical statements rather than simulations. We present very few numerical experiments to illustrate some key (and new) algorithmic ideas and methods, but the paper is otherwise a tour through the numerical analysis of adaptive approximation of linear elliptic PDEs.

By design, this paper goes deep into some foundational aspects of AFEMs theory, provides full discussions and proofs, as well as pointers to the main literature. We consider the following model problem on a polyhedral domain $\Omega \subset \mathbb{R}^d$ with $d \geq 2$:

$$L[u] := -\operatorname{div}(\boldsymbol{A}\nabla u) + cu = f, \tag{1.1}$$

with general variable coefficients $(\boldsymbol{A}, c)$, forcing $f \in H^{-1}(\Omega)$ and homogeneous Dirichlet boundary conditions $u = 0$ on $\partial\Omega$ mostly, but not exclusively. If $\mathbb{V} := H_0^1(\Omega)$ and $\mathcal{B}\colon \mathbb{V}\times\mathbb{V} \to \mathbb{R}$ is the bilinear form associated with (1.1), the weak form reads

$$u \in \mathbb{V}: \quad \mathcal{B}[u, v] = \langle f, v\rangle \quad \text{for all } v \in \mathbb{V}. \tag{1.2}$$

Given a conforming and shape-regular partition $\mathcal{T}$ of Ω, created by successive refinement of a coarse mesh $\mathcal{T}_0$, let $\mathbb{V}_\mathcal{T}$ denote the space of continuous piecewise polynomial functions of degree $n \geq 1$ over $\mathcal{T}$ vanishing on $\partial\Omega$. The Galerkin approximation $u_\mathcal{T}$ of u solves

$$u_\mathcal{T} \in \mathbb{V}_\mathcal{T}: \quad \mathcal{B}[u_\mathcal{T}, v] = \langle f, v\rangle \quad \text{for all } v \in \mathbb{V}_\mathcal{T}. \tag{1.3}$$

This is a conforming approximation because $\mathbb{V}_\mathcal{T} \subset \mathbb{V}$. The aim of this paper is as follows.

- To design and analyse practical ways to estimate the error $|u - u_\mathcal{T}|_{H_0^1(\Omega)} := \|\nabla(u - u_\mathcal{T})\|_{L^2(\Omega)}$ in terms of so-called *a posteriori* error estimators, which are computable quantities depending on the discrete solution $u_\mathcal{T}$ and data $\mathcal{D} = (\boldsymbol{A}, c, f)$.
- To design adaptive algorithms that equidistribute the local errors $\|\nabla(u - u_\mathcal{T})\|_{L^2(T)}$ for all elements $T \in \mathcal{T}$, thereby optimizing the computational effort; this is a key step that makes complex three-dimensional situations accessible computationally.
- To show that this strategy delivers a performance comparable with the best possible in terms of degrees of freedom, which is a measure of computational complexity. This is a delicate matter because it entails dealing with approximation classes and their relation to regularity classes in terms of Besov and Lipschitz spaces.
- To present and analyse the bisection method for mesh refinement, one of the most versatile techniques for local mesh refinement that guarantees shape regularity and optimal complexity; the latter is instrumental to the previous point. Our study includes conforming meshes as well as certain non-conforming meshes.
- To extend the theory to a range of important problems that fail to be conforming or coercive. The first class is that of discontinuous Galerkin methods and the second class is inf-sup stable FEMs. The former is a notorious example of non-conforming approximation, whereas the latter is non-coercive.

In achieving these goals we provide several new ideas and methods. We also refer to the pertinent literature but we do not give a full list of references or get into comparisons of various approaches. It is not our intention to be comprehensive but rather to cover basic aspects of adaptivity in depth at the expense of important topics we do not touch upon. Some of them are:

- adaptive eigenvalue approximation,
- goal-oriented error analysis,
- non-conforming discretizations (except for discontinuous Galerkin),
- coarsening or aggregation,

- anisotropic refinements,
- hp-adaptivity,
- tree approximation,
- other PDEs, e.g. convection–diffusion equations, nonlinear and evolution equations.

We devote the rest of this introduction to providing a roadmap for the rest of the paper. In doing so, we introduce notation that will be used later, and present some topics in their most primitive form to provide an early idea about how they fit and interrelate.

A posteriori error analysis. We refer to the books by Ainsworth and Oden (2000) and Verfürth (2013) for the classical theory. However, in contrast to most of the existing literature, the current theory deals with forcing $f \in H^{-1}(\Omega)$. This allows for rough data useful in applications, such as line Dirac masses, but also encompasses a new approach to error estimation that leads to an error-dominated estimator and oscillation, and prevents error overestimation; this extends Kreuzer and Veeser (2021) to (1.1) and polynomial degree $n \geq 1$. The new twist is the construction of a projection operator $P_{\mathcal{T}}\colon H^{-1}(\Omega) \to \mathbb{F}_{\mathcal{T}}$ into a space of piecewise polynomials in $\mathcal{T}$ and on its skeleton $\mathcal{F}$, namely the set of all internal faces. Such an operator happens to be locally stable on stars (or patches) ω_z of $\mathcal{T}$ for all vertices $z \in \mathcal{V}$ of $\mathcal{T}$:

$$\|P_{\mathcal{T}}\ell\|_{H^{-1}(\omega_z)} \leq C_{\mathrm{lStb}}\|\ell\|_{H^{-1}(\omega_z)} \quad \text{for all } \ell \in H^{-1}(\omega_z). \tag{1.4}$$

An important property of $P_{\mathcal{T}}$ and its range $\mathbb{F}_{\mathcal{T}}$ is that for piecewise polynomial coefficients $(\boldsymbol{A}, c)$, or in short discrete coefficients, $P_{\mathcal{T}}$ is invariant in the subspace $L[\mathbb{V}_{\mathcal{T}}]$ of $H^{-1}(\Omega)$ or equivalently

$$P_{\mathcal{T}}(L[v]) = L[v] \in \mathbb{F}_{\mathcal{T}} \quad \text{for all } v \in \mathbb{V}_{\mathcal{T}}. \tag{1.5}$$

It is worth realizing that $L[v]$ is made of two distinct parts. The first one is absolutely continuous relative to the Lebesgue measure, namely $-\operatorname{div}(\boldsymbol{A}\nabla v) + cv$ in every element $T \in \mathcal{T}$. The second part is singular and supported in the skeleton $\mathcal{F}$, namely $[[\boldsymbol{A}\nabla v]] \cdot \boldsymbol{n}|_F \delta_F$ for every face $F \in \mathcal{F}$, where $[[\cdot]]$ is the jump across F, $\boldsymbol{n}$ is a unit normal to F, and δ_F is the Dirac mass on F.

These two properties of $P_{\mathcal{T}}$ have the following crucial consequences. Let

$$R_{\mathcal{T}} := L[u - u_{\mathcal{T}}] = f - L[u_{\mathcal{T}}] \in H^{-1}(\Omega) \tag{1.6}$$

be the residual of the Galerkin approximation of (1.2). Using (1.5) yields

$$R_{\mathcal{T}} - P_{\mathcal{T}}R_{\mathcal{T}} = f - P_{\mathcal{T}}f.$$

This shows that $R_{\mathcal{T}}$ decomposes into a discrete, thus finite-dimensional and computable, PDE part $P_{\mathcal{T}}R_{\mathcal{T}} = P_{\mathcal{T}}f - L[u_{\mathcal{T}}]$ and an infinite-dimensional component $f - P_{\mathcal{T}}f$, the so-called data oscillation that depends on f and can only be evaluated with additional knowledge of f.

The non-local H^{-1}-norm of $R_\mathcal{T}$ splits into local contributions on stars, whence[1]

$$|u - u_\mathcal{T}|^2_{H^1_0(\Omega)} \approx \|R_\mathcal{T}\|^2_{H^{-1}(\Omega)} \approx \sum_{z\in\mathcal{V}} \|R_\mathcal{T}\|^2_{H^{-1}(\omega_z)}.$$

The discrete nature of $P_\mathcal{T} R_\mathcal{T}$ allows us to derive a computable L^2-weighted PDE estimator $\eta_\mathcal{T}(u_\mathcal{T}, z)$ equivalent to $\|R_\mathcal{T}\|_{H^{-1}(\omega_z)}$, which together with the remaining data oscillation $\mathrm{osc}_\mathcal{T}(f, z)_{-1} := \|f - P_\mathcal{T} f\|_{H^{-1}(\omega_z)}$ gives the upper bound

$$|u - u_\mathcal{T}|^2_{H^1_0(\Omega)} \lesssim \sum_{z\in\mathcal{V}} \big(\eta_\mathcal{T}(u_\mathcal{T}, z)^2 + \mathrm{osc}_\mathcal{T}(f, z)^2_{-1}\big).$$

It turns out that this estimate is sharp or, in other words, that there is no overestimation of the error. To see this important and unique property of these new estimators, we invoke (1.4) to write the local lower bounds

$$\begin{aligned}\eta_\mathcal{T}(u_\mathcal{T}, z) &\approx \|P_\mathcal{T} R_\mathcal{T}\|_{H^{-1}(\omega_z)} \le C_{\mathrm{lStb}}\|R_\mathcal{T}\|_{H^{-1}(\omega_z)},\\ \mathrm{osc}_\mathcal{T}(f, z)_{-1} &= \|R_\mathcal{T} - P_\mathcal{T} R_\mathcal{T}\|_{H^{-1}(\omega_z)} \le (1 + C_{\mathrm{lStb}})\|R_\mathcal{T}\|_{H^{-1}(\omega_z)}.\end{aligned}$$

Section 4 constructs the operator $P_\mathcal{T}$, and derives several important properties such as its local quasi-best approximation and the above error-dominated *a posteriori* bounds. The former guarantees the inequality for the local L^2-projection $\Pi_\mathcal{T}$, that is,

$$\|f - P_\mathcal{T} f\|_{H^{-1}(\omega_z)} \lesssim \|f - \Pi_\mathcal{T} f\|_{H^{-1}(\omega_z)} \lesssim \|h(f - \Pi_\mathcal{T} f)\|_{L^2(\omega_z)},$$

which is the typical form of data oscillation provided $f \in L^2(\Omega)$. However, this L^2-weighted oscillation is not bounded above by the error and is thus responsible for potential overestimation. Section 4 proves further properties of $\eta_\mathcal{T}(u_\mathcal{T})$ such as its reduction upon refinement and its localized discrete upper bound, as well as quasi-monotonicity of $\mathrm{osc}_\mathcal{T}(f)_{-1}$ upon refinement. These properties, known for the standard L^2-weighted estimator and oscillation, are thus retained by the new construction.

Section 4 also deals with the alternative error estimators that result from solving local problems, using hierarchy, or imposing flux equilibration. We show that all of them lead, essentially, to estimators equivalent to $\|R_\mathcal{T}\|_{H^{-1}(\omega_z)}$. Moreover, we present an optimal framework to deal with non-homogeneous Dirichlet boundary conditions as well as with Robin and Neumann boundary conditions.

Linear convergence of AFEMs. Local *a posteriori* error indicators are usually employed to mark elements (or sets of elements) with largest indicators for refinement. We are concerned with the most popular *Dörfler marking* (or bulk chasing): given a parameter $\theta \in (0, 1]$, select a set $\mathcal{M} \subset \mathcal{T}$ such that

$$\eta_\mathcal{T}(u_\mathcal{T}, \mathcal{M}) \ge \theta\, \eta_\mathcal{T}(u_\mathcal{T}); \tag{1.7}$$

[1] Throughout this work $A \lesssim B$ signifies $A \le CB$ with a constant C independent of the discretization parameters, and $A \approx B$ stands for $A \lesssim B$ and $B \lesssim A$.

hereafter we define

$$\eta_{\mathcal{T}}(u_{\mathcal{T}},\mathcal{M})^2 := \sum_{T\in\mathcal{M}} \eta_{\mathcal{T}}(u_{\mathcal{T}},T)^2,$$

where $\eta_{\mathcal{T}}(u_{\mathcal{T}},T)$ is the PDE indicator associated with a generic element $T \in \mathcal{T}$. Note that $\theta = 1$ corresponds to uniform refinement. In Section 5 we present three AFEMs in increasing order of complexity regarding data $\mathcal{D} = (\boldsymbol{A}, c, f)$ and prove their linear convergence.

The simplest algorithm, so-called GALERKIN, works for discrete $\mathcal{D}$ and is the usual adaptive loop

$$\text{SOLVE} \longrightarrow \text{ESTIMATE} \longrightarrow \text{MARK} \longrightarrow \text{REFINE}\,.$$

We assume SOLVE computes the exact Galerkin solution $u_{\mathcal{T}}$, so we refrain from addressing linear algebra issues. The module ESTIMATE computes the *a posteriori* error indicator and the module MARK implements (1.7); in most of the paper we deal with weighted L^2-error indicators $\eta_{\mathcal{T}}(u_{\mathcal{T}})$ but we also address linear convergence for alternative estimators. The module REFINE bisects marked elements and perhaps a few more to keep meshes conforming (or Λ-admissible if they are non-conforming). We let $|||u - u_{\mathcal{T}}|||_\Omega$ denote the energy error associated with the bilinear form $\mathcal{B}$. This error is monotone with refinement but may stagnate. We thus exploit the estimator reduction property with refinement, typical of $\eta_{\mathcal{T}}(u_{\mathcal{T}})$, to show that the combined quantity

$$\zeta_{\mathcal{T}}(u_{\mathcal{T}})^2 := |||u - u_{\mathcal{T}}|||_\Omega^2 + \gamma\,\eta_{\mathcal{T}}(u_{\mathcal{T}})^2 \tag{1.8}$$

contracts in every iteration of GALERKIN for a suitable scaling parameter $\gamma > 0$. This readily leads to linear convergence of both $|u - u_{\mathcal{T}}|_{H_0^1(\Omega)}$ and $\eta_{\mathcal{T}}(u_{\mathcal{T}})$.

We next keep the coefficients $(\boldsymbol{A}, c)$ discrete but allow for a general $f \in H^{-1}(\Omega)$. This is to prevent the *multiplicative interaction* between $(\boldsymbol{A}, c)$ and u that occurs in (1.1) if we were to approximate $(\boldsymbol{A}, c)$. In contrast, the effect of f is *linear* in (1.1). We show examples where $|||u - u_{\mathcal{T}}|||_\Omega$ may stagnate because the adaptive process is dominated by oscillation $\operatorname{osc}_{\mathcal{T}}(f)_{-1}$ (pre-asymptotic regime). To compensate for this fact, we design a one-step AFEM with switch as in Kreuzer, Veeser and Zanotti (2024), the so-called AFEM-SW, that proceeds like GALERKIN provided $\eta_{\mathcal{T}}(u_{\mathcal{T}})$ dominates and otherwise reduces $\operatorname{osc}_{\mathcal{T}}(f)_{-1}$ separately. We show that for a suitable parameter $\gamma > 0$, the combined quantity

$$\zeta_{\mathcal{T}}(u_{\mathcal{T}})^2 := |||u - u_{\mathcal{T}}|||_\Omega^2 + \gamma\eta_{\mathcal{T}}(u_{\mathcal{T}})^2 + \operatorname{osc}_{\mathcal{T}}(f)_{-1}^2 \tag{1.9}$$

contracts in every loop of AFEM-SW. This yields linear convergence of the error $|u - u_{\mathcal{T}}|_{H1_0(\Omega)}$ and the estimator $\mathcal{E}_{\mathcal{T}}(u_{\mathcal{T}}, f) = \eta_{\mathcal{T}}(u_{\mathcal{T}}) + \operatorname{osc}_{\mathcal{T}}(f)_{-1}$.

The third algorithm is the two-step AFEM, the so-called AFEM-TS, which allows for general data $\mathcal{D} = (\boldsymbol{A}, c, f)$. To handle the aforementioned nonlinear effect of $(\boldsymbol{A}, c)$ and also deal with general f, all data $\mathcal{D}$ are first approximated by a routine DATA to a desired level of accuracy, which is adjusted at every step of AFEM-TS, and then fed to GALERKIN which handles discrete data. Suitably combining the accuracies of each intermediate module leads to linear convergence and optimal complexity of GALERKIN within each loop. The structure of AFEM-TS is flexible enough to easily handle discontinuous coefficients $(\boldsymbol{A}, c)$ with discontinuities that may not be aligned with the mesh. This is because the approximation of $(\boldsymbol{A}, c)$ by discontinuous polynomials takes place in $L^p(\Omega)$ for $p < \infty$.

It is worth stressing two important points. First, the approximation of $\mathcal{D}$ is carried out by a GREEDY algorithm, which is shown to perform optimally starting from any refinement of $\mathcal{T}_0$. Second, the discontinuous piecewise polynomial approximations $(\widehat{\boldsymbol{A}}, \widehat{c})$ of $(\boldsymbol{A}, c)$ may not respect, for polynomial degree ≥ 1, the positivity bounds associated with the coefficients. This requires a nonlinear correction of the output of GREEDY that restores positivity and does not reduce accuracy beyond a modest multiplicative constant. We postpone the discussion of these two delicate and technical processes to Section 7, which can be omitted in a first reading.

Rate-optimality of AFEMs. Showing that AFEMs outperform classical FEMs is a difficult but important matter. This reduces to proving a superior relation between the required degrees of freedom (or number of elements) for a desired accuracy; the former is in fact an acceptable measure of complexity. Showing that AFEMs deliver performance comparable with the best entails the following basic ingredients.

- *Nonlinear approximation classes.* These classify functions in terms of the best possible algebraic decay rate of approximation $e_N(v)_X$ of a given function v in a given norm X with N number of elements; roughly speaking, we say $v \in \mathbb{A}_s$ if $e_N(v)_X \lesssim N^{-s}$. These classes are related to regularity classes (Sobolev, Besov and Lipschitz) along Sobolev embedding lines.
- *Dörfler marking.* If the oscillation $\operatorname{osc}_{\mathcal{T}}(f)_{-1}$ is dominated by the PDE estimator $\eta_{\mathcal{T}}(u_{\mathcal{T}})$ for a given mesh $\mathcal{T}$, then any conforming refinement $\mathcal{T}_* \geq \mathcal{T}$ of $\mathcal{T}$ that reduces $\eta_{\mathcal{T}}(u_{\mathcal{T}})$ by a substantial amount induces a refined set $\mathcal{R} := \mathcal{T} \setminus \mathcal{T}_*$ to modify $\mathcal{T}$ into $\mathcal{T}_*$ satisfying (1.7), namely $\eta_{\mathcal{T}}(u_{\mathcal{T}}, \mathcal{R}) \geq \theta\, \eta_{\mathcal{T}}(u_{\mathcal{T}})$.
- *Minimality of $\mathcal{M}$.* If the subset $\mathcal{M} \subset \mathcal{T}$ in (1.7) is minimal, then the cardinality of $\mathcal{M}$ compares favourably to the cardinality of the best mesh with a comparable error accuracy, thereby leading to rate-optimality of AFEM.

Together, these comprise the topic of Section 6. It is important to notice that membership of $\mathbb{A}_s$ is never used explicitly by AFEM to learn about problem (1.1) and improve its resolution. The fundamental reason for the superior performance of AFEM relative to FEM lies in nonlinear approximation theory. We now illustrate this point with the following insightful approximation example for $d = 1$ and $X = L^\infty(0, 1)$ (DeVore 1998, Kahane 1961): let $\Omega = (0, 1)$, $\mathcal{T}_N = \{[x_{j-1}, x_j]\}_{j=1}^N$ be a partition of Ω, with

$$0 = x_0 < x_1 < \cdots < x_j < \cdots < x_N = 1,$$

and let $v\colon \Omega \to \mathbb{R}$ be an absolutely continuous function to be approximated by a piecewise constant function v_N over $\mathcal{T}_N$. To quantify the difference between v and v_N we resort to the *maximum norm*, and study two cases depending on the regularity of v. We define $v_N(x) := v(x_{j-1})$ for all $x_{j-1} \le x < x_j$ and note that

$$|v(x) - v_N(x)| = |v(x) - v(x_{j-1})| \le \int_{x_{j-1}}^{x} |v'(t)|\,\mathrm{d}t.$$

- *Case 1: W^1_∞-regularity.* If $u \in W^1_\infty(0,1)$ and $x_{j-1} \le x < x_j$, then

  $$|v(x) - v_N(x)| \le h_j \|v'\|_{L^\infty(x_{j-1},x_j)} \quad \Rightarrow \quad \|v - v_N\|_{L^\infty(\Omega)} \le \frac{1}{N}\|v'\|_{L^\infty(\Omega)}$$

  for a *uniform* mesh. We thus deduce a rate N^{-1} using the same integrability L^∞ on both sides of the error estimate.
- *Case 2: W^1_1-regularity.* Let $\|v'\|_{L^1(\Omega)} = 1$ and $\mathcal{T}_N$ be a *graded* partition so that

  $$\int_{x_{j-1}}^{x_j} |v'(t)|\,\mathrm{d}t = \frac{1}{N}.$$

  Then, for $x \in [x_{j-1}, x_j]$,

  $$|v(x) - v(x_{j-1})| \le \int_{x_{j-1}}^{x_j} |v'(t)|\,\mathrm{d}t = \frac{1}{N} \quad \Rightarrow \quad \|v - v_N\|_{L^\infty(\Omega)} \le \frac{1}{N}\|v'\|_{L^1(\Omega)}.$$

We thus conclude that we could achieve the same rate of convergence N^{-1} for rougher functions with just $\|u'\|_{L^1(\Omega)} < \infty$. Three comments are now in order. First, the contrast between Cases 1 and 2 is more dramatic for $v(x) = x^\alpha$ with $\alpha \in (0,1)$ because Case 1 only yields the suboptimal rate $\|v - v_N\|_{L^\infty(\Omega)} \le N^{-\alpha}$. Second, $\mathcal{T}_N$ in Case 2 *equidistributes* the max-error, a concept that will permeate our discussions later. Third, the optimal rate of Case 2 is due to the exchange of differentiability with integrability along the critical Sobolev embedding line between the left- and right-hand sides of the error estimate (nonlinear Sobolev scale), while Case 1 relies on the linear Sobolev scale with constant integrability.

We exploit and further elaborate these concepts in Section 6 to show rate-optimality of the three algorithms GALERKIN, AFEM-SW and AFEM-TS, discussed in Section 5, provided u and $\mathcal{D}$ belong to suitable approximation classes. We also investigate the relation between these approximation classes with regularity classes, allowing for discontinuous coefficients, and present a fairly complete discussion.

Mesh refinement. A key component of any adaptive algorithm, such as the three AFEMs already described, is the routine REFINE which refines a current mesh $\mathcal{T}$ into $\mathcal{T}_*$ to improve resolution. In Section 8 we study the *bisection method*, which is the most popular method to refine simplicial meshes in $\mathbb{R}^d$ for $d \ge 1$. For simplicity we focus our attention on this method, but most results apply to other refinement strategies such as quadtrees (for quadrilaterals), octrees (for hexagons) and red–green (for simplicial meshes). We do not insist on these extensions but refer to Bonito and Nochetto (2010) for details.

Given an initial grid $\mathcal{T}_0$ with a suitable labelling, the bisection method splits a given simplex into two children. The rules for successive cutting of simplices, for instance newest vertex bisection for $d = 2$, are such that the ensuing meshes are shape-regular (with a uniform constant depending only on $\mathcal{T}_0$ and d). However, bisection may not be completely local to keep conformity. The analysis of propagation of refinement is a delicate combinatorial problem. It is easy to see by example that bisecting one element of large generation (i.e. the number of bisections needed to produce it) may require a chain of elements with length similar to the generation. Therefore the number of refined elements in one step cannot be bounded by the number of marked elements. The following amazing estimate by Binev, Dahmen and DeVore (2004) for $d = 2$ and Stevenson (2008) for $d > 2$ shows that the cumulative effect of bisection, counting all the marked elements $\mathcal{M}_j$ from $\mathcal{T}_0$, is quasi-optimal: there exists a constant $D > 0$, depending on $\mathcal{T}_0$ and d, such that

$$\#\mathcal{T}_k - \#\mathcal{T}_0 \leq D \sum_{j=0}^{k-1} \#\mathcal{M}_j. \tag{1.10}$$

This estimate is crucial for the study of rate-optimality of AFEM and is proved in Section 8 for $d = 2$ and for both conforming refinement and Λ-admissible refinement. The latter is a systematic way to handle non-conforming meshes that goes back to Beirão da Veiga *et al.* (2023). It associates a computable global index to hanging nodes and imposes a restriction on them not to exceed a preassigned value $\Lambda \geq 0$; if $\Lambda = 0$ then the mesh is conforming. Section 8 also discusses several interesting geometric properties of Λ-admissible meshes which turn out to be crucial for discontinuous Galerkin methods. Since Section 8 is quite technical, it can be skipped in a first reading.

Discontinuous Galerkin methods. These methods, so-called dG, are the natural first step in investigating the role of non-conformity in adaptivity, namely that the discrete space of discontinuous piecewise polynomials $\mathbb{V}_\mathcal{T}$ is no longer a subspace of $H_0^1(\Omega)$. To this end, we study the symmetric interior penalty dG method in Section 9 on Λ-admissible partitions $\mathcal{T}$ of $\mathcal{T}_0$. Such dG methods exhibit some characteristic and novel features with respect to conforming FEMs: the most notable one is the presence of weighted jumps that stabilize the method and compensate for the lack of H^1-conformity. We consider the formulation with lifting, which allows for minimal regularity $u \in H_0^1(\Omega)$, and forcing $f \in H^{-1}(\Omega)$ despite the fact that $\mathbb{V}_\mathcal{T}$ is not a subspace of $H_0^1(\Omega)$. The latter is possible because, within the framework of AFEM-TS, f is approximated by a piecewise polynomial $P_\mathcal{T} f$ for which the pairing with functions in $\mathbb{V}_\mathcal{T}$ is meaningful.

The fact that jumps are not monotone upon refinement constitutes one of the main obstructions to studying adaptivity for dG. To circumvent this issue we follow Bonito and Nochetto (2010), who in turn modified the original approach of Karakashian and Pascal (2007), and introduce the largest conforming subspace $\mathbb{V}_{\mathcal{T}}^0$ of $\mathbb{V}_{\mathcal{T}}$. It turns out that, despite being coarser, $\mathbb{V}_{\mathcal{T}}^0$ exhibits a local resolution comparable with $\mathbb{V}_{\mathcal{T}}$ because of key geometric properties of Λ-admissible meshes that control the degree of non-conformity of $\mathcal{T}$. In addition, $\mathbb{V}_{\mathcal{T}}^0$ is responsible for the scaled jumps being bounded by the PDE estimator $\eta_{\mathcal{T}}(u_{\mathcal{T}})$. Exploiting properties of $\eta_{\mathcal{T}}(u_{\mathcal{T}})$, similar to the conforming case, leads to a quasi-orthogonality estimate for the dG norm, a dG variant of the Pythagoras equality. This is instrumental in proving a contraction property for the error plus scaled estimator and deducing convergence for both GALERKIN and AFEM-TS. Moreover, we derive rate-optimality for both algorithms provided u and $\mathcal{D}$ belong to suitable approximation classes. Such classes are the same as for conforming AFEMs: in fact we prove that the approximation classes for u using continuous and discontinuous piecewise polynomials on Λ-admissible meshes coincide.

Inf-sup stable AFEMs. The convergence and optimality theories developed in Sections 5 and 6 rely on the bilinear form in (1.2) being *coercive*. We remove this strong restriction in Section 10 and consider uniformly *inf-sup stable* FEMs on conforming refinements $\mathcal{T}_j$ of $\mathcal{T}_0$. The lack of an energy norm and its monotone behaviour upon refinement has been an obstacle to the study of this class of problems. We follow the recent work by Feischl (2022), who introduced the following form of *quasi-orthogonality* between consecutive Galerkin solutions $u_j \in \mathbb{V}_j$, originally proposed in Carstensen, Feischl, Page and Praetorius (2014) as part of an abstract set of axioms of adaptivity:

$$\sum_{k=j}^{j+N} \|u_{k+1} - u_k\|_{\mathbb{V}}^2 \le C(N)\|u - u_j\|_{\mathbb{V}}^2, \quad j \ge 0, \tag{1.11}$$

where $C(N)/N \to 0$ as $N \to \infty$. This is our departure point for developing a *variational approach* to prove linear convergence of u_j provided data $\mathcal{D}$ is discrete; the latter is reflected in an equivalence property between error and estimator (without oscillation). This is the context of a GALERKIN routine, which is next used as a building block together with DATA for an AFEM-TS that handles general data $\mathcal{D}$. Moreover, we prove rate-optimality for both algorithms, thereby extending Sections 5 and 6.

This discussion is fairly abstract. We specialize it to the Stokes equations for viscous incompressible fluids and mixed formulations of (1.1) using Raviart–Thomas–Nédélec and Brezzi–Douglas–Marini elements. We thereby obtain convergence and rate-optimality for AFEM-SW for the Stokes equations and AFEM-TS for mixed methods with variable and possibly discontinuous coefficients $(\boldsymbol{A}, c)$.

We conclude with a complete proof of (1.11) following Feischl (2022). This is a *tour de force* in applied linear algebra and is rather technical. It can be omitted in a first reading.

2. Linear elliptic boundary value problems: basic theory

In this section we examine the variational formulation of elliptic partial differential equations (PDEs). We start with a brief review of Sobolev spaces and their properties, and continue with two model boundary value problems that are instrumental to our subsequent analysis. We next present the so-called inf-sup theory, which characterizes the existence, uniqueness and stability of variational problems, and apply it to coercive and saddle point problems. These two classes will play essential roles later.

2.1. *Sobolev spaces: scaling and embedding*

Let $\Omega \subset \mathbb{R}^d$ with $d > 1$ be a Lipschitz and bounded domain, and let $k \in \mathbb{N}$, $1 \le p \le \infty$. The Sobolev space $W_p^k(\Omega)$ is defined by

$$W_p^k(\Omega) := \{v\colon \Omega \to \mathbb{R} |\ D^\alpha v \in L^p(\Omega) \text{ for all } |\alpha| \le k\},$$

and is a Banach space with the norm

$$\|v\|_{W_p^k(\Omega)} = \left(\sum_{j=1}^{k} |v|^p_{W_p^j(\Omega)}\right)^{1/p}, \quad |v|_{W_p^j(\Omega)} = \left(\int_\Omega |D^j v|^p\right)^{1/p}.$$

The space $W_p^k(\Omega; \mathbb{R}^m)$ is the space $W_p^k(\Omega)$ of vector- or matrix-valued functions. If $p = 2$ we write $H^k(\Omega) = W_2^k(\Omega)$ and note that this is a Hilbert space. We let $H_0^1(\Omega) \subset H^1(\Omega)$ denote the completion of $C_0^\infty(\Omega)$ within $H^1(\Omega)$.

Sobolev spaces $W_p^k(\Omega)$ of fractional order $k > 0$ can be defined as well, by applying the real interpolation method between $W_p^{[k]}(\Omega)$ and $W_p^{[k]+1}(\Omega)$; see Bergh and Löfström (1976) and Adams and Fournier (2003) for the details. The subsequent definitions and properties hold for Sobolev spaces of integer or fractional order.

The *Sobolev number* of $W_p^k(\Omega)$ is given by

$$\operatorname{sob}(W_p^k) := k - \frac{d}{p}. \tag{2.1}$$

This number governs the scaling properties of the seminorm $|v|_{W_p^k(\Omega)}$, because rescaling variables $\widehat{x} = x/h$ for all $x \in \Omega$, transforms Ω into $\widehat{\Omega}$ and v into $\widehat{v}$, while the corresponding norms scale as

$$|\widehat{v}|_{W_p^k(\widehat{\Omega})} = h^{\operatorname{sob}(W_p^k)} |v|_{W_p^k(\Omega)}.$$

2.2. *Properties of Sobolev spaces*

We now summarize, but do not prove, several important properties of Sobolev spaces which play a key role later. We refer to Evans (2010), Gilbarg and Trudinger (2001) and Grisvard (2011) for details. We use the notation $\hookrightarrow$ to denote a continuous embedding.

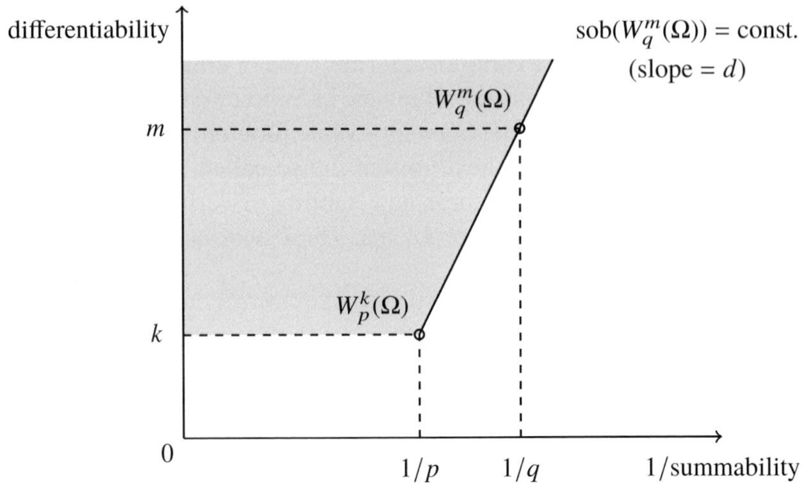

Figure 2.1. DeVore diagram (DeVore 1998). The space $W_p^k(\Omega)$ is represented by the point $(1/p, k)$ in the first quadrant. The line $\mathrm{sob}(W_q^m) = \text{const.} = \mathrm{sob}(W_p^k)$, with slope d, may be called the (critical) Sobolev embedding line for $W_p^k(\Omega)$. It represents all Sobolev spaces having the same Sobolev number as $W_p^k(\Omega)$. Sobolev spaces corresponding to points inside the grey region and on its boundary on the vertical axis are compactly embedded in $W_p^k(\Omega)$. Spaces on the oblique and horizontal lines emanating from $W_p^k(\Omega)$ are generally continuously but not compactly embedded in $W_p^k(\Omega)$ with exceptions such as $p = \infty$. Note that indices k and m may take non-integer values, corresponding to fractional Sobolev spaces.

Lemma 2.1 (Sobolev embedding). *Let $m > k \geq 0$ and assume $\mathrm{sob}(W_q^m) > \mathrm{sob}(W_p^k)$. Then the embedding $W_q^m(\Omega) \hookrightarrow W_p^k(\Omega)$ is compact and*

$$\|v\|_{W_p^k(\Omega)} \leq C\|v\|_{W_q^m(\Omega)} \quad \text{for all } v \in W_q^m(\Omega),$$

where $C = C(m, k, q, p, \Omega, d)$.

We say that two Sobolev spaces are in the same nonlinear Sobolev scale if they have the same Sobolev number; see Figure 2.1. We thus note that for compactness the space $W_q^m(\Omega)$ must be above the Sobolev scale of $W_p^k(\Omega)$, i.e. $\mathrm{sob}(W_q^m) > \mathrm{sob}(W_p^k)$.

The assumption on the Sobolev number cannot be relaxed. To see this, consider Ω to be the unit ball of $\mathbb{R}^d$ for $d \geq 2$ and set $v(x) = \log\log(|x|/2)$ for $x \in \Omega \setminus \{0\}$. Then we have $v \in W_d^1(\Omega) \setminus L^\infty(\Omega)$, but

$$\mathrm{sob}(W_d^1) = 1 - d/d = 0 = 0 - d/\infty = \mathrm{sob}(L^\infty).$$

Therefore equality cannot be expected in the embedding theorem. On the other hand, consider $d = 1$ and the spaces $W_1^1(\Omega)$ and $L^\infty(\Omega)$. We see that $\mathrm{sob}(W_1^1) =$

$\text{sob}(L^\infty) = 0$ but $W_1^1(\Omega)$ is compactly embedded in $L^\infty(\Omega)$ in this case. This shows that these two spaces are in the same nonlinear Sobolev scale and that the above inequality between Sobolev numbers for a compact embedding is only sufficient.

Moreover, if $0 < \alpha = \text{sob}(W_p^k) \le 1$, then functions of $W_p^k(\Omega)$ are Hölder-α and

$$|v|_{C^{0,\alpha}(\overline{\Omega})} \le C|v|_{W_p^k(\Omega)} \quad \text{for all } v \in W_p^k(\Omega).$$

This allows for the use of the standard Lagrange interpolation operator. We will exploit this fact later in Section 3.

Lemma 2.2 (first Poincaré inequality). *Let $\Omega \subset \mathbb{R}^d$ be a bounded and Lipschitz domain. Then there exists a constant $C_P = C_d|\Omega|^{1/d}$ such that*

$$\|v\|_{L^2(\Omega)} \le C_P\|\nabla v\|_{L^2(\Omega)} \quad \textit{for all } v \in H_0^1(\Omega). \tag{2.2}$$

The same inequality is valid in $W_p^1(\Omega)$ for any $1 \le p \le \infty$ provided v has vanishing trace (Gilbarg and Trudinger 2001, page 158).

Lemma 2.3 (second Poincaré inequality). *There exists C_P depending on Ω such that*

$$\|v - \overline{v}\|_{L^2(\Omega)} \le C_P\|\nabla v\|_{L^2(\Omega)} \quad \textit{for all } v \in H^1(\Omega), \tag{2.3}$$

where $\overline{v} := |\Omega|^{-1}\int_\Omega v$. The best constant within the class of convex domains is $C_P = \pi^{-1}\,\text{diam}(\Omega)$ (Payne and Weinberger 1960). The same inequality is valid in $W_p^1(\Omega)$ for $1 \le p \le \infty$ (Gilbarg and Trudinger 2001).

Lemma 2.4 (traces). *Let Ω be Lipschitz. There exists a unique linear operator $T\colon H^1(\Omega) \to L^2(\partial\Omega)$ such that*

$$\begin{aligned} \|Tv\|_{L^2(\partial\Omega)} &\le c(\Omega)\|v\|_{H^1(\Omega)} && \textit{for all } v \in H^1(\Omega), \\ Tv &= v_{|\partial\Omega} && \textit{for all } v \in C^0(\overline{\Omega}) \cap H^1(\Omega). \end{aligned}$$

The operator T is also well-defined on $W_p^1(\Omega)$ for $1 \le p \le \infty$ (Evans 2010, Grisvard 1985).

Since $Tv = v_{|\partial\Omega}$ for continuous functions, we write v for Tv. The image of T is a strict subspace of $L^2(\partial\Omega)$, the so-called $H^{1/2}(\partial\Omega)$. This is a Hilbert space if equipped with the norm $\|g\|_{H^{1/2}(\partial\Omega)} := \inf\{\|v\|_{H^1(\Omega)} \mid Tv = g\}$, and T is continuous with respect to this norm, since

$$\|Tv\|_{H^{1/2}(\partial\Omega)} \le \|v\|_{H^1(\Omega)} \quad \text{for all } v \in H^1(\Omega). \tag{2.4}$$

The definition of $H_0^1(\Omega)$ can be reconciled with that of traces because

$$H_0^1(\Omega) = \{v \in H^1(\Omega) \mid v = 0 \text{ on } \partial\Omega\}.$$

We point out that, in view of Lemma 2.2, the seminorm $|v|_{H_0^1(\Omega)} := \|\nabla v\|_{L^2(\Omega)}$ is a norm in $H_0^1(\Omega)$. We let $H^{-1}(\Omega)$ be the dual of $H_0^1(\Omega)$, with corresponding norm

$$\|f\|_{H^{-1}(\Omega)} := \sup_{v \in H_0^1(\Omega)} \frac{\langle f, v\rangle}{|v|_{H_0^1(\Omega)}}.$$

These definitions extend to any $p \in (1, \infty)$.

Lemma 2.5 (Gauss divergence theorem). *If* $\boldsymbol{n}$ *is the outer unit normal to* Ω, *then*

$$\int_\Omega \operatorname{div} \boldsymbol{w} = \int_{\partial\Omega} \boldsymbol{w} \cdot \boldsymbol{n} \quad \textit{for all } \boldsymbol{w} \in W_1^1(\Omega; \mathbb{R}^d).$$

Lemma 2.6 (Green's formula). *We have the integration by parts formula*

$$\int_\Omega \operatorname{div} \boldsymbol{w}\, v = -\int_\Omega \boldsymbol{w} \cdot \nabla v + \int_{\partial\Omega} v\, \boldsymbol{w} \cdot \boldsymbol{n} \quad \textit{for all } v \in H^1(\Omega),\ \boldsymbol{w} \in H^1(\Omega; \mathbb{R}^d).$$

2.3. Examples of boundary value problems

We consider two model elliptic problems in this paper. We start with the *scalar diffusion–reaction equation* with variable coefficients

$$\begin{aligned} L[u] := -\operatorname{div}(\boldsymbol{A}\nabla u) + cu &= f \quad \text{in } \Omega, \\ u &= 0 \quad \text{on } \partial\Omega, \end{aligned} \tag{2.5}$$

where $\Omega \subset \mathbb{R}^d$ is a bounded domain with Lipschitz boundary, $\boldsymbol{A} \in L^\infty(\Omega; \mathbb{R}^{d\times d})$ is a diffusion tensor, uniformly symmetric positive definite (SPD) over Ω, that is, there exist constants $0 < \alpha_1 \le \alpha_2$ such that

$$\alpha_1|\boldsymbol{\xi}|^2 \le \boldsymbol{\xi}^\top \boldsymbol{A}(x)\boldsymbol{\xi} \le \alpha_2|\boldsymbol{\xi}|^2 \quad \text{for all } x \in \Omega,\ \boldsymbol{\xi} \in \mathbb{R}^d, \tag{2.6}$$

$c \in L^\infty(\Omega)$, $c \ge 0$ is a reaction term, and $f \in L^2(\Omega)$ is a scalar load term.

To derive the variational formulation of (2.5) we let $\mathbb{V} = H_0^1(\Omega)$ and $\mathbb{V}^* = H^{-1}(\Omega)$. Since $H_0^1(\Omega)$ is the subspace of $H^1(\Omega)$ of functions with vanishing trace, asking for $u \in \mathbb{V}$ accounts for the homogeneous Dirichlet boundary values in (2.5). We next multiply (2.5) by a test function $v \in H_0^1(\Omega)$, integrate over Ω and use Lemma 2.6 (Green's formula), provided $\boldsymbol{w} = -\boldsymbol{A}\nabla u \in H^1(\Omega; \mathbb{R}^d)$, to obtain

$$u \in \mathbb{V}: \quad \mathcal{B}[u, v] = \langle f, v\rangle \quad \text{for all } v \in \mathbb{V}. \tag{2.7}$$

Here, $\mathcal{B}\colon \mathbb{V} \times \mathbb{V} \to \mathbb{R}$ stands for the bilinear form

$$\mathcal{B}[w, v] := \int_\Omega \nabla v \cdot \boldsymbol{A}\nabla w + cvw \quad \text{for all } v, w \in \mathbb{V} \tag{2.8}$$

and $\langle\cdot, \cdot\rangle$ stands for the $L^2(\Omega)$-scalar product and also for a duality pairing $H^{-1}(\Omega)$–$H_0^1(\Omega)$. Since $\boldsymbol{A}$ is assumed to be symmetric, the bilinear form $\mathcal{B}$ is also symmetric; however, $\boldsymbol{A}$ does not have to be symmetric in general. Note that the weak form

(2.7) allows for fluxes $\boldsymbol{w} \in L^2(\Omega; \mathbb{R}^d)$ and forcing $f \in H^{-1}(\Omega)$. We examine the existence, uniqueness and stability of (2.7) in Section 2.4 below.

We assume *homogeneous Dirichlet* boundary conditions in (2.5) for simplicity and because this will be our basic setting later. However, we could allow a *non-homogeneous Dirichlet* condition in the sense of traces, that is,

$$Tu = g \quad \text{on } \partial\Omega, \tag{2.9}$$

for any given function $g \in H^{1/2}(\partial\Omega)$. To write the companion variational formulation to (2.7), we first introduce the subspace $\mathbb{V}(g) \subset H^1(\Omega)$ of functions v with trace $Tv = g$ on $\partial\Omega$, and then rewrite (2.7) as follows:

$$u \in \mathbb{V}(g): \quad \mathcal{B}[u, v] = \langle f, v\rangle \quad \text{for all } v \in \mathbb{V}(0). \tag{2.10}$$

Moreover, we could consider a *Robin* boundary condition for given functions g and p on $\partial\Omega$:

$$\boldsymbol{A}\nabla u \cdot \boldsymbol{n} + pu = g \quad \text{on } \partial\Omega, \tag{2.11}$$

where $\boldsymbol{n}$ is the outer unit normal to Ω. To figure out the variational formulation, we now multiply the PDE in (2.5) by a test function $v \in H^1(\Omega)$ and integrate by parts to find the following variant of (2.7):

$$u \in H^1(\Omega): \quad \mathcal{B}[u, v] = \ell(v) \quad \text{for all } v \in H^1(\Omega), \tag{2.12}$$

where for all $v, w \in H^1(\Omega)$,

$$\mathcal{B}[w, v] := \int_\Omega \nabla v \cdot \boldsymbol{A}\nabla w + cvw + \int_{\partial\Omega} pvw, \quad \ell(v) := \langle f, v\rangle + \int_{\partial\Omega} gv. \tag{2.13}$$

We realize that (2.13) makes sense for $p \in L^\infty(\partial\Omega)$, $p \geq 0$ and $g \in H^{-1/2}(\partial\Omega)$, the dual space of $H^{1/2}(\partial\Omega)$, whence the last integral in (2.13) means a duality pairing. If $p = 0$, then (2.13) reduces to the weak form of the *Neumann* boundary condition.

The second model problem is the *Stokes system*, which is the simplest model of a stationary viscous incompressible fluid. Given an external force $\boldsymbol{f} \in L^2(\Omega; \mathbb{R}^d)$, let the velocity–pressure pair $(\boldsymbol{u}, p)$ satisfy the momentum and incompressibility equations with no-slip boundary condition:

$$\begin{aligned} -\Delta \boldsymbol{u} + \nabla p &= \boldsymbol{f} && \text{in } \Omega, \\ \operatorname{div} \boldsymbol{u} &= 0 && \text{in } \Omega, \\ \boldsymbol{u} &= 0 && \text{on } \partial\Omega. \end{aligned} \tag{2.14}$$

For the variational formulation we consider two Hilbert spaces $\mathbb{V} = H_0^1(\Omega; \mathbb{R}^d)$ and $\mathbb{Q} = L_0^2(\Omega)$, where $L_0^2(\Omega)$ is the space of L^2 functions with zero mean value. The space $H_0^1(\Omega; \mathbb{R}^d)$ takes care of the no-slip boundary values of the velocity. We first multiply the momentum equation in (2.14) by $\boldsymbol{v} \in \mathbb{V}$, assume $\boldsymbol{u} \in H^2(\Omega; \mathbb{R}^d)$ and use Lemma 2.6 (Green's formula) component-wise. We next multiply the incompressibility equation in (2.14) by $q \in \mathbb{Q}$ and integrate over Ω. We end up with the following variational formulation: find $(\boldsymbol{u}, p) \in \mathbb{V} \times \mathbb{Q}$ such that

$$\begin{aligned} a[\boldsymbol{u}, \boldsymbol{v}] + b[p, \boldsymbol{v}] &= \langle \boldsymbol{f}, \boldsymbol{v} \rangle \quad \text{for all } \boldsymbol{v} \in \mathbb{V}, \\ b[q, \boldsymbol{u}] &= 0 \qquad\quad\ \text{for all } q \in \mathbb{Q}. \end{aligned} \tag{2.15}$$

Here the bilinear forms $a\colon \mathbb{V} \times \mathbb{V} \to \mathbb{R}$ and $b\colon \mathbb{Q} \times \mathbb{V} \to \mathbb{R}$ read

$$a[\boldsymbol{w}, \boldsymbol{v}] := \int_\Omega \nabla \boldsymbol{v} : \nabla \boldsymbol{w} = \sum_{i=1}^d \int_\Omega \nabla v_i \cdot \nabla w_i \quad \text{for all } \boldsymbol{v}, \boldsymbol{w} \in \mathbb{V}$$

and

$$b[q, \boldsymbol{v}] := -\int_\Omega q \operatorname{div} \boldsymbol{v} \quad \text{for all } q \in \mathbb{Q},\ \boldsymbol{v} \in \mathbb{V}.$$

We observe that $a[\boldsymbol{w}, \boldsymbol{v}]$ does not require $\boldsymbol{w} \in H^2(\Omega; \mathbb{R}^d)$ and that (2.15) makes sense for $\boldsymbol{f} \in H^{-1}(\Omega; \mathbb{R}^d)$; note that the second equation in (2.15) is always satisfied for constant q due to Gauss's theorem, which explains the choice $q \in \mathbb{Q}$. Furthermore, (2.15) can be reformulated as (2.7), namely

$$(\boldsymbol{u}, p) \in \mathbb{V} \times \mathbb{Q}\colon \quad \mathcal{B}[(\boldsymbol{u}, p), (\boldsymbol{v}, q)] = \langle \boldsymbol{f}, \boldsymbol{v} \rangle \quad \text{for all } (\boldsymbol{v}, q) \in \mathbb{V} \times \mathbb{Q}, \tag{2.16}$$

with

$$\mathcal{B}[(\boldsymbol{u}, p), (\boldsymbol{v}, q)] := a[\boldsymbol{u}, \boldsymbol{v}] + b[p, \boldsymbol{v}] + b[q, \boldsymbol{u}].$$

We discuss the existence, uniqueness and stability of (2.16) in Section 2.4.

We could formulate the Stokes system with other boundary conditions. First, we may allow a non-homogeneous Dirichlet condition $\boldsymbol{u} = \boldsymbol{g}$ for a given function $\boldsymbol{g} \in H^{1/2}(\partial\Omega; \mathbb{R}^d)$ satisfying the compatibility condition $\int_\Omega \boldsymbol{g} \cdot \boldsymbol{n} = 0$ imposed by Gauss's theorem, and proceed as in the scalar case (2.10). Second, to deal with a Neumann boundary condition, we introduce the stress tensor $\boldsymbol{\sigma}(\boldsymbol{u}, p) := \frac{1}{2}(\nabla \boldsymbol{u} + \nabla \boldsymbol{u}^\top) - p\boldsymbol{I}$ and the symmetric part of the velocity gradient $\boldsymbol{\varepsilon}(\boldsymbol{u}) := \frac{1}{2}(\nabla \boldsymbol{u} + \nabla \boldsymbol{u}^\top)$. Then instead of (2.14) we could write the strong form of the Neumann problem as

$$\begin{aligned} -\operatorname{div} \boldsymbol{\sigma}(\boldsymbol{u}, p) = \boldsymbol{f}, \quad \operatorname{div} \boldsymbol{u} &= 0 \quad \text{in } \Omega, \\ \boldsymbol{\sigma}(\boldsymbol{u}, p)\boldsymbol{n} &= \boldsymbol{g} \quad \text{on } \partial\Omega, \end{aligned} \tag{2.17}$$

and its weak form as (2.15) but with

$$\mathbb{V} := \left\{ \boldsymbol{v} \in H^1(\Omega; \mathbb{R}^d) \mid \int_\Omega \boldsymbol{v} = \boldsymbol{0} \right\}, \quad \mathbb{Q} := L^2(\Omega), \tag{2.18}$$

as well as bilinear form a and right-hand side

$$a[\boldsymbol{u}, \boldsymbol{v}] := \int_\Omega \boldsymbol{\varepsilon}(\boldsymbol{u}) : \boldsymbol{\varepsilon}(\boldsymbol{v}), \quad \ell(\boldsymbol{v}) := \langle \boldsymbol{f}, \boldsymbol{v} \rangle + \int_{\partial\Omega} \boldsymbol{g} \cdot \boldsymbol{v} \tag{2.19}$$

for all $\boldsymbol{v} \in \mathbb{V}$. Again, the last integral in (2.19) is to be interpreted as a duality pairing for $\boldsymbol{g} \in H^{-1/2}(\partial\Omega; \mathbb{R}^d)$.

2.4. *Inf-sup theory*

We present a functional framework for the existence, uniqueness and stability of variational problems of the form (2.7) or (2.16). Throughout this section we let $(\mathbb{V}, \langle\cdot,\cdot\rangle_{\mathbb{V}})$ and $(\mathbb{W}, \langle\cdot,\cdot\rangle_{\mathbb{W}})$ be a pair of Hilbert spaces with induced norms $\|\cdot\|_{\mathbb{V}}$ and $\|\cdot\|_{\mathbb{W}}$. We let $\mathbb{V}^*$ and $\mathbb{W}^*$ denote their respective dual spaces equipped with norms

$$\|f\|_{\mathbb{V}^*} = \sup_{v\in\mathbb{V}} \frac{\langle f, v\rangle}{\|v\|_{\mathbb{V}}} \quad \text{and} \quad \|g\|_{\mathbb{W}^*} = \sup_{v\in\mathbb{W}} \frac{\langle g, v\rangle}{\|v\|_{\mathbb{W}}}.$$

We write $L(\mathbb{V}; \mathbb{W})$ for the space of all linear and continuous operators from $\mathbb{V}$ into $\mathbb{W}$ with operator norm

$$\|B\|_{L(\mathbb{V};\mathbb{W})} = \sup_{v\in\mathbb{V}} \frac{\|Bv\|_{\mathbb{W}}}{\|v\|_{\mathbb{V}}}.$$

The following result relates a continuous bilinear form $\mathcal{B}\colon \mathbb{V}\times\mathbb{W} \to \mathbb{R}$ with an operator $B \in L(\mathbb{V}; \mathbb{W})$ (Nečas 1962, Babuška 1971).

Theorem 2.7 (Banach–Nečas). *Let $\mathcal{B}\colon \mathbb{V}\times\mathbb{W} \to \mathbb{R}$ be a continuous bilinear form with norm*

$$\|\mathcal{B}\| := \sup_{v\in\mathbb{V}} \sup_{w\in\mathbb{W}} \frac{\mathcal{B}[v,w]}{\|v\|_{\mathbb{V}}\|w\|_{\mathbb{W}}}. \tag{2.20}$$

Then there exists a unique linear operator $B \in L(\mathbb{V}, \mathbb{W})$ such that

$$\langle Bv, w\rangle_{\mathbb{W}} = \mathcal{B}[v,w] \quad \textit{for all } v \in \mathbb{V},\ w \in \mathbb{W}$$

with operator norm

$$\|B\|_{L(\mathbb{V};\mathbb{W})} = \|\mathcal{B}\|.$$

Moreover, the bilinear form $\mathcal{B}$ satisfies the following two conditions:

$$\textit{there exists } \alpha > 0 \textit{ such that } \alpha\|v\|_{\mathbb{V}} \le \sup_{w\in\mathbb{W}} \frac{\mathcal{B}[v,w]}{\|w\|_{\mathbb{W}}} \textit{ for all } v \in \mathbb{V}; \tag{2.21a}$$

$$\textit{for every } 0 \neq w \in \mathbb{W} \textit{ there exists } v \in \mathbb{V} \textit{ such that } \mathcal{B}[v,w] \neq 0, \tag{2.21b}$$

if and only if $B\colon \mathbb{V} \to \mathbb{W}$ is an isomorphism with

$$\|B^{-1}\|_{L(\mathbb{W},\mathbb{V})} \le \alpha^{-1}. \tag{2.22}$$

We now consider the abstract variational problem

$$u \in \mathbb{V}\colon \quad \mathcal{B}[u,v] = \langle f, v\rangle \quad \text{for all } v \in \mathbb{W}. \tag{2.23}$$

The following result, due to Nečas (1962, Theorem 3.3), characterizes properties of the bilinear form $\mathcal{B}$ that imply that (2.23) is well-posed.

Theorem 2.8 (Nečas). *Let $\mathcal{B}\colon \mathbb{V}\times\mathbb{W} \to \mathbb{R}$ be a continuous bilinear form. Then the variational problem* (2.23) *admits a unique solution $u \in \mathbb{V}$ for all $f \in \mathbb{W}^*$, which depends continuously on f, if and only if the bilinear form $\mathcal{B}$ satisfies one of the following equivalent* inf-sup conditions.

(1) *There exists* $\alpha > 0$ *such that*

$$\sup_{w\in\mathbb{W}} \frac{\mathcal{B}[v,w]}{\|w\|_{\mathbb{W}}} \geq \alpha\|v\|_{\mathbb{V}} \quad \textit{for some } \alpha > 0; \tag{2.24a}$$

$$\textit{for every } 0 \neq w \in \mathbb{W} \textit{ there exists } v \in \mathbb{V} \textit{ such that } \mathcal{B}[v,w] \neq 0. \tag{2.24b}$$

(2) *We have*

$$\inf_{v\in\mathbb{V}}\sup_{w\in\mathbb{W}} \frac{\mathcal{B}[v,w]}{\|v\|_{\mathbb{V}}\|w\|_{\mathbb{W}}} > 0, \qquad \inf_{w\in\mathbb{W}}\sup_{v\in\mathbb{V}} \frac{\mathcal{B}[v,w]}{\|v\|_{\mathbb{V}}\|w\|_{\mathbb{W}}} > 0. \tag{2.25}$$

(3) *There exists* $\alpha > 0$ *such that*

$$\inf_{v\in\mathbb{V}}\sup_{w\in\mathbb{W}} \frac{\mathcal{B}[v,w]}{\|v\|_{\mathbb{V}}\|w\|_{\mathbb{W}}} = \inf_{w\in\mathbb{W}}\sup_{v\in\mathbb{V}} \frac{\mathcal{B}[v,w]}{\|v\|_{\mathbb{V}}\|w\|_{\mathbb{W}}} = \alpha. \tag{2.26}$$

In addition, the solution u *of* (2.23) *satisfies the stability estimate*

$$\|u\|_{\mathbb{V}} \leq \alpha^{-1}\|f\|_{\mathbb{W}^*}. \tag{2.27}$$

The equality in (2.26) might at first seem surprising but is just a consequence of $\|B^{-*}\|_{L(\mathbb{V};\mathbb{W})} = \|B^{-1}\|_{L(\mathbb{W};\mathbb{V})}$. In general, (2.24) is simpler to verify than (2.26) and α of (2.26) is the largest possible α in (2.24a). Since (2.22) shows that $\|B^{-1}\|_{L(\mathbb{W},\mathbb{V})}$ is the best inf-sup constant α in (2.21a), we readily obtain the following result.

Corollary 2.9 (well-posedness implies inf-sup). *If the variational problem* (2.23) *admits a unique solution* $u \in \mathbb{V}$ *for all* $f \in \mathbb{W}^*$, *so that*

$$\|u\|_{\mathbb{V}} \leq C\|f\|_{\mathbb{W}^*},$$

then $\mathcal{B}$ *satisfies the inf-sup condition* (2.26) *with* $\alpha \geq C^{-1}$.

We next apply these abstract results to two special but important cases. The first class comprises problems with coercive bilinear form and the second class comprises problems of saddle point type.

Coercive problems. An existence and uniqueness result for coercive bilinear forms was established by Lax and Milgram eight years prior to the result by Nečas (Lax and Milgram 1954). Coercivity of $\mathcal{B}$ is a sufficient condition for existence and uniqueness but it is not necessary. In this case, we assume $\mathbb{V} = \mathbb{W}$.

Corollary 2.10 (Lax–Milgram). *Let* $\mathcal{B}\colon \mathbb{V}\times\mathbb{V} \to \mathbb{R}$ *be a continuous bilinear form that is* coercive, *namely there exists* $\alpha > 0$ *such that*

$$\mathcal{B}[v,v] \geq \alpha\|v\|_{\mathbb{V}}^2 \quad \textit{for all } v \in \mathbb{V}. \tag{2.28}$$

Then (2.23) *has a unique solution that satisfies the stability estimate* (2.27).

If the bilinear form $\mathcal{B}$ is also symmetric, that is,

$$\mathcal{B}[v,w] = \mathcal{B}[w,v] \quad \text{for all } v,w \in \mathbb{V},$$

then $\mathcal{B}$ is a scalar product on $\mathbb{V}$. The norm induced by $\mathcal{B}$ is the so-called *energy norm*

$$|||v|||_{\Omega} := \mathcal{B}[v,v]^{1/2}.$$

For the reaction–diffusion equation (2.5), the bilinear form given in (2.8) satisfies

$$0 < \alpha_1 \leq \alpha \leq \|\mathcal{B}\| \leq \alpha_2 + \|c\|_{L^\infty(\Omega)} C_P^2, \tag{2.29}$$

where C_P is the constant in Lemma 2.2 (first Poincaré inequality). Coercivity and continuity of $\mathcal{B}$, with constants $c_{\mathcal{B}} = \alpha_1$ and $C_{\mathcal{B}} = \|\mathcal{B}\|$, in turn imply that the *energy norm* $|||\cdot|||_{\Omega}$ is equivalent to the natural norm $\|\cdot\|_{\mathbb{V}}$ in $\mathbb{V} = H_0^1(\Omega)$:

$$c_{\mathcal{B}}\|v\|_{\mathbb{V}}^2 \leq |||v|||_{\Omega}^2 \leq C_{\mathcal{B}}\|v\|_{\mathbb{V}}^2 \quad \text{for all } v \in \mathbb{V}. \tag{2.30}$$

Moreover, it is fairly easy to show that for symmetric and coercive $\mathcal{B}$ the solution u of (2.23) is the unique minimizer of the quadratic energy

$$J[v] := \frac{1}{2}\mathcal{B}[v,v] - \langle f, v\rangle \quad \text{for all } v \in \mathbb{V},$$

that is, $u = \arg\min_{v\in\mathbb{V}} J[v]$. In particular, the energy norm and the quadratic energy play a relevant role in Sections 5, 6 and 9.

This framework applies to the scalar diffusion–reaction equation (2.7) with *homogeneous Dirichlet* condition. Since the full H^1-norm $\|\cdot\|_{H^1(\Omega)}$ and the seminorm $|\cdot|_{H_0^1(\Omega)}$ are equivalent in the space $\mathbb{V} = H_0^1(\Omega)$, according to Lemma 2.2 (first Poincaré inequality), the bilinear form $\mathcal{B}$ in (2.8) is coercive and continuous in view of (2.6) and $c \geq 0$. This framework also applies to the *non-homogeneous Dirichlet* problem (2.10), upon extending g to a function $g \in H^1(\Omega)$, rewriting the problem in terms of $w = u - g \in H_0^1(\Omega)$ and forcing $\ell = f - L[g] \in H^{-1}(\Omega)$.

On the other hand, the bilinear form $\mathcal{B}$ in (2.13) associated with a *Robin* boundary condition is coercive provided $p \geq p_0$ on $\partial\Omega$ (or at least on an open subset of $\partial\Omega$) with some constant $p_0 > 0$. This is a consequence of the norm equivalence

$$\|v\|_{H^1(\Omega)}^2 \approx |v|_{H_0^1(\Omega)}^2 + \|v\|_{L^2(\partial\Omega)}^2 \quad \text{for all } v \in H^1(\Omega). \tag{2.31}$$

For the *Neumann* problem, instead, we have $p = 0$ in $\partial\Omega$, whence $\mathcal{B}$ is coercive whenever $c > 0$ in Ω (or at least in an open subset of Ω). If $c = 0$ in Ω, then $\mathcal{B}$ is only coercive in the subspace of $H^1(\Omega)$ of functions with vanishing mean value, according to Lemma 2.3 (second Poincaré inequality). Existence, uniqueness and stability is thus guaranteed by Corollary 2.10 (Lax–Milgram) provided the forcing in (2.13) satisfies the compatibility condition $\ell(1) = 0$.

Saddle point problems. We now consider an abstract problem a bit more general than (2.15), so the following results apply to the Stokes system (2.14).

Given a pair of Hilbert spaces $(\mathbb{V}, \mathbb{Q})$, we consider two continuous bilinear forms $a\colon \mathbb{V}\times\mathbb{V} \to \mathbb{R}$ and $b\colon \mathbb{Q}\times\mathbb{V} \to \mathbb{R}$. If $f \in \mathbb{V}^*$ and $g \in \mathbb{Q}^*$, then we seek a pair

$(u, p) \in \mathbb{V} \times \mathbb{Q}$ solving the *saddle point problem*

$$a[u, v] + b[p, v] = \langle f, v \rangle \quad \text{for all } v \in \mathbb{V}, \tag{2.32a}$$

$$b[q, u] = \langle g, q \rangle \quad \text{for all } q \in \mathbb{Q}. \tag{2.32b}$$

Problem (2.32) is variational and can be rewritten in the form (2.23)

$$(u, p) \in \mathbb{V} \times \mathbb{Q}: \quad \mathcal{B}[(u, p), (v, q)] = \langle f, v \rangle + \langle g, q \rangle \quad \text{for all } (v, q) \in \mathbb{V} \times \mathbb{Q}, \tag{2.33}$$

where $\mathcal{B}$ is the bilinear form

$$\mathcal{B}[(u, p), (v, q)] := a[u, v] + b[p, v] + b[q, u]. \tag{2.34}$$

Therefore the saddle point problem (2.32) is well-posed if and only if $\mathcal{B}$ satisfies the inf-sup condition (2.26). Since $\mathcal{B}$ is defined via the bilinear forms a and b, and (2.32) has a degenerate structure, it is not that simple to show (2.26) directly. However, the result is a consequence of the inf-sup theorem for saddle point problems given by Brezzi (1974).

Theorem 2.11 (Brezzi). *The saddle point problem* (2.32) *has a unique solution* $(u, p) \in \mathbb{V} \times \mathbb{Q}$ *for all data* $(f, g) \in \mathbb{V}^* \times \mathbb{Q}^*$, *that depends continuously on data, if and only if there exist constants* $\alpha, \beta > 0$ *such that*

$$\inf_{v \in \mathbb{V}_0} \sup_{w \in \mathbb{V}_0} \frac{a[v, w]}{\|v\|_{\mathbb{V}} \|w\|_{\mathbb{V}}} = \inf_{w \in \mathbb{V}_0} \sup_{v \in \mathbb{V}_0} \frac{a[v, w]}{\|v\|_{\mathbb{V}} \|w\|_{\mathbb{V}}} = \alpha > 0, \tag{2.35a}$$

$$\inf_{q \in \mathbb{Q}} \sup_{v \in \mathbb{V}} \frac{b[q, v]}{\|q\|_{\mathbb{Q}} \|v\|_{\mathbb{V}}} = \beta > 0, \tag{2.35b}$$

where

$$\mathbb{V}_0 := \{v \in \mathbb{V} \mid b[q, v] = 0 \textit{ for all } q \in \mathbb{Q}\}.$$

In addition, there exists $\gamma = \gamma(\alpha, \beta, \|a\|)$ *such that the solution* (u, p) *is stable, that is,*

$$\left(\|u\|_{\mathbb{V}}^2 + \|p\|_{\mathbb{Q}}^2\right)^{1/2} \le \gamma \left(\|f\|_{\mathbb{V}^*}^2 + \|g\|_{\mathbb{Q}^*}^2\right)^{1/2}. \tag{2.36}$$

Combining Theorem 2.11 (Brezzi) with Corollary 2.9 (well-posedness implies inf-sup), we infer the inf-sup condition for the bilinear form $\mathcal{B}$ in (2.34).

Corollary 2.12 (inf-sup of $\mathcal{B}$). *Let the bilinear form* $\mathcal{B}\colon \mathbb{W} \to \mathbb{W}$ *be defined by* (2.34). *Then*

$$\inf_{(v,q) \in \mathbb{W}} \sup_{(w,r) \in \mathbb{W}} \frac{\mathcal{B}[(v, q), (w, r)]}{\|(v, q)\|_{\mathbb{W}} \|(w, r)\|_{\mathbb{W}}} = \inf_{(w,r) \in \mathbb{W}} \sup_{(v,q) \in \mathbb{W}} \frac{\mathcal{B}[(v, q), (w, r)]}{\|(v, q)\|_{\mathbb{W}} \|(w, r)\|_{\mathbb{W}}} \ge \gamma^{-1},$$

where γ *is the stability constant from Theorem 2.11.*

Assume that $a\colon \mathbb{V} \times \mathbb{V} \to \mathbb{R}$ is symmetric and let (u, p) be the solution to (2.32). Then u is the unique minimizer of the energy $J[v] := \frac{1}{2} a[v, v] - \langle f, v \rangle$ under the constraint $b[\cdot, u] = g$ in $\mathbb{Q}^*$. In view of this, p is the corresponding Lagrange multiplier and the pair (u, p) is the unique saddle point of the Lagrangian

$$L[v, q] := J[v] + b[q, v] - \langle g, q\rangle \quad \text{for all } v \in \mathbb{V}, q \in \mathbb{Q}.$$

Stokes system. Theorem 2.11 (Brezzi) applies to the Stokes system (2.14) and (2.15) once we verify the inf-sup property (2.35b) for the bilinear form $b[q, \boldsymbol{v}] = -\int_\Omega q \operatorname{div} \boldsymbol{v}$. This turns out to be equivalent to the following problem: for any $q \in L^2_0(\Omega)$ there exists a $\boldsymbol{w} \in H^1_0(\Omega; \mathbb{R}^d)$ such that

$$-\operatorname{div} \boldsymbol{w} = q \quad \text{in } \Omega \quad \text{and} \quad |\boldsymbol{w}|_{H^1(\Omega;\mathbb{R}^d)} \leq C(\Omega)\|q\|_{L^2(\Omega)}. \tag{2.37}$$

This non-trivial result goes back to Nečas (Carroll *et al.* 1966) and a proof can be found in Galdi (1994, Theorem III.3.1), for example. This implies

$$\sup_{\boldsymbol{v}\in H^1_0(\Omega;\mathbb{R}^d)} \frac{b[q, \boldsymbol{v}]}{|\boldsymbol{v}|_{H^1(\Omega;\mathbb{R}^d)}} \geq \frac{b[q, \boldsymbol{w}]}{|\boldsymbol{w}|_{H^1(\Omega;\mathbb{R}^d)}} = \frac{\|q\|^2_{L^2(\Omega)}}{|\boldsymbol{w}|_{H^1(\Omega;\mathbb{R}^d)}} \geq C(\Omega)^{-1}\|q\|_{L^2(\Omega)}.$$

Therefore (2.35b) holds with $\beta \geq C(\Omega)^{-1}$.

The inf-sup condition is also satisfied for the spaces $\mathbb{V}$ and $\mathbb{Q}$ defined in (2.18), which are appropriate for the weak formulation (2.17) of the Neumann boundary value problem. Indeed, given any $q \in L^2(\Omega)$, we can split it as $q = (q - \hat{q}) + \hat{q}$ with $\hat{q} := |\Omega|^{-1} \int_\Omega q$. Let $\boldsymbol{w}_0 \in H^1_0(\Omega; \mathbb{R}^d)$ be the function defined as in (2.37) with q replaced by $q - \hat{q}$, and let $\hat{\boldsymbol{w}} = \frac{1}{2}(\hat{q}x, \hat{q}y)$. Then it is easily checked that the function $\boldsymbol{w} = \bar{\boldsymbol{w}} - |\Omega|^{-1} \int_\Omega \bar{\boldsymbol{w}}$ with $\bar{\boldsymbol{w}} = \boldsymbol{w}_0 + \hat{\boldsymbol{w}}$ belongs to $\mathbb{V}$ and satisfies

$$-\operatorname{div} \boldsymbol{w} = q \quad \text{in } \Omega \quad \text{and} \quad \|\boldsymbol{w}\|_{H^1(\Omega;\mathbb{R}^d)} \leq C\|q\|_{L^2(\Omega)}.$$

2.5. W^1_p-regularity of reaction–diffusion equation

It will be useful later in Lemma 5.20 to know whether L^∞-coefficients $(\boldsymbol{A}, c)$ allow for enhanced regularity beyond $H^1_0(\Omega)$ for solutions u of (2.7). We can reformulate this question as an extension of the Lax–Milgram theory, which states that the solution operator L^{-1} of (2.5) is an isomorphism between $H^{-1}(\Omega)$ and $H^1_0(\Omega)$; see Corollary 2.10 (Lax–Milgram).

This issue is well understood for the Laplace operator, i.e. $\boldsymbol{A} = \boldsymbol{I}$ and $c = 0$. It is known that for Lipschitz domains $\Omega \subset \mathbb{R}^d$, there exists $p_0 = p_0(\Omega) > 2$ such that

$$\|\nabla u\|_{L^p(\Omega)} \leq K\|f\|_{W^{-1}_p(\Omega)} \quad \text{for all } p \in [2, p_0], \tag{2.38}$$

where K depends on p (see e.g. Jerison and Kenig 1995); in particular, $p_0 > 4$ for $d = 2$ and $p_0 > 3$ for $d = 3$. Hereafter, $W^{-1}_p(\Omega)$ denotes the dual space of $\mathring{W}^1_q(\Omega)$, i.e. functions in $W^1_q(\Omega)$ with zero trace and $q = p/(p-1)$. For $\boldsymbol{A} \in L^\infty(\Omega, \mathbb{R}^{d\times d})$ and $c = 0$, estimate (2.38) was first derived by Meyers (1963) as a perturbation result for the Laplacian; see also Brenner and Scott (2008). We now present a simple proof for $p > 2$ following Bonito, DeVore and Nochetto (2013*b*). Let

$$\theta(p) := \frac{1/2 - 1/p}{1/2 - 1/p_0} \quad \text{for all } p \in [2, p_0], \tag{2.39}$$

and note that $\theta(p)$ increases strictly from 0 at $p = 2$ to 1 at $p = p_0$. Let K_0 be the constant K in (2.38) for $p = p_0$ and, for any $t \in (0, 1)$, define

$$p_*(t) := \max\bigl\{p \in [2, p_0] \mid K_0^{\theta(p)}(1 - t) \leq 1\bigr\}. \tag{2.40}$$

Lemma 2.13 (W^1_p-regularity). *Let $A \in L^\infty(\Omega, \mathbb{R}^{d\times d})$ satisfy (2.6) with $\alpha_1 \leq \alpha_2$, let $c \in L^\infty(\Omega)$ be non-negative, and let Ω be Lipschitz. If $f \in W^{-1}_p(\Omega)$ for some*

$$p \in \left[2, \min\left\{p_*\left(\frac{\alpha_1}{\alpha_2}\right), \frac{2d}{d-2}\right\}\right),$$

then the solution $u \in H^1_0(\Omega)$ of (2.7) satisfies

$$\|\nabla u\|_{L^p(\Omega)} \leq C(p)\bigl(1 + c_{\mathcal{B}}^{-1} C(\Omega)\|c\|_{L^\infty(\Omega)}\bigr)\|f\|_{W^{-1}_p(\Omega)} \tag{2.41}$$

with constants

$$C(p) = \frac{1}{\alpha_2} \frac{K_0^{\theta(p)}}{1 - K_0^{\theta(p)}\bigl(1 - \frac{\alpha_1}{\alpha_2}\bigr)}$$

and $C(\Omega) = C_s C_P$, where C_s is the constant in Lemma 2.1 (Sobolev embedding) and C_P is the constant in Lemma 2.2 (first Poincaré inequality).

Proof. We first consider the principal part of the operator L in (2.5), namely we take $c = 0$. In fact, let the operator $S\colon \mathring{W}^1_p(\Omega) \longrightarrow W^{-1}_p(\Omega)$ be defined by $Sv := -\operatorname{div}(\alpha_2{}^{-1} A\nabla v)$. In order to prove (2.41) for S, we resort to a perturbation argument for the Laplace operator $Tv := -\Delta v$.

The first task is to bound K in (2.38) in terms of K_0 and p. To this end, we recall that the operator $T\colon H^1_0(\Omega) \longrightarrow H^{-1}(\Omega)$ is an isomorphism with norm $\|T^{-1}\|_2 = 1$. Moreover, $T\colon \mathring{W}^1_p(\Omega) \longrightarrow W^{-1}_p(\Omega)$ is also an isomorphism with norm $\|T^{-1}\|_{p_0} = K_0$ according to (2.38) for $p = p_0$, provided we adopt the norm $\|\nabla \cdot\|_{L^p(\Omega)}$ in $\mathring{W}^1_p(\Omega)$. By the real method of interpolation, we know that $\mathring{W}^1_p(\Omega) = \bigl[H^1_0(\Omega), \mathring{W}^1_{p_0}(\Omega)\bigr]_{\theta(p),p}$ is the interpolation space between $H^1_0(\Omega)$ and $\mathring{W}^1_{p_0}(\Omega)$ with parameter $\theta(p)$ given by (2.39). Hence operator interpolation theory implies that $T\colon \mathring{W}^1_p(\Omega) \longrightarrow W^{-1}_p(\Omega)$ is an isomorphism with

$$K = \|T^{-1}\|_p = K_0^{\theta(p)}.$$

We regard S as a perturbation of T, define the operator $Q := T - S\colon \mathring{W}^1_p(\Omega) \longrightarrow W^{-1}_p(\Omega)$, and observe that $\|Q\|_p \leq 1 - \alpha_1/\alpha_2$ because

$$\langle Qv, w\rangle = \int_\Omega \left(I - \frac{1}{\alpha_2} A\right)\nabla v \nabla w \leq \left(1 - \frac{\alpha_1}{\alpha_2}\right)\|\nabla v\|_{L^p(\Omega)}\|\nabla w\|_{L^q(\Omega)}, \quad w \in \mathring{W}^1_q(\Omega).$$

Therefore the operator $T^{-1}Q\colon \mathring{W}^1_p(\Omega) \to \mathring{W}^1_p(\Omega)$ satisfies

$$\|T^{-1}Q\|_p \le \|T^{-1}\|_p\|Q\|_p \le K_0^{\theta(p)}\left(1 - \frac{\alpha_1}{\alpha_2}\right)$$

as well as $\|T^{-1}Q\|_p < 1$ for any $p \in \big[2, p_*(\alpha_1/\alpha_2)\big)$ in view of definition (2.40) of $p_*(t)$. We conclude by the Neumann theorem that the operator $S = T(I - T^{-1}Q)\colon \mathring{W}^1_p(\Omega) \to W^{-1}_p(\Omega)$ is invertible and its norm is bounded by

$$\|S^{-1}\|_p \le \|T^{-1}\|_p\|I - T^{-1}Q\|_p \le \frac{\|T^{-1}\|_p}{1 - \|T^{-1}Q\|_p} \le \frac{K_0^{\theta(p)}}{1 - K_0^{\theta(p)}(1 - \alpha_1/\alpha_2)}.$$

This yields the asserted estimate for $S = -\operatorname{div}({\alpha_2}^{-1}\boldsymbol{A}\nabla v)$.

Finally, we consider the operator L in (2.5) with $c \ne 0$. If $u \in H^1_0(\Omega)$ is the solution of (2.7) given by Corollary 2.10 (Lax–Milgram), rewrite (2.7) as

$$Su = -\operatorname{div}\left(\frac{1}{\alpha_2}\boldsymbol{A}\nabla u\right) = \frac{1}{\alpha_2}(f - cu) = \frac{1}{\alpha_2}g,$$

and apply the preceding estimate for S to infer that

$$\|S^{-1}g\|_{\mathring{W}^1_p(\Omega)} = \|\nabla u\|_{L^p(\Omega)} \le C(p)\|g\|_{W^{-1}_p(\Omega)} \le C(p)\big(\|f\|_{W^{-1}_p(\Omega)} + \|cu\|_{W^{-1}_p(\Omega)}\big).$$

It remains to estimate the last term on the right-hand side. Using the Cauchy Schwarz inequality in conjunction with Lemma 2.2 (first Poincaré inequality), i.e. $\|u\|_{L^2(\Omega)} \le C_P\|\nabla u\|_{L^2(\Omega)}$, and Lemma 2.1 (Sobolev embedding), i.e. $\|w\|_{L^2(\Omega)} \le C\|w\|_{\mathring{W}^1_q(\Omega)} \le C_s\|\nabla w\|_{L^q(\Omega)}$ for $q = p/(p-1) \ge 2d/(d+2)$, we see that

$$\begin{aligned}\langle cu, w\rangle &\le C_P C_s\|c\|_{L^\infty(\Omega)}\|\nabla u\|_{L^2(\Omega)}\|\nabla w\|_{L^q(\Omega)}\\ &\le \frac{C_P C_s}{C_B}\|c\|_{L^\infty(\Omega)}\|f\|_{H^{-1}(\Omega)}\|\nabla w\|_{L^q(\Omega)}\end{aligned}$$

because of the stability estimate (2.27) with constant $\alpha = c_{\mathcal{B}}$. Since $\|f\|_{H^{-1}(\Omega)} \le |\Omega|^{(p-2)/(2p)}\|f\|_{W^{-1}_p(\Omega)}$, the asserted estimate (2.41) for $c \ne 0$ follows immediately. □

3. *A priori* approximation theory

We devote this section to discussing basic concepts about piecewise polynomial approximation in Sobolev spaces over graded meshes in any dimension d. We start by introducing the Petrov–Galerkin method in an abstract setting (Section 3.1); this motivates our interest in approximation results in Sobolev spaces. Hence we briefly discuss the construction of finite element spaces in Section 3.2, along with polynomial interpolation of functions in Sobolev spaces in Section 3.3. This provides local estimates suitable for the comparison of quasi-uniform and graded meshes for $d > 1$. We exploit them in developing the so-called error equidistribution principle in Section 3.4 and the construction of suitably graded meshes via a greedy algorithm in Section 3.6. We conclude that graded meshes can deliver optimal interpolation rates for certain classes of singular functions, and thus supersede quasi-uniform refinement.

In the second part of the section, we explore the geometric aspects of mesh refinement for conforming meshes in Section 3.5 and non-conforming meshes in Section 3.7, but postpone a full and rather technical discussion to Section 8. We include a statement about complexity of the refinement procedure, which turns out to be instrumental later and will be proved in Section 8.

3.1. The Galerkin method: best approximation

In order to make the variational problem (2.23) computable, we let $\mathbb{V}_N \subset \mathbb{V}$ and $\mathbb{W}_N \subset \mathbb{W}$ be subspaces with the same dimension $N < \infty$ and consider the Petrov–Galerkin approximation

$$u_N \in \mathbb{V}_N: \quad \mathcal{B}[u_N, w] = \langle f, w \rangle \quad \text{for all } w \in \mathbb{W}_N. \tag{3.1}$$

If $\mathbb{V}_N = \mathbb{W}_N$, this is called Galerkin approximation. Since (3.1) is a square algebraic system, the existence and uniqueness of $u_N \in \mathbb{V}_N$ are equivalent to the kernel of the corresponding linear discrete operator being trivial. This leads to the following equivalent conditions for unique solvability (Nečas 1962, Babuška 1971); see also Nochetto *et al.* (2009, Proposition 1).

Lemma 3.1 (discrete inf-sup condition). *The following statements are equivalent.*

(1) *For every* $0 \neq v \in \mathbb{V}_N$ *there exists* $w \in \mathbb{W}_N$ *such that* $\mathcal{B}[v, w] \neq 0$.

(2) *For every* $0 \neq w \in \mathbb{W}_N$ *there exists* $v \in \mathbb{V}_N$ *such that* $\mathcal{B}[v, w] \neq 0$.

(3) *The following discrete inf-sup condition holds with a constant* $\beta_N > 0$:

$$\inf_{v \in \mathbb{V}_N} \sup_{w \in \mathbb{W}_N} \frac{\mathcal{B}[v, w]}{\|v\|_{\mathbb{V}} \|w\|_{\mathbb{W}}} = \inf_{w \in \mathbb{W}_N} \sup_{v \in \mathbb{V}_N} \frac{\mathcal{B}[v, w]}{\|v\|_{\mathbb{V}} \|w\|_{\mathbb{W}}} = \beta_N. \tag{3.2}$$

(4) $\displaystyle \inf_{v \in \mathbb{V}_N} \sup_{w \in \mathbb{W}_N} \frac{\mathcal{B}[v, w]}{\|v\|_{\mathbb{V}} \|w\|_{\mathbb{W}}} > 0.$

(5) $\displaystyle \inf_{w \in \mathbb{W}_N} \sup_{v \in \mathbb{V}_N} \frac{\mathcal{B}[v, w]}{\|v\|_{\mathbb{V}} \|w\|_{\mathbb{W}}} > 0.$

This is a discrete version of Theorem 2.8 (Nečas) and leads to the stability bound

$$\|u_N\|_{\mathbb{V}} \leq \frac{1}{\beta_N} \|f\|_{\mathbb{W}^*}. \tag{3.3}$$

Therefore β_N^{-1} acts as a stability constant for (3.1), and it is thus desirable for it to be uniformly bounded below away from zero. This is always the case for *coercive* problems because (2.28) is inherited within the subspaces $\mathbb{V}_N = \mathbb{W}_N \subset \mathbb{V}$, whence $\beta_N \geq \alpha > 0$. In contrast, a uniform lower bound for *saddle point* problems,

$$\beta_N \geq \beta > 0, \tag{3.4}$$

requires compatibility between the subspaces $\mathbb{V}_N$ and $\mathbb{W}_N$ (Boffi, Brezzi and Fortin 2013).

If we now subtract (3.1) from (2.23), we obtain *Galerkin orthogonality*:

$$\mathcal{B}[u - u_N, w] = 0 \quad \text{for all } w \in \mathbb{W}_N. \tag{3.5}$$

This relation is instrumental in deriving the following best approximation property as well as developing *a posteriori* error estimates in Section 4.

Proposition 3.2 (quasi-best-approximation property). *Let* $\mathcal{B}\colon \mathbb{V} \times \mathbb{W} \to \mathbb{R}$ *be continuous and satisfy* (3.2). *Then the error* $u - u_N$ *satisfies the bound*

$$\|u - u_N\|_{\mathbb{V}} \leq \frac{\|\mathcal{B}\|}{\beta_N} \min_{v \in \mathbb{V}_N} \|u - v\|_{\mathbb{V}}. \tag{3.6}$$

Proof. We give a simplified proof, which follows Babuška (1971) and Babuška and Aziz (1972), and yields the constant $1 + \|\mathcal{B}\|/\beta_N$. The asserted constant is due to Xu and Zikatanov (2003).

Combining (3.2), (3.5) and the continuity of $\mathcal{B}$, we derive for all $v \in \mathbb{V}_N$

$$\beta_N \|u_N - v\|_{\mathbb{V}} \leq \sup_{w \in \mathbb{W}_N} \frac{\mathcal{B}[u_N - v, w]}{\|w\|_{\mathbb{W}}} = \sup_{W \in \mathbb{W}_N} \frac{\mathcal{B}[u - v, w]}{\|w\|_{\mathbb{W}}} \leq \|\mathcal{B}\| \|u - v\|_{\mathbb{V}},$$

whence

$$\|u_N - v\|_{\mathbb{V}} \leq \frac{\|\mathcal{B}\|}{\beta_N} \|u - v\|_{\mathbb{V}}.$$

Using the triangle inequality yields

$$\|u - u_N\|_{\mathbb{V}} \leq \|u - v\|_{\mathbb{V}} + \|v - u_N\|_{\mathbb{V}} \leq \left(1 + \frac{\|\mathcal{B}\|}{\beta_N}\right) \|u - v\|_{\mathbb{V}}$$

for all $v \in \mathbb{V}_N$. It just remains to minimize in $\mathbb{V}_N$. □

Corollary 3.3 (quasi-monotonicity). *Let* $\mathcal{B}\colon \mathbb{V} \times \mathbb{W} \to \mathbb{R}$ *be continuous and satisfy* (3.2). *If* $\mathbb{V}_M$ *is a subspace of* $\mathbb{V}_N$, *then for all* $v \in \mathbb{V}_M$

$$\|u - u_N\|_{\mathbb{V}} \leq \frac{\|\mathcal{B}\|}{\beta_N} \|u - v\|_{\mathbb{V}}. \tag{3.7}$$

Moreover, if $\mathbb{V} = \mathbb{W}$ *and* $\mathcal{B}$ *is symmetric and coercive with constants* $c_{\mathcal{B}} \leq C_{\mathcal{B}}$, *then for all* $v \in \mathbb{V}_M$

$$|||u - u_N|||_{\Omega} \leq |||u - v|||_{\Omega}, \quad \|u - u_N\|_{\mathbb{V}} \leq C_{\text{Céa}} \|u - v\|_{\mathbb{V}}, \tag{3.8}$$

where $C_{\text{Céa}} := \sqrt{C_{\mathcal{B}}/c_{\mathcal{B}}}$.

Proof. Inequality (3.7) is a consequence of the previous bound (3.6) upon taking $v \in \mathbb{V}_M$ instead of $\mathbb{V}_N$. To show the left inequality in (3.8), we combine (2.28) and (3.5):

$$|||u - u_N|||_\Omega^2 = \mathcal{B}[u - u_N, u - v] \leq |||u - u_N|||_\Omega |||u - v|||_\Omega \quad \text{for all } v \in \mathbb{V}_M.$$

This together with the norm equivalence (2.30) gives the remaining inequality. □

The significance of (3.6) is that we need to construct discrete spaces $\mathbb{V}_N$ with good approximation properties. Next we introduce piecewise polynomial approximation, which gives rise to the finite element method.

3.2. *Finite element spaces*

In this section we focus on the construction of the discrete spaces $\mathbb{V}_N$ and $\mathbb{W}_N$. We consider the bilinear forms $\mathcal{B}$ introduced in Section 2.3 with emphasis on the diffusion–reaction case (2.8). We assume that Ω is a bounded polyhedral domain $\Omega \subset \mathbb{R}^d$ and is partitioned into a conforming or non-conforming mesh $\mathcal{T}$ made of simplices T, which are assumed to be closed with non-overlapping interiors; thus

$$\overline{\Omega} = \bigcup \{T \mid T \in \mathcal{T}\}.$$

The reference element is denoted by

$$T_d := \left\{ x = (x_1, \ldots, x_d) \in \mathbb{R}^d \mid 0 \leq x_i \leq 1,\ i = 1, \ldots, d,\ \sum_{i=1}^d x_i \leq 1 \right\}.$$

We will discuss the construction of conforming meshes in Section 3.5 by the bisection method and that of non-conforming meshes (constrained to the fulfilment of an *admissibility condition*) in Section 3.7, both for $d = 2$. We will embark on a thorough discussion in Section 8. We assume for the moment that $\mathcal{T}$ is an element of a (possibly infinite) class $\mathbb{T}$ of conforming *shape-regular* meshes. To define this geometric concept, we let $\overline{h}_T$ denote the diameter of $T \in \mathcal{T}$, let $\underline{h}_T$ denote the diameter of the largest ball contained in T, and impose the restriction

$$\sigma := \sup_{\mathcal{T} \in \mathbb{T}} \sup_{T \in \mathcal{T}} \frac{\overline{h}_T}{\underline{h}_T} < \infty. \tag{3.9}$$

The constant σ is referred to as the shape regularity constant of $\mathbb{T}$.

Given a shape-regular mesh $\mathcal{T} \in \mathbb{T}$, we define the finite element space of *discontinuous* piecewise polynomials of total degree up to $n \geq 1$,

$$\mathbb{S}_\mathcal{T}^{n,-1} := \{v \in L^2(\Omega) \mid v|_T \in \mathbb{P}_n(T) \text{ for all } T \in \mathcal{T}\},$$

and its globally *continuous* counterpart

$$\mathbb{S}_\mathcal{T}^{n,0} := \mathbb{S}_\mathcal{T}^{n,-1} \cap C^0(\overline{\Omega}).$$

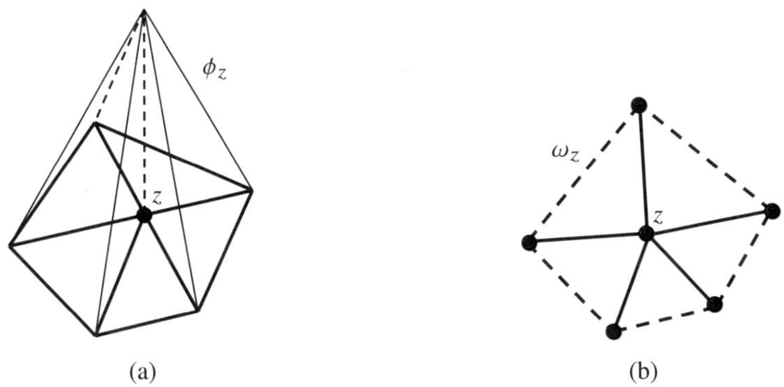

Figure 3.1. (a) Piecewise linear basis function ϕ_z corresponding to interior node z. (b) Support ω_z of ϕ_z and skeleton γ_z (solid line)

Note that $\mathbb{S}^{n,0}_{\mathcal{T}} \subset H^1(\Omega)$, which makes it adequate for (2.7)–(2.8). We refer to Braess (2007), Brenner and Scott (2008), Ciarlet (2002) and Siebert (2012) for a discussion of the local construction of this space (i.e. Lagrange elements of degree $n \geq 1$) along with its properties. We let

$$\mathbb{V}_{\mathcal{T}} := \mathbb{S}^{n,0}_{\mathcal{T}} \cap H^1_0(\Omega) \tag{3.10}$$

denote the subspace of finite element functions which vanish on $\partial\Omega$. Note that we do not explicitly refer to the polynomial degree, which will be clear in each context.

We focus on the conforming piecewise linear case $n = 1$ (Courant elements), but most results extend to non-conforming meshes or $n > 1$. In this vein, global continuity can be simply enforced by imposing continuity at the set $\mathcal{V}$ of vertices z of $\mathcal{T}$, the so-called *nodal values*. However, the following local construction leads to global continuity. If T is a generic simplex of $\mathcal{T}$, namely the convex hull of $\{z_i\}_{i=0}^d$, then we associate to each vertex z_i a *barycentric coordinate* λ_i^T, which is the linear function in T with nodal value 1 at z_i and 0 at the other vertices of T. Upon pasting together the barycentric coordinates λ_z^T of all simplices T containing the vertex $z \in \mathcal{V}$, we obtain a continuous piecewise linear function $\phi_z \in \mathbb{S}^{1,0}_{\mathcal{T}}$, as depicted in Figure 3.1 for $d = 2$.

The set $\{\phi_z\}_{z\in\mathcal{V}}$ of all such functions is the nodal basis of $\mathbb{S}^{1,0}_{\mathcal{T}}$, or Courant basis. We let $\omega_z := \operatorname{supp}(\phi_z)$ denote the support of ϕ_z, from now on called the *star* associated to z, and let γ_z be the interior skeleton of ω_z, namely the union of all the faces containing z.

In view of the definition of ϕ_z, we have the following unique representation of any function $v \in \mathbb{S}^{1,0}(\mathcal{T})$:

$$v(x) = \sum_{z\in\mathcal{V}} v(z)\phi_z(x).$$

The functions ϕ_z are non-negative and satisfy the *partition of unity* property

$$\sum_{z\in\mathcal{V}} \phi_z(x) = 1 \quad \text{for all } x \in \Omega. \tag{3.11}$$

If we further impose $v(z) = 0$ for all $z \in \partial\Omega \cap \mathcal{V}$, then $v \in H_0^1(\Omega)$.

For each simplex $T \in \mathcal{T}$, generated by vertices $\{z_i\}_{i=0}^d$, the *dual functions* $\{\lambda_i^*\}_{i=0}^d \subset \mathbb{P}_1(T)$ to the barycentric coordinates $\{\lambda_i\}_{i=0}^d$ satisfy the bi-orthogonality relation $\int_T \lambda_i^* \lambda_j = \delta_{ij}$, and are given by

$$\lambda_i^* = \frac{(1+d)^2}{|T|}\lambda_i - \frac{1+d}{|T|}\sum_{j\neq i}\lambda_j \quad \text{for all } 0 \le i \le d.$$

The *Courant dual basis* $\phi_z^* \in \mathbb{S}^{1,-1}(\mathcal{T})$ is formed by the functions over $\mathcal{T}$ given by

$$\phi_z^* = \frac{1}{V_z}\sum_{T\ni z}(\lambda_z^T)^*\chi_T \quad \text{for all } z \in \mathcal{V},$$

where $V_z \in \mathbb{N}$ is the valence of z (number of elements of $\mathcal{T}$ containing z) and χ_T is the characteristic function of T. These functions have the same support ω_z as the nodal basis ϕ_z and satisfy the global bi-orthogonality relation

$$\int_\Omega \phi_z^*\phi_y = \delta_{zy} \quad \text{for all } z, y \in \mathcal{V}.$$

Finally, we let $\mathcal{N}$ denote the Lagrange nodes of order n of a mesh $\mathcal{T}$, and let $\psi_z \in \mathbb{S}_\mathcal{T}^{n,0}$ be the corresponding Lagrange basis of $\mathbb{S}_\mathcal{T}^{n,0}$; hence $\mathbb{S}_\mathcal{T}^{n,0} = \operatorname{span}\{\psi_z\}_{z\in\mathcal{N}}$.

3.3. *Polynomial interpolation in Sobolev spaces*

We wish to use the space $\mathbb{V}_\mathcal{T}$ defined in (3.10) as the discrete space $\mathbb{V}_N$ in the Galerkin method (3.1). If the bilinear form $\mathcal{B}$ satisfies an inf-sup condition with constant $\beta_\mathcal{T} > 0$, we find a discrete solution $u_\mathcal{T} \in \mathbb{V}_\mathcal{T}$ which satisfies the error bound (3.6), that is,

$$\|u - u_\mathcal{T}\|_\mathbb{V} \le \frac{\|\mathcal{B}\|}{\beta_\mathcal{T}} \min_{v\in\mathbb{V}_\mathcal{T}} \|u - v\|_\mathbb{V}.$$

In turn, the minimum on the right-hand side can be bounded from above by the quantity $\|u - v\|_\mathbb{V}$ for any chosen $v \in \mathbb{V}_\mathcal{T}$. This motivates the search for quasi-best approximations of u in the norm of $\mathbb{V}$. One classical tool to generate approximations to any given function is *interpolation*. Interpolation in $\mathbb{V}_\mathcal{T}$ is discussed next.

If $v \in C^0(\overline{\Omega})$, we define the *Lagrange interpolant* $I_\mathcal{T} v$ of v as follows,

$$I_\mathcal{T} v(x) = \sum_{z\in\mathcal{N}} v(z)\psi_z(x),$$

and note that $I_{\mathcal{T}}v = v$ for all $v \in \mathbb{S}^{n,0}_{\mathcal{T}}$ (i.e. $I_{\mathcal{T}}$ is invariant in $\mathbb{S}^{n,0}_{\mathcal{T}}$). For functions without point values, such as functions in $H^1(\Omega)$ for $d > 1$, we need to determine nodal values by averaging. For any conforming mesh $\mathcal{T} \in \mathbb{T}$, the averaging process extends beyond nodes and so gives rise to the discrete neighbourhood

$$\omega_{\mathcal{T}}(T) := \bigcup_{\substack{T' \in \mathcal{T} \\ T' \cap T \neq \emptyset}} T'$$

for each element $T \in \mathcal{T}$, along with the *local quasi-uniformity* properties

$$\max_{T \in \mathcal{T}} \#\omega_{\mathcal{T}}(T) \leq C(\sigma), \qquad \max_{T' \subset \omega_{\mathcal{T}}(T)} \frac{|T|}{|T'|} \leq C(\sigma),$$

where σ is the shape regularity coefficient defined in (3.9). We will often write ω_T if there is no confusion about the underlying mesh $\mathcal{T}$. We shall also need a smaller subset, namely the set of elements sharing a face with a given element T:

$$\widetilde{\omega}_T := \widetilde{\omega}_{\mathcal{T}}(T) := \bigcup_{\substack{T' \in \mathcal{T} \\ T' \cap T \in \mathcal{F}}} T', \tag{3.12}$$

where $\mathcal{F}$ is the set of all $(d-1)$-dimensional faces of the mesh $\mathcal{T}$.

We now introduce one such operator $I_{\mathcal{T}}$ due to Scott and Zhang (Brenner and Scott 2008, Scott and Zhang 1990), from now on called a *quasi-interpolation operator*. We focus on polynomial degree $n = 1$, but the construction is valid for any n; see Brenner and Scott (2008) and Scott and Zhang (1990) for details. We recall that $\{\phi_z\}_{z \in \mathcal{V}}$ is the global Lagrange basis of $\mathbb{S}^{1,0}_{\mathcal{T}}$, $\{\phi^*_z\}_{z \in \mathcal{V}}$ is the global dual basis, and $\operatorname{supp} \phi^*_z = \operatorname{supp} \phi_z$ for all $z \in \mathcal{V}$. We thus define $I_{\mathcal{T}} \colon L^1(\Omega) \to \mathbb{S}^{1,0}_{\mathcal{T}}$ to be

$$I_{\mathcal{T}}v := \sum_{z \in \mathcal{V}} \langle v, \phi^*_z \rangle \phi_z, \tag{3.13}$$

If $0 \leq s \leq 2$ (integer) is a regularity index and $1 \leq p \leq \infty$ is an integrability index, then we would like to prove the *quasi-local error estimate*

$$|v - I_{\mathcal{T}}v|_{W^t_q(T)} \lesssim h_T^{\operatorname{sob}(W^s_p) - \operatorname{sob}(W^t_q)} |v|_{W^s_p(\omega_T)} \tag{3.14}$$

for all $T \in \mathcal{T}$, provided $0 \leq t \leq s$, $1 \leq q \leq \infty$ are such that $\operatorname{sob}(W^s_p) > \operatorname{sob}(W^t_q)$.

We first recall that $I_{\mathcal{T}}$ is invariant in $\mathbb{S}^{1,0}_{\mathcal{T}}$, namely,

$$I_{\mathcal{T}}w = w \quad \text{for all } w \in \mathbb{S}^{1,0}_{\mathcal{T}}.$$

Since the averaging process giving rise to the values of $I_{\mathcal{T}}v$ for each element $T \in \mathcal{T}$ takes place in the neighbourhood ω_T, we also deduce the local invariance

$$I_{\mathcal{T}}w|_T = w \quad \text{for all } w \in \mathbb{P}_1(\omega_T),$$

as well as the local stability estimate for any $1 \leq q \leq \infty$,

$$\|I_{\mathcal{T}}v\|_{L^q(T)} \lesssim \|v\|_{L^q(\omega_T)}.$$

We may thus write

$$v - I_{\mathcal{T}}v|_T = (v - w) - I_{\mathcal{T}}(v - w)|_T \quad \text{for all } T \in \mathcal{T},$$

where $w \in \mathbb{P}_{s-1}$ is arbitrary ($w = 0$ if $s = 0$). It now suffices to prove (3.14) in the reference element $\widehat{T}$ and scale back and forth to T; the definition (2.1) of Sobolev number accounts precisely for this scaling. We keep the notation T for $\widehat{T}$, apply the inverse estimate for linear polynomials $|I_{\mathcal{T}}v|_{W^t_q(T)} \lesssim \|I_{\mathcal{T}}v\|_{L^q(T)}$ to $v - w$ instead of v, and use the above local stability estimate, to infer that

$$|v - I_{\mathcal{T}}v|_{W^t_q(T)} \lesssim \|v - w\|_{W^t_q(\omega_T)} \lesssim \|v - w\|_{W^s_p(\omega_T)}.$$

The last inequality is a consequence of the inclusion $W^s_p(\omega_T) \subset W^t_q(\omega_T)$ because $\mathrm{sob}(W^s_p) > \mathrm{sob}(W^t_q)$ and $t \leq s$. Estimate (3.14) now follows from the Bramble–Hilbert lemma (see Brenner and Scott (2008, Lemma 4.3.8), Ciarlet (2002, Theorem 3.1.1), Dupont and Scott (1980)) or Proposition 6.34 below:

$$\inf_{w \in \mathbb{P}_{s-1}(\omega_T)} \|v - w\|_{W^s_p(\omega_T)} \lesssim |v|_{W^s_p(\omega_T)}. \tag{3.15}$$

This proves (3.14) for $n = 1$. The construction of $I_{\mathcal{T}}$ and ensuing estimate (3.14) is still valid for any $n > 1$ (Brenner and Scott 2008, Scott and Zhang 1990).

Proposition 3.4 (quasi-interpolant without boundary values). *Let s, t be regularity indices with $0 \leq t \leq s \leq n+1$, and let $1 \leq p, q \leq \infty$ be integrability indices so that $\mathrm{sob}(W^s_p) > \mathrm{sob}(W^t_q)$. Then there exists a quasi-interpolation operator $I_{\mathcal{T}}\colon L^1(\Omega) \to \mathbb{S}^{n,0}_{\mathcal{T}}$, which is invariant in $\mathbb{S}^{n,0}_{\mathcal{T}}$ and satisfies*

$$|v - I_{\mathcal{T}}v|_{W^t_q(T)} \lesssim h_T^{\mathrm{sob}(W^s_p)-\mathrm{sob}(W^t_q)} |v|_{W^s_p(\omega_T)} \quad \textit{for all } T \in \mathcal{T}. \tag{3.16}$$

The hidden constant in (3.14) *depends on the shape coefficient of $\mathcal{T}_0$ and d.*

To impose a vanishing trace on $I_{\mathcal{T}}v$ we may suitably modify the averaging process for boundary nodes. We thus define a set of dual functions with respect to an L^2-scalar product over $(d-1)$-subsimplices contained on $\partial\Omega$; see again Brenner and Scott (2008) and Scott and Zhang (1990) for details. This retains the invariance property of $I_{\mathcal{T}}$ on $\mathbb{S}^{n,0}(\mathcal{T})$ and guarantees that $I_{\mathcal{T}}v$ has a zero trace if $v \in \mathring{W}^1_1(\Omega)$ does. Hence the above argument applies, and (3.16) follows provided $s \geq 1$.

Proposition 3.5 (quasi-interpolant with boundary values). *Let s, t, p, q be as in Proposition 3.4. There exists a quasi-interpolation operator $I_{\mathcal{T}}\colon W^1_1(\Omega) \to \mathbb{S}^{n,0}_{\mathcal{T}}$ invariant in $\mathbb{S}^{n,0}_{\mathcal{T}}$ which satisfies* (3.16) *for $s \geq 1$ and preserves the boundary values of v provided they are piecewise polynomial of degree $\leq n$. In particular, if $v \in W^1_1(\Omega)$ has a vanishing trace on $\partial\Omega$, then so does $I_{\mathcal{T}}v$.*

Remark 3.6 (fractional regularity). We observe that (3.16) does not require the regularity indices t and s to be integer. The proof follows along the same lines but replaces the polynomial degree n with the greatest integer smaller than s; the generalization of (3.15) can be taken from Dupont and Scott (1980).

Remark 3.7 (local error estimate for Lagrange interpolant). Let the regularity index s and integrability index $1 \le p \le \infty$ satisfy $s - d/p > 0$. This implies that $\mathrm{sob}(W_p^s) > \mathrm{sob}(L^\infty)$, whence $W_p^s(\Omega) \subset C(\overline{\Omega})$ and the Lagrange interpolation operator $I_\mathcal{T}\colon W_p^s(\Omega) \to \mathbb{S}_\mathcal{T}^{n,0}$ is well-defined and satisfies the *local error estimate*

$$|v - I_\mathcal{T} v|_{W_q^t(T)} \lesssim h_T^{\mathrm{sob}(W_p^s)-\mathrm{sob}(W_q^t)} |v|_{W_p^s(T)}, \tag{3.17}$$

provided $0 \le t \le s$, $1 \le q \le \infty$ are such that $\mathrm{sob}(W_p^s) > \mathrm{sob}(W_q^t)$. We point out that ω_T in (3.14) is now replaced by T in (3.17). We also remark that if v vanishes on $\partial\Omega$ then so does $I_\mathcal{T} v$. The proof of (3.17) proceeds along the same lines as that of Proposition 3.4, except that the nodal evaluation does not extend beyond the element $T \in \mathcal{T}$, and the inverse and stability estimates over the reference element are replaced by

$$|I_\mathcal{T} v|_{W_q^t(\widehat{T})} \lesssim \|I_\mathcal{T} v\|_{L^q(\widehat{T})} \lesssim \|v\|_{L^\infty(\widehat{T})} \lesssim \|v\|_{W_p^s(\widehat{T})}.$$

The following global interpolation error estimate builds on Proposition 3.4 and relates to Figure 2.1 (DeVore diagram).

Theorem 3.8 (global interpolation error estimate). *Let* $1 \le s \le n+1$, $t = 0, 1$, $t < s$ *and* $1 \le p \le q$ *satisfy* $r := \mathrm{sob}(W_p^s) - \mathrm{sob}(W_q^t) > 0$. *If* $v \in W_p^s(\Omega)$, *then*

$$|v - I_\mathcal{T} v|_{W_q^t(\Omega)} \lesssim \left(\sum_{T \in \mathcal{T}} h_T^{rp} |v|_{W_p^s(\omega_T)}^p \right)^{1/p}. \tag{3.18}$$

Proof. Use Proposition 3.4 along with the elementary property of series $\sum_n a_n \le (\sum_n a_n^{p/q})^{q/p}$ for $0 < p/q \le 1$. □

Continuous vs. discontinuous approximation of gradients. The preceding discussion might induce us to believe that when dealing with Sobolev functions $v \in W_p^1(\Omega)$ without point values, namely $1 \le p \le d$, global continuity of the quasi-interpolant $I_\mathcal{T} v$ might degrade the approximation quality relative to discontinuous approximations. The following instrumental result shows that this is not the case (Veeser 2016, Theorem 2). It hinges on a new geometric concept: we say that a star ω_z is $(d-1)$-*face-connected* if, for any element $T \subset \omega_z$ and $(d-1)$-face $F \subset \omega_z$ containing z, there exists a sequence $(T_i)_{i=0}^m$ such that

- any T_i is an element of ω_z for $0 \le i \le m$,
- any intersection $T_i \cap T_{i+1}$ is a $(d-1)$-face of ω_z for $0 \le i \le m-1$,
- T_0 contains F and $T_m = T$.

Note that a star ω_z is $(d-1)$-face-connected if the set $\omega_z \cap \Omega$ is connected.

Proposition 3.9 (approximation of gradients). *Let* $v \in W_p^1(\Omega)$ *for* $1 \le p \le d$. *Let* $\mathcal{T}$ *be a conforming mesh such that its stars are* $(d-1)$-*face-connected. Then there exists a constant* $C(\sigma)$ *depending on the shape regularity coefficient* σ *of*

(3.9), *the dimension d and the polynomial degree $n \geq 1$, such that*

$$1 \leq \frac{\min_{w \in \mathbb{S}_{\mathcal{T}}^{n,0}} \|\nabla(v-w)\|_{L^p(\Omega)}}{\min_{w \in \mathbb{S}_{\mathcal{T}}^{n,-1}} \|\nabla(v-w)\|_{L^p(\Omega;\mathcal{T})}} \leq C(\sigma), \tag{3.19}$$

where $\|\nabla w\|_{L^p(\Omega;\mathcal{T})}$ stands for the broken norm over $\mathcal{T}$.

The left inequality in (3.19) is obvious because $\mathbb{S}_{\mathcal{T}}^{n,0} \subset \mathbb{S}_{\mathcal{T}}^{n,-1}$. In contrast, the right inequality is delicate and relies on examining the quasi-interpolant (3.13) (Veeser 2016, Theorems 1 and 22). An important consequence of (3.19) is the following localized version of (3.18).

Proposition 3.10 (localized quasi-interpolation estimate). *Let $1 < s \leq n+1$, $1 \leq p \leq d$ and $r = \mathrm{sob}(W_p^s) - \mathrm{sob}(W_q^1) > 0$. If $v \in W_q^1(\Omega)$, then*

$$\|\nabla(v - I_{\mathcal{T}}v)\|_{L^q(\Omega)} \lesssim \left(\sum_{T \in \mathcal{T}} h_T^{pr} |v|_{W_p^s(T)}^p\right)^{1/p}, \tag{3.20}$$

Proof. Since $v - I_{\mathcal{T}}v = (v-w) - I_{\mathcal{T}}(v-w)$ for any $w \in \mathbb{S}_{\mathcal{T}}^{n,0}$, combine Proposition 3.9 with Proposition 6.34 (Bramble–Hilbert for Sobolev spaces) to write

$$\|\nabla(v - I_{\mathcal{T}}v)\|_{L^q(\Omega)} \lesssim \min_{w \in \mathbb{S}_{\mathcal{T}}^{n,-1}} \|\nabla(v-w)\|_{L^q(\Omega)} \lesssim \left(\sum_{T \in \mathcal{T}} h_T^{pr} |v|_{W_p^s(T)}^p\right)^{1/p}.$$

This concludes the proof. □

The crucial difference between (3.20) and (3.18) is that the function $v \in W_q^1(\Omega)$ does not have to belong to $W_p^s(\Omega)$ globally but rather locally, namely $v \in W_p^s(T)$ for every $T \in \mathcal{T}$, to get optimal *a priori* error estimates. This property will find several applications later. A special case of (3.20) for $p = q = 2$ and $s = 2$ reads

$$|v - I_{\mathcal{T}}v|_{H^1(\Omega)}^2 \lesssim \sum_{T \in \mathcal{T}} h_T^2 |v|_{H^2(T)}^2,$$

for $v \in H^2(\Omega;\mathcal{T}) := \{w \in H^1(\Omega) \mid w_{|T} \in H^2(T) \text{ for all } T \in \mathcal{T}\}$.

Quasi-uniform meshes. We now apply Theorem 3.8 to *quasi-uniform* meshes, namely meshes $\mathcal{T} \in \mathbb{T}$ for which all its elements are of comparable size h, whence

$$h \approx (\#\mathcal{T})^{-1/d} |\Omega|^{1/d} \approx (\#\mathcal{T})^{-1/d}.$$

Corollary 3.11 (quasi-uniform meshes). *Let $1 < s \leq n+1$ and $v \in H^s(\Omega)$. If $\mathcal{T} \in \mathbb{T}$ is quasi-uniform, then*

$$\|\nabla(v - I_{\mathcal{T}}v)\|_{L^2(\Omega)} \lesssim |v|_{H^s(\Omega)} (\#\mathcal{T})^{-(s-1)/d}. \tag{3.21}$$

Remark 3.12 (optimal rate). If $s = n+1$, and so v has the maximal regularity $v \in H^{n+1}(\Omega)$, then we obtain the optimal convergence rate in a linear Sobolev scale

$$\|\nabla(v - I_{\mathcal{T}}v)\|_{L^2(\Omega)} \lesssim |v|_{H^{n+1}(\Omega)} (\#\mathcal{T})^{-n/d}. \tag{3.22}$$

Table 3.1. Rate of convergence s in term of uniform mesh size h. We observe an asymptotic error decay of about $h^{2/3}$ (i.e. $s = 2/3$), or equivalently $(\#\mathcal{T})^{-1/3}$, irrespective of the polynomial degree n. This provides a lower bound for $\|v - I_{\mathcal{T}} v\|_{L^2(\Omega)}$ and thus shows that (3.21) is sharp.

h	linear ($n = 1$)	quadratic ($n = 2$)	cubic ($n = 3$)
1/4	1.14	9.64	9.89
1/8	0.74	0.67	0.67
1/16	0.68	0.67	0.67
1/32	0.66	0.67	0.67
1/64	0.66	0.67	0.67
1/128	0.66	0.67	0.67

The order $-n/d$ is just dictated by the polynomial degree n and cannot be improved upon assuming either higher regularity than $H^{n+1}(\Omega)$ or a graded mesh $\mathcal{T}$. The presence of d in the exponent is referred to as the *curse of dimensionality*.

Example (corner singularity in two dimensions). To explore the effect of a geometric singularity on (3.21), we let $\Omega = (-1,1)^2 \setminus [0,1]^2$ be an L-shaped domain and let $v \in H^1(\Omega)$ be

$$v(r,\theta) = r^{2/3}\sin(2\theta/3) - r^2/4.$$

This function $v \in H^1(\Omega)$ exhibits the typical corner singularity of the solution of $-\Delta v = f$ with suitable Dirichlet boundary condition: $v \in H^s(\Omega)$ for $s < 5/3$. Table 3.1 displays the best approximation error for polynomial degree $n = 1,2,3$ and a sequence of *uniform* refinements in the seminorm $|\cdot|_{H^1(\Omega)} = \|\nabla\cdot\|_{L^2(\Omega)}$. This gives a *lower* bound for the interpolation error in (3.21).

Even though s is fractional, the error estimate (3.21) is still valid, as stated in Remark 3.6. In fact, for uniform refinement, (3.21) can be derived by space interpolation between $H^1(\Omega)$ and $H^{n+1}(\Omega)$. The asymptotic rate $(\#\mathcal{T})^{-1/3}$ reported in Table 3.1 is consistent with (3.21) and independent of the polynomial degree n; this shows that (3.21) is sharp. It is also suboptimal as compared with the optimal rate $(\#\mathcal{T})^{-n/2}$ of Remark 3.12.

The question arises whether the rate $(\#\mathcal{T})^{-1/3} \approx h^{2/3}$ in Table 3.1 is just a consequence of uniform refinement or unavoidable. It is important to realize that $v \notin H^s(\Omega)$ for $s \geq 5/3$ and thus (3.21) is not applicable. However, the problem is not that second-order derivatives of v do not exist but rather that they are not square-integrable. In particular, it is true that $v \in W^2_p(\Omega)$ if $1 \leq p < 3/2$. We may therefore apply Theorem 3.8 with, for example, $n = 1$, $s = 2$ and $p \in [1, 3/2)$, and then ask whether the structure of (3.18) can be exploited, for example by compensating the local singular behaviour of v with the local mesh size h. This enterprise naturally leads to *graded* meshes adapted to v.

3.4. *Principle of error equidistribution.*

We investigate the relation between local interpolation error and regularity for the design of optimal graded meshes adapted to a given function $v \in H^1(\Omega)$ for $d = 2$. We recall that $W^2_1(\Omega)$ is in the same nonlinear Sobolev scale of $H^1(\Omega)$, namely $\mathrm{sob}(W^2_1) = \mathrm{sob}(H^1)$, but $W^2_1(\Omega) \subset C^0(\overline{\Omega})$ (Brenner and Scott 2008, Lemma 4.3.4), and the *Lagrange interpolant* $I_\mathcal{T} v$ is well-defined and satisfies

$$\|\nabla(v - I_\mathcal{T} v)\|_{L^2(T)} \le C|v|_{W^2_1(T)} =: e_\mathcal{T}(v,T) \quad \text{for all } T \in \mathcal{T}. \tag{3.23}$$

We formulate a discrete minimization problem on the surrogate quantity $\boldsymbol{e} := (e_\mathcal{T}(v,T))_{T\in\mathcal{T}} \in \mathbb{R}^N$ with $N = \#\mathcal{T}$: minimize the square of the total H^1-error $E_\mathcal{T}(v)$,

$$E_\mathcal{T}(v)^2 := \sum_{T\in\mathcal{T}} e_\mathcal{T}(v,T)^2,$$

subject to the constraint

$$\sum_{T\in\mathcal{T}} e_\mathcal{T}(v,T) = C|v|_{W^2_1(\Omega)}.$$

We idealize the problem upon allowing $e_\mathcal{T}(v,T)$ to attain any non-negative real value, despite the fact that shape regularity of $\mathcal{T}$ entails geometric restrictions between adjacent elements. We next form the Lagrangian

$$\mathcal{L}[\boldsymbol{e},\lambda] := \sum_{T\in\mathcal{T}} e_\mathcal{T}(v,T)^2 - \lambda\left(\sum_{T\in\mathcal{T}} e_\mathcal{T}(v,T) - C|v|_{W^2_1(\Omega)}\right),$$

with Lagrange multiplier $\lambda \in \mathbb{R}$. We thus realize that the optimality condition reads

$$e_\mathcal{T}(v,T) = \frac{\lambda}{2} \quad \text{for all } T \in \mathcal{T}$$

or that $e_\mathcal{T}(v,T)$ is constant over $\mathcal{T}$. We rewrite this insightful conclusion as follows:

A graded mesh is quasi-optimal if the local error is equidistributed. (3.24)

This calculation yields

$$E_\mathcal{T}(v)^2 = \frac{\lambda^2}{4}N, \quad C|v|_{W^2_1(\Omega)} = \frac{\lambda}{2}N,$$

whence

$$E_\mathcal{T}(v) = C|v|_{W^2_1(\Omega)} N^{-1/2} \tag{3.25}$$

is the optimal decay rate but with regularity $v \in W^2_1(\Omega)$ rather than $H^2(\Omega)$. This is the second instance of nonlinear approximation, namely mesh design tailored to the specific function v at hand; the first one was in Section 1. The principle of error equidistribution (3.24) was originally derived by Babuška and Rheinboldt (1978) for $d = 1$, and extended to $d = 2$ by Nochetto and Veeser (2012, Section 1.6), using an idealized continuous minimization problem involving a mesh size function. The current formulation is closer to applications and does not require a positive power of h_T in (3.23).

Remark 3.13 (point singularities). Corner singularities (Grisvard 1985) as well as singularities due to intersecting interfaces (Kellogg 1974/75) are of the form

$$v(x) \approx r(x)^\gamma, \quad 0 < \gamma < 1, \tag{3.26}$$

for $d = 2$. This implies $v \in W_1^2(\Omega)$ for all γ and the decay rate (3.25) provided $\mathcal{T}$ equidistributes the H^1-error. Babuška, Kellogg and Pitkäranta (1979) and Grisvard (1985) designed such meshes for corner singularities using weighted H^2-regularity. The preceding approach is more powerful in that it does not require any characterization of the singularities, rather than $v \in W_1^2(\Omega)$, and also applies to line discontinuities for $d = 2$. We will come back to this point in Section 6.

We now consider an important abstract variant of the discrete minimization process leading to (3.24), which will be instrumental in understanding the success of greedy algorithms later. Suppose that $0 < q, p \leq \infty$, $v \in L^q(\Omega)$, $X_p^t(\Omega)$ is an abstract regularity space with $t = 1/p - 1/q > 0$, and $E_\mathcal{T}(v)_q$ and $e_\mathcal{T}(v,T)_q$ are global and local L^q-interpolation error indicators of v that satisfy the following two abstract properties.

- *Summability in ℓ^q.* There exists a constant $C_1 > 0$ such that

$$E_\mathcal{T}(v)_q^q \leq C_1^q \sum_{T \in \mathcal{T}} e_\mathcal{T}(v,T)_q^q. \tag{3.27}$$

- *Summability in ℓ^p.* There exists a constant $C_2 > 0$ such that

$$\sum_{T \in \mathcal{T}} e_\mathcal{T}(v,T)_q^p = C_2^p |v|_{X_p^t(\Omega)}^p. \tag{3.28}$$

We intend to find conditions on a mesh $\mathcal{T}$ that minimize the global L^q-error $E_\mathcal{T}(v)_q$ of v subject to the constraint (3.28). We again propose a Lagrangian

$$\mathcal{L}[\boldsymbol{e}, \lambda] := \sum_{T \in \mathcal{T}} e_\mathcal{T}(v,T)_q^q - \lambda\biggl(\sum_{T \in \mathcal{T}} e_\mathcal{T}(v,T)_q^p - C_2^p |v|_{X_p^t(\Omega)}^p\biggr).$$

The optimality condition for $\boldsymbol{e}$ reads

$$e_\mathcal{T}(v,T)_q = \left(\lambda \frac{p}{q}\right)^{1/(q-p)} \quad \text{for all } T \in \mathcal{T},$$

which is a third instance of error equidistribution and thus consistent with (3.24).

We now resort to (3.27) and (3.28), to arrive at

$$\sum_{T\in\mathcal{T}} e_{\mathcal{T}}(v,T)_q^q = N\left(\lambda\frac{p}{q}\right)^{q/(q-p)}, \quad C_2^p|v|^p_{X_p^t(\Omega)} = N\left(\lambda\frac{p}{q}\right)^{p/(q-p)},$$

whence

$$E_{\mathcal{T}}(v)_q \le C_1C_2|v|_{X_p^t(\Omega)}N^{1/q-1/p}. \tag{3.29}$$

We see that the decay rate in (3.29) is $-t = 1/q - 1/p < 0$ and is just dictated by the different summabilities of (3.27) and (3.28). In the applications of (3.29) below, $t = s/d$ will be proportional to a differentiability index s, and the condition

$$0 = t - \frac{1}{p} + \frac{1}{q} = \frac{s}{d} - \frac{1}{p} + \frac{1}{q}$$

will correspond to the spaces $L^q(\Omega)$ and $X_p^s(\Omega)$ being on the same nonlinear Sobolev scale. This minimization process is an idealization that does not account for mesh regularity of $\mathcal{T}$, which in turn entails some geometric constraints in the construction of $\mathcal{T}$. A key question is whether estimates of the form (3.29) can be achieved under conditions that are practical but weaker than (3.24). In Section 3.5 we will study the bisection method, a flexible technique for conforming mesh refinement with optimal complexity. In Section 3.6 we will present and analyse GREEDY, a practical algorithm that implements these ideas and constructs quasi-optimal conforming bisection meshes under the slightly stronger assumption

$$s - \frac{d}{p} + \frac{d}{q} > 0. \tag{3.30}$$

Moreover, in Section 3.7 we will extend this analysis to non-conforming meshes.

We realize from (3.29) that in order to maximize the error decay rate we would like to have p as small as possible, even $0 < p < 1$. The range of q, p does not matter in the argument above and, despite the fact that $q \ge 1$ in all applications below, the range of p is only limited by that of s, which in turn depends on the polynomial degree $n \ge 1$ in that $0 < s \le n + 1$.

We now return to the special case (3.23), namely $q = 2$, $p = 1$ and $\nabla v \in L^2(\Omega)$. As already shown in (3.23), in the nonlinear Sobolev scale

$$\operatorname{sob}(W_1^2) - \operatorname{sob}(H^1) = \left(2 - \frac{2}{1}\right) - \left(1 - \frac{2}{2}\right) = 0,$$

we expect the best error decay

$$\|\nabla(v - I_{\mathcal{T}}v)\|_{L^2(\Omega)} \lesssim |v|_{W_1^2(\Omega)}(\#\mathcal{T})^{-1/2},$$

whereas the linear Sobolev scale yields the reduced order

$$\|\nabla(v - I_{\mathcal{T}}v)\|_{L^2(\Omega)} \lesssim |v|_{H^s(\Omega)}(\#\mathcal{T})^{-(s-1)/2}$$

for $s < 1 + \gamma < 2$ and v satisfying (3.26), where $I_{\mathcal{T}}$ is the Lagrange interpolation

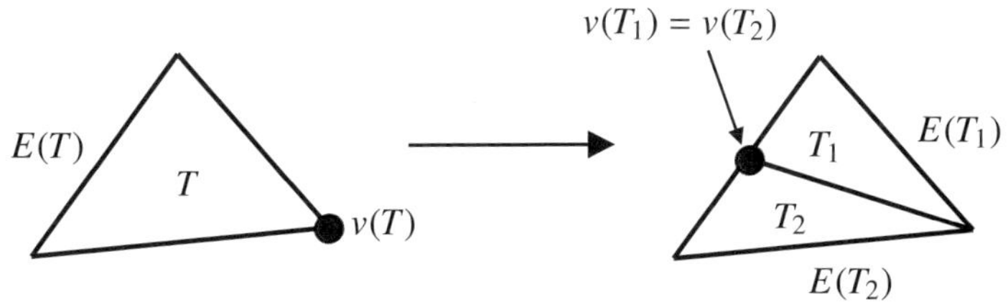

Figure 3.2. Triangle $T \in \mathcal{T}$ with vertex $v(T)$ and opposite refinement edge $E(T)$. The bisection rule for $d = 2$ consists of connecting $v(T)$ with the midpoint of $E(T)$, thereby giving rise to children T_1, T_2 with common vertex $v(T_1) = v(T_2)$, the newly created vertex, and opposite refinement edges $E(T_1), E(T_2)$.

operator. The nonlinear Sobolev scale entails a trade of differentiability with integrability: we gain up to differentiability $s = 2$ at the expense of lower integrability $p = 1$ for polynomial degree $n = 1$. This trade-off is at the heart of the optimal estimate (3.25) and is represented in the so-called *DeVore diagram* in Figure 2.1.

If the polynomial degree is $n \geq 2$, then the largest differentiability index is $s = n + 1$, which for $d = 2$ leads to integrability index $p < 1$:

$$\left(s - \frac{2}{p}\right) - \left(1 - \frac{2}{2}\right) = 0 \quad \Rightarrow \quad p = \frac{2}{n+1} < 1. \tag{3.31}$$

To measure regularity of v, the corresponding Sobolev space must be replaced by the Besov space $B^{n+1}_{p,p}(\Omega)$ or the Lipschitz space $\mathrm{Lip}^{n+1}_p(\Omega)$. We will introduce and study these spaces in Section 6.8.

3.5. *Conforming meshes: the bisection method*

In order to approximate functions in $W^k_p(\Omega)$ by piecewise polynomials, we decompose Ω into simplices. We briefly discuss the *bisection* method, an elegant and versatile technique for subdividing Ω in any dimension into a conforming mesh. We also briefly discuss non-conforming meshes in Section 3.7. We present complete proofs, especially of the complexity of bisection, later in Section 8.

We focus on $d = 2$ and follow Binev *et al.* (2004), but the results carry over to any dimension $d > 2$ (Stevenson 2008). We refer to Nochetto *et al.* (2009) for a fairly complete discussion for $d \geq 2$.

Let $\mathcal{T}$ denote a *mesh* (triangulation or grid) made of simplices T, and let $\mathcal{T}$ be *conforming* (edge-to-edge). Each element is labelled, namely it has an edge $E(T)$ assigned for refinement (and an opposite vertex $v(T)$ for $d = 2$); see Figure 3.2.

The bisection method consists of a suitable *labelling* of the initial mesh $\mathcal{T}_0$ and a rule to assign the refinement edge to the two children. For $d = 2$ we consider the *newest vertex bisection* as depicted in Figure 3.2. For $d > 2$ the situation is more complicated and we need the concepts of type and vertex order (Nochetto *et al.* 2009, Stevenson 2008). More precisely, we identify a simplex T with the set of its

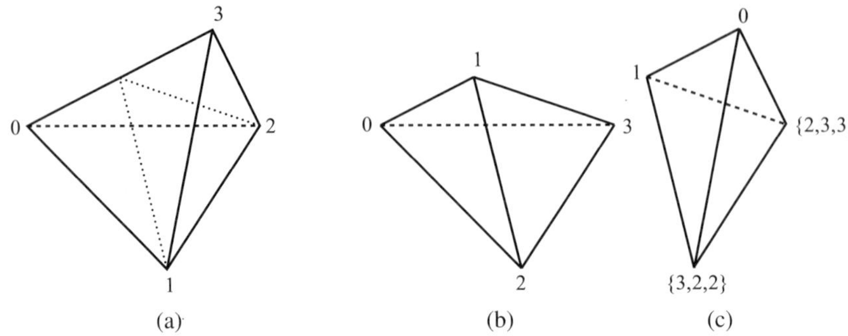

Figure 3.3. Refinement of a single tetrahedron T of type t. The child T_1 in (b) has the same node ordering regardless of type. In contrast, for the child T_2 in (c) a triple is appended to two nodes. The local vertex index is given for these nodes by the tth component of the triple.

ordered vertices and its *type t* by

$$T = \{z_0, z_1, \ldots, z_d\}_t,$$

with $t \in \{0, \ldots, d-1\}$. Given such a d-simplex T, we use the following bisection rule to split it in a unique fashion and to impose both vertex order and type on its children. The edge $\overline{z_0 z_d}$ connecting the first and last vertex of T is the *refinement edge* of T, and its midpoint $\bar{z} = (z_0 + z_d)/2$ becomes the new vertex. Connecting the new vertex $\bar{z}$ to the vertices of T other than z_0, z_d determines the common face $S = \{\bar{z}, z_1, \ldots, z_{d-1}\}$ shared by the two children T_1, T_2 of T. The *bisection rule* dictates the following vertex order and type for T_1, T_2:

$$\begin{aligned} T_1 &:= \{z_0, \bar{z}, \underbrace{z_1, \ldots, z_t}_{\rightarrow}, \underbrace{z_{t+1}, \ldots, z_{d-1}}_{\rightarrow}\}_{(t+1) \bmod d}, \\ T_2 &:= \{z_d, \bar{z}, \underbrace{z_1, \ldots, z_t}_{\rightarrow}, \underbrace{z_{d-1}, \ldots, z_{t+1}}_{\leftarrow}\}_{(t+1) \bmod d}, \end{aligned} \tag{3.32}$$

with the convention that arrows point in the direction of increasing indices and $\{z_1, \ldots, z_0\} = \emptyset$, $\{z_d, \ldots, z_{d-1}\} = \emptyset$. For instance, in three dimensions the children of $T = \{z_0, z_1, z_2, z_3\}_t$ are (see Figure 3.3)

$$\begin{aligned} t = 0: &\quad T_1 = \{z_0, \bar{z}, z_1, z_2\}_1 \quad \text{and} \quad T_2 = \{z_3, \bar{z}, z_2, z_1\}_1, \\ t = 1: &\quad T_1 = \{z_0, \bar{z}, z_1, z_2\}_2 \quad \text{and} \quad T_2 = \{z_3, \bar{z}, z_1, z_2\}_2, \\ t = 2: &\quad T_1 = \{z_0, \bar{z}, z_1, z_2\}_0 \quad \text{and} \quad T_2 = \{z_3, \bar{z}, z_1, z_2\}_0. \end{aligned}$$

Note that the vertex labelling of T_1 is type-independent, whereas that of T_2 is the same for type $t = 1$ and $t = 2$. To account for this fact, the vertices z_1 and z_2 of T are tagged $\{3, 2, 2\}$ and $\{2, 3, 3\}$ in Figure 3.3. The type of T then dictates which component of the triple is used to label the vertex.

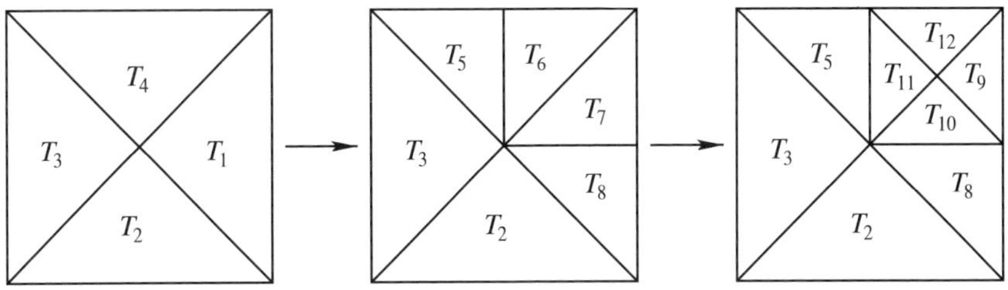

Figure 3.4. Sequence of bisection meshes $\{\mathcal{T}_k\}_{k=0}^2$ starting from the initial mesh $\mathcal{T}_0 = \{T_i\}_{i=1}^4$ with longest edges labelled for bisection. Mesh $\mathcal{T}_1$ is created from $\mathcal{T}_0$ upon bisecting T_1 and T_4, whereas mesh $\mathcal{T}_2$ arises from $\mathcal{T}_1$ upon refining T_6 and T_7. The bisection rule is described in Figure 3.2.

Bisection creates a *unique* master forest $\mathbb{F}$ of binary trees with infinite depth, where each node is a simplex (a triangle in two dimensions), its two successors are the two children created by bisection, and the roots of the binary trees are the elements of the initial conforming partition $\mathcal{T}_0$. It is important to realize that, no matter how an element arises in the subdivision process, its associated newest vertex is unique and depends only on the labelling of $\mathcal{T}_0$, so the edge $E(T)$ assigned for refinement (and the opposite vertex $v(T)$ for $d = 2$) are independent of the order of the subdivision process for all $T \in \mathbb{F}$; see Lemma 8.1. Therefore $\mathbb{F}$ is unique.

A finite subset $\mathcal{F} \subset \mathbb{F}$ is called a *forest* if $\mathcal{T}_0 \subset \mathcal{F}$ and the nodes of $\mathcal{F}$ satisfy

- all nodes of $\mathcal{F} \setminus \mathcal{T}_0$ have a predecessor;
- all nodes in $\mathcal{F}$ have either two successors or none.

Any node $T \in \mathcal{F}$ is thus uniquely connected with a node T_0 of the initial triangulation $\mathcal{T}_0$, that is, T belongs to the infinite tree $\mathbb{F}(T_0)$ emanating from T_0. Furthermore, any forest may have *interior nodes*, i.e. nodes with successors, as well as *leaf nodes*, i.e. nodes without successors. The set of leaves corresponds to a mesh (or triangulation, grid, partition) $\mathcal{T} = \mathcal{T}(\mathcal{F})$ of $\mathcal{T}_0$, which may not be conforming or edge-to-edge.

We thus introduce the set $\mathbb{T}$ of all *conforming* refinements of $\mathcal{T}_0$:

$$\mathbb{T} := \{\mathcal{T} = \mathcal{T}(\mathcal{F}) \mid \mathcal{F} \subset \mathbb{F} \text{ is finite and } \mathcal{T}(\mathcal{F}) \text{ is conforming}\}.$$

If $\mathcal{T}_* = \mathcal{T}(\mathcal{F}_*) \in \mathbb{T}$ is a conforming refinement of $\mathcal{T} = \mathcal{T}(\mathcal{F}) \in \mathbb{T}$, we write $\mathcal{T}_* \geq \mathcal{T}$ and understand this inequality in the sense of trees, namely $\mathcal{F} \subset \mathcal{F}_*$.

Example. Consider $\mathcal{T}_0 = \{T_i\}_{i=1}^4$ and the longest edge to be the refinement edge. Figure 3.4 displays a sequence of conforming meshes $\mathcal{T}_k \in \mathbb{T}$ created by bisection. Each element T_i of $\mathcal{T}_0$ is a root of a finite tree emanating from T_i, which together form the forest $\mathcal{F}_2$ corresponding to mesh $\mathcal{T}_2 = \mathcal{T}(\mathcal{F}_2)$. Figure 3.5 displays $\mathcal{F}_2$, whose leaf nodes are the elements of $\mathcal{T}_2$.

Figure 3.5. Forest $\mathcal{F}_2$ corresponding to the grid sequence $\{\mathcal{T}_k\}_{k=0}^2$ of Figure 3.4. The roots of $\mathcal{F}_2$ form the initial mesh $\mathcal{T}_0$ and the leaves of $\mathcal{F}_2$ constitute the conforming bisection mesh $\mathcal{T}_2$. Moreover, each level of $\mathcal{F}_2$ corresponds to all elements with generation equal to the level.

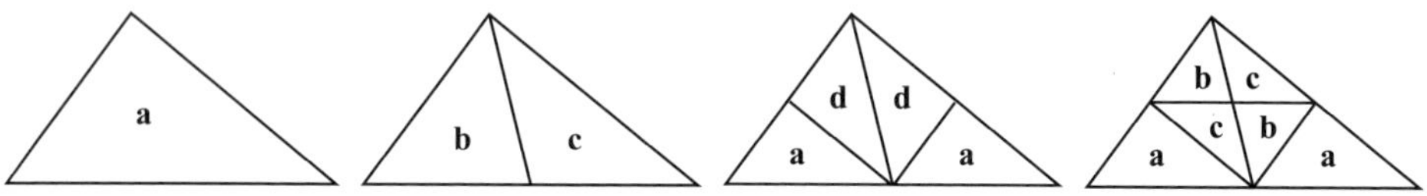

Figure 3.6. Bisection produces at most four similarity classes for any triangle.

Properties of bisection. We now discuss several crucial geometric properties of bisection. We start by recalling the concept of shape regularity. For any $T \in \mathcal{T}$, we define

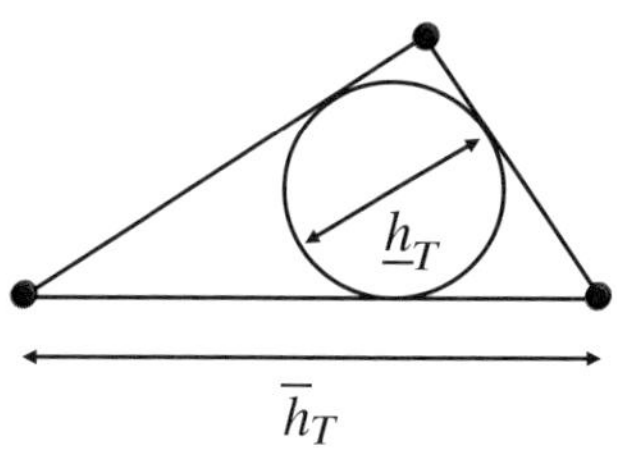

$$\overline{h}_T := \operatorname{diam}(T)$$
$$h_T := |T|^{1/d}$$
$$\underline{h}_T := 2 \sup\{r > 0 \mid B(x,r) \subset T \text{ for } x \in T\}.$$

Then

$$\underline{h}_T \le h_T \le \overline{h}_T \le \sigma \underline{h}_T \quad \text{for all } T \in \mathcal{T},$$

where $\sigma > 1$ is the shape regularity constant of (3.9). The next lemma guarantees that bisection keeps σ bounded.

Lemma 3.14 (shape regularity). *The partitions $\mathcal{T}$ generated by newest vertex bisection satisfy a uniform minimal angle condition, or equivalently σ is uniformly bounded, depending only on the initial partition $\mathcal{T}_0$.*

Proof. Each $T \in \mathcal{T}_0$ gives rise to a fixed number of similarity classes, namely four for $d = 2$ according to Figure 3.6. This, combined with the fact that $\#\mathcal{T}_0$ is finite, yields the assertion. □

We define the *generation (or level)* $g(T)$ of an element $T \in \mathcal{T}$ as the number of bisections needed to create T from its ancestor $T_0 \in \mathcal{T}_0$. Since bisection splits an element into two children with equal measure, we realize that

$$h_T = 2^{-g(T)/2} h_{T_0} \quad \text{for all } T \in \mathcal{T}. \tag{3.33}$$

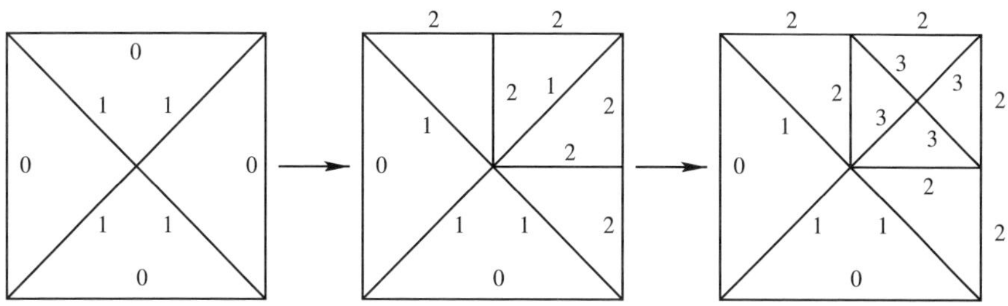

Figure 3.7. Initial labelling and its evolution for the sequence of conforming refinements $\mathcal{T}_0 \leq \mathcal{T}_1 \leq \mathcal{T}_2$ of Figure 3.4.

Referring to Figure 3.5, we observe that the leaf nodes $T_9, T_{10}, T_{11}, T_{12}$ have generation 2, whereas T_5, T_8 have generation 1 and T_2, T_3 have generation 0.

The following geometric property is a simple consequence of (3.33).

Lemma 3.15 (element size vs. generation). *There exist constants* $0 < D_1 < D_2$*, depending only on* $\mathcal{T}_0$*, such that*

$$D_1 2^{-g(T)/2} \leq h_T < \overline{h}_T \leq D_2 2^{-g(T)/2} \quad \textit{for all } T \in \mathcal{T}. \tag{3.34}$$

Labelling and bisection rule. Whether the recursive application of bisection does not lead to inconsistencies depends on a suitable initial labelling of edges and a bisection rule. For $d = 2$ they are simple to state (Binev *et al.* 2004). Given $T \in \mathcal{T}$ with generation $g(T) = i$, we assign the label $(i+1, i+1, i)$ to T with i corresponding to the refinement edge $E(T)$. The following rule dictates how the labelling changes with refinement: the side i is bisected and both new sides as well as the bisector are labelled $i + 2$ whereas the remaining labels do not change. To guarantee that the label of an edge is independent of the elements sharing this edge, we need a special labelling for $\mathcal{T}_0$, due to Mitchell (1989, Theorem 2.9) and Binev *et al.* (2004, Lemma 2.1), for $d = 2$:

Edges of $\mathcal{T}_0$ *have labels* 0 *or* 1 *and all elements* $T \in \mathcal{T}$ *have exactly two edges with label* 1 *and one with label* 0. (3.35)

There is a variant for $d > 2$ due to Stevenson (2008, Section 4). It is not obvious that labelling (3.35) exists, but if it does then all elements of $\mathcal{T}_0$ can be split into pairs of compatibly divisible elements. We refer to Figure 3.7 for an example of initial labelling of $\mathcal{T}_0$ satisfying (3.35) and the way it evolves for two successive refinements $\mathcal{T}_2 \geq \mathcal{T}_1 \geq \mathcal{T}_0$ corresponding to Figure 3.4.

To guarantee (3.35) we can proceed as follows: given a coarse mesh of elements T, we can bisect each T twice and label the four grandchildren as indicated in Figure 3.8, for the resulting mesh $\mathcal{T}_0$ to satisfy the initial labelling (Binev *et al.* 2004).

For $d \geq 3$ a general strategy of initial labelling is due to Stevenson (2008, Section 4, Condition (b)), who in turn improves upon Maubach (1995) and Traxler (1997) and shows how to impose it upon further refining each element of $\mathcal{T}_0$.

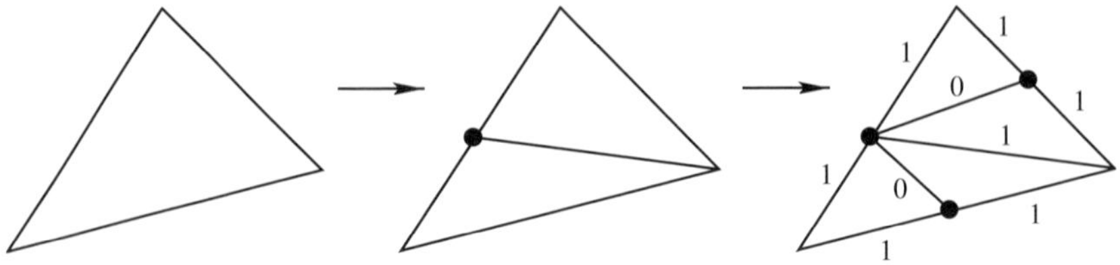

Figure 3.8. Bisecting each triangle of $\mathcal{T}_0$ twice and labelling edges in such a way that all boundary edges have label 1 yields an initial mesh satisfying (3.35).

We refer to the survey by Nochetto *et al.* (2009) for a discussion of this condition. A key consequence is:

$$\textit{Every uniform refinement of } \mathcal{T}_0 \textit{ gives a conforming bisection mesh.} \tag{3.36}$$

Condition (3.35) is still valid, and can be fulfilled by a construction by successive bisections, similar to but much trickier than the one described for $d = 2$; yet for $d = 3$ the number of elements increases by an order of magnitude, which indicates that (3.35) is a severe restriction in practice. Finding alternative, more practical conditions is an important problem.

Initialization of arbitrary triangulations. A novel initialization procedure that can be applied to *any* conforming triangulation $\mathcal{T}_0$ has recently been proposed by Diening, Gehring and Storn (2023); hereafter we present a short account of it.

The key concept is that of *colouring* the vertices of $\mathcal{T}_0$. A coloured initial triangulation in $\mathbb{R}^d$ is a pair $(\mathcal{T}_0, c)$, where $c\colon \mathcal{V}_{\mathcal{T}_0} \to \{0, \dots, d\}$ is such that the colours of all vertices of each $T \in \mathcal{T}_0$ are distinct. The colour map c allows us to sort the vertices of each initial element $T = \{z_0, \dots, z_d\}_t \in \mathcal{T}_0$ so that

$$c(z_j) = j, \quad j \in \{0, \dots, d\}.$$

To refine a marked $T \in \mathcal{T}_0$, we apply the Maubach bisection rule leading to (3.32), and possibly add a recursive closure, which is proved to terminate, to guarantee the conformity of the final triangulation. The colouring property is conserved in this process, and the conclusion of Theorem 3.16 below holds true, starting from any initially coloured triangulation $\mathcal{T}_0$.

Unfortunately, not every initial triangulation can be coloured. For instance, consider in dimension $d = 2$ a patch of triangles sharing a common vertex. If colour 0 is assigned to such an inner vertex, then the outer vertices must take colours 1 and 2 successively, but if the number of triangles in the patch is odd, there will be a vertex that is not colourable.

To overcome this obstruction, Diening *et al.* (2023) propose using more colours, and introduce the concept of *generalized colouring*: a pair $(\mathcal{T}_0, c)$ is an $(N + 1)$-coloured triangulation if there exists an integer $N \geq d$ and a mapping $c\colon \mathcal{V}_{\mathcal{T}_0} \to \{0, \dots, N\}$ such that the colours of all vertices of each $T \in \mathcal{T}_0$ are distinct. Any initial $\mathcal{T}_0$ can be coloured in this generalized sense: indeed, after the initialization $c(z) = +\infty$ for all $z \in \mathcal{V}_{\mathcal{T}_0}$, we define

$$c(z) := \min(\mathbb{N}_0 \setminus \{c(w) \mid [z, w] \text{ is an edge of } \mathcal{T}_0\}), \quad z \in \mathcal{V}_{\mathcal{T}_0},$$

as the smallest colour not already attained by a neighbouring vertex. Then, we set $N := \max\{c(z) \mid z \in \mathcal{V}_{\mathcal{T}_0}\}$, and note that N is bounded by the maximal number of edges connected to a vertex of $\mathcal{T}_0$.

A generalized $(N+1)$-coloured triangulation $(\mathcal{T}_0, c)$ in $\mathbb{R}^d$ can be seen as a collection of d-simplices contained in a virtual, coloured triangulation $\mathcal{T}_0^+$ in $\mathbb{R}^N$. It suffices to add $N - d$ virtual nodes to each simplex in $\mathcal{T}_0$, so that it becomes a N-simplex, and attribute to these nodes the remaining $N - d$ colours. Note that these virtual simplices are only connected via their d-subsimplices belonging to $\mathcal{T}_0$. In the example mentioned above of a patch of triangles sharing a vertex, a $(3+1)$-coloured triangulation is defined as follows: a tetrahedron is built on top of each triangle; the previously uncolourable vertex takes the new colour 3, whereas colour 2 is attributed to the new vertices of the two tetrahedra sharing that vertex; the new vertex of any other tetrahedron takes the colour 3.

With the new triangulation $\mathcal{T}_0^+$ at hand, one could apply the Mauback bisection rule to it, which as a by-product would refine the initial triangulation $\mathcal{T}_0$. However, Diening, Gehring and Storn (2023) suggest a short-cut that directly refines $\mathcal{T}_0$ by invoking an algorithm that bisects a k-simplex in dimension $m > k$. A further round of recursive refinements may be needed to guarantee conformity. Diening *et al.* prove that the recursion terminates. In addition, for any $(N+1)$-coloured initial triangulation, the conclusion of Theorem 3.16 below holds true in this case too, with a constant D satisfying $D \lesssim N^d$.

The procedure REFINE. Given $\mathcal{T} \in \mathbb{T}$ and a selected subset $\mathcal{M} \subset \mathcal{T}$ (the set of *marked* elements), the procedure

$$[\mathcal{T}_*] = \mathsf{REFINE}(\mathcal{T}, \mathcal{M})$$

creates a new conforming refinement $\mathcal{T}_*$ of $\mathcal{T}$ by bisecting all elements of $\mathcal{M}$ at least once and perhaps additional elements to keep conformity.

Conformity is a constraint in the refinement procedure that prevents it from being completely local. The propagation of refinement beyond the set of marked elements $\mathcal{M}$ is a rather delicate matter, which we discuss later in Section 8. For instance, we show that a naive estimate of the form

$$\#\mathcal{T}_* - \#\mathcal{T} \leq D\,\#\mathcal{M}$$

is *not* valid with an absolute constant D independent of the refinement level. This can be repaired upon considering the cumulative effect for a sequence of conforming bisection meshes $\{\mathcal{T}_k\}_{k=0}^{\infty}$. This is expressed in the following crucial complexity result due to Binev *et al.* (2004) for $d = 2$ and Stevenson (2008) for $d > 2$. We present a complete proof later in Section 8.

Theorem 3.16 (complexity of REFINE). *If $\mathcal{T}_0$ satisfies the initial labelling* (3.35) *for $d = 2$, or that in Stevenson (2008, Section 4) for $d > 2$, then there exists a constant $D > 0$ depending only on $\mathcal{T}_0$ and d such that, for all $k \geq 1$,*

$$\#\mathcal{T}_k - \#\mathcal{T}_0 \leq D \sum_{j=0}^{k-1} \#\mathcal{M}_j.$$

If elements $T \in \mathcal{M}$ are to be bisected $b \geq 1$ times, then the procedure REFINE can be applied recursively, and Theorem 3.16 remains valid with D also depending on b.

Mesh overlay. For the subsequent discussion it will be convenient to merge (or superpose) two conforming meshes $\mathcal{T}_1, \mathcal{T}_2 \in \mathbb{T}$, thereby giving rise to the so-called *overlay* $\mathcal{T}_1 \oplus \mathcal{T}_2$. This operation corresponds to the union in the sense of trees (Cascón, Kreuzer, Nochetto and Siebert 2008, Stevenson 2007). We next bound the cardinality of $\mathcal{T}_1 \oplus \mathcal{T}_2$ in terms of that of $\mathcal{T}_1$ and $\mathcal{T}_2$.

Lemma 3.17 (mesh overlay). *Let $\mathcal{T}_1, \mathcal{T}_2 \in \mathbb{T}$. The overlay $\mathcal{T} = \mathcal{T}_1 \oplus \mathcal{T}_2 \in \mathbb{T}$ is conforming and*

$$\#\mathcal{T} \leq \#\mathcal{T}_1 + \#\mathcal{T}_2 - \#\mathcal{T}_0. \tag{3.37}$$

For a proof we refer to Cascón *et al.* (2008, Lemma 3.7), and to Proposition 8.15 below for a more general situation.

3.6. *Constructive approximation*

We now construct graded bisection meshes $\mathcal{T}$ for $n = 1, d = 2$ that achieve the optimal decay rate $(\#\mathcal{T})^{-1/2}$ of (3.25) under the global regularity assumption

$$v \in W^2_p(\Omega), \quad p > 1. \tag{3.38}$$

Therefore $W^2_p(\Omega)$ is strictly above the Sobolev line for the space $H^1(\Omega)$: $\operatorname{sob}(W^2_p) = 2 - 2/p > 0 = \operatorname{sob}(H^1)$. Note that $s = 1$, $p > 1$ and $q = 2$ obey the restriction (3.30) for $\nabla v \in L^2(\Omega)$. In particular, $W^2_p(\Omega)$ is compactly embedded into $H^1(\Omega)$ according to Lemma 2.1 (Sobolev embedding).

Following Binev, Dahmen, DeVore and Petrushev (2002) and Gaspoz and Morin (2014), we use a greedy algorithm that is based on the knowledge of the element errors and on bisection. The algorithm hinges on (3.24): if $\delta > 0$ is a given tolerance, the element error is equidistributed and within tolerance $e_{\mathcal{T}}(v,T) \approx \delta$, and the global error decays with maximum rate $(\#\mathcal{T})^{-1/2}$, then

$$\delta^2 \#\mathcal{T} \approx \sum_{T \in \mathcal{T}} e_{\mathcal{T}}(v,T)^2 = |v - I_{\mathcal{T}} v|^2_{H^1(\Omega)} \lesssim (\#\mathcal{T})^{-1}$$

whence $\#\mathcal{T} \lesssim \delta^{-1}$; here $I_{\mathcal{T}}$ stands for the Lagrange interpolation operator. With this in mind, we impose $e_T(v) \leq \delta$ as a threshold to stop refining and expect $\#\mathcal{T} \lesssim \delta^{-1}$. The following algorithm implements this idea.

Algorithm 3.18 (greedy algorithm). Given a tolerance $\delta > 0$ and a conforming mesh $\mathcal{T}_0$, GREEDY finds a conforming refinement $\mathcal{T} \geq \mathcal{T}_0$ of $\mathcal{T}_0$ by bisection such that $e_{\mathcal{T}}(v, T) \leq \delta$ for all $T \in \mathcal{T}$: let $\mathcal{T} = \mathcal{T}_0$ and

```
[𝒯] = GREEDY(𝒯, δ, v)
   while ℳ := {T ∈ 𝒯 | e_𝒯(v,T) > δ} ≠ ∅
      𝒯 := REFINE(𝒯, ℳ)
   return 𝒯
```

Since $W_p^2(\Omega) \subset C^0(\overline{\Omega})$, because $p > 1$, we can use the Lagrange interpolant and local estimate (3.17) with $r = \mathrm{sob}(W_p^2) - \mathrm{sob}(H^1) = 2 - 2/p > 0$. We deduce

$$e_{\mathcal{T}}(v, T) \lesssim h_T^r \, \|D^2 v\|_{L^p(T)}. \tag{3.39}$$

We assess the quality of the resulting mesh in a slightly more general setting, following Bonito *et al.* (2016, Proposition 1 and Corollary 1), needed later in Sections 6 and 7 for solution and data approximation for any polynomial degree.

An abstract greedy algorithm. We consider a generic (possibly vector-valued) function $v \in L^q(\Omega, \mathbb{R}^M)$, with $M \geq 1$ and $1 \leq q \leq \infty$, let $e_{\mathcal{T}}(v, T) = \|v - \Pi_{\mathcal{T}} v\|_{L^q(T)}$ denote the abstract L^q-local error for $T \in \mathcal{T}$ used in the GREEDY procedure, and let $E_{\mathcal{T}}(v) = \|v - \Pi_{\mathcal{T}} v\|_{L^q(\Omega)}$ denote the global L^q-interpolation error by either continuous or discontinuous piecewise polynomials (the definition of $\Pi_{\mathcal{T}} v$ is irrelevant now). We formulate the following assumptions.

- *Summability in ℓ^q.* The errors $\{e_{\mathcal{T}}(v, T)\}_{T \in \mathcal{T}}$ satisfy

$$\|v\|_{L^q(\Omega)}^q \lesssim \sum_{T \in \mathcal{T}} e_{\mathcal{T}}(v, T)^q. \tag{3.40}$$

Rather than (3.38), we assume that v belongs to an abstract space $X_p^s(\Omega; \mathcal{T}_0)$ of functions with differentiability index $s \in (0, n]$ and integrability index $p \in (0, \infty]$ piecewise over $\mathcal{T}_0$ with two crucial properties.

- *Local error estimate.* For $r = s - d/p + d/q > 0$ and all $T \in \mathcal{T}$,

$$e_{\mathcal{T}}(v, T) \lesssim h_T^r \, |v|_{X_p^s(T)}. \tag{3.41}$$

- *Norm subadditivity.* For $p < \infty$, and obvious modification for $p = \infty$,

$$\sum_{T \in \mathcal{T}} |v|_{X_p^s(T)}^p \lesssim |v|_{X_p^s(\Omega; \mathcal{T}_0)}^p. \tag{3.42}$$

The space $X_p^s(\Omega; \mathcal{T}_0)$ will later be either a Sobolev space $W_p^s(\Omega; \mathcal{T}_0)$, a Besov space $B_{p,p}^s(\Omega; \mathcal{T}_0)$ or a Lipschitz space $\mathrm{Lip}_p^s(\Omega; \mathcal{T}_0)$, with piecewise regularity over $\mathcal{T}_0$; the latter two will allow $0 < p < 1$. For the moment we do not need to be specific and just rely on the two properties above.

Proposition 3.19 (abstract greedy error). *Let $\mathcal{T}_0$ be an initial subdivision of $\Omega \subset \mathbb{R}^d$ satisfying the initial labelling property* (3.35) *for $d = 2$, or its variant for*

$d > 2$. Let $M \geq 1$, $0 < q, p \leq \infty$ and $s - d/p + d/q > 0$. Let $v \in L^q(\Omega, \mathbb{R}^M)$ satisfy (3.40), (3.41) *and* (3.42). *Then* $\mathsf{GREEDY}(\mathcal{T}_0, \delta, v)$ *terminates in a finite number of iterations with local errors verifying $e_{\mathcal{T}}(v, T) \leq \delta$ for all $T \in \mathcal{T}$, and there is a constant $C = C(p, q, s, d, \Omega, \mathcal{T}_0)$ such that the output $\mathcal{T} \in \mathbb{T}$ satisfies*

$$\|v - \Pi_{\mathcal{T}} v\|_{L^q(\Omega)} \leq C |v|_{X_p^s(\Omega;\mathcal{T}_0)} (\#\mathcal{T} - \#\mathcal{T}_0)^{-s/d}. \tag{3.43}$$

Proof. We proceed in several steps.

[1] *Termination.* Since h_T decreases monotonically to 0 with bisection, so does $e_{\mathcal{T}}(v, T)$ in view of (3.41). Consequently, GREEDY terminates in a finite number $k \geq 1$ of iterations. Upon termination, the local errors satisfy $e_{\mathcal{T}}(v, T) \leq \delta$ for all $T \in \mathcal{T}$ by construction, whence (3.40) implies

$$\|v - \Pi_{\mathcal{T}} v\|_{L^q(\Omega)} \lesssim \delta (\#\mathcal{T})^{1/q}.$$

[2] *Counting.* Let $\mathcal{M} = \mathcal{M}_0 \cup \cdots \cup \mathcal{M}_{k-1}$ be the set of marked elements. We organize the elements in $\mathcal{M}$ by size in a way that allows for a counting argument. Let $\mathcal{P}_j$ be the set of elements T of $\mathcal{M}$ with size

$$2^{-(j+1)} \leq |T| < 2^{-j} \quad \Rightarrow \quad 2^{-(j+1)/d} \leq h_T < 2^{-j/d},$$

because $h_T = |T|^{1/d}$ for shape-regular meshes $\mathcal{T} \in \mathbb{T}$.

We first observe that all the T in $\mathcal{P}_j$ are *disjoint*. This is because if $T_1, T_2 \in \mathcal{P}_j$ and $\mathring{T}_1 \cap \mathring{T}_2 \neq \emptyset$, then one of them is contained in the other, say $T_1 \subset T_2$, due to the bisection procedure which works in any dimension $d \geq 1$; see Section 8. Hence

$$|T_1| \leq \frac{1}{2} |T_2|,$$

contradicting the definition of $\mathcal{P}_j$. This implies the first bound

$$2^{-(j+1)} \#\mathcal{P}_j \leq |\Omega| \quad \Rightarrow \quad \#\mathcal{P}_j \leq |\Omega| \, 2^{j+1}. \tag{3.44}$$

In light of (3.41), we have for $T \in \mathcal{P}_j$

$$\delta \leq e_{\mathcal{T}}(v, T) \lesssim 2^{-jr/d} |v|_{X_p^s(T)}.$$

Therefore, accumulating these quantities in ℓ^p and invoking (3.42) yields

$$\delta^p \, \#\mathcal{P}_j \lesssim 2^{-jrp/d} \sum_{T \in \mathcal{P}_j} |v|^p_{X_p^s(T)} \lesssim 2^{-jrp/d} \, |v|^p_{X_p^s(\Omega;\mathcal{T}_0)}$$

and gives rise to the second bound

$$\#\mathcal{P}_j \lesssim \delta^{-p} \, 2^{-jrp/d} \, |v|^p_{X_p^s(\Omega;\mathcal{T}_0)}. \tag{3.45}$$

[3] *Cardinality.* The two bounds for $\#\mathcal{P}$ in (3.44) and (3.45) are complementary. The first one is good for j small whereas the second is suitable for j large (think of $\delta \ll 1$). The crossover takes place for j_0 such that

$$2^{j_0+1} |\Omega| \approx \delta^{-p} \, 2^{-j_0 rp/d} |v|^p_{X_p^s(\Omega;\mathcal{T}_0)} \quad \Rightarrow \quad 2^{j_0} \approx \left(|\Omega|^{-1} \delta^{-p} |v|^p_{X_p^s(\Omega;\mathcal{T}_0)} \right)^{d/(d+rp)}.$$

We now compute

$$\#\mathcal{M} = \sum_j \#\mathcal{P}_j \lesssim \sum_{j\le j_0} 2^j|\Omega| + \delta^{-p}\,|v|^p_{X^s_p(\Omega)} \sum_{j>j_0}(2^{-rp/2})^j.$$

Since

$$\sum_{j\le j_0} 2^j \approx 2^{j_0}, \quad \sum_{j>j_0}(2^{-rp/d})^j \lesssim 2^{-rpj_0/d},$$

we can write

$$\#\mathcal{M} \lesssim |\Omega|^{1-d/(d+rp)}\big(\delta^{-1}|v|_{X^s_p(\Omega;\mathcal{T}_0)}\big)^{dp/(d+rp)}.$$

We finally apply Theorem 3.16 (complexity of REFINE), to arrive at

$$\#\mathcal{T} - \#\mathcal{T}_0 \lesssim \#\mathcal{M} \lesssim |\Omega|^{rp/(d+rp)}\big(\delta^{-1}|v|_{X^s_p(\Omega;\mathcal{T}_0)}\big)^{dp/(d+rp)},$$

or equivalently

$$\delta \lesssim |\Omega|^{r/d}|v|_{X^s_p(\Omega;\mathcal{T}_0)}(\#\mathcal{T} - \#\mathcal{T}_0)^{-(d+rp)/(dp)}.$$

[4] *Total error.* Since

$$\frac{d+rp}{dp} = \frac{s}{d} + \frac{1}{q},$$

we deduce from step [1] that

$$\|v - \Pi_\mathcal{T} v\|_{L^q(\Omega)} \lesssim \delta(\#\mathcal{T})^{1/q} \lesssim |\Omega|^{r/d}|v|_{X^s_p(\Omega;\mathcal{T}_0)}(\#\mathcal{T} - \#\mathcal{T}_0)^{-s/d},$$

which is the desired estimate. □

The output mesh $\mathcal{T}$ of GREEDY($\mathcal{T}_0, \delta, v$) starting from $\mathcal{T}_0$ satisfies $\#\mathcal{T} \ge c_0\#\mathcal{T}_0$ for some $c_0 > 1$, whence $\#\mathcal{T} - \#\mathcal{T}_0 \ge (1 - 1/c_0)\#\mathcal{T}$ and (3.43) yields

$$\|v - \Pi_\mathcal{T} v\|_{L^q(\Omega)} \lesssim C|v|_{X^s_p(\Omega;\mathcal{T}_0)}(\#\mathcal{T})^{-s/d}, \tag{3.46}$$

where C depends on c_0. In many applications of GREEDY, to be discussed later in Sections 6 and 7, it will be convenient for the starting mesh to be a conforming refinement of $\mathcal{T}_0$ to enhance its efficiency. We will prove in Section 7.1 that (3.46) remains valid.

It is instructive to realize that GREEDY is a practical algorithm that hinges on the different summabilities of (3.40) and (3.42), and delivers a global L^q-error consistent with (3.29) of Section 3.4. Moreover, the outcome graded grid $\mathcal{T}$ is quasi-optimal but may not equidistribute the error, not even approximately.

We are now in a position to show that GREEDY constructs optimal graded meshes for the interpolation error in $H^1(\Omega)$ alluded to at the beginning of this section. To this end, we let $I_\mathcal{T}$ be the Lagrange interpolation operator for $d = 2$.

Corollary 3.20 (optimal H^1-convergence rate). *If $v \in H^1(\Omega) \cap W^2_p(\Omega)$ for $1 < p \le 2$ and $d = 2$, then* GREEDY *yields graded bisection meshes $\mathcal{T}$ so that*

$$|v - I_\mathcal{T} v|_{H^1(\Omega)} \lesssim |\Omega|^{1-1/p}\|D^2 v\|_{L^p(\Omega)}(\#\mathcal{T})^{-1/2}.$$

Proof. We invoke Proposition 3.19 (abstract greedy error) and equation (3.46) for $\nabla v \in L^2(\Omega, \mathbb{R}^2)$ with $\Pi_{\mathcal{T}} \nabla v = \nabla I_{\mathcal{T}} v$ and $s = 1$, $q = 2$, $p > 1$, whence $s - d/p + d/q > 0$. □

Remark 3.21 (piecewise W^2_p-smoothness). Since (3.39) is completely local for $d = 2$, we see from (3.42) that it suffices for $v \in H^1(\Omega)$ to be piecewise in W^2_p over the initial partition $\mathcal{T}_0$, namely $W^2_p(\Omega; \mathcal{T}_0)$. It turns out that this statement is valid for any dimension $d \geq 2$ in view of Proposition 3.9 (approximation of gradients). We will revisit this issue in Section 6.8.

Remark 3.22 (case $p < 1$). We now consider polynomial degree $n \geq 1$. The integrability p corresponding to differentiability $n+1$ results from equating Sobolev numbers:

$$n + 1 - \frac{d}{p} = \operatorname{sob}(H^1) = 1 - \frac{d}{2} \quad \Rightarrow \quad p = \frac{2d}{2n + d}.$$

Depending on $d \geq 2$ and $n \geq 1$, this may lead to $0 < p < 1$, in which case $W^{n+1}_p(\Omega)$ is to be replaced by the Besov space $B^s_{p,p}(\Omega)$ for $s < n + 1$ or the Lipschitz space $\mathrm{Lip}^{n+1}_p(\Omega)$ (DeVore 1998). We will discuss this matter in Section 6.8 and make the abstract greedy setting precise.

Remark 3.23 (isotropic vs. anisotropic elements). Since geometric singularities are of the form (3.26) for $d = 2$, Corollary 3.20 (optimal H^1-convergence rate) shows that isotropic graded meshes are able to deliver optimal convergence rates for $d = 2$. Unfortunately this is no longer the case for $d > 2$, and anisotropic meshes are necessary for optimal meshes. This topic is delicate in several respects. Deriving reliable and efficient *a posteriori* error estimators is largely open for anisotropic meshes; this is the subject of Section 4 for isotropic meshes. Even having such estimators, building a theory of adaptivity is open; this is the subject of Sections 5, 6 and 10 for isotropic meshes. Finally, constructing anisotropic meshes based on *a posteriori* information alone and that easily allow for refinement and coarsening is problematic. For these reasons we do not dwell on anisotropic refinement in this survey.

3.7. Non-conforming meshes

More general subdivisions of Ω than those in Section 3.5 are used in practice. If the elements of $\mathcal{T}_0$ are quadrilaterals for $d = 2$, or their multidimensional variant for $d > 2$, then it is natural to allow for improper or *hanging nodes* for the resulting refinements $\mathcal{T}$ to be graded; see Figure 3.9(a). On the other hand, if $\mathcal{T}_0$ is made of triangles for $d = 2$, or simplices for $d > 2$, then red refinement without green completion also gives rise to graded meshes with hanging nodes; see Figure 3.9(b). In both cases, the presence of hanging nodes is inevitable to enforce mesh grading. Finally, bisection may produce meshes with hanging nodes, as depicted in Figure 3.9(c), if the completion process is incomplete. All three

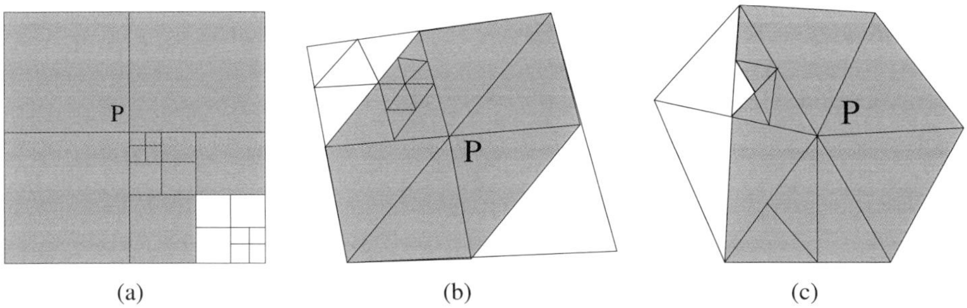

Figure 3.9. Non-conforming meshes made of quadrilaterals (a), triangles with red refinement (b) and triangles with bisection (c). The shaded regions depict the domain of influence of a proper or conforming node P.

refinements maintain shape regularity, but for both practice and theory, they cannot be arbitrary: we need to restrict the level of non-conformity. We discuss this next, starting with the case of polynomial degree $n = 1$ (Bonito and Nochetto 2010, Beirão da Veiga *et al.* 2023).

We say that a node P of $\mathcal{T}$ is a *proper* (or conforming) node if it is a vertex of all elements containing P; otherwise, we say that P is an *improper* (non-conforming or hanging) node. The set $\mathcal{N}$ of all nodes of $\mathcal{T}$ is thus partitioned into the set $\mathcal{P}$ of proper nodes, and the set $\mathcal{H} = \mathcal{N} \setminus \mathcal{P}$ of hanging nodes.

A useful notion in dealing with hanging nodes is the *global index* of a node, introduced in Beirão da Veiga *et al.* (2023): it measures the number of non-conforming refinements needed to generate a hanging node from proper nodes. To define it, for any $x \in \mathcal{H}$ which has been generated by the bisection of an edge $[x', x'']$, let us set $\mathcal{B}(x) = \{x', x''\}$.

Definition 3.24 (global index of a node). The global index $\lambda(x)$ of a node $x \in \mathcal{N}$ is defined recursively as follows:

- if $x \in \mathcal{P}$, set $\lambda(x) = 0$;
- if $x \in \mathcal{H}$ and $\mathcal{B}(x) = \{x', x''\}$, set $\lambda(x) = \max(\lambda(x'), \lambda(x'')) + 1$.

The set of all nodes of $\mathcal{T}$ is thus partitioned according to the value of the global index: for any integer $l \geq 0$, we set $\mathcal{H}_l = \{x \in \mathcal{N} \mid \lambda(x) = l\}$. Note that $\mathcal{H}_0 = \mathcal{P}$. An example of distribution of global indices for $d = 2$ is shown in Figure 3.10.

We define the global index of the triangulation $\mathcal{T}$ by $\lambda(\mathcal{T}) := \max_{x \in \mathcal{N}} \lambda(x)$. The level of non-conformity of the triangulations we are dealing with is controlled by the following condition of admissibility.

Definition 3.25 (Λ-admissibility). Let $\Lambda \geq 0$ be an integer. A refinement $\mathcal{T}$ of $\mathcal{T}_0$ is Λ-admissible if

$$\lambda(\mathcal{T}) \leq \Lambda. \tag{3.47}$$

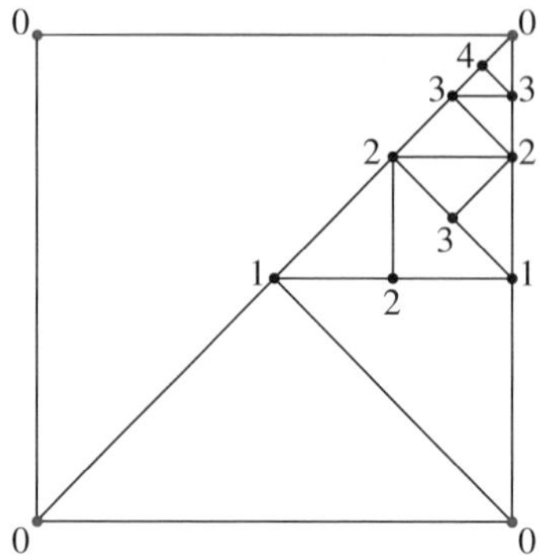

Figure 3.10. Example of distributions of proper nodes (red) and hanging nodes (black), with associated global indices λ.

If $\lambda(\mathcal{T}) \geq 1$, then $\mathcal{T}$ is non-conforming, but otherwise $\mathcal{T} \in \mathbb{T}$ is conforming if $\lambda(\mathcal{T}) = 0$. The collection of all Λ-admissible partitions is denoted by $\mathbb{T}^\Lambda$.

Λ-admissibility has the following basic implications.

Proposition 3.26 (properties of Λ-admissible partitions). *Let T be any element of a Λ-admissible partition $\mathcal{T}$.*

(i) *If $e \subset \partial T$ is an edge of T, then e may contain at most $2^\Lambda - 1$ hanging nodes.*

(ii) *If $e \subset \partial T$ is an edge of some other element T', then $h_{T'} \simeq h_T$, where the hidden constants depend only on the shape of the initial triangulation $\mathcal{T}_0$ and possibly on Λ.*

Proof. Case (i) stems from the fact that the edge may contain at most 2^{k-1} hanging nodes of level k for $1 \leq k \leq \Lambda$. To prove (ii) we observe that the length ratio $|\bar{e}|/|e|$, where $\bar{e}$ is the edge of T containing e, is at most 2^Λ, and we conclude by invoking the shape regularity of the partition. $\square$

In the space $\mathbb{V}_\mathcal{T}$ of continuous piecewise linear maps over $\mathcal{T}$, functions are uniquely defined by their values at the proper nodes of $\mathcal{T}$. So it is natural to introduce the canonical continuous piecewise linear basis functions ϕ_P associated with any proper node P. They satisfy

$$v = \sum_{P\in\mathcal{P}} v(P)\phi_P \quad \text{for all } v \in \mathbb{V}_\mathcal{T}, \tag{3.48}$$

and are defined by the conditions $\phi_P \in \mathbb{V}_\mathcal{T}$ and

- $\phi_P(z) = 1$ if $z = P$, $\phi_P(z) = 0$ if $z \in \mathcal{P} \setminus \{P\}$.

The values of ϕ_P at the hanging nodes, hence everywhere in the domain, can be reconstructed by linear interpolation as follows: assuming that ϕ_P has been defined

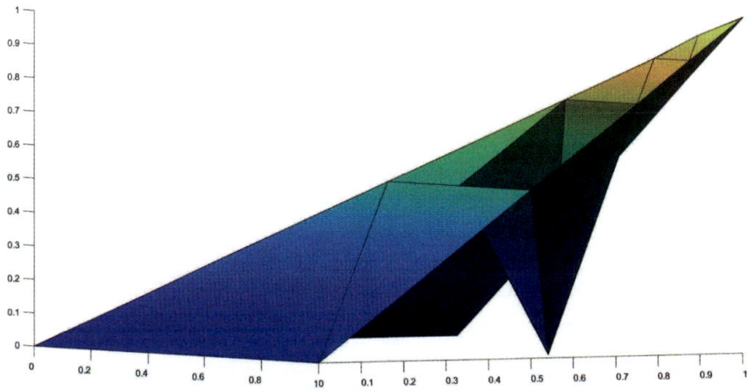

Figure 3.11. Plot of the basis function ϕ_P on the non-conforming triangulation shown in Figure 3.10, for P equal to the upper right corner of the domain, after using bisection to convert the lowest hanging node with global index 3 into a proper node.

at all nodes of global index $< l$, if $z \in \mathcal{H}_l$ and $\mathcal{B}(z) = \{z', z''\}$, then

$$\phi_P(z) = \frac{1}{2}(\phi_P(z') + \phi_P(z'')).$$

An example of basis function ϕ_P on a non-conforming triangulation is provided in Figure 3.11.

The *domain of influence* of a proper node P is the set

$$\omega_{\mathcal{T}}(P) = \operatorname{supp}(\phi_P),$$

highlighted in grey in Figure 3.9; this notion was introduced in Babuška and Miller (1987) in the context of K-meshes; see also Bonito and Nochetto (2010). To identify elements $T \in \mathcal{T}$ contained in $\omega_{\mathcal{T}}(P)$, we introduce for any node $x \in \mathcal{N}$ the set $\mathcal{P}(x)$ of the *proper nodes influencing* x, which is defined recursively as follows:

- initialize $\mathcal{P}(x) = \{x\}$;
- while $\mathcal{P}(x) \cap \mathcal{H} \neq \emptyset$, if $y \in \mathcal{P}(x) \cap \mathcal{H}$ replace $\mathcal{P}(x)$ with $(\mathcal{P}(x) \setminus \{y\}) \cup \mathcal{B}(y)$.

Then $T \subseteq \omega_{\mathcal{T}}(P)$ if and only if P influences some vertex of T, that is, there exists a vertex v of T such that $P \in \mathcal{P}(v)$.

One of the consequences of the Λ-admissibility assumption of $\mathcal{T}$ is the following result, which says that all elements T contained in $\omega_{\mathcal{T}}(P)$ have comparable size.

Proposition 3.27 (size of the domain of influence). *There exists a positive constant $C = C(\mathcal{T}_0, \Lambda)$, depending only on the shape of the initial triangulation $\mathcal{T}_0$ and possibly on Λ, such that for any $P \in \mathcal{P}$*

$$\operatorname{diam} \omega_{\mathcal{T}}(P) \leq C\, h_T \quad \textit{for all } T \in \mathcal{T},\ T \subseteq \omega_{\mathcal{T}}(P).$$

Proof. Elements in $\omega_{\mathcal{T}}(P)$ having P as a vertex share in pairs an edge or a portion of an edge, hence – as noted above – Λ-admissibility implies the existence of a characteristic size, say h_P, which is comparable to the diameter of each of them. On the other hand, any $T \subset \omega_{\mathcal{T}}(P)$ not containing P has at least one vertex $v_T \in \mathcal{H}$ such that $P \in \mathcal{P}(v_T)$. Thus there exists a sequence $\{y_k \mid 0 \le k \le K\}$ of vertices satisfying $y_0 = v_T$, $y_K = P$ and $y_{k+1} \in \mathcal{B}(y_k)$ for $0 \le k < K$; since $\lambda(y_k) \ge \lambda(y_{k+1}) + 1$, necessarily $K \le \Lambda$. Correspondingly, we can find a chain of at most K elements, starting at T and ending at an element containing P, which share in pairs an edge or a portion of an edge. We deduce that $h_T \simeq h_P$, and $\operatorname{dist}(T, P) \simeq h_T$, where the hidden constants may depend only on the shape of the initial triangulation and on Λ. The conclusion easily follows from these results. □

We now turn to the case of polynomial degree $n > 1$; we refer to Canuto and Fassino (2023) for more details. The concept of hanging node is no longer solely related to the geometry of the mesh, but also to the distribution of degrees of freedom along the edges of the elements. For instance, consider a full edge e shared by two triangles T and T', and bisect T' to create two new elements T_1 and T_2 having e as a vertex. If we use quadratic Lagrangian elements, the midpoint x of e carries a degree of freedom for the three elements that share it, so we do not consider it as a hanging node; on the other hand, the nodes at distance $\frac{1}{4}|e|$ and $\frac{3}{4}|e|$ from an endpoint of e are hanging nodes (despite being vertices of no triangle) since they do not carry a degree of freedom for the element T. If we move to cubic Lagrangian elements, then x becomes a hanging node, together with the nodes at distance $\frac{1}{6}|e|$ and $\frac{5}{6}|e|$ from an endpoint of e, whereas the nodes at distance $\frac{1}{3}|e|$ and $\frac{2}{3}|e|$ are not hanging nodes, since they carry a degree of freedom for each triangle they belong to.

In general, for a partition $\mathcal{T}$ made of classical affine Lagrangian or Hermitian elements, the hanging nodes are defined as follows.

- Given an element $T \in \mathcal{T}$, the set $\mathcal{P}_T$ of the *proper nodes of* T is made of all images of the reference n-lattice via the affine transformation. The set $\mathcal{H}_T$ of the *hanging nodes of* T collects the points of ∂T that are not proper nodes of T, but are proper nodes of some other contiguous element T'. The set of all nodes of T is $\mathcal{N}_T := \mathcal{P}_T \cup \mathcal{H}_T$.
- At the global level, if $\mathcal{N} = \bigcup_{T \in \mathcal{T}} \mathcal{N}_T$ is the set of all nodes of $\mathcal{T}$, the set $\mathcal{P} \subseteq \mathcal{N}$ of the *proper nodes of* $\mathcal{T}$ contains those nodes that are proper nodes for all elements they belong to. The complementary set $\mathcal{H} := \mathcal{N} \setminus \mathcal{P}$ is the set of the *hanging nodes of* $\mathcal{T}$.

In other words, a hanging node of $\mathcal{T}$ is a point that carries a degree of freedom for some but not all elements it belongs to. With this definition of proper nodes, representation (3.48) of continuous piecewise linear maps extends to $n > 1$.

The *global index* $\lambda(x)$ of a node $x \in \mathcal{N}$ is precisely defined as in Definition 3.24. The set $\mathcal{B}(x) \subset \mathcal{N}$ collects the endpoints of an interval $[x', x'']$, contained in the skeleton of $\mathcal{T}$, that has been bisected when x has been created, and contains no other node inside.

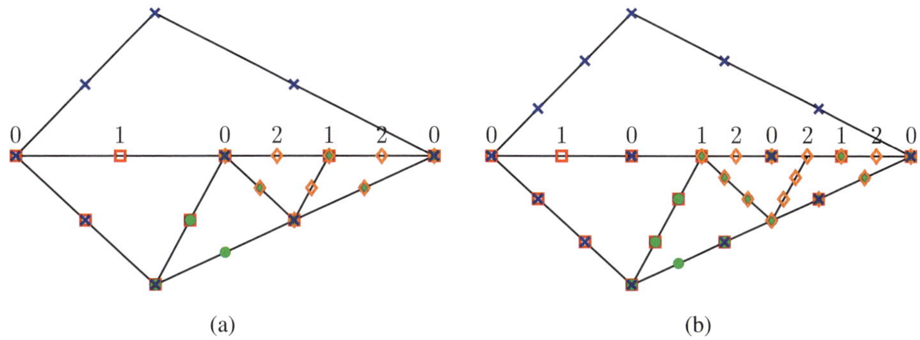

Figure 3.12. Triangulation after the three refinements in the case $n = 2$ (a) and in the case $n = 3$ (b). Blue crosses represent the original degrees of freedom on the initial conforming mesh. Red squares, green circles and orange triangles are used for the degrees of freedom of the first, second and third refinement, respectively. All nodes are proper, except those on the horizontal line, whose global index is reported.

Figure 3.12 provides two examples, for $n = 2$ and $n = 3$, of distributions of hanging nodes and corresponding global indices, created by successive bisections starting from an initial conforming partition.

The concept of Λ-admissibility, given in Definition 3.25, remains unchanged for $n > 1$. The statements in Proposition 3.26, too, extend to the higher-order case; the maximum number of hanging nodes on an edge now being $O(n2^{\Lambda})$. Consequently, the conclusion of Proposition 3.27 remains valid when $n > 1$ as well: there is a constant $C = C(\mathcal{T}_0, \Lambda)$ such that

$$\operatorname{diam} \omega_{\mathcal{T}}(P) \le C\, h_T \quad \text{for all } T \in \mathcal{T},\ T \subseteq \omega_{\mathcal{T}}(P). \tag{3.49}$$

Remark 3.28 (quadrilateral and hexahedral partitions). It is readily seen that the definitions of global index of a node and Λ-admissible partition extend seamlessly to shape-regular meshes made of quadrilaterals refined by a quadtree strategy ($d = 2$) (see Figure 3.9(a) for an example), or by hexahedra refined by an octree strategy ($d = 3$). The same holds for heterogeneous partitions made of a combination of simplices and hexahedra. All results reported above are valid for such partitions. We refer to Bonito and Nochetto (2010) for details.

Λ-admissible meshes under refinement. Given a Λ-admissible grid $\mathcal{T}$, a subset $\mathcal{M}$ of elements marked for refinement, and a desired number $b \ge 1$ of subdivisions to be performed in each marked element, the procedure

$$\mathcal{T}_* = \mathsf{REFINE}(\mathcal{T}, \mathcal{M}, \Lambda)$$

creates a minimal Λ-admissible mesh $\mathcal{T}_* \ge \mathcal{T}$ such that all the elements of $\mathcal{M}$ are subdivided at least b times. In order for $\mathcal{T}_*$ to be Λ-admissible, perhaps other elements not in $\mathcal{M}$ must be partitioned. Despite the fact that admissibility is a constraint on the refinement procedure weaker than conformity, it cannot avoid the propagation of refinements beyond $\mathcal{M}$. The complexity of REFINE is again an issue which we discuss in Section 8.2: we show that Theorem 3.16 extends to this case.

Theorem 3.29 (complexity of REFINE for Λ-admissible meshes). *Let $\mathcal{T}_0$ be an arbitrary conforming partition of Ω, except for the bisection algorithm in which case $\mathcal{T}_0$ satisfies the labelling (3.35) for $d = 2$ or its higher-dimensional counterpart (Stevenson 2008). Then the estimate*

$$\#\mathcal{T}_k - \#\mathcal{T}_0 \leq D \sum_{j=0}^{k-1} \#\mathcal{M}_j \quad \textit{for all } k \geq 1$$

holds with a constant D depending on $\mathcal{T}_0$, d, n and Λ.

The following result about uniform refinements of Λ-admissible partitions will be used below. *The uniform refinement $\mathcal{T}_*$ of a partition $\mathcal{T} \in \mathbb{T}^\Lambda$ is the partition obtained by bisecting each element of $\mathcal{T}$ d times.* This implies, in particular, that each edge of $\mathcal{T}$ is bisected once.

Proposition 3.30 (Λ-admissibility of uniform refinements). *If $\mathcal{T} \in \mathbb{T}^\Lambda$ is a Λ-admissible partition and $\mathcal{T}_*$ is its uniform refinement, then $\mathcal{T}_*$ is Λ-admissible.*

Proof. A simple recursion argument on the global index of the hanging nodes of $\mathcal{T}$ shows that after refinement each such node either becomes a proper node or its global index is reduced by 1. At the same time, new nodes are created by the refinement, whose global index is at most 1 plus the maximal global index of the pre-existing nodes. In both cases, the maximal global index of $\mathcal{T}_*$ cannot exceed Λ. □

A simple consequence is the following result, which is useful for controlling the mesh size between consecutive refinements.

Corollary 3.31 (bound on the refinements). REFINE *with $b = 1$ never refines an element of a Λ-admissible partition $\mathcal{T}$ more than d times.*

Proof. REFINE gives the smallest Λ-admissible mesh $\mathcal{T}_*$ such that all the marked elements of $\mathcal{T}$ have been refined. Since the uniform refinement of $\mathcal{T}$ remains Λ-admissible, the minimality of $\mathcal{T}_*$ implies that no element of the marked set can be refined more than d times. □

We conclude by emphasizing that the polynomial interpolation and adaptive approximation theories of Sections 3.3 and 3.6 extend to non-conforming meshes with fixed level of incompatibility as well.

4. *A posteriori* error analysis

Numerical solutions to a boundary value problem serve to approximate its unknown exact solution. In such a context, it is of interest

- to quantify the error of the numerical solution,
- to gain information for *adapting* the discretization to the exact solution

in a computationally accessible manner. These are the two goals of an *a posteriori error analysis*, where the term *a posteriori* hints at the fact that the numerical solution itself can be involved. To achieve the two goals, the *a posteriori* analysis individuates so-called *error estimators* that, ideally, are computable, split into local contributions called *indicators*, and bound the error from above and below.

This section exemplifies such an analysis, considering the numerical solution of the boundary value problem (2.5), that is,

$$-\operatorname{div}(\boldsymbol{A}\nabla u) + cu = f \quad \text{in } \Omega, \quad u = 0 \quad \text{on } \partial\Omega,$$

with Lagrange elements of arbitrary fixed order $n \geq 1$. Throughout this section we adopt the notation and assumptions of previous sections for this model setting. In particular, the *exact solution* $u \in H_0^1(\Omega)$ solves the variational problem (2.7) and, given a simplicial conforming mesh $\mathcal{T} \in \mathbb{T}$ of Ω and finite element space

$$\mathbb{V}_{\mathcal{T}} := \left\{ v \in \mathbb{S}_{\mathcal{T}}^{n,0} \mid v|_{\Omega} = 0 \right\} \subset H_0^1(\Omega),$$

the *Galerkin approximation* solves

$$u_{\mathcal{T}} \in \mathbb{V}_{\mathcal{T}}: \quad \mathcal{B}[u_{\mathcal{T}}, w] = \langle f, w \rangle \quad \text{for all } w \in \mathbb{V}_{\mathcal{T}}, \tag{4.1}$$

with the bilinear form $\mathcal{B}$ from (2.8).

We stress that the analysis will be conducted under the regularity assumptions

$$\boldsymbol{A} \in L^{\infty}(\Omega; \mathbb{R}^{d\times d}), \quad c \in L^{\infty}(\Omega), \quad f \in H^{-1}(\Omega) = H_0^1(\Omega)^*, \tag{4.2}$$

used in Section 2.4 to establish existence and uniqueness of the exact solution. This fact distinguishes the approach below, which builds on Kreuzer and Veeser (2019), from most other approaches requiring additional regularity; see e.g. Verfürth (2013). Notably, this difference not only allows for covering more examples but is also related to strengthening the relationship between error and estimator to a true equivalence on any admissible mesh $\mathcal{T} \in \mathbb{T}$.

It is useful to recall two differences between the forcing f and the coefficients $(\boldsymbol{A}, c)$. First, while the exact solution u depends linearly on the forcing f, it depends nonlinearly on the diffusion tensor $\boldsymbol{A}$ and the reaction coefficient c. To state the second difference, let $u \in H_0^1(\Omega)$ and note that the assumptions (4.2) on $(\boldsymbol{A}, c)$ imply the 'missing' $f \in H^{-1}(\Omega)$. On the other hand, the assumptions on (A, f) imply only $cu \in H^{-1}(\Omega)$, while the assumptions on (c, f) imply only $-\operatorname{div}(\boldsymbol{A}\nabla u) \in H^{-1}(\Omega)$. These conditions are weaker than the 'missing' $u \in H_0^1(\Omega)$, and are due to the multiplicative role of $(\boldsymbol{A}, c)$ in the differential equation.

In order to elucidate the new twists allowing for (4.2), this section is organized as follows. We start with steps of the *a posteriori* analysis that are common to the 'classical' and new approaches. We then illustrate the classical approach with the *standard residual estimator*, and afterwards develop the new approach resulting in a modification of the standard residual estimator, called the *modified residual estimator*. Finally, we conclude by adapting the new approach to other techniques of *a posteriori* error estimation and boundary conditions.

In what follows, the notation may or may not indicate the dependences of a given quantity. We shall balance readability and the importance of the dependence in the given context. For example, in (4.1), the discrete solution depends not only on the mesh $\mathcal{T}$ but also on the data Ω, $\boldsymbol{A}$, c and f in problem (2.5). We write $\mathcal{T}$ explicitly because of its more prominent role in the *a posteriori* analysis. Let $\mathcal{F} := \mathcal{F}_{\mathcal{T}}$ denote the set of all interior $(d-1)$-dimensional faces of $\mathcal{T}$. The letter C will be used for a generic constant, with possibly different values at each occurrence. If not stated otherwise, it may depend on the shape regularity coefficient σ from (3.9), the dimension d and the polynomial degree n in $\mathbb{V}_{\mathcal{T}}$.

4.1. Error, residual and localization of residual norm

This section starts the *a posteriori* analysis by establishing that a suitable norm of the so-called *residual*

- is equivalent to the error $\|\nabla(u - u_{\mathcal{T}})\|_{L^2(\Omega)}$, and
- admits a *localization* in the sense that it splits into suitable local contributions depending on accessible quantities, i.e. on data $\mathcal{D} = (\boldsymbol{A}, c, f)$ and the discrete solution $u_{\mathcal{T}}$.

We do not yet consider computability: this important aspect will be addressed in the following sections.

Replacing the exact solution u in the weak form (2.7) with its approximation $u_{\mathcal{T}}$, we define the *residual* $R_{\mathcal{T}} \in H^{-1}(\Omega)$:

$$\langle R_{\mathcal{T}}, w\rangle = \langle f, w\rangle - \mathcal{B}[u_{\mathcal{T}}, w] \quad \text{for all } w \in H_0^1(\Omega).$$

We thus have a quantity that depends only on data $\mathcal{D}$ and the approximate solution $u_{\mathcal{T}}$ and relates to the error function $u - u_{\mathcal{T}}$ as follows:

$$\langle R_{\mathcal{T}}, w\rangle = \mathcal{B}[u - u_{\mathcal{T}}, w] \quad \text{for all } w \in H_0^1(\Omega). \tag{4.3}$$

Continuity and coercivity of the bilinear form $\mathcal{B}$ then provide a quantitative relationship between error and residual.

Lemma 4.1 (error and residual). *The error of the approximation $u_{\mathcal{T}}$ is equivalent to the residual norm. More precisely,*

$$\frac{1}{\|B\|}\|R_{\mathcal{T}}\|_{H^{-1}(\Omega)} \le \|\nabla(u - u_{\mathcal{T}})\|_{L^2(\Omega)} \le \frac{1}{\alpha}\|R_{\mathcal{T}}\|_{H^{-1}(\Omega)},$$

where $\|\mathcal{B}\| \geq \alpha > 0$ *are, respectively, the continuity and coercivity constants of the bilinear form* $\mathcal{B}$*.*

Proof. The error–residual relationship (4.3) yields the lower bound,

$$\begin{aligned}\|R_{\mathcal{T}}\|_{H^{-1}(\Omega)} &= \sup_{w \in H_0^1(\Omega)} \frac{\langle R_{\mathcal{T}}, w\rangle}{\|\nabla w\|_{L^2(\Omega)}} = \sup_{w \in H_0^1(\Omega)} \frac{\mathcal{B}[u - u_{\mathcal{T}}, w]}{\|\nabla w\|_{L^2(\Omega)}} \\ &\leq \|\mathcal{B}\| \|\nabla(u - u_{\mathcal{T}})\|_{L^2(\Omega)},\end{aligned}$$

while the choice $w = u - u_{\mathcal{T}}$ therein gives

$$\begin{aligned}\alpha\|\nabla(u - u_{\mathcal{T}})\|^2_{L^2(\Omega)} &\leq \mathcal{B}[u - u_{\mathcal{T}}, u - u_{\mathcal{T}}] = \langle R_{\mathcal{T}}, u - u_{\mathcal{T}}\rangle \\ &\leq \|R_{\mathcal{T}}\|_{H^{-1}(\Omega)} \|\nabla(u - u_{\mathcal{T}})\|_{L^2(\Omega)}\end{aligned}$$

and thus the upper bound. □

Remark 4.2 (role of forcing vs. role of coefficients). In addition to the two differences between right-hand side f and coefficients $(\boldsymbol{A}, c)$ mentioned in the introduction of this section, a third one implicitly arises in the proof of Lemma 4.1: the coefficients defining the bilinear form $\mathcal{B}$ are fixed, while the right-hand side f is replaced by the residual $R_{\mathcal{T}}$ in (4.3), which varies with the mesh $\mathcal{T}$.

Remark 4.3 (local lower estimate for the error). The proof of Lemma 4.1 shows that the lower bound of the error hinges on the continuity of the bilinear form $\mathcal{B}$. Since the evaluation of $\mathcal{B}$ involves only local operators, one might expect that there are also local lower bounds. This, however, depends on the interplay of the underlying differential operator and the choice of the test space norm. Indeed, in the case of the Poisson problem, i.e. $\boldsymbol{A} = \boldsymbol{I}$, $c = 0$, and the test space norm $\|\nabla \cdot \|_{L^2(\Omega)}$, we easily see that

$$\|R_{\mathcal{T}}\|_{H^{-1}(\omega)} \leq \|\nabla(u - u_{\mathcal{T}})\|_{L^2(\omega)}$$

for any subdomain $\omega \subset \Omega$. This local lower bound, however, does not carry over to the general case with $c \neq 0$, as the error function itself is bounded by its gradient only through the global inequality in Lemma 2.2 (first Poincaré inequality). On the other hand, endowing the test space $H_0^1(\Omega)$ with the full H^1-norm $\|\cdot\|_{H^1(\Omega)}$ yields

$$\|R_{\mathcal{T}}\|_{(H^1(\omega))^*} \leq \max\{\alpha_1, \|c\|_{L^\infty(\Omega)}\} \|u - u_{\mathcal{T}}\|_{H^1(\omega)}$$

for any subdomain $\omega \subset \Omega$.

In line with Carstensen *et al.* (2014), we shall not invoke local lower bounds to derive convergence and rate-optimality for the error of AFEM, although they might appear useful or even crucial in other settings.

Remark 4.4 (constants in error–residual relationship). Given the norm measuring the error, i.e. the norm of the trial space, the choice of the test space norm is important for the ensuing constants in the error–residual relationship; see e.g. Verfürth (2013, Sections 4.3, 4.6). To avoid related additional technicalities and difficulties, the test space is endowed with the straightforward norm $\|\nabla\cdot\|_{L^2(\Omega)}$.

Lemma 4.1 establishes the first goal that was set out at the beginning of this section. We now turn to the second one, that is, we split the residual norm $\|R_{\mathcal{T}}\|_{H^{-1}(\Omega)}$ into local contributions. Note that the nature of the dual norm $\|\cdot\|_{H^{-1}(\Omega)}$ makes this task less obvious than for integral norms as in the error $\|\nabla(u-u_{\mathcal{T}})\|_{L^2(\Omega)}$.

We start by recalling that the definition of the Galerkin approximation $u_{\mathcal{T}}$ implies that its residual is orthogonal to the discrete trial space $\mathbb{V}_{\mathcal{T}} = \mathbb{S}^{n,0}_{\mathcal{T}} \cap H^1_0(\Omega)$:

$$\langle R_{\mathcal{T}}, w\rangle = 0 \quad \text{for all } w \in \mathbb{V}_{\mathcal{T}}.$$

Let $\mathcal{V}$ denote the set of vertices of $\mathcal{T}$ and let $\phi_z \in \mathbb{S}^{1,0}_{\mathcal{T}}$ be the hat function with $\phi_z(y) = \delta_{yz}$ for all vertices $y \in \mathcal{V}$. In what follows, the *partial* orthogonality

$$\langle R_{\mathcal{T}}, \phi_z\rangle = 0 \quad \text{for all } z \in \mathcal{V} \tag{4.4}$$

will be crucial for splitting the non-local norm of the residual into local contributions. The latter ones will be formulated in terms of the supports of the hat functions, and thus to each vertex $z \in \mathcal{V}$ we associate the following subset and submesh:

$$\omega_z := \operatorname{supp}\phi_z = \bigcup_{T\in\mathcal{T}_z} T \quad \text{with} \quad \mathcal{T}_z := \{T \in \mathcal{T} \mid T \ni z\}. \tag{4.5}$$

These subsets, called *stars*, form a *subdomain covering* of Ω, that is, each interior $\mathring{\omega}_z$ is a domain and $\overline{\Omega} = \cup_{z\in\mathcal{V}}\,\omega_z$. The *overlapping index* $\operatorname{ess\,sup}_{x\in\Omega} \#\{z \in \mathcal{V} \mid \omega_z \ni x\}$ of this covering is bounded by $(d+1)$.

Lemma 4.5 (localization of H^{-1}-norm). *Let $\ell \in H^{-1}(\Omega)$ be an arbitrary linear functional on $H^1_0(\Omega)$.*

(i) *If $\langle \ell, \phi_z\rangle = 0$ for all interior vertices $z \in \mathcal{V}\cap\Omega$, then*

$$\|\ell\|^2_{H^{-1}(\Omega)} \le (d+1)C^2_{\mathrm{loc}} \sum_{z\in\mathcal{V}} \|\ell\|^2_{H^{-1}(\omega_z)},$$

where C_{loc} depends only on the shape regularity coefficient σ from (3.9) *and d.*

(ii) *For any subdomain covering $(\omega_i)_{i\in I}$ of Ω with finite overlapping index $C_{\mathrm{ovrl}} :=$ $\operatorname{ess\,sup}_{x\in\Omega} \#\{i \in I \mid \omega_i \ni x\}$, we have*

$$\sum_{i\in I} \|\ell\|^2_{H^{-1}(\omega_i)} \le C_{\mathrm{ovrl}} \|\ell\|^2_{H^{-1}(\Omega)}.$$

Theorem 3.5 of Blechta, Málek and Vohralík (2020) generalizes Lemma 4.5 to the W^{-1}_p-norm, $1 < p < \infty$. Lemma 4.66 below provides an alternative localization with different local norms.

Proof. [1] We start by showing statement (i). Thanks to the orthogonality of ℓ, we may write

$$\langle \ell, w\rangle = \Big\langle \ell, w - \sum_{z\in\mathcal{V}} c_z\phi_z\Big\rangle,$$

where any c_z is an arbitrary constant if $z \in \mathcal{V}\cap\Omega$ is an interior vertex and 0 if $z \in \mathcal{V}\cap\partial\Omega$ is a boundary vertex. Using the partition of unity $\sum_{z\in\mathcal{V}}\phi_z = 1$ on Ω, we split the new test function

$$w - \sum_{z\in\mathcal{V}} c_z\phi_z = \sum_{z\in\mathcal{V}} (w - c_z)\phi_z$$

into local contributions $(w - c_z)\phi_z \in H^1_0(\omega_z)$, $z \in \mathcal{V}$. The constant c_z allows us to counter the gradient generated by the cut-off with ϕ_z. Indeed, the product rule, $0 \le \phi_z \le 1$ and $|\nabla\phi_z| \le C(d)\sigma h_T^{-1}$ on an element $T \in \mathcal{T}$ lead to

$$\begin{aligned}
\|\nabla((w - c_z)\phi_z)\|_{L^2(\omega_z)} &\le \|\phi_z\nabla(w - c_z)\|_{L^2(\omega_z)} + \|(w - c_z)\nabla\phi_z\|_{L^2(\omega_z)}\\
&\le \|\phi_z\|_{L^\infty(\omega_z)}\|\nabla w\|_{L^2(\omega_z)} + \|\nabla\phi_z\|_{L^\infty(\omega_z)}\|w - c_z\|_{L^2(\omega_z)}\\
&\le \|\nabla w\|_{L^2(\omega_z)} + C(d)\sigma\Big(\max_{T\subset\omega_z} h_T^{-1}\Big)\|w - c_z\|_{L^2(\omega_z)}.
\end{aligned}$$

If we choose $c_z = \fint_{\omega_z} w$ for interior vertices $z \in \mathcal{V}\cap\Omega$, then Lemma 2.3 (second Poincaré inequality) on reference stars implies

$$\|w - c_z\|_{L^2(\omega_z)} \lesssim \operatorname{diam}\omega_z\|\nabla w\|_{L^2(\omega_z)}.$$

The same inequality follows for boundary vertices $z \in \mathcal{V}\cap\partial\Omega$ thanks to the fact that w vanishes on at least one face of $\partial\omega_z \cap \partial\Omega$. Combining this with $\operatorname{diam}\omega_z \lesssim h_T$ for $T \subset \omega_z$, we thus obtain, for all local contributions, the stability bound

$$\|\nabla((w - c_z)\phi_z)\|_{L^2(\omega_z)} \le C_{\mathrm{loc}}\|\nabla w\|_{L^2(\omega_z)}, \tag{4.6}$$

where the constant C_{loc} depends only on d and σ. Hence

$$\langle \ell, w\rangle = \Big\langle \ell, w - \sum_{z\in\mathcal{V}} c_z\phi_z\Big\rangle = \sum_{z\in\mathcal{V}}\langle \ell, (w - c_z)\phi_z\rangle$$

gives

$$\begin{aligned}
|\langle \ell, w\rangle| &\le C_{\mathrm{loc}}\sum_{z\in\mathcal{V}}\|\ell\|_{H^{-1}(\omega_z)}\|\nabla w\|_{L^2(\omega_z)}\\
&\le \sqrt{d+1}\,C_{\mathrm{loc}}\Big(\sum_{z\in\mathcal{V}}\|\ell\|^2_{H^{-1}(\omega_z)}\Big)^{1/2}\|\nabla w\|_{L^2(\Omega)}
\end{aligned}$$

and (i) is proved.

2 We verify statement (ii). For each index $i \in I$, define $v_i \in H_0^1(\omega_i) \subset H_0^1(\Omega)$ by

$$\int_{\omega_i} \nabla v_i \cdot \nabla w = \langle \ell, w \rangle \quad \text{for all } w \in H_0^1(\omega_i).$$

We obtain

$$\langle \ell, v_i \rangle = \|\nabla v_i\|_{L^2(\omega_i)}^2 = \|\ell\|_{H^{-1}(\omega_z)}^2$$

by arguments similar to those in the proof of Lemma 4.1 (error and residual). The sum $v := \sum_{i\in I} v_i$ is in $H_0^1(\Omega)$ with

$$\begin{aligned}\|\nabla v\|_{L^2(\Omega)}^2 &\le \int_\Omega \Big| \sum_{i\in I_x} \nabla v_i(x) \Big|^2 \, dx \le \int_\Omega \# I_x \sum_{i\in I_x} |\nabla v_i(x)|^2 \, dx \\ &\le C_{\mathrm{ovrl}} \sum_{i\in I} \|\nabla v_i\|_{L^2(\omega_i)}^2 = C_{\mathrm{ovrl}} \sum_{i\in I} \|\ell\|_{H^{-1}(\omega_z)}^2,\end{aligned}$$

where we denote the set of active indices in $x \in \Omega$ by $I_x := \{i \in I \mid \omega_i \ni x\}$. Inserting this in

$$\sum_{i\in I} \|\ell\|_{H^{-1}(\omega_z)}^2 = \sum_{i\in I} \langle \ell, v_i \rangle = \langle \ell, v \rangle \le \|\ell\|_{H^{-1}(\omega_z)} \|\nabla v\|_{L^2(\Omega)}$$

establishes the desired inequality. □

Thanks to the partial orthogonality (4.4) and the properties of the star covering ω_z, $z \in \mathcal{V}$, we readily obtain the following statement.

Corollary 4.6 (star localization of residual norm). *The H^{-1}-norm of the residual can be split into local contributions on stars:*

$$\frac{1}{d+1} \sum_{z\in\mathcal{V}} \|R_\mathcal{T}\|_{H^{-1}(\omega_z)}^2 \le \|R_\mathcal{T}\|_{H^{-1}(\Omega)}^2 \le (d+1) C_{\mathrm{loc}} \sum_{z\in\mathcal{V}} \|R_\mathcal{T}\|_{H^{-1}(\omega_z)}^2,$$

where C_{loc} depends only on d and the shape regularity coefficient σ.

The upper bound of the global residual norm in Corollary 4.6 employs the stars ω_z, $z \in \mathcal{V}$, as local domains. The next remark assesses this choice by discussing conceivable alternatives in terms of elements and domains of the type

$$\omega_F := \bigcup_{T\in\mathcal{T}_F} T \quad \text{with } \mathcal{T}_F := \{T \in \mathcal{T} \mid T \supset F\}, \tag{4.7}$$

where $F \in \mathcal{F}$ is an interior face of $\mathcal{T}$.

Remark 4.7 (star localization is minimal for $d \ge 2$). The use of stars in the localization of the global residual norm is a sort of minimal choice, except for the special case $d = 1$ where elements can be used.

- If $d = 1$, point values are defined for functions in $H^1(\Omega)$. This allows an upper bound with elements instead of stars as local domains. In fact, choosing

$c_z = w(z)$ for all interval endpoints, the function $\sum_{z\in\mathcal{V}} c_z\phi_z$ amounts to the Lagrange interpolant $I_{\mathcal{T}}w \in \mathbb{S}^{1,0}_{\mathcal{T}} \cap H^1_0(\Omega)$, and we have $(w - I_{\mathcal{T}}w)|_I \in H^1_0(I)$ with $\|\nabla(w - I_{\mathcal{T}}w)\|_{L^2(I)} \lesssim \|\nabla w\|_{L^2(I)}$ for any interval I of the mesh $\mathcal{T}$. Arguing as in the proof of Lemma 4.5(i) then gives

$$\|R\|^2_{H^{-1}(\Omega)} \lesssim \sum_{I\in\mathcal{T}} \|R\|^2_{H^{-1}(I)}.$$

- An upper bound where the stars are replaced by elements $T \in \mathcal{T}$ cannot hold in general because it does not account for face-supported residual contributions. For example, consider our setting with

$$d \geq 2, \quad \boldsymbol{A} = \boldsymbol{I}, \quad c = 0, \quad \langle f, w\rangle = \int_F qw, \ w \in H^1_0(\Omega),$$

where $F \in \mathcal{F}$ and $q \neq 0$ is L^2-orthogonal to $\mathbb{P}_n(F)$. Then we have $u \neq 0 = u_{\mathcal{T}}$ and therefore $\|R_{\mathcal{T}}\|_{H^{-1}(\Omega)} > 0$ but $\|R_{\mathcal{T}}\|_{H^{-1}(T)} = 0$ for any $T \in \mathcal{T}$.

- An upper bound with pairs ω_F, $F \in \mathcal{F}$, instead of stars cannot hold in general. This can be shown by considering our setting with

$$d = 2, \quad \boldsymbol{A} = \boldsymbol{I}, \quad c = 0, \quad \langle f_\varepsilon, w\rangle = \frac{1}{\pi\varepsilon^2}\int_{B_\varepsilon(z)} w, \ w \in H^1_0(\Omega),$$

where $z \in \mathcal{V}$ is a vertex of a suitable triangulation $\mathcal{T}$, for $\varepsilon \searrow 0$. The limiting right-hand side is the Dirac measure in z, which, formally, is not seen by any $\|\cdot\|_{H^{-1}(\omega_F)}$, $F \in \mathcal{F}$. We thus have $\|R_{\mathcal{T}}\|_{H^{-1}(\Omega)} \to \infty$ but $\sum_{F\in\mathcal{F}} \|R_{\mathcal{T}}\|_{H^{-1}(\omega_F)} \lesssim 1$; see Tantardini, Veeser and Verfürth (2024).

The bisection method for mesh refinement is element-oriented. It is therefore advantageous to dispose of an element-indexed reformulation of the localization in Corollary 4.6. For that purpose, we recall the notion of *patches*,

$$\omega_T := \bigcup_{\substack{T'\in\mathcal{T} \\ T'\cap T\neq\emptyset}} T', \tag{4.8}$$

and may use the following equivalence.

Lemma 4.8 (localization re-indexing). *For any functional $\ell \in H^{-1}(\Omega)$, we have*

$$\sum_{z\in\mathcal{V}} \|\ell\|^2_{H^{-1}(\omega_z)} \approx \sum_{T\in\mathcal{T}} \|\ell\|^2_{H^{-1}(\omega_T)},$$

where the hidden constants depend on d and the shape regularity coefficient σ.

Note that, in contrast to the localization itself, its re-indexing does not require any orthogonality such as $\langle \ell, \phi_z\rangle = 0$ for all $z \in \mathcal{V} \cap \Omega$.

Proof. For any vertex $z \in \mathcal{V}$, there is an element $T \in \mathcal{T}$ containing z. Then the inclusion $\omega_z \subset \omega_T$ yields the inequality $\|\ell\|_{H^{-1}(\omega_z)} \le \|\ell\|_{H^{-1}(\omega_T)}$. Hence

$$\sum_{z\in\mathcal{V}} \|\ell\|^2_{H^{-1}(\omega_z)} \le \sum_{T\in\mathcal{T}} \|\ell\|^2_{H^{-1}(\omega_T)}.$$

To show the converse inequality, let $T \in \mathcal{T}$ be any element and $w \in H^1_0(\omega_T)$. Given any vertex $z \in \mathcal{V} \cap \omega_T$, Lemma 2.2 (first Poincaré inequality) on ω_T implies the stability bound

$$\begin{aligned}
\|\nabla(w\phi_z)\|_{L^2(\omega_z\cap\omega_T)} &\le \|\phi_z \nabla w\|_{L^2(\omega_z\cap\omega_T)} + \|w\nabla\phi_z\|_{L^2(\omega_z\cap\omega_T)} \\
&\le \|\nabla w\|_{L^2(\omega_z\cap\omega_T)} + C(d,\sigma) \max_{T\subset\omega_z\cap\omega_T} h_T^{-1} \|w\|_{L^2(\omega_T)} \\
&\lesssim \|\nabla w\|_{L^2(\omega_T)}.
\end{aligned}$$

We thus derive

$$\begin{aligned}
\langle \ell, w\rangle &= \sum_{z\in\mathcal{V}\cap\omega_T} \langle \ell, w\phi_z\rangle \\
&\le \sum_{z\in\mathcal{V}\cap\omega_T} \|\ell\|_{H^{-1}(\omega_z\cap\omega_T)} \|\nabla(w\phi_z)\|_{L^2(\omega_z\cap\omega_T)} \\
&\lesssim \Bigg(\sum_{z\in\mathcal{V}\cap\omega_T} \|\ell\|_{H^{-1}(\omega_z)}\Bigg) \|\nabla w\|_{L^2(\omega_T)}
\end{aligned}$$

and, since $\#(\mathcal{V} \cap \omega_T)$ is bounded in terms of the shape regularity coefficient σ,

$$\|\ell\|^2_{H^{-1}(\omega_T)} \lesssim \sum_{z\in\mathcal{V}\cap\omega_T} \|\ell\|^2_{H^{-1}(\omega_z)}.$$

Summing over $T \in \mathcal{T}$, and taking into account that $\#\{T \in \mathcal{T} \mid \omega_T \ni z\}$ is again bounded in terms of σ, concludes the proof. □

4.2. *Standard residual estimator and its flaws*

Exploiting the results of Section 4.1, we derive an *a posteriori upper bound* of the error in terms of the standard residual estimator and discuss its flawed sharpness. This discussion will serve as the starting point for an improved *a posteriori* analysis in the following sections.

The *standard residual estimator* needs the additional regularity

$$f \in L^2(\Omega) \quad \text{and} \quad \boldsymbol{A} \in W^1_\infty(\Omega; \mathbb{R}^{d\times d}) \tag{4.9}$$

for the data in our model problem (2.5). Given the Galerkin approximation $u_\mathcal{T}$ from (4.1), it may be defined as follows (see e.g. Verfürth 2013):

$$\mathcal{E}^{\text{std}}_\mathcal{T} := \mathcal{E}^{\text{std}}_\mathcal{T}(u_\mathcal{T}, \mathcal{D}) := \Bigg(\sum_{T\in\mathcal{T}} \mathcal{E}^{\text{std}}_\mathcal{T}(u_\mathcal{T}, \mathcal{D}, T)^2\Bigg)^{1/2} \tag{4.10a}$$

with the *local indicators*

$$\mathcal{E}^{\mathrm{std}}_{\mathcal{T}}(u_{\mathcal{T}},\mathcal{D},T)^2 := h_T\,\|j(u_{\mathcal{T}})\|^2_{L^2(\partial T\setminus\partial\Omega)} + h_T^2\,\|r(u_{\mathcal{T}})\|^2_{L^2(T)}, \tag{4.10b}$$

where

- the *scaling factor* $h_T = |T|^{1/d}$ measures the size of the element $T \in \mathcal{T}$,
- $j(v) = j_{\mathcal{T}}(v)$ is the *jump residual* given face-wise for $v \in \mathbb{V}_{\mathcal{T}}$ by

$$\begin{aligned} j(v)|_F &:= ([[\boldsymbol{A}\nabla v]]\cdot \boldsymbol{n}_{T_1})|_F := ((\boldsymbol{A}\nabla v)|_{T_1} - (\boldsymbol{A}\nabla v)|_{T_2})\cdot \boldsymbol{n}_{T_1} \\ &= (\boldsymbol{A}\nabla v)|_{T_1}\cdot \boldsymbol{n}_{T_1} + (\boldsymbol{A}\nabla v)|_{T_2}\cdot \boldsymbol{n}_{T_2}, \end{aligned}$$

  where $F \in \mathcal{F}$, $T_1, T_2 \in \mathcal{T}$ are such that $F = T_1 \cap T_2$, $\boldsymbol{n}_{T_i}$ denotes the outer normal of ∂T_i, $i = 1, 2$, and
- $r(v) = r_{\mathcal{T}}(v)$ is the *element residual*, a function given for $v \in \mathbb{V}_{\mathcal{T}}$ by

$$r(v)|_T := (f - cv + \operatorname{div}(\boldsymbol{A}\nabla v))|_T$$

  on any element $T \in \mathcal{T}$.

Note that the definition itself already uses the extra regularity (4.9). For notational simplicity, we shall write j and r instead of $j(u_{\mathcal{T}})$ and $r(u_{\mathcal{T}})$ for the rest of this section. Also, for any interior face $F \in \mathcal{F}$, we have $F = T_1 \cap T_2$ with $T_1, T_2 \in \mathcal{T}$. If $\boldsymbol{n}_{T_i}$ denotes the outer normal of ∂T_i, $i = 1, 2$, we set $\boldsymbol{n}_F = \boldsymbol{n}_{T_1}$. This particular choice of $\boldsymbol{n}_F$ is irrelevant as it does not affect the following definition of normal jump of any vector-valued field $\boldsymbol{g}$ with well-defined trace on F:

$$[[\boldsymbol{g}]]\cdot \boldsymbol{n}_F := \boldsymbol{g}|_{T_1}\cdot \boldsymbol{n}_{T_1} + \boldsymbol{g}|_{T_2}\cdot \boldsymbol{n}_{T_2}.$$

Theorem 4.9 (upper bound with standard residual estimator). *Suppose the additional regularity* (4.9) *holds. Then the error is bounded by the standard residual estimator:*

$$\|\nabla(u - u_{\mathcal{T}})\|_{L^2(\Omega)} \lesssim \mathcal{E}^{\mathrm{std}}_{\mathcal{T}},$$

where the hidden constant depends on the coefficients $(\boldsymbol{A}, c)$*, the shape regularity coefficient* σ*, and* d*.*

Proof. As Lemma 4.1 (error and residual) and Corollary 4.6 (star localization of residual norm) imply

$$\|\nabla(u - u_{\mathcal{T}})\|^2_{L^2(\Omega)} \lesssim \|R_{\mathcal{T}}\|^2_{H^{-1}(\Omega)} \lesssim \sum_{z\in\mathcal{V}} \|R_{\mathcal{T}}\|^2_{H^{-1}(\omega_z)}$$

and $\#\{z \in \mathcal{V} \mid \omega_z \supset T\} = d + 1$, it suffices to establish

$$\|R_{\mathcal{T}}\|^2_{H^{-1}(\omega_z)} \lesssim \sum_{T\subset\omega_z} \mathcal{E}^{\mathrm{std}}_{\mathcal{T}}(u_{\mathcal{T}}, f, T)^2 \tag{4.11}$$

for any vertex $z \in \mathcal{V}$. To this end, let $w \in H_0^1(\omega_z)$. The extra regularity $f \in L^2(\Omega)$ and $\boldsymbol{A} \in W^1_\infty(\Omega; \mathbb{R}^{d\times d})$ allows for piecewise integration by parts, which leads to the following L^2-representation of the residual:

$$\langle R_{\mathcal{T}}, w\rangle = \sum_{F\ni z} \int_F jw + \sum_{T\ni z} \int_T rw.$$

In order to bound the right-hand side suitably, we use the scaled trace theorem

$$\|w\|^2_{L^2(F)} \leq \frac{|F|}{|T|}\|w\|^2_{L^2(T)} + \frac{2}{d}\frac{|F|\operatorname{diam} T}{|T|}\|w\|_{L^2(T)}\|\nabla w\|_{L^2(T)} \tag{4.12}$$

for any face $F \subset \partial T$ (see e.g. Veeser and Verfürth 2009, Corollary 4.5), the inequality

$$\|w\|_{L^2(\omega_z)} \leq \operatorname{diam} \omega_z \|\nabla w\|_{L^2(\omega_z)}$$

from Lemma 2.2 (first Poincaré inequality), and the two geometric relationships

$$\operatorname{diam} \omega_z \lesssim h_T \quad \text{whenever } T \subset \omega_z, \qquad |F| \operatorname{diam} T \lesssim |T| \quad \text{for } F \subset \partial T.$$

We thus obtain

$$|\langle R_{\mathcal{T}}, w\rangle| \lesssim \left(\sum_{T\ni z} h_T \|r\|_{L^2(T)} + h_T^{1/2} \sum_{F\ni z, F\subseteq T} \|j\|_{L^2(F)}\right)\|\nabla w\|_{L^2(\omega_z)}.$$

As the number of faces and elements in the star ω_z is bounded in terms of the shape regularity coefficient σ, we arrive at the desired bound (4.11), and the proof is finished. □

Remark 4.10 (alternative derivation of upper bound). The upper bound for the standard residual estimator in Theorem 4.9 is often derived with a suitable interpolation operator, bypassing the localization of the H^{-1}-norm in Lemma 4.5. That approach is useful for the proof of Theorem 4.48 below and is presented therein. Here we opted for using the localization of the H^{-1}-norm in order to facilitate the comparison with thc following subsections. The approach at hand is also convenient to keep the ensuing constants small; see Veeser and Verfürth (2009).

An important question is the sharpness of the upper bound in Theorem 4.9. The so-called *a posteriori lower bounds* provide some answer by trying to bound the estimator in terms of the error. For many estimators, however, there arise additional terms of an *oscillatory* nature. The following remark justifies the presence of such terms for the case at hand.

Remark 4.11 (non-asymptotic overestimation). The lower bound

$$\mathcal{E}_{\mathcal{T}}^{\text{std}} \lesssim \|\nabla(u - u_{\mathcal{T}})\|_{L^2(\Omega)},$$

which would imply equivalence of error and estimator, *cannot hold* in general for the following reason.

Fix a mesh $\mathcal{T}$ and a functional $f \in H^{-1}(\Omega) \setminus L^2(\Omega)$ and consider a sequence $(f_n)_n$ of functions in $L^2(\Omega)$ with $\lim_{n\to\infty} \|f - f_n\|_{H^{-1}(\Omega)} = 0$. Then the sequences $(u_n)_n$ and $(u_{\mathcal{T},n})_n$ of exact and Galerkin solutions on a fixed mesh $\mathcal{T}$ remain bounded. The error sequence $\big(\|\nabla(u_n - u_{\mathcal{T},n})\|_{L^2(\Omega)}\big)_n$ is therefore also bounded, while the standard residual estimator $\mathcal{E}_{\mathcal{T}}^{\mathrm{std}}(u_{\mathcal{T},n}, f_n) \to \infty$ becomes unbounded. Note that in the special case $f = -\operatorname{div}(\boldsymbol{A}\nabla v) + cv$ with $v \in \mathbb{V}_{\mathcal{T}}$, we even have for the error $\lim_{n\to\infty} \|\nabla(u_n - u_{\mathcal{T},n})\|_{L^2(\Omega)} = 0$.

In other words, in certain cases, the standard residual estimator bounds almost 0 by almost ∞ and a lower bound has to involve an additional term that cannot be bounded by the error in general.

We shall define these additional terms with the help of the following local best approximations. Let K be an element or face of $\mathcal{T}$ and $m \in \mathbb{N}_0$ a polynomial degree. Given $v \in L^2(K)$, let $\Pi_K v := \Pi_K^m v$ denote the best approximation in $\mathbb{P}_m(K)$ with respect to the norm $\|\cdot\|_{L^2(K)}$. It is convenient to allow also for $m = -1$ with $\mathbb{P}_{-1}(K) = \{0\}$ and $\Pi_K^{(-1)} v = 0$. Writing $\mathcal{D} = (\boldsymbol{A}, c, f)$ for the data in problem (2.5), the (m_1, m_2)-oscillation for the standard residual estimator is then given by

$$\operatorname{osc}_{\mathcal{T}}^{\mathrm{std}}(u_{\mathcal{T}}, \mathcal{D})^2 := \sum_{T \in \mathcal{T}} \operatorname{osc}_{\mathcal{T}}^{\mathrm{std}}(u_{\mathcal{T}}, \mathcal{D}, T)^2, \tag{4.13a}$$

with the local indicators

$$\operatorname{osc}_{\mathcal{T}}^{\mathrm{std}}(u_{\mathcal{T}}, \mathcal{D}, T)^2 := h_T^2 \, \|r - \Pi_T^{m_2} r\|_{L^2(T)}^2 + h_T \sum_{F \subset \partial T \setminus \partial\Omega} \|j - \Pi_F^{m_1} j\|_{L^2(F)}^2. \tag{4.13b}$$

Proposition 4.12 (partial lower bound). *If $f \in L^2(\Omega)$ and $\boldsymbol{A} \in W^1_\infty(\Omega; \mathbb{R}^{d\times d})$, the standard residual estimator is bounded by error and oscillation:*

$$\mathcal{E}_{\mathcal{T}}^{\mathrm{std}} \lesssim \|\nabla(u - u_{\mathcal{T}})\|_{L^2(\Omega)} + \operatorname{osc}_{\mathcal{T}}^{\mathrm{std}}(u_{\mathcal{T}}, \mathcal{D}),$$

where the hidden constant depends on d, the coefficients $\boldsymbol{A}$ and c, the shape regularity coefficient σ as well as the oscillation degrees (m_1, m_2).

Proof. In light of Lemma 4.1 (error and residual) and Corollary 4.6 (star localization of residual norm), we may establish the claimed bound by bounding each indicator with a corresponding local residual norm. To this end, we shall consider here only the case of the oscillation degrees $(m_1, m_2) = (0, 0)$. The general case can be verified along the same lines with additional technicalities, and is treated in the proof of Lemma 4.28 below in a slightly different context.

[1] We start by bounding an arbitrary element residual $h_T \|r\|_{L^2(T)}$, $T \in \mathcal{T}$, in terms of some local residual norm. To this end, we may try to invert the following consequence of Lemma 2.2 (first Poincaré inequality):

$$\|R_{\mathcal{T}}\|_{H^{-1}(T)} = \sup_{w \in H_0^1(T)} \frac{|\langle R_{\mathcal{T}}, w\rangle|}{\|\nabla w\|_{L^2(T)}} = \sup_{w \in H_0^1(T)} \frac{\int_T rw}{\|\nabla w\|_{L^2(T)}} \lesssim h_T \|r\|_{L^2(T)},$$

the residual norm of which avoids involving the jump residual. We thus actually ask for an equivalence of two different smoothness norms. Such an equivalence can hold only for special r, e.g. from a finite-dimensional space. Furthermore, writing $\|r\|^2_{L^2(T)} = \int_T r(r\chi_T)$ suggests the choice $w = r\chi_T$, which, however, is not admissible for the residual $R_\mathcal{T}$ as neither r nor the characteristic function χ_T belong to $H^1_0(\Omega)$. We shall overcome these issues by replacing r with its mean value $\Pi^0_T r$ and χ_T with the element bubble

$$\phi_T := (d+1)^{(d+1)} \prod_{z\in\mathcal{V}\cap T} \phi_z. \tag{4.14}$$

Thanks to $\int_T \phi_T = C_d|T|$ and the inverse estimate $\|\nabla w\|_{L^2(T)} \lesssim h_T^{-1}\|w\|_{L^2(T)}$ for $w = (\Pi^0_T r)\phi_T \in H^1_0(T) \cap \mathbb{P}_{d+1}(T)$, we derive

$$\begin{aligned}\|\Pi^0_T r\|_{L^2(T)} &\lesssim \int_T (\Pi^0_T r)w \le \|(\Pi^0_T r)\chi_T\|_{H^{-1}(T)}\|\nabla w\|_{L^2(T)}\\ &\lesssim h_T^{-1}\|(\Pi^0_T r)\chi_T\|_{H^{-1}(T)}\|w\|_{L^2(T)}\\ &\le h_T^{-1}\|(\Pi^0_T r)\chi_T\|_{H^{-1}(T)}\|\Pi^0_T r\|_{L^2(T)},\end{aligned}$$

whence

$$h_T\|\Pi^0_T r\|_{L^2(T)} \lesssim \|(\Pi^0_T r)\chi_T\|_{H^{-1}(T)}. \tag{4.15}$$

This implies the desired partial lower bound for the element residual by a perturbation argument and the inequality $\|r - \Pi^0_T r\|_{H^{-1}(T)} \lesssim h_T\|r - \Pi^0_T r\|_{L^2(T)}$, which follows from another application of Lemma 2.2 (first Poincaré inequality):

$$\begin{aligned}h_T\|r\|_{L^2(T)} &\le h_T\|\Pi^0_T r\|_{L^2(T)} + h_T\|r - \Pi^0_T r\|_{L^2(T)}\\ &\lesssim \|(\Pi^0_T r)\chi_T\|_{H^{-1}(T)} + h_T\|r - \Pi^0_T r\|_{L^2(T)}\\ &\lesssim \|r\chi_T\|_{H^{-1}(T)} + h_T\|r - \Pi^0_T r\|_{L^2(T)}\\ &= \|R_\mathcal{T}\|_{H^{-1}(T)} + h_T\|r - \Pi^0_T r\|_{L^2(T)}.\end{aligned} \tag{4.16}$$

2 We bound an arbitrary jump residual $\|j\|_{L^2(F)}$, $F \in \mathcal{F}$, in a similar manner. Note that here an interference of the element residual is unavoidable because the support of non-trivial test functions has non-empty interior. We thus may try to insert

$$\|R_\mathcal{T}\|_{H^{-1}(\omega_F)} = \sup_{w\in H^1_0(\omega_F)} \frac{\int_F jw + \int_{\omega_F} rw}{\|\nabla w\|_{L^2(\omega_F)}} \lesssim h_F^{1/2}\|j\|_{L^2(F)} + \sum_{T\subset\omega_F} h_T\|r\|_{L^2(T)},$$

where the scaled trace theorem (4.12) is also used. To this end, we write δ_F for the Dirac measure of the face F,

$$\phi_F := d^d \prod_{z\in\mathcal{V}\cap F} \phi_z \tag{4.17}$$

for the face bubble of F, and choose the test function $w = (\Pi^0_F j)\phi_F \in H^1_0(\omega_F)$.

Using in addition $\|w\|_{L^2(\omega_F)} \lesssim h_F^{1/2}\|w\|_{L^2(F)}$ and (4.15), we deduce

$$\begin{aligned}
\|\Pi_F^0 j\|_{L^2(F)}^2 &\lesssim \int_F (\Pi_F^0 j)w + \sum_{T\subset\omega_F}\int_T (\Pi_T^0 r)w - \sum_{T\subset\omega_F}\int_T (\Pi_T^0 r)w \\
&\le \left\|(\Pi_F^0 j)\delta_F + \sum_{T\subset\omega_F}(\Pi_T^0 r)\chi_T\right\|_{H^{-1}(\omega_F)} \|\nabla w\|_{L^2(\omega_F)} \\
&\quad + \sum_{T\subset\omega_F} \|\Pi_F^0 r\|_{L^2(T)}\|w\|_{L^2(T)} \\
&\lesssim \left\|(\Pi_F^0 j)\delta_F + \sum_{T\subset\omega_F}(\Pi_T^0 r)\chi_T\right\|_{H^{-1}(\omega_F)} h_F^{-1/2}\|\Pi_F^0 j\|_{L^2(F)},
\end{aligned}$$

whence

$$h_F^{1/2}\|\Pi_F^0 j\|_{L^2(F)}^2 \lesssim \left\|(\Pi_F^0 j)\delta_F + \sum_{T\subset\omega_F}(\Pi_T^0 r)\chi_T\right\|_{H^{-1}(\omega_F)}. \tag{4.18}$$

Passing to the proper jump residual j, we arrive at the partial lower bound for the jump residual:

$$\begin{aligned}
h_F^{1/2}\|\Pi_F^0 j\|_{L^2(F)} &\lesssim \|R_{\mathcal{T}}\|_{H^{-1}(\omega_F)} \\
&\quad + h_F^{1/2}\|j - \Pi_F^0 j\|_{L^2(F)} + \sum_{T\subset\omega_F} h_T\|r - \Pi_T^0 r\|_{L^2(T)}.
\end{aligned} \tag{4.19}$$

[3] We square the bounds (4.16) and (4.19) from the previous steps and sum them, respectively, over all elements and faces to conclude the claimed partial lower bound with the help of Lemma 4.5(ii) (localization of H^{-1}-norm) and Lemma 4.1 (error and residual). □

The significance of Proposition 4.12 (partial lower bound) strongly depends on the choice of the polynomial degrees (m_1, m_2) in the oscillation from (4.13). The following two remarks address this important aspect.

Remark 4.13 (oscillation degrees: asymptotics). It is desirable that, under refinement, the oscillation in Proposition 4.12 (partial lower bound) converges to 0 at least as fast as the error. The maximal convergence order of the error under uniform refinement is $\|\nabla(u - u_{\mathcal{T}})\|_{L^2(\Omega)} = O(h^n)$ as $h \to 0$. In view of the scaling factors and derivative orders appearing in jump and element residual, we are thus led to require

$$m_1 \ge n-1 \quad \text{and} \quad m_2 \ge n-2.$$

One might hope that strict inequalities lead to higher order. Note, however, that since $\operatorname{osc}^{\mathrm{std}}$ involves in general both discrete solution $u_{\mathcal{T}}$ and data $\mathcal{D} = (\boldsymbol{A}, c, f)$, this will not be guaranteed without additional assumptions. Furthermore, increasing m_1 and m_2 entails bigger hidden constants in the lower bounds (4.15) and (4.18), as these bounds cannot hold for arbitrary L^2-functions. Consequently, a potentially higher asymptotic speed of the oscillation $\mathrm{osc}^{\mathrm{std}}$ comes with a bigger constant in front of it and therefore with diminished non-asymptotic significance.

Remark 4.14 (oscillation degrees: data oscillation reduction). In the particular case of the Poisson equation, i.e. $\boldsymbol{A} = \boldsymbol{I}$ and $c = 0$, and linear elements, i.e. $n = 1$, the oscillation with the degrees $(m_1, m_2) = (0, 0)$ reduces to the *data oscillation*

$$\mathrm{osc}^{\mathrm{std}}_{\mathcal{T}}(u_{\mathcal{T}}, \mathcal{D})^2 = \sum_{T \in \mathcal{T}} h_T^2 \|f - \Pi_T^0 f\|^2_{L^2(T)};$$

it depends only on the data, here the right-hand side f. Note also that here the regularity of f is determined by the regularity of the exact solution u.

For elements with degree $n \geq 2$, the choices $(m_1, m_2) = (n-1, n-2)$ ensure that for $F \in \mathcal{F}$,

$$\Pi_F^{n-1}([\![\nabla u_{\mathcal{T}}]\!]|_F \cdot \boldsymbol{n}_F) = [\![\nabla u_{\mathcal{T}}]\!]|_F \cdot \boldsymbol{n}_F \quad \text{and} \quad \Pi_T^{n-2}(\Delta u_{\mathcal{T}}|_T) = \Delta u_{\mathcal{T}}|_T, \tag{4.20}$$

and so, again, oscillation reduces to data oscillation in f:

$$\mathrm{osc}^{\mathrm{std}}_{\mathcal{T}}(u_{\mathcal{T}}, \mathcal{D})^2 = \sum_{T \in \mathcal{T}} h_T^2 \|f - \Pi_T^{n-2} f\|^2_{L^2(T)}.$$

If we add a reaction term, i.e. we consider $\boldsymbol{A} = \boldsymbol{I}$ and $c = 1$, we can again obtain the reduction to data oscillation by increasing m_2 to n.

For a more general operator with piecewise polynomial coefficients

$$\boldsymbol{A} \in \left(\mathbb{S}_{\mathcal{T}}^{n_A,-1}\right)^{d \times d} \quad \text{and} \quad c \in \mathbb{S}_{\mathcal{T}}^{n_c,-1},$$

the choice

$$(m_1, m_2) = (n_A + n - 1, \max\{n_c + n, n_A + n - 2\}) \tag{4.21}$$

again reduces $\mathrm{osc}^{\mathrm{std}}$ to data oscillation in f.

Finally, for a general operator without piecewise polynomial coefficients $(\boldsymbol{A}, c)$, a reduction to data oscillation with piecewise polynomial best approximations as before is not possible. The argument in Remark 4.13 suggests approximating the general coefficients with piecewise polynomial coefficients satisfying

$$n_A = n - 1 \quad \text{and} \quad n_c = n - 1.$$

As we shall see below in Section 4.8, the choice (4.21) with these values allows us to bound $\mathrm{osc}^{\mathrm{std}}$ in terms of $\|\nabla u_{\mathcal{T}}\|_{L^2(\Omega)}$, which is controlled by stability, and data oscillation terms involving f and the coefficients $\boldsymbol{A}$ and c. Note, however, that the nature of these data oscillation terms differs from the preceding reductions: for example, the regularity of the coefficients $\boldsymbol{A}$ and c is not determined by the exact solution u.

In light of Remark 4.13 (oscillation degrees: asymptotics), one might hope that the overestimation described in Remark 4.11 (non-asymptotic overestimation) disappears under refinement. This can be ensured under suitable regularity assumption but is not guaranteed in general, as the following remark reveals.

Remark 4.15 (asymptotic overestimation). Considering a variant of the standard residual estimator that allows for $f \in H^{-1}(\Omega)$ and adaptive refinement, Cohen, DeVore and Nochetto (2012, Section 6.4) give an example where the error converges asymptotically faster than the estimator; see also Kreuzer and Veeser (2021, Lemma 21).

Having recognized the above flaws of the standard residual estimator, let us conclude with an observation that will be the departure point of an improved analysis.

Corollary 4.16 (equivalence for discrete data). *Suppose all data* $\mathcal{D} = (\boldsymbol{A}, f, c)$ *of problem* (2.5) *are piecewise polynomial, that is, there are* $n_{\boldsymbol{A}}, n_c, n_f \in \mathbb{N}_0$ *such that*

$$\boldsymbol{A} \in \left(\mathbb{S}_{\mathcal{T}}^{n_{\boldsymbol{A}},-1}\right)^{d\times d}, \quad c \in \mathbb{S}_{\mathcal{T}}^{n_c,-1} \quad \text{and} \quad f \in \mathbb{S}_{\mathcal{T}}^{n_f,-1}.$$

Then error and standard residual estimator are equivalent:

$$\|\nabla(u - u_{\mathcal{T}})\|_{L^2(\Omega)} \approx \mathcal{E}_{\mathcal{T}}^{\text{std}},$$

where the hidden constants depend only on d, the coefficients $\boldsymbol{A}$ *and c, the shape regularity coefficient* σ*, and the degrees* $n_{\boldsymbol{A}}$*,* n_c *and* n_f*.*

Proof. The upper bound follows from Theorem 4.9 (upper bound with standard residual estimator), while Proposition 4.12 (partial lower bound) with

$$(m_1, m_2) = (n_{\boldsymbol{A}} + n - 1, \max\{n_{\boldsymbol{A}} + n - 2, n_c + n, n_f\})$$

yields the lower bound. □

Motivated by the above discussion, one may define a variant of the standard residual estimator, characterized by a splitting into two different parts. More precisely, choosing (m_1, m_2) according to Remark 4.14 (oscillation degrees: data oscillation reduction), one may replace the local indicators in (4.10b) with

$$\mathcal{E}_{\mathcal{T}}^{\text{std}}(u_{\mathcal{T}}, f, T)^2 := \eta_{\mathcal{T}}^{\text{std}}(u_{\mathcal{T}}, T)^2 + \operatorname{osc}_{\mathcal{T}}^{\text{std}}(u_{\mathcal{T}}, \mathcal{D}, T)^2, \tag{4.22a}$$

where the first part, the so-called *PDE indicator*, is given by

$$\eta_{\mathcal{T}}^{\text{std}}(u_{\mathcal{T}}, T)^2 := h_T^2 \|\Pi_T^{m_2} r\|_{L^2(T)}^2 + h_T \sum_{F \subset \partial T \setminus \partial\Omega} \|\Pi_F^{m_1} j\|_{L^2(F)}^2, \tag{4.22b}$$

while the second part corresponds to the local oscillation from (4.13b); compare with Verfürth (2013, Theorems 1.5 and 4.7). In this way,

- the PDE indicators are computable (in terms of the Galerkin approximation $u_{\mathcal{T}}$ and the local projections),
- the oscillation indicators typically have to be approximated by numerical quadrature,
- both types of indicators are, in general, not dominated by the error.

4.3. *Discrete functionals and* a posteriori *error analysis*

This section introduces the notion of *discrete functionals* and individuates properties in their approximation that are useful in *a posteriori* error analysis. The realization of these properties distinguishes the subsequent approach, which is adapted from Kreuzer and Veeser (2021) and Kreuzer *et al.* (2024).

The notion of discrete functionals and its local counterparts are of interest for at least two reasons. The first one is that their H^{-1}-norm can be rather easily quantified, as we shall see in Corollary 4.30 below. This property is related to Corollary 4.16 (equivalence for discrete data), which can be read in the following way: the standard residual estimator is equivalent to the error whenever the residual is a discrete functional. The second reason lies in the observation that an important part of the residual, namely the application of the differential operator to a discrete function, is itself of discrete nature. This feature is partially captured by the following definition of discrete functionals with polynomial densities and is discussed in Remark 4.18.

Definition 4.17 (discrete functionals and meshed subdomains). For $m_1 \in \mathbb{N}_0$, $m_2 \in \mathbb{N}_0 \cup \{-1\}$, let $\mathbb{F}_{\mathcal{T}} := \mathbb{F}(\mathcal{T}) := \mathbb{F}_{m_1,m_2}(\mathcal{T})$ denote the subspace

$$\Big\{\ell \in H^{-1}(\Omega) \mid \text{for all } w \in H_0^1(\Omega),\ \langle \ell, w\rangle = \sum_{F\in\mathcal{F}} \int_F q_F w + \sum_{T\in\mathcal{T}} \int_T q_T w \\ \text{with fixed } q_F \in \mathbb{P}_{m_1}(F),\ q_T \in \mathbb{P}_{m_2}(T)\Big\}$$

of *discrete functionals*, i.e. functionals that are given by piecewise polynomial densities over elements and interior faces. We call (m_1, m_2) the *degrees of the discrete functionals*.

A set ω is a $\mathcal{T}$*-meshed subdomain* if it is a subdomain of Ω and it is triangulated by a submesh $\mathcal{T}_\omega \subset \mathcal{T}$, that is, we have $\overline{\omega} = \cup_{T\in\mathcal{T}_\omega} T$. A functional $\ell \in H^{-1}(\Omega)$ is then *discrete in the meshed subdomain* ω whenever $\ell|_{H_0^1(\omega)} \in \mathbb{F}(\mathcal{T}_\omega)$. Here the faces

$$\mathcal{F}_\omega := \{F \in \mathcal{F} \mid F \cap \omega \neq \emptyset,\ F \not\subset \partial\omega\}$$

involved in $\mathbb{F}(\mathcal{T}_\omega)$ are interior to ω; for example, the subspaces $\mathbb{F}(\{T\})$, $T \in \mathcal{T}$, do not involve any faces. In accordance with (4.5), we use the abbreviations $\mathcal{T}_z$ and $\mathcal{F}_z$ for $\mathcal{T}_{\omega_z}$ and $\mathcal{F}_{\omega_z}$.

Alternatively, the local space $\mathbb{F}(\mathcal{T}_\omega)$ can be obtained from the global space $\mathbb{F}(\mathcal{T})$ by restriction:

$$\mathbb{F}(\mathcal{T}_\omega) = \mathbb{F}(\mathcal{T})|_{H_0^1(\omega)} := \big\{\ell|_{H_0^1(\omega)} \mid \ell \in \mathbb{F}(\mathcal{T})\big\}. \tag{4.23}$$

Remark 4.18 (differential operator and discrete functionals). The image of the finite element space $\mathbb{V}_{\mathcal{T}}$ under the linear differential operator $-\operatorname{div}(\boldsymbol{A}\nabla\cdot) + c\cdot$ is again a finite-dimensional space. For differential operators with piecewise polynomial coefficients $\boldsymbol{A}$ and c, the above notion captures this by the property that the application of such operators to discrete functions $v \in \mathbb{V}_{\mathcal{T}}$ yields discrete functionals. Indeed, if

$$\boldsymbol{A} \in \left(\mathbb{S}_{\mathcal{T}}^{n_{\boldsymbol{A}},-1}\right)^{d\times d} \quad \text{and} \quad c \in \mathbb{S}_{\mathcal{T}}^{n_c,-1}$$

with $n_{\boldsymbol{A}}, n_c \in \mathbb{N}_0$, piecewise integration by parts gives the representation

$$\int_\Omega \boldsymbol{A}\nabla v \cdot \nabla w + cvw = \sum_{F\in\mathcal{F}} \int_F [[\boldsymbol{A}\nabla v]] \cdot \boldsymbol{n}_F w + \sum_{T\in\mathcal{T}} \int_T (cv - \operatorname{div}(\boldsymbol{A}\nabla v))w, \tag{4.24}$$

where, for any interior face $F \in \mathcal{F}$ and any element $T \in \mathcal{T}$,

$$[[\boldsymbol{A}\nabla v]] \cdot \boldsymbol{n}_F \in \mathbb{P}_{m_1}(F), \quad cv - \operatorname{div}(\boldsymbol{A}\nabla v) \in \mathbb{P}_{m_2}(T)$$

with $m_1 = n_{\boldsymbol{A}} + n - 1$ and $m_2 = \max\{n_{\boldsymbol{A}} + n - 2, n_c + n\}$. Note, however, that not every functional in $\mathbb{F}_{m_1,m_2}(\mathcal{T})$ can be written in the form of (4.24). In fact, as the representation of a discrete functional is made up of L^2-scalar products on domains that are mutually disjoint or of different dimension, we have

$$\dim \mathbb{F}_{m_1,m_2}(\mathcal{T}) = \#\mathcal{F} \dim \mathbb{P}_{m_1} + \#\mathcal{T} \dim \mathbb{P}_{m_2}, \tag{4.25}$$

which is strictly greater than $\dim \mathbb{V}_{\mathcal{T}}$. This enlargement, which is implicitly used in the proof of Proposition 4.12 (partial lower bound), turns out to be convenient also in the constructive approximation of discrete functionals.

In view of the aforementioned properties of discrete functionals, we may split the residual into a discrete and a non-discrete part. Splitting the standard residual estimator in the alternative local indicators (4.22) is in a similar spirit. To see this, we introduce $\Pi_{\mathcal{T}}\ell \in H^{-1}(\Omega)$ given by

$$\langle \Pi_{\mathcal{T}}\ell, w\rangle := \sum_{F\in\mathcal{F}} \int_F \left(\Pi_F^{m_1} g\right) w + \sum_{T\in\mathcal{T}} \int_T \left(\Pi_T^{m_2} f\right) w, \quad w \in H_0^1(\Omega), \tag{4.26}$$

for all $\ell \in H^{-1}(\Omega)$ admitting the representation

$$\langle \ell, w\rangle = \sum_{F\in\mathcal{F}} \int_F gw + \sum_{T\in\mathcal{T}} \int_T fw, \quad w \in H_0^1(\Omega),$$

with suitable density functions g and f. Then the splitting of the alternative indicators (4.22) corresponds to writing

$$R_{\mathcal{T}} = \Pi_{\mathcal{T}} R_{\mathcal{T}} + (I - \Pi_{\mathcal{T}}) R_{\mathcal{T}}. \tag{4.27}$$

Moreover, Remark 4.14 (oscillation degrees: data oscillation reduction) discusses conditions for the identity

$$(I - \Pi_{\mathcal{T}}) R_{\mathcal{T}} = f - \Pi_{\mathcal{T}} f, \tag{4.28}$$

which follows from the property that $\Pi_{\mathcal{T}}$ reproduces the functionals in Remark 4.18 (differential operator and discrete functionals); compare with (4.20), which in terms of $\Pi_{\mathcal{T}}$ reads $\Pi_{\mathcal{T}}(\Delta u_{\mathcal{T}}) = \Delta u_{\mathcal{T}}$, where Δ is now the distributional Laplacian.

The fact that the definition of $\Pi_{\mathcal{T}}$ requires the extra regularity $f \in L^2(\Omega)$ and $A \in W^1_\infty(\Omega; \mathbb{R}^{d\times d})$ not only excludes applications but, in light of Remark 4.11 (non-asymptotic overestimation), entails overestimation. To circumvent this flaw, we therefore aim to construct a new approximation operator $P_{\mathcal{T}}$ that is defined for all functionals $\ell \in H^{-1}(\Omega)$. Furthermore, we want this operator to be a projection onto $\mathbb{F}_{\mathcal{T}}$ so that the counterpart

$$(I - P_{\mathcal{T}})R_{\mathcal{T}} = f - P_{\mathcal{T}} f$$

of (4.28) holds under the same conditions.

To summarize, our plan is to develop a quasi-optimal *a posteriori* error analysis by constructing a locally computable linear projection

$$P_{\mathcal{T}} \colon H^{-1}(\Omega) \to \mathbb{F}_{\mathcal{T}} \subset H^{-1}(\Omega)$$

onto the discrete functionals that induces a splitting

$$R_{\mathcal{T}} = P_{\mathcal{T}} R_{\mathcal{T}} + (I - P_{\mathcal{T}})R_{\mathcal{T}} \tag{4.29}$$

of the residual into a *discretized residual* $P_{\mathcal{T}} R_{\mathcal{T}}$, which can be easily quantified, as well as an *oscillatory residual* $(I - P_{\mathcal{T}})R_{\mathcal{T}}$, which under the conditions of Remark 4.18 (differential operator and discrete functionals) reduces to an oscillation of the right-hand side f.

The proof of an upper bound of the error will then involve a triangle inequality applied to the right-hand side of (4.29). The following remark provides criteria to prevent overestimation in such a context, and is followed by a comparison of the two approaches represented by (4.27) and (4.29).

Remark 4.19 (avoiding overestimation). Overestimation can often be avoided by ensuring two relatively simple conditions. In order to discuss them informally, consider the model inequality

$$|\cdot| \le |\cdot|_1 + |\cdot|_2, \tag{4.30}$$

where $|\cdot|, |\cdot|_i$, $i = 1, 2$, are seminorms and denote the domain and kernel of $|\cdot|$, respectively, by $\operatorname{dom}|\cdot|$ and $\ker|\cdot|$, etc.

The first condition, the *kernel condition*, is that zero is not overestimated:

$$\ker|\cdot| \subset \ker|\cdot|_1 \cap \ker|\cdot|_2. \tag{4.31a}$$

The second condition, the *domain condition*, is that a finite value is never bounded by ∞, or in other words, still informal, if the evaluation of the left-hand side is (or can be uniquely defined to be) a finite value, the same holds for the right-hand side:

$$\operatorname{dom}|\cdot| \subset \operatorname{dom}|\cdot|_1 \cap \operatorname{dom}|\cdot|_2. \tag{4.31b}$$

Kreuzer *et al.* (2024) provide a precise version of the domain condition (4.31b), showing that, given inequality (4.30), the two conditions (4.31) are also sufficient for equivalence, and discuss further applications of this viewpoint.

In order to illustrate the application of Remark 4.19, let us consider only the special case of the Poisson equation, i.e. $\boldsymbol{A} = \boldsymbol{I}$, $c = 0$, and linear elements, i.e. $n = 1$. We start with the upper bound in terms of the standard residual estimator in Theorem 4.9 and view it as a function of the right-hand side f. Then the domain condition is violated as the left-hand side is defined for any $f \in H^{-1}(\Omega)$, while the right-hand side is defined only for $f \in L^2(\Omega)$. Also, the kernel condition is not verified: the left-hand side vanishes whenever $f = -\Delta v$ for some $v \in \mathbb{V}_{\mathcal{T}}$, while the right-hand side vanishes only for $f = 0$. The splitting in the alternative local indicators (4.22) does not worsen this situation, that is, it does not add further instances in which kernel and domain condition are missed. Note, however, that the oscillation indicators alone are in conflict with the domain condition and therefore another PDE indicator cannot cure the overestimation. Finally, for the outlined approach, the splitting (4.29) and the required properties for the operator $P_{\mathcal{T}}$ ensure both kernel and domain condition.

4.4. *Testing discrete functionals*

The H^{-1}-projection $P_{\mathcal{T}}$ onto the discrete functionals $\mathbb{F}_{\mathcal{T}}$ will be defined by means of a Petrov–Galerkin-type approach. This section prepares its definition by individuating a suitable *test space* $\mathbb{V}_{\mathcal{T}}^{+}$. The key property of $\mathbb{V}_{\mathcal{T}}^{+}$ is that the dual pairing $\langle \cdot, \cdot \rangle$ in $H^{-1}(\Omega)$ is non-degenerate on the product $\mathbb{F}_{\mathcal{T}} \times \mathbb{V}_{\mathcal{T}}^{+}$. Doing so, the degrees (m_1, m_2) of the discrete functionals will be parameters that are omitted in the notation. The construction of the test space $\mathbb{V}_{\mathcal{T}}^{+}$ proceeds in two steps. First, we locally associate to the degrees of freedom in $\mathbb{F}_{\mathcal{T}}$ certain functions on Ω. For the degrees of freedom on the skeleton, this will involve a suitable *extension operator*. Second, we turn the ensuing functions into admissible test functions with the help of a *cut-off*.

The degrees of freedom in $\mathbb{F}_{\mathcal{T}}$ are given by density polynomials over element and faces. For an element $T \in \mathcal{T}$, if we extend such a density polynomial q_T by 0 off T, it is already a function on Ω. For a polynomial q_F associated with a face $F \in \mathcal{F}$, we employ the following extension operator E_F mapping a function v on F to a function on ω_F, the union of all elements T containing F.

Given such an element $T \subset \omega_F$, write $z_0, \ldots, z_d$ for its vertices, z_d being the one opposite to F, let

$$b_F := \frac{1}{d} \sum_{i=0}^{d-1} z_i$$

denote the barycentre of F, and set

$$(E_F v)(x) := v\left(\phi_d(x) b_F + \sum_{i=0}^{d-1} \phi_{z_i}(x) z_i \right), \quad x \in T,$$

and extend by 0 off ω_F. Note that the definition of E_F is affine-invariant and does not depend on the enumeration of the vertices of F. The next lemma collects two useful properties of this extension operator.

Lemma 4.20 (extending from faces). *Let $F \in \mathcal{F}$ be a face. For any function $v \in L^2(F)$, we have*

$$\|E_F v\|_{L^2(\omega_F)} \lesssim h_F^{1/2} \|v\|_{L^2(F)},$$

where h_F stands for the diameter of F and the hidden constant depends only on d and the shape regularity coefficient σ. Furthermore, if v is a polynomial, then $E_F v$ is a continuous piecewise polynomial of the same degree.

Proof. 1 In view of $(E_F v)^2 = E_F(v^2)$, we may show the inequality by verifying

$$\int_{\omega_F} E_F w \lesssim h_F \int_F w \tag{4.32}$$

for any positive function $w\colon F \to \mathbb{R}$, which amounts to L^1-stability. To this end, we shall use a standard argument involving the following reference situation, which slightly differs from the common one with

$$T_d = \left\{ x = (x_1, \dots, x_d) \in \mathbb{R}^d \mid 0 \le x_i \le 1,\ \sum_{i=1}^d x_i \le 1 \right\}$$

and $b_d := (d+1)^{-1}(1, \dots, 1) \in \mathbb{R}^d$. Let the reference face $\widehat{F} := T_{d-1} - b_{d-1} \subset \mathbb{R}^{d-1}$ be a translation of T_{d-1} and let the reference simplex $\widehat{T} \subset \mathbb{R}^d$ be the convex hull of $\widehat{F} \times \{0\}$ and the canonical basis vector $e_d = (0, \dots, 0, 1) \in \mathbb{R}^d$. The barycentre of $\widehat{F} \times \{0\}$ is then the origin in $\mathbb{R}^d$ and the barycentric coordinate of the vertex e_d of $\widehat{T}$ is x_d. Fixing an element T with $T \subset \omega_F$, let $G_T : \widehat{T} \to T$ denote a bi-affine map sending vertices of $\widehat{F} \times \{0\}$ into vertices of F and e_d into the vertex of T opposite to F, and write $G_F : \widehat{F} \times \{0\} \to F$ for the restriction $G_T|_{\widehat{F}\times\{0\}}$. The pullbacks of $E_F w$ and w satisfy

$$G_T^*(E_F w)(x', x_d) = G_F^* w(x', 0)$$

for all $x = (x', x_d) \in \widehat{T} = \{y = (y', y_d) \in \widehat{F} \times \mathbb{R} \mid 0 \le y_d \le 1 - |y' + b_{d-1}|_1\}$, where $|z'|_1 = \sum_{i=1}^{d-1} |z_i'|$ stands for the ℓ_1-norm in $\mathbb{R}^{d-1}$. Consequently, the transformation rule, the fact that the Jacobians of G_T and G_F are constant, the Fubini theorem, $w \ge 0$, and $|\widehat{F}|/|\widehat{T}| = |T_{d-1}|/|T_d| = d$ yield

$$\begin{aligned}
\int_T E_F w &= \frac{|T|}{|\widehat{T}|} \int_{\widehat{T}} G_T^*(E_F w) = \frac{|T|}{|\widehat{T}|} \int_{\widehat{F}} \int_0^{1-|x'+b_{d-1}|_1} G_F^* w(x', 0)\, \mathrm{d}x_d\, \mathrm{d}x' \\
&\le \frac{|T|}{|\widehat{T}|} \int_{\widehat{F}} G_F^* w(\cdot, 0) = \frac{|T||\widehat{F}|}{|\widehat{T}||F|} \int_F w = d \frac{|T|}{|F|} \int_F w.
\end{aligned}$$

Since the hidden constant in $|T|/|F| \lesssim h_F$ depends only on the shape regularity coefficient σ, this implies the L^1-stability bound (4.32) and so also the claimed L_2-stability is proved.

2 The second statement for polynomial arguments of E_F is a direct consequence of its definition. □

In view of the above different treatment of elements and faces, we need two types of cut-off functions: one for elements denoted by ϕ_T and another for faces denoted by ϕ_F. Possible choices are the element and face bubbles from (4.14) and (4.17). Since other choices will be useful in Section 4.8 below, we shall henceforth rely only on the following properties.

Assumption 4.21 (abstract cut-off). The *cut-off functions* ϕ_T, $T \in \mathcal{T}$, and ϕ_F, $F \in \mathcal{F}$, satisfy

$$\operatorname{supp} \phi_T = T, \quad 0 \le \phi_T \le 1, \quad \operatorname{supp} \phi_F = \omega_F, \quad 0 \le \phi_F \le 1,$$

and act in an affine-equivalent manner on the element level: there exists a finite-dimensional linear space $\mathbb{S}^+ \subset L^\infty(T_d)$ of functions defined on the reference element T_d such that $G_T^* \phi_T$ does not depend on T, $G_F^* \phi_F$ does not depend on F, and

$$\begin{aligned} &\text{for all } T \in \mathcal{T} \text{ and } q \in \mathbb{P}_{m_2}(T), && G_T^*(q\phi_T) \in \mathbb{S}^+, \\ &\text{for all } F \in \mathcal{F},\ q \in \mathbb{P}_{m_1}(F) \text{ and } T \in \mathcal{T}, && G_T^*(((E_F q)\phi_F)|_T) \in \mathbb{S}^+, \end{aligned}$$

where G_T is a bi-affine map from the reference element T_d to the generic element T, and $G_T^*(v) = v \circ G_T$ denotes the pullback of a function $v\colon T \to \mathbb{R}$ via G_T.

In the case of the bubble functions (4.14) and (4.17), Assumption 4.21 holds with $\mathbb{S}^+ = \mathbb{P}_{\max\{m_1+d-1, m_2+d\}}(T_d)$ as the extension operators E_F, $F \in \mathcal{F}$, preserve the polynomial degree.

Lemma 4.22 (properties of cut-off). *If the cut-off functions ϕ_T, $T \in \mathcal{T}$, and ϕ_F, $F \in \mathcal{F}$, satisfy Assumption 4.21, then we have*

$$\|q\|_{L^2(T)} \lesssim \|q\phi_T^{1/2}\|_{L^2(T)} \quad \text{and} \quad \|\nabla(q\phi_T)\|_{L^2(T)} \lesssim h_T^{-1}\|q\phi_T\|_{L^2(T)}$$

for all $q \in \mathbb{P}_{m_2}(T)$, as well as

$$\|q\|_{L^2(F)} \lesssim \|q\phi_F^{1/2}\|_{L^2(F)} \quad \text{and} \quad \|\nabla((E_F q)\phi_F)\|_{L^2(\omega_F)} \lesssim h_F^{-1}\|(E_F q)\phi_F\|_{L^2(\omega_F)}$$

for all $q \in \mathbb{P}_{m_1}(F)$. The hidden constants depend only on d, the shape regularity coefficient σ, the degrees (m_1, m_2) of the discrete functionals, and the space $\mathbb{S}^+$.

Proof. 1 To verify the first claimed inequality, we start by noting that, thanks to $\operatorname{supp} \phi_T = T$, we have $\phi_{T_d} := G_T^*(\phi_T) > 0$ in the interior of T_d. Hence $\|\cdot\|_{L^2(T_d)}$ and $\|\cdot\, \phi_{T_d}^{1/2}\|_{L^2(T_d)}$ are norms on $\mathbb{P}_{m_2}(T_d)$ and, thanks to $\dim \mathbb{P}_{m_2}(T_d) < \infty$, are equivalent. A standard round trip to the reference element and $G_T^* q\, G_T^* \phi_T^{1/2} = G_T^*(q\phi_T^{1/2})$ thus yields

$$\begin{aligned}\|q\|_{L^2(T)} &\lesssim h_T^{d/2}\|G_T^* q\|_{L^2(T_d)} \lesssim h_T^{d/2}\|G_T^* q\, G_T^* \phi_T^{1/2}\|_{L^2(T_d)} \\ &\lesssim \|q\phi_T^{1/2}\|_{L^2(T)},\end{aligned}$$

and the first claimed inequality is established. The third one is proved along the same lines, but with a round trip to the reference face.

2 For the other claimed inequalities, note that $\|\nabla \cdot\|_{L^2(T_d)}$ and $\inf_{c\in\mathbb{R}} \|\cdot - c\|_{L^2(T_d)}$ are equivalent norms on the finite-dimensional quotient space $\mathbb{S}^+/\mathbb{R}$. Consequently, further round trips to the reference element give

$$\begin{aligned}\|\nabla(q\phi_T)\|_{L^2(T)} &\lesssim h_T^{-1+d/2}\|\nabla G_T^*(q\phi_T)\|_{L^2(T_d)} \lesssim h_T^{-1+d/2} \inf_{c\in\mathbb{R}} \|G_T^*(q\phi_T) - c\|_{L^2(T_d)} \\ &\lesssim h_T^{-1+d/2}\|G_T^*(q\phi_T)\|_{L^2(T_d)} \lesssim h_T^{-1}\|q\phi_T\|_{L^2(T)}\end{aligned}$$

and

$$\begin{aligned}\|\nabla\big(E_F(q)\phi_T\big)\|^2_{L^2(\omega_F)} &= \sum_{T\subset\omega_F} \|\nabla\big(E_F(q)\phi_T\big)\|^2_{L^2(T)} \\ &\lesssim h_F^{-1} \sum_{T\subset\omega_F} \|E_F(q)\phi_T\|^2_{L^2(T)} = h_F^{-1}\|E_F(q)\phi_T\|^2_{L^2(\omega_F)},\end{aligned}$$

and the proof is completed. □

These preparations lead to the following test space for discrete functionals.

Definition 4.23 (test space for discrete functionals). Using the cut-off functions from Assumption 4.21, we associate to the space $\mathbb{F}_{\mathcal{T}}$ of discrete functionals the following *test space*:

$$\begin{aligned}\mathbb{V}_{\mathcal{T}}^+ := \mathbb{V}^+(\mathcal{T}) := \operatorname{span}\Big(&\{q_T\phi_T \mid q_T \in \mathbb{P}_{m_2}(T),\ T\in\mathcal{T}\} \\ &\bigcup\{E_F(q_F)\phi_F \mid q_F \in \mathbb{P}_{m_1}(F),\ F\in\mathcal{F}\}\Big).\end{aligned}$$

If ω is a subdomain of Ω meshed by $\mathcal{T}_\omega$, then $\mathbb{V}^+(\mathcal{T}_\omega)$ is the test space for $\mathbb{F}(\mathcal{T}_\omega)$.

Similarly as for the space $\mathbb{F}_{\mathcal{T}}$ of discrete functionals, the test space over a subdomain ω meshed by $\mathcal{T}_\omega$ can be obtained from the global test space, namely

$$\mathbb{V}^+(\mathcal{T}_\omega) = \mathbb{V}^+(\mathcal{T}) \cap H_0^1(\omega). \tag{4.33}$$

4.5. *A projection onto discrete functionals*

Having the test space $\mathbb{V}_{\mathcal{T}}^+$ from Definition 4.23 at our disposal, we are now ready to construct a H^{-1}*-projection* $P_{\mathcal{T}}$, as suggested in Section 4.3. As in the previous section, the degrees (m_1, m_2) of the discrete functionals in $\mathbb{F}_{\mathcal{T}}$ are hidden parameters.

Definition 4.24 (projection onto discrete functionals). Given the discrete functionals $\mathbb{F}_\mathcal{T}$ and the test space $\mathbb{V}_\mathcal{T}^+$, we define a *projection* $P_\mathcal{T}\colon H^{-1}(\Omega) \to \mathbb{F}_\mathcal{T}$ by

$$\langle P_\mathcal{T}\ell, w\rangle = \langle \ell, w\rangle \quad \text{for all } w \in \mathbb{V}_\mathcal{T}^+. \tag{4.34}$$

The well-posedness of this definition and algebraic properties of $P_\mathcal{T}$ are verified in the following Lemma 4.25. Moreover, a representation of $P_\mathcal{T}$ in the form of a quasi-interpolation operator is given in Corollary 4.61 below.

The *polynomial densities* of $P_\mathcal{T}\ell$ are denoted by $P_T\ell := P_{\mathcal{T},T}\ell$, $T \in \mathcal{T}$, and $P_F\ell := P_{\mathcal{T},F}$, $F \in \mathcal{F}$, so that

$$\langle P_\mathcal{T}\ell, w\rangle = \sum_{T\in\mathcal{T}} \int_T P_T\ell\, w + \sum_{F\in\mathcal{F}} \int_F P_F\ell\, w. \tag{4.35}$$

In the next lemma we show in particular that $P_\mathcal{T}$ is a local operator. In order to formulate this, we shall use *$\mathcal{T}$-meshed local subdomains*, i.e. $\mathcal{T}$-meshed subdomains ω for which there exists a mesh element $T \in \mathcal{T}$ with $\omega \subset \omega_T$. For the next lemma, addressing algebraic properties of the operator $P_\mathcal{T}$, recall the notation $\omega_F = \cup_{T\in\mathcal{T}_F} T$ from (4.7) for an interior face $F \in \mathcal{F}$.

Lemma 4.25 (algebraic properties). *The operator $P_\mathcal{T}$ is a local linear projection onto the subspace $\mathbb{F}_\mathcal{T}$ of discrete functionals. More precisely, for any local subdomain ω meshed by $\mathcal{T}$, there is a linear projection $P_\omega\colon H^{-1}(\omega) \to \mathbb{F}(\mathcal{T}_\omega)$ such that*

$$P_\mathcal{T}\ell|_{H^{-1}(\omega)} = P_\omega\big(\ell|_{H_0^1(\omega)}\big) \in \mathbb{F}(\mathcal{T}_\omega)$$

for all $\ell \in H^{-1}(\Omega)$.

Proof. [1] We first show that the degrees of freedom in Definition 4.23 of $\mathbb{V}_\mathcal{T}^+$ are linearly independent. To this end, we fix an element $T \in \mathcal{T}$, let $F_1, \dots, F_l$, $l \le d+1$ denote its faces that are in $\mathbb{F}_\mathcal{T}$ and write $\phi_0 := \phi_T$, $\phi_i := \phi_{F_i}$ and $E_i := E_{F_i}$ for $i = 1, \dots, l$. We then claim that, for all $q_0 \in \mathbb{P}_{m_2}(T)\setminus\{0\}$ and all $q_i \in \mathbb{P}_{m_1}(F_i)\setminus\{0\}$, $i = 1, \dots, l$, we have

$$\alpha_0 q_0 \phi_0 + \sum_{i=1}^{l} \alpha_i E_i(q_i)\phi_i = 0 \text{ in } T \quad\Rightarrow\quad \alpha_0 = \dots = \alpha_l = 0. \tag{4.36}$$

In light of $\operatorname{supp}\phi_T = T$ and $\operatorname{supp}\phi_F = \omega_F$, we observe $\phi_0|_{\partial T} = 0$ and $\phi_i|_{\partial T\setminus F_i} = 0$ for $i = 1, \dots, l$. We thus evaluate the hypothesis of (4.36) first on the faces $F_1, \dots, F_l$ and then in the element T. This gives $\alpha_i = 0$ for $i = 0, \dots, l$ and (4.36) is verified.

[2] Next, we discuss the well-posedness of (4.34). The linear independence (4.36) and (4.25) lead to

$$\dim \mathbb{V}_\mathcal{T}^+ = \#\mathcal{T} \dim \mathbb{P}_{m_2} + \#\mathcal{F} \dim \mathbb{P}_{m_1} = \dim \mathbb{F}_\mathcal{T}. \tag{4.37}$$

Thus, it suffices to show the implication

$$\ell \in \mathbb{F}_{\mathcal{T}} \colon \langle \ell, w \rangle = 0 \quad \text{for all } w \in \mathbb{V}^+_{\mathcal{T}} \quad \Rightarrow \quad \ell = 0. \tag{4.38}$$

For that purpose, let $\ell \in \mathbb{F}_{\mathcal{T}}$ and let $q_T, T \in \mathcal{T}$ and $q_F, F \in \mathcal{F}$ be the determining polynomials. For any element $T \in \mathcal{T}$, the choice $w = q_T \phi_T \in H^1_0(T)$ implies

$$0 = \langle \ell, w \rangle = \int_T (q_T)^2 \phi_T, \quad \text{i.e. } q_T = 0.$$

Thus ℓ does not have contributions from elements. Regarding faces, any choice $w = (E_F q_F)\phi_F \in H^1_0(\omega_F)$, $F \in \mathcal{F}$, therefore gives

$$0 = \langle \ell, w \rangle = \int_F (q_F)^2 \phi_F, \quad \text{i.e. } q_F = 0,$$

and implication (4.38) is established. Combining (4.37) and (4.38), we have that the dual pairing in $H^{-1}(\Omega)$ is non-degenerate on $\mathbb{F}_{\mathcal{T}} \times \mathbb{V}^+_{\mathcal{T}}$, which in turn ensures that $P_{\mathcal{T}}$ is well-defined. The Petrov–Galerkin character of the definition (4.34) then ensures that $P_{\mathcal{T}}$ is a linear projection onto $\mathbb{F}_{\mathcal{T}}$.

[3] It remains to show that $P_{\mathcal{T}}$ is a local operator. Given any local subdomain ω meshed by $\mathcal{T}_\omega$, we can apply the preceding proof to $\mathcal{T}_\omega$ instead of $\mathcal{T}$. This shows that the dual pairing in $H^{-1}(\omega)$ is non-degenerate on $\mathbb{F}(\mathcal{T}_\omega) \times \mathbb{V}^+(\mathcal{T}_\omega)$ and ensures a local projection operator $P_\omega \colon H^{-1}(\omega) \to \mathbb{F}(\mathcal{T}_\omega)$. Taking into account (4.23) and (4.33), we note that

$$P_{\mathcal{T}}|_{H^{-1}(\omega)} = P_\omega,$$

which completes the proof. □

The verification of (4.38), which amounts to a proof of uniqueness, suggests the following approach to computing $P_{\mathcal{T}}\ell$, $\ell \in H^{-1}(\Omega)$.

Remark 4.26 (local computation). Let $\ell \in H^{-1}(\Omega)$. Recalling (4.35), the polynomials $P_T\ell, T \in \mathcal{T}$, and $P_F\ell, F \in \mathcal{F}$ can be computed by solving first

$$\int_T P_T\ell\, q\phi_T = \langle \ell, q\phi_T \rangle \quad \text{for all } T \in \mathcal{T}, q \in \mathbb{P}_{m_2}(T), \tag{4.39}$$

and then

$$\int_F P_F\ell\, q\phi_F = \langle \ell, q\phi_F \rangle - \sum_{T \subset \omega_F} \int_T P_T\ell\, q\phi_F \quad \text{for all } F \in \mathcal{F}, q \in \mathbb{P}_{m_1}(F). \tag{4.40}$$

This amounts to two block diagonal linear systems with, respectively, $\#\mathcal{T}$ blocks of size $\dim \mathbb{P}_{m_2}$ and $\#\mathcal{F}$ blocks of size $\dim \mathbb{P}_{m_1}$. Each block and each corresponding right-hand side arises from local computations.

Remark 4.27 (star localization vs. locality of $P_{\mathcal{T}}$). Stars ω_z, $z \in \mathcal{V}$, are meshed local subdomains. Lemma 4.25 thus shows that, for any vertex $z \in \mathcal{V}$, there is a linear projection $P_z\colon H^{-1}(\omega_z) \to \mathbb{F}(\mathcal{T}_z)$ such that

$$P_\mathcal{T}\ell|_{H^{-1}(\omega_z)} = P_z\big(\ell|_{H_0^1(\omega_z)}\big) \in \mathbb{F}(\mathcal{T}_z)$$

for all $\ell \in H^{-1}(\Omega)$. The stars also appear in the localizing upper bound of the global residual norm in Corollary 4.6. As they are minimal subdomains therein (see Remark 4.7), it may appear that a finer localization with smaller domains cannot be exploited in *a posteriori* analysis. Although this is true in the context of upper bounds, the increased locality of $P_\mathcal{T}$ is useful in the context of lower bounds; see the reduced lower bound (5.17), which follows from the interior vertex property introduced in Definition 4.50, and is crucial for deriving the contraction result (5.26).

We have already mentioned that we shall use $P_\mathcal{T}$ to split the residual. In light of the bounds for the residual norm in Corollary 4.6 (star localization of residual norm), this should be done in a locally stable manner. In order to formulate and employ the local stability properties of $P_\mathcal{T}$, the following notation is useful. Given a local subdomain ω meshed by $\mathcal{T}_\omega$, we define the $\mathbb{V}^+(\mathcal{T}_\omega)$-*discrete dual norm* by

$$\|\ell\|_{\mathbb{V}^+(\mathcal{T}_\omega)^*} := \sup_{w\in\mathbb{V}^+(\mathcal{T}_\omega),\,\|\nabla w\|_{L^2(\omega)}=1} \langle \ell, w\rangle, \quad \ell \in H^{-1}(\omega). \tag{4.41}$$

In view of $\mathbb{V}^+(\mathcal{T}_\omega) \subset H_0^1(\omega)$, we have $\|\ell\|_{\mathbb{V}^+(\mathcal{T}_\omega)^*} \le \|\ell\|_{H^{-1}(\omega)}$ for all $\ell \in H^{-1}(\omega)$.

Lemma 4.28 (local H^{-1}-stability). *The projection $P_\mathcal{T}$ is locally H^{-1}-stable: for any local subdomain ω meshed by $\mathcal{T}_\omega$, we have*

$$\|P_\mathcal{T}\|_{\mathcal{L}(H^{-1}(\omega))} = \sup_{\ell\in\mathbb{F}(\mathcal{T}_\omega)} \frac{\|\ell\|_{H^{-1}(\omega)}}{\|\ell\|_{\mathbb{V}^+(\mathcal{T}_\omega)^*}} \le C_{\mathrm{lStb}},$$

where $C_{\mathrm{lStb}} = C_{\mathrm{lStb}}(d, \sigma, m_1, m_2)$ depends only on d, the shape regularity coefficient σ from (3.9)*, the degrees (m_1, m_2) of the discrete functionals and the space $\mathbb{S}^+$.*

Proof. [1] We start by verifying the '≤'-part of the claimed identity for the operator norm. The definition of the operator norm leads to

$$\|P_\omega\|_{\mathcal{L}(H^{-1}(\omega))} = \sup_{\ell\in H^{-1}(\omega)} \frac{\|P_\omega\ell\|_{H^{-1}(\omega)}}{\|\ell\|_{H^{-1}(\omega)}} \le \sup_{\ell\in H^{-1}(\omega)} \frac{\|P_\omega\ell\|_{H^{-1}(\omega)}}{\|\ell\|_{\mathbb{V}^+(\mathcal{T}_\omega)^*}}.$$

We now notice that $\|\ell\|_{\mathbb{V}^+(\mathcal{T}_\omega)^*} = \|P_\omega\ell\|_{\mathbb{V}^+(\mathcal{T}_\omega)^*}$ in view of (4.41) and (4.34). Hence

$$\|P_\omega\|_{\mathcal{L}(H^{-1}(\omega))} \le \sup_{\ell\in H^{-1}(\omega)} \frac{\|P_\omega\ell\|_{H^{-1}(\omega)}}{\|P_\omega\ell\|_{\mathbb{V}^+(\mathcal{T}_\omega)^*}} = \sup_{\ell\in\mathbb{F}(\mathcal{T}_\omega)} \frac{\|\ell\|_{H^{-1}(\omega)}}{\|\ell\|_{\mathbb{V}^+(\mathcal{T}_\omega)^*}},$$

because the projection P_ω is onto $\mathbb{F}(\mathcal{T}_\omega)$.

2 Next, we show that $\|P_{\mathcal{T}}\|_{\mathcal{L}(H^{-1}(\omega))}$ is uniformly bounded. Let $\ell \in \mathbb{F}(\mathcal{T}_\omega)$ be a discrete functional, namely

$$\langle \ell, w\rangle = \sum_{T\in\mathcal{T}_\omega} \int_T q_T w + \sum_{F\in\mathcal{F}_\omega} \int_F q_F w \quad \text{for all } w \in H_0^1(\omega).$$

We proceed in two steps that are quite similar to the classical standard residual estimates. Arguing as in (4.11), and using Lemma 2.2 (first Poincaré inequality) in the domain ω of diameter about h_T for $w \in H_0^1(\omega)$, we obtain

$$|\langle \ell, w\rangle| \lesssim \Bigg(\sum_{T\in\mathcal{T}_\omega} h_T^2 \|q_T\|_{L^2(T)}^2 + \sum_{F\in\mathcal{F}_\omega} h_F \|q_F\|_{L^2(F)}^2\Bigg)^{1/2} \|\nabla w\|_{L^2(\omega)},$$

whence

$$\|\ell\|_{H^{-1}(\omega)}^2 \lesssim \sum_{T\in\mathcal{T}_\omega} h_T^2 \|q_T\|_{L^2(T)}^2 + \sum_{F\in\mathcal{F}_\omega} h_F \|q_F\|_{L^2(F)}^2. \tag{4.42}$$

Here we do not exploit the fact that ℓ is discrete in ω; this will be crucial in the second step, when we bound each term on the right-hand side, in a manner recalling the derivation of classical lower bounds. For any $T \in \mathcal{T}_\omega$, we write $\mathbb{V}^+(T)^*$ as shorthand for $\mathbb{V}^+(\{T\})^*$, and exploit Lemma 4.22 (properties of cut-off) to deduce

$$\begin{aligned}\|q_T\|_{L^2(T)}^2 &\lesssim \int_T (q_T)^2\phi_T = \langle \ell, q_T\phi_T\rangle \le \|\ell\|_{\mathbb{V}^+(T)^*}\|\nabla(q_T\phi_T)\|_{L^2(T)}\\ &\lesssim \|\ell\|_{\mathbb{V}^+(T)^*} h_T^{-1}\|q_T\phi_T\|_{L^2(T)} \le \|\ell\|_{\mathbb{V}^+(T)^*} h_T^{-1}\|q_T\|_{L^2(T)},\end{aligned}$$

whence

$$h_T\|q_T\|_{L^2(T)} \lesssim \|\ell\|_{\mathbb{V}^+(T)^*}. \tag{4.43}$$

For an interior face $F \in \mathcal{F}_\omega$, we proceed similarly, also taking into account Lemma 4.20 and that $\mathbb{V}^+(T) \subset \mathbb{V}^+(\mathcal{T}_F)$ entails $\|\ell\|_{\mathbb{V}^+(T)^*} \le \|\ell\|_{\mathbb{V}^+(\mathcal{T}_F)}$ for $T \in \mathcal{T}_F$. We thus obtain

$$\begin{aligned}&\|q_F\|_{L^2(F)}^2\\ &\quad\lesssim \int_F (q_F)^2\phi_F = \langle \ell, (E_F q_F)\phi_F\rangle - \sum_{T\subset\omega_F}\int_T q_T(E_F q_F)\phi_F\\ &\quad\lesssim \|\ell\|_{\mathbb{V}^+(\mathcal{T}_F)^*}\|\nabla((E_F q_F)\phi_T)\|_{L^2(\omega_F)} + \sum_{T\subset\omega_F}\|q_T\|_{L^2(T)}\|(E_F q_F)\phi_T\|_{L^2(T)}\\ &\quad\lesssim \|\ell\|_{\mathbb{V}^+(\mathcal{T}_F)^*} h_T^{-1}\|(E_F q_F)\phi_T\|_{L^2(\omega_F)} \lesssim \|\ell\|_{\mathbb{V}^+(\mathcal{T}_F)^*} h_T^{-1/2}\|q_F\|_{L^2(F)},\end{aligned}$$

that is,

$$h_F^{1/2}\|q_F\|_{L^2(F)} \lesssim \|\ell\|_{\mathbb{V}^+(\mathcal{T}_F)^*}. \tag{4.44}$$

The number of elements and interior faces in the local subdomain ω is uniformly bounded by d and the shape regularity coefficient σ. Hence inequalities (4.43) and

(4.44) together with the inclusions $\mathbb{V}^+(T) \subset \mathbb{V}^+(\mathcal{T}_\omega)$ for $T \in \mathcal{T}_\omega$ and $\mathbb{V}^+(\mathcal{T}_F) \subset \mathbb{V}^+(\mathcal{T}_\omega)$ for $F \in \mathcal{F}_\omega$ imply

$$\sum_{T\in\mathcal{T}_\omega} h_T^2 \|q_T\|_{L^2(T)}^2 + \sum_{F\in\mathcal{F}_\omega} h_F \|q_F\|_{L^2(F)}^2 \lesssim \|\ell\|_{\mathbb{V}^+(\mathcal{T}_\omega)^*}^2. \tag{4.45}$$

Combining (4.42) and (4.45) shows that the ratio $\|\ell\|_{H^{-1}(\omega)}/\|\ell\|_{\mathbb{V}^+(\mathcal{T}_\omega)^*}$ for $\ell \in \mathbb{F}(\mathcal{T}_\omega)$ is bounded by a universal constant depending on d, σ, m_1, m_2 and $\mathbb{S}^+$.

[3] It remains to complete the proof of the claimed identity for the operator norm. To this end, we first introduce an operator $Q_\omega\colon H_0^1(\omega) \to \mathbb{V}^+(\mathcal{T}_\omega)$ by

$$\langle \ell, Q_\omega w\rangle = \langle \ell, w\rangle \quad \text{for all } \ell \in \mathbb{F}(\mathcal{T}_\omega).$$

Like the one for P_ω, this definition is well-posed because the pair $(\mathbb{F}(\mathcal{T}_\omega), \mathbb{V}^+(\mathcal{T}_\omega))$ is non-degenerate for the dual pairing of $H^{-1}(\omega)$; see (4.37) and (4.38) in the proof of Lemma 4.25 (algebraic properties). By the Petrov–Galerkin character of the definition, Q_ω is a linear projection onto $\mathbb{V}^+(\mathcal{T}_\omega)$. Given arbitrary $\ell \in H^{-1}(\omega)$ and $w \in H_0^1(\omega)$, the definitions of Q_ω and P_ω imply

$$\langle P_\omega \ell, w\rangle = \langle P_\omega \ell, Q_\omega w\rangle = \langle \ell, Q_\omega w\rangle,$$

that is, $Q_\omega = P_\omega^*$ is the (Hilbert) adjoint to P_ω. In other words, the adjoint P_ω^* is a projection onto $\mathbb{V}^+(\mathcal{T}_\omega)$. With this, we can prove the missing inequality. Let $\ell \in \mathbb{F}(\mathcal{T}_\omega)$ be discrete. In fact,

$$\langle \ell, w\rangle = \langle P_\omega \ell, w\rangle = \langle \ell, P_\omega^* w\rangle \quad\Rightarrow\quad \|\ell\|_{H^{-1}(\omega)} = \sup_{w\in H_0^1(\omega)} \frac{\langle \ell, P_\omega^* w\rangle}{\|\nabla w\|_{L^2(\omega)}}$$

leads to

$$\|\ell\|_{H^{-1}(\omega)} \le \|\ell\|_{\mathbb{V}^+(\mathcal{T}_\omega)^*} \|P_\omega^*\|_{\mathcal{L}(H_0^1(\omega))} = \|\ell\|_{\mathbb{V}^+(\mathcal{T}_\omega)^*} \|P_\omega\|_{\mathcal{L}(H^{-1}(\omega))}.$$

This concludes the proof. □

Remark 4.29 (failing global H^{-1}-stability). For Lebesgue norms, local stability of linear operators in terms of shape regularity entails that their respective global stability is uniform under mesh refinement. The fact that part (i) of Lemma 4.5 (localization of H^{-1}-norm) needs a condition to be true, may lead us to suspect that this implication might not be true in general for the H^{-1}-norm. This suspicion is confirmed by Example 4.63 below, where we show that $\|P_\mathcal{T}\|_{\mathcal{L}(H^{-1}(\Omega))}$ can tend to ∞ under mesh refinement.

The proof of Lemma 4.28 provides all non-trivial ingredients to allow the approximate computation of $\|\ell\|_{H^{-1}(\omega)}$ whenever $\ell \in H^{-1}(\Omega)$ is discrete in ω.

Corollary 4.30 (quantifying H^{-1}-norms of discrete functionals). *Let $\omega \subset \Omega$ be a local subdomain meshed by $\mathcal{T}_\omega$ and let $\ell \in H^{-1}(\Omega)$ be discrete in ω, given by the polynomials q_T for $T \in \mathcal{T}_\omega$ and q_F for $F \in \mathcal{F}_\omega$, where $F \in \mathcal{F}_\omega$ are the interior* *faces in ω. We then have*

$$\|\ell\|^2_{H^{-1}(\omega)} \approx \sum_{T\in\mathcal{T}_\omega} h_T^2 \|q_T\|^2_{L^2(T)} + \sum_{F\in\mathcal{F}_\omega} h_F \|q_F\|^2_{L^2(F)},$$

where the hidden constants depend on d, the shape regularity coefficient σ, the degrees (m_1, m_2) of the discrete functionals, and the space $\mathbb{S}^+$ appearing in Assumption 4.21.

Proof. This is a consequence of (4.42) and (4.45), the latter requiring ℓ to be discrete, along with the fact that $\|\ell\|_{\mathbb{V}^+(\mathcal{T}_\omega)^*} \le \|\ell\|_{H^{-1}\omega)^*}$ for any $\ell \in H^{-1}(\omega)$. □

Corollary 4.31 (local near-best approximation). *The projection $P_\mathcal{T}$ yields local near-best approximations: for any functional $\ell \in H^{-1}(\Omega)$ and any local subdomain ω meshed by $\mathcal{T}_\omega$, we have*

$$\|\ell - P_\mathcal{T}\ell\|_{H^{-1}(\omega)} \le C_{\mathrm{lStb}} \inf_{\chi\in\mathbb{F}(\mathcal{T}_\omega)} \|\ell - \chi\|_{H^{-1}(\omega)},$$

where C_{lStb} is the constant of Lemma 4.28 (local H^{-1}-stability).

Proof. Fix a local subdomain ω meshed by $\mathcal{T}_\omega$ and let $\chi \in \mathbb{F}(\mathcal{T}_\omega)$ be arbitrary. Thanks to Lemma 4.25 (algebraic properties), we have $P_\omega\chi = \chi$ and

$$(\ell - P_\mathcal{T}\ell)|_{H_0^1(\omega)} = (I - P_\omega)\ell|_{H_0^1(\omega)} = (I - P_\omega)(\ell - \chi)|_{H_0^1(\omega)}.$$

As P_ω is a non-trivial projection on the Hilbert space $H^{-1}(\omega)$, Szyld (2006) ensures

$$\|I - P_\omega\|_{\mathcal{L}(H^{-1}(\omega))} = \|P_\omega\|_{\mathcal{L}(H^{-1}(\omega))} \le C_{\mathrm{lStb}}. \tag{4.46}$$

Hence

$$\|\ell - P_\mathcal{T}\ell\|_{H^{-1}(\omega)} \le C_{\mathrm{lStb}}\|\ell - \chi\|_{H^{-1}(\omega)}$$

concludes the proof because $\chi \in \mathbb{F}(\mathcal{T}_\omega)$ is arbitrary. □

We illustrate the approximation of possible parts of the residual with the projection $P_\mathcal{T}$ in a series of three remarks. For that purpose, the approximation quality is to be measured with a local $H^{-1}(\omega)$-norm, and it is instructive to compare with the operator $\Pi_\mathcal{T}$ from (4.26). Recall that the operator $\Pi_\mathcal{T}$ is used implicitly in the standard approach (see Section 4.2) to approximate the discrete functionals $\mathbb{F}_\mathcal{T}$.

Remark 4.32 (approximating functions). For functions, the local error with $P_\mathcal{T}$ is uniformly dominated by the one with $\Pi_\mathcal{T}$. More precisely, if $m_2 \ge 0$ and $\ell \in H^{-1}(\Omega)$ satisfies $\langle \ell, w\rangle = \int_\Omega f w$ where $f \in L^p(\Omega)$ with $p > 2d/(2+d)$, then Corollary 4.31 (local near-best approximation) and $\Pi_\mathcal{T} f = \sum_{T\in\mathcal{T}}(\Pi_T^{m_2} f)\chi_T \in \mathbb{F}(\mathcal{T}_\omega)$ imply, for any local meshed subdomain ω,

$$\|\ell - P_\mathcal{T}\ell\|_{H^{-1}(\omega)} \lesssim \|f - \Pi_\mathcal{T} f\|_{H^{-1}(\omega)}.$$

Observe that although ℓ is a function, $P_\mathcal{T}\ell$ is typically not a function. This property might look undesirable but it is crucial for an advantage of $P_\mathcal{T}$ over $\Pi_\mathcal{T}$ and closely related to the fact that the opposite inequality does not hold; see Remark 4.34 about stability.

Furthermore, supposing $f \in L^2(\Omega)$ and combining the preceding inequality with Lemma 2.2 (first Poincaré inequality) gives

$$\|\ell - P_{\mathcal{T}}\ell\|^2_{H^{-1}(\omega_T)} \lesssim \sum_{T' \subset \omega_T} h_T^2 \|f - \Pi^{m_2}_{T'} f\|^2_{L^2(\omega_{T'})}, \tag{4.47}$$

which establishes that the local $P_{\mathcal{T}}$-oscillation of functions is uniformly dominated by its classical $\Pi_{\mathcal{T}}$-counterpart but not vice versa.

In Section 7.3.1 this case is considered in the context of adaptive approximation.

Remark 4.33 (approximating admissible functionals). For functionals allowing for the application of $\Pi_{\mathcal{T}}$, the local error with $P_{\mathcal{T}}$ is again uniformly dominated by the one with $\Pi_{\mathcal{T}}$. In view of the previous remark, let us consider only $\ell \in H^{-1}(\Omega)$ such that $\langle \ell, w\rangle = \int_\Sigma g w$, where $g \in L^p(\Sigma)$ with $\Sigma := \cup_{F\in\mathcal{F}} F$ and $p > 2(d-1)/d$. Note that we again have $\Pi_{\mathcal{T}}\ell \in \mathbb{F}(\mathcal{T}_\omega)$ as $\langle \Pi_{\mathcal{T}}\ell, w\rangle = \sum_{F\in\mathcal{F}} \int_F (\Pi^{m_1}_F g)\, w$ for all $w \in H^1_0(\Omega)$. Corollary 4.31 (local near-best approximation) thus ensures, for any local meshed subdomain ω,

$$\|\ell - P_{\mathcal{T}}\ell\|_{H^{-1}(\omega)} \lesssim \|\ell - \Pi_{\mathcal{T}}\ell\|_{H^{-1}(\omega)}.$$

Moreover, supposing $g \in L^2(\Sigma)$ and combining the scaled trace theorem (4.12) with Lemma 2.2 (first Poincaré inequality) yields

$$\|\ell - P_{\mathcal{T}}\ell\|^2_{H^{-1}(\omega_T)} \lesssim \sum_{F \subseteq \overline{\omega}_T, F \not\subseteq \partial\omega_T} h_F \|g - \Pi^{m_1}_F g\|^2_{L^2(F)}.$$

Also, this case will be revisited in the context of adaptive approximation, namely in Section 7.3.3.

Remark 4.34 (stability of approximation). The error with $P_{\mathcal{T}}$ is stable, while the one with $\Pi_{\mathcal{T}}$ is not. To see this by example, we restrict to $(m_1, m_2) = (0, 0)$, fix some interior face $F \in \mathcal{F}$ and, for $\varepsilon > 0$ sufficiently small, consider

$$\langle \ell_\varepsilon, w\rangle := \int_\Omega f_\varepsilon w = \frac{1}{2\varepsilon} \int_{-\varepsilon}^{\varepsilon} \int_F w(y + s\boldsymbol{n}_F)\, \mathrm{d}y\, \mathrm{d}s, \quad w \in H^1_0(\Omega),$$

where $f_\varepsilon = (2\varepsilon)^{-1}\chi_{F_\varepsilon}$ is a multiple of the characteristic function of $F_\varepsilon := \{x + s\boldsymbol{n}_F \mid x \in F, -\varepsilon < s < \varepsilon\}$. As

$$\begin{aligned}\langle \ell_\varepsilon - \delta_F, w\rangle &= \frac{1}{2\varepsilon} \int_{-\varepsilon}^{\varepsilon} \int_F \int_0^s \partial_{\boldsymbol{n}_F} w(y + t\boldsymbol{n}_F)\, \mathrm{d}t\, \mathrm{d}y\, \mathrm{d}s \\ &\le \frac{1}{2\varepsilon} \int_{-\varepsilon}^{\varepsilon} \int_{F_\varepsilon} |\nabla w(x)|\, \mathrm{d}x\, \mathrm{d}s \le |F_\varepsilon|^{1/2} \|\nabla w\|_{L^2(\Omega)},\end{aligned}$$

the functions $(\ell_\varepsilon)_{\varepsilon>0}$ tend to the proper functional δ_F:

$$\ell_\varepsilon \to \delta_F \quad \text{in } H^{-1}(\Omega).$$

Combining the convergence with the local stability of $P_\mathcal{T}$ (see (4.46)) yields

$$\big| \|\ell_\varepsilon - P_\mathcal{T}\ell_\varepsilon\|_{H^{-1}(\omega_F)} - \|\delta_F - P_\mathcal{T}\delta_F\|_{H^{-1}(\omega_F)} \big| \le \|(I - P_\mathcal{T})(\ell_\varepsilon - \delta_F)\|_{H^{-1}(\omega_F)} \\ \lesssim \|\ell_\varepsilon - \delta_F\|_{H^{-1}(\omega_F)} \to 0,$$

the stability of the error with $P_\mathcal{T}$. Furthermore, since $P_\mathcal{T}\delta_F = \delta_F$, the stability entails here $\|\ell_\varepsilon - P_\mathcal{T}\ell_\varepsilon\|_{H^{-1}(\omega_F)} \to 0$. For $\Pi_\mathcal{T}$, however, the approximation on the skeleton and in the volume are independent of each other. Hence, combining $\Pi_\mathcal{T}\delta_F = \delta_F$, which follows from (4.26), with $\lim_{\varepsilon\to 0}\Pi^0_T f_\varepsilon = |F|/(2|T|)$ for the two elements $T \in \mathcal{T}$ containing F, leads to

$$\|\delta_F - \Pi_\mathcal{T}\delta_F\|_{H^{-1}(\omega_F)} = 0 < \lim_{\varepsilon\to 0} \|\ell_\varepsilon - \Pi_\mathcal{T}\ell_\varepsilon\|_{H^{-1}(\omega_F)}.$$

Measuring the error in weighted L^2-norms instead of the H^{-1}-norm results in a more dramatic instability. Indeed, letting 1_K and 0_K denote the constant functions on a simplex K equal to 1 or 0, the two sides translate to

$$h_F^{1/2}\|1_F - \Pi^0_F 1_F\|_{L^2(F)} + \sum_{T\in\mathcal{T}_F} h_T\|0_T - \Pi^0_T 0_T\|_{L^2(T)} = 0$$

and

$$\lim_{\varepsilon\to 0}\bigg(h_F^{1/2}\|0_F - \Pi^0_F 0_F\|_{L^2(F)} + \sum_{T\in\mathcal{T}_F} h_T\|f_\varepsilon - \Pi^0_T f_\varepsilon\|_{L^2(T)}\bigg) = \infty.$$

Note that such a transformation of volume contributions into contributions on the skeleton may occur by perturbation in the right-hand side or, in the opposite direction, by an improvement of the Galerkin approximation thanks to refinement.

In view of this instability of $\Pi_\mathcal{T}$, the inequalities in the preceding Remarks 4.32 and 4.33 cannot be reversed – a fact that can also be inferred from Remark 4.19.

The above perturbations of δ_F are in the domain of $\Pi_\mathcal{T}$. For the functionals

$$\langle \widehat{\ell}_\varepsilon, w\rangle := \int_F w(y + \varepsilon \boldsymbol{n}_F)\,\mathrm{d}y, \quad w \in H^1_0(\Omega),$$

however, it is not clear how to directly apply $\Pi_\mathcal{T}$ for $\varepsilon \neq 0$. To the contrary, the approximations $P_\mathcal{T}\widehat{\ell}_\varepsilon$ are defined and stable around 0. Notably, $P_\mathcal{T}\widehat{\ell}_\varepsilon$ uses volume contributions to compensate for the displacement in the representation of the singular contribution.

4.6. Discretized and oscillatory residual

We now turn to the proper *a posteriori* analysis, that is, we shall derive *upper* and *lower bounds of the error*, implementing the following plan, which is motivated in Section 4.3. We use the projection $P_\mathcal{T}$ onto discrete functionals $\mathbb{F}_\mathcal{T}$ to split the residual into discretized and oscillatory parts. Then the quantification of the *oscillatory residual* is reduced to *data* oscillation through suitable choices of the degrees (m_1, m_2) of the discrete functionals. The *discretized residual* can be quantified in various ways; see Sections 4.2 and 4.9 below.

We start by introducing indicators reflecting the stated splitting of the residual into discretized and oscillatory parts. They are vertex-indexed and, given $z \in \mathcal{V}$, defined by

$$\begin{aligned} \mathcal{E}_{\mathcal{T}}^{\mathrm{abs}}(z)^2 &:= \eta_{\mathcal{T}}^{\mathrm{abs}}(z)^2 + \operatorname{osc}_{\mathcal{T}}(R_{\mathcal{T}}, z)^2 \quad \text{with} \\ \eta_{\mathcal{T}}^{\mathrm{abs}}(z) &:= \|P_{\mathcal{T}} R_{\mathcal{T}}\|_{H^{-1}(\omega_z)} \quad \text{and} \quad \operatorname{osc}_{\mathcal{T}}(R_{\mathcal{T}}, z) := \|(I - P_{\mathcal{T}}) R_{\mathcal{T}}\|_{H^{-1}(\omega_z)}. \end{aligned} \tag{4.48}$$

Note that these quantities are not proper indicators: they still need to be quantified in a computable manner. By using 'abs' (shorthand for 'abstract'), we hint at the fact that $\eta_{\mathcal{T}}^{\mathrm{abs}}$ can be quantified by various approaches.

Lemma 4.35 (splitting of local residual norm). *For any vertex $z \in \mathcal{V}$, the local residual norm is equivalent to the abstract indicator from* (4.48)*:*

$$\frac{1}{\sqrt{2} C_{\mathrm{lStb}}} \mathcal{E}_{\mathcal{T}}^{\mathrm{abs}}(z) \le \|R_{\mathcal{T}}\|_{H^{-1}(\omega_z)} \le \sqrt{2} \mathcal{E}_{\mathcal{T}}^{\mathrm{abs}}(z),$$

where C_{lStb} is the constant of Lemma 4.28 (local H^{-1}-stability).

Proof. As announced, we use the linear projection $P_{\mathcal{T}}$ in order to split the residual into a discretized and an oscillatory part.

$$R_{\mathcal{T}} = P_{\mathcal{T}} R_{\mathcal{T}} + (I - P_{\mathcal{T}}) R_{\mathcal{T}}. \tag{4.49}$$

The upper bound of the local residual norm then readily follows from the triangle inequality:

$$\|R_{\mathcal{T}}\|_{H^{-1}(\omega_z)} \le \eta_{\mathcal{T}}^{\mathrm{abs}}(z) + \operatorname{osc}_{\mathcal{T}}(R_{\mathcal{T}}, z) \le \sqrt{2} \mathcal{E}_{\mathcal{T}}^{\mathrm{abs}}(z).$$

To show the lower bound, we exploit the local stability of $P_{\mathcal{T}}$ (see (4.46)) to obtain

$$\eta_{\mathcal{T}}^{\mathrm{abs}}(z) = \|P_{\mathcal{T}} R_{\mathcal{T}}\|_{H^{-1}(\omega_z)} \le C_{\mathrm{lStb}} \|R_{\mathcal{T}}\|_{H^{-1}(\omega_z)}$$

and

$$\operatorname{osc}_{\mathcal{T}}(R_{\mathcal{T}}, z) = \|(I - P_{\mathcal{T}}) R_{\mathcal{T}}\|_{H^{-1}(\omega_z)} \le C_{\mathrm{lStb}} \|R_{\mathcal{T}}\|_{H^{-1}(\omega_z)}.$$

Squaring both inequalities, summing them, and then taking the square-root finishes the proof. □

In many *a posteriori* analyses, this lemma is replaced by steps breaking a possible true equivalence between error and estimator. Therefore the following remark points out the key ingredients.

Remark 4.36 (ensuring proper equivalence). The fact that the projection $P_{\mathcal{T}}$ and so also $I - P_{\mathcal{T}}$ are linear and locally bounded operators precludes overestimation; see also Remark 4.19. Comparing with Section 4.2 and $\Pi_{\mathcal{T}}$ in (4.27), we see that the local stability in H^{-1} is crucial to that end and, in view of Remark 4.34, requires discrete functionals with contributions on the skeleton.

Next, we want to simplify the residual oscillations $\operatorname{osc}_{\mathcal{T}}(R_{\mathcal{T}}, z)$, $z \in \mathcal{V}$, in the spirit of Remark 4.14. This will be dependent on the coefficients $\boldsymbol{A}$ and c of the differential operator and involve the following 'polynomial degrees':

$$n_{\boldsymbol{A}} := \min\{k \in \mathbb{N}_0 \mid \boldsymbol{A} \in (\mathbb{S}_{\mathcal{T}}^{k,-1})^{d\times d}\}, \tag{4.50a}$$

$$n_c := \min\{k \in \mathbb{N}_0 \cup \{-1\} \mid c \in \mathbb{S}_{\mathcal{T}}^{k,-1}\}, \tag{4.50b}$$

where we use the convention $\min \emptyset = \infty$. We shall say that the differential operator $-\operatorname{div}(\boldsymbol{A}\nabla\cdot) + c(\cdot)$ in (2.5) *has discrete coefficients* whenever $\max\{n_{\boldsymbol{A}}, n_c\} < \infty$; otherwise it has *non-discrete coefficients.*

Lemma 4.37 (data oscillation reduction for discrete coefficients). *If the coefficients A and c are discrete, the choices*

$$m_1 = n_{\boldsymbol{A}} + n - 1,$$

$$m_2 = \max\{n - 2 + n_{\boldsymbol{A}}, \widetilde{m}_c\} \quad \text{with} \quad \widetilde{m}_c = \begin{cases} n + n_c, & \text{if } c \neq 0 \\ 0, & \text{otherwise} \end{cases}$$

ensure that the oscillatory residual reduces to data oscillation of the right-hand side:

$$(I - P_{\mathcal{T}})R_{\mathcal{T}} = f - P_{\mathcal{T}} f.$$

Proof. The choices for m_1 and m_2 yield, for any face $F \in \mathcal{F}$ and any element $T \in \mathcal{T}$,

$$[[\boldsymbol{A}\nabla u_{\mathcal{T}}]] \cdot \boldsymbol{n}_F \in \mathbb{P}_{m_1}(F) \quad \text{and} \quad \operatorname{div}(\boldsymbol{A}\nabla u_{\mathcal{T}})|_T \in \mathbb{P}_{m_2}(T).$$

Furthermore, if $c \neq 0$, we also have $cu_{\mathcal{T}}|_T \in \mathbb{P}_{m_2}(T)$, and the claimed identity follows from $-\operatorname{div}(\boldsymbol{A}\nabla u_{\mathcal{T}}) + cu_{\mathcal{T}} \in \mathbb{F}_{\mathcal{T}}$. □

Remark 4.38 (Poisson equation with linear elements). In the case of the Poisson equation with linear elements, the choices in Lemma 4.37 lead to $m_1 = 0$ and $m_2 = 0$. Alternatively, one may use $m_1 = 0$ and $m_2 = -1$ (recall we have set $\mathbb{P}_{-1}(T) = \{0\}$); see Diening, Kreuzer and Stevenson (2016) or Siebert and Veeser (2007). The choice here leads to an oscillation for which the standard oscillation indicators $h_T \|f - \Pi_T f\|_T$, $T \in \mathcal{T}$, can be used as a surrogate; see also Remark 4.43 about surrogates.

If one of the coefficients, $\boldsymbol{A}$ or c, is non-discrete, the range of the finite element space $\mathbb{V}_{\mathcal{T}}$ under the differential operator $-\operatorname{div}(\boldsymbol{A}\nabla\cdot) + c(\cdot)$ consists of functionals whose densities are not piecewise polynomial. Consequently, the oscillatory residual cannot be reduced to the oscillation $f - P_{\mathcal{T}} f$, or to any other oscillation of f involving discrete functionals with piecewise polynomial densities. The next result illustrates the idea of a non-perfect remedy, namely bounding the residual oscillation defined in (4.48) in terms of data oscillation and discrete stability. For its formulation, we define global H^{-1}-oscillations by

$$\mathrm{osc}_{\mathcal{T}}(\ell)^2 := \sum_{z\in\mathcal{V}} \mathrm{osc}_{\mathcal{T}}(\ell,z)^2, \quad \ell \in H^{-1}(\Omega), \tag{4.51}$$

which, in contrast to $\|(I-P_{\mathcal{T}})\ell\|_{H^{-1}(\Omega)}$, is bounded in terms of ℓ; see Remark 4.29 (failing global H^{-1}-stability).

Lemma 4.39 (surrogate data oscillation reduction). *Let $m_{\boldsymbol{A}} := \min\{n_{\boldsymbol{A}}, n-1\}$ and $m_c := \min\{n_c, n-1\}$, and define m_1 and m_2 as in Lemma 4.37, but replacing $n_{\boldsymbol{A}}$ and n_c, respectively, with $m_{\boldsymbol{A}}$ and m_c. Given any approximations*

$$\widehat{\boldsymbol{A}} \in \left(\mathbb{S}_{\mathcal{T}}^{m_{\boldsymbol{A}},-1}\right)^{d\times d} \quad \text{and} \quad \widehat{c} \in \mathbb{S}_{\mathcal{T}}^{m_c,-1},$$

we then have, for all vertices $z\in\mathcal{V}$,

$$\begin{aligned}\mathrm{osc}_{\mathcal{T}}(R_{\mathcal{T}},z) \le \mathrm{osc}_{\mathcal{T}}(f,z) &+ C_{\mathrm{lStb}}C(d,\sigma)\|\boldsymbol{A}-\widehat{\boldsymbol{A}}\|_{L^\infty(\omega_z)}\|\nabla u_{\mathcal{T}}\|_{L^2(\omega_z)}\\ &+ C_{\mathrm{lStb}}C(d,\sigma)\|h(c-\widehat{c})\|_{L^\infty(\omega_z)}\|u_{\mathcal{T}}\|_{L^2(\omega_z)},\end{aligned}$$

and thus

$$\begin{aligned}\mathrm{osc}_{\mathcal{T}}(R_{\mathcal{T}})^2 &\le 3\,\mathrm{osc}_{\mathcal{T}}(f)^2\\ &+ \frac{3(d+1)}{\alpha^2}\|f\|^2_{H^{-1}(\Omega)}\left(\|\boldsymbol{A}-\widehat{\boldsymbol{A}}\|^2_{L^\infty(\Omega)} + C_P^2C(d,\sigma)\|h(c-\widehat{c})\|^2_{L^\infty(\Omega)}\right)\end{aligned}$$

where C_{lStb} is the constant from Lemma 4.28 (local H^{-1}-stability), h is the mesh size function defined by $h|_T = h_T$ for all $T\in\mathcal{T}$, α is the coercivity constant from (2.29), and C_P is the constant in Lemma 2.2 (first Poincaré inequality).

The bounds of Lemma 4.39 are obviously not convenient if $\boldsymbol{A}$ or c are not continuous. We therefore implement the underlying idea in Section 5.4 differently.

Proof. [1] To verify the local bound, let $z\in\mathcal{V}$ be any vertex. By linearity of $P_{\mathcal{T}}$, we obtain

$$\begin{aligned}\mathrm{osc}_{\mathcal{T}}(R_{\mathcal{T}},z) &\le \mathrm{osc}_{\mathcal{T}}(f,z)\\ &+ \|(I-P_{\mathcal{T}})(-\operatorname{div}(\boldsymbol{A}\nabla u_{\mathcal{T}}))\|_{H^{-1}(\omega_z)} + \|(I-P_{\mathcal{T}})(cu_{\mathcal{T}})\|_{H^{-1}(\omega_z)}\end{aligned}$$

and it remains to bound appropriately the two terms involving the coefficients $\boldsymbol{A}$ and c. As the definitions of m_1 and m_2 ensure $-\operatorname{div}(\widehat{\boldsymbol{A}}\nabla u_{\mathcal{T}}) \in \mathbb{F}_{\mathcal{T}}$, Corollary 4.31 (local near-best approximation), the scaled trace theorem (4.12) and Lemma 2.2 (first Poincaré inequality) give

$$\begin{aligned}\|(I-P_{\mathcal{T}})(-\operatorname{div}(\boldsymbol{A}\nabla u_{\mathcal{T}}))\|_{H^{-1}(\omega_z)} &\le C_{\mathrm{lStb}}\|-\operatorname{div}((\boldsymbol{A}-\widehat{\boldsymbol{A}})\nabla u_{\mathcal{T}})\|_{H^{-1}(\omega_z)}\\ &\le C C_{\mathrm{lStb}}\|\boldsymbol{A}-\widehat{\boldsymbol{A}}\|_{L^\infty(\omega_z)}\|\nabla u_{\mathcal{T}}\|_{L^2(\omega_z)}.\end{aligned}$$

As $\widehat{c}u_{\mathcal{T}} \in \mathbb{F}_{\mathcal{T}}$ thanks to the definition of m_2, a similar argument again using

Lemma 2.2 (first Poincaré inequality) and $\operatorname{diam}\omega_z \le Ch_z$ on ω_z provides

$$
\begin{aligned}
\|(I - P_{\mathcal{T}})(cu_{\mathcal{T}})\|_{H^{-1}(\omega_z)} &\le C_{\mathrm{lStb}}\|(c - \widehat{c})u_{\mathcal{T}}\|_{H^{-1}(\omega_z)} \\
&\le CC_{\mathrm{lStb}}\|h(c - \widehat{c})\|_{L^\infty(\omega_z)}\|u_{\mathcal{T}}\|_{L^2(\omega_z)},
\end{aligned}
$$

and the local bound is verified.

[2] To show the global bound, we square the local bound and sum it over all vertices $z \in \mathcal{V}$ to obtain

$$
\begin{aligned}
\operatorname{osc}_{\mathcal{T}}(R_{\mathcal{T}})^2 \le 3\sum_{z\in\mathcal{V}} \operatorname{osc}_{\mathcal{T}}(f, z)^2 &+ 3(d+1)C\|\boldsymbol{A} - \widehat{\boldsymbol{A}}\|^2_{L^\infty(\Omega)}\|\nabla u_{\mathcal{T}}\|^2_{L^2(\Omega)} \\
&+ 3(d+1)C\|h(c - \widehat{c})\|^2_{L^\infty(\Omega)}\|u_{\mathcal{T}}\|^2_{L^2(\Omega)}.
\end{aligned}
$$

Hence Lemma 2.2 (first Poincaré inequality) on Ω and discrete stability,

$$
\|u_{\mathcal{T}}\|_{L^2(\Omega)} \le C_P\|\nabla u_{\mathcal{T}}\|_{L^2(\Omega)} \le \frac{C_P}{\alpha}\|f\|_{H^{-1}(\Omega)},
$$

finish the proof. □

The following remarks set Lemma 4.35 (splitting of local residual norm) and the accompanying results Lemma 4.37 and Lemma 4.39 on the reduction to data oscillation in the context of adaptive algorithms.

Remark 4.40 (structure of splitting). Combining Lemma 4.35 (splitting of local residual norm) with Lemma 4.37 or Lemma 4.39 about reduction to data oscillation thus provides an abstract estimator with the following two global parts:

$$
\eta^{\mathrm{abs}}_{\mathcal{T}}(u_{\mathcal{T}})^2 := \sum_{z\in\mathcal{V}} \eta^{\mathrm{abs}}_{\mathcal{T}}(u_{\mathcal{T}}, z)^2
$$

and, writing $\mathcal{D} = (\boldsymbol{A}, c, f)$ for the data of the partial differential equation,

$$
\begin{aligned}
&\operatorname{osc}^{\mathrm{abs}}_{\mathcal{T}}(\mathcal{D})^2 := \operatorname{osc}_{\mathcal{T}}(f)^2, \\
&\operatorname{osc}^{\mathrm{abs}}_{\mathcal{T}}(\mathcal{D})^2 := \operatorname{osc}_{\mathcal{T}}(f)^2 + C_1 \max_{z\in\mathcal{V}} \|\boldsymbol{A} - \widehat{\boldsymbol{A}}\|^2_{L^\infty(\omega_z)} + C_2 \max_{z\in\mathcal{V}} \|h(c - \widehat{c})\|^2_{L^\infty(\omega_z)},
\end{aligned}
$$

the latter provided $(\boldsymbol{A}, c)$ are not discrete. It is important to note the different nature of these two parts. The first part $\eta^{\mathrm{abs}}_{\mathcal{T}}(u_{\mathcal{T}})$, the abstract *PDE indicator*,

- is strictly related to the structure of the underlying PDE,
- involves only discrete functionals from $\mathbb{F}_z$, and
- the evaluation of its local indicators $\eta^{\mathrm{abs}}_{\mathcal{T}}(u_{\mathcal{T}}, z)$ requires the global computation of the discrete solution $u_{\mathcal{T}}$.

In contrast, the second part $\operatorname{osc}^{\mathrm{abs}}_{\mathcal{T}}(\mathcal{D})$, the *oscillation (indicator)*,

- depends only on the data $\mathcal{D}$ of the differential operator,
- involves non-discrete functionals, and

- the evaluation of its local indicators $\operatorname{osc}_{\mathcal{T}}(f, z)$, $\|\boldsymbol{A} - \widehat{\boldsymbol{A}}\|_{L^\infty(\omega_z)}$, $\|h(c - \widehat{c})\|_{L^\infty(\omega_z)}$, $z \in \mathcal{V}$, is completely local.

The respective properties 'discrete nature' and 'local dependence' of the two parts are the key advantage over the whole local residual indicators $\|R_{\mathcal{T}}\|_{H^{-1}(\omega_z)}$, $z \in \mathcal{V}$, and will be instrumental to the algorithmic design in Sections 5 and 6 below.

Remark 4.41 (minimal regularity and regularizing $P_{\mathcal{T}}$). It is worth noting that the results in this section do not involve any regularity beyond (4.2) and that the projection $P_{\mathcal{T}}$ has a regularizing effect. In particular, we have

$$\operatorname{Im} P_{\mathcal{T}} = \mathbb{F}_{\mathcal{T}} \subset H^{-1/2-\varepsilon}(\Omega) \quad \text{for any small } \varepsilon > 0,$$

thanks to the trace theorem in fractional Sobolev spaces. As a consequence, most techniques for *a posteriori* error estimation can be directly applied to the discretized residual $P_{\mathcal{T}} R_{\mathcal{T}}$, without any special twisting and under natural regularity assumptions.

Remark 4.42 (reduction vs. surrogate reduction). The kernel condition of Remark 4.19 (avoiding overestimation) is not verified for the bounds in Lemma 4.39 (surrogate data oscillation reduction). These bounds may thus exhibit overestimation and cannot be reversed. If we use the right-hand side of an overestimating bound as a part of an estimator, we shall call that part a *surrogate*. This label marks a crucial difference between the cases represented by Lemma 4.39 (surrogate data oscillation reduction) and Lemma 4.37 (data oscillation reduction for discrete coefficients), which is free of any overestimation.

Remark 4.43 (surrogate data oscillation). Surrogates for data oscillation indicators can be useful for providing more direct access for computation. For example, if $f \in L^2(\Omega)$, the bound (4.47) by the classical $\Pi_{\mathcal{T}}$-oscillation can be approximated by numerical integration. In such a context, it is useful to take the following points into account.

- Computable surrogates, i.e. computable upper bounds, for data oscillation indicators are in general impossible. In fact, generic data from an infinite-dimensional space will not be completely seen by the finite information available at any stage of a computation; see also Kreuzer and Veeser (2021, Lemma 2 and Corollary 5) illustrating this fact for $\operatorname{osc}_{\mathcal{T}}(f)$ with the help of orthogonality. Hence computable surrogates will hinge on additional *a priori* information on the given data. We postpone a discussion of examples to Section 7.3.
- As a general rule, surrogates should be applied last. This avoids other parts of the estimators being affected by overestimation; see Remark 4.46 (modified vs. standard residual estimator) below.

4.7. Modified residual estimation

In view of the splitting into PDE and oscillation indicators and the discussion of the computability of the latter, it remains to quantify the abstract PDE indicators $\eta_{\mathcal{T}}^{\text{abs}}(u_{\mathcal{T}}, z)$, $z \in \mathcal{V}$. To this end, we can employ Corollary 4.30 (quantifying H^{-1}-norms of discrete functionals), resulting in a modification of the standard residual estimator $\mathcal{E}_{\mathcal{T}}^{\text{std}}(u_{\mathcal{T}}, \mathcal{D})$ from Section 4.2. Alternative quantifications by other techniques of *a posteriori* error estimation are discussed in Section 4.9 below. In so doing, for simplicity, we consider only the case given by the following assumption.

Assumption 4.44 (discrete coefficients and discrete functionals). Suppose that the coefficients $\boldsymbol{A}$ and c in (2.5) are discrete, and choose the degrees (m_1, m_2) of the discrete functionals in $\mathbb{F}_{\mathcal{T}}$ according to Lemma 4.37 (data oscillation reduction for discrete coefficients).

For non-discrete coefficients, we essentially have to invoke Lemma 4.39 (surrogate data oscillation reduction) instead of Lemma 4.37 in order to reduce to data oscillation.

We shall employ the bisection method in order to refine the mesh. Since this method is based upon the subdivision of elements, it is convenient to split the estimator into contributions associated with elements and not with vertices as in Section 4.6.

To define the modified residual estimator, we recall the representation (4.35) of the H^{-1}-projection $P_{\mathcal{T}}$, and we use Assumption 4.44 (discrete coefficients and discrete functionals) to set

$$
\begin{aligned}
\mathcal{E}_{\mathcal{T}}^2 &:= \sum_{T \in \mathcal{T}} \mathcal{E}_{\mathcal{T}}(T)^2 \quad \text{with} \\
\mathcal{E}_{\mathcal{T}}(T)^2 &:= \mathcal{E}_{\mathcal{T}}(u_{\mathcal{T}}, f, T)^2 := \eta_{\mathcal{T}}(u_{\mathcal{T}}, T)^2 + \operatorname{osc}_{\mathcal{T}}(f, T)^2, \\
\eta_{\mathcal{T}}(u_{\mathcal{T}}, T)^2 &:= h_T \sum_{F \subset \partial T \setminus \partial \Omega} \| [[\boldsymbol{A} \nabla u_{\mathcal{T}}]] \cdot \boldsymbol{n}_F - P_F f \|_{L^2(F)}^2 \\
&\quad + h_T^2 \| P_T f - c u_{\mathcal{T}} + \operatorname{div}(\boldsymbol{A} \nabla u_{\mathcal{T}}) \|_{L^2(T)}^2, \\
\operatorname{osc}_{\mathcal{T}}(f, T)^2 &:= \| f - P_{\mathcal{T}} f \|_{H^{-1}(\omega_T)}^2.
\end{aligned} \tag{4.52}
$$

Clearly, this is a variant of the standard residual estimator in (4.10), where the main differences are given by the corrections $P_F f$, $F \in \mathcal{F}$, of the jump residual and the replacement of $f|_T$ by $P_T f$, $T \in \mathcal{T}$, in the PDE indicator. As shown by the following theorem and remarks, the modification leads to more accurate *a posteriori* bounds.

Theorem 4.45 (modified residual estimator). *Under Assumption 4.44, the modified residual estimator* (4.52) *is equivalent to the error: more precisely, we have*

$$
C_L \mathcal{E}_{\mathcal{T}} \le \| \nabla (u - u_{\mathcal{T}}) \|_{L^2(\Omega)} \le C_U \mathcal{E}_{\mathcal{T}},
$$

where the constants $C_U \geq C_L > 0$ depend only on the coefficients $(\boldsymbol{A}, c)$, the shape regularity coefficient σ from (3.9)*, the polynomial degree n, and d.*

Proof. To derive the upper bound, we use those of Lemma 4.1 (error and residual), Corollary 4.6 (star localization of residual norm), Lemma 4.35 (splitting of local residual norm) and Corollary 4.30 (quantifying H^{-1}-norms of discrete functionals) with stars, and obtain

$$\begin{aligned}\|\nabla(u-u_{\mathcal{T}})\|^2_{L^2(\Omega)} &\lesssim \|R_{\mathcal{T}}\|^2_{H^{-1}(\Omega)} \lesssim \sum_{z\in\mathcal{V}} \|R_{\mathcal{T}}\|^2_{H^{-1}(\omega_z)}\\ &\lesssim \sum_{z\in\mathcal{V}} \mathcal{E}^{\mathrm{abs}}_{\mathcal{T}}(z)^2 = \sum_{z\in\mathcal{V}} \eta^{\mathrm{abs}}_{\mathcal{T}}(u_{\mathcal{T}},z)^2 + \sum_{z\in\mathcal{V}} \operatorname{osc}_{\mathcal{T}}(f,z)^2\\ &\lesssim \sum_{z\in\mathcal{V}}\sum_{T\in\mathcal{T}_z} \eta_{\mathcal{T}}(u_{\mathcal{T}},T)^2 + \sum_{z\in\mathcal{V}} \operatorname{osc}_{\mathcal{T}}(f,z)^2,\end{aligned}$$

with $\mathcal{T}_z = \{T \in \mathcal{T} \mid T \ni z\}$. As a given mesh element appears in the star meshes $\mathcal{T}_z$ for at most $d+1$ vertices, we have

$$\sum_{z\in\mathcal{V}}\sum_{T\in\mathcal{T}_z} \eta_{\mathcal{T}}(u_{\mathcal{T}},T)^2 \leq (d+1)\sum_{T\in\mathcal{T}} \eta_{\mathcal{T}}(u_{\mathcal{T}},T)^2$$

for the first sum, and Lemma 4.8 (localization re-indexing) yields

$$\sum_{z\in\mathcal{V}} \operatorname{osc}_{\mathcal{T}}(f,z)^2 \lesssim \sum_{T\in\mathcal{T}} \operatorname{osc}_{\mathcal{T}}(f,T)^2$$

for the second sum. Inserting the last two inequalities in the previous one, we conclude the upper bound:

$$\|\nabla(u-u_{\mathcal{T}})\|^2_{L^2(\Omega)} \lesssim \sum_{T\in\mathcal{T}} \eta_{\mathcal{T}}(u_{\mathcal{T}},T)^2 + \sum_{T\in\mathcal{T}} \operatorname{osc}_{\mathcal{T}}(f,T)^2 = \sum_{T\in\mathcal{T}} \mathcal{E}_{\mathcal{T}}(T)^2.$$

To show the lower bound, fix a mesh element $T \in \mathcal{T}$. Applying the local lower bounds in Corollary 4.30, Lemma 4.28 (local H^{-1}-stability) and Lemma 4.1 on the local meshed subdomain $\widetilde{\omega}_T$ defined in (3.12) yields for the PDE indicator

$$\eta_{\mathcal{T}}(u_{\mathcal{T}},T) \lesssim \|P_{\mathcal{T}}R_{\mathcal{T}}\|_{H^{-1}(\widetilde{\omega}_T)} \lesssim \|R_{\mathcal{T}}\|_{H^{-1}(\widetilde{\omega}_T)}. \tag{4.53}$$

In the case of the oscillation indicator, we exploit $-cu_{\mathcal{T}} + \operatorname{div}(\boldsymbol{A}\nabla u_{\mathcal{T}}) \in \mathbb{F}_{\mathcal{T}}$ with the help of Lemma 4.25 (algebraic properties) and apply Lemma 4.28 on the local meshed subdomain ω_T:

$$\operatorname{osc}_{\mathcal{T}}(f,T) = \|f - P_{\mathcal{T}}f\|_{H^{-1}(\omega_T)} = \|(I-P_{\mathcal{T}})R_{\mathcal{T}}\|_{H^{-1}(\omega_T)} \lesssim \|R_{\mathcal{T}}\|_{H^{-1}(\omega_T)}.$$

Thanks to $\widetilde{\omega}_T \subset \omega_T$, combining the last two inequalities gives the desired local lower bound:

$$\begin{aligned}\mathcal{E}_{\mathcal{T}}(T)^2 &= \eta_{\mathcal{T}}(u_{\mathcal{T}},T)^2 + \operatorname{osc}_{\mathcal{T}}(f,T)^2\\ &\lesssim \|R_{\mathcal{T}}\|_{H^{-1}(\widetilde{\omega}_T)} + \|R_{\mathcal{T}}\|_{H^{-1}(\omega_T)} \lesssim \|R_{\mathcal{T}}\|_{H^{-1}(\omega_T)}.\end{aligned} \tag{4.54}$$

As the number of patches $\omega_T, T \in \mathcal{T}$, containing a given mesh element is uniformly bounded by d and the shape regularity coefficient σ, summing this bound over all mesh elements yields the global lower bound

$$\sum_{T\in\mathcal{T}} \mathcal{E}_{\mathcal{T}}(T)^2 \lesssim \sum_{T\in\mathcal{T}} \|R_{\mathcal{T}}\|^2_{H^{-1}(\omega_T)} \lesssim \|\nabla(u-u_{\mathcal{T}})\|^2_{L^2(\Omega)},$$

with the help of Lemma 4.5 (localization of H^{-1}-norm) and Lemma 4.1. Thus the equivalence of error and estimator is established. □

A detailed comparison of the modified residual estimator with the standard estimator is in order.

Remark 4.46 (modified vs. standard residual estimator). We compare the modified residual estimator (4.52) with the standard one given by (4.10a) and the local split indicators (4.22). As a common characterizing feature, both residual estimators use properly scaled L^2-norms of jump and element residual, ready for numerical integration. However, we observe the following differences.

- While the modified estimator $\mathcal{E}_{\mathcal{T}}$ is defined under the natural regularity assumptions (4.2), the standard estimator $\mathcal{E}^{\mathrm{std}}_{\mathcal{T}}$ requires $\boldsymbol{A} \in W^1_\infty(\Omega;\mathbb{R}^{d\times d})$ and $f \in L^2(\Omega)$ in addition.
- While the modified estimator $\mathcal{E}_{\mathcal{T}}$ is truly equivalent to the error, the standard estimator $\mathcal{E}^{\mathrm{std}}_{\mathcal{T}}$ may overestimate it, limited, however, by Proposition 4.12 (partial lower bound).

By the domain test in Remark 4.19 (avoiding overestimation), we know that these two points are interrelated. However, the kernel test is also at play in the overestimation. Indeed, revisiting the proof of Theorem 4.9 (upper bound with standard residual estimator), we can replace the scaled L^2-norms of the element residuals on a star ω_z with $\|r\|_{H^{-1}(\omega_z)}$, and the resulting vertex-oriented variant of the residual estimator with unsplit local indicators is defined for all $f \in H^{-1}(\Omega)$. Overestimation can, however, still occur non-asymptotically as well as asymptotically; see Cohen *et al.* (2012). Indeed, in the case of the Poisson equation and linear finite elements, the kernel test is obviously not satisfied. This shows that the splitting in jump and element residual is quite delicate and highlights the crucial role of the modifications of the standard residual estimator: not only do they allow for stability in line with Remark 4.34 (stability of approximation) but they also imply the kernel test.

To conclude this comparison, let us illustrate the second point of Remark 4.43 (surrogate data oscillation), namely that surrogates should be applied last. Using (4.47) in Remark 4.32 (approximating functions), in the modified residual estimator we may replace the H^{-1}-oscillation $\mathrm{osc}_{\mathcal{T}}(f)$ with the standard oscillation $\mathrm{osc}^{\mathrm{std}}_{\mathcal{T}}(f)$, which can be readily approximated with numerical integration. In so doing, we first split the residual with $P_{\mathcal{T}}$ and then apply $\Pi_{\mathcal{T}}$ to obtain the surrogate. Note, however, that if we apply $\Pi_{\mathcal{T}}$ earlier to split the residual, the crucial modifications will not appear and, therefore, the PDE indicator of the standard residual estimator also exhibits overestimation.

4.8. *Bounds for corrections and reduction of PDE estimator*

In the following sections we shall use the modified residual estimator $\mathcal{E}_{\mathcal{T}}$ from (4.52) in adaptive algorithms. In their convergence analyses, not only is its relationship with the error important, but also its relationship with the norm $\|\nabla(u_{\mathcal{T}_*} - u_{\mathcal{T}})\|_{L^2(\Omega)}$ of (possible) *corrections*, where $u_{\mathcal{T}_*}$ is the Galerkin approximation to u over some refinement $\mathcal{T}_*$ of $\mathcal{T}$. This section establishes corresponding upper and lower bounds, as well as related results about the *global PDE indicator*

$$\eta_{\mathcal{T}}(u_{\mathcal{T}}, f)^2 := \sum_{T \in \mathcal{T}} \eta_{\mathcal{T}}(u_{\mathcal{T}}, T)^2 \tag{4.55}$$

and the *global oscillation*

$$\operatorname{osc}_{\mathcal{T}}(f)^2 := \sum_{T \in \mathcal{T}} \operatorname{osc}_{\mathcal{T}}(f, T)^2. \tag{4.56}$$

When it is important to indicate that the oscillations are measured in H^{-1}, we use the notation

$$\operatorname{osc}_{\mathcal{T}}(f)_{-1} \quad \text{and} \quad \operatorname{osc}_{\mathcal{T}}(f, T)_{-1}.$$

Let $\mathcal{T}_*$ be a conforming mesh that is a *refinement* of $\mathcal{T}$, that is, for any element $T \in \mathcal{T}$, there exists a submesh $\mathcal{T}_{*,T}$ of $\mathcal{T}_*$ such that $T = \cup\{T_* \mid T_* \in \mathcal{T}_{*,T}\}$. The Galerkin approximation in $\mathbb{V}_{\mathcal{T}_*}$ is characterized by

$$u_{\mathcal{T}_*} \in \mathbb{V}_{\mathcal{T}_*}: \quad \mathcal{B}[u_{\mathcal{T}_*}, w] = \langle f, w \rangle \quad \text{for all } w \in \mathbb{V}_{\mathcal{T}_*}.$$

Hence the discrete solution $u_{\mathcal{T}}$ on the original mesh $\mathcal{T}$ is not only a Galerkin approximation to the exact solution u satisfying (2.7) but also to $u_{\mathcal{T}_*}$. The norm $\|\nabla(u_{\mathcal{T}_*} - u_{\mathcal{T}})\|_{L^2(\Omega)}$ of the correction therefore can be viewed as the error in approximating $u_{\mathcal{T}_*}$ on the mesh $\mathcal{T}$. This viewpoint suggests considering the variant

$$\langle R_{\mathcal{T}}, w \rangle = \mathcal{B}[u_{\mathcal{T}_*} - u_{\mathcal{T}}, w] \quad \text{for all } w \in \mathbb{V}_{\mathcal{T}_*} \tag{4.57}$$

of the error–residual identity (4.3) and introducing the discrete dual norm

$$\|R_{\mathcal{T}}\|_{(\mathbb{V}_{\mathcal{T}_*})^*} := \sup_{w \in \mathbb{V}_{\mathcal{T}_*}} \frac{\langle R_{\mathcal{T}}, w \rangle}{\|\nabla w\|_{L^2(\Omega)}} \tag{4.58}$$

of the residual as a counterpart of $\|R_{\mathcal{T}}\|_{H^{-1}(\Omega)}$. Arguing as in the proof of Lemma 4.1 (error and residual), we thus readily obtain the following quantitative relationship between the correction and the residual.

Lemma 4.47 (correction and residual). *If $\mathcal{T}_*$ is a refinement of the mesh $\mathcal{T}$, the norm of the correction $u_{\mathcal{T}_*} - u_{\mathcal{T}}$ is equivalent to the discrete residual norm. More* *precisely,*

$$\frac{1}{\|\mathcal{B}\|}\|R_{\mathcal{T}}\|_{(\mathbb{V}_{\mathcal{T}_*})^*} \le \|\nabla(u_{\mathcal{T}_*} - u_{\mathcal{T}})\|_{L^2(\Omega)} \le \frac{1}{\alpha}\|R_{\mathcal{T}}\|_{(\mathbb{V}_{\mathcal{T}_*})^*},$$

where $\|\mathcal{B}\| \ge \alpha > 0$*, are, respectively, the continuity and coercivity constant of the bilinear form* $\mathcal{B}$*.*

We first exploit the upper bound in Lemma 4.47. As the inclusion $\mathbb{V}_{\mathcal{T}_*} \subset H_0^1(\Omega)$ implies

$$\|R_{\mathcal{T}}\|_{(\mathbb{V}_{\mathcal{T}_*})^*} \le \|R_{\mathcal{T}}\|_{H^{-1}(\Omega)}, \tag{4.59}$$

Theorem 4.45 (modified residual estimator) immediately yields the upper bound

$$\|\nabla(u_{\mathcal{T}_*} - u_{\mathcal{T}})\|_{L^2(\Omega)} \le C_U \mathcal{E}_{\mathcal{T}}(u_{\mathcal{T}}, f). \tag{4.60}$$

This bound, however, appears not to be accurate in view of the use of (4.59). We will sharpen it by following the lines of its proof but exploiting the *full* orthogonality

$$\langle R_{\mathcal{T}}, w\rangle = 0 \quad \text{for all } w \in \mathbb{V}_{\mathcal{T}}, \tag{4.61}$$

with suitably tuned Scott–Zhang interpolation (Scott and Zhang 1990).

In order to prepare the use of this interpolation, let $\mathcal{N}$ denote the Lagrange nodes of order n of the mesh $\mathcal{T}$ and let $\mathcal{F}$ and $\mathcal{F}_*$, respectively, denote the $(d-1)$-dimensional faces of $\mathcal{T}$ and $\mathcal{T}_*$, including boundary faces. Given a node $z \in \mathcal{N}$, fix a face $F_z \in \mathcal{F}$ such that F_z contains z and the following conditions are met:

$$\begin{aligned} z \in \partial\Omega \quad &\Rightarrow \quad F_z \subset \partial\Omega, \\ \{F \in \mathcal{F} \cap \mathcal{F}_* \mid F \ni z\} \ne \emptyset \quad &\Rightarrow \quad F_z \in \mathcal{F}_*. \end{aligned}$$

Furthermore, let ψ_z^* denote the polynomial in $\mathbb{P}_n(F_z)$ satisfying

$$\int_{F_z} \psi_z^* \psi_y = \delta_{yz} \quad \text{for all } y \in \mathcal{N},$$

where $\{\psi_y\}_{y\in\mathcal{N}}$ is the Lagrange basis of $\mathbb{S}_{\mathcal{T}}^{n,0}$, and define

$$I_{\mathcal{T}} w = \sum_{z\in\mathcal{N}} \left(\int_{F_z} \psi_z^* w\right) \psi_z. \tag{4.62}$$

The two conditions on the fixed face F_z then ensure, respectively,

$$w \in H_0^1(\Omega) \quad \Rightarrow \quad I_{\mathcal{T}} w \in \mathbb{V}_{\mathcal{T}}, \tag{4.63a}$$

$$w \in \mathbb{V}_{\mathcal{T}_*} \text{ and } T \in \mathcal{T} \cap \mathcal{T}_* \quad \Rightarrow \quad I_{\mathcal{T}} w = w \text{ on } T. \tag{4.63b}$$

In particular, if $w \in \mathbb{V}_{\mathcal{T}_*}$, its approximation $I_{\mathcal{T}} w \in \mathbb{V}_{\mathcal{T}}$ is an admissible test function and coincides with w whenever possible. Finally, $I_{\mathcal{T}}$ has the following stability and approximation properties, where the hidden constants depend only on d, n and the shape regularity coefficient σ: for any element $T \in \mathcal{T}$ and any face $F \in \mathcal{F}$,

$$\|\nabla I_{\mathcal{T}} w\|_{L^2(T)} \lesssim \|\nabla w\|_{L^2(\omega_T)}, \tag{4.64a}$$

$$\|w - I_{\mathcal{T}} w\|_{L^2(T)} \lesssim h_T \|\nabla w\|_{L^2(\omega_T)}, \tag{4.64b}$$

$$\|w - I_{\mathcal{T}} w\|_{L^2(F)} \lesssim h_F^{1/2} \|\nabla w\|_{L^2(\omega_F)}. \tag{4.64c}$$

The sharpening of the simple upper bound (4.60) lies in the fact that only a part of the estimator in (4.52) will be invoked. To formulate this, we define

$$\mathcal{E}_{\mathcal{T}}(u_{\mathcal{T}}, f, \widetilde{\mathcal{T}}) := \Big(\sum_{T \in \widetilde{\mathcal{T}}} \mathcal{E}_{\mathcal{T}}(u_{\mathcal{T}}, f, T)^2 \Big)^{1/2}, \tag{4.65}$$

where $\widetilde{\mathcal{T}} \subset \mathcal{T}$ is a subset of elements in $\mathcal{T}$. In the same vein, we shall denote $\eta_{\mathcal{T}}(u_{\mathcal{T}}, f, \widetilde{\mathcal{T}})$ and $\operatorname{osc}_{\mathcal{T}}(f, \widetilde{\mathcal{T}})$.

Theorem 4.48 (upper bound for corrections). *Let Assumption 4.44 hold and let $\mathcal{T}_*$ be a refinement of the mesh $\mathcal{T}$. The correction $u_{\mathcal{T}_*} - u_{\mathcal{T}}$ is bounded in terms of the indicators of the refined elements $\mathcal{T} \setminus \mathcal{T}_*$:*

$$\|\nabla(u_{\mathcal{T}} - u_{\mathcal{T}_*})\|_{L^2(\Omega)} \le \widetilde{C}_U \mathcal{E}_{\mathcal{T}}(u_{\mathcal{T}}, f, \mathcal{T} \setminus \mathcal{T}_*),$$

where the constant $\widetilde{C}_U > 0$ depends only on the dimension d, the coefficients $\mathbf{A}$ and c, the polynomial degree n, and the shape regularity coefficient σ from (3.9).

Proof. [1] *Localization and splitting of the residual norm.* In light of Lemma 4.47, it suffices to bound the discrete residual norm $\|R_{\mathcal{T}}\|_{(\mathbb{V}_{\mathcal{T}_*})^*}$. Given $w \in \mathbb{V}_{\mathcal{T}_*}$, we prepare the localization of the residual by full orthogonality (4.61) and split it with help of the projection $P_{\mathcal{T}}$ on discrete functionals:

$$\begin{aligned} |\langle R_{\mathcal{T}}, w\rangle| &= |\langle R_{\mathcal{T}}, w - I_{\mathcal{T}} w\rangle| \\ &\le |\langle P_{\mathcal{T}} R_{\mathcal{T}}, w - I_{\mathcal{T}} w\rangle| + |\langle f - P_{\mathcal{T}} f, w - I_{\mathcal{T}} w\rangle|, \end{aligned}$$

where we used the identity $R - P_{\mathcal{T}} R = f - P_{\mathcal{T}} f$ in the last step. In light of

$$\mathcal{E}_{\mathcal{T}}(u_{\mathcal{T}}, f, \mathcal{T} \setminus \mathcal{T}_*)^2 = \eta_{\mathcal{T}}(u_{\mathcal{T}}, f, \mathcal{T} \setminus \mathcal{T}_*)^2 + \operatorname{osc}_{\mathcal{T}}(f, \mathcal{T} \setminus \mathcal{T}_*)^2,$$

it remains to bound the two terms with discretized residual $P_{\mathcal{T}} R_{\mathcal{T}}$ and the oscillation of f appropriately.

[2] *Bounding the discretized residual.* We adopt the notation (4.35) for the densities of $P_{\mathcal{T}}$, and exploit the piecewise nature of the discretized residual and the local invariance (4.63b) of $I_{\mathcal{T}}$ to deduce

$$\begin{aligned} &\langle P_{\mathcal{T}} R_{\mathcal{T}}, w - I_{\mathcal{T}} w\rangle \\ &= \sum_{T \in \mathcal{T} \setminus \mathcal{T}_*} \Big(\int_T (P_{\mathcal{T}} R_{\mathcal{T}})(w - I_{\mathcal{T}} w) + \frac{1}{2} \sum_{F \subset \partial T \setminus \partial\Omega} \int_F (P_F R_{\mathcal{T}})(w - I_{\mathcal{T}} w) \Big). \end{aligned}$$

Invoking the local approximation properties (4.64b) and (4.64c) of $I_{\mathcal{T}}$ leads to the desired bound for the discretized residual:

$$
\begin{aligned}
|\langle P_{\mathcal{T}} R_{\mathcal{T}}, w - I_{\mathcal{T}} w\rangle| &\lesssim \sum_{T\in\mathcal{T}\setminus\mathcal{T}_*} \eta_{\mathcal{T}}(u_{\mathcal{T}}, f, T)\, \|\nabla w\|_{L^2(\omega_T)} \\
&\lesssim \eta_{\mathcal{T}}\big(u_{\mathcal{T}}, f, \mathcal{T}\setminus\mathcal{T}_*\big)\, \|\nabla w\|_{L^2(\Omega)}.
\end{aligned}
$$

3 *Bounding the oscillation.* We need to split the oscillation into suitable local contributions and first proceed similarly to the proof of Lemma 4.5(i) (localization of H^{-1}-norm). Writing

$$
w - I_{\mathcal{T}} w = \sum_{z\in\mathcal{V}} (w - I_{\mathcal{T}} w)\phi_z \quad \text{and} \quad \Omega_0 := \bigcup_{T\in\mathcal{T}\setminus\mathcal{T}_*} T,
$$

we have $(w - I_{\mathcal{T}} w)\phi_z \in H^1_0(\omega_z \cap \Omega_0)$ thanks to (4.63b) and, for any $T \subset \omega_z \cap \Omega_0$,

$$
\begin{aligned}
\|\nabla((w - I_{\mathcal{T}} w)\phi_z)\|_{L^2(T)} &\le \|\phi_z \nabla(w - I_{\mathcal{T}} w)\|_{L^2(T)} + \|(w - I_{\mathcal{T}} w)\nabla\phi_z\|_{L^2(T)} \\
&\le \|\nabla(w - I_{\mathcal{T}} w)\|_{L^2(T)} + C(d)\sigma \|\nabla w\|_{L^2(\omega_T)} \\
&\lesssim \|\nabla w\|_{L^2(\omega_T)}
\end{aligned}
$$

by means of $0 \le \phi_z \le 1$, $|\nabla\phi_z| \le C(d)\sigma h_T^{-1}$, (4.64a) and (4.64b). Hence we get

$$
\begin{aligned}
&|\langle f - P_{\mathcal{T}} f, w - I_{\mathcal{T}} w\rangle| \\
&\quad \le \sum_{z\in\mathcal{V}} |\langle f - P_{\mathcal{T}} f, (w - I_{\mathcal{T}} w)\phi_z\rangle| \\
&\quad \le \sum_{z\in\mathcal{V}} \|f - P_{\mathcal{T}} f\|_{H^{-1}(\omega_z\cap\Omega_0)} \|\nabla((w - I_{\mathcal{T}} w)\phi_z)\|_{L^2(\omega_z\cap\Omega_0)} \\
&\quad \lesssim \sum_{z\in\mathcal{V}} \|f - P_{\mathcal{T}} f\|_{H^{-1}(\omega_z\cap\Omega_0)} \|\nabla w\|_{L^2(\cup_{T\subseteq\omega_z\cap\Omega_0}\omega_T)} \\
&\quad \lesssim \bigg(\sum_{z\in\mathcal{V}} \|f - P_{\mathcal{T}} f\|^2_{H^{-1}(\omega_z\cap\Omega_0)}\bigg)^{1/2} \|\nabla w\|_{L^2(\Omega)}.
\end{aligned}
$$

Since

$$
\sum_{z\in\mathcal{V}} \|f - P_{\mathcal{T}} f\|^2_{H^{-1}(\omega_z\cap\Omega_0)} \lesssim \sum_{T\in\mathcal{T}\setminus\mathcal{T}_*} \|f - P_{\mathcal{T}} f\|^2_{H^{-1}(\omega_T)},
$$

the oscillation of f is therefore bounded by

$$
|\langle f - P_{\mathcal{T}} f, w - I_{\mathcal{T}} w\rangle| \le \operatorname{osc}_{\mathcal{T}}(f, \mathcal{T}\setminus\mathcal{T}_*)\|\nabla w\|_{L^2(\Omega)}
$$

and the proof is complete. □

Proposition 4.12 (partial lower bound) as well as Lemma 4.35 (splitting of local residual norm) illustrate that the test space $\mathbb{V}^+_{\mathcal{T}}$ is closely related to lower bounds for the error. This observation suggests establishing lower bounds for the correction

$\|\nabla(u_{\mathcal{T}} - u_{\mathcal{T}_*})\|_{L^2(\Omega)}$ by ensuring conditions such as

$$\mathbb{V}^+(\mathcal{T}_\omega) \subset \mathbb{V}(\mathcal{T}_*),$$

where ω is a $\mathcal{T}$-mesh subdomain. Inspecting the construction of $\mathbb{V}^+_{\mathcal{T}}$, we realize that such conditions can be achieved if $\max\{m_1, m_2\} \le n - 1$ and the cut-off is implemented with hat functions of a virtual refinement of $\mathcal{T}$.

Lemma 4.49 (cut-off by refined hat functions). *Let $\mathcal{T}_+$ be the minimal bisection refinement of $\mathcal{T}$ such that the relative interior of each element $T \in \mathcal{T}$ and each face $F \in \mathcal{F}$ of the original mesh $\mathcal{T}$ contains at least one vertex from $\mathcal{T}_+$. Then there exist hat functions ϕ_T, $T \in \mathcal{T}$, and ϕ_F, $F \in \mathcal{F}$, in $\mathbb{S}^{1,0}(\mathcal{T}_+)$ satisfying Assumption 4.21 if* $\max\{m_1, m_2\} \le n - 1$.

Proof. The details of the proof depend on bisection and we therefore restrict to the case $d = 2$; for $d > 2$, the following reference situation used to define the hat functions is replaced by several ones with 'tagged' reference simplices. Let $\widehat{T} = T_2$ be the reference element in $\mathbb{R}^2$ with the standard enumeration of its vertices $\widehat{z}_0 = 0$, $\widehat{z}_1 = e_1$ and $\widehat{z}_2 = e_2$. Furthermore, let $\widehat{\mathcal{T}_+}$ be the mesh obtained by applying five bisections so that vertices in the interiors of $\widehat{T}$ and of its faces are generated. Let $\widehat{\phi}_{\widehat{T}}$, $\widehat{\phi}_{F'}$, $F' \subset \widehat{T}$ denote the four hat functions in $\mathbb{S}^{1,0}(\widehat{\mathcal{T}_+})$ associated with these generated vertices. Given an arbitrary element $T \in \mathcal{T}$, let H_T denote the bi-affine map $T \longrightarrow \widehat{T}$ preserving the numbering of the vertices for bisection, and define the pullbacks

$$\phi_T := H_T^*(\widehat{\phi}_{\widehat{T}}), \quad \phi_F|_T := H_T^*(\widehat{\phi}_{H_T(F)}), \quad F \subset T,$$

and extend by 0 off T or ω_F. As the extension operators E_F preserve the polynomial degree (see Lemma 4.20 (extending from faces)), $\max\{m_1, m_2\} \le n - 1$, and

$$\{H_T^{-1}(\widehat{T}_+) \mid \widehat{T}_+ \in \widehat{\mathcal{T}_+}\} = \{T_+ \in \mathcal{T}_+ \mid T_+ \subset T\},$$

the hat functions ϕ_T, $T \in \mathcal{T}$, and ϕ_F, $F \in \mathcal{F}$, then satisfy Assumption 4.21 with $G_T = H_T^{-1}$ and $\mathbb{S}^+ = \mathbb{S}^{n,0}(\widehat{\mathcal{T}_+})$. □

Definition 4.50 (interior vertex property). A mesh element $T \in \mathcal{T}$ satisfies the *interior vertex property* with respect to $\mathcal{T}_* \ge \mathcal{T}$ whenever each interior face $F \subset \partial T \setminus \partial\Omega$ of T and each element in $\widetilde{\omega}_T$ (defined in (3.12)) have in their relative interiors at least one vertex from $\mathcal{T}_*$.

A set $\mathcal{M} \subset \mathcal{T}$ satisfies the interior vertex property with respect to a refinement $\mathcal{T}_* \ge \mathcal{T}$ if each element $T \in \mathcal{M}$ satisfies the interior vertex property.

The interior vertex property is valid upon enforcing a fixed number b of bisections ($b = 3, 6$ for $d = 2, 3$). An immediate consequence is the following lower bound for corrections.

Theorem 4.51 (lower bound for corrections). *Suppose A is piecewise constant over $\mathcal{T}$ and $c = 0$, define $P_{\mathcal{T}}$ with the help of the cut-off functions in Lemma 4.49,* *and let $\mathcal{M}$ denote the subset of elements in $\mathcal{T}$ satisfying the interior vertex property with respect to $\mathcal{T}_*$. Then*

$$\sum_{T\in\mathcal{M}} \mathcal{E}_{\mathcal{T}}(u_{\mathcal{T}}, f, T)^2 \le C\|\nabla(u_{\mathcal{T}} - u_{\mathcal{T}_*})\|^2_{L^2(\Omega)} + \sum_{T\in\mathcal{M}} \|f - P_{\mathcal{T}} f\|^2_{H^{-1}(\omega_T)},$$

where C depends on d, the shape regularity coefficient σ, the coefficients $(\boldsymbol{A}, c)$, and the polynomial degree n.

Proof. [1] We first show a local bound with the PDE indicator $\eta_{\mathcal{T}}(u_{\mathcal{T}}, T)$. In view of $\boldsymbol{A} \in \mathbb{S}^{0,-1}_{\mathcal{T}}$ and $c = 0$, we choose $m_1 = n - 1$ and $m_2 = n - 2$ as the degrees for the discrete functionals. We can thus apply Lemma 4.49 and construct $P_{\mathcal{T}}$ with the refined hat functions. Let $T \in \mathcal{M}$ and so, using the interior vertex property and the notation associated with $\widetilde{\omega}_T$ in (3.12), we deduce $\mathbb{V}^+(\widetilde{\mathcal{T}}_T) \subset \mathbb{V}(\mathcal{T}_*, T) := \mathbb{V}(\mathcal{T}_*) \cap H^1_0(\widetilde{\omega}_T)$, where $\widetilde{\mathcal{T}}_T = \{T \in \mathcal{T} \mid T \subseteq \widetilde{\omega}_T\}$. Combining this with inequality (4.45) and Definition 4.24 (projection onto discrete functionals), we conclude

$$\eta_{\mathcal{T}}(u_{\mathcal{T}}, T) \lesssim \|P_{\mathcal{T}} R_{\mathcal{T}}\|_{\mathbb{V}^+(\widetilde{\mathcal{T}}_T)^*} = \|R_{\mathcal{T}}\|_{\mathbb{V}^+(\widetilde{\mathcal{T}}_T)^*} \le \|R_{\mathcal{T}}\|_{\mathbb{V}(\mathcal{T}_*, T)^*}.$$

[2] To collect the local bounds of the first step, we first show that, for any $\ell \in \mathbb{V}(\mathcal{T}_*)^*$,

$$\sum_{T\in\mathcal{T}} \|\ell\|^2_{\mathbb{V}(\mathcal{T}_*, T)^*} \le (d+2)\|\ell\|^2_{\mathbb{V}(\mathcal{T}_*)^*}.$$

To this end, we just repeat the proof of Lemma 4.5 (localization of H^{-1}-norm), replacing the spaces $H^1_0(\omega_i)$ and $H^1_0(\Omega)$, respectively, with $\mathbb{V}(\mathcal{T}_*, T)$ and $\mathbb{V}(\mathcal{T}_*)$. Hence, squaring and summing the bound of the first step as well as using Lemma 4.1 (error and residual) yield

$$\sum_{T\in\mathcal{M}} \eta_{\mathcal{T}}(u_{\mathcal{T}}, T)^2 \lesssim \sum_{T\in\mathcal{M}} \|R_{\mathcal{T}}\|^2_{\mathbb{V}(\mathcal{T}_*, T)^*} \lesssim \|\nabla(u_{\mathcal{T}_*} - u_{\mathcal{T}})\|^2_{L^2(\Omega)}. \tag{4.66}$$

[3] We finally prove the claimed bound by simply inserting (4.66):

$$\begin{aligned}\sum_{T\in\mathcal{M}} \mathcal{E}_{\mathcal{T}}(u_{\mathcal{T}}, f, T)^2 &= \sum_{T\in\mathcal{M}} \big(\eta_{\mathcal{T}}(u_{\mathcal{T}}, T)^2 + \operatorname{osc}_{\mathcal{T}}(f, T)^2\big)\\ &\le C\|\nabla(u_{\mathcal{T}_*} - u_{\mathcal{T}})\|^2_{L^2(\Omega)} + \sum_{T\in\mathcal{M}} \operatorname{osc}_{\mathcal{T}}(f, T)^2,\end{aligned}$$

and the proof is finished. □

Remark 4.52 (oscillation and correction). In general, by first fixing the finer mesh $\mathcal{T}_*$, it is impossible to bound oscillation indicators $\operatorname{osc}_{\mathcal{T}}(f, T)$ by some suitable correction. Indeed, these indicators can contain contributions to f and so to $R_{\mathcal{T}}$ of 'arbitrarily high frequency', while the correction can control only contributions of the residual $R_{\mathcal{T}}$ with frequencies representable over $\mathcal{T}_*$; see (4.58).

Monotonicity properties of the error $|u - u_{\mathcal{T}}|_{H^1_0(\Omega)} = \|\nabla(u - u_{\mathcal{T}})\|_{L^2(\Omega)}$ and the PDE error estimator $\eta_{\mathcal{T}}(u_{\mathcal{T}}) = \eta_{\mathcal{T}}(u_{\mathcal{T}}, f)$ with respect to $\mathcal{T}$ would be useful but fail to hold. To investigate this issue, we consider two admissible meshes $\mathcal{T}, \mathcal{T}_* \in \mathbb{T}$, the latter being a refinement of the former $\mathcal{T}_* \geq \mathcal{T}$, and a third admissible mesh $\widehat{\mathcal{T}} \leq \mathcal{T}$. We further assume that data $\mathcal{D} = (\boldsymbol{A}, c, f)$ is discrete over $\widehat{\mathcal{T}}$ in the sense that $\mathcal{D} \in \mathbb{D}_{\widehat{\mathcal{T}}}$, where

$$\mathbb{D}_{\widehat{\mathcal{T}}} := \left[\mathbb{S}_{\widehat{\mathcal{T}}}^{n-1,-1}\right]^{d\times d} \times \mathbb{S}_{\widehat{\mathcal{T}}}^{n-1,-1} \times \mathbb{F}_{\widehat{\mathcal{T}}}$$

and $\mathcal{D}$ does not change in the transition from $\mathcal{T}$ to $\mathcal{T}_*$ irrespective of the degree of local refinement; in particular, $f = P_{\widehat{\mathcal{T}}} f \in \mathbb{F}_{\widehat{\mathcal{T}}}$. We will later denote discrete data as $\widehat{\mathcal{D}} = (\widehat{\boldsymbol{A}}, \widehat{c}, \widehat{f})$ to distinguish it from exact data $\mathcal{D}$, and to study their discrepancy, but we prefer to keep the simple notation $\mathcal{D} = \widehat{\mathcal{D}}$ now because there is no reason for confusion. In particular, this implies that the bilinear form in (2.8) and the forcing function are the same for both Galerkin solutions $u_{\mathcal{T}} \in \mathbb{V}_{\mathcal{T}}$ and $u_{\mathcal{T}_*} \in \mathbb{V}_{\mathcal{T}_*}$, whence the energy errors are monotone according to (3.8),

$$|\!|\!| u - u_{\mathcal{T}_*} |\!|\!|_\Omega \leq |\!|\!| u - u_{\mathcal{T}} |\!|\!|_\Omega,$$

but not $|u - u_{\mathcal{T}}|_{H_0^1(\Omega)}$. Moreover, $\eta_{\mathcal{T}}(u_{\mathcal{T}})$ is not monotone because the discrete solution $u_{\mathcal{T}} \in \mathbb{V}_{\mathcal{T}}$ changes with the mesh. It is thus useful to quantify the behaviour of $\eta_{\mathcal{T}}(u_{\mathcal{T}})$ in terms of $\mathcal{T}$ and $u_{\mathcal{T}}$ following Cascón *et al.* (2008); see also Morin, Siebert and Veeser (2008). We do this next.

The first lemma exploits the structure of the PDE residual estimator, namely the presence of a positive power of the local mesh size, and expresses the *reduction* of $\eta_{\mathcal{T}_*}(v, f)$ relative to $\eta_{\mathcal{T}}(v, f)$ for fixed functions $v \in \mathbb{V}_{\mathcal{T}}$ and $f \in \mathbb{F}_{\mathcal{T}}$. This quantitative property is instrumental in studying convergence of AFEMs for coercive problems in Section 6 as well as discontinuous Galerkin methods in Section 9 and inf-sup stable problems in Section 10.

Lemma 4.53 (reduction property of the estimator). *If the elements of $\mathcal{M} \subset \mathcal{T}$ are bisected at least $b \geq 1$ times to refine $\mathcal{T}$ into $\mathcal{T}_*$, and $\lambda = 1 - 2^{-b/d}$, then*

$$\eta_{\mathcal{T}_*}(v, f, \mathcal{T}_*)^2 \leq \eta_{\mathcal{T}}(v, f, \mathcal{T})^2 - \lambda\, \eta_{\mathcal{T}}(v, f, \mathcal{M})^2 \quad \textit{for all } v \in \mathbb{V}_{\mathcal{T}},\ f \in \mathbb{F}_{\mathcal{T}}. \tag{4.67}$$

Proof. Given $T \in \mathcal{T}$, we rewrite (4.52) as follows:

$$\eta_{\mathcal{T}}(v, T)^2 = h_T\, j_{\mathcal{T}}(v, T)^2 + h_T^2\, r_{\mathcal{T}}(v, T)^2,$$

with $\eta_{\mathcal{T}}(v, T) = \eta_{\mathcal{T}}(v, f, T)$ and

$$j_{\mathcal{T}}(v, T)^2 = j_{\mathcal{T}}(v, f, T)^2 = \sum_{\substack{F \in \mathcal{F} \\ F \subset \partial T}} \| [\![\boldsymbol{A} \nabla v]\!] \cdot \boldsymbol{n}_F - P_{\mathcal{T}} f \|_{L^2(F)}^2,$$

$$r_{\mathcal{T}}(v, T) = r_{\mathcal{T}}(v, f, T) = \| P_{\mathcal{T}} f + \operatorname{div}(\boldsymbol{A} \nabla v) - cv \|_{L^2(T)},$$

where $f = P_{\mathcal{T}} f = P_{\mathcal{T}_*} f \in \mathbb{F}_{\mathcal{T}}$ does not change from $\mathcal{T}$ to $\mathcal{T}_*$. We readily have

$$\sum_{T_* \in \mathcal{T}_*, T_* \subseteq T} \eta_{\mathcal{T}_*}(v, T_*)^2 \leq \eta_{\mathcal{T}}(v, T)^2,$$

because $h_{T_*} \le h_T$ for all $T_* \subset T$ and $T_* \in \mathcal{T}_*$. If, in addition, T is bisected at least b times, then any such T_* satisfies $h_{T_*} \le 2^{-b/d} h_T$, whence

$$\sum_{T_* \in \mathcal{T}_*, T_* \subseteq T} \eta_{\mathcal{T}_*}(v, T_*)^2 \le 2^{-b/d} \eta_{\mathcal{T}}(v, T)^2.$$

Therefore, adding over $T \in \mathcal{T}$, we obtain

$$\eta_{\mathcal{T}_*}(v)^2 = \sum_{T \in \mathcal{T}} \sum_{T_* \in \mathcal{T}_*, T_* \subset T} \eta_{\mathcal{T}_*}(v, T)^2 \le 2^{-b/d} \sum_{T \in \mathcal{M}} \eta_{\mathcal{T}}(v, T)^2 + \sum_{T \in \mathcal{T} \setminus \mathcal{M}} \eta_{\mathcal{T}}(v, T)^2,$$

which implies the assertion (4.67). □

The next result complements Lemma 4.53 in that it expresses the Lipschitz continuity of $\eta_{\mathcal{T}}(v, f)$ with respect to the argument $v \in \mathbb{V}_{\mathcal{T}}$ for fixed $\mathcal{T}$ and $f \in \mathbb{F}_{\mathcal{T}}$.

Lemma 4.54 (Lipschitz property of the estimator). *Let $\mathcal{T}$ and $f \in \mathbb{F}_{\mathcal{T}}$ be fixed. There exists a constant C_{Lip} proportional to $\|A\|_{L^\infty(\Omega)} + \|c\|_{L^\infty(\Omega)}$ such that*

$$|\eta_{\mathcal{T}}(v, f) - \eta_{\mathcal{T}}(w, f)| \le C_{\mathrm{Lip}} |v - w|_{H_0^1(\Omega)} \quad \textit{for all } v, w \in \mathbb{V}_{\mathcal{T}}. \tag{4.68}$$

Proof. Since $\eta_{\mathcal{T}}(v) = \eta_{\mathcal{T}}(v, f)$ is the ℓ^2-norm of the vector $(\eta_{\mathcal{T}}(v, T))_{T \in \mathcal{T}} \in \mathbb{R}^{\#\mathcal{T}}$, applying the triangle inequality gives

$$\begin{aligned} |\eta_{\mathcal{T}}(v) - \eta_{\mathcal{T}}(w)|^2 &\le \sum_{T \in \mathcal{T}} |\eta_{\mathcal{T}}(v, T) - \eta_{\mathcal{T}}(w, T)|^2 \\ &\le \sum_{T \in \mathcal{T}} h_T |j_{\mathcal{T}}(v, T) - j_{\mathcal{T}}(w, T)|^2 + h_T^2 |r_{\mathcal{T}}(v, T) - r_{\mathcal{T}}(w, T)|^2. \end{aligned}$$

We first consider the jump terms and again apply the triangle inequality followed by an inverse estimate to find that

$$\begin{aligned} |j_{\mathcal{T}}(v, T) - j_{\mathcal{T}}(w, T)|^2 &\le \sum_{\substack{F \in \mathcal{F} \\ F \subset \partial T}} \|[[\boldsymbol{A} \nabla (v - w)]] \cdot \boldsymbol{n}_F\|_{L^2(F)}^2 \\ &\lesssim h_T^{-1} \|\boldsymbol{A}\|_{L^\infty(\Omega)}^2 \|\nabla (v - w)\|_{L^2(\omega_T)}^2, \end{aligned}$$

where ω_T is the patch of T. A similar reasoning for the element residuals yields

$$\begin{aligned} |r_{\mathcal{T}}(v, T) - r_{\mathcal{T}}(w, T)|^2 &\lesssim \|c(v - w)\|_{L^2(T)}^2 + \|\operatorname{div}(\boldsymbol{A} \nabla (v - w))\|_{L^2(T)}^2 \\ &\lesssim \|c\|_{L^\infty(\Omega)}^2 \|(v - w)\|_{L^2(T)}^2 + h_T^{-2} \|\boldsymbol{A}\|_{L^\infty(\Omega)}^2 \|\nabla (v - w)\|_{L^2(T)}^2, \end{aligned}$$

because A is piecewise polynomial. Finally, adding over $T \in \mathcal{T}$ and applying Lemma 2.2 (first Poincaré inequality) concludes the proof. □

Since the estimator $\eta_{\mathcal{T}}(v, f)$ depends explicitly on $P_{\mathcal{T}} f$, and $P_{\mathcal{T}} f$ may change with $\mathcal{T}$, it is crucial to account for the variations of $\eta_{\mathcal{T}}(v, f)$ while keeping $\mathcal{T}$ and $v \in \mathbb{V}_{\mathcal{T}}$ fixed. This is the purpose of our next result.

Lemma 4.55 (estimator dependence on discrete forcing). *Let $\mathcal{T}$ and $v \in \mathbb{V}_{\mathcal{T}}$ be fixed. Then there exists a constant C_{Lip} such that*

$$|\eta_{\mathcal{T}}(v,f)-\eta_{\mathcal{T}}(v,g)| \le C_{\mathrm{Lip}}\Big(\sum_{T\in\mathcal{T}} \|f-g\|^2_{H^{-1}(\omega_T)}\Big)^{1/2} \quad \textit{for all } f,g \in \mathbb{F}_{\mathcal{T}}. \tag{4.69}$$

Proof. We proceed elementwise, as in Lemma 4.54, except that after applying the triangle inequality we end up with the weighted L^2-norms

$$h_T^2\|f-g\|^2_{L^2(T)} + h_T\|f-g\|^2_{L^2(\partial T)} \quad \text{for all } T\in\mathcal{T}.$$

Extending these norms to patches ω_T and its interior faces σ_T, and appealing to Corollary 4.30 (quantifying H^{-1}-norms of discrete functionals), we deduce

$$h_T^2\|f-g\|^2_{L^2(\omega_T)} + h_T\|f-g\|^2_{L^2(\sigma_T)} \approx \|f-g\|^2_{H^{-1}(\omega_T)}.$$

Adding over $T\in\mathcal{T}$ finishes the proof. □

In the subsequent applications of Lemma 4.54 the discrete coefficients $(\boldsymbol{A}, c)$ may change with the change of the supporting mesh $\widehat{\mathcal{T}}$, but they will always be uniformly bounded in $L^\infty(\Omega)$; hence the constant C_{Lip} is uniformly bounded as well. Upon combining Lemmas 4.53, 4.54 and 4.55, we obtain the following crucial property.

Proposition 4.56 (estimator reduction). *Given $\mathcal{T}\in\mathbb{T}$ and a subset $\mathcal{M}\subset\mathcal{T}$ of elements marked for refinement, let* REFINE *be the procedure discussed in Section 3.5 that bisects the elements of $\mathcal{M}$ at least b times, and let $\mathcal{T}_* = \mathsf{REFINE}(\mathcal{T},\mathcal{M})$ be the resulting conforming mesh. Let the coefficients $(\boldsymbol{A}, c)$ be discrete and fixed. Then, for $\lambda = 1-2^{-b/d}$, for all $v\in\mathbb{V}_{\mathcal{T}}$, $v_*\in\mathbb{V}_{\mathcal{T}_*}$, $f\in\mathbb{F}_{\mathcal{T}}$, $f_*\in\mathbb{F}_{\mathcal{T}_*}$, and any $\delta>0$,*

$$\begin{aligned}\eta_{\mathcal{T}_*}(v_*,f_*,\mathcal{T}_*)^2 &\le (1+\delta)\big(\eta_{\mathcal{T}}(v,f,\mathcal{T})^2 - \lambda\,\eta_{\mathcal{T}}(v,f,\mathcal{M})^2\big)\\ &\quad + 2(1+\delta^{-1})\,C_{\mathrm{Lip}}^2\left(|v_*-v|^2_{H^1_0(\Omega)} + \sum_{T_*\in\mathcal{T}_*}\|f_*-f\|^2_{H^{-1}(\omega_{T_*})}\right),\end{aligned}$$

where C_{Lip} is the constant in Lemmas 4.54 and 4.55.

Proof. For any $\delta>0$, write

$$\eta_{\mathcal{T}_*}(v_*,f_*,\mathcal{T}_*)^2 \le (1+\delta)\eta_{\mathcal{T}_*}(v,f,\mathcal{T}_*)^2 + (1+\delta^{-1})(\eta_{\mathcal{T}_*}(v_*,f_*,\mathcal{T}_*) - \eta_{\mathcal{T}_*}(v,f,\mathcal{T}_*))^2$$

and apply Lemma 4.53 to the first term and Lemmas 4.54 and 4.55 to the second term combined with a triangle inequality. □

We finish this section by investigating the behaviour of the global oscillation under refinement.

Lemma 4.57 (quasi-monotonicity of oscillation). *If $f\in H^{-1}(\Omega)$ and $\mathcal{T},\mathcal{T}_*\in\mathbb{T}$ with $\mathcal{T}_*\ge\mathcal{T}$, then*

$$\mathrm{osc}_{\mathcal{T}_*}(f) \le C_{\mathrm{osc}}\,\mathrm{osc}_{\mathcal{T}}(f),$$

where C_{osc} depends only on the shape regularity coefficient σ and d.

Proof. Given $T \in \mathcal{T}$, let $T_* \in \mathcal{T}_*$ such that $T_* \subset T$. Since $\mathcal{T}_*$ is a refinement of $\mathcal{T}$, this implies that the patch $\omega_{\mathcal{T}_*}(T_*)$ in $\mathcal{T}_*$ around T_* is contained in the patch $\omega_{\mathcal{T}}(T)$ in $\mathcal{T}$ around T. Thanks to Lemma 4.31 (local near-best approximation), we derive

$$\operatorname{osc}_{\mathcal{T}_*}(f, T_*)^2 = \|(I - P_{\mathcal{T}_*})f\|^2_{H^{-1}(\omega_{\mathcal{T}_*}(T_*))} \le C^2_{\text{lStb}} \|(I - P_{\mathcal{T}})f\|^2_{H^{-1}(\omega_{\mathcal{T}_*}(T_*))}$$

and therefore, with the help of (ii) of Lemma 4.5 (localization of H^{-1}-norm),

$$\sum_{T_* \subset T} \operatorname{osc}_{\mathcal{T}_*}(f, T_*)^2 \le C^2_{\text{lStb}} C_{\text{ovrl}} \|(I - P_{\mathcal{T}})f\|^2_{H^{-1}(\omega_{\mathcal{T}}(T))} = C^2_{\text{lStb}} C_{\text{ovrl}} \operatorname{osc}_{\mathcal{T}}(f, T)^2,$$

where C_{ovrl} is bounded in terms of the shape regularity coefficient σ and d. Hence summing over $T \in \mathcal{T}$ yields

$$\operatorname{osc}_{\mathcal{T}_*}(f)^2 = \sum_{T \in \mathcal{T}} \sum_{T_* \subset T} \operatorname{osc}_{\mathcal{T}_*}(f, T_*)^2 \le C^2_{\text{lStb}} C_{\text{ovrl}} \operatorname{osc}_{\mathcal{T}}(f)^2,$$

and the proof is finished. □

4.9. *Alternative estimators*

In Section 4.7 we used the H^{-1}-projection $P_{\mathcal{T}}$ to derive *a posteriori* bounds for the error in the spirit of the standard residual estimator. The goal of this section is to illustrate that the approach with $P_{\mathcal{T}}$ can also be combined with other techniques of *a posteriori* error estimation, generalizing and expanding the discussion in Kreuzer and Veeser (2021, Section 4) with the H^{-1}-projection $P_{\mathcal{T}}$.

Alternative techniques have been developed with the desire to reduce or even circumvent the fact that constants spoil the relationship between error and estimator. In the framework of the aforementioned approach, we shall see that the various techniques based upon

- local (discrete) problems,
- hierarchy,
- flux equilibration

amount to different ways of quantifying a local norm of the discretized residual $P_{\mathcal{T}} R_{\mathcal{T}}$. This observation is useful for comparing the techniques and for a common treatment in the following sections about adaptive algorithms.

As in Section 4.7 on modified residual estimation, we shall consider only the case given by Assumption 4.44 (discrete coefficients and discrete functionals). For the hidden constants in the results of this section, it is useful to keep in mind Remark 4.4 (constants in error–residual relationship).

Theorem 4.45 (modified residual estimator) analysed an element-indexed version of the residual estimator. For the sake of simplicity, we shall refrain here from such an element-indexed setting and remain in the vertex-indexed setting of the abstract analysis of Section 4.6. In order to facilitate the comparison with the other estimators below, we offer the following vertex-indexed variant of Theorem 4.45. For $z \in \mathcal{V}$, we set $\mathcal{F}_z := \{F \in \mathcal{F} \mid z \in F\}$ and $\mathcal{T}_z := \{T \in \mathcal{T} \mid z \in T\}$. Given the Galerkin approximation $u_\mathcal{T}$ from (4.1), define the PDE indicator by

$$\begin{aligned} \eta_\mathcal{T}^{\text{res}}(u_\mathcal{T}) &:= \sum_{z\in\mathcal{V}} \eta_\mathcal{T}^{\text{res}}(u_\mathcal{T}, z)^2 \quad \text{with} \\ \eta_\mathcal{T}^{\text{res}}(u_\mathcal{T}, z)^2 &:= \sum_{F\in\mathcal{F}_z} h_F \|P_F R_\mathcal{T}\|^2_{L^2(F)} + \sum_{T\in\mathcal{T}_z} h_T^2 \|P_T R_\mathcal{T}\|^2_{L^2(T)}, \end{aligned} \tag{4.70a}$$

where $R_\mathcal{T} = f + \operatorname{div}(\boldsymbol{A}\nabla u_\mathcal{T}) - cu_\mathcal{T} \in H^{-1}(\Omega)$ is the residual and P_F, $F \in \mathcal{F}$ and P_T, $T \in \mathcal{T}$ yield the polynomial densities of $P_\mathcal{T}$; see Definition 4.24. The vertex-indexed modified residual estimator is then

$$\mathcal{E}_\mathcal{T}^{\text{res}} := \mathcal{E}_\mathcal{T}^{\text{res}}(u_\mathcal{T}, f)^2 := \eta_\mathcal{T}^{\text{res}}(u_\mathcal{T})^2 + \operatorname{osc}_\mathcal{T}(f)^2. \tag{4.70b}$$

Theorem 4.58 (vertex-indexed modified residual estimator). *Suppose Assumption 4.44. The modified residual estimator* (4.70b) *is equivalent to the error:*

$$\frac{\min\{1, C_{L,\text{res}}\}}{C_{\text{lStb}}^\star C_d} \mathcal{E}_\mathcal{T}^{\text{res}} \lesssim \|\nabla(u - u_\mathcal{T})\|_{L^2(\Omega)} \lesssim \max\{1, C_{U,\text{res}}\} C_d C_{\text{loc}} \mathcal{E}_\mathcal{T}^{\text{res}},$$

while its PDE indicator (4.70a) *is locally equivalent to the discretized residual: for all vertices $z \in \mathcal{V}$,*

$$C_{L,\text{res}} \eta_\mathcal{T}^{\text{res}}(u_\mathcal{T}, z) \le \|P_\mathcal{T} R_\mathcal{T}\|_{H^{-1}(\omega_z)} \le C_{U,\text{res}} \eta_\mathcal{T}^{\text{res}}(u_\mathcal{T}, z).$$

Here, $C_{L,\text{res}}$ and $C_{U,\text{res}}$ are the hidden constants of Corollary 4.30 on stars, $C_{\text{lStb}}^\star$ is the stability constant of $P_\mathcal{T}$ on stars from Lemma 4.28, C_{loc} is the constant from Corollary 4.6, $C_d = \sqrt{2(d+1)}$, and the hidden constants depend only on the error–residual relationship in Lemma 4.1.

Proof. Local equivalence is a reformulation of Corollary 4.30 (quantifying H^{-1}-norms of discrete functionals) on stars. The global bounds follow by combining local equivalence with Lemma 4.1 (error and residual), Corollary 4.6 (star localization of residual norm) and Lemma 4.35 (splitting of local residual norm). □

Adjoint projection

The projection $P_\mathcal{T}$ relates residual $R_\mathcal{T}$ and discretized residual $P_\mathcal{T} R_\mathcal{T}$. In order to exploit this relationship on the test space $H_0^1(\Omega)$, we shall need the adjoint $P_\mathcal{T}^*$ to the projection $P_\mathcal{T}$. Curiously, operators employed in this vein appeared first; see e.g. Morin, Nochetto and Siebert (2003) and Veeser (2002).

Given $w \in H_0^1(\Omega)$, the function $P_\mathcal{T}^* w$ can be directly defined by requiring

$$P_\mathcal{T}^* w \in \mathbb{V}_\mathcal{T}^+ : \quad \langle \ell, P_\mathcal{T}^* w\rangle = \langle \ell, w\rangle \quad \text{for all } \ell \in \mathbb{F}_\mathcal{T}. \tag{4.71}$$

This definition is well-posed thanks to Lemma 3.1 (discrete inf-sup condition) and Lemma 4.25 (algebraic properties), especially (4.37) and (4.38). Clearly, $P^*_{\mathcal{T}}$ is a linear projection onto the finite-dimensional subspace $\mathbb{V}^+_{\mathcal{T}} \subset H^1_0(\Omega)$. A representation as interpolation operator will be derived in Corollary 4.61 below. Using both definitions of $P_{\mathcal{T}}$ and $P^*_{\mathcal{T}}$, we see that they are actually adjoint:

$$\langle P_{\mathcal{T}}\ell, w\rangle = \langle P_{\mathcal{T}}\ell, P^*_{\mathcal{T}}w\rangle = \langle \ell, P^*_{\mathcal{T}}w\rangle \quad \text{for all } \ell \in H^{-1}(\Omega),\ w \in H^1_0(\Omega). \tag{4.72}$$

Consequently, Lemmas 4.25 and 4.28 (local H^{-1}-stability) show that $P^*_{\mathcal{T}}$ is a local operator with

$$\|P^*_{\mathcal{T}}\|_{\mathcal{L}(H^1_0(\omega))} = \|P_{\mathcal{T}}\|_{\mathcal{L}(H^{-1}(\omega))} \leq C_{\mathrm{lStb}}. \tag{4.73}$$

The choice $\ell = R_{\mathcal{T}}$ in (4.72) leads to

$$\langle P_{\mathcal{T}}R_{\mathcal{T}}, w\rangle = \langle P_{\mathcal{T}}R_{\mathcal{T}}, P^*_{\mathcal{T}}w\rangle = \langle R_{\mathcal{T}}, P^*_{\mathcal{T}}w\rangle \quad \text{for all } w \in H^1_0(\Omega),$$

where the two identities show that the discretized residual $P_{\mathcal{T}}R_{\mathcal{T}}$ can be analysed with discrete test functions in $\mathbb{V}^+_{\mathcal{T}}$ only; see the norm equivalence in Lemma 4.28. Restricting to discrete test functions in $\mathbb{V}^+_{\mathcal{T}} = \operatorname{Im} P^*_{\mathcal{T}}$, we find the definition of $P_{\mathcal{T}}$:

$$\langle P_{\mathcal{T}}R_{\mathcal{T}}, w\rangle = \langle R_{\mathcal{T}}, w\rangle \quad \text{for all } w \in \mathbb{V}^+_{\mathcal{T}}. \tag{4.74}$$

An estimator based upon local problems

Local dual norms can be quantified by solving local problems. Requiring computability of these solutions leads to *finite-dimensional* or *discrete* local problems. In other words, we lift the residual to local and finite-dimensional extensions of the finite element space. Starting with Babuška and Rheinboldt (1978), this idea was used to soften the impact of constants in the relationship between error and estimator; see Verfürth (2013, Remark 1.22) for more references.

Within the approach of Sections 4.1 and 4.6, we can use local discrete problems to quantify the local H^{-1}-norms of the discretized residual $P_{\mathcal{T}}R_{\mathcal{T}}$. In this manner, constants arise only due to the localization of the residual norm and to the splitting into discretized and oscillatory residual by the H^{-1}-projection $P_{\mathcal{T}}$.

We start by introducing the vertex-oriented PDE indicator. Given the Galerkin approximation $u_{\mathcal{T}}$ from (4.1), set

$$\eta^{\mathrm{lpb}}_{\mathcal{T}}(u_{\mathcal{T}}) := \sum_{z\in\mathcal{V}} \eta^{\mathrm{lpb}}_{\mathcal{T}}(u_{\mathcal{T}}, z)^2 \quad \text{with} \quad \eta^{\mathrm{lpb}}_{\mathcal{T}}(u_{\mathcal{T}}, z) := \|\nabla v_z\|_{L^2(\omega_z)}, \tag{4.75a}$$

where $v_z \in \mathbb{V}^+(\mathcal{T}_z)$ is the solution of the local problem

$$\int_{\omega_z} \nabla v_z \cdot \nabla w = \langle R_{\mathcal{T}}, w\rangle \quad \text{for all } w \in \mathbb{V}^+(\mathcal{T}_z). \tag{4.75b}$$

Note that this problem is discrete for $\dim \mathbb{V}^+(\mathcal{T}_z) < \infty$ and can therefore be solved up to machine precision. The resulting estimator is then

$$\mathcal{E}^{\mathrm{lpb}}_{\mathcal{T}} := \mathcal{E}^{\mathrm{lpb}}_{\mathcal{T}}(u_{\mathcal{T}}, f)^2 := \eta^{\mathrm{lpb}}_{\mathcal{T}}(u_{\mathcal{T}})^2 + \mathrm{osc}_{\mathcal{T}}(f)^2. \tag{4.75c}$$

Theorem 4.59 (estimator based on local problems). *Under Assumption 4.44 the estimator* (4.75) *based on local problems is equivalent to the error, while its PDE indicator is locally equivalent to the discretized residual with constant* 1 *in the lower bound, so that*

$$\frac{1}{C^{\star}_{1\mathrm{Stb}} C_d} \mathcal{E}^{\mathrm{lpb}}_{\mathcal{T}} \lesssim \|\nabla(u - u_{\mathcal{T}})\|_{L^2(\Omega)} \lesssim C^{\star}_{1\mathrm{Stb}} C_d C_{\mathrm{loc}} \mathcal{E}^{\mathrm{lpb}}_{\mathcal{T}},$$

and, for all vertices $z \in \mathcal{V}$,

$$\eta^{\mathrm{lpb}}_{\mathcal{T}}(u_{\mathcal{T}}, z) \le \|P_{\mathcal{T}} R_{\mathcal{T}}\|_{H^{-1}(\omega_z)} \le C^{\star}_{1\mathrm{Stb}} \eta^{\mathrm{lpb}}_{\mathcal{T}}(u_{\mathcal{T}}, z)$$

where $C^{\star}_{1\mathrm{Stb}}$ *is the stability constant of* $P_{\mathcal{T}}$ *on stars from Lemma 4.28,* C_{loc} *from Corollary 4.6,* $C_d = \sqrt{2(d+1)}$ *and the hidden constants depend only on the error–residual relationship in Lemma 4.1.*

Proof. It suffices to show the local equivalence for the PDE indicator; see Theorem 4.58 (vertex-indexed modified residual estimator) and note that $C^{\star}_{1\mathrm{Stb}} \ge 1$. Let $z \in \mathcal{V}$ be any vertex. In view of (4.74), the definition of $v_z \in \mathbb{V}^+(\mathcal{T})$ readily implies

$$\eta^{\mathrm{lpb}}_{\mathcal{T}}(u_{\mathcal{T}}, z) = \|\nabla v_z\|_{L^2(\omega_z)} = \|P_{\mathcal{T}} R_{\mathcal{T}}\|_{\mathbb{V}^+(\mathcal{T})^*}.$$

Hence Lemma 4.28 on the local H^{-1}-stability of $P_{\mathcal{T}}$ yields the asserted local equivalence of PDE indicator and discretized residual $P_{\mathcal{T}} R_{\mathcal{T}}$. □

A stable biorthogonal system for $\mathbb{F}_{\mathcal{T}} \times \mathbb{V}^+_{\mathcal{T}}$

Stable biorthogonal systems induce linear bounded projections, which enjoy near-best approximation thanks to the Lebesgue lemma. Supposing Assumption 4.21 (abstract cut-off), we now outline the construction of such a system for the finite-dimensional product $\mathbb{F}_{\mathcal{T}} \times \mathbb{V}^+_{\mathcal{T}}$. The constructed system will induce both projections $P_{\mathcal{T}}$ and its adjoint $P^*_{\mathcal{T}}$. This generalizes the bi-orthogonal system in Kreuzer and Veeser (2021, Section 3.4) to arbitrary degrees of the discrete functionals and provides an alternative approach to $P_{\mathcal{T}}$, its local stability as well as its computation. Furthermore, we use it for devising a hierarchical estimator.

The construction is implemented in an affine equivalent manner and our first step consists in setting up a suitable *reference biorthogonal system.* Let $\widehat{T} := T_d$ be the reference element, let $\widehat{F} := T_{d-1} \times \{0\} \subset \widehat{T}$ be the reference face, and denote the polynomial degrees in $\mathbb{F}_{\mathcal{T}}$ by $m_1 \in \mathbb{N}_0$ and $m_2 \in \mathbb{N}_0$. Writing

$$K_1 := \dim \mathbb{P}_{m_1}(\widehat{F}), \quad K_2 := \dim \mathbb{P}_{m_2}(\widehat{T}),$$

assume that we are given orthonormal bases $q_{(\widehat{F},1)}, \ldots, q_{(\widehat{F},K_1)} \in \mathbb{P}_{m_1}(\widehat{F})$ and $q_{(\widehat{T},1)}, \ldots, q_{(\widehat{T},K_2)} \in \mathbb{P}_{m_2}(\widehat{T})$ in the sense that

$$\int_{\widehat{F}} q_{(\widehat{F},k)} q_{(\widehat{F},l)} \phi_{\widehat{F}} = \delta_{kl}, \quad \int_{\widehat{T}} q_{(\widehat{T},k)} q_{(\widehat{T},l)} \phi_{\widehat{T}} = \delta_{kl}. \tag{4.76}$$

for all admissible k, l, i.e. $k, l \in \{1, \dots, K_1\}$ or $\{1, \dots, K_2\}$ depending on the underlying domain. These bases induce the *reference functionals*

$$\widehat{\ell}_{(\widehat{F},k)}(w) := \int_{\widehat{F}} q_{(\widehat{F},k)} w, \quad \widehat{\ell}_{(\widehat{T},k)}(w) := \int_{\widehat{T}} q_{(\widehat{T},k)} w,$$

on $H^1(\widehat{T})$, which in turn span the reference space $\widehat{\mathbb{F}}$. In order to define complementary test functions, let $\widehat{E}$ be the extension operator (4.4) associated with the reference face $\widehat{F}$ adapted to the current situation with the only element $\widehat{T}$, and, given some $v \in L^2(\widehat{T})$, define $\widehat{Q}v \in \mathbb{P}_{m_2}$ by

$$\int_{\widehat{T}} q\big(\widehat{Q}v\big)\phi_{\widehat{T}} = \int_{\widehat{T}} qv \quad \text{for all } q \in \mathbb{P}_{m_2}. \tag{4.77}$$

We thus define the *reference test functions*

$$\begin{aligned} \widehat{w}_{(\widehat{T},k)} &:= q_{(\widehat{T},k)}\phi_{\widehat{T}}, \quad k = 1, \dots, K_2, \\ \widehat{w}_{(\widehat{F},k)} &:= \widehat{v}_{(\widehat{F},k)} - \big(\widehat{Q}\widehat{v}_{(\widehat{F},k)}\big)\phi_{\widehat{T}}, \quad k = 1, \dots, K_1, \end{aligned} \tag{4.78}$$

with $\widehat{v}_{(\widehat{F},k)} := (\widehat{E}q_{(\widehat{F},k)})\phi_{\widehat{F}}$. Note that $\widehat{w}_{(\widehat{F},k)} \neq 0$. Writing

$$\widehat{I} := \{(\widehat{F}, k) \mid k = 1, \dots, K_1\} \cup \{(\widehat{T}, k) \mid k = 1, \dots, K_2\},$$

we then have

$$\widehat{w}_i \in \widehat{\mathbb{V}}^+ := \{(\widehat{E}q_1)\phi_{\widehat{F}} + q_2\phi_{\widehat{T}} \mid q_1 \in \mathbb{P}_{m_1}, q_2 \in \mathbb{P}_{m_2}\} \quad \text{for all } i \in \widehat{I}$$

and the biorthogonality

$$\langle \widehat{\ell}_i, \widehat{w}_j \rangle = \delta_{ij} \quad \text{for all } i, j \in \widehat{I}, \tag{4.79}$$

thanks to (4.76) and (4.77). We thus dispose of a biorthogonal system in the reference product $\widehat{\mathbb{F}} \times \widehat{\mathbb{V}}^+$.

Using pullbacks with some minor tweaks, this reference biorthogonal system induces a global biorthogonal system. To this end, we employ bi-affine maps G_F, G_T, and $G_{(T,F)}$. Here, for example, given a pair $(T, F) \in \mathcal{T} \times \mathcal{F}$ with $F \subset T$, the map $G_{(T,F)}$ is bi-affine and sends vertices into vertices such that $G_{(T,F)}(\widehat{T}) = T$ and $G_{(T,F)}(\widehat{F}) = F$. The fact that these maps are only unique up to some renumbering of the vertices is irrelevant, as all objects in the reference situation on $(\widehat{T}, \widehat{F})$ are invariant under such renumberings. We denote the respective inverse maps of G_F, G_T and $G_{(T,F)}$ by H_F, H_T and $H_{(T,F)}$. Motivated by the transformation rule, we introduce the scaled pullbacks, for $F \in \mathcal{F}, T \in \mathcal{T}$, k admissible,

$$q_{(F,k)} := \left(\frac{|\widehat{F}|}{|F|}\right)^{1/2} H_F^* q_{(\widehat{F},k)}, \quad q_{(T,k)} := \left(\frac{|\widehat{T}|}{|T|}\right)^{1/2} H_T^* q_{(\widehat{T},k)}, \tag{4.80}$$

of the reference orthonormal bases in (4.76). These lead to the basis

$$\ell_{(F,k)}(w) := \int_F q_{(F,k)} w, \quad \ell_{(T,k)}(w) := \int_T q_{(T,k)} w \tag{4.81}$$

of $\mathbb{F}_{\mathcal{T}}$, while the associated test functions are again given via pullbacks:

$$w_{(T.k)} := \left(\frac{|\widehat{T}|}{|T|}\right)^{1/2} H_T^* \widehat{w}_{(\widehat{T},k)}, \quad w_{(F.k)}|_T := \left(\frac{|\widehat{F}|}{|F|}\right)^{1/2} H_{(T,F)}^* \widehat{w}_{(\widehat{F},k)} \tag{4.82}$$

for all $T \in \mathcal{T}$, $F \in \mathcal{F}$ with $F \subset T$ and all admissible k. Note that $w_{(F.k)} \in H_0^1(\Omega)$. Finally, we introduce the index set

$$I := (\mathcal{F} \times \{1, \ldots, K_1\}) \cup (\mathcal{T} \times \{1, \ldots, K_2\}),$$

and observe that $w_i \in \mathbb{V}_{\mathcal{T}}^+$ for all $i \in I$.

Lemma 4.60 (biorthogonal system). *The pairs (ℓ_i, w_i), $i \in I$, provide a stable biorthogonal system of the product $\mathbb{F}_{\mathcal{T}} \times \mathbb{V}_{\mathcal{T}}^+$: indeed, $\langle \ell_i, w_j \rangle = \delta_{ij}$ for all $i, j \in I$ and, writing $I_z := \{(S, k) \in I \mid S \ni z\}$ for all $z \in \mathcal{V}$,*

$$\sum_{i \in I_z} \|\ell_i\|_{H^{-1}(\omega_z)} \|\nabla w_i\|_{L^2(\omega_z)} \le C_{\mathrm{bOS}}^{\star},$$

where the constant $C_{\mathrm{bOS}}^{\star}$ depends only on d, m_1, m_2 and the shape regularity coefficient σ from (3.9).

Proof. [1] We first establish the biorthogonality. Thanks to the transformation rule, the scaled pullbacks (4.80) indeed form local orthonormal bases of $P_{m_1}(F)$, $F \in \mathcal{F}$, and $P_{m_2}(T)$, $T \in \mathcal{T}$:

$$\int_F q_{(F,k)} q_{(F,l)} \phi_F = \int_{\widehat{F}} q_{(\widehat{F},k)} q_{(\widehat{F},l)} \phi_{\widehat{F}} = \delta_{kl}, \quad \int_T q_{(T,k)} q_{(T,l)} \phi_T = \delta_{kl} \tag{4.83}$$

for all admissible k and l. This orthonormality, combined with the local supports of the pairs (ℓ_i, w_i), $i \in I$, shows the biorthogonality, except for the cases when $(T, F) \in \mathcal{T} \times \mathcal{F}$ with $F \subset T$ and k, l are admissible. Here the transformation rule and the definition of $\widehat{Q}$ imply

$$\begin{aligned}
\langle \ell_{(T,k)}, w_{(F,l)} \rangle &= \int_T q_{(T,k)} w_{(F,l)} = \left(\frac{|T|\,|\widehat{F}|}{|\widehat{T}|\,|F|}\right)^{1/2} \int_{\widehat{T}} q_{(\widehat{T},k)} \widehat{w}_{(\widehat{F},l)} \\
&= \left(\frac{|T|\,|\widehat{F}|}{|\widehat{T}|\,|F|}\right)^{1/2} \int_{\widehat{T}} q_{(\widehat{T},k)} \big(\widehat{v}_{(\widehat{F},l)} - (\widehat{Q}\widehat{v}_{(\widehat{F},l)}) \phi_{\widehat{T}}\big) = 0,
\end{aligned}$$

and biorthogonality is verified.

[2] It remains to show the stability bound for any vertex $z \in \mathcal{V}$. Given $i \in I_z$, we have either $i = (T, k)$ with $T \in \mathcal{T}$ or $i = (F, k)$ with $F \in \mathcal{F}$. On the one hand, the functional ℓ_i satisfies

$$\|\ell_{(T,k)}\|_{H^{-1}(\omega_z)} \lesssim h_T \quad \text{or} \quad \|\ell_{(F,k)}\|_{H^{-1}(\omega_z)} \lesssim h_F^{1/2}$$

since, by passing to the reference element and using orthonormality (4.83), (a variant of the) Poincaré inequality (2.2) and the trace inequality on $\widehat{T}$, we have

$$\begin{aligned}\langle \ell_{(T,k)}, w\rangle &= \int_T q_{(T,k)} w = \left(\frac{|T|}{|\widehat{T}|}\right)^{1/2} \int_{\widehat{T}} q_{(\widehat{T},k)} G_T^* w \\ &\leq \left(\frac{|T|}{|\widehat{T}|}\right)^{1/2} \|q_{(\widehat{T},k)}\|_{L^2(\widehat{T})} \|G_T^* w\|_{L^2(\widehat{T})} \\ &\lesssim \left(\frac{|T|}{|\widehat{T}|}\right)^{1/2} \left(\int_{\widehat{T}} |q_{(\widehat{T},k)}|^2 \phi_{\widehat{T}}\right) \|\nabla(G_T^* w)\|_{L^2(\widehat{T})} \lesssim h_T \|\nabla w\|_{L^2(T)}\end{aligned}$$

or, with $T \in \mathcal{T}$ such that $T \supset F$,

$$\begin{aligned}\langle \ell_{(F,k)}, w\rangle &= \int_F q_{(F,k)} w = \left(\frac{|F|}{|\widehat{F}|}\right)^{1/2} \int_{\widehat{F}} q_{(\widehat{F},k)} G_{(T,F)}^* w \\ &\leq \left(\frac{|F|}{|\widehat{F}|}\right)^{1/2} \|q_{(\widehat{F},k)}\|_{L^2(\widehat{F})} \|G_{(T,F)}^* w\|_{L^2(\widehat{F})} \\ &\lesssim \left(\frac{|F|}{|\widehat{F}|}\right)^{1/2} \left(\int_{\widehat{F}} |q_{(\widehat{F},k)}|^2 \phi_{\widehat{F}}\right) \|\nabla\big(G_{(T,F)}^* w\big)\|_{L^2(\widehat{T})} \\ &\lesssim \left(\frac{|F|\,|\widehat{T}|}{|\widehat{F}|\,|T|}\right)^{1/2} h_T \|\nabla w\|_{L^2(T)} \lesssim h_F^{1/2} \|\nabla w\|_{L^2(T)}.\end{aligned}$$

On the other hand, we obtain that the function w_i verifies

$$\|\nabla w_{(T,k)}\|_{L^2(T)} = \left(\frac{|\widehat{T}|}{|T|}\right)^{1/2} \|\nabla\big(H_T^* \widehat{w}_{(\widehat{T},k)}\big)\|_{L^2(T)} \lesssim h_T^{-1} \|\widehat{w}_{(\widehat{T},k)}\|_{L^2(\widehat{T})} \lesssim h_T^{-1}$$

or

$$\begin{aligned}\|\nabla w_{(F,k)}\|_{L^2(T)} &= \left(\frac{|\widehat{F}|}{|F|}\right)^{1/2} \|\nabla\big(H_{(T,F)}^* w_{(\widehat{T},k)}\big)\|_{L^2(T)} \\ &\lesssim h_T^{-1} \left(\frac{|T|}{|F|}\right)^{1/2} \|\widehat{w}_{(\widehat{T},k)}\|_{L^2(\widehat{T})} \lesssim h_F^{-1/2}.\end{aligned}$$

Using these inequalities to bound $\|\ell_i\|_{H^{-1}(\omega_z)} \|\nabla w_i\|_{L^2(\omega_z)}$, summing over all $i \in I_z$ then establishes the stability bound as the cardinality $\#I_z$ is uniformly bounded in terms of the shape regularity coefficient σ. □

Corollary 4.61 (projections as interpolation operators). *The biorthogonal system (ℓ_i, w_i), $i \in I$, induces the H^{-1}-projection $P_{\mathcal{T}}$ from Definition 4.24 and its adjoint $P_{\mathcal{T}}^*$. Indeed, we have*

$$P_{\mathcal{T}}\ell = \sum_{i\in I} \langle \ell, w_i\rangle \ell_i \quad \textit{and} \quad P_{\mathcal{T}}^* w = \sum_{i\in I} \langle \ell_i, w\rangle w_i$$

for all $\ell \in H^{-1}(\Omega)$ and $w \in H_0^1(\Omega)$. The stability of the biorthogonal system then *provides an alternative proof of the H^{-1}-stability on stars of both projection $P_{\mathcal{T}}$ and $P^*_{\mathcal{T}}$, entailing $C^\star_{\mathrm{lStb}} \le C^\star_{\mathrm{bOS}}$.*

Proof. [1] We only show the identity for $P_{\mathcal{T}}$; the one for $P^*_{\mathcal{T}}$ can be verified along the same lines. The biorthogonality in Lemma 4.60 readily implies

$$\left\langle \sum_{i\in I} \langle \ell, w_i\rangle \ell_i, w_j \right\rangle = \sum_{i\in I} \langle \ell, w_i\rangle \langle \ell_i, w_j\rangle = \langle \ell, w_j\rangle \quad \text{for all } j \in I.$$

As w_j, $j \in I$, is a basis of $\mathbb{V}^+_{\mathcal{T}}$, we conclude the claimed identity for $P_{\mathcal{T}}$.

[2] To verify the stability statement, we again restrict ourselves to the case of the projection $P_{\mathcal{T}}$. Observe first that the proof of the stability of the biorthogonal system does invoke the local stability of $P_{\mathcal{T}}$. Thanks to the representation of $P_{\mathcal{T}}$ and the stability of the biorthogonal system in Lemma 4.60, we have

$$\langle P_{\mathcal{T}}\ell, w\rangle = \sum_{i\in I_z} \langle \ell, w_i\rangle \langle \ell_i, w\rangle \le C^\star_{\mathrm{bOS}} \|\ell\|_{H^{-1}(\omega_z)} \|\nabla w\|_{L^2(\omega_z)}$$

for any $w \in H^1_0(w_z)$, $z \in \mathcal{V}$, and the proof is finished. □

The following two remarks illustrate the practical and theoretical usefulness of the representation formulae.

Remark 4.62 (alternative computation of $P_{\mathcal{T}}$). A by-product of Corollary 4.61 is a way of computing $P_{\mathcal{T}}\ell$ for a given functional $\ell \in H^{-1}(\Omega)$ that 'diagonalizes' the approach in Remark 4.26. In fact, given reference orthonormal bases as in (4.76), we can compute the functionals ℓ_i, $i \in I$, and test functions w_i, $i \in I$, by means of the formulae (4.78), (4.80), (4.81) and (4.82), whence, evaluating $\langle \ell, w_i\rangle$, $i \in I$, everything in the representation of $P_{\mathcal{T}}\ell$ in Corollary 4.61 is at our disposal.

Example 4.63 (global instability of $P_{\mathcal{T}}$ and $P^*_{\mathcal{T}}$). While the projections $P_{\mathcal{T}}$ and $P^*_{\mathcal{T}}$ are locally stable, both may become globally unbounded under mesh refinement. To see this, recall (4.73), note that $\|P_{\mathcal{T}}\|_{\mathcal{L}(H^{-1}(\Omega))} = \|P^*_{\mathcal{T}}\|_{\mathcal{L}(H^1_0(\Omega))}$ and, following the spirit of an example in Tantardini *et al.* (2024), consider

$$w := \sum_{z\in\mathcal{V}\cap\Omega} \phi_z \in H^1_0(\Omega).$$

Then, for all *quasi-uniform* meshes $\mathcal{T}$ with shape regularity coefficient σ, there is a constant C depending on σ and quasi-uniformity such that

$$\|P^*_{\mathcal{T}}\|^2_{\mathcal{L}(H^1_0(\Omega))} \ge \frac{\|\nabla(P^*_{\mathcal{T}}w)\|^2_{L^2(\Omega)}}{\|\nabla w\|^2_{L^2(\Omega)}} \ge C\frac{\#\{T \in \mathcal{T} \mid T \cap \partial\Omega = \emptyset\}}{\#\{T \in \mathcal{T} \mid T \cap \partial\Omega \ne \emptyset\}}. \tag{4.84}$$

Obviously, the last term tends to ∞ under uniform refinement.

To prove (4.84), we proceed in several steps, mostly hiding constants depending on quasi-uniformity of $\mathcal{T}$ and, as usual, the shape regularity coefficient σ and d.

1 We first bound $\|\nabla w\|_{L^2(\Omega)}$ from above. Noting that

$$w = 1 \quad \text{on} \quad \bigcup_{T \cap \partial\Omega = \emptyset} T,$$

the bound $\|\nabla \phi_z\|_{L^\infty(T)} \lesssim h_T^{-1}$ readily implies

$$\|\nabla w\|_{L^2(\Omega)}^2 = \sum_{T \cap \partial\Omega \neq \emptyset} \|\nabla w\|_{L^2(T)}^2 \lesssim \#\{T \in \mathcal{T} \mid T \cap \partial\Omega \neq \emptyset\}\, h_{\mathcal{T}}^{d-2}, \tag{4.85}$$

where $h_{\mathcal{T}}$ stands for the mesh size of $\mathcal{T}$.

2 The lower bound for $\|\nabla(P_{\mathcal{T}}^* w)\|_{L^2(\Omega)}$ is more involved. We start by showing the following representation for any $T \in \mathcal{T}$ with $T \cap \partial\Omega = \emptyset$:

$$P_{\mathcal{T}}^* w|_T = H_T^* \widehat{v} \tag{4.86}$$

with the fixed function

$$\widehat{v} := \sum_{(\widehat{T},k)\in\widehat{I}} \left(\int_{\widehat{T}} q_{(\widehat{T},k)}\right) \widehat{w}_{(\widehat{T},k)} + \sum_{(F',k)} \left(\frac{|F'|}{|\widehat{F}|}\right)^{1/2} \left(\int_{\widehat{F}} q_{(\widehat{F},k)}\right) \widehat{w}_{(F',k)} \notin \mathbb{P}_0,$$

where the indices of the second sum vary according to $F' \subset \widehat{T}$, $k = 1, \dots, K_1$ and $\widehat{w}_{(F',k)}$ is given by (4.82) with the transformation $H_{(\widehat{T},F')}$. Note first that, thanks to $w = 1$ on T and (4.80), the coefficients in the expansion of $P_{\mathcal{T}}^* w|_T$ satisfy

$$\langle \ell_{(T,k)}, w\rangle = \int_T q_{(T,k)} = \left(\frac{|T|}{|\widehat{T}|}\right)^{1/2} \int_{\widehat{T}} q_{(\widehat{T},k)}$$

and, for any $F \subset T$,

$$\langle \ell_{(F,k)}, w\rangle = \int_F q_{(F,k)} = \left(\frac{|F|}{|\widehat{F}|}\right)^{1/2} \int_{\widehat{F}} q_{(\widehat{F},k)}.$$

Combining these identities with (4.82) yields the claimed identity (4.86), and it remains to verify $\widehat{v} \notin \mathbb{P}_0$. Suppose $\widehat{v} = c \in \mathbb{R}$. As a consequence, for any face $F' \subset \widehat{T}$ and $k \in \{1, \dots, K_1\}$, we have

$$c = \widehat{w}_{(F',k)} = |\widehat{F}|^{1/2}|F'|^{-1/2}\widehat{w}_{(\widehat{F},k)} = |\widehat{F}|^{1/2}|F'|^{-1/2}c.$$

As not all faces of the reference simplex have the same volume, this yields $c = 0$. From (4.78) and (4.76), we infer that the coefficients in the definition of $\widehat{v}$ vanish. In particular, $\int_{\widehat{T}} q_{(\widehat{T},k)} = 0$ for all $k = 1, \dots, K_2$ means $\widehat{Q}1 = 0$, where $\widehat{Q}$ is the operator given in (4.77). This, however, is a contradiction because the restriction of $\widehat{Q}$ to $\mathbb{P}_{m_2}$ is injective. Hence $\widehat{v} \notin \mathbb{P}_0$ is proved.

3 We are ready to show the bound for $\|\nabla(P_{\mathcal{T}}^* w)\|_{L^2(\Omega)}$. Given any element $T \in \mathcal{T}$ with $T \cap \partial\Omega = \emptyset$, we pass to the reference element to exploit the previous step, and obtain

$$\|\nabla(P_{\mathcal{T}}^* w)\|_{L^2(T)} = \|\nabla(H_T^* \widehat{v})\|_{L^2(T)} \gtrsim h_T^{d/2-1} \|\nabla \widehat{v}\|_{L^2(\widehat{T})}$$

with $\|\nabla \widehat{v}\|_{L^2(\widehat{T})} > 0$ independent of T. Consequently,

$$\|\nabla(P_{\mathcal{T}}^* w)\|_{L^2(\Omega)}^2 \geq \sum_{T \cap \partial\Omega = \emptyset} \|\nabla(P_{\mathcal{T}}^* w)\|_{L^2(T)}^2 \gtrsim \#\{T \in \mathcal{T} \mid T \cap \partial\Omega = \emptyset\}\, h_{\mathcal{T}}^{d-2},$$

because $\mathcal{T}$ is quasi-uniform. Combining this lower bound with the upper bound (4.85) of the first step, we conclude (4.84).

A hierarchical estimator

Like estimators based upon local problems, hierarchical estimators aim at softening the impact of constants in the lower bound, with the difference that they are explicit. While global higher-order extensions were used originally, Bornemann, Erdmann and Kornhuber (1996) use an extension tailored to the residual structure and derive an upper bound with indicators testing the residual with a basis of the extension. One may expect that such explicit indicators come at the price of increased constants in the upper bound. For the following example, this expectation is confirmed by the inequality $C_{\mathrm{lStb}}^{\star} \leq C_{\mathrm{bOS}}^{\star}$.

Given the Galerkin approximation $u_{\mathcal{T}}$ from (4.1), the hierarchical PDE indicator is defined by

$$\eta_{\mathcal{T}}^{\mathrm{hier}}(u_{\mathcal{T}}) := \sum_{z \in \mathcal{V}} \eta_{\mathcal{T}}^{\mathrm{hier}}(u_{\mathcal{T}}, z)^2 \quad \text{with} \quad \eta_{\mathcal{T}}^{\mathrm{hier}}(u_{\mathcal{T}}, z) := \max_{i \in I_z} \frac{|\langle R_{\mathcal{T}}, w_i \rangle|}{\|\nabla w_i\|_{L^2(\omega_z)}}, \tag{4.87a}$$

with I and I_z as in Lemma 4.60 (biorthogonal system). Note that the test functions w_i, $i \in I$, are available (see Remark 4.62), and therefore $\eta_{\mathcal{T}}^{\mathrm{hier}}(u_{\mathcal{T}})$ is explicit. The resulting estimator is then

$$\mathcal{E}_{\mathcal{T}}^{\mathrm{hier}} := \mathcal{E}_{\mathcal{T}}^{\mathrm{hier}}(u_{\mathcal{T}}, f)^2 := \eta_{\mathcal{T}}^{\mathrm{hier}}(u_{\mathcal{T}})^2 + \mathrm{osc}_{\mathcal{T}}(f)^2. \tag{4.87b}$$

Theorem 4.64 (hierarchical estimator). *Suppose the coefficients* $\mathbf{A}$ *and* c *are discrete. The hierarchical estimator* (4.87) *is equivalent to the error, while its PDE indicator is locally equivalent to the discretized residual with constant* 1 *in the lower bound, so that*

$$\frac{1}{C_{\mathrm{lStb}}^{\star} C_d} \mathcal{E}_{\mathcal{T}}^{\mathrm{hier}} \lesssim \|\nabla(u - u_{\mathcal{T}})\|_{L^2(\Omega)} \lesssim C_{\mathrm{bOS}}^{\star} C_d C_{\mathrm{loc}} \mathcal{E}_{\mathcal{T}}^{\mathrm{hier}},$$

and, for all vertices $z \in \mathcal{V}$,

$$\eta_{\mathcal{T}}^{\mathrm{hier}}(u_{\mathcal{T}}, z) \leq \|P_{\mathcal{T}} R_{\mathcal{T}}\|_{H^{-1}(\omega_z)} \leq C_{\mathrm{bOS}}^{\star} \eta_{\mathcal{T}}^{\mathrm{hier}}(u_{\mathcal{T}}, z),$$

where $C_{\mathrm{lStb}}^{\star}$ *is the stability constant of* $P_{\mathcal{T}}$ *on stars from Lemma 4.5,* C_{loc} *from Corollary 4.6,* $C_d = \sqrt{2(d+1)}$*, and the hidden constants depend only on the error–residual relationship in Lemma 4.1.*

Proof. It suffices to verify the local equivalence for the PDE indicator; see Theorem 4.58 (vertex-indexed modified residual estimator). Its lower bound simply follows from (4.74): for all $i \in I_z$, we have

$$|\langle R_{\mathcal{T}}, w_i \rangle| = |\langle P_{\mathcal{T}} R_{\mathcal{T}}, w_i \rangle| \leq \|P_{\mathcal{T}} R_{\mathcal{T}}\|_{H^{-1}(\omega_z)} \|\nabla w_i\|_{L^2(\omega_z)}.$$

To show its upper bound, let $w \in H_0^1(\omega_z)$ and, with the help of Corollary 4.61 (projections as interpolation operators) and Lemma 4.60 (biorthogonal system), we derive

$$\begin{aligned}
\langle P_{\mathcal{T}} R_{\mathcal{T}}, w \rangle &= \sum_{i \in I_z} \langle R_{\mathcal{T}}, w_i \rangle \langle \ell_i, w \rangle \\
&\leq \sum_{i \in I_z} \frac{\langle R_{\mathcal{T}}, w_i \rangle}{\|\nabla w_i\|_{L^2(\omega_z)}} \|\nabla w_i\|_{L^2(\omega_z)} \|\ell_i\|_{H^{-1}(\omega_z)} \|\nabla w\|_{L^2(\omega_z)} \\
&\leq C_{\mathrm{bOS}}^{\star}\, \eta_{\mathcal{T}}^{\mathrm{hier}}(u_{\mathcal{T}}, z)\, \|\nabla w\|_{L^2(\omega_z)}
\end{aligned}$$

and the local equivalence is established. □

Remark 4.65 (different test functions). The hierarchical estimator (4.87) does not generalize the one in Bornemann *et al.* (1996) as it uses slightly different test functions for edges. The given framework, however, applies to their variant too; see Kreuzer and Veeser (2021, Section 4.1).

Alternative localization and residual splitting

Lemma 4.5 (localization of H^{-1}-norm) is not well suited to reducing or avoiding constants in the upper bounds. The following modification, however, allows this.

To this end, we replace the local spaces $H_0^1(\omega_z)$, $z \in \mathcal{V}$, with

$$\mathbb{W}_z := \begin{cases} \{w \in H^1(\omega_z) \mid \int_{\omega_z} w = 0\}, & \text{if } z \in \mathcal{V} \cap \Omega, \\ \{w \in H^1(\omega_z) \mid w = 0 \text{ on } \partial\omega_z \cap \partial\Omega\}, & \text{if } z \in \mathcal{V} \cap \partial\Omega, \end{cases}$$

endow them with the norm $\|\nabla \cdot\|_{L^2(\omega_z)}$, and let $\mathbb{W}_z^*$ denote the respective dual spaces endowed in turn with

$$\|\ell\|_{\mathbb{W}_z^*} := \sup\{\langle \ell, w \rangle \mid w \in \mathbb{W}_z,\ \|\nabla w\|_{L^2(\omega_z)} \leq 1\}. \tag{4.88}$$

Lemma 4.66 (alternative localization of H^{-1}-norm). *Let $\ell \in H^{-1}(\Omega)$ be any linear functional.*

(i) *If $\langle \ell, \phi_z \rangle = 0$ for all interior vertices $z \in \mathcal{V} \cap \Omega$, then*

$$\|\ell\|_{H^{-1}(\Omega)}^2 \leq (d+1) \sum_{z \in \mathcal{V}} \|\phi_z \ell\|_{\mathbb{W}_z^*}^2.$$

(ii) *We have*

$$\sum_{z \in \mathcal{V}} \|\phi_z \ell\|_{\mathbb{W}_z^*}^2 \leq (d+1) C_{\mathrm{loc}}^2 \|\ell\|_{H^{-1}(\Omega)}^2,$$

where C_{loc} is the constant in Lemma 4.5(i).

Proof. The proof is essentially a regrouping of the arguments in Lemma 4.5, where the constant C_{loc} in the stability bound (4.6) now arises in the proof of the lower bound from the following argument: we have

$$\|\phi_z \ell\|_{\mathbb{W}_z^*} \leq C_{\mathrm{loc}} \|\ell\|_{H^{-1}(\omega_z)}, \tag{4.89}$$

thanks to

$$\begin{aligned}\langle \phi_z \ell, w\rangle = \langle \ell, \phi_z w\rangle &\leq \|\ell\|_{H^{-1}(\omega_z)} \|\nabla(w\phi_z)\|_{L^2(\omega_z)} \\ &\leq C_{\mathrm{loc}} \|\ell\|_{H^{-1}(\omega_z)} \|\nabla w\|_{L^2(\omega_z)}\end{aligned}$$

for all $w \in \mathbb{W}_z$. □

The question arises whether the inequality (4.89) between the two local dual norms can be reversed. The following lemma reveals that this is only partially possible, covering discrete functionals as arguments.

Lemma 4.67 (partial equivalence for local dual norms). *If $z \in \mathcal{V} \cap \Omega$ is an interior vertex, the functional $\ell = \phi_z^{-1}$ satisfies*

$$\|\phi_z \ell\|_{\mathbb{W}_z^*} = 0 \quad \textit{and} \quad \|\ell\|_{H^{-1}(\omega_z)} > 0.$$

Furthermore, for any vertex $z \in \mathcal{V}$,

$$\|\ell\|_{H^{-1}(\omega_z)} \leq C_{\mathbb{F}} \|\phi_z \ell\|_{\mathbb{W}_z^*} \quad \textit{for all } \ell \in \mathbb{F}(\mathcal{T}_z),$$

where the constant $C_{\mathbb{F}}$ depends only on d, the shape regularity coefficient σ, and the degrees m_1 and m_2 of the discrete functionals.

Proof. [1] We show the claims on the functional $\ell = \phi_z^{-1}$ for an interior vertex $z \in \mathcal{V} \cap \Omega$. By the definition of $\mathbb{W}_z$, we have, for all $w \in \mathbb{W}_z$,

$$\langle \phi_z \ell, w\rangle = \int_{\omega_z} w = 0,$$

whence $\phi_z \ell \in \mathbb{W}_z^*$ with $\|\phi_z \ell\|_{\mathbb{W}_z^*} = 0$.

To verify that $\ell = \phi_z^{-1} \in H^{-1}(\omega_z)$, we write $d_z := \operatorname{dist}(\cdot, \partial\omega_z)$ for the distance function of the star boundary and shall use the weighted Poincaré inequality

$$\|w d_z^{-1}\|_{L^2(\omega_z)} \lesssim \|\nabla w\|_{L^2(\omega_z)} \quad \text{for all } w \in H_0^1(\omega_z),$$

which follows from the Hardy inequality; see Sacchi and Veeser (2006, Lemma 3.6). Consequently, exploiting $d_z \leq \phi_z$ on ω_z as well, we obtain, for all $w \in H_0^1(\Omega)$,

$$\langle \ell, w\rangle = \int_{\omega_z} (\phi_z^{-1} d_z)(w d_z^{-1}) \leq |\omega_z|^{1/2} \|w d_z^{-1}\|_{L^2(\omega_z)} \lesssim |\omega_z|^{1/2} \|\nabla w\|_{L^2(\omega_z)}.$$

This and $\langle \ell, \phi_z\rangle = |\omega_z|$ ensure $\ell \in H^{-1}(\omega_z)$ with $\|\ell\|_{H^{-1}(\omega_z)} > 0$.

[2] We start the proof of the asserted inequality by checking that $\|\cdot\|_{H^{-1}(\omega_z)}$ is a norm on $\mathbb{F}(\mathcal{T}_z)$. To this end, consider $q_F \in \mathbb{P}_{m_1}(F)$, $F \in \mathcal{F}_z$ and $q_T \in \mathbb{P}_{m_2}(T)$,

$T \in \mathcal{T}_z$ such that, for all $w \in H_0^1(\omega_z)$,

$$0 = \langle \ell, w \rangle := \sum_{F \in \mathcal{F}_z} \int_F q_F w + \sum_{T \in \mathcal{T}_z} \int_T q_T w.$$

We need to show $\ell = 0$. Testing with $w \in H_0^1(T)$, $T \in \mathcal{T}_z$, the fundamental lemma of the calculus of variations yields $q_T = 0$ for all $T \in \mathcal{T}_z$. Similarly, now testing with $w \in H_0^1(\omega_F)$, $F \in \mathcal{F}_z$, gives $q_F = 0$ for all $F \in \mathcal{T}_z$. Thus $\ell = 0$ holds.

[3] Next, we check that $\|\phi_z \cdot \|_{\mathbb{W}_z^*}$ is also a norm on $\mathbb{F}(\mathcal{T}_z)$. This time, consider $q_F \in \mathbb{P}_{m_1}(F)$, $F \in \mathcal{F}_z$ and $q_T \in \mathbb{P}_{m_2}(T)$, $T \in \mathcal{T}_z$ such that, for all $w \in \mathbb{W}_z$,

$$0 = \langle \phi_z \ell, w \rangle := \sum_{F \in \mathcal{F}_z} \int_F \phi_z q_F w + \sum_{T \in \mathcal{T}_z} \int_T \phi_z q_T w,$$

and again, we need to conclude $\ell = 0$. If $z \in \mathcal{V} \cap \partial\Omega$ is a boundary node, we obtain $\ell = 0$ by the arguments of the previous step. We are thus left with the case $z \in \mathcal{V} \cap \Omega$ of interior nodes. Given $w \in H^1(\omega_z)$, we set $c_w := \fint_{\omega_z} w$ and $c_\ell = |\omega_z|^{-1} \langle \phi_z \ell, 1 \rangle$, and observe

$$0 = \langle \phi_z \ell, w - c_w \rangle = \langle \phi_z \ell - c_\ell, w - c_w \rangle = \langle \phi_z \ell - c_\ell, w \rangle.$$

Hence, testing with $w \in H_0^1(T)$, $T \in \mathcal{T}_z$, we deduce $\phi_z q_T = c_\ell$ on each $T \in \mathcal{T}_z$. However, this is only possible if $c_\ell = 0$ and $q_T = 0$ for all $T \in \mathcal{T}_z$. Therefore, testing with $w \in H_0^1(\omega_F)$, $F \in \mathcal{F}_z$, yields $q_F = 0$ for all $F \in \mathcal{F}_z$, and $\ell = 0$ is established in general.

[4] To conclude the asserted inequality, note that $\mathbb{F}(\mathcal{T}_z)$ has finite dimension and, for fixed polynomial degrees m_1 and m_2, is invariant under continuous piecewise affine transformations. Furthermore, both norms scale in the same manner. We can therefore pass to reference stars and use the equivalence of norms in finite-dimensional spaces. Transforming the inequality back from the reference star then finishes the proof. □

The alternative localization entails that we need to adapt Lemma 4.35 (splitting of local residual norm). Relying on the local H^{-1}-stability of $P_\mathcal{T}$, Lemma 4.67 (partial equivalence for local dual norms) reveals that the adaptation has to be global.

Lemma 4.68 (alternative splitting). *Using the local norms $\| \cdot \|_{\mathbb{W}_z^*}$, $z \in \mathcal{V}$, the residual can be split into discretized and oscillatory residuals:*

$$\frac{1}{(C_{\mathrm{lStb}}^\star)^2 C_d^2 C_{\mathrm{loc}}^2} \sum_{z \in \mathcal{V}} \left(\|\phi_z P_\mathcal{T} R_\mathcal{T}\|_{\mathbb{W}_z^*}^2 + \|\phi_z (I - P_\mathcal{T}) R_\mathcal{T}\|_{\mathbb{W}_z^*}^2 \right)$$
$$\leq \|R_\mathcal{T}\|_{H^{-1}(\Omega)}^2 \leq C_d^2 \sum_{z \in \mathcal{V}} \left(\|\phi_z P_\mathcal{T} R_\mathcal{T}\|_{\mathbb{W}_z^*}^2 + \|\phi_z (I - P_\mathcal{T}) R_\mathcal{T}\|_{\mathbb{W}_z^*}^2 \right),$$

where $C^{\star}_{\mathrm{lStb}}$ *is the stability constant of* $P_{\mathcal{T}}$ *on stars from Lemma 4.28, and* $C_d = \sqrt{2(d+1)}$.

Proof. Combine the localization in Lemma 4.66 with the proof of Lemma 4.35, replacing the local norm $\|\cdot\|_{H^{-1}(\omega_z)}$ in most places, but apply (4.89) before using the local H^{-1}-stability of $P_{\mathcal{T}}$. □

An estimator based on flux equilibration

Estimators based on flux equilibration have been designed with the goal to obtain constant 1 in the upper bound. The principal obstruction that computation can access only a finite-dimensional part of infinite-dimensional objects such as the residual norm is overcome by means of the *Prager–Synge theorem*. Realizations of this approach can be found, for example, in Ainsworth (2010), Braess, Pillwein and Schöberl (2009), Ern, Smears and Vohralík (2017) and Luce and Wohlmuth (2004).

The definition of the PDE indicator needs some preparation. Let $d \in \{2,3\}$, as in the aforementioned works, and let $z \in \mathcal{V}$ be a vertex. Given the operator $\pi_z \colon \{\phi_z \ell \mid \ell \in H^{-1}(\Omega)\} \to \mathbb{W}_z^*$ defined by

$$\pi_z(\phi_z \ell) := \begin{cases} \phi_z \ell - \dfrac{\langle \ell, \phi_z \rangle}{|\omega_z|}, & \text{if } z \in \mathcal{V} \cap \Omega, \\ \phi_z \ell, & \text{if } z \in \mathcal{V} \cap \partial\Omega, \end{cases}$$

and

$$\gamma_z := \begin{cases} \partial\omega_z, & \text{if } z \in \mathcal{V} \cap \Omega, \\ \partial\omega_z \setminus \partial\Omega, & \text{if } z \in \mathcal{V} \cap \partial\Omega, \end{cases}$$

we introduce the local space $\mathbb{D}_z \neq \emptyset$,

$$\mathbb{D}_z := \{\boldsymbol{\xi} \in L^2(\omega_z; \mathbb{R}^d) \mid \operatorname{div} \boldsymbol{\xi} = \pi_z(\phi_z P_{\mathcal{T}} R_{\mathcal{T}}) \text{ and } \boldsymbol{\xi} \cdot \boldsymbol{n}_F = 0 \text{ on } F \text{ for all } F \subseteq \gamma_z\},$$

and its discretization

$$\mathbb{D}_z(\mathcal{T}) := \{\boldsymbol{\xi} \in \mathbb{D}_z \mid \boldsymbol{\xi} \in RTN_m(T) \text{ for all } T \in \mathcal{T}_z\},$$

with the Raviart–Thomas–Nédélec elements

$$RTN_m(T) = \{\boldsymbol{\xi} \colon T \to \mathbb{R}^d \mid \boldsymbol{\xi}(x) = \boldsymbol{q}(x) + q(x)x \text{ with } \boldsymbol{q} \in (\mathbb{P}_m)^d, q \in \mathbb{P}_m\}$$

of order $m := \max\{m_1, m_2\} + 1$. Given the Galerkin approximation $u_{\mathcal{T}}$ from (4.1), the PDE indicator is then given by

$$\eta_{\mathcal{T}}^{\mathrm{feq}}(u_{\mathcal{T}})^2 := \sum_{z \in \mathcal{V}} \eta_{\mathcal{T}}^{\mathrm{feq}}(u_{\mathcal{T}})^2 \quad \text{with} \quad \eta_{\mathcal{T}}^{\mathrm{feq}}(u_{\mathcal{T}}) := \min_{\boldsymbol{\xi} \in \mathbb{D}_z(\mathcal{T})} \|\boldsymbol{\xi}\|_{L^2(\omega_z)} \tag{4.90a}$$

and the total estimator by

$$\mathcal{E}_{\mathcal{T}}^{\mathrm{feq}} := \mathcal{E}_{\mathcal{T}}^{\mathrm{feq}}(u_{\mathcal{T}}, f)^2 := \eta_{\mathcal{T}}^{\mathrm{feq}}(u_{\mathcal{T}})^2 + \|\phi_z(f - P_{\mathcal{T}} f)\|^2_{\mathbb{W}_z^*}. \tag{4.90b}$$

Note that the local PDE indicators $\eta_{\mathcal{T}}^{\mathrm{feq}}(u_{\mathcal{T}}, z)$ are computable up to machine precision.

Theorem 4.69 (estimator based on flux equilibration). *Suppose that the coefficients $\mathbf{A}$ and c are discrete and that $d \in \{2, 3\}$. The estimator* (4.90) *based on flux equilibration is equivalent to the error, while its PDE indicator is locally equivalent to the discretized residual with constant* 1 *in the upper bound for the $\|\cdot\|_{\mathbb{W}_z^*}$-norm, so that*

$$\frac{C_{\mathbb{D}}}{C_{\mathrm{lStb}}^{\star} C_d C_{\mathrm{loc}}} \mathcal{E}_{\mathcal{T}}^{\mathrm{feq}} \lesssim \|\nabla(u - u_{\mathcal{T}})\|_{L^2(\Omega)} \lesssim C_d \mathcal{E}_{\mathcal{T}}^{\mathrm{feq}},$$

and, for all vertices $z \in \mathcal{V}$,

$$C_{\mathbb{D}} \eta_{\mathcal{T}}^{\mathrm{feq}}(u_{\mathcal{T}}, z) \le \|\phi_z P_{\mathcal{T}} R_{\mathcal{T}}\|_{\mathbb{W}_z^*} \le \eta_{\mathcal{T}}^{\mathrm{feq}}(u_{\mathcal{T}}, z)$$

as well as

$$\frac{C_{\mathbb{D}}}{C_{\mathrm{loc}}} \eta_{\mathcal{T}}^{\mathrm{feq}}(u_{\mathcal{T}}, z) \le \|P_{\mathcal{T}} R_{\mathcal{T}}\|_{H^{-1}(\omega_z)} \le C_{\mathbb{F}}\, \eta_{\mathcal{T}}^{\mathrm{feq}}(u_{\mathcal{T}}, z),$$

where $C_{\mathbb{D}}$ depends on d and the shape regularity coefficient σ, $C_{\mathrm{lStb}}^{\star}$ is the stability constant of $P_{\mathcal{T}}$ on stars from Lemma 4.28, C_{loc} comes from Lemma 4.5, $C_d = \sqrt{2(d+1)}$, $C_{\mathbb{F}}$ comes from Lemma 4.67, and the hidden constants depend only on the error–residual relationship in Lemma 4.1.

Proof. [1] We start by verifying the local equivalence for the $\|\cdot\|_{\mathbb{W}_z^*}$-norm. Let $z \in \mathcal{V}$ be any vertex. The Prager–Synge theorem on the star ω_z implies

$$\|\phi_z P_{\mathcal{T}} R_{\mathcal{T}}\|_{\mathbb{W}_z^*} = \|\pi_z(\phi_z P_{\mathcal{T}} R_{\mathcal{T}})\|_{\mathbb{W}_z^*} = \min_{\xi \in \mathbb{D}_z} \|\xi\|_{L^2(\omega_z)};$$

see e.g. Verfürth (2013, Proposition 1.40). Hence the upper bound with constant 1 readily follows the inclusion $\mathbb{D}_z(\mathcal{T}) \subset \mathbb{D}_z$, while the lower bound is a consequence of the non-trivial inequality

$$C_{\mathbb{D}} \min_{\xi \in \mathbb{D}_z(\mathcal{T})} \|\xi\|_{L^2(\omega_z)} \le \min_{\xi \in \mathbb{D}_z} \|\xi\|_{L^2(\omega_z)},$$

where $C_{\mathbb{D}}$ depends only on d and the shape regularity coefficient σ; see e.g. Braess *et al.* (2009, Theorem 7) and Ern *et al.* (2017, Theorem 1.1).

[2] We verify the local equivalence for the $\|\cdot\|_{H^{-1}(\omega_z)}$-norm. On the one hand, combining the first equivalence with (4.89), we obtain

$$C_{\mathbb{D}} \eta_{\mathcal{T}}^{\mathrm{feq}}(u_{\mathcal{T}}, z) \le \|\phi_z P_{\mathcal{T}} R_{\mathcal{T}}\|_{\mathbb{W}_z^*} \le C_{\mathrm{loc}} \|P_{\mathcal{T}} R_{\mathcal{T}}\|_{H^{-1}(\omega_z)}.$$

On the other hand, using Lemma 4.67 instead of (4.89) yields

$$\|P_{\mathcal{T}} R_{\mathcal{T}}\|_{H^{-1}(\omega_z)} \le C_{\mathbb{F}} \|\phi_z P_{\mathcal{T}} R_{\mathcal{T}}\|_{\mathbb{W}_z^*} \le C_{\mathbb{F}} \eta_{\mathcal{T}}^{\mathrm{feq}}(u_{\mathcal{T}}, z),$$

and the equivalence for the $\|\cdot\|_{H^{-1}(\omega_z)}$ is verified, too.

3 The global bounds follow by combining Lemmas 4.1 (error and residual), 4.68 (alternative splitting) and 4.37 (data oscillation reduction for discrete coefficients), as well as the first local equivalence. □

Remark 4.70 (improved upper bound). Applying the Prager–Synge theorem on Ω, we can improve the upper bound in Theorem 4.69 to

$$\|\nabla(u-u_{\mathcal{T}})\|_{L^2(\Omega)} \lesssim \|\xi_\Omega\|_{L^2(\Omega)} + \sqrt{d+1}\left(\sum_{z\in\mathcal{V}} \|\pi_z(\phi_z(P_{\mathcal{T}}f-f))\|^2_{\mathbb{W}_z^*}\right)^{1/2}, \quad (4.91)$$

with $\xi_\Omega := \sum_{z\in\mathcal{V}} \xi_z$, where $\xi_z := \arg\min_{\xi\in\mathbb{D}_z(\mathcal{T})} \|\xi\|_{L^2(\omega_z)}$ are the minimizing vector fields associated with the PDE indicators, extended by 0 off ω_z.

To see this, we derive, thanks to the partial orthogonality (4.4) of the residual and Lemma 4.37 (data oscillation reduction for discrete coefficients),

$$\begin{aligned}\operatorname{div}\xi_\Omega &= \sum_{z\in\mathcal{V}} \pi_z(\phi_z P_{\mathcal{T}} R_{\mathcal{T}}) = \sum_{z\in\mathcal{V}} \pi_z(\phi_z R_{\mathcal{T}}) + \sum_{z\in\mathcal{V}} \pi_z(\phi_z(P_{\mathcal{T}}R_{\mathcal{T}} - R_{\mathcal{T}}))\\ &= \sum_{z\in\mathcal{V}} \phi_z R_{\mathcal{T}} + \sum_{z\in\mathcal{V}} \pi_z(\phi_z(P_{\mathcal{T}}f-f)) = R_{\mathcal{T}} + \delta_{\mathcal{T}},\end{aligned}$$

with $\delta_{\mathcal{T}} := \sum_{z\in\mathcal{V}} \pi_z(\phi_z(P_{\mathcal{T}}f-f))$. Hence

$$\|\nabla(u-u_{\mathcal{T}})\|_{L^2(\Omega)} \lesssim \|R_{\mathcal{T}}\|_{H^{-1}(\Omega)} \leq \|R_{\mathcal{T}}+\delta_{\mathcal{T}}\|_{H^{-1}(\Omega)} + \|\delta_{\mathcal{T}}\|_{H^{-1}(\Omega)},$$

inserting ξ_Ω in the Prager–Synge theorem on Ω and Lemma 4.66 (alternative localization of H^{-1}-norm) establish the claimed bound.

In view of the bound (4.91), the alternative local PDE indicators $\|\phi_z^{1/2}\xi_\Omega\|_{L^2(\omega_z)}$, $z\in\mathcal{V}$, may be used in an adaptive context. Note, however, that this alternative does not necessarily strengthen the link with the local residual, as the definition of ξ_Ω suggests an increased overlapping in the lower bound.

4.10. *Other boundary conditions*

This section illustrates that the preceding analysis of homogeneous Dirichlet conditions can be adapted to other boundary conditions. In particular, we discuss

- *Robin and Neumann boundary conditions*, as an example for variationally formulated boundary conditions,
- the *pure Neumann problem*, with its global solvability constraint,
- *non-homogeneous Dirichlet boundary conditions*, formulated in an essential manner.

Mixed boundary conditions, suitably discretized, give rise to *a posteriori* error estimators combining in a straightforward manner the indicators of, for instance, the first and third of the above groups. We therefore omit further details of such a setting.

Robin and Neumann boundary conditions

The *Robin* bilinear form in (2.13) is coercive and continuous in $\mathbb{V} := H^1(\Omega)$ provided its coefficient $p \geq p_0$ on an open subset of $\partial\Omega$ for some constant $p_0 > 0$, according to the norm equivalence (2.31). Consequently, (2.12) admits a unique solution $u \in \mathbb{V}$. If $\mathbb{V}_\mathcal{T} = \mathbb{S}^{n,0}_\mathcal{T}$ is the subspace of $\mathbb{V}$ of continuous piecewise polynomial functions of degree $\leq n$, then the Galerkin counterpart of (4.1) reads

$$u_\mathcal{T} \in \mathbb{V}_\mathcal{T}: \quad \mathcal{B}[u_\mathcal{T}, v] = \ell(v) \quad \text{for all } v \in \mathbb{V}_\mathcal{T},$$

with $\ell = f + g\delta_{\partial\Omega} \in \mathbb{V}^*$; see (2.13). Its residual $R_\mathcal{T} \in \mathbb{V}^*$ is defined as

$$\langle R_\mathcal{T}, w\rangle := \ell(w) - \mathcal{B}[u_\mathcal{T}, w], \quad w \in \mathbb{V},$$

and $\|R_\mathcal{T}\|_{\mathbb{V}^*}$ is equivalent to the error $\|u - u_\mathcal{T}\|_{H^1(\Omega)}$ due to Lemma 4.1 (error and residual), whose proof easily extends to $\mathbb{V}$.

The global norm $\|R_\mathcal{T}\|_{\mathbb{V}^*}$ also localizes to all stars ω_z because Galerkin orthogonality $\langle R_\mathcal{T}, \phi_z\rangle = 0$ is now valid also for boundary vertices $z \in \mathcal{V} \cap \partial\Omega$. Indeed, the proof of Lemma 4.5 (localization of H^{-1}-norm) extends with minor modifications, where the local spaces for boundary vertices $z \in \mathcal{V} \cap \partial\Omega$ are now $\{v \in H^1(\omega_z) \mid v = 0 \text{ on } \partial\omega_z \setminus \partial\Omega\}$. Also, the proof of Lemma 4.66 (alternative localization of H^{-1}-norm) is easily modified, using the local space $\left\{v \in H^1(\omega_z) \mid \int_{\omega_z} v = 0\right\}$ at the boundary, too.

The next key step is the construction of a projection $P_\mathcal{T}\colon \mathbb{V}^* \to \mathbb{F}_\mathcal{T}$ that mimics the projection operator $P_\mathcal{T}$ of Section 4.4. For that purpose, the space of discrete functionals $\mathbb{F}_\mathcal{T}$ has to include boundary face Dirac masses $q_F\delta_F$ with densities $q_F \in \mathbb{P}_{m_1}(F)$ for $F \subset \partial\Omega$. Consequently, g can be approximated on $\partial\Omega$ similarly to the forcing f in Ω, while the coefficient p is at play like the coefficient c. Indeed, considering for simplicity only the case of discrete coefficients $(\boldsymbol{A}, c, p)$, the condition $m_1 \geq n_p + n$ arises in addition to those in Remark 4.14. With these caveats, the tools developed in Sections 4.3, 4.4 and 4.5 give rise to a suitably adapted projection $P_\mathcal{T}$ to split the residual $R_\mathcal{T}$ into a discretized residual $P_\mathcal{T} R_\mathcal{T}$ and an oscillatory residual $(f - P_\mathcal{T} f) + (g\delta_{\partial\Omega} - P_\mathcal{T}(g\delta_{\partial\Omega}))$. Here $g\delta_{\partial\Omega} - P_\mathcal{T}(g\delta_{\partial\Omega})$ is supported only on $\partial\Omega$ and therefore contributes only to the oscillation indicators based upon the aforementioned new local spaces for boundary stars. This modified oscillation $\operatorname{osc}^{\mathrm{Rob}}_\mathcal{T}(\mathcal{D})$ with $\mathcal{D} = (\boldsymbol{A}, c, p, f, g)$ can be combined with any of the presented PDE indicators, but we focus on residual estimation. In fact, the new discrete residual $P_\mathcal{T} R_\mathcal{T}$ leads to a definition of the PDE estimator $\eta^{\mathrm{Rob}}_\mathcal{T}(u_\mathcal{T})$ as in (4.52), but with additional contributions related to the boundary faces. Given any boundary face F of $\mathcal{T}$, such a contribution reads

$$h_F \| [[\boldsymbol{A}\nabla_\mathcal{T} u_\mathcal{T}]] \cdot \boldsymbol{n}_F + p\, u_\mathcal{T} - P_F g\|^2_{L^2(F)}$$

and measures the discretized Robin residual. Combining as usual the PDE estimator $\eta^{\mathrm{Rob}}_\mathcal{T}(u_\mathcal{T})$ and oscillation $\operatorname{osc}^{\mathrm{Rob}}_\mathcal{T}(\ell)$ yields the total estimator $\mathcal{E}^{\mathrm{Rob}}_\mathcal{T}(u_\mathcal{T}, \ell)$, whence the following variant of Theorem 4.45 (modified residual estimator) follows: for discrete coefficients $(\boldsymbol{A}, c, p)$, the H^1-error and $\mathcal{E}_{\mathcal{T}}^{\text{Rob}}(u_{\mathcal{T}}, \ell)$ are equivalent, that is,

$$C_L \mathcal{E}_{\mathcal{T}}^{\text{Rob}}(u_{\mathcal{T}}, \ell) \leq \|u - u_{\mathcal{T}}\|_{H^1(\Omega)} \leq C_U \mathcal{E}_{\mathcal{T}}^{\text{Rob}}(u_{\mathcal{T}}, \ell).$$

The estimates in Section 4.8 for corrections and estimator reduction extend as well.

Pure Neumann problem

Neumann conditions are already covered by the previous section, except for the case of the pure Neumann problem with $p = 0$ on $\partial\Omega$ in (2.12) requiring, as key novelty, the solvability constraint $\ell(1_\Omega) = 0$, that is, the right-hand side applied to the constant function equal to 1 gives 0. For such problems, unique exact and discrete solutions exist provided we choose $\mathbb{V}$ to be the subspace of $H^1(\Omega)$ of functions with zero mean value, and $\mathbb{V}_{\mathcal{T}}$ to be its natural finite element counterpart of degree n.

The residual $R_{\mathcal{T}}$ is defined on all $H^1(\Omega)$ and satisfies $\langle R_{\mathcal{T}}, 1_\Omega \rangle = 0$. Combining this fact with Lemma 2.3 (second Poincaré inequality) and $\inf_{c\in\mathbb{R}} \|v - c\|_{L^2(\Omega)} = \|w\|_{L^2(\Omega)}$ with $w = v - \fint_\Omega v \in \mathbb{V}$, we derive

$$\begin{aligned}\|R_{\mathcal{T}}\|_{\mathbb{V}^*} &= \sup_{w\in\mathbb{V}} \frac{\langle R_{\mathcal{T}}, w\rangle}{\|\nabla w\|_{L^2(\Omega)}} \approx \sup_{w\in\mathbb{V}} \frac{\langle R_{\mathcal{T}}, w\rangle}{\|w\|_{H^1(\Omega)}} \\ &= \sup_{v\in H^1(\Omega)} \frac{\langle R_{\mathcal{T}}, v\rangle}{\|v\|_{H^1(\Omega)}} = \|R_{\mathcal{T}}\|_{H^1(\Omega)^*}.\end{aligned}$$

Consequently, localizing $\|R_{\mathcal{T}}\|_{H^1(\Omega)^*}$ as in the previous section, we can derive *a posteriori* error estimators with suitable contributions from the boundary $\partial\Omega$.

However, the projection $P_{\mathcal{T}}$ from the previous section cannot be used to generate discrete data in some auxiliary problem because $\langle \ell, 1_\Omega\rangle = 0$ does not imply $\langle P_{\mathcal{T}}\ell, 1_\Omega\rangle = 0$ in general. Further, a simple modification like $P_{\mathcal{T}}\ell - \langle P_{\mathcal{T}}\ell, 1_\Omega\rangle\langle 1_\Omega, 1_\Omega\rangle^{-1} 1_\Omega$ with a global correction destroys the crucial local approximation properties.

To address this issue, we modify the projection $P_{\mathcal{T}}$ such that the new projection $\widetilde{P}_{\mathcal{T}}$ enforces locally $\langle \widetilde{P}_{\mathcal{T}}\ell, 1_\Omega\rangle = \langle \ell, 1_\Omega\rangle$ in the spirit of the construction of the Lagrange multiplier in Fierro and Veeser (2003). To this end, recall that $P_{\mathcal{T}}$ is now defined on $H^1(\Omega)$, and its range, the discrete functionals $\mathbb{F}(\mathcal{T})$, also includes boundary face Dirac masses, and that the first localization involves the local spaces $\mathbb{V}_z := \{v \in H^1(\omega_z) \mid v = 0 \text{ on } \partial\omega_z \setminus \partial\Omega\}$, $z \in \mathcal{V}$. Given $\ell \in H^1(\Omega)^*$, set

$$\widetilde{P}_{\mathcal{T}}\ell := \sum_{z\in\mathcal{V}} \phi_z \widetilde{P}_z \ell \quad \text{with} \quad \widetilde{P}_z\ell := P_{\mathcal{T}}\ell - \frac{\langle P_{\mathcal{T}}\ell - \ell, \phi_z\rangle}{\int_{\omega_z}\phi_z} 1_{\omega_z}. \tag{4.92}$$

Lemma 4.71 (new projection). *The operator* (4.92) *is linear, local, and satisfies*

$$\langle \widetilde{P}_z\ell, \phi_z\rangle = \langle \ell, \phi_z\rangle \quad \textit{for all } z \in \mathcal{V} \quad \textit{and} \quad \langle \widetilde{P}_{\mathcal{T}}\ell, 1\rangle = \langle \ell, 1\rangle.$$

Furthermore, $\widetilde{P}_z$ provides near-best approximation in $\mathbb{F}(\mathcal{T})|_{\mathbb{V}_z}$ and

$$\|\ell - \widetilde{P}_\mathcal{T}\ell\|^2_{H^1(\Omega)^*} \le C_{\text{loc}} \sum_{z\in\mathcal{V}} \|\ell - \widetilde{P}_z\ell\|^2_{\mathbb{V}_z^*}.$$

Proof. 1 We start with the algebraic properties. By the definition of $\widetilde{P}_z$, we have the local relationships $\langle \ell - \widetilde{P}_z\ell, \phi_z\rangle = 0$, i.e. $\langle \phi_z(\ell - \widetilde{P}_z\ell), 1\rangle = 0$ for all vertices $z \in \mathcal{V}$. Summing over all vertices immediately yields the global $\langle \ell - \widetilde{P}_\mathcal{T}\ell, 1\rangle = 0$.

2 To show that $\widetilde{P}_z$ is near-best approximating in $\mathbb{F}(\mathcal{T})|_{\mathbb{V}_z}$, we bound its error in terms of that of $P_\mathcal{T}$. The triangle inequality readily gives

$$\|\ell - \widetilde{P}_z\ell\|_{\mathbb{V}_z^*} \le \|\ell - P_\mathcal{T}\ell\|_{\mathbb{V}_z^*} + \left\| \frac{\langle P_\mathcal{T}\ell - \ell, \phi_z\rangle}{\int_\omega \phi_z} 1_{\omega_z} \right\|_{\mathbb{V}_z^*},$$

while a variant of Lemma 2.2 (first Poincaré inequality) and the properties of ϕ_z deliver

$$\begin{aligned}\left\| \frac{\langle P_\mathcal{T}\ell - \ell, \phi_z\rangle}{\int_\omega \phi_z} 1_{\omega_z} \right\|_{\mathbb{V}_z^*} &\lesssim |\omega_z|^{-1} h_z \|\langle P_\mathcal{T}\ell - \ell, \phi_z\rangle 1_{\omega_z}\|_{L^2(\omega_z)} \\ &\lesssim |\omega_z|^{-1/2} h_z \|\ell - P_\mathcal{T}\ell\|_{\mathbb{V}_z^*} \|\nabla\phi_z\|_{L^2(\omega_z)} \\ &\lesssim \|\ell - P_\mathcal{T}\ell\|_{\mathbb{V}_z^*}.\end{aligned}$$

Hence the error of $\widetilde{P}_z$ is dominated by that of $P_\mathcal{T}$,

$$\|\ell - \widetilde{P}_z\ell\|_{\mathbb{V}_z^*} \lesssim \|\ell - P_\mathcal{T}\ell\|_{\mathbb{V}_z^*}, \tag{4.93}$$

and the near-best approximation of $\widetilde{P}_z$ follows from Corollary 4.31 (local near-best approximation), adapted to the setting at hand.

3 It remains to prove the claimed inequality. Given $w \in H^1(\Omega)$, the definition of $\widetilde{P}_\mathcal{T}$ and the first step yield the following identity:

$$\begin{aligned}\langle \ell - \widetilde{P}_\mathcal{T}\ell, w\rangle &= \sum_{z\in\mathcal{V}} \langle \ell, w\phi_z\rangle - \langle \phi_z \widetilde{P}_z\ell, w\rangle \\ &= \sum_{z\in\mathcal{V}} \langle \ell - \widetilde{P}_z\ell, w\phi_z\rangle = \sum_{z\in\mathcal{V}} \langle \ell - \widetilde{P}_z\ell, (w - c_z)\phi_z\rangle,\end{aligned}$$

with $c_z = \fint_{\omega_z} w$. Proceeding as in Lemma 4.5 (localization of H^{-1}-norm) establishes the desired inequality and concludes the proof. □

The operator $\widetilde{P}_\mathcal{T}$ possesses additional enhanced global properties, which are not needed here. Lemma 4.71 and (4.93) allow us to solve auxiliary pure Neumann problems with discrete data $\widetilde{P}_\mathcal{T}\ell$, with the option of replacing the restrictions of $P_\mathcal{T}$ in the local indicators with $\widetilde{P}_z$.

Non-homogeneous Dirichlet boundary conditions
Let $\mathbb{V} := H^1(\Omega)$ and $\mathbb{V}_\mathcal{T} := \mathbb{S}^{n,0}_\mathcal{T}$ be the subspace of $\mathbb{V}$ of continuous piecewise polynomials of degree $\leq n$. Given Dirichlet boundary data $g \in H^{1/2}(\Gamma)$, where $\Gamma := \partial\Omega$ for simplicity, recall that $u \in \mathbb{V}(g) = \{v \in \mathbb{V} \mid v = g \text{ on } \Gamma\}$ satisfies (2.10). Let $g_\mathcal{T} \in \mathbb{S}^{n,0}_\mathcal{T}$ be a continuous finite element approximation of g on Γ and let $\mathbb{V}_\mathcal{T}(g_\mathcal{T})$ be the subspace of $\mathbb{V}_\mathcal{T}$ of discrete functions with trace $g_\mathcal{T}$. The Galerkin approximation of u satisfies

$$u_\mathcal{T} \in \mathbb{V}_\mathcal{T}(g_\mathcal{T}): \quad \mathcal{B}[u_\mathcal{T}, v] = \langle f, v\rangle \quad \text{for all } v \in \mathbb{V}_\mathcal{T}(0).$$

The error $e_\mathcal{T} = u - u_\mathcal{T}$ obviously satisfies Galerkin orthogonality

$$\mathcal{B}[e_\mathcal{T}, v] = 0 \quad \text{for all } v \in \mathbb{V}_\mathcal{T}(0),$$

but in general $e_\mathcal{T} = g - g_\mathcal{T} \neq 0$ on Γ. We follow Sacchi and Veeser (2006) to derive *a posteriori* bounds of $\|e_\mathcal{T}\|_{H^1(\Omega)}$ using minimal regularity $g \in H^{1/2}(\Gamma)$.

We start with an orthogonal decomposition of the error $e_\mathcal{T}$ arising from the two equations of the problem. Let $R_G = R_G(u_\mathcal{T}, f) \in H^{-1}(\Omega)$ be the Galerkin residual already introduced in Section 4.1, namely

$$\langle R_G, v\rangle = \langle f, v\rangle - \mathcal{B}[u_\mathcal{T}, v] \quad \text{for all } v \in \mathbb{V}(0) = H^1_0(\Omega),$$

and define the Galerkin error e_G as its representation in $H^1_0(\Omega)$:

$$e_G \in H^1_0(\Omega): \quad \mathcal{B}[e_G, v] = \langle R_G, v\rangle \quad \text{for all } v \in \mathbb{V}(0).$$

Furthermore, let $R_D = R_D(g) = g - g_\mathcal{T} \in H^{1/2}(\Gamma)$ be the *Dirichlet residual*, represented by the *Dirichlet error* e_D defined by

$$e_D \in \mathbb{V}(g - I_\mathcal{T} g): \quad \mathcal{B}[e_D, v] = 0 \quad \text{for all } v \in \mathbb{V}(0).$$

Then $e_\mathcal{T} = e_G + e_D$ and the orthogonality $\mathcal{B}[e_D, e_G] = 0$ yields

$$|||e_\mathcal{T}|||^2_\Omega = |||e_G|||^2_\Omega + |||e_D|||^2_\Omega,$$

while the derivation for homogeneous Dirichlet conditions readily provides

$$|||e_G|||_\Omega \approx \|\nabla e_G\|_{L^2(\Omega)} \approx \mathcal{E}_\mathcal{T}(u_\mathcal{T}, f),$$

where the Galerkin estimator $\mathcal{E}_\mathcal{T}(u_\mathcal{T}, f)$ is defined by (4.52), or any other estimator from Section 4.9. It thus remains to clarify whether $|||e_\mathcal{T}|||_\Omega$ is definite in the sense of $|||e_\mathcal{T}|||_\Omega = 0 \Rightarrow e_\mathcal{T} = 0$ and to derive suitable lower and upper bounds for $|||e_D|||_\Omega$.

To this end, we need to be more specific about the choice of $g_\mathcal{T}$. Let $g_\mathcal{T} = I_\mathcal{T} g$ be the Scott–Zhang quasi-interpolant of g, which is defined locally using boundary values of g exclusively (Scott and Zhang 1990, Brenner and Scott 2008) and satisfies

$$v \in \mathbb{V}_\mathcal{T}|_\Gamma \Rightarrow I_\mathcal{T} v = v \quad \text{on } \Gamma, \qquad \text{(invariance)} \tag{4.94a}$$

$$\|I_\mathcal{T} v\|_{L^2(\Gamma)} \lesssim \|v\|_{L^2(\Gamma)} \quad \text{for all } v \in \mathbb{V}. \qquad \text{(stability)} \tag{4.94b}$$

These two properties ensure a variant of the equivalence $\|\cdot\|_{H^1(\Omega)} \approx \|\nabla \cdot\|_{L^2(\Omega)}$ for functions with zero trace on Γ.

Lemma 4.72 (equivalence for vanishing discretized trace). *There exists a constant C depending only on the shape regularity of $\mathbb{T}$ and Ω such that*

$$\|v\|_{H^1(\Omega)} \le C\|\nabla v\|_{L^2(\Omega)} \quad \textit{for all } v \in \mathbb{V} \textit{ with } I_{\mathcal{T}} v = 0 \textit{ on } \Gamma.$$

Proof. Note that the core of the claimed inequality amounts to a variant of the first Poincaré inequality. In view of the norm equivalence (2.31), it suffices to prove that $\|v\|_{L^2(\Gamma)} \lesssim \|\nabla v\|_{L^2(\Omega)}$. Letting $\bar{v}_\Omega := \fint_\Omega v$, and using (4.94a) yields $v = v - I_{\mathcal{T}} v = (v - \bar{v}_\Omega) - I_{\mathcal{T}}(v - \bar{v}_\Omega)$ on Γ. Consequently, (4.94b) implies

$$\|v\|_{L^2(\Gamma)} \lesssim \|v - \bar{v}_\Omega\|_{L^2(\Gamma)} \lesssim \|v - \bar{v}_\Omega\|_{H^1(\Omega)} \lesssim \|\nabla v\|_{L^2(\Omega)}, \tag{4.95}$$

because of Lemma 2.4 (traces) and Lemma 2.3 (second Poincaré inequality). □

Observing $I_{\mathcal{T}} e_{\mathcal{T}} = I_{\mathcal{T}} g - I_{\mathcal{T}}^2 g = 0$ on Γ, we can apply Lemma 4.72 to get

$$\|e_{\mathcal{T}}\|_{H^1(\Omega)} \le C\|\nabla e_{\mathcal{T}}\|_{L^2(\Omega)} \le \frac{C}{\alpha_1} |||e_{\mathcal{T}}|||_\Omega \le \frac{C \max\{\alpha_2, \|c\|_{L^\infty(\Omega)}\}}{\alpha_1} \|e_{\mathcal{T}}\|_{H^1(\Omega)},$$

establishing in particular that $|||e_{\mathcal{T}}|||_\Omega$ is definite. In the same vein, we derive

$$|||e_D|||_\Omega \approx \|e_D\|_{H^1(\Omega)} \approx \|\nabla e_D\|_{L^2(\Omega)}$$

for the Dirichlet error.

With the intent to achieve directly computable bounds for the Dirichlet error, we next establish the equivalence $\|e_D\|_{H^1(\Omega)} \approx \|g - I_{\mathcal{T}} g\|_{H^{1/2}(\Gamma)}$, where the *intrinsic* $H^{1/2}$-norm combines the $L^2(\Gamma)$-norm with the seminorm

$$|v|^2_{H^{1/2}(\Gamma)} = \int_\Gamma \int_\Gamma \frac{|v(x) - v(y)|^2}{|x-y|^d} \, \mathrm{d}x \, \mathrm{d}y.$$

This equivalence follows with the help of the trace and extension theorems for $H^{1/2}(\Gamma)$; see e.g. Hackbusch (1992, Theorem 6.2.40). In fact, on the one hand, that trace theorem immediately gives $\|g - I_{\mathcal{T}} g\|_{H^{1/2}(\Gamma)} \lesssim \|e_D\|_{H^1(\Omega)}$. On the other hand, let $\chi \in H^1(\Omega)$ denote the extension of $g - I_{\mathcal{T}} g$ from Hackbusch (1992, Theorem 6.2.40). Then

$$\|e_D\|_{H^1(\Omega)} \lesssim |||e_D|||_\Omega \le |||\chi|||_\Omega \lesssim \|\chi\|_{H^1(\Omega)} \lesssim \|g - I_{\mathcal{T}} g\|_{H^{1/2}(\Gamma)},$$

where the second inequality is thanks to $\mathcal{B}[e_D, e_D - \chi] = 0$.

We are left with the issue that the $H^{1/2}(\Gamma)$-seminorm is *non-local*. To handle this delicate matter, we invoke its localization due to Faermann (2000, 2002), that is,

$$|v|^2_{H^{1/2}(\Gamma)} \le \sum_{F \in \mathcal{F}_\Gamma} \left(\int_F \int_{\omega_F} \frac{|v(x) - v(y)|^2}{|x-y|^d} \, \mathrm{d}x \, \mathrm{d}y + \frac{C}{h_F} \|v\|^2_{L^2(F)} \right),$$

where $F \in \mathcal{F}_\Gamma$ is a generic face of $\mathcal{T}$ lying on Γ and ω_F is the patch on Γ associated with F. The last term seems problematic. However, applied to $v = g - I_{\mathcal{T}} g$, we can mimic the steps of (4.95) with local variants of (4.94), but using in the last step the second Poincaré inequality in $H^{1/2}$ (see e.g. Sacchi and Veeser 2006, Lemma 3.2):

$$\|v\|^2_{L^2(F)} = \|v - I_{\mathcal{T}} v\|^2_{L^2(F)} \lesssim \|v - \bar{v}_F\|^2_{L^2(\omega_F)} \lesssim h_F |v|^2_{H^{1/2}(\omega_F)},$$

where $\bar{v}_F$ is the mean value of v on ω_F. Note that this bound also means that the L^2-part in $\|g - I_{\mathcal{T}} g\|_{H^{1/2}(\Gamma)}$ is (locally) controlled by its seminorm. Altogether, this leads to defining the *Dirichlet oscillation* with the following local indicators:

$$\begin{aligned} \operatorname{osc}_{\mathcal{T}}(g)^2_{1/2} &:= \sum_{F \in \mathcal{F}_\Gamma} \operatorname{osc}_{\mathcal{T}}(g, F)^2_{1/2}, \\ \operatorname{osc}_{\mathcal{T}}(g, F)^2_{1/2} &:= \int_{\omega_F} \int_{\omega_F} \frac{|(g - I_{\mathcal{T}} g)(x) - (g - I_{\mathcal{T}} g)(y)|^2}{|x - y|^d} \,\mathrm{d}x \,\mathrm{d}y. \end{aligned} \tag{4.96}$$

We observe that $\operatorname{osc}_{\mathcal{T}}(g, F)$ is a double singular integral but computationally accessible, for instance, by using suitable quadrature provided g is continuous (Sacchi and Veeser 2006, Section 4.1).

Proposition 4.73 (Dirichlet oscillation). *There exist constants $D_1 \geq D_2 > 0$ depending on the shape regularity of $\mathbb{T}$ and geometry of Γ, such that*

$$D_2 \operatorname{osc}_{\mathcal{T}}(g)_{1/2} \leq \|\nabla e_D\|_{L^2(\Omega)} \leq D_1 \operatorname{osc}_{\mathcal{T}}(g)_{1/2}.$$

Proof. The preceding derivation verifies the upper bound. For the lower bound, note that for any $v \in H^1(\Omega)$ such that $v = g - I_{\mathcal{T}} g$ on Γ

$$\begin{aligned} \operatorname{osc}_{\mathcal{T}}(g)^2_{1/2} &= \sum_{F \in \mathcal{F}_\Gamma} \operatorname{osc}_{\mathcal{T}}(g, F)^2_{1/2} \\ &\leq \sum_{F \in \mathcal{F}_\Gamma} \int_{\omega_F} \int_{\Gamma} \frac{|v(x) - v(y)|^2}{|x - y|^d} \,\mathrm{d}x \,\mathrm{d}y \\ &\lesssim \int_{\Gamma} \int_{\Gamma} \frac{|v(x) - v(y)|^2}{|x - y|^d} \,\mathrm{d}x \,\mathrm{d}y = |v|^2_{H^{1/2}(\Gamma)}, \end{aligned}$$

because the patches ω_F, $F \in \mathcal{F}_\Gamma$, possess a uniform overlapping property due to shape regularity of $\mathbb{T}$. Applying this to $v = e_D$ finishes the proof. □

For suitable settings, local lower *a posteriori* estimates for the Dirichlet error e_D can be derived; see Sacchi and Veeser (2006, Theorem 3.2).

Combining the Dirichlet oscillation with some Galerkin estimator $\mathcal{E}_{\mathcal{T}}(u_{\mathcal{T}}, f)$ by

$$\mathcal{E}^{\mathrm{Dir}}_{\mathcal{T}}(u_{\mathcal{T}}, f, g)^2 := \mathcal{E}_{\mathcal{T}}(u_{\mathcal{T}}, f)^2 + \operatorname{osc}_{\mathcal{T}}(g)^2_{1/2},$$

the preceding discussion is summarized by the following result.

Theorem 4.74 (estimators for general Dirichlet condition). *If Assumption 4.44 (discrete coefficients and discrete functionals) is valid, then there exist constants $C_L \leq C_U$ depending on $(\boldsymbol{A}, c)$, Ω, Γ, and the shape regularity of $\mathbb{T}$ such that*

$$C_L \mathcal{E}^{\mathrm{Dir}}_{\mathcal{T}}(u_{\mathcal{T}}, f, g) \leq \|\nabla(u - u_{\mathcal{T}})\|_{L^2(\Omega)} \leq C_U \mathcal{E}^{\mathrm{Dir}}_{\mathcal{T}}(u_{\mathcal{T}}, f, g).$$

5. Convergence of AFEM for coercive problems

In this section we consider the coercive problem (2.5) with the intent to design and analyse three AFEMs in increasing order of complexity and applicability, depending on properties of data $\mathcal{D}$. Our basic regularity assumption on data reads $\mathcal{D} = (\boldsymbol{A}, c, f) \in \mathbb{D}$, where

$$\mathbb{D} := L^\infty(\Omega; \mathbb{R}^{d\times d}) \times L^\infty(\Omega) \times H^{-1}(\Omega). \tag{5.1}$$

We approximate $\mathcal{D}$ with discrete data $\widehat{\mathcal{D}} = (\widehat{\boldsymbol{A}}, \widehat{c}, \widehat{f}) \in \mathbb{D}_{\widehat{\mathcal{T}}}$, where

$$\mathbb{D}_{\widehat{\mathcal{T}}} := \big[\mathbb{S}^{n-1,-1}_{\widehat{\mathcal{T}}}\big]^{d\times d} \times \mathbb{S}^{n-1,-1}_{\widehat{\mathcal{T}}} \times \mathbb{F}_{\widehat{\mathcal{T}}} \tag{5.2}$$

is subordinate to a partition $\widehat{\mathcal{T}} \in \mathbb{T}$. We will often assume that *data is discrete*, meaning precisely that $\mathcal{D} = \widehat{\mathcal{D}}$.

We start with the *one-step* AFEM, hereafter called GALERKIN, which is the standard SEMR loop

$$\mathsf{SOLVE} \longrightarrow \mathsf{ESTIMATE} \longrightarrow \mathsf{MARK} \longrightarrow \mathsf{REFINE}$$

introduced by Dörfler (1996) and further developed by Morin, Nochetto and Siebert (2000, 2002) and Cascón *et al.* (2008). This is the simplest algorithm in that it requires data $\mathcal{D} = (\boldsymbol{A}, c, f)$ to be discrete, but it is a building block for the other two methods. After reviewing a few crucial properties of error and estimator in Section 5.1, we fully discuss GALERKIN in Section 5.2.

The second algorithm is the *one-step* AFEM *with switch*, which still assumes the coefficients $(\boldsymbol{A}, c)$ to be discrete but allows for general forcing $f \in H^{-1}(\Omega)$. This is a new contribution of this survey that, similarly to Kreuzer *et al.* (2024), exploits the structure of the error estimator $\mathcal{E}_{\mathcal{T}}(u_{\mathcal{T}}, f)$ of Section 4,

$$\mathcal{E}_{\mathcal{T}}(u_{\mathcal{T}}, f)^2 = \eta_{\mathcal{T}}(u_{\mathcal{T}})^2 + \operatorname{osc}_{\mathcal{T}}(f)^2_{-1},$$

and its equivalence to the energy error. The PDE estimator $\eta_{\mathcal{T}}(u_{\mathcal{T}})$ relies on the discrete forcing $P_{\mathcal{T}} f \in \mathbb{F}_{\mathcal{T}}$ and is fully computable, whereas the data oscillation $\operatorname{osc}_{\mathcal{T}}(f)_{-1}$ encodes the infinite-dimensional nature of f and could be estimated in important cases of practical interest further discussed in Section 7.3. The quantity $\operatorname{osc}_{\mathcal{T}}(f)_{-1}$ measures the deviation of f from being discrete and may dictate the pre-asymptotic regime of AFEM. Therefore $\operatorname{osc}_{\mathcal{T}}(f)_{-1}$ must be handled separately from $\eta_{\mathcal{T}}(u_{\mathcal{T}})$; hence the name of the new method, hereafter called AFEM-SW. Assuming that $\operatorname{osc}_{\mathcal{T}}(f)_{-1}$ is computable, the module

$$[\widehat{\mathcal{T}}] = \mathsf{DATA}(\mathcal{T}, f, \tau)$$

deals with $\operatorname{osc}_{\mathcal{T}}(f)_{-1}$ whenever it is large relative to $\mathcal{E}_{\mathcal{T}}(u_{\mathcal{T}}, f)$. In fact, it creates an admissible refinement $\widehat{\mathcal{T}}$ of the input mesh $\mathcal{T}$ such that $\operatorname{osc}_{\widehat{\mathcal{T}}}(f)_{-1}$ is below the desired tolerance τ, i.e. $\operatorname{osc}_{\widehat{\mathcal{T}}}(f)_{-1} \le \tau$. We explain the role of data oscillation for error analysis, design AFEM-SW and prove its linear convergence in Section 5.3.

The third algorithm deals with variable data $\mathcal{D}$ and various degrees of regularity of $\mathcal{D}$, and is able to handle discontinuous coefficients $(\boldsymbol{A}, c)$ not aligned with admissible meshes $\mathcal{T} \in \mathbb{T}$ emanating from $\mathcal{T}_0$. To handle the multiplicative structure of $(\boldsymbol{A}, c)$ in the model problem (2.5), we consider the following *two-step* AFEM.

Algorithm 5.1 (AFEM-TS). Given an initial mesh $\mathcal{T}_0$, an initial tolerance ε_0, and a parameter ω sufficiently small to be determined later, iterate

AFEM-TS($\mathcal{T}_0, \varepsilon_0, \omega$)
  $k = 0$
  $[\widehat{\mathcal{T}}_k, \widehat{\mathcal{D}}_k]$ = DATA($\mathcal{T}_k, \mathcal{D}, \omega\, \varepsilon_k$)
  $[\mathcal{T}_{k+1}, u_{k+1}]$ = GALERKIN($\widehat{\mathcal{T}}_k, \widehat{\mathcal{D}}_k, \varepsilon_k$)
  $\varepsilon_{k+1} = \frac{1}{2}\varepsilon_k;\ k \leftarrow k+1$

This structure was first proposed by Stevenson (2008) and further explored by Bonito *et al.* (2013*b*), Cohen *et al.* (2012), Bonito *et al.* (2016), Bonito, Cascón, Morin and Nochetto (2013*a*) and Bonito and Devaud (2015). The three components of data $\mathcal{D} = (\boldsymbol{A}, c, f) \in \mathbb{D}$ are first approximated by discrete data $\widehat{\mathcal{D}} = (\widehat{\boldsymbol{A}}, \widehat{c}, \widehat{f}) \in \mathbb{D}_{\widehat{\mathcal{T}}}$, as defined in (5.1) and (5.2), within the module

$$[\widehat{\mathcal{T}}, \widehat{\mathcal{D}}] = \mathsf{DATA}(\mathcal{T}, \mathcal{D}, \tau)$$

to accuracy $\tau = \omega\varepsilon$ significantly smaller than ε. This is achieved by an algorithm similar to Algorithm 3.18 (greedy algorithm), which is fully discussed along with applications to $\mathcal{D}$ in Section 7. The resulting admissible refinement $\widehat{\mathcal{T}}$ of $\mathcal{T}$ and discrete data $\widehat{\mathcal{D}}$ over $\widehat{\mathcal{T}}$ are next taken by GALERKIN to reduce the PDE error to the desired tolerance ε, namely the module

$$[\mathcal{T}, u_{\mathcal{T}}] = \mathsf{GALERKIN}(\widehat{\mathcal{T}}, \widehat{\mathcal{D}}, \varepsilon)$$

constructs a refinement $\mathcal{T}$ of $\widehat{\mathcal{T}}$ with discrete data $\widehat{\mathcal{D}}$ over $\widehat{\mathcal{T}}$ such that $\eta_{\mathcal{T}}(u_{\mathcal{T}}) \le \varepsilon$. We point out that if the data is discrete, i.e. $\mathcal{D} = \widehat{\mathcal{D}}$, then DATA is skipped and AFEM-TS reduces to GALERKIN. We tackle AFEM-TS in Section 5.4, where we prove a perturbation estimate with respect to $\mathcal{D}$ and next discuss convergence properties of AFEM-TS. We will extend this approach to discontinuous FEMs in Section 9 and to mixed FEMs for (2.5) as well as the Stokes system (2.14) in Section 10.

5.1. Properties of error and estimator

We follow Cascón, Kreuzer, Nochetto and Siebert (2008) and summarize some basic properties of GALERKIN that emanate from the symmetry of the differential operator (i.e. of $\boldsymbol{A}$) and features of the modules. In doing this, any explicit constant or hidden constant in $\lesssim$ will depend only on the uniform shape regularity of $\mathbb{T}$, the dimension d, the polynomial degree n and the (global) eigenvalues of $\boldsymbol{A}$, but not on a specific grid $\mathcal{T} \in \mathbb{T}$, unless explicitly stated.

We recall that the bilinear form $\mathcal{B}$ in (2.8) with continuous coefficients $(\boldsymbol{A}, c)$ is symmetric, coercive and continuous in the space $H_0^1(\Omega)$ (see (2.30)), namely $|\!|\!| v |\!|\!|_\Omega = \mathcal{B}[v,v]^{1/2}$ is a norm equivalent to $|v|_{H_0^1(\Omega)}$ with equivalence constants $0 < c_\mathcal{B} \le C_\mathcal{B}$

$$c_\mathcal{B} |v|^2_{H_0^1(\Omega)} \le |\!|\!| v |\!|\!|^2_\Omega \le C_\mathcal{B} |v|^2_{H_0^1(\Omega)} \quad \text{for all } v \in H_0^1(\Omega). \tag{5.3}$$

The module DATA approximates $(\boldsymbol{A}, c)$ over a mesh $\mathcal{T}$ by piecewise polynomial coefficients $(\widehat{\boldsymbol{A}}, \widehat{c})$ obeying side constraints so that the corresponding perturbed bilinear form $\widehat{\mathcal{B}}$ still defines a uniform scalar product in $H_0^1(\Omega)$,

$$|\!|\!| v |\!|\!|_\Omega = \widehat{\mathcal{B}}[v,v]^{1/2} \quad \text{for all } v \in H_0^1(\Omega), \tag{5.4}$$

which satisfies (5.3) with constants $0 < c_{\widehat{\mathcal{B}}} \le C_{\widehat{\mathcal{B}}}$ independent of $\mathcal{T}$. We hope this slight abuse of notation will not create confusion because we will always refer to the energy norm in (5.4) when dealing with $\widehat{\mathcal{B}}$. We let $\widehat{u} = u(\widehat{\mathcal{D}}) \in H_0^1(\Omega)$ denote the solution of (2.7) with coefficients $(\widehat{\boldsymbol{A}}, \widehat{c})$ and forcing function either $\widehat{f} = f \in H^{-1}(\Omega)$ or its projection $\widehat{f} = P_\mathcal{T} f \in \mathbb{F}_\mathcal{T}$ defined in (4.35), namely

$$\widehat{\mathcal{B}}[\widehat{u}, v] = \langle \widehat{f}, v \rangle \quad \text{for all } v \in H_0^1(\Omega). \tag{5.5}$$

In what follows, we will often compare discrete functions on different meshes. Given $\mathcal{T} \in \mathbb{T}$, we let $\mathcal{T}_* \in \mathbb{T}$ denote an admissible refinement of $\mathcal{T}$, and write

$$\mathcal{T} \le \mathcal{T}_* \quad \Leftrightarrow \quad \mathbb{T}(\mathcal{T}) \subset \mathbb{T}(\mathcal{T}_*), \tag{5.6}$$

in the sense that the supporting tree of $\mathcal{T}$ is contained in the tree of $\mathcal{T}_*$. For any $\mathcal{T}_* \ge \mathcal{T}$, we have the following crucial property.

Lemma 5.2 (Pythagoras). *Let $\mathcal{T}_* \ge \mathcal{T} \ge \widehat{\mathcal{T}}$ and let $\widehat{u} \in H_0^1(\Omega)$ be the solution of* (5.5) *with discrete coefficients $(\widehat{\boldsymbol{A}}, \widehat{c})$ over $\widehat{\mathcal{T}}$. The corresponding Galerkin solutions $u_\mathcal{T} \in \mathbb{V}_\mathcal{T}$ and $u_{\mathcal{T}_*} \in \mathbb{V}_{\mathcal{T}_*}$ with coefficients $(\widehat{\boldsymbol{A}}, \widehat{c})$ and forcing $f \in H^{-1}(\Omega)$ satisfy the orthogonality property*

$$|\!|\!| \widehat{u} - v_\mathcal{T} |\!|\!|^2_\Omega = |\!|\!| \widehat{u} - u_{\mathcal{T}_*} |\!|\!|^2_\Omega + |\!|\!| u_{\mathcal{T}_*} - v_\mathcal{T} |\!|\!|^2_\Omega \quad \textit{for all } v_\mathcal{T} \in \mathbb{V}_\mathcal{T}. \tag{5.7}$$

Proof. Exploit the nestedness property $\mathbb{V}_\mathcal{T} \subset \mathbb{V}_{\mathcal{T}_*}$ and the Galerkin orthogonality property $\widehat{\mathcal{B}}[\widehat{u} - u_{\mathcal{T}_*}, v_\mathcal{T} - u_{\mathcal{T}_*}] = 0$ in $\mathbb{V}_{\mathcal{T}_*}$ for the scalar product induced by $\widehat{\mathcal{B}}$. □

Property (5.7) is very restrictive: it relies on space nestedness and is valid exclusively for the energy norm. However, it is instrumental to the subsequent analysis in the energy norm or the equivalent norm $|\cdot|_{H_0^1(\Omega)}$, but it does not extend to other, perhaps more practical, norms such as the maximum norm. This is an important open problem and a serious limitation of this theory.

We recall that the residual *a posteriori* error analysis of Section 4 relies on the projection operator $P_\mathcal{T}\colon H^{-1}(\Omega) \to \mathbb{F}_\mathcal{T}$, with element and face components

$P_\mathcal{T} f|_T = P_T f$ for $T \in \mathcal{T}$ and $P_\mathcal{T} f|_F = P_F f$. The *full local error indicator*

$$\mathcal{E}_\mathcal{T}(u_\mathcal{T}, f, T)^2 = \eta_\mathcal{T}(u_\mathcal{T}, T)^2 + \operatorname{osc}_\mathcal{T}(f, T)_{-1}^2$$

splits into a computable *PDE error indicator* with discrete coefficients $(\widehat{\boldsymbol{A}}, \widehat{c})$,

$$\eta_\mathcal{T}(v, T)^2 = h_T^2 \, \|r(v)\|_T^2 + h_T \, \|j(v)\|_{\partial T}^2 \quad \text{for all } T \in \mathcal{T}, \tag{5.8}$$

where the *interior* and *jump residuals* are given by

$$\begin{aligned} r(v)|_T &= P_T f + \operatorname{div}(\widehat{\boldsymbol{A}} \nabla v) - \widehat{c} v && \text{for all } T \in \mathcal{T}, \\ j(v)|_F &= [\widehat{\boldsymbol{A}} \nabla v] \cdot \boldsymbol{n} \,|_F - P_F f && \text{for all } F \in \mathcal{F}, \end{aligned} \tag{5.9}$$

and $j(v)|_F = 0$ for boundary faces F, and *data oscillation*

$$\operatorname{osc}_\mathcal{T}(f, T)_{-1}^2 = \|f - P_\mathcal{T} f\|_{H^{-1}(\omega_T)}^2 \quad \text{for all } T \in \mathcal{T}, \tag{5.10}$$

where ω_T is the patch associated with T. The corresponding global quantities are

$$\begin{gathered} \mathcal{E}_\mathcal{T}(u_\mathcal{T}, f)^2 = \sum_{T \in \mathcal{T}} \mathcal{E}(u_\mathcal{T}, f, T)^2, \\ \eta_\mathcal{T}(u_\mathcal{T})^2 = \sum_{T \in \mathcal{T}} \eta_\mathcal{T}(u_\mathcal{T}, T)^2, \quad \operatorname{osc}_\mathcal{T}(f)_{-1}^2 = \sum_{T \in \mathcal{T}} \operatorname{osc}_\mathcal{T}(f, T)_{-1}^2, \end{gathered} \tag{5.11}$$

and have the following *a posteriori* error estimates proved in Theorem 4.45 (modified residual estimator) for the H_0^1-norm.

Proposition 5.3 (*a posteriori* error estimates). *Let $\widehat{u} \in H_0^1(\Omega)$ be the solution of* (5.5) *with discrete coefficients $(\widehat{\boldsymbol{A}}, \widehat{c})$ over $\mathcal{T} \in \mathbb{T}$ but general forcing $f \in H^{-1}(\Omega)$. Then there exist constants $0 < C_L \le C_U$, depending on the shape regularity of $\mathbb{T}$, such that the Galerkin solution $u_\mathcal{T} \in \mathbb{V}_\mathcal{T}$ satisfies*

$$C_L \, \mathcal{E}_\mathcal{T}(u_\mathcal{T}, f) \le |\widehat{u} - u_\mathcal{T}|_{H_0^1(\Omega)} \le C_U \, \mathcal{E}_\mathcal{T}(u_\mathcal{T}, f). \tag{5.12}$$

Moreover, if $|\!|\!| \cdot |\!|\!|_\Omega$ stands for the energy norm in (5.4) with equivalence constants $c_{\widehat{\mathcal{B}}} \le C_{\widehat{\mathcal{B}}}$ satisfying (5.3), then (5.12) yields

$$C_2 \, \mathcal{E}_\mathcal{T}(u_\mathcal{T}, f) \le |\!|\!| \widehat{u} - u_\mathcal{T} |\!|\!|_\Omega \le C_1 \, \mathcal{E}_\mathcal{T}(u_\mathcal{T}, f), \tag{5.13}$$

with $C_1 = \sqrt{C_{\widehat{\mathcal{B}}}} C_U$ and $C_2 = \sqrt{c_{\widehat{\mathcal{B}}}} C_L$.

There is a fundamental difference between (5.12) and earlier versions of *a posteriori* error estimates, going back to the seminal paper of Babuška and Miller (1987); see also Ainsworth and Oden (2000), Braess (2007), Nochetto *et al.* (2009) and Verfürth (2013). It is about the role of data oscillation $\operatorname{osc}_\mathcal{T}(f)_{-1}$, which is now dominated by the error $|\widehat{u} - u_\mathcal{T}|_{H_0^1(\Omega)}$ and does not spoil the lower bound. This is due to the fact that $\operatorname{osc}_\mathcal{T}(f)_{-1}$ is evaluated in the natural space $H^{-1}(\Omega)$ and quantifies the discrepancy between f and a suitable projection $P_\mathcal{T} f$ which gives rise to a quasi-best local approximation of f. We refer to Nochetto *et al.* (2009) and Kreuzer and Veeser (2021) for a discussion of data oscillation.

Suppose now that we have two conforming meshes $\mathcal{T}, \mathcal{T}_* \in \mathbb{T}$ with $\mathcal{T}_* \geq \mathcal{T}$. Let

$$\mathcal{R} := \mathcal{R}_{\mathcal{T}\to\mathcal{T}_*} := \mathcal{T}\setminus\mathcal{T}_* \tag{5.14}$$

be the subset of refined elements of $\mathcal{T}$, namely those elements in $\mathcal{T}$ that are no longer in $\mathcal{T}_*$. We stress that the upper bound in (5.12) cannot be local due to the non-local nature of the error $|\widehat{u} - u_{\mathcal{T}}|_{H_0^1(\Omega)}$. However, in view of Theorem 4.48 (upper bound for corrections), the following remarkable local upper bound for Galerkin solutions $u_{\mathcal{T}} \in V_{\mathcal{T}}, u_{\mathcal{T}_*} \in V_{\mathcal{T}_*}$ holds:

$$|\!|\!| u_{\mathcal{T}} - u_{\mathcal{T}_*} |\!|\!|_\Omega \leq C_1 \mathcal{E}_{\mathcal{T}}(u_{\mathcal{T}}, f, \mathcal{R}), \tag{5.15}$$

where for $\mathcal{S} \subset \mathcal{T}$,

$$\mathcal{E}_{\mathcal{T}}(u_{\mathcal{T}}, f, \mathcal{S}) := \left(\sum_{T \in \mathcal{S}} \mathcal{E}_{\mathcal{T}}(u_{\mathcal{T}}, f, T)^2 \right)^{1/2}$$

is the error estimator restricted to $\mathcal{S}$. Consequently, only the elements where $\mathcal{T}$ and $\mathcal{T}_*$ differ, namely the set $\mathcal{R}$, account for the discrepancy between $u_{\mathcal{T}}$ and $u_{\mathcal{T}_*}$. This turns out to be consistent with (5.13) because $\mathcal{T}$ has to be refined everywhere to get to $\widehat{u}$, whence $\mathcal{R} = \mathcal{T}$.

In contrast to the upper bound in (5.12), the corresponding lower bound is local according to Theorem 4.45 (modified residual estimator). This is due to the local nature of the PDE (2.5). However, when comparing $u_{\mathcal{T}}$ and $u_{\mathcal{T}_*}$, this bound is not valid unless the *interior vertex property* (given in Definition 4.50) is satisfied (Morin *et al.* 2000); in fact, we present a counterexample later in Example 5.7 taken from Morin *et al.* (2000).

The interior vertex property is valid upon enforcing a fixed number b of bisections ($b = 3, 6$ for $d = 2, 3$). An immediate consequence, proved in Theorem 4.51 (lower bound for corrections), is the *discrete* lower *a posteriori* bound for piecewise constant diffusion coefficient $\boldsymbol{A}$ and reaction coefficient $c = 0$ on $\mathcal{T}_0$,

$$C_{L,1}\, \mathcal{E}_{\mathcal{T}}(u_{\mathcal{T}}, f, \mathcal{M}) \leq |\!|\!| u_{\mathcal{T}} - u_{\mathcal{T}_*} |\!|\!|_\Omega + C_{L,2}\, \mathrm{osc}_{\mathcal{T}}(f, \omega(\mathcal{M}))_{-1}, \tag{5.16}$$

where $\omega(\mathcal{M}) := \cup\{\omega_T \mid T \subset \mathcal{M}\}$ is the union of all patches of elements in $\mathcal{M}$ and $\mathrm{osc}_{\mathcal{T}}(f, \omega(\mathcal{M}))_{-1}^2 = \sum_{T \in \omega(\mathcal{M})} \mathrm{osc}_{\mathcal{T}}(f, T)_{-1}^2$; we refer to Morin *et al.* (2000, 2002). We stress that if $f = P_{\mathcal{T}} f$ is discrete, then $\mathrm{osc}_{\mathcal{T}}(f)_{-1} = 0$ and (5.16) reduces to

$$C_2\, \eta_{\mathcal{T}}(u_{\mathcal{T}}, \mathcal{M}) \leq |\!|\!| u_{\mathcal{T}} - u_{\mathcal{T}_*} |\!|\!|_\Omega. \tag{5.17}$$

One serious difficulty in dealing with AFEM is that we have access to the energy error $|\!|\!| \widehat{u} - u_{\mathcal{T}} |\!|\!|_\Omega$, or equivalently to $|\widehat{u} - u_{\mathcal{T}}|_{H_0^1(\Omega)}$, only through the full error estimator $\mathcal{E}_{\mathcal{T}}(u_{\mathcal{T}}, f)$. Lemma 5.2 (Pythagoras) implies monotonicity of the energy error with respect to $\mathcal{T}$, namely, for $\mathcal{T}_* \geq \mathcal{T}$,

$$|\!|\!| \widehat{u} - u_{\mathcal{T}_*} |\!|\!|_\Omega \leq |\!|\!| \widehat{u} - u_{\mathcal{T}} |\!|\!|_\Omega.$$

However, the PDE estimator $\eta_{\mathcal{T}}(u_{\mathcal{T}})$ fails to be monotone for fixed discrete coefficients $(\widehat{\boldsymbol{A}}, \widehat{c})$ because it depends on the discrete solution $u_{\mathcal{T}} \in \mathbb{V}_{\mathcal{T}}$ that changes with the mesh. The following estimate, proved in Proposition 4.56 (estimator reduction), quantifies the deviation of $\eta_{\mathcal{T}}(u_{\mathcal{T}})$ from monotonicity: there exists $\lambda > 0$ such that for any $\delta > 0$, $v \in \mathbb{V}_{\mathcal{T}}$ and $v_* \in \mathbb{V}_{\mathcal{T}_*}$,

$$\eta_{\mathcal{T}_*}(v_*, \mathcal{T}_*)^2 \leq (1+\delta)\big(\eta_{\mathcal{T}}(v,\mathcal{T})^2 - \lambda\, \eta_{\mathcal{T}}(v,\mathcal{M})^2\big) \\ + 2(1+\delta^{-1})\, C_{\mathrm{Lip}} \left(\|v_* - v\|_{\Omega}^2 + \sum_{T \in \mathcal{T}_*} \|P_{\mathcal{T}} f - P_{\mathcal{T}_*} f\|_{H^{-1}(\omega_T)}^2 \right),$$

where C_{Lip} depends on $(\boldsymbol{A}, c)$ and the shape regularity constant of $\mathbb{T}$. We refer to Cascón *et al.* (2008) and Morin *et al.* (2008) for the case $P_{\mathcal{T}} f = P_{\mathcal{T}_*} f = f \in L^2(\Omega)$.

5.2. *Convergence for discrete data: one-step* AFEM

We now present the four basic modules of GALERKIN, the one-step AFEM within Algorithm 5.1 (AFEM-TS), namely

$$\text{SOLVE} \longrightarrow \text{ESTIMATE} \longrightarrow \text{MARK} \longrightarrow \text{REFINE}, \tag{5.18}$$

discuss their main properties, and prove a contraction property between consecutive iterates of GALERKIN. According to Algorithm 5.1, given discrete data $\widehat{\mathcal{D}}$ over a conforming mesh $\widehat{\mathcal{T}}$, created by DATA, and a desired tolerance $\varepsilon > 0$, the module

$$[\mathcal{T}, u_{\mathcal{T}}] - \text{GALERKIN}(\widehat{\mathcal{T}}, \widehat{\mathcal{D}}, \varepsilon) \tag{5.19}$$

stops the loop (5.18) as soon as the error tolerance ε is reached, i.e. as soon as

$$\eta_{\mathcal{T}}(u_{\mathcal{T}}) \leq \varepsilon. \tag{5.20}$$

Since the data never changes within GALERKIN and is always discrete, we assume in this section that $\mathcal{D} \in \mathbb{D}_{\mathcal{T}}$ and do not use the hat symbol to indicate quantities defined using the (discrete) data.

5.2.1. *Modules of* GALERKIN

Module SOLVE. If $\mathcal{T} \in \mathbb{T}$ is a conforming refinement of $\mathcal{T}_0$, and $\mathbb{V}_{\mathcal{T}}$ is the finite element space of C^0 piecewise polynomials of degree $\leq n$, then

$$[u_{\mathcal{T}}] = \text{SOLVE}(\mathcal{T})$$

determines the Galerkin FEM solution *exactly*, namely without algebraic error,

$$u_{\mathcal{T}} \in \mathbb{V}_{\mathcal{T}}: \quad \mathcal{B}[u_{\mathcal{T}}, v] = \int_{\Omega} \nabla v \cdot \boldsymbol{A} \nabla u_{\mathcal{T}} + c v u = \langle f, v \rangle, \tag{5.21}$$

where $f \in H^{-1}(\Omega)$. However, if $f \in \mathbb{F}_{\mathcal{T}}$ is discrete as defined in (4.35), then

$$\langle f, v \rangle = \sum_{T \in \mathcal{T}} \int_T q_T v + \sum_{F \in \mathcal{F}} \int_F q_F v \quad \text{for all } v \in \mathbb{V}_{\mathcal{T}}.$$

The assumption of exact solvability is made for simplicity. The algebraic error committed in solving (5.21) by iterative solvers can be accommodated within the forthcoming theory. We refer to Stevenson (2007) and Daniel and Vohralík (2023) for details about how to relate the algebraic and PDE errors.

Module ESTIMATE. Given a conforming mesh $\mathcal{T} \in \mathbb{T}$ and the Galerkin solution $u_{\mathcal{T}} \in \mathbb{V}_{\mathcal{T}}$, the output of

$$[\{\eta_{\mathcal{T}}(u_{\mathcal{T}}, T), \operatorname{osc}_{\mathcal{T}}(f, T)_{-1}\}_{T \in \mathcal{T}}] = \mathsf{ESTIMATE}(u_{\mathcal{T}}, \mathcal{T}, \mathcal{D})$$

gives the element error indicators $\eta_{\mathcal{T}}(u_{\mathcal{T}}, T)$ defined in (5.8) with the discrete data $\mathcal{D}$, namely

$$\eta_{\mathcal{T}}(u_{\mathcal{T}}, T)^2 = h_T^2 \|r(u_{\mathcal{T}})\|_T^2 + h_T \|j(u_{\mathcal{T}})\|_{\partial T}^2, \quad T \in \mathcal{T},$$

and element data oscillation $\operatorname{osc}_{\mathcal{T}}(f, T)_{-1}$ defined in (5.10), namely

$$\operatorname{osc}_{\mathcal{T}}(f, T)_{-1} = \|f - P_{\mathcal{T}} f\|_{H^{-1}(\omega_T)}.$$

We observe that for discrete forcing $f = P_{\mathcal{T}} f$, global data oscillation vanishes, that is,

$$\operatorname{osc}_{\mathcal{T}}(f)_{-1} = \|f - P_{\mathcal{T}} f\|_{H^{-1}(\Omega)} = 0; \tag{5.22}$$

this property is always valid within GALERKIN. In this case, the output of ESTIMATE reduces to just the PDE error indicators. Given $\mathcal{S} \subset \mathcal{T}$, we denote

$$\eta_{\mathcal{T}}(v, \mathcal{S})^2 := \sum_{T \in \mathcal{S}} \eta_{\mathcal{T}}(v, T)^2, \quad \eta_{\mathcal{T}}(v) = \eta_{\mathcal{T}}(v, \mathcal{T}), \quad v \in \mathbb{V}_{\mathcal{T}}.$$

Module MARK. Given $\mathcal{T} \in \mathbb{T}$, the Galerkin solution $u_{\mathcal{T}} \in \mathbb{V}_{\mathcal{T}}$, and element error indicators $\{\eta_{\mathcal{T}}(u_{\mathcal{T}}, T)\}_{T \in \mathcal{T}}$, the module MARK selects elements for refinement using *Dörfler marking* (or bulk chasing) (Dörfler 1996, Morin *et al.* 2000, Nochetto *et al.* 2009, Nochetto and Veeser 2012), that is, given a parameter $\theta \in (0, 1]$, the output $\mathcal{M}$ of

$$[\mathcal{M}] = \mathsf{MARK}(\{\eta_{\mathcal{T}}(u_{\mathcal{T}}, T)\}_{T \in \mathcal{T}}, \mathcal{T}, \theta)$$

satisfies

$$\eta_{\mathcal{T}}(u_{\mathcal{T}}, \mathcal{M}) \geq \theta\, \eta_{\mathcal{T}}(u_{\mathcal{T}}, \mathcal{T}). \tag{5.23}$$

This marking guarantees that $\mathcal{M}$ contains a substantial part of the total (or bulk) error, hence its name. The choice of $\mathcal{M}$ does not have to be minimal at this stage, that is, the marked elements $T \in \mathcal{M}$ do not necessarily have to be those with largest indicators.

Module REFINE. Let $b \in \mathbb{N}$ be the number of desired bisections per marked element. Given $\mathcal{T} \in \mathbb{T}$ and a subset $\mathcal{M}$ of marked elements, the output $\mathcal{T}_* \in \mathbb{T}$ of

$$[\mathcal{T}_*] = \mathsf{REFINE}(\mathcal{T}, \mathcal{M})$$

is the smallest admissible refinement $\mathcal{T}_*$ of $\mathcal{T}$ such that all elements of $\mathcal{M}$ are bisected at least b times. Therefore we have $h_{\mathcal{T}_*} \leq h_{\mathcal{T}}$ and the strict reduction property

$$h_{\mathcal{T}_*}|_T \le 2^{-b/d} h_{\mathcal{T}}|_T \quad \text{for all } T \in \mathcal{M}, \tag{5.24}$$

where $h_{\mathcal{T}} : \Omega \to \mathbb{R}^+$ is a piecewise constant mesh size function that coincides with $h_T = |T|^{1/d}$ on every $T \in \mathcal{T}$. We finally let

$$\mathcal{R} := \mathcal{R}_{\mathcal{T}\to\mathcal{T}_*} := \mathcal{T}\backslash\mathcal{T}_*$$

be the subset of refined elements of $\mathcal{T}$ and note that $\mathcal{M} \subseteq \mathcal{R}$.

Concatenating these four modules, we get the standard SEMR one-step AFEM.

Algorithm 5.4 (GALERKIN). Let $\mathcal{T} \ge \mathcal{T}_0$ be a conforming refinement of a suitable initial mesh $\mathcal{T}_0$. Let data $\mathcal{D} = (\boldsymbol{A}, c, f) \in D_{\mathcal{T}}$ be discrete on $\mathcal{T}$ and let $\varepsilon > 0$ be a stopping tolerance. The following one-step AFEM creates a conforming refinement $\mathcal{T}_* \ge \mathcal{T}$ and Galerkin solution $u_{\mathcal{T}_*} \in \mathbb{V}_{\mathcal{T}_*}$ for data $\mathcal{D}$ such that $\eta_{\mathcal{T}_*}(u_{\mathcal{T}_*}) \le \varepsilon$:

```
[T_*, u_{T_*}] = GALERKIN(T, D, ε)
    set j = 0, T_0 = T
    do
        [u_j] = SOLVE(T_j)
        [{η_j(u_j, T)}_{T∈T_j}] = ESTIMATE(u_j, T_j, D)
        if η_j(u_j) ≤ ε
            return T_j, u_j
        [M_j] = MARK({η_j(u_j, T)}_{T∈T_j}, T_j, θ)
        [T_{j+1}] = REFINE(T_j, M_j)
        j ← j + 1
    while true
```

5.2.2. *Contraction property of* GALERKIN

A key question to ask is what is (are) the quantity(ies) that GALERKIN may contract. In light of (5.7), an obvious candidate is the energy error $|||u - u_j|||_\Omega$, where $u_j \in \mathbb{V}_j = \mathbb{V}_{\mathcal{T}_j}$ solves the problem

$$\mathcal{B}[u_j, w] = \langle f, w\rangle \quad \text{for all } w \in \mathbb{V}_j. \tag{5.25}$$

We now show that this is in fact the case for discrete data $\mathcal{D} \in \mathbb{D}_{\mathcal{T}}$ provided the *discrete local estimate* (5.17) holds. The latter is a consequence of the *interior vertex property* of Definition 4.50 whenever $\boldsymbol{A}$ is piecewise constant, $c = 0$ in $\mathcal{T}$, and data oscillation vanishes, i.e. $\mathrm{osc}_{\mathcal{T}}(f)_{-1} = 0$ (Morin *et al.* 2000, 2002).

Lemma 5.5 (contraction property with discrete lower bound). *Let data $\mathcal{D} \in \mathbb{D}_{\mathcal{T}}$ be discrete and let $u = u(\mathcal{D}) \in H_0^1(\Omega)$ be the corresponding exact solution. If the subset $\mathcal{M}_j \subset \mathcal{T}_j$ of elements marked by* MARK *satisfies the discrete local estimate* (5.17) *with respect to $\mathcal{T}_{j+1} \ge \mathcal{T}_j$, then for*

$$\alpha := \left(1 - \left(\theta\frac{C_2}{C_1}\right)^2\right)^{1/2} < 1,$$

the Galerkin solutions $u_j \in \mathbb{V}_j, u_{j+1} \in \mathbb{V}_{j+1}$ *of* (5.25) *satisfy*

$$|||u - u_{j+1}|||_\Omega \le \alpha |||u - u_j|||_\Omega, \tag{5.26}$$

where $0 < \theta < 1$ *is the parameter in* (5.23) *and* $C_1 \ge C_2$ *are the constants in* (5.13) *and* (5.17) *respectively.*

Proof. For convenience, we use the notation

$$e_j = |||u - u_j|||_\Omega, \quad E_j = |||u_{j+1} - u_j|||_\Omega, \quad \eta_j = \eta_j(u_j, \mathcal{T}_j), \quad \eta_j(\mathcal{M}_j) = \eta_j(u_j, \mathcal{M}_j)$$

and recall that $\mathcal{E}_{\mathcal{T}_j}(u_j, f) = \eta_j$ because $\mathrm{osc}_{\mathcal{T}_j}(f)_{-1} = 0$. The key idea is to use the Pythagoras equality (5.7), namely $e_{j+1}^2 = e_j^2 - E_j^2$, and show that E_j is a significant portion of e_j. Since (5.17) implies

$$C_2 \eta_j(\mathcal{M}_j) \le E_j,$$

applying Dörfler marking (5.23) and the upper bound in (5.13), we deduce

$$E_j^2 \ge C_2^2 \theta^2 \eta_j^2 \ge \left(\theta \frac{C_2}{C_1}\right)^2 e_j^2.$$

This is the desired property of E_j and leads to (5.26). □

The contraction property (5.26) is very special and only valid for the energy norm. For the H_0^1-norm we have the following simple but useful consequence.

Corollary 5.6 (linear convergence). *If* $c_\mathcal{B} \le C_\mathcal{B}$ *are the constants in* (5.3)*, then*

$$|u - u_k|_{H_0^1(\Omega)} \le \sqrt{\frac{C_\mathcal{B}}{c_\mathcal{B}}}\, \alpha^{k-j}\, |u - u_j|_{H_0^1(\Omega)}, \quad k \ge j \ge 0.$$

We wonder whether or not the interior vertex property is necessary for (5.17), and thus for (5.26). We present an example, introduced by Morin *et al.* (2000, 2002) to justify such a property for constant data and $n = 1$.

Example 5.7 (lack of strict error monotonicity). Let $\Omega = (0,1)^2$, $\boldsymbol{A} = \boldsymbol{I}$, $c = 0$, $f = 1$ (constant data), and consider the sequences of meshes depicted in Figure 5.1. If ϕ_0 denotes the basis function associated with the only interior vertex of the initial mesh $\mathcal{T}_0$, then $u_0 = u_1 = \frac{1}{12}\phi_0$ and $u_2 \ne u_1$.

The mesh $\mathcal{T}_1 \ge \mathcal{T}_0$ is produced by a standard two-step bisection ($b = 2$) in two dimensions. Since $u_0 = u_1$, we conclude that the energy error does not change $|||u - u_0|||_\Omega = |||u - u_1|||_\Omega$, whence (5.17) fails, between two consecutive steps of GALERKIN for $b = d = 2$. This is no longer true provided an interior vertex in each marked element is created, because then Lemma 5.5 (contraction property with discrete lower bound) holds.

Circumventing the discrete lower bound. Enforcing (5.17) requires a minimal number b_* of bisections, say $b_* = 3, 6$ for $d = 2, 3$, to guarantee the interior vertex property. This can be quite taxing, especially for $d = 3$, and relies on the strong

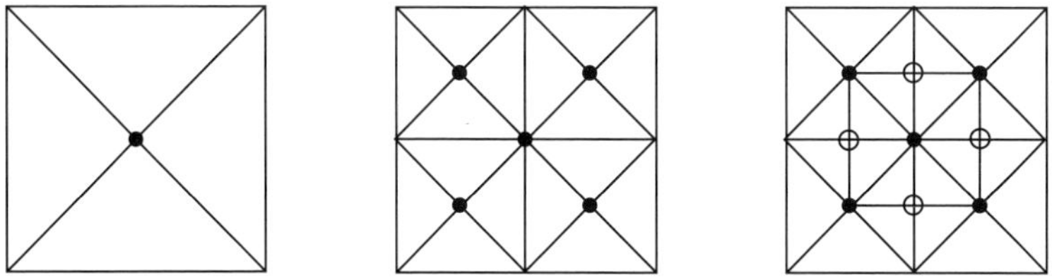

Figure 5.1. Grids $\mathcal{T}_0$, $\mathcal{T}_1$ and $\mathcal{T}_2$ of Example 5.7. The mesh $\mathcal{T}_1$ has nodes in the middle of edges of $\mathcal{T}_0$, but only $\mathcal{T}_2$ has nodes in the interior of elements of $\mathcal{T}_0$. Hence $\mathcal{T}_2$ satisfies the interior vertex property of Definition 4.50 with respect to $\mathcal{T}_0$ whereas $\mathcal{T}_1$ does not.

assumption of $\boldsymbol{A}$ being piecewise constant and $c = 0$ on $\mathcal{T}$. It is clear from the preceding discussion that the energy error alone cannot be expected to contract between consecutive iterates. We explore next what quantity to monitor instead of the energy error in the analysis, with the aim of avoiding (5.17) and building a theory applicable to general discrete coefficients $(\boldsymbol{A}, c)$. This exploits the special structure of residual estimators and does not directly extend to non-residual estimators.

Heuristics. According to (5.7), the energy error is monotone $|||u - u_{j+1}|||_\Omega \le |||u - u_j|||_\Omega$, but the previous example shows that strict inequality may fail. However, if $u_{j+1} = u_j$, estimate (4.67) reveals a strict estimator reduction $\eta_{j+1}(u_{j+1}) < \eta_j(u_j)$. We thus expect that, for a suitable scaling factor $\gamma > 0$, the so-called *quasi-error*

$$\zeta_j^2(u_j) := |||u - u_j|||_\Omega^2 + \gamma\, \eta_j^2(u_j) \tag{5.27}$$

may contract. This heuristic illustrates a distinct aspect of AFEM theory, the interplay between continuous quantities, such as the energy error $|||u - u_j|||_\Omega$, and discrete quantities, such as the estimator $\eta_j(u_j)$: neither one alone has the requisite properties to yield a contraction between consecutive adaptive steps. This result was originally proved by Cascón *et al.* (2008).

Theorem 5.8 (general contraction property). *Let $\mathcal{D} \in \mathbb{D}_\mathcal{T}$ be discrete data. Let $\theta \in (0, 1]$ be the Dörfler marking parameter, and let $\{\mathcal{T}_j, \mathbb{V}_j, u_j\}_{j=0}^\infty$ be a sequence of conforming meshes, finite element spaces and discrete solutions $u_j \in \mathbb{V}_j$ created by* GALERKIN *for the model problem* (5.25)*. If $u = u(\mathcal{D}) \in H_0^1(\Omega)$ is the exact solution of* (5.5)*, then there exist constants $\gamma > 0$ and $0 < \alpha < 1$, additionally depending on the number $b \ge 1$ of bisections and θ, such that for all $j \ge 0$*

$$|||u - u_{j+1}|||_\Omega^2 + \gamma\, \eta_{j+1}^2(u_{j+1}) \le \alpha^2 \left(|||u - u_j|||_\Omega^2 + \gamma\, \eta_j^2(u_j) \right). \tag{5.28}$$

Proof. We split the proof into four steps and use the notation in Lemma 5.5 (contraction property with discrete lower bound).

1 The error orthogonality (5.7) reads

$$e_{j+1}^2 = e_j^2 - E_j^2. \tag{5.29}$$

Employing Proposition 4.56 (estimator reduction) with $\mathcal{T} = \mathcal{T}_j$, $\mathcal{T}_* = \mathcal{T}_{j+1}$, $v = u_j$, $v_* = u_{j+1}$ and $f = f_* \in \mathbb{F}_j$ gives

$$\eta_{j+1}^2 \le (1+\delta)\big(\eta_j^2 - \lambda\,\eta_j^2(\mathcal{M}_j)\big) + (1+\delta^{-1})\,C_{\mathrm{Lip}}^2\,E_j^2. \tag{5.30}$$

After multiplying (5.30) by $\gamma > 0$, to be determined later, we add (5.29) and (5.30) to obtain

$$e_{j+1}^2 + \gamma\,\eta_{j+1}^2 \le e_j^2 + \big(\gamma\,(1+\delta^{-1})\,C_{\mathrm{Lip}}^2 - 1\big)\,E_j^2 + \gamma\,(1+\delta)\,\big(\eta_j^2 - \lambda\,\eta_j^2(\mathcal{M}_j)\big).$$

2 We now choose the parameters δ, γ: let δ satisfy

$$(1+\delta)(1-\lambda\theta^2) = 1 - \frac{\lambda\theta^2}{2},$$

and let γ verify

$$\gamma\,(1+\delta^{-1})\,C_{\mathrm{Lip}}^2 = 1.$$

Note that this choice of γ yields

$$e_{j+1}^2 + \gamma\,\eta_{j+1}^2 \le e_j^2 + \gamma\,(1+\delta)\,\big(\eta_j^2 - \lambda\,\eta_j^2(\mathcal{M}_j)\big). \tag{5.31}$$

3 We next employ Dörfler marking (5.23), namely $\eta_j(\mathcal{M}_j) \ge \theta\eta_j$, to deduce

$$e_{j+1}^2 + \gamma\,\eta_{j+1}^2 \le e_j^2 + \gamma(1+\delta)(1-\lambda\theta^2)\eta_j^2.$$

This, in conjunction with the choice of δ, gives

$$e_{j+1}^2 + \gamma\,\eta_{j+1}^2 \le e_j^2 + \gamma\left(1 - \frac{\lambda\theta^2}{2}\right)\eta_j^2, \tag{5.32}$$

which we write as

$$e_{j+1}^2 + \gamma\,\eta_{j+1}^2 \le e_j^2 - \frac{\gamma\lambda\theta^2}{4}\eta_j^2 + \gamma\left(1 - \frac{\lambda\theta^2}{4}\right)\eta_j^2.$$

4 Finally, the upper bound in (5.13), namely $e_j \le C_1\,\eta_j$, implies that

$$e_{j+1}^2 + \gamma\,\eta_{j+1}^2 \le \left(1 - \frac{\gamma\lambda\theta^2}{4C_1^2}\right)e_j^2 + \gamma\left(1 - \frac{\lambda\theta^2}{4}\right)\eta_j^2.$$

This in turn leads to

$$e_{j+1}^2 + \gamma\,\eta_{j+1}^2 \le \alpha^2\big(e_j^2 + \gamma\,\eta_j^2\big),$$

with

$$\alpha^2 := \max\left\{1 - \frac{\gamma\lambda\theta^2}{4C_1^2}, 1 - \frac{\lambda\theta^2}{4}\right\} < 1,$$

and thus concludes the proof of the theorem. □

Remark 5.9 (basic ingredients). This proof solely uses Dörfler marking (5.23), the Pythagoras identity (5.7), the *a posteriori* upper bound in (5.13), and Proposition 4.56 (estimator reduction). The proof altogether circumvents use of the lower bound in (5.13) and the discrete lower bound (5.17).

The contraction property (5.28) is valid for a suitable combination of the energy norm $|||u - u_j|||_\Omega$ and the PDE estimator $\eta_j(u_j)$. We cannot expect this type of result for the underlying space norm $|u - u_j|_{H_0^1(\Omega)}$. We instead have the following statement, whose structure reflects the possible stagnation of $|u - u_j|_{H_0^1(\Omega)}$ during the refinement process, as documented in Example 5.7.

Corollary 5.10 (linear convergence of error). *If the assumptions of Theorem 5.8 are valid, and* $0 < \alpha < 1, \gamma > 0$ *are the constants in* (5.28), *then*

$$|u - u_k|_{H_0^1(\Omega)} \leq C_* \alpha^{k-j} |u - u_j|_{H_0^1(\Omega)} \quad \textit{for all } k \geq j \geq 0, \tag{5.33}$$

with

$$C_* = \left(\frac{C_\mathcal{B}}{c_\mathcal{B}} \left(1 + \frac{\gamma}{C_2^2} \right) \right)^{1/2} > 1$$

and constants $C_\mathcal{B} \geq c_\mathcal{B} > 0$ *and* $C_2 > 0$ *given in* (5.3) *and* (5.13) *respectively.*

Proof. Simply concatenate (5.3), (5.28) and (5.13) to obtain

$$\begin{aligned} c_\mathcal{B} |u - u_k|^2_{H_0^1(\Omega)} &\leq |||u - u_k|||^2_\Omega + \gamma\, \eta_k(u_k)^2 \\ &\leq \alpha^{2(k-j)} \left(|||u - u_j|||^2_\Omega + \gamma\, \eta_j(u_j)^2 \right) \\ &\leq \alpha^{2(k-j)} \left(C_\mathcal{B} \left(1 + \frac{\gamma}{C_2^2} \right) \right) |u - u_j|^2_{H_0^1(\Omega)}. \end{aligned}$$

This implies (5.33) and concludes the proof. □

We stress that, in contrast to (5.28), (5.33) relies on the lower bound in (5.13). This is not the case if we express linear convergence in terms of the PDE estimator. The proof is similar to the preceding one and is omitted.

Corollary 5.11 (linear convergence of estimator). *If the assumptions of Theorem 5.8 are valid, and* $0 < \alpha < 1, \gamma > 0$ *are the constants in* (5.28), *then*

$$\eta_k(u_k) \leq C_\# \alpha^{k-j} \eta_j(u_j) \quad \textit{for all } k \geq j \geq 0, \tag{5.34}$$

with $C_\# = (1 + C_1^2/\gamma)^{1/2} > 1$ *and* C_1 *given in* (5.13).

Remark 5.12 (stopping). In view of (5.34), (5.12), we realize that GALERKIN requires $j \leq J$ iterations until the stopping criterion $\eta_j \leq \varepsilon$ is satisfied and delivers the error $|u - u_j|_{H_0^1(\Omega)} \leq C_U \varepsilon$, where

$$J \leq 1 + \frac{\log\left(\varepsilon / (C_\# \eta_0)\right)}{\log \alpha}.$$

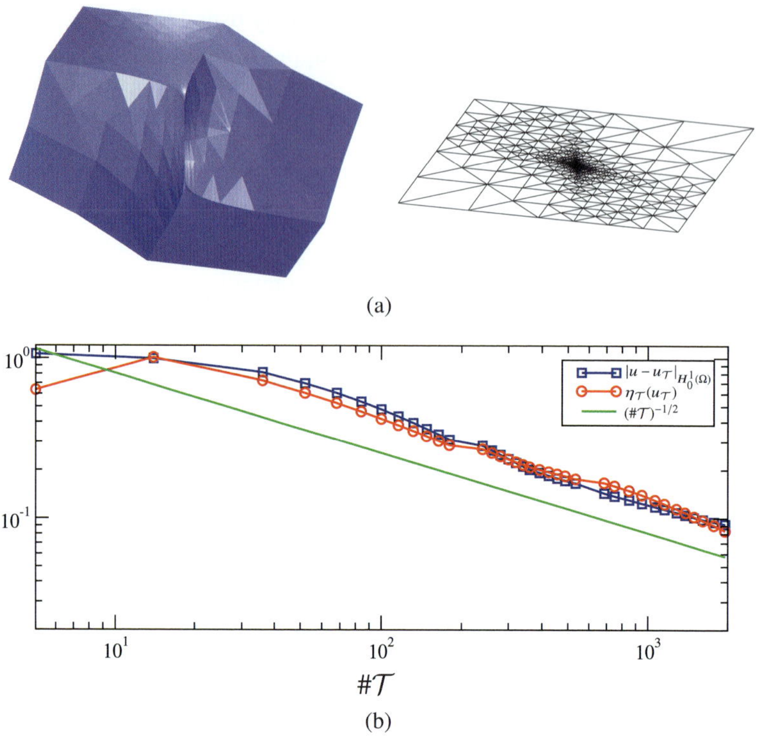

Figure 5.2. Discontinuous coefficients in a checkerboard pattern: (a) graph of the discrete solution u, which is $u \approx r^{0.1}$, and underlying strongly graded grid $\mathcal{T}$ towards the origin (notice the steep gradient of u at the origin); (b) estimate and true error in terms of $\#\mathcal{T}$ (the optimal decay for piecewise linear elements in two dimensions is indicated by the green line with slope $-1/2$).

5.2.3. *Discontinuous coefficients: Kellogg's example*

We examine a simple yet quite demanding example with piecewise constant coefficients in a checkerboard pattern for $d = 2$ due to Kellogg (1974/75), and used by Morin *et al.* (2000, 2002) as a benchmark for GALERKIN. We consider $\Omega = (-1, 1)^2$, $\boldsymbol{A} = a_1\boldsymbol{I}$ in the first and third quadrants, and $\boldsymbol{A} = a_2\boldsymbol{I}$ in the second and fourth quadrants. This checkerboard pattern is the worst for the regularity of the solution u at the origin. For $f = c = 0$, a function of the form $u(r, \theta) = r^\gamma \mu(\theta)$ in polar coordinates solves (2.5) with non-vanishing Dirichlet condition for suitable $0 < \gamma < 2$ and μ (Morin *et al.* 2000, 2002, Nochetto *et al.* 2009). We choose $\gamma = 0.1$, which leads to $u \in H^s(\Omega)$ for $1 \le s < 1.1$ and piecewise in W^2_p for some $p > 1$. This corresponds to diffusion coefficients $a_1 \approx 161.44$ and $a_2 = 1$, which can be computed via Newton's method; the closer γ is to 0, the larger is the ratio a_1/a_2. The solution u and a sample mesh are depicted in Figure 5.2(a).

Figure 5.2(b) documents the optimal performance of GALERKIN: both the energy error and estimator exhibit optimal decay $(\#\mathcal{T})^{-1/2}$ in terms of the cardinality $\#\mathcal{T}$

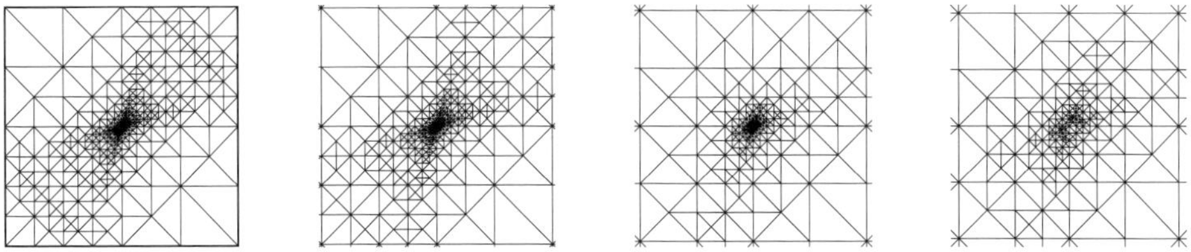

(a) (b) (c) (d)

Figure 5.3. Discontinuous coefficients in a checkerboard pattern: (a) final grid $\mathcal{T}$ highly graded towards the origin with cardinality $\#\mathcal{T} \approx 2000$; (b) zoom to $(-10^{-3}, 10^{-3})^2$; (c) zoom to $(-10^{-6}, 10^{-6})^2$; (d) zoom to $(-10^{-9}, 10^{-9})^2$. For a similar resolution, a uniform grid $\mathcal{T}$ would require cardinality $\#\mathcal{T} \approx 10^{20}$.

of the underlying mesh $\mathcal{T}$ for piecewise linear finite elements. On the other hand, Figure 5.3 displays a strongly graded mesh $\mathcal{T}$ towards the origin generated by GALERKIN using bisection, and three zooms which reveal a self-similar structure. It is worth stressing that the mesh size is of order 10^{-10} at the origin and that $\#\mathcal{T} \approx 2 \times 10^3$, whereas to reach a similar resolution with a uniform mesh $\mathcal{T}$ we would need $\#\mathcal{T} \approx 10^{20}$. This example clearly reveals that adaptivity can restore optimal performance even with modest computational resources.

Classical FEMs with quasi-uniform meshes $\mathcal{T}$ require regularity $u \in H^2(\Omega)$ to deliver an optimal convergence rate $(\#\mathcal{T})^{-1/2}$ with polynomial degree $n = 1$. Since $u \notin H^s(\Omega)$ for any $s > 1.1$, this is not possible for the example above. However, the problem is not quite the lack of second derivatives, but rather the fact that they are not square integrable. In fact, the function u is in W_p^2 for $p > 1$ in each quadrant, and so over the initial mesh $\mathcal{T}_0$, namely $u \in W_p^2(\Omega; \mathcal{T}_0)$. The computational rate of convergence $(\#\mathcal{T})^{-1/2}$ is consistent with Corollary 3.20. We will prove that GALERKIN delivers this rate in Section 6.

5.3. *Data oscillation: one-step* AFEM *with switch*

In Section 5.2 we assumed that the full data $\mathcal{D} = (\boldsymbol{A}, c, f) \in \mathbb{D}_{\mathcal{T}}$ is discrete, and in particular $f = P_{\mathcal{T}} f \in \mathbb{F}_{\mathcal{T}}$. The finite-dimensional nature of $\widehat{\mathcal{D}}$ allowed us to develop a rather simple theory of convergence for GALERKIN, the one-step AFEM, that hinges exclusively on the PDE local error indicator $\eta_{\mathcal{T}}(u_{\mathcal{T}}, T)$ defined in (5.8). We now keep $(\boldsymbol{A}, c)$ discrete, whence the elliptic operator in (2.5) includes the Laplacian, but explore the role of a general forcing $f \neq P_{\mathcal{T}} f$. Therefore, in contrast to (5.22), we now investigate the effect of data oscillation (5.11),

$$\operatorname{osc}_{\mathcal{T}}(f)_{-1}^2 = \sum_{T \in \mathcal{T}} \|f - P_{\mathcal{T}} f\|_{H^{-1}(\omega_T)}^2$$

for any $\mathcal{T} \in \mathbb{T}$, and present a linear convergence theory. We recall from Theorem 4.45 (modified residual estimator) that the total error estimator $\mathcal{E}_{\mathcal{T}}(u_{\mathcal{T}}, f)^2 =$

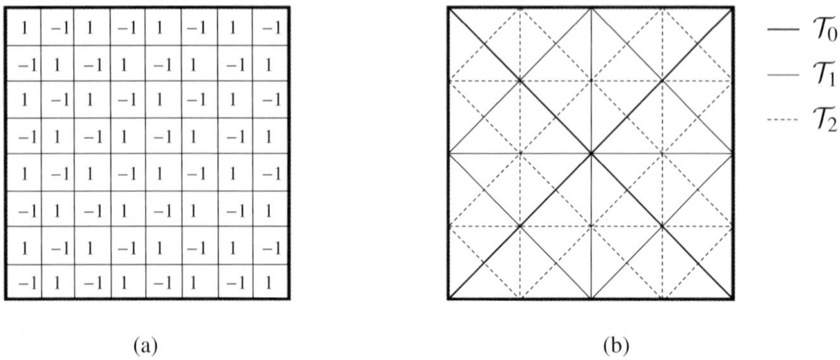

Figure 5.4. Representation of the checkerboard function f of Example 5.13 for $m = 3$ (a), and grids $\mathcal{T}_k$ for $k = 0, 1, 2$ (b).

$\eta_\mathcal{T}(u_\mathcal{T})^2 + \operatorname{osc}_\mathcal{T}(f)^2_{-1}$ is equivalent to the H^1-error, namely

$$C_L \mathcal{E}_\mathcal{T}(u_\mathcal{T}, f) \le \|\nabla(u - u_\mathcal{T})\|_{L^2(\Omega)} \le C_U \mathcal{E}_\mathcal{T}(u_\mathcal{T}, f). \tag{5.35}$$

As in the previous section, to simplify notation we do not use the hat symbol to indicate quantities defined with the discrete data $(\boldsymbol{A}, c)$.

5.3.1. *Role of data oscillation*

At first sight, it might seem that Example 5.7 (lack of strict error monotonicity) is very special and can only occur at the beginning of the refinement process. We now show that this situation can happen at any stage and that even an interior vertex property may not guarantee error or data oscillation decrease.

Example 5.13 (interior vertex). Let the polynomial degree be $n = 1$, fix $m \in \mathbb{N}$ and consider (5.21) with $\boldsymbol{A} = \boldsymbol{I}$ the identity matrix, $c = 0$, $\Omega = (0,1)^2$ and checkerboard f given by the following expression and depicted in Figure 5.4(a):

$$f(x) = \begin{cases} 1, & \text{if } x \in (i\,2^{-m}, (i+1)\,2^{-m}) \times (j\,2^{-m}, (j+1)\,2^{-m}) \text{ and } i+j \text{ odd}, \\ -1, & \text{otherwise.} \end{cases}$$

We start with the same mesh $\mathcal{T}_0$ with four elements as in Example 5.7, and construct recursively grids $\mathcal{T}_{k+1} \in \mathbb{T}$, $k \ge 0$, as a conforming refinement of $\mathcal{T}_k \in \mathbb{T}$ via two newest-vertex bisections of every triangle of $\mathcal{T}_k$; see Figure 5.4(b). Since f is L^2-orthogonal to every piecewise linear basis function of the space $\mathbb{V}_{\mathcal{T}_k} = \mathbb{S}^{1,0}_{\mathcal{T}_k}$ for $0 \le k \le m-1$, we deduce that $u_{\mathcal{T}_k} = 0$ and the energy error does not change

$$|\!|\!| u - u_{\mathcal{T}_k} |\!|\!|_\Omega = |\!|\!| u - u_{\mathcal{T}_0} |\!|\!|_\Omega, \quad 0 \le k \le m-1. \tag{5.36}$$

We see that this procedure creates three interior vertices in every triangle of $\mathcal{T}_k$ after two refinement steps, namely in $\mathcal{T}_{k+2}$ as long as $k + 2 \le m$. Since the error does not change, we conclude that the interior vertex property is necessary for error reduction but is not sufficient in the presence of data oscillation $\operatorname{osc}_{\mathcal{T}_k}(f)_{-1} \neq 0$. We conclude that

$$\text{\textit{Data oscillation} } \operatorname{osc}_{\mathcal{T}}(f)_{-1} \text{ \textit{is not generally of higher order than the error, especially in the early stages of the adaptive process.}} \tag{5.37}$$

On the other hand, for $k = m$ the discrete solution $u_{\mathcal{T}_m}$ no longer vanishes globally, but is still zero along the lines where f changes sign due to the symmetry of the problem, and the same happens with $u_{\mathcal{T}_{m+1}}$. Therefore the behaviour of $u_{\mathcal{T}m}$ and $u_{\mathcal{T}_{m+1}}$ in a fixed square, where f is constant, is exactly the same as in Example 5.7. This implies that $u_{\mathcal{T}_m} = u_{\mathcal{T}_{m+1}}$, and illustrates that the rather special situation of Example 5.7 can occur at any stage of the refinement process.

Example 5.14 (vanishing of $P_{\mathcal{T}}f$ for $n = 1$). Since $P_{\mathcal{T}}f$ is constructed locally upon testing f against cubic and quadratic bubbles (see Remark 4.26 (local computation)), and f of Example 5.13 is highly oscillatory, we realize that $P_{\mathcal{T}_k}f$ is rather small relative to f in $H^{-1}(\Omega)$, but it is not zero. This is due to the lack of complete symmetry of the checkerboard pattern and the triangular grid. Suppose that each square of Figure 5.4, where $f = \pm 1$, is further split across the diagonals into four triangles, and that f is assigned the alternating values ± 1 and ∓ 1 in each triangle depending on whether f was originally 1 or -1 in that square; this configuration is displayed in Figure 5.5. Suppose further that the coefficients $(\boldsymbol{A}, c)$ of the operator (2.5) are piecewise constant, as happens for the Laplacian, the polynomial degree is $n = 1$, and the definition (4.39) of P_T over a triangle $T \in \mathcal{T}$ uses $q \in \mathbb{P}_0$ rather than $\mathbb{P}_1$. In light of (4.39) and (4.40), symmetry yields, for all $T \in \mathcal{T}$ and $F \in \mathcal{F}$,

$$\int_T f\phi_T = 0 \quad \Rightarrow \quad P_T f = 0, \quad \int_F f\phi_F = 0 \quad \Rightarrow \quad P_F f = 0, \tag{5.38}$$

whence $P_{\mathcal{T}}f = 0$. Since also $u_{\mathcal{T}} = 0$ because f is orthogonal to all basis functions of $\mathbb{V}_{\mathcal{T}}$, we deduce $\mathcal{E}_{\mathcal{T}}(u_{\mathcal{T}}, f) = 0$, and all the information about the error $\|u - u_{\mathcal{T}}\|_\Omega \neq 0$ resides in the data oscillation $\operatorname{osc}_{\mathcal{T}}(f)_{-1} \neq 0$. Moreover, the fact that $P_{\mathcal{T}}f = 0$ for several iterations reveals the important property that $\operatorname{osc}_{\mathcal{T}}(f)_{-1}$ may not change upon refinement because

$$\operatorname{osc}_{\mathcal{T}}(f)_{-1}^2 = \sum_{T \in \mathcal{T}} \|f\|_{H^{-1}(\omega_T)}^2. \tag{5.39}$$

Since $\|u - u_{\mathcal{T}}\|_\Omega \approx \operatorname{osc}_{\mathcal{T}}(f)_{-1}$, according to (4.45), special care must be exercised to reduce data oscillation when it dominates. This justifies the structure of Algorithm 5.16 (AFEM-SW) below.

Example 5.15 (vanishing of $P_{\mathcal{T}}f$ for $n > 1$). Given $n \geq 1$ a polynomial degree and $\mathcal{T}_k$, $k = 1, \ldots, m$, uniform refinements of $\mathcal{T}_0$, there are finitely many conditions to verify for $f \in H^{-1}(\Omega)$ to be orthogonal to $\mathbb{V}_{\mathcal{T}_k}$ and to $\mathbb{F}_{\mathcal{T}_k}$. Since $\dim H^{-1}(\Omega) = \dim L^2(\Omega) = \infty$, there are infinitely many loads $f \in H^{-1}(\Omega)$ as well as in $L^2(\Omega)$ that yield $u_{\mathcal{T}_k} = P_{\mathcal{T}_k}f = 0$, which implies (5.36). Moreover, $\eta_{\mathcal{T}_k}(u_k) = 0$ and $\mathcal{E}_{\mathcal{T}_k}(u_{\mathcal{T}_k}) = \operatorname{osc}_{\mathcal{T}_k}(f)_{-1}$ satisfies (5.39). One explicit example is as follows.

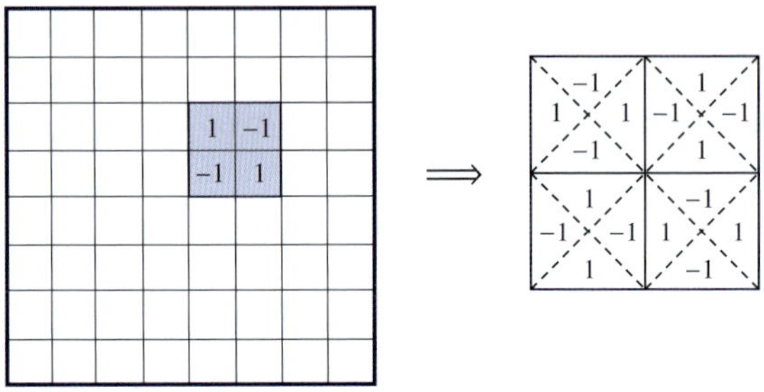

Figure 5.5. Refinement of the shaded area according to the process described in Example 5.14.

Given an initial mesh $\mathcal{T}_0$, suppose that f consists of line Dirac masses supported on the skeleton of $\mathcal{T}_0$ with densities g_F on $F \in \mathcal{F}_{\mathcal{T}_0}$ made of piecewise polynomials of degree $2n + 1$. We further assume that the g_F are orthogonal to $\mathbb{P}_{2n}$ over F as well as over all sub-faces obtained from $m \geq 1$ uniform refinements of $\mathcal{T}_0$; see Figure 5.4. In such a situation, (5.38) applies and $u_{\mathcal{T}_k} = P_{\mathcal{T}_k} f = 0$, whence (5.36) and (5.39) are valid for $0 \leq k \leq m$.

These three examples reveal the following crucial and novel feature about the interplay of the energy error $|\!|\!| u - u_\mathcal{T} |\!|\!|_\Omega$ and data oscillation $\operatorname{osc}_\mathcal{T}(f)_{-1}$:

Data oscillation $\operatorname{osc}_\mathcal{T}(f)_{-1}$ *may be responsible for the energy error* $|\!|\!| u - u_\mathcal{T} |\!|\!|_\Omega$ *to stagnate, even with the interior vertex property, and may entirely dominate it relative to the error estimator* $\mathcal{E}_\mathcal{T}(u_\mathcal{T}, f)$ *over many mesh refinements unless it is reduced.* (5.40)

5.3.2. Reducing data oscillation

The PDE error estimator $\eta_\mathcal{T}(u_\mathcal{T})$ in (5.8) is fully discrete and thus computable. In contrast, the computation, or rather estimation, of $\operatorname{osc}_\mathcal{T}(f)_{-1}$ hinges on *a priori* knowledge of f and cannot be assessed in general. Assuming that the local indicators introduced in Lemma 4.8 (localization re-indexing),

$$\operatorname{osc}_\mathcal{T}(f, T)_{-1} = \|f - P_\mathcal{T} f\|_{H^{-1}(\omega_T)}, \quad T \in \mathcal{T}, \tag{5.41}$$

are computable without further regularity than $f \in H^{-1}(\Omega)$, it is natural to think of *tree approximation* as the algorithm of choice to reduce $\operatorname{osc}_\mathcal{T}(f)_{-1}$ (Binev and DeVore 2004, Binev, Fierro and Veeser 2023, Binev 2018). However, this optimal algorithm is not readily applicable because of the lack of a suitable sub-additivity property.

On the other hand, greedy algorithms, such as that in Section 3.6 (constructive approximation), do not work under minimal regularity. In Section 7.3 we present practical examples of rough f for which $\operatorname{osc}_\mathcal{T}(f)_{-1}$ can be replaced by a larger computable surrogate estimator $\widetilde{\operatorname{osc}}_\mathcal{T}(f)_{-1}$. The latter splits into element contributions and is amenable to a greedy strategy. Since this is specialized and technical, we prefer to postpone the full discussion to Section 7.3 and now assume the existence of a module DATA with the following property: given a tolerance $\tau > 0$ and a conforming mesh $\mathcal{T} \in \mathbb{T}$, DATA constructs a conforming refinement $\mathcal{T}_* \in \mathbb{T}$,

$$[\mathcal{T}_*] = \mathsf{DATA}(\mathcal{T}, f, \tau),$$

such that $\operatorname{osc}_{\mathcal{T}_*}(f)_{-1} \leq \tau$. The complexity of DATA depends on the decay rate of the best approximation error $\min_{\mathcal{T}\in\mathbb{T}_N} \widetilde{\operatorname{osc}}_{\mathcal{T}}(f)_{-1}$ of f with N degrees of freedom. We address this important issue in Section 7.3 for each example separately.

5.3.3. *Linear convergence*

The following algorithm, AFEM-SW, a one-step AFEM with switch, is a minor, but essential, modification of GALERKIN in that the call to the modules MARK and REFINE is conditional on the size of $\operatorname{osc}_{\mathcal{T}}(f)_{-1}$ relative to $\mathcal{E}_{\mathcal{T}}(u_{\mathcal{T}}, f)$. This structure is consistent with the heuristic discussion by Cascón *et al.* (2008, Section 6) to avoid separate marking. A similar algorithm is being developed in Kreuzer *et al.* (2024).

Algorithm 5.16 (AFEM-SW). Let $\mathcal{T}_0$ be a suitable initial mesh, let the coefficients $(\boldsymbol{A}, c)$ be discrete over $\mathcal{T}_0$, and let $\varepsilon > 0$ be a stopping tolerance. Given parameters $0 < \theta, \omega, \xi < 1$, AFEM-SW iterates the following loop until $\mathcal{E}_{\mathcal{T}}(u_{\mathcal{T}}, f) \leq \varepsilon$:

```
[T, u_T] = AFEM-SW(T_0, D, ε)
    set j = 0
    do
        [u_{T_j}] = SOLVE(T_j)
        [η_{T_j}(u_{T_j}), osc_{T_j}(f)_{-1}] = ESTIMATE(u_{T_j}, T_j, D)
        if E_{T_j}(u_{T_j}, f) ≤ ε
            return T_j, u_{T_j}
        else if osc_{T_j}(f)_{-1} ≤ σ_j := ω E_{T_j}(u_{T_j}, f)
            [M_j] = MARK({η_{T_j}(u_{T_j}, T)}_{T∈T_j}, T_j, θ)
            [T_{j+1}] = REFINE(T_j, M_j)
        else
            [T_{j+1}] = DATA(T_j, f, ξσ_j)
        j ← j + 1
    while true
```

Note that SOLVE computes the Galerkin approximation using the exact right-hand side $f \in H^{-1}(\Omega)$ (not necessarily in $\mathbb{F}_{\mathcal{T}_j}$), thereby preserving the Galerkin orthogonality property. Moreover, ESTIMATE is now responsible for computing the PDE estimator

$$\eta_{\mathcal{T}_j}(u_{\mathcal{T}_j}) = \eta_{\mathcal{T}_j}(u_{\mathcal{T}_j}, f, \mathcal{T}_j)$$

using $P_{\mathcal{T}_j} f \in \mathbb{F}_{\mathcal{T}_j}$, as well as data oscillation $\operatorname{osc}_{\mathcal{T}_j}(f)_{-1}$, which together give

$$\mathcal{E}_{\mathcal{T}_j}(u_{\mathcal{T}_j}, f) = \left(\eta_{\mathcal{T}_j}(u_{\mathcal{T}_j})^2 + \operatorname{osc}_{\mathcal{T}_j}(f)_{-1}^2\right)^{1/2},$$

and MARK consists of Dörfler marking (5.23) with parameter θ.

We proceed as in Section 5.2.2 to prove linear convergence of AFEM-SW. We first show a contraction property for the *quasi-error*, which instead of (5.27) reads

$$\zeta_{\mathcal{T}_j}(u_{\mathcal{T}_j}, f)^2 := \|u - u_{\mathcal{T}_j}\|_\Omega^2 + \gamma \eta_{\mathcal{T}_j}(u_{\mathcal{T}_j})^2 + \operatorname{osc}_{\mathcal{T}_j}(f)_{-1}^2, \tag{5.42}$$

where $u = u(\boldsymbol{A}, c, f)$ is the Galerkin solution with $(\boldsymbol{A}, c)$ discrete but f exact and the scaling parameter satisfies $0 < \gamma \leq 1$.

Theorem 5.17 (contraction property of AFEM-SW). *Let $(\boldsymbol{A}, c)$ be discrete coefficients over $\mathcal{T}_0$ and let $f \in H^{-1}(\Omega)$. Let $\theta \in (0, 1]$ be the Dörfler parameter and let $(\mathcal{T}_j, \mathbb{V}_j, u_j)$ be the sequence of conforming meshes $\mathcal{T}_j$, finite element spaces $\mathbb{V}_j$, and Galerkin solutions $u_j \in \mathbb{V}_j$ produced by* AFEM-SW. *There exist parameters $0 < \omega_0 < 1$ sufficiently small and $0 < \gamma \leq 1$ and $0 < \alpha < 1$ such that for any $\omega \leq \omega_0$ and $\xi \leq 1/2$, the quasi-error $\zeta_{\mathcal{T}_j}$ in* (5.42) *contracts*

$$\zeta_{\mathcal{T}_{j+1}}(u_{\mathcal{T}_{j+1}}, f) \leq \alpha \zeta_{\mathcal{T}_j}(u_{\mathcal{T}_j}, f), \quad j \geq 0. \tag{5.43}$$

Proof. We argue as in Theorem 5.8 (general contraction property) upon distinguishing the two possible cases within Algorithm 5.16. But first we must account for a crucial difference: the discrete forcing function $P_{\mathcal{T}_j} f$ used in the definition of the estimator $\mathcal{E}_{\mathcal{T}_j}(u_{\mathcal{T}_j}, f)$ changes in each iteration. We use the same notation as in Theorem 5.8 along with $\operatorname{osc}_j := \operatorname{osc}_{\mathcal{T}_j}(f)_{-1}$, $\mathcal{E}_j^2 := \eta_j^2 + \operatorname{osc}_j^2$ and $P_j := P_{\mathcal{T}_j}$.

[1] *Estimator reduction property.* In view of Proposition 4.56 (estimator reduction), we need to estimate the discrepancy between discrete forcing functions

$$\sum_{T \in \mathcal{T}_{j+1}} \|P_{j+1} f - P_j f\|_{H^{-1}(\omega_T)}^2 \leq 2 \sum_{T \in \mathcal{T}_{j+1}} \big(\|f - P_{j+1} f\|_{H^{-1}(\omega_T)}^2 + \|f - P_j f\|_{H^{-1}(\omega_T)}^2 \big).$$

For the first term we recall Lemma 4.57 (quasi-monotonicity of oscillation) to write

$$\sum_{T \in \mathcal{T}_{j+1}} \|f - P_{j+1} f\|_{H^{-1}(\omega_T)}^2 = \operatorname{osc}_{j+1}^2 \leq C_{\mathrm{osc}}^2 \operatorname{osc}_j^2 .$$

For the second term, instead, we combine the projection property $P_{j+1}(P_j f) = P_j f$ with Lemma 4.5 (localization of H^{-1}-norm) and Corollary 4.31 (local near-best approximation), and the fact that $\mathcal{T}_{j+1}$ is a refinement of $\mathcal{T}_j$, to see that

$$\begin{aligned} \sum_{T' \subset \omega_T} \|f - P_{j+1}(P_j f)\|_{H^{-1}(\omega_{T'})}^2 &\leq C_{\mathrm{lStb}}^2 \sum_{T' \subset \omega_T} \|f - P_j f\|_{H^{-1}(\omega_{T'})}^2 \\ &\leq C_{\mathrm{lStb}}^2 C_{\mathrm{ovrl}}^2 \operatorname{osc}_{\mathcal{T}_j}(f, \omega_T)_{-1}^2 \quad \text{for all } T \in \mathcal{T}_j. \end{aligned}$$

Adding over T and recalling Proposition 4.56, we end up with

$$\begin{aligned} \eta_{\mathcal{T}_{j+1}}(u_{j+1}, f, \mathcal{T}_{j+1})^2 \leq{}& (1 + \delta)\big(\eta_{\mathcal{T}_j}(u_j, f, \mathcal{T}_j)^2 - \lambda \, \eta_{\mathcal{T}_j}(u_j, f, \mathcal{M}_j)^2\big) \\ &+ (1 + \delta^{-1}) \, C_{\mathrm{Lip}}^2 \left(|u_j - u_{j+1}|_{H_0^1(\Omega)}^2 + \operatorname{osc}_j^2 \right), \end{aligned} \tag{5.44}$$

for a constant C_{Lip} large enough to absorb all preceding constants, and any $\delta > 0$.

[2] *Case* $\mathrm{osc}_j \le \omega \mathcal{E}_j$. We first observe that $\eta_j^2 \ge (1-\omega^2)\mathcal{E}_j^2$ and $\mathrm{osc}_j^2 \le \omega^2(1-\omega^2)^{-1}\eta_j^2$. We then proceed as in Theorem 5.8 with the quantity $e_j^2 + \gamma\eta_j^2$, and observe that the choices of δ and γ,

$$\delta \le -1 + \frac{1-\lambda\theta^2/2}{1-\lambda\theta^2} = \frac{\lambda\theta^2}{2(1-\lambda\theta^2)}, \qquad \gamma \le \frac{\delta}{4C_{\mathrm{Lip}}^2} \le \frac{1}{2(1+\delta^{-1})C_{\mathrm{Lip}}^2}, \tag{5.45}$$

imply $\gamma(1+\delta^{-1})C_{\mathrm{Lip}}^2 \le 1/2$. This, together with (5.7) and (5.44), leads to

$$e_{j+1}^2 + \gamma\eta_{j+1}^2 \le e_j^2 + \gamma\left(1 - \frac{\lambda\theta^2}{2}\right)\eta_j^2 + \frac{1}{2}\,\mathrm{osc}_j^2;$$

compare with (5.32). We invoke the upper bound in (5.13) to write

$$\eta_j^2 \ge (1-\omega^2)\mathcal{E}_j^2 \ge (1-\omega^2)\frac{e_j^2}{C_1^2} \ge \frac{e_j^2}{2C_1^2}$$

provided $\omega^2 \le 1/2$, whence

$$e_{j+1}^2 + \gamma\eta_{j+1}^2 \le \left(1 - \frac{\gamma\lambda\theta^2}{8C_1^2}\right)e_j^2 + \gamma\left(1 - \frac{\lambda\theta^2}{8}\right)\eta_j^2 - \frac{\gamma\lambda\theta^2}{8}\eta_j^2 + \frac{1}{2}\,\mathrm{osc}_j^2.$$

We next consider the data oscillation, for which we invoke Lemma 4.57 (quasi-monotonicity of oscillation):

$$\mathrm{osc}_{j+1} \le C_{\mathrm{osc}}\,\mathrm{osc}_j, \quad \mathrm{osc}_j^2 \le C_{\mathrm{osc}}^2 \frac{\omega^2}{1-\omega^2}\eta_j^2 \le 2C_{\mathrm{osc}}^2\omega^2\eta_j^2.$$

Adding the two preceding inequalities yields

$$\begin{aligned}\zeta_{j+1}^2 = e_{j+1}^2 + \gamma\eta_{j+1}^2 + \mathrm{osc}_{j+1}^2 \le{}& \left(1 - \frac{\gamma\lambda\theta^2}{8C_1^2}\right)e_j^2 \\ &+ \left(1 - \frac{\lambda\theta^2}{8}\right)\left(\gamma\eta_j^2 + \mathrm{osc}_j^2\right) \\ &+ \left[-\gamma\frac{\lambda\theta^2}{8} + 2\left(C_{\mathrm{osc}}^2 - 1 + \frac{\lambda\theta^2}{8} + \frac{1}{2}\right)\omega^2\right]\eta_j^2.\end{aligned}$$

We drop the term $-1/2 + \lambda\theta^2/8 \le 0$ and let $\gamma = \delta/(4C_{\mathrm{Lip}}^2)$, which is consistent with (5.45). We seek conditions on ω that make the factor of η_j^2 non-positive. Imposing

$$\omega^2 \le \frac{\gamma\lambda\theta^2}{16C_{\mathrm{osc}}^2} = \frac{\lambda\theta^2}{64C_{\mathrm{osc}}^2 C_{\mathrm{Lip}}^2}\delta \tag{5.46}$$

yields

$$\zeta_{j+1}^2 \le \alpha_1^2\zeta_j^2,$$

with

$$\alpha_1^2 := \max\left\{1 - \frac{\delta\lambda\theta^2}{32C_1^2C_{\mathrm{Lip}}^2}, 1 - \frac{\lambda\theta^2}{8}\right\} < 1.$$

[3] *Case* $\mathrm{osc}_j > \omega\mathcal{E}_j$. The module DATA with input parameter $\xi \le 1/2$ gives

$$\mathrm{osc}_{j+1} \le \xi\omega\mathcal{E}_j < \xi\,\mathrm{osc}_j\,.$$

We now exploit the contraction of osc_j to compensate the moderate increase of η_j^2 and presence of osc_j^2, both governed by (5.44). In fact, $\gamma(1+\delta^{-1})C_{\mathrm{Lip}}^2 \le 1/2$ yields

$$e_{j+1}^2 + \gamma\eta_{j+1}^2 \le e_j^2 + \gamma(1+\delta)\eta_j^2 + \frac{1}{2}\,\mathrm{osc}_j^2\,.$$

We add osc_{j+1}^2 to both sides and rewrite the right-hand side to arrive at

$$\begin{aligned}\zeta_{j+1}^2 = e_{j+1}^2 + \gamma\eta_{j+1}^2 + \mathrm{osc}_{j+1}^2 \le e_j^2 &- \frac{1-2\xi^2}{8}\,\mathrm{osc}_j^2 \\ &+ (1-\delta)\gamma\eta_j^2 + \left(\frac{1+2\xi^2}{4} + \frac{1}{2}\right)\mathrm{osc}_j^2 \\ &+ 2\delta\gamma\eta_j^2 - \frac{1-2\xi^2}{8}\,\mathrm{osc}_j^2\,.\end{aligned}$$

Our next task is to find conditions on ω for the last line to be non-positive. To this end, we resort to the upper bound $\eta_j^2 < \omega^{-2}\,\mathrm{osc}_j^2$ and $\xi \le 1/2$ to obtain

$$2\delta\gamma\eta_j^2 - \frac{1-2\xi^2}{8}\,\mathrm{osc}_j^2 < \left(\frac{2\delta\gamma}{\omega^2} - \frac{1}{16}\right)\mathrm{osc}_j^2 \le 0$$

provided we impose the relation

$$\omega^2 \ge 32\delta\gamma = \frac{8}{C_{\mathrm{Lip}}^2}\delta^2. \tag{5.47}$$

We next use the upper bound $e_j \le C_1\mathcal{E}_j \le C_1\omega^{-1}\,\mathrm{osc}_j$ to write

$$e_j^2 - \frac{1-2\xi^2}{8}\,\mathrm{osc}_j^2 \le \left(1 - \frac{\omega^2}{16C_1^2}\right)e_j^2,$$

whence we end up with

$$\zeta_{j+1}^2 \le \alpha_2^2\zeta_j^2$$

provided we define

$$\alpha_2^2 := \max\left\{1 - \frac{\omega^2}{16C_1^2}, 1-\delta, \frac{3+2\xi^2}{4}\right\} < 1.$$

[4] *Choosing the parameters.* We see that the asserted estimate (5.43) is valid with $\alpha = \max\{\alpha_1, \alpha_2\} < 1$ provided the constraints (5.46) and (5.47) are compatible, that is,

$$\frac{8}{C_{\text{Lip}}}\delta^2 \le \omega^2 \le \frac{\lambda\theta^2}{64C_{\text{osc}}^2 C_{\text{Lip}}^2}\delta.$$

We choose

$$\delta_0 = \frac{\lambda\theta^2}{512C_{\text{osc}}^2 C_{\text{Lip}}} \quad \text{and} \quad \omega_0 = \frac{\lambda\theta^2}{128C_{\text{osc}}^2 C_{\text{Lip}}\sqrt{2C_{\text{Lip}}}}.$$

Then, for all $\omega \le \omega_0$, there exists $\delta \le \delta_0$ that satisfies the previous inequalities as well as $\gamma = \delta/(4C_{\text{Lip}}^2) \le 1$, perhaps upon reducing δ_0. This completes the proof of Theorem 5.43. □

Note that we could replace the conditional $\operatorname{osc}_{\mathcal{T}_j}(f)_{-1} \le \omega\,\mathcal{E}_{\mathcal{T}_j}(u_{\mathcal{T}_j}, f)$ with $\operatorname{osc}_{\mathcal{T}_j}(f)_{-1} \le \omega\,\eta_{\mathcal{T}_j}(u_{\mathcal{T}_j})$, but the tolerance τ of DATA cannot be

$$\tau = \xi\,\omega\,\eta_{\mathcal{T}_j}(u_{\mathcal{T}_j})$$

because the algorithm might not terminate when $\eta_{\mathcal{T}_j}(u_{\mathcal{T}_j}) = 0$; see e.g. Examples 5.13–5.15. In fact, the tolerance $\tau = \xi\omega\mathcal{E}_{\mathcal{T}_j}(u_{\mathcal{T}_j}, f)$ is *dynamic* and relative to $\mathcal{E}_{\mathcal{T}_j}(u_{\mathcal{T}_j}, f)$. This avoids separate marking, which was shown by Cascón *et al.* (2008, Section 6) to give non-optimal convergence rates. In contrast, we will prove in Section 6 that Algorithm 5.16 is rate-optimal.

It turns out that Theorem 5.17 yields linear convergence of error and estimator.

Corollary 5.18 (linear convergence of error). *For $0 < \alpha < 1$ and $0 < \omega \le \omega_0$, $\xi \le 1/2$ as in Theorem 5.17, and $C_* = \left(1 + C_2^{-1}\right)^{1/2}$ with C_2 as in* (5.13), *we have*

$$|u - u_{\mathcal{T}_k}|_{H_0^1(\Omega)} \le C_*\,\alpha^{k-j}|u - u_{\mathcal{T}_j}|_{H_0^1(\Omega)} \quad \textit{for all } k \ge j \ge 0.$$

Proof. We again use the same notation as in Lemma 5.5 and Theorem 5.17. In view of the definition (5.42) of quasi-error $\zeta_j := \zeta_{\mathcal{T}_j}(u_{\mathcal{T}_j}, f)$, we thus have $e_j \le \zeta_j$ and

$$\zeta_j^2 \le e_j^2 + \eta_j^2 + \operatorname{osc}_j^2 \le \left(1 + C_2^{-1}\right)e_j^2$$

because $C_2\mathcal{E}_j \le e_j$ from (5.13). This implies

$$e_j \le \zeta_j \le C_* e_j \quad \text{for all } j \ge 0,$$

and invoking Theorem 5.17 (contraction property for AFEM-SW),

$$e_k^2 \le \zeta_k^2 \le \alpha^{2(k-j)}\zeta_j^2 \le \alpha^{2(k-j)}C_*^2 e_j$$

gives the desired estimate. □

We stress that Corollary 5.18 relies on the lower bound in (5.13) whereas Corollary 5.19 uses only the upper bound. Its proof is similar and is thus omitted.

Corollary 5.19 (linear convergence of estimator). *For $0 < \alpha < 1$ and $0 < \omega \le \omega_0$, $\xi \le 1/2$ as in Theorem 5.17, and $C_\# = \big((1+C_1^2)\gamma^{-1}\big)^{1/2}$ with C_1 as in (5.13), we have*

$$\mathcal{E}_{\mathcal{T}_k}(u_{\mathcal{T}_k}, f) \le C_\# \, \alpha^{k-j}\, \mathcal{E}_{\mathcal{T}_j}(u_{\mathcal{T}_j}, f) \quad \textit{for all } k \ge j \ge 0.$$

5.4. *Convergence for general data: two-step* AFEM

We now remove the restriction of Sections 5.2 and 5.3 to discrete data and allow for general data $\mathcal{D} = (\boldsymbol{A}, c, f) \in \mathbb{D}$ as defined in (5.2). The current goal is to study Algorithm 5.1 (AFEM-TS), which concatenates the modules DATA and GALERKIN. We start with the study of continuous dependence with respect to data $\mathcal{D}$. We next discuss the approximation of $\mathcal{D}$ within the module DATA, the computational cost of GALERKIN, and eventually the convergence of Algorithm 5.1.

5.4.1. *Perturbation theory*

We start with a brief discussion of data perturbation. Given constants $0 < \alpha_1 \le \alpha_2$ and $0 \le c_1 \le c_2$, we define the constrained spaces for the diffusion and reaction coefficients by

$$M(\alpha_1, \alpha_2) := \big\{\boldsymbol{A} \in L^\infty\big(\Omega; \mathbb{R}^{d\times d}_{\text{sym}}\big) \mid 0 < \alpha_1 \le \lambda_j(\boldsymbol{A}(x)) \le \alpha_2 \\ \text{for a.e. } x \in \Omega,\ 1 \le j \le d\big\}, \tag{5.48}$$

where $\lambda_j(\boldsymbol{A}(x))$ denotes the jth eigenvalue of $\boldsymbol{A}$ at $x \in \Omega$ and

$$R(c_1, c_2) := \{c \in L^\infty(\Omega) \mid c_1 \le c(x) \le c_2 \text{ for a.e. } x \in \Omega\}. \tag{5.49}$$

The coefficients $(\boldsymbol{A}, c)$ are assumed to satisfy the structural assumption

$$\boldsymbol{A} \in M(\alpha_1, \alpha_2), \qquad c \in R(c_1, c_2); \tag{5.50}$$

see (2.6). This guarantees coercivity and continuity of the bilinear form $\mathcal{B}$ in (2.8), and thus unique solvability of (2.7).

Regarding the discrete coefficients, $(\widehat{\boldsymbol{A}}, \widehat{c})$ will ultimately be piecewise polynomials in a grid $\widehat{\mathcal{T}} \in \mathbb{T}$. The side constraints in (5.48) and (5.49) are generally violated by any linear projection onto piecewise polynomials of degree $n-1 \ge 1$, e.g. the L^2-projection, and require a nonlinear correction maintaining high-order accuracy. This is a crucial but delicate matter addressed later in Section 7.2. For the moment, we simply assume that the discrete coefficients $(\widehat{\boldsymbol{A}}, \widehat{c})$ satisfy

$$\widehat{\boldsymbol{A}} \in M(\widehat{\alpha}_1, \widehat{\alpha}_2), \qquad \widehat{c} \in R(\widehat{c}_1, \widehat{c}_2), \tag{5.51}$$

with

$$\frac{\alpha_1}{2} \le \widehat{\alpha}_1 \le \widehat{\alpha}_2 \le C_{\text{ctr}}\alpha_2, \qquad -\frac{\alpha_1}{4C_P^2} \le \widehat{c}_1 \le \widehat{c}_2 \le C_{\text{ctr}}(\alpha_1 + c_2), \tag{5.52}$$

where $C_P > 0$ is the Poincaré constant in (2.2) and $C_{\text{ctr}} \ge 1$ is a constant; see (7.21) and (7.23). This implies coercivity and continuity of the perturbed bilinear form

$$\widehat{\mathcal{B}}[v,w] := \int_\Omega \nabla v \cdot \widehat{\boldsymbol{A}} \nabla w + \widehat{c} v w, \quad \text{for all } v, w \in H_0^1(\Omega), \tag{5.53}$$

because for all $v, w \in H_0^1(\Omega)$

$$\widehat{\mathcal{B}}[v,v] \geq \widehat{\alpha}_1 \int_\Omega |\nabla v|^2 - \frac{\alpha_1}{4C_P^2} \int_\Omega |v|^2 \geq \frac{\alpha_1}{4} |v|^2_{H_0^1(\Omega)}$$

and

$$|\widehat{\mathcal{B}}[v,w]| \leq \int_\Omega \widehat{\alpha}_2 |\nabla v|\,|\nabla w| + \widehat{c}_2 |v|\,|w| \leq \left(\widehat{\alpha}_2 + \widehat{c}_2 C_P^2\right) |v|_{H_0^1(\Omega)} |w|_{H_0^1(\Omega)}.$$

Therefore the energy norm $|\!|\!| v |\!|\!|_\Omega^2 = \widehat{\mathcal{B}}[v,v]$ is equivalent to the H_0^1-seminorm

$$c_{\widehat{\mathcal{B}}} |v|^2_{H_0^1(\Omega)} \leq |\!|\!| v |\!|\!|_\Omega^2 \leq C_{\widehat{\mathcal{B}}} |v|^2_{H_0^1(\Omega)}, \tag{5.54}$$

where $c_{\widehat{\mathcal{B}}} = \alpha_1/4$ and $C_{\widehat{\mathcal{B}}} = \widehat{\alpha}_2 + \widehat{c}_2 C_P^2$. Hence the Lax–Milgram theorem guarantees the existence of a unique solution $\widehat{u} = u(\widehat{\mathcal{D}}) \in H_0^1(\Omega)$ of the perturbed problem (5.5) defined using the discrete data $\widehat{\mathcal{D}} = (\widehat{\boldsymbol{A}}, \widehat{c}, \widehat{f})$.

We now quantify the effect of perturbing data from $\mathcal{D}$ to $\widehat{\mathcal{D}}$ in the space

$$\widehat{D}(\Omega) := L^r(\Omega; \mathbb{R}^{d\times d}) \times W_q^{-s}(\Omega) \times H^{-1}(\Omega), \tag{5.55}$$

where $2 \leq r \leq \infty$ and $0 \leq s \leq 1$, $d/(2-s) < q \leq \infty$; $W_q^{-s}(\Omega)$ is the dual of $W_{q*}^s(\Omega)$ with $q^* = q/(q-1)$. The use of $r = \infty$ for $\boldsymbol{A}$ entails the further assumption

$$\boldsymbol{A} \textit{ is piecewise uniformly continuous over a generic mesh } \mathcal{T} \in \mathbb{T}, \tag{5.56}$$

which turns out to be rather restrictive but customary in the theory of AFEM. Our present approach allows for $r < \infty$ and thus for discontinuous coefficients $(\boldsymbol{A}, c)$ not aligned with $\mathcal{T}$, which is important in practice. However, it requires the following slightly stronger regularity property of the solution $u \in H_0^1(\Omega)$ of (2.7):

$$\|\nabla u\|_{L^p(\Omega)} \leq C_p \|f\|_{W_p^{-1}(\Omega)}, \quad 2 < p \leq p_0. \tag{5.57}$$

We refer to Lemma 2.13 (W_p^1-regularity), which shows the existence of $C_p > 0$ and $p_0 > 2$ that depend only on $\Omega, \alpha_1, \alpha_2$ and c_2.

Lemma 5.20 (continuous dependence on data). *Let $\mathcal{D} = (\boldsymbol{A}, c, f) \in \mathbb{D}$ be such that $\boldsymbol{A} \in M(\alpha_1, \alpha_2)$ and $c \in R(c_1, c_2)$. Let $\widehat{\mathcal{D}} = (\widehat{\boldsymbol{A}}, \widehat{c}, \widehat{f}) \in \mathbb{D}$ be an approximation of $\mathcal{D}$ such that $\widehat{\boldsymbol{A}} \in M(\widehat{\alpha}_1, \widehat{\alpha}_2)$ and $\widehat{c} \in R(\widehat{c}_1, \widehat{c}_2)$. Let $2 \leq r \leq \infty, 2 \leq r_* = 2r/(r-2) \leq p_0$ be such that $f \in W_{r_*}^{-1}(\Omega)$. If $u = u(\mathcal{D}), \widehat{u} = u(\widehat{\mathcal{D}}) \in H_0^1(\Omega)$ are the solutions of* (2.7) *and* (5.5) *with data $\mathcal{D}, \widehat{\mathcal{D}}$, respectively, and u satisfies* (5.57) *with $p = r_*$ for $r < \infty$, then for any $0 \leq s \leq 1$ and $d/(2-s) < q \leq \infty$ we have*

$$\|\nabla(u - \widehat{u})\|_{L^2(\Omega)} \leq C(\mathcal{D}, \Omega)\, \|\mathcal{D} - \widehat{\mathcal{D}}\|_{\widehat{D}(\Omega)}, \tag{5.58}$$

where the constant $C(\mathcal{D},\Omega)$ depends on $\mathcal{D}$, Ω, p_0, q and s, and blows up as $q \to d/(2-s)$ for $d=2$ while it remains bounded for $d>2$.

Proof. Subtracting the weak formulations (2.7) for u and (5.5) for $\widehat{u}$, and reordering, we easily obtain for any $v \in H_0^1(\Omega)$

$$\int_\Omega \nabla v \cdot \widehat{\boldsymbol{A}}\nabla(u-\widehat{u}) + \widehat{c}v(u-\widehat{u}) = \int_\Omega \nabla v \cdot (\widehat{\boldsymbol{A}} - \boldsymbol{A})\nabla u + (\widehat{c}-c)vu + \langle f - \widehat{f}, v\rangle.$$

We choose $v = u - \widehat{u} \in H_0^1(\Omega)$ and invoke (5.54) to deduce

$$c_{\widehat{\mathcal{B}}}\|\nabla v\|_{L^2(\Omega)}^2 \le \int_\Omega \nabla v \cdot (\widehat{\boldsymbol{A}} - \boldsymbol{A})\nabla u + (\widehat{c}-c)vu + \langle f - \widehat{f}, v\rangle.$$

We estimate each term separately, starting with the first and last terms

$$\int_\Omega \nabla v \cdot (\widehat{\boldsymbol{A}} - \boldsymbol{A})\nabla u \le \|\widehat{\boldsymbol{A}} - \boldsymbol{A}\|_{L^r(\Omega)}\|\nabla u\|_{L^{r^*}(\Omega)}\|\nabla v\|_{L^2(\Omega)} \tag{5.59}$$

with $2 \le r_* = 2r/(r-2) \le p_0$, as well as

$$\langle f - \widehat{f}, v\rangle \le \|f - \widehat{f}\|_{H^{-1}(\Omega)}\|\nabla v\|_{H^1(\Omega)}.$$

For the reaction term, which is more delicate, we invoke the duality pairing $W_{q'}^s$–W_q^{-s} for any $0 \le s \le 1$ and $q' = q/(q-1) \ge 1$, to obtain

$$\int_\Omega (\widehat{c}-c)vu \le \|\widehat{c}-c\|_{W_q^{-s}(\Omega)}|vu|_{W_{q'}^s(\Omega)}.$$

We now estimate $|vu|_{W_{q'}^s(\Omega)} \lesssim |vu|_{W_{p'}^1(\Omega)}$, where $1/p' = \min\{1, (1-s)/d + 1/q'\}$ guarantees that $W_{p'}^1(\Omega) \subset W_{q'}^s(\Omega)$ (Leoni 2009, Theorem 14.32). Recalling that $q > d/(2-s)$, we deduce

$$\frac{1-s}{d} + \frac{1}{q'} = \frac{1-s}{d} + 1 - \frac{1}{q} > \frac{d-1}{d} \ge \frac{1}{2},$$

whence $1/p' > 1/2$, and there exists $t < \infty$ satisfying $1/t + 1/2 = 1/p'$ and

$$|vu|_{W_{q'}^s(\Omega)} \lesssim \|\nabla v\|_{L^2(\Omega)}\|u\|_{L^t(\Omega)} + \|v\|_{L^t(\Omega)}\|\nabla u\|_{L^2(\Omega)}.$$

Using the definition of p', we obtain the explicit expression $t = \max\{2, t_0\}$, where

$$t_0 = \frac{2dq}{q(2(1-s)+d) - 2d}.$$

Moreover, for the Sobolev embedding $H^1(\Omega) \hookrightarrow L^t(\Omega)$, we require

$$1 - d\left(\frac{1}{2} - \frac{1}{t}\right) > 0 \quad \Rightarrow \quad q > \frac{d}{2-s},$$

which is our assumption on q. Therefore, as $q \to d/(2-s)$, we see that $t_0 \to 2d/(d-2)$, and the limit is infinite for $d=2$ but finite and larger than 2 for

$d > 2$. Sobolev embedding together with the first Poincaré inequality (2.2) gives the estimate

$$\|vu\|_{W^s_{q'}(\Omega)} \leq C(\Omega, t)\|\nabla v\|_{L^2(\Omega)}\|\nabla u\|_{L^2(\Omega)},$$

where $C(\Omega, t)$ is proportional to t for $d = 2$.

We finally observe that the factors $\|\nabla u\|_{L^{r_*}(\Omega)}$ and $\|\nabla u\|_{L^2(\Omega)}$ appear in the estimates of the coefficients $\boldsymbol{A}$ and c, thereby reflecting the multiplicative nature of these terms. Since $2 \leq r_* \leq p_0$, they can be further bounded in terms of $\|f\|_{W^{-1}_{r_*}(\Omega)}$ according to (5.57). This in conjunction with the preceding estimates yields the assertion (5.58). □

A natural and rather popular choice of parameters (r, q, s) in Lemma 5.20 (continuous dependence on data) is $r = q = \infty$ and $s = 0$, but this would prevent the coefficients $(\boldsymbol{A}, c)$ from being discontinuous within elements; see (5.56). We will explore this matter further in Section 7 (data approximation).

Remark 5.21 (L^2**-approximation of** $\boldsymbol{A}$)**.** It is appealing to estimate the distortion $\boldsymbol{A} - \widehat{\boldsymbol{A}}$ in $L^2(\Omega)$ rather than in $L^r(\Omega)$ because it is a simpler norm to deal with. Since $\|\boldsymbol{A}\|_{L^\infty(\Omega)} \leq \alpha_2$, $\|\widehat{\boldsymbol{A}}\|_{L^\infty(\Omega)} \leq \widehat{\alpha}_2$ and $2 \leq r \leq \infty$, we deduce

$$\|\boldsymbol{A} - \widehat{\boldsymbol{A}}\|_{L^r(\Omega)} \leq \|\boldsymbol{A} - \widehat{\boldsymbol{A}}\|^{1-2/r}_{L^\infty(\Omega)}\|\boldsymbol{A} - \widehat{\boldsymbol{A}}\|^{2/r}_{L^2(\Omega)} \lesssim \|\boldsymbol{A} - \widehat{\boldsymbol{A}}\|^{2/r}_{L^2(\Omega)}.$$

However, this may be sub-optimal in general. One important situation where this is sharp corresponds to $\boldsymbol{A}$ being piecewise constant with jump discontinuities across a Lipschitz hypersurface γ and $\widehat{\boldsymbol{A}} = \boldsymbol{A}$ on every element $T \in \mathcal{T}$ not intersecting γ. In that case, the equivalence

$$\|\boldsymbol{A} - \widehat{\boldsymbol{A}}\|_{L^p(\Omega)} \approx |\{x \in \Omega \mid \boldsymbol{A}(x) \neq \widehat{\boldsymbol{A}}(x)\}|^{1/p}$$

is valid for $1 \leq p \leq \infty$, whence

$$\|\boldsymbol{A} - \widehat{\boldsymbol{A}}\|_{L^r(\Omega)} \approx \|\boldsymbol{A} - \widehat{\boldsymbol{A}}\|^{2/r}_{L^2(\Omega)}.$$

5.4.2. *Approximation of $\mathcal{D}$: module* DATA

In this section we briefly discuss the structure of DATA, which is the module of Algorithm 5.1 (AFEM-TS) responsible for data approximation.

Henceforth we will no longer rely on the Banach space $\widehat{D}(\Omega)$ defined in (5.55) and used in Lemma 5.20 (continuous dependence on data). Instead we restrict the error notion to the following stronger Banach space:

$$D(\Omega) := L^r(\Omega; \mathbb{R}^{d\times d}) \times L^q(\Omega) \times H^{-1}(\Omega), \tag{5.60}$$

where $q = 2$ for $d < 4$ or $q > d/2$ for $d \geq 4$; we justify the choice of q below. Let $\mathbb{D}$ and $\mathbb{D}_\mathcal{T}$ be the spaces defined in (5.1) and (5.2) for a conforming mesh $\mathcal{T} \in \mathbb{T}$. Given $\mathcal{D} = (\boldsymbol{A}, c, f) \in \mathbb{D}$, let $\delta_\mathcal{T}(\mathcal{D})$ be the best approximation error of $\mathcal{D}$ within

$\mathbb{D}_{\mathcal{T}}$ measured in the space $D(\Omega)$, namely

$$\delta_{\mathcal{T}}(\mathcal{D}) := \inf_{\mathcal{D}_{\mathcal{T}} \in \mathbb{D}_{\mathcal{T}}} \|\mathcal{D} - \mathcal{D}_{\mathcal{T}}\|_{D(\Omega)}. \tag{5.61}$$

This quantity characterizes the approximation quality of $\mathbb{D}_{\mathcal{T}}$, thereby having theoretical value. Since $\delta_{\mathcal{T}}(\mathcal{D})$ is hard to access in view of the norms involved in the definition of $D(\Omega)$, the module DATA computes the surrogate quantity

$$\operatorname{osc}_{\mathcal{T}}(\mathcal{D}) := \|\mathcal{D} - \widehat{\mathcal{D}}\|_{D(\Omega)}$$

for some approximation $\widehat{\mathcal{D}} \in \mathbb{D}_{\mathcal{T}}$ to be specified below.

Assumption 5.22 (properties of DATA). Given a conforming mesh $\mathcal{T} \in \mathbb{T}$ and a tolerance $\tau > 0$, the call

$$[\widehat{\mathcal{T}}, \widehat{\mathcal{D}}] = \mathsf{DATA}(\mathcal{T}, \mathcal{D}, \tau) \tag{5.62}$$

creates an admissible refinement $\widehat{\mathcal{T}}$ of $\mathcal{T}$ and discrete data $\widehat{\mathcal{D}} = \mathcal{D}_{\widehat{\mathcal{T}}} \in \mathbb{D}_{\widehat{\mathcal{T}}}$ such that for a constant C_{data},

$$\operatorname{osc}_{\widehat{\mathcal{T}}}(\mathcal{D}) := \|\mathcal{D} - \widehat{\mathcal{D}}\|_{D(\Omega)} \le C_{\text{data}}\tau, \tag{5.63}$$

as well as the structural conditions (5.51), are achieved in a finite number of iterations that depends on the regularity of $\mathcal{D}$, and such that

$$\operatorname{osc}_{\widehat{\mathcal{T}}}(\mathcal{D}) \le \Lambda_{\text{data}}\, \delta_{\widehat{\mathcal{T}}}(\mathcal{D}), \tag{5.64}$$

with $\Lambda_{\text{data}} \ge 1$ depending only on the shape regularity of $\mathbb{T}$, the polynomial degree n and the Lebesgue exponents in the space $D(\Omega)$.

In view of Lemma 5.20 (continuous dependence on data), there exists a constant $C_D > 0$ depending on $\mathcal{D}, \Omega$, and the shape regularity of $\mathbb{T}$, such that the exact solutions $u = u(\mathcal{D})$ and $\widehat{u} = u(\widehat{\mathcal{D}})$ of (2.5) and (5.5), corresponding to data $\mathcal{D}$ and $\widehat{\mathcal{D}}$ respectively, satisfy the error estimate

$$|u - \widehat{u}|_{H_0^1(\Omega)} \le C_D \tau. \tag{5.65}$$

A brief discussion follows about computing $\operatorname{osc}_{\widehat{\mathcal{T}}}(\mathcal{D})$, where $\widehat{\mathcal{T}}$ remains fixed and is replaced by $\mathcal{T}$ to simplify the notation. Specific details are given later in Assumptions 6.10 and 6.11 of Section 6.10 and especially in Section 7.

Approximating the coefficients. We now construct approximations $(\widehat{\boldsymbol{A}}, \widehat{c})$ using local L^2-projections, and emphasize that this does not enforce the side constraints in the structural assumption (5.51). In Section 7 we propose a nonlinear correction satisfying the side constraints without sacrificing the accuracy.

Given $T \in \mathcal{T}$, and $v \in L^p(T)$ with $1 \le p \le \infty$, we let $\Pi_T v := \Pi_T^{n-1} v$ denote the L^2-projection of v onto the space $\mathbb{P}_{n-1}$ of polynomials of degree $\le n-1$, namely

$$\Pi_T v \in \mathbb{P}_{n-1}: \qquad \int_T \Pi_T v\, w = \int_T v\, w \quad \text{for all } w \in \mathbb{P}_{n-1}. \tag{5.66}$$

Lemma 5.23 (L^p-stability of Π_T). *For every $1 \le p \le \infty$ and $v \in L^p(T)$, there exists a constant C depending on p, n and the shape regularity of $\mathbb{T}$ such that*

$$\|\Pi_T v\|_{L^p(T)} \le C\|v\|_{L^p(T)} \quad \text{for all } T \in \mathcal{T}. \tag{5.67}$$

Proof. It is trivial to see that $\|\Pi_T v\|_{L^2(T)} \le \|v\|_{L^2(T)}$. Let $2 < p \le \infty$ and combine an inverse estimate with a Hölder inequality to write

$$\|\Pi_T v\|_{L^p(T)} \le Ch_T^{d/p-d/2}\|\Pi_T v\|_{L^2(T)} \le Ch_T^{d/p-d/2}\|v\|_{L^2(T)} \le C\|v\|_{L^p(T)}.$$

For $1 \le p < 2$ we proceed by duality. Let $\varphi \in L^q(T)$ with $q = p/(p-1)$. Then

$$\int_T \Pi_T v\varphi = \int_T v\Pi_T\varphi \le \|v\|_{L^p(T)}\|\Pi_T\varphi\|_{L^q(T)} \le C\|v\|_{L^p(T)}\|\varphi\|_{L^q(T)},$$

which implies (5.67) and concludes the proof. □

We immediately have the following simple consequence of Lemma 5.23.

Corollary 5.24 (best approximation of Π_T). *For every $1 \le p \le \infty$ and $v \in L^p(T)$, there exists a constant $C_{BA} \ge 1$ depending on p, n and the shape regularity of $\mathbb{T}$ such that*

$$\|v - \Pi_T v\|_{L^p(T)} \le C_{BA} \inf_{w \in \mathbb{P}_{n-1}} \|v - w\|_{L^p(T)}. \tag{5.68}$$

Proof. We combine the invariance of Π_T on $\mathbb{P}_{n-1}$, i.e. $\Pi_T w = w$ for $w \in \mathbb{P}_{n-1}$, with (5.67) to see that

$$\|v - \Pi_T v\|_{L^p(T)} = \|(v - w) - \Pi_T(v - w)\|_{L^p(T)} \le C\|v - w\|_{L^p(T)}.$$

This implies (5.68) as asserted. □

The L^2-projection is easily computable because it entails solving the linear system (5.66). However, this flexibility comes at the expense of a best approximation constant $C_{BA} > 1$ in (5.68) for $p \ne 2$. The best L^p-approximation of v in T is also computable, because it boils down to a convex minimization problem, and would result in $C_{BA} = 1$. This excellent property is superseded by the simplicity of (5.66), which makes $\Pi_T v$ the approximation of choice.

Corollary 5.25 (quasi-monotonicity of Π_T). *Let $\mathcal{T}, \mathcal{T}_* \in \mathbb{T}$ be such that $\mathcal{T} \le \mathcal{T}_*$, and let $T \in \mathcal{T}, T_* \in \mathcal{T}_*$ satisfy $T_* \subset T$. If C_{BA} is the constant in* (5.68), *then*

$$\|v - \Pi_{T_*} v\|_{L^p(T_*)} \le C_{BA}\|v - \Pi_T v\|_{L^p(T)} \tag{5.69}$$

for all $1 \le p \le \infty$, and $C_{BA} = 1$ for $p = 2$.

Proof. Simply use (5.68) to write

$$\|v - \Pi_{T_*} v\|_{L^p(T_*)} \le C_{BA}\|v - \Pi_T v\|_{L^p(T_*)} \le C_{BA}\|v - \Pi_T v\|_{L^p(T)}.$$

This is the desired bound. □

We are now ready to define the discontinuous $\mathbb{P}_{n-1}$-approximation $\widehat{v}$ of $v \in L^p(\Omega)$. Inequality (5.69) with $C_{BA} > 1$ is fine for most instances except Lemma 7.5 below. Therefore we introduce a nonlinear modification of the obvious choice $\widehat{v}$ for $T \in \mathcal{T}$, namely $\widehat{v} = \Pi_T v$. We give a recursive (and computable) definition as follows: if $T \in \mathcal{T}_0$, then $\widehat{v}|_T := \Pi_T v$; if $T \in \mathcal{T}$, let $\widehat{v}|_{P(T)} \in \mathbb{P}_{n-1}$ be the approximation of v in the parent element $P(T)$ of T, and set

$$\widehat{v}|_T := \begin{cases} \Pi_T v, & \text{if } \|v - \Pi_T v\|_{L^p(T)} \leq \|v - \widehat{v}|_{P(T)}\|_{L^p(T)}, \\ \widehat{v}|_{P(T)}, & \text{if } \|v - \Pi_T v\|_{L^p(T)} > \|v - \widehat{v}|_{P(T)}\|_{L^p(T)}. \end{cases} \tag{5.70}$$

We then define

$$\operatorname{osc}_{\mathcal{T}}(v, T)_p := \|v - \widehat{v}\|_{L^p(T)} \quad \text{for all } T \in \mathcal{T}. \tag{5.71}$$

Since the chain of elements emanating from $\mathcal{T}_0$ and culminating with T is unique, the notion $\operatorname{osc}_{\mathcal{T}}(v, T)_p$ is well-defined and independent of $\mathcal{T}$. The following result is an immediate consequence of (5.70).

Lemma 5.26 (monotonicity of oscillation). *For all $1 \leq p \leq \infty$, $\mathcal{T}, \mathcal{T}_* \in \mathbb{T}$ with $\mathcal{T} \leq \mathcal{T}_*$, and $T_* \in \mathcal{T}_*, T \in \mathcal{T}$ so that $T_* \subset T$, we have*

$$\operatorname{osc}_{\mathcal{T}_*}(v, T_*)_p \leq \operatorname{osc}_{\mathcal{T}}(v, T)_p. \tag{5.72}$$

Consequently, for any $n \geq 1$ and $T \in \mathcal{T}$, let

$$\widehat{\boldsymbol{A}} \in \left[\mathbb{S}_{\mathcal{T}}^{n-1,-1}\right]^{d \times d}, \quad \widehat{c} \in \mathbb{S}_{\mathcal{T}}^{n-1,-1}$$

be defined locally via (5.70), and let the surrogate element error indicators of $(\boldsymbol{A}, c)$ be given by

$$\operatorname{osc}_{\mathcal{T}}(\boldsymbol{A}, T)_r := \|\boldsymbol{A} - \widehat{\boldsymbol{A}}\|_{L^r(T)}, \quad \operatorname{osc}_{\mathcal{T}}(c, T)_q := \|c - \widehat{c}\|_{L^q(T)}, \tag{5.73}$$

for some $2 \leq r \leq \infty$ and $d/2 < q \leq \infty$ according to (5.58) for $s = 0$. The simplest choice $q = 2$ yields $\widehat{c}_T = \Pi_T c$ in (5.70), but requires the restriction $d < 4$, which is fine in practice.

For $n = 1$ the situation is a bit special on two counts. First, $\Pi_T v$ reduces to mean values of v, namely

$$\Pi_T \boldsymbol{A} := \frac{1}{|T|} \int_T \boldsymbol{A}, \quad \Pi_T c := \frac{1}{|T|} \int_T c \quad \text{for all } T \in \mathcal{T}. \tag{5.74}$$

for $\boldsymbol{A} \in M(\alpha_1, \alpha_2)$, $c \in R(c_1, c_2)$ defined in (5.50). Hence $\widehat{\boldsymbol{A}} \in M(\widehat{\alpha}_1, \widehat{\alpha}_2)$ with $\widehat{\alpha}_1 = \alpha_1, \widehat{\alpha}_2 = \alpha_2$ and $\widehat{c} \in R(\widehat{c}_1, \widehat{c}_2)$ with $\widehat{c}_1 = c_1, \widehat{c}_2 = c_2$, that is, the L^2-projections (5.74) on piecewise constants over $\mathcal{T}$ as well as $\widehat{\boldsymbol{A}}$ and $\widehat{c}$ satisfy the side conditions in (5.51) without changing the original range of parameters. In addition, instead of (5.73), we can exploit *superconvergence* in $W_q^{-1}(\Omega)$ with $q > d/(2 - s) = d$ in (5.58). In fact we utilize the orthogonality of Π_T in conjunction with (5.68) and

(3.16), to obtain, for an arbitrary function $w \in W^1_{q^*}(\Omega)$ and $q^* = q/(q-1)$,

$$\int_T (c - \Pi_{\mathcal{T}} c) w = \int_T (c - \Pi_{\mathcal{T}} c)(w - \Pi_{\mathcal{T}} w) \lesssim h_T^t \|c - \Pi_{\mathcal{T}} c\|_{L^r(T)} |w|_{W^1_{q^*}(\omega_T)},$$

where $t = 1 - d/q^* + d/r^* = 1 + d/q - d/r > 0$ and $r^* = r/(r-1)$. We consider two cases: $r = 2, \infty$. If $r = 2$ and $s = 1$, then $q > d$ results in $0 < t < 2 - d/2$ and entails the restriction $d < 4$. This implies $\|c - \widehat{c}\|_{W_q^{-1}(\Omega)} \lesssim \operatorname{osc}_{\mathcal{T}}(c)_2$, where

$$\operatorname{osc}_{\mathcal{T}}(c, T)_2 := h_T^t \|c - \Pi_{\mathcal{T}} c\|_{L^2(T)}. \tag{5.75}$$

If $r = \infty$ and $s = 1$, then $q = \infty$ yields $t = 1$ and $\|c - \widehat{c}\|_{W_q^{-1}(\Omega)} \lesssim \operatorname{osc}_{\mathcal{T}}(c)_\infty$, where

$$\operatorname{osc}_{\mathcal{T}}(c, T)_\infty := h_T \|c - \Pi_{\mathcal{T}} c\|_{L^\infty(T)}. \tag{5.76}$$

Approximating the load. Dealing with $f \in H^{-1}(\Omega)$ is trickier for several reasons. First, the norm in $H^{-1}(\Omega)$ is non-local, so its localization is non-obvious. We recall the definition (4.52) of local oscillation $\operatorname{osc}_{\mathcal{T}}(f, T)_{-1}$ for $T \in \mathcal{T}$ and Corollary 4.31 (local near-best approximation), to deduce

$$\operatorname{osc}_{\mathcal{T}}(f, T)_{-1} := \|f - P_{\mathcal{T}} f\|_{H^{-1}(\omega_T)} \le C_{\mathrm{lStb}} \inf_{\chi \in \mathbb{F}_{\mathcal{T}_{\omega_T}}} \|f - \chi\|_{H^{-1}(\omega_T)}, \tag{5.77}$$

where C_{lStb} is the constant in Lemma 4.28 (local H^{-1}-stability); equivalently, $\operatorname{osc}_{\mathcal{T}}(f, T)_{-1}$ delivers a near-best approximation of f in $H^{-1}(\omega_T)$. The second issue at stake is that without further assumptions on f, it is not possible to evaluate or bound the left-hand side of (5.77). In Section 7 we will consider several classes of loads amenable to computation and yet relevant in practice.

A popular variant of this approach for $f \in L^2(\Omega)$ replaces χ in (5.77) with the L^2-projection $\Pi_{\mathcal{T}}$ onto discontinuous piecewise polynomials of degree $n-1$, and sets $\widehat{f} = \Pi_{\mathcal{T}} f$. This leads to the standard local weighted L^2-element error indicator

$$\widetilde{\operatorname{osc}}_{\mathcal{T}}(f, T)_{-1} := h_T \|f - \widehat{f}\|_{L^2(T)} \quad \text{for all } T \in \mathcal{T}. \tag{5.78}$$

Data error estimators. They are the following quantities for the coefficients $(\boldsymbol{A}, c)$:

$$\begin{aligned} \operatorname{osc}_{\mathcal{T}}(\boldsymbol{A})_r &:= \left(\sum_{T \in \mathcal{T}} \operatorname{osc}_{\mathcal{T}}(\boldsymbol{A}, T)_r^r \right)^{1/r}, \\ \operatorname{osc}_{\mathcal{T}}(c)_q &:= \left(\sum_{T \in \mathcal{T}} \operatorname{osc}_{\mathcal{T}}(c, T)_q^q \right)^{1/q}, \end{aligned} \tag{5.79}$$

which accumulate in ℓ^r and ℓ^q for $2 \le r \le \infty$ and $d/2 < q$; recall that $q = 2$ is an admissible choice provided $d < 4$. In contrast, the global error estimator for f,

$$\operatorname{osc}_{\mathcal{T}}(f)_{-1} := \left(\sum_{T \in \mathcal{T}} \operatorname{osc}_{\mathcal{T}}(f, T)_{-1}^2 \right)^{1/2}, \tag{5.80}$$

accumulates in ℓ^2. The total data error estimator satisfies (5.64) and reads

$$\operatorname{osc}_{\mathcal{T}}(\mathcal{D}) := \operatorname{osc}_{\mathcal{T}}(\boldsymbol{A})_r + \operatorname{osc}_{\mathcal{T}}(c)_q + \operatorname{osc}_{\mathcal{T}}(f)_{-1}. \tag{5.81}$$

The module DATA. This module reduces the oscillation of data $\mathcal{D} = (\boldsymbol{A}, c, f)$ sequentially. It consists of a linear approximation followed by a nonlinear correction.

Given a coefficient $v = \boldsymbol{A}, c$, a mesh $\mathcal{T} \in \mathbb{T}$, a tolerance τ, an accumulation index $1 \leq p \leq \infty$, and a number of bisections $b \geq 1$ per marked element, the call

$$[\widetilde{\mathcal{T}}, \widetilde{v}] = \mathsf{GREEDY}(v, \mathcal{T}, \tau, p, b)$$

returns a conforming refinement $\widetilde{\mathcal{T}}$ of $\mathcal{T}$ and a piecewise polynomial approximation $\widetilde{v}$ of v over $\widetilde{\mathcal{T}}$ such that the oscillation computed with $v - \widetilde{v}$ satisfies

$$\operatorname{osc}_{\widetilde{\mathcal{T}}}(v)_p \leq \tau.$$

For the load function f, since the computation of $\operatorname{osc}_{\mathcal{T}}(f)_{-1}$ is impossible without further assumptions on f, we will consider three surrogate estimators $\widetilde{\operatorname{osc}}_{\mathcal{T}}(f)_{-1}$ in Section 7.3 that also accumulate in ℓ^p such that, for all $\mathcal{T} \in \mathbb{T}$,

$$\operatorname{osc}_{\mathcal{T}}(f)_{-1} \leq C_{\text{data}} \widetilde{\operatorname{osc}}_{\mathcal{T}}(f)_{-1},$$

where $C_{\text{data}} \geq 1$. GREEDY applied to the surrogate estimator constructs $\widetilde{\mathcal{T}} \geq \mathcal{T}$ satisfying

$$\widetilde{\operatorname{osc}}_{\widetilde{\mathcal{T}}}(f)_{-1} \leq \tau \quad \Rightarrow \quad \operatorname{osc}_{\widetilde{\mathcal{T}}}(f)_{-1} \leq C_{\text{data}} \tau. \tag{5.82}$$

In all cases, the routine GREEDY is similar to that in Algorithm 3.18 (greedy algorithm) with several important distinctions: it accumulates the local error indicators in the ℓ^p-norm and starts from any mesh $\mathcal{T} \geq \mathcal{T}_0$ to save computational work.

Finally, the structure of the module DATA is as follows: it concatenates GREEDY with CONSTRAINT-A and CONSTRAINT-c in order to satisfy Assumption 5.22 (properties of DATA). The routine GREEDY deals with pure approximation without constraints: called with tolerance $\tau/3$, it sequentially reduces the oscillation for $\boldsymbol{A}, c, f$ with the most recent updated mesh to reduce their errors so that

$$\operatorname{osc}_{\widehat{\mathcal{T}}}(\boldsymbol{A})_r \leq \tau/3, \quad \operatorname{osc}_{\widehat{\mathcal{T}}}(c)_q \leq \tau/3, \quad \widetilde{\operatorname{osc}}_{\widehat{\mathcal{T}}}(f)_{-1} \leq \tau/3$$

on a conforming refinement $\widehat{\mathcal{T}} \geq \mathcal{T}$. This is discussed in detail in Section 7.1.

From (5.82) we get $\operatorname{osc}_{\widehat{\mathcal{T}}}(f)_{-1} \leq C_{\text{data}} \tau/3$. On the other hand, the resulting coefficients $(\widetilde{\boldsymbol{A}}, \widetilde{c})$ most likely do not satisfy the constraints (5.51) for $n > 1$. This requires a further nonlinear correction

$$[\widehat{\boldsymbol{A}}] = \mathsf{CONSTRAINT\text{-}A}(\widehat{\mathcal{T}}, \widetilde{\boldsymbol{A}}), \quad [\widehat{c}] = \mathsf{CONSTRAINT\text{-}c}(\widehat{\mathcal{T}}, \widetilde{c}),$$

that enforces (5.51) on the same grid $\widehat{\mathcal{T}}$ without compromising the accuracy gain produced by GREEDY: there exists a constant ≥ 1, still denoted by C_{data} for simplicity, such that

$$\operatorname{osc}_{\widehat{\mathcal{T}}}(\widehat{\boldsymbol{A}})_r \leq C_{\text{data}} \tau/3, \quad \operatorname{osc}_{\widehat{\mathcal{T}}}(\widehat{c})_q \leq C_{\text{data}} \tau/3 \quad \Rightarrow \quad \operatorname{osc}_{\widehat{\mathcal{T}}}(\mathcal{D}) \leq C_{\text{data}} \tau.$$

For instance, for a fixed parameter $L \geq 2$, we get $\widehat{\alpha}_1 = \frac{1}{2}\alpha_1$ and $\widehat{\alpha}_2 = (1+4L)(\alpha_2/2)$ for the parameters in (5.51). We give details in Sections 7.2, 7.3 and 7.4.

The optimality properties of DATA hinge on the performance of GREEDY and the regularity of $\mathcal{D}$. Since this is not necessary for the present convergence assessment, we discuss it later in Section 7.

5.4.3. *Computational cost of* GALERKIN

The output pair $(\widehat{\mathcal{T}}, \widehat{\mathcal{D}})$ of DATA is next taken by GALERKIN, the one-step AFEM of Algorithm 5.4 in Section 5.2.1, to run an inner loop of the form (5.18) with fixed discrete data $\widehat{\mathcal{D}}$ and initial mesh $\widehat{\mathcal{T}}$. The call (5.19) of GALERKIN stops as soon as the error tolerance ε is reached, which takes a finite number of iterations because GALERKIN is a contraction between consecutive iterates, and creates the next mesh-solution pair $(\mathcal{T}, u_{\mathcal{T}})$. It is worth noticing that, in the absence of this stopping test, the Galerkin solution $u_{\mathcal{T}}$ would converge to the solution $\widehat{u} = u(\widehat{\mathcal{D}})$ of (5.5), which is not the desired solution $u = u(\mathcal{D})$ of (2.5).

We stress that, in view of (5.63) and (5.65), the relative resolution of the modules DATA and GALERKIN is critical for the discrepancy between the exact and perturbed solutions u and $\widehat{u}$. This is ultimately responsible for the performance of AFEM-TS and is studied in Section 6.

We now investigate the number of iterations within GALERKIN, which dictate its computational cost. We point out that at iteration $k-1 \geq 0$ of AFEM-TS, the output $(\mathcal{T}_k, u_k)$ of GALERKIN, and thus of AFEM-TS, satisfies

$$\eta_k(u_k) = \eta_{\mathcal{T}_k}(u_k) \leq \varepsilon_{k-1} \quad \Rightarrow \quad |u_k - \widehat{u}_{k-1}|_{H_0^1(\Omega)} \leq C_U \varepsilon_{k-1} \tag{5.83}$$

according to (5.12). We recall that $\widehat{u}_{k-1} = \widehat{u}_{k-1}(\widehat{\mathcal{D}}_{k-1}) \in H_0^1(\Omega)$ is the exact solution with discrete data $\widehat{\mathcal{D}}_{k-1}$, and that $\mathcal{E}_{\mathcal{T}_k}(u_k, f)$ is defined with discrete data $\widehat{\mathcal{D}}_{k-1}$ and satisfies $\mathcal{E}_{\mathcal{T}_k}(u_k, f) = \eta_{\mathcal{T}_k}(u_k)$ because data oscillation $\operatorname{osc}_{\mathcal{T}_k}(f)_{-1} = 0$. The next iteration k of AFEM-TS calls DATA, which in turn refines the mesh $\mathcal{T}_k$ to $\widehat{\mathcal{T}}_k$ and updates the data approximation from $\widehat{\mathcal{D}}_{k-1}$ to $\widehat{\mathcal{D}}_k$ over $\widehat{\mathcal{T}}_k$. The pair $(\widehat{\mathcal{T}}_k, \widehat{\mathcal{D}}_k)$ determines the first Galerkin solution $u_{k,0} \in \mathbb{V}_{k,0} = \mathbb{V}_{\widehat{\mathcal{T}}_k}$ of GALERKIN and corresponding estimator $\eta_{k,0}(u_{k,0})$ with $\mathcal{T}_{k,0} = \widehat{\mathcal{T}}_k$, which must satisfy

$$\eta_{k,0}(u_{k,0}) > \varepsilon_k \tag{5.84}$$

for GALERKIN to be executed. The reduction of $\eta_{\mathcal{T}_{k,j}}(u_{k,j})$ for $j \geq 1$ dictates the number of iterations of GALERKIN. We examine this next.

Proposition 5.27 (computational cost of GALERKIN). *If the assumptions of Theorem 5.8 are valid, then for any $k \in \mathbb{N}$, the number of subiterations J_k inside a call to* GALERKIN *at iteration k of* AFEM-TS *is bounded independently of k.*

Proof. The jth error $e_{k,j} := |\widehat{u}_k - u_{k,j}|_{H_0^1(\Omega)}$ within GALERKIN converges linearly in view of Corollary 5.10 (linear convergence of error) because the discrete data $\widehat{\mathcal{D}}_k$

is fixed in these inner iterations. Exploiting the lower bound $C_L\eta_{k,j}(u_{k,j}) \le e_{k,j}$ stated in (5.12), we thus deduce

$$\eta_{k,j}(u_{k,j}) \le C_L^{-1} e_{k,j} \le C_L^{-1} C_* \alpha^{j-i} e_{k,i}, \quad j \ge i \ge 0,$$

whence $\eta_{k,j}(u_{k,j}) \le C_\# \alpha^j e_{k,0}$ with $C_\# := C_L^{-1} C_*$. The number of iterations of GALERKIN depends on the size of $\eta_{k,0}(u_{k,0})$ relative to ε_k. We assume that $\eta_{k,0}(u_{k,0}) > \varepsilon_k$ according to (5.84). We first prove that $\eta_{k,0}(u_{k,0}) \lesssim \varepsilon_k$, and next argue that J_k is bounded uniformly in k. We proceed in two steps.

[1] *Bound on* $|\widehat{u}_k - u_{k,0}|_{H_0^1(\Omega)}$. Since $u_k \in \mathbb{V}_k \subset \mathbb{V}_{k,0} = \mathbb{V}_{\widehat{\mathcal{T}}_k}$, and the Galerkin solution $u_{k,0} \in \mathbb{V}_{k,0}$ minimizes the error $|\!|\!| u_{k,0} - \widehat{u}_k |\!|\!|_\Omega$ in $\mathbb{V}_{k,0}$, relative to the energy norm induced by the bilinear form $\widehat{\mathcal{B}}$ with discrete data $\widehat{\mathcal{D}}_k$, we deduce

$$|\!|\!| u_{k,0} - \widehat{u}_k |\!|\!|_\Omega \le |\!|\!| u_k - \widehat{u}_k |\!|\!|_\Omega \le \sqrt{C_{\widehat{\mathcal{B}}}}\big(|u_k - \widehat{u}_{k-1}|_{H_0^1(\Omega)} + |\widehat{u}_{k-1} - \widehat{u}_k|_{H_0^1(\Omega)}\big),$$

where the last inequality uses (5.3) for $\widehat{\mathcal{B}}$. Invoking the *a posteriori* upper bound (5.13) and the termination condition of GALERKIN at step $k-1$, we obtain

$$|u_k - \widehat{u}_{k-1}|_{H_0^1(\Omega)} \le C_U \mathcal{E}_{\mathcal{T}_k}(u_k, f) = C_U \eta_k(u_k) \le C_U \varepsilon_{k-1} = 2C_U \varepsilon_k.$$

On the other hand, using (5.65) with $\tau = \omega\varepsilon_k$ and $0 < \omega \le 1$, we arrive at

$$|u - \widehat{u}_k|_{H_0^1(\Omega)} \le C_1 \varepsilon_k,$$

with $C_1 = \omega C_D$. The triangle inequality thus yields

$$|\widehat{u}_{k-1} - \widehat{u}_k|_{H_0^1(\Omega)} \le |u - \widehat{u}_{k-1}|_{H_0^1(\Omega)} + |u - \widehat{u}_k|_{H_0^1(\Omega)} \le C_1(\varepsilon_{k-1} + \varepsilon_k) = 3C_1\varepsilon_k,$$

whence

$$e_{k,0} = |u_{k,0} - \widehat{u}_k|_{H_0^1(\Omega)} \le \sqrt{\frac{C_{\widehat{\mathcal{B}}}}{c_{\widehat{\mathcal{B}}}}}(2C_U + 3C_1)\varepsilon_k =: C_2\varepsilon_k.$$

[2] *Bound on* J_k. We observe that GALERKIN stops once $\eta_{k,j}(u_{k,j}) \le \varepsilon_k$. Since the smallest such j is J_k, we see that

$$\varepsilon_k < \eta_{k,J_k-1}(u_{k,J_k-1}) \le C_\# \alpha^{J_k-1} e_{k,0} \le C_\# C_2 \varepsilon_k \alpha^{J_k-1}.$$

This implies the asserted bound

$$J_k \le 1 + \frac{\log(C_\# C_2)}{\log \alpha^{-1}}$$

uniform in k. □

5.4.4. Realization of AFEM-TS

We now make the two-step AFEM algorithm precise.

Algorithm 5.28 (AFEM-TS). Given an initial tolerance $\varepsilon_0 > 0$, a target tolerance `tol` and initial mesh $\mathcal{T}_0$, as well as a safety parameter $\omega \in (0, 1]$, AFEM consists of the two-step algorithm:

$[\mathcal{T}, u_\mathcal{T}] = \mathsf{AFEM\text{-}TS}(\mathcal{T}_0, \varepsilon_0, \omega, \mathtt{tol})$
  set $k = 0$ and do
    $[\widehat{\mathcal{T}}_k, \widehat{\mathcal{D}}_k] = \mathsf{DATA}(\mathcal{T}_k, \mathcal{D}, \omega\,\varepsilon_k)$
    $[\mathcal{T}_{k+1}, u_{k+1}] = \mathsf{GALERKIN}(\widehat{\mathcal{T}}_k, \widehat{\mathcal{D}}_k, \varepsilon_k)$
    $\varepsilon_{k+1} = \frac{1}{2}\varepsilon_k$
    $k \leftarrow k + 1$
  while $\varepsilon_{k-1} > \mathtt{tol}$
  return $\mathcal{T}_k, u_k$

Proposition 5.29 (convergence of AFEM-TS). *For each $k \geq 0$, the modules* DATA *and* GALERKIN *converge in a finite number of iterations, the latter independent of k. Moreover, there exists a constant C_* depending on $\mathcal{T}_0, \Omega, d, n$, the Lebesgue exponents r, q in $D(\Omega)$, the parameters $\alpha_1, \alpha_2, c_1, c_2$ in* (5.48) *and* (5.49), *and the shape regularity constant of $\mathbb{T}$, such that the output of the $(k + 1)$th iteration $[\mathcal{T}_{k+1}, u_{k+1}] = \mathsf{GALERKIN}(\widehat{\mathcal{T}}_k, \widehat{\mathcal{D}}_k, \varepsilon_k)$ satisfies $|u - u_{k+1}|_{H^1_0(\Omega)} \leq C_*\varepsilon_k$ for all $k \geq 0$. Therefore* AFEM-TS *stops after*

$$K < 2 + \frac{\log(\varepsilon_0/\mathtt{tol})}{\log 2}$$

iterations and delivers

$$|u - u_K|_{H^1_0(\Omega)} \leq C_*\,\mathtt{tol}\,.$$

Proof. In view of Assumption 5.22 (properties of DATA), the module DATA iterates a finite number of steps to reach tolerance $\tau = \omega\varepsilon_k$ for every $k \geq 0$. Moreover, the number of iterations of GALERKIN is independent of k due to Proposition 5.27 (computational cost of GALERKIN), whence we deduce that each loop of AFEM-TS requires finite iterations. Thus, the output u_{k+1} of the $(k + 1)$th loop satisfies

$$|u - u_{k+1}|_{H^1_0(\Omega)} \leq |u - \widehat{u}_k|_{H^1_0(\Omega)} + |\widehat{u}_k - u_{k+1}|_{H^1_0(\Omega)} \leq (\omega C_D + C_U)\varepsilon_k = C_*\varepsilon_k,$$

according to (5.65) with $\tau \leq \omega\varepsilon_k$ and (5.83) for all $k \geq 0$. Finally, AFEM-TS terminates after K loops, where K satisfies $\frac{1}{2}\,\mathtt{tol} < \varepsilon_{K-1} \leq \mathtt{tol}$, and the asserted estimate holds. □

This elementary proof gives no insight into whether the H^1_0-error decays optimally in terms of degrees of freedom. We assess this fundamental question in Sections 6 and 7, but investigate it computationally in Section 5.4.5.

A two-step algorithm similar to AFEM-TS was first proposed by Stevenson (2008), and further explored by Bonito *et al.* (2013*b*) and Cohen *et al.* (2012). Note that other quantities, such as the number of degrees of freedom, could be employed to stop AFEM-TS instead. It is also worth realizing that the structure of the algorithm is independent of the size of tolerance `tol`. In this vein, a user could take $\varepsilon_0 = \texttt{tol}$, provided `tol` is affordable by the computational resources at hand. With such a choice, the modules DATA and GALERKIN run only once, in sequence: data are approximated to the desired accuracy in one shot, then fed to the PDE solver which produces the approximate solution. Since the quasi-optimality theory in Section 6 would also hold for this choice of ε_0, one might wonder why we do not use this simpler strategy. We stress that iterating over ε_k has the following advantages.

- *Restarts.* Dynamical shrinking of `tol`, for instance to account for the user decision to improve the accuracy, does not entail a restart of AFEM-TS but rather a continuation from the previous computed solution. In this sense, the resulting iteration would be similar to the proposed structure of AFEM-TS.
- *Computational resources.* AFEM-TS allows for 'balanced investment' of computational resources between the modules DATA and GALERKIN. If the stopping criterion, either accuracy or number of degrees of freedom, is unrealistic for the problem at hand, AFEM-TS would still produce a discrete solution with equilibrated data and solution errors.
- *Nonlinear problems.* The interleaving approach of AFEM-TS appears to be better suited to treating nonlinear problems for which data $\mathcal{D}$ may depend on the solution. Therefore a call to GALERKIN, and corresponding solution update, must precede a call to DATA.
- *Iterative solvers.* If an efficient iterative solver is adopted within SOLVE, then the previous discrete solution of GALERKIN could be taken as initial iterate, thereby making SOLVE fast because $\epsilon_{k+1}/\epsilon_k = 1/2$. If instead we compute with DATA alone until the fixed tolerance `tol` is reached, then GALERKIN would work directly on fine meshes, which are not adapted to the geometric domain singularities, and without a good initial guess. This would lead to fewer but heavier iterations of GALERKIN, which is detrimental from a linear algebra perspective.

5.4.5. Computational assessment of AFEM-TS

In this section we explore computationally the relative performance of GALERKIN and DATA, for the two-step AFEM, and elucidate the behaviour of data and coefficient oscillations within DATA. Our observations motivate the rigorous study of Section 6, which provides theoretical support to our experiments. The numerical computations are made with the help of Funken, Praetorius and Wissgott (2011).

We consider problem (2.5) in the L-shaped domain $\Omega = (-1,1)^2 \setminus ([0,1] \times [-1,0])$, with diffusion term $\boldsymbol{A} = a\boldsymbol{I}$, where

$$a(x,y) = 1 + \exp(-50((x+0.5)^2 + (y+0.5)^2)) + \exp(-50((x+0.5)^2 + (y-0.5)^2))$$

and reaction term

$$c(x,y) = 1 + \exp(-50((x+0.5)^2 + y^2)) + \exp(-50(x^2 + (y-0.5)^2));$$

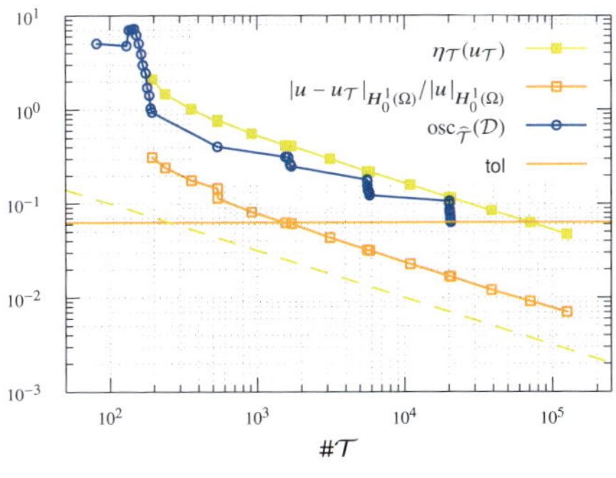

(a)

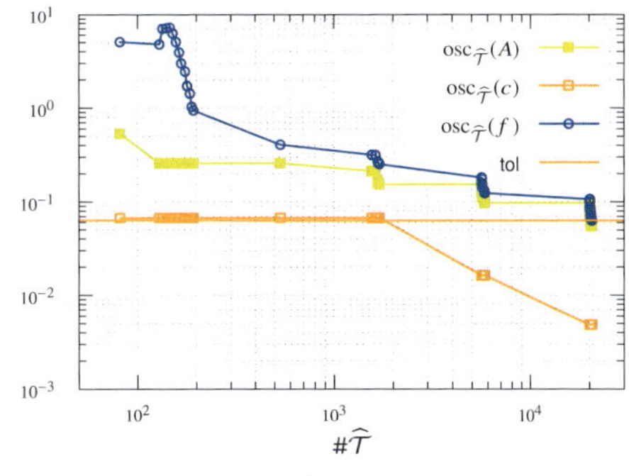

(b)

Figure 5.6. (a) Estimator $\eta_{\mathcal{T}}(u_{\mathcal{T}})$, data error $\mathrm{osc}_{\widehat{\mathcal{T}}}(\mathcal{D})$, and relative H^1-error obtained with the algorithm AFEM performing $b = 3$ bisections per marked element. The optimal decay is indicated by the dashed line with slope -0.5. (b) Diffusion error $\mathrm{osc}_{\widehat{\mathcal{T}}}(\boldsymbol{A})$, reaction error $\mathrm{osc}_{\widehat{\mathcal{T}}}(c)$, load error $\mathrm{osc}_{\widehat{\mathcal{T}}}(f)_{-1}$, obtained with the algorithm AFEM.

note that the Gaussians in the definition of a and c have the same intensity but are located in different places within Ω. The load term f and the Dirichlet boundary conditions are chosen in accordance with the analytical solution

$$u(x,y) = r^{2/3}\sin(2\alpha/3) + \exp(-1000((x-0.5)^2 + (y-0.5)^2)),$$

where (r,α) are the polar coordinates around the origin. Notice that the exact solution u is singular at the re-entrant corner: it belongs to the Sobolev spaces $H(\Omega)^{5/3-\varepsilon}$ with $\varepsilon > 0$ and $W_p^2(\Omega)$ with $p > 1$. It also exhibits a rapid transition of order $10^{-3/2}$ around the point $(0.5, 0.5)$ due to the presence of a very narrow Gaussian. The Gaussians are meant to test the performance of the module DATA, while in addition the corner singularity of the solution tests the execution of the module GALERKIN.

We utilize the following parameters in the numerical test:

$$\theta = 0.5, \quad \omega = 1, \quad \mathtt{tol} = 2^{-4}, \quad h_0 = 0.125, \quad \varepsilon_0 = 1.$$

Notice that the number of iterations of the algorithm AFEM is $K = \log_2(\epsilon_0/\mathtt{tol}) = 4$. We compute the relative H^1-error between the exact solution u and the FEM solution $u_{\mathcal{T}}$ and notice that its decay rate is $(\#\mathcal{T})^{-1/2}$ in Figure 5.6(a). This rate is consistent with that of the PDE estimator $\eta_{\mathcal{T}}(u_{\mathcal{T}})$ and data estimator $\mathrm{osc}_{\mathcal{T}}(\mathcal{D})$. In Figure 5.6(b) we display the component of the data error $\mathrm{osc}_{\widehat{\mathcal{T}}}(\boldsymbol{A})$, $\mathrm{osc}_{\widehat{\mathcal{T}}}(c)$, $\mathrm{osc}_{\widehat{\mathcal{T}}}(f)_{-1}$ defined in (5.79) and (5.80) with local contributions defined in (5.73) for $\boldsymbol{A}$ with $r = \infty$, in (5.76) for c with $t = 1$ and (5.78) for f. Recall that at each iteration k, DATA circles through $\mathrm{osc}_{\widehat{\mathcal{T}}}(\boldsymbol{A})$, $\mathrm{osc}_{\widehat{\mathcal{T}}}(c)$, and $\mathrm{osc}_{\widehat{\mathcal{T}}}(f)_{-1}$, reducing each of these oscillations to $\frac{1}{3}$ of the iteration tolerance $\varepsilon_k = 2^{-k}$. The presence of the weight h^t in $\mathrm{osc}_{\widehat{\mathcal{T}}}(c)$ considerably reduces the influence of the approximation of c,

Table 5.1. Number of marked elements to reduce the data and Galerkin errors at each iteration $k = 1, 2, 3, 4$ of AFEM-TS when using $b = 1$ and $b = 3$ refinements per marked element. Regardless of the value used for b, the reduction of the Galerkin error is driving most of the refinements followed by the error in the approximation of the diffusion coefficient $\boldsymbol{A}$. The approximation of f is subordinate to the approximation of u and $\boldsymbol{A}$ arising earlier in the adaptive loop, and thus does not generate any refinement except during the first iteration, when the Galerkin error has not yet been tackled by the algorithm. The approximation of c is below the final tolerance from the start and does not generate any refinement.

k	$\mathrm{osc}_{\widehat{\mathcal{T}}}(\boldsymbol{A})$		$\mathrm{osc}_{\widehat{\mathcal{T}}}(c)$		$\mathrm{osc}_{\widehat{\mathcal{T}}}(f)_{-1}$		$\eta_{\mathcal{T}}(u_{\mathcal{T}})$	
	$b=1$	$b=3$	$b=1$	$b=3$	$b=1$	$b=3$	$b=1$	$b=3$
1	32	16	0	0	26	13	363	308
2	16	16	0	0	0	0	1 636	1 138
3	120	43	0	0	0	0	7 447	4 227
4	123	62	0	0	0	0	42 792	15 268
5	82	138	0	0	0	0	144 345	102 350

which is below threshold from the start and thus never generates any refinement (see Table 5.1). The local oscillation for f also includes a weight vanishing as $h \to 0$ but $\mathrm{osc}_{\widehat{\mathcal{T}}}(f)_{-1}$ is above the desired tolerance, which would in principle generate refinements. However, since at each iteration DATA considers $\mathrm{osc}_{\widehat{\mathcal{T}}}(\boldsymbol{A})$ first and the regions refined to reduce $\mathrm{osc}_{\widehat{\mathcal{T}}}(f)_{-1}$ are included in the regions needed to be refined to reduce $\eta_{\mathcal{T}}(u_{\mathcal{T}})$ and $\mathrm{osc}_{\widehat{\mathcal{T}}}(\boldsymbol{A})$, the GREEDY routine applied to f does not refine any element except during the first iteration, when the Galerkin error has not yet been reduced by the algorithm. Overall, the reduction of the Galerkin error is driving most of the refinements. The number of marked elements to reduce the approximation errors of $\boldsymbol{A}$, c, f and the residual estimator are reported in Table 5.1 along with those when $b = 1$ refinement is used per marked element. In Figure 5.7 we provide the resulting meshes after the first iteration of DATA and GALERKIN.

5.5. *Convergence for other boundary conditions*

First we consider the variational problem (2.13) with *Robin* boundary condition. We approximate data $\mathcal{D} = (\boldsymbol{A}, c, p, f, g)$ by piecewise polynomials $\widehat{\mathcal{D}} = (\widehat{\boldsymbol{A}}, \widehat{c}, \widehat{p}, \widehat{f}, \widehat{g})$, The only difference with respect to (5.2) is that the new functions (p, g) are approximated on $\partial\Omega$ by discontinuous polynomials $(\widehat{p}, \widehat{g})$ of degree $n-1$ and $2n-1$. The projection operator $P_{\mathcal{T}}$ approximates $g\delta_{\partial\Omega}$ by $\widehat{g\delta_{\partial\Omega}} = P_{\mathcal{T}}(g\delta_{\partial\Omega})$ without component in the bulk because $g\delta_{\partial\Omega}$ is a line Dirac mass aligned with the mesh. Discrete functions $(\widehat{p}, \widehat{g})$ must be produced by DATA, subject to a sign constraint on p. The

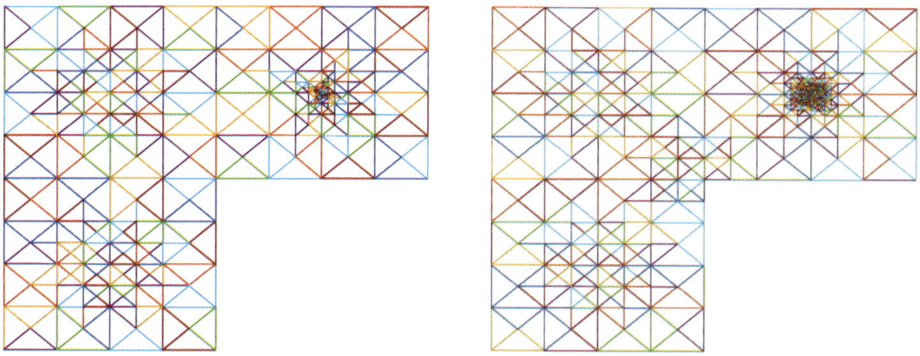

(a) (b)

Figure 5.7. Resulting meshes after the first iteration of DATA (a) and after the first iteration of GALERKIN (b). DATA marked 29 elements for refinement while GALERKIN marked 308 elements. Refer to Table 5.1 for more details.

approximate bilinear form $\widehat{\mathcal{B}}$ and linear functional $\widehat{\ell}$ read

$$\widehat{\mathcal{B}}[w,v] := \int_\Omega \nabla v \cdot \widehat{\mathbf{A}}\nabla w + \widehat{c}vw + \int_{\partial\Omega} \widehat{p}vw, \quad \widehat{\ell}(v) := \langle \widehat{f}, v\rangle + \int_{\partial\Omega} \widehat{g}\,v. \quad (5.85)$$

The *a posteriori* error estimates of Section 4 extend to this pair $(\widehat{\mathcal{B}}, \widehat{\ell})$. The algorithms GALERKIN, AFEM-SW and AFEM-TS are similar to those above and possess a similar supporting convergence theory. The *Neumann* boundary condition is a particular case with $p = 0$. We do not pursue this any further.

However, the *pure Neumann* boundary condition is special because of the global compatibility condition $\widehat{\ell}(1) = \langle \widehat{\ell}, 1\rangle = 0$. In Section 4.10 we introduce a new projection operator $\widetilde{P}_\mathcal{T}$, a modification of $P_\mathcal{T}$, with the requisite properties of local approximation and global compatibility $\langle \widetilde{P}_\mathcal{T}\ell, 1\rangle = 0$ provided $\ell \in H^1(\Omega)^*$ satisfies $\langle \ell, 1\rangle = 0$. We thus set $\widetilde{\ell} = \widetilde{P}_\mathcal{T}\ell$ to solve the Galerkin problems and use $\widetilde{P}_z$ in the local indicators. We do not explore this matter further.

For a *non-homogeneous Dirichlet* boundary data $g \in H^{1/2}(\partial\Omega)$, DATA must produce a continuous piecewise polynomial approximation $\widehat{g}$ of degree n, thereby consistent with the Galerkin solution $u_\mathcal{T}$. The Dirichlet oscillation $\mathrm{osc}_\mathcal{T}(g)_{1/2}$ is defined in (4.96) and is locally computable. Data oscillation now becomes

$$\mathrm{osc}_\mathcal{T}(\ell) = \mathrm{osc}_\mathcal{T}(f)_{-1} + \mathrm{osc}_\mathcal{T}(g)_{1/2},$$

and added to the PDE estimator $\eta_\mathcal{T}(u_\mathcal{T})$ for $g = 0$ gives a full estimator equivalent to the error, according to Theorem 4.74 (estimators for general Dirichlet conditions). With these minor modifications, the convergence theory for GALERKIN, AFEM-SW and AFEM-TS extends to this case. We do not provide any further details.

5.6. Convergence for alternative estimators

We have so far developed a convergence theory for the residual estimator $\mathcal{E}_{\mathcal{T}}(u_{\mathcal{T}}, f)$. The purpose of this section is to extend this theory to the three alternative estimators discussed in Section 4.9, namely

- $\mathcal{E}_{\mathcal{T}}^{\mathrm{lpb}}(u_{\mathcal{T}}, f)^2 = \eta_{\mathcal{T}}^{\mathrm{lpb}}(u_{\mathcal{T}})^2 + \operatorname{osc}_{\mathcal{T}}(f)_{-1}^2$: estimator based on *local problems*,
- $\mathcal{E}_{\mathcal{T}}^{\mathrm{hier}}(u_{\mathcal{T}}, f)^2 = \eta_{\mathcal{T}}^{\mathrm{hier}}(u_{\mathcal{T}})^2 + \operatorname{osc}_{\mathcal{T}}(f)_{-1}^2$: *hierarchical* estimator,
- $\mathcal{E}_{\mathcal{T}}^{\mathrm{feq}}(u_{\mathcal{T}}, f)^2 = \eta_{\mathcal{T}}^{\mathrm{feq}}(u_{\mathcal{T}})^2 + \operatorname{osc}_{\mathcal{T}}(f)_{-1}^2$: estimator based on *flux equilibration*.

They are all computed on stars ω_z with $z \in \mathcal{V}$ and possess a similar structure. The first term is the PDE estimator, from now on called $\zeta_{\mathcal{T}}(u_{\mathcal{T}})$ to refer to any of them, and is locally equivalent to the discrete residual $P_{\mathcal{T}} R_{\mathcal{T}}$

$$\zeta_{\mathcal{T}}(u_{\mathcal{T}}, z) \approx \|P_{\mathcal{T}} R_{\mathcal{T}}\|_{H^{-1}(\omega_z)} \quad \text{for all } z \in \mathcal{V}; \tag{5.86}$$

see Theorems 4.59, 4.64 and 4.69. In fact they are all different mechanisms to extract information from $P_{\mathcal{T}} R_{\mathcal{T}}$. Since the vertex-indexed residual PDE indicator $\eta_{\mathcal{T}}(u_{\mathcal{T}}, z) := \eta_{\mathcal{T}}^{\mathrm{res}}(u_{\mathcal{T}}, z)$, defined in (4.70a), is also proved to be equivalent to $\|P_{\mathcal{T}} R_{\mathcal{T}}\|_{H^{-1}(\omega_z)}$ in Theorem 4.58 (vertex-indexed modified residual estimator), we deduce the existence of two equivalence constants $C_L^{\mathrm{eq}} \le C_U^{\mathrm{eq}}$ such that

$$C_L^{\mathrm{eq}} \eta_{\mathcal{T}}(u_{\mathcal{T}}, z) \le \zeta_{\mathcal{T}}(u_{\mathcal{T}}, z) \le C_U^{\mathrm{eq}} \eta_{\mathcal{T}}(u_{\mathcal{T}}, z) \quad \text{for all } z \in \mathcal{V}. \tag{5.87}$$

Following Kreuzer and Siebert (2011), we will exploit this property to prove convergence of AFEM driven by $\zeta_{\mathcal{T}}(u_{\mathcal{T}})$. An obstruction to a direct convergence theory is that our preceding results rely heavily on Lemma 4.53 (reduction property of the estimator), which is not necessarily valid for any of the alternative estimators. We refer to Cascón and Nochetto (2012), who present a direct approach based on the local lower bound for discrete solutions of Theorem 4.51 (lower bound for corrections). The latter is guaranteed by Definition 4.50 (interior vertex property) for operators with coefficients $\boldsymbol{A}$ piecewise constant and $c = 0$, and any polynomial degree $n \ge 1$, but we do not know its validity for more general coefficients $(\boldsymbol{A}, c)$.

The key for convergence is imposing a Dörfler marking. We say that a set of vertices $\mathcal{M}_{\mathcal{V}}$ satisfies a Dörfler property with parameter $\theta \le 1$ if

$$\zeta_{\mathcal{T}}(u_{\mathcal{T}}, \mathcal{M}_{\mathcal{V}})^2 := \sum_{z \in \mathcal{M}_{\mathcal{V}}} \zeta_{\mathcal{T}}(u_{\mathcal{T}}, z)^2 \ge \theta^2 \sum_{z \in \mathcal{V}} \zeta_{\mathcal{T}}(u_{\mathcal{T}}, z)^2 =: \zeta_{\mathcal{T}}(u_{\mathcal{T}})^2. \tag{5.88}$$

Let $\mathcal{M}$ be the collection of elements contained in the stars ω_z with $z \in \mathcal{M}_{\mathcal{V}}$. Then MARK marks all elements in $\mathcal{M}$, and REFINE bisects them $b \ge 1$ times. This gives rise to a star-driven GALERKIN procedure.

Lemma 5.30 (Dörfler property). *If the set of vertices $\mathcal{M}_{\mathcal{V}}$ satisfies a Dörfler property with parameter θ for $\zeta_{\mathcal{T}}(u_{\mathcal{T}})$, then $\mathcal{M}$ satisfies a Dörfler property with parameter $\overline{\theta} = (C_L^{\mathrm{eq}}/C_U^{\mathrm{eq}})\theta$ for $\eta_{\mathcal{T}}(u_{\mathcal{T}})$.*

Proof. Simply use (5.87) to derive (5.88) for $\eta_{\mathcal{T}}(u_{\mathcal{T}})$ with parameter $\overline{\theta}$. □

Hence, star-driven procedures for $\zeta_{\mathcal{T}}(u_{\mathcal{T}})$ lead to the corresponding counterparts for $\eta_{\mathcal{T}}(u_{\mathcal{T}})$. It turns out that algorithms GALERKIN, AFEM-SW and AFEM-TS can be reformulated for vertex-indexed indicators $\{\eta_{\mathcal{T}}(u_{\mathcal{T}}, z)\}_{z\in\mathcal{V}}$ as defined in (4.70a), without changing their essential properties. We may thus wonder about them driven by $\{\zeta_{\mathcal{T}}(u_{\mathcal{T}}, z)\}_{z\in\mathcal{V}}$ instead. Since these algorithms hinge on the Dörfler property (5.88), Lemma 5.30 gives rise to similar convergence properties for $\zeta_{\mathcal{T}}(u_{\mathcal{T}})$-driven algorithms provided (5.88) is enforced. We state this next without proof.

Corollary 5.31 (convergence of GALERKIN). *If the coefficients $(\boldsymbol{A}, c, f) \in \mathbb{D}_{\mathcal{T}}$, then there exist $0 < \overline{\alpha} < 1$ and $C_*, C_\# > 0$ such that the solution–estimator pairs $(u_j, \zeta_j(u_j))$ of* GALERKIN *converge linearly, namely, for all $k \ge j \ge 0$,*

$$|u - u_k|_{H_0^1(\Omega)} \le C_* \overline{\alpha}^{k-j} |u - u_j|_{H_0^1(\Omega)}, \quad \zeta_k(u_k) \le C_\# \overline{\alpha}^{k-j} \zeta_j(u_j).$$

Corollary 5.32 (convergence of AFEM-SW). *If the coefficients $(\boldsymbol{A}, c)$ are discrete and $f \in H^{-1}(\Omega)$, then for $0 < \omega \le \omega_0$, $\xi \le \frac{1}{2}$ as in Theorem 5.17, there exist $0 < \overline{\alpha} < 1$ and $C_*, C_\# > 0$ such that the solution–estimator pairs $(u_j, \mathcal{E}_j(u_j, f))$ of* AFEM-SW*, where $\mathcal{E}_j(u_j, f)^2 = \zeta_j(u_j)^2 + \operatorname{osc}_j(f)_{-1}^2$, converge linearly: for all $k \ge j \ge 0$,*

$$|u - u_k|_{H_0^1(\Omega)} \le C_* \overline{\alpha}^{k-j} |u - u_j|_{H_0^1(\Omega)}, \quad \mathcal{E}_k(u_k, f) \le C_\# \overline{\alpha}^{k-j} \mathcal{E}_j(u_j, f).$$

Both GALERKIN and AFEM-SW converge under restrictions on $\mathcal{D} = (\boldsymbol{A}, c, f)$. For arbitrary data $\mathcal{D}$, AFEM-TS concatenates GALERKIN and DATA, the latter being unrelated to $\zeta_{\mathcal{T}}(u_{\mathcal{T}})$. Therefore Corollary 5.31 and Proposition 5.27 (computational cost of GALERKIN) yield the following extension of Proposition 5.29 (convergence of AFEM-TS).

Corollary 5.33 (convergence of AFEM-TS). *The algorithm* AFEM-TS *driven by $\zeta_{\mathcal{T}}(u_{\mathcal{T}})$ stops after*

$$K < 2 + \frac{\log(\varepsilon_0/\mathtt{tol})}{\log 2}$$

iterations and delivers the error

$$|u - u_K|_{H_0^1(\Omega)} \le C_* \,\mathtt{tol}\,.$$

The number of iterations of GALERKIN *is bounded uniformly for all outer loops.*

6. Convergence rates of AFEM for coercive problems

The ultimate goal of AFEM is to produce a *quasi-best* approximation $u_{\mathcal{T}} \in \mathbb{V}_{\mathcal{T}}$ to the solution $u \in \mathbb{V}$ of (2.7) with error measured in $\mathbb{V} = H_0^1(\Omega)$. The performance of AFEM is measured by the size of the error $|u - u_{\mathcal{T}}|_{H_0^1(\Omega)}$ relative to the cardinality $\#\mathcal{T}$ of $\mathcal{T}$. The latter usually reflects the total computational cost of implementing AFEM. As a benchmark, it is useful to compare the performance of AFEM with the best approximation of $u \in \mathbb{V}$ and $\mathcal{D} = (\boldsymbol{A}, c, f) \in \mathbb{D}$, provided we have full knowledge of them. This is the main purpose of this section.

Under suitable assumptions on the solution u and data $\mathcal{D}$, we prove the existence of constants $C(u, \mathcal{D}) > 0$ and $s \in (0, n/d]$ such that

$$|u - u_{\mathcal{T}_k}|_{H_0^1(\Omega)} \leq C(u, \mathcal{D})\, (\#\mathcal{T}_k)^{-s}, \tag{6.1}$$

provided s is the best decay rate with meshes in $\mathbb{T}$ with a comparable number of degrees of freedom. The upper bound n/d of s is dictated by the best decay rate with polynomials of degree $n \geq 1$ in dimension d unless u is degenerate (for instance, u belongs to a finite element space $\mathbb{V}_{\mathcal{T}}$ with $\mathcal{T} \in \mathbb{T}$). The dependence on $\mathcal{D}$ of the constant $C(u, \mathcal{D})$ accounts for the multiplicative structure of the interaction between the coefficients $(\boldsymbol{A}, c)$ and u, and cannot be avoided in general.

A crucial insight for the simplest scenario, the Laplacian and piecewise constant forcing f, is due to Stevenson (2007). It has been extended to operators with variable coefficients by Cascón *et al.* (2008) and later expressed in terms of the estimator by Carstensen *et al.* (2014). It reads as follows:

If a marking strategy reduces the PDE estimator $\eta_{\mathcal{T}}(u_{\mathcal{T}})$ to a fraction of its current value, then the refined set of elements $\mathcal{R}$ inherits an error indicator $\eta_{\mathcal{T}}(u_{\mathcal{T}}, \mathcal{R})$ comparable to $\eta_{\mathcal{T}}(u_{\mathcal{T}})$, hence a Dörfler marking. (6.2)

This allows us to compare meshes produced by AFEM with optimal meshes and to conclude a quasi-optimal error decay. To this end, in Section 6.1 we introduce approximation classes for functions in $\mathbb{V}$ and $\mathbb{D}$, tailored to the decomposition of Ω into conforming refinements of an initial conforming partition $\mathcal{T}_0$, the root of $\mathbb{T}$. We will assume that $u = u(\mathcal{D}) \in \mathbb{V}$ and $\mathcal{D} = (\boldsymbol{A}, c, f) \in \mathbb{D}$ belong to these classes which, however, are not characterized in terms of regularity of u and $\mathcal{D}$. In Section 6.2, we investigate the approximability properties of perturbations $\widehat{u} = u(\widehat{\mathcal{D}})$ of the exact solution u, namely exact solutions of (5.5) with perturbed data $\widehat{\mathcal{D}}$. Next, in Section 6.3, we consider a conforming refinement $\mathcal{T}_* \in \mathbb{T}$ of a partition $\mathcal{T} \in \mathbb{T}$, and give conditions under which an optimal Dörfler marking property holds. We first apply this in Section 6.4 to study and derive rate-optimality of GALERKIN and AFEM-SW, the one-step AFEMs. We then combine the quasi-optimal performances of GALERKIN and DATA to prove rate-optimality of the two-step AFEM in Section 6.5. We conclude in Section 6.8 upon bridging the gap between appproximation and regularity classes. In particular, we give sufficient conditions for functions in Besov, Sobolev and Lipschitz spaces to belong to the approximation classes.

6.1. Nonlinear approximation classes

In Section 6.1.1 we discuss approximation classes for functions in $\mathbb{V}$, which are applicable to the solution u of (2.7). In Section 6.1.2 we turn our attention to approximation classes for functions in $\mathbb{D}$, which are in turn applicable to data $\mathcal{D}$.

We refer to DeVore (1998), as well as DeVore and Lorentz (1993) and Binev *et al.* (2002), for a discussion within nonlinear approximation theory.

6.1.1. Nonlinear approximation classes for functions in $\mathbb{V}$

For any $N \in \mathbb{N}$, $N \geq \#\mathcal{T}_0$, we define the following collection of partitions within $\mathbb{T}$:

$$\mathbb{T}_N = \{\mathcal{T} \mid \mathcal{T} \in \mathbb{T} \text{ satisfies } \#\mathcal{T} \leq N\}.$$

This is the set of *conforming* meshes generated from $\mathcal{T}_0$ with at most $N - \#\mathcal{T}_0$ bisections. Given $v \in \mathbb{V}$, we let $\sigma_N(v)$ be the smallest approximation H_0^1-error incurred on v with *continuous* piecewise polynomial functions of degree $\leq n$ over meshes $\mathbb{T}_N$:

$$\sigma_N(v) := \inf_{\mathcal{T} \in \mathbb{T}_N} \inf_{v_\mathcal{T} \in \mathbb{V}_\mathcal{T}} |v - v_\mathcal{T}|_{H_0^1(\Omega)}. \tag{6.3}$$

This is a theoretical measure of performance, in that finding a mesh $\mathcal{T} \in \mathbb{T}_N$ that realizes $\sigma_N(v)$ has exponential complexity. Proving a bound $|v - v_\mathcal{T}|_{1,\Omega} \leq C_1 \sigma_{C_2 N}(v)$ for $\mathcal{T} \in \mathbb{T}_N$ with $C_2 \leq 1 \leq C_1$ independent of N, the so-called *instance optimality*, is rather difficult and beyond the scope of this survey. In fact, a function $v \in \mathbb{V}_\mathcal{T}$ with $\mathcal{T} \in \mathbb{T}_N$ could be the solution of our model problem (2.7), because we allow forcing $f \in H^{-1}(\Omega)$. Hence we see that $\sigma_N(v) = 0$, and AFEM should then capture v exactly on a finer mesh $\mathcal{T} \in \mathbb{T}_{C_2^{-1}N}$. We refer to Diening *et al.* (2016) for a proof of instance optimality for a forcing $f \in L^2(\Omega)$ and the Laplace operator, namely for coefficients $\boldsymbol{A} = \boldsymbol{I}$ and $c = 0$.

We will instead be able to prove that the error $|v - v_\mathcal{T}|_{H_0^1(\Omega)}$ for the Galerkin solution $v_\mathcal{T}$ for $\mathcal{T} \in \mathbb{T}_N$ decays in terms of N with the same rate N^{-s} as $\sigma_N(v)$; we thus say that AFEM is *rate-optimal*. We first note that for $v \in H^{n+1}(\Omega)$ and $\mathcal{T} \in \mathbb{T}_N$ quasi-uniform, we expect to have

$$\inf_{v_\mathcal{T} \in \mathbb{V}_\mathcal{T}} |v - v_\mathcal{T}|_{H_0^1(\Omega)} \lesssim N^{-n/d} |v|_{H^{n+1}(\Omega)} \tag{6.4}$$

because the global mesh size h and N satisfy $h \approx N^{-1/d}$. This error estimate within the linear Sobolev scale provides the largest possible decay rate $-n/d$.

Definition 6.1 (approximation class of u). Given $0 < s \leq n/d$, the class $\mathbb{A}_s := \mathbb{A}_s(H_0^1(\Omega); \mathcal{T}_0)$, relative to the partition $\mathcal{T}_0$ and approximation in the H_0^1-norm by continuous piecewise polynomials of degree $\leq n$ on the forest $\mathbb{T}$ emanating from $\mathcal{T}_0$, is the set of functions $v \in \mathbb{V} = H_0^1(\Omega)$ such that

$$|v|_{\mathbb{A}_s} := \sup_{N \geq \#\mathcal{T}_0} (N^s \sigma_N(v)) < \infty, \tag{6.5a}$$

whence

$$\sigma_N(v) \leq |v|_{\mathbb{A}_s} N^{-s} \quad \text{for all } N \geq \#\mathcal{T}_0. \tag{6.5b}$$

We also write $\mathbb{A}_s = \mathbb{A}_s^0$ to emphasize continuity of the discrete functions in $\mathbb{V}_\mathcal{T} = \mathbb{S}_\mathcal{T}^{n,0} \cap \mathbb{V}$ with $\mathcal{T} \in \mathbb{T}$. The quantity $|v|_{\mathbb{A}_s}$ is a quasi seminorm in $\mathbb{A}_s$, which is not a linear space but rather a nonlinear class of functions. Notice that as s increases, the cost of membership to be in $\mathbb{A}_s$ increases, namely $\mathbb{A}_{s_1} \subset \mathbb{A}_{s_2}$ for $s_1 \geq s_2$.

We may as well consider approximating $v \in \mathbb{V}$ with *discontinuous* piecewise polynomials $\mathbb{S}^{n,-1}_{\mathcal{T}}$ of degree $\leq n$, which is a richer space than $\mathbb{S}^{n,0}_{\mathcal{T}}$. We can likewise define the corresponding modulus of approximation

$$\sigma_N^{(-1)}(v) := \inf_{\mathcal{T} \in \mathbb{T}_N} \inf_{v_\mathcal{T} \in \mathbb{S}^{n,-1}_\mathcal{T}} |v - v_\mathcal{T}|_{H^1_0(\Omega;\mathcal{T})} \tag{6.6}$$

and approximation class $\mathbb{A}_s^{-1} := \mathbb{A}_s^{-1}(H^1_0(\Omega); \mathcal{T}_0)$ of functions $v \in H^1_0(\Omega)$ such that

$$|v|_{\mathbb{A}_s^{-1}} := \sup_{N \geq \#\mathcal{T}_0} \big(N^s \sigma_N^{(-1)}(v)\big) < \infty \quad \Rightarrow \quad \sigma_N^{(-1)}(v) \leq |v|_{\mathbb{A}_s^{-1}} N^{-s}. \tag{6.7}$$

It is obvious that $\sigma_N^{(-1)}(v) \leq \sigma_N(v)$ for all $v \in H^1_0(\Omega)$ because $\mathbb{S}^{n,0}_\mathcal{T} \subset \mathbb{S}^{n,-1}_\mathcal{T}$. However, we have the following equivalence result taken from Veeser (2016). The original proof, although more complicated and for a different notion of error relevant to discontinuous Galerkin approximations, can be traced back to Bonito and Nochetto (2010, Proposition 5.2); see Proposition 9.4.

Proposition 6.2 (equivalence of classes). *Assume that all stars of meshes $\mathcal{T} \in \mathbb{T}$ are $(d-1)$-face-connected. Then, there exists a constant C_{dG} that depends on the shape regularity of $\mathbb{T}$, the dimension d and the polynomial degree $n \geq 1$, such that*

$$\sigma_N(v) \leq C_{\mathrm{dG}}\, \sigma_N^{(-1)}(v) \quad \textit{for all } v \in H^1_0(\Omega), \quad N \geq \#\mathcal{T}_0.$$

Moreover, the approximation classes coincide, i.e. $\mathbb{A}_s^0 = \mathbb{A}_s^{-1}$.

Proof. We simply resort to (3.19) of Proposition 3.9 (approximation of gradients), namely, for $v \in H^1_0(\Omega)$,

$$1 \leq \frac{\min_{w \in \mathbb{S}^{n,0}_\mathcal{T}} |v - w|_{H^1_0(\Omega)}}{\min_{w \in \mathbb{S}^{n,-1}_\mathcal{T}} |v - w|_{H^1_0(\Omega;\mathcal{T})}} \leq C_{\mathrm{dG}},$$

and use the definitions (6.3) and (6.6). This completes the proof. □

In the rest of the paper we will make the following approximability assumption.

Assumption 6.3 (approximability of u). The exact solution $u \in H^1_0(\Omega)$ of problem (2.5) belongs to the approximation class $\mathbb{A}_s(H^1_0(\Omega); \mathcal{T}_0)$ with $s = s_u \in (0, n/d]$.

The following condition (6.8) is simpler to handle in practice than (6.5).

Lemma 6.4 (membership of $\mathbb{A}_s$). *Let $v \in \mathbb{A}_s$ and $\varepsilon_0 := \inf_{v_{\mathcal{T}_0} \in \mathbb{V}_{\mathcal{T}_0}} |v - v_{\mathcal{T}_0}|_{H^1_0(\Omega)}$. Then, for all $0 < \varepsilon \leq \varepsilon_0$, there exist $\mathcal{T}_\varepsilon \in \mathbb{T}$ and $v_\varepsilon \in \mathbb{V}_{\mathcal{T}_\varepsilon}$ such that*

$$|v - v_\varepsilon|_{H^1_0(\Omega)} \leq \varepsilon, \quad \#\mathcal{T}_\varepsilon \leq 1 + |v|_{\mathbb{A}_s}^{1/s} \varepsilon^{-1/s}. \tag{6.8}$$

Proof. Given $0 < \varepsilon \le \varepsilon_0$, let $\mathcal{T}_\varepsilon \in \mathbb{T}$ be a conforming refinement of $\mathcal{T}_0$ with minimal cardinality and $v_\varepsilon \in \mathbb{V}_{\mathcal{T}_\varepsilon}$ such that

$$|v - v_\varepsilon|_{H_0^1(\Omega)} \le \varepsilon.$$

Therefore, if $\varepsilon < \varepsilon_0$, we deduce from the minimal property of $\mathcal{T}_\varepsilon$ that

$$\inf_{v_\mathcal{T} \in \mathbb{V}_\mathcal{T}} |v - v_\mathcal{T}|_{H_0^1(\Omega)} > \varepsilon \quad \text{for all } \mathcal{T} \in \mathbb{T} \quad \text{such that} \quad \#\mathcal{T} \le \#\mathcal{T}_\varepsilon - 1.$$

If $N := \#\mathcal{T}_\varepsilon - 1$, definition (6.5) implies

$$\varepsilon < \inf_{\mathcal{T} \in \mathbb{T}_N} \inf_{v_\mathcal{T} \in \mathbb{V}_\mathcal{T}} |v - v_\mathcal{T}|_{H_0^1(\Omega)} = \sigma_N(v) \le |v|_{\mathbb{A}_s} N^{-s},$$

whence

$$\#\mathcal{T}_\varepsilon = 1 + N \le 1 + |v|_{\mathbb{A}_s}^{1/s} \varepsilon^{-1/s},$$

as asserted in (6.8). On the other hand, if $\varepsilon = \varepsilon_0$ we see that

$$\varepsilon_0 \le |v|_{\mathbb{A}_s} (\#\mathcal{T}_0)^{-s} \quad \Rightarrow \quad \#\mathcal{T}_0 \le |v|_{\mathbb{A}_s}^{1/s} \varepsilon_0^{-1/s} < 1 + |v|_{\mathbb{A}_s}^{1/s} \varepsilon_0^{-1/s}.$$

This completes the proof. □

Remark 6.5. If $d = 2$ and $n = 1$, then Corollary 3.20 (optimal H^1-convergence rate) shows that $W_p^2(\Omega) \subset \mathbb{A}^{1/2}$ for $p > 1$. The space $W_p^2(\Omega)$ is much larger than $H^2(\Omega)$, fits within the nonlinear Sobolev scale, and delivers the same decay rate as (6.4). We will investigate the connection between approximation classes $\mathbb{A}_s$ and regularity classes in any dimension d and for any polynomial degree $n \ge 1$ later in Section 6.8.

6.1.2. Nonlinear approximation classes for data in $\mathbb{D}$

Given data $\mathcal{D} = (\boldsymbol{A}, c, f) \in \mathbb{D}$ and a mesh $\mathcal{T} \in \mathbb{T}$, we consider the best approximation of $\mathcal{D}$ by discrete (piecewise polynomial) data $\widehat{\mathcal{D}} = (\widehat{\boldsymbol{A}}, \widehat{c}, \widehat{f}) \in \mathbb{D}_\mathcal{T}$, where $\mathbb{D}$ and $\mathbb{D}_\mathcal{T}$ are defined in (5.1) and (5.2). We measure the error in the space $D(\Omega)$ defined in (5.60) with $q = 2$ for $d < 4$ or $q > d/2$ for $d \ge 4$. We now discuss the best approximation errors for the components of data in $D(\Omega)$, which are used to define the corresponding approximation classes. For the coefficients $(\boldsymbol{A}, c)$, they are characterized by the quantities

$$\delta_\mathcal{T}(\boldsymbol{A})_r := \inf_{\widehat{\boldsymbol{A}} \in [\mathbb{S}_\mathcal{T}^{n-1,-1}]^{d\times d}} \|\boldsymbol{A} - \widehat{\boldsymbol{A}}\|_{L^r(\Omega)}, \quad \delta_\mathcal{T}(c)_q := \inf_{\widehat{c} \in \mathbb{S}_\mathcal{T}^{n-1,-1}} \|c - \widehat{c}\|_{L^q(\Omega)} \quad (6.9)$$

for $r, q \in [2, \infty]$ as above. Note that $\widehat{\boldsymbol{A}}$ and $\widehat{c}$ in (6.9) are unconstrained in the sense that they do not necessarily satisfy the structural assumption (5.51) and are thus not suited to the perturbed problem (5.5). We define the best constrained approximation errors for $\boldsymbol{A} \in M(\alpha_1, \alpha_2)$ and $c \in R(c_1, c_2)$ by

$$\begin{aligned} \widetilde{\delta}_{\mathcal{T}}(\boldsymbol{A})_r &:= \inf_{\widehat{\boldsymbol{A}} \in [\mathbb{S}_{\mathcal{T}}^{n-1,-1}]^{d\times d} \cap M(\widehat{\alpha}_1, \widehat{\alpha}_2)} \|\boldsymbol{A} - \widehat{\boldsymbol{A}}\|_{L^r(\Omega)}, \\ \widetilde{\delta}_{\mathcal{T}}(c)_q &:= \inf_{\widehat{c} \in \mathbb{S}_{\mathcal{T}}^{n-1,-1} \cap R(\widehat{c}_1, \widehat{c}_2)} \|c - \widehat{c}\|_{L^q(\Omega)}, \end{aligned} \tag{6.10}$$

where in view of (5.52)

$$\widehat{\alpha}_1 = \frac{\alpha_1}{2}, \quad \widehat{\alpha}_2 = C_{\mathrm{ctr}}\alpha_2, \quad \widehat{c}_1 = -\frac{\alpha_1}{4C_P^2}, \quad \widehat{c}_2 = C_{\mathrm{ctr}}(\alpha_1 + c_1). \tag{6.11}$$

We mention in anticipation that in Section 7.4 we prove the equivalences

$$\delta_{\mathcal{T}}(\boldsymbol{A})_r \le \widetilde{\delta}_{\mathcal{T}}(\boldsymbol{A})_r \le C_{\mathrm{data}}\delta_{\mathcal{T}}(\boldsymbol{A})_r, \quad \delta_{\mathcal{T}}(c)_q \le \widetilde{\delta}_{\mathcal{T}}(c)_q \le C_{\mathrm{data}}\delta_{\mathcal{T}}(c)_q \tag{6.12}$$

for all $\boldsymbol{A} \in M(\alpha_1, \alpha_2)$ and $c \in R(c_1, c_2)$; see Remarks 7.13 and 7.17. For the load function f, the definition (4.56) of $\mathrm{osc}_{\mathcal{T}}(f)_{-1}$ suggests considering

$$\delta_{\mathcal{T}}(f)_{-1} := \left(\sum_{T\in\mathcal{T}} \inf_{\widehat{f} \in \mathbb{F}_{\mathcal{T}_{\omega_T}}} \|f - \widehat{f}\|^2_{H^{-1}(\omega_T)} \right)^{1/2}.$$

All these best approximation errors are hard to evaluate and are thus replaced by the computable oscillations defined in (5.79) and (5.80) in practice. We recall that they rely on the local L^2-projection operator $\Pi_{\mathcal{T}}$ for $(\boldsymbol{A}, c)$ and the local H^{-1}-projection operator $P_{\mathcal{T}}$ for f to compute linear approximations $\widetilde{\mathcal{D}}$ of $\mathcal{D}$ to a desired accuracy. These projections are later modified nonlinearly to give rise to $\widehat{\mathcal{D}}$ satisfying the side constraints (5.51) without compromising accuracy. We recall that the DATA module is assumed to construct approximations so that

$$\mathrm{osc}_{\mathcal{T}}(\boldsymbol{A})_r \le \Lambda_{\mathrm{data}}\delta_{\mathcal{T}}(\boldsymbol{A})_r, \quad \mathrm{osc}_{\mathcal{T}}(c)_q \le \Lambda_{\mathrm{data}}\delta_{\mathcal{T}}(c)_q, \quad \mathrm{osc}_{\mathcal{T}}(f)_{-1} \le \Lambda_{\mathrm{data}}\delta_{\mathcal{T}}(f)_{-1}$$

with a mesh independent constant Λ_{data}; see Assumption 5.22. In Section 5.4.2 we discuss practical realizations of DATA.

For the purpose of assessing the cardinality of AFEM, we do not need the specific form of $\widehat{\mathcal{D}}$ but rather the decay of the best approximation errors in terms of degrees of freedom. Therefore we postpone to Section 7 the construction of $\widehat{\mathcal{D}}$ for $n \ge 1$, and to Section 6.8 the discussion of regularity properties of $\mathcal{D}$ that guarantee membership of the following approximation classes.

Definition 6.6 (approximation classes of $\boldsymbol{A}$). For $0 < \alpha_1 \le \alpha_2$, $2 \le r \le \infty$, let $\mathbb{M}_s := \mathbb{M}_s(L^r(\Omega; \mathbb{R}^{d\times d}); \mathcal{T}_0)$ be the set of matrix-valued functions $\boldsymbol{A} \in M(\alpha_1, \alpha_2)$ satisfying

$$|\boldsymbol{A}|_{\mathbb{M}_s} := \sup_{N \ge \#\mathcal{T}_0} \left(N^s \inf_{\mathcal{T}\in\mathbb{T}_N} \widetilde{\delta}_{\mathcal{T}}(\boldsymbol{A})_r \right) < \infty \Rightarrow \inf_{\mathcal{T}\in\mathbb{T}_N} \widetilde{\delta}_{\mathcal{T}}(\boldsymbol{A})_r \le |\boldsymbol{A}|_{\mathbb{M}_s} N^{-s}. \tag{6.13}$$

Definition 6.7 (approximation classes of c). The class $\mathbb{C}_s := \mathbb{C}_s(L^q(\Omega); \mathcal{T}_0)$ is the set of functions $c \in R(c_1, c_2)$ such that

$$|c|_{\mathbb{C}_s} := \sup_{N \geq \#\mathcal{T}_0} \Big(N^s \inf_{\mathcal{T} \in \mathbb{T}_N} \widetilde{\delta}_{\mathcal{T}}(c)_q \Big) < \infty \Rightarrow \inf_{\mathcal{T} \in \mathbb{T}_N} \widetilde{\delta}_{\mathcal{T}}(c)_q \leq |c|_{\mathbb{C}_s} N^{-s}. \quad (6.14)$$

Definition 6.8 (approximation classes of f). The class $\mathbb{F}_s := \mathbb{F}_s(H^{-1}(\Omega); \mathcal{T}_0)$ is the set of functions $f \in H^{-1}(\Omega)$ such that

$$|f|_{\mathbb{F}_s} := \sup_{N \geq \#\mathcal{T}_0} \Big(N^s \inf_{\mathcal{T} \in \mathbb{T}_N} \delta_{\mathcal{T}}(f)_{-1} \Big) < \infty \Rightarrow \inf_{\mathcal{T} \in \mathbb{T}_N} \delta_{\mathcal{T}}(f)_{-1} \leq |f|_{\mathbb{F}_s} N^{-s}. \quad (6.15)$$

Since the polynomial degree of discrete coefficients $(\widehat{\boldsymbol{A}}, \widehat{c})$ in definition (5.2) is $n-1$, we expect decay rates $s_A, s_c \leq n/d$ according to nonlinear approximation theory. The specific values of (s_A, s_c) depend on the regularity of $(\boldsymbol{A}, c)$, a delicate topic that we further investigate in Sections 6.8 and 7. However, because u and $\mathcal{D} = (\boldsymbol{A}, c, f)$ satisfy the elliptic problem (2.5), the above approximation classes are somewhat related. We now quantify this statement.

Lemma 6.9 (relation between approximation classes). *Let $2 \leq r, q \leq \infty$ be such that $d/2 < q$. If $u \in \mathbb{A}_{s_u}(H_0^1(\Omega); \mathcal{T}_0)$, $\boldsymbol{A} \in \mathbb{M}_{s_A}(L^r(\Omega; \mathbb{R}^{d \times d}); \mathcal{T}_0)$ and $c \in \mathbb{C}_{s_c}(L^q(\Omega); \mathcal{T}_0)$, with $0 < s_u, s_A, s_c < n/d$, then $f \in \mathbb{F}_{s_f}(H^{-1}(\Omega); \mathcal{T}_0)$ and*

$$|f|_{\mathbb{F}_{s_f}} \leq C\big(|u|_{\mathbb{A}_{s_u}} + |\boldsymbol{A}|_{\mathbb{M}_{s_A}} + |c|_{\mathbb{C}_{s_c}}\big), \quad s_f = \min\{s_u, s_A, s_c\}, \quad (6.16)$$

where the constant $C > 0$ depends on $\|u\|_{W_p^1(\Omega)}$, $p = 2r/(r-2)$ and $\alpha_1, \alpha_2, c_1, c_2$. In particular, if $(\boldsymbol{A}, c)$ are discrete in $\mathcal{T}_0$, then

$$|f|_{\mathbb{F}_{s_f}} \leq C|u|_{\mathbb{A}_{s_u}}, \quad s_f = s_u. \quad (6.17)$$

Proof. Let $L[u] := -\operatorname{div}(\boldsymbol{A}\nabla u) + cu$ be the operator in (2.5) and note that $f = L[u] \in H^{-1}(\Omega)$ can be approximated by $\widehat{f} = -\operatorname{div}(\widehat{\boldsymbol{A}}\nabla\widehat{v}) + \widehat{c}\widehat{v} \in \mathbb{F}_{\mathcal{T}}$, where the discrete space $\mathbb{F}_{\mathcal{T}}$ is given in Definition 4.17 and $\widehat{v} \in \mathbb{S}_{\mathcal{T}}^{n,0}$, $\widehat{\boldsymbol{A}} \in (\mathbb{S}_{\mathcal{T}}^{n-1,-1})^{d \times d}$, $\widehat{c} \in \mathbb{S}_{\mathcal{T}}^{n-1,-1}$. Let us now express $f - \widehat{f}$ as follows:

$$f - \widehat{f} = -\operatorname{div}((\boldsymbol{A} - \widehat{\boldsymbol{A}})\nabla u) + (c - \widehat{c})u - \operatorname{div}(\widehat{\boldsymbol{A}}\nabla(u - \widehat{v})) + \widehat{c}(u - \widehat{v}),$$

and recall that we have to estimate $\|f - \widehat{f}\|_{H^{-1}(\omega_T)}$ for every $T \in \mathcal{T}$, rather than a global norm in $H^{-1}(\Omega)$, to get an upper bound on $\delta_{\mathcal{T}}(f)_{-1}$. Therefore we proceed as in the proof of Lemma 5.20 (continuous dependence on data), to obtain

$$\sum_{T \in \mathcal{T}} \|f - \widehat{f}\|_{H^{-1}(\omega_T)}^2 \lesssim \|\nabla u\|_{L^p(\Omega)}^2 \|\boldsymbol{A} - \widehat{\boldsymbol{A}}\|_{L^r(\Omega)}^2 + \|\nabla u\|_{L^2(\Omega)}^2 \|c - \widehat{c}\|_{L^q(\Omega)}^2$$
$$+ \|\widehat{\boldsymbol{A}}\|_{L^\infty(\Omega)}^2 \|\nabla(u - \widehat{v})\|_{L^2(\Omega)}^2 + \|\widehat{c}\|_{L^\infty(\Omega)}^2 \|u - \widehat{v}\|_{L^2(\Omega)}^2,$$

where $p = 2r/(r-1)$ and $\|\nabla u\|_{L^p(\Omega)} < \infty$ according to (2.41) and $d/2 < q \leq \infty$.

Note that thanks to (6.11), $\|\widehat{A}\|_{L^\infty(\Omega)} \le \widehat{\alpha}_2 = C_{\mathrm{ctr}}\alpha_2$ and $\|\widehat{c}\|_{L^\infty(\Omega)} \le \widehat{c}_2 = C_{\mathrm{ctr}}(\alpha_1 + c_2)$. Moreover, since $\widehat{v}, \widehat{\boldsymbol{A}}, \widehat{c}$ can be chosen separately, invoking (6.5), (6.13), (6.14) and (6.15), we realize that

$$
\begin{aligned}
\inf_{\mathcal{T}\in\mathbb{T}_N} \delta_{\mathcal{T}}(f)_{-1} &\lesssim \inf_{\mathcal{T}\in\mathbb{T}_N} \inf_{\widehat{v}\in\mathbb{S}^{n,0}_{\mathcal{T}}} \|\nabla(u-\widehat{v})\|_{L^2(\Omega)} \\
&\quad + \inf_{\mathcal{T}\in\mathbb{T}_N} \inf_{\widehat{\boldsymbol{A}}\in[\mathbb{S}^{n-1,-1}_{\mathcal{T}}]^{d\times d}} \|\boldsymbol{A}-\widehat{\boldsymbol{A}}\|_{L^r(\Omega)} \\
&\quad + \inf_{\mathcal{T}\in\mathbb{T}_N} \inf_{\widehat{c}\in\mathbb{S}^{n-1,-1}_{\mathcal{T}}} \|c-\widehat{c}\|_{L^q(\Omega)} \\
&\le |u|_{\mathbb{A}_{s_u}} N^{-s_u} + |\boldsymbol{A}|_{\mathbb{M}_{s_A}} N^{-s_A} + |c|_{\mathbb{C}_{s_c}} N^{-s_c}
\end{aligned}
$$

gives (6.16) with $s_f = \min\{s_u, s_A, s_c\}$; (6.17) is a trivial consequence. □

Estimate (6.17) will be useful later in Theorem 6.20 (rate-optimality of one-step AFEMs). It is important to realize that the multiplicative structure between solution u and coefficients $(\boldsymbol{A}, c)$ is hidden in the constants C in (6.16) and (6.17). Moreover, these estimates are possible due to the fact that the space $H^{-1}(\Omega)$ is the range of the linear operator $L\colon H^1_0(\Omega) \to H^{-1}(\Omega)$ and that the discrete functions in $\mathbb{F}_{\mathcal{T}}$ are images by L of functions in $\mathbb{V}_{\mathcal{T}}$. This would not be true for L^2-weighted surrogates of $\delta_{\mathcal{T}}(f)_{-1}$ that typically overestimate the error in $H^{-1}(\Omega)$.

Assumption 6.10 (approximability of data). There exist $s_A, s_c, s_f \in (0, n/d]$ such that data $\mathcal{D} = (\boldsymbol{A}, c, f) \in \mathbb{D}$ satisfies $\boldsymbol{A} \in \mathbb{M}_{s_A}$, $c \in \mathbb{C}_{s_c}$, $f \in \mathbb{F}_{s_f}$.

We recall that if $\mathrm{osc}_{\mathcal{T}}(\mathcal{D}) = \|\mathcal{D}-\widehat{\mathcal{D}}\|_{D(\Omega)} > C_{\mathrm{data}}\tau$ over a conforming refinement $\mathcal{T} \in \mathbb{T}$ of $\mathcal{T}_0$, then the call

$$[\widehat{\mathcal{T}}, \widehat{\mathcal{D}}] = \mathsf{DATA}(\mathcal{T}, \mathcal{D}, \tau)$$

produces a conforming refinement $\widehat{\mathcal{T}}$ of $\mathcal{T}$ and approximate data $\widehat{\mathcal{D}} = (\widehat{\boldsymbol{A}}, \widehat{c}, \widehat{f}) \in \mathbb{D}_{\widehat{\mathcal{T}}}$ over $\widehat{\mathcal{T}}$ that satisfies $\mathrm{osc}_{\widehat{\mathcal{T}}}(\mathcal{D}) \le \Lambda_{\mathrm{data}}\delta_{\widehat{\mathcal{T}}}(\mathcal{D})$, and for $r, q \in [2, \infty]$,

$$\mathrm{osc}_{\widehat{\mathcal{T}}}(\mathcal{D}) = \mathrm{osc}_{\widehat{\mathcal{T}}}(\boldsymbol{A})_r + \mathrm{osc}_{\widehat{\mathcal{T}}}(c)_q + \mathrm{osc}_{\widehat{\mathcal{T}}}(f)_{-1} \le C_{\mathrm{data}}\tau,$$

as well as the constraints $\widehat{\boldsymbol{A}} \in M(\widehat{\alpha}_1, \widehat{\alpha}_2)$ and $\widehat{c} \in R(\widehat{c}_1, \widehat{c}_2)$ defined in (5.51). We will show in Section 7 that the routine responsible for reducing oscillations, namely GREEDY, exhibits optimal performance in the sense that the cardinalities $N_{\mathcal{T}}(\boldsymbol{A}), N_{\mathcal{T}}(c), N_{\mathcal{T}}(f)$ of the sets of elements necessary to reduce the individual oscillations of $(\boldsymbol{A}, c, f)$ below the threshold $C_{\mathrm{data}}(\tau/3)$ starting from any $\mathcal{T} \ge \mathcal{T}_0$ satisfy

$$N_{\mathcal{T}}(\boldsymbol{A}) \lesssim |\boldsymbol{A}|_{\mathbb{M}_{s_A}}^{1/s_A} \tau^{-1/s_A}, \quad N_{\mathcal{T}}(c) \lesssim |c|_{\mathbb{C}_{s_c}}^{1/s_c} \tau^{-1/s_c}, \quad N_{\mathcal{T}}(f) \lesssim |f|_{\mathbb{F}_{s_f}}^{1/s_f} \tau^{-1/s_f}. \tag{6.18}$$

Therefore the cost of one call to DATA can be quantified by the total number $N_{\mathcal{T}}(\mathcal{D})$

of elements marked, which obeys the relation

$$\begin{aligned} N_{\mathcal{T}}(\mathcal{D}) &= N_{\mathcal{T}}(\boldsymbol{A}) + N_{\mathcal{T}}(c) + N_{\mathcal{T}}(f) \\ &\lesssim |\boldsymbol{A}|_{\mathbb{M}_{s_A}}^{1/s_A} \tau^{-1/s_A} + |c|_{\mathbb{C}_{s_c}}^{1/s_c} \tau^{-1/s_c} + |f|_{\mathbb{F}_{s_f}}^{1/s_f} \tau^{-1/s_f} \\ &\le |\mathcal{D}|_{\mathbb{A}_{\mathcal{D}}}^{1/s_{\mathcal{D}}} \tau^{-1/s_{\mathcal{D}}}, \end{aligned}$$

with

$$s_{\mathcal{D}} := \min\{s_A, s_c, s_f\}, \quad |\mathcal{D}|_{\mathbb{A}_{\mathcal{D}}} := \left(|\boldsymbol{A}|_{\mathbb{M}_{s_A}}^{1/s_A} + |c|_{\mathbb{C}_{s_c}}^{1/s_c} + |f|_{\mathbb{F}_{s_f}}^{1/s_f} \right)^{s_{\mathcal{D}}}. \tag{6.19}$$

It is thus natural to make the following assumption on DATA.

Assumption 6.11 (quasi-optimality of DATA). The call $[\widehat{\mathcal{T}}, \widehat{\mathcal{D}}] = \mathsf{DATA}(\mathcal{T}, \mathcal{D}, \tau)$, from an arbitrary conforming refinement $\mathcal{T}$ of $\mathcal{T}_0$ with tolerance τ, marks the number of elements $N_{\mathcal{T}}(\mathcal{D})$ to produce an approximation $\widehat{\mathcal{D}}$ of $\mathcal{D}$ over $\widehat{\mathcal{T}}$ so that

$$\operatorname{osc}_{\widehat{\mathcal{T}}}(\mathcal{D}) = \|\mathcal{D} - \widehat{\mathcal{D}}\|_{D(\Omega)} \le C_{\text{data}} \tau, \quad N_{\mathcal{T}}(\mathcal{D}) \lesssim |\mathcal{D}|_{\mathbb{A}_{\mathcal{D}}}^{1/s_{\mathcal{D}}} \tau^{-1/s_{\mathcal{D}}}. \tag{6.20}$$

6.2. *ε-approximation of order s*

Inspection of the structure of algorithm AFEM-TS (Algorithm 5.28) reveals that the approximate data $\mathcal{D}_k$ is fixed inside GALERKIN. Therefore the performance of GALERKIN is dictated by the regularity of the exact solution $\widehat{u}_k = \widehat{u}_k(\mathcal{D}_k) \in H_0^1(\Omega)$ with data $\mathcal{D}_k$, rather than the exact solution $u = u(\mathcal{D})$ with data $\mathcal{D}$. We know that $u \in \mathbb{A}_s$, and wonder what regularity is inherited by $\widehat{u}_k$. This leads to the following concept introduced in Bonito *et al.* (2013*b*, Definition 3.1, Lemma 3.2).

Definition 6.12 (ε-approximation of order s). Given $u \in \mathbb{A}_s(H_0^1(\Omega); \mathcal{T}_0)$ and $\varepsilon > 0$, a function $v \in H_0^1(\Omega)$ is said to be an ε-approximation of order s to u if $|u - v|_{H_0^1(\Omega)} \le \varepsilon$ and there exists a constant $C > 0$ independent of ε, u and v such that for all $\delta \ge \varepsilon$ there exists $N \ge \#\mathcal{T}_0$ satisfying

$$\sigma_N(v) \le \delta, \quad N \le 1 + C|u|_{\mathbb{A}_s}^{1/s} \delta^{-1/s}. \tag{6.21}$$

Lemma 6.13 (ε-approximation of u of order s). *Let $u \in \mathbb{A}_s(H_0^1(\Omega); \mathcal{T}_0)$ and $v \in H_0^1(\Omega)$ satisfy $|u - v|_{H_0^1(\Omega)} \le \varepsilon$ for some $0 < \varepsilon \le \varepsilon_0$ with ε_0 defined in Lemma 6.4. Then v is a 2ε-approximation of order s to u.*

Proof. Let $\delta \ge 2\varepsilon$. By definition (6.3) of $\sigma_N(v)$, it suffices to invoke the triangle inequality to realize that

$$\sigma_N(v) \le |u - v|_{H_0^1(\Omega)} + \sigma_N(u) \le \frac{\delta}{2} + \sigma_N(u).$$

Since $u \in \mathbb{A}_s(H_0^1(\Omega))$, in view of Lemma 6.4, there exist $N \ge \#\mathcal{T}_0$ and $\mathcal{T} \in \mathbb{T}_N$:

$$\sigma_N(u) \le \frac{\delta}{2}, \quad N \le 1 + |u|_{\mathbb{A}_s}^{1/s} \left(\frac{\delta}{2}\right)^{-1/s}.$$

The estimate (6.21) thus follows with constant $C = 2^{1/s}$. □

This is a simple but crucial result for studying AFEM-TS. It says that any function v that is ε-close to a function $u \in \mathbb{A}_s(X;\mathcal{T}_0)$ in the norm of the space X defining the approximation class $\mathbb{A}_s(X;\mathcal{T}_0)$ can be approximated with a decay rate similar to u in X for as long as the desired accuracy does not exceed ε. In other words, the approximability of u is inherited by v up to scale ε. However, beyond the scale ε, the approximability of v may differ from that of u. Note that neither the definition (6.3) of $\sigma_N(v)$ nor Lemma 6.13 require $X = H_0^1(\Omega)$.

6.3. *Properties of Dörfler marking*

We follow the ideas of Stevenson (2007), Cascón *et al.* (2008) and Carstensen *et al.* (2014) to explore the insight (6.2) about Dörfler marking. Hereafter, we recall (5.6) and consider two admissible partitions $\mathcal{T}, \mathcal{T}_* \in \mathbb{T}$ such that $\mathcal{T} \le \mathcal{T}_*$, that is, the latter is a refinement of the former obtained by applying (newest-vertex) bisection to some of the elements of $\mathcal{T}$.

In what follows, we let $u = \widehat{u} \in H_0^1(\Omega)$ be the exact solution with discrete coefficients $(\widehat{\boldsymbol{A}}, \widehat{c})$ over a fixed mesh $\widehat{\mathcal{T}} \le \mathcal{T}$ and forcing function f that may or may not be discrete. We rewrite the *a posteriori* error estimates (5.12),

$$C_L\, \mathcal{E}_{\mathcal{T}}(u_{\mathcal{T}}, f) \le |\widehat{u} - u_{\mathcal{T}}|_{H_0^1(\Omega)} \le C_U\, \mathcal{E}_{\mathcal{T}}(u_{\mathcal{T}}, f), \tag{6.22}$$

where the total estimator $\mathcal{E}_{\mathcal{T}}(u_{\mathcal{T}}, f)$ consists of the PDE estimator $\eta_{\mathcal{T}}(u_{\mathcal{T}}, f)$ and the oscillation $\mathrm{osc}_{\mathcal{T}}(f)_{-1}$ and reads, according to (5.11),

$$\mathcal{E}_{\mathcal{T}}(u_{\mathcal{T}}, f)^2 = \eta_{\mathcal{T}}(u_{\mathcal{T}}, f)^2 + \mathrm{osc}_{\mathcal{T}}(f)_{-1}^2.$$

We also recall that when $f = P_{\mathcal{T}} f \in \mathbb{F}_{\mathcal{T}}$ is discrete, $\mathrm{osc}_{\mathcal{T}}(f)_{-1} = 0$ and $\mathcal{E}_{\mathcal{T}}(u_{\mathcal{T}}, f)$ reduces to $\eta_{\mathcal{T}}(u_{\mathcal{T}}, f)$, and that $P_{\mathcal{T}} f$ is used within $\eta_{\mathcal{T}}(u_{\mathcal{T}}, f)$ rather than f. The global nature of the elliptic boundary value problem (2.5) prevents upper *a posteriori* energy error estimates such as (6.22) between the continuous and discrete solution from being local. Remarkably, the situation for two Galerkin solutions $u_{\mathcal{T}} \in \mathbb{V}_{\mathcal{T}}$ and $u_{\mathcal{T}_*} \in \mathbb{V}_{\mathcal{T}_*}$ is different, as stated in Theorem 4.48 (upper bound for corrections):

$$|u_{\mathcal{T}_*} - u_{\mathcal{T}}|_{H_0^1(\Omega)} \le \widetilde{C}_U\, \mathcal{E}_{\mathcal{T}}(u_{\mathcal{T}}, f, \mathcal{R}), \tag{6.23}$$

where $\mathcal{R} = \mathcal{T} \backslash \mathcal{T}_*$ is the refined set defined in (5.14). It thus turns out that $|u_{\mathcal{T}_*} - u_{\mathcal{T}}|_{H_0^1(\Omega)}$ is controlled by the estimator $\mathcal{E}_{\mathcal{T}}(u_{\mathcal{T}}, f, \mathcal{R})$ on the set of elements $\mathcal{R}$ where the meshes differ. This crucial observation goes back to Stevenson (2007); see also Cascón *et al.* (2008) and Nochetto *et al.* (2009).

Henceforth we will impose restrictions on the ranges of the Dörfler marking parameter (5.23) and the threshold parameter ω for GALERKIN and AFEM-SW. We will impose a different restriction later on ω for AFEM-TS.

Assumption 6.14 (marking parameter). Let θ satisfy $\theta \in (0, \theta_0)$ with

$$\theta_0 := \min\{(2C_{\mathrm{Lip}}\widetilde{C}_U)^{-1}, 1\},$$

where $C_{\mathrm{Lip}}, \widetilde{C}_U$ are the constants in (4.68) and (6.23) respectively.

Assumption 6.15 (restriction on ω). We assume $0 \le \omega \le \omega_0 < 1$, with

$$\omega_0 := \sqrt{\frac{\theta_0^2 - \theta^2}{2 + \theta_0^2 - \theta^2}}.$$

We are now ready to make Stevenson's insight (6.2) precise.

Lemma 6.16 (Dörfler marking). *Let Assumptions 6.14 and 6.15 hold and $0 < \mu \le \frac{1}{2}$. Let $\mathcal{T} \in \mathbb{T}$, and let $\mathcal{T}_* \in \mathbb{T}$ be a refinement of $\mathcal{T}$ with respective Galerkin solutions $u_{\mathcal{T}} \in \mathbb{V}_{\mathcal{T}}$ and $u_{\mathcal{T}_*} \in \mathbb{V}_{\mathcal{T}_*}$; let $\mathcal{R} = \mathcal{T} \setminus \mathcal{T}_*$ be the refined set. Assume that the oscillation on $\mathcal{T}$ is dominated by the total estimator*

$$\mathrm{osc}_{\mathcal{T}}(f)_{-1} \le \omega\, \mathcal{E}_{\mathcal{T}}(u_{\mathcal{T}}, f), \tag{6.24}$$

and that

$$\eta_{\mathcal{T}_*}(u_{\mathcal{T}_*}, f) \le \mu\, \eta_{\mathcal{T}}(u_{\mathcal{T}}, f). \tag{6.25}$$

Then Dörfler marking is valid for any $0 < \theta < \theta_0$:

$$\theta\, \eta_{\mathcal{T}}(u_{\mathcal{T}}, f) \le \eta_{\mathcal{T}}(u_{\mathcal{T}}, f, \mathcal{R}). \tag{6.26}$$

Proof. We invoke Proposition 4.56 (estimator reduction) with $\delta = 1$ along with the localized upper bound (6.23) to write

$$\eta_{\mathcal{T}}(u_{\mathcal{T}}, f)^2 \le 2\, \eta_{\mathcal{T}_*}(u_{\mathcal{T}_*}, f)^2 + 2C_{\mathrm{Lip}}^2 \big(\widetilde{C}_U^2 \mathcal{E}_{\mathcal{T}}(u_{\mathcal{T}}, f, \mathcal{R})^2 + \mathrm{osc}_{\mathcal{T}}(f)_{-1}^2\big).$$

The last term accounts for the presence of $P_{\mathcal{T}} f$ and $P_{\mathcal{T}_*} f$ in the definitions of $\eta_{\mathcal{T}}(u_{\mathcal{T}}, f)$ and $\eta_{\mathcal{T}_*}(u_{\mathcal{T}_*}, f)$. In view of (6.24) and the definition (5.11) of the total estimator, we have

$$\mathrm{osc}_{\mathcal{T}}(f)_{-1} \le \sigma\, \eta_{\mathcal{T}}(u_{\mathcal{T}}), \quad \sigma^2 := \frac{\omega^2}{1 - \omega^2}$$

so that

$$\eta_{\mathcal{T}}(u_{\mathcal{T}}, f)^2 \le 2\eta_{\mathcal{T}_*}(u_{\mathcal{T}_*}, f)^2 + 2C_{\mathrm{Lip}}^2 \widetilde{C}_U^2 \big(\eta_{\mathcal{T}}(u_{\mathcal{T}}, f, \mathcal{R})^2 + 2\sigma^2 \eta_{\mathcal{T}}(u_{\mathcal{T}}, f)^2\big).$$

Using (6.25) and rearranging the above expression, we obtain

$$\big(\theta_0^2 - 2\sigma^2\big)\eta_{\mathcal{T}}(u_{\mathcal{T}}, f)^2 \le \left(\frac{1 - 2\mu^2}{2C_{\mathrm{Lip}}^2 \widetilde{C}_U^2} - 2\sigma^2\right)\eta_{\mathcal{T}}(u_{\mathcal{T}}, f)^2 \le \eta_{\mathcal{T}}(u_{\mathcal{T}}, f, \mathcal{R})^2$$

provided $0 < \mu \le \frac{1}{2}$, because of the definition of θ_0 in Assumption 6.14. Finally, for any $\theta < \theta_0$ we realize that ω_0 from Assumption 6.15 satisfies

$$\sigma_0^2 := \frac{\omega_0^2}{1 - \omega_0^2} = \frac{1}{2}\big(\theta_0^2 - \theta^2\big) \quad \Rightarrow \quad \theta^2 = \theta_0^2 - 2\sigma_0^2 \le \theta_0^2 - 2\sigma^2,$$

and Dörfler marking is valid for $\mathcal{R}$ with parameter θ. □

We remark that Lemma 6.16 requires that the oscillation on $\mathcal{T}$ is dominated by the total estimator to guarantee a Dörfler marking property. This is always the case when f is discrete as in Algorithm 5.4 (GALERKIN), because in that case $\operatorname{osc}_{\mathcal{T}}(f)_{-1} = 0$, or within Algorithm 5.16 (AFEM-SW), which marks elements for refinement only if this property holds.

We also see that θ_0 in Assumption 6.14 corresponds to the choices $\mu = \frac{1}{2}$ and $\delta = 1$. However, the proof reveals that for $\mu \to 0$ we could obtain the largest possible value $\theta_0 = (C_{\mathrm{Lip}}\widetilde{C}_U)^{-1}$, thereby the less restrictive. Since this is just twice the value of θ_0 in Assumption 6.14, the practical choice $\mu = \frac{1}{2}$ is justified.

Lemma 6.16 hinges on two ingredients: the Lipschitz property (4.68) and the localized upper bound (6.23) of the estimator. In particular, it does not rely on the lower *a posteriori* error estimate in (6.22), like the original proofs in Stevenson (2007) and Cascón *et al.* (2008), and easily extends to discontinuous Galerkin methods in Section 9 and inf-sup stable methods in Section 10. The original statement is, however, a bit more insightful: if $\theta_0^2 = C_2^2/2C_1^2$, then for all $0 < \theta < \theta_0$, $\omega^2 \le \theta_0^2 - \theta^2$,

$$|\!|\!| u - u_{\mathcal{T}_*} |\!|\!|_\Omega \le \mu |\!|\!| u - u_{\mathcal{T}} |\!|\!|_\Omega \quad \Rightarrow \quad \eta_{\mathcal{T}}(u_{\mathcal{T}}, \mathcal{R}) \ge \theta\, \eta_{\mathcal{T}}(u_{\mathcal{T}})$$

provided $0 < \mu \le 2^{-1/2}$. We see that the threshold θ_0 is related to the gap between *reliability* constant C_1 and *efficiency* constant C_2 in the *a posteriori* bounds (5.13) in the energy norm; hence the ratio $C_2/C_1 \le 1$ is a quality measure of the estimator $\eta_{\mathcal{T}}(u_{\mathcal{T}})$. It is thus reasonable to be cautious about marking decisions if the constants C_1 and C_2 are very disparate, and thus the ratio C_2/C_1 is far from 1. This justifies the constraint $\theta < \theta_0$.

6.4. *Rate-optimality of one-step* AFEMs

Recall that $\mathcal{M}_j$ is the output of the module MARK and that $\mathcal{T}_j, u_j$ are the meshes and associated Galerkin solutions generated within Algorithms 5.4 (GALERKIN) and 5.16 (AFEM-SW). To express the cardinality $N_j(u)$ of $\mathcal{M}_j$ in terms of $|u - u_j|_{H_0^1(\Omega)}$, we must relate the performance of these one-step AFEMs with the approximation classes $\mathbb{A}_s = \mathbb{A}_s(H_0^1(\Omega); \mathcal{T}_0)$ for u and $\mathbb{F}_s = \mathbb{F}_s(H^{-1}(\Omega); \mathcal{T}_0)$ for f, which are never used in the design of these algorithms. Even though this might appear infeasible, the key to unravel this connection is given by Lemma 6.16 (Dörfler marking) and the following assumption.

Assumption 6.17 (cardinality of $\mathcal{M}$). The module MARK selects a set $\mathcal{M}$ in (5.23) with *minimal* cardinality.

According to the equidistribution principle (3.24) and the local lower bound (4.54) in the proof of Theorem 4.45 (modified residual estimator) for discrete coefficients, that is,

$$C_L \eta_{\mathcal{T}_j}(u_j, T) \le C_L \mathcal{E}_{\mathcal{T}_j}(u_j, f, T) \le |u - u_j|_{H^1(\omega_T)},$$

it is natural to mark elements with largest error indicators. This explains the choice of a minimal set $\mathcal{M}$ in Assumption 6.17.

We are now ready to bound the cardinality of $\mathcal{M}_j$ in terms of $|u - u_j|_{H_0^1(\Omega)}$.

Proposition 6.18 (cardinality of $\mathcal{M}_j$). *Let Assumptions 6.14, 6.15 and 6.17 be valid. If $u \in \mathbb{A}_s$ and*

$$\operatorname{osc}_{\mathcal{T}_j}(f)_{-1} \leq \omega\, \mathcal{E}_{\mathcal{T}_j}(u_j, f),$$

then the cardinality $N_j(u)$ of $\mathcal{M}_j$ satisfies

$$N_j(u) \lesssim |u|_{\mathbb{A}_s}^{1/s} |u - u_j|_{H_0^1(\Omega)}^{-1/s} \quad \textit{for all } j \geq 0.$$

Proof. Let $C_{\text{Céa}} := \sqrt{C_{\mathcal{B}}/c_{\mathcal{B}}}$ be the quasi-monotonicity constant in (3.8). Let $\delta = \mu(C_L/C_{\text{Céa}})\eta_j(u_j)$ with $\mu \leq \frac{1}{2}$. We invoke (6.8) for $u \in \mathbb{A}_s$ to find a mesh $\mathcal{T}_\delta \in \mathbb{T}$ and a Galerkin solution $u_\delta \in \mathbb{V}_{\mathcal{T}_\delta}$, so that

$$|u - u_{\mathcal{T}_\delta}|_{H_0^1(\Omega)} \leq \delta, \quad \#\mathcal{T}_\delta \lesssim |u|_{\mathbb{A}_s}^{1/s} \delta^{-1/s}.$$

Since $\mathcal{T}_\delta$ may be totally unrelated to $\mathcal{T}_j$, we introduce the overlay $\mathcal{T}_* = \mathcal{T}_j \oplus \mathcal{T}_\delta$. We exploit that $\mathcal{T}_* \geq \mathcal{T}_\delta$, hence the space nestedness $\mathbb{V}_{\mathcal{T}_\delta} \subset \mathbb{V}_{\mathcal{T}_*}$, along with the property that the Galerkin solution $u_{\mathcal{T}_*} \in \mathbb{V}_{\mathcal{T}_*}$ minimizes the energy error in $\mathbb{V}_{\mathcal{T}_*}$,

$$\eta_{\mathcal{T}_*}(u_{\mathcal{T}_*}) \leq \frac{1}{C_L}|u - u_{\mathcal{T}_*}|_{H_0^1(\Omega)} \leq \frac{C_{\text{Céa}}}{C_L}|u - u_{\mathcal{T}_\delta}|_{H_0^1(\Omega)} \leq \frac{C_{\text{Céa}}}{C_L}\delta = \mu\,\eta_j(u_j),$$

because of the lower bound in (5.12) and (3.8). Therefore Lemma 6.16 (Dörfler marking) implies that the refined set $\mathcal{R} = \mathcal{T}\backslash\mathcal{T}_*$ satisfies Dörfler marking with parameter $\theta < \theta_0$. Since MARK delivers a minimal set $\mathcal{M}_j$ with this property, according to Assumption 6.17, we deduce

$$N_j(u) = \#\mathcal{M}_j \leq \#\mathcal{R} \leq \#\mathcal{T}_* - \#\mathcal{T} \leq \#\mathcal{T}_\varepsilon - \#\mathcal{T}_0 \lesssim |u|_{\mathbb{A}_s}^{1/s} \delta^{-1/s},$$

where we have used Lemma 3.17 (mesh overlay). The assertion follows from $\operatorname{osc}_j(f)_{-1} \leq \omega\mathcal{E}_j(u_j, f)$ in Assumption 6.15 and the upper bound in (5.12),

$$|u - u_j|_{H_0^1(\Omega)} \leq C_U \mathcal{E}_j(u_j, f) \leq \frac{C_U}{\sqrt{1 - \omega^2}}\eta_j(u_j),$$

and completes the proof. □

We next prove rate-optimality of the one-step AFEMs of Algorithm 5.4 and Algorithm 5.16. To this end, we need an additional assumption.

Assumption 6.19 (initial labelling). If the initial mesh $\mathcal{T}_0$ is made of simplices, then let the initial labelling (3.35) for $d = 2$, or that of Stevenson (2008, Section 4) for $d > 2$, be valid.

This assumption ensures the validity of Theorem 3.16 (complexity of REFINE): if $\mathcal{M}_j \subset \mathcal{T}_j$ is a set of marked elements for a sequence $\{\mathcal{T}_j\}_{j=0}^{k-1}$ of consecutive refinements of $\mathcal{T}_0$, then the cardinality of the kth mesh satisfies

$$\#\mathcal{T}_k - \#\mathcal{T}_0 \leq D \sum_{j=0}^{k-1} \#\mathcal{M}_j, \tag{6.27}$$

with a universal constant D depending only on $\mathcal{T}_0$ and d. We always assume that $\#\mathcal{T}_k \geq \frac{3}{2}\#\mathcal{T}_0$, whence $\#\mathcal{T}_k - \#\mathcal{T}_0 \geq \frac{1}{3}\#\mathcal{T}_k$ and, if $\widetilde{D} = 3D$, (6.27) instead reads

$$\#\mathcal{T}_k \leq \widetilde{D} \sum_{j=0}^{k-1} \#\mathcal{M}_j, \tag{6.28}$$

Theorem 6.20 (rate-optimality of one-step AFEMs). *For Algorithms 5.4 (with $\mathcal{T} = \mathcal{T}_0$) and 5.16, let Assumptions 6.14, 6.17 and 6.19 be valid, and in addition let the parameter $\omega > 0$ satisfy Assumption 6.15 for Algorithm 5.16. If $u \in \mathbb{A}_s$, then both one-step* AFEM*s give rise to sequences $\{\mathcal{T}_k, \mathbb{V}_k, u_k\}_{k=0}^{\infty}$ such that*

$$|u - u_k|_{H_0^1(\Omega)} \lesssim |u|_{\mathbb{A}_s} (\#\mathcal{T}_k)^{-s}. \tag{6.29}$$

Proof. We first consider Algorithm 5.4, for which the forcing $f \in \mathbb{F}_{\mathcal{T}_0}$ is discrete, whence $\operatorname{osc}_j(f)_{-1} = 0$ for all $j \geq 0$ and $\omega = \sigma = 0$ in Assumption 6.15. In view of (6.28), we apply Proposition 6.18 (cardinality of $\mathcal{M}_j$) to infer that

$$\#\mathcal{T}_k \leq \widetilde{D} \sum_{j=0}^{k-1} \#\mathcal{M}_j \lesssim |u|_{\mathbb{A}_s}^{1/s} \sum_{j=0}^{k-1} |u - u_j|_{H_0^1(\Omega)}^{-1/s}.$$

We now recall the inequality $|u-u_k|_{H_0^1(\Omega)} \leq C_* \alpha^{k-j} |u-u_j|_{H_0^1(\Omega)}$ from Corollary 5.6 (linear convergence), and replace the sum above with

$$\sum_{j=0}^{k-1} |u - u_j|_{H_0^1(\Omega)}^{-1/s} \leq |u - u_k|_{H_0^1(\Omega)}^{-1/s} \sum_{j=0}^{k} \alpha^{(k-j)/s} \leq \frac{\alpha^{1/s}}{1 - \alpha^{1/s}} |u - u_k|_{H_0^1(\Omega)}^{-1/s},$$

because $0 < \alpha < 1$ and the geometric series is summable.

We now deal with Algorithm 5.16. If the algorithm calls MARK, then $\operatorname{osc}_j(f)_{-1} \leq \omega \mathcal{E}_j(u_j, f)$ and the number of marked elements $N_j(u)$ obeys Proposition 6.18:

$$N_j(u) \lesssim |u|_{\mathbb{A}_s}^{1/s} |u - u_j|_{H_0^1(\Omega)}^{-1/s}.$$

Instead, if the algorithm calls DATA, then $\operatorname{osc}_j(f)_{-1} > \sigma_j = \omega \mathcal{E}_j(u_j, f)$ and DATA returns a mesh $\mathcal{T}_{j+1}$ and reduces the oscillation $\operatorname{osc}_{j+1}(f)_{-1} \leq \sigma_j$ with optimal complexity. To quantify the cost, we recall that $u \in \mathbb{A}_s$ yields $f \in \mathbb{F}_s$ according to Lemma 6.9 (relation between approximation classes) and $|f|_{\mathbb{F}_s} \lesssim |u|_{\mathbb{A}_s}$. Therefore the number of marked elements $N_j(f)$ to reduce $\operatorname{osc}_j(f)_{-1}$ to tolerance σ_j satisfies

$$N_j(f) \lesssim |f|_{\mathbb{F}_s}^{1/s} \sigma_j^{-1/s} \lesssim |u|_{\mathbb{A}_s}^{1/s} \mathcal{E}_j(u_j, f)^{-1/s} \lesssim |u|_{\mathbb{A}_s}^{1/s} |u - u_j|_{H_0^1(\Omega)}^{-1/s},$$

because of (5.12). It thus remains to sum over j, again apply (6.28),

$$\#\mathcal{T}_k \leq \widetilde{D} \sum_{j=0}^{k-1} (N_j(u) + N_j(f)) \lesssim |u|_{\mathbb{A}_s}^{1/s} \sum_{j=0}^{k-1} |u - u_j|_{H_0^1(\Omega)}^{-1/s},$$

and finally argue as before with the help of Corollary 5.18 (linear convergence of error). □

6.5. *Rate-optimality of two-step* AFEM

The output of $[\widehat{\mathcal{T}}_k, \widehat{\mathcal{D}}_k] = \mathsf{DATA}(\mathcal{T}_k, \mathcal{D}, \omega\varepsilon_k)$, in the k-step of AFEM-TS (Algorithm 5.28), is fed to $[\mathcal{T}_{k+1}, u_{k+1}] = \mathsf{GALERKIN}(\widehat{\mathcal{T}}_k, \widehat{\mathcal{D}}_k, \varepsilon_k)$, which in turn iterates J_k times. We let $(\mathcal{T}_{k,j}, \mathcal{M}_{k,j}, u_{\mathcal{T}_{k,j}})$ denote the triplets of grids, marked sets and discrete solutions computed within $\mathsf{GALERKIN}(\widehat{\mathcal{T}}_k, \widehat{\mathcal{D}}_k, \varepsilon_k)$ for $0 \leq j < J_k$. We further assume that

$$\widehat{\varepsilon}_k := \eta_{\mathcal{T}_{k,0}}(u_{\mathcal{T}_{k,0}}, \widehat{\mathcal{D}}_k) > \varepsilon_k,$$

for otherwise the module GALERKIN is skipped. In view of the lower *a posteriori* error estimate in (6.22) for discrete data $\widehat{\mathcal{D}}_k$, we infer that

$$|\widehat{u}_k - u_{\mathcal{T}_{k,0}}|_{H_0^1(\Omega)} \geq C_L \widehat{\varepsilon}_k > C_L \varepsilon_k,$$

where $\widehat{u}_k \in H_0^1(\Omega)$ is the exact solution of (5.5) with approximate data $\widehat{\mathcal{D}}_k$. The module DATA guarantees (5.63) and (5.65), namely

$$\|\mathcal{D} - \widehat{\mathcal{D}}_k\|_{D(\Omega)} \leq C_{\text{data}} \omega \varepsilon_k \quad \Rightarrow \quad |u - \widehat{u}_k|_{H_0^1(\Omega)} \leq C_D \omega \varepsilon_k, \tag{6.30}$$

where $u = u(\mathcal{D}) \in H_0^1(\Omega)$ is the exact solution of (2.7). We see that the parameter ω controls the discrepancy between u and $\widehat{u}_k = \widehat{u}_k(\mathcal{D}_k)$ relative to ε_k. We now make an assumption on the appropriate size of ω, which replaces Assumption 6.15 for AFEM-SW.

Assumption 6.21 (size of ω). The parameter ω in AFEM-TS satisfies $\omega \in (0, \omega_0]$, where

$$\omega_0 := \frac{\mu C_L}{2 C_D C_{\text{Céa}}}$$

with $C_{\text{Céa}}$ as in (3.8) and the parameter $0 < \mu \leq \frac{1}{2}$ appears in Lemma 6.16 (Dörfler marking).

Consequently, if Assumption 6.21 is valid then (6.30) yields for $\omega \leq \omega_0$

$$|u - \widehat{u}_k|_{H_0^1(\Omega)} \leq \frac{\mu C_L}{2 C_{\text{Céa}}} \varepsilon_k. \tag{6.31}$$

Corollary 6.22 (cardinality of marked sets). *Let Assumptions 6.14, 6.17 and 6.21 hold. If $u \in \mathbb{A}_s(H_0^1(\Omega); \mathcal{T}_0)$ and $\widehat{\varepsilon}_k > \varepsilon_k$, then* GALERKIN *is called and there* *exists a constant* C_0 *such that, for all* $0 \le j < J_k$,

$$\#\mathcal{M}_{k,j} \le C_0 |u|_{\mathbb{A}_s}^{1/s} |u - u_{\mathcal{T}_{k,j}}|_{H_0^1(\Omega)}^{-1/s} \tag{6.32}$$

and

$$\#\mathcal{M}_{k,j} \le C_0 |u|_{\mathbb{A}_s}^{1/s} \varepsilon_k^{-1/s}. \tag{6.33}$$

Proof. We argue as in Proposition 6.18. Fix $0 \le j < J_k$ and set

$$\delta := \mu \frac{C_L}{C_{\text{Céa}}} \eta_{\mathcal{T}_{k,j}}(u_{\mathcal{T}_{k,j}}) \quad \Rightarrow \quad \delta \ge \mu \frac{C_L}{C_{\text{Céa}}} \varepsilon_k.$$

Since $|u - \widehat{u}_k|_{H_0^1(\Omega)} \le \delta/2$, by virtue of (6.31), we deduce that $\widehat{u}_k$ is an δ-approximation of order s to u according to Lemma 6.13 (ε-approximation of u of order s). Thus there exists an admissible mesh $\mathcal{T}_\delta \in \mathbb{T}$ such that

$$|\widehat{u}_k - u_{\mathcal{T}_\delta}|_{H_0^1(\Omega)} \le \delta, \quad \#\mathcal{T}_\delta \lesssim |u|_{\mathbb{A}_s}^{1/s} \delta^{-1/s},$$

and we proceed exactly as in Proposition 6.18, to show that

$$\#\mathcal{M}_{k,j} \lesssim |u|_{\mathbb{A}_s}^{1/s} \delta^{-1/s} \approx |u|_{\mathbb{A}_s}^{1/s} |u - u_{\mathcal{T}_{k,j}}|_{H_0^1(\Omega)}^{-1/s} \lesssim |u|_{\mathbb{A}_s}^{1/s} \varepsilon_k^{-1/s}.$$

This concludes the proof. □

Corollary 6.23 (quasi-optimality of GALERKIN). *Let Assumptions 6.3, 6.14, 6.17 and 6.21 be valid. Then the number of marked elements* $N_k(u)$ *within the* k*th call to* GALERKIN *satisfies*

$$N_k(u) \le J C_0 |u|_{\mathbb{A}_s}^{1/s} \varepsilon_k^{-1/s},$$

where $J \ge J_k$ *is a uniform upper bound for the number of iterations of* GALERKIN.

Proof. Use $N_k(u) = \sum_{j=0}^{J_k-1} \#\mathcal{M}_{k,j}$ and combine Corollary 6.22 (cardinality of marked sets) and Proposition 5.27 (computational cost of GALERKIN). □

We finally address the rate-optimality of the two-step algorithm AFEM-TS, by proving the stated bound (6.1).

Theorem 6.24 (rate-optimality of AFEM-TS). *Let Assumptions 6.3 (approximability of u), 6.10 (approximability of data), 6.11 (quasi-optimality of* DATA*), 6.14 (marking parameter), 6.21 (size of* ω*) and 6.19 (initial labelling) be valid. Then* AFEM-TS *gives rise to a sequence* $(\mathcal{T}_k, \mathbb{V}_{\mathcal{T}_k}, u_{\mathcal{T}_k})_{k=0}^K$ *such that*

$$|u - u_k|_{H_0^1(\Omega)} \le C(u, \mathcal{D})(\#\mathcal{T}_k)^{-s}, \quad 1 \le k \le K,$$

where $0 < s = \min\{s_u, s_{\mathcal{D}}\} = \min\{s_u, s_A, s_c, s_f\} \le n/d$ *and*

$$C(u, \mathcal{D}) = C_* \left(|u|_{\mathbb{A}_{s_u}}^{1/s_u} + |\boldsymbol{A}|_{\mathbb{M}_{s_A}}^{1/s_A} + |c|_{\mathbb{C}_{s_c}}^{1/s_c} + |f|_{\mathbb{F}_{s_f}}^{1/s_f} \right)^s,$$

with constant $C_* > 0$ *independent of* u *and* $\mathcal{D}$.

Proof. In view of Assumption 6.3, Corollary 6.23 implies that the number of marked elements $N_k(u)$ within the $(k+1)$th call to GALERKIN satisfies

$$N_k(u) \leq C_3 |u|_{\mathbb{A}_{s_u}}^{1/s_u} \varepsilon_k^{-1/s_u},$$

with $s_u \leq n/d$ and $C_3 > 0$ a suitable constant. Moreover, by Assumption 6.11 the number of marked elements $N_k(\mathcal{D})$ within the $(k+1)$th call to DATA satisfies

$$N_k(\mathcal{D}) \leq C_3 |\mathcal{D}|_{\mathbb{A}_{s_\mathcal{D}}}^{1/s_\mathcal{D}} \varepsilon_k^{-1/s_\mathcal{D}},$$

with $s_\mathcal{D} \leq n/d$. The total number of marked elements in the $(k+1)$th loop of AFEM-TS is thus

$$N_k(\mathcal{D}) + N_k(u) \leq C_3 \left(|u|_{\mathbb{A}_{s_u}}^{1/s_u} + |\mathcal{D}|_{\mathbb{A}_{s_\mathcal{D}}}^{1/s_\mathcal{D}} \right) \varepsilon_k^{-1/s}.$$

Upon termination, DATA and GALERKIN give

$$|u - \widehat{u}_k|_{H_0^1(\Omega)} \leq \frac{\mu C_L}{2C_{\text{Céa}}} \varepsilon_k \leq \frac{C_L}{4C_{\text{Céa}}} \varepsilon_k,$$
$$|\widehat{u}_k - u_{k+1}|_{H_0^1(\Omega)} \leq C_U \eta_{\mathcal{T}_{k+1}}(u_{k+1}, \widehat{\mathcal{D}}_k) \leq C_U \varepsilon_k,$$

because of (6.31), (6.22) and the fact that $\mu < \frac{1}{2}$. This implies by the triangle inequality

$$|u - u_{k+1}|_{H_0^1(\Omega)} \leq \left(\frac{C_L}{4C_{\text{Céa}}} + C_U \right) \varepsilon_k = C_4 \varepsilon_k.$$

Therefore, applying Theorem 3.16 (complexity of REFINE), the total amount of elements created by $k+1$ iterations within AFEM-TS, besides those in $\mathcal{T}_0$, obeys the expression

$$\#\mathcal{T}_{k+1} \leq \widetilde{D} \sum_{j=0}^{k} (N_j(\mathcal{D}) + N_j(u)) \leq \widetilde{D} C_3 \left(|u|_{\mathbb{A}_{s_u}}^{1/s_u} + |\mathcal{D}|_{\mathbb{A}_{s_\mathcal{D}}}^{1/s_\mathcal{D}} \right) \sum_{j=0}^{k} \varepsilon_j^{-1/s}.$$

according to (6.28). Since $\varepsilon_j = 2^{-j}\varepsilon_0$ and

$$\sum_{j=0}^{k-1} (2^{-1/s})^j \leq \frac{1}{1 - 2^{-1/s}},$$

we obtain

$$\#\mathcal{T}_{k+1} \leq C \left(|u|_{\mathbb{A}_{s_u}}^{1/s_u} + |\mathcal{D}|_{\mathbb{A}_{s_\mathcal{D}}}^{1/s_\mathcal{D}} \right) \varepsilon_k^{-1/s},$$

with

$$C = \frac{\widetilde{D} C_3 \varepsilon_0}{1 - 2^{-1/s}}$$

provided $\#\mathcal{T}_{k+1} \geq \frac{3}{2}\#\mathcal{T}_0$. This together with $|u - u_{k+1}|_{H_0^1(\Omega)} \leq C_4 \varepsilon_k$ gives the asserted estimate after $1 \leq k+1 \leq K$ loops. □

Remark 6.25. The thresholds θ_0, ω_0 play no role in Proposition 5.29 (convergence of AFEM-TS) but are critical in Theorem 6.24 (rate-optimality of AFEM-TS). The former takes care of the discrepancy between error and estimator (Stevenson 2007, Cascón *et al.* 2008, Nochetto *et al.* 2009, Bonito and Nochetto 2010, Nochetto and Veeser 2012). The latter guarantees that the perturbation error (6.30) is much smaller than ε_k and enables GALERKIN to learn the regularity of u from $\widehat{u}_{\widehat{\mathcal{T}}_k}$ (Stevenson 2007, Bonito *et al.* 2013*b*).

Remark 6.26. We claim that the convergence rate $s = \min\{s_u, s_{\mathcal{D}}\}$ cannot be improved to s_u (the optimal rate for approximations of $u \in \mathbb{A}_{s_u}(H_0^1(\Omega); \mathcal{T}_0)$) when $s_{\mathcal{D}} < s_u$ by any algorithm that uses approximations $\widehat{\mathcal{D}} = (\widehat{\boldsymbol{A}}, \widehat{c}, \widehat{f})$ of data $\mathcal{D} = (\boldsymbol{A}, c, f)$. In fact, given any $\delta > 0$, consider the ball

$$B(\mathcal{D}, \delta) := \{\widehat{\mathcal{D}} \in \mathbb{D} \mid \|\mathcal{D} - \widehat{\mathcal{D}}\|_{D(\Omega)} \leq \delta\}, \tag{6.34}$$

where $D(\Omega)$ is defined in (5.60). If $u, \widehat{u} \in H_0^1(\Omega)$ are the exact solutions for data $\mathcal{D}, \widehat{\mathcal{D}}$, then there are constants $0 < c_* \leq C_*$ such that

$$c_*\delta \leq \sup_{\widehat{\mathcal{D}} \in B(\mathcal{D}, \delta)} |u - \widehat{u}|_{H_0^1(\Omega)} \leq C_*\delta.$$

The rightmost inequality is a consequence of Lemma 5.20 (continuous dependence on data). For the leftmost inequality, first consider a perturbation $\widehat{f} = (1+\delta)f$ of the source term with coefficients $(\widehat{\boldsymbol{A}}, \widehat{c}) = (\boldsymbol{A}, c)$, whence $\|\mathcal{D} - \widehat{\mathcal{D}}\|_{D(\Omega)} = \delta$. Proceeding as in (2.30), the coercivity and continuity of the bilinear form $\mathcal{B}$ imply

$$c_{\mathcal{B}}\, |u - \widehat{u}|_{H_0^1(\Omega)} \leq \|f - \widehat{f}\|_{H^{-1}(\Omega)} = \delta \leq C_{\mathcal{B}}\, |u - \widehat{u}|_{H_0^1(\Omega)}.$$

On the other hand, if $\widehat{f} = f$ and $(\widehat{\boldsymbol{A}}, \widehat{c}) = \alpha^{-1}(\boldsymbol{A}, c)$ with $\alpha = 1 + \delta/\|\mathcal{D}\|_{D(\Omega)}$, then

$$\|\mathcal{D} - \widehat{\mathcal{D}}\|_{D(\Omega)} < \delta, \quad |u - \widehat{u}|_{H_0^1(\Omega)} = \frac{|u|_{H_0^1(\Omega)}}{\|\mathcal{D}\|_{D(\Omega)}}\delta \geq \frac{\|f\|_{H^{-1}(\Omega)}}{C_{\mathcal{B}}\,\|\mathcal{D}\|_{D(\Omega)}}\delta.$$

This argument takes care of the multiplicative nature of $(\boldsymbol{A}, c)$ in (2.5), which makes $\widehat{u} = \alpha u$, and proves our claim.

6.6. *Rate-optimality of* AFEM *with other boundary conditions*

The key ingredient for rate-optimality of AFEM, regardless of boundary conditions, is the validity of Lemma 6.16 (Dörfler marking). This lemma provides a bridge between FEM meshes and optimal meshes and, in turn, hinges on three properties of the PDE residual estimator $\eta_{\mathcal{T}}(u_{\mathcal{T}}, f)$: Theorem 4.48 (upper bound for corrections), Lemma 4.54 (Lipschitz property of the estimator) and Lemma 4.55 (estimator dependence on discrete forcing) to account for the possible change in the discrete forcing $P_{\mathcal{T}} f$. Since their proofs are insensitive to boundary conditions, because they do not alter the structure of $\eta_{\mathcal{T}}(u_{\mathcal{T}}, f)$, we conclude their validity as well as for Robin, Neumann and non-homogeneous Dirichlet conditions.

Therefore our three AFEMs based on Dörfler marking deliver the same asymptotic convergence rates associated with the approximations classes $\mathbb{A}_{s_u}$ for the solution $u \in H^1(\Omega)$ and $\mathbb{A}_{s_\mathcal{D}}$ for data $\mathcal{D} = (\boldsymbol{A}, c, p, \ell)$ for Robin and Neumann boundary conditions, and $\mathcal{D} = (\boldsymbol{A}, c, f, g)$ for Dirichlet boundary conditions. We need three new approximation classes for $p \in \mathbb{P}_{s_p}(L^\infty(\partial\Omega); \mathcal{T}_0)$, $\ell \in \mathbb{L}_{s_\ell}(H^1(\Omega)^*; \mathcal{T}_0)$ for Robin or Neumann conditions and $g \in \mathbb{G}_{s_g}(H^{1/2}(\partial\Omega); \mathcal{T}_0)$.

If coefficients $(\boldsymbol{A}, c, p)$ are discrete for the Robin condition, then Lemma 6.9 (relation between approximation classes) extends and yields

$$\langle \ell - \widehat{\ell}, v\rangle = \mathcal{B}[u - \widehat{u}, v] \quad \Rightarrow \quad \|\ell - \widehat{\ell}\|_{\mathbb{V}^*} \le C\|u - \widehat{u}\|_{\mathbb{V}},$$

with $\mathbb{V} = H^1(\Omega)$ and $\mathcal{B} = \widehat{\mathcal{B}}, \widehat{\ell}$ given in (5.85). This in turn implies $|\ell|_{\mathbb{L}_{s_\ell}} \le C|u|_{\mathbb{A}_{s_u}}$, $s_\ell = s_u$ and the validity of Theorem 6.20 (rate-optimality of one-step AFEMs). In AFEM-TS, DATA approximates ℓ along with the other data and Theorem 6.24 (rate-optimality of AFEM-TS) is also valid for Robin and Neumann boundary conditions. We do not explore this matter any further.

For non-homogeneous Dirichlet boundary conditions the analysis is simpler. If g is discrete, then there is no difference to $g = 0$. If not, we note that the solution map $g \mapsto u$ (all other data being fixed) is affine and that the error and augmented total estimator $\mathcal{E}_\mathcal{T}(u_\mathcal{T}, f, g) := \mathcal{E}_\mathcal{T}(u_\mathcal{T}, f) + \text{osc}_\mathcal{T}(g)_{1/2}$ are equivalent (Theorem 4.74). This indicates that the role of g is similar to the role of f. Therefore it suffices to replace $\mathcal{E}_\mathcal{T}(u_\mathcal{T}, f)$ with $\mathcal{E}_\mathcal{T}(u_\mathcal{T}, f, g)$ and $\text{osc}_\mathcal{T}(f)_{-1}$ with $\text{osc}_\mathcal{T}(f)_{-1} + \text{osc}_\mathcal{T}(g)_{1/2}$ in AFEM-SW. For AFEM-TS, the approximation of g is handled by DATA along with the other data. Hence we again conclude that Theorems 6.20 and 6.24 extend to non-vanishing Dirichlet conditions.

6.7. *Rate-optimality of* AFEM *driven by alternative estimators*

We recall the notation $\zeta_\mathcal{T}(u_\mathcal{T})$ of Section 5.6 for any of the three alternative estimators in Section 4.9 and the crucial local properties (5.86) and (5.87).

As already alluded to in Section 6.6, the key instrument for rate-optimality is Lemma 6.16 (Dörfler marking). We now check the validity of its three main pillars: Theorem 4.48 (upper bound for corrections), Lemma 4.54 (Lipschitz property of the estimator) and Lemma 4.55 (estimator dependence on discrete forcing) to account for possible change in the discrete forcing $P_\mathcal{T} f$. It turns out that if they were valid for $\zeta_\mathcal{T}(u_\mathcal{T}) = \zeta_\mathcal{T}(u_\mathcal{T}, f)$, then statements about rates of convergence similar to those for $\eta_\mathcal{T}(u_\mathcal{T})$ would follow for $\zeta_\mathcal{T}(u_\mathcal{T})$.

Lemma 6.27 (localized discrete upper bound). *Let $\mathcal{T}, \mathcal{T}_* \in \mathbb{T}$ and $\mathcal{T}_*$ be a refinement of $\mathcal{T}$. Let the coefficients $(\widehat{\boldsymbol{A}}, \widehat{c})$ be discrete over $\mathcal{T}$ and $f \in H^{-1}(\Omega)$. Then the error between the corresponding Galerkin solutions $u_\mathcal{T} \in \mathbb{V}_\mathcal{T}$ and $u_{\mathcal{T}_*} \in \mathbb{V}_{\mathcal{T}_*}$ is bounded by the indicator in the refined set $\overline{\mathcal{R}}$ plus data oscillation*

$$|u_\mathcal{T} - u_{\mathcal{T}_*}|_{H_0^1(\Omega)} \le \overline{C}_U \big(\big(C_L^{\text{eq}}\big)^{-1} \zeta_\mathcal{T}(u_\mathcal{T}, \overline{\mathcal{R}})^2 + \text{osc}_\mathcal{T}(f)_{-1}^2 \big)^{1/2},$$

where $\overline{\mathcal{R}} := \{z \in \mathcal{V} \mid T \in \mathcal{T}\backslash\mathcal{T}_*, T \subset \omega_z\}$ *collects all vertices whose associated stars change from* $\mathcal{T}$ *to* $\mathcal{T}_*$.

Proof. It suffices to realize that $\mathcal{T}\backslash\mathcal{T}_* \subset \bigcup\{\omega_z \mid z \in \overline{\mathcal{R}}\}$ and appeal to Theorem 4.48 and (5.87) to arrive at

$$\begin{aligned}|u_{\mathcal{T}} - u_{\mathcal{T}_*}|^2_{H^1_0(\Omega)} &\le \overline{C}_U^2 \left(\eta_{\mathcal{T}}(u_{\mathcal{T}}, \mathcal{R})^2 + \mathrm{osc}_{\mathcal{T}}(f, \mathcal{R})^2_{-1}\right)\\ &\le \overline{C}_U^2 \left(\eta_{\mathcal{T}}(u_{\mathcal{T}}, \overline{\mathcal{R}})^2 + \mathrm{osc}_{\mathcal{T}}(f)^2_{-1}\right)\\ &\le \overline{C}_U^2 \left(\left(C_L^{\mathrm{eq}}\right)^{-2} \zeta_{\mathcal{T}}(u_{\mathcal{T}}, \overline{\mathcal{R}})^2 + \mathrm{osc}_{\mathcal{T}}(f)^2_{-1}\right).\end{aligned}$$

This is the desired estimate. □

Lemma 6.28 (Lipschitz property of the estimator). *Let the coefficients* $(\widehat{\mathbf{A}}, \widehat{c})$ *be discrete over* $\mathcal{T}$. *There exists* C_{Lip} *such that*

$$|\zeta_{\mathcal{T}}(v_1) - \zeta_{\mathcal{T}}(v_2)| \le C_{\mathrm{Lip}} |v_1 - v_2|_{H^1_0(\Omega)} \quad \textit{for all } v_1, v_2 \in \mathbb{V}_{\mathcal{T}}.$$

Proof. We resort to the star equivalence (5.86) between discrete residual and estimator. It thus suffices to derive the Lipschitz property for $\|P_{\mathcal{T}} R_{\mathcal{T}}(v)\|_{H^{-1}(\omega_z)}$ with respect to $v \in \mathbb{V}_{\mathcal{T}}$ for all $z \in \mathcal{V}$. Since $P_{\mathcal{T}} R_{\mathcal{T}}(v) = P_{\mathcal{T}} f - \widehat{\mathcal{B}}[v, \cdot]$, we get

$$\langle P_{\mathcal{T}} R_{\mathcal{T}}(v_1) - P_{\mathcal{T}} R_{\mathcal{T}}(v_2), w\rangle = \int_{\omega_z} \nabla w \cdot \widehat{\mathbf{A}} \nabla(v_1 - v_2) + \widehat{c}(v_1 - v_2) w$$

for all $w \in H^1_0(\omega_z)$. Therefore Lemma 2.2 (first Poincaré inequality) yields

$$\|P_{\mathcal{T}} R_{\mathcal{T}}(v) - P_{\mathcal{T}} R_{\mathcal{T}}(v_2)\|_{H^{-1}(\omega_z)} \le C(\widehat{\mathbf{A}}, \widehat{c}) \|v_1 - v_2\|_{H^1(\omega_z)},$$

where $C(\widehat{\mathbf{A}}, \widehat{c})$ depends on the L^∞-norms of $(\widehat{\mathbf{A}}, \widehat{c})$. Finally, using the triangle inequality to accumulate over $z \in \mathcal{V}$ together with (5.86) gives the assertion. □

Lemmas 6.27 and 6.28 lead to Lemma 6.16 (Dörfler marking) for $\zeta_{\mathcal{T}}(u_{\mathcal{T}})$. If we further choose a *minimal* set $\mathcal{M}$ of vertices that satisfies Dörfler property (5.88), then the previous rates of convergence for the three algorithms GALERKIN, AFEM-SW and AFEM-TS but now driven by $\zeta_{\mathcal{T}}(u_{\mathcal{T}})$ are valid provided $u \in \mathbb{A}_s$, the approximation class in Definition 6.1. We do not restate these results.

6.8. *Approximation vs. regularity classes*

The purpose of this section is to reconcile the notion of approximation classes, discussed above, with that of regularity classes. We recall the DeVore diagram of Figure 2.1, which depicts the Sobolev line for the energy space $H^1_0(\Omega)$, namely

$$\mathrm{sob}(H^1_0) = \mathrm{sob}(W^s_p) \quad \Rightarrow \quad s - \frac{d}{p} = 1 - \frac{d}{2}.$$

The differentiability $s \ge 1$ is only limited by the polynomial degree n, so $s \in [1, n+1]$. On the other hand, the integrability p is not restricted to be $p \ge 1$ as is customary with Sobolev spaces. For example, for $d = 2$ and $s = n+1$, we get $p = 2/(n+1) < 1$ provided $n \geq 2$. Therefore, to take full advantage of nonlinear approximation theory, we need to abandon the framework of Sobolev spaces $W_p^s(\Omega)$ and deal with *Besov spaces* $B_{p,q}^s(\Omega)$ (frequently denoted by $B_q^s(L^p(\Omega))$ or $B_q^s(L_p(\Omega))$ in the literature) with integrability index $p \in (0,\infty]$. The second index $q \in (0,\infty]$ is useful in characterizing special limiting cases; below we will provide a few interesting examples but take $p = q$ most of the time. At this point, we only mention that when s is non-integer and $1 \leq p \leq \infty$, $B_{p,p}^s(\Omega) = W_p^s(\Omega)$, while when r is integer $W_p^r(\Omega)$ for $p \neq 2$ is not a Besov space but it is slightly smaller than $B_{p,\infty}^r(\Omega)$. The case $p = 2$ is special since $B_{2,2}^s(\Omega) = H^s(\Omega)$ even when s is an integer.

This section is devoted to the definition and properties of Besov and Lipschitz spaces, including their close relation to approximation classes. Our presentation closely follows Binev *et al.* (2002) for $n = 1$ and Gaspoz and Morin (2014, 2017) for $n \geq 1$, but it adds a few new ingredients. Since our results involve three different type of spaces to account for the particular cases when the differentiability is integer, it is pertinent to introduce the following abstract space $X_p^s(\Omega)$ with differentiability index $s \in (0, n+1]$ and integrability index $p \in (0,\infty]$:

$$X_p^s(\Omega) := \begin{cases} B_{p,p}^s(\Omega), & s \in (0, n+1), p \in (0,\infty], \\ W_p^{n+1}(\Omega), & s = n+1, p \in [1,\infty], \\ \mathrm{Lip}_p^{n+1}(\Omega), & s = n+1, p \in (0,1). \end{cases} \tag{6.35}$$

Here $\mathrm{Lip}_p^s(\Omega) = \mathrm{Lip}(s, L^p(\Omega))$, $s \in \mathbb{N}$, are the Lipschitz spaces; see (6.58) below. For $s \in \mathbb{N}$ and $1 < p < \infty$ the Sobolev spaces coincide with the Lipschitz spaces (Leoni 2009, Theorem 10.55), that is,

$$\mathrm{Lip}_p^s(\Omega) = W_p^s(\Omega), \quad s \in \mathbb{N},\ 1 < p < \infty, \tag{6.36}$$

while for $p = 1$ we only have

$$W_1^s(\Omega) \hookrightarrow \mathrm{Lip}_1^s(\Omega), \quad s \in \mathbb{N}. \tag{6.37}$$

We use the following conventions: $X_p^s(\Omega) := L^p(\Omega)$ for $s = 0$; $X_p^s(\Omega;\mathcal{T})$ is the space of functions with piecewise regularity X_p^s over $\mathcal{T} \in \mathbb{T}$; $X_p^s(\Omega;\mathbb{R}^m)$ is the space $X_p^s(\Omega)$ of vector- or matrix-valued functions.

In Section 6.8.4 we will prove the following crucial approximation results for functions in $L^q(\Omega)$ by discontinuous piecewise polynomials $\mathbb{S}_{\mathcal{T}}^{n,-1}$ of degree $n \geq 1$ over conforming refinements $\mathcal{T}$ of $\mathcal{T}_0$. It turns out that this will also allow us to deal with approximations in $H_0^1(\Omega)$ by continuous piecewise polynomials $\mathbb{V}_{\mathcal{T}}$ of degree $n \geq 1$.

Theorem 6.29 (regularity yields approximation). *Let* $q \in [1,\infty]$, $p \in (0,\infty]$, $s \in (0, n+1]$ *and a function* $g \in L^q(\Omega)$ *satisfy* $g \in X_p^s(\Omega)$ *with* $s - d/p + d/q > 0$.

Then there exists a constant $C = C(p, q, s, t, d, \Omega, \mathcal{T}_0)$ *such that*

$$E_n(g, \Omega)_q := \inf_{\mathcal{T} \in \mathbb{T}_N} \inf_{v \in \mathbb{S}_{\mathcal{T}}^{n,-1}} |g - v|_{L^q(\Omega)} \leq C |g|_{X_p^s(\Omega)} N^{-s/d}. \tag{6.38}$$

Therefore $g \in \mathbb{A}_{s/d} = \mathbb{A}_{s/d}(L^q(\Omega); \mathcal{T}_0)$ *and*

$$|g|_{\mathbb{A}_{s/d}} \leq C |g|_{X_p^s(\Omega)}. \tag{6.39}$$

We see that the decay rate s/d in (6.38) is proportional to the difference of the differentiability indices between the space $X_p^s(\Omega)$ and $L^q(\Omega)$ provided the Sobolev numbers satisfy the relation

$$\mathrm{sob}(X_p^s(\Omega)) > \mathrm{sob}(L^q(\Omega)),$$

which implies that the embedding of $X_p^s(\Omega)$ into $L^q(\Omega)$ is compact. The factor d in the denominator is a manifestation of the so-called *curse of dimensionality*. The limiting case $s = n + 1$ entails dealing with Sobolev spaces $W_p^s(\Omega)$ and Lipschitz spaces $\mathrm{Lip}_p^s(\Omega)$ depending on whether $p \geq 1$ or $p < 1$.

6.8.1. *Modulus of smoothness*

Difference operators. Since we intend to allow $p \in (0, 1)$, the underlying functions in $B_{p,p}^s(\Omega)$ might not be locally integrable, whence they might not be distributions in Ω. Therefore the notion of weak derivative does not apply, which in turn has the disadvantage of being defined for integers and not for fractional numbers. This leads to the most standard definition of Besov spaces $B_{p,q}^s(\Omega)$ using difference operators, which only requires integrability in $L^p(\Omega)$ and is valid for any $s > 0$, $p, q \in (0, \infty]$. Other definitions, which provide equivalent results in the range $1 \leq p, q \leq \infty$, can be found in Adams and Fournier (2003) and Bergh and Löfström (1976).

Given a bounded Lipschitz domain $\Omega \subset \mathbb{R}^d$, and a vector $h \in \mathbb{R}^d$, we set

$$\Omega_h := \{x \in \Omega \mid [x, x + h] \subset \Omega\},$$

where $[x, x + h]$ denotes the closed segment connecting x and $x + h$, and define the first-order difference operator to be

$$\Delta_h^1 g(x) = \Delta_h^1(g, x, \Omega) := \begin{cases} g(x + h) - g(x), & x \in \Omega_h, \\ 0, & \text{otherwise.} \end{cases} \tag{6.40}$$

For $k \in \mathbb{N}$, $k \geq 1$, we define the kth difference operator by iteration,

$$\Delta_h^k g(x) := \Delta_h^1 \left(\Delta_h^{k-1}\right) g(x), \quad x \in \Omega_{kh}, \tag{6.41}$$

and observe that it has the explicit form

$$\Delta_h^k g(x) = \begin{cases} \displaystyle\sum_{j=0}^{k} (-1)^{k+j} \binom{k}{j} g(x + jh), & [x, x + kh] \subset \Omega, \\ 0, & \text{otherwise.} \end{cases}$$

Note the property

$$p \in \mathbb{P}_k \quad \Rightarrow \quad \Delta_h^{k+1} p = 0 \quad \text{for all } h. \tag{6.42}$$

Smoothness. Given $p \in (0, \infty]$ and $t > 0$, we define the *modulus of smoothness* of order k in $L^p(\Omega)$ to be

$$\omega_k(g,t)_p = \omega_k(g,t,\Omega)_p := \sup_{|h|\le t} \|\Delta_h^k g\|_{L^p(\Omega)}. \tag{6.43}$$

We note that if $\omega_k(g,t)_p = o(t^{n+1})$ as $t \to 0$, then g is a.e. a polynomial in $\mathbb{P}_n$ and

$$g \notin \mathbb{P}_n \quad \Rightarrow \quad \omega_k(g,t)_p \ge C t^{n+1}, \quad 0 < t \le 1, \tag{6.44}$$

for some $C > 0$ (DeVore and Lorentz 1993, Proposition 7.4). We also observe that the definition (6.43) only requires L^p-integrability of g and leads to the following celebrated Whitney estimate of the *best approximation error*

$$E_n(g,G)_p := \inf_{v \in \mathbb{P}_n} \|g - v\|_{L^p(G)}$$

of g by polynomials of degree $\le n$ in $G \subset \Omega$; see Binev *et al.* (2002), Dekel and Leviatan (2004, Theorem 1.4) and Gaspoz and Morin (2014, 2017, Lemma 4.4).

Lemma 6.30 (Whitney's lemma). *Let $\mathcal{T} \in \mathbb{T}$ be an admissible grid, and let $T \in \mathcal{T}$ be a generic element. If $0 < p \le \infty$ and $n \ge 0$, then*

$$E_n(g,T)_p \le C\omega_{n+1}(g, h_T, T)_p \quad \textit{for all } g \in L^p(T),$$

where $C = C(p,n,d,\mathcal{T}_0)$ but is independent of g and the size of T.

6.8.2. Besov spaces

Given $s > 0$ and $0 < p, q \le \infty$, the Besov space $B^s_{p,q}(\Omega)$ is the set of all functions $v \in L^p(\Omega)$ such that the following quantity is finite:

$$|v|_{B^s_{p,q}(\Omega)} := \begin{cases} \left(\displaystyle\int_0^\infty \left[t^{-s}\omega_k(v,t)_p\right]^q \frac{\mathrm{d}t}{t} \right)^{1/q}, & 0 < q < \infty, \\ \sup_{t>0} \left[t^{-s}\omega_k(v,t)_p\right], & q = \infty, \end{cases} \tag{6.45}$$

with $k = [s] + 1 \in \mathbb{N}$ and $[s]$ stands for the integer part of s. If we split the integral in (6.45) for $0 < q < \infty$ in dyadic intervals, we obtain the following equivalent expression for $|v|_{B^s_{p,q}(\Omega)}$:

$$|v|^q_{B^s_{p,q}(\Omega)} = \sum_{m\in\mathbb{Z}} \int_{2^{-m-1}}^{2^{-m}} t^{-sq}\omega_k(v,t)_p^q \frac{\mathrm{d}t}{t} \approx \sum_{m\in\mathbb{Z}} 2^{msq}\omega_k(v, 2^{-m})_p^q. \tag{6.46}$$

Here we have used that both $\omega_k(v,t)_p$ and t^{-s} are monotone functions of t. The hidden constants depend on s and q but are otherwise independent of v, k and p. Note that with obvious changes, (6.46) is also valid for $q = \infty$:

$$|v|_{B^s_{p,\infty}(\Omega)} \approx \sup_{m\in\mathbb{Z}} 2^{ms}\omega_k(v, 2^{-m})_p. \tag{6.47}$$

We point out that $|v|_{B^s_{p,q}(\Omega)}$ is a seminorm for $p, q \geq 1$ and is otherwise a semi-(quasi-)norm in that the triangle inequality is valid up to a constant larger than 1; note that $|1|_{B^s_{p,q}(\Omega)} = 0$. The quasi-norm of $B^s_{p,q}(\Omega)$ is defined to be

$$\|v\|_{B^s_{p,q}(\Omega)} := \|v\|_{L^p(\Omega)} + |v|_{B^s_{p,q}(\Omega)}.$$

If an integer $k' > k$ is chosen in (6.45), then the ensuing quasi-norms $\|v\|_{B^s_{p,q}(\Omega)}$ are equivalent. This hinges on the *Marchaud inequality* (see DeVore and Popov 1988, (2.6), and Ditzian 1988, Theorems 1 and 3)

$$\omega_k(v,t)_p \leq Ct^k\left(\|v\|_{L^p(\Omega)} + \left(\int_t^\infty \left(z^{-k}\omega_{k'}(v,z)_p\right)^p \frac{\mathrm{d}z}{z}\right)^{1/p}\right). \tag{6.48}$$

The following lemma characterizes the precise blow-up of $|v|_{B^s_{p,p}(\Omega)}$ as $s \longrightarrow n+1$.

Lemma 6.31 (blow-up of $|v|_{B^s_{p,p}(\Omega)}$). *Let $s \in (0, n+1)$, $p \in (0,\infty]$. Then*

$$|v|_{B^s_{p,p}(\Omega)} \leq (p(n+1-s))^{-1/p}\|v\|_{B^{n+1}_{p,p}(\Omega)} \quad \textit{for all } v \in B^{n+1}_{p,p}(\Omega).$$

Proof. We take $p < \infty$ and combine the definition (6.45) with (6.48), after replacing the upper limit of integration with $\mathrm{diam}(\Omega) \approx 1$, to write

$$|v|^p_{B^s_{p,p}(\Omega)} \approx \int_0^1 \left(t^{-s}\omega_{n+1}(v,t)_p\right)^p \frac{\mathrm{d}t}{t} \lesssim I + II,$$

with

$$I = \int_0^1 t^{(n+1-s)p} \int_t^1 \left(z^{-(n+1)}\omega_{n+2}(v,z)_p\right)^p \frac{\mathrm{d}z}{z}\frac{\mathrm{d}t}{t}, \quad II = \int_0^1 t^{(n+1-s)p}\|v\|^p_{L^p(\Omega)}\frac{\mathrm{d}t}{t}.$$

Exchanging the order of integration yields

$$\begin{aligned} I &= \int_0^1 \left(z^{-(n+1)}\omega_{n+2}(v,z)_p\right)^p \left(\int_0^z t^{(n+1-s)p-1}\,\mathrm{d}t\right)\frac{\mathrm{d}z}{z} \\ &= \frac{1}{p(n+1-s)}\int_0^1 \left(z^{-s}\omega_{n+2}(v,z)_p\right)^p \frac{\mathrm{d}z}{z} \leq \frac{1}{p(n+1-s)}|v|^p_{B^{n+1}_{p,p}(\Omega)}. \end{aligned}$$

Since $II = (p(n+1-s))^{-1}\|v\|^p_{L^p(\Omega)}$, the proof is thus complete. □

The following equivalence between Sobolev and Besov spaces is valid for fractional differentiability s (Leoni 2009, Proposition 14.40) (see also Bergh and Löfström 1976, Section 6.4.4, and Adams and Fournier 2003, Sections 7.33, 7.67): for all $s \geq 0, s \notin \mathbb{N}$ and $p \in [1, \infty]$,

$$B^s_{p,p}(\Omega) = W^s_p(\Omega). \tag{6.49}$$

However, if $s \in \mathbb{N}$ is an integer, then $B^s_{p,q}(\Omega)$ is defined using $k = s+1$ differences, whereas $W^s_p(\Omega)$ involves s weak derivatives in $L^p(\Omega)$ provided $p \in [1,\infty]$. It turns out that for integer values $s \in \mathbb{N}$ the spaces differ, that is,

$$B^s_{p,p}(\Omega) \neq W^s_p(\Omega), \quad p \neq 2, \tag{6.50}$$

except for the exceptional case $p = 2$ for which $B^s_{2,2}(\Omega) = H^s(\Omega)$ (DeVore 1998).

The Besov seminorm is *sub-additive* in the following sense: if $\{T_i\}_{i=1}^N$ is a disjoint collection of elements $T_i \in \mathcal{T}$ and $\mathcal{T} \in \mathbb{T}$, $p \in (0, \infty]$ and $s > 0$, then there exists a constant C depending on p, s, d and $\mathcal{T}_0$ but independent of N such that

$$\sum_{i=1}^N |v|^p_{B^s_{p,p}(T_i)} \le C|v|^p_{B^s_{p,p}(\Omega)} \quad \text{for all } v \in B^s_{p,p}(\Omega). \tag{6.51}$$

The localization of Besov norms is more general than (6.51). In fact, if $\omega_{\mathcal{T}}(T)$ denotes the patch of elements in $\mathcal{T}$ around $T \in \mathcal{T}$ (first ring), then the following is valid with equivalence constants depending on p, s, d and $\mathcal{T}_0$ but independent of N (Binev *et al.* 2002, Lemmas 4.3, 4.4):

$$\sum_{T \in \mathcal{T}} |v|^p_{B^s_{p,p}(\omega_{\mathcal{T}}(T))} \approx C|v|^p_{B^s_{p,p}(\Omega)} \quad \text{for all } v \in B^s_{p,p}(\Omega). \tag{6.52}$$

The following statements about embeddings between Besov spaces on bounded Lipschitz domains Ω will turn out to be useful below (Triebel 2010, Sections 3.2.4, 3.3.1): if $0 < p \le \infty$, $0 < q_1, q_2 \le \infty$ and $s_1, s_2, s > 0$, then

$$\begin{aligned} s_1 > s_2 \quad &\Rightarrow \quad B^{s_1}_{p,q_1}(\Omega) \hookrightarrow B^{s_2}_{p,q_1}(\Omega), \\ q_1 < q_2 \quad &\Rightarrow \quad B^{s}_{p,q_1}(\Omega) \hookrightarrow B^{s}_{p,q_2}(\Omega). \end{aligned} \tag{6.53}$$

Because of the second relation in (6.53), statements valid for all second index q are written for the largest space corresponding to $q = \infty$. In addition, for all $0 < p, q, r \le \infty$ and $s > 0$, the *discrepancy* between the spaces $B^s_{p,r}(\Omega)$ and $L^q(\Omega)$ is the quantity

$$\delta := s - \frac{d}{p} + \frac{d}{q}. \tag{6.54}$$

The discrepancy δ governs the embedding between these two spaces (Leoni 2009, Theorems 14.29, 14.32, DeVore 1998), namely

$$\delta > 0 \Rightarrow B^s_{p,\infty}(\Omega) \hookrightarrow L^q(\Omega), \quad \delta = 0 \Rightarrow B^s_{p,p}(\Omega) \hookrightarrow L^q(\Omega), \ q \ne \infty, \tag{6.55}$$

and the embedding is compact when $\delta > 0$. Notice that $\delta = 0$ determines the Sobolev embedding line of the DeVore diagram in Figure 2.1.

We stress that when $\delta > 0$, the third parameter r in $B^s_{p,r}(\Omega)$ plays no role in (6.55) involving the largest space $B^s_{p,\infty}(\Omega)$; see (6.53). However, it turns out to be useful to quantify regularity in extreme cases. For instance, the characteristic function χ_G of a smooth set $G \subsetneq \Omega$ satisfies

$$\chi_G \in B^{1/p}_{p,\infty}(\Omega) \backslash B^{1/p}_{p,r}(\Omega), \quad 0 < p, r < \infty.$$

Moreover, the Lagrange basis functions $\{\phi_z\}_{z\in\mathcal{N}}$ of $\mathbb{V}_\mathcal{T}$ satisfy for any $0 < p \leq \infty$ (Gaspoz and Morin 2014, Proposition 4.7)

$$\omega_{n+1}(\phi_z, t)_p = \begin{cases} |\operatorname{supp}\phi_z|^{(d-1-p)/(dp)} t^{1+1/p}, & 0 < t \leq |\operatorname{supp}\phi_z|^{1/d}, \\ |\operatorname{supp}\phi_z|, & t > |\operatorname{supp}\phi_z|^{1/d}. \end{cases}$$

This readily implies that for all $0 < s < 1 + 1/p$ and $0 < q < \infty$,

$$\mathbb{V}_\mathcal{T} \subset B^s_{p,q}(\Omega), \quad \mathbb{V}_\mathcal{T} \subset B^{1+1/p}_{p,\infty}(\Omega). \tag{6.56}$$

6.8.3. Local approximation

We are now in a position to prove a key approximation estimate. In finite element theory it goes by the name of Bramble–Hilbert lemma, whereas in nonlinear approximation theory it is called Jackson's theorem. We distinguish between the case $0 < s < n + 1$ and the limit integral case $s = n + 1$.

Proposition 6.32 (Bramble–Hilbert for Besov spaces). *Let $\mathcal{T} \in \mathbb{T}$ and $T \in \mathcal{T}$. Assume $0 < p, q \leq \infty$, $0 < s < n + 1$, and either $s - d/p + d/q \geq 0$, $q < \infty$ or $s > d/p$, $q = \infty$. Set $r = \infty$ when $s - d/p + d/q > 0$ and $r = p$ otherwise. Then we have*

$$\inf_{P\in\mathbb{P}_n} \|v - P\|_{L^q(T)} \leq C h_T^{s-d/p+d/q} |v|_{B^s_{p,r}(T)} \quad \textit{for all } v \in B^s_{p,r}(T), \tag{6.57}$$

where the constant $C = C(p, q, s, d, n, \mathcal{T}_0)$ is independent of v and T.

Proof. We first point out that we could use $k = n + 1 \geq [s] + 1$ in the definition of $|v|_{B^s_{p,r}(T)}$ according to (6.48). We next proceed in three steps.

$\boxed{1}$ Suppose first that T is the master element, namely $|T| \approx 1$. If $P \in \mathbb{P}_n$ is an arbitrary polynomial, using that the discrepancy $\delta = s - d/p + d/q \geq 0$ yields

$$E_n(v,T)_q \leq \|v - P\|_{L^q(T)} \lesssim \|v - P\|_{B^s_{p,r}(T)} = \|v - P\|_{L^p(T)} + |v - P|_{B^s_{p,r}(T)}$$

due to the embedding (6.55). Since the definition of $\omega_{n+1}(v,t)_p$ involves $n + 1$ differences, we deduce $\Delta_h^{n+1} P = 0$ in view of (6.42), whence $|v - P|_{B^s_{p,r}(T)} = |v|_{B^s_{p,r}(T)}$. We now take P to be the best approximation of v in $L^p(T)$, to derive

$$E_n(v,T)_q \lesssim E_n(v,T)_p + |v|_{B^s_{p,r}(T)}.$$

$\boxed{2}$ We perform a scaling argument from the element $T \in \mathcal{T}$ to the master element $\widehat{T}$. Let $\widehat{x} = |T|^{-1/d} x$ be the change of variables and note that

$$\begin{aligned}\omega_{n+1}(v,t,T)_p &= \sup_{|h|\leq t} \|\Delta_h^{n+1} v\|_{L^p(T)} \\ &= |T|^{1/p} \sup_{|h|\leq t} \|\Delta^{n+1}_{h|T|^{-1/d}} \widehat{v}\|_{L^p(\widehat{T})} = |T|^{1/p} \omega_{n+1}(\widehat{v}, \widehat{t}, \widehat{T})_p,\end{aligned}$$

with $\widehat{t} = t|T|^{-1/d}$, whence

$$|v|^r_{B^s_{p,r}(T)} = \int_0^\infty (t^{-s}\omega_{n+1}(v,t,T)_p)^r \frac{\mathrm{d}t}{t}$$
$$= |T|^{r/p-sr/d} \int_0^\infty (\widehat{t}^{-s}\omega_{n+1}(\widehat{v},\widehat{t},\widehat{T})_p)^r \frac{\mathrm{d}\widehat{t}}{\widehat{t}} = |T|^{r/p-sr/d}|\widehat{v}|^r_{B^s_{p,r}(\widehat{T})}.$$

Therefore, since $E_n(v,T)_q = |T|^{1/q} E_n(\widehat{v},\widehat{T})_q$, we obtain

$$E_n(v,T)_q \lesssim |T|^{1/q-1/p} E_n(v,T)_p + |T|^{s/d-1/p+1/q}|v|_{B^s_{p,r}(T)}.$$

$\boxed{3}$ It remains to estimate $E_n(v,T)_p$ which, in view of Lemma 6.30 (Whitney's lemma), satisfies $E_n(v,T)_p \le C\omega_{n+1}(v,h_T,T)_p$ with $h_T \approx |T|^{1/d} \approx 2^{-m}$ for some $m \in \mathbb{Z}$. Since $k = n+1 \ge [s]+1$, invoking the equivalent definition (6.46) of $|v|_{B^s_{p,r}(T)}$ yields

$$E_n(v,T)^r_p \lesssim \omega_{n+1}(v,2^{-m},T)^r_p \lesssim h^{sr}_T \sum_{m\in\mathbb{Z}} 2^{msr}\omega_{n+1}(v,2^{-m},T)^r_p = h^{sr}_T|v|^r_{B^s_{p,r}(T)}.$$

Inserting this estimate into that of step $\boxed{2}$ gives (6.57), as asserted. □

We now consider the integer case $s = n+1$. The first thing to notice is that (6.57) cannot possibly be valid: the definition (6.45) requires $k = [s]+1 = n+2$, whence any polynomial $g \in \mathbb{P}_{n+1} \setminus \mathbb{P}_n$ satisfies $E_n(g,T)_p > 0$ as well as $\omega_{n+2}(g,T)_p = 0$ according to (6.42). Lemma 6.31 (blow-up of $|v|_{B^s_{p,p}(\Omega)}$) reveals that replacing the seminorm $|v|_{B^s_{p,p}(\Omega)}$ with the full norm $\|v\|_{B^{n+1}_{p,p}(\Omega)}$ is not a good idea either. To overcome this problem, we now introduce the space $\mathrm{Lip}^s_p(\Omega) := \mathrm{Lip}(s, L^p(\Omega))$ of s-Lipschitz functions with values in $L^p(\Omega)$, $0 < p < \infty$ (DeVore 1998, page 92):

$$|g|_{\mathrm{Lip}^s_p(\Omega)} := \sup_{t>0}(t^{-s}\omega_{n+1}(g,t,\Omega)_p). \tag{6.58}$$

Comparing with (6.45), we realize that $\mathrm{Lip}^s_p(\Omega) = B^s_{p,\infty}(\Omega)$ provided $s \notin \mathbb{N}$ but $\mathrm{Lip}^s_p(\Omega) \ne B^s_{p,\infty}(\Omega)$ when $s \in \mathbb{N}$. Moreover,

$$\delta = s - \frac{d}{p} + \frac{d}{q} > 0,\ s \in \mathbb{N} \quad \Rightarrow \quad \mathrm{Lip}^s_p(\Omega) \hookrightarrow L^q(\Omega), \tag{6.59}$$

with compact embedding. If $\delta = 0$ and $p \ge 1$, $q \ne \infty$, the above embedding is continuous in view of (6.36) and (6.55).

Proposition 6.33 (Bramble–Hilbert for Lipschitz spaces). *Let $\mathcal{T} \in \mathbb{T}$ and $T \in \mathcal{T}$. If $p \in (0,\infty)$, $q \in (0,\infty]$, $k \ge 0$ integer, and $k+1-d/p+d/q \ge 0$, with strict inequality when $p < 1$ or $q = \infty$, then we have*

$$\inf_{P\in\mathbb{P}_k} \|v-P\|_{L^q(T)} \le Ch_T^{k+1-d/p+d/q}|v|_{\mathrm{Lip}^{k+1}_p(T)} \quad \textit{for all } v \in \mathrm{Lip}^{k+1}_p(T), \tag{6.60}$$

where the constant $C = C(p,q,d,k,\mathcal{T}_0)$ is independent of v and T.

Proof. In view of (6.59), we proceed as in the proof of Proposition 6.32, except for the following change in step [3]. For $h_T \approx 2^{-m}$, we instead have

$$E_k(v,T)_p \lesssim \omega_{k+1}(v,2^{-m},T)_p$$
$$\lesssim h_T^{k+1} \sup_{m\in\mathbb{Z}} \big(2^{m(k+1)}\omega_{k+1}(v,2^{-m},T)_p\big) = h_T^{k+1}|v|_{\mathrm{Lip}_p^{k+1}(T)}.$$

This concludes the proof. □

It is instructive to realize that Propositions 6.32 and 6.33 extend to Besov and Lipschitz spaces the usual Bramble–Hilbert lemma for Sobolev spaces (Brenner and Scott 2008, Lemma 4.3.8).

Proposition 6.34 (Bramble–Hilbert for Sobolev spaces). *Let $\mathcal{T} \in \mathbb{T}$ and $T \in \mathcal{T}$. For all $1 \le p,q, \le \infty$ and $0 < s \le n+1$ such that $s - d/p + d/q \ge 0$ with strict inequality when $q = \infty$, then*

$$\inf_{P\in\mathbb{P}_n} \|v - P\|_{L^q(T)} \le Ch_T^{s-d/p+d/q}|v|_{W_p^s(T)} \quad \text{for all } v \in W_p^s(T), \tag{6.61}$$

where the constant $C = C(p,q,s,d,n,\mathcal{T}_0)$ but is independent of v and T.

Proof. When s is fractional, $W_p^s(T) = B_{p,p}^s(T)$ in view of (6.49) and the result follows from Proposition 6.32. Instead, when s is integral, we invoke (6.36) and (6.37) to deduce the result from Proposition 6.33. □

6.8.4. *Global approximation: direct estimates*

We now collect local contributions from Propositions 6.32, 6.33 and 6.34, depending on the range of parameters p,q,s, to find global error estimates for the solution u as well as the coefficients $(\boldsymbol{A},c)$ of (2.7). They are trivial consequences of Theorem 6.29, which we prove first. The analysis of the forcing function f is somewhat different, due to the non-locality of the corresponding norm $H^{-1}(\Omega)$, and is postponed to Section 7.3.

Proof of Theorem 6.29. Since the discrepancy $\delta = s-d/p+d/q > 0$, the embedding $X_p^s(\Omega) \hookrightarrow L^q(\Omega)$ is compact according to (6.55) and (6.59). Given $g \in X_p^s(\Omega)$, we consider the surrogate quantity $e_{\mathcal{T}}(g,T) := Ch_T^\delta|g|_{X_p^s(T)}$, which satisfies

$$E_n(g,T)_q = \inf_{v\in\mathbb{P}_n} \|g - v\|_{L^q(T)} \le e_{\mathcal{T}}(g,T) \quad \text{for all } T \in \mathcal{T}$$

by virtue of Bramble–Hilbert Propositions 6.32, 6.33 and 6.34. We finally combine Proposition 3.19 (abstract greedy error) with the subadditivity property (6.51) to deduce the desired estimate (6.3). The remaining estimate (6.39) follows from the definition of $|g|_{\mathbb{A}_{s/d}}$. This concludes the proof. □

Inspection of this proof reveals that our estimate is stronger than (6.38). In fact, we need the weaker regularity

$$|g|^p_{X_p^s(\Omega;\mathcal{T})} = \sum_{T\in\mathcal{T}} |g|^p_{X_p^s(T)} < \infty,$$

which allows for piecewise Besov smoothness of g very much in the spirit of (3.20). This may accommodate singular behaviour of g aligned with the initial mesh $\mathcal{T}_0$.

Corollary 6.35 (approximation class of u). *Let the solution $u \in H_0^1(\Omega)$ of (2.7) satisfy $u \in X_p^s(\Omega)$ with $s \in (0, n+1]$, $p \in (0, \infty]$ and $s - 1 - d/p + d/2 > 0$, where $X_p^s(\Omega)$ is defined in (6.35). Then $u \in \mathbb{A}_{(s-1)/d}\big(H_0^1(\Omega); \mathcal{T}_0\big)$ and*

$$|u|_{\mathbb{A}_{(s-1)/d}} \lesssim |u|_{X_p^s(\Omega)}. \tag{6.62}$$

Equivalently, $\sigma_N(u)$ defined in (6.3) satisfies

$$\sigma_N(u) \lesssim |u|_{X_p^s(\Omega)} N^{-(s-1)/d}, \quad N \geq \#\mathcal{T}_0. \tag{6.63}$$

Proof. In view of (3.19) of Proposition 3.9 (approximation of gradients), namely

$$\inf_{v \in \mathbb{S}_{\mathcal{T}}^{n,0}} \|\nabla(u - v)\|_{L^2(\Omega)} \lesssim \inf_{v \in \mathbb{S}_{\mathcal{T}}^{n,-1}} \|\nabla(u - v)\|_{L^2(\Omega)},$$

we realize that it suffices to bound the element errors for $g = \nabla u \in L^2(\Omega; \mathbb{R}^d)$ by vector-valued discontinuous piecewise polynomials of degree $\leq n - 1$. Therefore, applying Theorem 6.29 (regularity yields approximation) with n replaced by $n - 1$ gives the desired estimates (6.62) and (6.63). □

We now turn our attention to the coefficients $(\boldsymbol{A}, c)$. Regarding $\boldsymbol{A}$, Lemma 5.20 (continuous dependence on data) shows that the natural function space for $\boldsymbol{A}$ is $L^\infty(\Omega, \mathbb{R}^{d\times d})$ provided $u \in H_0^1(\Omega)$. However, Lemma 5.20 also allows for $\boldsymbol{A} \in L^r(\Omega, \mathbb{R}^{d\times d})$, $2 \leq r < \infty$, provided $u \in W_p^1(\Omega)$ with $2 \leq p = 2r/(r-2) < p_1$, which in turn is guaranteed by Lemma 2.13 (W_p^1-regularity). The latter permits discontinuities of $\boldsymbol{A}$ within elements, which is of practical importance. Therefore we consider the most general situation $2 \leq r \leq \infty$ below.

Corollary 6.36 (approximation class of $\boldsymbol{A}$). *For $0 < \alpha_1 \leq \alpha_2$ and $2 \leq r \leq \infty$ let the diffusion coefficient $\boldsymbol{A} \in M(\alpha_1, \alpha_2)$ of (2.7) satisfy $\boldsymbol{A} \in X_p^s(\Omega; \mathbb{R}^{d\times d})$ with $s \in (0, n], p \in (0, \infty]$ and $s - d/p + d/r > 0$. Then $\boldsymbol{A} \in \mathbb{M}_{s/d} = \mathbb{M}_{s/d}((L^r(\Omega))^{d\times d}; \mathcal{T}_0)$ and*

$$|\boldsymbol{A}|_{\mathbb{M}_{s/d}} \lesssim |\boldsymbol{A}|_{X_p^s(\Omega)}. \tag{6.64}$$

Equivalently, $\widetilde{\delta}_{\mathcal{T}}(\boldsymbol{A})_r$ defined in Section 6.1.2 satisfies

$$\inf_{\mathcal{T} \in \mathbb{T}_N} \widetilde{\delta}_{\mathcal{T}}(\boldsymbol{A})_r \lesssim |\boldsymbol{A}|_{X_p^s(\Omega)} N^{-s/d} \quad \textit{for all } \mathcal{T} \in \mathbb{T}_N,\ N \geq \#\mathcal{T}_0, \tag{6.65}$$

and this error decay is achieved by Algorithm 3.18 (greedy algorithm).

Proof. Simply recall the relation (6.12) between the best constrained and unconstrained approximation errors and apply Theorem 6.29 (regularity yields approximation). □

Consider the special case $s = n$ and $r = \infty$ in Corollary 6.36. We readily see that $p > d/n$ which might be less than 1 for $n > d$, hence the need for Besov spaces.

We finally deal with the reaction coefficient $c \in L^\infty(\Omega)$. Given Lemma 5.20 (continuous dependence on data) and the discussion in Section 5.4.2, a natural space for c is $L^q(\Omega)$ with $d/2 < q \leq \infty$, $s = 0$; we could take $q = 2$ for $d < 4$. Section 5.4.2 also reveals that the case $n = 1$ is somewhat special in that we can exploit superconvergence in $W_q^{-1}(\Omega)$ with $q > d$. In fact, combining the argument following (5.75) with (5.68) yields

$$\inf_{\widehat{c} \in \mathbb{S}_{\mathcal{T}}^{n-1,-1}} \|c - \widehat{c}\|^2_{W_q^{-1}(\Omega)} \lesssim \sum_{T \in \mathcal{T}} h_T^{2t} \|c - \Pi_{\mathcal{T}} c\|^2_{L^2(T)} \lesssim \sum_{T \in \mathcal{T}} h_T^{2t} \delta_{\mathcal{T}}(c, T)_2^2 = \operatorname{osc}_{\mathcal{T}}(c)_2^2$$

with $0 < t = 1 - d/2 + d/q < 2 - d/2$ provided $d < 4$. This gives the following statement. We note that (5.76) could also be combined with (5.68) for $n = 1$ to obtain a similar result for $\operatorname{osc}_{\mathcal{T}}(c)_\infty$ with $t = 1$ and any $d \geq 2$; however, we do not elaborate further.

Corollary 6.37 (approximation class of c). *Let $0 \leq c_1 \leq c_2$ and the reaction coefficient $c \in R(c_1, c_2)$ satisfy $c \in X_p^s(\Omega)$ with $s \in (0, n]$, $p \in (0, \infty]$. If $n \geq 1$, $q > d/2$, and $s - d/p + d/q > 0$, then $c \in \mathbb{C}_{s/d} = \mathbb{C}_{s/d}(L^q(\Omega); \mathcal{T}_0)$ and*

$$|c|_{\mathbb{C}_{s/d}} \lesssim |c|_{X_p^s(\Omega)}. \tag{6.66}$$

If instead $n = 1$, $q > d$, $s - d/p + d/2 \geq 0$, $0 < t = 1 - d/2 + d/q < 2 - d/2$ and $d < 4$, then $c \in \mathbb{C}_{(s+t)/d} = \mathbb{C}_{(s+t)/d}(L^2(\Omega); \mathcal{T}_0)$ and

$$|c|_{\mathbb{C}_{(s+t)/d}} \lesssim |c|_{X_p^s(\Omega)}. \tag{6.67}$$

Equivalently, for all $n \geq 1$, $\widetilde{\delta}_{\mathcal{T}}(c)_q$ defined in Section 6.1.2 satisfies

$$\inf_{\mathcal{T} \in \mathbb{T}_N} \widetilde{\delta}_{\mathcal{T}}(c)_q \lesssim |c|_{X_p^s(\Omega)} N^{-(s+t)/d} \quad \text{for all } \mathcal{T} \in \mathbb{T}_N,\ N \geq \#\mathcal{T}_0,$$

with $t = 0$ when $n > 1$. This error decay is achieved by Algorithm 3.18 (greedy algorithm).

Proof. In view of (6.12), inequality (6.66) is a direct application of Theorem 6.29 (regularity yields approximation). The superconvergence rate in (6.67) is a consequence of (6.12) and the proof of Proposition 3.19 (abstract greedy error) with s replaced by $s + t$. □

We finally go back to the abstract space $X_p^s(\Omega)$, defined in (6.35), and introduce the corresponding abstract approximation class $\mathbb{X}_{s/d} = \mathbb{X}_{s/d}(L^q(\Omega); \mathcal{T}_0)$ of functions $v \in L^q(\Omega)$ such that

$$|v|_{\mathbb{X}_{s/d}} = \sup_{N \geq \#\mathcal{T}_0} \Big(N^{s/d} \inf_{\mathcal{T} \in \mathbb{T}_N} \operatorname{osc}_{\mathcal{T}}(v)_q \Big) < \infty \quad \Rightarrow \quad \inf_{\mathcal{T} \in \mathbb{T}_N} \operatorname{osc}_{\mathcal{T}}(v)_q \leq |v|_{\mathbb{X}_{s/d}} N^{-s/d}.$$

Consequently, Theorem 6.29 (regularity yields approximation) implies

$$X_p^s(\Omega) \subset \mathbb{X}_{s/d}, \quad |v|_{\mathbb{X}_{s/d}} \lesssim |v|_{X_p^s(\Omega)}. \tag{6.68}$$

We will utilize this abstract notation and estimates in Section 7 while discussing the approximation of data $\mathcal{D} = (\boldsymbol{A}, c, f)$ by a greedy algorithm.

6.8.5. Global approximation: inverse estimates

Theorem 6.29 gives sufficient regularity properties for a function $g \in L^q(\Omega)$ to belong to an approximation class $\mathbb{A}_{s/d}(L^q(\Omega); \mathcal{T}_0)$; this is called *direct estimate*. Such regularity is written in terms of a Besov space $B^s_{p,p}(\Omega)$, except in the limiting case $s = n+1$. The converse statement is also true and is called an *inverse estimate*: if g belongs to an approximation class $\mathbb{A}_{t/d}(L^q(\Omega); \mathcal{T}_0)$, then it is a member of a Besov space $\widehat{B}^s_{p,p}(\Omega)$ provided $t > s$ and $0 < s < n+1$, $s - d/p + d/q = 0$ (Binev *et al.* 2002, Gaspoz and Morin 2014).

Several comments are in order. The Besov space $\widehat{B}^s_{p,p}(\Omega)$ is defined via a multilevel decomposition of $L^p(\Omega)$ and coincides with $B^s_{p,p}(\Omega)$ only when $s < 1 + 1/p$. This restriction of s is natural because $\mathbb{V}_\mathcal{T} \subset \widehat{B}^s_{p,p}(\Omega)$ for all s, but $\mathbb{V}_\mathcal{T} \subset B^s_{p,p}(\Omega)$ requires $s < 1+1/p$ according to (6.56). The discrepancy between the spaces $\widehat{B}^s_{p,p}(\Omega)$ and $L^q(\Omega)$ is $\delta = s - d/p + d/q = 0$, but the decay rate t/d of $\mathbb{A}_{t/d}(L^q(\Omega); \mathcal{T}_0)$ is larger than s/d. This accounts for the embedding of $\mathbb{A}_{t/d}(L^q(\Omega); \mathcal{T}_0)$, that is,

$$\sum_{n\in\mathbb{N}} \big(\sigma_{2^n}(g) 2^{(s/d)n}\big)^p \le \sup_{n\in\mathbb{N}} \big(\sigma_{2^n}(g) 2^{(t/d)n}\big)^p \sum_{n\in\mathbb{N}} 2^{((s-t)/d)pn} \lesssim |g|^p_{\mathbb{A}_{t/d}}$$

into a space with decay s/d and summability ℓ^q that in turn embeds into $\widehat{B}^s_{p,p}(\Omega)$ (Binev *et al.* 2002, Gaspoz and Morin 2014). This reveals that there is no complete characterization of the approximation class $\mathbb{A}_{s/d}$ in terms of Besov regularity.

7. Data approximation

This section focuses on the module DATA of Algorithms 5.1 (AFEM-TS) and 5.16 (AFEM-SW). According to Assumption 6.11 (quasi-optimality of DATA), the call

$$[\widehat{\mathcal{T}}, \widehat{\mathcal{D}}] = \mathsf{DATA}(\mathcal{T}, \mathcal{D}, \tau) \tag{7.1}$$

is meant to construct a quasi-optimal conforming refinement $\widehat{\mathcal{T}}$ of $\mathcal{T} \in \mathbb{T}$ and approximate piecewise polynomial data $\widehat{\mathcal{D}} = (\widehat{\boldsymbol{A}}, \widehat{c}, \widehat{f}) \in \mathbb{D}_{\widehat{\mathcal{T}}}$ over $\widehat{\mathcal{T}}$ that satisfies

$$\|\mathcal{D} - \widehat{\mathcal{D}}\|_{D(\Omega)} \le C_{\mathrm{data}}\tau \tag{7.2}$$

as well as the constraints $\widehat{\boldsymbol{A}} \in M(\widehat{\alpha}_1, \widehat{\alpha}_2)$ and $\widehat{c} \in R(\widehat{c}_1, \widehat{c}_2)$ defined in (5.50).

Sections 7.2.2 and 7.2.3 are devoted to the construction of $(\widehat{\boldsymbol{A}}, \widehat{c})$. To approximate the coefficients $(\boldsymbol{A}, c)$, we proceed in two steps. First, we solve an *unconstrained* approximation problem upon computing the L^2-projection $(\widetilde{\boldsymbol{A}}, \widetilde{c})$ of $(\boldsymbol{A}, c)$ onto the space of discontinuous piecewise polynomials of degree $\le n-1$; this step is linear, easily achieves the desired accuracy, but does not guarantee the monotonicity of oscillations with respect to refinement (5.72) and violates the constraints in (5.50) unless $n = 1$. Second, we resort to the nonlinear selection (5.70) of the local L^2 approximation to force the resulting oscillations to be monotone. Third, we solve a *constrained* problem, which modifies $(\widetilde{\boldsymbol{A}}, \widetilde{c})$ locally into $(\widehat{\boldsymbol{A}}, \widehat{c})$ and restores

(5.50) without accuracy degradation; this is a delicate nonlinear procedure executed element by element, introduced and discussed in Section 7.2.

The approximation of the right-hand side $f \in H^{-1}(\Omega)$ is a conceptually different linear process. Without further structural assumptions on f it is not possible to evaluate $\operatorname{osc}_{\mathcal{T}}(f)_{-1}$ and reduce it. Hence we introduce surrogate estimators $\widetilde{\operatorname{osc}}_{\mathcal{T}}(f)_{-1}$, which are larger than $\operatorname{osc}_{\mathcal{T}}(f)_{-1}$, but computable, for several classes of forcing functions f relevant in practice. We discuss this in Section 7.3.

We start in Section 7.1 with a presentation and assessment of quasi-optimal GREEDY algorithms to reduce the data error. An important consideration is that the local error estimators $\{\operatorname{osc}_{\mathcal{T}}(v,T)\}_{T\in\mathcal{T}}$ may accumulate in ℓ^∞ as well as in ℓ^q for $q<\infty$. Both are handled via a GREEDY algorithm similar to Algorithm 3.18 but with different stopping criteria when the local errors accumulate in ℓ^q with $q<\infty$. The module DATA combines both: its structure is displayed in Algorithm 7.23 and its performance is elucidated in Corollary 7.24 below.

7.1. Quasi-optimal GREEDY *algorithms for data reduction*

Algorithm 3.18 (greedy algorithm) is well suited to dealing with local error estimators $\operatorname{osc}_{\mathcal{T}}(v,T)_q$ that accumulate with respect to $T\in\mathcal{T}$ in the space ℓ^∞. This is the framework for approximating coefficients $v = \boldsymbol{A}, c$ in $L^\infty(\Omega)$, in which case the local error estimators $\operatorname{osc}_{\mathcal{T}}(v,T)_\infty$ are defined in (5.73) for $r=q=\infty$. This requires $v=\boldsymbol{A},c$ to be piecewise uniformly continuous on $\mathcal{T}$ for $\operatorname{osc}_{\mathcal{T}}(v;T)_\infty \to 0$ as $h_{\mathcal{T}} \to 0$. However, for discontinuous $(\boldsymbol{A},c)$ and the forcing function f, the accumulation of $\operatorname{osc}_{\mathcal{T}}(v,T)_q$ for $v=\boldsymbol{A},c,f$ is in ℓ^q for $q<\infty$. In this case, Algorithm 3.18 does not provide a direct relation between a desired output tolerance τ for the total error

$$E_{\mathcal{T}}(v)_q := \|\{\operatorname{osc}_{\mathcal{T}}(v,T)_q\}_{T\in\mathcal{T}}\|_{\ell^q} = \left(\sum_{T\in\mathcal{T}} \operatorname{osc}_{\mathcal{T}}(v,T)_q^q\right)^{1/q}$$

and the threshold δ; recall that $\operatorname{osc}_{\mathcal{T}}(v,T)_q := \|v-\widehat{v}\|_{L^q(T)}$ for $T\in\mathcal{T}$.

Another subtle difference from Algorithm 3.18 is that the algorithm GREEDY below starts not from $\mathcal{T}_0$ but from any $\mathcal{T}\in\mathbb{T}$. Since DATA and thus GREEDY are called repeatedly within AFEM, it seems advantageous to exploit the mesh refinement already performed in the adaptive process rather than restart from scratch; this then improves the computational efficiency.

Algorithm 7.1 (GREEDY). Given a tolerance $\tau>0$, $0<q\le\infty$, a number of bisections $b\ge 1$ performed per element to be refined, and an arbitrary conforming grid $\mathcal{T}\in\mathbb{T}$, not necessarily $\mathcal{T}_0$, GREEDY finds a conforming refinement $\widehat{\mathcal{T}}\ge\mathcal{T}$ of $\mathcal{T}$ by bisection and $\widehat{v}\in\mathbb{S}^{n-1,-1}_{\mathcal{T}}$ such that $E_{\mathcal{T}}(v)_q\le\tau$:

```
[T̂, v̂] = GREEDY(T, τ, q, b, v)
  [v̂] = PROJECT(T, v)
  while E_T(v)_q > τ
```

$[\mathcal{M}] = \arg\max\{\text{osc}_{\mathcal{T}}(v,T)_q : T \in \mathcal{T}\}$
$[\mathcal{T}] = \text{REFINE}(\mathcal{T}, \mathcal{M}, b)$
$[\widehat{v}] = \text{PROJECT}(\mathcal{T}, v)$
return $\mathcal{T}, \widehat{v}$

In GREEDY above, the element T with largest error is refined as long as the total error $E_{\mathcal{T}}(v)_q$ exceeds the target tolerance τ. When the largest error is achieved by several elements, an *ad hoc* criterion such as lexical order is used to break ties. We also recall that the routine REFINE bisects all the elements in $\mathcal{M}$ (in this case only one) b times and performs additional refinements necessary to produce a conforming subdivision. PROJECT computes the local approximations $\widehat{v}$ of v needed to evaluate $\text{osc}_{\mathcal{T}}(v,T)_q$; refer to Section 5.4.2 and (5.70) for the definition of $\widehat{v}$. The dependence on $\widehat{v}$ in $\text{osc}_{\mathcal{T}}(v,T)_q$ and $E_{\mathcal{T}}(v)$ is not indicated.

To discuss the performances of the GREEDY algorithm, we recall that $X_p^s(\Omega;\mathcal{T}_0)$ is the abstract space defined in (6.35), which satisfies

$$\sum_{T\in\mathcal{T}} |v|^p_{X_p^s(T)} \lesssim |v|^p_{X_p^s(\Omega;\mathcal{T}_0)} \tag{7.3}$$

for all $\mathcal{T} \in \mathbb{T}$ and $v \in X_p^s(\Omega;\mathcal{T}_0)$.

The GREEDY algorithm analysed in Proposition 3.19 (abstract greedy error) relies on the abstract assumptions (3.40), (3.41) and (3.42). With the aim of reducing the data oscillations, we make these assumptions more concrete.

Assumption 7.2 (admissible set of parameters for GREEDY). We say that the set of parameters (v,s,t,p,q) is admissible for GREEDY with local oscillations $\{\text{osc}_{\mathcal{T}}(v,T)_q\}_{T\in\mathcal{T}}$ if $0 < p, q \le \infty$, $s,t \ge 0$ satisfy

(i) $v \in X_p^s(\Omega;\mathcal{T}_0)$,

(ii) $t + s > 0$, $s - d/p + d/q \ge 0$ with strict inequality when $q = \infty$ or $s = n+1$, $p < 1$,

(iii) for $r := t + s - d/p + d/q > 0$,

$$\text{osc}_{\mathcal{T}}(v,T)_q \lesssim h_T^r |v|_{X_p^s(T)} \quad \text{for all } T \in \mathcal{T},\ \mathcal{T} \in \mathbb{T}. \tag{7.4}$$

When the local oscillations considered are clear from the context, we say that (v,s,t,p,q) is admissible for GREEDY.

Relation (7.4) replaces (3.41) and is a regularity assumption guaranteeing a convergence rate when approximating v by $\widehat{v} = \text{PROJECT}(\mathcal{T}, v)$ (appearing in the definition of $\text{osc}_{\mathcal{T}}(v,T)_q$). We refer to Propositions 6.32, 6.33 and 6.34 for examples where Assumption 7.2 holds. Note that in view of (6.55) and (6.59), condition (ii) in Assumption 7.2 guarantees $v \in X_p^s(\Omega) \subset L^q(\Omega)$. The parameter $t \ge 0$ reflects a possible additional power of h in the oscillation term; see e.g. (5.75), (5.76) and (5.78). Furthermore, in view of (7.3), assumption (6.51) is always satisfied by the $X_p^s(\Omega;\mathcal{T}_0)$ seminorms, and (3.28) is no longer needed.

As alluded to above, the case $q < \infty$ is more complex to analyse and cannot rely solely on the decay property (7.4) as in the proof of Proposition 3.19. It requires the local oscillations to be monotone with respect to refinements.

Assumption 7.3 (monotonicity of local oscillations). We say that for $0 < q \leq \infty$, the local errors satisfy the monotonicity property in ℓ^q if, for any $v \in L^q(\Omega)$, any $\mathcal{T}, \mathcal{T}_* \in \mathbb{T}$ with $\mathcal{T}_* \geq \mathcal{T}$ and any $T_* \in \mathcal{T}_*, T \in \mathcal{T}$ with $T_* \subset T$, we have

$$\mathrm{osc}_{\mathcal{T}_*}(v, T_*)_q \leq \mathrm{osc}_{\mathcal{T}}(v, T)_q. \tag{7.5}$$

In view of Lemma 5.26 (monotonicity of oscillation), Assumption 7.3 holds for the oscillations on $\boldsymbol{A}$ and c given in (5.73) and (5.75), but not for the oscillations (5.77) of f. However, in Section 7.3 below we derive computable surrogates for the local error $\mathrm{osc}_{\mathcal{T}}(f, T)_{-1}$. These surrogates satisfy the monotonicity property and are used in turn to drive the GREEDY algorithm. In passing, we note that we refrain from using the right-hand side of inequality (7.4) as a surrogate for the local oscillation. In fact it is monotone with respect to refinements but at the expense of being difficult to evaluate because it involves the seminorm $|v|_{X_p^s(T)}$.

The following result is the counterpart of Proposition 3.19 (abstract greedy error) for GREEDY with errors accumulating in ℓ^q, $0 < q \leq \infty$, and still starting from $\mathcal{T}_0$. We address the case where $\mathcal{T} \neq \mathcal{T}_0$ in Lemma 7.5 below.

Proposition 7.4 (performance of GREEDY). *Let the initial subdivision $\mathcal{T}_0$ of $\Omega \subset \mathbb{R}^d$ satisfy Assumption 6.19 (initial labelling). Let $\tau > 0$ be the target tolerance and let $b \geq 1$ be the number of bisections performed on each marked element. Let (v, s, t, p, q) satisfy Assumption 7.2 (admissible set of parameters for* GREEDY*) with local errors $\{\mathrm{osc}_{\mathcal{T}}(v, T)_q\}_{T \in \mathcal{T}}$ which in turn verify Assumption 7.3 (monotonicity of local oscillations) in ℓ^q. Then* GREEDY$(\mathcal{T}_0, \tau, q, b, v)$ *terminates in a finite number of iterations and*

$$E_{\mathcal{T}}(v)_q \leq \tau \leq C|v|_{X_p^s(\Omega;\mathcal{T}_0)}(\#\mathcal{T})^{-(s+t)/d}, \tag{7.6}$$

with a constant $C = C(p, q, s, b, d, \Omega, \mathcal{T}_0)$. Furthermore, $v \in \mathbb{X}_{(s+t)/d}$ and $|v|_{\mathbb{X}_{(s+t)/d}} \lesssim |v|_{X_p^s(\Omega;\mathcal{T}_0)}$. Moreover, the estimate (7.6) is valid for tensor-valued functions v.

Proof. Since the proof is similar to that of Proposition 3.19 (abstract greedy error) with $e_{\mathcal{T}}(v, T)_q = \mathrm{osc}_{\mathcal{T}}(v, T)_q$, we only report the new ingredients. We recall that we use the convention $1/\infty = 0$. Let $\mathcal{T}_1, \ldots, \mathcal{T}_k$ be the sequence of refinements produced by GREEDY, and let $T_1, \ldots T_k$ be the sequence of marked elements. We need to estimate $\#\mathcal{M} = k$ with $\mathcal{M} = \{T_1, \ldots, T_k\}$. Set

$$\delta_i := \mathrm{osc}_{\mathcal{T}_i}(v, T_i)_q \quad (1 \leq i \leq k) \quad \text{and} \quad \delta := \delta_{k-1}.$$

Then

$$E_{\mathcal{T}_k}(v)_q \leq \tau < E_{\mathcal{T}_{k-1}}(v)_q \leq \delta\, (\#\mathcal{T}_{k-1})^{1/q} \leq \delta\, (\#\mathcal{T}_k)^{1/q}. \tag{7.7}$$

On the other hand, as REFINE does not increase the element estimators $\mathrm{osc}_{\mathcal{T}_i}(v, T_i)$ thanks to (7.5), we have $\delta_i \geq \delta$ for any i, whence

$$\mathrm{osc}_{\mathcal{T}_i}(v, T_i)_q = \delta_i \geq \delta \quad \text{for all } 1 \leq i < k.$$

Let us now partition $\mathcal{M}$ into disjoint subsets $\mathcal{P}_j$ as in the proof of Proposition 3.19. If $T_i \in \mathcal{P}_j$, (7.4) implies

$$\delta \leq \mathrm{osc}_{\mathcal{T}_i}(v, T_i)_q \lesssim h_T^r |v|_{X_p^s(T_i)} \leq 2^{-jr/2} |v|_{X_p^s(T_i)},$$

whence, exploiting the ℓ^p summability (7.3) gives

$$\#\mathcal{P}_j \lesssim \delta^{-p} 2^{-jrp/2} |v|^p_{X_p^s(\Omega;\mathcal{T}_0)}.$$

which is similar to (3.45). Recalling (3.44), and proceeding as in the proof of Proposition 3.19, yields

$$\delta \lesssim |v|_{X_p^s(\Omega;\mathcal{T}_0)} \, (\#\mathcal{T}_k - \#\mathcal{T}_0)^{-(s+t)/d-1/q}.$$

We conclude the proof using (7.7) and the bound $\#\mathcal{T}_k \geq c_0 \#\mathcal{T}_0$ for $c_0 > 1$. □

In contrast to Section 3, and most of the existing literature, Algorithm 7.1 starts from a refinement $\mathcal{T}$ of $\mathcal{T}_0$ rather than $\mathcal{T}_0$ and thus exploits the mesh refinement already performed in the adaptive process. We now give a simple argument, updated from Bonito *et al.* (2013*b*), that shows that the number of elements $N(\mathcal{T}, \tau, b, v)$ marked by GREEDY starting from $\mathcal{T}$ with target tolerance τ and refined $b \geq 1$ times is dominated by $N(\mathcal{T}_0, \tau, 1, v)$, namely

$$N(\mathcal{T}, \tau, b, v) \leq N(\mathcal{T}_0, \tau, 1, v). \tag{7.8}$$

Estimate (7.8) is crucial because it avoids studying the cardinality of GREEDY starting from $\mathcal{T} \neq \mathcal{T}_0$ directly, and simplifies the analysis. Even though (7.8) is plausible, the fact that the output of GREEDY($\mathcal{T}_0, \tau, q, 1, v$) is unrelated to $\mathcal{T}$ makes it non-obvious. In fact, note that we do not claim that $N(\mathcal{T}, \tau, b, v) \leq N(\mathcal{T}_0, \tau, b, v)$, which is unclear. The proof presented below hinges on the fact that all the elements refined within GREEDY($\mathcal{T}_0, \tau, q, 1, v$) are either refined because they are marked by GREEDY (and thus of largest oscillation) or because their refinement is necessary to guarantee conformity of the resulting subdivision. For our purposes, (7.8) suffices.

Lemma 7.5 (GREEDY starting from $\mathcal{T}$). *Let $\tau > 0$ be a target tolerance and let $b \geq 1$ be the number of bisections per marked element. Assume that the local errors employed by* GREEDY *satisfy Assumption 7.3 (monotonicity of local oscillations) in ℓ^q. Then the number of elements $N(\mathcal{T}, \tau, b, v)$ marked by* GREEDY($\mathcal{T}, \tau, q, b, v$) *satisfies* (7.8) *for any admissible refinement $\mathcal{T} \in \mathbb{T}$ of $\mathcal{T}_0$.*

Proof. We simply write GREEDY($\mathcal{T}_0, \tau, 1$) and GREEDY($\mathcal{T}, \tau, b$) because v and q are fixed. Let $N := N(\mathcal{T}_0, \tau, 1, v)$, and recall that the bisection rules define a unique forest $\mathbb{T}$ emanating from $\mathcal{T}_0$ and a unique sequence of elements $\{T_i\}_{i=1}^N$ marked by GREEDY($\mathcal{T}_0, \tau, 1$). We let $\{\mathcal{T}^i\}_{i=1}^N$ denote the sequence of intermediate subdivisions built within GREEDY($\mathcal{T}_0, \tau, 1$) starting with $\mathcal{T}^0 = \mathcal{T}_0$: $T_i \in \mathcal{T}^{i-1}$ is bisected once by REFINE, which also produces the smallest conforming refinement $\mathcal{T}^i$ of $\mathcal{T}^{i-1}$ containing the two children of T_i. We thus say that GREEDY$(\mathcal{T}_0, \tau, 1)$ satisfies the *minimality property* that all the elements refined are either marked elements because their error is largest or necessary to guarantee conforming subdivisions. Notice that this is not true for GREEDY$(\mathcal{T}_0, \tau, b)$ when $b > 1$.

For any $\mathcal{T} \in \mathbb{T}$, we let $\Lambda_{\mathcal{T}}$ be the set of indices $j \in \{1, \dots, N\}$ such that T_j is never refined in the process to create $\mathcal{T}$, that is, T_j is either an element of $\mathcal{T}$ or a successor of an element of $\mathcal{T}$. We show that

$$N(\mathcal{T}, \tau, b, v) \leq \#\Lambda_{\mathcal{T}} \tag{7.9}$$

by induction on $\#\Lambda_{\mathcal{T}}$. If $\#\Lambda_{\mathcal{T}} = 0$ then $\mathcal{T}$ is a refinement of $\mathcal{T}^N$, whence the monotonicity of the total error

$$E_{\mathcal{T}}(v)_q \leq E_{\mathcal{T}^N}(v)_q \leq \tau,$$

guaranteed by (7.5), implies that $N(\mathcal{T}, \tau, b, v) = 0$; this satisfies (7.9) as desired.

We now assume that (7.9) is valid for any $\mathcal{T} \in \mathbb{T}$ such that $\#\Lambda_{\mathcal{T}} \leq k$, a non-negative integer, and deduce that it must also hold for any $\mathcal{T} \in \mathbb{T}$ such that $\#\Lambda_{\mathcal{T}} \leq k + 1$. Let $\mathcal{T} \in \mathbb{T}$ be one such mesh, namely $\#\Lambda_{\mathcal{T}} = k + 1$. If $E_{\mathcal{T}}(v)_q \leq \tau$, then $N(\mathcal{T}, \tau, b, v) = 0$ and $N(\mathcal{T}, \tau, b, v) \leq \#\Lambda_{\mathcal{T}}$ holds trivially.

When instead $E_{\mathcal{T}}(v)_q > \tau$, we let j be the smallest index in $\Lambda_{\mathcal{T}}$ and show that $T_j \in \mathcal{T}$ using the minimality property of GREEDY$(\mathcal{T}_0, \tau, 1)$. Assume by contradiction that $T_j \notin \mathcal{T}$ but T_j belongs to a refinement $\widetilde{\mathcal{T}}$ of $\mathcal{T}$ and is thus a successor of an element $T \in \mathcal{T}$. Note that T is refined by GREEDY$(\mathcal{T}_0, \tau, 1)$ to produce T_j but was not marked, because otherwise $T = T_i$ for some $i < j$ and $i \in \Lambda_{\mathcal{T}}$, which would contradict the minimality of j. Hence T must have been refined by the REFINE routine to guarantee conformity when bisecting a marked element T_ℓ, $\ell < j$. Invoking the minimality of j again yields that $\ell \notin \Lambda_{\mathcal{T}}$ and T_ℓ cannot be in $\mathcal{T}$ because T_ℓ has been refined to get to $\mathcal{T}$ by definition of $\Lambda_{\mathcal{T}}$. Since REFINE refines the minimal number of non-marked elements to guarantee conformity, and $\mathcal{T}$ is conforming, T must have been refined as well when refining T_ℓ in the process of constructing $\mathcal{T}$ and therefore cannot be in $\mathcal{T}$. This is a contradiction and $T_j \in \mathcal{T}$.

Therefore $\mathcal{T}$ is a refinement of $\mathcal{T}^{j-1}$ because all the elements marked or refined to ensure conformity by GREEDY$(\mathcal{T}_0, \tau, 1)$ have been refined in the process of creating $\mathcal{T}$. Moreover, $T_j \in \mathcal{T}$ is the element with largest error $\operatorname{osc}_{\mathcal{T}}(v, T_j)$ within $\mathcal{T}$ (with *ad hoc* criteria to break ties), because $\operatorname{osc}_{\mathcal{T}^{j-1}}(v, T_j)$ is largest in $\mathcal{T}^{j-1}$ by definition of T_j and monotonicity of the local error (7.5); hence T_j must be the first element marked by GREEDY$(\mathcal{T}, \tau, b)$. Let $\mathcal{T}^*$ be the subdivision obtained from $\mathcal{T}$ upon bisecting b times T_j. Notice that $\Lambda_{\mathcal{T}^*}$ is a strict subset of $\Lambda_{\mathcal{T}}$, because $j \notin \Lambda_{\mathcal{T}^*}$, so that the induction assumption yields

$$N(\mathcal{T}, \tau, b, v) = 1 + N(\mathcal{T}^*, \tau, b, v) \leq 1 + \#\Lambda_{\mathcal{T}^*} \leq \#\Lambda_{\mathcal{T}}.$$

This proves (7.9), and (7.8) follows immediately since $\#\Lambda_{\mathcal{T}} \leq N(\mathcal{T}_0, \tau, 1, v)$. □

Estimate (7.8) is critical to analysing the performances of GREEDY starting from any admissible subdivision $\mathcal{T} \in \mathbb{T}$. We emphasize that the complexity estimate provided by Corollary 7.6 is expressed in terms of the number of marked elements $N(\mathcal{T},\tau,q,b,v)$ and tolerance τ instead of error and cardinality of $\mathcal{T}$. This is why GREEDY can start from any mesh $\mathcal{T} \in \mathbb{T}$.

Corollary 7.6 (performance of GREEDY). *Let the initial subdivision $\mathcal{T}_0$ of $\Omega \subset \mathbb{R}^d$ satisfy Assumption 6.19 (initial labelling) and let $\mathcal{T} \in \mathbb{T}$ be any admissible refinement of $\mathcal{T}_0$. Let $\tau > 0$ be the target tolerance and let $b \geq 1$ be the number of bisections performed on each marked element. Let (v,s,t,p,q) satisfy Assumption 7.2 (admissible set of parameters for* GREEDY*) with local errors $\{\mathrm{osc}_{\mathcal{T}}(v,T)_q\}_{T\in\mathcal{T}}$ which in turn verify Assumption 7.3 (monotonicity of local oscillations) in ℓ^q. The number of marked elements $N(\mathcal{T},\tau,q,b,v)$ by* GREEDY$(\mathcal{T},\tau,q,b,v)$ *satisfies*

$$N(\mathcal{T},\tau,q,b,v) \leq C|v|_{X_p^s(\Omega)}^{d/(s+t)}\tau^{-d/(s+t)}, \tag{7.10}$$

with a constant $C = C(p,q,s,b,d,\Omega,\mathcal{T}_0)$. Moreover, the estimate (7.10) *is valid for tensor-valued functions v.*

Proof. Invoking Proposition 7.4 (performance of GREEDY), which gives rise to a mesh $\widehat{\mathcal{T}}$, and Lemma 7.5 (GREEDY starting from $\mathcal{T}$), we readily deduce

$$N(\mathcal{T},\tau,q,b,v) \leq N(\mathcal{T}_0,\tau,q,1,v) \leq \#\widehat{\mathcal{T}} \leq C|v|_{X_p^s(\Omega)}^{d/(s+t)}\tau^{-d/(s+t)},$$

which is the desired inequality (7.10). $\square$

7.2. *Constrained approximations*

We discuss how the approximations produced by GREEDY (see Corollary 7.6) can be modified to satisfy the structural assumption (5.51) without sacrificing their accuracy.

7.2.1. *Constrained approximations of scalar functions*

The approximate data $\widetilde{\mathcal{D}} = (\widetilde{\boldsymbol{A}},\widetilde{c},\widetilde{f})$ constructed in the previous sections using the GREEDY algorithm are not guaranteed to satisfy the necessary conditions for perturbed problem (5.5) with $\widehat{\mathcal{D}} = \widetilde{\mathcal{D}}$ to have a solution $\widehat{u} = \widehat{u}(\widehat{\mathcal{D}}) \in H_0^1(\Omega)$. Recall that the data $\mathcal{D} = (\boldsymbol{A},c,f) \in D(\Omega)$ is assumed to satisfy the structural assumption (5.50), i.e. $\boldsymbol{A} \in M(\alpha_1,\alpha_2)$ and $c \in R(c_1,c_2)$ with $0 < \alpha_1 \leq \alpha_2$ and $0 \leq c_1 \leq c_2$. It turns out that constructing approximate data $\widehat{\mathcal{D}}$ with the same constraints is a difficult task. We follow Bonito *et al.* (2013*b*) and modify the data $\widetilde{\mathcal{D}}$ to obtain $\widehat{\mathcal{D}} = (\widehat{\boldsymbol{A}},\widehat{c},\widehat{f})$ in such a way that the approximation property of $\widetilde{\mathcal{D}}$ is preserved, that is,

$$\|\mathcal{D} - \widehat{\mathcal{D}}\|_{D(\Omega)} \leq C_{\mathrm{data}}\|\mathcal{D} - \widetilde{\mathcal{D}}\|_{D(\Omega)},$$

while ensuring that

$$\widehat{A} \in M\left(\frac{\alpha_1}{2}, C\widehat{\alpha}_2\right), \quad \widehat{c} \in R\left(-\frac{\alpha_1}{4C_P^2}, C\widehat{c}_2\right). \tag{7.11}$$

Here C is a constant independent of relevant quantities (we make this more precise below). In particular, the data $\widehat{\mathcal{D}}$ satisfies the structural assumption (7.11) which guarantees that the perturbed problem (5.5) has a unique solution. Note that the general case is more subtle than when the data are approximated by piecewise constant approximations (5.74), which are directly satisfying the structural assumption and used as motivation in Section 5.4.2.

We start by discussing a process modifying the approximation of a strictly positive scalar function $v \in L^\infty(\Omega)$, i.e. $v \in R(c_1, c_2)$ for some $0 < c_1 \le c_2$; see (5.49). Because the polynomial degree used to approximate the data might differ depending on the application, we use $m \in \mathbb{N}$ to denote a generic polynomial degree. We think of $\widetilde{v} \in \mathbb{S}^{m,-1}_{\mathcal{T}}$ as an approximation to v not necessarily strictly positive. The following process modifies $\widetilde{v}$ locally to construct $\widehat{v} \in \mathbb{S}^{m,-1}_{\mathcal{T}}$. It involves a parameter $L > 2$ responsible for the truncation of $\widetilde{v}$ whenever it is too large, i.e. $\widetilde{v} \ge Lc_2$. For $T \in \mathcal{T}$, we set $\widehat{v}|_T := \widehat{v}_T$, where

$$\widehat{v}_T := \begin{cases} c_2, & \text{when } \|\widetilde{v}\|_{L^\infty(T)} \ge Lc_2, \\ \widetilde{v}|_T - \min_{x\in T} \widetilde{v}(x) + \dfrac{c_1}{2}, & \text{when otherwise } \min_{x\in T} \widetilde{v}(x) < \dfrac{c_1}{2}, \\ \widetilde{v}|_T, & \text{otherwise.} \end{cases} \tag{7.12}$$

Corollary 7.9 below is in essence Proposition 3 of Bonito *et al.* (2013*b*), and states that the constructed $\widehat{v}$ satisfies

$$0 < \frac{c_1}{2} \le \widehat{v} \le \left(\frac{1}{2} + 2L\right) c_2 \quad \text{a.e. in } \Omega.$$

This is at the expense of inflating the approximation error in L^q, $1 \le q \le \infty$, by a multiplicative constant C depending only on d, m, c_2/c_1, q, L and the shape regularity of $\mathbb{T}$,

$$\|v - \widehat{v}\|_{L^q(T)} \le C\|v - \widetilde{v}\|_{L^q(T)}.$$

In preparation for this result, we introduce the following notation. We let C_I denote the smallest constant such that for any $T \in \mathcal{T}$ and any polynomial $P \in \mathbb{P}_m(T)$, we have the inverse inequality

$$\|\nabla P\|_{L^\infty(T)} \le C_I \|P\|_{L^\infty(T)} |T|^{-1/d}. \tag{7.13}$$

The inverse inequality constant C_I depends only on the shape regularity of $\mathbb{T}$, m and d. Note that for such a polynomial $P \in \mathbb{P}_m(T)$, we have

$$|P(x) - P(y)| \le C_I \|P\|_{L^\infty(T)} |T|^{-1/d} |x - y|, \quad \text{for all } x, y \in T.$$

Consequently, for any $\rho > 0$ and $x \in T$, we define

$$T(x,\rho) := T \cap \overline{B(x, \rho|T|^{1/d}/C_I)},$$

which is motivated by the fact that for $x \in T$ and $y \in T(x,\rho)$ we have

$$|P(x) - P(y)| \le C_I \|P\|_{L^\infty(T)} |T|^{-1/d} |x-y| \le \rho \|P\|_{L^\infty(T)}. \tag{7.14}$$

Critical for the analysis below is the existence of a constant $0 < C_S(\rho) \le 1$ depending on ρ but also on d, m and the shape regularity of $\mathbb{T}$, such that

$$|T(x,\rho)| \ge C_S(\rho)|T| \quad \text{for all } x \in T,\ T \in \mathcal{T}. \tag{7.15}$$

This constant $C_S(\rho)$ assesses the area of a subset of T where the polynomial P varies no more than $\rho\|P\|_{L^\infty(T)}$ away from $P(x)$.

We are now in a position to analyse the effect of the nonlinear correction (7.12). We proceed locally over each $T \in \mathcal{T}$ and start with the case where $\|\widetilde{v}\|_{L^\infty(T)}$ is large (Lemma 7.7). We then discuss the case where $\widetilde{v}(x)$ is small on T (Lemma 7.8), while for the remaining case the function $\widetilde{v}$ does not need to be modified on T. These three cases are collected in Corollary 7.9 for scalar-valued functions and in Corollary 7.11 for matrix-valued functions. In all the arguments below we used the convention $a^{1/\infty} = 1$ for any $a > 0$.

Lemma 7.7 (locally enforcing constraints for large approximations). *Let $\mathcal{T} \in \mathbb{T}$ be any conforming refinement of $\mathcal{T}_0$ satisfying Assumption 6.19 (initial labelling). Let $c_2 > 0$, $T \in \mathcal{T}$ and $v_T \in L^\infty(T)$ satisfying $0 < v_T \le c_2$ a.e. in T. Furthermore, for $m \ge 0$ and $L > 2$, assume that $\widetilde{v}_T \in \mathbb{P}_m(T)$ satisfies*

$$\|\widetilde{v}_T\|_{L^\infty(T)} \ge Lc_2. \tag{7.16}$$

Then, for the constant function $\widehat{v}_T := c_2 \in \mathbb{P}_m(T)$,

$$\frac{c_1}{2} < \widehat{v}_T < Lc_2 < \left(\frac{1}{2} + 2L\right)c_2.$$

Moreover, for $1 \le q \le \infty$, we have

$$\|v_T - \widehat{v}_T\|_{L^q(T)} \le C_2^+ \|v_T - \widetilde{v}_T\|_{L^q(T)},$$

where

$$C_2^+ := \frac{4C_S^{-1/q}}{L-2}$$

and $C_S = C_S(1/2)$ is the constant appearing in (7.15) with $\rho = 1/2$.

Proof. Let $x_0 \in T$ and $\widetilde{c}_{2,T}$ defined by the relation

$$\widetilde{c}_{2,T} := |\widetilde{v}_T(x_0)| := \|\widetilde{v}_T\|_{L^\infty(T)}.$$

In view of the Lipschitz property (7.14) applied to $P = \widetilde{v}_T$ and with $\rho = \frac{1}{2}$, we have

$$|\widetilde{v}_T(x) - \widetilde{v}_T(x_0)| \le \frac{\widetilde{c}_{2,T}}{2}$$

for $x \in T_0 := T(x_0, \frac{1}{2}) \subset T$. Recall (7.15), which implies that $|T_0| \geq \widetilde{C}_S |T|$ for some constant $\widetilde{C}_S := C_S(1/2)$ depending only on d, n and the shape regularity of $\mathbb{T}$. On the one hand, this implies that $\widetilde{v}_T|_{T_0}$ is bounded below with $|\widetilde{v}_T(x)| \geq \widetilde{c}_{2,T}/2$ for $x \in T_0$ and, on the other hand, v_T is bounded from above by

$$0 \leq v_T(x) \leq c_2 \leq L^{-1}\widetilde{c}_{2,T}, \quad x \in T.$$

Consequently, for $x \in T_0$ and since $L > 2$, we have

$$0 \leq v_T(x) \leq L^{-1}\widetilde{c}_{2,T} \leq \frac{\widetilde{c}_{2,T}}{2} \leq |\widetilde{v}_T(x)|$$

and thus

$$|v_T(x) - \widetilde{v}_T(x)| \geq |\widetilde{v}_T(x)| - v_T(x) \geq \left(\frac{1}{2} - \frac{1}{L}\right)\widetilde{c}_{2,T} = \frac{L-2}{2L}\widetilde{c}_{2,T},$$

which indicates that v_T and $\widetilde{v}_T$ are sufficiently far apart on a substantial portion T_0 of T. This is responsible for the L^q-bound below. In fact, we have

$$\|v_T - \widetilde{v}_T\|_{L^q(T)} \geq \|v_T - \widetilde{v}_T\|_{L^q(T_0)} \geq \frac{L-2}{2L}\widetilde{c}_{2,T}|T_0|^{1/q}, \tag{7.17}$$

whence, from the definition $\widehat{v}_T := c_2$ and using (7.15), we deduce

$$\|v_T - \widehat{v}_T\|_{L^q(T)} \leq 2c_2|T|^{1/q} \leq 2L^{-1}\widetilde{c}_{2,T}|T|^{1/q} \leq \frac{4\widetilde{C}_S^{-1/q}}{L-2}\|v_T - \widetilde{v}_T\|_{L^q(T)}$$

as desired. □

Lemma 7.8 (locally enforcing constraints for small approximations). *Let $\mathcal{T} \in \mathbb{T}$ be any conforming refinement of $\mathcal{T}_0$ satisfying Assumption 6.19 (initial labelling). Let $0 < c_1 \leq c_2$, $T \in \mathcal{T}$ and $v_T \in L^\infty(T)$ satisfying $c_1 < v_T \leq c_2$ a.e. in T. Furthermore, for $m \geq 0$ and $L > 2$ assume that $\widetilde{v}_T \in \mathbb{P}_m(T)$ satisfies*

$$\|\widetilde{v}_T\|_{L^\infty(T)} \leq Lc_2 \tag{7.18}$$

and

$$\min_{x \in T} \widetilde{v}_T(x) < \frac{c_1}{2}. \tag{7.19}$$

Then the function $\widehat{v}_T := c_1/2 + \widetilde{v}_T - \min_{x\in T}\widetilde{v}_T(x) \in \mathbb{P}_n(T)$ is such that

$$\frac{c_1}{2} \leq \widehat{v}_T \leq 2Lc_2 + \frac{c_1}{2} \leq \left(2L + \frac{1}{2}\right)c_2$$

and

$$\|v_T - \widehat{v}_T\|_{L^q(T)} \leq C_1^+\|v_T - \widetilde{v}_T\|_{L^q(T)},$$

where

$$C_1^+ := \left(1 + C_S^{-1/q}(\rho)\right)$$

and $C_S(\rho)$ is the constant appearing in (7.15) with $\rho = c_1/(2Lc_2)$.

Proof. We define $x_0 \in T$, $\widetilde{c}_{1,T} \in \mathbb{R}$ by the relations

$$\widetilde{c}_{1,T} := \widetilde{v}_T(x_0) := \min_{x \in T} \widetilde{v}_T(x).$$

From the Lipschitz property (7.14) and the assumption (7.18), we find that

$$|\widetilde{v}_T(x) - \widetilde{v}_T(x_0)| \leq \frac{c_1}{2}$$

for $x \in T_0 := T(x_0, \rho)$ with $\rho := c_1/(2Lc_2)$. Recall (7.15), which implies that $|T_0| \geq \widetilde{C}_S|T|$ for some constant $\widetilde{C}_S := C_S(\rho)$ depending only on d, m, c_2/c_1, L and the shape regularity of $\mathbb{T}$.

For $x \in T_0$, we proceed by estimating the difference

$$v_T(x) - \widetilde{v}_T(x) = v_T(x) - (\widetilde{v}_T(x) - \widetilde{v}_T(x_0)) - \widetilde{v}_T(x_0) \geq c_1 - \frac{c_1}{2} - \widetilde{c}_{1,T} = \frac{c_1}{2} - \widetilde{c}_{1,T} > 0,$$

because $\widetilde{c}_{1,T} < c_1/2$ by assumption (7.19). This implies that

$$|T_0|^{1/q}\left(\frac{c_1}{2} - \widetilde{c}_{1,T}\right) \leq \|v_T - \widetilde{v}_T\|_{L^q(T)},$$

and v_T and $\widetilde{v}_T$ are uniformly far apart in the substantial part T_0 of T. Therefore $\widehat{v}_T := \widetilde{v}_T + (c_1/2 - \widetilde{c}_{1,T})$ satisfies

$$\frac{c_1}{2} \leq \widehat{v}_T \leq 2Lc_2 + \frac{c_1}{2}$$

because $\widetilde{c}_{1,T} \geq -\|\widetilde{v}\|_{L^\infty(T)} \geq -Lc_2$ by assumption (7.18), and

$$\begin{aligned}
\|v_T - \widehat{v}_T\|_{L^q(T)} &\leq \|v_T - \widetilde{v}_T\|_{L^q(T)} + \left(\frac{c_1}{2} - \widetilde{c}_{1,T}\right)|T|^{1/q} \\
&\leq \|v_T - \widetilde{v}_T\|_{L^q(T)} + \left(\frac{c_1}{2} - \widetilde{c}_{1,T}\right)\widetilde{C}_S^{-1/q}|T_0|^{1/q} \\
&\leq \left(1 + \widetilde{C}_S^{-1/q}\right)\|v_T - \widetilde{v}_T\|_{L^q(T)}.
\end{aligned}$$

This proves the assertions. □

Corollary 7.9 (locally enforcing constraints). *Let $\mathcal{T} \in \mathbb{T}$ be any conforming refinement of $\mathcal{T}_0$ satisfying Assumption 6.19 (initial labelling). Let $0 < c_1 \leq c_2$, $T \in \mathcal{T}$ and $v_T \in L^\infty(T)$ satisfying $c_1 \leq v_T \leq c_2$ a.e. in T. Then, for $m \geq 0$, $L > 2$, and $\widetilde{v}_T \in \mathbb{P}_m(T)$, the function $\widehat{v}_T \in \mathbb{P}_m(T)$ defined in (7.12) satisfies*

$$\frac{c_1}{2} \leq \widehat{v}_T \leq \left(\frac{1}{2} + 2L\right)c_2 \quad \textit{a.e. in } T.$$

Moreover, for $1 \leq q \leq \infty$, we have

$$\|v_T - \widehat{v}_T\|_{L^q(T)} \leq \max(C_1^+, C_2^+)\|v_T - \widetilde{v}_T\|_{L^q(T)} \quad \textit{for all } T \in \mathcal{T},$$

where C_1^+ and C_2^+ are the constants appearing in Lemmas 7.8 and 7.7, which depend only on d, m, c_2/c_1, L and the shape regularity of $\mathbb{T}$.

Proof. The desired results follow from Lemma 7.7 when

$$\|\widetilde{v}_T\|_{L^\infty(T)} \geq Lc_2,$$

and from Lemma 7.8 when

$$\|\widetilde{v}_T\|_{L^\infty(T)} < Lc_2 \quad \text{and} \quad \min_{x\in T} \widetilde{v}_T < \frac{c_1}{2}.$$

In the remaining case,

$$\|\widetilde{v}_T\|_{L^\infty(T)} < Lc_2 \quad \text{and} \quad \min_{x\in T} \widetilde{v}_T(x) \geq \frac{c_1}{2},$$

since $\widehat{v}_T = \widetilde{v}_T$ satisfies the desired constraints, there is nothing to prove. □

7.2.2. Constrained approximation of the diffusion coefficients

For matrix-valued functions, the constraints are on the eigenvalues of the matrix rather than on the coefficients themselves. Although this requires a few adjustments, the process is similar to the scalar case. We recall that for $0 < \alpha_1 \leq \alpha_2$, $M(\alpha_1, \alpha_2) \subset L^\infty(\Omega; \mathbb{R}^{d\times d}_{\text{sym}})$ denotes the class of symmetric matrix-valued functions whose eigenvalues lie between α_1 and α_2; see (5.48).

Algorithm CONSTRAINT-A is based on (7.12) and modifies approximations $\widetilde{\boldsymbol{A}} \in (\mathbb{S}^{n_A,-1}_{\mathcal{T}})^{d\times d}$ of $\boldsymbol{A} \in M(\alpha_1, \alpha_2)$ to produce uniformly positive definite approximations $\widehat{\boldsymbol{A}} \in (\mathbb{S}^{n_A,-1}_{\mathcal{T}})^{d\times d}$ of $\boldsymbol{A}$.

Algorithm 7.10 (CONSTRAINT-A). Given a threshold parameter $L > 2$, $0 < \alpha_1 \leq \alpha_2$, a conforming refinement $\mathcal{T} \in \mathbb{T}$ of $\mathcal{T}_0$, and $\widetilde{\boldsymbol{A}} \in (\mathbb{S}^{n_A,-1}_{\mathcal{T}})^{d\times d}$, this routine constructs a positive definite $\widehat{\boldsymbol{A}} \in (\mathbb{S}^{n_A,-1}_{\mathcal{T}})^{d\times d}$.

$[\widehat{\boldsymbol{A}}]$ = CONSTRAINT-A$(\mathcal{T}, \alpha_1, \alpha_2, L, \widetilde{\boldsymbol{A}})$
  For $T \in \mathcal{T}$
    $\widetilde{\alpha}_{1,T} = \inf\{y^t\widetilde{\boldsymbol{A}}(x)y,\ x \in T,\ |y| = 1\}$
    $\widetilde{\alpha}_{2,T} = \sup\{|y^t\widetilde{\boldsymbol{A}}(x)y|,\ x \in T,\ |y| = 1\}$
    if $\widetilde{\alpha}_{2,T} \geq L\alpha_2$
      $\widehat{\boldsymbol{A}}|_T = \alpha_2 \boldsymbol{I}_d$
    else if $\widetilde{\alpha}_{1,T} < \alpha_1/2$
      $\widehat{\boldsymbol{A}}|_T = \widetilde{\boldsymbol{A}}|_T - (\alpha_1/2 - \widetilde{\alpha}_{1,T})\boldsymbol{I}_d$
    else
      $\widehat{\boldsymbol{A}}|_T = \widetilde{\boldsymbol{A}}|_T$
  return $\widehat{\boldsymbol{A}}$

Notice that CONSTRAINT-A preserves symmetry, that is, if $\widetilde{\boldsymbol{A}}$ is symmetric then so is the output $\widehat{\boldsymbol{A}}$. In addition, when $n = 1$ and $\widetilde{\boldsymbol{A}} \in (\mathbb{S}^{0,-1}_{\mathcal{T}})^{d\times d}$ is the piecewise constant local average of $\boldsymbol{A}$, the output $\widehat{\boldsymbol{A}}$ of CONSTRAINT-A is $\widehat{\boldsymbol{A}} = \widetilde{\boldsymbol{A}}$ since in that case the parameters $\widetilde{\alpha}_{1,T}$ and $\widetilde{\alpha}_{2,T}$ satisfy

$$\widetilde{\alpha}_{1,T} \geq \alpha_1 > \frac{\alpha_1}{2} \quad \text{and} \quad \widetilde{\alpha}_{2,T} \leq \alpha_2 < L\alpha_2 \quad \text{for all } T \in \mathcal{T}.$$

This is consistent with the observation made in Section 5.4.2.

The next corollary hinges on Corollary 7.9 (locally enforcing constraints) to derive properties of CONSTRAINT-A. In passing, we recall that for $\boldsymbol{A} \in L^p(\Omega; \mathbb{R}^{d\times d})$ we write

$$\|\boldsymbol{A}\|_{L^p(\Omega)} := \||\boldsymbol{A}|\|_{L^p(\Omega)},$$

where for $x \in \Omega$, $|\boldsymbol{A}(x)|$ is the spectral norm of $\boldsymbol{A}(x)$.

Corollary 7.11 (locally enforcing constraints for matrices). *Let the threshold be $L > 2$, $0 < \alpha_1 \leq \alpha_2$ and $\boldsymbol{A} \in M(\alpha_1, \alpha_2)$. Let $\mathcal{T} \in \mathbb{T}$ be any conforming refinement of $\mathcal{T}_0$ and let $\widetilde{\boldsymbol{A}} \in \left(\mathbb{S}_{\mathcal{T}}^{n_A,-1}\right)^{d\times d}$ be a symmetric approximation of $\boldsymbol{A}$. Then the output $[\widehat{\boldsymbol{A}}] =$ CONSTRAINT-A$(\mathcal{T}, \alpha_1, \alpha_2, L, \widetilde{\boldsymbol{A}})$ is symmetric and satisfies*

$$\frac{\alpha_1}{2} \leq \lambda_j(\widehat{\boldsymbol{A}}) \leq \left(\frac{1}{2} + 2L\right)\alpha_2 \quad \text{a.e. in } \Omega,\ 1 \leq j \leq d.$$

Moreover, for $1 \leq q \leq \infty$, we have

$$\|\boldsymbol{A} - \widehat{\boldsymbol{A}}\|_{L^q(T)} \leq C_{\text{data}} \|\boldsymbol{A} - \widetilde{\boldsymbol{A}}\|_{L^q(T)} \quad \text{for all } T \in \mathcal{T},$$

where $C_{\text{data}} := \max(C_1^+, C_2^+)$ and C_1^+ and C_2^+ are the constants appearing in Lemmas 7.8 and 7.7, which depend only on d, n_A, α_2/α_1, L and the shape regularity of $\mathbb{T}$.

Proof. We observe that $\widetilde{\boldsymbol{A}}$ is not assumed to be positive semidefinite. We argue locally and fix $T \in \mathcal{T}$. Let $\widetilde{\alpha}_{2,T} > 0$ and $y_0 \in \mathbb{R}^d$ be such that $|y_0| = 1$ and

$$\widetilde{\alpha}_{2,T} := \sup_{x\in T} |y_0^t \widetilde{\boldsymbol{A}}(x) y_0| := \sup_{x\in T} \sup_{y\in\mathbb{R}^d, |y|=1} |y^t \widetilde{\boldsymbol{A}}(x) y|.$$

We first consider the case $\widetilde{\alpha}_{2,T} \geq L\alpha_2$ for which $\widehat{\boldsymbol{A}}|_T := \alpha_2 \boldsymbol{I}_d$. For $x \in T$, we set

$$a(x) := y_0^t \boldsymbol{A}(x) y_0 \quad \text{and} \quad \widetilde{a}_T(x) = y_0^t \widetilde{\boldsymbol{A}}(x) y_0 \in \mathbb{P}_{n_A}(T).$$

This notation allows us to reduce to the scalar case upon noting that

$$\|a - \widetilde{a}_T\|_{L^q(T)} \leq \|\boldsymbol{A} - \widetilde{\boldsymbol{A}}\|_{L^q(T)}$$

and $\alpha_1 \leq a \leq \alpha_2$ a.e. in Ω. Because $\widetilde{\alpha}_{2,T} \geq L\alpha_2$, Lemma 7.7 with $m = n_A$ guarantees that $\widehat{a}_T := \alpha_2$ satisfies

$$\|a - \widehat{a}_T\|_{L^q(T)} \leq C_2^+ \|a - \widetilde{a}_T\|_{L^q(T)} \leq C_2^+ \|\boldsymbol{A} - \widetilde{\boldsymbol{A}}\|_{L^q(T)}.$$

Consequently, the matrix-valued approximation $\widehat{\boldsymbol{A}}|_T := \alpha_2 \boldsymbol{I}_d$ satisfies

$$\|\boldsymbol{A} - \widehat{\boldsymbol{A}}\|_{L^q(T)} = \|a - \widehat{a}_T\|_{L^q(T)} \leq C_2^+ \|\boldsymbol{A} - \widetilde{\boldsymbol{A}}\|_{L^q(T)}.$$

This proves the desired result when $\widetilde{\alpha}_{2,T} \geq L\alpha_2$.

We now consider the case where $\widetilde{\alpha}_{2,T} < L\alpha_2$ and define $\widetilde{\alpha}_{1,T} \in \mathbb{R}$, $y_1 \in \mathbb{R}^d$ with $|y_1| = 1$ by the relations

$$\widetilde{\alpha}_{1,T} = \inf_{x\in T} y_1^t \widetilde{\boldsymbol{A}}(x) y_1 = \inf_{x\in T} \inf_{|y|=1} y^t \widetilde{\boldsymbol{A}}(x) y.$$

We also redefine the associated scalar functions for $x \in T$ using y_1 instead of y_0:

$$a(x) := y_1^t \boldsymbol{A}(x) y_1 \quad \text{and} \quad \widetilde{a}_T(x) = y_1^t \widetilde{\boldsymbol{A}}(x) y_1 \in \mathbb{P}_{n_A}(T).$$

If $\widetilde{\alpha}_{1,T} < \alpha_1/2$ then $\widehat{\boldsymbol{A}}|_T = \widetilde{\boldsymbol{A}}|_T + (\alpha_1/2 - \widetilde{\alpha}_{1,T})\boldsymbol{I}_d$. Lemma 7.8 with $m = n_A$ ensures that $\widehat{a}_T = \widetilde{a}_T + \alpha_1/2 - \widetilde{\alpha}_{1,T}$ satisfies

$$\frac{\alpha_1}{2} \leq \widehat{a}_T \leq \left(\frac{1}{2} + 2M\right)\alpha_2$$

and

$$\|a - \widehat{a}_T\|_{L^q(T)} \leq C_1^+ \|a - \widetilde{a}_T\|_{L^q(T)} \leq C_1^+ \|\boldsymbol{A} - \widetilde{\boldsymbol{A}}\|_{L^q(T)}.$$

Thus $\widehat{\boldsymbol{A}}|_T$ satisfies the desired properties provided $\widetilde{\alpha}_{2,T} \geq L\alpha_2$ as well.

It the remaining case $\widetilde{\alpha}_{2,T} < L\alpha_2$ and $\widetilde{\alpha}_{1,T} \geq \alpha_1/2$, the function $\widehat{\boldsymbol{A}}|_T = \widetilde{\boldsymbol{A}}|_T$ satisfies the desired properties and there is nothing to prove. □

As a corollary, we report the complexity of an algorithm that concatenates the linear approximation of GREEDY with the nonlinear correction into the constraint of CONSTRAINT-A. We recall from Corollary 6.36 (approximation class of $\boldsymbol{A}$) that the admissible set of parameters of $\boldsymbol{A}$ for GREEDY are $n_A \leq n - 1$:

$$s_A \in (0, n_A], \quad p_A \in (0, \infty], \quad q_A \in [2, \infty], \quad s_A - \frac{d}{p_A} + \frac{d}{q_A} > 0, \quad t_A = 0.$$

Corollary 7.12 (complexity of constrained GREEDY for $\boldsymbol{A}$). *Let the initial mesh $\mathcal{T}_0$ of $\Omega \subset \mathbb{R}^d$ satisfy Assumption 6.19 (initial labelling) and let $\mathcal{T} \in \mathbb{T}$ be any admissible refinement of $\mathcal{T}_0$. Let $\tau > 0$ be the target tolerance, let $b \geq 1$ be the number of bisections performed on each marked element, and let $L > 2$ be a threshold parameter. Furthermore, assume that $(\boldsymbol{A}, s_A, t_A, p_A, q_A)$ satisfies Assumption 7.2 (admissible set of parameters for* GREEDY*) with local oscillations $\{\|\boldsymbol{A} - \widehat{\boldsymbol{A}}\|_{L^{q_A}(T)}\}_{T\in\mathcal{T}}$ and, in addition, $\boldsymbol{A} \in M(\alpha_1, \alpha_2)$ for some $0 < \alpha_1 \leq \alpha_2$. The algorithm*

$$[\widehat{\mathcal{T}}, \widetilde{\boldsymbol{A}}] = \mathsf{GREEDY}(\mathcal{T}, \tau, q_A, b, \boldsymbol{A})$$
$$[\widehat{\boldsymbol{A}}] = \mathsf{CONSTRAINT\text{-}A}(\widehat{\mathcal{T}}, \alpha_1, \alpha_2, L, \widetilde{\boldsymbol{A}})$$

where GREEDY *is applied to the $d(d+1)/2$ distinct components of $\boldsymbol{A}$, marks N elements of $\mathcal{T}$ for refinement with*

$$N \leq C |\boldsymbol{A}|_{X_{p_A}^{s_A}(\Omega;\mathcal{T}_0)}^{d/s_A} \tau^{-d/s_A} \tag{7.20}$$

and $C = C(p_A, q_A, s_A, b, d, n_A, \alpha_2/\alpha_1, L, \Omega, \mathcal{T}_0)$. *Moreover,* $\widehat{\boldsymbol{A}} \in \left(\mathbb{S}_{\mathcal{T}}^{n_A,-1}\right)^{d\times d}$ *satisfies*

$$\widehat{\boldsymbol{A}} \in M(\widehat{\alpha}_1, \widehat{\alpha}_2): \quad \widehat{\alpha}_1 = \frac{\alpha_1}{2}, \quad \widehat{\alpha}_2 = (1+4L)\frac{\alpha_2}{2}, \tag{7.21}$$

and there is a constant $C_{\mathrm{data}} > 0$ *such that*

$$\|\boldsymbol{A} - \widehat{\boldsymbol{A}}\|_{L^q(\Omega)} \le C_{\mathrm{data}}\tau.$$

Proof. This result follows upon invoking Corollary 7.6 (performance of GREEDY) and Corollary 7.11 (locally enforcing constraints for matrices). □

Remark 7.13 (constrained approximation class of matrices). As a consequence of Corollary 7.12, we realize that for $A \in M(\alpha_1, \alpha_2)$,

$$\delta_{\mathcal{T}}(\boldsymbol{A})_r \le \widetilde{\delta}_{\mathcal{T}}(\boldsymbol{A})_r \le C_{\mathrm{data}}\delta_{\mathcal{T}}(\boldsymbol{A})_r,$$

where the best approximation error $\delta_{\mathcal{T}}(\boldsymbol{A})_r$ and best constrained approximation error $\widetilde{\delta}_{\mathcal{T}}(\boldsymbol{A})_r$ are defined in (6.9) and (6.10).

7.2.3. *Constrained approximation of the reaction coefficients*

If the reaction coefficient $c \in R(c_1, c_2)$ is strictly positive ($c_1 > 0$), then Corollary 7.9 (locally enforcing constraints) with $m = n_c$ directly applies to $v_T = c|_T$, $T \in \mathcal{T}$, and guarantees that the approximate coefficient $\widehat{c} \in \mathbb{S}_{\mathcal{T}}^{n_c,-1}$ defined on $T \in \mathcal{T}$ by $\widehat{c}|_T := \widehat{v}_T$ satisfies

$$\widehat{c} \in R(\widehat{c}_1, \widehat{c}_2): \quad \widehat{c}_1 = \frac{c_1}{2}, \quad \widehat{c}_2 = (1+4L)\frac{c_2}{2}.$$

However, reaction coefficients are not necessarily strictly positive on $\overline{\Omega}$, and Corollary 7.9 cannot be invoked directly. Instead, we take advantage of the fact that the perturbed problem (5.5) is still well-posed provided $\widehat{c} \ge -\widehat{\alpha}_1/(2C_P^2)$ and the approximate diffusion coefficient $\widehat{\boldsymbol{A}} \in M(\widehat{\alpha}_1, \widehat{\alpha}_2)$ of $\boldsymbol{A} \in M(\alpha_1, \alpha_2)$ satisfies $\widehat{\alpha}_1 \ge \alpha_1/2$ according to (5.52); hence $\widehat{c} \ge -\alpha_1/(4C_P^2)$. Therefore we apply Corollary 7.9 to the shifted reaction coefficient $v = c + \widehat{\alpha}_1/C_P^2$, which satisfies

$$v_1 := c_1 + \frac{\widehat{\alpha}_1}{C_P^2} \le v \le c_2 + \frac{\widehat{\alpha}_1}{C_P^2} =: v_2. \tag{7.22}$$

Below is the proposed algorithm for the construction of $\widehat{c}$ in the general case $c_1 \ge 0$.

Algorithm 7.14 (CONSTRAINT-c). Given $L > 2$, $\widehat{\alpha}_1 > 0$, a conforming refinement $\mathcal{T} \in \mathbb{T}$ of $\mathcal{T}_0$, and $\widetilde{c} \in \mathbb{S}_{\mathcal{T}}^{n_c,-1}$, this routine constructs $\widehat{c} \in \mathbb{S}_{\mathcal{T}}^{n_c,-1}$ as follows:

$[\widehat{c}]$ = CONSTRAINT-c($\mathcal{T}, \widehat{\alpha}_1, L, \widetilde{c}$)
  $\widetilde{v} = \widetilde{c} + \widehat{\alpha}_1/C_P^2$
  For $T \in \mathcal{T}$
    if $\|\widetilde{v}\|_{L^\infty(T)} \ge Lv_2$
      $\widehat{v}|_T = v_2$

else if $\min_{x\in T} \widetilde{v}(x) < v_1/2$
$\widehat{v}|_T = \widetilde{v}|_T - \min_{x\in T} \widetilde{v}(x) + v_1/2$
else
$\widehat{v}|_T = \widetilde{v}|_T$
$\widehat{c} = \widehat{v} - \widehat{\alpha}_1/C_P^2$
return $\widehat{c}$

We note that if $n_c = 0$, then $\widetilde{c}$ is the piecewise average of c and CONSTRAINT-c does not modify $\widetilde{c}$, which already satisfies the structural assumption (7.11).

The next result shows that the output $\widehat{c}$ of CONSTRAINT-c is a modification of $\widetilde{c}$ which satisfies $\widehat{c} \in R(\widehat{c}_1, \widehat{c}_2)$, with

$$\widehat{c}_1 := \frac{c_1}{2} - \frac{\widehat{\alpha}_1}{2C_P^2} \quad \text{and} \quad \widehat{c}_2 := (1+4L)\frac{c_2}{2} + (4L-1)\frac{\widehat{\alpha}_1}{2C_P^2} \tag{7.23}$$

without affecting the approximation of c in L^q, $1 \le q \le \infty$ (up to a multiplicative constant). In particular, $\widehat{c}_1 \ge -\widehat{\alpha}_1/(2C_P^2)$, which is necessary for the well-posedness of the perturbed problem (5.5) when $\widehat{A} \in M(\widehat{\alpha}_1, \widehat{\alpha}_2)$.

Corollary 7.15 (locally enforcing constraints for non-negative scalar functions). *Let $A \in M(\widehat{\alpha}_1, \widehat{\alpha}_2)$ with $0 < \widehat{\alpha}_1 \le \widehat{\alpha}_2$, and $c \in R(c_1, c_2)$ with $0 \le c_1 \le c_2$. Let $L > 2$ and $v_1 \le v_2$ be defined in* (7.22). *Let $\mathcal{T} \in \mathbb{T}$ be any conforming refinement of $\mathcal{T}_0$ and $\widetilde{c} \in \mathbb{S}_{\mathcal{T}}^{n_c,-1}$. Then the output $[\widehat{c}]$ =* CONSTRAINT-c$(\mathcal{T}, \widehat{\alpha}_1, L, \widetilde{c})$ *satisfies*

$$\widehat{c}_1 \le \widehat{c} \le \widehat{c}_2 \quad \text{a.e. in } \Omega,$$

where $\widehat{c}_1$ and $\widehat{c}_2$ are given by (7.23). *Moreover, for $0 < q \le \infty$, we have*

$$\|c - \widehat{c}\|_{L^q(T)} \le C_{\text{data}} \|c - \widetilde{c}\|_{L^q(T)} \quad \text{for all } T \in \mathcal{T},$$

where C_{data} is a constant depending only on d, n, v_2/v_1, Ω, L and the shape regularity of $\mathbb{T}$.

Proof. Set $\kappa := \widehat{\alpha}_1/C_P^2$ and $v := c + \kappa \in R(c_1 + \kappa, c_2 + \kappa)$ so that $c_1 + \kappa > 0$. On each $T \in \mathcal{T}$, we invoke Corollary 7.9 (locally enforcing constraints) with $m = n_c$, $\widetilde{v}_T = \widetilde{c}|_T + \kappa$ and where c_1, c_2 are replaced by $c_1 + \kappa$, $c_2 + \kappa$ respectively. Hence we deduce that the function $\widehat{v}$ constructed within CONSTRAINT-c satisfies

$$\frac{c_1 + \kappa}{2} \le \widehat{v} \le (1+4L)\frac{c_2+\kappa}{2}.$$

and

$$\|v - \widehat{v}\|_{L^q(T)} \le C_{\text{data}} \|v - \widetilde{v}\|_{L^q(T)} \quad \text{for all } T \in \mathcal{T}, \tag{7.24}$$

with a constant C_{data} depending on $d, n, v_2/v_1, L$, and the shape regularity of $\mathbb{T}$. Shifting back, $c = v - \kappa$ and $\widehat{c} := \widehat{v} - \kappa$, we find that the approximation $\widehat{c}$ constructed by CONSTRAINT-c satisfies

$$\frac{c_1+\kappa}{2} - \kappa \le \widehat{c} \le (1+4L)\frac{c_2+\kappa}{2} - \kappa$$

or equivalently

$$\frac{c_1}{2} - \frac{k}{2} \le \widehat{c} \le (1+4L)\frac{c_2}{2} + (4L-1)\frac{\kappa}{2}.$$

In view of (7.23) and $\kappa = \widehat{\alpha}_1/C_P^2$, this is the first desired inequality in disguise.

Furthermore, the second desired inequality follows from (7.24) because for $T \in \mathcal{T}$ we have $c - \widehat{c} = v - \widehat{v}$ and $c - \widetilde{c} = v - \widetilde{v}$. □

The next corollary combines the linear approximation of GREEDY together with the nonlinear correction into the constraint of CONSTRAINT-c. We recall from Corollary 6.37 (approximation class of c) that the admissible set of parameters of c for GREEDY are $n_c \le n-1$, $s_c \in (0, n_c]$, $p_c \in (0, \infty]$, where

$$n_c > 0 \quad \Rightarrow \quad q_c > \frac{d}{2}, \quad s_c - \frac{d}{p_c} + \frac{d}{q_c} > 0, \quad t_c = 0,$$

$$n_c = 0 \quad \Rightarrow \quad q_c = 2, \quad s_c - \frac{d}{p_c} + \frac{d}{2} > 0, \quad 0 < t_c < 2 - \frac{d}{2}.$$

Corollary 7.16 (complexity of constrained GREEDY for c). *Let the initial subdivision $\mathcal{T}_0$ of $\Omega \subset \mathbb{R}^d$ satisfy Assumption 6.19 (initial labelling) and $\mathcal{T} \in \mathbb{T}$ be any admissible refinement of $\mathcal{T}_0$. Let $\tau > 0$ be the target tolerance, $b \ge 1$ be the number of bisections performed on each marked element, let $L > 2$ be the threshold parameter, and $\widehat{\alpha}_1 > 0$. Furthermore, assume that (c, s_c, t_c, p_c, q_c) satisfies Assumption 7.2 (admissible set of parameters for* GREEDY*) with local oscillations $\{\|c - \widehat{c}\|_{L^{q_c}(T)}\}_{T \in \mathcal{T}}$ and that $c \in R(c_1, c_2)$ for some $0 \le c_1 \le c_2$. The algorithm*

$[\widehat{\mathcal{T}}, \widehat{c}] = \mathsf{GREEDY}(\mathcal{T}, \tau, q_c, b, c)$
$[\widetilde{c}] = \mathsf{CONSTRAINT\text{-}c}(\widehat{\mathcal{T}}, \widehat{\alpha}_1, L, \widehat{c})$

marks N elements of $\mathcal{T}$ for refinement with

$$N \le C|c|_{X_{p_c}^{s_c}(\Omega;\mathcal{T}_0)}^{d/(s_c+t_c)} \tau^{-d/(s_c+t_c)} \tag{7.25}$$

and a constant $C = C(p_c, q_c, s_c, b, d, n_c, v_2/v_1, L, \Omega, \mathcal{T}_0)$ with $v_1 \le v_2$ defined in (7.22) to construct $\widehat{\mathcal{T}}$. The function $\widetilde{c} \in \mathbb{S}_{\widehat{\mathcal{T}}}^{n_c,-1}$ is a piecewise polynomial of degree $\le n_c$ over $\widehat{\mathcal{T}}$ and satisfies

$$\widetilde{c} \in R(\widehat{c}_1, \widehat{c}_2),$$

where $\widehat{c}_1 \le \widehat{c}_2$ are given by (7.23). Moreover, for $1 < q_c \le \infty$, there is a constant C_{data} depending only on d, n_c, v_2/v_1, Ω, L and the shape regularity of $\mathbb{T}$ such that

$$\|c - \widetilde{c}\|_{L^{q_c}(\Omega)} \le C_{\mathrm{data}} \tau.$$

Proof. Simply apply Corollary 7.6 (performance of GREEDY) and Corollary 7.15 (locally enforcing constraints for non-negative scalar functions). □

Remark 7.17 (constrained approximation class of scalars). Corollary 7.16 implies that for $c \in R(c_1, c_2)$

$$\delta_{\mathcal{T}}(c)_q \le \widetilde{\delta}_{\mathcal{T}}(c)_q \le C_{\mathrm{data}} \delta_{\mathcal{T}}(c)_q,$$

where the best approximation error $\delta_{\mathcal{T}}(c)_q$ and best constrained approximation error $\widetilde{\delta}_{\mathcal{T}}(c)_q$ are defined in (6.9) and (6.10).

7.3. Approximation of the load term f

We now turn our attention to the question of designing a practical algorithm for reducing the global oscillation

$$E_{\mathcal{T}}(f)_{-1}^2 := \sum_{T \in \mathcal{T}} \|f - P_{\mathcal{T}} f\|_{H^{-1}(\omega_T)}^2 \approx \sum_{z \in \mathcal{V}} \|f - P_{\mathcal{T}} f\|_{H^{-1}(\omega_z)}^2, \tag{7.26}$$

where the projection $P_{\mathcal{T}}$ is defined in (4.34). The approximation of functionals in $H^{-1}(\Omega)$ is rather intricate and out of reach without assuming additional structure enabling practical evaluation of their actions on polynomial functions.

We examine three cases of independent interest. In Section 7.3.1 we consider $f \in L^q(\Omega)$ for q satisfying $2d/(d+2) < q \le \infty$, which includes the most common setting $f \in L^2(\Omega)$. Sections 7.3.2 and 7.3.3 present examples of right-hand sides not in L^1. In Section 7.3.2 we treat the case $f = g\delta_\Gamma$, where Γ is a hypersurface not necessarily captured by the faces of the subdivisions and $g \in L^q(\Gamma)$, $q \ge 2$, while in Section 7.3.3 we consider $f = \operatorname{div} \boldsymbol{g}$ for some $\boldsymbol{g} \in L^2(\Omega; \mathbb{R}^d)$. In all cases, the total error $E_{\mathcal{T}}(f)_{-1}$ is estimated by a surrogate $\widetilde{E}_{\mathcal{T}}(f)_{-1}$, namely $E_{\mathcal{T}}(f)_{-1} \le C_{\mathrm{data}} \widetilde{E}_{\mathcal{T}}(f)_{-1}$,

$$\widetilde{E}_{\mathcal{T}}(f)_{-1}^2 := \sum_{T \in \mathcal{T}} \widetilde{\operatorname{osc}}_{\mathcal{T}}(f, T)_q^2,$$

with a definition of $\widetilde{\operatorname{osc}}_{\mathcal{T}}(f, T)_q$ depending on the situation but local to $T \in \mathcal{T}$ (and not on stars). This allows Algorithm 7.1 (GREEDY) to reduce $\widetilde{E}_{\mathcal{T}}(f)_{-1}$.

Before starting, we recall relevant definitions and results from Section 4 (*a posteriori* error analysis). For $z \in \mathcal{V}$, we let $\mathcal{T}_z \subset \mathcal{T}$ denote all the elements in ω_z and $\mathcal{F}_z \subset \mathcal{F}$ all the faces in ω_z. For $\ell \in H^{-1}(\Omega)$, the restriction $P_{\mathcal{T}} \ell|_{\omega_z}$ belongs to the space $\mathbb{F}(\mathcal{T}_z) = \mathbb{F}_{m_1,m_2}(\mathcal{T}_z)$ of functionals whose action against $w \in H_0^1(\omega_z)$ reads

$$\langle \ell, w \rangle = \sum_{T \in \mathcal{T}_z} \int_T q_T\, w + \sum_{F \in \mathcal{F}_z} \int_F q_F\, w \tag{7.27}$$

for some $q_F \in P_{m_1}(F)$, $F \in \mathcal{F}_z$ and $q_T \in P_{m_2}(T)$, $T \in \mathcal{T}_z$. The polynomial degrees are chosen to be $m_1 = n - 1$ and $m_2 = n - 2$ but can be general in this discussion.

Corollary 4.31 (local near-best approximation) guarantees that $P_{\mathcal{T}} \ell|_{\omega_z}$ is the quasi-best discrete functional in $\mathbb{F}(\mathcal{T}_z)$, namely

$$\|\ell - P_{\mathcal{T}} \ell\|_{H^{-1}(\omega_z)} \le C_P \inf_{\chi \in \mathbb{F}(\mathcal{T}_z)} \|\ell - \chi\|_{H^{-1}(\omega_z)}. \tag{7.28}$$

This will be used repeatedly to replace $P_{\mathcal{T}}\ell$ with more tractable quantities and justify the use of GREEDY algorithms to reduce (7.26).

7.3.1. The case $f \in L^q(\Omega)$

In this section we show how to reduce the oscillation error (7.26) when $f \in L^q(\Omega)$, with $q > 2d/(d+2)$ to guarantee that $L^q(\Omega)$ compactly embeds in $H^{-1}(\Omega)$. Note that this not only includes the most treated case in the literature $f \in L^2(\Omega)$ but also the more intricate cases $q < 2$ originally analysed by Cohen *et al.* (2012).

If $\Pi_{\mathcal{T}} f$ is the L^2-projection of f into the space $\mathbb{S}^{n_f,-1}_{\mathcal{T}}$ of discontinuous piecewise polynomials of degree n_f, let $\widehat{f} \in \mathbb{S}^{n_f,-1}_{\mathcal{T}}$ be defined by (5.70); $n_f = n-1$ in some applications but not always. Since $\widehat{f}|_{\omega_z} \in \mathbb{F}(\mathcal{T}_z)$ by taking $q_F = 0$ and $q_T = \widehat{f}|_T$ in (7.27), the local near-best approximation property (7.28) of $P_{\mathcal{T}}$ implies

$$\|f - P_{\mathcal{T}} f\|_{H^{-1}(\omega_z)} \leq C_P \|f - \widehat{f}\|_{H^{-1}(\omega_z)}.$$

Furthermore, for $v \in H^1_0(\omega_z)$ we have

$$\langle f - \widehat{f}, v\rangle \leq \|f - \widehat{f}\|_{L^q(\omega_z)} \|v\|_{L^{\widetilde{q}}(\omega_z)}$$

where $1/q + 1/\widetilde{q} = 1$. Note that the restriction $q > 2d/(d+2)$ guarantees that $1 \leq \widetilde{q} < 2d/(d-2)$ and thus $\mathrm{sob}(H^1) > \mathrm{sob}(L^{\widetilde{q}})$. Therefore Lemma 2.2 (first Poincaré inequality) yields

$$\|f - \widehat{f}\|_{H^{-1}(\omega_z)} \lesssim \mathrm{diam}(\omega_z)^{1+d(1/2-1/q)} \|f - \widehat{f}\|_{L^q(\omega_z)}.$$

Returning to (7.26), after rearranging the terms element-wise and invoking the shape regularity of $\mathbb{T}$, we obtain $E_{\mathcal{T}}(f)_{-1} \leq C_{\mathrm{data}} \widetilde{E}_{\mathcal{T}}(f)_{-1}$, where

$$\widetilde{E}_{\mathcal{T}}(f)^2_{-1} := \sum_{T \in \mathcal{T}} \widetilde{\mathrm{osc}}_{\mathcal{T}}(f,T)^2_q, \tag{7.29}$$

and $\widetilde{\mathrm{osc}}_{\mathcal{T}}(f,T)_q := h^t_T \|f - \widehat{f}\|_{L^q(T)}$ with $t := 1 + d(1/2 - 1/q) > 0$.

In view of the definition (5.70), the local oscillations $\widetilde{\mathrm{osc}}(f,T)_q$ satisfy Assumption 7.3 (monotonicity of local oscillations) in ℓ^2 and we can now employ Algorithm 7.1 (GREEDY) with local errors $\widetilde{\mathrm{osc}}_{\mathcal{T}}(f,T)_q$ accumulating in ℓ^2. Recall that we use the convention $X^0_q(\Omega;\mathcal{T}_0) = L^q(\Omega)$.

Corollary 7.18 (approximation class of $f \in L^q(\Omega)$). *Let the initial subdivision $\mathcal{T}_0$ of $\Omega \subset \mathbb{R}^d$ satisfy Assumption 6.19 (initial labelling) and let $\mathcal{T} \in \mathbb{T}$ be any admissible refinement of $\mathcal{T}_0$. Let $\tau > 0$ be the target tolerance and let $b \geq 1$ be the number of bisections performed on each marked element. Let $2d/(d+2) < q \leq \infty$ and set $t = 1 + d(1/2 - 1/q)$. Let $(f,s,t,p,2)$ satisfy Assumption 7.2 (admissible set of parameters for* GREEDY*) with local oscillations $\{\widetilde{\mathrm{osc}}(f,T)_q\}_{T\in\mathcal{T}}$. Then $[\widehat{\mathcal{T}},\widehat{f}] =$ GREEDY$(\mathcal{T},\tau,2,b,f)$ terminates in a finite number of steps with $\widetilde{E}_{\widehat{\mathcal{T}}}(f)_{-1} \leq \tau$, whence*

$$E_{\widehat{\mathcal{T}}}(f)_{-1} \leq C_{\mathrm{data}}\tau.$$

Moreover, the number N of marked elements by GREEDY *satisfies*

$$N \lesssim |f|_{X_p^s(\Omega;\mathcal{T}_0)}^{d/(s+t)} \tau^{-d/(s+t)}. \tag{7.30}$$

In particular, $f \in \mathbb{F}_{d/(s+t)}$ *with* $|f|_{\mathbb{F}_{d/(s+t)}} \lesssim \|f\|_{X_p^s(\Omega;\mathcal{T}_0)}$.

Proof. Directly apply Corollary 7.6 (performance of GREEDY). □

7.3.2. The case $f = g\delta_{\mathcal{C}}$

We now consider the case where the right-hand side data f is a density supported on a Lipschitz hyper-surface $\mathcal{C} \subset \Omega$ in $\mathbb{R}^d$ with $(d-1)$-measure $|\mathcal{C}| < \infty$.

The intricate interactions between bulk and interface contributions on $P_{\mathcal{T}}$ makes it difficult to analyse when $f = g\delta_{\mathcal{C}}$ with density $g \in L^q(\mathcal{C})$. We take a simpler approach, likely suboptimal when $n > 1$ and $d > 2$, which discards $P_{\mathcal{T}}$ in view of the near-best approximation property (7.28):

$$\|f - P_{\mathcal{T}} f\|_{H^{-1}(\omega_z)} \lesssim \|f\|_{H^{-1}(\omega_z)}. \tag{7.31}$$

The right-hand side of the above estimate is the starting point of the analysis by Cohen *et al.* (2012) assuming $n = 1$ and $d = 2$.

We start with the derivation of a first upper bound for the local error $\|f\|_{H^{-1}(\omega_z)}$.

Lemma 7.19 (local oscillation). *Let $\mathcal{T} \in \mathbb{T}$, $z \in \mathcal{N}$, and $q > 2(d-1)/d$. If $g \in L^q(\mathcal{C})$ and $t := d/2 - (d-1)/q > 0$, then*

$$\|f\|_{H^{-1}(\omega_z)} \lesssim |\omega_z \cap \mathcal{C}|^{t/(d-1)} \|g\|_{L^q(\omega_z \cap \mathcal{C})} \lesssim \sum_{T \subset \omega_z} h_T^t \|g\|_{L^q(T \cap \mathcal{C})}. \tag{7.32}$$

Proof. For $v \in H_0^1(\omega_z)$ and $1/q + 1/\widetilde{q} = 1$, we have

$$\langle f, v\rangle = \int_{\omega_z \cap \mathcal{C}} gv \le \|g\|_{L^q(\omega_z \cap \mathcal{C})} \|v\|_{L^{\widetilde{q}}(\omega_z \cap \mathcal{C})}. \tag{7.33}$$

We realize that $H^{1/2}(\omega_z \cap \mathcal{C})$ compactly embeds in $L^{\widetilde{q}}(\omega_z \cap \mathcal{C})$ because

$$\begin{aligned} t &:= \operatorname{sob}(H^{1/2}(\omega_z \cap \mathcal{C})) - \operatorname{sob}(L^{\widetilde{q}}(\omega_z \cap \mathcal{C})) \\ &= \frac{1}{2} - (d-1)\left(\frac{1}{2} - \frac{1}{\widetilde{q}}\right) \\ &= \frac{d}{2} - \frac{1}{q}(d-1) > 0, \end{aligned}$$

provided $q > 2(d-1)/d$. Consequently, we find that

$$\|v\|_{L^{\widetilde{q}}(\omega_z \cap \mathcal{C})} \lesssim |\omega_z \cap \mathcal{C}|^{t/(d-1)} \|v\|_{H^{1/2}(\omega_z \cap \mathcal{C})}.$$

It remains to invoke the continuity (2.4) of the trace operator to write

$$\|v\|_{L^{\widetilde{q}}(\omega_z \cap \mathcal{C})} \lesssim |\omega_z \cap \mathcal{C}|^{t/(d-1)} \|v\|_{H^1(\omega_z)},$$

which, together with (7.33), yields the first estimate in (7.32). To deduce the second estimate, it suffices to note that $|\omega_z \cap \mathcal{C}| \lesssim \operatorname{diam}(\omega_z)^{d-1} \lesssim h_T^{d-1}$ for $T \subset \omega_z$ and that $\|g\|_{L^q(\omega_z\cap\mathcal{C})} \le \sum_{T\subset\omega_z} \|g\|_{L^q(T\cap\mathcal{C})}$. □

Estimate (7.31) and Lemma 7.19 provide a surrogate for data oscillation

$$\widetilde{E}_{\mathcal{T}}(f)_{-1}^2 := \sum_{T\in\mathcal{T}} \widetilde{\operatorname{osc}}_{\mathcal{T}}(g,T)_q^2, \quad \widetilde{\operatorname{osc}}_{\mathcal{T}}(g,T)_q := h_T^t \|g\|_{L^q(T\cap\mathcal{C})}, \tag{7.34}$$

where $t = d/2 - (d-1)/q$. The quantity $\widetilde{\operatorname{osc}}_{\mathcal{T}}(g,T)_q$ verifies Assumption 7.3 (monotonicity of local oscillations) with Ω replaced by $\mathcal{C}$ because of its element-wise structure. Therefore Proposition 7.4 (performance of GREEDY) states that Algorithm 7.1 (GREEDY) can reduce $\widetilde{E}_{\mathcal{T}}(f)_{-1}$. This is in contrast to the star-wise GREEDY algorithm analysed in Cohen *et al.* (2012), which requires that all marked stars are refined d times to ensure all the faces in the marked stars are refined.

We now discuss the performance of GREEDY with local indicators $\widetilde{\operatorname{osc}}_{\mathcal{T}}(g,T)_q$.

Lemma 7.20 (approximation class of $f = g\delta_{\mathcal{C}}$). *Let $\mathcal{C} \subset \Omega$ be a Lipschitz hypersurface. Let the initial subdivision $\mathcal{T}_0$ of $\Omega \subset \mathbb{R}^d$ satisfy Assumption 6.19 (initial labelling) and let $\mathcal{T} \in \mathbb{T}$ be any admissible refinement of $\mathcal{T}_0$. Let $\tau > 0$ be the target tolerance and let $b \ge 1$ be the number of bisections performed on each marked element, and $2(d-1)/d < q \le \infty$. Then $[\widehat{\mathcal{T}}, \widehat{f}] =$ GREEDY$(\mathcal{T}, \tau, 2, b, f)$ terminates in a finite number of steps with surrogate estimator $\widetilde{E}_{\widehat{\mathcal{T}}}(f)_{-1} \le \tau$ defined in (7.34), whence*

$$E_{\widehat{\mathcal{T}}}(f)_{-1} \le C_{\text{data}}\tau.$$

Moreover, the number N of marked elements by GREEDY satisfies

$$N \lesssim \|g\|_{L^q(\mathcal{C})}^{2(d-1)} \tau^{-2(d-1)}. \tag{7.35}$$

In particular, $f = g\delta_{\mathcal{C}} \in \mathbb{F}_{1/(2(d-1))}$ with $|f|_{\mathbb{F}_{1/(2(d-1))}} \lesssim \|g\|_{L^q(\mathcal{C})}$.

Proof. This proof mainly follows the proof of Proposition 7.4 (performance of GREEDY) but requires a few modifications to account for the geometry of the problem. Since in turn the proof of Proposition 7.4 describes modifications to the proof of Proposition 3.19 (abstract greedy error), we now provide a complete proof. We proceed in several steps. We first consider the call GREEDY$(\mathcal{T}_0, \tau, 2, 1, f)$ from $\mathcal{T}_0$ with one bisection $b = 1$ and accumulation in ℓ^2, and discuss the general call from $\mathcal{T}$ with $b \ge 1$ in the last step of this proof.

[1] *Termination.* Since h_T decreases monotonically to 0 with bisection, so does $\widetilde{\operatorname{osc}}_{\mathcal{T}}(g,T)_q$. Consequently, GREEDY terminates in a finite number $k \ge 1$ of iterations. Let $T_1, \dots T_k$ be the sequence of marked elements, with $\mathcal{M} = \{T_1, \dots, T_k\}$ and let $\mathcal{T}_1, \dots, \mathcal{T}_k$ be the sequence of refinements produced by GREEDY starting from $\mathcal{T}_0$. Upon termination, the surrogate error satisfies $\widetilde{E}_{\mathcal{T}_k}(f)_{-1} \le \tau$, whence $E_{\mathcal{T}_k}(f)_{-1} \le C_{\text{data}}\tau$.

[2] *Counting.* To estimate the cardinality of $\mathcal{T}_k$, we need to count $\#\mathcal{M}$. Set

$$\delta_i := \widetilde{\operatorname{osc}}_{\mathcal{T}_i}(g,T_i)_q, \quad 1 \le i \le k \quad \text{and} \quad \delta := \delta_{k-1}.$$

Then we obtain

$$\widetilde{E}_{\mathcal{T}_k}(f)_{-1} \le \tau < \widetilde{E}_{\mathcal{T}_{k-1}}(f)_{-1} \le \delta\,(\#\mathcal{T}_{k-1})^{1/2} \le \delta\,(\#\mathcal{T}_k)^{1/2}. \tag{7.36}$$

We organize the elements in $\mathcal{M}$ by size in such a way that allows for a counting argument. Let $\mathcal{P}_j$ be the set of elements T of $\mathcal{M}$ with size

$$2^{-(j+1)} \le |T| < 2^{-j} \quad \Rightarrow \quad 2^{-(j+1)/d} \le h_T < 2^{-j/d}.$$

We first observe that all the T in $\mathcal{P}_j$ are *disjoint*. This is because if $T_1, T_2 \in \mathcal{P}_j$ and $\mathring{T}_1 \cap \mathring{T}_2 \neq \emptyset$, then one of them is contained in the other, say $T_1 \subset T_2$, due to the bisection procedure which works in any dimension $d \ge 1$; see Section 3.5. Hence

$$|T_1| \le \frac{1}{2}\,|T_2|,$$

contradicting the definition of $\mathcal{P}_j$. On the one hand, this implies the first bound

$$2^{-(j+1)(d-1)/d}\,\#\mathcal{P}_j \lesssim |\mathcal{C}| \quad \Rightarrow \quad \#\mathcal{P}_j \lesssim |\mathcal{C}|\,2^{(j+1)(d-1)/d}, \tag{7.37}$$

where we used that $h_T^{d-1} \approx |\omega_T \cap \mathcal{C}|$ since $T \cap \mathcal{C} \neq \emptyset$ for all marked elements. Recall that ω_T stands for the patch of elements around T.

On the other hand, the monotonicity of the local error indicators $\widetilde{\operatorname{osc}}_{\mathcal{T}_i}(g,T)_q = h_T^t\,\|g\|_{L^q(T\cap\mathcal{C})}$ implies that REFINE does not increase $\widetilde{\operatorname{osc}}_{\mathcal{T}_i}(g,T)_q$ and thus

$$\delta \le \delta_i = \widetilde{\operatorname{osc}}_{\mathcal{T}_i}(g,T_i)_q, \qquad 1 \le i \le k-1,$$

where $t = d/2 - (d-1)/q$. In view of (7.34), if $T_i \in \mathcal{P}_j$, then we obtain

$$\delta \le \widetilde{\operatorname{osc}}_{\mathcal{T}_i}(g,T_i)_q \lesssim 2^{-jt/d}\|g\|_{L^q(T_i\cap\mathcal{C})}.$$

Therefore, accumulating these quantities in ℓ^q yields

$$\delta^q\,\#\mathcal{P}_j \lesssim 2^{-jtq/d}\|g\|_{L^q(\mathcal{C})}^q$$

and gives rise to the second bound

$$\#\mathcal{P}_j \lesssim \delta^{-q}\,2^{-jtq/d}\,\|g\|_{L^q(\mathcal{C})}^q. \tag{7.38}$$

[3] *Cardinality.* The two bounds for $\#\mathcal{P}$ in (7.37) and (7.38) are complementary. The first one is good for j small whereas the second is suitable for j large (think of $\delta \ll 1$). The crossover takes place for j_0 such that

$$2^{(j_0+1)(d-1)/d}|\mathcal{C}| \approx \delta^{-q}\,2^{-j_0tq/d}\|g\|_{L^q(\mathcal{C})}^q \quad \Rightarrow \quad 2^{j_0} \approx |\mathcal{C}|^{-2/q}\delta^{-2}\|g\|_{L^q(\mathcal{C})}^2,$$

upon using the expression for t. We now compute

$$k = \#\mathcal{M} = \sum_j \#\mathcal{P}_j \lesssim \sum_{j\le j_0} 2^{j(d-1)/d}|\mathcal{C}| + \delta^{-q}\,\|g\|_{L^q(\mathcal{C})}^q \sum_{j>j_0} 2^{-(tq/d)j}.$$

Since

$$\sum_{j\le j_0} 2^{j(d-1)/d} \approx 2^{j_0(d-1)/d}, \quad \sum_{j>j_0} (2^{-tq/d})^j \lesssim 2^{-tqj_0/d},$$

we can write

$$\#\mathcal{M} \lesssim |\mathcal{C}|^{1-2(d-1)/(qd)} \big(\delta^{-1}\|g\|_{L^q(\mathcal{C})}\big)^{2(d-1)/d}.$$

We finally apply Theorem 3.16 (complexity of REFINE), to arrive at

$$\#\mathcal{T}_k - \#\mathcal{T}_0 \lesssim \#\mathcal{M} \lesssim |\mathcal{C}|^{1-2(d-1)/(qd)} \big(\delta^{-1}\|g\|_{L^q(\mathcal{C})}\big)^{2(d-1)/d},$$

or equivalently

$$\delta \lesssim |\mathcal{C}|^{d/(d-1)-2/q}\|g\|_{L^q(\mathcal{C})}(\#\mathcal{T} - \#\mathcal{T}_0)^{-d/(2(d-1))}.$$

We deduce from (7.36) that

$$\tau \lesssim \delta(\#\mathcal{T})^{1/2} \lesssim |\mathcal{C}|^{d/(d-1)-2/q}\|g\|_{L^q(\mathcal{C})}(\#\mathcal{T} - \#\mathcal{T}_0)^{-d/(2(d-1))+1/2}$$

or equivalently

$$\#\mathcal{T}_k - \#\mathcal{T}_0 \lesssim \|g\|_{L^q(\mathcal{C})}^{2(d-1)} \tau^{-2(d-1)}, \tag{7.39}$$

From this we conclude that $f = g\delta_{\mathcal{C}} \in \mathbb{F}_{1/(2(d-1))}$ with $|f|_{\mathbb{F}_{1/(2(d-1))}} \lesssim \|g\|_{L^q(\mathcal{C})}$, as desired.

[4] *Starting from $\mathcal{T}$.* To derive similar properties for GREEDY starting from $\mathcal{T} \in \mathbb{T}$, we proceed as in the proof of Corollary 7.6 (performance of GREEDY). We distinguish the output $[\widetilde{\mathcal{T}}, \widetilde{f}] = \mathsf{GREEDY}(\mathcal{T}_0, \tau, 2, 1, f)$ starting from $\mathcal{T}_0$ and performing $b = 1$ bisection per marked element with $[\widehat{\mathcal{T}}, \widehat{f}] = \mathsf{GREEDY}(\mathcal{T}, \tau, 2, b, f)$ starting from $\mathcal{T} \in \mathbb{T}$ and performing $b \geq 1$ bisections per marked element. Lemma 7.5 (GREEDY starting from $\mathcal{T}$) guarantees that $\mathsf{GREEDY}(\mathcal{T}, \tau, 2, b, f)$ terminates with $\widetilde{E}_{\widehat{\mathcal{T}}}(f)_{-1} \leq \tau$. Moreover, Lemma 7.5 also ensures that the number of marked elements satisfies

$$N \leq \#\widetilde{\mathcal{T}} - \#\mathcal{T}_0 \lesssim \|g\|_{L^q(\mathcal{C})}^{2(d-1)} \tau^{-2(d-1)},$$

where we used (7.39) to derive the last inequality. This ends the proof. □

7.3.3. The case $f = \operatorname{div} \boldsymbol{g}$ with $\boldsymbol{g} \in L^2(\Omega; \mathbb{R}^d)$

A characterization of distributions in $H^{-1}(\Omega)$ is given in Evans (2010, Section 5.9.1): they are of the form

$$f = f_0 + \operatorname{div} \boldsymbol{g}$$

with $f_0 \in L^2(\Omega), \boldsymbol{g} \in L^2(\Omega; \mathbb{R}^d)$. Since we have already treated separately the ubiquitous case $\boldsymbol{g} = \mathbf{0}$ in Section 7.3.1, we now consider the case $f_0 = 0$. Therefore

$$\langle f, v \rangle = -\int_\Omega \boldsymbol{g} \cdot \nabla v \quad \text{for all } v \in H_0^1(\Omega) \tag{7.40}$$

gives the action of f on v, and its norm is (Evans 2010, Section 5.9.1)

$$\|f\|_{H^{-1}(\Omega)} = \inf\{\|\boldsymbol{g}\|_{L^2(\Omega)} \mid \boldsymbol{g} \in L^2(\Omega; \mathbb{R}^d) \text{ satisfies (7.40)}\}. \tag{7.41}$$

Since adding the curl of a smooth vector field to $\boldsymbol{g}$ does not change (7.40), we realize that the actual computation of (7.41) is problematic. We assume here that $\boldsymbol{g}$ is given and simply deal directly with $\boldsymbol{g}$, thereby exploiting the relation

$$\|f\|_{H^{-1}(\Omega)} \leq \|\boldsymbol{g}\|_{L^2(\Omega)}; \tag{7.42}$$

this leads to a surrogate estimator. We first approximate $\boldsymbol{g}$ by discontinuous piecewise polynomials of degree $n_f \leq n-1$, namely, we compute the L^2-projection $\boldsymbol{g}_{\mathcal{T}} = \Pi_{\mathcal{T}}\boldsymbol{g}$ onto $\left[\mathbb{S}_{\mathcal{T}}^{n_f,-1}\right]^d$, then we let $f_{\mathcal{T}} := \operatorname{div}\boldsymbol{g}_{\mathcal{T}} \in \mathbb{F}_{\mathcal{T}} \subset H^{-1}(\Omega)$ be the approximation of f:

$$\langle f_{\mathcal{T}}, v\rangle = -\sum_{T\in\mathcal{T}}\int_T \operatorname{div}\boldsymbol{g}_{\mathcal{T}} v - \sum_{F\in\mathcal{F}}\int_F [[\boldsymbol{g}_{\mathcal{T}}]]\cdot\boldsymbol{n}_F v \quad \text{for all } v\in H_0^1(\Omega).$$

We see that for $z\in\mathcal{V}$, $f_{\mathcal{T}}|_{\omega_z}$ has the form of a functional in $\mathbb{F}(\mathcal{T}_z)$ (see (7.27)) with $q_T = \operatorname{div}\boldsymbol{g}_{\mathcal{T}}|_T \in \mathbb{P}_{n_f-1}, q_F = [[\boldsymbol{g}_{\mathcal{T}}]]\cdot\boldsymbol{n}_F \in \mathbb{P}_{n_f}$ for all $T\in\mathcal{T}, F\in\mathcal{F}$, but with smaller polynomial degree than functions in $\mathbb{F}(\mathcal{T}_z)$. We next exploit the local near-best approximation (7.28) to replace $P_{\mathcal{T}}f$ by $f_{\mathcal{T}}$,

$$\|f - P_{\mathcal{T}}f\|_{H^{-1}(\omega_z)} \leq C_P\|f - f_{\mathcal{T}}\|_{H^{-1}(\omega_z)} \leq C_P\|\boldsymbol{g}-\boldsymbol{g}_{\mathcal{T}}\|_{L^2(\omega_z)}, \tag{7.43}$$

by virtue of (7.42) with Ω replaced by ω_z. This leads to the surrogate element-wise oscillation $\widetilde{\operatorname{osc}}_{\mathcal{T}}(\boldsymbol{g},T)_2 := \|\boldsymbol{g}-\boldsymbol{g}_{\mathcal{T}}\|_{L^2(T)}$, which satisfies Assumption 7.3 (monotonicity of local oscillations). We thus have the global surrogate

$$\widetilde{E}_{\mathcal{T}}(f)_{-1}^2 := \sum_{T\in\mathcal{T}}\widetilde{\operatorname{osc}}_{\mathcal{T}}(\boldsymbol{g},T)_2^2.$$

Corollary 7.21 (approximation class of $\operatorname{div}\boldsymbol{g}$**).** *Let the initial subdivision $\mathcal{T}_0$ of $\Omega\subset\mathbb{R}^d$ satisfy Assumption 6.19 (initial labelling) and let $\mathcal{T}\in\mathbb{T}$ be any admissible refinement of $\mathcal{T}_0$. Let $\tau>0$ be the target tolerance and let $b\geq 1$ be the number of bisections performed on each marked element. Let $(\boldsymbol{g},s,0,p,2)$ satisfy Assumption 7.2 (admissible set of parameters for* GREEDY*) with local oscillations $\{\widetilde{\operatorname{osc}}(\boldsymbol{g},T)_2\}_{T\in\mathcal{T}}$. Then $[\widehat{\mathcal{T}},\widehat{f}] =$ GREEDY$(\mathcal{T},\tau,2,b,f)$ terminates in a finite number of steps with $\widetilde{E}_{\widehat{\mathcal{T}}}(f)_{-1}\leq\tau$, whence $E_{\widehat{\mathcal{T}}}(f)_{-1}\leq C_{\text{data}}\tau$. Moreover, the number N of marked elements by* GREEDY *satisfies*

$$N \lesssim \|\boldsymbol{g}\|_{X_p^s(\Omega)}^{d/s}\tau^{-d/s}.$$

In particular, $f = \operatorname{div}\boldsymbol{g}\in\mathbb{F}_{s/d}$ with

$$|f|_{\mathbb{F}_{s/d}} \lesssim \|\boldsymbol{g}\|_{X_p^s(\Omega)}.$$

Proof. Apply Corollary 7.6 (performance of GREEDY) with $q=2$ to $\boldsymbol{g}$. □

7.4. DATA *module*

We now summarize in one single algorithm, called DATA, all the developments in Sections 7.2.2, 7.2.3 and 7.3. We first recall that Corollaries 7.12 (complexity of constrained GREEDY for $\boldsymbol{A}$) and 7.16 (complexity of constrained GREEDY for c) deliver piecewise polynomial approximations $(\widehat{\boldsymbol{A}}, \widehat{c})$ of the coefficients $(\boldsymbol{A}, c)$ over an admissible mesh $\widehat{\mathcal{T}}$ that satisfies both the global errors estimates

$$E_{\widehat{\mathcal{T}}}(\boldsymbol{A})_{q_A} \leq C_{\text{data}}\tau, \quad E_{\widehat{\mathcal{T}}}(c)_{q_c} \leq C_{\text{data}}\tau,$$

where $2 \leq q_A, q_c \leq \infty$ are the corresponding integrability indices, as well as the structural constraint (5.51).

The situation for the load f is more intricate due to the evaluation of the non-local norm $H^{-1}(\Omega)$, which requires further structure of f besides regularity. Section 7.3 provides three examples of practical significance that allow for computable surrogate errors $\widetilde{E}_{\mathcal{T}}(f)_{-1}$ larger than the desired oscillations $E_{\mathcal{T}}(f)_{-1}$. Since these examples have different requirements for the approximation procedure to work, we gather the salient structural points in the following assumption.

Assumption 7.22 (structure of f). Let (s_f, p_f) denote the additional regularity–integrability indices of f beyond the basic H^{-1}-regularity, which are required by Assumption 7.2 (admissible set of parameters for GREEDY). Let $|f|_{\widetilde{X}^{s_f}_{p_f}(\Omega;\mathcal{T}_0)}$ be a measure of piecewise regularity of f in $\mathcal{T}_0$ expressed below in terms of surrogates. Assume that exactly one of the following cases holds, and note that all accumulate local oscillations in ℓ^2.

- $f \in L^q(\Omega)$, with $2d/(d+2) < q \leq \infty$. Let $\widetilde{\operatorname{osc}}_{\mathcal{T}}(f, T)_q = h_T^{t_f} \|f - \widehat{f}\|_{L^q(T)}$ be the local oscillation with $t_f = 1 + d(1/2 - 1/q) \geq 0$ and $(f, s_f, t_f, p_f, 2)$ satisfy Assumption 7.2, and set $|f|_{\widetilde{X}^{s_f}_{p_f}(\Omega;\mathcal{T}_0)} := |f|_{X^{s_f}_{p_f}(\Omega;\mathcal{T}_0)}$.
- $f = g\delta_{\mathcal{C}}$ where $\mathcal{C} \subset \Omega$ is a Lipschitz hyper-surface and $g \in L^q(\mathcal{C})$ with $2(d-1)/d < q \leq \infty$. Let $\widetilde{\operatorname{osc}}_{\mathcal{T}}(g, T)_q = h_T^r \|g\|_{L^q(T\cap\mathcal{C})}$ be the local oscillation with $r = d/2 - (d-1)/q > 0$. Set $s_f = 0$, $t_f = d/(2(d-1))$, $p_f = q$, and $|f|_{\widetilde{X}^{s_f}_{p_f}(\Omega;\mathcal{T}_0)} := \|g\|_{L^q(\mathcal{C})}$.
- $f = \operatorname{div} \boldsymbol{g}$ with $\boldsymbol{g} \in L^2(\Omega; \mathbb{R}^d)$. Let $\widetilde{\operatorname{osc}}_{\mathcal{T}}(f, T)_2 = \|\boldsymbol{g} - \Pi_{\mathcal{T}}\boldsymbol{g}\|_{L^2(T)}$ be the local oscillation, $t_f = 0$, and $(\boldsymbol{g}, s_f, t_f, p_f, 2)$ satisfy Assumption 7.3, and set $|f|_{\widetilde{X}^{s_f}_{p_f}(\Omega;\mathcal{T}_0)} := \|\boldsymbol{g}\|_{X^{s_f}_{p_f}(\Omega;\mathcal{T}_0)}$.

In all these cases, GREEDY algorithms with tolerance $\tau > 0$ reduce the surrogate error $\widetilde{E}_{\mathcal{T}}(f)^2_{-1}$ and eventually guarantee that

$$E_{\mathcal{T}}(f)_{-1} \leq C_{\text{data}}\tau,$$

where $C_{\text{data}} \geq 1$ is the constant appearing in Corollary 7.18, Lemma 7.20 or Corollary 7.21 depending on Assumption 7.22 (structure of f).

Algorithm 7.23 (DATA). Given a tolerance $\tau > 0$ and an arbitrary conforming grid $\mathcal{T} \in \mathbb{T}$, not necessarily $\mathcal{T}_0$, DATA finds a conforming refinement $\widehat{\mathcal{T}} \geq \mathcal{T}$ of $\mathcal{T}$

and approximate data $\widehat{\mathcal{D}} = (\widehat{\boldsymbol{A}}, \widehat{c}, \widehat{f}) \in \mathbb{D}_{\widehat{\mathcal{T}}}$ over $\widehat{\mathcal{T}}$ such that

$$\|\mathcal{D} - \widehat{\mathcal{D}}\|_{D(\Omega)} = E_{\widehat{\mathcal{T}}}(A)_{q_A} + E_{\widehat{\mathcal{T}}}(c)_{q_c} + E_{\widehat{\mathcal{T}}}(f)_{-1} \le C_{\mathrm{data}}\tau.$$

$[\widehat{\mathcal{T}}, \widehat{\mathcal{D}}] = \mathsf{DATA}(\mathcal{T}, \tau, \mathcal{D})$
    $[\mathcal{T}_{\boldsymbol{A}}, \widetilde{\boldsymbol{A}}] = \mathsf{GREEDY}(\mathcal{T}, \tau/3, q_A, b, \boldsymbol{A})$
    $\widehat{\boldsymbol{A}} = \mathsf{CONSTRAINT\text{-}A}(\mathcal{T}_A, \alpha_1, \alpha_2, L, \widetilde{\boldsymbol{A}})$
    Set $\widehat{\alpha}_1 = \frac{1}{2}\alpha_1$ and $\widehat{\alpha}_2 = (1 + 4L)\alpha_2/2$
    $[\mathcal{T}_c, \widetilde{c}] = \mathsf{GREEDY}(\mathcal{T}_A, \tau/3, q_c, b, c)$
    $\widehat{c} = \mathsf{CONSTRAINT\text{-}c}(\mathcal{T}_c, \widehat{\alpha}_1, L, \widetilde{c})$
    $[\widehat{\mathcal{T}}, \widehat{f}] = \mathsf{GREEDY}(\mathcal{T}_c, \tau/3, 2, b, f)$
    return $\widehat{\mathcal{T}}, \widehat{\mathcal{D}}$

Note that DATA depends on the threshold parameter $L > 2$ used in CONSTRAINT-A and CONSTRAINT-c, although for simplicity it is not listed among the input parameters.

The next result summarizes the properties of DATA.

Corollary 7.24 (performance of DATA). *Let the initial subdivision $\mathcal{T}_0$ of $\Omega \subset \mathbb{R}^d$ satisfy Assumption 6.19 (initial labelling) and let $\mathcal{T} \in \mathbb{T}$ be any admissible refinement of $\mathcal{T}_0$. Let $b \ge 1$ be the number of bisections performed on each marked element. Let the assumptions of Corollaries 7.12 and 7.16 for the coefficients $(\boldsymbol{A}, c)$ be valid, and let f satisfy Assumption 7.22.*

For any target tolerance $\tau > 0$ and any threshold parameter $L > 2$, $[\widehat{\mathcal{T}}, \widehat{\mathcal{D}}] =$ DATA$(\mathcal{T}, \tau, \mathcal{D})$ terminates in a finite number of iterations and outputs $\widehat{\mathcal{D}}$, $\widehat{\mathcal{T}} \in \mathbb{T}$ such that $\widehat{\boldsymbol{A}}$ is symmetric and

$$\widehat{\boldsymbol{A}} \in M(\widehat{\alpha}_1, \widehat{\alpha}_2), \quad \widehat{c} \in R(\widehat{c}_1, \widehat{c}_2),$$

where $\widehat{\alpha}_1, \widehat{\alpha}_2$ are given by (7.21) while $\widehat{c}_1, \widehat{c}_2$ are given by (7.23). Moreover, there is a constant $C_{\mathrm{data}} \ge 1$ such that DATA terminates with

$$\|\mathcal{D} - \widehat{\mathcal{D}}\|_{D(\Omega)} \le C_{\mathrm{data}}\tau,$$

and the number N of elements marked to construct $\widehat{\mathcal{T}}$ satisfies

$$N \lesssim |\mathcal{D}|_{\mathbb{X}_{s_{\mathcal{D}}/d}}^{d/s_{\mathcal{D}}} \tau^{-d/s_{\mathcal{D}}}, \tag{7.44}$$

with $s_{\mathcal{D}} := \min\{s_A, s_c + t_c, s_f + t_f\}$, and

$$|\mathcal{D}|_{\mathbb{X}_{s_{\mathcal{D}}/d}} = \left(|\boldsymbol{A}|^{d/s_A}_{X^{s_A}_{p_A}(\Omega,\mathcal{T}_0)} + |c|^{d/s_c}_{X^{s_c}_{p_c}(\Omega,\mathcal{T}_0)} + |f|^{d/s_f}_{\widetilde{X}^{s_f}_{p_f}(\Omega,\mathcal{T}_0)} \right)^{s_{\mathcal{D}}/d}.$$

Proof. Since the local oscillations for $\boldsymbol{A}$ and c satisfy Assumption 7.3 (monotonicity of local oscillations), we deduce that global oscillations do not increase upon refinement, namely, for $\widehat{\mathcal{T}} \ge \mathcal{T}_c \ge \mathcal{T}_{\boldsymbol{A}}$,

$$E_{\widehat{\mathcal{T}}}(\boldsymbol{A})_{q_A} + E_{\widehat{\mathcal{T}}}(c)_{q_c} \le E_{\mathcal{T}_A}(\boldsymbol{A})_{q_A} + E_{\mathcal{T}_c}(c)_{q_c}.$$

In view of Corollaries 7.12 and 7.16, this in turn implies

$$E_{\widehat{\mathcal{T}}}(\boldsymbol{A})_{q_A} + E_{\widehat{\mathcal{T}}}(c)_{q_c} \le C_{\text{data}} \frac{2}{3}\tau.$$

For the load term f, we invoke Corollary 7.18, Lemma 7.20 or Corollary 7.21, depending on Assumption 7.22 (structure of f), to infer that

$$E_{\widehat{\mathcal{T}}}(f)_{-1} \le C_{\text{data}} \frac{1}{3}\tau.$$

Hence

$$\|\mathcal{D} - \widehat{\mathcal{D}}\|_{D(\Omega)} = E_{\widehat{\mathcal{T}}}(\boldsymbol{A})_{q_A} + E_{\widehat{\mathcal{T}}}(c)_{q_c} + E_{\widehat{\mathcal{T}}}(f)_{-1} \le C_{\text{data}}\tau$$

as desired. The complexity estimate (7.44) directly follows from the complexity estimates given in Corollaries 7.12 and 7.16 for $(\boldsymbol{A}, c)$, and Corollary 7.18, Lemma 7.20 or Corollary 7.21 for f depending on its structure. □

Similar ideas apply to approximate non-vanishing Dirichlet data or boundary flux conditions for Robin or Neumann problems, but we do not elaborate on this.

8. Mesh refinement: the bisection method

This section is devoted to the complexity analysis of REFINE for Λ-admissible triangulations. More precisely, we prove the existence of a constant $D > 0$ such that

$$\#\mathcal{T}_k - \#\mathcal{T}_0 \le D \sum_{j=0}^{k-1} \#\mathcal{M}_j, \quad k \ge 0.$$

This kind of result holds for conforming meshes ($\Lambda = 0$) and was stated in Theorem 3.16, and for non-conforming meshes ($\Lambda > 1$) as anticipated in Theorem 3.29. The results of Sections 8.1 and 8.2 are valid for $d = 2$ but the proofs of the cited theorems extend to $d > 2$. We refer to the survey by Nochetto *et al.* (2009) for a full discussion for $d \ge 2$.

8.1. *Conforming meshes*

8.1.1. *Chains and labelling for $d = 2$*

In order to study non-local effects of bisection for $d = 2$, we now introduce the concept of chain (Binev *et al.* 2004); this concept is inadequate for $d > 2$ (Nochetto *et al.* 2009, Stevenson 2008). Recall that $E(T)$ denotes the edge of T assigned for refinement. To each $T \in \mathcal{T}$ we associate the element $F(T) \in \mathcal{T}$ sharing the edge $E(T)$ if $E(T)$ is interior and $F(T) = \emptyset$ if $E(T)$ is on $\partial\Omega$. A *chain* $\mathcal{C}(T, \mathcal{T})$, with starting element $T \in \mathcal{T}$, is a sequence $\{T, F(T), \ldots, F^m(T)\}$ with no repetitions of elements and with

$$F^{m+1}(T) = F^k(T) \text{ for some } k \in \{0, \ldots, m-1\} \text{ or } F^{m+1}(T) = \emptyset;$$

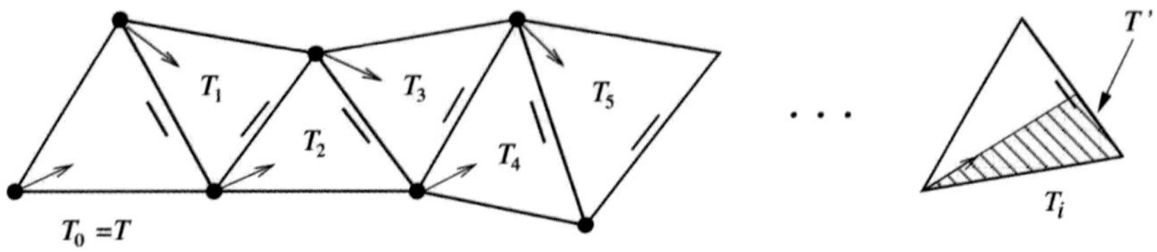

Figure 8.1. Typical chain $\mathcal{C}(T,\mathcal{T}) = \{T_j\}_{j=0}^{i}$ emanating from $T = T_0 \in \mathcal{T}$ with $T_j = F(T_{j-1})$, $j \geq 1$.

see Figure 8.1. We observe that if an element T belongs to two different grids, then the corresponding chains may be different as well. Two adjacent elements $T, T' = F(T)$ are *compatibly divisible* (or equivalently T, T' form a *compatible bisection patch*) if $F(T') = T$. Hence $\mathcal{C}(T,\mathcal{T}) = \{T, T'\}$, and a bisection of either T or T' does not propagate outside the patch.

Example (chains). Let $\mathcal{F} = \{T_i\}_{i=1}^{12}$ be the forest of Figure 3.5. Then $\mathcal{C}(T_6,\mathcal{T}) = \{T_6, T_7\}, \mathcal{C}(T_9,\mathcal{T}) = \{T_9\}$ and $\mathcal{C}(T_{10},\mathcal{T}) = \{T_{10}, T_8, T_2\}$ are chains, but only $\mathcal{C}(T_6,\mathcal{T})$ is a compatible bisection patch.

To study the structure of chains we rely on the initial labelling (3.35) and the bisection rule of Section 3.5 (see Figure 3.7):

> *Every triangle $T \in \mathcal{T}$ with generation $g(T) = i$ receives the label $(i+1, i+1, i)$ with i corresponding to the refinement edge $E(T)$, its side i is bisected and both new sides as well as the bisector are labelled $i+2$ whereas the remaining labels do not change.* (8.1)

We first show that once the initial labelling and bisection rule are set, the resulting master forest $\mathbb{F}$ is uniquely determined: the label of an edge is independent of the elements sharing this edge and no ambiguity arises in the recursion process.

Lemma 8.1 (labelling). *Let the initial labelling* (3.35) *for $\mathcal{T}_0$ and the above bisection rule be enforced. If $\mathcal{T}_0 \leq \mathcal{T}_1 \leq \cdots \leq \mathcal{T}_n$ are generated according to* (8.1), *then each side in $\mathcal{T}_k$ has a unique label independent of the two triangles sharing this edge.*

Proof. We argue by induction over $\mathcal{T}_k$. For $k = 0$ the assertion is valid due to the initial labelling. Suppose the statement is true for $\mathcal{T}_k$. An edge S in $\mathcal{T}_{k+1}$ can be obtained in two ways. The first is that S is a bisector and so a new edge, in which case there is nothing to prove about its label being unique. The second possibility is that S was obtained by bisecting an edge $S' \in \mathcal{S}_k$. Let T, $T' \in \mathcal{T}_k$ be the elements sharing S', and let us assume that $E(T') = S'$. Let $(i+1, i+1, i)$ be the label of T', which means that S is assigned the label $i+2$. By induction assumption over $\mathcal{T}_k$, the label of S' as an edge of T is also i. There are two possible cases for the label of T.

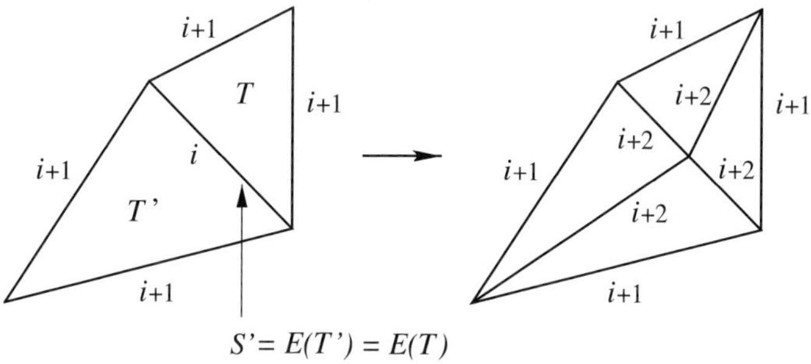

Figure 8.2. T and T' form a compatible patch, as they share the generation.

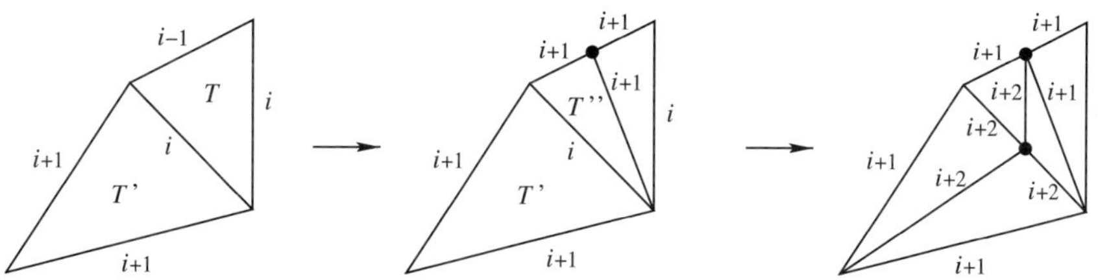

Figure 8.3. T' form a compatible patch with the child T'' of T; indeed T has a lower generation than T'.

- Label $(i+1, i+1, i)$: this situation is symmetric, $E(T) = S'$, and S' is bisected with both halves getting the label $i + 2$. This is depicted in Figure 8.2.
- Label $(i, i, i-1)$: a bisection of side $E(T)$ with label $i-1$ creates a child T'' with label $(i+1, i+1, i)$ that is compatibly divisible with T'. Joining the new node of T with the midpoint of S' creates a conforming partition with level $i+2$ assigned to S. This is depicted in Figure 8.3.

Therefore, in both cases the label $i + 2$ assigned to S is the same from both sides, as asserted. □

The two possible configurations displayed in the two figures above lead readily to the following statement about generations.

Corollary 8.2 (generation of consecutive elements). *For any $\mathcal{T} \in \mathbb{T}$ and $T, T' \in \mathcal{T}$ with $T = F(T')$, we have either*

(a) *$g(T) = g(T')$ and T, T' are compatibly divisible, or*

(b) *$g(T) = g(T') - 1$ and T' is compatibly divisible with a child of T.*

Corollary 8.3 (generations within a chain). *For all $\mathcal{T} \in \mathbb{T}$ and $T \in \mathcal{T}$, its chain $\mathcal{C}(T, \mathcal{T}) = \{T_k\}_{k=0}^{m}$ with $T_k = F^k(T)$ has the property*

$$g(T_k) = g(T) - k, \quad 0 \le k \le m - 1,$$

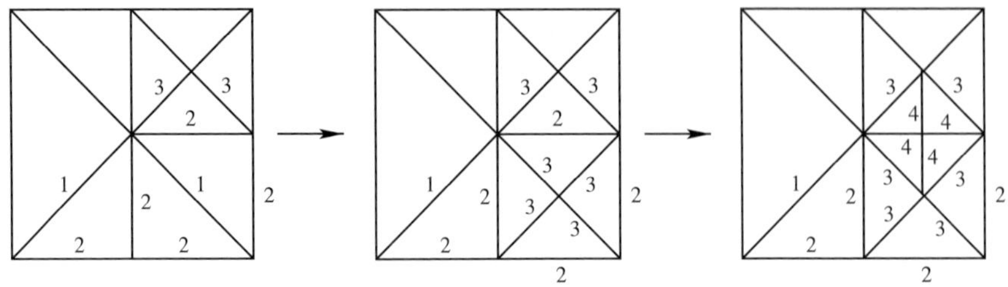

Figure 8.4. The recursive refinement of $T_{10} \in \mathcal{T}$ in Figure 3.4 using REFINE_RECURSIVE. This entails refining the chain $\mathcal{C}(T_{10}, \mathcal{T}) = \{T_{10}, T_8, T_2\}$, starting from the last element $T_2 \in \mathcal{T}$, which forms a compatible bisection patch on its own because its refinement edge is on the boundary, and continuing with $T_8 \in \mathcal{T}$ and finally $T_{10} \in \mathcal{T}$. Note that the successive meshes are always conforming and that REFINE_RECURSIVE bisects elements in $\mathcal{C}(T_{10}, \mathcal{T})$ twice before getting back to T_{10}.

and $T_m = F^m(T)$ has generation $g(T_m) = g(T_{m-1})$ or it is a boundary element with lowest labelled edge on $\partial\Omega$. In the first case, T_{m-1} and T_m are compatibly divisible.

Proof. Apply Corollary 8.2 repeatedly to consecutive elements of $\mathcal{C}(T, \mathcal{T})$. ☐

8.1.2. *Recursive bisection*

Given an element $T \in \mathcal{M}$ to be refined, the routine REFINE_RECURSIVE($\mathcal{T}, T$) recursively refines the chain $\mathcal{C}(T, \mathcal{T})$ of T, from the end back to T, and creates a minimal conforming partition $\mathcal{T}_* \geq \mathcal{T}$ such that T is bisected once. This procedure reads as follows:

```
[T*] = REFINE_RECURSIVE(T, T)
  if g(F(T)) < g(T)
     [T] = REFINE_RECURSIVE(T, F(T))
  else
     bisect the compatible bisection patch C(T, T)
     update T
  return T
```

We let $\mathcal{C}_*(T, \mathcal{T}) \subset \mathcal{T}_*$ denote the recursive refinement of $\mathcal{C}(T, \mathcal{T})$ (or completion of $\mathcal{C}(T, \mathcal{T})$) caused by bisection of T. Since REFINE_RECURSIVE refines solely compatible bisection patches, intermediate meshes are always conforming.

We refer to Figure 8.4 for an example of recursive bisection $\mathcal{C}_*(T_{10}, \mathcal{T})$ of $\mathcal{C}(T_{10}, \mathcal{T}) = \{T_{10}, T_8, T_2\}$ in Figure 3.4: REFINE_RECURSIVE starts bisecting from the end of $\mathcal{C}(T_{10}, \mathcal{T})$, namely T_2, which is a boundary element, and goes back the chain bisecting elements twice until it gets to T_{10}.

We now establish a fundamental property of REFINE_RECURSIVE($\mathcal{T}, T$) relating the generation of elements within $\mathcal{C}_*(T, \mathcal{T})$ (Binev *et al.* 2004).

Lemma 8.4 (recursive refinement). *Let $\mathcal{T}_0$ satisfy the labelling* (3.35), *and let $\mathcal{T} \in \mathbb{T}$ be a conforming refinement of $\mathcal{T}_0$. A call to* REFINE_RECURSIVE($\mathcal{T}, T$) *terminates, for all T in the set $\mathcal{M}$ of marked elements, and outputs the smallest conforming refinement $\mathcal{T}_*$ of $\mathcal{T}$ such that T is bisected. In addition, all newly created $T' \in \mathcal{C}_*(T, \mathcal{T})$ satisfy*

$$g(T') \leq g(T) + 1. \tag{8.2}$$

Proof. We first observe that T has maximal generation within $\mathcal{C}(T, \mathcal{T})$. So recursion is applied to elements with generation $\leq g(T)$, whence the recursion terminates. We also note that this procedure creates children of T and either children or grandchildren of triangles $T_k \in \mathcal{C}(T, \mathcal{T}) = \{T_i\}_{i=0}^m$ with $k \geq 1$. If T' is a child of T there is nothing to prove. If not, we first consider $m = 1$, in which case T' is a child of T_1 because T_0 and T_1 are compatibly divisible and so have the same generation; thus $g(T') = g(T_1) + 1 = g(T_0) + 1$. Finally, if $m > 1$, then $g(T_k) < g(T)$ and we apply Corollary 8.3 to deduce

$$g(T') \leq g(T_k) + 2 \leq g(T) + 1,$$

as asserted. □

The following crucial lemma links generation and distance between T and $T' \in \mathcal{C}_*(T, \mathcal{T})$, the latter being defined as (Binev *et al.* 2004)

$$\operatorname{dist}(T', T) := \inf_{x' \in T', x \in T} |x' - x|.$$

Lemma 8.5 (distance and generation). *Let $T \in \mathcal{M}$. Any newly created $T' \in \mathcal{C}_*(T, \mathcal{T})$ by* REFINE_RECURSIVE($\mathcal{T}, T$) *satisfies*

$$\operatorname{dist}(T', T) \leq D_2 \frac{2}{\sqrt{2} - 1} 2^{-g(T')/2}, \tag{8.3}$$

where $D_2 > 0$ is the constant in (3.34).

Proof. Suppose $T' \subset T_i \in \mathcal{C}(T, \mathcal{T})$ has been created by subdividing T_i (see Figure 8.1). If $i \leq 1$ then $\operatorname{dist}(T', T) = 0$ and there is nothing to prove. If $i > 1$, then we observe that $\operatorname{dist}(T', T_{i-1}) = 0$, whence

$$\begin{aligned}\operatorname{dist}(T', T) &\leq \operatorname{dist}(T_{i-1}, T) + \operatorname{diam}(T_{i-1}) \leq \sum_{k=1}^{i-1} \operatorname{diam}(T_k)\\ &\leq D_2 \sum_{k=1}^{i-1} 2^{-g(T_k)/2} < D_2 \frac{1}{1 - 2^{-1/2}} 2^{-g(T_{i-1})/2},\end{aligned}$$

because the generations decrease exactly by 1 along the chain $\mathcal{C}(T)$ according to Corollary 8.2(b). Since T' is a child or grandchild of T_i, we deduce

$$g(T') \leq g(T_i) + 2 = g(T_{i-1}) + 1,$$

whence

$$\operatorname{dist}(T', T) < D_2 \frac{2^{1/2}}{1 - 2^{-1/2}} 2^{-g(T')/2}.$$

This is the desired estimate. □

The recursive procedure REFINE_RECURSIVE is the core of the routine REFINE of Section 3.5: given a conforming mesh $\mathcal{T} \in \mathbb{T}$ and a subset $\mathcal{M} \subset \mathcal{T}$ of marked elements, REFINE creates a conforming refinement $\mathcal{T}_* \geq \mathcal{T}$ of $\mathcal{T}$ such that all elements of $\mathcal{M}$ are bisected at least once:

$[\mathcal{T}_*]$ = REFINE($\mathcal{T}, \mathcal{M}$)
  for all $T \in \mathcal{M} \cap \mathcal{T}$ do
    $[\mathcal{T}]$ = REFINE_RECURSIVE($\mathcal{T}, T$)
  return $\mathcal{T}$

It may happen that an element $T' \in \mathcal{M}$ is scheduled prior to T for refinement and $T \in \mathcal{C}(T', \mathcal{T})$. Since the call REFINE_RECURSIVE($\mathcal{T}, T'$) bisects T, its two children replace T in $\mathcal{T}$. This implies that $T \notin \mathcal{M} \cap \mathcal{T}$, which prevents further refinement of T.

In practice, we often like to bisect selected elements several times: for instance, each marked element is scheduled for $b \geq 1$ bisections. This can be done by assigning the number $b(T) = b$ of bisections that have to be executed for each marked element T. If T is bisected then we assign $b(T) - 1$ as the number of pending bisections to its children and the set of marked elements is $\mathcal{M} := \{T \in \mathcal{T} \mid b(T) > 0\}$.

8.1.3. *Complexity of bisection for conforming meshes*

Figure 8.4 reveals that the issue of propagation of mesh refinement to keep conformity is rather delicate. In particular, an estimate of the form

$$\#\mathcal{T}_k - \#\mathcal{T}_{k-1} \leq C\, \#\mathcal{M}_{k-1}$$

is not valid with a constant C independent of k; in fact the constant can be proportional to k according to Figure 8.4.

Binev, Dahmen and DeVore (2004) for $d = 2$ and Stevenson (2008) for $d > 2$ show that control of the propagation of refinement by bisection is possible when considering the collective effect:

$$\#\mathcal{T}_k - \#\mathcal{T}_0 \leq D \sum_{j=0}^{k-1} \#\mathcal{M}_j. \tag{8.4}$$

This can be heuristically motivated as follows. Consider the set $\mathcal{M} := \bigcup_{j=0}^{k-1} \mathcal{M}_j$ used to generate the sequence $\mathcal{T}_0 \leq \mathcal{T}_1 \leq \cdots \leq \mathcal{T}_k =: \mathcal{T}$. Suppose that each element $T_* \in \mathcal{M}$ is assigned a fixed amount C_1 of money to spend on refined elements in $\mathcal{T}$, i.e. on $T \in \mathcal{T} \setminus \mathcal{T}_0$. Assume further that $\lambda(T, T_*)$ is the portion of money spent by T_* on T. Then it must hold that

$$\sum_{T \in \mathcal{T} \setminus \mathcal{T}_0} \lambda(T, T_*) \leq C_1 \quad \text{for all } T_* \in \mathcal{M}. \tag{8.5a}$$

In addition, we suppose that the investment of all elements in $\mathcal{M}$ is fair in the sense that each $T \in \mathcal{T} \setminus \mathcal{T}_0$ gets at least a fixed amount C_2, whence

$$\sum_{T_* \in \mathcal{M}} \lambda(T, T_*) \geq C_2 \quad \text{for all } T \in \mathcal{T} \setminus \mathcal{T}_0. \tag{8.5b}$$

Therefore, summing up (8.5b) and using the upper bound (8.5a), we readily obtain

$$C_2(\#\mathcal{T} - \#\mathcal{T}_0) \leq \sum_{T \in \mathcal{T} \setminus \mathcal{T}_0} \sum_{T_* \in \mathcal{M}} \lambda(T, T_*) = \sum_{T_* \in \mathcal{M}} \sum_{T \in \mathcal{T} \setminus \mathcal{T}_0} \lambda(T, T_*) \leq C_1 \, \#\mathcal{M},$$

which proves (8.4) for $\mathcal{T}$ and $\mathcal{M}$. In the remainder of this section we design such an allocation function $\lambda \colon \mathcal{T} \times \mathcal{M} \to \mathbb{R}^+$ in several steps and prove that recurrent refinement by bisection yields (8.5) provided $\mathcal{T}_0$ satisfies (3.35), thereby establishing Theorem 3.16 (complexity of REFINE).

Construction of the allocation function. The function $\lambda(T, T_*)$ is defined with the help of two sequences $(a(\ell))_{\ell=-1}^{\infty}, (b(\ell))_{\ell=0}^{\infty} \subset \mathbb{R}^+$ of positive numbers satisfying

$$\sum_{\ell \geq -1} a(\ell) = A < \infty, \quad \sum_{\ell \geq 0} 2^{-\ell/2} \, b(\ell) = B < \infty, \quad \inf_{\ell \geq 1} b(\ell) \, a(\ell) = c_* > 0,$$

and $b(0) \geq 1$. Valid instances are $a(\ell) = (\ell + 2)^{-2}$ and $b(\ell) = 2^{\ell/3}$.

With these settings we are prepared to define $\lambda \colon \mathcal{T} \times \mathcal{M} \to \mathbb{R}^+$ by

$$\lambda(T, T_*) := \begin{cases} a(g(T_*) - g(T)), & \operatorname{dist}(T, T_*) < D_3 \, B \, 2^{-g(T)/d} \text{ and } g(T) \leq g(T_*) + 1, \\ 0, & \text{else}, \end{cases}$$

where $D_3 := D_2(1 + 2(\sqrt{2} - 1)^{-1})$. Therefore the investment of money by $T_* \in \mathcal{M}$ is restricted to cells T that are sufficiently close and are of generation $g(T) \leq g(T_*) + 1$. Only elements of these generations can be created during refinement of T_* according to Lemma 8.4. We stress that except for the definition of B, this construction is multidimensional, and we refer to Nochetto *et al.* (2009) and Stevenson (2008) for details.

The following lemma shows that the total amount of money spent by the allocation function $\lambda(T, T_*)$ per marked element T_* is bounded.

Lemma 8.6 (upper bound). *There exists a constant $C_1 > 0$ depending only on $\mathcal{T}_0$ such that λ satisfies* (8.5a), *that is,*

$$\sum_{T \in \mathcal{T} \setminus \mathcal{T}_0} \lambda(T, T_*) \leq C_1 \quad \textit{for all } T_* \in \mathcal{M}.$$

Proof. We proceed in two steps.

[1] Given $T_* \in \mathcal{M}$, we set $g_* = g(T_*)$ and we let $0 \le g \le g_* + 1$ be a generation of interest in the definition of λ. We claim that for such g the cardinality of the set

$$\mathcal{T}(T_*, g) = \{T \in \mathcal{T} \mid \operatorname{dist}(T, T_*) < D_3\, B\, 2^{-g/2} \text{ and } g(T) = g\}$$

is uniformly bounded, i.e. $\#\mathcal{T}(T_*, g) \le C$ with C depending solely on D_1, D_2, D_3, B.

From (3.34) we learn that $\operatorname{diam}(T_*) \le D_2 2^{-g_*/2} \le 2D_2 2^{-(g_*+1)/2} \le 2D_2 2^{-g/2}$ as well as $\operatorname{diam}(T) \le D_2 2^{-g/2}$ for any $T \in \mathcal{T}(T_*, g)$. Hence all elements of the set $\mathcal{T}(T_*, g)$ lie inside a ball centred at the barycentre of T_* with radius $(D_3 B + 3D_2)2^{-g/2}$. Again relying on (3.34), we thus conclude that

$$\#\mathcal{T}(T_*, g) D_1 2^{-g} \le \sum_{T \in \mathcal{T}(T_*, g)} |T| \le c(D_3 B + 3D_2)^2 2^{-g},$$

whence $\#\mathcal{T}(T_*, g) \le c\, D_1^{-1}\, (D_3 B + 3D_2)^2 =: C$.

[2] Accounting only for non-zero contributions $\lambda(T, T_*)$, we deduce

$$\sum_{T \in \mathcal{T} \setminus \mathcal{T}_0} \lambda(T, T_*) = \sum_{g=0}^{g_*+1} \sum_{T \in \mathcal{T}(T_*, g)} a(g_* - g) \le C \sum_{\ell=-1}^{\infty} a(\ell) = CA =: C_1,$$

which is the desired upper bound. □

The definition of λ also implies that each refined element receives a fixed amount of money. We show this next.

Lemma 8.7 (lower bound). *There exists a constant $C_2 > 0$ depending only on $\mathcal{T}_0$ such that λ satisfies* (8.5b), *that is,*

$$\sum_{T_* \in \mathcal{M}} \lambda(T, T_*) \ge C_2 \quad \text{for all } T \in \mathcal{T} \setminus \mathcal{T}_0.$$

Proof. We proceed in several steps.

[1] Fix an arbitrary $T_0 \in \mathcal{T} \setminus \mathcal{T}_0$. Then there is an iteration count $1 \le k_0 \le k$ such that $T_0 \in \mathcal{T}_{k_0}$ and $T_0 \notin \mathcal{T}_{k_0-1}$. Therefore there exists an $T_1 \in \mathcal{M}_{k_0-1} \subset \mathcal{M}$ such that T_0 is generated during REFINE_RECURSIVE($\mathcal{T}_{k_0-1}, T_1$). Iterating this process, we construct a sequence $\{T_j\}_{j=1}^J \subset \mathcal{M}$ with corresponding iteration counts $\{k_j\}_{j=1}^J$ such that T_j is created by REFINE_RECURSIVE($\mathcal{T}_{k_j-1}, T_{j+1}$). The sequence is finite since the iteration counts are strictly decreasing and thus $k_J = 0$ for some $J > 0$, or equivalently $T_J \in \mathcal{T}_0$.

Since T_j is created during refinement of T_{j+1}, we infer from (8.2) that

$$g(T_{j+1}) \ge g(T_j) - 1.$$

Accordingly, $g(T_{j+1})$ can decrease the previous value of $g(T_j)$ by at most 1. Since $g(T_J) = 0$, there exists a smallest value s such that $g(T_s) = g(T_0) - 1$. Note that for $j = 1, \dots, s$ we have $\lambda(T_0, T_j) > 0$ if $\operatorname{dist}(T_0, T_j) \le D_3 B g^{-g(T_0)/d}$.

2 We next estimate the distance $\operatorname{dist}(T_0, T_j)$. For $1 \le j \le s$ and $\ell \ge 0$ we define the set

$$\mathcal{T}(T_0, \ell, j) := \{T \in \{T_0, \dots, T_{j-1}\} \mid g(T) = g(T_0) + \ell\}$$

and denote its cardinality by $m(\ell, j)$. The triangle inequality combined with an induction argument yields

$$\begin{aligned}\operatorname{dist}(T_0, T_j) &\le \operatorname{dist}(T_0, T_1) + \operatorname{diam}(T_1) + \operatorname{dist}(T_1, T_j)\\ &\le \sum_{i=1}^{j} \operatorname{dist}(T_{i-1}, T_i) + \sum_{i=1}^{j-1} \operatorname{diam}(T_i).\end{aligned}$$

We apply (8.3) for the terms of the first sum and (3.34) for the terms of the second sum, to obtain

$$\begin{aligned}\operatorname{dist}(T_0, T_j) &< D_2 \frac{2}{\sqrt{2}-1} \sum_{i=1}^{j} 2^{-g(T_{i-1})/2} + D_2 \sum_{i=1}^{j-1} 2^{-g(T_i)/2}\\ &\le D_2 \left(1 + \frac{2}{\sqrt{2}-1}\right) \sum_{i=0}^{j-1} 2^{-g(T_i)/2}\\ &= D_3 \sum_{\ell=0}^{\infty} m(\ell, j)\, 2^{-(g(T_0)+\ell)/2}\\ &= D_3 2^{-g(T_0)/2} \sum_{\ell=0}^{\infty} m(\ell, j)\, 2^{-\ell/2}.\end{aligned}$$

To establish the lower bound we distinguish two cases depending on the size of $m(\ell, s)$. This is done next.

3 *Case 1: $m(\ell, s) \le b(\ell)$ for all $\ell \ge 0$.* From this we conclude

$$\operatorname{dist}(T_0, T_s) < D_3 2^{-g(T_0)/2} \sum_{\ell=0}^{\infty} b(\ell)\, 2^{-\ell/2} = D_3 B\, 2^{-g(T_0)/2},$$

and the definition of λ then readily implies

$$\sum_{T_* \in \mathcal{M}} \lambda(T_0, T_*) \ge \lambda(T_0, T_s) = a(g(T_s) - g(T_0)) = a(-1) > 0.$$

4 *Case 2: there exists $\ell \ge 0$ such that $m(\ell, s) > b(\ell)$.* For each of these ℓ there exists a smallest $j = j(\ell)$ such that $m(\ell, j(\ell)) > b(\ell)$. We let ℓ^* be the index ℓ that gives rise to the smallest $j(\ell)$, and set $j^* = j(\ell^*)$. Consequently

$$m(\ell, j^* - 1) \le b(\ell) \quad \text{for all } \ell \ge 0,\; m(\ell^*, j^*) > b(\ell^*).$$

As in Case 1, we see $\operatorname{dist}(T_0, T_i) < D_3 B\, 2^{-g(T_0)/2}$ for all $i \le j^* - 1$, or equivalently

$$\operatorname{dist}(T_0, T_i) < D_3 B\, 2^{-g(T_0)/2} \quad \text{for all } T_i \in \mathcal{T}(T_0, \ell, j^*).$$

We next show that the elements in $\mathcal{T}(T_0, \ell^*, j^*)$ spend enough money on T_0. We first consider $\ell^* = 0$ and note that $T_0 \in \mathcal{T}(T_0, 0, j^*)$. Since $m(0, j^*) > b(0) \ge 1$ we discover $j^* \ge 2$. Hence there is an $T_i \in \mathcal{T}(T_0, 0, j^*) \cap \mathcal{M}$, which yields the estimate

$$\sum_{T_* \in \mathcal{M}} \lambda(T_0, T_*) \ge \lambda(T_0, T_i) = a(g(T_i) - g(T_0)) = a(0) > 0.$$

For $\ell^* > 0$ we see that $T_0 \notin \mathcal{T}(T_0, \ell^*, j^*)$, whence $\mathcal{T}(T_0, \ell^*, j^*) \subset \mathcal{M}$. In addition, $\lambda(T_0, T_i) = a(\ell^*)$ for all $T_i \in \mathcal{T}(T_0, \ell^*, j^*)$. From this we conclude

$$\begin{aligned}\sum_{T_* \in \mathcal{M}} \lambda(T_0, T_*) \ge \sum_{T_* \in \mathcal{T}(T_0, \ell^*, j^*)} \lambda(T_0, T_*) &= m(\ell^*, j^*)\, a(\ell^*) \\ &> b(\ell^*)\, a(\ell^*) \ge \inf_{\ell \ge 1} b(\ell)\, a(\ell) = c_* > 0.\end{aligned}$$

5 In summary, we have proved the assertion, since for any $T_0 \in \mathcal{T} \setminus \mathcal{T}_0$

$$\sum_{T_* \in \mathcal{M}} \lambda(T_0, T_*) \ge \min\{a(-1), a(0), c_*\} =: C_2 > 0. \tag{8.6}$$

This completes the proof. □

Remark 8.8 (complexity with $b > 1$ bisections). To show the complexity estimate when REFINE performs $b > 1$ bisections, the set $\mathcal{M}_k$ is to be understood as a sequence of *single* bisections recorded in sets $\{\mathcal{M}_k(j)\}_{j=1}^{b}$, which belong to intermediate triangulations between $\mathcal{T}_k$ and $\mathcal{T}_{k+1}$ with $\#\mathcal{M}_k(j) \le 2^{j-1} \#\mathcal{M}_k$, $j = 1, \ldots, b$. Then we also obtain Theorem 3.16 because

$$\sum_{j=1}^{b} \#\mathcal{M}_k(j) \le \sum_{j-1}^{b} 2^{j-1} \#\mathcal{M}_k = (2^b - 1) \#\mathcal{M}_k.$$

In practice, it is customary to take $b = d$ (Siebert 2012).

8.2. *Non-conforming meshes*

In this subsection we consider two kinds of non-conforming meshes undergoing a refinement process: (a) quadrilateral meshes with at most one hanging node per edge ($\Lambda = 1$ in the definition of Λ-admissible meshes), and (b) triangular meshes having global index bounded by a fixed, but arbitrary $\Lambda > 1$.

8.2.1. *Complexity of bisection for non-conforming quadrilateral meshes*

We briefly examine the refinement process for quadrilaterals with one hanging node per edge, which gives rise to the so-called 1-*meshes*. The refinement of $T \in \mathcal{T}$

might affect four elements of $\mathcal{T}$ for $d = 2$ (or 2^d elements for any dimension $d \geq 2$), all contained in the *refinement patch* $R(T, \mathcal{T})$ of T in $\mathcal{T}$. The latter is defined as

$$R(T, \mathcal{T}) := \{T' \in \mathcal{T} \mid T' \text{ and } T \text{ share an edge and } g(T') \leq g(T)\},$$

and is called *compatible* provided $g(T') = g(T)$ for all $T' \in R(T, \mathcal{T})$. The generation gap between elements sharing an edge, in particular those in $R(T, \mathcal{T})$, is always ≤ 1 for 1-meshes, and is 0 if $R(T, \mathcal{T})$ is compatible. The element size satisfies

$$h_T = 2^{-g(T)} h_{T_0} \quad \text{for all } T \in \mathcal{T},$$

where $T_0 \in \mathcal{T}_0$ is the ancestor of T in the initial mesh $\mathcal{T}_0$. Lemma 3.15 is thus valid:

$$h_T < \bar{h}_T \leq D_2 2^{-g(T)} \quad \text{for all } T \in \mathcal{T}. \tag{8.7}$$

Given an element $T \in \mathcal{M}$ to be refined, the routine REFINE_RECURSIVE($\mathcal{T}, T$) refines $R(T, \mathcal{T})$ recursively in such a way that the intermediate meshes are always 1-meshes, and reads as follows:

```
[𝒯*] = REFINE_RECURSIVE(𝒯, T)
   if g = min{g(T''): T'' ∈ R(T, 𝒯)} < g(T)
      let T' ∈ R(T, 𝒯) satisfy g(T') = g
      [𝒯] = REFINE_RECURSIVE(𝒯, T')
   else
      subdivide T
      update 𝒯 upon replacing T with its children
   return 𝒯
```

The conditional prevents the generation gap within $R(T, \mathcal{T})$ from getting larger than 1. If it fails, then the refinement patch $R(T, \mathcal{T})$ is compatible, and refining T increases the generation gap from 0 to 1 without violating the 1-mesh structure. This implies a variant of Lemma 8.4: REFINE_RECURSIVE($\mathcal{T}, T$) creates a minimal 1-mesh $\mathcal{T}_* \geq \mathcal{T}$ refinement of $\mathcal{T}$ so that for all newly created elements $T' \in \mathcal{T}_*$,

$$g(T') \leq g(T) + 1 \tag{8.8}$$

and T is subdivided only *once*. This yields Lemma 8.5: there exist a geometric constant $D_g > 0$ such that for all newly created elements $T' \in \mathcal{T}_*$

$$\operatorname{dist}(T, T') \leq D_g 2^{g(T')}. \tag{8.9}$$

The procedure REFINE_RECURSIVE is the core of REFINE, which is conceptually identical to that in Section 8.1.2. Suppose that each marked element $T \in \mathcal{M}$ is to be subdivided $b \geq 1$ times. We assign a flag $q(T)$ to each element T which is initialized $q(T) = b$ if $T \in \mathcal{M}$ and $q(T) = 0$ otherwise. The marked set $\mathcal{M}$ is then the set of elements T with $q(T) > 0$, and every time T is subdivided it is removed from $\mathcal{T}$ and replaced by its children, which inherit the flag $q(T) - 1$. This

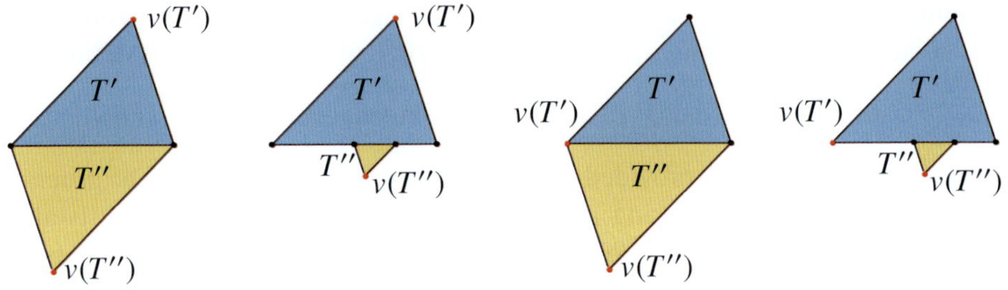

Figure 8.5. The elements T' and T'' are adjacent in cases A to D. They are compatible in cases A and B, and non-compatible in cases C and D.

avoids the conflict of again subdividing an element that has been previously refined by REFINE_RECURSIVE. The procedure REFINE($\mathcal{T}, \mathcal{M}$) reads

```
[T_*] = REFINE(T, M)
   for all T ∈ M ∩ T do
      [T] = REFINE_RECURSIVE(T, T);
   end
   return T
```

and its output is a minimal 1-mesh $\mathcal{T}_* \geq \mathcal{T}$ refinement of $\mathcal{T}$, so that all marked elements of $\mathcal{M}$ are refined at least b times. Since $\mathcal{T}_*$ has one hanging node per side it is thus admissible in the sense of (3.47). However, the refinement may spread outside $\mathcal{M}$ and the issue of complexity of REFINE again becomes non-trivial.

With the above ingredients in place, a statement similar to Theorem 3.16 (complexity of REFINE) for non-conforming quadrilateral meshes follows along the lines of Section 8.1.3.

8.2.2. *Complexity of bisection for Λ-admissible triangular meshes*

Let $\mathcal{T} \in \mathbb{T}^{\Lambda}$ be a Λ-admissible simplicial mesh. Given any $T \in \mathcal{T}$, let us again denote by $E(T)$ the edge of T assigned for refinement, i.e. the edge opposite to the newest vertex $v(T)$. Let $x(T)$ denote the midpoint of the edge $E(T)$.

Two elements $T', T'' \in \mathcal{T}$ are said to be *adjacent* if $E = T' \cap T''$ is an edge for at least one element, and are said to be *compatible* if they are adjacent and both $E(T')$ and $E(T'')$ belong to the same line (see Figure 8.5, cases A and B).

The following technical results will be helpful in the design of the refinement procedure.

Lemma 8.9 (global index of a hanging node). *Consider an edge $E = [x', x'']$ of the partition $\mathcal{T}$. If $x \in \mathcal{H} \cap \operatorname{int} E$ is generated by $m \geq 1$ bisections of E, then its global index $\lambda(x)$ satisfies*

$$\lambda(x) = \max(\lambda(x'), \lambda(x'')) + m.$$

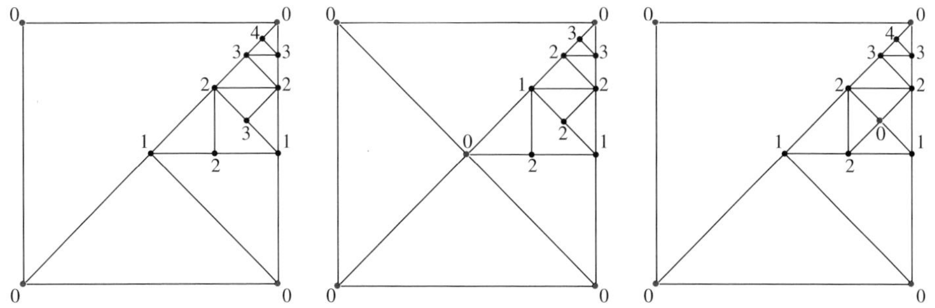

Figure 8.6. Three examples of distributions of proper nodes (red) and hanging nodes (black), with associated global indices λ. The bisection added in (b) converts the centre node into a proper node, and induces non-local changes of global indices on chains associated with it; if $\Lambda = 3$, then mesh (a) is not admissible and this procedure is instrumental in restoring admissibility. Mesh (c) illustrates the creation of a proper node without non-local effects on global indices.

Proof. If $m = 1$, then $x = x_M$ is the midpoint of E, and the formula is just Definition 3.24 (global index of a node). If $m > 1$, then x is generated by bisecting some interval $[z', z''] \subset E$, and $\lambda(x) = \max(\lambda(z'), \lambda(z'')) + 1$. Exactly one between z', z'' has been generated by $m - 1$ bisections, whereas the other one has been generated by less than $m - 1$ bisections. Hence we conclude by induction. □

Lemma 8.10 (reducing the global index of hanging nodes). *Let $\mathcal{H} \cap \operatorname{int} E$ contain at least the midpoint x_M of E. Assume that a bisection of some element in $\mathcal{T}$ transforms x_M into a proper node, and let λ_{new} denote the new global-index mapping of the nodes in $\mathcal{H} \cap \operatorname{int} E$ after the bisection. Then we have*

$$\lambda_{\text{new}}(x) \leq \lambda(x) - 1 \quad \textit{for all } x \in \mathcal{H} \cap \operatorname{int} E.$$

Proof. If $x = x_M$, then trivially $\lambda_{\text{new}}(x) = 0 \leq \lambda(x) - 1$. If $x \in \mathcal{H} \cap \operatorname{int} E$ is contained, say, in (x', x_M) and has been generated by $m > 1$ successive bisections of E, then it is generated by $m - 1$ successive bisections of $[x', x_M]$. Thus, by applying Lemma 8.9, we get

$$\begin{aligned}\lambda_{\text{new}}(x) &\leq \max(\lambda_{\text{new}}(x'), \lambda_{\text{new}}(x_M)) + m - 1 \\ &= \max(\lambda(x'), 0) + m - 1 = \lambda(x') + m - 1 \\ &\leq \max((\lambda(x'), \lambda(x'')) + m - 1 = \lambda(x) - 1.\end{aligned}$$

This gives the desired estimate. □

The result just established is the motivation for the proposed refinement strategy, introduced by Beirão da Veiga *et al.* (2024). Indeed, it ensures that in order to reduce the global index of a hanging node sitting on an edge, it is enough to transform the midpoint of the edge into a proper node. The situation is well represented in Figure 8.6.

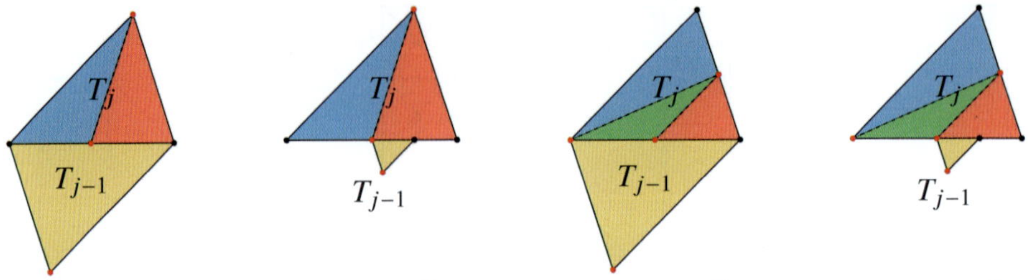

Figure 8.7. Two elements T_{j-1} and T_j in the chain $\mathcal{C}(\mathcal{T}, T)$: T_{j-1} can be bisected in a Λ-admissible way, only after T_j is refined once (cases A and B) or twice (cases C and D).

The following remark will be useful below.

Remark 8.11 (facing element). Given a Λ-admissible mesh $\mathcal{T}$ and $T \in \mathcal{T}$, let $x(T)$ be the midpoint of $E(T)$, and suppose that $\lambda(x(T)) > \Lambda$. Then $x(T)$ is not a node of $\mathcal{T}$, whence the edge $E(T)$ cannot contain any hanging node in its interior. We conclude that there exists a unique adjacent element $\widetilde{T} \in \mathcal{T}, \widetilde{T} \neq T$, such that $T \cap \widetilde{T} = E(T)$. This element will be called the element *facing* T, and denoted by $F(T)$.

Given an element $T \in \mathcal{T}$ which has been marked for refinement, we are ready to identify those elements in $\mathcal{T}$ that need be bisected with T in order to create a Λ-admissible refinement of $\mathcal{T}$. Figure 8.7 illustrates the possible situations.

Definition 8.12 (chain of elements to be refined). Define by recurrence the chain of elements starting at T,

$$\mathcal{C}(T, \mathcal{T}) = \{T_0, T_1, \dots, T_k\},$$

for some $k \geq 0$, as follows. First set $T_0 = T$. Assuming we have defined T_j for $j \geq 0$, then

(i) if $\lambda(x(T_j)) \leq \Lambda$, set $k = j$ and stop;

(ii) if $\lambda(x(T_j)) = \Lambda + 1$ and the facing element $F(T_j)$ is compatible with T_j, set $T_{j+1} = F(T_j)$, $k = j + 1$ and stop;

(iii) if $\lambda(x(T_j)) = \Lambda + 1$ and the facing element $F(T_j)$ is not compatible with T_j, set $T_{j+1} = F(T_j)$ and continue.

Lemma 8.13 (properties of the chain of refinement). *The chain $\mathcal{C}(T, \mathcal{T})$ has finite length; precisely, we have $k \leq g(T) + 1$, where $g(T)$ is the generation of T, defined in Section 3.5. Furthermore, the sequence of element generations $\{g(T_j)\}_{j=0}^k$ is not increasing.*

Proof. We claim that step (iii) in Definition 8.12 reduces the generation by at least one. In fact T_j coincides with or is a refinement of a triangle $\hat{T} \in \mathbb{T}$ sharing a full edge with T_{j+1}; thus $g(T_j) \geq g(\hat{T})$. Such a triangle $\hat{T}$ satisfies $g(\hat{T}) = g(T_{j+1}) + 1$, whence

$$g(T_{j+1}) = g(\hat{T}) - 1 \leq g(T_j) - 1. \tag{8.10}$$

Therefore, for as long as case (iii) is active, i.e. for all $j < k$, we have $g(T_j) \leq g(T_0) - j$ and

$$0 \leq g(T_{k-1}) \leq g(T_0) - (k-1),$$

which gives the first statement of the lemma. The monotonicity of $\{g(T_j)\}_{j=0}^k$ follows from (8.10) and the fact that $g(T_{k-1}) = g(T_k)$ in case (ii). □

Once the chain $\mathcal{C}(\mathcal{T}, T)$ is defined, all its elements are refined, starting from the last one and proceeding backwards. This is accomplished in the following procedure.

```
[𝒯*] = REFINE_RECURSIVE(𝒯, T, Λ)
   if λ(x(T)) ≤ Λ
      bisect T
      update 𝒯
   else if F(T) is compatible with T
      bisect F(T) and T
      update 𝒯
   else
      [𝒯] = REFINE_RECURSIVE(𝒯, F(T), Λ)
   return 𝒯
```

Proposition 8.14 (properties of REFINE_RECURSIVE). *If $\mathcal{T}$ is Λ-admissible, the call $[\mathcal{T}_*]$ = REFINE_RECURSIVE$(\mathcal{T}, T, \Lambda)$ outputs the smallest Λ-admissible refinement $\mathcal{T}_*$ of $\mathcal{T}$ such that T is bisected. In addition, every element $T' \in \mathcal{T}_*$ generated by this call satisfies*

$$g(T') \leq g(T) + 1. \tag{8.11}$$

Proof. Let $\mathcal{C}(T, \mathcal{T}) = \{T_j\}_{j=0}^k$ and observe that, for $j \geq 1$, one or two bisections of T_j convert the midpoint of the edge E of T_j shared with T_{j-1} into a proper node. Therefore Lemma 8.10 (reducing the global index of hanging nodes) implies that the global indices of all interior nodes to E decrease by at least 1, and makes the bisection of T_{j-1} Λ-admissible as desired.

To prove (8.11) we take $j \geq 1$ and consider the following two mutually exclusive cases. If T_j and T_{j-1} are compatible, then T_j is replaced by two elements $T' \in \mathcal{T}_*$ of generation

$$g(T') = g(T_j) + 1 \leq g(T) + 1,$$

according to Lemma 8.13 (properties of the chain of refinement). On the other hand, if T_j and T_{j-1} are not compatible, then T_j is replaced by one element of generation $g(T_j) + 1$ and two elements $T' \in \mathcal{T}_*$ of generation

$$g(T') = g(T_j) + 2 \le g(T_{j-1}) + 1 \le g(T) + 1$$

because of (8.10). Finally, the element $T_0 = T$ is replaced by two elements of generation $g(T) + 1$. □

If we consider the chains starting at any element $T \in \mathcal{M}$, we obtain the procedure REFINE($\mathcal{T}, \mathcal{M}, \Lambda$), which reads

```
[T*] = REFINE(T, M, Λ)
    for all T ∈ M ∩ T do
        [T] = REFINE_RECURSIVE(T, T, Λ)
    return T
```

and outputs a minimal Λ-admissible mesh $\mathcal{T}_* \ge \mathcal{T}$, refinement of $\mathcal{T}$, so that all marked elements of $\mathcal{M}$ are refined.

Proof of Theorem 3.29 (complexity of REFINE *for Λ-admissible meshes).* The arguments given in Section 8.1.3 for the conforming case can be easily adapted to the current situation. The two crucial properties needed are the relation (8.3) between the distance of two elements in a chain and their generation, which is valid for bisection grids regardless of Λ-admissibility, and the relation (8.11) between generations of elements. □

8.2.3. Mesh overlay and Λ-admissibility

Given two partitions $\mathcal{T}_A$ and $\mathcal{T}_B$, let $\mathcal{T}_A \oplus \mathcal{T}_B$ denote the *overlay* of $\mathcal{T}_A$ and $\mathcal{T}_B$, i.e. the partition whose associated tree is the union of the trees of $\mathcal{T}_A$ and $\mathcal{T}_B$. The following property holds.

Proposition 8.15 (mesh overlay is Λ-admissible). *If $\mathcal{T}_A$ and $\mathcal{T}_B$ are Λ-admissible, then $\mathcal{T}_A \oplus \mathcal{T}_B$ remains Λ-admissible.*

Proof. Let $\mathcal{N}$ denote the set of all nodes obtained by newest-vertex bisection from the root partition $\mathcal{T}_0$. Let $\mathcal{N}_0, \mathcal{N}_A, \mathcal{N}_B, \mathcal{N}_{A+B}$, respectively, be the set of nodes of the partitions $\mathcal{T}_0, \mathcal{T}_A, \mathcal{T}_B, \mathcal{T}_A \oplus \mathcal{T}_B$. It is easily seen that for each $x \in \mathcal{N} \setminus \mathcal{N}_0$ there exists a unique set $\mathcal{B}(x) = \{x', x''\} \subset \mathcal{N}$ such that x is generated by the bisection of the segment $[x', x'']$. Furthermore, if $x \in \mathcal{N}_{A+B}$ is a proper node of $\mathcal{T}_A$ (resp. $\mathcal{T}_B$), then it is also a proper node of $\mathcal{T}_A \oplus \mathcal{T}_B$.

Let $\lambda_A, \lambda_B, \lambda_{A+B}$, respectively, denote the global-index mappings defined on $\mathcal{N}_A, \mathcal{N}_B, \mathcal{N}_{A+B}$. It is convenient to extend the definition of λ_A and λ_B to the whole $\mathcal{N}_{A+B}$ by setting

$$\lambda_A(x) = +\infty \quad \text{if } x \in \mathcal{N}_{A+B} \setminus \mathcal{N}_A, \quad \lambda_B(x) = +\infty \quad \text{if } x \in \mathcal{N}_{A+B} \setminus \mathcal{N}_B.$$

With this notation at hand, we are going to prove the inequality

$$\lambda_{A+B}(x) \le \min(\lambda_A(x), \lambda_B(x)) \quad \text{for all } x \in \mathcal{N}_{A+B}, \tag{8.12}$$

from which the thesis immediately follows.

We proceed by induction on $k = \lambda_{A+B}(x)$, $x \in \mathcal{N}_{A+B}$. If $k = 0$, the inequality is trivial since $\lambda_A(x), \lambda_B(x) \geq 0$. So suppose (8.12) holds up to some $k \geq 0$. If $x \in \mathcal{N}_{A+B}$ satisfies $\lambda_{A+B}(x) = k + 1 > 0$, then it is a hanging node of $\mathcal{T}_A \oplus \mathcal{T}_B$ by the definition of global index, so it is a hanging node of $\mathcal{T}_A$ or $\mathcal{T}_B$; without loss of generality, suppose it is a hanging node of $\mathcal{T}_A$. If x is generated by the bisection of the segment $[x', x'']$, then again by the definition of global index it holds that

$$k + 1 = \lambda_{A+B}(x) = \max(\lambda_{A+B}(x'), \lambda_{A+B}(x'')) + 1,$$

which implies

$$\lambda_{A+B}(x') \leq k, \quad \lambda_{A+B}(x'') \leq k.$$

By induction,

$$\lambda_{A+B}(x') \leq \min(\lambda_A(x'), \lambda_B(x')), \quad \lambda_{A+B}(x'') \leq \min(\lambda_A(x''), \lambda_B(x'')),$$

from which we obtain

$$\lambda_{A+B}(x) \leq \max(\lambda_A(x'), \lambda_A(x'')) + 1 = \lambda_A(x)$$

since x is a hanging node of $\mathcal{T}_A$. On the other hand, either $x \in \mathcal{N}_B$ or $x \notin \mathcal{N}_B$. In the latter case $\lambda_B(x) = +\infty$, and (8.12) is proved. In the former case, necessarily x is a hanging node of $\mathcal{T}_B$, hence as above

$$\lambda_{A+B}(x) \leq \max(\lambda_B(x'), \lambda_B(x'')) + 1 = \lambda_B(x),$$

and the thesis is proved. □

9. Discontinuous Galerkin methods

So far we have studied conforming finite element approximations. In this section we present and analyse a two-step AFEM for discontinuous Galerkin methods (dG). The core PDE routine GALERKIN is thereby replaced by GALERKIN-DG, which hinges on the interior penalty discontinuous FEM. We regard dG as a prototype non-conforming method of practical importance and thus the natural first step to investigate the effects of non-conformity within adaptivity.

Finite element functions, being discontinuous, allow for non-conforming meshes to support them. We consider Λ-admissible subdivisions, according to Definition 3.25, where $\Lambda \geq 0$ restricts the level of non-conformity, and let $\mathbb{T}^\Lambda$ denote the collection of all Λ-admissible refinements of an initial subdivision $\mathcal{T}_0$; we refer to Section 8 for details. However, we further assume that $\mathcal{T}_0$ is conforming, to limit the level of technicalities.

There are several novel but characteristic aspects of dG. The most notable one is the appearance of jumps in its formulation, to compensate for the lack of H^1-conformity, as well as in the *a posteriori* upper bounds and the comparison of Galerkin solutions on different meshes. The lack of monotonicity of these jumps presents a formidable obstruction to the available proof techniques in adaptivity. However, we show in Lemma 9.11 that they are controlled by the residual estimator, thereby enabling us to loosely follow the roadmap of the conforming method, namely Sections 4, 5 and 6. Our approach is based on Bonito and Nochetto (2010) for the one-step AFEM.

The extra flexibility provided by non-conforming meshes, and corresponding discontinuous functions, does not yield a better asymptotic rate in H^1. An early manifestation of this fact, although written for conforming subdivisions, is Proposition 6.2 (equivalence of classes). We extend this result below for general Λ-admissible partitions.

One advantage of the two-step AFEM is that its design and analysis allows for $f \in H^{-1}(\Omega)$ without added difficulties: the function f is replaced by the discrete functional $\widehat{f} = P_{\mathcal{T}} f \in \mathbb{F}_{\mathcal{T}}$, which applies to functions in $\mathbb{S}^{n,-1}_{\mathcal{T}}$. This is in contrast to f, which cannot be applied to functions in $\mathbb{S}^{n,-1}_{\mathcal{T}}$. We exploit this property and thereby extend the applicability of dG to load functions in $H^{-1}(\Omega)$.

Our intention is to analyse the following algorithm for the approximation of the solution $u \in H^1_0(\Omega)$ to the coercive problem (2.7).

Algorithm 9.1 (AFEM-DG-TS). Given an initial tolerance $\varepsilon_0 > 0$, a target tolerance `tol` and initial mesh $\mathcal{T}_0$, as well as a safety parameter $\omega \in (0, 1]$, AFEM-DG-TS is a two-step algorithm alternating between the resolution of data $\mathcal{D}$ and the Galerkin solution $u_{\mathcal{T}}$:

$[\mathcal{T}, u_{\mathcal{T}}] = \text{AFEM-DG-TS}(\mathcal{T}_0, \varepsilon_0, \omega, \texttt{tol})$
 set $k = 0$
 do
  $[\widehat{\mathcal{T}}_k, \widehat{\mathcal{D}}_k] = \text{DATA}(\mathcal{T}_k, \mathcal{D}, \omega\, \varepsilon_k)$
  $[\mathcal{T}_{k+1}, u_{k+1}] = \text{GALERKIN-DG}(\widehat{\mathcal{T}}_k, \widehat{\mathcal{D}}_k, \varepsilon_k)$
  $\varepsilon_{k+1} = \frac{1}{2}\varepsilon_k$
  $k \leftarrow k + 1$
 while $\varepsilon_{k-1} > \texttt{tol}$
 return $\mathcal{T}_k, u_k$

In AFEM-DG-TS, the module DATA$(\mathcal{T}, \mathcal{D}, \tau)$ is the same as described in Section 5.4.2, except that it produces approximate data $\widehat{\mathcal{D}} \in \mathbb{D}_{\widehat{\mathcal{T}}}$, defined in (5.2), subordinate to a Λ-admissible refinement $\widehat{\mathcal{T}}$ of $\mathcal{T}_0$ for $\Lambda \geq 0$, rather than $\Lambda = 0$ (conforming). The discrete data $\widehat{\mathcal{D}}$ also satisfies the structural assumption (5.51) as discussed in Section 7. It is worth pointing out that the projection $P_{\mathcal{T}}$ used to approximate the right-hand side $f \in H^{-1}(\Omega)$ as well as all the results and algorithms presented in Section 7.4 are restricted to conforming subdivisions $\mathbb{T}$. We briefly discuss in Section 9.7 the extension of $P_{\mathcal{T}}$ and DATA to Λ-admissible subdivisions. Algorithm 9.17 describes the module GALERKIN-DG, the counterpart of GALERKIN for dG formulations.

In Section 9.1 we introduce notation and tools relevant for the characterization of discontinuous finite elements. Among them is the operator $\mathcal{I}_{\mathcal{T}}^{\mathrm{dG}}$, which projects piecewise polynomial functions onto globally continuous piecewise polynomial functions. It is instrumental in deriving a Poincaré inequality on the discontinuous spaces and guarantees that the approximation classes $\mathbb{A}_s^{-1}$ for the solution u using discontinuous approximation on Λ-admissible subdivisions are equivalent to their conforming counterparts $\mathbb{A}_s^0$ introduced in Section 6. We present the discontinuous Galerkin method in Section 9.2. We start with the standard symmetric interior penalty, discuss its drawbacks regarding the unnecessary regularity beyond $H_0^1(\Omega)$ imposed on the exact solution u, and describe a reformulation valid in $H_0^1(\Omega)$. The latter suffers from lack of consistency that needs to be accounted for. The *a posteriori* estimates for the perturbed problem (5.5) are derived in Section 9.3. Because the data is polynomial within GALERKIN-DG, the *a posteriori* estimators are oscillation-free. The GALERKIN-DG module is analysed in Section 9.4 while the discussion of rate-optimality of AFEM-DG-TS is reserved for Section 9.5.

9.1. *Discontinuous Galerkin setting*

We start with an initial conforming subdivision $\mathcal{T}_0$ made of simplices or hexahedra satisfying Assumption 6.19 (initial labelling). Given $\Lambda > 0$, the refinement procedure REFINE is designed to produce a Λ-admissible sequence of meshes $\mathbb{T}^\Lambda$ obeying Theorem 3.29 (complexity of REFINE for Λ-admissible meshes). From now on, we do not specify the dependence on Λ in the constants.

9.1.1. *Basic setting*

For $\mathcal{T} \in \mathbb{T}^\Lambda$, we let

$$\mathbb{V}_{\mathcal{T}}^{-1} := \mathbb{S}_{\mathcal{T}}^{n,-1} := \prod_{T \in \mathcal{T}} \mathbb{P}_n(T)$$

denote the space of piecewise polynomials of degree at most $n \geq 1$ subordinate to a partition $\mathcal{T}$. In contrast to the conforming spaces

$$\mathbb{V}_{\mathcal{T}}^{0} := \mathbb{S}_{\mathcal{T}}^{n,0} \cap H_0^1(\Omega)$$

considered earlier, the space $\mathbb{V}_{\mathcal{T}}^{-1}$ consists of (possibly) discontinuous functions across the elements $T \in \mathcal{T}$ and do not necessarily satisfy the vanishing boundary condition. Continuity across elements and vanishing boundary condition will be weakly imposed in the discontinuous Galerkin formulations.

We recall from Section 3.7 that for a proper (interior) node $P \in \mathcal{P}$, the *domain of influence* $\omega_{\mathcal{T}}(P) = \operatorname{supp}(\psi_P)$ is the support of the Lagrange basis function $\psi_P \in \mathbb{V}_{\mathcal{T}}^0$ associated with the node P; we refer to Figure 3.11. Since the sequence of meshes is Λ-admissible, Proposition 3.27 (size of the domain of influence) shows that the number of elements $T \in \mathcal{T}$ such that $T \subset \omega_{\mathcal{T}}(P)$ is uniformly bounded for $\mathcal{T} \in \mathbb{T}^\Lambda$.

The set of faces associated with a subdivision $\mathcal{T} \in \mathbb{T}^\Lambda$ is denoted $\mathcal{F}^+ := \mathcal{F}^+(\mathcal{T})$, and it contains boundary faces as well as interior faces. The set of interior faces is denoted $\mathcal{F}$. For a face $F \in \mathcal{F}^+$, we let $\{\{v\}\}|_F$ and $[[.]]|_F$ denote the average and jump operators across a face F. To define them precisely, we associate for each face $F \in \mathcal{F}^+$ one of the two unit normals $\boldsymbol{n}_F$. The choice of $\boldsymbol{n}_F$ is fixed but irrelevant as long as the outward pointing normal to Ω is chosen for boundary faces. Let $T_\pm \in \mathcal{T}$ be the elements that share the interior face F, namely $F = T_- \cap T_+$, and let $\mp\boldsymbol{n}_F$ be their outward pointing normals. Now, given $v \in \mathbb{V}_{\mathcal{T}}^{-1}$, let $v_\pm := v|_{T_\pm}$ and define for an interior face F

$$\{\{v\}\}|_F := \frac{1}{2}(v_- + v_+)|_F, \quad [[v]]|_F := (v_- - v_+)|_F, \tag{9.1}$$

By convention, we set $\{\{v\}\}|_F := v_-$ and $[[v]]|_F := v_-$ whenever F is a boundary face. These definitions extend readily to vector-valued functions.

We use the subscript $\mathcal{T}$ to denote the piecewise version of differential operators. For instance, the *broken gradient* $\nabla_{\mathcal{T}}$ is the piecewise gradient $\nabla_{\mathcal{T}} v|_{\mathring{T}} = \nabla v|_{\mathring{T}}$ for $T \in \mathcal{T}$ and $v \in \mathbb{V}_{\mathcal{T}}^{-1}$. For simplicity, we write

$$\|v\|_{L^2(\tau)}^2 := \sum_{T \in \tau} \|v\|_{L^2(T)}^2$$

for any subset $\tau \subset \mathcal{T}$ of elements, and

$$\|v\|_{L^2(\sigma)}^2 := \sum_{F \in \sigma} \|v\|_{L^2(F)}^2$$

for any subset $\sigma \subset \mathcal{F}^+$ of faces. We also define a mesh size function $h := h_{\mathcal{T}} : \overline{\Omega} \to (0, \infty)$ such that $h|_{\mathring{T}} \approx \operatorname{diam}(T)$ for $T \in \mathcal{T}$ and $h|_F \approx \operatorname{diam}(F)$ for $F \in \mathcal{F}^+$. With this notation at hand, the broken H^1 space

$$\mathbb{E}_{\mathcal{T}} := H^1(\Omega; \mathcal{T}) = \prod_{T \in \mathcal{T}} H^1(T)$$

is endowed with the mesh-dependent seminorm

$$\|v\|_{a,\mathcal{T}}^2 := \|\nabla_{\mathcal{T}} v\|_{L^2(\mathcal{T})}^2 + a\|h^{-1/2}[[v]]\|_{L^2(\mathcal{F}^+)}^2, \tag{9.2}$$

where a is some positive parameter. We will prove below that this is indeed a norm.

With this notation we can extend functionals $\widehat{f} \in \mathbb{F}_{\widehat{\mathcal{T}}}$ in Definition 4.17 to $\mathbb{V}_{\mathcal{T}}^{-1}$ for $\mathcal{T} \geq \widehat{\mathcal{T}}$. Before doing so, recall that for $\widehat{f} \in \mathbb{F}_{\widehat{\mathcal{T}}}$ and $v \in H_0^1(\Omega)$ we have

$$\langle \widehat{f}, v\rangle = \sum_{\widehat{T} \in \widehat{\mathcal{T}}} \int_{\widehat{T}} \widehat{f} v + \sum_{\widehat{F} \in \mathcal{F}(\widehat{\mathcal{T}})} \int_{\widehat{F}} \widehat{f} v,$$

where, compared to Definition 4.17, we slightly abused the notation

$$\widehat{f}|_{\widehat{T}} = \widehat{f}_T \in \mathbb{P}_{2n-2}(\widehat{T}) \quad \widehat{f}|_{\widehat{F}} = \widehat{f}_{\widehat{F}} \in \mathbb{P}_{2n-1}(\widehat{F}).$$

In view of this, we can extend the duality pairing to $\mathbb{V}_{\mathcal{T}}^{-1}$ by setting

$$\langle \widehat{f}, v\rangle_{\mathcal{T}} := \sum_{T \in \mathcal{T}} \int_T \widehat{f} v + \sum_{F \in \mathcal{F}} \int_F \widehat{f}\{\{v\}\} \tag{9.3}$$

so that consistency in $H_0^1(\Omega)$ is preserved, that is,

$$\langle \widehat{f}, v\rangle_{\mathcal{T}} = \langle \widehat{f}, v\rangle \quad \text{for all } v \in H_0^1(\Omega). \tag{9.4}$$

9.1.2. *Interpolation operator* $\mathcal{I}_{\mathcal{T}}^{\mathrm{dG}}$

We shall need the interpolation operator $\mathcal{I}_{\mathcal{T}}^{\mathrm{dG}} : \mathbb{E}_{\mathcal{T}} \to \mathbb{V}_{\mathcal{T}}^0$ from Bonito and Nochetto (2010). Its construction is based on an original idea of Clément (1975); see also Bernardi and Girault (1998) and other alternatives (Brenner 2003, Bonito, Nochetto and Ntogkas 2021).

Before embarking on the construction of $\mathcal{I}_{\mathcal{T}}^{\mathrm{dG}}$, we introduce some notation. For an interior or boundary proper node $P \in \mathcal{P}$ of the subdivision $\mathcal{T}$, we let

$$\mathbb{V}^0_{\omega_{\mathcal{T}}(P)} := H_0^1(\Omega) \cap \prod_{T \subset \omega_{\mathcal{T}}(P)} \mathbb{P}_n(T) \tag{9.5}$$

denote the space of continuous piecewise polynomial with support on the domain of influence $\omega_{\mathcal{T}}(P)$ of P and vanishing on $\partial\Omega$. When the underlying grid $\mathcal{T}$ is clear from the context, we will simplify the notation and write $\mathbb{V}_P^0 := \mathbb{V}^0_{\omega_{\mathcal{T}}(P)}$ and $\omega_P := \omega_{\mathcal{T}}(P)$; we refer to Figures 3.9 and 8.6.

We now construct $\mathcal{I}_{\mathcal{T}}^{\mathrm{dG}}$ in two steps. First, we define $V_P \in \mathbb{V}_P^0$ locally as satisfying

$$\int_{\omega_P} (v - V_P)w = 0 \quad \text{for all } w \in \mathbb{V}_P^0. \tag{9.6}$$

The value $V_P(P)$ is then used as the nodal value of $\mathcal{I}_{\mathcal{T}}^{\mathrm{dG}} v$, namely

$$\mathcal{I}_{\mathcal{T}}^{\mathrm{dG}} v := \sum_{P \in \mathcal{P}} V_P(P)\psi_P, \tag{9.7}$$

and we recall that $\{\psi_P\}_{P\in\mathcal{P}}$ is a basis of $\mathbb{V}_{\mathcal{T}}^0$ (see Section 3); note that $\mathcal{I}_{\mathcal{T}}^{\mathrm{dG}} v = 0$ on $\partial\Omega$ for all $v \in \mathbb{E}_{\mathcal{T}}$. Moreover, including boundary proper nodes in the definition (9.6)–(9.7) and replacing $H_0^1(\Omega)$ with $H^1(\Omega)$ in the definition (9.5), $\mathcal{I}_{\mathcal{T}}^{\mathrm{dG}}$ easily extends to $\mathbb{S}_{\mathcal{T}}^{n,0}$ without zero trace; we denote this operator by $\mathcal{I}_{\mathcal{T},+}^{\mathrm{dG}} : \mathbb{E}_{\mathcal{T}} \to \mathbb{S}_{\mathcal{T}}^{n,0}$. An immediate property of $\mathcal{I}_{\mathcal{T}}^{\mathrm{dG}}$ is *local invariance*,

$$v \in \mathbb{V}^0_{\omega_T} \quad \Rightarrow \quad v = \mathcal{I}_{\mathcal{T}}^{\mathrm{dG}} v \quad \text{in } T, \tag{9.8}$$

where $\omega_T := \bigcup\{\omega_P \mid P \in \mathcal{P}, T \subset \omega_P\}$; a similar property is valid for $\mathcal{I}_{\mathcal{T},+}^{\mathrm{dG}}$. We next gather a few more properties satisfied by $\mathcal{I}_{\mathcal{T}}^{\mathrm{dG}}$.

Lemma 9.2 (interpolation operator). *Let Assumption 6.19 (initial labelling) hold and let $\mathcal{T} \in \mathbb{T}^\Lambda$. For $v \in H_0^1(\Omega)$,*

$$\|v - \mathcal{I}_{\mathcal{T}}^{\mathrm{dG}} v\|_{L^2(T)} \lesssim \|h\nabla v\|_{L^2(\omega_T)}, \quad \|\nabla \mathcal{I}_{\mathcal{T}}^{\mathrm{dG}} v\|_{L^2(T)} \lesssim \|\nabla v\|_{L^2(\omega_T)}, \tag{9.9}$$

where ω_T is defined above. Instead, for $v \in \mathbb{E}_{\mathcal{T}}$,

$$\|v - \mathcal{I}_{\mathcal{T}}^{\mathrm{dG}} v\|_{L^2(T)} + \|h\nabla_{\mathcal{T}}(v - \mathcal{I}_{\mathcal{T}}^{\mathrm{dG}} v)\|_{L^2(T)} \\ \lesssim \|h^{1/2}[[v]]\|_{L^2(\mathcal{F}^+\cap\omega_T)} + \|h\nabla_{\mathcal{T}}(v - \Pi_{\mathcal{T}} v)\|_{L^2(\omega_T)}, \tag{9.10}$$

where $\Pi_{\mathcal{T}}$ is the L^2 projection operator onto $\mathbb{V}_{\mathcal{T}}^{-1} = \mathbb{S}_{\mathcal{T}}^{n,-1}$.

Proof. We start with (9.9) and let $v \in H_0^1(\Omega)$. The definition (9.6) of the local projection $V_P \in \mathbb{V}_P^0$ yields for all $P \in \mathcal{P}$

$$\|V_P\|_{L^2(\omega_P)} \le \|v\|_{L^2(\omega_P)} \quad \Rightarrow \quad \|V_P\|_{L^\infty(\omega_P)} \lesssim \operatorname{diam}(\omega_P)^{-d/2}\|v\|_{L^2(\omega_P)}.$$

Proposition 3.27 (size of the domain of influence) gives $\operatorname{diam}\omega_T \le Ch_T$, whence the number of ω_P containing T is uniformly bounded. Combining this with the definition (9.7) of $\mathcal{I}_{\mathcal{T}}^{\mathrm{dG}}$ implies

$$\|\mathcal{I}_{\mathcal{T}}^{\mathrm{dG}} v\|_{L^2(T)} \lesssim \sum_{P\in\mathcal{P}:\, T\subset\omega_P} |V_P(P)|\,\|\psi_P\|_{L^2(T)} \lesssim \|v\|_{L^2(\omega_T)} \quad \text{for all } T \in \mathcal{T}. \tag{9.11}$$

Since $\mathcal{I}_{\mathcal{T}}^{\mathrm{dG}}$ reproduces constants exactly locally, according to (9.8), the first relation in (9.9) follows from invoking the local L^2-stability property (9.11) together with Proposition 6.34 (Bramble–Hilbert for Sobolev spaces). The second relation is proved using the same arguments and an inverse inequality

$$\|\nabla\mathcal{I}_{\mathcal{T}}^{\mathrm{dG}} v\|_{L^2(T)} \lesssim h_T^{-1} \inf_{v_0\in\mathbb{R}} \|\mathcal{I}_{\mathcal{T}}^{\mathrm{dG}}(v - v_0)\|_{L^2(T)} \lesssim \|\nabla v\|_{L^2(\omega_T)}. \tag{9.12}$$

We now consider $v \in \mathbb{E}_{\mathcal{T}}$ and let $\widehat{v} = \Pi_{\mathcal{T}} v \in \mathbb{V}_{\mathcal{T}}^{-1}$. We intend to prove (9.10) by dealing with $v - \widehat{v}$ and $\widehat{v}$ separately and applying the triangle inequality. Since $v - \widehat{v}$ has zero mean in T according to (5.66), we apply Lemma 2.3 (second Poincaré inequality) to deduce

$$\|v - \widehat{v}\|_{L^2(T)} \lesssim h_T\,\|\nabla(v - \widehat{v})\|_{L^2(T)},$$

whence, combining an inverse estimate with (9.11), we further infer that

$$h_T\,\|\nabla\mathcal{I}_{\mathcal{T}}^{\mathrm{dG}}(v-\widehat{v})\|_{L^2(T)} \lesssim \|\mathcal{I}_{\mathcal{T}}^{\mathrm{dG}}(v-\widehat{v})\|_{L^2(T)} \lesssim \|v-\widehat{v}\|_{L^2(\omega_T)} \lesssim h_T\,\|\nabla(v-\widehat{v})\|_{L^2(\omega_T)}.$$

This argument yields the inequality (9.10) for $v - \widehat{v}$. It remains to deal with $\widehat{v}$.

We scale ω_T to a reference domain with unit diameter. Estimate (3.49) on the size of the domains of influence guarantees that the number of such reference patches is uniformly finite over $\mathbb{T}^\Lambda$. We relabel $\widehat{v}$ as v and examine the seminorm $\|\,[[v]]\,\|_{L^2(\mathcal{F}^+\cap\omega_P)}$ on the space of discontinuous piecewise polynomials

$$\{v \in \Pi_{T\subset\omega_P}\mathbb{P}_n(T)|\ V_P = 0\},$$

where V_P is defined by (9.6). If this seminorm vanishes then v is continuous in ω_P and thus $v \in \mathbb{V}_P^0$, whence the seminorm dominates any norm in this finite-dimensional space. Consequently, scaling back gives

$$\|v - V_P\|_{L^2(\omega_P)} + \|h\nabla_{\mathcal{T}}(v - V_P)\|_{L^2(\omega_P)} \lesssim \|h^{1/2}[[v]]\|_{L^2(\mathcal{F}^+\cap\omega_P)}. \tag{9.13}$$

We now deduce corresponding estimates for $\mathcal{I}_{\mathcal{T}}^{\mathrm{dG}}$. For $T \in \mathcal{T}$ and $P, Q \in \omega_T \cap \mathcal{P}$, (9.13) implies that

$$\|V_P - V_Q\|_{L^2(T)} + \|h\nabla_{\mathcal{T}}(V_P - V_Q)\|_{L^2(T)} \lesssim \|h^{1/2}[[v]]\|_{L^2(\mathcal{F}^+\cap\omega_T)}. \tag{9.14}$$

Consequently, the definition (9.7) of $\mathcal{I}_{\mathcal{T}}^{\mathrm{dG}}$ yields

$$v - \mathcal{I}_{\mathcal{T}}^{\mathrm{dG}} v = v - \sum_{P\in\omega_T\cap\mathcal{P}} V_P\psi_P = (v - V_Q) - \sum_{P\in\omega_T\cap\mathcal{P}} (V_P - V_Q)\psi_P,$$

which, combined with (9.13) and (9.14), implies

$$\|v - \mathcal{I}_{\mathcal{T}}^{\mathrm{dG}} v\|_{L^2(T)} + \|h\nabla_{\mathcal{T}}\big(v - \mathcal{I}_{\mathcal{T}}^{\mathrm{dG}} v\big)\|_{L^2(T)} \lesssim \|h^{1/2}[[v]]\|_{L^2(\mathcal{F}^+\cap\omega_T)}.$$

This is the desired estimate (9.10) for $v = \widehat{v} \in \mathbb{V}_{\mathcal{T}}^{-1}$. To finish the proof we still need to express the right-hand side of the last inequality in terms of $v \in \mathbb{E}_{\mathcal{T}}$. Applying the triangle inequality, we are left with estimating $\|[[v - \widehat{v}]]\|_{L^2(F)}$ for any $F \in \mathcal{F}^+ \cap \omega_T$. If $T_F \in \mathcal{T}$ is an element within ω_T that contains F in its boundary, we employ the scaled trace inequality to arrive at

$$h_T^{1/2}\|v - \widehat{v}\|_{L^2(F)} \lesssim h_T\|\nabla(v - \widehat{v})\|_{L^2(T_F)} + \|v - \widehat{v}\|_{L^2(T_F)} \lesssim h_T\|\nabla(v - \widehat{v})\|_{L^2(T_F)}.$$

Finally, collecting all the estimates completes the proof. □

We now discuss consequences of Lemma 9.2. The first one is that jumps are solely responsible for controlling the discrepancy between $v \in \mathbb{V}_{\mathcal{T}}^{-1}$ and $\mathcal{I}_{\mathcal{T}}^{\mathrm{dG}} v \in \mathbb{V}_{\mathcal{T}}^{0}$:

$$\|v - \mathcal{I}_{\mathcal{T}}^{\mathrm{dG}} v\|_{L^2(T)} + \|h\nabla_{\mathcal{T}}\big(v - \mathcal{I}_{\mathcal{T}}^{\mathrm{dG}} v\big)\|_{L^2(T)} \lesssim \|h^{1/2}[[v]]\|_{L^2(\mathcal{F}^+\cap\omega_T)}, \tag{9.15}$$

because $\Pi_{\mathcal{T}} v = v$ in ω_T. We next observe that (9.10) is also valid for $\mathcal{I}_{\mathcal{T},+}^{\mathrm{dG}}$ with the same proof. We can thus apply (9.10) for $\mathcal{I}_{\mathcal{T},+}^{\mathrm{dG}}$ to $w = v - \mathcal{I}_{\mathcal{T},+}^{\mathrm{dG}} v$, use the invariance of $\mathcal{I}_{\mathcal{T},+}^{\mathrm{dG}}$ in $\mathbb{S}_{\mathcal{T}}^{n,0}$, and its continuity across internal faces in $\mathcal{F}$, to deduce

$$\begin{aligned}\|v - \mathcal{I}_{\mathcal{T},+}^{\mathrm{dG}} v\|_{L^2(T)} + \|h\nabla_{\mathcal{T}}\big(v - \mathcal{I}_{\mathcal{T},+}^{\mathrm{dG}} v\big)\|_{L^2(T)} &\lesssim \|h\nabla_{\mathcal{T}}(v - \Pi_{\mathcal{T}} v)\|_{L^2(\omega_T)} \\ &+ \|h^{1/2}[[v]]\|_{L^2(\mathcal{F}\cap\omega_T)} + \|h^{1/2}\big(v - \mathcal{I}_{\mathcal{T},+}^{\mathrm{dG}} v\big)\|_{L^2(\partial\Omega\cap\omega_T)}.\end{aligned} \tag{9.16}$$

A third consequence of (9.10) is the following Poincaré-type inequality on $\mathbb{E}_{\mathcal{T}}$.

Lemma 9.3 (Poincaré-type inequality on $\mathbb{E}_{\mathcal{T}}$). *Let $\mathcal{T} \in \mathbb{T}^{\Lambda}$ be a Λ-admissible refinement of $\mathcal{T}_0$ satisfying Assumption 6.19 (initial labelling). There exists $C_P = C_P(\Omega, \mathcal{T}_0)$, such that, for all $v \in \mathbb{E}_{\mathcal{T}}$,*

$$\|v\|_{L^2(\Omega)} \le C_P\big(\|\nabla_{\mathcal{T}} v\|_{L^2\mathcal{T})} + \|h^{-1/2}[[v]]\|_{L^2(\mathcal{F})} + \|h^{-1/2} v\|_{L^2(\partial\Omega)}\big). \tag{9.17}$$

In particular, if $v = 0$ on $\partial\Omega$ then (9.17) *is a dG version of* (2.2).

Proof. We argue locally with (9.10). First we realize that an argument similar to (9.12) yields $\|\nabla_{\mathcal{T}}(v - \Pi_{\mathcal{T}} v)\|_{L^2(\omega_T)} \lesssim \|\nabla_{\mathcal{T}} v\|_{L^2(\omega_T)}$, whence adding over $T \in \mathcal{T}$ we obtain

$$\|v - \mathcal{I}_{\mathcal{T}}^{\mathrm{dG}} v\|_{L^2(\Omega)} + \|\nabla \mathcal{I}_{\mathcal{T}}^{\mathrm{dG}} v\|_{L^2(\Omega)} \lesssim \|\nabla_{\mathcal{T}} v\|_{L^2(\Omega)} + \|h^{-1/2}[[v]]\|_{L^2(\mathcal{F}^+)}.$$

It thus suffices to write

$$\|v\|_{L^2(\Omega)} \le \|\mathcal{I}_{\mathcal{T}}^{\mathrm{dG}} v\|_{L^2(\Omega)} + \|v - \mathcal{I}_{\mathcal{T}}^{\mathrm{dG}} v\|_{L^2(\Omega)},$$

and invoke (2.2) for $\mathcal{I}_{\mathcal{T}}^{\mathrm{dG}} v \in H_0^1(\Omega)$ together with the preceding inequality. □

Another important property obtained using the interpolation operator $\mathcal{I}_{\mathcal{T}}^{\mathrm{dG}}$ is that the approximation classes $\mathbb{A}_s^0 := \mathbb{A}_s(H_0^1(\Omega); \mathcal{T}_0)$ defined using globally continuous piecewise polynomial approximations of degree $\le n$ on conforming subdivisions are equivalent to those without global continuity on Λ-admissible subdivisions $\mathcal{T} \in \mathbb{T}^\Lambda$, provided $\|\cdot\|_{1,\mathcal{T}}$ (defined in (9.2)) is used as norm on $\mathbb{E}_{\mathcal{T}}$. We define

$$\sigma_N^{n,-1}(v) := \inf_{\mathcal{T} \in \mathbb{T}_N^\Lambda} \inf_{v_{\mathcal{T}} \in \mathbb{S}_{\mathcal{T}}^{n,-1}} \|v - v_{\mathcal{T}}\|_{1,\mathcal{T}} \tag{9.18}$$

and $\mathbb{A}_s^{-1} := \mathbb{A}_s^{-1}(H_0^1(\Omega); \mathcal{T}_0)$ to be the class of functions $v \in H_0^1(\Omega)$ such that

$$|v|_{\mathbb{A}_s^{-1}} := \sup_{N \ge \#\mathcal{T}_0} \left(N^s \sigma_N^{n,-1}(v)\right) < \infty \quad \Rightarrow \quad \sigma_N^{n,-1}(v) \le |v|_{\mathbb{A}_s^{-1}} N^{-s}.$$

Note that the scaling parameter a for jumps in the definition of $\sigma_N^{n,-1}$ is just $a = 1$.

The following result can be traced back to Bonito and Nochetto (2010).

Proposition 9.4 (equivalence of classes for u). *Let $\mathcal{T}_0$ be an initial conforming subdivision satisfying Assumption 6.19 (initial labelling). There are two constants $m \in \mathbb{N}$ and $C \ge 1$ such that, for all $N \ge \#\mathcal{T}_0$ and all $v \in H_0^1(\Omega)$,*

$$\sigma_N^{n,-1}(v) \le \sigma_N^{n,0}(v) \quad \textit{and} \quad \sigma_{mN}^{n,0}(v) \le C\sigma_N^{n,-1}(v).$$

In particular, the approximation classes coincide: $\mathbb{A}_s^0 \equiv \mathbb{A}_s^{-1}$, $s \ge 0$.

Proof. We start with the first inequality. For $v \in H_0^1(\Omega)$ and $N \ge \#\mathcal{T}_0$, we let $\mathcal{T} \in \mathbb{T}_N$ be a conforming subdivision of $\mathcal{T}_0$ and let $v_{\mathcal{T}}^0 \in \mathbb{V}_{\mathcal{T}}^0 \subset \mathbb{V}_{\mathcal{T}}^{-1}$ be such that

$$\sigma_N^{n,0}(v) = |v - v_{\mathcal{T}}^0|_{H_0^1(\Omega)}.$$

Because $v - v_{\mathcal{T}}^0 \in H_0^1(\Omega)$, we have $\|v - v_{\mathcal{T}}^0\|_{1,\mathcal{T}} = |v - v_{\mathcal{T}}^0|_{H_0^1(\Omega)}$ and thus

$$\sigma_N^{n,-1}(v) \le |v - v_{\mathcal{T}}^0|_{H_0^1(\Omega)} = \sigma_N^{n,0}(v).$$

We now prove the second inequality. For $v \in H_0^1(\Omega)$ and $N \ge \#\mathcal{T}_0$, let $\mathcal{T} \in \mathbb{T}_N^\Lambda$ be a Λ-admissible mesh with N elements and let $v_{\mathcal{T}} \in \mathbb{V}_{\mathcal{T}}^{-1}$ be such that

$$\|v - v_{\mathcal{T}}\|_{1,\mathcal{T}} = \sigma_N^{n,-1}(v).$$

We first show that $\mathcal{I}_{\mathcal{T}}^{\mathrm{dG}} v_{\mathcal{T}} \in \mathbb{V}_{\mathcal{T}}^0$ satisfies

$$|v - \mathcal{I}_{\mathcal{T}}^{\mathrm{dG}} v_{\mathcal{T}}|_{H_0^1(\Omega)} \lesssim \sigma_N^{n,-1}(v).$$

Indeed, using the triangle inequality we obtain

$$\|v - \mathcal{I}_{\mathcal{T}}^{\mathrm{dG}} v_{\mathcal{T}}\|_{1,\mathcal{T}} \le \|v - v_{\mathcal{T}}\|_{1,\mathcal{T}} + \|v_{\mathcal{T}} - \mathcal{I}_{\mathcal{T}}^{\mathrm{dG}} v_{\mathcal{T}}\|_{1,\mathcal{T}}.$$

Interpolation estimate (9.15) yields

$$\|v_{\mathcal{T}} - \mathcal{I}_{\mathcal{T}}^{\mathrm{dG}} v_{\mathcal{T}}\|_{1,\mathcal{T}} \lesssim \|h^{-1/2} [[v_{\mathcal{T}}]]\|_{L^2(\mathcal{F}^+)}, \tag{9.19}$$

because $v_{\mathcal{T}} - \mathcal{I}_{\mathcal{T}}^{\mathrm{dG}} v_{\mathcal{T}} \in \mathbb{V}_{\mathcal{T}}^{-1}$, whence

$$|v - \mathcal{I}_{\mathcal{T}}^{\mathrm{dG}} v_{\mathcal{T}}|_{H_0^1(\Omega)} = \|v - \mathcal{I}_{\mathcal{T}}^{\mathrm{dG}} v_{\mathcal{T}}\|_{1,\mathcal{T}} \leq C\sigma_N^{n,-1}(v)$$

as claimed for a constant $C \geq 1$ independent of v and N. To assert an estimate on $\sigma_N^{n,0}(v)$, we now exhibit a conforming refinement $\overline{\mathcal{T}}$ of $\mathcal{T}$ with a comparable number of elements. To do this, we note that because $\mathcal{T} \in \mathbb{T}^{\Lambda}$ is Λ-admissible, it is the product of successive calls $[\mathcal{T}_j] = \mathsf{REFINE}(\mathcal{T}_{j-1}, T_{j-1})$, $j = 1, \ldots, J$, where $\mathcal{T}_j$ is the smallest Λ-*admissible* refinement of $\mathcal{T}_{j-1}$ such that the element $T_{j-1} \in \mathcal{T}_{j-1}$ is bisected once. We now let $\overline{\mathcal{T}} \in \mathbb{T}$ be the conforming subdivision obtained from the successive calls $[\overline{\mathcal{T}}_j] = \mathsf{REFINE}(\overline{\mathcal{T}}_{j-1}, \{T_{j-1}\} \cap \overline{\mathcal{T}}_{j-1})$ with $\overline{\mathcal{T}}_0 = \mathcal{T}_0$, but where this time REFINE produces the smallest *conforming* refinement of $\overline{\mathcal{T}}_{j-1}$, where the element of T_{j-1} is bisected once if $T_{j-1} \in \overline{\mathcal{T}}_{j-1}$ or otherwise $\overline{\mathcal{T}}_j = \overline{\mathcal{T}}_{j-1}$. A simple induction argument, exploiting the minimality of the meshes generated by REFINE, reveals that $\overline{\mathcal{T}}_j \geq \mathcal{T}_j$ for $0 \leq j \leq J$. Consequently, Theorem 3.16 (complexity of REFINE) guarantees that

$$\#\overline{\mathcal{T}} - \#\mathcal{T}_0 \leq D \sum_{j=0}^{J-1} \#(\{T_{j-1}\} \cap \overline{\mathcal{T}}_{j-1}) \leq DJ \leq D(\#\mathcal{T} - \#\mathcal{T}_0),$$

whence $\#\overline{\mathcal{T}} \leq D\#\mathcal{T} \leq mN$ with $m := \lceil D \rceil$ because $D \geq 1$.

Therefore $\mathbb{V}_{\mathcal{T}}^0 \subset \mathbb{V}_{\overline{\mathcal{T}}}^0$ because $\overline{\mathcal{T}}$ is a conforming refinement of $\mathcal{T}$. Since $\#\overline{\mathcal{T}} \leq mN$ and $\mathcal{I}_{\mathcal{T}}^{\mathrm{dG}} v_{\mathcal{T}} \in \mathbb{V}_{\mathcal{T}}^0$, we deduce

$$\sigma_{mN}^{n,0}(v) \leq |v - \mathcal{I}_{\mathcal{T}}^{\mathrm{dG}} v_{\mathcal{T}}|_{H_0^1(\Omega)} \leq C\sigma_N^{n,-1}(v),$$

which is the desired inequality. Finally, the equivalence of moduli of approximation yields $\mathbb{A}_s^0 \equiv \mathbb{A}_s^{-1}$ and completes the proof. □

Remark 9.5 (equivalence of classes for $\mathcal{D}$). The approximation classes for data $\mathcal{D} = (\boldsymbol{A}, c, f)$, namely $\mathbb{M}_s((L^r(\Omega))^{d\times d}); \mathcal{T}_0, \mathbb{C}_s(L^q(\Omega); \mathcal{T}_0)$ and $\mathbb{F}_s(H^{-1}(\Omega); \mathcal{T}_0)$, are defined for conforming subdivisions in Section 6. However, repeating the construction of the smallest conforming refinement $\overline{\mathcal{T}}$ of any Λ-admissible subdivision $\mathcal{T}$, and using the fact that $\#\overline{\mathcal{T}} \approx \#\mathcal{T}$ proved above, we deduce that these classes are equivalent to their counterparts on non-conforming meshes. Therefore we do not repeat the proof here, and from now on we use the same notation to denote the approximation classes on Λ-admissible subdivisions.

9.2. *Discontinuous Galerkin formulation*

This section discusses the SOLVE routine at the core of the module $\mathsf{GALERKIN\text{-}DG}$. Recall that within the two-step method $\mathsf{AFEM\text{-}DG\text{-}TS}$, data $\mathcal{D} = (\boldsymbol{A}, c, f)$ is approximated by $\widehat{\mathcal{D}} = (\widehat{\boldsymbol{A}}, \widehat{c}, \widehat{f}) \in \mathbb{D}_{\widehat{\mathcal{T}}}$ subordinate to a partition $\widehat{\mathcal{T}} \in \mathbb{T}^{\Lambda}$. For a subdivision

$\mathcal{T} \in \mathbb{T}^\Lambda$, $\mathcal{T} \geq \widehat{\mathcal{T}}$, the Galerkin solution $[u_\mathcal{T}] = \mathsf{SOLVE}(\mathcal{T})$ is constructed to approximate $\widehat{u} = u(\widehat{\mathcal{D}}) \in H_0^1(\Omega)$, the exact weak solution of the perturbed problem (5.5) with approximate data $\widehat{\mathcal{D}} = (\widehat{\boldsymbol{A}}, \widehat{c}, \widehat{f})$ constructed using Algorithm 7.23 (DATA). Corollary 7.24 (performance of DATA) guarantees that the output $[\widehat{\mathcal{D}}, \widehat{\mathcal{T}}]$ of DATA satisfies the structural assumption

$$\widehat{\boldsymbol{A}} \in M(\widehat{\alpha}_1, \widehat{\alpha}_2), \qquad \widehat{c} \in R(\widehat{c}_1, \widehat{c}_2) \tag{9.20}$$

with $0 < \widehat{\alpha}_1 \leq \widehat{\alpha}_2$ and $-\widehat{\alpha}_1/(2C_P^2) \leq \widehat{c}_1 \leq \widehat{c}_2$ upon replacing the Poincaré constant C_P with the larger constant C_P appearing in Lemma 9.3 in Algorithm 7.14 (CONSTRAINT-c). We do not specify the dependence on $\widehat{\alpha}_1$, $\widehat{\alpha}_2$, $\widehat{c}_1$ and $\widehat{c}_2$ of the constants appearing in the analysis below. We also emphasize that the constants involved in (9.20) do not depend on $\widehat{\mathcal{T}}$ and are thus uniform among all the discrete data constructed within AFEM-DG-TS.

Relation (9.20) not only ensures the existence and uniqueness of a solution $\widehat{u} \in H_0^1(\Omega)$ satisfying the perturbed problem (5.5) but also, as we shall see in Corollary 9.8, the existence and uniqueness of its discontinuous Galerkin approximation. We first present the standard symmetric interior penalty method and point out that its consistency requires the exact solution $u \in H^s(\Omega)$, $s > 3/2$. To circumvent this rather restrictive assumption, we introduce lifting operators allowing a reformulation valid in $H^1(\Omega)$. However, this reformulation is only consistent on the conforming subspace $\mathbb{V}_\mathcal{T}^0 = \mathbb{V}_\mathcal{T}^{-1} \cap H_0^1(\Omega)$, and requires our analysis to decompose the discrete space $\mathbb{V}_\mathcal{T}^{-1}$ into $\mathbb{V}_\mathcal{T}^0$ and its complement $\mathbb{V}_\mathcal{T}^\perp$ with respect to an appropriate scalar product.

9.2.1. *The symmetric interior penalty method*

The symmetric interior penalty (SIP) formulation is the most standard discontinuous Galerkin method. For $\mathcal{T} \in \mathbb{T}^\Lambda$, it consists in finding $u_\mathcal{T} \in \mathbb{V}_\mathcal{T}^{-1}$ satisfying

$$\mathcal{B}_\mathcal{T}[u_\mathcal{T}, v] = \langle \widehat{f}, v \rangle_\mathcal{T} \quad \text{for all } v \in \mathbb{V}_\mathcal{T}^{-1}, \tag{9.21}$$

where $\mathcal{B}_\mathcal{T} \colon \mathbb{V}_\mathcal{T}^{-1} \times \mathbb{V}_\mathcal{T}^{-1} \to \mathbb{R}$ is the bilinear form defined by

$$\begin{aligned}\mathcal{B}_\mathcal{T}[w, v] := \int_\Omega (\nabla_\mathcal{T} v \cdot \widehat{\boldsymbol{A}} \nabla_\mathcal{T} w + \widehat{c} w v) - \sum_{F \in \mathcal{F}^+} \int_F [[v]] \boldsymbol{n}_F \cdot \{\{\widehat{\boldsymbol{A}} \nabla_\mathcal{T} w\}\} \\ - \sum_{F \in \mathcal{F}^+} \int_F [[w]] \boldsymbol{n}_F \cdot \{\{\widehat{\boldsymbol{A}} \nabla_\mathcal{T} v\}\} + \kappa \sum_{F \in \mathcal{F}^+} \int_F h_F^{-1} [[w]] [[v]].\end{aligned} \tag{9.22}$$

The parameter $\kappa > 0$ is responsible for keeping the discontinuity of the Galerkin solution under control and its value is discussed below. Unless specified otherwise, all the constants appearing in the discussion below are independent of κ, and the notation $A \lesssim B$ signifies $A \leq CB$ with a constant C independent of the discretization parameters and κ.

A few comments regarding the weak formulation (9.21) are in order. An integration by parts reveals that the method is consistent whenever the exact solution satisfies the additional regularity $u \in H^s(\Omega)$, $s > 3/2$. However, we do not make this assumption in the analysis below but rather extend the formulation to the energy space $\mathbb{E}_{\mathcal{T}} \supset \mathbb{V}_{\mathcal{T}}^{-1}$ using lifting operators. The same integration by parts also indicates that the term $\sum_{F\in\mathcal{F}^+}\int_F [[w]]\boldsymbol{n}_F \cdot \{\{\widehat{\boldsymbol{A}}\nabla v\}\}$ is not necessary but included to achieve a symmetric formulation. Recall that $\widehat{\boldsymbol{A}}$ constructed by DATA is symmetric. In addition, the presence of $\langle \widehat{f}, v\rangle_{\mathcal{T}}$ is not standard but allows for right-hand sides $\widehat{f} \in \mathbb{F}_{\widehat{\mathcal{T}}}$ and in turn for $f \in H^{-1}(\Omega)$ within the AFEM-DG-TS algorithm.

9.2.2. Lifting operators

The interior penalty bilinear form (9.22) includes inter-element terms

$$\sum_{F\in\mathcal{F}^+}\int_F [[v]]\boldsymbol{n}_F \cdot \{\{\widehat{\boldsymbol{A}}\nabla_{\mathcal{T}} w\}\} + \sum_{F\in\mathcal{F}^+}\int_F [[w]]\boldsymbol{n}_F \cdot \{\{\widehat{\boldsymbol{A}}\nabla_{\mathcal{T}} v\}\}, \tag{9.23}$$

which are not defined on $H^1(\Omega)$ but on $H^s(\Omega)$, $s > 3/2$. In turn, the method is consistent when $u \in H^s(\Omega)$, $s > 3/2$. The key ingredient to extending $\mathcal{B}_{\mathcal{T}}[w, v]$ to $\mathbb{E}_{\mathcal{T}} \times \mathbb{E}_{\mathcal{T}}$ without additional regularity is a lifting operator (Brezzi *et al.* 2000, Arnold, Brezzi, Cockburn and Marini 2002, Perugia and Schötzau 2003, Houston, Schötzau and Wihler 2004, 2007, Bonito and Nochetto 2010) introduced in this section.

For $n' > 0$, we define $\mathcal{L}_{\mathcal{T}}^{n'}: \mathbb{E}_{\mathcal{T}} \to [\mathbb{S}_{\mathcal{T}}^{n',-1}]^d$ by the relations

$$\int_\Omega \mathcal{L}_{\mathcal{T}}^{n'}[v]\cdot \widehat{\boldsymbol{A}}\boldsymbol{w} = \sum_{F\in\mathcal{F}^+}\int_F [[v]]\boldsymbol{n}_F \cdot \{\{\widehat{\boldsymbol{A}}\boldsymbol{w}\}\} \quad \text{for all } \boldsymbol{w} \in \left[\mathbb{S}_{\mathcal{T}}^{n',-1}\right]^d. \tag{9.24}$$

From this definition, we easily deduce an L^2-stability estimate.

Lemma 9.6 (stability of lift). *Let $\mathcal{T} \in \mathbb{T}^\Lambda$ be a Λ-admissible subdivision of $\mathcal{T}_0$ satisfying Assumption 6.19 (initial labelling). Assume $\widehat{\boldsymbol{A}} \in M(\widehat{\alpha}_1, \widehat{\alpha}_2)$ with $0 < \widehat{\alpha}_1 \le \widehat{\alpha}_2$. For $n' \ge 0$ and all $v \in \mathbb{S}_{\mathcal{T}}^{n',-1}$,*

$$\|\mathcal{L}_{\mathcal{T}}^{n'}[v]\|_{L^2(\Omega)} \le C\|h^{-1/2}[[v]]\|_{L^2(\mathcal{F}^+)}, \tag{9.25}$$

where $C = C(\widehat{\alpha}_2/\widehat{\alpha}_1, \mathcal{T}_0, n')$.

Proof. Let $v \in \mathbb{S}_{\mathcal{T}}^{n',-1}$ and set $\boldsymbol{w} = \mathcal{L}_{\mathcal{T}}^{n'}[v]$ in (9.24) to write

$$\begin{aligned}
\|\widehat{\boldsymbol{A}}^{1/2}\mathcal{L}_{\mathcal{T}}^{n'}[v]\|_{L^2(\Omega)}^2 &= \int_\Omega \mathcal{L}_{\mathcal{T}}^{n'}[v]\cdot \widehat{\boldsymbol{A}}\mathcal{L}_{\mathcal{T}}^{n'}[v] \\
&= \sum_{F\in\mathcal{F}^+}\int_F h^{-1/2}[[v]]\boldsymbol{n}_F \cdot h^{1/2}\{\{\widehat{\boldsymbol{A}}\mathcal{L}_{\mathcal{T}}^{n'}[v]\}\} \\
&\le \|h^{-1/2}[[v]]\|_{L^2(\mathcal{F}^+)}\|h^{1/2}\{\{\widehat{\boldsymbol{A}}\mathcal{L}_{\mathcal{T}}^{n'}[v]\}\}\|_{L^2(\mathcal{F}^+)}.
\end{aligned}$$

A local inverse estimate along with the eigenvalue bounds for $\widehat{\boldsymbol{A}} \in M(\widehat{\alpha}_1, \widehat{\alpha}_2)$ yields

$$\|h^{1/2}\{\{\widehat{\boldsymbol{A}}\mathcal{L}_{\mathcal{T}}^{n'}[v]\}\}\|_{L^2(\mathcal{F}^+)} \le C\widehat{\alpha}_2\|\mathcal{L}_{\mathcal{T}}^{n'}[v]\|_{L^2(\Omega)},$$

where C depends only on the shape regularity constant of $\mathcal{T}_0$ and n'. Combining the above two inequalities and again taking advantage of the assumption $\widehat{A} \in M(\widehat{\alpha}_1, \widehat{\alpha}_2)$ implies (9.25). □

We record two estimates based on (9.25) and used multiple times in the analysis below. Combining the estimate (9.25) on the lifting operator with assumption (9.20) and a Cauchy–Schwarz inequality, we find that

$$\int_\Omega \mathcal{L}_\mathcal{T}^{n'}[v] \cdot \widehat{A}\nabla_\mathcal{T} w \leq C_{\text{lift}} \|h^{-1/2}[[v]]\|_{L^2(\mathcal{F}^+)} \|\nabla_\mathcal{T} w\|_{L^2(\mathcal{T})} \quad \text{for all } v, w \in \mathbb{E}_\mathcal{T}, \tag{9.26}$$

for a constant $C_{\text{lift}} = C_{\text{lift}}(\widehat{\alpha}_1, \widehat{\alpha}_2, \mathcal{T}_0, n')$ and in particular independent of the discretization parameters and κ. This, together with a Young inequality, yields for any $\epsilon > 0$ the second estimate for all $v, w \in \mathbb{E}_\mathcal{T}$:

$$\int_\Omega \mathcal{L}_\mathcal{T}^{n'}[v] \cdot \widehat{A}\nabla_\mathcal{T} w \leq \frac{C_{\text{lift}}^2}{2\epsilon} \|h^{-1/2}[[v]]\|_{L^2(\mathcal{F}^+)}^2 + \frac{\epsilon}{2} \|\nabla_\mathcal{T} w\|_{L^2(\mathcal{T})}^2. \tag{9.27}$$

We now return to the SIP weak formulation (9.21) and take advantage of the lifting operators to deduce an equivalent expression of the bilinear form $\mathcal{B}_\mathcal{T}$ on $\mathbb{V}_\mathcal{T}^{-1}$, which is well-defined on $\mathbb{E}_\mathcal{T}$. The problematic inter-element terms (9.23) are equivalently rewritten as

$$\int_\Omega \mathcal{L}_\mathcal{T}^{n'}[v] \cdot \widehat{A}\nabla_\mathcal{T} w + \int_\Omega \mathcal{L}_\mathcal{T}^{n'}[w] \cdot \widehat{A}\nabla_\mathcal{T} v, \tag{9.28}$$

provided

$$\nabla_\mathcal{T} \mathbb{V}_\mathcal{T}^{-1} \subset \left[\mathbb{S}_\mathcal{T}^{n',-1}\right]^d.$$

The above condition is satisfied when $n' \geq n - 1$ for subdivisions $\mathcal{T}$ made of simplices and $n' \geq n$ for hexahedra. To continue with an analysis incorporating both cases, we set $n' = n$ and write $\mathcal{L}_\mathcal{T} := \mathcal{L}_\mathcal{T}^n$. With this choice, the bilinear form $\mathcal{B}_\mathcal{T}$ in the symmetric interior penalty method (9.21) reads

$$\begin{aligned} \mathcal{B}_\mathcal{T}[w, v] = a_\mathcal{T}[w, v] &- \int_\Omega \mathcal{L}_\mathcal{T}[v] \cdot \widehat{A}\nabla_\mathcal{T} w \\ &- \int_\Omega \mathcal{L}_\mathcal{T}[w] \cdot \widehat{A}\nabla_\mathcal{T} v + \kappa \sum_{F \in \mathcal{F}^+} \int_F h_F^{-1} [[w]]\,[[v]], \end{aligned} \tag{9.29}$$

for all $w, v \in \mathbb{V}_\mathcal{T}^{-1}$ and where we used

$$a_\mathcal{T}[w, v] := \int_\Omega \nabla_\mathcal{T} v \cdot \widehat{A}\nabla_\mathcal{T} w + \widehat{c} w v \tag{9.30}$$

to denote the bilinear form related to the conforming method.

Expression (9.29) is well-defined for $w, v \in \mathbb{E}_\mathcal{T}$ and the weak formulation (9.21) is well-posed. These two claims follow from Corollary 9.8 below, which in turn is a consequence of the next result focusing on the bilinear form $a_\mathcal{T}$; we recall (5.52).

Lemma 9.7 (properties of $a_{\mathcal{T}}$). *Let $\mathcal{T} \in \mathbb{T}^{\Lambda}$ be a Λ-admissible refinement of $\mathcal{T}_0$ satisfying Assumption 6.19 (initial labelling). Furthermore, assume that $\widehat{\mathbf{A}}$ and $\widehat{c}$ satisfy the structural assumption* (9.20)*. Then we have*

$$a_{\mathcal{T}}[w, v] \le \left(\widehat{\alpha}_2 + |\widehat{c}_2| C_P^2\right) \|v\|_{1,\mathcal{T}} \|w\|_{1,\mathcal{T}} \quad \textit{for all } v, w \in \mathbb{E}_{\mathcal{T}} \tag{9.31}$$

and

$$a_{\mathcal{T}}[v, v] \ge \frac{\widehat{\alpha}_1}{2} \|\nabla_{\mathcal{T}} v\|^2_{L^2(\Omega)} + \min(0, \widehat{c}_1) C_P^2 \|h^{-1/2} [\![v]\!]\|^2_{L^2(\mathcal{F}^+)} \quad \textit{for all } v \in \mathbb{E}_{\mathcal{T}}, \tag{9.32}$$

where C_P is the constant in Lemma 9.3 (Poincaré-type inequality in $\mathbb{E}_{\mathcal{T}}$).

Proof. We start with the continuity estimate (9.31). The assumption on the discretized coefficients implies that for $v, w \in \mathbb{E}_{\mathcal{T}}$ we have

$$a_{\mathcal{T}}[w, v] \le \alpha_2 \|\nabla_{\mathcal{T}} w\|_{L^2(\mathcal{T})} \|\nabla_{\mathcal{T}} v\|_{L^2(\mathcal{T})} + |c_2| \|w\|_{L^2(\Omega)} \|v\|_{L^2(\Omega)}.$$

It remains to invoke Lemma 9.3 (Poincaré-type inequality on $\mathbb{E}_{\mathcal{T}}$) to deduce (9.31).

Similarly, for the partial coercivity estimate (9.32) we have

$$a_{\mathcal{T}}[v, v] \ge \widehat{\alpha}_1 \|\nabla_{\mathcal{T}} v\|^2_{L^2(\Omega)} + \widehat{c}_1 \|v\|^2_{L^2(\Omega)} \ge \widehat{\alpha}_1 \|\nabla_{\mathcal{T}} v\|^2_{L^2(\Omega)} + \min(0, \widehat{c}_1) C_P^2 \|v\|^2_{1,\mathcal{T}},$$

and the desired estimate follows from the assumption $-\widehat{\alpha}_1/(2C_P^2) \le \widehat{c}_1$. □

For the next result, we recall that the discrete norm $\|\cdot\|_{\kappa,\mathcal{T}}$ is defined in (9.2).

Corollary 9.8 (properties of $\mathcal{B}_{\mathcal{T}}$). *Let $\mathcal{T} \in \mathbb{T}^{\Lambda}$ be a Λ-admissible refinement of $\mathcal{T}_0$ satisfying Assumption 6.19 (initial labelling). Furthermore, assume that $\widehat{\mathbf{A}}$ and $\widehat{c}$ satisfy the structural assumption* (9.20)*. There exists a constant C_{cont} such that*

$$\mathcal{B}_{\mathcal{T}}[w, v] \le C_{\mathrm{cont}} \|v\|_{\kappa,\mathcal{T}} \|w\|_{\kappa,\mathcal{T}} \quad \textit{for all } v, w \in \mathbb{E}_{\mathcal{T}}. \tag{9.33}$$

Moreover, there are constants $\overline{\kappa}_{\mathrm{stab}}, C_{\mathrm{coer}} > 0$ such that for all $\kappa > \overline{\kappa}_{\mathrm{stab}}$ we have

$$C_{\mathrm{coer}} \|v\|^2_{\kappa,\mathcal{T}} \le \mathcal{B}_{\mathcal{T}}[v, v] \quad \textit{for all } v \in \mathbb{E}_{\mathcal{T}}. \tag{9.34}$$

In particular, the Galerkin formulation (9.21) *has a unique solution $u_{\mathcal{T}} \in \mathbb{V}_{\mathcal{T}}^{-1}$.*

Proof. The continuity estimate is a direct consequence of the continuity estimate (9.31), estimate (9.26) for the lifting terms, and Cauchy–Schwarz inequality

$$\kappa \sum_{F \in \mathcal{F}^+} \int_F h_F^{-1} [\![w]\!] [\![v]\!] \le \kappa \|h^{-1/2} [\![w]\!]\|_{L^2(\mathcal{F}^+)} \|h^{-1/2} [\![v]\!]\|_{L^2(\mathcal{F}^+)},$$

which holds for all $v, w \in \mathbb{E}_{\mathcal{T}}$.

We now focus on the coercivity estimate (9.34) and start from (9.32), which we write for $v \in \mathbb{E}_{\mathcal{T}}$ as

$$\frac{\widehat{\alpha}_1}{2} \|\nabla_{\mathcal{T}} v\|^2_{L^2(\mathcal{T})} - \max\left(0, -c_1 \widehat{C}_P^2\right) \|h^{-1/2} [\![v]\!]\|^2_{L^2(\mathcal{F}^+)} \le a_{\mathcal{T}}[v, v]. \tag{9.35}$$

Furthermore, the terms involving the lifting operators in the definition (9.29) of the bilinear form $\mathcal{B}_{\mathcal{T}}$ reduce to $-2\int_\Omega \mathcal{L}_{\mathcal{T}}[v]\cdot\widehat{\boldsymbol{A}}\nabla_{\mathcal{T}}v$ when $w=v$. Hence the estimate (9.27) with $\epsilon=\widehat{\alpha}_1/4$ implies that

$$2\left|\int_\Omega \mathcal{L}_{\mathcal{T}}[v]\cdot\widehat{\boldsymbol{A}}\nabla_{\mathcal{T}}v\right| \le \frac{\widehat{\alpha}_1}{4}\|\nabla_{\mathcal{T}}v\|^2_{L^2(\mathcal{T})} + \frac{4C^2_{\mathcal{L}}}{\widehat{\alpha}_1}\|h^{-1/2}[[v]]\|^2_{L^2(\mathcal{F}^+)}.$$

Gathering the above inequalities and recalling definition (9.29) of $\mathcal{B}_{\mathcal{T}}$, we find that

$$\frac{\alpha_1}{4}\|\nabla_{\mathcal{T}}v\|^2_{L^2(\mathcal{T})} + (\kappa-\overline{\kappa}_{\text{stab}})\|h^{-1/2}[[v]]\|^2_{L^2(\mathcal{F}^+)} \le \mathcal{B}_{\mathcal{T}}[v,v],$$

with

$$\overline{\kappa}_{\text{stab}} := \frac{4C^2_{\mathcal{L}}}{\alpha_1} + \max\big(0,-\widehat{c}_1C_P^2\big).$$

The desired coercivity estimate directly follows provided $\kappa>\overline{\kappa}_{\text{stab}}$. □

9.2.3. Partial consistency and role of the conforming Galerkin solution

From now on we shall use the expression of $\mathcal{B}_{\mathcal{T}}$ in (9.29) extending $\mathcal{B}_{\mathcal{T}}$ to $\mathbb{E}_{\mathcal{T}}\times\mathbb{E}_{\mathcal{T}}$. This reformulation comes at the price of partial consistency. Since $\mathcal{L}_{\mathcal{T}}[v]=0$ whenever $v\in H^1_0(\Omega)$ and the duality product $\langle\cdot,\cdot\rangle_{\mathcal{T}}$ satisfies the consistency (9.4), we have

$$\mathcal{B}_{\mathcal{T}}[\widehat{u},v] = a_{\mathcal{T}}[\widehat{u},v] = \langle\widehat{f},v\rangle_{\mathcal{T}} \quad \text{for all } v\in H^1_0(\Omega), \tag{9.36}$$

which indicates that the reformulation (9.29) using lifts is consistent on $H^1_0(\Omega)$. However, (9.36) does not hold for all $v\in\mathbb{V}^{-1}_{\mathcal{T}}$.

This suggests splitting $\mathbb{V}^{-1}_{\mathcal{T}}$ into a conforming space where the consistency holds and its orthogonal complement. We decompose the discontinuous space as

$$\mathbb{V}^{-1}_{\mathcal{T}} = \mathbb{V}^0_{\mathcal{T}}\oplus\mathbb{V}^\perp_{\mathcal{T}}, \tag{9.37}$$

where $\mathbb{V}^0_{\mathcal{T}}=\mathbb{V}^{-1}_{\mathcal{T}}\cap H^1_0(\Omega)$ is the finest conforming subspace of $\mathbb{V}^{-1}_{\mathcal{T}}$ and $\mathbb{V}^\perp_{\mathcal{T}}$ is the orthogonal complement with respect to the $\mathcal{B}_{\mathcal{T}}[\cdot,\cdot]$ scalar product. Note that the latter is well-defined provided the assumption on the penalty parameter $\kappa>\overline{\kappa}_{\text{stab}}$, required by Corollary 9.8, is satisfied. From now on we assume this is the case, and point out that although the constants appearing in the analysis below do not depend on κ, they may depend on $\overline{\kappa}_{\text{stab}}$.

We also emphasize that there might not be a conforming subdivision associated with $\mathbb{V}^0_{\mathcal{T}}$. The latter is the span of the basis functions associated with proper nodes; see Figure 3.11 for an illustration and refer to Section 8 for more details. Consequently, the analysis provided below relies on the decomposition (9.37) of the space $\mathbb{V}_{\mathcal{T}}$ rather than on a subdivision $\mathcal{T}$. It is also worth pointing out that the conforming part $u^0_{\mathcal{T}}\in\mathbb{V}^0_{\mathcal{T}}$ of the Galerkin solution $u_{\mathcal{T}}\in\mathbb{V}^{-1}_{\mathcal{T}}$ satisfies

$$\mathcal{B}_{\mathcal{T}}\big[u^0_{\mathcal{T}},v\big] = \langle\widehat{f},v\rangle \quad \text{for all } v\in\mathbb{V}^0_{\mathcal{T}}. \tag{9.38}$$

Hence $u_{\mathcal{T}}^0$ is the conforming Galerkin approximation on $\mathbb{V}_{\mathcal{T}}^0$. As we shall see, this finest coarser conforming Galerkin solution plays a critical role in the convergence of AFEM-DG-TS. This justifies the orthogonal decomposition (9.37) associated with the $\mathcal{B}_{\mathcal{T}}$ scalar product.

Another advantage of using the $\mathcal{B}_{\mathcal{T}}$-orthogonal decomposition (9.37) is that it offers a control on the non-conforming component of $v \in \mathbb{V}_{\mathcal{T}}^{-1}$ by its scaled jumps. To achieve this, the operator $\mathcal{I}_{\mathcal{T}}^{\mathrm{dG}}$ defined by (9.7) is instrumental.

Lemma 9.9 (control of non-conformity). *Let $\mathcal{T} \in \mathbb{T}^{\Lambda}$ be a Λ-admissible refinement of $\mathcal{T}_0$ satisfying Assumption 6.19 (initial labelling). Assume that $\widehat{\mathbf{A}}$ and $\widehat{c}$ satisfy the structural assumption* (9.20). *For $\kappa > \overline{\kappa}_{\mathrm{stab}}$, if $v = v^0 + v^{\perp} \in \mathbb{V}_{\mathcal{T}}^{-1}$ according to* (9.37), *then*

$$\|v^{\perp}\|_{\kappa,\mathcal{T}} \lesssim \kappa^{1/2}\|h^{-1/2}[[v^{\perp}]]\|_{L^2(\mathcal{F}^+)} = \kappa^{1/2}\|h^{-1/2}[[v]]\|_{L^2(\mathcal{F}^+)}.$$

Proof. Because $\mathcal{I}_{\mathcal{T}}^{\mathrm{dG}} v \in \mathbb{V}_{\mathcal{T}}^0$, the orthogonal decomposition (9.37) implies that

$$\mathcal{B}_{\mathcal{T}}[v^{\perp}, v^{\perp}] \leq \mathcal{B}_{\mathcal{T}}\big[v - \mathcal{I}_{\mathcal{T}}^{\mathrm{dG}} v, v - \mathcal{I}_{\mathcal{T}}^{\mathrm{dG}} v\big].$$

The desired result follows from the coercivity (9.34) and continuity (9.33) of $\mathcal{B}_{\mathcal{T}}$ along with the interpolation estimate (9.10). □

9.3. *A posteriori* error estimates

We derive a residual error estimate for the discontinuous Galerkin method. Because the data $\widehat{\mathcal{D}} \in \mathbb{D}_{\mathcal{T}}$ is discrete, the analysis is free from data oscillation. In the notation introduced in Section 4, this means $\mathcal{E}_{\mathcal{T}} = \eta_{\mathcal{T}}$, where for $v \in \mathbb{E}_{\mathcal{T}}$

$$\eta_{\mathcal{T}}^2(v) := \sum_{T\in\mathcal{T}} \eta_{\mathcal{T}}(v,T)^2,$$

and

$$\eta_{\mathcal{T}}(v,T)^2 := h_T \sum_{F\subset\partial T\setminus\partial\Omega} \|j_{\mathcal{T}}(v) - \widehat{f}\|_{L^2(F)}^2 + h_T^2\|r_{\mathcal{T}}(v)\|_{L^2(T)}^2,$$

with $j_{\mathcal{T}}(v)|_F := \boldsymbol{n}_F \cdot [[\widehat{\mathbf{A}}\nabla_{\mathcal{T}} v]]$ and $r_{\mathcal{T}}(v)|_T := \widehat{f} - \widehat{c}v + \mathrm{div}_{\mathcal{T}}(\widehat{\mathbf{A}}\nabla_{\mathcal{T}} v)$.

We start with a result mimicking the conforming argument, and then discuss its drawbacks and remedies.

Lemma 9.10 (*a posteriori* error estimates). *Let $\mathcal{T} \in \mathbb{T}^{\Lambda}$ be a Λ-admissible refinement of $\mathcal{T}_0$ satisfying Assumption 6.19 (initial labelling). Assume that $\widehat{\mathbf{A}}$ and $\widehat{c}$ satisfy the structural assumption* (9.20). *If $\kappa > \overline{\kappa}_{\mathrm{stab}}$, then*

$$\|\widehat{u} - u_{\mathcal{T}}\|_{\kappa,\mathcal{T}}^2 \lesssim \eta_{\mathcal{T}}(u_{\mathcal{T}})^2 + \kappa\|h^{-1/2}[[u_{\mathcal{T}}]]\|_{L^2(\mathcal{F}^+)}^2 \tag{9.39}$$

and

$$C_L\,\eta_{\mathcal{T}}(u_{\mathcal{T}}) \leq \|\widehat{u} - u_{\mathcal{T}}\|_{\kappa,\mathcal{T}}, \tag{9.40}$$

for some constant C_L.

Proof. We start with the upper bound (9.39). To exploit the consistency (9.36) in $\mathbb{V}^0_{\mathcal{T}}$, we decompose the error $e := \widehat{u} - u_{\mathcal{T}} \in \mathbb{E}_{\mathcal{T}}$ into a conforming part $e^0 := \widehat{u} - u^0_{\mathcal{T}} \in H^1_0(\Omega)$ and a non-conforming part $e^\perp := -u^\perp_{\mathcal{T}} \in \mathbb{V}^{-1}_{\mathcal{T}}$ according to (9.37). The proof thus relies on techniques used in the conforming theory coupled with Lemma 9.9 (control of non-conformity). We let C denote a generic constant independent of the discretization and κ but possibly depending on $\overline{\kappa}_{\text{stab}}$.

Using the coercivity (9.34) and partial consistency (9.36), we get

$$C_{\text{coer}}\|e\|^2_{\kappa,\mathcal{T}} \le \mathcal{B}_{\mathcal{T}}[e,e] = \mathcal{B}_{\mathcal{T}}[e, e^0 - \mathcal{I}_{\mathcal{T}}e^0] - \mathcal{B}_{\mathcal{T}}[e, u^\perp_{\mathcal{T}}], \tag{9.41}$$

where $I_{\mathcal{T}}$ is the Scott–Zhang interpolant provided in Proposition 3.5. For the first term, note that since $e^0 - I_{\mathcal{T}}e^0 \in H^1_0(\Omega)$ we have

$$\mathcal{B}_{\mathcal{T}}\big[e, e^0 - I_{\mathcal{T}}e^0\big] = a_{\mathcal{T}}[e, e^0 - I_{\mathcal{T}}e^0] - \int_\Omega \mathcal{L}_{\mathcal{T}}[e]\cdot \boldsymbol{A}\nabla(e^0 - I_{\mathcal{T}}e^0).$$

For the term involving the bilinear form $a_{\mathcal{T}}$, we proceed as in the conforming case, to arrive at

$$a_{\mathcal{T}}[e, e^0 - I_{\mathcal{T}}e^0] \lesssim \eta_{\mathcal{T}}(u_{\mathcal{T}})\|\nabla e^0\|_{L^2(\Omega)}.$$

This, combined with estimate (9.26) on the lifting operators and the H^1-stability of the Scott–Zhang interpolant, yields

$$\mathcal{B}_{\mathcal{T}}\big[e, e^0 - I_{\mathcal{T}}e^0\big] \lesssim \big(\eta_{\mathcal{T}}(u_{\mathcal{T}})^2 + \|h^{-1/2}[\![u_{\mathcal{T}}]\!]\|^2_{L^2(\mathcal{F}^+)}\big)^{1/2}\|\nabla e^0\|_{L^2(\Omega)}.$$

We rewrite $e^0 = e + u^\perp_{\mathcal{T}}$ and use the estimate on the non-conforming component provided by Lemma 9.9 along with a Young inequality, to write

$$\mathcal{B}_{\mathcal{T}}\big[e, e^0 - I_{\mathcal{T}}e^0\big] \le C\big(\eta_{\mathcal{T}}(u_{\mathcal{T}})^2 + \kappa\|h^{-1/2}[\![u_{\mathcal{T}}]\!]\|^2_{L^2(\mathcal{F}^+)}\big) + \frac{C_{\text{coer}}}{4}\|\nabla e\|^2_{L^2(\Omega)}.$$

For the second term in (9.41), the continuity (9.33) of the bilinear form $\mathcal{B}_{\mathcal{T}}$, Lemma 9.9 again, and a Young inequality yield

$$\mathcal{B}_{\mathcal{T}}[e, u^\perp_{\mathcal{T}}] \le \frac{C_{\text{coer}}}{4}\|e\|^2_{\kappa,\mathcal{T}} + C\kappa\|h^{-1/2}[\![u_{\mathcal{T}}]\!]\|^2_{L^2(\mathcal{F}^+)}.$$

Returning to (9.41), we find that

$$\|e\|^2_{\mathcal{T}} \lesssim \eta_{\mathcal{T}}(u_{\mathcal{T}})^2 + \kappa\|h^{-1/2}[\![u_{\mathcal{T}}]\!]\|^2_{L^2(\mathcal{F}^+)},$$

which is the desired upper bound.

We finally deal with the lower bound (9.40). For $T \in \mathcal{T}$ and $v \in H^1_0(T)$, we get

$$\int_T (-\operatorname{div}_{\mathcal{T}}(\widehat{\boldsymbol{A}}\nabla_{\mathcal{T}}u_{\mathcal{T}}) + \widehat{c}u_{\mathcal{T}} - \widehat{f})v = \int_T \nabla v\cdot\widehat{\boldsymbol{A}}\nabla_{\mathcal{T}}(\widehat{u} - u_{\mathcal{T}})v + \widehat{c}(\widehat{u} - u_{\mathcal{T}})v.$$

For an interior face $F \in \mathcal{F}$, $v \in H_0^1(\omega_F)$ with $\omega_F := \{T \in \mathcal{T} \mid T \cap F \neq \emptyset\}$, we have

$$\int_F ([[\widehat{\mathbf{A}}\nabla_{\mathcal{T}} u_{\mathcal{T}}]] - \widehat{f})v = \int_{\omega_F} (-\operatorname{div}_{\mathcal{T}}(\widehat{\mathbf{A}}\nabla_{\mathcal{T}} u_{\mathcal{T}}) + \widehat{c}u_{\mathcal{T}} - \widehat{f})v$$
$$- \int_{\omega_F} \nabla v \cdot \widehat{\mathbf{A}}\nabla_{\mathcal{T}}(\widehat{u} - u_{\mathcal{T}}) - \widehat{c}(\widehat{u} - u_{\mathcal{T}})v.$$

The desired lower bound follows from the same arguments as in the conforming case; we refer to Proposition 4.12 (partial lower bound). □

Upper bound (9.39) may suggest adding the jump term $\kappa^{1/2}\|h^{-1/2}[[u_{\mathcal{T}}]]\|_{L^2(\mathcal{F}^+)}$ to the residual estimator $\eta_{\mathcal{T}}(u_{\mathcal{T}})$. This would result in a clean upper bound but, because of the presence of negative powers of the mesh size, it would be at the expense of destroying the monotonicity property of the estimator; see e.g. Proposition 4.56 (estimator reduction). The latter is instrumental to the analysis provided below.

The next result mitigates the effect of the additional jump term by showing that $\|h^{-1/2}[[u_{\mathcal{T}}]]\|_{L^2(\mathcal{F}^+)}$ can be bounded by $\eta_{\mathcal{T}}(u_{\mathcal{T}})/\kappa$ and can thus be absorbed by the estimator in the upper bound provided κ is sufficiently large. We follow the proof provided in Bonito and Nochetto (2010) and refer to Karakashian and Pascal (2007, (3.20)) for an alternative (original) proof.

Lemma 9.11 (discontinuity control). *Let $\mathcal{T} \in \mathbb{T}^\Lambda$ be a Λ-admissible refinement of $\mathcal{T}_0$ satisfying Assumption 6.19 (initial labelling). Assume that $\widehat{\mathbf{A}}$ and $\widehat{c}$ satisfy the structural assumption* (9.20). *There exists a constant $\overline{\kappa}_{\text{jump}} \geq \overline{\kappa}_{\text{stab}} > 0$ such that if $\kappa \geq \overline{\kappa}_{\text{jump}}$, then*

$$\|h^{-1/2}[[u_{\mathcal{T}}]]\|_{L^2(\mathcal{F}^+)} \lesssim \kappa^{-1}\eta_{\mathcal{T}}(u_{\mathcal{T}}).$$

Proof. For $v_{\mathcal{T}}^0 \in \mathbb{V}_{\mathcal{T}}^0$ we realize that because $[[v_{\mathcal{T}}^0]] = 0$, the coercivity estimate (9.34) implies that

$$C_{\text{coerc}}\kappa\|h^{-1/2}[[u_{\mathcal{T}}]]\|_{L^2(\mathcal{F}^+)}^2 \leq \mathcal{B}_{\mathcal{T}}\big[u_{\mathcal{T}} - v_{\mathcal{T}}^0, u_{\mathcal{T}} - v_{\mathcal{T}}^0\big]. \tag{9.42}$$

We now rewrite the right-hand side of (9.42) to produce residual terms. Since $u_{\mathcal{T}}$ solves (9.21), we have

$$\mathcal{B}_{\mathcal{T}}\big[u_{\mathcal{T}} - v_{\mathcal{T}}^0, u_{\mathcal{T}} - v_{\mathcal{T}}^0\big] = \langle \widehat{f}, u_{\mathcal{T}} - v_{\mathcal{T}}^0\rangle_{\mathcal{T}} - \mathcal{B}_{\mathcal{T}}\big[v_{\mathcal{T}}^0, u_{\mathcal{T}} - v_{\mathcal{T}}^0\big]. \tag{9.43}$$

We concentrate for the moment on the second term. Since $[[v_{\mathcal{T}}^0]] = 0$, the stabilization term vanishes as well:

$$\kappa \sum_{F \in \mathcal{F}^+} \int_F [[v_{\mathcal{T}}^0]]\,[[u_{\mathcal{T}} - v_{\mathcal{T}}^0]] = 0.$$

Hence, writing $v_{\mathcal{T}}^0 = u_{\mathcal{T}} + (v_{\mathcal{T}}^0 - u_{\mathcal{T}})$, we deduce that

$$\begin{aligned}\mathcal{B}_{\mathcal{T}}\big[v_{\mathcal{T}}^0, u_{\mathcal{T}} - v_{\mathcal{T}}^0\big] &= a_{\mathcal{T}}[u_{\mathcal{T}}, u_{\mathcal{T}} - v_{\mathcal{T}}^0] \\ &\quad - a_{\mathcal{T}}[u_{\mathcal{T}} - v_{\mathcal{T}}^0, u_{\mathcal{T}} - v_{\mathcal{T}}^0] - \int_\Omega \nabla v_{\mathcal{T}}^0 \cdot \widehat{\boldsymbol{A}} \mathcal{L}_{\mathcal{T}}[u_{\mathcal{T}}],\end{aligned}$$

where we again invoked the property $[\![v_{\mathcal{T}}^0]\!] = 0$ to infer that $\mathcal{L}_{\mathcal{T}}[v_{\mathcal{T}}^0] = 0$. Integrating the first term on the right-hand side by parts, adding it to the first term on the right-hand side of (9.43), and using the extended definition (9.3) of the duality pairing leads to the following expression involving the residuals $r_{\mathcal{T}}(u_{\mathcal{T}}), j_{\mathcal{T}}(u_{\mathcal{T}})$:

$$\begin{aligned}&\mathcal{B}_{\mathcal{T}}\big[u_{\mathcal{T}} - v_{\mathcal{T}}^0, u_{\mathcal{T}} - v_{\mathcal{T}}^0\big] \\ &\quad = \int_\Omega r_{\mathcal{T}}(u_{\mathcal{T}})(u_{\mathcal{T}} - v_{\mathcal{T}}^0) + \int_{\mathcal{F}^+} (j_{\mathcal{T}}(u_{\mathcal{T}}) - \widehat{f})\{\!\{u_{\mathcal{T}} - v_{\mathcal{T}}^0\}\!\} \\ &\quad\quad + a_{\mathcal{T}}[u_{\mathcal{T}} - v_{\mathcal{T}}^0, u_{\mathcal{T}} - v_{\mathcal{T}}^0] - \int_\Omega \nabla(u_{\mathcal{T}} - v_{\mathcal{T}}^0) \cdot \widehat{\boldsymbol{A}} \mathcal{L}_{\mathcal{T}}[u_{\mathcal{T}}].\end{aligned}$$

We point out that we have also employed the definition (9.24) of lift to rewrite the resulting face terms. Inserting this estimate back in (9.42), together with the bound (9.26) for lifts and the continuity estimate (9.31) of $a_{\mathcal{T}}$, gives

$$\begin{aligned}\kappa \|h^{-1/2}[\![u_{\mathcal{T}}]\!]\|_{L^2(\mathcal{F}^+)}^2 &\lesssim \|u_{\mathcal{T}} - v_{\mathcal{T}}^0\|_{1,\mathcal{T}}^2 \\ &\quad + \eta_{\mathcal{T}}(u_{\mathcal{T}})\big(\|h^{-1}(u_{\mathcal{T}} - v_{\mathcal{T}}^0)\|_{L^2(\Omega)} + \|h^{-1/2}\{\!\{u_{\mathcal{T}} - v_{\mathcal{T}}^0\}\!\}\|_{L^2(\mathcal{F}^+)}\big).\end{aligned}$$

Note that the presence of $\|\cdot\|_{1,\mathcal{T}}$ rather than $\|\cdot\|_{\kappa,\mathcal{T}}$ on the right-hand side of the above estimate is critical for the argument below. The former is independent of κ and can thus be absorbed on the left-hand side for sufficiently large κ provided $v_{\mathcal{T}}^0 = \mathcal{I}_{\mathcal{T}}^{\mathrm{dG}} u_{\mathcal{T}}$. In fact the interpolation estimates (9.15) in turn imply

$$\begin{aligned}&\|h^{-1}(u_{\mathcal{T}} - \mathcal{I}_{\mathcal{T}}^{\mathrm{dG}} u_{\mathcal{T}})\|_{L^2(\Omega)} + \|h^{-1/2}\{\!\{u_{\mathcal{T}} - \mathcal{I}_{\mathcal{T}}^{\mathrm{dG}} u_{\mathcal{T}}\}\!\}\|_{L^2(\mathcal{F}^+)} \\ &\quad + \|u_{\mathcal{T}} - \mathcal{I}_{\mathcal{T}}^{\mathrm{dG}} u_{\mathcal{T}}\|_{1,\mathcal{T}} \lesssim \|h^{-1/2}[\![u_{\mathcal{T}}]\!]\|_{L^2(\mathcal{F}^+)}.\end{aligned}$$

Hence a Young's inequality yields

$$(\kappa - \overline{\kappa}_{\mathrm{stab}})\|h^{-1/2}[\![u_{\mathcal{T}}]\!]\|_{L^2(\mathcal{F}^+)}^2 \lesssim \kappa^{-1} \eta_{\mathcal{T}}(u_{\mathcal{T}})^2 + \|h^{-1/2}[\![u_{\mathcal{T}}]\!]\|_{L^2(\mathcal{F}^+)}^2,$$

and the desired estimate follows provided κ is sufficiently large. □

As a direct consequence of the previous lemma, we obtain a simpler practical upper bound.

Corollary 9.12 (stabilization-free *a posteriori* upper bound). *If we make the same assumptions as Lemma 9.11, there exists a constant C_U such that for all $\kappa \geq \overline{\kappa}_{\mathrm{jump}}$ we have*

$$\|\widehat{u} - u_{\mathcal{T}}\|_{\kappa,\mathcal{T}} \leq C_U\, \eta_{\mathcal{T}}(u_{\mathcal{T}}). \tag{9.44}$$

Proof. Combine the upper bound (9.39) and Lemma 9.11. □

The partial consistency (9.36) leads to *partial Galerkin orthogonality*

$$\mathcal{B}_{\mathcal{T}}[\widehat{u} - u_{\mathcal{T}}, v] = 0 \quad \text{for all } v \in \mathbb{V}^0_{\mathcal{T}}. \tag{9.45}$$

This would suggest that a quasi-best approximation (Céa's lemma) result in the full space $\mathbb{V}^{-1}_{\mathcal{T}}$ is questionable. However, the lack of consistency is built into the jump terms, which are in turn controlled by the estimator weighted by a negative power of the penalty parameter κ. It thus remains to resort to the lower bound to return to the error and derive a quasi-best approximation estimate for sufficiently large κ. We prove this result next, which expresses the important fact that dG is quasi-optimal with respect to the norm $\|\cdot\|_{\kappa,\mathcal{T}}$ defined in (9.2). This has two significant consequences: first it leads to quasi-monotonicity of the error upon refinement (see Corollary 9.14 below), and second it dictates the approximation class for dG already alluded to in Proposition 9.4 (equivalence of classes for u).

Corollary 9.13 (Céa's lemma). *Under the assumptions of Lemma 9.11, there is* $\overline{\kappa}_{\text{Céa}} \geq \overline{\kappa}_{\text{jump}}$ *such that, for* $\kappa \geq \overline{\kappa}_{\text{Céa}}$,

$$\|\widehat{u} - u_{\mathcal{T}}\|_{\kappa,\mathcal{T}} \leq C_{\text{Céa}} \inf_{v_{\mathcal{T}} \in \mathbb{V}^{-1}_{\mathcal{T}}} \|\widehat{u} - v_{\mathcal{T}}\|_{\kappa,\mathcal{T}}. \tag{9.46}$$

Proof. We combine the orthogonal decomposition (9.37) and the partial Galerkin orthogonality (9.45) to write

$$\mathcal{B}_{\mathcal{T}}[\widehat{u} - u_{\mathcal{T}}, \widehat{u} - u_{\mathcal{T}}] = \mathcal{B}_{\mathcal{T}}[\widehat{u} - u_{\mathcal{T}}, \widehat{u} - v_{\mathcal{T}}] - \mathcal{B}_{\mathcal{T}}[\widehat{u} - u_{\mathcal{T}}, u^{\perp}_{\mathcal{T}}] + \mathcal{B}_{\mathcal{T}}[\widehat{u} - u_{\mathcal{T}}, v^{\perp}_{\mathcal{T}}]$$

for all $v_{\mathcal{T}} \in \mathbb{V}_{\mathcal{T}}$. Invoking the coercivity and continuity of $\mathcal{B}_{\mathcal{T}}$ in Lemma 9.8 (properties of $\mathcal{B}_{\mathcal{T}}$) in conjunction with Lemma 9.9 (control of non-conformity) yields

$$\|\widehat{u} - u_{\mathcal{T}}\|_{\kappa,\mathcal{T}} \lesssim \|\widehat{u} - v_{\mathcal{T}}\|_{\kappa,\mathcal{T}} + \kappa^{1/2}\|h^{-1/2}[\![u_{\mathcal{T}}]\!]\|_{L^2(\mathcal{F}^+)} + \kappa^{1/2}\|h^{-1/2}[\![v_{\mathcal{T}}]\!]\|_{L^2(\mathcal{F}^+)}.$$

Now applying Lemma 9.11 (discontinuity control) results in

$$\|h^{-1/2}[\![u_{\mathcal{T}}]\!]\|_{L^2(\mathcal{F}^+)} \lesssim \kappa^{-1}\eta_{\mathcal{T}}(u_{\mathcal{T}}) \lesssim \kappa^{-1}\|\widehat{u} - u_{\mathcal{T}}\|_{\kappa,\mathcal{T}}$$

because of the lower bound (9.40). We thus end up with

$$\|\widehat{u} - u_{\mathcal{T}}\|_{\kappa,\mathcal{T}} \lesssim \|\widehat{u} - v_{\mathcal{T}}\|_{\kappa,\mathcal{T}} + \kappa^{-1/2}\|\widehat{u} - u_{\mathcal{T}}\|_{\kappa,\mathcal{T}},$$

which for κ sufficiently large gives the desired bound. □

With this best approximation result, we deduce the following crucial property.

Corollary 9.14 (quasi-monotonicity). *Under the assumptions of Lemma 9.11, there is a constant* C_{Mo} *independent of the discretization parameters and* κ *such that for all* $\kappa \geq \overline{\kappa}_{\text{jump}}$ *and* $\mathcal{T}_* \geq \mathcal{T}$ *we have*

$$\|\widehat{u} - u_{\mathcal{T}_*}\|_{\kappa,\mathcal{T}_*} \leq C_{\text{Mo}}\|\widehat{u} - u_{\mathcal{T}}\|_{\kappa,\mathcal{T}}. \tag{9.47}$$

Proof. We rely on the orthogonal decomposition (9.37) to write $u_{\mathcal{T}} = u_{\mathcal{T}}^0 + u_{\mathcal{T}}^\perp$ and on Corollary 9.13 (Céa's lemma). Since $u_{\mathcal{T}}^0 \in \mathbb{V}_{\mathcal{T}}^0 \subset \mathbb{V}_{\mathcal{T}_*}^{-1}$, we see that

$$C_{\text{Céa}}^{-1} \|\widehat{u} - u_{\mathcal{T}_*}\|_{\kappa,\mathcal{T}_*} \le \|\widehat{u} - u_{\mathcal{T}}^0\|_{\kappa,\mathcal{T}_*} = |\widehat{u} - u_{\mathcal{T}}^0|_{H_0^1(\Omega)} = \|\widehat{u} - u_{\mathcal{T}}^0\|_{\kappa,\mathcal{T}}.$$

Therefore, adding and subtracting $u_{\mathcal{T}}^\perp$ and making use of Lemma 9.9 (control of non-conformity) together with Lemma 9.11 (discontinuity control) implies

$$\|\widehat{u} - u_{\mathcal{T}_*}\|_{\kappa,\mathcal{T}_*} \lesssim \|\widehat{u} - u_{\mathcal{T}}\|_{\kappa,\mathcal{T}} + \kappa^{-1/2}\eta_{\mathcal{T}}(u_{\mathcal{T}}).$$

It remains to invoke the lower bound (9.40) to deduce the desired result. □

Corollary 9.14 assumes the same data. In estimating the cost of GALERKIN-DG we need a variant of this result that allows for different data. We establish this next.

Corollary 9.15 (quasi-monotonicity with different data). *Let $\mathcal{T}_* \ge \mathcal{T}$ and $\widehat{\mathcal{D}}_*, \widehat{\mathcal{D}}$ be discrete data on these meshes. Let $\widehat{u}_* = u(\widehat{\mathcal{D}}_*)$, $\widehat{u} = u(\widehat{\mathcal{D}}) \in H_0^1(\Omega)$ and let $u_{\mathcal{T}_*} \in \mathbb{V}_{\mathcal{T}_*}^{-1}$, $u_{\mathcal{T}} \in \mathbb{V}_{\mathcal{T}}^{-1}$ be the corresponding exact and Galerkin solutions. Under the assumptions of Lemma 9.11, for $\kappa \ge \overline{\kappa}_{\text{jump}}$ we have*

$$\|\widehat{u}_* - u_{\mathcal{T}_*}\|_{\kappa,\mathcal{T}_*} \le C_{\text{Mo}}\big(\|\widehat{u} - u_{\mathcal{T}}\|_{\kappa,\mathcal{T}} + |\widehat{u}_* - \widehat{u}|_{H_0^1(\Omega)}\big). \tag{9.48}$$

Proof. We proceed as in the proof of Corollary 9.14 with $\widehat{u}_*$, but in the last step use the fact that $u_{\mathcal{T}}$ and $\widehat{u}$ are the functions associated with the same data $\widehat{\mathcal{D}}$ and thereby satisfy $C_L\eta_{\mathcal{T}}(u_{\mathcal{T}}) \le \|\widehat{u} - u_{\mathcal{T}}\|_{\kappa,\mathcal{T}}$ according to (9.40). Applying the triangle inequality and the property $\|\widehat{u}_* - \widehat{u}\|_{\kappa,\mathcal{T}} = |\widehat{u}_* - \widehat{u}|_{H_0^1(\Omega)}$ concludes the proof. □

We end this section with the dG counterpart of Theorem 4.48 (upper bound for corrections). One striking difference is that the lack of consistency prevents the discrete lower bound in the dG context from localizing to the refined set $\mathcal{T} \setminus \mathcal{T}_*$ for $\mathcal{T}_* \ge \mathcal{T}$. Rather, it contains a global jump term that expresses the lack of conformity and vanishes as $\kappa \to \infty$ in view of Lemma 9.11 (discontinuity control). This is consistent with the upper bound (9.39). We use the notation $\omega_{\mathcal{T}}(\tau)$ for a set of elements $\tau \in \mathcal{T}$ to denote τ augmented by one layer of elements

$$\omega(\tau) := \omega_{\mathcal{T}}(\tau) = \bigcup_{T\in\tau} \omega_{\mathcal{T}}(T).$$

Lemma 9.16 (quasi-localized discrete upper bound). *Assume $\mathcal{T}, \mathcal{T}_* \in \mathbb{T}^\Lambda$, with $\mathcal{T}_* \ge \mathcal{T}$, are two Λ-admissible refinements of $\mathcal{T}_0$ satisfying Assumption 6.19 (initial labelling). Assume that $\widehat{\mathbf{A}}$ and $\widehat{c}$ satisfy the structural assumption (9.20), $\widehat{f} \in \mathbb{F}_{\mathcal{T}}$, and let $u_{\mathcal{T}} \in \mathbb{V}_{\mathcal{T}}^{-1}$, $u_{\mathcal{T}_*} \in \mathbb{V}_{\mathcal{T}_*}^{-1}$ denote the two Galerkin solutions associated with $\mathcal{T}$, $\mathcal{T}_*$ respectively. There is a constant C_{LU} such that for all $\kappa > \overline{\kappa}_{\text{stab}}$ we have*

$$\|u_{\mathcal{T}_*}^0 - u_{\mathcal{T}}\|_{\kappa,\mathcal{T}}^2 \le C_{LU}^2\big(\eta_{\mathcal{T}}^2(u_{\mathcal{T}}, \omega(\mathcal{T} \setminus \mathcal{T}_*)) + \kappa\|h^{-1/2}[\![u_{\mathcal{T}}]\!]\|_{L^2(\mathcal{F}^+)}^2\big),$$

where $u_{\mathcal{T}_*} = u_{\mathcal{T}_*}^0 + u_{\mathcal{T}_*}^\perp$ is the orthogonal decomposition according to (9.37).

Moreover, if $\kappa \ge \overline{\kappa}_{\text{jump}}$, then

$$\|u^0_{\mathcal{T}_*} - u_{\mathcal{T}}\|^2_{\kappa,\mathcal{T}} \le C^2_{LU}\big(\eta^2_{\mathcal{T}}(u_{\mathcal{T}}, \omega(\mathcal{T}\setminus\mathcal{T}_*)) + \kappa^{-1}\eta^2_{\mathcal{T}}(u_{\mathcal{T}})\big). \tag{9.49}$$

Proof. We decompose $u_{\mathcal{T}_*} = u^0_{\mathcal{T}_*} + u^\perp_{\mathcal{T}_*}$ according to (9.37), exploit the partial consistency (9.36) for $u^0_{\mathcal{T}_*}$ with $v^0 \in \mathbb{V}^0_{\mathcal{T}}$, and $\widehat{f} \in \mathbb{F}_{\mathcal{T}} \subset \mathbb{F}_{\mathcal{T}_*}$ to obtain

$$\mathcal{B}_{\mathcal{T}_*}[u_{\mathcal{T}_*}, v^0] = a_{\mathcal{T}_*}[u_{\mathcal{T}_*}, v^0] = a_{\mathcal{T}}[u_{\mathcal{T}_*}, v^0] = \mathcal{B}_{\mathcal{T}}[u_{\mathcal{T}_*}, v^0] = \langle \widehat{f}, v^0\rangle_{\mathcal{T}}.$$

Since $\mathcal{B}_{\mathcal{T}}[u_{\mathcal{T}}, v^0] = \langle \widehat{f}, v^0\rangle_{\mathcal{T}}$, we readily see that

$$\mathcal{B}_{\mathcal{T}}[u^0_{\mathcal{T}_*} - u_{\mathcal{T}}, v^0] = 0 \quad \text{for all } v^0 \in \mathbb{V}^0_{\mathcal{T}}.$$

We rely on this reduced form of Galerkin orthogonality to prove the assertions. To this end, we write $u^0_{\mathcal{T}_*} - u_{\mathcal{T}} = e^0_* - u^\perp_{\mathcal{T}}$, with $e^0_* := u^0_{\mathcal{T}_*} - u^0_{\mathcal{T}}$. Using the coercivity estimate (9.34) for $v = u^0_{\mathcal{T}_*} - u_{\mathcal{T}} \in \mathbb{E}_{\mathcal{T}}$ yields for $\kappa \ge \overline{\kappa}_{\text{stab}}$

$$\begin{aligned}\|u^0_{\mathcal{T}_*} - u_{\mathcal{T}}\|^2_{\kappa,\mathcal{T}} &\lesssim \mathcal{B}_{\mathcal{T}}[u^0_{\mathcal{T}_*} - u_{\mathcal{T}}, u^0_{\mathcal{T}_*} - u_{\mathcal{T}}] \\ &= \mathcal{B}_{\mathcal{T}}[u^0_{\mathcal{T}_*} - u_{\mathcal{T}}, e^0_*] - \mathcal{B}_{\mathcal{T}}[u^0_{\mathcal{T}_*} - u_{\mathcal{T}}, u^\perp_{\mathcal{T}}].\end{aligned} \tag{9.50}$$

Note that the last term cannot be localized, and accounts for the lack of consistency of the dG method. However, it can be made arbitrarily small by increasing the penalty parameter κ. In fact, combining the continuity (9.33) with Lemma 9.9 (control of non-conformity) gives

$$\mathcal{B}_{\mathcal{T}}[u^0_{\mathcal{T}_*} - u_{\mathcal{T}}, u^\perp_{\mathcal{T}}] \lesssim \kappa^{1/2}\|h^{-1/2}[[u_{\mathcal{T}}]]\|_{L^2(\mathcal{F}^+)}\|u^0_{\mathcal{T}_*} - u_{\mathcal{T}}\|_{\kappa,\mathcal{T}}.$$

To localize $\mathcal{B}_{\mathcal{T}}[u^0_{\mathcal{T}_*} - u_{\mathcal{T}}, e^0_*]$, we choose $v^0 = \mathcal{I}^{\text{dG}}_{\mathcal{T}} e^0_*$, where the interpolation operator $\mathcal{I}^{\text{dG}}_{\mathcal{T}}$ is given by (9.7), and exploit the reduced Galerkin orthogonality. Since $e^0_* - \mathcal{I}^{\text{dG}}_{\mathcal{T}} e^0_* \in H^1_0(\Omega)$, the decomposition (9.29) of the bilinear form $\mathcal{B}_{\mathcal{T}}$ reads

$$\begin{aligned}&\mathcal{B}_{\mathcal{T}}\big[u^0_{\mathcal{T}_*} - u_{\mathcal{T}}, e^0_*\big] \\ &\quad= \mathcal{B}_{\mathcal{T}}\big[u^0_{\mathcal{T}_*} - u_{\mathcal{T}}, e^0_* - \mathcal{I}^{\text{dG}}_{\mathcal{T}} e^0_*\big] \\ &\quad= a_{\mathcal{T}}\big[u^0_{\mathcal{T}_*} - u_{\mathcal{T}}, e^0_* - \mathcal{I}^{\text{dG}}_{\mathcal{T}} e^0_*\big] - \int_\Omega \mathcal{L}_{\mathcal{T}}\big[u^0_{\mathcal{T}_*} - u_{\mathcal{T}}\big] \cdot \widehat{\mathbf{A}}\nabla_{\mathcal{T}}(e^0_* - \mathcal{I}^{\text{dG}}_{\mathcal{T}} e^0_*).\end{aligned}$$

We handle the first term as in the conforming case (Theorem 4.48), namely

$$a_{\mathcal{T}}(u^0_{\mathcal{T}_*} - u_{\mathcal{T}}, e^0_*) \lesssim \eta_{\mathcal{T}}(u_{\mathcal{T}}, \omega(\mathcal{T}\setminus\mathcal{T}_*))\,|u^0_{\mathcal{T}_*} - u^0_{\mathcal{T}}|_{H^1_0(\Omega)}.$$

Note that the interpolation estimate (9.9) for $\mathcal{I}^{\text{dG}}_{\mathcal{T}}$ is responsible for the appearance of $\omega(\mathcal{T}\setminus\mathcal{T}_*)$ rather than the smaller set $\mathcal{T}\setminus\mathcal{T}_*$.

For the second term, we use the lift estimate (9.26) along with the H^1-stability of $\mathcal{I}^{\text{dG}}_{\mathcal{T}}$ and $[[u^0_{\mathcal{T}_*}]] = 0$ to write

$$\int_\Omega \mathcal{L}_{\mathcal{T}}[u^0_{\mathcal{T}_*} - u_{\mathcal{T}}] \cdot \widehat{\mathbf{A}}\nabla_{\mathcal{T}} e^0_* \lesssim \|h^{-1/2}[[u_{\mathcal{T}}]]\|_{L^2(\mathcal{F}^+)}\,|u^0_{\mathcal{T}_*} - u^0_{\mathcal{T}}|_{H^1_0(\Omega)}.$$

Inserting the estimates into (9.50), and recalling that $1 \lesssim \overline{\kappa}_{\text{stab}} \leq \kappa$, we find that

$$\|u^0_{\mathcal{T}_*} - u_{\mathcal{T}}\|^2_{\kappa,\mathcal{T}} \lesssim \kappa^{1/2}\|h^{-1/2}[\![u_{\mathcal{T}}]\!]\|_{L^2(\mathcal{F}^+)}\|u^0_{\mathcal{T}_*} - u_{\mathcal{T}}\|_{\kappa,\mathcal{T}} + \big(\eta_{\mathcal{T}}(u_{\mathcal{T}}, \mathcal{T}\setminus\mathcal{T}_*) + \kappa^{1/2}\|h^{1/2}[\![u_{\mathcal{T}}]\!]\|_{L^2(\mathcal{F}^+)}\big)|u^0_{\mathcal{T}_*} - u^0_{\mathcal{T}}|_{H^1_0(\Omega)}.$$

Notice that $u^0_{\mathcal{T}_*} - u^0_{\mathcal{T}} = u^0_{\mathcal{T}_*} - u_{\mathcal{T}} + u^{\perp}_{\mathcal{T}}$, so that in view of Lemma 9.9 (control of non-conformity), we have

$$|u^0_{\mathcal{T}_*} - u^0_{\mathcal{T}}|_{H^1_0(\Omega)} = \|u^0_{\mathcal{T}_*} - u^0_{\mathcal{T}}\|_{\kappa,\mathcal{T}} \lesssim \|u^0_{\mathcal{T}_*} - u_{\mathcal{T}}\|_{\kappa,\mathcal{T}} + \kappa^{1/2}\|h^{-1/2}[\![u_{\mathcal{T}}]\!]\|_{L^2(\mathcal{F}^+)}.$$

The first desired inequality follows from the last two estimates. For the second inequality, it suffices to further invoke Lemma 9.11 (discontinuity control). □

9.4. *Module* GALERKIN-DG

The main ingredients for the *a posteriori* estimation have been derived in the previous section and we can now turn our attention to the adaptive method. In essence, it is the same as in the conforming case (Algorithm 5.4) but accounting for the perturbation arising from the non-conforming setting. Compared to Algorithm 5.4, SOLVE($\mathcal{T}$) determines the discontinuous Galerkin solution to (9.21) and REFINE($\mathcal{T}, \mathcal{M}$) produces the smallest Λ-admissible refinement of $\mathcal{T}$ where all the marked elements $\mathcal{M}$ are refined at least $b \geq 1$ times.

Algorithm 9.17 (GALERKIN-DG). Let $\widehat{\mathcal{T}} \geq \mathcal{T}_0$ be a Λ-admissible refinement, $\Lambda \geq 0$, of a suitable initial mesh $\mathcal{T}_0$. Let data $\widehat{\mathcal{D}} = (\widehat{\boldsymbol{A}}, \widehat{c}, \widehat{f}) \in \mathbb{D}_{\widehat{\mathcal{T}}}$ be discrete on $\widehat{\mathcal{T}}$ and let $\varepsilon > 0$ be a stopping tolerance. The following routine creates a Λ-admissible refinement $\mathcal{T} \geq \widehat{\mathcal{T}}$ and discontinuous Galerkin solution $u_{\mathcal{T}} \in \mathbb{V}^{-1}_{\mathcal{T}}$ for data $\widehat{\mathcal{D}}$ such that $\eta_{\mathcal{T}}(u_{\mathcal{T}}) \leq \varepsilon$.

$[\mathcal{T}, u_{\mathcal{T}}]$ = GALERKIN-DG($\widehat{\mathcal{T}}, \widehat{\mathcal{D}}, \varepsilon$)
  set $j = 0$, $\mathcal{T}_0 = \widehat{\mathcal{T}}$ and do
    $[u_j]$ = SOLVE($\mathcal{T}_j$);
    $[\{\eta_j(u_j, T)\}_{T\in\mathcal{T}_j}]$ = ESTIMATE($u_j, \mathcal{T}_j, \widehat{\mathcal{D}}$);
    if $\eta_j(u_j) \leq \varepsilon$;
      return $(\mathcal{T}_j, u_j)$
    $[\mathcal{M}_j]$ = MARK($\{\eta_j(u_j, T)\}_{T\in\mathcal{T}_j}, \mathcal{T}_j, \theta$);
    $[\mathcal{T}_{j+1}]$ = REFINE($\mathcal{T}_j, \mathcal{M}_j$);
    $j \leftarrow j + 1$;
  while true

We start the analysis of GALERKIN-DG by investigating how the energy norm $\mathcal{B}_{\mathcal{T}}[v, v]^{1/2}$ changes upon refining $\mathcal{T}$. Note that in the conforming case, Lemma 5.2 (Pythagoras) directly provides the relation $|\!|\!|\widehat{u} - u_{\mathcal{T}_*}|\!|\!|_{\Omega} \leq |\!|\!|\widehat{u} - u_{\mathcal{T}}|\!|\!|_{\Omega}$. In the non-conforming setting, the constant on the right-hand side is no longer 1 and jump terms are present in the estimate. Regardless, it is possible to assess the effect of refinement in the energy norm and, in turn, compare two consecutive Galerkin solutions $u_{\mathcal{T}}$ and $u_{\mathcal{T}_*}$, where $\mathcal{T} \in \mathbb{T}^\Lambda$ and $\mathcal{T}_* = \mathsf{REFINE}(\mathcal{T}, \mathcal{M})$ for some $\mathcal{M} \subset \mathcal{T}$. This is the subject of the next three results, but before embarking on this path, we mention a key ingredient for this comparison to hold: *the routine* REFINE *does not refine elements in* $\mathcal{T}$ *more than d times for* $b = 1$; see Corollary 3.31. This implies that for any $F \in \mathcal{F}^+$ and $F_* \in \mathcal{F}_*^+$ with $F_* \subset F$, we have

$$h_F \lesssim h_{F_*}. \tag{9.51}$$

Lemma 9.18 (mesh perturbation). *Let $\mathcal{T} \in \mathbb{T}^\Lambda$ be a Λ-admissible refinement of $\mathcal{T}_0$ satisfying Assumption 6.19 (initial labelling), $\mathcal{M} \subset \mathcal{T}$ and $\mathcal{T}_* = \mathsf{REFINE}(\mathcal{T}, \mathcal{M})$. Assume that $\widehat{\mathbf{A}}$ and $\widehat{c}$ satisfy the structural assumption* (9.20). *There is a constant C such that, for $0 < \varepsilon < 1$ and all $v \in \mathbb{E}_{\mathcal{T}}$,*

$$\mathcal{B}_{\mathcal{T}_*}[v,v] \le (1+\varepsilon)\mathcal{B}_{\mathcal{T}}[v,v] + C\varepsilon^{-1}\kappa \|h^{-1/2}[\![v]\!]\|^2_{L^2(\mathcal{F}^+)}. \tag{9.52}$$

Proof. Because $\nabla_{\mathcal{T}_*} v = \nabla_{\mathcal{T}} v$ when $v \in \mathbb{E}_{\mathcal{T}}$, we directly deduce that

$$\begin{aligned}\mathcal{B}_{\mathcal{T}_*}[v,v] = \mathcal{B}_{\mathcal{T}}[v,v] &+ 2\int_\Omega (\mathcal{L}_{\mathcal{T}}[v] - \mathcal{L}_{\mathcal{T}_*}[v]) \cdot \widehat{\mathbf{A}} \nabla_{\mathcal{T}} v \\ &+ \kappa \|h_*^{-1/2}[\![v]\!]\|^2_{L^2(\mathcal{F}_*^+)} - \kappa \|h^{-1/2}[\![v]\!]\|^2_{L^2(\mathcal{F}^+)},\end{aligned} \tag{9.53}$$

where $h_* := h_{\mathcal{T}_*}$ denotes the mesh size function of $\mathcal{T}_*$.

Unlike the broken gradients, the lifting operators are affected by refinements. However, this effect is controlled by the scaled jumps, as we now show. Using estimate (9.27) twice with $\epsilon = \varepsilon C_{\text{coer}}/2$ yields

$$\begin{aligned}&2\int_\Omega (\mathcal{L}_{\mathcal{T}}[v] - \mathcal{L}_{\mathcal{T}_*}[v]) \cdot \widehat{\mathbf{A}} \nabla_{\mathcal{T}} v \\ &\quad\le \varepsilon C_{\text{coer}} \|\nabla_{\mathcal{T}} v\|^2_{L^2(\mathcal{T})} + \frac{2C_{\text{lift}}^2}{\varepsilon C_{\text{coer}}} \big(\|h_*^{-1/2}[\![v]\!]\|^2_{L^2(\mathcal{F}_*^+)} + \|h^{-1/2}[\![v]\!]\|^2_{L^2(\mathcal{F}^+)}\big).\end{aligned}$$

Hence the coercivity estimate (9.34) gives

$$\begin{aligned}&2\int_\Omega (\mathcal{L}_{\mathcal{T}}[v] - \mathcal{L}_{\mathcal{T}_*}[v]) \cdot \widehat{\mathbf{A}} \nabla_{\mathcal{T}} v \\ &\quad\le \epsilon \mathcal{B}_{\mathcal{T}}[v,v] + \frac{2C_{\text{lift}}^2}{\varepsilon C_{\text{coer}}} \big(\|h^{-1/2}[\![v]\!]\|^2_{L^2(\mathcal{F}^+)} + \|h_*^{-1/2}[\![v]\!]\|^2_{L^2(\mathcal{F}_*^+)}\big).\end{aligned}$$

Inserting this back into (9.53), and using the fact that the jumps of v occur only on $\mathcal{F}^+ \subset \mathcal{F}_*^+$, the mesh size relation (9.51) proves the desired estimate. □

Lemma 9.19 (comparison of solutions). *Let $\mathcal{T} \in \mathbb{T}^\Lambda$ be a Λ-admissible refinement of $\mathcal{T}_0$ satisfying Assumption 6.19 (initial labelling), $\mathcal{M} \subset \mathcal{T}$, and $\mathcal{T}_* =$ $\mathsf{REFINE}(\mathcal{T}, \mathcal{M})$. Assume that $\widehat{\mathbf{A}}$ and $\widehat{c}$ satisfy the structural assumption* (9.20). *Let $\widehat{u} = u(\widehat{\mathcal{D}}) \in H_0^1(\Omega)$ be the solution of the perturbed problem* (5.5) *with discrete data $\widehat{\mathcal{D}}$ and let $u_{\mathcal{T}} \in \mathbb{V}_{\mathcal{T}}^{-1}$, $u_{\mathcal{T}_*} \in \mathbb{V}_{\mathcal{T}_*}^{-1}$ denote the Galerkin solutions associated to $\mathcal{T}$, $\mathcal{T}_*$ respectively with data $\widehat{\mathcal{D}}$. Let $\overline{\kappa}_{\text{jump}}$ be as in Lemma 9.11. There exists a* *constant* C_{comp} *such that for all* $\kappa \geq \overline{\kappa}_{\mathrm{jump}}$ *and all* $0 < \epsilon < 1$ *we have*

$$\begin{aligned}&\mathcal{B}_{\mathcal{T}_*}[\widehat{u} - u_{\mathcal{T}_*}, \widehat{u} - u_{\mathcal{T}_*}]\\ &\quad\leq (1+\epsilon)\mathcal{B}_{\mathcal{T}}[\widehat{u} - u_{\mathcal{T}}, \widehat{u} - u_{\mathcal{T}}]\\ &\qquad - \frac{C_{\mathrm{coer}}}{2}\|\nabla_{\mathcal{T}_*}(u_{\mathcal{T}_*} - u_{\mathcal{T}})\|^2_{L^2(\mathcal{T}_*)} + \frac{C_{\mathrm{comp}}}{\epsilon\kappa}\big(\eta_{\mathcal{T}}(u_{\mathcal{T}})^2 + \eta_{\mathcal{T}_*}(u_{\mathcal{T}_*})^2\big).\end{aligned}$$

Proof. We invoke the partial Galerkin orthogonality (9.45) of $\widehat{u} - u_{\mathcal{T}_*}$ upon testing with $v^0 := u^0_{\mathcal{T}_*} - u^0_{\mathcal{T}} \in \mathbb{V}^0_{\mathcal{T}_*}$:

$$\mathcal{B}_{\mathcal{T}_*}[\widehat{u} - u_{\mathcal{T}_*}, \widehat{u} - u_{\mathcal{T}_*}] = \mathcal{B}_{\mathcal{T}_*}[\widehat{u} - u_{\mathcal{T}_*} + v^0, \widehat{u} - u_{\mathcal{T}_*} + v^0] - \mathcal{B}_{\mathcal{T}_*}[v^0, v^0].$$

Note that

$$\widehat{u} - u_{\mathcal{T}_*} + v^0 = \widehat{u} - u_{\mathcal{T}} + u^{\perp}_{\mathcal{T}} - u^{\perp}_{\mathcal{T}_*}$$

and $\|v^0\|_{\kappa,\mathcal{T}_*} = \|\nabla_{\mathcal{T}_*} v^0\|_{L^2(\mathcal{T}_*)}$, which is critical to the argument below. Hence, from the coercivity and continuity of $\mathcal{B}_{\mathcal{T}_*}$ (Corollary 9.8), we deduce that

$$\begin{aligned}&\mathcal{B}_{\mathcal{T}_*}[\widehat{u} - u_{\mathcal{T}_*}, \widehat{u} - u_{\mathcal{T}_*}]\\ &\quad\leq \mathcal{B}_{\mathcal{T}_*}[\widehat{u} - u_{\mathcal{T}}, \widehat{u} - u_{\mathcal{T}}] + 2C^{1/2}_{\mathrm{cont}}\mathcal{B}_{\mathcal{T}_*}[\widehat{u} - u_{\mathcal{T}}, \widehat{u} - u_{\mathcal{T}}]^{1/2}\|u^{\perp}_{\mathcal{T}} - u^{\perp}_{\mathcal{T}_*}\|_{\kappa,\mathcal{T}_*}\\ &\qquad + C_{\mathrm{cont}}\|u^{\perp}_{\mathcal{T}} - u^{\perp}_{\mathcal{T}_*}\|^2_{\kappa,\mathcal{T}_*} - C_{\mathrm{coer}}\|\nabla_{\mathcal{T}_*}(u^0_{\mathcal{T}_*} - u^0_{\mathcal{T}})\|^2_{L^2(\mathcal{T}_*)}.\end{aligned}$$

We now apply the reverse triangle inequality and Young's inequality

$$\|\nabla_{\mathcal{T}_*}(u^0_{\mathcal{T}_*} - u^0_{\mathcal{T}})\|^2_{L^2(\mathcal{T}_*)} \geq \frac{1}{2}\|\nabla_{\mathcal{T}_*}(u_{\mathcal{T}_*} - u_{\mathcal{T}})\|^2_{L^2(\mathcal{T}_*)} - \|\nabla_{\mathcal{T}_*}(u^{\perp}_{\mathcal{T}_*} - u^{\perp}_{\mathcal{T}})\|^2_{L^2(\mathcal{T}_*)}$$

to deduce that for any $0 < \varepsilon < 1$

$$\begin{aligned}\mathcal{B}_{\mathcal{T}_*}[\widehat{u} - u_{\mathcal{T}_*}, \widehat{u} - u_{\mathcal{T}_*}] \leq{}& (1+\varepsilon)\mathcal{B}_{\mathcal{T}_*}[\widehat{u} - u_{\mathcal{T}}, \widehat{u} - u_{\mathcal{T}}]\\ & - \frac{C_{\mathrm{coer}}}{2}\|\nabla_{\mathcal{T}_*}(u_{\mathcal{T}_*} - u_{\mathcal{T}})\|^2_{L^2(\mathcal{T}_*)} + \frac{C}{\varepsilon}\|u^{\perp}_{\mathcal{T}_*} - u^{\perp}_{\mathcal{T}}\|^2_{\kappa,\mathcal{T}_*},\end{aligned}$$

where C is for the remainder of this proof a constant independent of the discretization parameters and κ.

To bound the last term we recall Lemma 9.9 (control of non-conformity),

$$\|u^{\perp}_{\mathcal{T}_*} - u^{\perp}_{\mathcal{T}}\|_{\kappa,\mathcal{T}_*} \lesssim \kappa^{1/2}\|h_*^{-1/2}[[u_{\mathcal{T}_*}]]\|_{L^2(\mathcal{F}^+_*)} + \kappa^{1/2}\|h_*^{-1/2}[[u_{\mathcal{T}}]]\|_{L^2(\mathcal{F}^+_*)},$$

and notice that the last integral over $\mathcal{F}^+_*$ has weights relative to the local mesh size of $\mathcal{T}_* \geq \mathcal{T}$. Since for consecutive meshes the local mesh sizes are comparable, according to (9.51), we can write $\|h^{-1/2}[[u_{\mathcal{T}}]]\|_{L^2(\mathcal{F}^+)}$ instead. Inserting these expressions in the preceding estimate, and using Lemma 9.18 (mesh perturbation) to replace $\mathcal{B}_{\mathcal{T}_*}$ with $\mathcal{B}_{\mathcal{T}}$ on the right-hand side, yields

$$\begin{aligned}\mathcal{B}_{\mathcal{T}_*}[\widehat{u} - u_{\mathcal{T}_*}, \widehat{u} - u_{\mathcal{T}_*}] \leq{}& (1+\varepsilon)\mathcal{B}_{\mathcal{T}}[\widehat{u} - u_{\mathcal{T}}, \widehat{u} - u_{\mathcal{T}}] - \frac{C_{\mathrm{coer}}}{2}\|\nabla_{\mathcal{T}_*}(u_{\mathcal{T}_*} - u_{\mathcal{T}})\|^2_{L^2(\mathcal{T}_*)}\\ & + C\frac{\kappa}{\varepsilon}\big(\|h_*^{-1/2}[[u_{\mathcal{T}_*}]]\|^2_{L^2(\mathcal{F}^+_*)} + \|h^{-1/2}[[u_{\mathcal{T}}]]\|^2_{L^2(\mathcal{F}^+)}\big),\end{aligned}$$

where 2ε has been relabelled ε. Finally, to derive the desired estimate, it remains to invoke Lemma 9.11 (discontinuity control). □

Combining Lemma 9.19 (comparison of solutions) with Lemma 9.10 (*a posteriori* error estimates), we derive the following dG version of Lemma 5.2 (Pythagoras).

Corollary 9.20 (quasi-orthogonality of dG errors). *If we make the same assumptions as Lemma 9.19, then for all*

$$\kappa \geq \kappa_{QO} := \frac{C_{\text{comp}}}{\varepsilon^2 C_L} \quad \textit{and} \quad 0 < \varepsilon \leq \frac{1}{4}$$

we obtain

$$\|\widehat{u} - u_{\mathcal{T}_*}\|^2_{\kappa,\mathcal{T}_*} \leq (1+4\varepsilon)\, \|\widehat{u} - u_{\mathcal{T}}\|^2_{\kappa,\mathcal{T}} - \frac{C_{\text{coer}}}{2}\, \|\nabla_{\mathcal{T}_*}(u_{\mathcal{T}_*} - u_{\mathcal{T}})\|^2_{L^2(\mathcal{T}_*)}.$$

Proof. We make use of the lower bound (9.40), and set

$$D := \frac{C_{\text{comp}}}{\varepsilon C_L},$$

to rewrite the estimate of Lemma 9.19 as follows:

$$\left(1 - \frac{D}{\kappa}\right) \|\widehat{u} - u_{\mathcal{T}_*}\|^2_{\kappa,\mathcal{T}_*} \leq \left(1 + \varepsilon + \frac{D}{\kappa}\right) \|\widehat{u} - u_{\mathcal{T}}\|^2_{\kappa,\mathcal{T}} - \frac{C_{\text{coer}}}{2} \|\nabla_{\mathcal{T}_*}(u_{\mathcal{T}_*} - u_{\mathcal{T}})\|^2_{L^2(\mathcal{T}_*)}.$$

For $\kappa \geq \kappa_{QO} = D/\varepsilon$ this inequality implies

$$\|\widehat{u} - u_{\mathcal{T}_*}\|^2_{\kappa,\mathcal{T}_*} \leq \frac{1+2\varepsilon}{1-\varepsilon} \|\widehat{u} - u_{\mathcal{T}}\|^2_{\kappa,\mathcal{T}} - \frac{C_{\text{coer}}}{2(1-\varepsilon)} \|\nabla_{\mathcal{T}_*}(u_{\mathcal{T}_*} - u_{\mathcal{T}})\|^2_{L^2(\mathcal{T}_*)}.$$

It remains to realize that $(1+2\varepsilon)/(1-\varepsilon) \leq 1 + 4\varepsilon$ provided $\varepsilon \leq \frac{1}{4}$. □

The last ingredient to prove convergence of GALERKIN-DG is a dG version of Proposition 4.56 (estimator reduction) with $f = f_* \in \mathbb{F}_{\mathcal{T}}$. It turns out that the same estimate and proof are valid for dG except that the H^1_0-seminorm is to be replaced by the broken H^1_0-seminorm. We thus state the result without proof.

Proposition 9.21 (estimator reduction). *Given $\mathcal{T} \in \mathbb{T}^\Lambda$ and a subset $\mathcal{M} \subset \mathcal{T}$ of elements marked for refinement, let $\mathcal{T}_* =$ REFINE$(\mathcal{T}, \mathcal{M})$. If $f = P_{\mathcal{T}} f \in \mathbb{F}_{\mathcal{T}}$, then there is a constant C_{Lip} such that, for all $v \in \mathbb{V}_{\mathcal{T}}$, $v_* \in \mathbb{V}_{\mathcal{T}_*}$ and any $\delta > 0$,*

$$\eta_{\mathcal{T}_*}(v_*, \mathcal{T}_*)^2 \leq (1+\delta)\big(\eta_{\mathcal{T}}(v, \mathcal{T})^2 - \lambda\, \eta_{\mathcal{T}}(v, \mathcal{M})^2\big) + (1+\delta^{-1})\, C^2_{\text{Lip}}\, \|\nabla_{\mathcal{T}_*}(v_* - v)\|^2_{L^2(\mathcal{T}_*)}.$$

We are now in a position to prove a contraction property between two consecutive iterations of the adaptive loop GALERKIN-DG.

Theorem 9.22 (contraction property). *Let $\widehat{\mathcal{T}}$ be a Λ-admissible refinement of $\mathcal{T}_0$ satisfying Assumption 6.19 (initial labelling). Let $\widehat{\mathcal{D}} \in \mathbb{D}_{\widehat{\mathcal{T}}}$ be such that $\widehat{\mathbf{A}}$ and $\widehat{c}$ satisfy the structural assumption* (9.20). *Let $\theta \in (0, 1]$ be the Dörfler marking parameter used in the* MARK *module and let $\{\mathcal{T}_j, \mathbb{V}_j, u_j\}_{j=0}^J$ be a sequence of* *conforming meshes, finite element spaces and discrete solutions* $u_j = u_j(\widehat{\mathcal{D}}) \in \mathbb{V}_j$ *created within* GALERKIN-DG. *If* $\widehat{u} \in H_0^1(\Omega)$ *is the exact solution of* (5.5) *with discrete data* $\widehat{\mathcal{D}}$, *then there exist constants* $\overline{\kappa}_{\text{conv}} \geq 0$, $\gamma > 0$, *and* $0 < \alpha < 1$ *independent of the discretization parameters and* κ, *such that, for all* $\kappa \geq \overline{\kappa}_{\text{conv}}$ *and* $0 \leq j < J$,

$$B_{j+1}^2 + \gamma \eta_{\mathcal{T}_{j+1}}^2(u_{j+1}) \leq \alpha^2 \big(B_j^2 + \gamma \eta_{\mathcal{T}_j}^2(u_j) \big), \tag{9.54}$$

where $B_j := (\mathcal{B}_{\mathcal{T}_j}[\widehat{u} - u_j, \widehat{u} - u_j])^{1/2}$ *is the dG norm of* $\widehat{u} - u_j$.

Proof. In essence, we proceed as in Theorem 5.8 (general contraction property) for the conforming case but with minor changes that account for non-conformity. We only explain the differences below. For $j \geq 0$, we shorten the notation and write $\eta_j := \eta_{\mathcal{T}_j}(u_j)$ and $E_j := \|\nabla_{\mathcal{T}_{j+1}}(u_{j+1} - u_j)\|_{L^2(\mathcal{T}_{j+1})}$.

Corollary 9.20 (quasi-orthogonality of dG errors) gives for any $0 < \varepsilon \leq \frac{1}{4}$

$$B_{j+1}^2 \leq (1 + 4\varepsilon) B_j^2 - \frac{C_{\text{coer}}}{2} E_j^2.$$

Combining Proposition 9.21 (estimator reduction), written in terms of $\mathcal{T} = \mathcal{T}_j$, $\mathcal{T}_* = \mathcal{T}_{j+1}$, $v = u_j$ and $v_* = u_{j+1}$, with Dörfler marking $\eta_j(u_j, \mathcal{M}_j) \geq \theta \eta_j$, yields

$$\eta_{j+1}^2 \leq (1 + \delta)(1 - \lambda\theta^2)\eta_j^2 + (1 + \delta^{-1}) C_{\text{Lip}}^2 E_j^2$$

for any $\delta > 0$. We now multiply this inequality by $\gamma > 0$ and add it to the previous one with the following choice of parameters:

$$\delta = \frac{1 - \lambda\theta^2/2}{1 - \lambda\theta^2} - 1, \quad \gamma = \frac{C_{\text{coer}}}{2C_{\text{Lip}}^2(1 + \delta^{-1})}.$$

Consequently, the terms involving E_j^2 cancel out and we end up with

$$\begin{aligned} B_{j+1}^2 + \gamma \eta_{j+1}^2 &\leq (1 + 4\varepsilon) B_j^2 + \gamma (1 + \delta)(1 - \lambda\theta^2)\eta_j^2 \\ &= \left(1 + 4\varepsilon - \frac{\gamma\lambda\theta^2}{4}\right) B_j^2 + \gamma \left(1 - \frac{\lambda\theta^2}{4}\right)\eta_j^2. \end{aligned}$$

We finally choose

$$\varepsilon := \frac{\gamma\lambda\theta^2}{32}$$

to obtain (9.54) with

$$\alpha^2 = \max\left\{1 - \frac{\gamma\lambda\theta^2}{8}, 1 - \frac{\lambda\theta^2}{4}\right\},$$

and conclude the proof. □

Corollary 9.23 (linear convergence). *Under the assumptions of Theorem 9.22, and if* $0 < \alpha < 1$, $\gamma > 0$, $\overline{\kappa}_{\text{conv}} > 0$ *are the constants in* (9.54), *then for all* $\kappa \geq \overline{\kappa}_{\text{conv}}$

we obtain

$$\|\widehat{u} - u_k\|_{\kappa,\mathcal{T}_k} \le C_*\alpha^{k-j}\|\widehat{u} - u_j\|_{\kappa,\mathcal{T}_j},$$

for some constant C_* independent of the discretization parameters and κ.

Proof. Let $e_k^2 = \|\widehat{u} - u_k\|^2_{\kappa,\mathcal{T}_k}$ and $B_k^2 = \mathcal{B}_{\mathcal{T}_k}[\widehat{u} - u_k, \widehat{u} - u_k]^2$, and use the coercivity estimate (9.34), the contraction property (9.54), the continuity estimate (9.33) and the lower bound (9.40), to arrive at

$$C_{\mathrm{coer}} e_k^2 \le B_k^2 \le \alpha^{2(k-j)}\big(B_j^2 + \gamma\eta_j^2\big) \le \alpha^{2(k-j)}\left(C_{\mathrm{cont}} + \frac{\gamma}{C_L^2}\right)e_j^2.$$

This is the desired estimate in disguise with

$$C_* := \frac{1}{C_{\mathrm{coer}}}\left(C_{\mathrm{cont}} + \frac{\gamma}{C_L^2}\right)^{1/2}. \qquad \square$$

We end the discussion of GALERKIN-DG by deriving the optimality property of the Dörfler marking strategy. We mimic the proof of Lemma 6.16 (Dörfler marking) but directly use the optimal parameter $\mu = \frac{1}{2}$ to simplify the argument. We refer to the discussion after Lemma 6.16 for the role of μ and its influence on θ_0. Notice that θ_0 depends on κ^{-1} because of its appearance in the perturbed localized upper bound (9.49). It plays a similar role to σ in Assumption 6.15 (restriction on ω) in the presence of oscillations (one-step method with switch).

Lemma 9.24 (Dörfler marking). *Let $\mathcal{T}_* \ge \mathcal{T}$ be two Λ-admissible refinements of $\mathcal{T}_0$ satisfying Assumption 6.19 (initial labelling). Let $\widehat{\mathcal{D}} \in \mathbb{D}_{\widehat{\mathcal{T}}}$ be such that $\widehat{\mathbf{A}}$ and $\widehat{c}$ satisfy the structural assumption* (9.20). *Let $u_{\mathcal{T}} \in \mathbb{V}_{\mathcal{T}}^{-1}$, $u_{\mathcal{T}_*} \in \mathbb{V}_{\mathcal{T}_*}^{-1}$ denote the Galerkin solutions associated with $\mathcal{T}$, $\mathcal{T}_*$, respectively, and let $\widehat{u} \in H_0^1(\Omega)$ denotes the solution to* (5.5) *with discrete data $\widehat{\mathcal{D}}$. Assume*

$$\kappa > \overline{\kappa}_D := \max\big(\overline{\kappa}_{\mathrm{stab}}, 4C_{\mathrm{Lip}}^2 C_{LU}^2\big).$$

If

$$\eta_{\mathcal{T}_*}(u_{\mathcal{T}_*}) \le \frac{1}{2}\eta_{\mathcal{T}}(u_{\mathcal{T}}), \tag{9.55}$$

then the refined set $\mathcal{T} \setminus \mathcal{T}_*$ satisfies the Dörfler property

$$\eta_{\mathcal{T}}(u_{\mathcal{T}}, \omega(\mathcal{T} \setminus \mathcal{T}_*)) \ge \theta_0\, \eta_{\mathcal{T}}(u_{\mathcal{T}}), \tag{9.56}$$

with

$$0 < \theta_0^2 := \theta_0^2(\kappa) := \frac{1 - 4C_{\mathrm{Lip}}^2 C_{LU}^2 \kappa^{-1}}{4C_{\mathrm{Lip}}^2 C_{LU}^2} < \frac{1}{4C_{\mathrm{Lip}}^2 C_{LU}^2}.$$

Proof. To relate $\eta_{\mathcal{T}}$ to $\eta_{\mathcal{T}_*}$, we invoke Proposition 9.21 (estimator reduction) with $\delta = 1$, along with the localized upper bound (9.49), to write

$$\eta_{\mathcal{T}}(u_{\mathcal{T}})^2 \le 2\eta_{\mathcal{T}_*}(u_{\mathcal{T}_*})^2 + 2C_{\mathrm{Lip}}^2 C_{LU}^2\big(\eta_{\mathcal{T}}(u_{\mathcal{T}}, \omega(\mathcal{T} \setminus \mathcal{T}_*))^2 + \kappa^{-1}\eta_{\mathcal{T}}(u_{\mathcal{T}})^2\big).$$

This, combined with (9.55), yields

$$\left(\frac{1}{2}-2C_{\mathrm{Lip}}^2C_{LU}^2\kappa^{-1}\right)\eta_{\mathcal{T}}(u_{\mathcal{T}})^2 \le 2C_{\mathrm{Lip}}^2C_{LU}^2\eta_{\mathcal{T}}(u_{\mathcal{T}},\omega(\mathcal{T}\setminus\mathcal{T}_*))^2$$

for $\kappa \ge \overline{\kappa}_{\mathrm{D}}$. This is the desired result in disguise. □

9.5. *Convergence of* AFEM-DG-TS

Algorithm 9.1 (AFEM-DG-TS) relies on two modules: GALERKIN-DG and DATA. We have analysed the performance of GALERKIN-DG in the previous section and showed in Section 7 that the output $[\widehat{\mathcal{T}}_k,\widehat{\mathcal{D}}_k] = \mathsf{DATA}(\mathcal{T}_k,\mathcal{D},\omega\varepsilon_k)$ satisfies

$$\|\mathcal{D}-\widehat{\mathcal{D}}_k\|_{D(\Omega)} \le \omega\varepsilon_k \quad\Rightarrow\quad |u-\widehat{u}_k|_{H_0^1(\Omega)} \le C_{D}\omega\varepsilon_k. \tag{9.57}$$

Recall that $u = u(\mathcal{D}), \widehat{u}_k = u(\widehat{\mathcal{D}}_k) \in H_0^1(\Omega)$ are the exact solutions to (2.7) with exact data $\mathcal{D}$ and discrete data $\widehat{\mathcal{D}}_k$, respectively. We also recall that $\widehat{\mathcal{D}}_k$ satisfies the structural assumption (9.20) uniformly in k and thus C_D does not depend on k.

We start with a result guaranteeing that the cost of GALERKIN-DG does not depend on the iteration counter k within AFEM-DG-TS.

Lemma 9.25 (computational cost of GALERKIN-DG). *For any $\kappa \ge \overline{\kappa}_{\mathrm{conv}}$ and any $k\in\mathbb{N}$, the number of sub-iterations J_k inside a call of* GALERKIN-DG *at iteration k of Algorithm 9.1* (AFEM-DG-TS) *is bounded independently of k.*

Proof. We proceed as in the proof of Proposition 5.27 (computational cost of GALERKIN) for the conforming case, and focus on the essential differences. We fix the iteration counter $k \ge 1$, recall that the output of the $(k-1)$th loop of AFEM-DG-TS is $[\mathcal{T}_k, u_k] = \mathsf{GALERKIN\text{-}DG}(\widehat{\mathcal{T}}_{k-1},\widehat{\mathcal{D}}_{k-1},\varepsilon_{k-1})$, and let $\mathcal{T}_{k,j}$ and $\widehat{u}_{k,j}\in\mathbb{V}_{\mathcal{T}_{k,j}}^{-1}$ denote the jth mesh and Galerkin solution to (9.21) with data $\widehat{\mathcal{D}}_k$ in the kth loop of AFEM-DG-TS. The exact solution to the perturbed problem (5.5) with discrete coefficient $\widehat{\mathcal{D}}_k$ is $\widehat{u}_k = u(\widehat{\mathcal{D}}_k)$.

We recall that $\mathcal{T}_{k,0} = \widehat{\mathcal{T}}_k$ is the mesh produced by DATA, and assume that $u_{k,0}\in\mathbb{V}_{k,0}$ satisfies $\eta_{\mathcal{T}_{k,0}}(u_{k,0}) > \varepsilon_k$, because otherwise $J_k = 0$ and there is nothing to prove. In view of Corollary 9.23 (linear convergence), all we need to prove is that the error $\|u_{k,0}-\widehat{u}_k\|_{\kappa,\mathcal{T}_{k,0}}$ entering GALERKIN-DG is bounded by ε_k. We resort to Corollary 9.15 (quasi-monotonicity with different data) to write

$$\|u_{k,0}-\widehat{u}_k\|_{\kappa,\mathcal{T}_{k,0}} \le C_{\mathrm{Mo}}\big(\|u_k-\widehat{u}_{k-1}\|_{\kappa,\mathcal{T}_k} + |\widehat{u}_k-\widehat{u}_{k-1}|_{H_0^1(\Omega)}\big).$$

The appearance of the last term is the only difference with respect to Proposition 5.27. However, in view of property (9.57) of DATA, we infer that

$$|\widehat{u}_k-\widehat{u}_{k-1}|_{H_0^1(\Omega)} \le |\widehat{u}_k-u|_{H_0^1(\Omega)} + |\widehat{u}_{k-1}-u|_{H_0^1(\Omega)} \le C_D\omega(\varepsilon_k+\varepsilon_{k-1}) = 3C_D\omega\varepsilon_k.$$

Moreover, the stabilization-free upper bound (9.44) implies

$$\|u_k-\widehat{u}_{k-1}\|_{\kappa,\mathcal{T}_k} \le C_U\,\eta_{\mathcal{T}_k}(u_k) \le C_U\varepsilon_{k-1} = 2C_U\varepsilon_k,$$

which, combined with the lower bound (9.40), further yields

$$\|u_{k,0} - \widehat{u}_k\|_{\kappa,\mathcal{T}_{k,0}} \le (3C_D\omega + 2C_U)\varepsilon_k = \Lambda\varepsilon_k \quad \Rightarrow \quad \eta_{\mathcal{T}_{k,0}} \le C_L^{-1}\Lambda\varepsilon_k.$$

This is the requisite estimate. In fact, recalling Corollary 9.23, we see that

$$\eta_{\mathcal{T}_{k,j}}(u_{k,j}) \le C_L^{-1}\|u_{k,j} - \widehat{u}_k\|_{\kappa,\mathcal{T}_{k,j}} \le C_L^{-1}C_*\alpha^j\|u_{k,0} - \widehat{u}_k\|_{\kappa,\mathcal{T}_{k,0}} \le C_L^{-1}C_*\Lambda\varepsilon_k\alpha^j.$$

Since GALERKIN-DG stops when $\eta_{\mathcal{T}_{J_k,k}}(u_{J_k,k}) \le \varepsilon_k$, we finally conclude as in the proof of Proposition 5.27 that J_k is independent of k. □

The proof of convergence of AFEM-DG-TS is identical to the proof of Proposition 5.29 upon replacing the seminorm $|.|_{H_0^1(\Omega)}$ with the appropriate dG norm. It is therefore not repeated here.

Proposition 9.26 (convergence of AFEM-DG-TS). *For any $\kappa \ge \overline{\kappa}_{\text{conv}}$ and $k \ge 0$, the $(k+1)$th iteration of* AFEM-DG-TS *terminates and requires a finite number of inner iterations of* GALERKIN-DG *independent of k. Moreover, if $u \in H_0^1(\Omega)$ denotes the solution to (2.7), there exists a constant C_* such that the output of $[\mathcal{T}_{k+1}, u_{k+1}] =$* GALERKIN-DG*$(\widehat{\mathcal{T}}_k, \widehat{\mathcal{D}}_k, \varepsilon_k)$ satisfies*

$$\|u - u_{k+1}\|_{\kappa,\mathcal{T}_{k+1}} \le C_*\varepsilon_k \quad \textit{for all } k \ge 0.$$

Therefore AFEM-DG-TS *stops after*

$$K < 2 + \frac{\log \varepsilon_0/\texttt{tol}}{\log 2}$$

iterations and delivers

$$\|u - u_{K+1}\|_{\kappa,\mathcal{T}_K} \le C_*\,\texttt{tol}.$$

9.6. *Rate-optimality of* AFEM-DG-TS

To derive rates of convergence for the discontinuous Galerkin method, we proceed similarly to Section 6 for the conforming case. Recall that in the kth step of Algorithm 9.1 (AFEM-DG-TS), the output of $[\widehat{\mathcal{T}}_k, \widehat{\mathcal{D}}_k] =$ DATA$(\mathcal{T}_k, \mathcal{D}, \omega\varepsilon_k)$ is fed to $[\mathcal{T}_{k+1}, u_{k+1}] =$ GALERKIN-DG$(\widehat{\mathcal{T}}_k, \widehat{\mathcal{D}}_k, \varepsilon_k)$, which in turn iterates J_k times. Lemma 9.25 shows that J_k is uniformly bounded in k, and we assume that $J_k \ge 1$, for otherwise the module GALERKIN-DG is skipped altogether. We let $(\mathcal{T}_{k,j}, \mathcal{M}_{k,j}, u_{\mathcal{T}_{k,j}})$ denote the triplets of grids, marked sets and discrete solutions computed within GALERKIN-DG$(\widehat{\mathcal{T}}_k, \widehat{\mathcal{D}}_k, \varepsilon_k)$ for $0 \le j < J_k$. Note that

$$\widehat{\varepsilon}_{k,j} := \eta_{\mathcal{T}_{k,j}}(u_{\mathcal{T}_{k,j}}, \widehat{\mathcal{D}}_k) > \varepsilon_k, \quad 0 \le j < J_k$$

so that together with the lower *a posteriori* error estimate (9.40), we infer that

$$\|\widehat{u}_k - u_{\mathcal{T}_{k,j}}\|_{\kappa,\mathcal{T}_{k,j}} \ge C_L\widehat{\varepsilon}_{k,j} > C_L\varepsilon_k,$$

where $\widehat{u}_k = u(\widehat{\mathcal{D}}_k) \in H_0^1(\Omega)$ is the exact solution with approximate data $\widehat{\mathcal{D}}_k$.

The module DATA guarantees (9.57), and the parameter ω modulates the discrepancy between u and $\widehat{u}_k$ relative to ε_k. The error due to data approximation can be made small relative to the finite element approximation by choosing ω much smaller than 1. In addition, we have established Lemma 9.24 (Dörfler marking) for $\theta_0 < 1$, which implies a Dörfler property for any $0 < \theta \leq \theta_0$. The restrictions on the parameters κ, ω and θ are gathered in the following assumption.

Assumption 9.27 (restrictions on κ, ω and θ). Assume that $\kappa > \max(\overline{\kappa}_D, \overline{\kappa}_{\text{conv}})$, that $0 < \omega \leq \frac{1}{4} C_{\text{Mo}}^{-1} C_L C_D^{-1}$ and that $0 < \theta \leq \theta_0(\kappa)$, where κ_D and θ_0 are defined in Lemma 9.24 (Dörfler marking).

Note that if Assumption 9.27 is valid then

$$|u - \widehat{u}_k|_{H_0^1(\Omega)} \leq \frac{1}{4} C_{\text{Mo}}^{-1} C_L \varepsilon_k. \tag{9.58}$$

The next results rely on Assumption 6.3 (approximability of u) and Assumption 6.10 (approximability of data). They are stated and proved for conforming meshes and continuous approximations of u. However, Proposition 9.4 (equivalence of classes for u) and Remark 9.5 (equivalence of classes for $\mathcal{D}$) show that these classes coincide with the conforming case.

Proposition 9.28 (cardinality of marked sets). *Let Assumptions 6.3 (approximability of u), 6.17 (cardinality of $\mathcal{M}$) and 9.27 (restrictions on κ, ω and θ) hold. If $\widehat{\varepsilon}_{k,0} > \varepsilon_k$, then* GALERKIN-DG *at iteration k of* AFEM-DG-TS *is called and the cardinality $N_{k,j}(u)$ of the marked set $\mathcal{M}_{k,j}$ satisfies*

$$N_{k,j}(u) \lesssim |u|_s^{1/s} \, \|u - u_{\mathcal{T}_{k,j}}\|_{\kappa,\mathcal{T}_{k,j}}^{-1/s} \quad \textit{for all } 0 \leq j < J_k. \tag{9.59}$$

Proof. Fix $0 \leq j < J_k$ and set

$$\delta := \frac{1}{2} C_{\text{Mo}}^{-1} C_L \, \eta_{\mathcal{T}_{k,j}}(u_{\mathcal{T}_{k,j}}) \geq \frac{1}{2} C_{\text{Mo}}^{-1} C_L \varepsilon_k,$$

because $\eta_{\mathcal{T}_{k,j}}(u_{\mathcal{T}_{k,j}}) > \varepsilon_k$ for $j < J_k$. Thanks to (9.58), $\widehat{u}_k$ is an $(\frac{1}{2} C_{\text{Mo}}^{-1} C_L \varepsilon_k)$-approximation of order s to u according to Lemma 6.13 (ε-approximation of u of order s). Hence there exists a conforming mesh $\mathcal{T}_\delta \in \mathbb{T}^\Lambda$ and $u_{\mathcal{T}_\delta}^0 \in \mathbb{V}_{\mathcal{T}_\delta}$ such that

$$\|\widehat{u}_k - u_{\mathcal{T}_\delta}^0\|_{\kappa,\mathcal{T}_\delta} = |\widehat{u}_k - u_{\mathcal{T}_\delta}^0|_{H_0^1(\Omega)} \leq \delta, \quad \#\mathcal{T}_\delta \lesssim |u|_{\mathbb{A}_s}^{1/s} \delta^{-1/s}.$$

To compare $\mathcal{T}_\delta$ with $\mathcal{T}_{k,j}$ we consider the overlay $\mathcal{T}_* = \mathcal{T}_{k,j} \oplus \mathcal{T}_\delta$, which satisfies

$$\#\mathcal{T}_* \leq \#\mathcal{T}_{k,j} + \#\mathcal{T}_\delta - \#\mathcal{T}_0;$$

see Proposition 8.15 (mesh overlay is Λ-admissible). Let $u_{\mathcal{T}_*} \in \mathbb{V}_{\mathcal{T}_*}^{-1}$ be the Galerkin solution on the subspace $\mathbb{V}_{\mathcal{T}_*}^{-1}$ and invoke Corollary 9.13 (Céa's lemma), to write

$$\eta_{\mathcal{T}_*}(u_{\mathcal{T}_*}) \leq C_L^{-1} \|\widehat{u}_k - u_{\mathcal{T}_*}\|_{\kappa,\mathcal{T}_*} \leq C_L^{-1} C_{\text{Mo}} \|\widehat{u}_k - u_{\mathcal{T}_\delta}^0\|_{\kappa,\mathcal{T}_*} = C_L^{-1} C_{\text{Mo}} |\widehat{u}_k - u_{\mathcal{T}_\delta}^0|_{H_0^1(\Omega)},$$

whence $\eta_{\mathcal{T}_*}(u_{\mathcal{T}_*}) \leq C_L^{-1} C_{\mathrm{Mo}} \delta$ and

$$\eta_{\mathcal{T}_*}(u_{\mathcal{T}_*}) \leq \frac{1}{2} \eta_{\mathcal{T}_{k,j}}(u_{\mathcal{T}_{k,j}}).$$

Applying Lemma 9.24 (Dörfler marking) to $\mathcal{T}_*$ and $\mathcal{T}_{k,j}$, we infer that the enlarged refined set $\omega(\mathcal{T}_{k,j} \setminus \mathcal{T}_*)$ satisfies the Dörfler marking property

$$\eta_{\mathcal{T}_{k,j}}(u_{\mathcal{T}_{k,j}}, \omega(\mathcal{T}_{k,j} \setminus \mathcal{T}_*)) \geq \theta\, \eta_{\mathcal{T}_*}(u_{\mathcal{T}_*})$$

since $0 < \theta \leq \theta_0$ by Assumption 9.27. The Dörfler marking involves a minimal set $\mathcal{M}_{k,j}$ according to Assumption 6.17, which thus implies

$$N_{k,j}(u) \leq \#\omega(\mathcal{T}_{k,j} \setminus \mathcal{T}_*) \lesssim \#(\mathcal{T}_{k,j} \setminus \mathcal{T}_*) \leq \#\mathcal{T}_\delta - \#\mathcal{T}_0 \lesssim |u|_{\mathbb{A}_s}^{1/s} \delta^{-1/s} \lesssim |u|_{\mathbb{A}_s}^{1/s} \varepsilon_k^{-1/s},$$

because $\#(\mathcal{T}_{k,j} \setminus \mathcal{T}_*) \leq \#\mathcal{T}_* - \#\mathcal{T}_{k,j}$. This concludes the proof. □

Corollary 9.29 (quasi-optimality of GALERKIN-DG). *Let Assumptions 6.3 (approximability of u), 6.17 (cardinality of $\mathcal{M}$) and 9.27 (restrictions on κ, ω and θ) hold. Assume $\kappa \geq \max(\overline{\kappa}_{\mathrm{conv}}, \overline{\kappa}_{\mathrm{D}})$. Then the total number of marked elements $N_k(u)$ within a call to* GALERKIN-DG *satisfies*

$$N_k(u) \leq J C_0 |u|_{\mathbb{A}_s}^{1/s} \varepsilon_k^{-1/s},$$

where $J \geq J_k$ is a uniform upper bound for the number of iterations required by GALERKIN-DG *according to Lemma 9.25 (computational cost of* GALERKIN-DG*).*

Proof. Use the fact that $N_k(u) = \sum_{j=0}^{J_k-1} N_{k,j}(u)$ and combine Propositions 9.28 (cardinality of marked sets) and 9.25 (computational cost of GALERKIN-DG). □

We finally address the rate-optimality of the two-step algorithm AFEM-DG-TS.

Theorem 9.30 (rate-optimality of AFEM-DG-TS). *Let Assumptions 6.3 (approximability of u), 6.10 (approximability of data), 6.11 (quasi-optimality of* DATA*), 6.17 (cardinality of $\mathcal{M}$), 6.19 (initial labelling) and 9.27 (restrictions on κ, ω and θ) hold. Then* AFEM-DG-TS *gives rise to a sequence $\left(\mathcal{T}_k, \mathbb{V}_{\mathcal{T}_k}^{-1}, u_{\mathcal{T}_k}\right)_{k=0}^{K+1}$ such that*

$$\|u - u_{\mathcal{T}_k}\|_{\kappa, \mathcal{T}_k} \leq C(u, \mathcal{D})(\#\mathcal{T}_k)^{-s}, \quad 1 \leq k \leq K+1,$$

where $0 < s = \min\{s_u, s_{\mathcal{D}}\} = \min\{s_u, s_A, s_c, s_f\} \leq n/d$ and

$$C(u, \mathcal{D}) = C_* \left(|u|_{\mathbb{A}_{s_u}}^{1/s_u} + |\boldsymbol{A}|_{\mathbb{M}_{s_A}}^{1/s_A} + |c|_{\mathbb{C}_{s_c}}^{1/s_c} + |f|_{\mathbb{F}_{s_f}}^{1/s_f} \right)^s$$

with constant $C_* > 0$ independent of u and $\mathcal{D}$.

Proof. Assumptions 6.3, 6.17 and 9.27 combined with Corollary 9.29 for u, and Assumptions 6.10 and 6.11 for $\mathcal{D}$, imply the existence of a constant $C_\#$ such that the total number of marked elements within one loop of AFEM-DG-TS is

$$N_k(u) + N_k(\mathcal{D}) \leq C_\# \left(|u|_{\mathbb{A}_{s_u}}^{1/s_u} + |\mathcal{D}|_{\mathbb{A}_{s_{\mathcal{D}}}}^{1/s_{\mathcal{D}}} \right) \varepsilon_k^{-1/s},$$

with $s_u, s_{\mathcal{D}} \leq n/d$. Moreover, upon termination DATA and GALERKIN-DG give

$$|u - \widehat{u}_k|_{H_0^1(\Omega)} \leq \frac{1}{4} C_{\mathrm{Mo}}^{-1} C_L \varepsilon_k,$$

$$\|\widehat{u}_k - u_{\mathcal{T}_{k+1}}\|_{\kappa,\mathcal{T}_{k+1}} \leq C_U \eta_{\mathcal{T}_{k+1}}(u_{\mathcal{T}_{k+1}}) \leq C_U \varepsilon_k,$$

because of (9.58) and (9.44). This implies, by the triangle inequality,

$$\|u - u_{\mathcal{T}_{k+1}}\|_{\kappa,\mathcal{T}_{k+1}} \leq \left(\frac{1}{4} C_{\mathrm{Mo}}^{-1} C_L + C_U\right) \varepsilon_k.$$

We finally conclude as in Theorem 6.24 (rate-optimality of AFEM-TS). □

Remark 6.25 on the role of ω, θ^* and Remark 6.26 on the optimality of the result, written after Theorem 6.24 for the conforming case, remain valid for the non-conforming case and are not repeated here.

9.7. Operator $P_{\mathcal{T}}$ and routine DATA *on Λ-admissible partitions*

In this section we have extensively used the notion of Λ-admissible meshes for the design and study of dG methods, including forcing $f \in H^{-1}(\Omega)$. To this end, as well as for the design of the two-step AFEM for dG, namely AFEM-DG-TS, the construction of the local projection $P_{\mathcal{T}} f \in \mathbb{F}_{\mathcal{T}}$ is critical. We discuss this now.

Recall that for a conforming partition $\mathcal{T} \in \mathbb{T}$, $P_{\mathcal{T}} f$ is defined as a projection to $\mathbb{F}_{\mathcal{T}}$; see Definition 4.24 (projection onto discrete functionals). The definition and subsequent properties of $P_{\mathcal{T}}$ hinge on extensions E_F for $F \in \mathcal{F}$, studied in Lemma 4.20 (extending from faces), as well as on bubble functions ϕ_T, $T \in \mathcal{T}$, and ϕ_F, $F \in \mathcal{F}$ satisfying Assumption 4.21 (abstract cut-off).

The definition of the element bubble functions ϕ_T in (4.14) is local to T and is thus unchanged on non-conforming subdivisions. The situation is different for faces. If F is a conforming face, we have the conforming definitions of E_F and ϕ_F as in (4.17). Instead, if F is a non-conforming face, $F = T \cap T_*$ with $g(T_*) > g(T)$, we use a virtual conforming refinement of ω_F to define E_F and ϕ_F as in (4.17). Recall that $g(T)$ is the generation of $T \in \mathcal{T}$, and $\mathcal{T} \in \mathbb{T}$ is a uniform refinement of $\mathcal{T}_0$ if and only if g is constant on $\mathcal{T}$. Let $\overline{\mathcal{T}}$ be the uniform refinement of $\mathcal{T}_0$ containing T_*, whence $g(T) = g(T_*)$ for all $T \in \overline{\mathcal{T}}$; $\overline{\mathcal{T}}$ is conforming thanks to Assumption 6.19 (initial labelling) on $\mathcal{T}_0$. Let $\overline{T} \in \overline{\mathcal{T}}$ be the element sharing F with T_* (and thus contained in T) and let $\overline{\omega}_F := T_* \cup \overline{T}$ be the virtual conforming patch around F. We now proceed by defining E_F via (4.4) with ω_F replaced by $\overline{\omega}_F$ and ϕ_F as in (4.17) using the basis functions ϕ_z, $z \in \mathcal{V} \cap F$, associated with $\prod_{T \subset \overline{\omega}_F} \mathbb{P}_n(T) \cap H_0^1(\overline{\omega}_F)$. Note that because $\mathcal{T}$ is Λ-admissible, Proposition 3.27 guarantees that the diameters of $\overline{T}$, T, T_*, ω_F and $\overline{\omega}_F$ are all comparable with constants depending on the initial mesh $\mathcal{T}_0$ and Λ.

Assumption 4.21 is an important ingredient in the analysis of $P_{\mathcal{T}}$ and it holds true with ω_F replaced by $\overline{\omega}_F$ when F is a non-conforming face. Therefore Remark 4.26 (local computation) and Corollary 4.31 (local near-best approximation) are valid for Λ-admissible partitions as well. Consequently, all the algorithms and results presented in Section 7 (data approximation) readily extend to Λ-admissible subdivisions as well. We do not dwell on this matter any further.

10. AFEMs for inf-sup stable problems

We go back to the functional framework introduced in Section 2.4. Precisely, let the bilinear form $\mathcal{B}\colon \mathbb{V} \times \mathbb{W} \to \mathbb{R}$ be continuous and *inf-sup stable* (i.e. it satisfies one of the equivalent conditions stated in Theorem 2.8 (Nečas)). Given $f \in \mathbb{W}^*$, let $u \in \mathbb{V}$ be the unique solution of the variational problem

$$u \in \mathbb{V}: \quad \mathcal{B}[u, w] = \langle f, w \rangle \quad \text{for all } w \in \mathbb{W}. \tag{10.1}$$

Let $\mathbb{V}_j \subset \mathbb{V}$, $\mathbb{W}_j \subset \mathbb{W}$ be finite-dimensional subspaces depending on an integer parameter $j \geq 0$, such that

$$\dim \mathbb{V}_j = \dim \mathbb{W}_j = n_j, \quad \mathbb{V}_j \subset \mathbb{V}_{j+1}, \quad \mathbb{W}_j \subset \mathbb{W}_{j+1}.$$

(Note that the notation has changed with respect to Section 3.1, where $\mathbb{V}_N$ was a subspace of dimension N. Here $\mathbb{V}_j$ may stand for $\mathbb{V}_{\mathcal{T}_j}$, where $\mathcal{T}_j$ is the jth mesh generated by an adaptive algorithm.)

We assume $\mathcal{B}$ satisfies a *uniform* discrete inf-sup condition on any product of subspaces $\mathbb{V}_j \times \mathbb{W}_j$, that is, there exists a constant $\beta > 0$ such that for all j

$$\inf_{v \in \mathbb{V}_j} \sup_{w \in \mathbb{W}_j} \frac{\mathcal{B}[v, w]}{\|v\|_{\mathbb{V}} \|w\|_{\mathbb{W}}} \geq \beta. \tag{10.2}$$

Let $u_j \in \mathbb{V}_j$ be the solution of the (Petrov–)Galerkin problem

$$u_j \in \mathbb{V}_j: \quad \mathcal{B}[u_j, w] = \langle f, w \rangle \quad \text{for all } w \in \mathbb{W}_j. \tag{10.3}$$

The first part of this section, which is mostly based on the recent work by Feischl (2022), is devoted to studying the convergence of this approximation. Convergence and rate-optimality of different AFEMs will be discussed next in Section 10.3. Applications will be given to the Stokes problem (see Section 10.4) and the mixed formulation of a scalar diffusion problem (see Section 10.5).

10.1. Linear convergence of inf-sup stable methods

We make the following key assumptions that guarantee the convergence of the sequence u_j to u in the $\mathbb{V}$-norm, and comment about them afterwards. The first assumption is a relaxed form of the *general quasi-orthogonality* property introduced in Carstensen *et al.* (2014) as part of an abstract set of axioms of adaptivity.

Assumption 10.1 (relaxed quasi-orthogonality). For each $N \in \mathbb{N}$ there exists a non-decreasing constant $C = C(N)$ such that

$$\sum_{k=j}^{j+N} \|u_{k+1} - u_k\|_{\mathbb{V}}^2 \leq C(N) \|u - u_j\|_{\mathbb{V}}^2, \quad j \geq 0, \tag{10.4}$$

and

$$C(N) = o(N) \quad \text{as } N \to \infty.$$

Assumption 10.2 (equivalence of error and estimator). There exist constants $C_U \geq C_L > 0$ and, for each $j \geq 0$, an error estimator $\eta_j = \eta_j(u_j)$, such that

$$C_L \eta_j \leq \|u - u_j\|_{\mathbb{V}} \leq C_U \eta_j, \quad j \geq 0. \tag{10.5}$$

Assumption 10.3 (estimator reduction). There exist constants $0 < \rho_1 < 1$ and $C_1 > 0$ such that

$$\eta_{j+1}^2 \leq \rho_1 \eta_j^2 + C_1 \|u_{j+1} - u_j\|_{\mathbb{V}}^2, \quad j \geq 0. \tag{10.6}$$

Remark 10.4. Assumptions 10.2 and 10.3 are abstract and allow for a general convergence theory. In the context of our model problems of Section 2.3, they are valid for discrete data, that is, if the coefficients of the linear operator corresponding to the bilinear form $\mathcal{B}$ are piecewise polynomials on the adopted meshes, and if $f \in \mathbb{F}_{\mathcal{T}}$ (see Section 4.3). We make this concrete in Sections 10.4 and 10.5 below.

Remark 10.5. We comment on the significance of Assumption 10.1 upon considering two extreme cases.

(1) Assumption 10.1 with $C(N) = O(1)$ is precisely the general quasi-orthogonality property of Carstensen *et al.* (2014). It is valid with $C(N) = 1$ for $\mathbb{V} = \mathbb{W}$ and $\mathcal{B}$ symmetric and coercive. Indeed,

$$\mathcal{B}[u_{k+1} - u_k, u - u_{k+1}] = 0 \quad \text{(Galerkin orthogonality)},$$

whence

$$|||u_{k+1} - u_k|||_{\Omega}^2 + |||u - u_{k+1}|||_{\Omega}^2 = |||u - u_k|||_{\Omega}^2,$$

where $|||\cdot|||_{\Omega}$ is the energy norm induced by $\mathcal{B}$. Adding upon k and using telescopic cancellation yields

$$\begin{aligned}\sum_{k=j}^{j+N} |||u_{k+1} - u_k|||_{\Omega}^2 &= \sum_{k=j}^{j+N} |||u - u_k|||_{\Omega}^2 - |||u - u_{k+1}|||_{\Omega}^2 \\ &= |||u - u_j|||_{\Omega}^2 - |||u - u_{j+N+1}|||_{\Omega}^2 \leq |||u - u_j|||_{\Omega}^2.\end{aligned}$$

Finally, the equivalence (2.30) of the norms $\|\cdot\|_{\mathbb{V}}$ and $|||\cdot|||_{\Omega}$ yields the result.

(2) Assumption 10.1 trivially holds with $C(N) = O(N)$ for $\mathcal{B}$ continuous and inf-sup stable. Indeed, choosing in Corollary 3.3 (quasi-monotonicity) $\mathbb{V}_N = \mathbb{V}_k$ or $\mathbb{V}_{k+1}$ and $\mathbb{V}_M = \mathbb{V}_j$ for $j \leq k$, and using the triangle inequality gives

$$\|u_{k+1} - u_k\|_{\mathbb{V}}^2 \lesssim \|u_{k+1} - u\|_{\mathbb{V}}^2 + \|u_k - u\|_{\mathbb{V}}^2 \lesssim \|u - u_j\|_{\mathbb{V}}^2.$$

Adding, we get

$$\sum_{k=j}^{j+N} \|u_{k+1} - u_k\|_{\mathbb{V}}^2 \leq C \sum_{k=j}^{j+N} \|u - u_j\|_{\mathbb{V}}^2 = CN \|u - u_j\|_{\mathbb{V}}^2.$$

However, the relation $C(N) = O(N)$ is not enough for the subsequent analysis. In fact, we need $C(N) = o(N)$.

We now prove that the stated assumptions guarantee the linear convergence of the sequence of Petrov–Galerkin solutions (10.3). This result is similar to the convergence result for the estimators given in Feischl (2022), and exploits the equivalence (10.5) between errors and estimators.

Theorem 10.6 (linear convergence). *Under Assumptions 10.1, 10.2 and 10.3, the discretization* (10.3) *is convergent; precisely, there exist constants* $0 < \rho < 1$ *and* $c > 0$ *such that*

$$e_{j+i} \leq c\rho^i e_j \quad \textit{for all } i, j \in \mathbb{N}, \tag{10.7}$$

where $e_j := \|u - u_j\|_{\mathbb{V}}$.

Proof. The proof is divided into several steps. First, we set

$$E_k := \|u_k - u_{k-1}\|_{\mathbb{V}}.$$

[1] We start by iterating (10.6) $1 \leq n \leq k$ times to obtain

$$\begin{aligned}
\eta_k^2 &\leq \rho_1 \eta_{k-1}^2 + C_1 E_k^2 \\
&\leq \rho_1\left(\rho_1 \eta_{k-2}^2 + C_1 E_{k-1}^2\right) + C_1 E_k^2 \leq \rho_1^2 \eta_{k-2}^2 + C_1\left(E_k^2 + E_{k-1}^2\right) \\
&\leq \rho_1^n \eta_{k-n}^2 + C_1 \sum_{\ell=k-n+1}^{k} E_\ell^2.
\end{aligned}$$

We now invoke Assumption 10.2 to state the upper bound

$$e_k^2 \leq c_1 \eta_k^2$$

and the lower bound

$$\eta_k^2 \leq c_2 e_k^2$$

(with $c_1 = C_U^2$ and $c_2 = C_L^{-2}$). This yields

$$e_k^2 \leq c_1 \eta_k^2 \leq c_1 c_2 \rho_1^n e_{k-n}^2 + c_1 C_1 \sum_{\ell=k-n+1}^{k} E_\ell^2. \tag{10.8}$$

Let $n \in \mathbb{N}$ be sufficiently large that

$$\rho_2 = c_1 c_2 \rho_1^n < 1,$$

and let us relabel $c_1 C_1$ as C_1 to get

$$e_k^2 \leq \rho_2 e_{k-n}^2 + C_1 \sum_{\ell=k-n+1}^{k} E_\ell^2. \tag{10.9}$$

This shows that the reduction property (10.6) of the estimator is valid for the error after n iterations. We cannot expect (10.9) to hold for e_k with $n = 1$, even in the coercive case: see Example 5.7 (lack of strict error monotonicity) for $\boldsymbol{A} = \boldsymbol{I}$ and $f = 1$. It is thus convenient to rewrite (10.9) as follows:

$$e_{kn}^2 \leq \rho_2 e_{(k-1)n}^2 + C_1 \sum_{\ell=(k-1)n+1}^{kn} E_\ell^2. \tag{10.10}$$

[2] Sum up (10.10) from $k = j + 1$ to $k = j + N$, to get

$$\sum_{k=j+1}^{j+N} e_{kn}^2 \leq \rho_2 \sum_{k=j+1}^{j+N} e_{(k-1)n}^2 + C_1 \sum_{\ell=jn+1}^{(j+N)n} E_\ell^2.$$

Using (10.4) we see that

$$\sum_{\ell=jn+1}^{(j+N)n} E_\ell^2 = \sum_{\ell=jn}^{(j+N)n-1} E_{\ell+1}^2 \leq C(Nn-1)\, e_{jn}^2 \leq C(Nn)\, e_{jn}^2,$$

whence

$$\begin{aligned}\sum_{k=j+1}^{j+N} e_{kn}^2 &\leq \rho_2 \sum_{k=j+1}^{j+N} e_{(k-1)n}^2 + C_1 C(Nn)\, e_{jn}^2 \\ &\leq \rho_2 \left(\sum_{k=j+1}^{j+N} e_{kn}^2 + e_{jn}^2 \right) + C_1 C(Nn)\, e_{jn}^2.\end{aligned}$$

This implies

$$(1-\rho_2) \sum_{k=j+1}^{j+N} e_{kn}^2 \leq (\rho_2 + C_1 C(Nn))\, e_{jn}^2,$$

or equivalently

$$\frac{1-\rho_2}{\rho_2 + C_1 C(Nn)} \sum_{k=j+1}^{j+N} e_{kn}^2 \leq e_{jn}^2. \tag{10.11}$$

Let us add the quantity $\sum_{k=j+1}^{j+N} e_{kn}^2$ to both sides to arrive at

$$\left(1 + \frac{1-\rho_2}{\rho_2 + C_1 C(Nn)}\right) \sum_{k=j+1}^{j+N} e_{kn}^2 \leq \sum_{k=j}^{j+N} e_{kn}^2.$$

We can rewrite this inequality as follows:

$$\sum_{k=j+1}^{j+N} e_{kn}^2 \leq \rho(N) \sum_{k=j}^{j+N} e_{kn}^2, \tag{10.12}$$

where

$$\rho(N) = \frac{1}{1 + \frac{1-\rho_2}{\rho_2 + C_1 C(Nn)}} = \frac{\rho_2 + C_1 C(Nn)}{1 + C_1 C(Nn)}$$
$$= 1 - \frac{1-\rho_2}{1 + C_1 C(Nn)} = 1 - \frac{1}{D(N)}$$

with

$$D(N) = \frac{1 + C_1 C(Nn)}{1 - \rho_2} \to \infty, \quad N \to \infty,$$

whenever $C(N)$ diverges. Therefore (10.12) is a contraction for the quantity $\sum_{k=j+1}^{j+N} e_{kn}^2$ with a constant $\rho(N)$ uniform in j that may degenerate to 1 as $N \to \infty$.

[3] We iterate (10.12) and exploit the fact that the left-hand side has one fewer term than the right-hand side. Take

$$j \to j+1, \quad N \to N-1,$$

to get

$$\sum_{k=j+2}^{j+N} e_{kn}^2 \le \rho(N-1) \sum_{k=j+1}^{j+N} e_{kn}^2,$$

whence

$$\sum_{k=j+2}^{j+N} e_{kn}^2 \le \rho(N-1)\rho(N) \sum_{k=j}^{j+N} e_{kn}^2.$$

Iterating, we get

$$e_{(j+N)n}^2 = \sum_{k=j+N}^{j+N} e_{kn}^2 \le \rho(1)\ldots\rho(N-1)\rho(N) \sum_{k=j}^{j+N} e_{kn}^2.$$

We now need to bound the sum on the right-hand side by a single term. To this end, we resort to (10.11), that is,

$$\sum_{k=j+1}^{j+N} e_{kn}^2 \le \frac{\rho_2 + C_1 C(Nn)}{1-\rho_2} e_{jn}^2,$$

and add e_{jn}^2 to both sides:

$$\sum_{k=j}^{j+N} e_{kn}^2 \le \left(1 + \frac{\rho_2 + C_1 C(Nn)}{1-\rho_2}\right) e_{jn}^2 = D(N)\, e_{jn}^2.$$

Altogether, we arrive at

$$e^2_{(j+N)n} \le \rho(1) \ldots \rho(N) D(N)\, e^2_{jn} = D(N) \prod_{k=1}^{N} \left(1 - \frac{1}{D(k)}\right) e^2_{jn}.$$

4 We estimate the factor on the right-hand side. For $N_0 > 0$ to be chosen later, set

$$\rho_0 := D(N_0) \prod_{k=1}^{N_0} \left(1 - \frac{1}{D(k)}\right)$$

and compute the logarithm of ρ_0 via

$$\log(\rho_0) = \log(D(N_0)) + \sum_{k=1}^{N_0} \log\left(1 - \frac{1}{D(k)}\right) \le \log(D(N_0)) - \sum_{k=1}^{N_0} \frac{1}{D(k)},$$

because the log is concave and $\log(1 + x) \le x$. Since we assume in (10.4) that $D(k) \simeq C(kN_0) = o(k)$, the series diverges and we see that

$$\log(\rho_0) < 0$$

for N_0 sufficiently large. Summarizing, there exist $N_0 > 0$ and $0 < \rho_0 < 1$ such that

$$e^2_{(j+N_0)n} \le \rho_0\, e^2_{jn} \quad \text{for all } j \in \mathbb{N}. \tag{10.13}$$

5 For any $j, i \in \mathbb{N}$, we now find $c > 0$ and $0 < \rho < 1$ such that the inequality

$$e_{j+i} \le c\rho^i e_j$$

holds. We decompose j and $j + i$ in terms of integers k, m,

$$j = (k-1)n + \widehat{j}, \quad k \ge 1, \quad 0 \le \widehat{j} < n,$$
$$j + i = (k+m)n + \widehat{i}, \quad m \ge -1, \quad 0 \le \widehat{i} < n,$$

and first examine the case $m \ge 0$. We further decompose

$$m = aN_0 + b, \quad a, b \in \mathbb{N}, \quad 0 \le b < N_0 \quad \Rightarrow \quad a = \frac{m}{N_0} - \frac{b}{N_0}.$$

Note that

$$kn = j - \widehat{j} + n > j, \quad m = \frac{i}{n} - \left(\frac{\widehat{i} - \widehat{j}}{n} + 1\right), \quad a > \frac{i}{nN_0} - \frac{2 + N_0}{N_0}.$$

Therefore, invoking Corollary 3.3 (quasi-monotonicity),

$$e_{j_2} \le \frac{\|\mathcal{B}\|}{\beta} e_{j_1} = C_* e_{j_1}, \quad j_2 \ge j_1 \ge 0,$$

in conjunction with (10.13), yields

$$e_{j+i} \le C_* e_{(k+aN_0)n} \le C_* \rho_0^{a/2} e_{kn} \le C_*^2 \rho_0^{a/2} e_j < C_*^2 \rho_0^{-(2+N_0)/(2N_0)} \left(\rho_0^{1/(2nN_0)}\right)^i e_j.$$

This is the desired estimate with $c = C_*^2 \rho_0^{-(2+N_0)/(2N_0)}$ and $\rho = \rho_0^{1/(2nN_0)}$ for $m \geq 0$. We finally consider $m = -1$ and again use the error quasi-monotonicity to write

$$e_{j+i} \leq C_* e_j = \frac{C_*}{\rho^i} \rho^i e_j < \frac{C_*}{\rho^n} \rho^i e_j.$$

This concludes the proof with

$$c = \max\left\{ C_*^2 \rho_0^{-(2+N_0)/(2N_0)}, \frac{C_*}{\rho^n} \right\}. \qquad \square$$

Remark 10.7 (improving on Assumption 10.1). It is worth emphasizing that while linear convergence (10.7) is established in Theorem 10.6 using Assumption 10.1, the same property combined with uniform inf-sup stability tells us *a posteriori* that the constant $C(N)$ in (10.4) can be made independent of N, i.e. $C(N) = O(1)$. To see this, we apply the linear convergence bound

$$\|u - u_k\|_{\mathbb{V}} \leq c\rho^{k-j} \|u - u_j\|_{\mathbb{V}}$$

in conjunction with the triangle inequality

$$\|u_{k+1} - u_k\|_{\mathbb{V}} \leq \|u - u_{k+1}\|_{\mathbb{V}} + \|u - u_k\|_{\mathbb{V}} \leq 2c\rho^{k-j} \|u - u_j\|_{\mathbb{V}};$$

summation of a geometric series gives

$$\sum_{k=j}^{j+N} \|u_{k+1} - u_k\|_{\mathbb{V}}^2 \leq C \|u - u_j\|_{\mathbb{V}}^2$$

with $C = 4c^2 \sum_{\ell=0}^{\infty} \rho^{2\ell} < +\infty$. This suggests that Assumption 10.1 might be too pessimistic.

10.2. Inf-sup stability implies quasi-orthogonality

We aim at proving the following key result in this section.

Theorem 10.8 (sufficient condition for Assumption 10.1). *Assumption 10.1 (relaxed quasi-orthogonality) is valid if the bilinear form $\mathcal{B}\colon \mathbb{V} \times \mathbb{W} \to \mathbb{R}$ is continuous and uniformly inf-sup stable on the sequence of subspaces $\mathbb{V}_j \times \mathbb{W}_j$, $j \geq 0$.*

To accomplish this task, we proceed in two steps. Using variational techniques, we first establish an intermediate result formally similar to (10.4) (see Corollary 10.14), but involving the norm of a matrix $\boldsymbol{U}$ related to the form $\mathcal{B}$. Next, we rely on algebraic techniques to estimate such a norm (see Theorem 10.15) and complete the proof of the desired result.

In order to perform the first step, we introduce orthonormal bases of the finite element spaces $\mathbb{V}_j$, $\mathbb{W}_j$, $0 \leq j \leq N$, and next we biorthogonalize them. This procedure turns out to be crucial.

We start with some notation. Let

$$n_j = \dim \mathbb{V}_j = \dim \mathbb{W}_j.$$

Let $\mathbb{V}_{j-1}^{\perp}$ and $\mathbb{W}_{j-1}^{\perp}$ denote the orthogonal complements of $\mathbb{V}_{j-1}$ and $\mathbb{W}_{j-1}$ within $\mathbb{V}_j$ and $\mathbb{W}_j$, respectively. Let

$$d_j = n_j - n_{j-1} = \dim \mathbb{V}_{j-1}^{\perp} = \dim \mathbb{W}_{j-1}^{\perp}$$

be the dimension of the Galerkin update to augment the space $\mathbb{V}_{j-1}$ into the next space $\mathbb{V}_j$, and likewise with the space $\mathbb{W}_{j-1}$ and $\mathbb{W}_j$.

We consider orthonormal bases

$$\boldsymbol{v} = \{\boldsymbol{v}(j)\}_{j=0}^{N} \subset \mathbb{V}_N, \quad \boldsymbol{w} = \{\boldsymbol{w}(i)\}_{i=0}^{N} \subset \mathbb{W}_N \tag{10.14}$$

partitioned into blocks for $1 \leq j \leq N$

$$\boldsymbol{v}(j) = (v_k)_{k=n_{j-1}+1}^{n_j} \subset \mathbb{V}_{j-1}^{\perp}, \quad \boldsymbol{w}(i) = (w_k)_{k=n_{i-1}+1}^{n_i} \subset \mathbb{W}_{i-1}^{\perp}, \tag{10.15}$$

and $\boldsymbol{v}(0) \subset \mathbb{V}_0$, $\boldsymbol{w}(0) \subset \mathbb{W}_0$. In other words, $(\boldsymbol{v}(j), \boldsymbol{w}(j))$ represent the d_j new directions added by Galerkin to the current spaces $(\mathbb{V}_{j-1}, \mathbb{W}_{j-1})$ for $1 \leq j \leq N$.

We recall that the bilinear form $\mathcal{B}\colon \mathbb{V}_N \times \mathbb{W}_N \to \mathbb{R}$ satisfies the following uniform properties for all $0 \leq j \leq N$.

(P1) Continuity:

$$\left|\mathcal{B}[v, w]\right| \leq \|\mathcal{B}\| \|v\|_{\mathbb{V}} \|w\|_{\mathbb{W}} \quad \text{for all } v \in \mathbb{V}_j,\ w \in \mathbb{W}_j. \tag{10.16}$$

(P2) Inf-sup condition:

$$\beta \|v\|_{\mathbb{V}} \leq \sup_{w \in \mathbb{W}_j} \frac{\mathcal{B}[v, w]}{\|w\|_{\mathbb{W}}} \quad \text{for all } v \in \mathbb{V}_j. \tag{10.17}$$

The block bases $\boldsymbol{v}$ and $\boldsymbol{w}$ given in (10.14) induce a block matrix

$$\boldsymbol{B} := (\boldsymbol{B}(i, j))_{i,j=0}^{N} \in \mathbb{R}^{n_N \times n_N}$$

defined by

$$\boldsymbol{B}(i, j) = \mathcal{B}[\boldsymbol{v}(j), \boldsymbol{w}(i)]. \tag{10.18}$$

Note that the actual size of $\boldsymbol{B}$ is $n_N = \dim \mathbb{V}_N \gg N$, and that the following analysis entails expressing important quantities in terms of the number of blocks N rather than the dimension n_N.

We will use this block decomposition for a generic matrix

$$\boldsymbol{M} = (\boldsymbol{M}(i, j))_{i,j=0}^{N} \in \mathbb{R}^{n_N \times n_N},$$

and we let $\boldsymbol{M}[k] = (\boldsymbol{M}(i, j))_{i,j=0}^{k}$ denote the *principal kth block* of $\boldsymbol{M}$. Figure 10.1 shows schematically what this means.

We stress that (P2) implies that $\boldsymbol{B}[k]$ is uniformly invertible with

$$\|\boldsymbol{B}[k]^{-1}\|_2 \leq \frac{1}{\beta} \quad \text{for all } 0 \leq k \leq N. \tag{10.19}$$

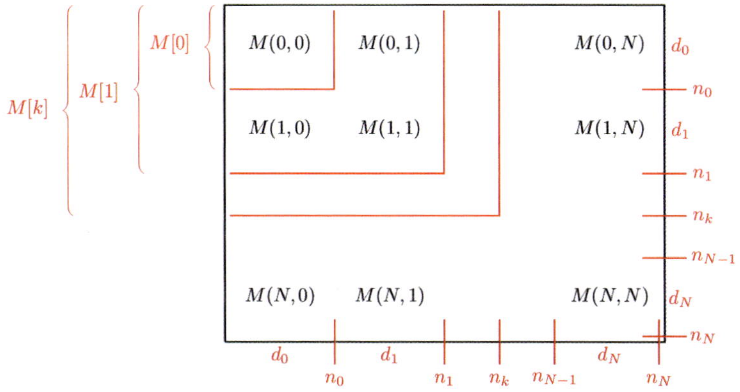

Figure 10.1. Block partition of a matrix $\boldsymbol{M} \in \mathbb{R}^{n_N \times n_N}$ with $(N+1)\times(N+1)$ blocks $\boldsymbol{M}(i, j) \in \mathbb{R}^{d_i \times d_j}$ and principal kth block $\boldsymbol{M}[k] \in \mathbb{R}^{n_k \times n_k}$ with $0 \le i, j, k \le N$.

In fact, (P2) with j replaced by k can be rephrased as follows in terms of the coordinates $\mathbf{v} \in \mathbb{R}^{n_k}$ relative to the orthonormal basis $\{\boldsymbol{v}(j)\}_{j=0}^{k}$ of $\mathbb{V}_k$ of a generic vector in $\mathbb{V}_k$:

$$\beta\|\mathbf{v}\|_2 \le \|\boldsymbol{B}[k]\mathbf{v}\|_2 \quad \text{for all } \mathbf{v} \in \mathbb{R}^{n_k},$$

that is, setting $\mathbf{z} = \boldsymbol{B}[k]\mathbf{v}$,

$$\beta\,\|\boldsymbol{B}[k]^{-1}\mathbf{z}\|_2 \le \|\mathbf{z}\|_2 \quad \text{for all } \mathbf{z} \in \mathbb{R}^{n_k},$$

which is precisely (10.19).

A fundamental linear algebra theorem of Gaussian elimination guarantees the existence of a unique normalized block $\boldsymbol{LU}$ decomposition of $\boldsymbol{B}$ without pivoting due to (10.19):

$$\boldsymbol{B} = \boldsymbol{LU}, \tag{10.20}$$

with block partitioning

$$\boldsymbol{L}(i, j) \in \mathbb{R}^{d_i \times d_j}, \quad \boldsymbol{L}(i, j) = \boldsymbol{0} \quad \text{for } j > i, \quad \boldsymbol{L}(i, i) = \boldsymbol{I}(i, i), \tag{10.21}$$

$$\boldsymbol{U}(i, j) \in \mathbb{R}^{d_i \times d_j}, \quad \boldsymbol{U}(i, j) = \boldsymbol{0} \quad \text{for } i > j. \tag{10.22}$$

10.2.1. Matrix representation

The kth Galerkin solution u_k satisfies

$$u_k \in \mathbb{V}_k: \quad \mathcal{B}[u_k, w] = \langle f, w\rangle \quad \text{for all } w \in \mathbb{W}_k.$$

Equivalently, if $\{\boldsymbol{\gamma}(j)\}_{j=0}^{k} \in \mathbb{R}^{n_k}$ are the coordinates of u_k with respect to the orthonormal basis $\{\boldsymbol{v}(j)\}_{j=0}^{k}$, that is,

$$u_k = \sum_{j=0}^{k} \boldsymbol{\gamma}(j) \cdot \boldsymbol{v}(j),$$

then

$$\sum_{j=0}^{k} \boldsymbol{\gamma}(j) \cdot \mathcal{B}[\boldsymbol{v}(j), \boldsymbol{w}(i)] = \langle f, \boldsymbol{w}(i) \rangle \quad \text{for all } 0 \leq i \leq k,$$

or using matrix notation

$$\sum_{j=0}^{k} \boldsymbol{B}(i,j)\,\boldsymbol{\gamma}(j) = \boldsymbol{f}(i) = \langle f, \boldsymbol{w}(i) \rangle \quad \text{for all } 0 \leq i \leq k. \tag{10.23}$$

If we further write

$$\boldsymbol{\gamma}_k = (\boldsymbol{\gamma}(j))_{j=0}^k \in \mathbb{R}^{n_k}, \quad \boldsymbol{f}_k = (\boldsymbol{f}(i))_{i=0}^k \in \mathbb{R}^{n_k},$$

then (10.23) reduces to

$$\boldsymbol{B}[k]\boldsymbol{\gamma}_k = \boldsymbol{f}_k. \tag{10.24}$$

In view of the definition of $\boldsymbol{f}_k$ we realize that the kth section $\boldsymbol{f}_N[k] \in \mathbb{R}^{n_k}$ of $\boldsymbol{f}_N$ coincides with $\boldsymbol{f}_k$:

$$\boldsymbol{f}_N[k] = (\boldsymbol{f}(i))_{i=0}^k = \boldsymbol{f}_k.$$

However, this statement is *not* true for the solution $\boldsymbol{\gamma}_k$ of (10.24), namely

$$\boldsymbol{\gamma}_N[k] \neq \boldsymbol{\gamma}_k.$$

10.2.2. Block biorthogonal bases

We define biorthogonal bases $\widetilde{\boldsymbol{v}} \subset \mathbb{V}_N$ and $\widetilde{\boldsymbol{w}} \subset \mathbb{W}_N$ as follows:

$$\widetilde{\boldsymbol{v}} := \boldsymbol{U}^{-\top}\boldsymbol{v} \quad \Rightarrow \quad \widetilde{\boldsymbol{v}}(j) = \sum_{m=0}^{j} \boldsymbol{U}^{-\top}(j,m)\boldsymbol{v}(m), \quad 0 \leq j \leq N, \tag{10.25}$$

$$\widetilde{\boldsymbol{w}} := \boldsymbol{L}^{-1}\boldsymbol{w} \quad \Rightarrow \quad \widetilde{\boldsymbol{w}}(i) = \sum_{m=0}^{i} \boldsymbol{L}^{-1}(i,m)\boldsymbol{w}(m), \quad 0 \leq i \leq N. \tag{10.26}$$

We will see below that these bases are convenient for representing the Galerkin solution $u_k \in \mathbb{V}_k$. We start with a list of properties.

Lemma 10.9 (span of new bases). *The vectors $\widetilde{\boldsymbol{v}}$ and $\widetilde{\boldsymbol{w}}$ are bases of $\mathbb{V}_N$ and $\mathbb{W}_N$, respectively, and satisfy*

$$\operatorname{span}\{\widetilde{\boldsymbol{v}}(j)\}_{j=0}^k = \operatorname{span}\{\boldsymbol{v}(j)\}_{j=0}^k,$$
$$\operatorname{span}\{\widetilde{\boldsymbol{w}}(i)\}_{i=0}^k = \operatorname{span}\{\boldsymbol{w}(i)\}_{i=0}^k.$$

Proof. This relies on the fact that $\boldsymbol{U}^{-\top}$ and $\boldsymbol{L}^{-1}$ are lower triangular and the diagonal blocks are non-singular (i.e. both $\boldsymbol{L}$ and $\boldsymbol{U}$ are invertible). □

Now consider the matrix $\widetilde{\boldsymbol{B}}$ induced by $(\widetilde{\boldsymbol{v}}, \widetilde{\boldsymbol{w}})$, namely

$$\widetilde{\boldsymbol{B}} := \mathcal{B}[\widetilde{\boldsymbol{v}}, \widetilde{\boldsymbol{w}}] \in \mathbb{R}^{n_N \times n_N}. \tag{10.27}$$

Lemma 10.10 (biorthogonality). *The block matrix $\widetilde{\boldsymbol{B}}$ is equal to the identity, namely*

$$\widetilde{\boldsymbol{B}}(i,j) = \boldsymbol{I}(i,j) \quad \textit{for all } 0 \le i, j \le N.$$

Proof. We simply combine the definition (10.27) with (10.25) and (10.26) to deduce, for all $0 \le i, j \le N$, that

$$\begin{aligned}
\widetilde{\boldsymbol{B}}(i,j) &= \mathcal{B}[\widetilde{\boldsymbol{v}}(j), \widetilde{\boldsymbol{w}}(i)] \\
&= \mathcal{B}\left[\sum_{m=0}^{j} \boldsymbol{U}^{-\top}(j,m)\boldsymbol{v}(m), \sum_{k=0}^{i} \boldsymbol{L}^{-1}(i,k)\boldsymbol{w}(k)\right] \\
&= \sum_{m=0}^{j}\sum_{k=0}^{i} \boldsymbol{L}^{-1}(i,k)\mathcal{B}[\boldsymbol{v}(m), \boldsymbol{w}(k)]\boldsymbol{U}^{-\top}(j,m) \\
&= \sum_{m=0}^{j}\sum_{k=0}^{i} \boldsymbol{L}^{-1}(i,k)\boldsymbol{B}(k,m)\boldsymbol{U}^{-1}(m,j) \\
&= (\boldsymbol{L}^{-1}\boldsymbol{B}\boldsymbol{U}^{-\top})(i,j) \\
&= (\boldsymbol{L}^{-1}(\boldsymbol{L}\boldsymbol{U})\boldsymbol{U}^{-1})(i,j) \\
&= \boldsymbol{I}(i,j),
\end{aligned}$$

as asserted. □

Generic functions $v \in \mathbb{V}_N$ and $w \in \mathbb{W}_N$ can be represented as follows in terms of the old and new bases:

$$v = \sum_{j=0}^{N} \gamma(j) \cdot \boldsymbol{v}(j) = \sum_{j=0}^{N} \widetilde{\gamma}(j) \cdot \widetilde{\boldsymbol{v}}(j), \tag{10.28}$$

$$w = \sum_{i=0}^{N} \alpha(i) \cdot \boldsymbol{w}(i) = \sum_{i=0}^{N} \widetilde{\alpha}(i) \cdot \widetilde{\boldsymbol{w}}(i). \tag{10.29}$$

The following lemma relates the coordinates in the two systems.

Lemma 10.11 (change of basis). *The coordinates*

$$\gamma = (\gamma(j))_{j=0}^{N}, \quad \alpha = (\alpha(i))_{i=0}^{N}$$

satisfy

$$\gamma = \boldsymbol{U}^{-1}\widetilde{\gamma}, \quad \alpha = \boldsymbol{L}^{-\top}\widetilde{\alpha}. \tag{10.30}$$

Proof. Write (10.28) in vector form and use (10.25) to obtain

$$v = \widetilde{\gamma}^{\top}\widetilde{\boldsymbol{v}} = \widetilde{\gamma}^{\top}(\boldsymbol{U}^{-\top}\boldsymbol{v}) = (\boldsymbol{U}^{-1}\widetilde{\gamma})^{\top}\boldsymbol{v} = \gamma^{\top}\boldsymbol{v},$$

whence

$$\gamma = \boldsymbol{U}^{-1}\widetilde{\gamma}.$$

Similarly, combining (10.29) and (10.26) yields

$$w = \widetilde{\boldsymbol{\alpha}}^\top \widetilde{\boldsymbol{w}} = \widetilde{\boldsymbol{\alpha}}^\top (\boldsymbol{L}^{-1}\boldsymbol{w}) = (\boldsymbol{L}^{-\top}\widetilde{\boldsymbol{\alpha}})^\top \boldsymbol{w} = \boldsymbol{\alpha}^\top \boldsymbol{w}$$

and

$$\boldsymbol{\alpha} = \boldsymbol{L}^{-\top}\widetilde{\boldsymbol{\alpha}}.$$

This concludes the proof. □

10.2.3. An intermediate inequality

We intend to prove the following crucial estimate. This result is in Feischl (2022), but we give a different proof based on variational arguments.

Proposition 10.12 (quasi-orthogonality I). *If $u_k \in \mathbb{V}_k$ denotes the kth Galerkin solution of* (10.3), *then*

$$\sum_{k=0}^{N-1} \|u_{k+1} - u_k\|_{\mathbb{V}}^2 \leq \frac{\|\boldsymbol{U}\|_2^2}{\beta^2} \|u_N - u_0\|_{\mathbb{V}}^2. \tag{10.31}$$

Proof. We proceed in several steps.

[1] *Estimate of $\|u_{k-1} - u_k\|_{\mathbb{V}}$.* Galerkin orthogonality yields

$$\mathcal{B}[u_{k+1} - u_k, w] = 0 \quad \text{for all } w \in \mathbb{W}_k. \tag{10.32}$$

The uniform discrete inf-sup property (P2) implies the existence of $\overline{w} \in \mathbb{W}_{k+1}$ with $\|\overline{w}\|_{\mathbb{W}} = 1$ such that

$$\beta\|u_{k+1} - u_k\|_{\mathbb{V}} \leq \mathcal{B}[u_{k+1} - u_k, \overline{w}]. \tag{10.33}$$

We decompose $\overline{w}$ orthogonally as follows:

$$\overline{w} = \overline{w}_k + \overline{w}_k^\perp, \quad \overline{w}_k \in \mathbb{W}_k, \quad \overline{w}_k^\perp \in \mathbb{W}_k^\perp,$$

with $\|\overline{w}_k^\perp\|_{\mathbb{W}} \leq 1$. In view of (10.32), (10.33) also reads

$$\beta\|u_{k+1} - u_k\|_{\mathbb{V}} \leq \mathcal{B}[u_{k+1} - u_k, \overline{w}_k^\perp] \leq \mathcal{B}\left[u_{k+1} - u_k, \frac{\overline{w}_k^\perp}{\|\overline{w}_k^\perp\|_{\mathbb{W}}}\right].$$

We now let

$$\widehat{w}_{k+1} := \frac{\overline{w}_k^\perp}{\|\overline{w}_k^\perp\|_{\mathbb{W}}} \in \mathbb{W}_k^\perp \subset \mathbb{W}_{k+1},$$

and decompose it along the oblique subspaces $\mathbb{W}_k = \operatorname{span}\{\widetilde{\boldsymbol{w}}(j)\}_{j=0}^k$ and $\operatorname{span}\{\widetilde{\boldsymbol{w}}(k+1)\}$, as illustrated in Figure 10.2. Since $\mathbb{W}_k^\perp = \operatorname{span}\{\boldsymbol{w}(k+1)\}$ and $\boldsymbol{w}(k+1) = (w_j)_{j=n_k+1}^{n_{k+1}}$ is an orthonormal basis, the function $\widehat{w}_{k+1} \in \mathbb{W}_k^\perp$ can be written uniquely as

$$\widehat{w}_{k+1} = \boldsymbol{\alpha}(k+1) \cdot \boldsymbol{w}(k+1),$$

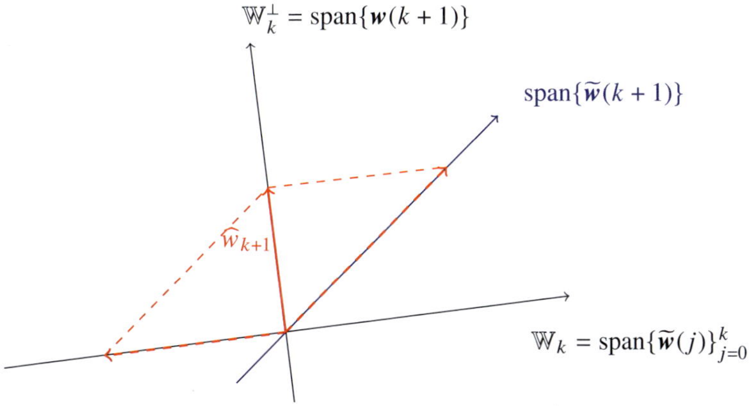

Figure 10.2. Oblique decomposition of the space $\mathbb{W}_{k+1}$ into the subspaces $\mathbb{W}_k = \mathrm{span}\{\widetilde{\boldsymbol{w}}(j)\}_{j=0}^k$ and $\mathrm{span}\{\widetilde{\boldsymbol{w}}(k+1)\}$.

with $\boldsymbol{\alpha}(k+1) \in \mathbb{R}^{d_{k+1}}$ satisfying

$$\|\boldsymbol{\alpha}(k+1)\|_2 = 1 = \|\widehat{w}_{k+1}\|_{\mathbb{W}}.$$

Invoking (10.26), we can express $\boldsymbol{w}(k+1)$ in terms of $\{\widetilde{\boldsymbol{w}}(j)\}_{j=0}^k$ as

$$\begin{aligned}\boldsymbol{w}(k+1) &= \sum_{j=0}^{k+1} \boldsymbol{L}(k+1,j)\,\widetilde{\boldsymbol{w}}(j)\\ &= \widetilde{\boldsymbol{w}}(k+1) + \sum_{j=0}^{k} \boldsymbol{L}(k+1,j)\,\widetilde{\boldsymbol{w}}(j)\end{aligned}$$

because $\boldsymbol{L}(k+1,k+1) = \boldsymbol{I}(k+1,k+1) \in \mathbb{R}^{d_{k+1}\times d_{k+1}}$. Consequently

$$\mathcal{B}[u_{k+1}-u_k, \widehat{w}_{k+1}] = \mathcal{B}[u_{k+1}-u_k, \boldsymbol{\alpha}(k+1)\cdot\widetilde{\boldsymbol{w}}(k+1)]$$

because (10.32) implies

$$\mathcal{B}[u_{k+1}-u_k, \widetilde{\boldsymbol{w}}(j)] = 0 \quad \text{for all } 0 \le j \le k.$$

In addition, the biorthogonality of $\widetilde{\boldsymbol{w}}(k+1)$ with respect to $\widetilde{\boldsymbol{v}}(j)$ for $0 \le j \le k$ translates into

$$\mathcal{B}[u_k, \widetilde{\boldsymbol{w}}(k+1)] = 0 = \mathcal{B}[u_0, \widetilde{\boldsymbol{w}}(k+1)].$$

Moreover, Galerkin orthogonality yields

$$\mathcal{B}[u_{k+1}, \widetilde{\boldsymbol{w}}_{k+1}] = \langle f, \widetilde{\boldsymbol{w}}_{k+1}\rangle = \mathcal{B}[u_N, \widetilde{\boldsymbol{w}}_{k+1}],$$

and collecting the preceding expressions we obtain

$$\|u_{k+1}-u_k\| \le \frac{1}{\beta}\boldsymbol{\alpha}(k+1)\cdot\mathcal{B}[u_N-u_0, \widetilde{\boldsymbol{w}}(k+1)]. \tag{10.34}$$

2 *Estimate of* $\mathcal{B}[u_N - u_0, \widetilde{\boldsymbol{w}}(k+1)]$. We exploit the biorthogonality between $\{\widetilde{\boldsymbol{v}}(j)\}_{j=0}^N$ and $\{\widetilde{\boldsymbol{w}}(j)\}_{j=0}^N$. In fact we write

$$u_N - u_0 = \sum_{j=0}^N \widetilde{\boldsymbol{\gamma}}(j) \cdot \widetilde{\boldsymbol{v}}(j),$$

and substitute into the right-hand side of (10.34) to arrive at

$$\mathcal{B}[u_N - u_0, \widetilde{\boldsymbol{w}}(k+1)] = \sum_{j=0}^N \widetilde{\boldsymbol{\gamma}}(j) \cdot \mathcal{B}[\widetilde{\boldsymbol{v}}(j), \widetilde{\boldsymbol{w}}(k+1)] = \widetilde{\boldsymbol{\gamma}}(k+1).$$

Therefore (10.34) gives

$$\|u_{k+1} - u_k\|_{\mathbb{V}} \leq \frac{1}{\beta} \boldsymbol{\alpha}(k+1) \cdot \widetilde{\boldsymbol{\gamma}}(k+1),$$

whence

$$\|u_{k+1} - u_k\|_{\mathbb{V}} \leq \frac{1}{\beta} \|\widetilde{\boldsymbol{\gamma}}(k+1)\|_2$$

because $\|\boldsymbol{\alpha}(k+1)\|_2 = 1$.

3 *Final estimate.* Compute

$$\sum_{k=0}^{N-1} \|u_{k+1} - u_k\|_{\mathbb{V}}^2 \leq \frac{1}{\beta^2} \sum_{k=0}^{N-1} \|\widetilde{\boldsymbol{\gamma}}(k+1)\|_2^2 \leq \frac{1}{\beta^2} \|\widetilde{\boldsymbol{\gamma}}\|_2^2 \leq \frac{\|\boldsymbol{U}\|_2^2}{\beta^2} \|\boldsymbol{\gamma}\|_2^2,$$

according to (10.30). Since $\{\boldsymbol{v}(j)\}_{j=0}^N$ are orthonormal, we get

$$u_N - u_0 = \sum_{j=0}^N \boldsymbol{\gamma}(j) \cdot \boldsymbol{v}(j) \quad \Rightarrow \quad \|u_N - u_0\|_{\mathbb{V}} = \|\boldsymbol{\gamma}\|_2$$

and

$$\sum_{k=0}^{N-1} \|u_{k+1} - u_k\|_{\mathbb{V}}^2 \leq \frac{\|\boldsymbol{U}\|_2^2}{\beta^2} \|u_N - u_0\|_{\mathbb{V}}^2,$$

as asserted. This concludes the proof. □

In order to get the quasi-orthogonality estimate, we still need to compare the errors $\|u_N - u_0\|_{\mathbb{V}}$ and $\|u - u_0\|_{\mathbb{V}}$. The following is a variant of (3.3).

Lemma 10.13 (stability). *We have*

$$\|u_N - u_0\|_{\mathbb{V}} \leq \frac{\|\mathcal{B}\|}{\beta} \|u - u_0\|_{\mathbb{V}}.$$

Proof. We use (10.17) and (10.16), in this order, together with Galerkin orthogonality, to deduce

$$\beta\|u_N - u_0\|_{\mathbb{V}} \leq \sup_{w\in\mathbb{W}_N} \frac{\mathcal{B}[u_N - u_0, w]}{\|w\|_{\mathbb{W}}} = \sup_{w\in\mathbb{W}_N} \frac{\mathcal{B}[u - u_0, w]}{\|w\|_{\mathbb{W}}} \leq \|\mathcal{B}\|\,\|u - u_0\|_{\mathbb{V}}.$$

This completes the proof. □

Corollary 10.14 (quasi-orthogonality II). *Let $u_k \in \mathbb{V}_k$ be the kth Galerkin solution of* (10.3). *Then, for all $0 \leq j \leq N$, we have*

$$\sum_{k=j}^{j+N-1} \|u_{k+1} - u_k\|_{\mathbb{V}}^2 \leq \frac{\|\mathcal{B}\|^2}{\beta^4}\|\boldsymbol{U}\|_2^2\|u - u_j\|_{\mathbb{V}}^2.$$

Proof. Combining Proposition 10.12 with Lemma 10.13 yields

$$\sum_{k=0}^{N-1} \|u_{k+1} - u_k\|_{\mathbb{V}}^2 \leq \frac{\|\mathcal{B}\|^2}{\beta^4}\|\boldsymbol{U}\|_2^2\|u - u_0\|_{\mathbb{V}}^2.$$

Finally, replacing $u_0 \in \mathbb{V}_0$ with the jth Galerkin solution $u_j \in \mathbb{V}_j$, we obtain the desired estimate. □

This corollary says that in order to prove Theorem 10.8, i.e. to check the validity of Assumption 10.1, it is enough to investigate the growth of the block triangular factor $\boldsymbol{U}$ introduced in (10.20), and more precisely to prove that

$$\|\boldsymbol{U}\|_2 = o(N^{1/2}).$$

This is the second step of our analysis. In fact we prove something more, which is expressed by the following result.

Theorem 10.15 (bound of block matrices $\boldsymbol{L}$ and $\boldsymbol{U}$). *There are constants $C_{LU} > 0$ and $p > 2$ such that the block $\boldsymbol{LU}$ factors of $\boldsymbol{B}$ satisfy*

$$\|\boldsymbol{U}\|_2 + \|\boldsymbol{L}\|_2 + \|\boldsymbol{U}^{-1}\|_2 + \|\boldsymbol{L}^{-1}\|_2 \leq C_{LU}\, N^{1/p}. \tag{10.35}$$

The proof of this theorem is lengthy and very technical; it involves subtle linear algebra arguments, which may not be familiar to many readers. For such reasons, we prefer to postpone it to the end of this section (see Section 10.6).

10.3. Convergence rates of AFEMs for inf-sup stable methods

In this section we discuss AFEMs to solve a boundary value problem admitting a variational formulation of the form

$$u \in \mathbb{V}: \quad \mathcal{B}[u, w] = \langle f, w\rangle \quad \text{for all } w \in \mathbb{W}, \tag{10.36}$$

in which the bilinear form $\mathcal{B}$ on $\mathbb{V} \times \mathbb{W}$ is continuous and inf-sup stable, with inf-sup constant $\beta > 0$, and $f \in \mathbb{W}^*$. We will consider the one-step AFEM given by Algorithm 5.4 (GALERKIN) when all data are discrete, the one-step AFEM with switch given by Algorithm 5.16 (AFEM-SW) when the operator coefficients are discrete but the forcing term is not (as in the Stokes problem), and the general two-step AFEM given by Algorithm 5.1 (AFEM-TS).

10.3.1. Algorithm 5.4 (GALERKIN)

For $j \geq 0$, let $(\mathcal{T}_j, \mathbb{V}_j, u_j)$, with $u_j \in \mathbb{V}_j = \mathbb{V}_{\mathcal{T}_j}$, denote the sequence of meshes, subspaces and Galerkin approximations to (10.36) generated by GALERKIN. Let

$$\eta_{\mathcal{T}_j}(v) = \eta_{\mathcal{T}_j}(v, f) = \left(\sum_{T \in \mathcal{T}_j} \eta_{\mathcal{T}}(v, T)^2 \right)^{1/2} \tag{10.37}$$

be the PDE error estimators used in the loop. If such estimators fulfil Assumptions 10.2 (equivalence of error and estimator) and 10.3 (estimator reduction), then Theorem 10.6 (linear convergence) applies and the following result holds.

Proposition 10.16 (convergence and termination of GALERKIN). *The module* GALERKIN *produces a sequence* $\{u_j\}$ *converging linearly to* $u \in \mathbb{V}$,

$$\|u - u_{j+i}\|_{\mathbb{V}} \leq C \rho^i \|u - u_j\|_{\mathbb{V}} \quad \text{for all } j, i \geq 0, \quad 0 < \rho < 1,$$

thereby reaching any prescribed accuracy $\|u - u_j\|_{\mathbb{V}} \leq \varepsilon$ *in a finite number of iterations.*

10.3.2. Algorithm 5.16 (AFEM-SW)

This algorithm applies to the situation in which the operator coefficients are discrete, whereas the forcing $f \in \mathbb{W}^*$ is not. Then the PDE estimator $\eta_{\mathcal{T}}(v, f)$ depends on f via a projection $P_{\mathcal{T}} f$ upon a finite-dimensional subspace of $\mathbb{W}^*$. Inspired by Lemma 4.5 (localization of H^{-1}-norm), we let $\mathbb{W}^*_{\mathcal{T}}$ denote a suitable decomposition of $\mathbb{W}^*$ subordinate to $\mathcal{T}$ with norm $\|f\|_{\mathbb{W}^*_{\mathcal{T}}}$. In this part of the discussion, we prefer to make the dependence of $\eta_{\mathcal{T}}$ upon $P_{\mathcal{T}} f$ explicit to avoid confusion, so we will write $\eta_{\mathcal{T}}(v, P_{\mathcal{T}} f)$ rather than $\eta_{\mathcal{T}}(v, f)$ as usual.

Let us begin by stating two assumptions on the estimator (10.37) to be used below.

Assumption 10.17 (Lipschitz continuity of estimator). There exists a constant $C_{\text{Lip}} > 0$ such that for any $\mathcal{T} \in \mathbb{T}$, any $v, w \in \mathbb{V}_{\mathcal{T}}$ and any $f, g \in \mathbb{W}^*$, we have

$$|\eta_{\mathcal{T}}(v, P_{\mathcal{T}} f) - \eta_{\mathcal{T}}(w, P_{\mathcal{T}} g)| \leq C_{\text{Lip}} \big(\|v - w\|_{\mathbb{V}} + \|P_{\mathcal{T}} f - P_{\mathcal{T}} g\|_{\mathbb{W}^*_{\mathcal{T}}} \big).$$

Assumption 10.18 (monotonicity of estimator). If $\mathcal{T} \in \mathbb{T}$ and $\mathcal{T}_*$ is a refinement of $\mathcal{T}$, then the projection operator satisfies $P_{\mathcal{T}_*} P_{\mathcal{T}} = P_{\mathcal{T}}$ and

$$\eta_{\mathcal{T}_*}(v, P_{\mathcal{T}} f, T) \leq \eta_{\mathcal{T}}(v, P_{\mathcal{T}} f, T) \quad \text{for all } T \in \mathcal{T} \text{ and } v \in \mathbb{V}_{\mathcal{T}},\ f \in \mathbb{W}^*.$$

It is useful for the subsequent applications to have explicit criteria that guarantee the fulfilment of Assumption 10.3. This is the purpose of the following result.

Proposition 10.19 (estimator reduction under Dörfler marking). *Let the estimator $\eta_{\mathcal{T}}(v, P_{\mathcal{T}} f)$ in (10.37) be used in* GALERKIN. *Let Assumptions 10.17 and 10.18 be valid. Let $\mathcal{T}_*$ be a refinement of $\mathcal{T}$, with estimator $\eta_{\mathcal{T}_*}(v, P_{\mathcal{T}_*} f)$, obtained by bisecting the elements $T \in \mathcal{M}$ marked in* MARK, *using a Dörfler condition on the estimator $\eta_{\mathcal{T}}(u_{\mathcal{T}}, P_{\mathcal{T}} f)$ for the Galerkin solution $u_{\mathcal{T}} \in \mathbb{V}_{\mathcal{T}}$.*

Suppose that there exists $\lambda \in (0, 1)$ such that

$$\eta_{\mathcal{T}_*}(u_{\mathcal{T}}, P_{\mathcal{T}} f, T)^2 \le \lambda\, \eta_{\mathcal{T}}(u_{\mathcal{T}}, P_{\mathcal{T}} f, T)^2 \quad \text{for all } T \in \mathcal{M}. \tag{10.38}$$

Then there exists $0 < \rho < 1$ and $C > 0$ such that, for all $v_{\mathcal{T}_*} \in \mathbb{V}_{\mathcal{T}_*}$,

$$\eta_{\mathcal{T}_*}(v_{\mathcal{T}_*}, P_{\mathcal{T}_*} f)^2 \le \rho\, \eta_{\mathcal{T}}(u_{\mathcal{T}}, P_{\mathcal{T}} f)^2 + C\big(\|v_{\mathcal{T}_*} - u_{\mathcal{T}}\|_{\mathbb{V}}^2 + \|P_{\mathcal{T}_*} f - P_{\mathcal{T}} f\|_{\mathbb{W}^*_{\mathcal{T}_*}}^2\big).$$

Proof. By Assumption 10.17 applied to $\mathcal{T}_*$, we have

$$\eta_{\mathcal{T}_*}(v_{\mathcal{T}_*}, P_{\mathcal{T}_*} f) \le \eta_{\mathcal{T}_*}(u_{\mathcal{T}}, P_{\mathcal{T}} f) + C_{\mathrm{Lip}}(\|v_{\mathcal{T}_*} - u_{\mathcal{T}}\|_{\mathbb{V}} + \|P_{\mathcal{T}_*} f - P_{\mathcal{T}} f\|_{\mathbb{W}^*_{\mathcal{T}_*}}).$$

Using Assumption 10.18 while extending Proposition 4.56 to the current abstract setting, we have for any $\delta > 0$

$$\begin{aligned}\eta_{\mathcal{T}_*}(v_{\mathcal{T}_*}, P_{\mathcal{T}_*} f)^2 \le (1+\delta)\big(&\eta_{\mathcal{T}}(u_{\mathcal{T}}, P_{\mathcal{T}} f)^2 - (1-\lambda)\, \eta_{\mathcal{T}}(u_{\mathcal{T}}, P_{\mathcal{T}} f, \mathcal{M})^2\big) \\ &+ 2(1+\delta^{-1})\, C_{\mathrm{Lip}}^2\big(\|v_{\mathcal{T}_*} - u_{\mathcal{T}}\|_{\mathbb{V}}^2 + \|P_{\mathcal{T}_*} f - P_{\mathcal{T}} f\|_{\mathbb{W}^*_{\mathcal{T}_*}}^2\big).\end{aligned}$$

We conclude using Dörfler condition $\eta_{\mathcal{T}}(u_{\mathcal{T}}, \mathcal{M}) \ge \theta \eta_{\mathcal{T}}(u_{\mathcal{T}})$ and choosing δ small enough. □

Before proceeding further, let us introduce the quantity

$$\mathrm{osc}_{\mathcal{T}}(f) := \|f - P_{\mathcal{T}} f\|_{\mathbb{W}^*_{\mathcal{T}}},$$

which is a measure of the oscillation of the data f. If $u_{\mathcal{T}} \in \mathbb{V}_{\mathcal{T}}$ is the solution of AFEM-SW, we let $\mathcal{E}_{\mathcal{T}}(u_{\mathcal{T}}, f)$ indicate the full estimator defined by

$$\mathcal{E}_{\mathcal{T}}(u_{\mathcal{T}}, f)^2 := \eta_{\mathcal{T}}(u_{\mathcal{T}}, P_{\mathcal{T}} f)^2 + \mathrm{osc}_{\mathcal{T}}(f)^2. \tag{10.39}$$

We formulate the following assumption on the data oscillation.

Assumption 10.20 (quasi-monotonicity of oscillation). There exists a constant $C_{\mathrm{osc}} > 0$ such that, for any $\mathcal{T} \in \mathbb{T}$ and any admissible refinement $\mathcal{T}_* \ge \mathcal{T}$, we have

$$\mathrm{osc}_{\mathcal{T}_*}(f) \le C_{\mathrm{osc}}\, \mathrm{osc}_{\mathcal{T}}(f).$$

A consequence of this assumption is the bound

$$\|P_{\mathcal{T}_*} f - P_{\mathcal{T}} f\|_{\mathbb{W}^*_{\mathcal{T}_*}} \le \|f - P_{\mathcal{T}_*} f\|_{\mathbb{W}^*_{\mathcal{T}_*}} + \|f - P_{\mathcal{T}} f\|_{\mathbb{W}^*_{\mathcal{T}_*}} \le (1 + C_{\mathrm{osc}})\, \mathrm{osc}_{\mathcal{T}}(f),$$

which, inserted into the reduction estimate of Proposition 10.19, gives the existence of $0 < \rho_0 < 1$ and $C_0 > 0$ independent of $\mathcal{T}$ such that

$$\eta_{\mathcal{T}_*}(u_{\mathcal{T}_*}, P_{\mathcal{T}_*} f)^2 \le \rho_0\, \eta_{\mathcal{T}}(u_{\mathcal{T}}, P_{\mathcal{T}} f)^2 + C_0\big(\|u_{\mathcal{T}_*} - u_{\mathcal{T}}\|_{\mathbb{V}}^2 + \mathrm{osc}_{\mathcal{T}}(f)^2\big). \tag{10.40}$$

We aim at establishing a linear convergence result similar to Theorem 10.6 for the sequence $\{u_j\}_{j=0}^{\infty}$ generated by AFEM-SW. To this end, we introduce as usual the short-hand notation $e_j = \|u - u_j\|_{\mathbb{V}}$, $E_{j+1} = \|u_{j+1} - u_j\|_{\mathbb{V}}$, $\eta_j = \eta_{\mathcal{T}_j}(u_j, P_{\mathcal{T}_j} f)$, $\operatorname{osc}_j = \operatorname{osc}_{\mathcal{T}_j}(f)$, $\mathcal{E}_j = \mathcal{E}_{\mathcal{T}_j}(u_j, f)$, and we also introduce the scaled estimator

$$\zeta_j^2 := \eta_j^2 + \gamma \operatorname{osc}_j^2, \tag{10.41}$$

where the parameter $\gamma > 0$ is to be found. Note that at this point we have three parameters $\omega \in (0,1)$, $\xi \in (0,1)$ and $\gamma > 0$ to play with, and the idea is to find conditions on them such that an inequality similar to (10.9) in the proof of Theorem 10.6 holds true. The following result is an intermediate step.

Lemma 10.21 (linear estimator reduction). *Let Assumptions 10.3 (estimator reduction), 10.17 (Lipschitz continuity of estimator), 10.18 (monotonicity of estimator) and 10.20 (quasi-monotonicity of oscillation) be valid. There exists $\omega_0 > 0$ such that, for any choice of parameters $0 < \omega \leq \omega_0$ and $0 < \xi \leq 1/\sqrt{2}$ in* AFEM-SW, *there exist constants $0 < \rho < 1$, $\Lambda > 0$, $\gamma \geq 1$ for which*

$$\zeta_k^2 \leq \rho^{k-j} \zeta_j^2 + \Lambda \sum_{\ell=j+1}^{k} E_\ell^2, \quad k \geq j \geq 0. \tag{10.42}$$

Proof. We discuss the two alternatives in Algorithm 5.16 (AFEM-SW) separately.

[1] *Case* $\operatorname{osc}_j \leq \omega \mathcal{E}_j$. We use (10.40) to get

$$\eta_{j+1}^2 \leq \rho_0 \eta_j^2 + C_0 E_{j+1}^2 + C_0 \operatorname{osc}_j^2$$

and Assumption 10.20 to write

$$\operatorname{osc}_{j+1} \leq C_{\mathrm{osc}} \operatorname{osc}_j .$$

From

$$\operatorname{osc}_j^2 \leq \omega^2 \mathcal{E}_j^2 = \omega^2 (\eta_j^2 + \operatorname{osc}_j^2) \leq \omega^2 (\eta_j^2 + \gamma \operatorname{osc}_j^2) = \omega^2 \zeta_j^2,$$

provided $\gamma \geq 1$, we deduce

$$\begin{aligned}
\zeta_{j+1}^2 = \eta_{j+1}^2 + \gamma \operatorname{osc}_{j+1}^2 &\leq \rho_0 \eta_j^2 + C_0 E_{j+1}^2 + \left(C_0 + \gamma C_{\mathrm{osc}}^2\right) \operatorname{osc}_j^2 \\
&\leq \rho_0 \eta_j^2 + C_0 E_{j+1}^2 + \left(C_0 + \gamma C_{\mathrm{osc}}^2\right) \omega^2 \left(\eta_j^2 + \gamma \operatorname{osc}_j^2\right) \\
&= \left[\rho_0 + \left(C_0 + \gamma C_{\mathrm{osc}}^2\right)\omega^2\right] \eta_j^2 + \left[\left(C_0 + \gamma C_{\mathrm{osc}}^2\right)\omega^2\right] \gamma \operatorname{osc}_j^2 + C_0 E_{j+1}^2 \\
&\leq \left[\rho_0 + \left(C_0 + \gamma C_{\mathrm{osc}}^2\right)\omega^2\right] \zeta_j^2 + C_0 E_{j+1}^2.
\end{aligned} \tag{10.43}$$

Below we will impose

$$\rho_1 := \rho_0 + \left(C_0 + \gamma C_{\mathrm{osc}}^2\right)\omega^2 < 1, \tag{10.44}$$

which will yield the desired bound

$$\zeta_{j+1}^2 \leq \rho_1 \zeta_j^2 + C_0 E_{j+1}^2. \tag{10.45}$$

[2] *Case* $\mathrm{osc}_j > \omega \mathcal{E}_j$. We use $\mathcal{E}_j^2 = \eta_j^2 + \mathrm{osc}_j^2 > \eta_j^2$ to get

$$\eta_j^2 < \frac{1}{\omega^2} \mathrm{osc}_j^2 .$$

Proceeding as in the proofs of Proposition 10.19 (now with $\mathcal{M} = \emptyset$) and (10.40), we obtain for any $\delta > 0$

$$\eta_{j+1}^2 \leq (1+\delta)\eta_j^2 + C_\delta \big(E_{j+1}^2 + \mathrm{osc}_j^2\big) \leq (1-\delta)\eta_j^2 + C_\delta E_{j+1}^2 + \left(\frac{2\delta}{\omega^2} + C_\delta\right) \mathrm{osc}_j^2$$

with $C_\delta = C_0^2(1+\delta^{-1})$. On the other hand, since osc_{j+1} is computed after a call to DATA, it satisfies

$$\mathrm{osc}_{j+1}^2 \leq \xi^2 \sigma_j^2 = \xi^2 \omega^2 \mathcal{E}_j^2 < \xi^2 \mathrm{osc}_j^2 = \frac{1+\xi^2}{2} \mathrm{osc}_j^2 - \frac{1-\xi^2}{2} \mathrm{osc}_j^2 .$$

Combining the last two equations, we obtain

$$\begin{aligned} \zeta_{j+1}^2 &= \eta_{j+1}^2 + \gamma \, \mathrm{osc}_{j+1}^2 \\ &\leq (1-\delta)\eta_j^2 + \gamma \frac{1+\xi^2}{2} \mathrm{osc}_j^2 + C_\delta E_{j+1}^2 + \left(\frac{2\delta}{\omega^2} + C_\delta - \gamma \frac{1-\xi^2}{2}\right) \mathrm{osc}_j^2 . \end{aligned}$$

Below we will enforce

$$\Gamma := \frac{2\delta}{\omega^2} + C_\delta - \gamma \frac{1-\zeta^2}{2} \leq 0, \tag{10.46}$$

which will guarantee

$$\zeta_{j+1}^2 \leq \rho_2 \zeta_j^2 + C_\delta E_{j+1}^2, \tag{10.47}$$

with $\rho_2 := \max(1-\delta, (1+\xi^2)/2) < 1$.

[3] *Choice of parameters.* Summarizing, in both cases [1] and [2] we have obtained

$$\zeta_{j+1}^2 \leq \rho \zeta_j^2 + \Lambda E_{j+1}^2, \tag{10.48}$$

with $\rho := \max(\rho_1, \rho_2) < 1$ and $\Lambda := \max(C_0, C_\delta)$, which holds under the conditions (10.44) and (10.46). Iterating (10.48), we obtain the desired bound (10.42).

To fulfil (10.44), we write ω^2 in the form $\omega^2 = \sigma_0/\gamma$, which gives

$$\rho_1 = \rho_0 + \big(C_0 + \gamma C_{\mathrm{osc}}^2\big)\frac{\sigma_0}{\gamma} = \rho_0 + \left(\frac{C_0}{\gamma} + C_{\mathrm{osc}}^2\right)\sigma_0 \leq \rho_0 + \big(C_0 + C_{\mathrm{osc}}^2\big)\sigma_0$$

since $\gamma \geq 1$, and we pick a $\sigma_0 > 0$ small enough to make $\rho_0 + \big(C_0 + C_{\mathrm{osc}}^2\big)\sigma_0 < 1$.

To fulfil (10.46), we use $\xi \leq 1/\sqrt{2}$ and again $\omega^2 = \sigma_0/\gamma$ to write

$$\Gamma = C_\delta + \left(\frac{2\delta}{\sigma_0} - \frac{1-\xi^2}{2}\right)\gamma \leq C_\delta + \left(\frac{2\delta}{\sigma_0} - \frac{1}{4}\right)\gamma.$$

Choosing $\delta = \delta_0 = \sigma_0/16$ yields

$$\Gamma \leq C_{\delta_0} - \frac{1}{8}\gamma \leq 0 \quad \text{provided} \quad \gamma \geq 8C_{\delta_0}.$$

In conclusion, setting $\gamma_0 = \max(1, 8C_{\delta_0})$ and $\omega_0 = \sqrt{\sigma_0/\gamma_0}$, we fulfil both conditions (10.44) and (10.46) for any $0 < \omega \leq \omega_0$, by choosing the scaling parameter $\gamma = \sigma_0/\omega^2 \geq \gamma_0$. This completes the proof. □

Before establishing the linear convergence result for Algorithm 5.16 (AFEM-SW), we need to extend Assumption 10.2 (equivalence of error and estimator) to the present situation, in which the estimator $\eta_{\mathcal{T}}$ is replaced by the full estimator $\mathcal{E}_{\mathcal{T}}$ defined in (10.39); see Theorem 4.45 (modified residual estimator).

Assumption 10.22 (equivalence of error and full estimator). There exist constants $C_U \geq C_L > 0$ such that

$$C_L \mathcal{E}_j \leq \|u - u_j\|_{\mathbb{V}} \leq C_U \mathcal{E}_j, \quad j \geq 0, \tag{10.49}$$

where $\mathcal{E}_j = \mathcal{E}_{\mathcal{T}_j}(u_{\mathcal{T}_j}, f)$.

Theorem 10.23 (linear convergence for AFEM-SW). *Suppose Assumptions 10.22 (equivalence of error and full estimator), 10.3 (estimator reduction), 10.17 (Lipschitz continuity of estimator) and 10.20 (quasi-monotonicity of oscillation) are valid. There exists* $\omega_0 \in (0, 1]$ *such that, for any choice of parameters* $0 < \omega < \omega_0$ *and* $0 < \xi, \theta < 1$ *in* AFEM-SW, *constants* $0 < \rho < 1$ *and* $c > 0$ *exist for which*

$$e_{j+1} \leq c\rho^i e_j \quad \textit{for all } i, j \in \mathbb{N}, \tag{10.50}$$

where $e_j := \|u - u_j\|_{\mathbb{V}}$.

Proof. By Assumption 10.22 and $\gamma \geq 1$ in (10.41), we get the equivalence of error and scaled estimator

$$\frac{C_L^2}{\gamma}\zeta_j^2 = \frac{C_L^2}{\gamma}(\eta_j^2 + \gamma\,\mathrm{osc}_j^2) \leq C_L^2\mathcal{E}_j^2 \leq e_j^2 \leq C_U^2\mathcal{E}_j^2 \leq C_U^2(\eta_j^2 + \gamma\,\mathrm{osc}_j^2) = C_U^2\zeta_j^2.$$

Invoking (10.42) yields

$$\begin{aligned} e_k^2 \leq C_U^2\zeta_k^2 &\leq C_U^2\rho^{k-j}\zeta_j^2 + C_U^2\Lambda\sum_{\ell=j+1}^{k} E_\ell^2 \\ &\leq \gamma\frac{C_U^2}{C_L^2}\rho^{k-j}e_j^2 + C_U^2\Lambda\sum_{\ell=j+1}^{k} E_\ell^2. \end{aligned}$$

This inequality is similar to the expression (10.8) obtained in step [1] in the proof of Theorem 10.6. Therefore we can finally proceed as in that proof and obtain the desired result. □

10.3.3. Algorithm 5.28 (AFEM-TS)

As usual, DATA produces discrete data $\widehat{\mathcal{D}}_k$ on a mesh $\widehat{\mathcal{T}}_k \geq \mathcal{T}_k$, whereas GALERKIN produces an approximation u_{k+1} on a mesh $\mathcal{T}_{k+1} \geq \widehat{\mathcal{T}}_k$ to the exact solution $\widehat{u}_k$ of the boundary value problem of interest with data $\widehat{\mathcal{D}}_k$. Its kernel is given in Algorithm 5.4 (GALERKIN).

In order to proceed, we need some notation and some assumptions. Let $D(\Omega)$ denote the space of admissible data $\mathcal{D}$ for the boundary value problem at hand; let $\|\mathcal{D}\|_{D(\Omega)}$ be a (quasi-)norm on $D(\Omega)$. If $\mathcal{D}$ collects all data of problem (10.36), we write $\mathcal{B} = \mathcal{B}(\mathcal{D})$ and $\mathcal{F} = \mathcal{F}(\mathcal{D}) = \langle f(\mathcal{D}), \cdot\rangle$ to highlight the dependence of the bilinear and linear forms on the chosen data; similarly, we write $u = u(\mathcal{D})$ for the corresponding solution. A perturbation $\widehat{\mathcal{D}}$ of $\mathcal{D}$ generates perturbed bilinear and linear forms $\widehat{\mathcal{B}} = \mathcal{B}(\widehat{\mathcal{D}})$ and $\widehat{\mathcal{F}} = \mathcal{F}(\widehat{\mathcal{D}}) = \langle f(\widehat{\mathcal{D}}), \cdot\rangle$, and a perturbation $\widehat{u} = u(\widehat{\mathcal{D}})$ of u, which satisfies

$$\widehat{u} \in \mathbb{V}: \quad \widehat{\mathcal{B}}[\widehat{u}, w] = \widehat{\mathcal{F}}[w] \quad \text{for all } w \in \mathbb{W}. \tag{10.51}$$

We assume, as in Section 5.4.2, that a call $[\widehat{\mathcal{T}}, \widehat{\mathcal{D}}] = \mathsf{DATA}(\mathcal{T}, \mathcal{D}, \tau)$ generates an admissible refinement $\widehat{\mathcal{T}}$ of $\mathcal{T}$ and discrete data $\widehat{\mathcal{D}}$ over $\widehat{\mathcal{T}}$, such that

$$\|\mathcal{D} - \widehat{\mathcal{D}}\|_{D(\Omega)} \le C_{\text{data}}\tau, \tag{10.52}$$

where $C_{\text{data}} > 0$ depends on data (see Section 7.2). Finally, we associate to any admissible refinement $\mathcal{T}$ of $\mathcal{T}_0$ two finite-dimensional spaces $\mathbb{V}_{\mathcal{T}} \subset \mathbb{V}$ and $\mathbb{W}_{\mathcal{T}} \subset \mathbb{W}$ of equal dimension, made of piecewise polynomial functions on $\mathcal{T}$ (typically this is accomplished by choosing a type of finite element compatible with the pair $(\mathbb{V}, \mathbb{W})$ and adopting it in any $\mathcal{T} \in \mathbb{T}$).

We are ready to state the assumptions which will rule our forthcoming analysis of AFEM-TS.

Assumption 10.24 (perturbation estimate). For any $\widetilde{\mathcal{T}} \in \mathbb{T}$ and $\varepsilon \le \varepsilon_0$, let $[\widehat{\mathcal{T}}, \widehat{\mathcal{D}}] = \mathsf{DATA}(\widetilde{\mathcal{T}}, \mathcal{D}, \varepsilon)$ and let $\widehat{u} = u(\widehat{\mathcal{D}})$ be the solution of (10.51). There exists a constant $C_D > 0$, independent of $\widetilde{\mathcal{T}}$ and ε, such that

$$\|u - \widehat{u}\|_{\mathbb{V}} \le C_D \|\mathcal{D} - \widehat{\mathcal{D}}\|_{D(\Omega)}. \tag{10.53}$$

Note that concatenating this inequality with (10.52) for $\tau = \varepsilon$, we can quantify the effect of a call to DATA on the perturbation of the exact solution; we indeed have

$$\|u - \widehat{u}\|_{\mathbb{V}} \le C_D C_{\text{data}} \varepsilon. \tag{10.54}$$

Assumption 10.25 (uniform continuity constant). For any $\widetilde{\mathcal{T}} \in \mathbb{T}$ and $\varepsilon \le \varepsilon_0$, let $[\widehat{\mathcal{T}}, \widehat{\mathcal{D}}] = \mathsf{DATA}(\widetilde{\mathcal{T}}, \mathcal{D}, \varepsilon)$ and let $\widehat{\mathcal{B}} = \mathcal{B}(\widehat{\mathcal{D}})$ be the associated bilinear form. There exists a constant $C_B \ge \|\mathcal{B}\|$, independent of $\widetilde{\mathcal{T}}$ and ε, such that

$$\|\widehat{\mathcal{B}}\| \le C_B. \tag{10.55}$$

Assumption 10.26 (uniform inf-sup constant). For any $\widetilde{\mathcal{T}} \in \mathbb{T}$ and $\varepsilon \le \varepsilon_0$, let $[\widehat{\mathcal{T}}, \widehat{\mathcal{D}}] = \mathsf{DATA}(\widetilde{\mathcal{T}}, \mathcal{D}, \varepsilon)$, let $\widehat{\mathcal{B}} = \mathcal{B}(\widehat{\mathcal{D}})$ be the associated bilinear form, let $\mathcal{T}$ be either $\widehat{\mathcal{T}}$ or an admissible refinement of $\widehat{\mathcal{T}}$, and finally let $\mathbb{V}_{\mathcal{T}} \subset \mathbb{V}$, $\mathbb{W}_{\mathcal{T}} \subset \mathbb{W}$ be the subspaces built on $\mathcal{T}$ as above. There exists a constant $0 < \bar{\beta} \le \beta$, independent of $\widetilde{\mathcal{T}}$, ε and $\mathcal{T}$, such that

$$\inf_{w\in\mathbb{W}_{\mathcal{T}}}\sup_{v\in\mathbb{V}_{\mathcal{T}}}\frac{\widehat{\mathcal{B}}[v,w]}{\|v\|_{\mathbb{V}}\|w\|_{\mathbb{W}}}\geq\bar{\beta}. \tag{10.56}$$

The last assumption guarantees the well-posedness of all the discrete variational problems

$$u_{\mathcal{T}}\in\mathbb{V}_{\mathcal{T}}:\quad \widehat{\mathcal{B}}[u_{\mathcal{T}},w_{\mathcal{T}}]=\widehat{\mathcal{F}}[w_{\mathcal{T}}]\quad\text{for all } w_{\mathcal{T}}\in\mathbb{W}_{\mathcal{T}}, \tag{10.57}$$

associated with the successive refinements of the initial mesh $\mathcal{T}_0$ performed by AFEM-TS.

We want to prove, as in the coercive case (see Proposition 5.27), that the number of iterations performed in any call to GALERKIN inside AFEM-TS (which is finite by Proposition 10.16) is indeed uniformly bounded.

Proposition 10.27 (computational cost of GALERKIN). *Let Assumptions 10.2, 10.3, 10.24, 10.25 and 10.26 be valid. For any $k\in\mathbb{N}$, the number of subiterations J_k inside a call to* GALERKIN *at iteration k of* AFEM-TS *is bounded by a constant J independent of k.*

Proof. Let $\mathcal{T}_{k,j}$ denote the successive refinements of $\widehat{\mathcal{T}}_k$ defined in GALERKIN at iteration k, and let $u_{k,j}\in\mathbb{V}_{k,j}=\mathbb{V}_{\mathcal{T}_{k,j}}$ be the corresponding Galerkin solutions, which are approximations of the solution $\widehat{u}_k\in\mathbb{V}$ of the perturbed problem (10.51) with forms $\widehat{\mathcal{B}}=\widehat{\mathcal{B}}_k=\mathcal{B}(\widehat{\mathcal{D}}_k)$ and $\widehat{f}=\widehat{f}_k=f(\widehat{\mathcal{D}}_k)$. Note also that we use *a posteriori* estimators $\eta_{k,j}=\eta_{k,j}(v)$ defined on $\mathbb{V}_{k,j}$, which depend on $\widehat{\mathcal{D}}_k$ via the coefficients of the equation. However, in reference to Assumptions 10.2 and 10.3, we always suppose the constants in the bounds (10.5) and (10.6) to be independent of k and j.

Let us pick $j:=J_k-1$. By definition of stopping criterion in GALERKIN, and by (10.5) and (10.7), we get

$$\varepsilon_k<\eta_{k,j}(u_{k,j})\leq\frac{1}{C_L}\|\widehat{u}_k-u_{k,j}\|_{\mathbb{V}}\leq\frac{c}{C_L}\rho^j\|\widehat{u}_k-u_{k,0}\|_{\mathbb{V}}. \tag{10.58}$$

The norm on the right-hand side can be bounded via Corollary 3.3 (quasi-monotonicity), applied to $\mathcal{B}:=\widehat{\mathcal{B}}_k$, $u:=\widehat{u}_k\in\mathbb{V}$, $u_N:=u_{k,0}\in\mathbb{V}_N:=\mathbb{V}_{\widehat{\mathcal{T}}_k}$ and $v:=u_k\in\mathbb{V}_M:=\mathbb{V}_{\mathcal{T}_k}\subseteq\mathbb{V}_{\widehat{\mathcal{T}}_k}$ (the output of GALERKIN at iteration $k-1$). Using Assumptions 10.25 (uniform continuity constant) and 10.26 (uniform inf-sup constant), we thus have

$$\|\widehat{u}_k-u_{k,0}\|_{\mathbb{V}}\leq\lambda\|\widehat{u}_k-u_k\|_{\mathbb{V}}, \tag{10.59}$$

with $\lambda=C_B/\bar{\beta}$.

Finally, we again use the triangle inequality to get

$$\begin{aligned}\|\widehat{u}_k-u_k\|_{\mathbb{V}}&\leq\|\widehat{u}_k-\widehat{u}_{k-1}\|_{\mathbb{V}}+\|\widehat{u}_{k-1}-u_k\|_{\mathbb{V}}\\&\leq\|u-\widehat{u}_k\|_{\mathbb{V}}+\|u-\widehat{u}_{k-1}\|_{\mathbb{V}}+\|\widehat{u}_{k-1}-u_k\|_{\mathbb{V}};\end{aligned}$$

then, Assumption 10.24 (perturbation estimate) yields $\|u - \widehat{u}_k\|_{\mathbb{V}} \le C_D C_{\text{data}}\, \omega\varepsilon_k$ and $\|u - \widehat{u}_{k-1}\|_{\mathbb{V}} \le C_D C_{\text{data}}\, \omega\varepsilon_{k-1}$, whereas the termination test for GALERKIN at iteration $k-1$ yields $\|\widehat{u}_{k-1} - u_k\|_{\mathbb{V}} \le C_U\, \varepsilon_{k-1}$. Hence, recalling $\varepsilon_{k-1} = 2\varepsilon_k$ and $\omega \le 1$, we get

$$\|\widehat{u}_k - u_k\|_{\mathbb{V}} \le \sigma\, \varepsilon_k \tag{10.60}$$

with $\sigma = 3C_D C_{\text{data}} + 2C_U$. Finally, concatenating (10.58), (10.59) and (10.60), we obtain

$$\rho^j \ge \frac{C_L}{c\lambda\sigma},$$

which implies

$$J_k \le 1 + \left(\log \frac{C_L}{c\lambda\sigma}\right)(\log \rho)^{-1} =: J. \qquad \square$$

The remainder of this section is devoted to investigating the rate-optimality of AFEMs for inf-sup stable problems. Precisely, we aim at establishing the analogue of bound (6.1) for such problems, that is,

$$\|u - u_{\mathcal{T}}\|_{\mathbb{V}} \le C(u, \mathcal{D})\, (\#\mathcal{T})^{-s}. \tag{10.61}$$

To this end, we have to introduce approximation classes for the solution and the data, and to study the quasi-optimality properties of mesh refinement and GALERKIN.

10.3.4. Nonlinear approximation classes

The definition of the approximation class $\mathbb{A}_s = \mathbb{A}_s(\mathbb{V}; \mathcal{T}_0)$ for functions in $\mathbb{V}$ is identical to that given in Section 6.1.1 for functions in $H_0^1(\Omega)$ (see Definition 6.1), provided the norm $|v|_{H_0^1(\Omega)}$ is replaced by the norm $\|v\|_{\mathbb{V}}$ at all occurrences.

In the rest of the section we will make the following regularity assumption.

Assumption 10.28 (approximability of u). The exact solution $u \in \mathbb{V}$ of problem (10.36) belongs to the approximation class $\mathbb{A}_s(\mathbb{V}; \mathcal{T}_0)$ for some $s = s_u \in (0, n/d]$.

The approximation classes of data $\mathcal{D} \in D(\Omega)$ are defined via discrete approximations $\mathcal{D}_{\mathcal{T}} \in \mathbb{D}_{\mathcal{T}}$ subordinate to a partition $\mathcal{T} \in \mathbb{T}$, which produce the oscillation

$$\operatorname{osc}_{\mathcal{T}}(\mathcal{D}) = \inf_{\mathcal{D}_{\mathcal{T}} \in \mathbb{D}_{\mathcal{T}}} \|\mathcal{D} - \mathcal{D}_{\mathcal{T}}\|_{D(\Omega)}.$$

Definition 10.29 (approximation classes of $\mathcal{D}$). Let $\mathbb{D}_s := \mathbb{D}_s(D(\Omega); \mathcal{T}_0)$ be the set of data $\mathcal{D} \in D(\Omega)$ satisfying

$$|\mathcal{D}|_{\mathbb{D}_s} := \sup_{N \ge \#\mathcal{T}_0} \left(N^s \inf_{\mathcal{T} \in \mathbb{T}_N} \operatorname{osc}_{\mathcal{T}}(\mathcal{D})\right) < \infty \;\Rightarrow\; \inf_{\mathcal{T} \in \mathbb{T}_N} \operatorname{osc}_{\mathcal{T}}(\mathcal{D}) \le |\mathcal{D}|_{\mathbb{D}_s} N^{-s}. \tag{10.62}$$

The following assumptions on the data of our boundary value problem will be valid in the rest of the section.

Assumption 10.30 (approximability of $\mathcal{D}$). The data $\mathcal{D} \in D(\Omega)$ of problem (10.36) belongs to the approximation class $\mathbb{D}_s(D(\Omega); \mathcal{T}_0)$ for some $s = s_{\mathcal{D}} \in (0, n/d]$.

Assumption 10.31 (quasi-optimality of DATA). A call $[\widehat{\mathcal{T}}, \widehat{\mathcal{D}}] = \mathsf{DATA}(\mathcal{T}, \mathcal{D}, \varepsilon)$ marks a set of elements $\mathcal{M}_{\mathcal{D}}$ whose cardinality $N(\mathcal{D}) = \#\mathcal{M}_{\mathcal{D}}$ obeys

$$N(\mathcal{D}) \lesssim |\mathcal{D}|_{\mathbb{D}_s}^{1/s_{\mathcal{D}}} \varepsilon^{-1/s_{\mathcal{D}}}. \tag{10.63}$$

The concept of *ε-approximation of order s* of $u \in \mathbb{A}_s(\mathbb{V}; \mathcal{T}_0)$ is identical to the one given in Definition 6.12, and so is the proof of the following result.

Lemma 10.32 (ε-approximation of u of order s). *Let $u \in \mathbb{A}_s(\mathbb{V}; \mathcal{T}_0)$ and $v \in \mathbb{V}$ satisfy $\|u - v\|_{\mathbb{V}} \le \varepsilon$ for some $0 < \varepsilon \le \varepsilon_0$. Then v is a 2ε-approximation of order s to u.*

10.3.5. Rate-optimality of GALERKIN

To estimate the growth of the cardinality of the meshes produced inside a call to GALERKIN, which always deals with discrete data, and to relate it to the approximation class of the exact solution u, we need an additional assumption of the estimators $\eta_{\mathcal{T}}$. Henceforth, for any subset $\mathcal{S} \subset \mathcal{T}$, we define $\eta_{\mathcal{T}}(v, \mathcal{S})$ by

$$\eta_{\mathcal{T}}(v, \mathcal{S})^2 = \sum_{T \in \mathcal{S}} \eta_{\mathcal{T}}(v, T)^2.$$

Assumption 10.33 (discrete reliability of the estimator). There exists a constant $c_2 > 0$ such that for any $\mathcal{T} \in \mathbb{T}$ and any refinement $\mathcal{T}_* \ge \mathcal{T}$, if $\mathcal{R} = \mathcal{R}_{\mathcal{T} \to \mathcal{T}_*} = \mathcal{T} \setminus \mathcal{T}_*$ is the set of refined elements of $\mathcal{T}$, then

$$\|u_{\mathcal{T}_*} - u_{\mathcal{T}}\|_{\mathbb{V}} \le c_2 \eta_{\mathcal{T}}(u_{\mathcal{T}}, \mathcal{R}),$$

where $u_{\mathcal{T}}$ (resp. $u_{\mathcal{T}_*}$) are the Galerkin solutions in $\mathbb{V}_{\mathcal{T}}$ (resp. $\mathbb{V}_{\mathcal{T}_*}$).

We recall that the module MARK in GALERKIN implements Dörfler's strategy, that is, for a fixed $\theta \in (0, 1]$, it identifies a subset $\mathcal{M} \subseteq \mathcal{T}$ of elements undergoing bisection by the condition

$$\eta_{\mathcal{T}}(u_{\mathcal{T}}, \mathcal{M}) \ge \theta\, \eta_{\mathcal{T}}(u_{\mathcal{T}}). \tag{10.64}$$

The following property is the analogue of the one stated in Lemma 6.16 for coercive problems. Since the proof is similar, we omit it.

Lemma 10.34 (Dörfler marking). *Let Assumptions 10.17 and 10.33 be valid. For all $0 < \mu < \frac{1}{2}$ there exists $0 < \theta_0 < 1$ such that, if $\mathcal{T} \in \mathbb{T}$ and $\mathcal{T}_*$ is a refinement of $\mathcal{T}$ with refined set $\mathcal{R} = \mathcal{T} \setminus \mathcal{T}_*$, and if the Galerkin solutions $u_{\mathcal{T}} \in \mathbb{V}_{\mathcal{T}}$ and $u_{\mathcal{T}_*} \in \mathbb{V}_{\mathcal{T}_*}$ satisfy*

$$\eta_{\mathcal{T}_*}(u_{\mathcal{T}_*}) \le \mu\, \eta_{\mathcal{T}}(u_{\mathcal{T}}),$$

then

$$\theta_0\, \eta_{\mathcal{T}}(u_{\mathcal{T}}) \le \eta_{\mathcal{T}}(u_{\mathcal{T}}, \mathcal{R}).$$

We are ready to investigate the rate-optimality of the kth call to GALERKIN in the two-step AFEM (see Definition 5.1). We let $\mathcal{M}_{k,j} \subseteq \mathcal{T}_{k,j}$ denote the marked set at the jth iteration inside GALERKIN (hereafter we refer to the notation in the proof of Proposition 10.27). To achieve quasi-optimality, the following assumption is fundamental.

Assumption 10.35 (minimality of marked sets). The module MARK selects a set $\mathcal{M}_{k,j}$ with *minimal* cardinality among those satisfying Dörfler's condition

$$\eta_{k,j}(u_{k,j}, \mathcal{M}) \geq \theta\, \eta_{k,j}(u_{k,j}) \quad \text{for all } k, j.$$

Proposition 10.36 (cardinality of marked sets). *Let Assumptions 10.2, 10.24, 10.25, 10.26, 10.28, 10.17, 10.33 and 10.35 hold true. There exists a constant $C_0 > 0$ independent of k and j such that the cardinality $N_{k,j}(u)$ of $\mathcal{M}_{k,j}$ satisfies*

$$N_{k,j}(u) \leq C_0\, |u|_{\mathbb{A}_s}^{1/s} \varepsilon_k^{-1/s} \tag{10.65}$$

and

$$N_{k,j}(u) \leq C_0\, |u|_{\mathbb{A}_s}^{1/s} \|u - u_{k,j}\|_{\mathbb{V}}^{-1/s}. \tag{10.66}$$

Proof. The proof can be easily obtained by slightly adapting to the current abstract setting the proof of Corollary 6.22, also taking into account Proposition 6.18. □

Let $\mathcal{M}_k$ denote the set of marked elements in GALERKIN at iteration k of AFEM. Since the cardinality $N_k(u) = \#\mathcal{M}_k$ of $\mathcal{M}_k$ satisfies $N_k(u) = \sum_{j=0}^{J_k-1} N_{k,j}(u)$, we can estimate its cardinality by combining Propositions 10.27 and 10.36.

Corollary 10.37 (rate-optimality of GALERKIN). *Under the assumptions of Propositions 10.27 and 10.36, the total number of marked elements $\mathcal{M}_k$ in* GALERKIN *at iteration k of* AFEM *satisfies*

$$N_k(u) \leq J\, C_0\, |u|_{\mathbb{A}_s}^{1/s} \varepsilon_k^{-1/s}.$$

10.3.6. Rate-optimality of AFEM-TS

At last, we focus on the two-step AFEM in Definition 5.1 (AFEM-TS), and prove its rate-optimality, in relation to the nonlinear approximation classes of the solution u and the problem data $\mathcal{D}$.

Theorem 10.38 (rate-optimality of AFEM-TS). *Under the same assumptions as Proposition 10.36, plus Assumptions 10.30 and 10.31, there exists a constant C_* independent of u and $\mathcal{D}$ such that the sequence $(\mathcal{T}_k, \mathbb{V}_{\mathcal{T}_k}, u_{\mathcal{T}_k})$, $k \geq 0$, produced by* AFEM-TS *satisfies*

$$\|u - u_{\mathcal{T}_k}\|_{\mathbb{V}} \leq C_* \left(|u|_{\mathbb{A}_{s_u}}^{1/s_u} + |\mathcal{D}|_{\mathbb{D}_{s_\mathcal{D}}}^{1/s_\mathcal{D}} \right)^s (\#\mathcal{T}_k)^{-s},$$

with $0 < s = \min(s_u, s_\mathcal{D}) \leq n/d$.

Proof. Let $\mathcal{M}_\ell^u$ (resp. $\mathcal{M}_\ell^{\mathcal{D}}$) denote the set of elements marked by GALERKIN (resp. DATA) at iteration ℓ of AFEM. By Corollary 10.37 and Assumption 10.31, there exist constants D_1, D_2 independent of u, $\mathcal{D}$ and k such that

$$N_\ell(u) \le D_1 \, |u|_{\mathbb{A}_{s_u}}^{1/s_u} \varepsilon_\ell^{-1/s_u}, \quad N_\ell(\mathcal{D}) \le D_2 \, |\mathcal{D}|_{\mathbb{D}_{s_\mathcal{D}}}^{1/s_\mathcal{D}} \varepsilon_\ell^{-1/s_\mathcal{D}}.$$

Then we conclude as in the proof of Theorem 6.24. □

10.4. The Stokes problem

Here we consider the Stokes problem

$$\begin{aligned} -\Delta \boldsymbol{u} + \nabla p &= \boldsymbol{f} \quad \text{in } \Omega, \\ \operatorname{div} \boldsymbol{u} &= 0 \quad \text{in } \Omega, \\ \boldsymbol{u} &= 0 \quad \text{on } \partial\Omega, \end{aligned} \tag{10.67}$$

already introduced in Section 2.3. Assuming $\boldsymbol{f} \in H^{-1}(\Omega;\mathbb{R}^d) = \mathbb{V}^*$, its weak formulation is given in (2.15) or, equivalently, in (2.16), where the bilinear form $\mathcal{B}$ is continuous and inf-sup stable, as a consequence of Theorem 2.11 (Brezzi); see Section 2.4.

A Galerkin discretization of this problem, based on finite-dimensional subspaces $\mathbb{V}_\mathcal{T} \subset \mathbb{V} = H_0^1(\Omega;\mathbb{R}^d)$ and $\mathbb{Q}_\mathcal{T} \subset \mathbb{Q} = L_0^2(\Omega)$, reads as follows: find $(\boldsymbol{u}_\mathcal{T}, p_\mathcal{T}) \in \mathbb{V}_\mathcal{T} \times \mathbb{Q}_\mathcal{T}$ such that

$$\begin{aligned} a[\boldsymbol{u}_\mathcal{T}, \boldsymbol{v}] + b[p_\mathcal{T}, \boldsymbol{v}] &= \langle \boldsymbol{f}, \boldsymbol{v}\rangle \quad \text{for all } \boldsymbol{v} \in \mathbb{V}_\mathcal{T}, \\ b[q, \boldsymbol{u}_\mathcal{T}] &= 0 \qquad\quad \text{for all } q \in \mathbb{Q}_\mathcal{T}, \end{aligned} \tag{10.68}$$

or equivalently

$$(\boldsymbol{u}_\mathcal{T}, p_\mathcal{T}) \in \mathbb{V}_\mathcal{T} \times \mathbb{Q}_\mathcal{T}: \quad \mathcal{B}[(\boldsymbol{u}_\mathcal{T}, p_\mathcal{T}), (\boldsymbol{v}, q)] = \langle \boldsymbol{f}, \boldsymbol{v}\rangle \quad \text{for all } (\boldsymbol{v}, q) \in \mathbb{V}_\mathcal{T} \times \mathbb{Q}_\mathcal{T}.$$

We assume that the pair $(\mathbb{V}_\mathcal{T}, \mathbb{Q}_\mathcal{T})$ is *uniformly inf-sup stable* for the form b, that is, there exists a constant $\beta > 0$, independent of the refinement $\mathcal{T}$, such that

$$\inf_{q \in \mathbb{Q}_\mathcal{T}} \sup_{\boldsymbol{v} \in \mathbb{V}_\mathcal{T}} \frac{b[q, \boldsymbol{v}]}{\|\boldsymbol{v}\|_\mathbb{V} \|w\|_\mathbb{W}} \ge \beta. \tag{10.69}$$

This condition is equivalent to the uniform inf-sup stability of the bilinear form $\mathcal{B}$ on the product space $\mathbb{X}_\mathcal{T} := \mathbb{V}_\mathcal{T} \times \mathbb{Q}_\mathcal{T}$. Then, applying a discrete form of Brezzi's theorem, we obtain the existence and uniqueness of the solution of (10.68), which satisfies the stability bound

$$\|\boldsymbol{u}_\mathcal{T}\|_\mathbb{V} + \|p_\mathcal{T}\|_\mathbb{Q} \le C \, \|\boldsymbol{f}\|_{\mathbb{V}^*}, \tag{10.70}$$

where C depends only on the continuity constant $\|a\|$ and the coercivity constant α of the form a, and the inf-sup constant β. Furthermore, we have the quasi-best approximation bounds (Boffi *et al.* 2013, Proposition 8.2.1)

$$\|\boldsymbol{u} - \boldsymbol{u}_{\mathcal{T}}\|_{\mathbb{V}} \le C_{11} \min_{\boldsymbol{v}\in\mathbb{V}_{\mathcal{T}}} \|\boldsymbol{u} - \boldsymbol{v}\|_{\mathbb{V}} + C_{12} \min_{q\in\mathbb{Q}_{\mathcal{T}}} \|p - q\|_{\mathbb{Q}}, \tag{10.71}$$

$$\|p - p_{\mathcal{T}}\|_{\mathbb{V}} \le C_{21} \min_{\boldsymbol{v}\in\mathbb{V}_{\mathcal{T}}} \|\boldsymbol{u} - \boldsymbol{v}\|_{\mathbb{V}} + C_{22} \min_{q\in\mathbb{Q}_{\mathcal{T}}} \|p - q\|_{\mathbb{Q}}, \tag{10.72}$$

where the constants C_{ij}, $1 \le i, j \le 2$, depend only on the quantities $\|a\|$, $\|b\|$, α and β.

There are many families of finite element spaces that are uniformly inf-sup stable for the Stokes problem; see Boffi *et al.* (2013, Chapter 8). Among them, we consider here the Taylor–Hood element (Taylor and Hood 1973) and its higher-order versions. They all use continuous discrete pressures, so they fit into the general form

$$\mathbb{V}_{\mathcal{T}} = \{\boldsymbol{v} \in H^1_0(\Omega;\mathbb{R}^d) \mid \boldsymbol{v}|_T \in \boldsymbol{V}_T,\ T \in \mathcal{T}\},$$
$$\mathbb{Q}_{\mathcal{T}} = \{q \in L^2_0(\Omega) \cap C^0(\bar{\Omega}) \mid q|_T \in Q_T,\ T \in \mathcal{T}\},$$

where $\boldsymbol{V}_T$ and Q_T are spaces of polynomials on the element T. Considering simplicial elements, we have for $n \ge 2$

$$\boldsymbol{V}_T = (\mathbb{P}_n(T))^d, \quad Q_T = \mathbb{P}_{n-1}(T). \tag{10.73}$$

The convergence and optimality of an adaptive algorithm for the Stokes problem based on Taylor–Hood elements was first established by Feischl (2019) (see also Feischl 2022, Section 6). We aim at deriving a similar result using the abstract framework presented in this section.

We start by developing the *a posteriori* error analysis, and for this we introduce the weak residual

$$\langle \boldsymbol{\mathcal{R}}_{\mathcal{T}}, (\boldsymbol{v}, q)\rangle := \langle \boldsymbol{f}, \boldsymbol{v}\rangle - \mathcal{B}[(\boldsymbol{u}_{\mathcal{T}}, p_{\mathcal{T}}), (\boldsymbol{v}, q)] \quad \text{for all } (\boldsymbol{v}, q) \in \mathbb{V} \times \mathbb{Q},$$

which we represent as $\boldsymbol{\mathcal{R}}_{\mathcal{T}} = (\boldsymbol{\mathcal{R}}^m_{\mathcal{T}}, \mathcal{R}^c_{\mathcal{T}})$ according to the momentum and continuity equations; note that $\mathcal{R}^c_{\mathcal{T}} = \operatorname{div} \boldsymbol{u}_{\mathcal{T}}$. The continuity and inf-sup stability properties of the exact Stokes form $\mathcal{B}$ yield the equivalence

$$\|\boldsymbol{u} - \boldsymbol{u}_{\mathcal{T}}\|_{\mathbb{V}} + \|p - p_{\mathcal{T}}\|_{\mathbb{Q}} \approx \|\boldsymbol{\mathcal{R}}_{\mathcal{T}}\|_{\mathbb{V}^*\times\mathbb{Q}^*} \approx \|\boldsymbol{\mathcal{R}}^m_{\mathcal{T}}\|_{\mathbb{V}^*} + \|\operatorname{div} \boldsymbol{u}_{\mathcal{T}}\|_{L^2(\Omega)}. \tag{10.74}$$

We now apply Corollary 4.6 (star localization of residual norm) to each component of the momentum residual $\boldsymbol{\mathcal{R}}^m_{\mathcal{T}}$, to get

$$\|\boldsymbol{\mathcal{R}}^m_{\mathcal{T}}\|^2_{\mathbb{V}^*} \approx \sum_{z\in\mathcal{V}} \|\boldsymbol{\mathcal{R}}^m_{\mathcal{T}}\|^2_{(H^{-1}(\omega_z))^d},$$

whereas Lemma 4.35 (splitting of local residual norm) yields the equivalence

$$\|\boldsymbol{\mathcal{R}}^m_{\mathcal{T}}\|^2_{(H^{-1}(\omega_z))^d} \approx \|P_{\mathcal{T}}\boldsymbol{f} + \Delta\boldsymbol{u}_{\mathcal{T}} - \nabla p_{\mathcal{T}}\|^2_{(H^{-1}(\omega_z))^d} + \|\boldsymbol{f} - P_{\mathcal{T}}\boldsymbol{f}\|^2_{(H^{-1}(\omega_z))^d}.$$

In view of mesh refinement, we recall Lemma 4.8 (localization re-indexing), and we express the error indicator in terms of elements $T \in \mathcal{T}$ rather than stars ω_z, in analogy with the scalar case (4.52). To this end, define

$$\begin{aligned} \eta_{\mathcal{T}}^m(\boldsymbol{u}_{\mathcal{T}}, p_{\mathcal{T}}, T)^2 &:= h_T^2 \, \|P_{\mathcal{T}} \boldsymbol{f} + \Delta \boldsymbol{u}_{\mathcal{T}} - \nabla p_{\mathcal{T}}\|^2_{(L^2(T))^d} \\ &\quad + h_T \sum_{F \subset \partial T \setminus \partial\Omega} \|[[(\nabla \boldsymbol{u}_{\mathcal{T}})\boldsymbol{n}_F]] - P_F \boldsymbol{f}\|^2_{(L^2(F))^d}, \\ \operatorname{osc}_{\mathcal{T}}(\boldsymbol{f}, T)^2_{-1} &:= \|\boldsymbol{f} - P_{\mathcal{T}} \boldsymbol{f}\|^2_{(H^{-1}(\omega_T))^d}. \end{aligned} \tag{10.75}$$

Note that the jump term $[[(\nabla \boldsymbol{u}_{\mathcal{T}})\boldsymbol{n}_F]]$ does not contain the pressure contribution, since discrete pressures in $\mathbb{Q}_{\mathcal{T}}$ are globally continuous. We thus have

$$\|\mathcal{R}_{\mathcal{T}}^m\|^2_{\mathbb{V}^*} \approx \sum_{T \in \mathcal{T}} \big(\eta_{\mathcal{T}}^m(\boldsymbol{u}_{\mathcal{T}}, p_{\mathcal{T}}, T)^2 + \operatorname{osc}_{\mathcal{T}}(\boldsymbol{f}, T)^2_{-1}\big). \tag{10.76}$$

Recalling (10.74), the full local PDE residual indicator could be defined as

$$\eta_{\mathcal{T}}(\boldsymbol{u}_{\mathcal{T}}, p_{\mathcal{T}}, T)^2 := \eta_{\mathcal{T}}^m(\boldsymbol{u}_{\mathcal{T}}, p_{\mathcal{T}}, T)^2 + \|\operatorname{div} \boldsymbol{u}_{\mathcal{T}}\|^2_{L^2(T)}.$$

However, such a quantity is not guaranteed to strictly reduce under mesh refinement, due to the presence of the divergence term, which is not scaled by a positive power of the mesh size. The following result provides an equivalent expression of the last term, which does reduce. We recall the definition (9.1) of jumps across faces.

Lemma 10.39 (norm equivalence for divergence). *We have*

$$\|\operatorname{div} \boldsymbol{u}_{\mathcal{T}}\|^2_{L^2(\Omega)} \approx \sum_{T \in \mathcal{T}} \sum_{F \subset \partial T \setminus \partial\Omega} h_F \, \|[[\operatorname{div} \boldsymbol{u}_{\mathcal{T}}]]\|^2_{L^2(F)}.$$

Proof. The result follows from applying to $\varphi = \operatorname{div} \boldsymbol{u}_{\mathcal{T}}$ the equivalence

$$\|\varphi - \Pi_{\mathcal{T}}\varphi\|^2_{L^2(\Omega)} \approx \sum_{T \in \mathcal{T}} \sum_{F \subset \partial T \setminus \partial\Omega} h_F \, \|[[\varphi]]\|^2_{L^2(F)} \quad \text{for all } \varphi \in \mathbb{S}_{\mathcal{T}}^{n-1,-1}$$

(where $\Pi_{\mathcal{T}}$ is the L^2-orthogonal projection upon $\mathbb{S}_{\mathcal{T}}^{n-1,0}$), after observing that $\Pi_{\mathcal{T}}\varphi = 0$ since $\boldsymbol{u}_{\mathcal{T}}$ is discretely divergence-free, that is, it satisfies the second set of equations in (10.68)). To prove the equivalence for arbitrary $\varphi \in \mathbb{S}_{\mathcal{T}}^{n-1,-1}$, we use the quasi-interpolation operator $\mathcal{I}_{\mathcal{T}}^{\mathrm{dG}}$ introduced in Section 9.1.2, which leaves $\mathbb{S}_{\mathcal{T}}^{n-1,0}$ invariant. Then it is easily seen that

$$\|\varphi - \Pi_{\mathcal{T}}\varphi\|^2_{L^2(\Omega)} \approx \|\varphi - \mathcal{I}_{\mathcal{T}}^{\mathrm{dG}}\varphi\|^2_{L^2(\Omega)},$$

so it is enough to prove the equivalence with $\Pi_{\mathcal{T}}$ replaced by $\mathcal{I}_{\mathcal{T}}^{\mathrm{dG}}$. But this calculation can be done on patches ω_T since $\mathcal{I}_{\mathcal{T}}^{\mathrm{dG}}$ is quasi-local:

$$\begin{aligned} \|\varphi - \mathcal{I}_{\mathcal{T}}^{\mathrm{dG}}\varphi\|^2_{L^2(T)} &\lesssim \sum_{F \subset \omega_T} h_F \, \|[[\varphi]]\|^2_{L^2(F)} \\ &\lesssim \|\varphi - \mathcal{I}_{\mathcal{T}}^{\mathrm{dG}}\varphi\|^2_{L^2(\omega_T)} \quad \text{for all } \varphi \in \mathbb{S}_{\mathcal{T}}^{n-1,-1}. \end{aligned}$$

The first inequality follows from (9.10) (see also Bänsch, Morin and Nochetto 2002,

Proposition 5.4). The second inequality results from the fact that if the rightmost term vanishes on ω_T then $\varphi = \mathcal{I}_{\mathcal{T}}^{\mathrm{dG}}\varphi$, whence φ is continuous in ω_T. This yields $[[\varphi]]|_F = 0$ for all internal faces F of ω_T and the middle term vanishes. □

Applying Lemma 10.39, we are led to define the elemental residual indicator

$$\eta_{\mathcal{T}}(\boldsymbol{u}_{\mathcal{T}}, p_{\mathcal{T}}, T)^2 := \eta_{\mathcal{T}}^m(\boldsymbol{u}_{\mathcal{T}}, p_{\mathcal{T}}, T)^2 + h_T \sum_{F \subset \partial T \setminus \partial\Omega} \|[[\operatorname{div} \boldsymbol{u}_{\mathcal{T}}]]\|_{L^2(F)}^2. \tag{10.77}$$

Concatenating (10.74), (10.76) and Lemma 10.39, we fulfil Assumption 10.22 (equivalence of error and full estimator). The precise result is as follows.

Proposition 10.40 (*a posteriori* error analysis for Stokes). *There exist constants* $C_U \geq C_L > 0$ *such that*

$$C_L \mathcal{E}_{\mathcal{T}}(\boldsymbol{u}_{\mathcal{T}}, p_{\mathcal{T}}, \boldsymbol{f}) \leq \|\boldsymbol{u} - \boldsymbol{u}_{\mathcal{T}}\|_{\mathbb{V}} + \|p - p_{\mathcal{T}}\|_{\mathbb{Q}} \leq C_U \mathcal{E}_{\mathcal{T}}(\boldsymbol{u}_{\mathcal{T}}, p_{\mathcal{T}}, \boldsymbol{f}),$$

where the full estimator is defined by

$$\mathcal{E}_{\mathcal{T}}(\boldsymbol{u}_{\mathcal{T}}, p_{\mathcal{T}}, \boldsymbol{f})^2 := \sum_{T \in \mathcal{T}} \mathcal{E}_{\mathcal{T}}(\boldsymbol{u}_{\mathcal{T}}, p_{\mathcal{T}}, \boldsymbol{f}, T)^2,$$

with $\mathcal{E}_{\mathcal{T}}(\boldsymbol{u}_{\mathcal{T}}, p_{\mathcal{T}}, \boldsymbol{f}, T)^2 := \eta_{\mathcal{T}}(\boldsymbol{u}_{\mathcal{T}}, p_{\mathcal{T}}, T)^2 + \operatorname{osc}_{\mathcal{T}}(\boldsymbol{f}, T)_{-1}^2$ *introduced in* (10.77) *and* (10.75) *and* $\operatorname{osc}_{\mathcal{T}}(f, T)_{-1} = \|f - P_{\mathcal{T}} f\|_{H^{-1}(\omega_T)}$.

Since the Stokes problem has constant coefficients but variable forcing, it is natural to resort to Algorithm 5.16 (AFEM-SW), the one-step AFEM with switch, for its adaptive discretization. With respect to the functional setting of Section 10.3.2, the ambient space $\mathbb{W}$ is $\mathbb{V} \times \mathbb{Q}$ and the data projection operator $P_{\mathcal{T}}$ is

$$\boldsymbol{P}_{\mathcal{T}} := \left((P_{\mathcal{T}})^d, \Pi_{\mathcal{T}}^{n-1}\right) : \mathbb{W}^* \to (\mathbb{F}_{\mathcal{T}})^d \times \mathbb{S}^{n-1,-1},$$

where $\mathbb{F}_{\mathcal{T}}$ is the scalar discrete space introduced in Definition 4.17, $P_{\mathcal{T}}$ is here the scalar projection operator introduced in Definition 4.24, and $\Pi_{\mathcal{T}}^{n-1}$ is the L^2-orthogonal projection upon $\mathbb{S}^{n-1,-1}$. Furthermore, the norm used to measure data perturbations is

$$\|(\boldsymbol{f}, g)\|_{\mathbb{W}_{\mathcal{T}}^*}^2 = \sum_{T \in \mathcal{T}} \left(\|\boldsymbol{f}\|_{(H^{-1}(\omega_T))^d}^2 + \|g\|_{L^2(T)}^2\right).$$

It is easily seen that $\eta_{\mathcal{T}}(\boldsymbol{u}_{\mathcal{T}}, p_{\mathcal{T}}, T)$ satisfies Assumptions 10.17 (Lipschitz continuity of estimator) and 10.18 (monotonicity of estimator) as well as the hypotheses of Proposition 10.19 (estimator reduction under Dörfler marking): the estimator is clearly Lipschitz-continuous and monotone, and it satisfies condition (10.38) since all its addends appear multiplied by a positive power of the mesh size. In addition, the oscillation $\operatorname{osc}_{\mathcal{T}}((\boldsymbol{f}, g)) = \|(\boldsymbol{f}, g) - \boldsymbol{P}_{\mathcal{T}}(\boldsymbol{f}, g)\|_{\mathbb{W}_{\mathcal{T}}^*}$ fulfils Assumption 10.20 (quasi-monotonicity of oscillation).

Theorem 10.23 provides sufficient conditions for the linear convergence of the algorithm, and these conditions have been verified according to the previous discussion. Therefore we may summarize our findings in the following theorem.

Theorem 10.41 (linear convergence for Stokes). *Consider the Galerkin discretization* (10.68) *of the Stokes problem which uses Taylor–Hood elements of order* $n \geq 2$, *and let the* a posteriori *estimator be given in Proposition 10.40. Then Theorem 10.6 guarantees the linear convergence of Algorithm 5.16 (*AFEM-SW*) applied to this problem, that is, for some* $c > 0$ *and* $0 \leq \rho < 1$,

$$e_{j+i} \leq c\rho^i e_j \quad \text{for all } i, j \in \mathbb{N},$$

with $e_j := \|\nabla(\boldsymbol{u} - \boldsymbol{u}_j)\|_\Omega + \|p - p_j\|_\Omega$.

In order to assess the optimality of the discretization, we specify the definition of approximation classes for the solution of the Stokes problem. Precisely, given $(\boldsymbol{v}, q) \in \mathbb{V} \times \mathbb{Q}$, we let $\sigma_N(\boldsymbol{v}, q)$ be the smallest approximation error incurred on $(\boldsymbol{v}, q)$ with elements in $\mathbb{V}_\mathcal{T} \times \mathbb{Q}_\mathcal{T}$ over meshes belonging to $\mathbb{T}_N$:

$$\sigma_N(\boldsymbol{v}, q) := \inf_{\mathcal{T}\in\mathbb{T}_N} \inf_{(\boldsymbol{v}_\mathcal{T}, q_\mathcal{T})\in\mathbb{V}_\mathcal{T}\times\mathbb{Q}_\mathcal{T}} \left(\|\nabla(\boldsymbol{v} - \boldsymbol{v}_\mathcal{T})\|_\Omega^2 + \|q - q_\mathcal{T}\|_\Omega^2\right)^{1/2}. \tag{10.78}$$

For $0 < s \leq n/d$, the class $\mathbb{A}_s = \mathbb{A}_s(\mathbb{V} \times \mathbb{Q}; \mathcal{T}_0)$, relative to the partition $\mathcal{T}_0$ is the set of functions $(\boldsymbol{v}, q) \in \mathbb{V} \times \mathbb{Q}$ such that

$$|(\boldsymbol{v}, q)|_{\mathbb{A}_s} := \sup_{N\geq\#\mathcal{T}_0} (N^s \sigma_N(\boldsymbol{v}, q)) < \infty. \tag{10.79}$$

By adapting the arguments used in the proof of Theorem 6.20 (rate-optimality of one-step AFEMs), we can prove the following result.

Theorem 10.42 (rate-optimality of AFEM-SW for Stokes). *Let the assumptions of Theorem 10.41 be valid. If* $(\boldsymbol{u}, p) \in \mathbb{A}_s$, *then the sequence* $\{\mathcal{T}_k, \mathbb{V}_k, (\boldsymbol{u}_k, p_k)\}_{k\geq 0}$ *generated by* AFEM-SW *satisfies*

$$\|\nabla(\boldsymbol{u} - \boldsymbol{u}_k)\|_{L^2(\Omega)} + \|p - p_k\|_{L^2(\Omega)} \lesssim |(\boldsymbol{u}, p)|_{\mathbb{A}_s} (\#\mathcal{T}_k)^{-s}, \quad k \geq 0. \tag{10.80}$$

Remark 10.43 (limits of the analysis). Other inf-sup stable elements, such as the Mini element or the Crouzeix–Raviart element (see e.g. Boffi *et al.* 2013), do not fit into the present setting of the analysis, since their velocities contain element-wise bubble components (which are indeed responsible for the stability of the elements). Unfortunately, a bubble on an element does not restrict to two bubbles when the element is bisected, preventing the nestedness condition $\mathbb{V}_\mathcal{T} \subset \mathbb{V}_{\mathcal{T}_*}$ from being satisfied when $\mathcal{T}_*$ is a refinement of $\mathcal{T}$.

10.5. Mixed FEMs for scalar elliptic PDEs

The diffusion–reaction problem (2.5) can be formulated in mixed form as follows:

$$\begin{aligned} \boldsymbol{A}^{-1}\boldsymbol{\sigma} &= \nabla u && \text{in } \Omega, \\ -\operatorname{div} \boldsymbol{\sigma} + cu &= f && \text{in } \Omega, \\ u &= 0 && \text{on } \partial\Omega. \end{aligned} \tag{10.81}$$

Introducing the porosity matrix $\boldsymbol{K} := \boldsymbol{A}^{-1}$, we assume hereafter that data

$$\mathcal{D} = (\boldsymbol{K}, c, f) \in D(\Omega) := M(\alpha_1, \alpha_2) \times R(c_1, c_2) \times L^2(\Omega), \tag{10.82}$$

where $M(\alpha_1, \alpha_2)$ and $R(c_1, c_2)$ are defined in (5.48) and (5.49), respectively. Note that the current parameters α_1, α_2 are the reciprocals of α_2, α_1 in (5.48), but to avoid complicating the notation further, we relabel them hereafter.

Weak formulation. To write the weak formulation of these equations, we introduce the Hilbert space

$$H(\mathrm{div}; \Omega) := \{\boldsymbol{\tau} \in L^2(\Omega; \mathbb{R}^d) \mid \mathrm{div}\, \boldsymbol{\tau} \in L^2(\Omega)\} \tag{10.83}$$

equipped with the norm $\|\boldsymbol{\tau}\|^2_{H(\mathrm{div};\Omega)} := \|\boldsymbol{\tau}\|^2_\Omega + \|\mathrm{div}\, \boldsymbol{\tau}\|^2_\Omega$. Then we multiply the first equation in (10.81) by $\boldsymbol{\tau} \in H(\mathrm{div}; \Omega)$ and the second equation by $v \in L^2(\Omega)$, we integrate over Ω and apply partial integration to the term containing ∇u, taking into account the Dirichlet boundary condition. In this way we obtain the following variational problem: find $(\boldsymbol{\sigma}, u) \in \mathbb{V} := H(\mathrm{div}; \Omega) \times L^2(\Omega)$ such that

$$\begin{aligned} \int_\Omega \boldsymbol{K\sigma} \cdot \boldsymbol{\tau} + \int_\Omega u\, \mathrm{div}\, \boldsymbol{\tau} &= 0 && \text{for all } \boldsymbol{\tau} \in H(\mathrm{div}; \Omega), \\ \int_\Omega v\, \mathrm{div}\, \boldsymbol{\sigma} - \int_\Omega c\, u\, v &= -\int_\Omega f\, v && \text{for all } v \in L^2(\Omega). \end{aligned} \tag{10.84}$$

This can be written as follows: find $(\boldsymbol{\sigma}, u) \in V \times Q$ such that

$$\begin{aligned} a[\boldsymbol{\sigma}, \boldsymbol{\tau}] + b[u, \boldsymbol{\tau}] &= 0 && \text{for all } \boldsymbol{\sigma} \in V, \\ b[v, \boldsymbol{\sigma}] + m[u, v] &= -\langle f, v\rangle && \text{for all } v \in Q, \end{aligned} \tag{10.85}$$

if we set $V := H(\mathrm{div}; \Omega)$, $Q := L^2(\Omega)$, and we introduce the continuous bilinear forms $a\colon V \times V \to \mathbb{R}$, $b\colon Q \times V \to \mathbb{R}$ and $m\colon Q \times Q \to \mathbb{R}$ by

$$a[\boldsymbol{\sigma}, \boldsymbol{\tau}] = \int_\Omega \boldsymbol{K\sigma} \cdot \boldsymbol{\tau}, \quad b[v, \boldsymbol{\tau}] = \int_\Omega v\, \mathrm{div}\, \boldsymbol{\tau}, \quad m[u, v] = -\int_\Omega c\, u\, v,$$

and the linear form $\langle f, v\rangle = \int_\Omega f v$. An equivalent formulation, similar to (2.16), is as follows:

$$(\boldsymbol{\sigma}, u) \in V \times Q\colon \quad \mathcal{B}[(\boldsymbol{\sigma}, u), (\boldsymbol{\tau}, v)] = -\langle f, v\rangle \quad \text{for all } (\boldsymbol{\tau}, v) \in V \times Q, \tag{10.86}$$

with

$$\mathcal{B}[(\boldsymbol{\sigma}, u), (\boldsymbol{\tau}, v)] := a[\boldsymbol{\sigma}, \boldsymbol{\tau}] + b[u, \boldsymbol{\tau}] + b[v, \boldsymbol{\sigma}] + m[u, v].$$

Formulation (10.85) is a generalization of the classical saddle point problem considered in Section 2.4, given by the presence of the third bilinear form m. According to Boffi *et al.* (2013, Theorem 4.3.1), the well-posedness of such a problem can be derived from the following three conditions:

(i) the form a is coercive on $V_0 = \{\boldsymbol{\tau} \in V \mid b[v, \boldsymbol{\tau}] = 0 \text{ for all } v \in Q\}$,

(ii) the form b satisfies an inf-sup condition on $V \times Q$,

(iii) the form m is non-positive on Q, i.e. $m[v,v] \leq 0$ for all $v \in Q$.

These conditions are easily checked for our mixed formulation of the Dirichlet problem.

Discretization. To define a finite element discretization of this problem, we consider partitions $\mathcal{T} \in \mathbb{T}$ obtained by conforming bisection refinements of an initial partition $\mathcal{T}_0$, and let $V_{\mathcal{T}} \subset V$ and $Q_{\mathcal{T}} \subset Q$ be finite-dimensional subspaces made of piecewise polynomial functions on $\mathcal{T}$. Among the families of *uniformly inf-sup stable* finite element spaces for this problem, we consider the Raviart–Thomas–Nédélec family (Raviart and Thomas 1977, Nédélec 1980), and the Brezzi–Douglas–Marini family (Brezzi, Douglas Jr and Marini 1985) on simplicial elements. They fit into the general definition

$$
\begin{aligned}
V_{\mathcal{T}} &= \{\boldsymbol{\tau} \in H(\mathrm{div};\Omega) \mid \boldsymbol{\tau}|_T \in \boldsymbol{V}_T,\ T \in \mathcal{T}\},\\
Q_{\mathcal{T}} &= \{q \in L^2(\Omega) \mid q|_T \in Q_T,\ T \in \mathcal{T}\}.
\end{aligned}
$$

For the Raviart–Thomas–Nédélec (RTN) family we have

$$
\boldsymbol{V}_T = (\mathbb{P}_{n-1}(T))^d + \boldsymbol{x}\,\mathbb{P}_{n-1}(T), \quad Q_T = \mathbb{P}_{n-1}(T), \quad n \geq 1,
$$

where $\boldsymbol{x} = (x_1, \ldots, x_d)$ is the coordinate vector, whereas for the Brezzi–Douglas–Marini (BDM) family we have

$$
\boldsymbol{V}_T = (\mathbb{P}_n(T))^d, \quad Q_T = \mathbb{P}_{n-1}(T), \quad n \geq 1.
$$

Note that for any face F of the triangulation we have $\boldsymbol{\tau}|_F \cdot \boldsymbol{n}_F \in \mathbb{P}_{n-1}(F)$ for the RTN family, and $\boldsymbol{\tau}|_F \cdot \boldsymbol{n}_F \in \mathbb{P}_n(F)$ for the BDM family; furthermore, $\mathrm{div}\,\boldsymbol{V}_T = Q_T$. We refer to Boffi *et al.* (2013, Sections 2.3.1, 7.1.2) for more details.

Due to the presence of variable data, it is natural to perform the adaptive discretization of the problem by adopting Algorithm 5.28 (AFEM-TS), the two-step AFEM. The procedure $[\widehat{\mathcal{T}}, \widehat{\mathcal{D}}] = \mathsf{DATA}(\mathcal{T}, \mathcal{D}, \tau)$ generates an admissible refinement $\widehat{\mathcal{T}}$ of $\mathcal{T}$ and discrete data

$$
\widehat{\mathcal{D}} = (\widehat{\boldsymbol{K}}, \widehat{c}, \widehat{f}) \in \mathbb{D}_{\widehat{\mathcal{T}}} := \big[\mathbb{S}^{n-1,-1}_{\widehat{\mathcal{T}}}\big]^{d\times d} \times \mathbb{S}^{n-1,-1}_{\widehat{\mathcal{T}}} \times \mathbb{S}^{n-1,-1}_{\widehat{\mathcal{T}}}
$$

over $\widehat{\mathcal{T}}$, such that $\widehat{\boldsymbol{K}} \in M(\widehat{\alpha}_1, \widehat{\alpha}_2)$, $\widehat{c} \in R(\widehat{c}_1, \widehat{c}_2)$ (see Sections 7.2.2 and 7.2.3), and

$$
\|\mathcal{D} - \widehat{\mathcal{D}}\|_{\widehat{D}(\Omega)} \leq C_{\mathrm{data}}\tau,
$$

where the space $\widehat{D}(\Omega)$ is defined in (5.55).

The Galerkin discretization with these discrete data reads: find $(\boldsymbol{\sigma}_{\mathcal{T}}, u_{\mathcal{T}}) \in V_{\mathcal{T}} \times Q_{\mathcal{T}}$ such that

$$
\begin{aligned}
\widehat{a}[\boldsymbol{\sigma}_{\mathcal{T}}, \boldsymbol{\tau}] + b[u_{\mathcal{T}}, \boldsymbol{\tau}] &= 0 && \text{for all } \boldsymbol{\tau} \in V_{\mathcal{T}},\\
b[v, \boldsymbol{\sigma}_{\mathcal{T}}] + \widehat{m}[u_{\mathcal{T}}, v] &= -\langle \widehat{f}, v\rangle && \text{for all } v \in Q_{\mathcal{T}},
\end{aligned}
\tag{10.87}
$$

with

$$\widehat{a}[\boldsymbol{\sigma},\boldsymbol{\tau}] = \int_\Omega \widehat{\boldsymbol{K}}\boldsymbol{\sigma}\cdot\boldsymbol{\tau}, \quad \widehat{m}[u,v] = -\int_\Omega \widehat{c}\,u\,v, \quad \langle\widehat{f},v\rangle = \int_\Omega \widehat{f}\,v,$$

or equivalently

$$(\boldsymbol{\sigma}_\mathcal{T},u_\mathcal{T}) \in V_\mathcal{T}\times Q_\mathcal{T}: \quad \widehat{\mathcal{B}}[(\boldsymbol{\sigma}_\mathcal{T},u_\mathcal{T}),(\boldsymbol{\tau},v)] = -\langle\widehat{f},\,v\rangle \quad \text{for all } (\boldsymbol{\tau},v)\in V_\mathcal{T}\times Q_\mathcal{T}.$$

A posteriori *error estimator.* Let $(\widehat{\boldsymbol{\sigma}},\widehat{u}) \in V\times Q$ denote the exact solution of the perturbed problem

$$\widehat{\mathcal{B}}[(\widehat{\boldsymbol{\sigma}},\widehat{u}),(\boldsymbol{\tau},v)] = -\langle\widehat{f},\,v\rangle \quad \text{for all } (\boldsymbol{\tau},v)\in V\times Q;$$

note that the forcing $\widehat{f}$ appears with a negative sign. Then, by continuity and uniform inf-sup stability of the form $\widehat{\mathcal{B}}$, we know that the error

$$\|\widehat{\boldsymbol{\sigma}}-\boldsymbol{\sigma}_\mathcal{T}\|_{H(\mathrm{div};\Omega)} + \|\widehat{u}-u_\mathcal{T}\|_{L^2(\Omega)}$$

is equivalent to the quantity

$$\sup_{(\boldsymbol{\tau},v)\in V\times Q} \frac{\widehat{\mathcal{B}}[(\widehat{\boldsymbol{\sigma}}-\boldsymbol{\sigma}_\mathcal{T},\widehat{u}-u_\mathcal{T}),(\boldsymbol{\tau},v)]}{\|\boldsymbol{\tau}\|_{H(\mathrm{div};\Omega)}+\|v\|_{L^2(\Omega)}} = \sup_{(\boldsymbol{\tau},v)\in V\times Q} \frac{\langle\widehat{f},v\rangle + \widehat{\mathcal{B}}[(\boldsymbol{\sigma}_\mathcal{T},u_\mathcal{T}),(\boldsymbol{\tau},v)]}{\|\boldsymbol{\tau}\|_{H(\mathrm{div};\Omega)}+\|v\|_{L^2(\Omega)}}.$$

By Galerkin orthogonality, the numerator is equal to

$$\langle\widehat{f},v-v_\mathcal{T}\rangle + \widehat{\mathcal{B}}[(\boldsymbol{\sigma}_\mathcal{T},u_\mathcal{T}),(\boldsymbol{\tau}-\boldsymbol{\tau}_\mathcal{T},v-v_\mathcal{T})] \quad \text{for all } (\boldsymbol{\tau}_\mathcal{T},v_\mathcal{T})\in V_\mathcal{T}\times Q_\mathcal{T},$$

which we now proceed to estimate. The term

$$\widehat{\mathcal{B}}[(\boldsymbol{\sigma}_\mathcal{T},u_\mathcal{T}),(\boldsymbol{\tau}-\boldsymbol{\tau}_\mathcal{T},0)]$$

can be analysed as in Carstensen (1997) (see also Verfürth 2013, Section 4.8), by resorting to a stable decomposition of $H(\mathrm{div};\Omega)$: precisely, given $\boldsymbol{\tau}\in H(\mathrm{div};\Omega)$, there exist $\boldsymbol{\Phi}\in(H^1(\Omega))^d$ and $\boldsymbol{u}\in(H^1(\Omega))^d$ such that

$$\boldsymbol{\tau} = \boldsymbol{\Phi} + \mathrm{curl}\,\boldsymbol{u} \tag{10.88}$$

with $\|\boldsymbol{\Phi}\|_{(H^1(\Omega))^d} + \|\boldsymbol{u}\|_{(H^1(\Omega))^d} \lesssim \|\boldsymbol{\tau}\|_{H(\mathrm{div};\Omega)}$ (see Xu, Chen and Nochetto 2009, Section 5.1.3). Note that if Ω is convex, then (10.88) is the Helmholtz decomposition of $\boldsymbol{\tau}$, with $\boldsymbol{\Phi} = \nabla G$ for some $G\in(H^2(\Omega))^d$. Using (10.88) and a suitable choice of $\boldsymbol{\tau}_\mathcal{T}$, one can show that

$$|\widehat{\mathcal{B}}[(\boldsymbol{\sigma}_\mathcal{T},u_\mathcal{T}),(\boldsymbol{\tau}-\boldsymbol{\tau}_\mathcal{T},0)]| \lesssim \eta_{\mathcal{T},1}((\boldsymbol{\sigma}_\mathcal{T},u_\mathcal{T}))\,\|\boldsymbol{\tau}\|_{H(\mathrm{div};\Omega)}, \tag{10.89}$$

with

$$\eta_{\mathcal{T},1}((\boldsymbol{\sigma}_\mathcal{T},u_\mathcal{T}))^2 = \sum_{T\in\mathcal{T}} \eta_{\mathcal{T},1}((\boldsymbol{\sigma}_\mathcal{T},u_\mathcal{T}),T)^2$$

and

$$\begin{aligned}
&\eta_{\mathcal{T},1}((\boldsymbol{\sigma}_{\mathcal{T}}, u_{\mathcal{T}}), T)^2 \\
&\quad := h_T^2 \|\widehat{\boldsymbol{K}}\boldsymbol{\sigma}_{\mathcal{T}} - \nabla u_{\mathcal{T}}\|_{L^2(T)}^2 + h_T^2 \|\operatorname{curl}(\widehat{\boldsymbol{K}}\boldsymbol{\sigma}_{\mathcal{T}})\|_{L^2(T)}^2 \\
&\qquad + h_T \sum_{F \subset \partial T \setminus \partial\Omega} \|[[(\widehat{\boldsymbol{K}}\boldsymbol{\sigma}_{\mathcal{T}})_t]]\|_{L^2(F)}^2 + h_T \sum_{F \subset \partial T \cap \partial\Omega} \|(\widehat{\boldsymbol{K}}\boldsymbol{\sigma}_{\mathcal{T}})_t\|_{L^2(F)}^2,
\end{aligned} \tag{10.90}$$

where $\boldsymbol{\phi}_t = \boldsymbol{\phi} - (\boldsymbol{\phi} \cdot \boldsymbol{n}_F)\boldsymbol{n}_F$ denotes the tangential component of the vector field $\boldsymbol{\phi}$ on F. On the other hand, the term

$$\langle \widehat{f}, v - v_{\mathcal{T}} \rangle + \widehat{\mathcal{B}}[(\boldsymbol{\sigma}_{\mathcal{T}}, u_{\mathcal{T}}), (0, v - v_{\mathcal{T}})] = \int_\Omega (\widehat{f} + \operatorname{div} \boldsymbol{\sigma}_{\mathcal{T}} - \widehat{c}\, u_{\mathcal{T}})(v - v_{\mathcal{T}})$$

can be bounded as follows. For any $T \in \mathcal{T}$, let $\Pi_T = \Pi_T^{n-1}$ be the L^2-orthogonal projection upon $Q_T = \mathbb{P}_{n-1}(T)$, and let us choose $(v_{\mathcal{T}})|_T = \Pi_T v$. Then, noticing that $\widehat{f} + \operatorname{div} \boldsymbol{\sigma}_{\mathcal{T}} \in Q_{\mathcal{T}}$, we have

$$\begin{aligned}
\int_\Omega (\widehat{f} + \operatorname{div} \boldsymbol{\sigma}_{\mathcal{T}} - \widehat{c}\, u_{\mathcal{T}})(v - v_{\mathcal{T}}) &= \sum_{T \in \mathcal{T}} \int_T (\widehat{f} + \operatorname{div} \boldsymbol{\sigma}_{\mathcal{T}} - \Pi_T(\widehat{c}\, u_{\mathcal{T}}))(v - \Pi_T v) \\
&\quad - \sum_{T \in \mathcal{T}} \int_T (\widehat{c}\, u_{\mathcal{T}} - \Pi_T(\widehat{c}\, u_{\mathcal{T}}))(v - \Pi_T v) \\
&= - \sum_{T \in \mathcal{T}} \int_T (\widehat{c}\, u_{\mathcal{T}} - \Pi_T(\widehat{c}\, u_{\mathcal{T}}))v,
\end{aligned} \tag{10.91}$$

whence

$$|\langle \widehat{f}, v - v_{\mathcal{T}} \rangle + \widehat{\mathcal{B}}[(\boldsymbol{\sigma}_{\mathcal{T}}, u_{\mathcal{T}}), (0, v - v_{\mathcal{T}})]| \le \sum_{T \in \mathcal{T}} \|\widehat{c}\, u_{\mathcal{T}} - \Pi_T(\widehat{c}\, u_{\mathcal{T}})\|_{L^2(T)} \|v\|_{L^2(T)}.$$

Conversely, it is easily checked that (10.91) implies the bound

$$\|\widehat{c}\, u_{\mathcal{T}} - \Pi_T(\widehat{c}\, u_{\mathcal{T}})\|_{L^2(T)} \lesssim \|\operatorname{div} \boldsymbol{\sigma} - \operatorname{div} \boldsymbol{\sigma}_{\mathcal{T}}\|_{L^2(T)} + \|\widehat{c}\|_{L^\infty(T)} \|\widehat{u} - u_{\mathcal{T}}\|_{L^2(T)}.$$

The choice $n = 1$ yields $\widehat{c}\, u_{\mathcal{T}} \in \mathbb{P}_0(T)$, hence $\widehat{c}\, u_{\mathcal{T}} - \Pi_T(\widehat{c}\, u_{\mathcal{T}}) = 0$. For $n \ge 2$, we could define as a (squared) local error indicator the quantity

$$\eta_{\mathcal{T},1}((\boldsymbol{\sigma}_{\mathcal{T}}, u_{\mathcal{T}}), T)^2 + \|\widehat{c}\, u_{\mathcal{T}} - \Pi_T(\widehat{c}\, u_{\mathcal{T}})\|_{L^2(T)}^2,$$

but the second addend is not guaranteed to reduce under refinement, since it is not scaled by a positive power of the mesh size. However, there is an equivalent quantity which does reduce, as stated in the following result.

Lemma 10.44 (equivalence of local error indicators). *Assume $\widehat{c}, u_{\mathcal{T}} \in \mathbb{P}_{n-1}(T)$, for $n \ge 2$. Let Π_T^j be the L^2-orthogonal projection upon $\mathbb{P}_j(T)$. Then*

$$\|\widehat{c}\, u_{\mathcal{T}} - \Pi_T(\widehat{c}\, u_{\mathcal{T}})\|_{L^2(T)} \approx h_T \sum_{j=1}^{n} \|u_{\mathcal{T}} - \Pi_T^{n-1-j} u_{\mathcal{T}}\|_{L^2(T)} \|\nabla \widehat{c} - \Pi_T^{j-2} \nabla \widehat{c}\|_{L^\infty(T)}, \tag{10.92}$$

where the constants hidden in the symbol $\approx$ are independent of $\widehat{c}$, $u_\mathcal{T}$ and T.

Proof. By the Bramble–Hilbert theorem,

$$\|\widehat{c}\,u_\mathcal{T} - \Pi_T(\widehat{c}\,u_\mathcal{T})\|_{L^2(T)} \lesssim h_T^n |\widehat{c}\,u_\mathcal{T}|_{H^n(T)}$$

and

$$|\widehat{c}\,u_\mathcal{T}|_{H^n(T)} \lesssim \sum_{j=1}^{n-1} |\widehat{c}|_{W^j_\infty(T)} |u_\mathcal{T}|_{H^{n-j}(T)}.$$

Moreover,

$$|\widehat{c}|_{W^j_\infty(T)} = |\nabla\widehat{c}|_{W^{j-1}_\infty(T)} = |\nabla\widehat{c} - \Pi_T^{j-2}\nabla\widehat{c}|_{W^{j-1}_\infty(T)}$$

and

$$|u_\mathcal{T}|_{H^{n-j}(T)} = |u_\mathcal{T} - \Pi_T^{n-1-j} u_\mathcal{T}|_{H^{n-j}(T)}.$$

Applying inverse estimates for seminorms, we obtain the $\lesssim$ inequality in (10.92).

To get the opposite inequality, it is enough to check that the vanishing of the left-hand side implies the vanishing of the right-hand side, since both quantities are defined on finite-dimensional spaces and their scaling with respect to the element size is the same. Now, $\widehat{c}\,u_\mathcal{T} = \Pi_T(\widehat{c}\,u_\mathcal{T})$ implies $\widehat{c}\,u_\mathcal{T} \in \mathbb{P}_{n-1}(T)$. Let us assume that $u_\mathcal{T} \in \mathbb{P}_{n-1-k}(T)$ for some $0 \le k \le n-1$, and consequently $\widehat{c} \in \mathbb{P}_k(T)$, i.e. $\nabla\widehat{c} \in \mathbb{P}_{k-1}(T)$. Then $\Pi_T^{n-1-j} u_\mathcal{T} = u_\mathcal{T}$ for any $j \le k$, and hence the corresponding differences in the summation on the right-hand side of (10.92) vanish. Conversely, for $j > k$ we have $j-2 \ge k-1$, which implies $\Pi_T^{j-2}\nabla\widehat{c} = \nabla\widehat{c}$, that is, the corresponding differences in the summation on the right-hand side vanish. In conclusion, all terms in the summation in (10.92) vanish, and the thesis is proved. $\square$

Summarizing, we have obtained the following result.

Proposition 10.45 (*a posteriori* error estimator for mixed methods). *For every $T \in \mathcal{T}$, the local quantity*

$$\begin{aligned}
&\eta_\mathcal{T}((\boldsymbol{\sigma}_\mathcal{T}, u_\mathcal{T}), T)^2 \\
&\quad := h_T^2 \|\widehat{\boldsymbol{K}}\boldsymbol{\sigma}_\mathcal{T} - \nabla u_\mathcal{T}\|^2_{L^2(T)} + h_T^2 \|\operatorname{curl}(\widehat{\boldsymbol{K}}\boldsymbol{\sigma}_\mathcal{T})\|^2_{L^2(T)} \\
&\qquad + h_T \sum_{F \subset \partial T \setminus \partial\Omega} \|[(\widehat{\boldsymbol{K}}\boldsymbol{\sigma}_\mathcal{T})_t]\|^2_{L^2(F)} + h_T \sum_{F \subset \partial T \cap \partial\Omega} \|(\widehat{\boldsymbol{K}}\boldsymbol{\sigma}_\mathcal{T})_t\|^2_{L^2(F)} \\
&\qquad + h_T^2 \sum_{j=1}^{n} \|u_\mathcal{T} - \Pi_T^{n-1-j} u_\mathcal{T}\|^2_{L^2(T)} \|\nabla\widehat{c} - \Pi_T^{j-2}\nabla\widehat{c}\|^2_{L^\infty(T)}
\end{aligned} \tag{10.93}$$

is a (squared) a posteriori *error indicator for the mixed problem* (10.87)*, which gives rise to a global* a posteriori *error estimator $\eta_\mathcal{T}(\boldsymbol{\sigma}_\mathcal{T}, u_\mathcal{T})$ that satisfies Assumption 10.2.*

Finally, Assumption 10.3 follows from Proposition 10.19 (estimator reduction under Dörfler marking), since the estimator is clearly Lipschitz-continuous and monotone, and it satisfies condition (10.38) since all its addends are scaled by positive powers of the mesh size.

As a consequence, the GALERKIN step in AFEM-TS converges linearly by Theorem 10.6, and the number of sub-iterations in the kth call to GALERKIN is bounded by a constant J independent of k (Proposition 10.27). Furthermore, Theorem 10.38 guarantees the quasi-optimality of the two-step AFEM.

Theorem 10.46 (quasi-optimality of AFEM-TS for mixed methods). *Let the exact solution $(\boldsymbol{\sigma}, u)$ of the mixed problem* (10.84) *belong to the approximation class $\mathbb{A}_{s_u}(V;\mathcal{T}_0)$, and let the data $(\boldsymbol{K}, c, f)$ belong to the approximation class $\mathbb{D}_{s_\mathcal{D}}(D(\Omega);\mathcal{T}_0)$. Let Assumptions 6.14 (marking parameter), 6.21 (size of ω) and 6.19 (initial labelling) be valid. Consider the Galerkin discretization* (10.87) *based on one of the Raviart–Thomas–Nédélec or Brezzi–Douglas–Marini finite element pairs. There exists a constant C_* independent of the exact solution $(\boldsymbol{\sigma}, u)$ and the data $\mathcal{D} = (\boldsymbol{K}, c, f)$ such that the sequence $\{(\mathcal{T}_k, \mathbb{V}_{\mathcal{T}_k}\times\mathbb{Q}_{\mathcal{T}_k}, (\boldsymbol{\sigma}_{\mathcal{T}_k}, u_{\mathcal{T}_k}))\}_{k\geq 0}$ produced by* AFEM-TS *satisfies for $k \geq 0$*

$$\|\boldsymbol{\sigma} - \boldsymbol{\sigma}_{\mathcal{T}_k}\|_{H(\mathrm{div};\Omega)} + \|u - u_{\mathcal{T}_k}\|_{L^2(\Omega)} \leq C_* \Big(|(\boldsymbol{\sigma}, u)|_{\mathbb{A}_{s_u}}^{1/s_u} + |\mathcal{D}|_{\mathbb{D}_{s_\mathcal{D}}}^{1/s_\mathcal{D}} \Big)^s (\#\mathcal{T}_k)^{-s},$$

with $0 < s = \min(s_u, s_\mathcal{D}) \leq n/d$.

Remark 10.47. Another family of uniformly inf-sup stable spaces is that of Brezzi, Douglas Jr, Fortin and Marini (1987), where $\mathbb{V}_T = \{\boldsymbol{\tau} \in (\mathbb{P}_n(T))^d \mid \boldsymbol{\tau}\cdot\boldsymbol{n}_F \in \mathbb{P}_{n-1}(F) \text{ for all } F \in \mathcal{F}\cap\partial T\}$ and $Q_T = \mathbb{P}_n(T)$, $n \geq 1$. However, the imposed condition on the normal component of vector fields on each face of $\mathcal{T}$ prevents the inclusion of $\mathbb{V}_\mathcal{T}$ into $\mathbb{V}_{\mathcal{T}_*}$ from holding if $\mathcal{T}_*$ is a bisection refinement of $\mathcal{T}$.

10.6. Proof of Theorem 10.15

This section is devoted to establishing Theorem 10.15, which in turn contributes with Corollary 10.14 to the proof of Theorem 10.6.

It is important to notice that the growth of $\|\boldsymbol{U}\|_2$ is dictated by the number of blocks N rather than the actual dimension $n_N \gg N$ of $\boldsymbol{U}$. Therefore we again use the block notation from Section 10.2,

$$\boldsymbol{B} = (\boldsymbol{B}(i,j))_{i,j=0}^N \in \mathbb{R}^{n_N\times n_N},$$

with lower and upper triangular factors

$$\boldsymbol{L} = (\boldsymbol{L}(i,j))_{i,j=0}^N \in \mathbb{R}^{n_N\times n_N}, \quad \boldsymbol{U} = (\boldsymbol{U}(i,j))_{i,j=0}^N \in \mathbb{R}^{n_N\times n_N}.$$

We also set

$$\boldsymbol{A} = (\boldsymbol{A}(i,j))_{i,j=0}^N := \sqrt{\alpha}\boldsymbol{B}$$

for a suitable parameter $\alpha > 0$ defined below.

10.6.1. Representation of block inverse matrices

We first show that it suffices to derive the estimates

$$\|U^{-1}\|_2 \lesssim N^{-1/p}, \quad p > 2, \tag{10.94}$$

$$\|\tilde{U}^{-1}\|_2 \lesssim N^{-1/p}, \quad p > 2, \tag{10.95}$$

$$\|D\|_2 \lesssim 1, \tag{10.96}$$

where $B^\top = \tilde{L}\tilde{U}$ is the normalized block triangular decomposition of $B^\top$, and $D \in \mathbb{R}^{n_N \times n_N}$ stands for the block diagonal part of U,

In fact, in view of property (P1) (continuity of $\mathcal{B}$), we see that

$$\|B\|_2 = \|B^\top\|_2 \leq \|\mathcal{B}\|,$$

whence

$$L = BU^{-1} \Rightarrow \|L\|_2 \leq \|B\|_2\|U^{-1}\|_2 \leq \|\mathcal{B}\|\|U^{-1}\|_2$$

and, similarly,

$$\|\tilde{L}\|_2 \leq \|\mathcal{B}\|\|\tilde{U}^{-1}\|_2.$$

On the other hand, from

$$B = LD(D^{-1}U) \quad \Rightarrow \quad B^\top = (U^\top D^{-1})DL^\top$$

we infer that

$$\begin{aligned} \tilde{L} = U^\top D^{-1} \quad &\Rightarrow \quad U = D\tilde{L}^\top, \\ \tilde{U} = DL^\top \quad &\Rightarrow \quad L^{-1} = D\tilde{U}^{-\top}, \end{aligned} \tag{10.97}$$

which implies

$$\begin{aligned} \|U\|_2 &\leq \|D\|_2\|\tilde{L}^\top\|_2, \\ \|L^{-1}\|_2 &\leq \|D\|_2\|\tilde{U}^{-\top}\|_2. \end{aligned}$$

Therefore we can focus on proving (10.94) and (10.96), since the proof of (10.95) is identical to that of (10.94). We proceed in several steps. The most delicate estimate is (10.94).

$\boxed{1}$ *jth column of* U^{-1}. To prove (10.94), it turns out to be convenient to first get an explicit expression for the jth column of U^{-1}. We achieve this next.

Lemma 10.48 (representation of the jth column of U^{-1}). *We have*

$$U^{-1}(i,j) = B[j]^{-1}(i,j) \quad \textit{for all } 0 \leq i \leq j \leq N. \tag{10.98}$$

Proof. We compute the (i,j) block of $B[j]^{-1} = U[j]^{-1}L[j]^{-1}$,

$$B[j]^{-1}(i,j) = \sum_{k=0}^{j} U^{-1}[j](i,k)L^{-1}[j](k,j) = U[j]^{-1}(i,j),$$

because $\boldsymbol{L}^{-1}(k,j) = \boldsymbol{0}$ for $k < j$ and $\boldsymbol{L}^{-1}(j,j) = \boldsymbol{I}(j,j)$. Moreover, we claim that

$$\boldsymbol{U}^{-1}[j](i,j) = \boldsymbol{U}^{-1}(i,j), \quad i \le j,$$

because $\boldsymbol{U}^{-1}$ is block upper triangular. To see this, let

$$\boldsymbol{x}(:,j) = \boldsymbol{U}^{-1}(:,j) \in \mathbb{R}^{n_N \times d_j}$$

be the jth block column of $\boldsymbol{U}^{-1}$, which satisfies

$$\boldsymbol{U}\boldsymbol{x}(:,j) = \boldsymbol{I}(:,j) \in \mathbb{R}^{n_N \times d_j}.$$

Since $\boldsymbol{I}(i,j) = \boldsymbol{0}$ for $i > j$ and $\boldsymbol{U}$ is block upper triangular, we have $\boldsymbol{x}(i,j) = \boldsymbol{0}$ for $i > j$. Therefore the matrix

$$\tilde{\boldsymbol{x}}(:,j) = (\boldsymbol{x}(i,j))_{i=0}^{j} \in \mathbb{R}^{n_j \times d_j},$$

with the first j blocks of $\boldsymbol{x}(:,j)$, satisfies the reduced system

$$\begin{aligned} \boldsymbol{U}[j](j,j)\tilde{\boldsymbol{x}}(j,j) &= \boldsymbol{I}(j,j), \\ \sum_{k=i}^{j} \boldsymbol{U}[j](i,k)\tilde{\boldsymbol{x}}(k,j) &= \boldsymbol{0}, \quad 0 \le i \le j. \end{aligned}$$

We thus deduce that $\tilde{\boldsymbol{x}}(:,j) = \boldsymbol{U}[j]^{-1}(:,j)$, as asserted. □

This lemma justifies dealing with $\boldsymbol{B}[j]^{-1}$.

[2] *Representation of* $\boldsymbol{B}[j]^{-1}$. We resort to the Neumann series expansion. We first consider the uniform SPD matrix

$$\boldsymbol{B}[j]\boldsymbol{B}[j]^\top \in \mathbb{R}^{n_j \times n_j},$$

for which there exists $\alpha > 0$ such that

$$\|\boldsymbol{I}[j] - \alpha \boldsymbol{B}[j]\boldsymbol{B}[j]^\top\|_2 < 1$$

uniformly in j. In fact, note that for $\boldsymbol{x} \in \mathbb{R}^{n_j \times n_j}$

$$\|\boldsymbol{x} - \alpha\boldsymbol{B}[j]\boldsymbol{B}[j]^\top\boldsymbol{x}\|_2^2 = \|\boldsymbol{x}\|_2^2 - 2\alpha\langle\boldsymbol{x},\, \boldsymbol{B}[j]\boldsymbol{B}[j]^\top\boldsymbol{x}\rangle + \alpha^2\|\boldsymbol{B}[j]\boldsymbol{B}[j]^\top\boldsymbol{x}\|_2^2,$$

as well as

$$\langle\boldsymbol{x},\, \boldsymbol{B}[j]\boldsymbol{B}[j]^\top\boldsymbol{x}\rangle = \|\boldsymbol{B}[j]^\top\boldsymbol{x}\|_2^2 \ge \beta^2\|\boldsymbol{x}\|_2^2$$

in view of property (P2) (discrete inf-sup) and (3.2), and

$$\|\boldsymbol{B}[j]\boldsymbol{B}[j]^\top\boldsymbol{x}\|_2 \le \|\boldsymbol{B}[j]\|_2^2\|\boldsymbol{x}\|_2 \le \|\mathcal{B}\|^2\|\boldsymbol{x}\|_2.$$

Consequently

$$\|\boldsymbol{x} - \alpha\boldsymbol{B}[j]\boldsymbol{B}[j]^\top\boldsymbol{x}\|_2^2 \le (1 - 2\alpha\beta^2 + \alpha^2\|\mathcal{B}\|^4)\|\boldsymbol{x}\|_2^2.$$

The quadratic polynomial in α on the right-hand side is minimized by $\alpha = \beta^2/\|\mathcal{B}\|^4$, and gives

$$\|\boldsymbol{I}[j] - \alpha \boldsymbol{B}[j]\boldsymbol{B}[j]^\top\|_2^2 \leq 1 - \frac{\beta^4}{\|\mathcal{B}\|^4} =: \rho^2. \tag{10.99}$$

From now on, we fix this value of α and assume the uniform bound (10.99). Let

$$\boldsymbol{A}[j] = \sqrt{\alpha}\boldsymbol{B}[j] \in \mathbb{R}^{n_j \times n_j}$$

be the jth principal section of the matrix $\boldsymbol{A}$ introduced previously, and let

$$\boldsymbol{G}[j] := \boldsymbol{I}[j] - \boldsymbol{A}[j]\boldsymbol{A}[j]^\top \in \mathbb{R}^{n_j \times n_j}. \tag{10.100}$$

Lemma 10.49 (representation of $\boldsymbol{B}[j]^{-1}$). *The following expression is valid:*

$$\boldsymbol{B}[j]^{-1} = \alpha \boldsymbol{B}[j]^\top \sum_{m=0}^{\infty} \boldsymbol{G}[j]^m \quad \textit{for all } 0 \leq j \leq N. \tag{10.101}$$

Proof. Since $\|\boldsymbol{G}[j]\|_2 \leq \rho < 1$ according to (10.99), the Neumann series theorem guarantees that

$$\boldsymbol{A}[j]\boldsymbol{A}[j]^\top = \boldsymbol{I}[j] - \boldsymbol{G}[j]$$

is invertible, and the inverse reads

$$\boldsymbol{A}[j]^{-\top}\boldsymbol{A}[j]^{-1} = \sum_{m=0}^{\infty} \boldsymbol{G}[j]^m,$$

where $\boldsymbol{G}[j]^0 = \boldsymbol{I}[j]$. Multiplying on the left by $\boldsymbol{A}[j]^\top$, we obtain

$$\frac{1}{\sqrt{\alpha}}\boldsymbol{B}[j]^{-1} = \sqrt{\alpha}\boldsymbol{B}[j] \sum_{m=0}^{\infty} \boldsymbol{G}[j]^m,$$

which yields the assertion. □

[3] *Representation of* $\boldsymbol{U}^{-1}$. In order to obtain a representation of $\boldsymbol{U}^{-1}$, we now build on (10.98), which gives a formula for the jth column of $\boldsymbol{U}^{-1}$ in terms of $\boldsymbol{B}[j]^{-1}$, and (10.101), which provides a series representation of $\boldsymbol{B}[j]^{-1}$. To this end, we introduce the block upper triangular matrix $\boldsymbol{G}_m \in \mathbb{R}^{n_N \times n_N}$ given by

$$\boldsymbol{G}_m(i,j) := \begin{cases} \boldsymbol{G}[j]^m(i,j), & i \leq j, \\ 0, & i > j, \end{cases}$$

for $m \geq 1$ and $\boldsymbol{G}_0 = \boldsymbol{I}$. Hence

$$\boldsymbol{U}^{-1}(i,j) = \begin{cases} \alpha\left(\boldsymbol{B}[j]^\top \displaystyle\sum_{m=0}^{\infty} \boldsymbol{G}_m[j]\right)(i,j), & i \leq j, \\ 0, & i > j. \end{cases}$$

To write this expression in compact form, it is convenient to introduce the *block upper triangular truncation operator* $\mathcal{U}\colon \mathbb{R}^{n_N\times n_N} \to \mathbb{R}^{n_N\times n_N}$ defined by

$$\mathcal{U}(\boldsymbol{M})(i,j) := \begin{cases} \boldsymbol{M}(i,j), & i \le j, \\ 0, & i > j, \end{cases} \quad \text{for all } \boldsymbol{M} \in \mathbb{R}^{n_N\times n_N}.$$

Lemma 10.50 (representation of U^{-1}). *We have*

$$\boldsymbol{U}^{-1} = \alpha\, \mathcal{U}\biggl(\boldsymbol{B}^\top \sum_{m=0}^{\infty} \boldsymbol{G}_m\biggr). \tag{10.102}$$

Proof. Since $\boldsymbol{G}_m$ is block upper triangular for all $m \ge 0$, so is the series $\sum_{m=0}^{\infty} \boldsymbol{G}_m$. It thus suffices to check that

$$\biggl(\boldsymbol{B}^\top \sum_{m=0}^{\infty} \boldsymbol{G}_m\biggr)(i,j) = \biggl(\boldsymbol{B}[j]^\top \sum_{m=0}^{\infty} \boldsymbol{G}_m[j]\biggr)(i,j), \quad i \le j.$$

This shows the desired relation (10.102). □

4 *Recursion.* In order to estimate $\boldsymbol{G}_m$, it is useful to relate $\boldsymbol{G}_m$ to $\boldsymbol{G}_{m-1}$. We start with a simple property of the operator $\mathcal{U}$: for $\boldsymbol{A}, \boldsymbol{B} \in \mathbb{R}^{n\times n}$ and $1 \le i \le j \le n$, we have

$$(\boldsymbol{A}\,\mathcal{U}(\boldsymbol{B}))_{ij} = \sum_{k=1}^{n} \boldsymbol{A}_{ik}\,\mathcal{U}(\boldsymbol{B})_{kj} = \sum_{k=1}^{j} \boldsymbol{A}_{ik}\,\boldsymbol{B}_{kj} = (\boldsymbol{A}[j]\boldsymbol{B}[j])(i,j).$$

Lemma 10.51 (recursion). *The following is valid for all $m \ge 1$:*

$$\boldsymbol{G}_m = \boldsymbol{G}_{m-1} - \mathcal{U}(\boldsymbol{A}\,\mathcal{U}(\boldsymbol{A}^\top \boldsymbol{G}_{m-1})), \tag{10.103}$$

with $\boldsymbol{G}_0 = \boldsymbol{I}$. Therefore the jth column of $\boldsymbol{G}_m$ reads

$$\boldsymbol{G}_m(0\colon j, j) = \boldsymbol{G}[j]\boldsymbol{G}_{m-1}(0\colon j, j), \quad 0 \le j \le N. \tag{10.104}$$

Proof. First take $m = 1$ and apply the proceeding relation for $0 \le i \le j \le N$, to obtain

$$\begin{aligned}(\boldsymbol{G}_0 - \mathcal{U}(\boldsymbol{A}\,\mathcal{U}(\boldsymbol{A}^\top \boldsymbol{G}_0)))(i,j) &= \boldsymbol{I}(i,j) - (\boldsymbol{A}\,\mathcal{U}(\boldsymbol{A}^\top))(i,j) \\ &= \boldsymbol{I}(i,j) - \boldsymbol{A}[j]\boldsymbol{A}[j]^\top(i,j) \\ &= \boldsymbol{G}[j](i,j) = \boldsymbol{G}_1(i,j),\end{aligned}$$

in light of (10.100). Then take $m > 1$ and $0 \le i \le j$, to arrive at

$$\begin{aligned}&(\boldsymbol{G}_{m-1} - \mathcal{U}(\boldsymbol{A}\,\mathcal{U}(\boldsymbol{A}^\top \boldsymbol{G}_{m-1})))(i,j) \\ &\quad = \boldsymbol{G}_{m-1}(i,j) - \sum_{k,\ell=1}^{j} \boldsymbol{A}(i,k)\,\boldsymbol{A}^\top(k,\ell)\,\boldsymbol{G}_{m-1}(\ell,j) \\ &\quad = \sum_{\ell=1}^{j} (\boldsymbol{I}[j] - \boldsymbol{A}[j]\boldsymbol{A}[j]^\top)(i,\ell)\,\boldsymbol{G}_{m-1}(\ell,j)\end{aligned}$$

$$= \sum_{\ell=1}^{j} \boldsymbol{G}[j](i,\ell)\,\boldsymbol{G}[j]^{m-1}(\ell,j)$$
$$= \boldsymbol{G}[j]^m(i,j) = \boldsymbol{G}_m(i,j).$$

This is the asserted equality (10.103). The remaining relation (10.104) follows from the last equality upon realizing that

$$\boldsymbol{G}_m(0\colon j,j) = (\boldsymbol{G}_m(i,j))_{i=0}^{j} = \boldsymbol{G}[j]\boldsymbol{G}[j]^{m-1}(0\colon j,j) = \boldsymbol{G}[j]\boldsymbol{G}_{m-1}(0\colon j,j).$$

This completes the proof. □

5 *Schatten norms.* In the view of Lemmas 10.51 (recursion) and 10.50 (representation of $\boldsymbol{U}^{-1}$), we intend to estimate $\|\boldsymbol{U}^{-1}\|_2$ in terms of suitable norms of $\boldsymbol{G}_m$ that depend on the number N of blocks rather than the dimension n_N, because $n_N \gg N$. These special norms are called *block Schatten norms.*

However, for the sake of clarity, we start with the definition and properties of the usual Schatten norms. They include the operator 2-norm, the Frobenius norm, and satisfy a Hölder inequality.

Definition 10.52 (Schatten norms). Given $\boldsymbol{M} \in \mathbb{R}^{n\times n}$, let

$$\sigma_1(\boldsymbol{M}) \geq \sigma_2(\boldsymbol{M}) \geq \cdots \geq \sigma_n(\boldsymbol{M}) \geq 0$$

be the singular values of $\boldsymbol{M}$. Given $1 \leq p \leq \infty$, let the p-Schatten norm be

$$|\boldsymbol{M}|_p := \left(\sum_{m=1}^{n} \sigma_m(\boldsymbol{M})^p\right)^{1/p}.$$

Remark 10.53. Note that if $p = \infty$, the Schatten norm reduces to the 2-norm, that is,

$$|\boldsymbol{M}|_\infty = \sigma_1(\boldsymbol{M}) = \|\boldsymbol{M}\|_2,$$

and if $p = 2$ it is equivalent to the Frobenius norm,

$$|\boldsymbol{M}|_2 = \left(\sum_{m=1}^{n} \sigma_m(\boldsymbol{M})^2\right)^{1/2} = \left(\sum_{m=1}^{n} \boldsymbol{M}_{ij}^2\right)^{1/2} = \|\boldsymbol{M}\|_F.$$

We now list a number of useful properties of these norms.

Lemma 10.54 (properties of $|\cdot|_p$). *The following properties hold for* $1 \leq p \leq \infty$*:*

(i) $\sigma_i(\boldsymbol{M}^\top\boldsymbol{M}) = \sigma_i(\boldsymbol{M})^2 \quad \Rightarrow \quad |\boldsymbol{M}^\top\boldsymbol{M}|_p = |\boldsymbol{M}|_{2p}^2$,

(ii) $\sigma_i(\boldsymbol{M}) = \sigma_i(\boldsymbol{M}^\top) \quad \Rightarrow \quad |\boldsymbol{M}|_p = |\boldsymbol{M}^\top|_p$,

(iii) *Hölder inequality: for* $1/r = 1/p + 1/q$*, with* $r, p, q \in [1,\infty]$*,*

$$|\boldsymbol{M}_1\,\boldsymbol{M}_2|_r \leq |\boldsymbol{M}_1|_p\,|\boldsymbol{M}_2|_q,$$

(iv) $|\mathcal{U}(\boldsymbol{M})|_\infty \lesssim \log(n)\, |\boldsymbol{M}|_\infty$,

(v) $|\mathcal{U}(\boldsymbol{M})|_{2^j} \leq 2^{j-1} |\boldsymbol{M}|_{2^j}$.

Remark 10.55. Properties (i) and (ii) are trivial. We refer to Dunford and Schwartz (1988, Lemma XI.9.20) for property (iii), to Bhatia (2000, (15)) for property (iv), and to Davies (1988) and Feischl (2022, Lemma 17) for property (v).

To define the block Schatten norms, we consider the subspace $\mathcal{D}_b$ of $\mathbb{R}^{n_N \times (N+1)}$ of matrices of the form

$$\boldsymbol{X} = \begin{bmatrix} \boldsymbol{X}_0 & & & \\ & \boldsymbol{X}_1 & & \\ & & \ddots & \\ & & & \boldsymbol{X}_N \end{bmatrix}, \qquad \boldsymbol{X}_j \in \mathbb{R}^{d_j}, \quad 0 \leq j \leq N,$$

or equivalently

$$\boldsymbol{X} \in \mathcal{D}_b \quad \Longleftrightarrow \quad \boldsymbol{X}_{ij} = \boldsymbol{0} \quad \text{for all } i \neq n_{j-1} + 1, \ldots, n_j.$$

We can represent $\boldsymbol{X}$ using block notation as follows:

$$\boldsymbol{X} = (\boldsymbol{X}(i,j))_{i,j=0}^N, \qquad \boldsymbol{X}(i,j) \in \mathbb{R}^{d_j \times 1},$$

where

$$\boldsymbol{X}(i,j) = \begin{cases} \boldsymbol{X}_j, & i = j, \\ \boldsymbol{0}, & i \neq j. \end{cases}$$

Given a block matrix $\boldsymbol{M} = (\boldsymbol{M}(i,j))_{i,j=0}^N \in \mathbb{R}^{n_N \times n}$, we consider

$$\boldsymbol{M}\boldsymbol{X} = (\boldsymbol{M}(i,j)\boldsymbol{X}_j)_{i,j=0}^N \in \mathbb{R}^{n_N \times (N+1)},$$

namely the jth block column of $\boldsymbol{M}\boldsymbol{X}$ is

$$(\boldsymbol{M}(i,j)\boldsymbol{X}_j)_{i=0}^N \in \mathbb{R}^{n_N}.$$

Definition 10.56 (block Schatten norms). For $1 \leq p \leq \infty$, let

$$|\boldsymbol{M}|_{b,p} := \sup_{\boldsymbol{X} \in \mathcal{D}_b,\ |\boldsymbol{X}|_\infty \leq 1} |\boldsymbol{M}\boldsymbol{X}|_p \quad \text{for all } \boldsymbol{M} \in \mathbb{R}^{n_N \times n_N}.$$

Note the unusual norm $|\boldsymbol{X}|_\infty$ instead of $|\boldsymbol{X}|_p$ in this definition of operator norm $|\boldsymbol{M}|_{b,p}$. This choice is deliberate and will be useful later; see Remark 10.58. We now list important properties of the block Schatten norms; see Feischl (2022, Lemmas 15, 16, 17) for proofs.

Lemma 10.57 (properties of $|\cdot|_{b,p}$). *The following properties hold for all $\boldsymbol{M}$, $\boldsymbol{M}_1$, $\boldsymbol{M}_2 \in \mathbb{R}^{n_N \times n_N}$ and $1 \leq p \leq \infty$:*

(i) $|\boldsymbol{M}|_{b,p} \leq (N+1)^{1/p} |\boldsymbol{M}|_\infty = (N+1)^{1/p} \|\boldsymbol{M}\|_2$,

(ii) $|\boldsymbol{M}_1 \boldsymbol{M}_2|_{b,p} \leq |\boldsymbol{M}_1|_\infty |\boldsymbol{M}_2|_{b,p}$,

(iii) $|\boldsymbol{M}|_\infty \le |\boldsymbol{M}|_{b,p}$, $|\boldsymbol{M}|_{b,\infty} = |\boldsymbol{M}|_\infty$,

(iv) *if* $\boldsymbol{M}_1 \in \mathbb{R}^{n_N \times n_N}$ *is block triangular with* j*th block column*

$$\boldsymbol{M}_1(0\colon j, j) = \boldsymbol{P}_j \boldsymbol{M}_2(0\colon j, j), \quad \boldsymbol{P}_j \in \mathbb{R}^{n_j \times n_j}$$

for $0 \le j \le N$, *then*

$$|\boldsymbol{M}_1|_{b,2} \le \max_{0\le j\le N} |\boldsymbol{P}_j|_\infty \, |\boldsymbol{M}_2|_{b,2},$$

(v) $|\mathcal{U}(\boldsymbol{M})|_{b,2^k} \le 2^{k-1} |\boldsymbol{M}|_{b,2^k}$, $k = 1, 2$,

(vi) $|\mathcal{U}(\boldsymbol{M})|_\infty \le (\lceil \log_2(N) \rceil + 1) |\boldsymbol{M}|_\infty$.

Remark 10.58. To understand the significance of Definition 10.56, we examine the growth of the usual and block p-Schatten norm relative to the ∞-Schatten norm for $1 \le p < \infty$. Given $\boldsymbol{M} \in \mathbb{R}^{n_N \times n_N}$, we have for the usual p-norm

$$|\boldsymbol{M}|_p = \left(\sum_{i=1}^{n_N} \sigma_i(\boldsymbol{M})^p \right)^{1/p} \le n_N^{1/p} \sigma_1(\boldsymbol{M}) = n_N^{1/p} |\boldsymbol{M}|_\infty = n_N^{1/p} \|\boldsymbol{M}\|_2,$$

whereas for the block p-norm we get

$$|\boldsymbol{M}|_{b,p} \le (N+1)^{1/p} |\boldsymbol{M}|_\infty = (N+1)^{1/p} \|\boldsymbol{M}\|_2,$$

according to Lemma 10.57(i). In fact, given $\boldsymbol{X} \in \mathcal{D}_b$ with $|\boldsymbol{X}|_\infty = \|\boldsymbol{X}\|_2 = 1$, we first note that

$$|\boldsymbol{M}\boldsymbol{X}|_p = \left(\sum_{j=0}^{N} \sigma_j(\boldsymbol{M}\boldsymbol{X})^p \right)^{1/p} \le \sigma_0(\boldsymbol{M}\boldsymbol{X})(N+1)^{1/p} = \|\boldsymbol{M}\boldsymbol{X}\|_2 (N+1)^{1/p},$$

and also that

$$\|\boldsymbol{M}\boldsymbol{X}\|_2 = \sup_{\boldsymbol{x}\in\mathbb{R}^{N+1}} \frac{\|\boldsymbol{M}\boldsymbol{X}\boldsymbol{x}\|_2}{\|\boldsymbol{x}\|_2} \le \|\boldsymbol{M}\|_2 \sup_{\boldsymbol{x}\in\mathbb{R}^{N+1}} \frac{\|\boldsymbol{X}\boldsymbol{x}\|_2}{\|\boldsymbol{x}\|_2} \le \|\boldsymbol{M}\|_2,$$

because $\|\boldsymbol{M}\|_2 = |\boldsymbol{M}|_\infty = 1$. On the one hand, this explains why it is convenient to have the norm $|\boldsymbol{X}|_\infty$ rather than $|\boldsymbol{X}|_p$ in Definition 10.56. On the other hand, this calculation reveals the key point that

$$|\boldsymbol{M}|_{b,p} \ll |\boldsymbol{M}|_p,$$

because the growth of $|\boldsymbol{M}|_{b,p}$ is dictated by the number of blocks $N+1$ whereas that of $|\boldsymbol{M}|_p$ is proportional to the dimension n_N of $\boldsymbol{M}$ and $n_N \gg N$. This property is essential to the estimate of $\|\boldsymbol{U}^{-1}\|_2$ below.

[6] *Estimate of* $\|\boldsymbol{U}^{-1}\|_2$. We are now in a position to prove the desired bound (10.94).

Proposition 10.59 (estimate of $\|U^{-1}\|_2$). *Let* $\boldsymbol{B} \in \mathbb{R}^{n_N \times n_N}$ *be a block matrix such that*

$$\|\boldsymbol{B}\|_2 \le \|\mathcal{B}\|, \quad \max_{0\le j\le N} \|\boldsymbol{B}[j]^{-1}\|_2 \le \frac{1}{\beta}.$$

Then there exist constants C_{LU} and $p > 2$ such that the block upper triangular factor $\boldsymbol{U}$ of $\boldsymbol{B}$ satisfies

$$\|\boldsymbol{U}^{-1}\|_2 \le C_{LU} N^{1/p}. \tag{10.105}$$

Proof. We recall (10.102) of Lemma 10.50 (representation of $\boldsymbol{U}^{-1}$),

$$\boldsymbol{U}^{-1} = \alpha\, \mathcal{U}\left(\boldsymbol{B}^\top \sum_{m=0}^{\infty} \boldsymbol{G}_m\right),$$

along with (10.103) of Lemma 10.51 (recursion),

$$\boldsymbol{G}_m = \boldsymbol{G}_{m-1} - \mathcal{U}(\boldsymbol{A}\, \mathcal{U}(\boldsymbol{A}^\top \boldsymbol{G}_{m-1})), \quad m \ge 1,$$

and (10.104) of Lemma 10.51,

$$\boldsymbol{G}(0\colon j, j) = \boldsymbol{G}[j] \boldsymbol{G}_{m-1}(0\colon j, j), \quad 0 \le j \le N,$$

with $\boldsymbol{G}_0 = \boldsymbol{I}$. We use these expressions in conjunction with Lemma 10.57 (properties of $|\cdot|_{b,p}$) to prove (10.105). We proceed in several steps.

(i) *Bound for* $|\boldsymbol{G}_m|_{b,2}$. In light of (10.99) and (10.100),

$$|\boldsymbol{G}[j]|_\infty = \|\boldsymbol{G}[j]\|_2 \le \rho = \sqrt{1 - \frac{\beta^4}{\|\mathcal{B}\|^4}} < 1, \quad 0 \le j \le N.$$

Applying Lemma 10.57(iv) to $\boldsymbol{G}_m$ yields

$$\begin{aligned}|\boldsymbol{G}_m|_{b,2} &\le \max_{0 \le j \le N} |\boldsymbol{G}[j]|_\infty \, |\boldsymbol{G}_{m-1}|_{b,2} \\ &\le \rho\, |\boldsymbol{G}_{m-1}|_{b,2} \le \rho^m\, |\boldsymbol{I}|_{b,2}.\end{aligned}$$

Recalling Lemma 10.57(i),

$$|\boldsymbol{I}|_{b,2} \le (N+1)^{1/2} \|\boldsymbol{I}\|_2 = (N+1)^{1/2},$$

whence

$$|\boldsymbol{G}_m|_{b,2} \le \rho^m\, (N+1)^{1/2}.$$

We observe that this bound is not good enough for our purposes because it scales like $N^{1/2}$ instead of $N^{1/p}$ for $p > 2$. We next improve upon this bound.

(ii) *Bound for* $|\boldsymbol{G}_m|_{b,4}$. We take $k = 2$ in Lemma 10.57(v) and use the triangle inequality to arrive at

$$\begin{aligned}|\boldsymbol{G}_m|_{b,4} &\le |\boldsymbol{G}_{m-1}|_{b,4} + |\mathcal{U}(\boldsymbol{A}\, \mathcal{U}(\boldsymbol{A}^\top \boldsymbol{G}_{m-1}))|_{b,4} \\ &\le |\boldsymbol{G}_{m-1}|_{b,4} + 2|\boldsymbol{A}\, \mathcal{U}(\boldsymbol{A}^\top \boldsymbol{G}_{m-1})|_{b,4}.\end{aligned}$$

We further apply parts (ii) and (v) of Lemma 10.57 to obtain

$$|\boldsymbol{A}\, \mathcal{U}(\boldsymbol{A}^\top \boldsymbol{G}_{m-1})|_{b,4} \le 2|\boldsymbol{A}|_\infty\, |\boldsymbol{A}^\top \boldsymbol{G}_{m-1}|_{b,4} \le 2|\boldsymbol{A}|_\infty^2\, |\boldsymbol{G}_{m-1}|_{b,4}.$$

Therefore

$$|\boldsymbol{G}_m|_{b,4} \le (1 + 4|\boldsymbol{A}|_\infty^2)|\boldsymbol{G}_{m-1}|_{b,4},$$

but the prefactor on the right-hand side is greater than 1 and thus not suitable for iteration. We still have

$$|\boldsymbol{G}_m|_{b,4} \le (1 + 4|\boldsymbol{A}|_\infty^2)^m |\boldsymbol{I}|_{b,4}.$$

(iii) *Bound for* $|\boldsymbol{G}_m|_\infty$. We combine the estimates from steps (i) and (ii) to exploit their relative merits. Recall from Lemma 10.57(iii) that

$$|\boldsymbol{G}_m|_\infty \le |\boldsymbol{G}_m|_{b,p} \quad \text{for all } 1 \le p \le \infty.$$

Take $p = 2, 4$ and $0 < t < 1$ to be chosen later, and write

$$\begin{aligned} |\boldsymbol{G}_m|_\infty &\le |\boldsymbol{G}_m|_{b,2}^{1-t} |\boldsymbol{G}_m|_{b,4}^t \\ &\le \left[\rho^{1-t}\left(1 + 4|\boldsymbol{A}|_\infty^2\right)^t\right]^m |\boldsymbol{I}|_{b,2}^{1-t} |\boldsymbol{I}|_{b,4}^t. \end{aligned}$$

Consequently, there exists $0 < t_0 < 1$ such that

$$q := \rho^{1-t}\left(1 + 4|\boldsymbol{A}|_\infty^2\right)^t < 1, \quad 0 < t < t_0$$

and

$$|\boldsymbol{G}_m|_\infty \le q^m |\boldsymbol{I}|_{b,2}^{1-t} |\boldsymbol{I}|_{b,4}^t.$$

We now estimate the two terms on the right-hand side via Lemma 10.57(i), namely

$$\begin{aligned} |\boldsymbol{I}|_{b,2} &\le (N+1)^{1/2} \|\boldsymbol{I}\|_2 = (N+1)^{1/2}, \\ |\boldsymbol{I}|_{b,4} &\le (N+1)^{1/4} \|\boldsymbol{I}\|_2 = (N+1)^{1/4}. \end{aligned}$$

We thus obtain

$$|\boldsymbol{G}_m|_\infty \le q^m (N+1)^{1/\widetilde{p}},$$

with $1/\widetilde{p} = (1-t)/2 + t/4 < 1/2$ for $0 < t < t_0$.

(iv) *Estimate of* $\|\boldsymbol{U}^{-1}\|_2$. Recalling the expression

$$\boldsymbol{U}^{-1} = \alpha\, \mathcal{U}\left(\boldsymbol{B}^\top \sum_{m=0}^{\infty} \boldsymbol{G}_m\right),$$

and applying parts (vi), (ii) and (iii) of Lemma 10.57, we see that

$$\begin{aligned} \|\boldsymbol{U}^{-1}\|_2 = |\boldsymbol{U}^{-1}|_\infty &\lesssim \log(N)\, |\boldsymbol{B}|_\infty \sum_{m=0}^{\infty} |\boldsymbol{G}_m|_\infty \\ &\lesssim |\boldsymbol{B}|_\infty (N+1)^{1/\widetilde{p}} \log(N) \sum_{m=0}^{\infty} q^m \\ &\lesssim \|\boldsymbol{B}\|_2 (N+1)^{1/\widetilde{p}} \log(N). \end{aligned}$$

Finally, for any $2 < p < \widetilde{p}$, we can absorb the logarithm, thereby getting

$$\|\boldsymbol{U}^{-1}\|_2 \lesssim \|\boldsymbol{B}\|_2 (N+1)^{1/p},$$

which is the desired estimate (10.105).

This concludes the proof. □

7 *Estimate of block diagonal* $\boldsymbol{D}$. We recall that $\boldsymbol{D} = \operatorname{diag} \boldsymbol{U} \in \mathbb{R}^{n_N \times n_N}$ is the block diagonal of $\boldsymbol{U}$. We consider the block partitioning of $\boldsymbol{B}[j]$,

$$\boldsymbol{B}[j] = \begin{bmatrix} \boldsymbol{B}[j-1] & \boldsymbol{R}_1 \\ \boldsymbol{R}_2^\top & \boldsymbol{R}_3 \end{bmatrix} \in \mathbb{R}^{n_j \times n_j},$$

where

$$\begin{aligned} \boldsymbol{R}_1 &= \boldsymbol{B}[j](1\colon j-1, j) \in \boldsymbol{R}^{n_{j-1} \times d_j}, \\ \boldsymbol{R}_2^\top &= \boldsymbol{B}[j](j, 1\colon j-1) \in \boldsymbol{R}^{d_j \times n_{j-1}}, \\ \boldsymbol{R}_3 &= \boldsymbol{B}[j](j,j) \in \mathbb{R}^{d_j \times d_j}. \end{aligned}$$

Lemma 10.60 (bound of $\|\boldsymbol{D}\|_2$). *We have*

$$\|\boldsymbol{D}\|_2 \le \|\mathcal{B}\| + \frac{\|\mathcal{B}\|^2}{\beta} = C_D. \tag{10.106}$$

Proof. Compute the $\boldsymbol{LU}$ factorization of $\boldsymbol{B}[j]$,

$$\boldsymbol{B}[j] = \begin{bmatrix} \boldsymbol{I}[j-1] & 0 \\ \boldsymbol{R}_2^\top \boldsymbol{B}[j-1]^{-1} & 1 \end{bmatrix} \begin{bmatrix} \boldsymbol{B}[j-1] & \boldsymbol{R}_1 \\ 0 & \boldsymbol{R}_3 - \boldsymbol{R}_2^\top \boldsymbol{B}[j-1]^{-1} \boldsymbol{R}_1 \end{bmatrix},$$

and realize that

$$\boldsymbol{U}(j,j) = \boldsymbol{D}(j,j) = \boldsymbol{R}_3 - \boldsymbol{R}_2^\top \boldsymbol{B}[j-1]^{-1} \boldsymbol{R}_1 \in \mathbb{R}^{d_j \times d_j}.$$

Since

$$\begin{aligned} |\boldsymbol{R}_i|_\infty &= \|\boldsymbol{R}_i\|_2 \le \|\boldsymbol{B}[j]\|_2 \le \|\boldsymbol{B}\|_2 = \|\mathcal{B}\|, \quad i = 1,2, \\ |\boldsymbol{R}_3|_\infty &= \|\boldsymbol{R}_3\|_2 \le \|\boldsymbol{B}[j]\|_2 \le \|\mathcal{B}\|, \end{aligned}$$

and

$$|\boldsymbol{B}[j-1]^{-1}|_\infty = \|\boldsymbol{B}[j-1]\|_2 \le \frac{1}{\beta},$$

according to properties (P1) and (P2) of the bilinear form $\mathcal{B}$, we deduce

$$|\boldsymbol{D}(j,j)|_\infty = \|\boldsymbol{D}(j,j)\|_2 \le \|\mathcal{B}\| + \frac{\|\mathcal{B}\|^2}{\beta}$$

as asserted. □

8 *Bound of* $\boldsymbol{LU}$ *factors.* We are finally in the position to prove Theorem 10.15. We combine Proposition 10.59 (estimate of $\|\boldsymbol{U}^{-1}\|_2$) and $\boldsymbol{L} = \boldsymbol{B}\boldsymbol{U}^{-1}$ to obtain

$$\|\boldsymbol{L}\|_2 \le \|\boldsymbol{B}\|_2 \|\boldsymbol{U}^{-1}\|_2 \le \|\mathcal{B}\| C_{LU} N^{1/p}.$$

Then, invoking (10.97) in conjunction with Proposition 10.59 and Lemma 10.60

(bound of $\|G\|_2$), as well as the bounds of $\|U^{-1}\|_2$ and $\|L\|_2$, yields

$$\begin{aligned} \|U\|_2 &\le \|D\|_2 \|\tilde{L}^{\top}\|_2 \le C_D \|\mathcal{B}\| C_{LU} N^{1/p}, \\ \|L^{-1}\|_2 &\le \|D\|_2 \|\tilde{U}^{-\top}\|_2 \le C_D C_{LU} N^{1/p}, \end{aligned} \tag{10.107}$$

with C_D being the constant in (10.106). This completes the proof. □

Acknowledgements

We are grateful to R. DeVore, P. Morin and A. Salgado for discussions about regularity classes, and to Ch. Kreuzer for pointing out a gap in an earlier version of the proof of Theorem 5.17. We are also thankful to G. Vacca for providing simulations and D. Fassino for assistance with the manuscript.

This work was partially supported by NSF grants DMS-2110811 (AB) and DMS-1908267 (RHN), and MUR-PRIN grants 20227K44ME (CC) and 201752HKM (AV), and INdAM research group GNCS (CC and AV). RHN was also partially supported by NSF grant DMS-1929284 while he was in residence at the Institute for Computational and Experimental Research in Mathematics in Providence, RI, during the Numerical PDEs: Analysis, Algorithms, and Data Challenges semester programme.

Index

Algorithms

AFEM-SW: AFEM with switch, 303
AFEM-TS: two-step AFEM successively approximating the data and the Galerkin solution with approximate data, 287, 319
AFEM-DG-TS: interior penalty version of AFEM-TS, 398
CONSTRAINT-A: modify an approximation of $\boldsymbol{A}$ to satisfy the structural constraints, 366
CONSTRAINT-c: modify an approximation of $\boldsymbol{A}$ to satisfy the structural constraints, 369
DATA: procedure to approximate the data $\mathcal{D} = (\boldsymbol{A}, c, f)$, 287, 379
ESTIMATE: compute the element error indicators and element data oscillations, 292
GALERKIN: procedure that iterates SOLVE, ESTIMATE, MARK, REFINE, 287, 293
GALERKIN-DG: discontinuous Galerkin version of GALERKIN, 418
GREEDY: abstract greedy algorithm for DATA, 356
MARK: Dörfler marking, 292
REFINE: refine all marked elements b times and others necessary to produce a conforming mesh, 292, 392
REFINE: refine marked elements and others necessary to produce a Λ-admissible mesh, 396
SOLVE: construct FEM approximation, 291

Assumptions

Abstract cut-off, 237
Admissible set of parameters for GREEDY, 357
Approximability of u, 328
Approximability of data, 332
Cardinality of the marked set, 336
Discrete coefficients and discrete functionals, 252
Equivalence of error and estimator, 430
Equivalence of error and full estimator, 448
Estimator reduction, 430
Initial labelling, 337
Lipschitz continuity of estimator, 444
Marking parameter θ, 334
Monotonicity of estimator, 444
Monotonicity of local oscillations, 358
Properties of DATA, 312
Quasi-monotonicity of oscillation, 445
Quasi-optimality of DATA, 333
Relaxed quasi-orthogonality, 429
Restriction on ω, 335
Restrictions on κ, ω and θ, 426
Size of ω, 339
Structural assumption for discrete data, 308
Structural assumption for exact data, 308
Structure of f, 379

Constants

(C_L, C_U): *a posteriori* lower and upper bounds constants, 252, 289, 334
$(C_L^{\mathrm{eq}}, C_U^{\mathrm{eq}})$: lower and upper estimators equivalence constant, 324
(α_1, α_2): lower and upper bounds of the diffusion coefficient spectrum, 176, 308
$(\widehat{\alpha}_1, \widehat{\alpha}_2)$: lower and upper bounds of the approximate diffusion coefficient spectrum, 308
$(\widehat{\alpha}_1, \widehat{\alpha}_2)$: lower and upper bounds on the spectrum of $\widehat{\boldsymbol{A}}$, 369
(c_1, c_2): lower and upper bounds of the reaction coefficient, 308
$(\widehat{c}_1, \widehat{c}_2)$: lower and upper bounds of the approximate reaction coefficient, 308
$(c_{\mathcal{B}}, C_{\mathcal{B}})$: norm equivalence constants, 181, 288
$(c_{\widehat{\mathcal{B}}}, C_{\widehat{\mathcal{B}}})$: norm equivalence constants for the perturbed problem, 309
C_D: DATA constant, 312
C_P: Poincaré constant, 175
C_{loc}: localization constant, 220
C_{ovrl}: overlay constant, 220
$C_{\mathrm{Céa}}$: best approximation constant, 187
C_{Lip}: estimator Lipschitz property constant, 291

C_{osc}: oscillation quasi-monotonicity constant, 263
L: threshold parameter for constrained approximation, 362
C_{BA}: best approximation constant of $\Pi_{\mathcal{T}}$, 313
C_{BA}: quasi-monotonicity constant of $\Pi_{\mathcal{T}}$, 313
D: complexity of REFINE constant, 206, 216, 381
$\widetilde{D}$: modified complexity of REFINE constant, 338
C_{ctr}: constrain upper bound amplification constant, 308
C_{data}: DATA approximation constant, 312, 316
C_{Lip}: estimator Lipschitz property constant, 262, 263
Λ: Λ-admissibility constant, 211
Λ_{data}: DATA quasi-optimality constant, 312, 330
α: inf sup constant, 179
α: contraction constant, 294, 304
σ: shape regularity constant of $\mathbb{T}$, 188
θ: Dörfler marking parameter, 292
$\widetilde{C}_U$: localized upper bound constant, 257, 334

Definitions
ε-approximation of order s, 333
$\mathcal{T}$-meshed subdomain, 232
Face-connected, 193
Global index of a node, 211
Interior vertex property, 259
Sobolev number sob(W_p^k), 173

Error estimators
$E_{\mathcal{T}}(v)_q$: generic total error, 356
$R_{\mathcal{T}}$: residual in $H^{-1}(\Omega)$, 218
$\mathcal{E}_{\mathcal{T}}^{\text{abs}}(z)$: abstract total estimator, 247
$\mathcal{E}_{\mathcal{T}}$: total estimator, 252
$\mathcal{E}_{\mathcal{T}}(T)$, $\mathcal{E}_{\mathcal{T}}(u_{\mathcal{T}}, f, T)$: local total estimator, 252
$\mathcal{E}_{\mathcal{T}}^{\text{std}}(u_{\mathcal{T}}, \mathcal{D})$: standard residual estimator, 224
$\mathcal{E}_{\mathcal{T}}^{\text{std}}(u_{\mathcal{T}}, \mathcal{D}, T)$: standard local indicators, 225
$\mathcal{E}_{\mathcal{T}}^{\text{std}}(u_{\mathcal{T}}, f, T)$: standard local indicators, 231
$\eta_{\mathcal{T}}^{\text{abs}}(z)$: abstract PDE estimator, 247
$\eta_{\mathcal{T}}(u_{\mathcal{T}}, T)$: PDE local estimator, 252
$\eta_{\mathcal{T}}(u_{\mathcal{T}}, f)$: PDE estimator, 255
$\eta_{\mathcal{T}}^{\text{std}}(u_{\mathcal{T}}, T)$: standard local PDE indicators, 231
$\text{osc}_{\mathcal{T}}^{\text{abs}}(R_{\mathcal{T}}, z)$: abstract oscillation, 247
$\text{osc}_{\mathcal{T}}(\boldsymbol{A}, T)_r$: local surrogate for the diffusion coefficient approximation error, 314
$\text{osc}_{\mathcal{T}}(\mathcal{D})$: surrogate for the data error, 312
$\text{osc}_{\mathcal{T}}(c, T)_2$: super-convergent local surrogate for the reaction coefficient approximation error, 315

$\mathrm{osc}_{\mathcal{T}}(c,T)_\infty$: super-convergent local surrogate for the reaction coefficient approximation error, 315
$\mathrm{osc}_{\mathcal{T}}(c,T)_q$: local surrogate for the reaction coefficient approximation error, 314
$\mathrm{osc}_{\mathcal{T}}(f)$, $\mathrm{osc}_{\mathcal{T}}(f)_{-1}$: oscillation for the load function, 255
$\mathrm{osc}_{\mathcal{T}}(f,T)$, $\mathrm{osc}_{\mathcal{T}}(f,T)_{-1}$: local oscillation for the load function, 252, 255, 315
$\mathrm{osc}_{\mathcal{T}}(v,T)_p$: generic surrogate for data error, 314
$\mathrm{osc}_{\mathcal{T}}^{\mathrm{std}}(u_{\mathcal{T}},\mathcal{D})$: standard oscillation, 227
$\mathrm{osc}_{\mathcal{T}}^{\mathrm{std}}(u_{\mathcal{T}},\mathcal{D},T)$: standard local oscillation, 227
$\mathrm{osc}_{\mathcal{T}}(\mathcal{D})$: total data error estimator, 316
$\mathrm{osc}_{\mathcal{T}}(\boldsymbol{A})_r$: oscillation for the diffusion coefficient, 315
$\mathrm{osc}_{\mathcal{T}}(c)_q$: oscillation for the reaction coefficient, 315
$\mathrm{osc}_{\mathcal{T}}(f)$, $\mathrm{osc}_{\mathcal{T}}(f)_{-1}$: oscillation for the load function, 315
$\widetilde{E}_{\mathcal{T}}(f)^2_{-1}$: generic surrogate estimator for the approximation of the load term, 372
$j(u_{\mathcal{T}})$: jump residual, 225
$r_{\mathcal{T}}(u_{\mathcal{T}})$, $r(u_{\mathcal{T}})$: element residual, 225

Functional spaces

$B^s_{p,q}(\Omega)$: Besov spaces, 347
$D(\Omega)$: metric space for the data perturbation, 311
$\widehat{D}(\Omega)$: temporary metric space for the data perturbation, 309
$M(\alpha_1,\alpha_2)$: admissible set for $\boldsymbol{A}$, 308
$R(c_1,c_2)$: admissible set for c, 308
$W^k_p(\Omega)$: Sobolev spaces, 173
$W^{-s}_q(\Omega)$: dual of $W^s_{q*}(\Omega)$ with $q^* = q/(q-1)$, 309
$X^s_p(\Omega)$: abstract functional spaces, 345
$\mathbb{M}_s$: approximation classes of $\boldsymbol{A}$, 330
$\mathbb{C}_s$: approximation classes of c, 331
$\mathbb{F}_s$: approximation classes of f, 331
$\mathbb{D}$: data, 286
$\mathbb{D}_{\widehat{\mathcal{T}}}$: discrete data subordinate to $\widehat{\mathcal{T}}$, 286
$\mathbb{F}(\mathcal{T}_\omega)$: local discrete functionals, 232
$\mathbb{F}_{\mathcal{T}}$, $\mathbb{F}(\mathcal{T})$: discrete functionals, 232
$\mathbb{V}^+(\mathcal{T}_\omega)$: local test space for discrete functionals, 238
$\mathbb{V}_{\mathcal{T}}$: conforming finite element space, 189
$\mathbb{V}^+(\mathcal{T})$, $\mathbb{V}^+_{\mathcal{T}}$: test space for discrete functionals, 238
$\mathbb{A}_s$: approximation class for u, 327
$\mathbb{A}^{-1}_s$: approximation classes for v for the discontinuous Galerkin norm, 404
$\mathbb{E}_{\mathcal{T}}$: broken H^1 space, 400
$\mathbb{V}^{-1}_{\mathcal{T}}$: non-conforming finite element space, 399
$\mathrm{Lip}^{n+1}_p(\Omega)$: Lipschitz spaces, 345
$\mathbb{S}^{n,-1}_{\mathcal{T}}$: piecewise polynomials of degree $\le n$, 188
$\mathbb{S}^{n,0}_{\mathcal{T}}$: globally continuous piecewise polynomials of degree $\le n$, 188

Functions

ϕ_z: Lagrange basis of $\mathbb{S}^{1,0}_{\mathcal{T}}$, 189
ψ_z: Lagrange basis of $\mathbb{S}^{n,0}_{\mathcal{T}}$, 190
$\widehat{u}$: solution to the perturbed problem (5.5), 288
u: solution to weak formulation (2.7), 177
$u_{\mathcal{T}}$: Galerkin approximation, 217

Meshes

T_d: reference element, 188
$\mathcal{P}$: proper nodes, 211, 214
$\mathcal{F}, \mathcal{F}_{\mathcal{T}}$: interior faces, 218
$\mathcal{F}_\omega$: faces interior to ω, 232
$\mathcal{F}_z, \mathcal{F}_{\omega_z}$: faces interior to ω_z, 232
γ_z: skeleton of ω_z, 189
$\mathcal{T} \le \mathcal{T}_*$: refinement relation, 288
$\mathcal{T}_1 \oplus \mathcal{T}_2$: mesh overlay, 206
$\mathcal{T}_\omega$: triangulated submesh, 232
$\mathcal{T}_z, \mathcal{T}_{\omega_z}$: elements forming ω_z, 232
$\mathcal{T}_z$: star of elements sharing the vertex z, 220
$\mathbb{T}$: set of all conforming refinements of $\mathcal{T}_0$, 201
$\mathbb{T}^\Lambda$: set of all Λ-admissibility refinements of $\mathcal{T}_0$, 212
$\mathbb{T}_N$: set of all conforming refinement of $\mathcal{T}_0$ with no more than N elements, 327
$[[\boldsymbol{g}]] \cdot \boldsymbol{n}_F$: normal jump across F, 225
$[[\cdot]]$: jump across faces, 400
$\{\{\cdot\}\}$: average on faces, 400
$\mathcal{N}$: Lagrange nodes of order n, 190
ω_F: region of elements containing the face F, 222
$\omega_T, \omega_{\mathcal{T}}(T)$: region of elements intersecting T, 191, 223
$\widetilde{\omega}_T, \widetilde{\omega}_{\mathcal{T}}(T)$: elements sharing a face with T, 191
$\omega_{\mathcal{T}}(P)$: domain of influence of a proper node P, 213
ω_z: region made of elements sharing the vertex z, 189, 220
$\boldsymbol{n}_F$: normal to the face F, 225
$\mathcal{V}$: set of vertices, 189
$g(T)$: generation of T, 202
$\lambda(x)$: global index of a node $x \in \mathcal{N}$, 211

Norms

$|||\cdot|||_\Omega$: energy norm with exact coefficients, 288
$|||\cdot|||_\Omega$: energy norm with perturbed coefficients, 288
$\|v\|_{a,\mathcal{T}}$: discontinuous Galerkin norm, 400

Operators
$I_{\mathcal{T}}$: quasi-interpolation operator, 191
P_T, P_F: polynomial densities of $P_{\mathcal{T}}$, 239
$P_{\mathcal{T}}$: projection operator from $H^{-1}(\Omega)$ into $\mathbb{F}_{\mathcal{T}}$, 239
$\mathcal{I}_{\mathcal{T}}^{\mathrm{dG}}$: discontinuous Galerkin quasi-interpolant, 401
Π_K, Π_K^m: L^2 projection onto $\mathbb{P}_m(K)$, 227, 312

References

R. A. Adams and J. J. F. Fournier (2003), *Sobolev Spaces*, Vol. 140 of Pure and Applied Mathematics (Amsterdam), second edition, Elsevier/Academic Press.

M. Ainsworth (2010), A framework for obtaining guaranteed error bounds for finite element approximations, *J. Comput. Appl. Math.* **234**, 2618–2632.

M. Ainsworth and J. T. Oden (2000), *A Posteriori Error Estimation in Finite Element Analysis*, Pure and Applied Mathematics (New York), Wiley-Interscience.

D. N. Arnold, F. Brezzi, B. Cockburn and L. D. Marini (2002), Unified analysis of discontinuous Galerkin methods for elliptic problems, *SIAM J. Numer. Anal.* **39**, 1749–1779.

I. Babuška (1971), Error-bounds for finite element method, *Numer. Math.* **16**, 322–333.

I. Babuška (1971), The finite element method for elliptic differential equations, in *Numerical Solution of Partial Differential Equations, II (SYNSPADE 1970)* (B. Hubbard, ed.), Academic Press, pp. 69–106.

I. Babuška and A. K. Aziz (1972), Survey lectures on the mathematical foundations of the finite element method, in *The Mathematical Foundations of the Finite Element Method with Applications to Partial Differential Equations* (A. K. Aziz, ed.), Academic Press, pp. 1–359.

I. Babuška and A. Miller (1987), A feedback finite element method with *a posteriori* error estimation, I: The finite element method and some basic properties of the *a posteriori* error estimator, *Comput. Methods Appl. Mech. Engrg* **61**, 1–40.

I. Babuška and W. C. Rheinboldt (1978), Error estimates for adaptive finite element computations, *SIAM J. Numer. Anal.* **15**, 736–754.

I. Babuška, R. B. Kellogg and J. Pitkäranta (1979), Direct and inverse error estimates for finite elements with mesh refinements, *Numer. Math.* **33**, 447–471.

E. Bänsch, P. Morin and R. H. Nochetto (2002), An adaptive Uzawa FEM for the Stokes problem: Convergence without the inf-sup condition, *SIAM J. Numer. Anal.* **40**, 1207–1229.

L. Beirão da Veiga, C. Canuto, R. H. Nochetto, G. Vacca and M. Verani (2023), Adaptive VEM: Stabilization-free *a posteriori* error analysis and contraction property, *SIAM J. Numer. Anal.* **61**, 457–494.

L. Beirão da Veiga, C. Canuto, R. H. Nochetto, G. Vacca and M. Verani (2024), Adaptive VEM for variable data: Convergence and optimality, *IMA J. Numer. Anal.* Available at https://doi.org/10.1093/imanum/drad085.

J. Bergh and J. Löfström (1976), *Interpolation Spaces: An Introduction*, Vol. 223 of Grundlehren der mathematischen Wissenschaften, Springer.

C. Bernardi and V. Girault (1998), A local regularization operator for triangular and quadrilateral finite elements, *SIAM J. Numer. Anal.* **35**, 1893–1916.

R. Bhatia (2000), Pinching, trimming, truncating, and averaging of matrices, *Amer. Math. Monthly* **107**, 602–608.

P. Binev (2018), Tree approximation for hp-adaptivity, *SIAM J. Numer. Anal.* **56**, 3346–3357.

P. Binev and R. DeVore (2004), Fast computation in adaptive tree approximation, *Numer. Math.* **97**, 193–217.

P. Binev, W. Dahmen and R. DeVore (2004), Adaptive finite element methods with convergence rates, *Numer. Math.* **97**, 219–268.

P. Binev, W. Dahmen, R. DeVore and P. Petrushev (2002), Approximation classes for adaptive methods, *Serdica Math. J.* **28**, 391–416.

P. Binev, F. Fierro and A. Veeser (2023), Near-best adaptive approximation on conforming meshes, *Constr. Approx.* **57**, 327–349.

J. Blechta, J. Málek and M. Vohralík (2020), Localization of the W_q^{-1}-norm for local *a posteriori* efficiency, *IMA J. Numer. Anal.* **40**, 914–950.

D. Boffi, F. Brezzi and M. Fortin (2013), *Mixed Finite Element Methods and Applications*, Vol. 44 of Springer Series in Computational Mathematics, Springer.

A. Bonito and D. Devaud (2015), Adaptive finite element methods for the Stokes problem with discontinuous viscosity, *Math. Comp.* **84**, 2137–2162.

A. Bonito and R. H. Nochetto (2010), Quasi-optimal convergence rate of an adaptive discontinuous Galerkin method, *SIAM J. Numer. Anal.* **48**, 734–771.

A. Bonito, J. M. Cascón, K. Mekchay, P. Morin and R. H. Nochetto (2016), High-order AFEM for the Laplace–Beltrami operator: Convergence rates, *Found. Comput. Math.* **16**, 1473–1539.

A. Bonito, J. M. Cascón, P. Morin and R. H. Nochetto (2013*a*), AFEM for geometric PDE: The Laplace–Beltrami operator, in *Analysis and Numerics of Partial Differential Equations* (F. Brezzi *et al.*, eds), Springer, pp. 257–306.

A. Bonito, R. A. DeVore and R. H. Nochetto (2013*b*), Adaptive finite element methods for elliptic problems with discontinuous coefficients, *SIAM J. Numer. Anal.* **51**, 3106–3134.

A. Bonito, R. H. Nochetto and D. Ntogkas (2021), DG approach to large bending plate deformations with isometry constraint, *Math. Models Methods Appl. Sci.* **31**, 133–175.

F. Bornemann, B. Erdmann and R. Kornhuber (1996), *A posteriori* error estimates for elliptic problems in two and three space dimensions, *SIAM J. Numer. Anal.* **33**, 1188–1204.

D. Braess (2007), *Finite Elements: Theory, Fast solvers, and Applications in Elasticity Theory*, third edition, Cambridge University Press.

D. Braess, V. Pillwein and J. Schöberl (2009), Equilibrated residual error estimates are p-robust, *Comput. Methods Appl. Mech. Engrg* **198**, 1189–1197.

S. C. Brenner (2003), Poincaré–Friedrichs inequalities for piecewise H 1 functions, *SIAM J. Numer. Anal.* **41**, 306–324.

S. C. Brenner and L. R. Scott (2008), *The Mathematical Theory of Finite Element Methods*, Vol. 15 of Texts in Applied Mathematics, third edition, Springer.

F. Brezzi (1974), On the existence, uniqueness and approximation of saddle-point problems arising from Lagrangian multipliers, *Rev. Fr. Autom. Inform. Rech. Opér. Anal. Numér.* **8**, 129–151.

F. Brezzi, J. Douglas Jr and L. D. Marini (1985), Two families of mixed finite elements for second order elliptic problems, *Numer. Math.* **47**, 217–235.

F. Brezzi, J. Douglas Jr, M. Fortin and L. D. Marini (1987), Efficient rectangular mixed finite elements in two and three space variables, *RAIRO Modél. Math. Anal. Numér.* **21**, 581–604.

F. Brezzi, G. Manzini, D. Marini, P. Pietra and A. Russo (2000), Discontinuous Galerkin approximations for elliptic problems, *Numer. Methods Partial Differential Equations* **16**, 365–378.

C. Canuto and D. Fassino (2023), Higher-order adaptive virtual element methods with contraction properties, *Math. Engrg* **5**, 1–33.

R. Carroll, G. Duff, J. Friberg, J. Gobert, P. Grisvard, J. Nečas and R. Seeley (1966), *Equations aux Dérivées Partielles*, Vol. 19 of Séminaire de mathématiques supérieures, Les Presses de l'Université de Montréal.

C. Carstensen (1997), *A posteriori* error estimate for the mixed finite element method, *Math. Comp.* **66**, 465–476.

C. Carstensen, M. Feischl, M. Page and D. Praetorius (2014), Axioms of adaptivity, *Comput. Math. Appl.* **67**, 1195–1253.

J. M. Cascón and R. H. Nochetto (2012), Quasioptimal cardinality of AFEM driven by nonresidual estimators, *IMA J. Numer. Anal.* **32**, 1–29.

J. M. Cascón, C. Kreuzer, R. H. Nochetto and K. G. Siebert (2008), Quasi-optimal convergence rate for an adaptive finite element method, *SIAM J. Numer. Anal.* **46**, 2524–2550.

P. G. Ciarlet (2002), *The Finite Element Method for Elliptic Problems*, Vol. 40 of Classics in Applied Mathematics, SIAM. Reprint of the 1978 original.

P. Clément (1975), Approximation by finite element functions using local regularization, *Rev. Fr. Autom. Inform. Rech. Opér. Anal. Numér.* **9**, 77–84.

A. Cohen, R. DeVore and R. H. Nochetto (2012), Convergence rates of AFEM with H^{-1} data, *Found. Comput. Math.* **12**, 671–718.

P. Daniel and M. Vohralík (2023), Guaranteed contraction of adaptive inexact hp-refinement strategies with realistic stopping criteria, *ESAIM Math. Model. Numer. Anal.* **57**, 329–366.

E. B. Davies (1988), Lipschitz continuity of functions of operators in the Schatten classes, *J. London Math. Soc. (2)* **37**, 148–157.

S. Dekel and D. Leviatan (2004), Whitney estimates for convex domains with applications to multivariate piecewise polynomial approximation, *Found. Comput. Math.* **4**, 345–368.

R. A. DeVore (1998), Nonlinear approximation, *Acta Numer.* **7**, 51–150.

R. A. DeVore and G. G. Lorentz (1993), *Constructive Approximation*, Vol. 303 of Grundlehren der mathematischen Wissenschaften [Fundamental Principles of Mathematical Sciences], Springer.

R. A. DeVore and V. A. Popov (1988), Interpolation of Besov spaces, *Trans. Amer. Math. Soc.* **305**, 397–414.

L. Diening, L. Gehring and J. Storn (2023), Adaptive mesh refinement for arbitrary initial triangulations. Available at arXiv:2304.02674.

L. Diening, C. Kreuzer and R. Stevenson (2016), Instance optimality of the adaptive maximum strategy, *Found. Comput. Math.* **16**, 33–68.

Z. Ditzian (1988), On the Marchaud-type inequality, *Proc. Amer. Math. Soc.* **103**, 198–202.

W. Dörfler (1996), A convergent adaptive algorithm for Poisson's equation, *SIAM J. Numer. Anal.* **33**, 1106–1124.

N. Dunford and J. T. Schwartz (1988), *Linear Operators*, part II, *Spectral Theory: Self-adjoint Operators in Hilbert Space*, Wiley Classics Library, Wiley. Reprint of the 1963 original.

T. Dupont and R. Scott (1980), Polynomial approximation of functions in Sobolev spaces, *Math. Comp.* **34**, 441–463.

A. Ern, I. Smears and M. Vohralík (2017), Discrete p-robust $H(\mathrm{div})$-liftings and *a posteriori* estimates for elliptic problems with H^{-1} source terms, *Calcolo* **54**, 1009–1025.

L. C. Evans (2010), *Partial Differential Equations*, Vol. 19 of Graduate Studies in Mathematics, second edition, American Mathematical Society.

B. Faermann (2000), Localization of the Aronszajn–Slobodeckij norm and application to adaptive boundary element methods I: The two-dimensional case, *IMA J. Numer. Anal.* **20**, 203–234.

B. Faermann (2002), Localization of the Aronszajn–Slobodeckij norm and application to adaptive boundary element methods II: The three-dimensional case, *Numer. Math.* **92**, 467–499.

M. Feischl (2019), Optimality of a standard adaptive finite element method for the Stokes problem, *SIAM J. Numer. Anal.* **57**, 1124–1157.

M. Feischl (2022), Inf-sup stability implies quasi-orthogonality, *Math. Comp.* **91**, 2059–2094.

F. Fierro and A. Veeser (2003), *A posteriori* error estimators for regularized total variation of characteristic functions, *SIAM J. Numer. Anal.* **41**, 2032–2055.

S. Funken, D. Praetorius and P. Wissgott (2011), Efficient implementation of adaptive P1-FEM in Matlab, *Comput. Methods Appl. Math.* **11**, 460–490.

G. P. Galdi (1994), *An Introduction to the Mathematical Theory of the Navier–Stokes Equations*, Vol. I, *Linearized Steady Problems*, Vol. 38 of Springer Tracts in Natural Philosophy, Springer.

F. D. Gaspoz and P. Morin (2014), Approximation classes for adaptive higher order finite element approximation, *Math. Comp.* **83**, 2127–2160.

F. D. Gaspoz and P. Morin (2017), Errata to 'Approximation classes for adaptive higher order finite element approximation', *Math. Comp.* **86**, 1525–1526.

D. Gilbarg and N. S. Trudinger (2001), *Elliptic Partial Differential Equations of Second Order*, Classics in Mathematics, Springer. Reprint of the 1998 edition.

P. Grisvard (1985), *Elliptic Problems in Nonsmooth Domains*, Vol. 24 of Monographs and Studies in Mathematics, Pitman (Advanced Publishing Program).

P. Grisvard (2011), *Elliptic Problems in Nonsmooth Domains*, Vol. 69 of Classics in Applied Mathematics, SIAM. Reprint of the 1985 original.

W. Hackbusch (1992), *Elliptic Differential Equations: Theory and Numerical Treatment*, Vol. 18 of Springer Series in Computational Mathematics, Springer.

P. Houston, D. Schötzau and T. P. Wihler (2004), Mixed hp-discontinuous Galerkin finite element methods for the Stokes problem in polygons, in *Numerical Mathematics and Advanced Applications: Proceedings of the 5th European Conference on Numerical Mathematics and Advanced Applications (ENUMATH 2003)*, Springer, pp. 493–501.

P. Houston, D. Schötzau and T. P. Wihler (2007), Energy norm *a posteriori* error estimation of hp-adaptive discontinuous Galerkin methods for elliptic problems, *Math. Models Methods Appl. Sci.* **17**, 33–62.

D. Jerison and C. E. Kenig (1995), The inhomogeneous Dirichlet problem in Lipschitz domains, *J. Funct. Anal.* **130**, 161–219.

J.-P. Kahane (1961), *Teoria Constructiva de Funciones*, Universidad de Buenos Aires.

O. A. Karakashian and F. Pascal (2007), Convergence of adaptive discontinuous Galerkin approximations of second-order elliptic problems, *SIAM J. Numer. Anal.* **45**, 641–665.

R. B. Kellogg (1974/75), On the Poisson equation with intersecting interfaces, *Applicable Anal.* **4**, 101–129.

C. Kreuzer and K. G. Siebert (2011), Decay rates of adaptive finite elements with Dörfler marking, *Numer. Math.* **117**, 679–716.

C. Kreuzer and A. Veeser (2019), Convergence of adaptive finite element methods with error-dominated oscillation, in *Numerical Mathematics and Advanced Applications (ENUMATH 2017)*, Vol. 126 of Lecture Notes in Computational Science and Engineering, Springer, pp. 471–479.

C. Kreuzer and A. Veeser (2021), Oscillation in *a posteriori* error estimation, *Numer. Math.* **148**, 43–78.

C. Kreuzer, A. Veeser and P. Zanotti (2024), Accurate error bounds for finite element methods. In preparation.

P. D. Lax and A. N. Milgram (1954), Parabolic equations, in *Contributions to the Theory of Partial Differential Equations*, Vol. 33 of Annals of Mathematics Studies, Princeton University Press, pp. 167–190.

G. Leoni (2009), *A First Course in Sobolev Spaces*, Vol. 105 of Graduate Studies in Mathematics, American Mathematical Society.

R. Luce and B. Wohlmuth (2004), A local *a posteriori* error estimator based on equilibrated fluxes, *SIAM J. Numer. Anal.* **42**, 1394–1414.

J. M. Maubach (1995), Local bisection refinement for n-simplicial grids generated by reflection, *SIAM J. Sci. Comput.* **16**, 210–227.

N. G. Meyers (1963), An L^p-estimate for the gradient of solutions of second order elliptic divergence equations, *Ann. Scuola Norm. Sup. Pisa Cl. Sci. (3)* **17**, 189–206.

W. F. Mitchell (1989), A comparison of adaptive refinement techniques for elliptic problems, *ACM Trans. Math. Software* **15**, 326–347.

P. Morin, R. Nochetto and K. Siebert (2003), Local problems on stars: *A posteriori* error estimators, convergence, and performance, *Math. Comp.* **72**, 1067–1097.

P. Morin, R. H. Nochetto and K. G. Siebert (2000), Data oscillation and convergence of adaptive FEM, *SIAM J. Numer. Anal.* **38**, 466–488.

P. Morin, R. H. Nochetto and K. G. Siebert (2002), Convergence of adaptive finite element methods, *SIAM Rev.* **44**, 631–658. Revised reprint of 'Data oscillation and convergence of adaptive FEM'.

P. Morin, K. G. Siebert and A. Veeser (2008), A basic convergence result for conforming adaptive finite elements, *Math. Models Methods Appl. Sci.* **18**, 707–737.

J.-C. Nédélec (1980), Mixed finite elements in $\mathbb{R}^3$, *Numer. Math.* **35**, 315–341.

J. Nečas (1962), Sur une méthode pour résoudre les équations aux dérivées partielles du type elliptique, voisine de la variationnelle, *Ann. Scuola Norm. Sup. Pisa Cl. Sci. (3)* **16**, 305–326.

R. H. Nochetto and A. Veeser (2012), Primer of adaptive finite element methods, in *Multiscale and Adaptivity: Modeling, Numerics and Applications*, Vol. 2040 of Lecture Notes in Mathematics, Springer, pp. 125–225.

R. H. Nochetto, K. G. Siebert and A. Veeser (2009), Theory of adaptive finite element methods: An introduction, in *Multiscale, Nonlinear and Adaptive Approximation* (R. DeVore and A. Kunoth, eds), Springer, pp. 409–542.

L. E. Payne and H. F. Weinberger (1960), An optimal Poincaré inequality for convex domains, *Arch. Rational Mech. Anal.* **5**, 286–292.

I. Perugia and D. Schötzau (2003), The hp-local discontinuous Galerkin method for low-frequency time-harmonic Maxwell equations, *Math. Comp.* **72**, 1179–1214.

P.-A. Raviart and J. M. Thomas (1977), A mixed finite element method for 2-nd order elliptic problems, in *Mathematical Aspects of Finite Element Methods*, Vol. 606 of Lecture Notes in Mathematics, Springer, pp. 292–315.

R. Sacchi and A. Veeser (2006), Locally efficient and reliable *a posteriori* error estimators for Dirichlet problems, *Math. Models Methods Appl. Sci.* **16**, 319–346.

L. R. Scott and S. Zhang (1990), Finite element interpolation of nonsmooth functions satisfying boundary conditions, *Math. Comp.* **54**, 483–493.

K. Siebert and A. Veeser (2007), A unilaterally constrained quadratic minimization with adaptive finite elements, *SIAM J. Optim.* **18**, 260–289.

K. G. Siebert (2012), Mathematically founded design of adaptive finite element software, in *Multiscale and Adaptivity: Modeling, Numerics and Applications*, Vol. 2040 of Lecture Notes in Mathematics, Springer, pp. 227–309.

R. Stevenson (2007), Optimality of a standard adaptive finite element method, *Found. Comput. Math.* **7**, 245–269.

R. Stevenson (2008), The completion of locally refined simplicial partitions created by bisection, *Math. Comp.* **77**, 227–241.

D. B. Szyld (2006), The many proofs of an identity on the norm of oblique projections, *Numer. Algorithms* **42**, 309–323.

F. Tantardini, A. Veeser and R. Verfürth (2024), Best error localization in the approximation of functionals with piecewise polynomials. In preparation.

C. Taylor and P. Hood (1973), A numerical solution of the Navier–Stokes equations using the finite element technique, *Int. J. Comput. Fluids* **1**, 73–100.

C. T. Traxler (1997), An algorithm for adaptive mesh refinement in n dimensions, *Computing* **59**, 115–137.

H. Triebel (2010), *Theory of Function Spaces*, Modern Birkhäuser Classics, Birkhäuser/Springer. Reprint of 1983 edition.

A. Veeser (2002), Convergent adaptive finite elements for the nonlinear Laplacian, *Numer. Math.* **92**, 743–770.

A. Veeser (2016), Approximating gradients with continuous piecewise polynomial functions, *Found. Comput. Math.* **16**, 723–750.

A. Veeser and R. Verfürth (2009), Explicit upper bounds for dual norms of residuals, *SIAM J. Numer. Anal.* **47**, 2387–2405.

R. Verfürth (2013), *A Posteriori Error Estimation Techniques for Finite Element Methods*, Numerical Mathematics and Scientific Computation, Oxford University Press.

J. Xu and L. Zikatanov (2003), Some observations on Babuška and Brezzi theories, *Numer. Math.* **94**, 195–202.

J. Xu, L. Chen and R. H. Nochetto (2009), Optimal multilevel methods for $H(\text{grad})$, $H(\text{curl})$, and $H(\text{div})$ systems on graded and unstructured grids, in *Multiscale, Nonlinear and Adaptive Approximation* (R. DeVore and A. Kunoth, eds), Springer, pp. 599–659.

Acta Numerica (2024), pp. 487–632 Printed in the United Kingdom
doi:10.1017/S0962492923000065

The geometry of monotone operator splitting methods

Patrick L. Combettes*
North Carolina State University, Department of Mathematics,
Raleigh, NC 27695-8205, USA
E-mail: plc@math.ncsu.edu

We propose a geometric framework to describe and analyse a wide array of operator splitting methods for solving monotone inclusion problems. The initial inclusion problem, which typically involves several operators combined through monotonicity-preserving operations, is seldom solvable in its original form. We embed it in an auxiliary space, where it is associated with a surrogate monotone inclusion problem with a more tractable structure and which allows for easy recovery of solutions to the initial problem. The surrogate problem is solved by successive projections onto half-spaces containing its solution set. The outer approximation half-spaces are constructed by using the individual operators present in the model separately. This geometric framework is shown to encompass traditional methods as well as state-of-the-art asynchronous block-iterative algorithms, and its flexible structure provides a pattern to design new ones.

2020 Mathematics Subject Classification: Primary 47H05, 90C48, 90C46
Secondary 47H09, 90C25

CONTENTS

1	Introduction	488
2	Monotone operators	492
3	Structured monotone inclusions	510
4	Two geometric convergence principles	517
5	The proximal point algorithm	529
6	Douglas–Rachford splitting	544
7	Tseng's forward–backward–forward splitting	553
8	Forward–backward splitting	564
9	Block-iterative Kuhn–Tucker projective splitting	593
10	Block-iterative saddle projective splitting	602
11	Extensions and variants	610
	References	613

* This work was supported by the National Science Foundation under grant CCF-2211123.

1. Introduction

Throughout, $\mathcal{H}$ is a real Hilbert space with scalar product $\langle \cdot \mid \cdot \rangle$ and $2^{\mathcal{H}}$ stands for the power set of $\mathcal{H}$. Our main focus is on the following monotone inclusion problem.

Problem 1.1. Let $M\colon \mathcal{H} \to 2^{\mathcal{H}}$ be a monotone operator, that is,

$$(\forall x \in \mathcal{H})(\forall y \in \mathcal{H})(\forall x^* \in Mx)(\forall y^* \in My) \quad \langle x - y \mid x^* - y^* \rangle \geq 0. \tag{1.1}$$

The task is to find $x \in \mathcal{H}$ such that $0 \in Mx$.

Monotone inclusion problems are intimately linked to the birth of nonlinear analysis. They first appeared as powerful models to establish existence, uniqueness and stability results for various nonlinear problems (Browder 1968/1976, Ghizzetti 1969, Kačurovskiĭ 1960, Zarantonello 1960, 1971). Over the past six decades, monotone inclusion models have penetrated almost all areas of mathematics and its applications. Nowadays, Problem 1.1 models a broad range of equilibria in areas such as dynamical systems (Adly, Hantoute and Le 2017), ill-posed problems (Alber and Ryazantseva 2006), domain decomposition methods (Alduncin 2023, Attouch, Briceño-Arias and Combettes 2016, Attouch, Cabot, Frankel and Peypouquet 2011), circuit theory (Anderson Jr and Trapp 1976, Chaffey and Sepulchre 2024, Chaffey, Banert, Giselsson and Pates 2023*a*, Chaffey, Forni and Sepulchre 2023*b*, Goeleven 2017), machine learning (Argyriou, Foygel and Srebro 2012, Combettes, Salzo and Villa 2018, Jenatton, Mairal, Obozinski and Bach 2011, Vaiter, Peyré and Fadili 2018), evolution equations (Attouch, Briceño-Arias and Combettes 2010, Brézis 1973, Showalter 1997), partial differential equations (Barbu 2010, Brézis and Browder 1998, Clason and Valkonen 2017, Ghoussoub 2009, Pascali and Sburlan 1978, Showalter 1997, Zeidler 1990), signal processing (Beck and Teboulle 2010, Combettes and Pesquet 2011, Combettes and Wajs 2005, Potter and Arun 1993), image processing (Bednarczuk, Jezierska and Rutkowski 2018, Chambolle and Pock 2016, Combettes and Woodstock 2022, Glowinski, Osher and Yin 2016, Pesquet, Repetti, Terris and Wiaux 2021), game theory (Belgioioso, Nedich and Grammatico 2021, Börgens and Kanzow 2021, Briceño-Arias and Combettes 2013, Bùi and Combettes 2022*a*, Cohen 1987, Facchinei and Pang 2003, Facchinei, Fischer and Piccialli 2007, Gautam, Sahu, Dixit and Som 2021), network flow problems (Bertsekas 1998, Bùi 2022*a*, Rockafellar 1984, 1995), equilibrium theory (Briceño-Arias 2012, Combettes and Hirstoaga 2005, Moudafi and Théra 1999), mean-field games (Briceño-Arias, Deride, López-Rivera and Silva 2023, Briceño-Arias, Kalise and Silva 2018), control theory (Brogliato and Tanwani 2020, Brogliato, Lozano, Maschke and Egeland 2007, Camlibel and Schumacher 2016, Doležal 1979*a*, Singh, Weiss and Tucsnak 2022), data science (Chan, Wang and Elgendy 2017, Combettes and Pesquet 2021, Wright and Recht 2022), optimization (Combettes 2018, Eckstein and Bertsekas 1992, Gol'shtein and Tret'yakov 1996, Tseng 1990, 1991), statistics (Combettes and Müller 2020, Yan and Bien 2021), neural networks (Combettes and Pesquet 2020, Winston and

Kolter 2020, Yi and Ching 2020), traffic equilibrium (Dafermos 1980, Fukushima 1996), systems theory (Desoer and Vidyasagar 1975, Doležal 1979*b*), mechanics (Fortin and Glowinski 1983, Mercier 1980), optimal transportation (Papadakis, Peyré and Oudet 2014) and minimax theory (Rockafellar 1970*b*).

Early numerical solution methods to solve Problem 1.1 can be found in Antipin (1976), Bruck (1973, 1974), Korpelevič (1976), Lions (2010), Petryshyn (1966), Sibony (1970), Vaĭnberg (1960, 1961) and Zarantonello (1960, 1964). These methods are of the explicit Euler type, meaning that, at iteration n, the update x_{n+1} is determined by finding a point in Mx_n. An alternative method, which first appeared in Lieutaud (1969*a*) and then in more detail in Rockafellar (1976*b*), is the proximal point algorithm, where the update is obtained through the implicit relation $x_n - x_{n+1} \in Mx_{n+1}$. Such approaches have limited potential since they can be directly implemented only in specific situations. For instance, the Euler step methods of Bruck (1973, 1974, 1975) impose certain properties on M and asymptotically vanishing step sizes, which is detrimental to numerical stability and speed of convergence. On the other hand, the proximal point algorithm requires explicit expressions for the resolvent of M, which is seldom possible. In most problems, however, M has a complex structure and it is typically expressed in terms of monotonicity-preserving operations involving simpler operators. The principle governing *splitting methods* is to devise algorithms in which each of the elementary operators arising in the decomposition of M are used individually, hence breaking up Problem 1.1 into tasks that are more manageable.

The first monotone operator splitting methods arose in the late 1970s and were motivated by applications in mechanics and partial differential equations (Fortin and Glowinski 1983, Glowinski and Le Tallec 1989, Mercier 1980). The three main algorithms that dominated the field were designed for problems in which

$$M = A + B, \tag{1.2}$$

where $A\colon \mathcal{H} \to 2^{\mathcal{H}}$ and $B\colon \mathcal{H} \to 2^{\mathcal{H}}$ are maximally monotone: the forward–backward method (Mercier 1979), the Douglas–Rachford method (Lions and Mercier 1979) and Tseng's forward–backward–forward method (Tseng 2000). In recent years, the field of monotone operator splitting algorithms has benefited from a new impetus, fuelled by the emerging application areas mentioned above and their demand for solving efficiently increasingly complex large-dimensional problems. Thus, duality techniques have arisen to address composite models of the form

$$M = A + L^* \circ B \circ L, \tag{1.3}$$

where L is a linear operator from $\mathcal{H}$ to a Hilbert space $\mathcal{G}$ and $A\colon \mathcal{H} \to 2^{\mathcal{H}}$ and $B\colon \mathcal{G} \to 2^{\mathcal{G}}$ are maximally monotone (Briceño-Arias and Combettes 2011). These techniques have been further developed to devise splitting algorithms for the more structured model (Boţ and Hendrich 2014, Combettes and Pesquet 2012, Vũ 2013)

$$M = A + \sum_{k=1}^{p} L_k^* \circ \left(B_k^{-1} + D_k^{-1}\right)^{-1} \circ L_k + C, \tag{1.4}$$

where each linear operator L_k maps $\mathcal{H}$ to a Hilbert space $\mathcal{G}_k$, and the operators $A\colon \mathcal{H} \to 2^{\mathcal{H}}$, $B_k\colon \mathcal{G}_k \to 2^{\mathcal{G}_k}$, $D_k\colon \mathcal{G}_k \to 2^{\mathcal{G}_k}$ and $C\colon \mathcal{H} \to \mathcal{H}$ are maximally monotone. Splitting algorithms for models which are more finely structured than (1.4) have also been proposed, as well as multivariate versions that capture coupled systems of monotone inclusions; see Bùi and Combettes (2022*b*) and the references therein. On a different front, block-iterative algorithms, which allow for the activation of only a subgroup of operators present in the model at a given iteration, have also been developed (Bùi 2022*b*, Bùi and Combettes 2022*b*, Combettes and Eckstein 2018, Johnstone and Eckstein 2022). At the same time, a multitude of splitting algorithms tailored to specific models have been elaborated. For instance, if $A\colon \mathcal{H} \to 2^{\mathcal{H}}$ and $B\colon \mathcal{H} \to 2^{\mathcal{H}}$ are maximally monotone and $C\colon \mathcal{H} \to \mathcal{H}$ is cocoercive, splitting algorithms have been proposed in Davis and Yin (2017) and Raguet (2019) for the decomposition $M = A + B + C$, and in particular in Briceño-Arias and Davis (2018) if $B\colon \mathcal{H} \to \mathcal{H}$ is Lipschitzian and in Latafat and Patrinos (2017) if $B\colon \mathcal{H} \to \mathcal{H}$ is linear and bounded.

Given the abundance of activity in monotone operator splitting techniques, it is important to identify general structures and principles, as well as possible bonds between algorithm design methodologies, in order not only to simplify and clarify the state of the art but also to facilitate the developments of new methods in the future. From the outset, fixed point theory has been a tool of choice to achieve this goal. For instance, it has played an important role in the analysis of the proximal point algorithm (Kryanev 1973, Martinet 1972, Rockafellar 1976*b*). Combettes (2004) showed that fixed point iterations of averaged operators provide a convenient framework to investigate the asymptotic behaviour of classical splitting algorithms such as the forward–backward, backward–backward, Douglas–Rachford and Peaceman–Rachford algorithms. Further applications of averaged operator iterations to design and analyse splitting methods can be found in Briceño-Arias and Roldán (2023), Chambolle and Pock (2016), Combettes and Glaudin (2017), Combettes and Pesquet (2021), Combettes and Yamada (2015), Condat, Kitahara, Contreras and Hirabayashi (2023), Davis and Yin (2017), Raguet (2019), Raguet and Landrieu (2015), Raguet, Fadili and Peyré (2013), Ryu, Taylor, Bergeling and Giselsson (2020) and Xue (2023*b*). Fixed point modelling is also a central algorithmic development tool in recent works such as those of Aragón-Artacho, Boţ and Torregrosa-Belén (2023), Briceño-Arias and Davis (2018) and Malitsky and Tam (2023). In spite of these achievements, fixed point methods seem less well suited to capturing in simple terms the most flexible splitting methods, such as the block-iterative asynchronous methods of Bùi (2022*b*), Bùi and Combettes (2022*b*), Combettes and Eckstein (2018) and Johnstone and Eckstein (2022), which were built using geometric arguments. The purpose of the present paper is to provide a standardized pattern for building and analysing splitting methods around the following geometric framework. It comprises an embedding step, where the initial Problem 1.1 is replaced by a more tractable surrogate inclusion problem in an auxiliary space X from which the solutions to the original problem can be easily recovered. The second step is an iterative process in which the current iterate is projected onto a closed half-space that serves as an outer approximation to the surrogate solution set.

Framework 1.2. Geometric algorithmic template for solving Problem 1.1.

(i) *Embedding*. Find a real Hilbert space X, a maximally monotone operator $\boldsymbol{\mathcal{M}}\colon \mathsf{X} \to 2^{\mathsf{X}}$ and an operator $\boldsymbol{\mathcal{T}}\colon \mathsf{X} \to \mathcal{H}$ such that $\boldsymbol{\mathcal{T}}(\operatorname{zer}\boldsymbol{\mathcal{M}}) \subset \operatorname{zer} M$. We call $(\mathsf{X}, \boldsymbol{\mathcal{M}}, \boldsymbol{\mathcal{T}})$ an *embedding* of Problem 1.1.

(ii) *Iterations.*

$$
\begin{array}{l}
\text{for } n = 0, 1, \ldots \\
\left\lfloor
\begin{array}{l}
\mathsf{H}_n \text{ is a closed half-space of } \mathsf{X} \text{ such that } \operatorname{zer}\boldsymbol{\mathcal{M}} \subset \mathsf{H}_n \\
\mathsf{x}_{n+1} \text{ is a relaxed projection of } \mathsf{x}_n \text{ onto } \mathsf{H}_n.
\end{array}
\right.
\end{array}
\tag{1.5}
$$

In optimization, the use of half-spaces as outer approximations to the solution set goes back to the cutting plane methods of Cheney and Goldstein (1959*a*), Kelley (1960) and Levitin and Polyak (1966); see also Laurent and Martinet (1970), Veinott (1967) and Zangwill (1969). In monotone inclusion problems, modelling iterations as successive projections onto separating half-spaces occurs in several papers (Bauschke and Combettes 2001, Combettes 2001*a*, Solodov and Svaiter 1999*a*,*b*). We aim at showing that Framework 1.2 is sufficiently broad and flexible to encompass a wide array of existing methods while providing a template to create new ones. It will allow us to derive in a unified fashion simple proofs of existing convergence results. It will also make it possible to establish seamlessly strongly convergent variants of these algorithms. The proofs we provide are new, and so are some of the results.

The remainder of the paper is organized as follows. To make our presentation self-contained, Section 2 covers the necessary mathematical background on monotone operator theory. It also contains various examples of maximally monotone operators and a detailed history of the field. In Section 3 we present several models for decomposing M in Problem 1.1. These decompositions will generate the embeddings required in Framework 1.2 and form the backbone of the splitting methods discussed in the paper. The geometric principles underlying our approach are presented in Section 4, where the main convergence theorems are laid out. In Section 5 we study the proximal point algorithm and explore several of its facets. In Sections 6, 7 and 8 we study, respectively, the Douglas–Rachford, forward–backward–forward and forward–backward methods through the lens of Framework 1.2 and capture a broad range of algorithms and applications by embedding them in bigger spaces. Block-iterative Kuhn–Tucker and saddle projective splitting methods are addressed in Sections 9 and 10, respectively. Finally, several extensions and variants of the results are discussed in Section 11.

2. Monotone operators

2.1. *Notation and basic definitions*

The material of this section can be found in Bauschke and Combettes (2017).

2.1.1. *General notation*

$\mathcal{H}$ and $\mathcal{G}$ are real Hilbert spaces, $\mathcal{B}(\mathcal{H},\mathcal{G})$ is the space of bounded linear operators from $\mathcal{H}$ to $\mathcal{G}$, $\mathcal{B}(\mathcal{H}) = \mathcal{B}(\mathcal{H},\mathcal{H})$, and $\mathcal{H}\oplus\mathcal{G}$ denotes the Hilbert direct sum of $\mathcal{H}$ and $\mathcal{G}$. The identity operator of $\mathcal{H}$ is denoted by $\mathrm{Id}_{\mathcal{H}}$, its scalar product by $\langle\cdot\mid\cdot\rangle_{\mathcal{H}}$, and the associated norm by $\|\cdot\|_{\mathcal{H}}$ (the subscripts will be omitted when the context is clear). The weak convergence of a sequence $(x_n)_{n\in\mathbb{N}}$ to x is denoted by $x_n \rightharpoonup x$, whereas $x_n \to x$ denotes its strong convergence; the set of weak sequential cluster points of $(x_n)_{n\in\mathbb{N}}$ is denoted by $\mathfrak{W}(x_n)_{n\in\mathbb{N}}$.

2.1.2. *Sets*

Let C be a subset of $\mathcal{H}$. The interior of C is $\operatorname{int} C$, the *indicator function* of C is

$$\iota_C\colon \mathcal{H}\to\,]-\infty,+\infty] : x\mapsto \begin{cases} 0, & \text{if } x\in C,\\ +\infty, & \text{otherwise,}\end{cases} \tag{2.1}$$

the *support function* of C is

$$\sigma_C\colon \mathcal{H}\to[-\infty,+\infty] : x^*\mapsto \sup_{x\in C}\langle x\mid x^*\rangle, \tag{2.2}$$

and the *distance function* to C is

$$d_C\colon \mathcal{H}\to\,]-\infty,+\infty] : x\mapsto \inf_{y\in C}\|x-y\|. \tag{2.3}$$

Suppose that C is convex. We let $\operatorname{cone} C$ denote the smallest cone that contains C and let $\operatorname{sri} C$ denote the *strong relative interior* of C, that is,

$$\operatorname{sri} C = \{x\in C \mid \operatorname{cone}(-x+C) \text{ is a closed vector subspace of } \mathcal{H}\}. \tag{2.4}$$

If $\mathcal{H}$ is finite-dimensional, $\operatorname{sri} C$ coincides with the *relative interior* $\operatorname{ri} C$ of C, i.e. the interior of C relative to the smallest affine subspace of $\mathcal{H}$ containing C. Suppose that C is nonempty, closed and convex. For every $x\in\mathcal{H}$,

$$\operatorname{proj}_C x \text{ is the unique point in } C \text{ such that } d_C(x) = \|x-\operatorname{proj}_C x\|. \tag{2.5}$$

This process defines the *projection operator* $\operatorname{proj}_C\colon \mathcal{H}\to\mathcal{H}$ of C. The simple case of a closed half-space is central to our approach.

Example 2.1 (Bauschke and Combettes 2017, Example 29.20). Let $u^* \in \mathcal{H}$, let $\eta \in \mathbb{R}$, and suppose that $H = \{z \in \mathcal{H} \mid \langle z \mid u^* \rangle \leq \eta\} \neq \varnothing$. Let $x \in \mathcal{H}$ and set

$$d = \begin{cases} \dfrac{\langle x \mid u^* \rangle - \eta}{\|u^*\|^2} u^*, & \text{if } \langle x \mid u^* \rangle > \eta, \\ 0, & \text{otherwise.} \end{cases} \tag{2.6}$$

Then $\operatorname{proj}_H x = x - d$.

2.1.3. Functions

The set of minimizers of a function $f\colon \mathcal{H} \to \left]-\infty, +\infty\right]$ is denoted by $\operatorname{Argmin} f$ and, if it is a singleton, its unique element is denoted by $\operatorname{argmin}_{x \in \mathcal{H}} f(x)$. The *infimal convolution* of $f\colon \mathcal{H} \to \left]-\infty, +\infty\right]$ and $h\colon \mathcal{H} \to \left]-\infty, +\infty\right]$ is

$$f \,\square\, h\colon \mathcal{H} \to [-\infty, +\infty] : x \mapsto \inf_{y \in \mathcal{H}} (f(y) + h(x - y)). \tag{2.7}$$

Let $\Gamma_0(\mathcal{H})$ denote the class of functions $f\colon \mathcal{H} \to \left]-\infty, +\infty\right]$ which are lower semicontinuous, convex, and such that $\operatorname{dom} f = \{x \in \mathcal{H} \mid f(x) < +\infty\} \neq \varnothing$. Let $f \in \Gamma_0(\mathcal{H})$. The *conjugate* of f is

$$\Gamma_0(\mathcal{H}) \ni f^*\colon x^* \mapsto \sup_{x \in \mathcal{H}} (\langle x \mid x^* \rangle - f(x)). \tag{2.8}$$

For every $x \in \mathcal{H}$,

$$\operatorname{prox}_f x \text{ is the unique minimizer over } \mathcal{H} \text{ of } y \mapsto f(y) + \frac{1}{2}\|x - y\|^2. \tag{2.9}$$

This process defines the *proximity operator* $\operatorname{prox}_f\colon \mathcal{H} \to \mathcal{H}$ of f. We have

$$(\forall \gamma \in \left]0, +\infty\right[)(\forall x \in \mathcal{H}) \quad x = \operatorname{prox}_{\gamma f} x + \gamma \operatorname{prox}_{f^*/\gamma}(x/\gamma). \tag{2.10}$$

The *Moreau envelope* of f of parameter $\gamma \in \left]0, +\infty\right[$ is

$$^{\gamma}f = f \,\square\, \left(\frac{1}{2\gamma}\|\cdot\|^2\right). \tag{2.11}$$

2.1.4. Set-valued operators

Let $M\colon \mathcal{H} \to 2^{\mathcal{H}}$. The *graph* of M is

$$\operatorname{gra} M = \{(x, x^*) \in \mathcal{H} \times \mathcal{H} \mid x^* \in Mx\}. \tag{2.12}$$

The *inverse* of M is the operator $M^{-1}\colon \mathcal{H} \to 2^{\mathcal{H}}$ defined through the relation

$$(\forall (x, x^*) \in \mathcal{H} \times \mathcal{H}) \quad x^* \in Mx \quad \Leftrightarrow \quad x \in M^{-1}x^*. \tag{2.13}$$

Thus,

$$\operatorname{gra} M^{-1} = \{(x^*, x) \in \mathcal{H} \times \mathcal{H} \mid (x, x^*) \in \operatorname{gra} M\}. \tag{2.14}$$

The set of *fixed points* of M is

$$\operatorname{Fix} M = \{x \in \mathcal{H} \mid x \in Mx\}, \tag{2.15}$$

the set of *zeros* of M is

$$\operatorname{zer} M = M^{-1}0 = \{x \in \mathcal{H} \mid 0 \in Mx\}, \tag{2.16}$$

and the *resolvent* of M is the operator

$$J_M = (\mathrm{Id} + M)^{-1}. \tag{2.17}$$

In other words,

$$(\forall x \in \mathcal{H})(\forall p \in \mathcal{H}) \quad p \in J_M x \Leftrightarrow (p, x - p) \in \operatorname{gra} M, \tag{2.18}$$

and therefore

$$\operatorname{zer} M = \operatorname{Fix} J_M. \tag{2.19}$$

We have

$$(\forall \gamma \in \left]0, +\infty\right[)(\forall x \in \mathcal{H}) \quad x - J_{\gamma M} x = \gamma J_{M^{-1}/\gamma}(x/\gamma). \tag{2.20}$$

The *Yosida approximation* of index $\gamma \in \left]0, +\infty\right[$ of M is

$${}^{\gamma}M = \frac{\mathrm{Id} - J_{\gamma M}}{\gamma} = (\gamma \mathrm{Id} + M^{-1})^{-1} = (J_{\gamma^{-1}M^{-1}}) \circ \gamma^{-1}\mathrm{Id} \tag{2.21}$$

and it satisfies

$$\operatorname{zer} M = \operatorname{zer} {}^{\gamma}M. \tag{2.22}$$

The *domain* of M is

$$\operatorname{dom} M = \{x \in \mathcal{H} \mid Mx \neq \varnothing\} \tag{2.23}$$

and the *range* of M is

$$\operatorname{ran} M = \bigcup_{x \in \operatorname{dom} M} Mx = \{x^* \in \mathcal{H} \mid (\exists x \in \operatorname{dom} M)\ x^* \in Mx\}. \tag{2.24}$$

We have

$$\operatorname{dom} M^{-1} = \operatorname{ran} M \quad \text{and} \quad \operatorname{ran} M^{-1} = \operatorname{dom} M. \tag{2.25}$$

If, for some $x \in \mathcal{H}$, Mx is a singleton, we let Mx denote its single element. We say that M is *injective* if $(\forall x \in \mathcal{H})(\forall y \in \mathcal{H})\ Mx \cap My \neq \varnothing \Rightarrow x = y$. Finally, given $A\colon \mathcal{H} \to 2^{\mathcal{H}}$, $B\colon \mathcal{G} \to 2^{\mathcal{G}}$, $L \in \mathcal{B}(\mathcal{H}, \mathcal{G})$ and $\alpha \in \mathbb{R}$, we set

$$A + \alpha L^* \circ B \circ L\colon \begin{array}{l} \mathcal{H} \to 2^{\mathcal{H}} \\ x \mapsto \{x^* + \alpha L^* y^* \mid x^* \in Ax \text{ and } y^* \in B(Lx)\}. \end{array} \tag{2.26}$$

2.1.5. Monotone operators

Let $M\colon \mathcal{H} \to 2^{\mathcal{H}}$. Then M is *monotone* if

$$(\forall (x, x^*) \in \operatorname{gra} M)(\forall (y, y^*) \in \operatorname{gra} M) \quad \langle x - y \mid x^* - y^* \rangle \geq 0 \tag{2.27}$$

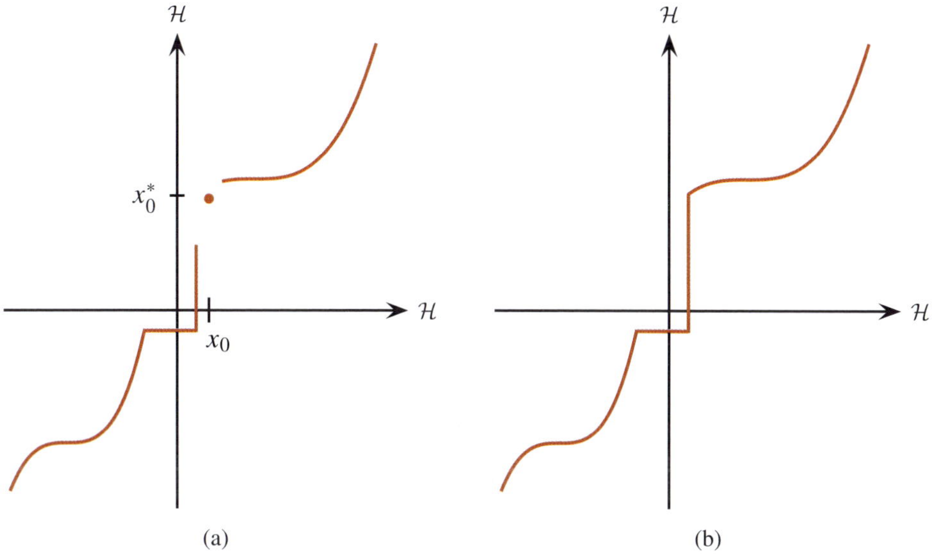

Figure 2.1. (a) Graph of a monotone, but not maximally monotone, operator: the point (x_0, x_0^*) can be added to the graph and the resulting graph remains monotone. (b) Graph of a maximally monotone operator: adding any point to the graph does not preserve its monotonicity.

and *maximally monotone* if, further, there exists no monotone operator $A\colon \mathcal{H} \to 2^{\mathcal{H}}$ such that $\operatorname{gra} M \subset \operatorname{gra} A \neq \operatorname{gra} M$, that is (see Figure 2.1),

$$(\forall (x, x^*) \in \mathcal{H} \times \mathcal{H})$$
$$[(x, x^*) \in \operatorname{gra} M \iff (\forall (y, y^*) \in \operatorname{gra} M)\ \langle x - y \mid x^* - y^* \rangle \geq 0]. \tag{2.28}$$

We have

$$M \text{ maximally monotone} \implies \operatorname{zer} M \text{ is closed and convex.} \tag{2.29}$$

Let $\beta \in \,]0, +\infty[$. Then M is *β-strongly monotone* if $M - \beta\mathrm{Id}$ is monotone, that is,

$$(\forall (x, x^*) \in \operatorname{gra} M)(\forall (y, y^*) \in \operatorname{gra} M) \quad \langle x - y \mid x^* - y^* \rangle \geq \beta \|x - y\|^2. \tag{2.30}$$

Now let D be a nonempty subset of $\mathcal{H}$, let $\alpha \in \,]0, +\infty[$, and let $M\colon D \to \mathcal{H}$. Then M is *nonexpansive* if

$$(\forall x \in D)(\forall y \in D) \quad \|Mx - My\| \leq \|x - y\|, \tag{2.31}$$

α-averaged if $\alpha \leq 1$ and $\mathrm{Id} + \alpha^{-1}(M - \mathrm{Id})$ is nonexpansive, *α-cocoercive* if M^{-1} is α-strongly monotone, that is,

$$(\forall x \in D)(\forall y \in D) \quad \langle x - y \mid Mx - My \rangle \geq \alpha \|Mx - My\|^2, \tag{2.32}$$

and *firmly nonexpansive* if it is 1-cocoercive. Alternatively,

$$M \text{ is firmly nonexpansive} \iff 2M - \mathrm{Id} \text{ is nonexpansive.} \tag{2.33}$$

The following result is known as the Baillon–Haddad theorem.

Lemma 2.2 (Baillon and Haddad 1977). Let $\alpha \in \,]0,+\infty[$ and let $f\colon \mathcal{H} \to \mathbb{R}$ be convex, Fréchet-differentiable, and such that ∇f is $1/\alpha$-Lipschitzian. Then ∇f is α-cocoercive.

2.2. History

Monotonicity goes back to classical calculus and the notion of an increasing real-valued function defined on an interval $D \subset \mathbb{R}$, i.e. a function $f\colon D \to \mathbb{R}$ that satisfies

$$(\forall x \in D)(\forall y \in D) \quad (x-y)(f(x)-f(y)) \geq 0. \tag{2.34}$$

The special properties enjoyed by such functions have long been recognized; see for instance Darboux (1875), Froda (1929) and Hahn (1921). The monotonicity condition (2.34) is also tied to the infancy of the theory of convex functions. Thus Jensen (1906) showed that, if D is open and $g\colon D \to \mathbb{R}$ is a twice differentiable function with derivative f, then (2.34) implies that g is convex. On the numerical side, (2.34) is an important property in connection with solving iteratively the root finding problem (Papakonstantinou and Tapia 2013)

$$\text{find } x \in D \text{ such that } f(x) = 0. \tag{2.35}$$

Monotone operators on $\mathbb{R}$ also appeared in nonlinear circuit theory in the 1940s in the form of quasi-linear resistors (Duffin 1946, 1947, 1948). A quasi-linear resistor is a two-pole circuit element characterized by the property that the current going through it increases smoothly with the voltage across it. In other words, the transformation underlying its current–voltage characteristic is differentiable and increasing. Dipoles with monotonic characteristics were further investigated in Millar (1951). To study networks involving a broader range of devices, Minty (1960, 1961) extended this concept to maximally monotone set-valued transformations on $\mathbb{R}$; see Figure 2.2 and Cederbaum (1962) for examples. Interestingly, as will be discussed shortly, Minty turned out to be one of the founders of monotone operator theory. For further relevant early work on the connections between monotone operators and network theory, see Berge and Ghouila-Houri (1962) and Desoer and Wu (1974) and, for more abstract ramifications, see Doležal (1979*b*) and Rockafellar (1984).

Another precursor of monotonicity is found in linear functional analysis, where a linear operator $M\colon \mathcal{H} \supset D \to \mathcal{H}$ is declared accretive if (Kato 1980)

$$(\forall x \in D) \quad \langle x \mid Mx \rangle \geq 0. \tag{2.36}$$

In this context, the notion of a maximally accretive operator was introduced in Phillips (1959). Accretive operators are also central to passive linear network theory (Beltrami 1972, Zames and Falb 1968). One of the first instances of (2.36) in electrical networks is the current–voltage transformation of the four-pole circuit element known as an ideal gyrator (Tellegen 1948).

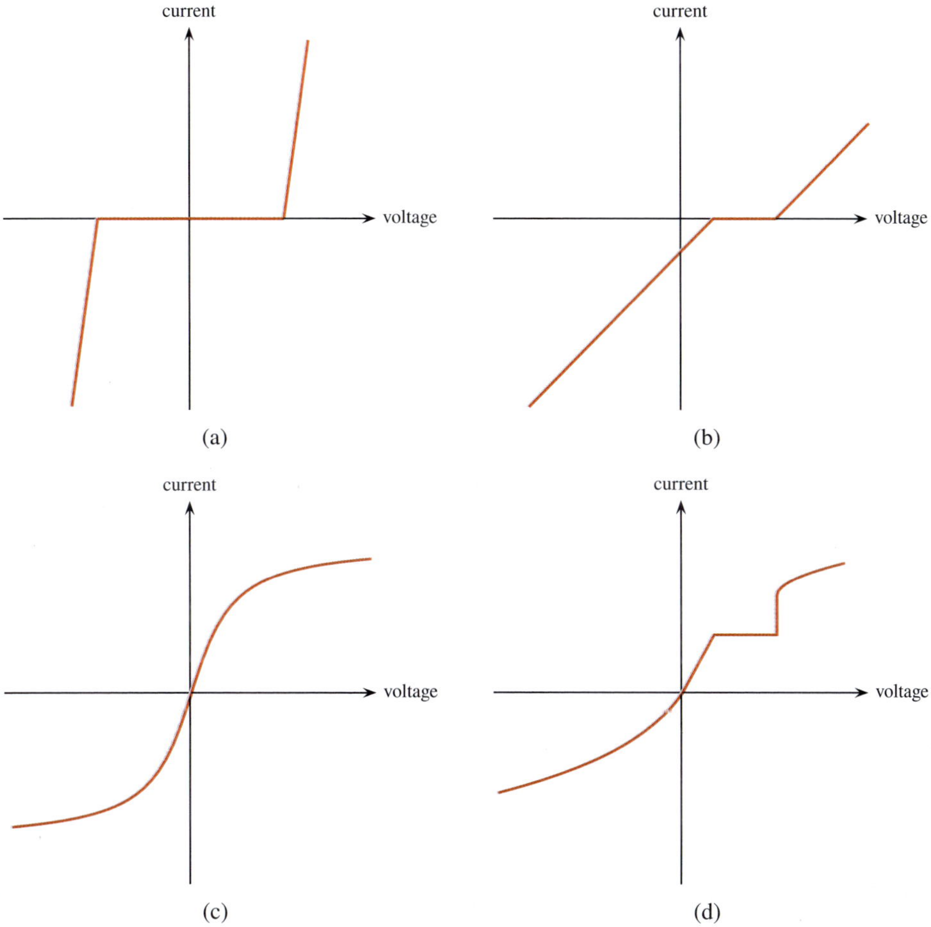

Figure 2.2. Current–voltage characteristics of quasi-linear resistors as monotone operators from $\mathbb{R}$ to $2^{\mathbb{R}}$. (a) Breakdown diodes in series (Reich 1961). (b) Breakdown diode and resistance in series (Reich 1961). (c) Anode–dynode beam-deflection tube (Reich 1961). (d) The maximally monotone current–voltage characteristic of Minty (1961).

The above notions of increasing functions and positive operators can be brought together by considering an operator $M\colon \mathcal{H} \supset D \to \mathcal{H}$ such that

$$(\forall x \in D)(\forall x \in D) \quad \langle x - y \mid Mx - My \rangle \geq 0. \tag{2.37}$$

Instances of (2.37) appear implicitly in Golomb (1935) and, more explicitly, in Vaĭnberg (1956, 1959) in connection with the existence of solutions to Hammerstein integral equations; see also Golomb (1936) for more general types of equations. Another instance, which corresponds to what is now called strict monotonicity, appears in Buck (1956), where $\mathcal{H}$ is the standard Euclidean space. The systematic study of operators satisfying (2.37) started in 1960 and opened an important new chapter of nonlinear functional analysis. Three independent papers submitted that year are associated with the birth of monotone operator theory.

- Kačurovskiĭ (1960), in an article submitted in February 1960, used the term *monotone* to describe an operator that satisfies (2.37). This paper concerned the monotonicity of the gradient of a differentiable convex function (see also Vaĭnberg and Kačurovskiĭ 1959) and the existence of solutions to certain nonlinear equations. It also introduced strongly monotone operators.
- Zarantonello (1960), in a technical report completed in June 1960, called (2.37) an (isotonically) monotonicity property and discussed supra-unitary (in modern language, strongly monotone) operators. In connection with the solution of nonlinear equations, an important result of Zarantonello (1960) is that, if $M\colon \mathcal{H} \to \mathcal{H}$ is monotone and Lipschitzian, then $\mathrm{Id} + M$ is surjective.
- Minty (1962), in an article submitted in December 1960, also called $M\colon D \to \mathcal{H}$ monotone if it satisfies (2.37). In addition, he introduced the fundamental concept of maximal monotonicity and established key connections with nonexpansive operators. Although, strictly speaking, his definitions dealt with single-valued operators, he established results on monotone relations that naturally suggest extensions to the set-valued case (1.1). According to Browder (1965), who initiated the study of set-valued monotone operators in Banach spaces, the Hilbertian setting was worked out by Minty in unpublished notes.

Accounts of the history of the development of monotone operator theory in the 1960s can be found in Borwein (2010), Browder (1968/1976), Kačurovskiĭ (1968), Lions (1969, Section 2.12), Minty (1969) and Vaĭnberg (1972, Chapter VI). In that period, the main mathematical areas of applications were nonlinear equations, partial differential equations, boundary-value problems, nonexpansive semigroups, convex analysis, evolution equations and variational inequalities; see Brézis (1966), Browder (1963, 1968/1976), Ghizzetti (1969), Kōmura (1967), Leray and Lions (1965), Moreau (1966–1967), Vishik (1961) and Zarantonello (1971), and their bibliographies. At the same time, monotonicity continued to be used in the analysis of networks and systems, for instance in Zames (1966*a*,*b*), where it is known as incremental positiveness; see also Desoer and Vidyasagar (1975), where monotonicity is called incremental passivity. The main use of monotone operators was to establish existence, uniqueness or stability results in a variety of nonlinear problems in analysis.

2.3. *Examples of maximally monotone operators*

The following example concerns single-valued operators; Examples 2.4–2.10 follow from it (Bauschke and Combettes 2017, Chapter 20).

Example 2.3 (Minty 1963). Let $A\colon \mathcal{H} \to \mathcal{H}$ be monotone and *hemicontinuous* (in particular, continuous) in the sense that

$$(\forall (x,y,z) \in \mathcal{H}^3) \quad \lim_{0<\alpha\downarrow 0} \langle z \mid A(x+\alpha y)\rangle = \langle z \mid Ax\rangle. \tag{2.38}$$

Then A is maximally monotone.

Example 2.4. Let $T\colon \mathcal{H} \to \mathcal{H}$ be nonexpansive and let $\alpha \in [-1,1]$. Then $\mathrm{Id}+\alpha T$ is maximally monotone. In particular, set $A = \mathrm{Id}-T$. Then A is maximally monotone and $\operatorname{zer} A = \operatorname{Fix} T$.

Example 2.5. Let $A\colon \mathcal{H} \to \mathcal{H}$ be cocoercive. Then A is maximally monotone.

Example 2.6. Let $M\colon \mathcal{H} \to 2^{\mathcal{H}}$ be maximally monotone and set $A = J_M$. Then A is maximally monotone and $\operatorname{zer} A = \operatorname{zer} M^{-1}$.

Example 2.7. Let $M\colon \mathcal{H} \to 2^{\mathcal{H}}$ be maximally monotone, let $\gamma \in \left]0,+\infty\right[$, and set $A = {}^{\gamma}M$ (see (2.21)). Then A is γ-cocoercive, hence maximally monotone, and $\operatorname{zer} A = \operatorname{zer} M$.

Example 2.8. Let $f \in \Gamma_0(\mathcal{H})$ and set $A = \operatorname{prox}_f$. Then A is maximally monotone.

Example 2.9. Let C be a nonempty closed convex subset of $\mathcal{H}$ and set $A = \operatorname{proj}_C$. Then A is maximally monotone.

Example 2.10. Let $A \in \mathcal{B}(\mathcal{H})$ be a *skew* operator, i.e. $A^* = -A$. Then A is maximally monotone.

Here is an elementary example of a maximally monotone set-valued operator on the real line.

Example 2.11. Let $a \in \mathbb{R}$ and $b \in \mathbb{R}$ be such that $a < b$, let $f\colon [a,b] \to \mathbb{R}$ be increasing (see (2.34)), and define

$$(\forall x \in \mathbb{R}) \quad Ax = \begin{cases} \varnothing, & \text{if } x \notin [a,b],\\ \left]-\infty, f(a)\right], & \text{if } x = a,\\ \left[f(b),+\infty\right[, & \text{if } x = b,\\ \left[\sup f([a,x[),\, \inf f(]x,b])\right], & \text{if } x \in \left]a,b\right[. \end{cases} \tag{2.39}$$

Then A is maximally monotone.

The following example is a central result in variational methods; for a special case see Minty (1964, corollary on page 244).

Example 2.12 (Moreau 1965). Let $f\colon \mathcal{H} \to]-\infty,+\infty]$ be proper. Then the *subdifferential*

$$\partial f\colon \mathcal{H} \to 2^{\mathcal{H}}\colon x \mapsto \{x^* \in \mathcal{H} \mid (\forall y \in \mathcal{H})\ \langle y - x \mid x^*\rangle + f(x) \leq f(y)\} \tag{2.40}$$

of f is monotone and (*Fermat's rule*) $\operatorname{zer} \partial f = \operatorname{Argmin} f$. If $f \in \Gamma_0(\mathcal{H})$, then ∂f is maximally monotone and $(\partial f)^{-1} = \partial f^*$.

Example 2.13 (Rockafellar 1970*a*, Theorem 24.3). Let $A\colon \mathbb{R} \to 2^{\mathbb{R}}$ be maximally monotone. Then there exists $f \in \Gamma_0(\mathbb{R})$ such that $A = \partial f$.

Example 2.14. Let C be a nonempty convex subset of $\mathcal{H}$. Then, setting $f = \iota_C$ in Example 2.12, we conclude that the *normal cone* operator

$$N_C = \partial \iota_C\colon\ \mathcal{H} \to 2^{\mathcal{H}}\\ x \mapsto \begin{cases} \{x^* \in \mathcal{H} \mid (\forall y \in C)\ \langle y - x \mid x^*\rangle \leq 0\}, & \text{if } x \in C,\\ \varnothing, & \text{otherwise} \end{cases} \tag{2.41}$$

of C is monotone and that it is maximally monotone if C is closed, in which case $(N_C)^{-1} = \partial \sigma_C$.

Example 2.15. Let V be a closed vector subspace of $\mathcal{H}$. Then it follows from Example 2.14 that

$$N_V\colon \mathcal{H} \to 2^{\mathcal{H}}\colon x \mapsto \begin{cases} V^{\perp}, & \text{if } x \in V,\\ \varnothing, & \text{otherwise} \end{cases} \tag{2.42}$$

is maximally monotone and $(N_V)^{-1} = N_{V^{\perp}}$.

The next two examples involve the Laplacian operator and are central to partial differential equations (Attouch, Buttazzo and Michaille 2014, Barbu 2010, Brézis 1971, Ghoussoub 2009, Zeidler 1990).

Example 2.16 (Attouch *et al.* 2014, Theorem 17.2.10). Let Ω be a nonempty bounded open subset of $\mathbb{R}^N$, suppose that $\mathcal{H} = L^2(\Omega)$, and set

$$A\colon \mathcal{H} \to 2^{\mathcal{H}}\colon x \mapsto \begin{cases} -\Delta x, & \text{if } x \in H_0^1(\Omega) \text{ and } \Delta x \in \mathcal{H},\\ \varnothing, & \text{otherwise.} \end{cases} \tag{2.43}$$

Then it follows from Example 2.12 that A is maximally monotone as the subdifferential of the function

$$f\colon \mathcal{H} \to]-\infty,+\infty]\colon x \mapsto \begin{cases} \dfrac{1}{2}\displaystyle\int_{\Omega} \|\nabla x(\omega)\|^2\, \mathrm{d}\omega, & \text{if } x \in H_0^1(\Omega),\\ +\infty, & \text{otherwise,} \end{cases} \tag{2.44}$$

which is in $\Gamma_0(\mathcal{H})$. In addition, if bdry Ω is of class $\mathscr{C}^2$, then $\operatorname{dom}\partial f = H^2(\Omega) \cap H_0^1(\Omega)$.

Example 2.17 (Attouch *et al.* 2014, Section 17.2.9). Let Ω be a nonempty bounded open subset of $\mathbb{R}^N$ such that bdry Ω is of class $\mathscr{C}^2$, let $\partial/\partial\nu$ denote the outward normal derivative to bdry Ω, suppose that $\mathcal{H} = L^2(\Omega)$, let $h \in \mathcal{H}$, and set

$$A\colon \quad \mathcal{H} \to 2^{\mathcal{H}}$$
$$x \mapsto \begin{cases} -\Delta x - h, & \text{if } x \in H^2(\Omega) \text{ and } \partial x/\partial\nu = 0 \text{ a.e. on bdry } \Omega, \\ \varnothing, & \text{otherwise.} \end{cases} \tag{2.45}$$

Then it follows from Example 2.12 that A is maximally monotone as the subdifferential of the function

$$f\colon \quad \mathcal{H} \to \left]-\infty, +\infty\right]$$
$$x \mapsto \begin{cases} \dfrac{1}{2}\displaystyle\int_\Omega \|\nabla x(\omega)\|^2 \,\mathrm{d}\omega - \int_\Omega x(\omega)h(\omega)\,\mathrm{d}\omega, & \text{if } x \in H^1(\Omega), \\ +\infty, & \text{otherwise,} \end{cases} \tag{2.46}$$

which is in $\Gamma_0(\mathcal{H})$.

The next scenario arises in the study of evolution equations by monotonicity methods (Brézis 1971, 1973, Showalter 1997, Zeidler 1990).

Example 2.18 (Brézis 1971, Example 4; Showalter 1997, Chapter IV; Zeidler 1990, Chapter 32). Let H be a separable real Hilbert space, let $T \in \left]0, +\infty\right[$, and suppose that $\mathcal{H} = L^2([0,T];\mathsf{H})$. For every $y \in \mathcal{H}$, the function $x\colon [0,T] \to \mathsf{H}\colon t \mapsto \int_0^t y(s)\,\mathrm{d}s$ is differentiable a.e. on $\left]0,T\right[$ with $x' = y$ a.e. Define

$$H^1([0,T];\mathsf{H}) = \{x \in \mathcal{H} \mid x' \in L^2([0,T];\mathsf{H})\}, \tag{2.47}$$

let $\mathsf{x}_0 \in \mathsf{H}$, and set

$$A\colon \mathcal{H} \to 2^{\mathcal{H}}\colon x \mapsto \begin{cases} \{x'\}, & \text{if } x \in H^1([0,T];\mathsf{H}) \text{ and } x(0) = \mathsf{x}_0, \\ \varnothing, & \text{otherwise} \end{cases} \tag{2.48}$$

and

$$B\colon \mathcal{H} \to 2^{\mathcal{H}}\colon x \mapsto \begin{cases} \{x'\}, & \text{if } x \in H^1([0,T];\mathsf{H}) \text{ and } x(0) = x(T), \\ \varnothing, & \text{otherwise.} \end{cases} \tag{2.49}$$

Then A and B are maximally monotone.

Example 2.19 (Brézis 1973, Exemple 2.3.3). Let $(\Omega, \mathcal{F}, \mu)$ be a measure space, let H be a separable real Hilbert space, let $\mathsf{A}\colon \mathsf{H} \to 2^{\mathsf{H}}$ be maximally monotone, and set $\mathcal{H} = L^2((\Omega,\mathcal{F},\mu);\mathsf{H})$. Define $A\colon \mathcal{H} \to 2^{\mathcal{H}}$ via

$$(\forall x \in \mathcal{H})(\forall x^* \in \mathcal{H}) \quad (x, x^*) \in \operatorname{gra} A \Leftrightarrow$$
$$\text{for } \mu\text{-almost every } \omega \in \Omega,\ (x(\omega), x^*(\omega)) \in \operatorname{gra}\mathsf{A}, \tag{2.50}$$

and suppose that one of the following holds:

(i) $\mu(\Omega) < +\infty$.

(ii) $0 \in A0$.

Then A is maximally monotone.

We now turn to an equilibrium problem in the sense of Blum and Oettli (1994).

Example 2.20 (Aoyama, Kimura and Takahashi 2008). Let C be a nonempty closed convex subset of $\mathcal{H}$ and suppose that $F: C \times C \to \mathbb{R}$ satisfies the following:

(i) $(\forall x \in C)\ F(x,x) = 0$.

(ii) $(\forall x \in C)(\forall y \in C)\ F(x,y) + F(y,x) \leq 0$.

(iii) For every $x \in C$, $F(x,\cdot): C \to \mathbb{R}$ is lower semicontinuous and convex.

(iv) $(\forall x \in C)(\forall y \in C)(\forall z \in C)\ \overline{\lim}_{0<\varepsilon\to 0} F((1-\varepsilon)x + \varepsilon z, y) \leq F(x,y)$.

Set

$$A: \mathcal{H} \to 2^{\mathcal{H}}$$
$$x \mapsto \begin{cases} \{x^* \in \mathcal{H} \mid (\forall y \in C)\ F(x,y) + \langle x - y \mid x^* \rangle \geq 0\}, & \text{if } x \in C, \\ \varnothing, & \text{otherwise}. \end{cases} \tag{2.51}$$

Then A is maximally monotone and $\operatorname{zer} A = \{x \in C \mid (\forall y \in C)\ F(x,y) \geq 0\}$ is the set of *equilibria* of F.

We conclude with an example in the theory of saddle functions.

Example 2.21 (Rockafellar 1970*b*). Let $F: \mathcal{H} \oplus \mathcal{G} \to [-\infty, +\infty]$ be a *saddle function*, i.e. a convex–concave function which is proper and closed in the sense of Rockafellar (1970*b*, 1971) (e.g. for every $x \in \mathcal{H}$ and every $y \in \mathcal{G}$, $-F(x,\cdot) \in \Gamma_0(\mathcal{G})$ and $F(\cdot,y) \in \Gamma_0(\mathcal{H})$). Set

$$(\forall x \in \mathcal{H})(\forall y \in \mathcal{G}) \quad A(x,y) = \partial F(\cdot,y)(x) \times \partial(-F(x,\cdot))(y). \tag{2.52}$$

Then A is maximally monotone and

$$\operatorname{zer} A = \{(x,y) \in \mathcal{H} \oplus \mathcal{G} \mid F(x,y) = \inf F(\mathcal{H},y) = \sup F(x,\mathcal{G})\} \tag{2.53}$$

is the set of *saddle points* of F.

The following illustration is set in the powerful perturbation framework of Rockafellar (1969, 1970*b*, 1974) (see also Joly and Laurent 1971), which provides a systematic tool to construct duality frameworks in minimization problems.

Example 2.22. Let $\mathcal{V}$ be a real Hilbert space, let $f: \mathcal{H} \to\ ]-\infty, +\infty]$ be a proper function, and consider the primal problem

$$\underset{x \in \mathcal{H}}{\operatorname{minimize}}\ f(x). \tag{2.54}$$

Let $F\colon \mathcal{H}\oplus\mathcal{V}\to\,]-\infty,+\infty]$ be a *perturbation* of f, i.e. $(\forall x\in\mathcal{H})\ f(x)=F(x,0)$. The associated *Lagrangian* is

$$\mathscr{L}_F\colon \mathcal{H}\oplus\mathcal{V}\mapsto[-\infty,+\infty]:(x,v^*)\mapsto\inf_{v\in\mathcal{V}}(F(x,v)-\langle v\mid v^*\rangle), \tag{2.55}$$

the associated *dual problem* is

$$\underset{v^*\in\mathcal{V}}{\text{minimize}}\ \ \sup_{x\in\mathcal{H}}(-\mathscr{L}_F(x,v^*)), \tag{2.56}$$

and the associated *saddle operator* is

$$\boldsymbol{\mathcal{S}}_F\colon \mathcal{H}\oplus\mathcal{V}\to 2^{\mathcal{H}\oplus\mathcal{V}}\colon (x,v^*)\mapsto\partial(\mathscr{L}_F(\cdot,v^*))(x)\times\partial(-\mathscr{L}_F(x,\cdot))(v^*). \tag{2.57}$$

It follows from Example 2.21 that $\boldsymbol{\mathcal{S}}_F$ is maximally monotone. In addition, if $(x,v^*)\in\operatorname{zer}\boldsymbol{\mathcal{S}}_F$, then x solves (2.54) and v^* solves (2.56).

2.4. *Basic theory*

2.4.1. *Operations preserving maximal monotonicity*

The examples of Section 2.3 can be combined in various fashions to create maximally monotone operators.

Lemma 2.23 (Bauschke and Combettes 2017, Proposition 20.22). Let $A\colon\mathcal{H}\to 2^{\mathcal{H}}$ be maximally monotone, let $z\in\mathcal{H}$, let $u\in\mathcal{H}$, and let $\gamma\in\,]0,+\infty[$. Then A^{-1} and $x\mapsto u+\gamma A(x+z)$ are maximally monotone.

Lemma 2.24 (Bauschke and Combettes 2017, Proposition 23.18). Let $(\mathcal{H}_i)_{i\in I}$ be a finite family of real Hilbert spaces, set

$$\boldsymbol{\mathcal{H}}=\bigoplus_{i\in I}\mathcal{H}_i, \tag{2.58}$$

and, for every $i\in I$, let $A_i\colon\mathcal{H}_i\to 2^{\mathcal{H}_i}$ be maximally monotone. Set

$$\boldsymbol{A}\colon\boldsymbol{\mathcal{H}}\to 2^{\boldsymbol{\mathcal{H}}}\colon (x_i)_{i\in I}\mapsto\bigtimes_{i\in I}A_ix_i. \tag{2.59}$$

Then $\boldsymbol{A}$ is maximally monotone.

Lemma 2.25. Let $\beta\in\,]0,+\infty[$, let $A\colon\mathcal{H}\to 2^{\mathcal{H}}$, let $U\in\mathcal{B}(\mathcal{H})$ be self-adjoint and β-strongly monotone, and let $\mathcal{X}$ be the real Hilbert space obtained by endowing $\mathcal{H}$ with the scalar product $(x,y)\mapsto\langle Ux\mid y\rangle$. Then the following hold:

(i) $\operatorname{zer}(U^{-1}\circ A)=\operatorname{zer}A$.

(ii) Suppose that $A\colon\mathcal{H}\to 2^{\mathcal{H}}$ is maximally monotone. Then $U^{-1}\circ A\colon\mathcal{X}\to 2^{\mathcal{X}}$ is maximally monotone.

(iii) Let $\alpha\in\,]0,+\infty[$ and suppose that $A\colon\mathcal{H}\to\mathcal{H}$ is α-cocoercive. Then $U^{-1}\circ A\colon\mathcal{X}\to 2^{\mathcal{X}}$ is $\alpha\beta$-cocoercive.

Proof. We note that (i) is clear and (ii) is proved in Combettes and Vũ (2014, Lemma 3.7(i)).

(iii) Take $(x, y) \in \mathcal{H} \times \mathcal{H}$. Then

$$\begin{aligned}\langle x - y \mid (U^{-1} \circ A)x - (U^{-1} \circ A)y\rangle_{\mathcal{X}} &= \langle x - y \mid Ax - Ay\rangle_{\mathcal{H}} \\ &\geq \alpha \|Ax - Ay\|_{\mathcal{H}}^2. \end{aligned} \tag{2.60}$$

However, $\|U^{-1}x\|_{\mathcal{X}}^2 = \langle x \mid U^{-1}x\rangle_{\mathcal{H}} \leq \|U\|^{-1} \|x\|_{\mathcal{H}}^2$ and $\|U\|^{-1} \leq \beta^{-1}$ (Kato 1980, Section VI.2.6). □

Lemma 2.26 (Bauschke and Combettes 2017, Theorem 25.3; Boţ 2010, Section 24; Pennanen 2000, Corollary 4.2(a)). Let $A\colon \mathcal{H} \to 2^{\mathcal{H}}$ and $B\colon \mathcal{G} \to 2^{\mathcal{G}}$ be maximally monotone, let $L \in \mathcal{B}(\mathcal{H}, \mathcal{G})$, and suppose that

$$\text{cone}(L(\text{dom}\, A) - \text{dom}\, B) \text{ is a closed vector subspace of } \mathcal{G}. \tag{2.61}$$

Then $A + L^* \circ B \circ L$ is maximally monotone.

Lemma 2.27 (Bauschke and Combettes 2017, Corollary 25.5). Let $A\colon \mathcal{H} \to 2^{\mathcal{H}}$ and $B\colon \mathcal{H} \to 2^{\mathcal{H}}$ be maximally monotone and such that one of the following holds:

(i) cone (dom A − dom B) is a closed vector subspace of $\mathcal{H}$.

(ii) dom $B = \mathcal{H}$.

(iii) dom $A \cap$ int dom $B \neq \varnothing$.

Then $A + B$ is maximally monotone.

Lemma 2.28 (Alimohammady, Ramazannejad and Roohi 2014). Let $A\colon \mathcal{H} \to 2^{\mathcal{H}}$ be maximally monotone and let $B\colon \mathcal{H} \to 2^{\mathcal{H}}$ be monotone and such that dom $B = \mathcal{H}$ and $A - B$ is monotone. Then $A - B$ is maximally monotone.

Lemma 2.29. Let $A\colon \mathcal{H} \to 2^{\mathcal{H}}$ and $B\colon \mathcal{H} \to 2^{\mathcal{H}}$ be maximally monotone. Define the *parallel sum* of A and B as

$$A \,\square\, B = (A^{-1} + B^{-1})^{-1} \tag{2.62}$$

and suppose that cone (ran A − ran B) is a closed vector subspace of $\mathcal{H}$. Then $A \,\square\, B$ is maximally monotone.

Proof. This follows from (2.25), Lemma 2.23 and Lemma 2.27(i). □

Lemma 2.30 (Becker and Combettes 2014). Let $A\colon \mathcal{H} \to 2^{\mathcal{H}}$ and $B\colon \mathcal{G} \to 2^{\mathcal{G}}$ be maximally monotone, and let $L \in \mathcal{B}(\mathcal{H}, \mathcal{G})$. Define the *parallel composition* of A with L as

$$L \triangleright A = (L \circ A^{-1} \circ L^*)^{-1}. \tag{2.63}$$

Suppose that

$$\text{cone}(\text{ran}\, A - L^*(\text{ran}\, B)) \text{ is a closed vector subspace of } \mathcal{H}. \tag{2.64}$$

Then $(L \triangleright A) \,\square\, B$ is a maximally monotone operator from $\mathcal{G}$ to $2^{\mathcal{G}}$.

Example 2.31 (Combettes 2023). Let $L \in \mathcal{B}(\mathcal{H}, \mathcal{G})$ be such that $\|L\| \leq 1$ and let $B\colon \mathcal{G} \to 2^{\mathcal{G}}$ be maximally monotone. Define the *resolvent composition* of B with L as

$$L \diamond B = L^* \triangleright (B + \mathrm{Id}_{\mathcal{G}}) - \mathrm{Id}_{\mathcal{H}} \tag{2.65}$$

and the *resolvent cocomposition* of B with L as $L \blacklozenge B = (L \diamond B^{-1})^{-1}$. Then $L \diamond B$ and $L \blacklozenge B$ are maximally monotone operators from $\mathcal{H}$ to $2^{\mathcal{H}}$.

Example 2.32. Let $0 < p \in \mathbb{N}$, let $(\omega_k)_{1 \leq k \leq p}$ be a family in $]0, 1]$ such that $\sum_{k=1}^{p} \omega_k = 1$, and let $(A_k)_{1 \leq k \leq p}$ be maximally monotone operators from $\mathcal{H}$ to $2^{\mathcal{H}}$. Then the *resolvent average*

$$\left(\sum_{k=1}^{p} \omega_k J_{A_k} \right)^{-1} - \mathrm{Id}_{\mathcal{H}} \tag{2.66}$$

is maximally monotone. This result was originally established in Bartz, Bauschke, Moffat and Wang (2016, Proposition 2.7) and derived from Example 2.31 in Combettes (2023, Remark 4.10(ii)).

Example 2.33. Let $A\colon \mathcal{H} \to 2^{\mathcal{H}}$ be a maximally monotone operator and let V be a closed vector subspace of $\mathcal{H}$. The *partial inverse* of A with respect to V is the operator $A_V : \mathcal{H} \to 2^{\mathcal{H}}$ with graph

$$\operatorname{gra} A_V = \{(\operatorname{proj}_V x + \operatorname{proj}_{V^\perp} x^*, \operatorname{proj}_V x^* + \operatorname{proj}_{V^\perp} x) \mid (x, x^*) \in \operatorname{gra} A\}. \tag{2.67}$$

This construction was introduced in Spingarn (1983), which contains the following (see Spingarn 1983, Section 2):

(i) A_V is maximally monotone.

(ii) Let $x \in \mathcal{H}$. Then $x \in \operatorname{zer} A_V \Leftrightarrow (\operatorname{proj}_V x, \operatorname{proj}_{V^\perp} x) \in \operatorname{gra} A$.

2.4.2. *Resolvent*

In terms of solving inclusion problems, the resolvent of (2.17) is the most important operator attached to a monotone operator A. First, as seen in (2.18), it can be employed as a device to generate points in the graph of A. Second, as seen in (2.19), its fixed point set coincides with the set of zeros of A. Third, resolvents provide an effective bridge between the theory of nonexpansive operators and that of monotone operators. This connection goes back to the theory of semigroups of linear nonexpansive operators. The following result, essentially due to Minty (1962), establishes such a connection in the nonlinear case. It states in particular that the resolvent of a maximally monotone operator is a firmly nonexpansive operator which is defined everywhere.

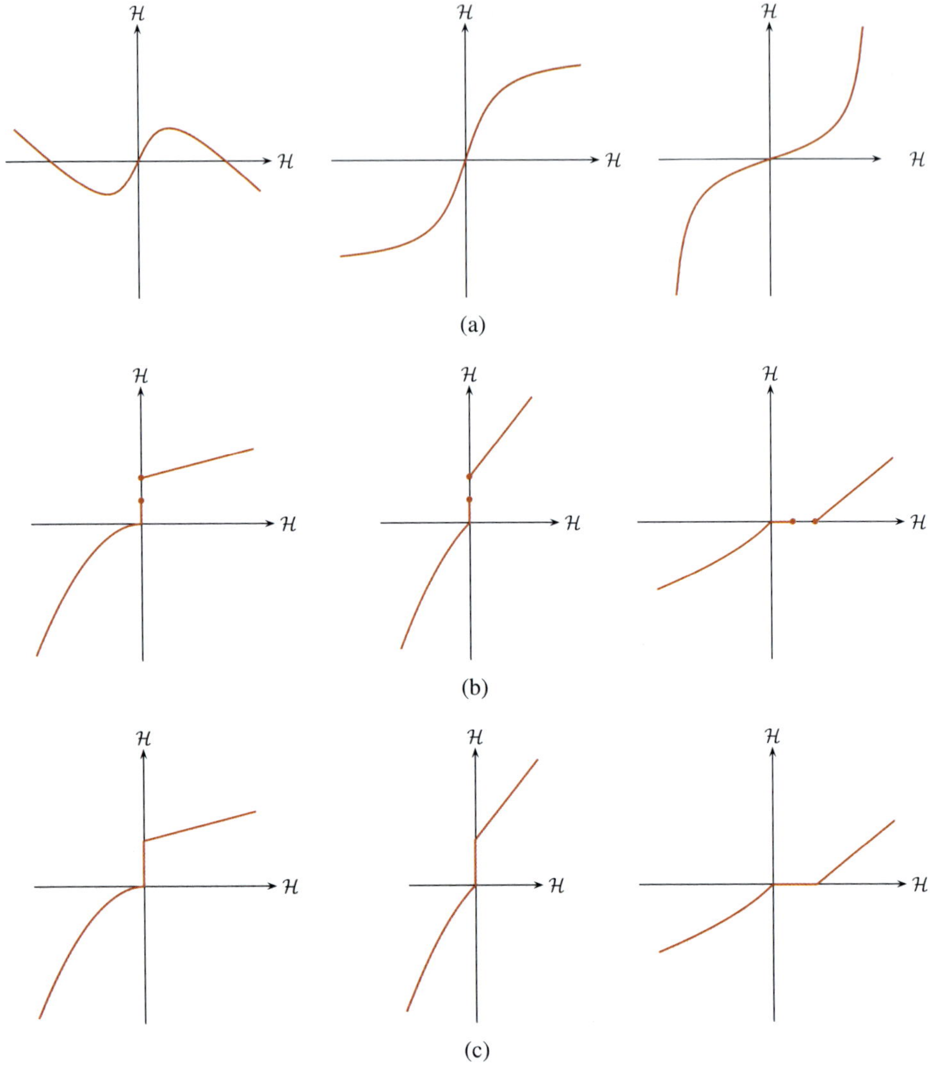

Figure 2.3. Illustration of Minty's theorem (Lemma 2.34). From left to right on each row: graph of A, graph of $\mathrm{Id} + A$ and graph of J_A. (a) A is not monotone: $\mathrm{ran}(\mathrm{Id}+A) = \mathrm{dom}\, J_A \neq \mathcal{H}$ and J_A is not firmly nonexpansive. (b) A is monotone but not maximally monotone: J_A is firmly nonexpansive but $\mathrm{ran}(\mathrm{Id}+A) = \mathrm{dom}\, J_A \neq \mathcal{H}$. (c) A is maximally monotone: J_A is firmly nonexpansive with $\mathrm{ran}(\mathrm{Id} + A) = \mathrm{dom}\, J_A = \mathcal{H}$.

Lemma 2.34 (Bauschke and Combettes 2017, Proposition 23.8). Let D be a nonempty subset of $\mathcal{H}$, let $T\colon D \to \mathcal{H}$, and set $A = T^{-1} - \mathrm{Id}$. Then the following hold (see Figure 2.3):

(i) $D = \mathrm{ran}(\mathrm{Id} + A)$ and $T = J_A$.

(ii) T is firmly nonexpansive if and only if A is monotone.

(iii) T is firmly nonexpansive and $D = \mathcal{H}$ if and only if A is maximally monotone.

Here are a few examples of resolvents that will be explicitly needed; see Bauschke and Combettes (2017), Chierchia, Chouzenoux, Combettes and Pesquet (2016) and Combettes and Pesquet (2011) for additional examples with closed-form expressions and, in particular, instances of proximity operators.

Example 2.35 (Moreau 1965). Let $f \in \Gamma_0(\mathcal{H})$. Then $J_{\partial f} = \mathrm{prox}_f$.

Example 2.36 (Moreau 1962). Let C be a nonempty closed convex subset of $\mathcal{H}$. Then $J_{N_C} = \mathrm{prox}_{\iota_C} = \mathrm{proj}_C$.

Example 2.37 (Bauschke and Combettes 2017, Proposition 23.18). Let $0 < m \in \mathbb{N}$, let $(\mathcal{H}_i)_{1\le i\le m}$ be real Hilbert spaces, set

$$\boldsymbol{\mathcal{H}} = \bigoplus_{i=1}^{m} \mathcal{H}_i, \tag{2.68}$$

and, for every $i \in \{1,\dots,m\}$, let $A_i\colon \mathcal{H}_i \to 2^{\mathcal{H}_i}$ be maximally monotone. Set

$$\boldsymbol{A}\colon \boldsymbol{\mathcal{H}} \to 2^{\boldsymbol{\mathcal{H}}}\colon (x_i)_{1\le i\le m} \mapsto \bigtimes_{1\le i\le m} A_i x_i. \tag{2.69}$$

Then $\boldsymbol{A}$ is maximally monotone (Lemma 2.24) and

$$J_{\boldsymbol{A}}\colon \boldsymbol{\mathcal{H}} \to \boldsymbol{\mathcal{H}}\colon (x_i)_{1\le i\le m} \mapsto (J_{A_i} x_i)_{1\le i\le m}. \tag{2.70}$$

Example 2.38. Let $A\colon \mathcal{H} \to 2^{\mathcal{H}}$ be maximally monotone, let V be a closed vector subspace of $\mathcal{H}$, and let A_V be the partial inverse of Example 2.33. In addition, let $x \in \mathcal{H}$ and $p \in \mathcal{H}$. Then

$$p = J_{A_V} x \quad \Leftrightarrow \quad \mathrm{proj}_V\, p + \mathrm{proj}_{V^\perp}(x - p) = J_A x. \tag{2.71}$$

Proof. This is implicit in Spingarn (1983, Section 4); see Alghamdi, Alotaibi, Combettes and Shahzad (2014, Lemma 2.2) for a proof. □

Example 2.39 (Combettes and Vũ 2014). As in Lemma 2.25(ii), $A\colon \mathcal{H} \to 2^{\mathcal{H}}$ is maximally monotone, $U \in \mathcal{B}(\mathcal{H})$ is self-adjoint and strongly monotone, and $\mathcal{X}$ is the real Hilbert space obtained by endowing $\mathcal{H}$ with the scalar product $(x, y) \mapsto \langle Ux \mid y\rangle$. Then $J_{U^{-1}\circ A} = (U + A)^{-1} \circ U$.

Example 2.40 (Combettes 2023). Let $L \in \mathcal{B}(\mathcal{H}, \mathcal{G})$ be such that $\|L\| \le 1$, let $B\colon \mathcal{G} \to 2^{\mathcal{G}}$ be maximally monotone, and consider the resolvent compositions of Example 2.31. Then

$$J_{L\diamond B} = L^* \circ J_B \circ L \quad \text{and} \quad J_{L\blacklozenge B} = \mathrm{Id}_{\mathcal{H}} - L^* \circ L + L^* \circ J_B \circ L. \tag{2.72}$$

2.4.3. Warped resolvents

A generalization of the notion of a resolvent is the following.

Definition 2.41 (Bùi and Combettes 2020*b*). Let D be a nonempty subset of $\mathcal{H}$, let $U: D \to \mathcal{H}$, and let $M: \mathcal{H} \to 2^{\mathcal{H}}$ be such that $\operatorname{ran} U \subset \operatorname{ran}(U+M)$ and $U+M$ is injective. The *warped resolvent* of M with kernel U is $J_M^U = (U+M)^{-1} \circ U: D \to D$.

The properties of warped resolvent generalize those of classical ones. In this respect, here is an extension of (2.18)–(2.19).

Lemma 2.42. Let D and E be nonempty subsets of $\mathcal{H}$, let $U: D \to \mathcal{H}$, let $C: E \to \mathcal{H}$, and let $W: \mathcal{H} \to 2^{\mathcal{H}}$ be such that $\operatorname{ran} U \subset \operatorname{ran}(U + W + C)$ and $U + W + C$ is injective. Then the following hold:

(i) Let $x \in D$ and $p \in D$. Then $p = J_{W+C}^U x \Leftrightarrow (p, Ux - Up - Cp) \in \operatorname{gra} W$.

(ii) $\operatorname{Fix} J_{W+C}^U = D \cap \operatorname{zer}(W + C)$.

Proof. Note that $J_{W+C}^U: D \to D$ is well defined.

(i) $p = J_{W+C}^U x \Leftrightarrow p = (U+W+C)^{-1}(Ux) \Leftrightarrow Ux \in Up+Wp+Cp \Leftrightarrow Ux-Up-Cp \in Wp$.

(ii) Let $x \in \mathcal{H}$. Then (i) yields $x = J_{W+C}^U x \Leftrightarrow [x \in D \text{ and } (x, -Cx) \in \operatorname{gra} W] \Leftrightarrow [x \in D \text{ and } x \in \operatorname{zer}(W + C)]$. □

An instance of a warped resolvent with a linear kernel appears in Example 2.39, where $D = \mathcal{H}$ and $U \in \mathcal{B}(\mathcal{H})$ is a self-adjoint strongly monotone operator. Self-adjoint monotone operators which are not strongly monotone have also been used as kernels; see Bredies, Chenchene, Lorenz and Naldi (2022) and Xue (2023*b*). The next example features a monotone kernel in $\mathcal{B}(\mathcal{H})$ which is not self-adjoint.

Example 2.43. Let $A: \mathcal{H} \to 2^{\mathcal{H}}$ and $B: \mathcal{G} \to 2^{\mathcal{G}}$ be maximally monotone, and suppose that $0 \neq L \in \mathcal{B}(\mathcal{H}, \mathcal{G})$. Set $\mathbf{X} = \mathcal{H} \oplus \mathcal{G}$ and

$$\begin{cases} \mathcal{K}: \mathbf{X} \to 2^{\mathbf{X}}: (x, y^*) \mapsto (Ax + L^*y^*) \times (B^{-1}y^* - Lx), \\ U: \mathbf{X} \to \mathbf{X}: (x, y^*) \mapsto (x - L^*y^*, Lx + y^*). \end{cases} \tag{2.73}$$

As will be seen in Lemma 3.8, $\mathcal{K}$ is the Kuhn–Tucker operator associated with the problem of finding a zero of $A + L^* \circ B \circ L$. It follows from (2.73) that

$$J_{\mathcal{K}}^U: \mathbf{X} \to \mathbf{X}: (x, y^*) \mapsto (J_A(x - L^*y^*), J_{B^{-1}}(Lx + y^*)), \tag{2.74}$$

whereas $J_{\mathcal{K}}$ is typically intractable.

The next examples employ nonlinear kernels.

Example 2.44. Let $M: \mathcal{H} \to 2^{\mathcal{H}}$ be maximally monotone and such that $\operatorname{zer} M \neq \varnothing$, let $f: \mathcal{H} \to]-\infty, +\infty]$ be a Legendre function such that $\operatorname{dom} M \subset \operatorname{int} \operatorname{dom} f$, and set $D = \operatorname{int} \operatorname{dom} f$ and $U = \nabla f$. Then it follows from Bauschke, Borwein and Combettes (2003, Corollary 3.14(ii)) that $J_M^U : D \to D$ is a well-defined warped resolvent, called the D-resolvent of M. It is an essential tool in the study of algorithms based on Bregman distances which goes back to Brègman (1967), Censor and Zenios (1992), Eckstein (1993) and Teboulle (1992).

Example 2.45. Let $A: \mathcal{H} \to 2^{\mathcal{H}}$ and $B: \mathcal{H} \to 2^{\mathcal{H}}$ be maximally monotone, and let $f \in \Gamma_0(\mathcal{H})$ be essentially smooth (Bauschke, Borwein and Combettes 2003). Suppose that $D = (\operatorname{int}\operatorname{dom} f) \cap \operatorname{dom} A$ is a nonempty subset of $\operatorname{int}\operatorname{dom} B$, that B is single-valued on $\operatorname{int}\operatorname{dom} B$, that ∇f is strictly monotone on D, and that $(\nabla f - B)(D) \subset \operatorname{ran}(\nabla f + A)$. Set $M = A + B$ and $U: D \to \mathcal{H}: x \mapsto \nabla f(x) - Bx$. Then the warped resolvent coincides with the Bregman forward–backward operator $J_M^U = (\nabla f + A)^{-1} \circ (\nabla f - B)$ investigated in Bùi and Combettes (2021), where it is shown to capture a construction found in Renaud and Cohen (1997) and known as the *auxiliary principle*. In the case when A and B are subdifferentials, J_M^U is the operator studied in Nguyen (2017) and, in Euclidean spaces, in Bauschke, Bolte and Teboulle (2017). Scenarios in which J_M^U is more manageable than J_M are discussed in Bauschke *et al.* (2017), Bùi and Combettes (2021), Lu, Freund and Nesterov (2018), Nguyen (2017), Renaud and Cohen (1997) and Teboulle (2018).

Example 2.46. Let $A: \mathcal{H} \to 2^{\mathcal{H}}$, let $C: \mathcal{H} \to \mathcal{H}$ be cocoercive, let $Q: \mathcal{H} \to \mathcal{H}$ be monotone and Lipschitzian, and let $\gamma \in \left]0, +\infty\right[$. The underlying problem is to find a point in $\operatorname{zer}(A + C + Q)$ and we recover the *nonlinear forward–backward operator* of Giselsson (2021) as a warped resolvent as follows. Set $M = \gamma(A+C+Q)$, let $K: \mathcal{H} \to \mathcal{H}$ be strongly monotone and Lipschitzian, and set $U = K - \gamma(C + Q)$. Then $J_M^U = (K+\gamma A)^{-1} \circ (K - \gamma(C+Q))$, which is the operator driving the algorithms of Giselsson (2021).

Remark 2.47. If B is cocoercive and $f = \|\cdot\|^2/2$ in Example 2.45, or if $K = \mathrm{Id}$ and $Q = 0$ and $C = B$ in Example 2.46, then $J_M^U = J_{\gamma A} \circ (\mathrm{Id} - \gamma B)$. This operator will arise in the forward–backward algorithm of Section 8.

Lemma 2.48. Let $Q: \mathcal{H} \to \mathcal{H}$ be Lipschitzian with constant $\beta \in \left]0, +\infty\right[$, let $K: \mathcal{H} \to \mathcal{H}$ be strongly monotone with constant $\alpha \in \left]0, +\infty\right[$, let $\varepsilon \in \left]0, \alpha\right[$, and set $U = K - \gamma Q$. Then the following hold:

(i) Let $\gamma \in \left]0, (\alpha - \varepsilon)/\beta\right]$. Then U is ε-strongly monotone (Bùi and Combettes 2020*b*, Lemma 5.1(i)).

(ii) Suppose that $\alpha = 1$ and $K = \mathrm{Id}$, and let $\gamma \in \left]0, (1 - \varepsilon)/\beta\right]$. Then U is cocoercive with constant $1/(2-\varepsilon)$ (Bùi and Combettes 2020*b*, Lemma 5.1(ii)).

(iii) Suppose that $\alpha = 1$, $K = \mathrm{Id}$, and Q is $1/\beta$-cocoercive, and let $\gamma \in \left]0, 2/\beta\right[$. Then U is $\gamma\beta/2$-averaged, hence nonexpansive (Combettes 2004, Lemma 2.3).

2.4.4. Topological properties

We record key properties of the graphs of monotone operators.

Lemma 2.49 (Bauschke and Combettes 2017, Proposition 20.38(ii)). Let $M\colon \mathcal{H} \to 2^{\mathcal{H}}$ be maximally monotone. Then $\operatorname{gra} M$ is sequentially closed in $\mathcal{H}^{\text{weak}} \times \mathcal{H}^{\text{strong}}$, that is, for every sequence $(x_n, x_n^*)_{n\in\mathbb{N}}$ in $\operatorname{gra} M$ and every $(x, x^*) \in \mathcal{H}\times\mathcal{H}$, if $x_n \rightharpoonup x$ and $x_n^* \to x^*$, then $(x, x^*) \in \operatorname{gra} M$.

Lemma 2.50 (Bauschke and Combettes 2017, Corollary 26.6). Let $A\colon \mathcal{H} \to 2^{\mathcal{H}}$ and $B\colon \mathcal{H} \to 2^{\mathcal{H}}$ be maximally monotone, let $(x_n, x_n^*)_{n\in\mathbb{N}}$ be a sequence in $\operatorname{gra} A$, let $(y_n, y_n^*)_{n\in\mathbb{N}}$ be a sequence in $\operatorname{gra} B$, let $x \in \mathcal{H}$, and let $x^* \in \mathcal{H}$. Suppose that

$$x_n \rightharpoonup x, \quad x_n^* \rightharpoonup x^*, \quad x_n - y_n \to 0 \quad \text{and} \quad x_n^* + y_n^* \to 0. \tag{2.75}$$

Then $x \in \operatorname{zer}(A+B)$, $-x^* \in \operatorname{zer}(-A^{-1}\circ(-\mathrm{Id}) + B^{-1})$, $(x, x^*) \in \operatorname{gra} A$ and $(x, -x^*) \in \operatorname{gra} B$.

2.4.5. Subdifferentials

The subdifferential operator of Example 2.12 is an essential tool in variational analysis.

Lemma 2.51 (Bauschke and Combettes 2017, Proposition 16.6 and Theorem 16.47(i)). Let $f \in \Gamma_0(\mathcal{H})$, $g \in \Gamma_0(\mathcal{G})$ and $L \in \mathcal{B}(\mathcal{H},\mathcal{G})$ be such that $(L(\operatorname{dom} f)) \cap \operatorname{dom} g \neq \varnothing$. Then the following hold:

(i) $\operatorname{zer}(\partial f + L^*\circ(\partial g)\circ L) \subset \operatorname{zer}\partial(f + g\circ L) = \operatorname{Argmin}(f + g\circ L)$.

(ii) Suppose that one of the following is satisfied:

  (a) $0 \in \operatorname{sri}(L(\operatorname{dom} f) - \operatorname{dom} g)$.

  (b) $L(\operatorname{dom} f) - \operatorname{dom} g$ is a closed vector subspace of $\mathcal{G}$.

  (c) $\operatorname{dom} g = \mathcal{G}$.

  (d) $\mathcal{G}$ is finite-dimensional and $(\operatorname{ri} L(\operatorname{dom} f)) \cap (\operatorname{ri}\operatorname{dom} g) \neq \varnothing$.

  Then $\partial(f + g\circ L) = \partial f + L^*\circ(\partial g)\circ L$.

3. Structured monotone inclusions

Our master problem is the following two-operator inclusion.

Problem 3.1. Let $A\colon \mathcal{H} \to 2^{\mathcal{H}}$ and $B\colon \mathcal{H} \to 2^{\mathcal{H}}$ be maximally monotone. The objective is to

$$\text{find } x \in \mathcal{H} \text{ such that } 0 \in Ax + Bx. \tag{3.1}$$

3.1. Two-operator formulations

We provide problem formulations which correspond to specific choices of the operators A and B in Problem 3.1 from the examples of Section 2.3.

Problem 3.2. In Problem 3.1, let $f \in \Gamma_0(\mathcal{H})$, set $A = \partial f$, and suppose that B is at most single-valued. Then (3.1) reduces to the variational inequality problem (Lions 1969)

$$\text{find } x \in \mathcal{H} \text{ such that } (\forall y \in \mathcal{H})\ \langle x - y \mid Bx\rangle + f(x) \leq f(y). \tag{3.2}$$

Problem 3.3. In Problem 3.2, let C be a nonempty closed convex subset of $\mathcal{H}$ and set $f = \iota_C$. Then (3.2) reduces to the standard variational inequality problem (Fichera 1963, Kinderlehrer and Stampacchia 1980)

$$\text{find } x \in C \text{ such that } (\forall y \in C)\ \langle x - y \mid Bx\rangle \leq 0. \tag{3.3}$$

Problem 3.4. In Problem 3.3, suppose that C is a cone with dual cone $C^{\oplus}$. Then (3.3) reduces to the *complementarity problem* (Facchinei and Pang 2003)

$$\text{find } x \in C \text{ such that } x \perp Bx \text{ and } Bx \in C^{\oplus}. \tag{3.4}$$

Problem 3.5. In Problem 3.1, let $f \in \Gamma_0(\mathcal{H})$ and $g \in \Gamma_0(\mathcal{H})$, and set $A = \partial f$ and $B = \partial g$. Suppose that one of the following holds:

(i) $0 \in \operatorname{sri}(\operatorname{dom} f - \operatorname{dom} g)$.

(ii) $g\colon \mathcal{H} \to \mathbb{R}$ is differentiable.

Then the objective is to

$$\underset{x\in\mathcal{H}}{\text{minimize}}\ \ f(x) + g(x). \tag{3.5}$$

Problem 3.6. In Problem 3.5, let C be a nonempty closed convex subset of $\mathcal{H}$ and set $f = \iota_C$. Suppose that one of the following holds:

(i) $0 \in \operatorname{sri}(C - \operatorname{dom} g)$.

(ii) $g\colon \mathcal{H} \to \mathbb{R}$ is differentiable.

Then the objective is to

$$\underset{x\in C}{\text{minimize}}\ \ g(x). \tag{3.6}$$

3.2. Composite problems

We start by presenting a duality framework for monotone inclusions introduced in Pennanen (2000) and Robinson (1999, 2001); for special cases, see Alduncin (2005), Attouch and Théra (1996), Eckstein and Ferris (1999), Fukushima (1996), Gabay (1983), Mercier (1980), Mosco (1972) and Robinson (1998).

Problem 3.7. Let $A\colon \mathcal{H} \to 2^{\mathcal{H}}$ and $B\colon \mathcal{G} \to 2^{\mathcal{G}}$ be maximally monotone, and let $L \in \mathcal{B}(\mathcal{H}, \mathcal{G})$. The objective is to solve the primal inclusion

$$\text{find } x \in \mathcal{H} \text{ such that } 0 \in Ax + L^*(B(Lx)) \tag{3.7}$$

together with the dual inclusion

$$\text{find } y^* \in \mathcal{G} \text{ such that } 0 \in -L(A^{-1}(-L^*y^*)) + B^{-1}y^*. \tag{3.8}$$

Lemma 3.8 (Briceño-Arias and Combettes 2011). In the setting of Problem 3.7, let $\mathbf{X} = \mathcal{H} \oplus \mathcal{G}$, let Z and Z^* be the sets of solutions to (3.7) and (3.8), respectively, and set

$$\begin{cases} \boldsymbol{M} : \mathbf{X} \to 2^{\mathbf{X}} : (x, y^*) \mapsto Ax \times B^{-1}y^*, \\ \boldsymbol{S} : \mathbf{X} \to \mathbf{X} : (x, y^*) \mapsto (L^*y^*, -Lx). \end{cases} \tag{3.9}$$

Define the *Kuhn–Tucker operator* of Problem 3.7 as

$$\boldsymbol{\mathcal{K}} = \boldsymbol{M} + \boldsymbol{S} \tag{3.10}$$

and the set of *Kuhn–Tucker points* as $\operatorname{zer} \boldsymbol{\mathcal{K}}$. Then the following hold:

(i) $\boldsymbol{M}$ is maximally monotone.

(ii) $\boldsymbol{S} \in \mathcal{B}(\mathbf{X})$ is skew and maximally monotone, with $\|\boldsymbol{S}\| = \|L\|$.

(iii) $\boldsymbol{\mathcal{K}}$ is maximally monotone.

(iv) $\operatorname{zer} \boldsymbol{\mathcal{K}}$ is a closed convex subset of $Z \times Z^*$ in $\mathbf{X}$.

(v) $Z \neq \varnothing \Leftrightarrow \operatorname{zer} \boldsymbol{\mathcal{K}} \neq \varnothing \Leftrightarrow Z^* \neq \varnothing$ (see also Eckstein and Ferris 1999, Pennanen 2000, Robinson 1999).

The best known instance for Problem 3.7 is the classical Fenchel–Rockafellar duality framework (Rockafellar 1967).

Problem 3.9. Let $f \in \Gamma_0(\mathcal{H})$, $g \in \Gamma_0(\mathcal{G})$ and $L \in \mathcal{B}(\mathcal{H}, \mathcal{G})$ be such that

$$0 \in \operatorname{sri}(L(\operatorname{dom} f) - \operatorname{dom} g). \tag{3.11}$$

Set $A = \partial f$ and $B = \partial g$ in Problem 3.7. Then it follows from Lemma 2.51 that (3.7) is the primal problem

$$\underset{x \in \mathcal{H}}{\text{minimize}} \;\; f(x) + g(Lx), \tag{3.12}$$

(3.8) is the *Fenchel–Rockafellar dual* problem

$$\underset{y^* \in \mathcal{G}}{\text{minimize}} \;\; f^*(-L^*y^*) + g^*(y^*), \tag{3.13}$$

and (3.10) yields the Kuhn–Tucker operator

$$\boldsymbol{\mathcal{K}} : (x, y^*) \mapsto (\partial f(x) + L^*y^*) \times (-Lx + \partial g^*(y^*)). \tag{3.14}$$

Problem 3.10. Let V be a closed vector subspace of $\mathcal{H}$ and let $A : \mathcal{H} \to 2^{\mathcal{H}}$ be maximally monotone. Then, in the case when $\mathcal{G} = \mathcal{H}$ and $L = \mathrm{Id}$, the Kuhn–Tucker operator (3.10) associated with the operators N_V and A is

$$\boldsymbol{\mathcal{K}} : \mathcal{H} \oplus \mathcal{H} \to 2^{\mathcal{H} \oplus \mathcal{H}} : (x, x^*) \mapsto (N_V x + x^*) \times (A^{-1}x^* - x). \tag{3.15}$$

In view of Example 2.15, the problem of finding a zero of the maximally monotone operator $\boldsymbol{\mathcal{K}}$ reduces to

$$\text{find } x \in V \text{ and } x^* \in V^{\perp} \text{ such that } x^* \in Ax. \tag{3.16}$$

This formulation was first considered by Spingarn (1983).

An extension of Problem 3.7 involving several linearly composed terms is the following.

Problem 3.11. Let $0 < p \in \mathbb{N}$, let $A\colon \mathcal{H} \to 2^{\mathcal{H}}$ be maximally monotone, and, for every $k \in \{1\dots,p\}$, let $\mathcal{G}_k$ be a real Hilbert space, let $B_k\colon \mathcal{G}_k \to 2^{\mathcal{G}_k}$ be maximally monotone, and let $L_k \in \mathcal{B}(\mathcal{H},\mathcal{G}_k)$. The objective is to solve the primal inclusion

$$\text{find } x \in \mathcal{H} \text{ such that } 0 \in Ax + \sum_{k=1}^{p} L_k^*(B_k(L_k x)) \tag{3.17}$$

together with the dual inclusion

$$\text{find } y_1^* \in \mathcal{G}_1,\dots,y_p^* \in \mathcal{G}_p \text{ such that}$$

$$\left(\exists x \in A^{-1}\left(-\sum_{k=1}^{p} L_k^* y_k^*\right)\right)\ (\forall k \in \{1,\dots,p\})\ L_k x \in B_k^{-1} y_k^*. \tag{3.18}$$

Lemma 3.12. In the setting of Problem 3.11, set $\mathbf{X} = \mathcal{H} \oplus \mathcal{G}_1 \oplus \cdots \oplus \mathcal{G}_p$ and let Z and Z^* be the sets of solutions to (3.17) and (3.18), respectively. Define the *Kuhn–Tucker operator* of Problem 3.11 as

$$\mathcal{K}\colon \mathbf{X} \to 2^{\mathbf{X}}\colon (x,y_1^*,\dots,y_p^*) \mapsto$$
$$\left(Ax + \sum_{k=1}^{p} L_k^* y_k^*\right) \times \left(-L_1 x + B_1^{-1} y_1^*\right) \times \cdots \times \left(-L_p x + B_p^{-1} y_p^*\right) \tag{3.19}$$

and the set of *Kuhn–Tucker points* as $\operatorname{zer}\mathcal{K}$. Then the following hold:

(i) $\mathcal{K}$ is maximally monotone.

(ii) $\operatorname{zer}\mathcal{K}$ is a closed convex subset of $Z \times Z^*$ in $\mathbf{X}$.

(iii) $Z \neq \varnothing \Leftrightarrow \operatorname{zer}\mathcal{K} \neq \varnothing \Leftrightarrow Z^* \neq \varnothing$.

Proof. The proof is similar to that of Lemma 3.8. □

An alternative angle on Problem 3.9 is provided by the Lagrangian approach of Example 2.22. Set $\boldsymbol{f}\colon \mathcal{H} \oplus \mathcal{G} \to\,]-\infty,+\infty]\colon \boldsymbol{x} = (x,y) \mapsto f(x) + g(y)$, $\boldsymbol{L}\colon \mathcal{H} \oplus \mathcal{G} \to \mathcal{G}\colon (x,y) \mapsto Lx - y$ and $\mathbf{X} = \mathcal{H} \oplus \mathcal{G} \oplus \mathcal{G}$. Then the primal problem (3.12) is equivalent to

$$\underset{\boldsymbol{x} \in \ker \boldsymbol{L}}{\text{minimize}}\ \boldsymbol{f}(\boldsymbol{x}) \tag{3.20}$$

and its standard perturbation function is (Rockafellar 1974, Example 4′; see also Bauschke and Combettes 2017, Proposition 19.21)

$$\boldsymbol{F}\colon \mathbf{X} \to\,]-\infty,+\infty]\colon (\boldsymbol{x},v) \mapsto \boldsymbol{f}(\boldsymbol{x}) + \iota_{\{0\}}(\boldsymbol{L}\boldsymbol{x} + v). \tag{3.21}$$

We derive from (2.55) that the associated Lagrangian is

$$\mathscr{L}_{\boldsymbol{F}}\colon \mathbf{X} \to\,]-\infty,+\infty]\colon (\boldsymbol{x},v^*) \mapsto \boldsymbol{f}(\boldsymbol{x}) + \langle \boldsymbol{L}\boldsymbol{x} \mid v^*\rangle, \tag{3.22}$$

from (2.56) that the associated dual problem is (3.13), and from (2.57) that the associated saddle operator is

$$\boldsymbol{\mathcal{S}}_{\boldsymbol{F}}\colon \mathbf{X} \to 2^{\mathbf{X}}\colon (\boldsymbol{x}, v^*) \mapsto (\partial \boldsymbol{f}(\boldsymbol{x}) + \boldsymbol{L}^* v^*) \times \{-\boldsymbol{L}\boldsymbol{x}\}, \tag{3.23}$$

that is,

$$\boldsymbol{\mathcal{S}}_{\boldsymbol{F}}\colon \begin{array}{ccl} \mathbf{X} & \to & 2^{\mathbf{X}} \\ (x, y, v^*) & \mapsto & (\partial f(x) + L^* v^*) \times (\partial g(y) - v^*) \times \{-Lx + y\}. \end{array} \tag{3.24}$$

We saw in Example 2.22 that, if $(x, y, v^*) \in \operatorname{zer} \boldsymbol{\mathcal{S}}_{\boldsymbol{F}}$, then x solves the primal problem (3.12) and v^* solves the dual problem (3.13). A version of this result for Problem 3.7 is the following where, although there is no notion of a Lagrangian, we can introduce a saddle operator.

Lemma 3.13. In the setting of Problem 3.7, set $\mathbf{X} = \mathcal{H} \oplus \mathcal{G} \oplus \mathcal{G}$ and let Z and Z^* be the sets of solutions to (3.7) and (3.8), respectively. Define the Kuhn–Tucker operator $\boldsymbol{\mathcal{K}}$ as in (3.10) and define the *saddle operator* of Problem 3.7 as

$$\boldsymbol{\mathcal{S}}\colon \begin{array}{ccl} \mathbf{X} & \to & 2^{\mathbf{X}} \\ (x, y, v^*) & \mapsto & (Ax + L^* v^*) \times (By - v^*) \times \{-Lx + y\}. \end{array} \tag{3.25}$$

Then the following hold:

(i) $\boldsymbol{\mathcal{S}}$ is maximally monotone.

(ii) $\operatorname{zer} \boldsymbol{\mathcal{S}}$ is closed and convex.

(iii) Suppose that $(x, y, v^*) \in \operatorname{zer} \boldsymbol{\mathcal{S}}$. Then $(x, v^*) \in \operatorname{zer} \boldsymbol{\mathcal{K}} \subset Z \times Z^*$.

(iv) $Z^* \neq \varnothing \Leftrightarrow \operatorname{zer} \boldsymbol{\mathcal{S}} \neq \varnothing \Leftrightarrow \operatorname{zer} \boldsymbol{\mathcal{K}} \neq \varnothing \Leftrightarrow Z \neq \varnothing$.

Proof. This is a special case of Bùi and Combettes (2022*b*, Proposition 1(i)–(v)(a)). ☐

3.3. Examples of embeddings in Framework 1.2

Example 3.14. Suppose that it is computationally feasible solve Problem 1.1 directly in the original space $\mathcal{H}$. Then an embedding of Problem 1.1 is just $(\mathcal{H}, M, \mathrm{Id})$.

Example 3.15. Let $M\colon \mathcal{H} \to 2^{\mathcal{H}}$ be a maximally monotone operator, let $U \in \mathcal{B}(\mathcal{H})$ be a self-adjoint strongly monotone operator, let $\mathbf{X}$ be the real Hilbert space obtained by endowing $\mathcal{H}$ with the scalar product $(x, y) \mapsto \langle Ux \mid y\rangle$, let $\boldsymbol{\mathcal{M}} = U^{-1} \circ M$, and set $\boldsymbol{\mathcal{T}} = \mathrm{Id}$. Then it follows from Lemma 2.25(i)–(ii) that $(\mathbf{X}, \boldsymbol{\mathcal{M}}, \boldsymbol{\mathcal{T}})$ is an embedding of Problem 1.1.

Example 3.16. Let $\alpha \in\,]0, 1]$ and let $T\colon \mathcal{H} \to \mathcal{H}$ be α-averaged. In Problem 1.1, suppose that $M = \mathrm{Id} - T$ (see Example 2.4) and set

$$\mathbf{X} = \mathcal{H}, \quad \boldsymbol{\mathcal{M}} = \left(\mathrm{Id} + \frac{1}{2\alpha}(T - \mathrm{Id})\right)^{-1} - \mathrm{Id} \quad \text{and} \quad \boldsymbol{\mathcal{T}} = \mathrm{Id}. \tag{3.26}$$

Then $(\mathbf{X}, \mathcal{M}, \mathcal{T})$ is an embedding of Problem 1.1. Indeed, since $\mathrm{Id} + \alpha^{-1}(T - \mathrm{Id})$ is nonexpansive, we derive from Bauschke and Combettes (2017, Proposition 4.4) that $\mathrm{Id} + (2\alpha)^{-1}(T - \mathrm{Id})$ is firmly nonexpansive, and hence from Lemma 2.34(iii) that $\mathcal{M}$ is maximally monotone, with $\operatorname{zer} \mathcal{M} = \operatorname{zer} M = \operatorname{Fix} T$.

Example 3.17. Let $A\colon \mathcal{H} \to 2^{\mathcal{H}}$ and $B\colon \mathcal{H} \to 2^{\mathcal{H}}$ be maximally monotone, and let $\gamma \in\,]0, +\infty[$. Let

$$\mathbf{X} = \mathcal{H}, \quad \mathcal{M} = (J_{\gamma A} \circ (2J_{\gamma B} - \mathrm{Id}) + \mathrm{Id} - J_{\gamma B})^{-1} - \mathrm{Id} \quad \text{and} \quad \mathcal{T} = J_{\gamma B}. \tag{3.27}$$

Then it follows from Eckstein and Bertsekas (1992, Section 4) that $(\mathbf{X}, \mathcal{M}, \mathcal{T})$ is an embedding of Problem 3.1. In this setting, we actually have $\mathcal{T}(\operatorname{zer} \mathcal{M}) = \operatorname{zer} M$ (Combettes 2004, Lemma 2.6(iii)).

Example 3.18. Let $A\colon \mathcal{H} \to 2^{\mathcal{H}}$ and $B\colon \mathcal{H} \to 2^{\mathcal{H}}$ be maximally monotone. Let $\mathbf{X} = \mathcal{H} \oplus \mathcal{H}$, $\mathcal{M}\colon \mathbf{X} \to 2^{\mathbf{X}}\colon (x, x^*) \mapsto (Ax + x^*) \times (-x + B^{-1}x^*)$ and $\mathcal{T}\colon \mathbf{X} \to \mathcal{H}\colon (x, x^*) \mapsto x$. Then applying Lemma 3.8 with $\mathcal{G} = \mathcal{H}$ and $L = \mathrm{Id}$ shows that $(\mathbf{X}, \mathcal{M}, \mathcal{T})$ is an embedding of Problem 3.1. This embedding is implicitly present in the projective splitting algorithm of Eckstein and Svaiter (2008), which is therefore an instance of Framework 1.2.

We now discuss structured inclusion problems that offer greater modelling flexibility by involving three or more operators. The principle of a splitting algorithm, which is to involve each operator individually, faces a serious challenge in the presence of such formulations. Indeed, since inclusion is a binary relation, for reasons discussed in Briceño-Arias and Combettes (2011) and Combettes (2013*a*), and analysed in more depth in Ryu (2020), it is not possible to split problems that involve more than two set-valued operators. A purpose of Framework 1.2 is to circumvent this fundamental limitation by seeking more tractable reformulations in bigger spaces.

Example 3.19. Let $0 < p \in \mathbb{N}$ and, for every $k \in \{1, \dots, p\}$, let $A_k\colon \mathcal{H} \to 2^{\mathcal{H}}$ be maximally monotone. The problem is to

$$\text{find } x \in \mathcal{H} \text{ such that } 0 \in \sum_{k=1}^{p} A_k x. \tag{3.28}$$

Let $\mathbf{X}$ be the p-fold Hilbert direct sum $\mathcal{H}^p$ and set

$$\begin{cases} \mathbf{V} = \{(x_1, \dots, x_p) \in \mathbf{X} \mid x_1 = \cdots = x_p\}, \\ \mathbf{A}\colon \mathbf{X} \to 2^{\mathbf{X}}\colon (x_1, \dots, x_p) \mapsto A_1 x_1 \times \cdots \times A_p x_p, \\ \mathcal{M} = \mathbf{A} + N_{\mathbf{V}}, \\ \mathcal{T}\colon \mathbf{X} \to \mathcal{H}\colon (x_1, \dots, x_p) \mapsto x_1. \end{cases} \tag{3.29}$$

Then

$$\mathbf{V}^{\perp} = \left\{ (x_1^*, \dots, x_p^*) \in \mathbf{X} \;\middle|\; \sum_{k=1}^{p} x_k^* = 0 \right\}, \tag{3.30}$$

and it follows from Example 2.15 that $(\mathbf{X}, \mathcal{M}, \mathcal{T})$ is an embedding of (3.28). This setting to split the sum of $p > 2$ monotone operators was introduced by Spingarn (1983, Section 5); see also Gol'shtein (1987). It reduces the p-operator problem (3.28) to the two-operator inclusion $\mathbf{0} \in \boldsymbol{A}\boldsymbol{x} + N_{\boldsymbol{V}}\boldsymbol{x}$. The idea of rephrasing multi-operator problems in product spaces finds its roots in convex feasibility problems (Pierra 1976, 1984), where the problem of finding a point in the intersection $\bigcap_{k=1}^{p} C_k$ of closed convex subsets $(C_k)_{1 \leq k \leq p}$ of $\mathcal{H}$ is associated with that of finding a point in $\boldsymbol{C} \cap \boldsymbol{V}$ in $\mathbf{X}$, where $\boldsymbol{C} = C_1 \times \cdots \times C_p$.

Example 3.20. In the setting of Problem 3.7, set $\mathbf{X} = \mathcal{H} \oplus \mathcal{G}$, define $\boldsymbol{M}$ and $\boldsymbol{S}$ as in (3.9), let $\mathcal{K} = \boldsymbol{M} + \boldsymbol{S}$ be the Kuhn–Tucker operator of (3.10), and let $\mathcal{T} \colon \mathbf{X} \to \mathcal{H} \colon (x, y^*) \mapsto x$. Then, in view of Lemma 3.8(iv), $(\mathbf{X}, \mathcal{K}, \mathcal{T})$ is an embedding of (3.7). This embedding, which underlies the *monotone+skew* framework of Briceño-Arias and Combettes (2011), reduces Problem 3.7, which involves three operators in the primal space $\mathcal{H}$ (namely, A, B and L), to a problem in $\mathbf{X}$ that involves the two operators $\boldsymbol{M}$ and $\boldsymbol{S}$.

Example 3.21. In the setting of Problem 3.11, set $\mathbf{X} = \mathcal{H} \oplus \mathcal{G}_1 \oplus \cdots \oplus \mathcal{G}_p$, let $\mathcal{K}$ be the Kuhn–Tucker operator of (3.19), and let

$$\mathcal{T} \colon \mathbf{X} \to \mathcal{H} \colon (x, y_1^*, \ldots, y_p^*) \mapsto x. \tag{3.31}$$

Then it follows from Lemma 3.12(ii) that $(\mathbf{X}, \mathcal{K}, \mathcal{T})$ is an embedding of (3.17).

Next, we consider an embedding for strongly monotone problems.

Example 3.22. Let $\rho \in \left]0, +\infty\right[$, let $0 < p \in \mathbb{N}$, let $z \in \mathcal{H}$, and let $A \colon \mathcal{H} \to 2^{\mathcal{H}}$ be maximally monotone. For every $k \in \{1, \ldots, p\}$, let $B_k \colon \mathcal{G}_k \to 2^{\mathcal{G}_k}$ and $D_k \colon \mathcal{G}_k \to 2^{\mathcal{G}_k}$ be maximally monotone, and suppose that $0 \neq L_k \in \mathcal{B}(\mathcal{H}, \mathcal{G}_k)$. The problem is to

$$\text{find } x \in \mathcal{H} \text{ such that } z \in Ax + \sum_{k=1}^{p} L_k^*((B_k \,\square\, D_k)(L_k x)) + \rho x. \tag{3.32}$$

Let $\mathbf{X} = \mathcal{G}_1 \oplus \cdots \oplus \mathcal{G}_p$, let

$$\begin{aligned} \mathcal{M} \colon \quad & \mathbf{X} \to 2^{\mathbf{X}} \\ & (y_1^*, \ldots, y_p^*) \mapsto \left(-L_1\left(J_{A/\rho}\left(\frac{1}{\rho}\left(z - \sum_{k=1}^{p} L_k^* y_k^*\right)\right)\right) + B_1^{-1} y_1^* + D_1^{-1} y_1^*\right) \\ & \times \cdots \times \left(-L_p\left(J_{A/\rho}\left(\frac{1}{\rho}\left(z - \sum_{k=1}^{p} L_k^* y_k^*\right)\right)\right) + B_p^{-1} y_p^* + D_p^{-1} y_p^*\right), \end{aligned} \tag{3.33}$$

and let

$$\mathcal{T} \colon \mathbf{X} \to \mathcal{H} \colon (y_1^*, \ldots, y_p^*) \mapsto J_{A/\rho}\left(\frac{1}{\rho}\left(z - \sum_{k=1}^{p} L_k^* y_k^*\right)\right). \tag{3.34}$$

Then it follows from Combettes and Vũ (2014, Proposition 5.2(iii)) that $(\mathbf{X}, \mathcal{M}, \mathcal{T})$ is an embedding of (3.32).

Our last example concerns an embedding based on a saddle operator.

Example 3.23. In the setting of Problem 3.7, set $\mathbf{X} = \mathcal{H} \oplus \mathcal{G} \oplus \mathcal{G}$, let $\boldsymbol{\mathcal{S}}$ be the saddle operator of (3.25), and let $\mathcal{T}\colon \mathbf{X} \to \mathcal{H}\colon (x, y, v^*) \mapsto x$. Then it follows from Lemma 3.13(iii) that $(\mathbf{X}, \boldsymbol{\mathcal{S}}, \mathcal{T})$ is an embedding of (3.7).

Additional examples of embeddings will be provided by Examples 7.9, 9.8 and 10.4.

4. Two geometric convergence principles

4.1. Overview

The methodology of Framework 1.2 is to identify a target set Z in a suitable Hilbert space in such a way that every point in Z yields a solution to the original problem of interest. The algorithms we shall consider are Fejérian in the sense that every iteration brings the current iterate closer to every point in Z.

4.2. Fejér monotone scheme

Let us first recall some basic facts about weak and strong convergence in Hilbert spaces.

Lemma 4.1 (Bauschke and Combettes 2017, Section 2.5). Let $(x_n)_{n\in\mathbb{N}}$ be a sequence in $\mathcal{H}$ and let $x \in \mathcal{H}$. Then the following hold:

(i) Let Z be a nonempty subset of $\mathcal{H}$. Suppose that $\mathfrak{W}(x_n)_{n\in\mathbb{N}} \subset Z$ and that, for every $z \in Z$, $(\|x_n - z\|)_{n\in\mathbb{N}}$ converges. Then $(x_n)_{n\in\mathbb{N}}$ converges weakly to a point in Z.

(ii) $x_n \rightharpoonup x \Leftrightarrow [(x_n)_{n\in\mathbb{N}}$ is bounded and $\mathfrak{W}(x_n)_{n\in\mathbb{N}} = \{x\}]$.

(iii) $x_n \to x \Leftrightarrow [x_n \rightharpoonup x$ and $\overline{\lim}\, \|x_n\| \leq \|x\|]$.

Theorem 4.2. Let Z be a nonempty closed convex subset of $\mathcal{H}$, let $(\lambda_n)_{n\in\mathbb{N}}$ be a sequence of relaxation parameters in $]0, 2[$, and let $x_0 \in \mathcal{H}$. Iterate (see Figure 4.1)

$$
\begin{array}{l}
\text{for } n = 0, 1, \ldots \\
\left\lfloor
\begin{array}{l}
H_n \text{ is a closed half-space such that } Z \subset H_n \\
p_n = \operatorname{proj}_{H_n} x_n \\
x_{n+1} = x_n + \lambda_n(p_n - x_n).
\end{array}
\right.
\end{array}
\tag{4.1}
$$

Then the following hold:

(i) Fejér monotonicity: $(\forall z \in Z)(\forall n \in \mathbb{N})\ \|x_{n+1} - z\| \leq \|x_n - z\|$.

(ii) $\sum_{n\in\mathbb{N}} \lambda_n(2 - \lambda_n)\|p_n - x_n\|^2 < +\infty$.

(iii) Suppose that $\sup_{n\in\mathbb{N}} \lambda_n < 2$. Then $\sum_{n\in\mathbb{N}} \|x_{n+1} - x_n\|^2 < +\infty$.

(iv) Suppose that $\mathfrak{W}(x_n)_{n\in\mathbb{N}} \subset Z$. Then $(x_n)_{n\in\mathbb{N}}$ converges weakly to a point in Z.

Proof. Let $z \in Z$. Then, for every $n \in \mathbb{N}$, $H_n = \{u \in \mathcal{H} \mid \langle u - p_n \mid x_n - p_n\rangle \leq 0\}$ and, since $z \in H_n$, (4.1) yields

$$\begin{aligned}\|x_{n+1} - z\|^2 &= \|x_n - z\|^2 + 2\lambda_n\langle x_n - z \mid p_n - x_n\rangle + \lambda_n^2\|p_n - x_n\|^2 \\ &= \|x_n - z\|^2 - \lambda_n(2 - \lambda_n)\|p_n - x_n\|^2 + 2\lambda_n\langle z - p_n \mid x_n - p_n\rangle \\ &\leq \|x_n - z\|^2 - \lambda_n(2 - \lambda_n)\|p_n - x_n\|^2 \qquad (4.2) \\ &= \|x_n - z\|^2 - \frac{2 - \lambda_n}{\lambda_n}\|x_{n+1} - x_n\|^2 \qquad (4.3) \\ &\leq \|x_n - z\|^2. \qquad (4.4)\end{aligned}$$

(i) See (4.4).

(ii) Fix $N \in \mathbb{N}$. Then (4.2) yields

$$\sum_{n=0}^{N} \lambda_n(2 - \lambda_n)\|p_n - x_n\|^2 \leq \|x_0 - z\|^2 \qquad (4.5)$$

and we conclude by letting $N \to +\infty$.

(ii)⇒(iii) This follows from (4.3).

(iv) In view of (i), $(\|x_n - z\|)_{n\in\mathbb{N}}$ converges. The claim therefore follows from Lemma 4.1(i). □

Remark 4.3. Fejér (1922) studied the following problem: given a nonempty closed set $Z \subset \mathbb{R}^N$ and a point $y \notin Z$, can one find a point $x \in \mathbb{R}^N$ such that

$$(\forall z \in Z) \quad \|x - z\| < \|y - z\|. \qquad (4.6)$$

This led Motzkin and Schoenberg (1954) to adopt the terminology *Fejér monotone* to describe sequences satisfying property (i) in Theorem 4.2. In their paper (see also Agmon 1954), an algorithm was developed to solve systems of linear inequalities in $\mathbb{R}^N$ by successive projections onto the half-spaces defining the polyhedral solution set Z, and Fejér monotonicity was shown to be an adequate tool to study the convergence of this algorithm. Further analysis of Fejér monotonicity was proposed in Brègman (1965), Eremin (1968*a*,*b*) and Raik (1967, 1969), and nowadays it constitutes a central tool to analyse the asymptotic behaviour of various algorithms (Bauschke and Combettes 2017).

Remark 4.4. In general, the convergence of $(x_n)_{n\in\mathbb{N}}$ to $x \in Z$ in Theorem 4.2(iv) is only weak and, even if it were strong, there exists no rate of convergence on $(\|x_n - x\|)_{n\in\mathbb{N}}$, even in Euclidean spaces (Bauschke, Deutsch and Hundal 2009, Gubin, Polyak and Raik 1967, Youla 1987). In particular, achieving a linear rate of

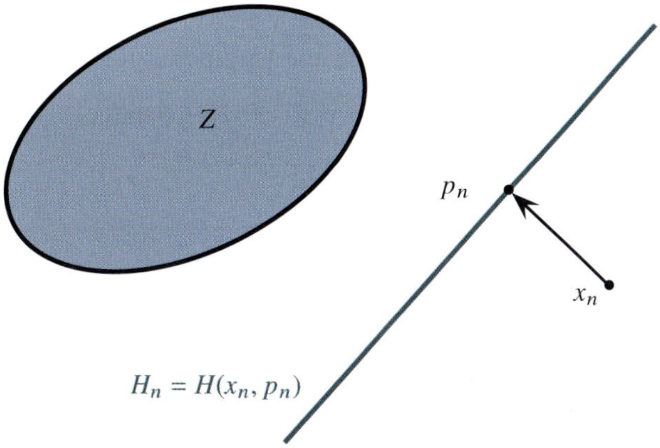

Figure 4.1. Iteration n of the Fejérian algorithm (4.1).

convergence, that is, securing the existence of $\kappa \in]0, +\infty[$ and $\rho \in]0, 1[$ such that

$$(\forall n \in \mathbb{N}) \quad \|x_n - x\| \leq \kappa\rho^n, \tag{4.7}$$

requires stringent additional assumptions on the problem. In our inclusion context, a typical assumption is strong monotonicity; see Bauschke and Combettes (2017, Proposition 26.16) for an example. In the broader context of Theorem 4.2(i), it is clear that $(d_C(x_n))_{n\in\mathbb{N}}$ decreases and that, for every $n \in \mathbb{N}$ and $m \in \mathbb{N}$, $\|x_n - x_{n+m}\| \leq \|x_n - \operatorname{proj}_C x_n\| + \|x_{n+m} - \operatorname{proj}_C x_n\| \leq 2d_C(x_n)$. Hence, (4.7) will hold with $\kappa = 2d_C(x_0)$ if the decreasing property can be strengthened to $(\forall n \in \mathbb{N})$ $d_C(x_{n+1}) \leq \rho d_C(x_n)$.

Remark 4.5. The implementation of (4.1) is said to be *unrelaxed* if $(\forall n \in \mathbb{N})$ $\lambda_n = 1$.

4.3. Haugazeau-like scheme

Theorem 4.2 guarantees only weak convergence to an unspecified point in Z and, as will be seen on several occasions later, strong convergence fails in general (many of these examples will be based on a scenario of Hundal (2004) concerning the method of alternating projections). However, in some infinite-dimensional applications in areas such as inverse problems, control, mechanics, PDEs, optics and analogue computing, weak convergence does not offer sufficient guarantees and strong convergence is required. The geometric approach described in this section emanates from ideas found in the work of Haugazeau (1967, 1968) on the convex feasibility problem. It will provide strong convergence to a specific point in Z, namely the projection of the initial point onto Z. This means that the resulting algorithm is also of interest, even in Euclidean spaces, as a best approximation method.

The following technical fact from Haugazeau (1968, Théorème 3-1) will be employed repeatedly; see also Bauschke and Combettes (2017, Corollary 29.25).

Lemma 4.6 (Haugazeau 1968). Let $(x, y, z) \in \mathcal{H}^3$. Define

$$H(x, y) = \{z \in \mathcal{H} \mid \langle z - y \mid x - y\rangle \leq 0\}, \tag{4.8}$$

$C = H(x, y) \cap H(y, z)$, and, if $C \neq \varnothing$,

$$\mathsf{Q}(x, y, z) = \operatorname{proj}_C x. \tag{4.9}$$

Set $\chi = \langle x - y \mid y - z\rangle$, $\mu = \|x - y\|^2$, $\nu = \|y - z\|^2$ and $\rho = \mu\nu - \chi^2$. Then exactly one of the following holds:

(i) $\rho = 0$ and $\chi < 0$, in which case $C = \varnothing$.

(ii) $[\rho = 0$ and $\chi \geq 0]$ or $\rho > 0$, in which case $C \neq \varnothing$ and

$$\mathsf{Q}(x, y, z) = \begin{cases} z, & \text{if } \rho = 0 \text{ and } \chi \geq 0,\\ x + (1 + \chi/\nu)(z - y), & \text{if } \rho > 0 \text{ and } \chi\nu \geq \rho,\\ y + (\nu/\rho)(\chi(x - y) + \mu(z - y)), & \text{if } \rho > 0 \text{ and } \chi\nu < \rho. \end{cases} \tag{4.10}$$

The essential components of the following theorem are found in the unpublished thesis of Haugazeau (1968) (see Haugazeau 1967 for a preliminary variant), where he considered the specific problem of projecting a point onto the intersection of finitely many sets using their individual projection operators cyclically.

Theorem 4.7. Let Z be a nonempty closed convex subset of $\mathcal{H}$, let $(\lambda_n)_{n\in\mathbb{N}}$ be a sequence of relaxation parameters in $]0, 1]$, and let $x_0 \in \mathcal{H}$. Iterate (see Figure 4.2)

$$\begin{array}{l} \text{for } n = 0, 1, \ldots\\ \left\lfloor \begin{array}{l} H_n \text{ is a closed half-space such that } Z \subset H_n\\ p_n = \operatorname{proj}_{H_n} x_n\\ r_n = x_n + \lambda_n(p_n - x_n)\\ x_{n+1} = \mathsf{Q}(x_0, x_n, r_n). \end{array} \right. \end{array} \tag{4.11}$$

Then the sequence $(x_n)_{n\in\mathbb{N}}$ is well defined and the following hold:

(i) $(\forall n \in \mathbb{N})\ Z \subset H(x_0, x_n) \cap H(x_n, r_n)$.

(ii) $(\exists \ell \in [0, +\infty[)\ \|x_n - x_0\| \uparrow \ell \leq d_Z(x_0)$.

(iii) $\sum_{n\in\mathbb{N}} \|x_{n+1} - x_n\|^2 < +\infty$.

(iv) $\sum_{n\in\mathbb{N}} \lambda_n^2 \|p_n - x_n\|^2 < +\infty$.

(v) Suppose that $\mathfrak{W}(x_n)_{n\in\mathbb{N}} \subset Z$. Then $(x_n)_{n\in\mathbb{N}}$ converges strongly to $\operatorname{proj}_Z x_0$.

Proof. First, recall that the projector onto a nonempty closed convex subset D of $\mathcal{H}$ is characterized by (Bauschke and Combettes 2017, Theorem 3.16)

$$(\forall x \in \mathcal{H}) \quad \operatorname{proj}_D x \in D \quad \text{and} \quad D \subset H(x, \operatorname{proj}_D x). \tag{4.12}$$

We also observe that (4.11) implies that

$$\begin{aligned}(\forall n \in \mathbb{N}) \quad & H(x_n, p_n) \\ & = \{z \in \mathcal{H} \mid \langle z - p_n \mid x_n - r_n \rangle \leq 0\} \\ & = \{z \in \mathcal{H} \mid \langle z - r_n \mid x_n - r_n \rangle \leq \langle p_n - r_n \mid x_n - r_n \rangle\} \\ & = \{z \in \mathcal{H} \mid \langle z - r_n \mid x_n - r_n \rangle \leq -\lambda_n(1 - \lambda_n)\|x_n - p_n\|^2\} \\ & \subset H(x_n, r_n). \end{aligned} \tag{4.13}$$

(i) Let $n \in \mathbb{N}$ be such that x_n exists. It follows from (4.11) and (4.13) that $Z \subset H_n = H(x_n, p_n) \subset H(x_n, r_n)$. It is therefore enough to show that $Z \subset H(x_0, x_n)$. This inclusion certainly holds for $n = 0$ since $H(x_0, x_0) = \mathcal{H}$. Furthermore, it follows from (4.12) and (4.11) that

$$\begin{aligned} Z \subset H(x_0, x_n) \quad & \Rightarrow \quad Z \subset H(x_0, x_n) \cap H(x_n, r_n) \\ & \Rightarrow \quad Z \subset H(x_0, \mathsf{Q}(x_0, x_n, r_n)) \\ & \Leftrightarrow \quad Z \subset H(x_0, x_{n+1}), \end{aligned} \tag{4.14}$$

which establishes the assertion by induction. This also shows that $H(x_0, x_n) \cap H(x_n, r_n) \neq \varnothing$ and hence that x_{n+1} is well defined.

(ii)–(iii) Let $n \in \mathbb{N}$. By construction, $x_{n+1} = \mathsf{Q}(x_0, x_n, r_n) \in H(x_0, x_n) \cap H(x_n, r_n)$. Consequently, since x_n is the projection of x_0 onto $H(x_0, x_n)$ and $x_{n+1} \in H(x_0, x_n)$, we have $\|x_0 - x_n\| \leq \|x_0 - x_{n+1}\|$. On the other hand, since $\operatorname{proj}_Z x_0 \in Z \subset H(x_0, x_n)$, we have $\|x_0 - x_n\| \leq \|x_0 - \operatorname{proj}_Z x_0\|$. It follows that $(\|x_0 - x_k\|)_{k \in \mathbb{N}}$ converges to some $\ell \in [0, \|x_0 - \operatorname{proj}_Z x_0\|]$, which establishes (ii), and that

$$\lim \|x_0 - x_k\| \leq \|x_0 - \operatorname{proj}_Z x_0\|. \tag{4.15}$$

However, since $x_{n+1} \in H(x_0, x_n)$, we have

$$\begin{aligned} \|x_{n+1} - x_n\|^2 & \leq \|x_{n+1} - x_n\|^2 + 2\langle x_{n+1} - x_n \mid x_n - x_0 \rangle \\ & = \|x_0 - x_{n+1}\|^2 - \|x_0 - x_n\|^2. \end{aligned} \tag{4.16}$$

Hence,

$$\sum_{k=0}^{n} \|x_{k+1} - x_k\|^2 \leq \|x_0 - x_{n+1}\|^2 \leq \|x_0 - \operatorname{proj}_Z x_0\|^2 \tag{4.17}$$

and therefore

$$\sum_{k \in \mathbb{N}} \|x_{k+1} - x_k\|^2 < +\infty. \tag{4.18}$$

(iv) For every $n \in \mathbb{N}$, we derive from the inclusion $x_{n+1} \in H(x_n, r_n)$ that

$$\begin{aligned} \|r_n - x_n\|^2 & \leq \|x_{n+1} - r_n\|^2 + \|x_n - r_n\|^2 \\ & \leq \|x_{n+1} - r_n\|^2 + 2\langle x_{n+1} - r_n \mid r_n - x_n \rangle + \|x_n - r_n\|^2 \\ & = \|x_{n+1} - x_n\|^2. \end{aligned} \tag{4.19}$$

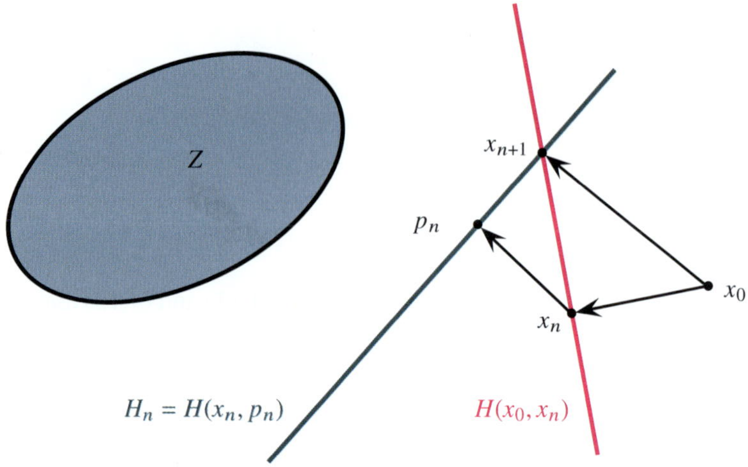

Figure 4.2. Iteration n of the Haugazeau-like algorithm (4.11) with $\lambda_n = 1$.

Hence, by (iii) and (4.11),

$$\sum_{n\in\mathbb{N}} \lambda_n^2 \|p_n - x_n\|^2 = \sum_{n\in\mathbb{N}} \|r_n - x_n\|^2 < +\infty. \tag{4.20}$$

(v) Let us note that (ii) implies that $(x_n)_{n\in\mathbb{N}}$ is bounded. Now let $x \in \mathfrak{W}(x_n)_{n\in\mathbb{N}}$, say $x_{k_n} \rightharpoonup x$. Then, by weak lower semicontinuity of $\|\cdot\|$ (Bauschke and Combettes 2017, Lemma 2.42) and (ii),

$$\|x_0 - x\| \leq \underline{\lim}\, \|x_0 - x_{k_n}\| \leq \|x_0 - \mathrm{proj}_Z\, x_0\| = \inf_{z\in Z} \|x_0 - z\|. \tag{4.21}$$

Hence, since $x \in Z$, $x = \mathrm{proj}_Z\, x_0$ is the only weak sequential cluster point of $(x_n)_{n\in\mathbb{N}}$ and it follows from Lemma 4.1(ii) that $x_n \rightharpoonup \mathrm{proj}_Z\, x_0$. In turn, (ii) yields

$$\|x_0 - \mathrm{proj}_Z\, x_0\| \leq \underline{\lim}\, \|x_0 - x_n\| = \lim \|x_0 - x_n\| \leq \|x_0 - \mathrm{proj}_Z\, x_0\|. \tag{4.22}$$

Thus, $x_0 - x_n \rightharpoonup x_0 - \mathrm{proj}_Z\, x_0$ and $\|x_0 - x_n\| \to \|x_0 - \mathrm{proj}_Z\, x_0\|$. We therefore derive from Lemma 4.1(iii) that $x_0 - x_n \to x_0 - \mathrm{proj}_Z\, x_0$, i.e. $x_n \to \mathrm{proj}_Z\, x_0$. □

4.4. Graph-based cuts

We consider the problem of finding a zero of a maximally monotone operator $M\colon \mathcal{H} \to 2^{\mathcal{H}}$ decomposed as $M = W + C$, where $W\colon \mathcal{H} \to 2^{\mathcal{H}}$ is maximally monotone and $C\colon \mathcal{H} \to \mathcal{H}$ is cocoercive, using the geometric principles of Theorems 4.2 and 4.7. To this end, we shall construct half-spaces by selecting points in the graph of W. Let us start with a weak convergence result.

Theorem 4.8. Let $\alpha \in \left]0, +\infty\right[$, let $W\colon \mathcal{H} \to 2^{\mathcal{H}}$ be maximally monotone, let $C\colon \mathcal{H} \to \mathcal{H}$ be α-cocoercive and such that $Z = \mathrm{zer}(W + C) \neq \varnothing$, let $x_0 \in \mathcal{H}$, and let $(\lambda_n)_{n\in\mathbb{N}}$ be a sequence in $]0,2[$. Iterate

$$
\begin{array}{l}
\text{for } n = 0, 1, \ldots \\
\left\lfloor
\begin{array}{l}
(w_n, w_n^*) \in \operatorname{gra} W, \;\; q_n \in \mathcal{H} \\
t_n^* = w_n^* + Cq_n \\
\delta_n = \langle x_n - w_n \mid t_n^* \rangle - \|w_n - q_n\|^2/(4\alpha) \\
d_n = \begin{cases} \dfrac{\delta_n}{\|t_n^*\|^2} t_n^*, & \text{if } \delta_n > 0, \\ 0, & \text{otherwise} \end{cases} \\
x_{n+1} = x_n - \lambda_n d_n.
\end{array}
\right.
\end{array}
\tag{4.23}
$$

Then the following hold:

(i) $(x_n)_{n\in\mathbb{N}}$ is bounded.

(ii) $\sum_{n\in\mathbb{N}} \lambda_n(2-\lambda_n)\|d_n\|^2 < +\infty$.

(iii) Suppose that $w_n - x_n \rightharpoonup 0$, $w_n - q_n \to 0$ and $t_n^* \to 0$. Then $(x_n)_{n\in\mathbb{N}}$ converges weakly to a point in Z.

Proof. We first observe that (4.23) is well defined since $(\forall n \in \mathbb{N})\ \delta_n > 0 \Rightarrow t_n^* \neq 0$. It follows from Example 2.5 and Lemma 2.27(ii) that

$$
W + C \text{ is maximally monotone,} \tag{4.24}
$$

and hence from (2.29) that Z is a nonempty closed convex subset of $\mathcal{H}$. Set

$$
(\forall n \in \mathbb{N}) \quad H_n = \left\{ z \in \mathcal{H} \;\middle|\; \langle z - w_n \mid t_n^* \rangle \leq \frac{\|w_n - q_n\|^2}{4\alpha} \right\} \tag{4.25}
$$

and let $z \in Z$. For every $n \in \mathbb{N}$, since $(z, -Cz) \in \operatorname{gra} W$ and $(w_n, w_n^*) \in \operatorname{gra} W$, it results from the monotonicity of W that $\langle w_n - z \mid w_n^* + Cz \rangle \geq 0$. Hence, since C is α-cocoercive,

$$
\begin{aligned}
(\forall n \in \mathbb{N}) \quad & \langle z - w_n \mid t_n^* \rangle \\
&= \langle z - w_n \mid w_n^* + Cq_n \rangle \\
&\leq \langle z - w_n \mid Cq_n - Cz \rangle && (4.26) \\
&= \langle q_n - w_n \mid Cq_n - Cz \rangle + \langle z - q_n \mid Cq_n - Cz \rangle \\
&\leq \langle q_n - w_n \mid Cq_n - Cz \rangle - \alpha\|Cq_n - Cz\|^2 && (4.27) \\
&= 2\left\langle \frac{q_n - w_n}{\sqrt{4\alpha}} \;\middle|\; \sqrt{\alpha}(Cq_n - Cz) \right\rangle - \left\|\sqrt{\alpha}(Cq_n - Cz)\right\|^2 \\
&= \frac{\|w_n - q_n\|^2}{4\alpha} - \left\| \sqrt{\alpha}(Cq_n - Cz) + \frac{w_n - q_n}{\sqrt{4\alpha}} \right\|^2 \\
&\leq \frac{\|w_n - q_n\|^2}{4\alpha}. && (4.28)
\end{aligned}
$$

This shows that $(\forall n \in \mathbb{N})\ Z \subset H_n$. In addition, it results from (4.23) and Example 2.1 that

$$(\forall n \in \mathbb{N}) \quad x_{n+1} = x_n + \lambda_n(\operatorname{proj}_{H_n} x_n - x_n), \tag{4.29}$$

which corresponds to the setting of Theorem 4.2.

(i) This follows from Theorem 4.2(i).

(ii) This follows from Theorem 4.2(ii).

(iii) Let $x \in \mathfrak{W}(x_n)_{n\in\mathbb{N}}$, say $x_{k_n} \rightharpoonup x$. Then $w_{k_n} = x_{k_n} + (w_{k_n} - x_{k_n}) \rightharpoonup x$. On the other hand, since C is $1/\alpha$-Lipschitzian,

$$\|w_n^* + Cw_n\| = \|t_n^* + Cw_n - Cq_n\| \leq \|t_n^*\| + \frac{\|w_n - q_n\|}{\alpha} \to 0. \tag{4.30}$$

In addition, since $(w_n, w_n^*)_{n\in\mathbb{N}}$ is in $\operatorname{gra} W$, $(w_n, w_n^* + Cw_n)_{n\in\mathbb{N}}$ is in $\operatorname{gra}(W + C)$. It then follows from (4.24) and Lemma 2.49 that $x \in Z$. We conclude by invoking Theorem 4.2(iv). □

We now turn to strong convergence.

Theorem 4.9. Let $\alpha \in \left]0, +\infty\right[$, let $W\colon \mathcal{H} \to 2^{\mathcal{H}}$ be maximally monotone, let $C\colon \mathcal{H} \to \mathcal{H}$ be α-cocoercive and such that $Z = \operatorname{zer}(W + C) \neq \varnothing$, let $x_0 \in \mathcal{H}$, and let $(\lambda_n)_{n\in\mathbb{N}}$ be a sequence in $\left]0, 1\right]$. Iterate

$$\begin{array}{l}
\text{for } n = 0, 1, \ldots \\
\left\lfloor
\begin{array}{l}
(w_n, w_n^*) \in \operatorname{gra} W, \ q_n \in \mathcal{H} \\
t_n^* = w_n^* + Cq_n \\
\delta_n = \langle x_n - w_n \mid t_n^* \rangle - \|w_n - q_n\|^2/(4\alpha) \\
d_n = \begin{cases} \dfrac{\delta_n}{\|t_n^*\|^2} t_n^*, & \text{if } \delta_n > 0, \\ 0, & \text{otherwise} \end{cases} \\
r_n = x_n - \lambda_n d_n \\
x_{n+1} = \mathsf{Q}(x_0, x_n, r_n),
\end{array}
\right.
\end{array} \tag{4.31}$$

where Q is defined in Lemma 4.6. Then the following hold:

(i) $(x_n)_{n\in\mathbb{N}}$ is bounded.

(ii) $\sum_{n\in\mathbb{N}} \|x_{n+1} - x_n\|^2 < +\infty$.

(iii) $\sum_{n\in\mathbb{N}} \lambda_n^2 \|d_n\|^2 < +\infty$.

(iv) Suppose that $w_n - x_n \rightharpoonup 0$, $w_n - q_n \to 0$ and $t_n^* \to 0$. Then $(x_n)_{n\in\mathbb{N}}$ converges strongly to $\operatorname{proj}_Z x_0$.

Proof. Define $(H_n)_{n\in\mathbb{N}}$ as in (4.25) and note that (4.28) yields $Z \subset \bigcap_{n\in\mathbb{N}} H_n$. Furthermore, we derive from (4.31) and Example 2.1 that $(\forall n \in \mathbb{N})\ r_n = x_n + \lambda_n(\operatorname{proj}_{H_n} x_n - x_n)$. This places us in the setting of Theorem 4.7.

(i) This follows from Theorem 4.7(ii).

(ii) See Theorem 4.7(iii).

(iii) This follows from Theorem 4.7(iv).

(iv) As in the proof of Theorem 4.8(iii), $\mathfrak{W}(x_n)_{n\in\mathbb{N}} \subset Z$. The claim follows from Theorem 4.7(v). □

In the absence of the cocoercive operator C, we can choose $(q_n)_{n\in\mathbb{N}} = (w_n)_{n\in\mathbb{N}}$ in (4.23) and (4.31), and Theorems 4.8 and 4.9 simplify as follows.

Proposition 4.10. Let $M\colon \mathcal{H} \to 2^{\mathcal{H}}$ be a maximally monotone operator such that $Z = \operatorname{zer} M \neq \varnothing$, let $x_0 \in \mathcal{H}$, and let $(\lambda_n)_{n\in\mathbb{N}}$ be a sequence in $]0,2[$. Iterate

$$
\begin{array}{l}
\text{for } n = 0, 1, \ldots \\
\left\lfloor
\begin{array}{l}
(m_n, m_n^*) \in \operatorname{gra} M \\
d_n = \begin{cases} \dfrac{\langle x_n - m_n \mid m_n^*\rangle}{\|m_n^*\|^2} m_n^*, & \text{if } \langle x_n - m_n \mid m_n^*\rangle > 0, \\ 0, & \text{otherwise} \end{cases} \\
x_{n+1} = x_n - \lambda_n d_n.
\end{array}
\right.
\end{array}
\tag{4.32}
$$

Then the following hold:

(i) $\sum_{n\in\mathbb{N}} \lambda_n(2-\lambda_n)\|d_n\|^2 < +\infty$.

(ii) Suppose that $m_n - x_n \rightharpoonup 0$ and $m_n^* \to 0$. Then $(x_n)_{n\in\mathbb{N}}$ converges weakly to a point in Z.

Proposition 4.11. Let $M\colon \mathcal{H} \to 2^{\mathcal{H}}$ be a maximally monotone operator such that $Z = \operatorname{zer} M \neq \varnothing$, let $x_0 \in \mathcal{H}$, and let $(\lambda_n)_{n\in\mathbb{N}}$ be a sequence in $]0,1]$. Iterate

$$
\begin{array}{l}
\text{for } n = 0, 1, \ldots \\
\left\lfloor
\begin{array}{l}
(m_n, m_n^*) \in \operatorname{gra} M \\
d_n = \begin{cases} \dfrac{\langle x_n - m_n \mid m_n^*\rangle}{\|m_n^*\|^2} m_n^*, & \text{if } \langle x_n - m_n \mid m_n^*\rangle > 0, \\ 0, & \text{otherwise} \end{cases} \\
r_n = x_n - \lambda_n d_n \\
x_{n+1} = \mathsf{Q}(x_0, x_n, r_n),
\end{array}
\right.
\end{array}
\tag{4.33}
$$

where Q is defined in Lemma 4.6. Then the following hold:

(i) $\sum_{n\in\mathbb{N}} \lambda_n^2\|d_n\|^2 < +\infty$.

(ii) Suppose that $m_n - x_n \rightharpoonup 0$ and $m_n^* \to 0$. Then $(x_n)_{n\in\mathbb{N}}$ converges strongly to $\operatorname{proj}_Z x_0$.

4.5. Warped resolvent cuts

Algorithms (4.23) and (4.31) are conceptual in the sense that they do not provide an explicit mechanism to find points in the graph of W. In this section, we propose implementable versions that pick points in $\operatorname{gra} W$ using the warped resolvents of Lemma 2.42.

Theorem 4.12. Let $\alpha \in\,]0,+\infty[$, let $W\colon \mathcal{H} \to 2^{\mathcal{H}}$ be maximally monotone, let $C\colon \mathcal{H} \to \mathcal{H}$ be α-cocoercive and such that $Z = \operatorname{zer}(W+C) \neq \varnothing$, let $x_0 \in \mathcal{H}$, and let $(\lambda_n)_{n\in\mathbb{N}}$ be a sequence in $]0,2[$. Further, for every $n \in \mathbb{N}$, let $U_n\colon \mathcal{H} \to \mathcal{H}$ be an operator such that $\operatorname{ran} U_n \subset \operatorname{ran}(U_n + W + C)$ and $U_n + W + C$ is injective. Iterate

$$
\begin{array}{l}
\text{for } n = 0, 1, \ldots \\
\left\lfloor
\begin{array}{l}
w_n = J^{U_n}_{W+C} x_n \\
w_n^* = U_n x_n - U_n w_n - C w_n \\
q_n \in \mathcal{H} \\
t_n^* = w_n^* + C q_n \\
\delta_n = \langle x_n - w_n \mid t_n^* \rangle - \|w_n - q_n\|^2/(4\alpha) \\
d_n = \begin{cases} \dfrac{\delta_n}{\|t_n^*\|^2} t_n^*, & \text{if } \delta_n > 0, \\ 0, & \text{otherwise} \end{cases} \\
x_{n+1} = x_n - \lambda_n d_n.
\end{array}
\right.
\end{array}
\tag{4.34}
$$

Then the following hold:

(i) $\sum_{n\in\mathbb{N}} \lambda_n(2-\lambda_n)\|d_n\|^2 < +\infty$.

(ii) Suppose that one of the following is satisfied:

(a) $\sum_{n\in\mathbb{N}} \lambda_n(2-\lambda_n) = +\infty$ and $(\|d_n\|)_{n\in\mathbb{N}}$ converges,

(b) $\inf_{n\in\mathbb{N}} \lambda_n > 0$ and $\sup \lambda_n < 2$,

together with one of the following:

(c) $w_n - x_n \rightharpoonup 0$, $U_n w_n - U_n x_n \to 0$ and $w_n - q_n \to 0$,

(d) $q_n - x_n \to 0$ and there exist $\beta_1 \in\,]1/(4\alpha),+\infty[$ and $\beta_2 \in\,]0,+\infty[$ such that the kernels $(U_n)_{n\in\mathbb{N}}$ are β_1-strongly monotone and β_2-Lipschitzian.

Then $(x_n)_{n\in\mathbb{N}}$ converges weakly to a point in Z.

Proof. Lemma 2.42(i) indicates that (4.34) is governed by the scenario of Theorem 4.8.

(i) See Theorem 4.8(ii).

(ii) A consequence of (i) under (iia) or (iib) is that

$$\|d_n\| \to 0. \tag{4.35}$$

Indeed, the claim is clear under (iib), whereas under (iia) we have $\underline{\lim}\, \|d_n\| = 0$ and therefore $\lim \|d_n\| = 0$. Next, let us assume that (iic) holds. Then it follows from (4.34) and (2.32) that

$$
\begin{aligned}
(\forall n \in \mathbb{N}) \quad \|t_n^*\| &= \|U_n w_n - U_n x_n + C w_n - C q_n\| \\
&\leq \|U_n w_n - U_n x_n\| + \|C w_n - C q_n\|
\end{aligned}
\tag{4.36}
$$

$$\begin{aligned}&\leq \|U_n w_n - U_n x_n\| + \frac{\|w_n - q_n\|}{\alpha}\\ &\to 0.\end{aligned} \tag{4.37}$$

In view of Theorem 4.8(iii), the claim is established. It remains to show that (iid)⇒(iic). Because the operators $(U_n + W + C)_{n\in\mathbb{N}}$ are β_1-strongly monotone, the operators $(U_n + W + C)^{-1}_{n\in\mathbb{N}}$ are β_1-cocoercive, hence $1/\beta_1$-Lipschitzian. Consequently, since the operators $(U_n)_{n\in\mathbb{N}}$ are β_2-Lipschitzian, the operators $\big(J^{U_n}_{W+C}\big)_{n\in\mathbb{N}}$ are β_2/β_1-Lipschitzian. Now let $z \in Z$. Then we derive from (4.34) and Lemma 2.42(ii) that

$$(\forall n \in \mathbb{N})\ \|w_n - z\| = \big\|J^{U_n}_{W+C} x_n - J^{U_n}_{W+C} z\big\| \leq \frac{\beta_2}{\beta_1}\|x_n - z\|. \tag{4.38}$$

Appealing to Theorem 4.8(i), we infer that $(w_n)_{n\in\mathbb{N}}$ is bounded. Thus, since $q_n - x_n \to 0$ and C is $1/\alpha$-Lipschitzian, the sequences

$$(\|w_n - x_n\|)_{n\in\mathbb{N}},\ (\|w_n - q_n\|)_{n\in\mathbb{N}} \text{ and } (\|Cw_n - Cq_n\|)_{n\in\mathbb{N}} \text{ are bounded.} \tag{4.39}$$

However, (4.36) entails that

$$(\forall n \in \mathbb{N})\quad \|t_n^*\| \leq \beta_2\|w_n - x_n\| + \frac{\|w_n - q_n\|}{\alpha}, \tag{4.40}$$

which verifies that $(\|t_n^*\|)_{n\in\mathbb{N}}$ is bounded. In turn, (4.34) and (4.35) imply that

$$\overline{\lim}\, \delta_n \leq \lim \|t_n^*\|\, \|d_n\| = 0. \tag{4.41}$$

Moreover, for every $n \in \mathbb{N}$, (4.34) yields

$$\begin{aligned}\delta_n &= \langle w_n - x_n \mid U_n w_n - U_n x_n\rangle + \langle w_n - x_n \mid Cw_n - Cq_n\rangle - \frac{\|w_n - q_n\|^2}{4\alpha}\\ &\geq \beta_1\|w_n - x_n\|^2 + \langle w_n - q_n \mid Cw_n - Cq_n\rangle + \langle q_n - x_n \mid Cw_n - Cq_n\rangle\\ &\quad - \frac{\|w_n - q_n\|^2}{4\alpha}\\ &\geq \beta_1(\|w_n - q_n\|^2 + 2\langle w_n - q_n \mid q_n - x_n\rangle + \|q_n - x_n\|^2)\\ &\quad + \alpha\|Cw_n - Cq_n\|^2 + \langle q_n - x_n \mid Cw_n - Cq_n\rangle - \frac{\|w_n - q_n\|^2}{4\alpha}\\ &\geq \left(\beta_1 - \frac{1}{4\alpha}\right)\|w_n - q_n\|^2 + \beta_1(2\langle w_n - q_n \mid q_n - x_n\rangle + \|q_n - x_n\|^2)\\ &\quad + \langle q_n - x_n \mid Cw_n - Cq_n\rangle\\ &\geq \left(\beta_1 - \frac{1}{4\alpha}\right)\|w_n - q_n\|^2\\ &\quad + \|q_n - x_n\|(\beta_1\|q_n - x_n\| - 2\beta_1\|w_n - q_n\| + \|Cw_n - Cq_n\|).\end{aligned} \tag{4.42}$$

Therefore, since $\|q_n - x_n\| \to 0$, it follows from (4.39) and (4.41) that $w_n - q_n \to 0$

and hence that $w_n - x_n \to 0$. Since

$$\|U_n w_n - U_n x_n\| \leq \beta_2 \|w_n - x_n\| \leq \beta_2(\|w_n - q_n\| + \|q_n - x_n\|) \to 0, \tag{4.43}$$

the proof is complete. □

Remark 4.13. In the special case when $C = 0$, $(q_n)_{n\in\mathbb{N}} = (w_n)_{n\in\mathbb{N}}$, and conditions (iib) and (iic) are satisfied, Theorem 4.12(ii) is closely related to Theorem 4.2(ii) of Bùi and Combettes (2020*b*).

We conclude this section with the strongly convergent best approximation companion algorithm resulting from Theorem 4.9.

Theorem 4.14. Let $\alpha \in \left]0,+\infty\right[$, let $W\colon \mathcal{H} \to 2^{\mathcal{H}}$ be maximally monotone, let $C\colon \mathcal{H} \to \mathcal{H}$ be α-cocoercive and such that $Z = \operatorname{zer}(W+C) \neq \varnothing$, let $x_0 \in \mathcal{H}$, and let $(\lambda_n)_{n\in\mathbb{N}}$ be a sequence in $\left]0,1\right]$. Further, for every $n \in \mathbb{N}$, let $U_n\colon \mathcal{H} \to \mathcal{H}$ be an operator such that $\operatorname{ran} U_n \subset \operatorname{ran}(U_n + W + C)$ and $U_n + W + C$ is injective. Iterate

$$\begin{array}{l}
\text{for } n = 0,1,\ldots\\
\left\lfloor
\begin{array}{l}
w_n = J^{U_n}_{W+C} x_n\\
w_n^* = U_n x_n - U_n w_n - C w_n\\
q_n \in \mathcal{H}\\
t_n^* = w_n^* + C q_n\\
\delta_n = \langle x_n - w_n \mid t_n^*\rangle - \|w_n - q_n\|^2/(4\alpha)\\
d_n = \begin{cases} \dfrac{\delta_n}{\|t_n^*\|^2} t_n^*, & \text{if } \delta_n > 0,\\ 0, & \text{otherwise}\end{cases}\\
r_n = x_n - \lambda_n d_n\\
x_{n+1} = \mathsf{Q}(x_0, x_n, r_n),
\end{array}
\right.
\end{array} \tag{4.44}$$

where Q is defined in Lemma 4.6. Then the following hold:

(i) $\sum_{n\in\mathbb{N}} \lambda_n^2 \|d_n\|^2 < +\infty$.

(ii) Suppose that one of the following is satisfied:

(a) $\sum_{n\in\mathbb{N}} \lambda_n^2 = +\infty$ and $(\|d_n\|)_{n\in\mathbb{N}}$ converges,

(b) $\inf_{n\in\mathbb{N}} \lambda_n > 0$,

together with one of the following:

(c) $w_n - x_n \rightharpoonup 0$, $U_n w_n - U_n x_n \to 0$ and $w_n - q_n \to 0$,

(d) $q_n - x_n \to 0$ and there exist $\beta_1 \in \left]1/(4\alpha),+\infty\right[$ and $\beta_2 \in \left]0,+\infty\right[$ such that the kernels $(U_n)_{n\in\mathbb{N}}$ are β_1-strongly monotone and β_2-Lipschitzian.

Then $(x_n)_{n\in\mathbb{N}}$ converges strongly to $\operatorname{proj}_Z x_0$.

Proof. In view of Lemma 2.42(i), (4.44) is an instance of (4.31) and we shall therefore employ Theorem 4.9.

(i) See Theorem 4.9(iii).

(ii) It follows from (i) and (4.44) that $d_n \rightharpoonup 0$. Indeed, this is evident under (iib), whereas under (iia) we have $\underline{\lim}\,\|d_n\| = 0$ and therefore $\lim \|d_n\| = 0$. Let us now assume that (iic) holds. Then (4.37) is satisfied and we obtain the assertion by invoking Theorem 4.9(iv). Finally, to show that (iid)⇒(iic), we remark that Theorem 4.9(i) asserts that $(x_n)_{n\in\mathbb{N}}$ is bounded. Hence, we follow the same pattern as in the proof of Theorem 4.12(iid) to conclude. □

5. The proximal point algorithm

5.1. Preview

The proximal point algorithm is an implicit method to construct a zero of a maximally monotone operator which goes back to a quadratic programming method proposed in Bellman, Kalaba and Lockett (1966, Section 5.8). In the nonlinear case, it first appeared in the work of Lieutaud (1969*a*) (this fact seems to have been overlooked in the literature: see Remark 6.1), then in Martinet (1970, 1972) for subdifferentials and Rockafellar (1976*b*) for the general case. Iteration n of the unrelaxed form of the algorithm can be interpreted as a backward Euler discretization of the Cauchy problem (Aubin and Cellina 1984, Section 3.2) (see Example 2.18)

$$\begin{cases} x(0) = x_0, \\ -x'(t) \in Mx(t), \quad \text{for a.e. } t \in \left]0, +\infty\right[, \end{cases} \tag{5.1}$$

with time step $\gamma_n \in \left]0, +\infty\right[$, that is,

$$\frac{x_n - x_{n+1}}{\gamma_n} \in Mx_{n+1} \tag{5.2}$$

or, equivalently, $x_{n+1} = J_{\gamma_n M} x_n$.

5.2. Fejérian algorithm

The following theorem, which brings together results from Brézis and Lions (1978), Eckstein and Bertsekas (1992), Gabay (1983), Gol'shtein and Tret'yakov (1979), Lemaire (1989), Martinet (1970, 1972) and Rockafellar (1976*b*), will be derived from Theorem 4.12.

Theorem 5.1. Let $M\colon \mathcal{H} \to 2^{\mathcal{H}}$ be a maximally monotone operator such that $Z = \operatorname{zer} M \neq \varnothing$, let $x_0 \in \mathcal{H}$, let $(\lambda_n)_{n\in\mathbb{N}}$ be a sequence in $\left]0, 2\right[$, and let $(\gamma_n)_{n\in\mathbb{N}}$ be a sequence in $\left]0, +\infty\right[$. Iterate

$$(\forall n \in \mathbb{N}) \quad x_{n+1} = x_n + \lambda_n (J_{\gamma_n M} x_n - x_n), \tag{5.3}$$

and suppose that one of the following holds:

(i) $\sum_{n\in\mathbb{N}} \lambda_n(2 - \lambda_n) = +\infty$ and $(\forall n \in \mathbb{N})\ \gamma_n = 1$.

(ii) $\sum_{n\in\mathbb{N}} \gamma_n^2 = +\infty$ and $(\forall n \in \mathbb{N})\ \lambda_n = 1$.

(iii) $\inf_{n\in\mathbb{N}} \lambda_n > 0$, $\sup_{n\in\mathbb{N}} \lambda_n < 2$ and $\inf_{n\in\mathbb{N}} \gamma_n > 0$.

Then $\|J_{\gamma_n M} x_n - x_n\|/\gamma_n \to 0$ and $(x_n)_{n\in\mathbb{N}}$ converges weakly to a point in Z.

Proof. Let us apply Theorem 4.12 with

$$C = 0 \quad \text{and} \quad (\forall n \in \mathbb{N}) \quad U_n = \gamma_n^{-1}\mathrm{Id} \quad \text{and} \quad q_n = w_n. \tag{5.4}$$

We derive from (2.19) that the variables of the iterations (4.34) satisfy

$$(\forall n \in \mathbb{N}) \quad t_n^* = \frac{x_n - w_n}{\gamma_n}, \qquad \delta_n = \gamma_n \|t_n^*\|^2 \quad \text{and} \quad d_n = x_n - w_n. \tag{5.5}$$

Thus, the sequence $(x_n)_{n\in\mathbb{N}}$ produced by (5.3) coincides with that of (4.34). In turn, Theorem 4.12(i) yields

$$\sum_{n\in\mathbb{N}} \lambda_n(2 - \lambda_n)\|d_n\|^2 < +\infty. \tag{5.6}$$

We now show that one of conditions (iia)–(iib) and one of conditions (iic)–(iid) of Theorem 4.12(ii) are fulfilled in each scenario. We also recall from (4.35) that (iia) and (iib) in Theorem 4.12 each imply that

$$d_n \to 0. \tag{5.7}$$

(i) Let us check that conditions (iia) and (iid) are fulfilled. For (iia), it is enough to show that $(\|d_n\|)_{n\in\mathbb{N}}$ decreases. To this end, set $T = 2J_M - \mathrm{Id}$. Then Lemma 2.34(iii) and (2.33) assert that T is nonexpansive. Therefore, (5.5) yields

$$\begin{aligned}(\forall n \in \mathbb{N}) \quad 2\|d_{n+1}\| &= \|Tx_{n+1} - x_{n+1}\| \\ &= \|Tx_{n+1} - Tx_n + (1 - \lambda_n/2)(Tx_n - x_n)\| \\ &\leq \|x_{n+1} - x_n\| + (1 - \lambda_n/2)\|Tx_n - x_n\| \\ &= (\lambda_n/2)\|Tx_n - x_n\| + (1 - \lambda_n/2)\|Tx_n - x_n\| \\ &= 2\|d_n\|, \end{aligned} \tag{5.8}$$

as desired. For (iid), note that (5.7) and (5.5) imply that $q_n - x_n = w_n - x_n = -d_n \to 0$. In addition, it is clear from (5.4) that $(U_n)_{n\in\mathbb{N}}$ satisfies the required conditions with $\beta_1 = \beta_2 = 1$.

(ii) Condition (iib) holds. To show that (iic) holds as well, we first infer from (5.5) and (5.6) that $\sum_{n\in\mathbb{N}} \gamma_n^2 \|t_n^*\|^2 < +\infty$ and hence that $w_n - x_n = -\gamma_n t_n^* \to 0$. Furthermore, since $\sum_{n\in\mathbb{N}} \gamma_n^2 = +\infty$, $\underline{\lim}\, \|t_n^*\| = 0$. On the other hand, $(\forall n \in \mathbb{N})$ $t_n^* = \gamma_n^{-1}(x_n - w_n) = \gamma_n^{-1}(x_n - x_{n+1})$. Hence, using (2.18), the monotonicity of M and the Cauchy–Schwarz inequality, we obtain

$$\begin{aligned}(\forall n \in \mathbb{N}) \quad 0 &\leq \langle w_n - w_{n+1} \mid t_n^* - t_{n+1}^*\rangle/\gamma_{n+1} \\ &= \langle x_{n+1} - x_{n+2} \mid t_n^* - t_{n+1}^*\rangle/\gamma_{n+1} \\ &= \langle t_{n+1}^* \mid t_n^* - t_{n+1}^*\rangle \end{aligned}$$

$$
\begin{aligned}
&= \langle t_{n+1}^* \mid t_n^* \rangle - \|t_{n+1}^*\|^2 \\
&\leq \|t_{n+1}^*\|(\|t_n^*\| - \|t_{n+1}^*\|),
\end{aligned}
\tag{5.9}
$$

which shows that $(\|t_n^*\|)_{n\in\mathbb{N}}$ decreases. Altogether, $U_n x_n - U_n w_n = t_n^* \to 0$.

(iii) Condition (iib) is assumed. Let us check (iic). Since (5.5) and (5.6) yield $\sum_{n\in\mathbb{N}} \gamma_n^2 \|t_n^*\|^2 < +\infty$, we have $x_n - w_n = \gamma_n t_n^* \to 0$. Finally, since $\inf_{n\in\mathbb{N}} \gamma_n > 0$, $U_n x_n - U_n w_n = t_n^* \to 0$.

We conclude the proof by noting that in all three cases above $\|J_{\gamma_n M} x_n - x_n\|/\gamma_n = \|t_n^*\| \to 0$. ☐

Remark 5.2. Let $f \in \Gamma_0(\mathcal{H})$ and suppose that $M = \partial f$ in Theorem 5.1. Then, as seen in Example 2.12, M is maximally monotone and $Z = \operatorname{Argmin} f$. In this case, the condition on $(\gamma_n)_{n\in\mathbb{N}}$ in Theorem 5.1(ii) can be improved to $\sum_{n\in\mathbb{N}} \gamma_n = +\infty$ (Brézis and Lions 1978, Théorème 9).

5.3. Haugazeau-like algorithm

We employ Theorem 4.14 to obtain a strongly convergent variant of the proximal point algorithm; see Bauschke and Combettes (2001) and Solodov and Svaiter (2000) for related results. Examples of proximal point iterations that fail to converge strongly are constructed in Bauschke, Matoušková and Reich (2004), Combettes (2018) and Güler (1991).

Theorem 5.3. Let $M\colon \mathcal{H} \to 2^{\mathcal{H}}$ be a maximally monotone operator such that $Z = \operatorname{zer} M \neq \varnothing$, let $x_0 \in \mathcal{H}$, let $(\lambda_n)_{n\in\mathbb{N}}$ be a sequence in $]0,1]$ such that $\inf_{n\in\mathbb{N}} \lambda_n > 0$, and let $(\gamma_n)_{n\in\mathbb{N}}$ be a sequence in $]0,+\infty[$ such that $\inf_{n\in\mathbb{N}} \gamma_n > 0$. Iterate

$$
(\forall n \in \mathbb{N}) \quad x_{n+1} = \mathsf{Q}(x_0, x_n, x_n + \lambda_n(J_{\gamma_n M} x_n - x_n)),
\tag{5.10}
$$

where Q is defined in Lemma 4.6. Then $(x_n)_{n\in\mathbb{N}}$ converges strongly to $\operatorname{proj}_Z x_0$.

Proof. In Theorem 4.14, set $C = 0$ and $(\forall n \in \mathbb{N})$ $U_n = \gamma_n^{-1}\mathrm{Id}$ and $q_n = w_n$. Then (5.5) holds and the sequence $(x_n)_{n\in\mathbb{N}}$ produced by (5.10) coincides with that of (4.44). In turn, Theorem 4.14(i) yields $\sum_{n\in\mathbb{N}} \lambda_n^2 \|d_n\|^2 < +\infty$. Therefore, $x_n - w_n = d_n \to 0$ and $U_n x_n - U_n w_n = \gamma_n^{-1} d_n \to 0$. This confirms that condition (iic) in Theorem 4.14(ii) is fulfilled. Since condition (iib) holds by assumption, the proof is complete. ☐

5.4. Special cases and variants

As mentioned in Section 1, direct implementations of the proximal point algorithm are limited due to the potential difficulty of evaluating the resolvents in (5.3) and (5.10). As we shall see in this section, the proximal point framework can nonetheless be an effective device to establish indirectly the convergence of algorithms that can be identified, possibly in a different space, as an instance of (5.3). Early examples in the context of inequality-constrained minimization problems are found in Rockafellar (1976*a*), where a dual application of an approximate proximal point algorithm was shown to yield a method of multipliers (also called the augmented Lagrangian method) that extends some classical ones from Hestenes (1969) and Powell (1969); see also Rockafellar (1973). A primal–dual quadratically perturbed variant of this algorithm, known as the proximal method of multipliers, was also introduced in Rockafellar (1976*a*) as an application of an approximate proximal point algorithm to find saddle points of the Lagrangian; see also Rockafellar (2024), Shefi and Teboulle (2014) and their bibliographies for recent work along these lines. The applications described below reduce to implementations of the proximal point algorithm that feature full operator splitting when several linear and nonlinear operators are present in the original problem.

5.4.1. The Euler method

We derive from the proximal point algorithm a (forward) Euler method to find a zero of a cocoercive operator.

Proposition 5.4. Let $\alpha \in]0,+\infty[$ and let $B\colon \mathcal{H} \to \mathcal{H}$ be α-cocoercive, with $\operatorname{zer} B \neq \varnothing$. Let $(\gamma_n)_{n\in\mathbb{N}}$ be a sequence in $]0, 2\alpha[$ such that $\sum_{n\in\mathbb{N}} \gamma_n(2\alpha - \gamma_n) = +\infty$ and let $x_0 \in \mathcal{H}$. Iterate

$$(\forall n \in \mathbb{N}) \quad x_{n+1} = x_n - \gamma_n B x_n. \tag{5.11}$$

Then $(x_n)_{n\in\mathbb{N}}$ converges weakly to a point in $\operatorname{zer} B$.

Proof. Set $M = (\mathrm{Id} - \alpha B)^{-1} - \mathrm{Id}$. Since αB is firmly nonexpansive with domain $\mathcal{H}$, $\mathrm{Id} - \alpha B$ is likewise and Lemma 2.34(iii) asserts that M is maximally monotone. On the other hand, $\operatorname{zer} M = \operatorname{zer} B$, $J_M = \mathrm{Id} - \alpha B$, and hence (5.11) becomes

$$(\forall n \in \mathbb{N}) \quad x_{n+1} = x_n + \lambda_n(J_M x_n - x_n), \quad \text{where } \lambda_n = \gamma_n/\alpha \in]0, 2[. \tag{5.12}$$

Thus, since $\sum_{n\in\mathbb{N}} \lambda_n(2 - \lambda_n) = +\infty$, the claim follows from Theorem 5.1(i). □

Remark 5.5. As just shown, the Euler method (5.11) is an instance of the proximal point algorithm (5.3). Conversely, we can interpret the proximal point iterations in the format

$$(\forall n \in \mathbb{N}) \quad x_{n+1} = x_n + \lambda_n(J_M x_n - x_n), \quad \text{where } \lambda_n \in]0, 2[, \tag{5.13}$$

as an instance of (5.11). Indeed, let $M\colon \mathcal{H} \to 2^{\mathcal{H}}$ be maximally monotone and set $B = {}^{1}M$ and $(\forall n \in \mathbb{N})\ \gamma_n = \lambda_n$. Then, as seen in Example 2.7, $\operatorname{zer} M = \operatorname{zer} B$ and B is 1-cocoercive, while (2.21) implies that (5.13) reduces to (5.11).

The following example is about the gradient method; see Cauchy (1847) and Curry (1944) for the premises of this algorithm.

Example 5.6. Let $\alpha \in]0,+\infty[$ and let $g\colon \mathcal{H} \to \mathbb{R}$ be convex, differentiable, and such that ∇g is $1/\alpha$-Lipschitzian, with $\operatorname{Argmin} g \neq \varnothing$. Let $(\gamma_n)_{n\in\mathbb{N}}$ be a sequence in $]0, 2\alpha[$ such that $\sum_{n\in\mathbb{N}} \gamma_n(2\alpha - \gamma_n) = +\infty$ and let $x_0 \in \mathcal{H}$. Iterate

$$(\forall n \in \mathbb{N}) \quad x_{n+1} = x_n - \gamma_n \nabla g(x_n). \tag{5.14}$$

Then $(x_n)_{n\in\mathbb{N}}$ converges weakly to a point in Argmin g.

Proof. Combine Lemma 2.2 and Proposition 5.4. □

As noted in Bauschke, Combettes and Reich (2005, Remark 4.8(ii)) in the context of Example 5.6, the convergence in Proposition 5.4 can fail to be strong. The next result, which guarantees strong convergence, is obtained by defining M and $(\lambda_n)_{n\in\mathbb{N}}$ as in the proof of Proposition 5.4 and using Theorem 5.3.

Proposition 5.7. Let $\alpha \in \,]0,+\infty[$ and let $B\colon \mathcal{H} \to \mathcal{H}$ be α-cocoercive, with $\operatorname{zer} B \neq \varnothing$. Let $(\gamma_n)_{n\in\mathbb{N}}$ be a sequence in $]0,\alpha]$ such that $\inf_{n\in\mathbb{N}} \gamma_n > 0$ and let $x_0 \in \mathcal{H}$. Iterate

$$(\forall n \in \mathbb{N}) \quad x_{n+1} = \mathsf{Q}(x_0, x_n, x_n - \gamma_n B x_n), \tag{5.15}$$

where Q is defined in Lemma 4.6. Then $(x_n)_{n\in\mathbb{N}}$ converges strongly to $\operatorname{proj}_{\operatorname{zer} B} x_0$.

5.4.2. *Fixed point problem*

We address the basic problem of constructing a fixed point of a nonexpansive operator $T\colon \mathcal{H} \to \mathcal{H}$. The following result is derived as an instance of the proximal point algorithm of Theorem 5.1 via the embedding of Example 3.16.

Proposition 5.8. Let $\alpha \in \,]0,1]$ and let $T\colon \mathcal{H} \to \mathcal{H}$ be *α-averaged.* Suppose that $\operatorname{Fix} T \neq \varnothing$, let $(\lambda_n)_{n\in\mathbb{N}}$ be a sequence in $]0,1/\alpha[$ such that $\sum_{n\in\mathbb{N}} \lambda_n(1-\alpha\lambda_n) = +\infty$, and let $x_0 \in \mathcal{H}$. Iterate

$$(\forall n \in \mathbb{N}) \quad x_{n+1} = x_n + \lambda_n(Tx_n - x_n). \tag{5.16}$$

Then $(x_n)_{n\in\mathbb{N}}$ converges weakly to a point in $\operatorname{Fix} T$.

Proof. We use the embedding of Example 3.16. Define $\boldsymbol{\mathcal{M}}$ as in (3.26) and note that $J_{\boldsymbol{\mathcal{M}}} = \mathrm{Id} + (2\alpha)^{-1}(T - \mathrm{Id})$. We therefore rewrite (5.16) as

$$(\forall n \in \mathbb{N}) \quad x_{n+1} = x_n + \mu_n(J_{\boldsymbol{\mathcal{M}}} x_n - x_n), \quad \text{where } \mu_n = 2\alpha\lambda_n \in \,]0,2[. \tag{5.17}$$

Then $\sum_{n\in\mathbb{N}} \mu_n(2 - \mu_n) = +\infty$ and, appealing to Theorem 5.1(i), we conclude that $(x_n)_{n\in\mathbb{N}}$ converges weakly to a point in $\operatorname{zer} \boldsymbol{\mathcal{M}} = \operatorname{Fix} T$. □

In the case when $\alpha = 1$, Proposition 5.8 is due to Groetsch (1972) and (5.16) is known as the *Krasnosel'skiĭ–Mann iteration*, owing to its connection with iterative schemes proposed in Krasnosel'skiĭ (1955) and Mann (1953), and it is a pillar of nonlinear numerical functional analysis (Bauschke and Combettes 2017, Cegielski 2012, Dong *et al.* 2022). Here is a strongly convergent variant derived from Theorem 5.3; see Genel and Lindenstrauss (1975) for an example of the failure of strong convergence in Proposition 5.8.

Proposition 5.9. Let $\alpha \in \,]0,1]$ and let $T\colon \mathcal{H} \to \mathcal{H}$ be α-averaged. Suppose that $\operatorname{Fix} T \neq \varnothing$, let $(\lambda_n)_{n\in\mathbb{N}}$ be a sequence in $]0,1/(2\alpha)]$ such that $\inf_{n\in\mathbb{N}} \lambda_n > 0$, and let $x_0 \in \mathcal{H}$. Iterate

$$(\forall n \in \mathbb{N}) \quad x_{n+1} = \mathsf{Q}(x_0, x_n, x_n + \lambda_n(Tx_n - x_n)), \tag{5.18}$$

where Q is defined in Lemma 4.6. Then $(x_n)_{n\in\mathbb{N}}$ converges strongly to $\operatorname{proj}_{\operatorname{Fix} T} x_0$.

Proof. Define $\boldsymbol{\mathcal{M}}$ as in (3.26), argue as in the proof of Proposition 5.8 to observe that (5.18) is an instance of (5.10), and conclude by invoking Theorem 5.3. □

5.4.3. Resolvent compositions

We focus on the inclusion problem of Combettes (2023, Section 6), which is modelled by resolvent compositions (see Example 2.40) and solvable via the proximal point algorithm.

Proposition 5.10. Suppose that $L \in \mathcal{B}(\mathcal{H}, \mathcal{G})$ satisfies $0 < \|L\| \leq 1$, let $B\colon \mathcal{G} \to 2^{\mathcal{G}}$ be maximally monotone, let $V \neq \{0\}$ be a closed vector subspace of $\mathcal{H}$, and let $\gamma \in \left]0, +\infty\right[$. Let S be the set of solutions to the problem

$$\text{find } x \in V \text{ such that } 0 \in B(Lx) \tag{5.19}$$

and let Z be the set of solutions to the problem

$$\text{find } x \in \mathcal{H} \text{ such that } 0 \in (\operatorname{proj}_V \diamond (L \blacklozenge (\gamma B)))x. \tag{5.20}$$

Then (5.20) is an exact relaxation of (5.19) in the sense that $S \neq \varnothing \Rightarrow Z = S$. Now assume that $Z \neq \varnothing$, let $(\lambda_n)_{n\in\mathbb{N}}$ be a sequence in $\left]0, 2\right[$ such that $\sum_{n\in\mathbb{N}} \lambda_n(2-\lambda_n) = +\infty$, and let $x_0 \in V$. Iterate

$$\begin{array}{l}
\text{for } n = 0, 1, \ldots \\
\left\lfloor \begin{array}{l}
y_n = Lx_n \\
q_n = J_{\gamma B} y_n - y_n \\
z_n = L^* q_n \\
x_{n+1} = x_n + \lambda_n \operatorname{proj}_V z_n.
\end{array} \right.
\end{array} \tag{5.21}$$

Then $(x_n)_{n\in\mathbb{N}}$ converges weakly to a point in Z.

Proof. The exact relaxation claim is established in Theorem 6.3(v) of Combettes (2023). Now set $M = \operatorname{proj}_V \diamond (L \blacklozenge (\gamma B))$ and note that $\|\operatorname{proj}_V\| = 1$ and $\operatorname{proj}_V^* = \operatorname{proj}_V$. Hence, it follows from Example 2.31 that M is maximally monotone, and from Example 2.40 that $J_M = \operatorname{proj}_V \circ (\operatorname{Id}_{\mathcal{H}} - L^* \circ L + L^* \circ J_{\gamma B} \circ L) \circ \operatorname{proj}_V$. Altogether, the convergence result follows from Theorem 5.1(i) □

Here is a strongly convergent algorithm based on the Haugazeau variant.

Proposition 5.11. Suppose that $L \in \mathcal{B}(\mathcal{H}, \mathcal{G})$ satisfies $0 < \|L\| \leq 1$, let $B\colon \mathcal{G} \to 2^{\mathcal{G}}$ be maximally monotone, let $V \neq \{0\}$ be a closed vector subspace of $\mathcal{H}$, and let

$\gamma \in]0, +\infty[$. Suppose that the set Z of solutions to the problem

$$\text{find } x \in \mathcal{H} \text{ such that } 0 \in (\mathrm{proj}_V \diamond (L \blacklozenge (\gamma B)))x \tag{5.22}$$

is not empty. Let $(\lambda_n)_{n\in\mathbb{N}}$ be a sequence in $]0, 1]$ such that $\inf_{n\in\mathbb{N}} \lambda_n > 0$, and let $x_0 \in V$. Iterate

$$\begin{array}{l} \text{for } n = 0, 1, \ldots \\ \left\lfloor \begin{array}{l} y_n = Lx_n \\ q_n = J_{\gamma B} y_n - y_n \\ z_n = L^* q_n \\ x_{n+1} = \mathsf{Q}(x_0, x_n, x_n + \lambda_n \,\mathrm{proj}_V\, z_n), \end{array} \right. \end{array} \tag{5.23}$$

where Q is defined in Lemma 4.6. Then $(x_n)_{n\in\mathbb{N}}$ converges strongly to $\mathrm{proj}_Z\, x_0$.

Proof. Arguing as in the proof of Proposition 5.10, this is an application of Theorem 5.3 with $M = \mathrm{proj}_V \diamond (L \blacklozenge (\gamma B))$ and $(\forall n \in \mathbb{N})\ \gamma_n = 1$. □

Below we recover the relaxation framework of Combettes and Woodstock (2022) for signal reconstruction in the presence of possibly inconsistent nonlinear observations.

Example 5.12. Let $0 < p \in \mathbb{N}$, let $\gamma \in]0, +\infty[$, and let $V \neq \{0\}$ be a closed vector subspace of $\mathcal{H}$. For every $k \in \{1, \ldots, p\}$, let $\mathcal{G}_k$ be a real Hilbert space, let $L_k \in \mathcal{B}(\mathcal{H}, \mathcal{G}_k)$, let $\omega_k \in]0, +\infty[$, let $F_k : \mathcal{G}_k \to \mathcal{G}_k$ be firmly nonexpansive, and let $r_k \in \mathcal{G}_k$. Consider the nonlinear reconstruction problem (Combettes and Woodstock 2022, Problem 1.1)

$$\text{find } x \in V \text{ such that } (\forall k \in \{1, \ldots, p\})\ F_k(L_k x) = r_k \tag{5.24}$$

and the relaxed variational inequality problem (Combettes and Woodstock 2022, Problem 1.3)

$$\text{find } x \in V \text{ such that } \sum_{k=1}^{p} \omega_k L_k^*(F_k(L_k x) - r_k) \in V^\perp. \tag{5.25}$$

Suppose that $0 < \sum_{k=1}^{p} \omega_k \|L_k\|^2 \leq 1$ and that (5.25) admits solutions. Let $x_0 \in V$, let $(\lambda_n)_{n\in\mathbb{N}}$ be a sequence in $]0, 2[$ such that $\sum_{n\in\mathbb{N}} \lambda_n(2 - \lambda_n) = +\infty$, and iterate

$$\begin{array}{l} \text{for } n = 0, 1, \ldots \\ \left\lfloor \begin{array}{l} \text{for } k = 1, \ldots, p \\ \left\lfloor \begin{array}{l} y_{k,n} = L_k x_n \\ q_{k,n} = r_k - F_k y_{k,n} \end{array} \right. \\ z_n = \sum_{k=1}^{p} \omega_k L_k^* q_{k,n} \\ x_{n+1} = x_n + \lambda_n \,\mathrm{proj}_V\, z_n. \end{array} \right. \end{array} \tag{5.26}$$

Then $(x_n)_{n\in\mathbb{N}}$ converges weakly to a solution to (5.25).

Proof. Let $\mathcal{G}$ be the standard product vector space $\mathcal{G}_1 \times \cdots \times \mathcal{G}_p$, with generic element $\boldsymbol{y} = (y_k)_{1 \le k \le p}$, and equipped with the scalar product $(\boldsymbol{y}, \boldsymbol{y}') \mapsto \sum_{k=1}^{p} \omega_k \langle y_k \mid y'_k \rangle$. Further, set $L\colon \mathcal{H} \to \mathcal{G}\colon x \mapsto (L_1 x, \ldots, L_p x)$ and

$$B\colon \mathcal{G} \to 2^{\mathcal{G}}\colon \boldsymbol{y} \mapsto ((\mathrm{Id}-F_1+r_1)^{-1}y_1-y_1)\times\cdots\times((\mathrm{Id}-F_p+r_p)^{-1}y_p-y_p). \quad (5.27)$$

In this setting, (5.24) is a realization of (5.19), (5.25) of (5.20), and (5.26) of (5.21); for details, see Combettes (2023, Example 6.10). The claim therefore results from Proposition 5.10. □

5.4.4. The method of partial inverses

We go back to a formulation already touched upon in Problem 3.10. Given a maximally monotone operator $A\colon \mathcal{H} \to 2^{\mathcal{H}}$ and a closed vector subspace V of $\mathcal{H}$, Spingarn (1983) considered the problem

$$\text{find } x \in V \text{ and } x^* \in V^{\perp} \text{ such that } x^* \in Ax, \quad (5.28)$$

and solved it by applying the proximal point algorithm to the partial inverse A_V (see Example 2.33). The resulting algorithm is called the *method of partial inverses*. The following is a relaxed version of the convergence result of Spingarn (1983, Theorem 4.1(i)) (see Alghamdi *et al.* 2014, Theorem 2.4).

Theorem 5.13. Let $A\colon \mathcal{H} \to 2^{\mathcal{H}}$ be a maximally monotone operator, let V be a closed vector subspace of $\mathcal{H}$, and let $(\lambda_n)_{n\in\mathbb{N}}$ be a sequence in $]0,2[$ such that $\sum_{n\in\mathbb{N}} \lambda_n(2-\lambda_n) = +\infty$. Suppose that (5.28) has solutions, let $x_0 \in V$, let $x_0^* \in V^{\perp}$, and iterate

$$\begin{array}{l}\text{for } n = 0,1,\ldots\\ \left\lfloor\begin{array}{l} p_n = J_A(x_n + x_n^*)\\ p_n^* = x_n + x_n^* - p_n\\ x_{n+1} = x_n - \lambda_n \operatorname{proj}_V p_n^*\\ x_{n+1}^* = x_n^* - \lambda_n \operatorname{proj}_{V^{\perp}} p_n. \end{array}\right.\end{array} \quad (5.29)$$

Then the following hold:

(i) $\operatorname{proj}_V p_n - x_n \to 0$ and $\operatorname{proj}_{V^{\perp}} p_n^* - x_n^* \to 0$.

(ii) There exists a solution (x, x^*) to (5.28) such that $x_n \rightharpoonup x$ and $x_n^* \rightharpoonup x^*$.

Proof. Set

$$(\forall n \in \mathbb{N}) \quad z_n = x_n + x_n^* \quad (5.30)$$

and note that, since $(x_n)_{n\in\mathbb{N}}$ lies in V and $(x_n^*)_{n\in\mathbb{N}}$ lies in $V^{\perp}$, (5.29) can be rewritten as

$$\begin{array}{l}\text{for } n = 0,1,\ldots\\ \left\lfloor\begin{array}{l} p_n = J_A(x_n + x_n^*)\\ p_n^* = x_n + x_n^* - p_n\\ x_{n+1} = x_n + \lambda_n(\operatorname{proj}_V p_n - x_n)\\ x_{n+1}^* = x_n^* + \lambda_n(\operatorname{proj}_{V^{\perp}} p_n^* - x_n^*). \end{array}\right.\end{array} \quad (5.31)$$

Thus,

$$
\begin{aligned}
(\forall n \in \mathbb{N}) \quad & \operatorname{proj}_V\left(\frac{z_{n+1}-z_n}{\lambda_n}+z_n\right)+\operatorname{proj}_{V^\perp}\left(z_n-\left(\frac{z_{n+1}-z_n}{\lambda_n}+z_n\right)\right)\\
&= \operatorname{proj}_V\left(\frac{z_{n+1}-z_n}{\lambda_n}+z_n\right)+\operatorname{proj}_{V^\perp}\left(\frac{z_n-z_{n+1}}{\lambda_n}\right)\\
&= \operatorname{proj}_V\left(\frac{x_{n+1}-x_n}{\lambda_n}+x_n\right)+\operatorname{proj}_{V^\perp}\left(\frac{x_n^*-x_{n+1}^*}{\lambda_n}\right)\\
&= \operatorname{proj}_V p_n+\operatorname{proj}_{V^\perp}(x_n^*-p_n^*)\\
&= \operatorname{proj}_V p_n+\operatorname{proj}_{V^\perp}(p_n-x_n)\\
&= p_n\\
&= J_A z_n.
\end{aligned}
\tag{5.32}
$$

Hence, it follows from (5.30), (5.31) and Example 2.38 that

$$(\forall n \in \mathbb{N}) \quad z_{n+1} = z_n + \lambda_n(J_{A_V} z_n - z_n). \tag{5.33}$$

Altogether, we derive from Theorem 5.1(i) that

$$J_{A_V} z_n - z_n \to 0 \tag{5.34}$$

and that there exists $z \in \operatorname{zer} A_V$ such that

$$z_n \rightharpoonup z. \tag{5.35}$$

(i) In view of (5.31), (5.30), Example 2.38 and (5.34), we have

$$\operatorname{proj}_V p_n - x_n = \operatorname{proj}_V(J_{A_V} z_n) - x_n = \operatorname{proj}_V(J_{A_V} z_n - z_n) \to 0 \tag{5.36}$$

and

$$x_n^* - \operatorname{proj}_{V^\perp} p_n^* = \operatorname{proj}_{V^\perp}(p_n - x_n) = \operatorname{proj}_{V^\perp} J_A z_n = \operatorname{proj}_{V^\perp}(z_n - J_{A_V} z_n) \to 0. \tag{5.37}$$

(ii) As seen above, $z \in \operatorname{zer} A_V$. Now set $(x, x^*) = (\operatorname{proj}_V z, \operatorname{proj}_{V^\perp} z)$. Then Example 2.33(ii) guarantees that (x, x^*) solves (5.28). In addition, since proj_V and $\operatorname{proj}_{V^\perp}$ are linear and continuous, they are weakly continuous. We conclude that $x_n = \operatorname{proj}_V z_n \rightharpoonup \operatorname{proj}_V z = x$ and $x_n^* = \operatorname{proj}_{V^\perp} z_n \rightharpoonup \operatorname{proj}_{V^\perp} z = x^*$. □

Example 5.14. In Theorem 5.13, let $f \in \Gamma_0(\mathcal{H})$ be such that $0 \in \operatorname{sri}(\operatorname{dom} f - V)$, set $A = \partial f$, and suppose that f admits minimizers over V. Then (5.28) amounts to finding a solution to the Fenchel dual pair

$$\underset{x \in V}{\operatorname{minimize}}\ f(x) \quad \text{and} \quad \underset{x^* \in V^\perp}{\operatorname{minimize}}\ f^*(x^*). \tag{5.38}$$

In this case, given $x_0 \in V$ and $x_0^* \in V^\perp$, the method of partial inverses (5.29) iterates

$$\begin{array}{l}\text{for } n = 0, 1, \ldots \\ \left\lfloor \begin{array}{l} p_n = \operatorname{prox}_f(x_n + x_n^*) \\ p_n^* = x_n + x_n^* - p_n \\ x_{n+1} = x_n - \lambda_n \operatorname{proj}_V p_n^* \\ x_{n+1}^* = x_n^* - \lambda_n \operatorname{proj}_{V^\perp} p_n, \end{array} \right.\end{array} \tag{5.39}$$

and Theorem 5.13(ii) guarantees that there exists a primal–dual solution (x, x^*) of (5.38) such that $x_n \rightharpoonup x$ and $x_n^* \rightharpoonup x^*$.

Algorithm (5.29) has many applications in convex optimization, for example Idrissi, Lefebvre and Michelot (1989), Lemaire (1989), Lenoir and Mahey (2017), Pennanen (2002) and Spingarn (1983, 1985, 1987). As shown in Rockafellar and Sun (2019), it also constitutes the basic building block of the *progressive hedging algorithm* in stochastic programming (Rockafellar and Wets 1991).

Although the method of partial inverses (5.29) is presented in the context of the simple problem (5.28), it has far-reaching ramifications. Below we present an application proposed in Alghamdi *et al.* (2014), where it is applied to Problem 3.11. In terms of Framework 1.2, this approach can be seen as a rephrasing of Problem 3.11 as an instance of (5.28) in $\mathbf{X} = \mathcal{H} \oplus \mathcal{G}_1 \oplus \cdots \oplus \mathcal{G}_p$.

Proposition 5.15. Let $0 < p \in \mathbb{N}$, let $A\colon \mathcal{H} \to 2^{\mathcal{H}}$ be maximally monotone, and, for every $k \in \{1 \ldots, p\}$, let $\mathcal{G}_k$ be a real Hilbert space, let $B_k\colon \mathcal{G}_k \to 2^{\mathcal{G}_k}$ be maximally monotone, and let $L_k \in \mathcal{B}(\mathcal{H}, \mathcal{G}_k)$. Suppose that the set Z of solutions to the inclusion

$$\text{find } x \in \mathcal{H} \text{ such that } 0 \in Ax + \sum_{k=1}^{p} L_k^*(B_k(L_k x)) \tag{5.40}$$

is not empty and let Z^* be the set of solutions to the dual inclusion

$$\begin{array}{l}\text{find } y_1^* \in \mathcal{G}_1, \ldots, y_p^* \in \mathcal{G}_p \text{ such that} \\ \left(\exists x \in A^{-1}\left(-\sum_{k=1}^{p} L_k^* y_k^*\right)\right)(\forall k \in \{1, \ldots, p\})\ L_k x \in B_k^{-1} y_k^*.\end{array} \tag{5.41}$$

Let $x_0 \in \mathcal{H}$ and let $(\lambda_n)_{n\in\mathbb{N}}$ be a sequence in $]0, 2[$ such that $\sum_{n\in\mathbb{N}} \lambda_n(2-\lambda_n) = +\infty$. Set

$$U = \left(\mathrm{Id} + \sum_{k=1}^{p} L_k^* \circ L_k\right)^{-1} \tag{5.42}$$

and, for every $k \in \{1, \ldots, p\}$, let $y_{k,0}^* \in \mathcal{G}_k$ and set $y_{k,0} = L_k x_0$. Additionally, set

$$x_0^* = -\sum_{k=1}^{p} L_k^* y_{k,0}^*, \tag{5.43}$$

and iterate

$$
\begin{array}{l}
\text{for } n = 0, 1, \ldots \\
\left\lfloor
\begin{array}{l}
p_n = J_A(x_n + x_n^*) \\
p_n^* = x_n + x_n^* - p_n \\
\text{for } k = 1, \ldots, p \\
\left\lfloor
\begin{array}{l}
q_{k,n} = J_{B_k}(y_{k,n} + y_{k,n}^*) \\
q_{k,n}^* = y_{k,n} + y_{k,n}^* - q_{k,n}
\end{array}
\right. \\
t_n = U\big(p_n^* + \sum_{k=1}^{p} L_k^* q_{k,n}^*\big) \\
w_n = U\big(p_n + \sum_{k=1}^{p} L_k^* q_{k,n}\big) \\
x_{n+1} = x_n - \lambda_n t_n \\
x_{n+1}^* = x_n^* + \lambda_n(w_n - p_n) \\
\text{for } k = 1, \ldots, p \\
\left\lfloor
\begin{array}{l}
y_{k,n+1} = y_{k,n} - \lambda_n L_k t_n \\
y_{k,n+1}^* = y_{k,n}^* + \lambda_n(L_k w_n - q_{k,n}).
\end{array}
\right.
\end{array}
\right.
\end{array}
\tag{5.44}
$$

Then there exist $x \in Z$ and $(y_k^*)_{1\leq k\leq p} \in Z^*$ such that $x_n \rightharpoonup x$ and, for every $k \in \{1, \ldots, p\}$, $y_{k,n}^* \rightharpoonup y_k^*$.

Proof. Define

$$
\begin{cases}
\mathcal{G} = \mathcal{G}_1 \oplus \cdots \oplus \mathcal{G}_p, \\
B\colon \mathcal{G} \to 2^{\mathcal{G}} : (y_1, \ldots, y_p) \mapsto B_1 y_1 \times \cdots \times B_p y_p, \\
L\colon \mathcal{H} \to \mathcal{G} : x \mapsto (L_1 x, \ldots, L_p x),
\end{cases}
\tag{5.45}
$$

and note that $L^*\colon \mathcal{G} \to \mathcal{H} : (y_1^*, \ldots, y_p^*) \mapsto L_1^* y_1^* + \cdots + L_p^* y_p^*$. Moreover set, for every $n \in \mathbb{N}$, $q_n = (q_{k,n})_{1\leq k\leq p}$, $q_n^* = (q_{k,n}^*)_{1\leq k\leq p}$, $y_n = (y_{k,n})_{1\leq k\leq p}$ and $y_n^* = (y_{k,n}^*)_{1\leq k\leq p}$. In this setting, B is maximally monotone and $J_B\colon (y_k)_{1\leq k\leq p} \mapsto (J_{B_k} y_k)_{1\leq k\leq p}$ (Example 2.37), so that (5.44) can be rewritten as

$$
\begin{array}{l}
\text{for } n = 0, 1, \ldots \\
\left\lfloor
\begin{array}{l}
p_n = J_A(x_n + x_n^*) \\
q_n = J_B(y_n + y_n^*) \\
p_n^* = x_n + x_n^* - p_n \\
q_n^* = y_n + y_n^* - q_n \\
t_n = U(p_n^* + L^* q_n^*) \\
w_n = U(p_n + L^* q_n) \\
x_{n+1} = x_n - \lambda_n t_n \\
y_{n+1} = y_n - \lambda_n L t_n \\
x_{n+1}^* = x_n^* + \lambda_n(w_n - p_n) \\
y_{n+1}^* = y_n^* + \lambda_n(L w_n - q_n).
\end{array}
\right.
\end{array}
\tag{5.46}
$$

Let us introduce

$$\begin{cases} \mathbf{X} = \mathcal{H} \oplus \mathcal{G}, \\ \boldsymbol{V} = \{(x, y) \in \mathbf{X} \mid Lx = y\}, \\ \boldsymbol{Z} = \{(x, y^*) \in \mathbf{X} \mid -L^*y^* \in Ax \text{ and } y^* \in B(Lx)\}, \\ \boldsymbol{A} \colon \mathbf{X} \to 2^{\mathbf{X}} \colon (x, y) \mapsto Ax \times By, \\ \boldsymbol{S} = \{(\boldsymbol{x}, \boldsymbol{x}^*) \in \boldsymbol{V} \times \boldsymbol{V}^{\perp} \mid \boldsymbol{x}^* \in \boldsymbol{A}\boldsymbol{x}\}, \end{cases} \tag{5.47}$$

and observe that

$$\begin{cases} \boldsymbol{V}^{\perp} = \{(x^*, y^*) \in \mathbf{X} \mid x^* = -L^*y^*\}, \\ \boldsymbol{S} = \{((x, Lx), (-L^*y^*, y^*)) \in \mathbf{X} \times \mathbf{X} \mid (x, y^*) \in \boldsymbol{Z}\}. \end{cases} \tag{5.48}$$

Then Lemma 3.12(iii) implies that

$$\text{(5.40) admits solutions} \;\Leftrightarrow\; \boldsymbol{Z} \neq \varnothing \;\Leftrightarrow\; \boldsymbol{S} \neq \varnothing. \tag{5.49}$$

Now define $(\forall n \in \mathbb{N})$ $\boldsymbol{p}_n = (p_n, q_n)$, $\boldsymbol{p}_n^* = (p_n^*, q_n^*)$, $\boldsymbol{x}_n = (x_n, y_n)$ and $\boldsymbol{x}_n^* = (x_n^*, y_n^*)$. Then $\boldsymbol{x}_0 \in \boldsymbol{V}$ and $\boldsymbol{x}_0^* \in \boldsymbol{V}^{\perp}$. Moreover, by Lemma 2.24 and Example 2.37, $\boldsymbol{A}$ is maximally monotone and

$$(\forall n \in \mathbb{N}) \quad J_{\boldsymbol{A}}(\boldsymbol{x}_n + \boldsymbol{x}_n^*) = (J_A(x_n + x_n^*), J_B(y_n + y_n^*)). \tag{5.50}$$

Furthermore, since $U = (\mathrm{Id} + L^* \circ L)^{-1}$, it follows from (5.47) and Bauschke and Combettes (2017, Example 29.19) that

$$(\forall n \in \mathbb{N}) \quad \mathrm{proj}_{\boldsymbol{V}^{\perp}} \boldsymbol{p}_n = (p_n - U(p_n + L^*q_n), q_n - L(U(p_n + L^*q_n))) \tag{5.51}$$

and

$$(\forall n \in \mathbb{N}) \quad \mathrm{proj}_{\boldsymbol{V}} \boldsymbol{p}_n^* = (U(p_n^* + L^*q_n^*), L(U(p_n^* + L^*q_n^*))). \tag{5.52}$$

Combining (5.50), (5.51) and (5.52), we rewrite (5.46) as

$$\begin{array}{l} \text{for } n = 0, 1, \ldots \\ \left\lfloor \begin{array}{l} \boldsymbol{p}_n = J_{\boldsymbol{A}}(\boldsymbol{x}_n + \boldsymbol{x}_n^*) \\ \boldsymbol{p}_n^* = \boldsymbol{x}_n + \boldsymbol{x}_n^* - \boldsymbol{p}_n \\ \boldsymbol{x}_{n+1} = \boldsymbol{x}_n - \lambda_n \,\mathrm{proj}_{\boldsymbol{V}} \boldsymbol{p}_n^* \\ \boldsymbol{x}_{n+1}^* = \boldsymbol{x}_n^* - \lambda_n \,\mathrm{proj}_{\boldsymbol{V}^{\perp}} \boldsymbol{p}_n. \end{array} \right. \end{array} \tag{5.53}$$

In turn, Theorem 5.13(ii) implies that there exists $(\boldsymbol{x}, \boldsymbol{x}^*) \in \boldsymbol{S}$ such that $\boldsymbol{x}_n \rightharpoonup \boldsymbol{x}$ and $\boldsymbol{x}_n^* \rightharpoonup \boldsymbol{x}^*$. We then derive from (5.48) that there exists $(x, y^*) \in \boldsymbol{Z}$ such that $(x_n, y_n^*) \rightharpoonup (x, y^*)$. We complete the proof by invoking Lemma 3.12(ii). □

5.4.5. Renorming

The potency of the proximal point algorithm can be further extended by setting it up in a renormed space. In terms of Framework 1.2, the guiding principle lies in the embedding of Example 3.15. Here is a weak convergence result.

Proposition 5.16. Let $M\colon \mathcal{H} \to 2^{\mathcal{H}}$ be a maximally monotone operator such that $Z = \operatorname{zer} M \neq \varnothing$, let $U \in \mathcal{B}(\mathcal{H})$ be a self-adjoint strongly monotone operator, and let $\mathcal{X}$ be the real Hilbert space obtained by endowing $\mathcal{H}$ with the scalar product $(x, y) \mapsto \langle Ux \mid y\rangle$. Let $x_0 \in \mathcal{H}$, let $(\lambda_n)_{n\in\mathbb{N}}$ be a sequence in $]0, 2[$, and let $(\gamma_n)_{n\in\mathbb{N}}$ be a sequence in $]0, +\infty[$. Iterate

$$
\begin{array}{l}
\text{for } n = 0, 1, \ldots \\
\left\lfloor
\begin{array}{l}
u_n = \gamma_n^{-1} U x_n \\
p_n = \big(\gamma_n^{-1} U + M\big)^{-1} u_n \\
x_{n+1} = x_n + \lambda_n (p_n - x_n),
\end{array}
\right.
\end{array}
\tag{5.54}
$$

and suppose that one of the following holds:

(i) $\sum_{n\in\mathbb{N}} \lambda_n(2 - \lambda_n) = +\infty$ and $(\forall n \in \mathbb{N})\ \gamma_n = 1$.

(ii) $\sum_{n\in\mathbb{N}} \gamma_n^2 = +\infty$ and $(\forall n \in \mathbb{N})\ \lambda_n = 1$.

(iii) $\inf_{n\in\mathbb{N}} \lambda_n > 0$, $\sup_{n\in\mathbb{N}} \lambda_n < 2$ and $\inf_{n\in\mathbb{N}} \gamma_n > 0$.

Then $(x_n)_{n\in\mathbb{N}}$ converges weakly to a point in Z.

Proof. In view of Lemma 2.25(ii) and Example 2.39, (5.54) is just the proximal point algorithm (5.3) applied to the maximally monotone operator $U^{-1} \circ M$ in $\mathcal{X}$. Since weak convergences in $\mathcal{H}$ and $\mathcal{X}$ coincide, the claims follow from Lemma 2.25(i) and Theorem 5.1. □

Remark 5.17. In terms of the warped resolvent of Section 2.4.3, the update in (5.54) can be written as $x_{n+1} = x_n + \lambda_n\big(J^{U}_{\gamma_n M} x_n - x_n\big)$.

Likewise, Theorem 5.3 leads to a strongly convergent algorithm.

Proposition 5.18. Let $M\colon \mathcal{H} \to 2^{\mathcal{H}}$ be a maximally monotone operator such that $Z = \operatorname{zer} M \neq \varnothing$, let $U \in \mathcal{B}(\mathcal{H})$ be a self-adjoint strongly monotone operator, and let $\mathcal{X}$ be the real Hilbert space obtained by endowing $\mathcal{H}$ with the scalar product $(x, y) \mapsto \langle Ux \mid y\rangle$. Let $x_0 \in \mathcal{H}$, let $(\lambda_n)_{n\in\mathbb{N}}$ be a sequence in $]0, 1]$ such that $\inf_{n\in\mathbb{N}} \lambda_n > 0$, and let $(\gamma_n)_{n\in\mathbb{N}}$ be a sequence in $]0, +\infty[$ such that $\inf_{n\in\mathbb{N}} \gamma_n > 0$. Iterate

$$
\begin{array}{l}
\text{for } n = 0, 1, \ldots \\
\left\lfloor
\begin{array}{l}
u_n = \gamma_n^{-1} U x_n \\
p_n = (\gamma_n^{-1} U + M)^{-1} u_n \\
x_{n+1} = \mathsf{Q}(x_0, x_n, x_n + \lambda_n (p_n - x_n)),
\end{array}
\right.
\end{array}
\tag{5.55}
$$

where Q is defined in Lemma 4.6. Then $(x_n)_{n\in\mathbb{N}}$ converges strongly to $\operatorname{proj}_Z x_0$.

Proof. It follows from Lemma 2.25(ii) and Example 2.39 that applying the algorithm (5.10) to the maximally monotone operator $U^{-1} \circ M$ in $\mathcal{X}$ yields (5.55). Since strong convergences in $\mathcal{H}$ and $\mathcal{X}$ coincide, the assertion follows from Lemma 2.25(i) and Theorem 5.3. □

Although the inversion of the operators $(\gamma_n^{-1}U + M)_{n\in\mathbb{N}}$ in (5.54) and (5.55) may be intimidating, we show below that the renormed proximal point algorithm leads to important instances of fully executable splitting algorithms. First, we revisit a classical minimization problem and recover an algorithm known as the *proximal Landweber method.*

Example 5.19. Let $\varphi \in \Gamma_0(\mathcal{H})$, let $\mu \in \left]0, +\infty\right[$, and let $y \in \mathcal{G}$. Suppose that $0 \neq L \in \mathcal{B}(\mathcal{H}, \mathcal{G})$ and that the set Z of solutions to the optimization problem

$$\underset{x\in\mathcal{H}}{\text{minimize}} \;\; \varphi(x) + \frac{\mu}{2}\|Lx - y\|^2 \tag{5.56}$$

is not empty. Without loss of generality (rescale), assume that $\mu\|L\|^2 < 1$. Let $x_0 \in \mathcal{H}$, let $(\lambda_n)_{n\in\mathbb{N}}$ be a sequence in $\left]0, 2\right[$ such that $\sum_{n\in\mathbb{N}} \lambda_n(2 - \lambda_n) = +\infty$, and iterate

$$\begin{array}{l}\text{for } n = 0, 1, \ldots\\ \left\lfloor \begin{array}{l} u_n = x_n - \mu L^*(Lx_n)\\ p_n = \operatorname{prox}_\varphi(u_n + \mu L^* y)\\ x_{n+1} = x_n + \lambda_n(p_n - x_n). \end{array}\right.\end{array} \tag{5.57}$$

Then $(x_n)_{n\in\mathbb{N}}$ converges weakly to a point in Z.

Proof. Set $f = \varphi - \mu\langle\cdot \mid L^*y\rangle$, $M = \partial(\varphi + \mu\|L\cdot - y\|^2/2) = \partial f + \mu L^* \circ L$ and $U = \operatorname{Id} - \mu L^* \circ L$. Then $f \in \Gamma_0(\mathcal{H})$, M is maximally monotone with $\operatorname{zer} M = Z$ by virtue of Example 2.12, $U \in \mathcal{B}(\mathcal{H})$ is self-adjoint and strongly monotone, and $(U+M)^{-1} = \operatorname{prox}_f = \operatorname{prox}_\varphi(\cdot + \mu L^* y)$. Consequently, (5.57) is the implementation of (5.54) with, for every $n \in \mathbb{N}$, $\gamma_n = 1$, and Proposition 5.16(i) brings the conclusion. □

Next, we return to the primal–dual composite inclusion framework of Problem 3.7 and approach it via Framework 1.2 where, as discussed in Example 3.20, the embedding is based on $\mathbf{X} = \mathcal{H} \oplus \mathcal{G}$ and the Kuhn–Tucker operator $\boldsymbol{\mathcal{K}}$ of Lemma 3.8.

Example 5.20. Let $A\colon \mathcal{H} \to 2^{\mathcal{H}}$ and $B\colon \mathcal{G} \to 2^{\mathcal{G}}$ be maximally monotone, and let $L \in \mathcal{B}(\mathcal{H}, \mathcal{G})$. Suppose that the set Z of solutions to the primal inclusion

$$\text{find } x \in \mathcal{H} \text{ such that } 0 \in Ax + L^*(B(Lx)) \tag{5.58}$$

is not empty and let Z^* be the set of solutions to the dual inclusion

$$\text{find } y^* \in \mathcal{G} \text{ such that } 0 \in -L(A^{-1}(-L^*y^*)) + B^{-1}y^*. \tag{5.59}$$

Let $(\lambda_n)_{n\in\mathbb{N}}$ be a sequence in $\left]0, 2\right[$ such that $\sum_{n\in\mathbb{N}} \lambda_n(2 - \lambda_n) = +\infty$, let $x_0 \in \mathcal{H}$, let $y_0^* \in \mathcal{G}$, and let $\sigma \in \left]0, +\infty\right[$ and $\tau \in \left]0, +\infty\right[$ be such that $\tau\sigma\|L\|^2 < 1$. Iterate

$$
\begin{array}{l}
\text{for } n = 0, 1, \ldots \\
\left\lfloor
\begin{array}{l}
x_n^* = \tau L^* y_n^* \\
p_n = J_{\tau A}(x_n - x_n^*) \\
y_n = \sigma L(2p_n - x_n) \\
q_n^* = J_{\sigma B^{-1}}(y_n^* + y_n) \\
x_{n+1} = x_n + \lambda_n(p_n - x_n) \\
y_{n+1}^* = y_n^* + \lambda_n(q_n^* - y_n^*).
\end{array}
\right.
\end{array}
\tag{5.60}
$$

Then there exist $x \in Z$ and $y^* \in Z^*$ such that $x_n \rightharpoonup x$ and $y_n^* \rightharpoonup y^*$.

Proof. Set $\mathbf{X} = \mathcal{H} \oplus \mathcal{G}$ and

$$
\begin{cases}
\boldsymbol{\mathcal{K}} \colon \mathbf{X} \to 2^{\mathbf{X}} \colon (x, y^*) \mapsto (Ax + L^*y^*) \times (-Lx + B^{-1}y^*), \\
\boldsymbol{U} \colon \mathbf{X} \to \mathbf{X} \colon (x, y^*) \mapsto (\tau^{-1}x - L^*y^*, -Lx + \sigma^{-1}y^*).
\end{cases}
\tag{5.61}
$$

As seen in Lemma 3.8(iii)–(iv), $\boldsymbol{\mathcal{K}}$ is the maximally monotone Kuhn–Tucker operator associated with (5.58)–(5.59), and to prove the claim it is enough to show that $(x_n, y_n^*)_{n\in\mathbb{N}}$ converges weakly to a point in $\operatorname{zer}\boldsymbol{\mathcal{K}}$, which we shall derive from Proposition 5.16(i). It is clear that $\boldsymbol{U} \in \mathcal{B}(\mathbf{X})$ is self-adjoint. Now set $\beta = 1 - \sqrt{\sigma\tau}\|L\|$. Then, since $\tau\sigma\|L\|^2 < 1$, $\beta \in \left]0, 1\right[$ and, for every $(x, y^*) \in \mathbf{X}$, the Cauchy–Schwarz inequality yields

$$
\begin{aligned}
\langle \boldsymbol{U}(x, y^*) \mid (x, y^*) \rangle_{\mathbf{X}} &= \tau^{-1}\|x\|^2 - 2\langle Lx \mid y^* \rangle + \sigma^{-1}\|y^*\|^2 \\
&\geq \tau^{-1}\|x\|^2 - 2\sqrt{\tau\sigma}\|L\| \left\|\frac{x}{\sqrt{\tau}}\right\| \left\|\frac{y^*}{\sqrt{\sigma}}\right\| + \sigma^{-1}\|y^*\|^2 \\
&= \tau^{-1}\|x\|^2 - 2(1-\beta) \left\|\frac{x}{\sqrt{\tau}}\right\| \left\|\frac{y^*}{\sqrt{\sigma}}\right\| + \sigma^{-1}\|y^*\|^2 \\
&= \left(\left\|\frac{x}{\sqrt{\tau}}\right\| - \left\|\frac{y^*}{\sqrt{\sigma}}\right\|\right)^2 + 2\beta \left\|\frac{x}{\sqrt{\tau}}\right\| \left\|\frac{y^*}{\sqrt{\sigma}}\right\| \\
&= (1-\beta)\left(\left\|\frac{x}{\sqrt{\tau}}\right\| - \left\|\frac{y^*}{\sqrt{\sigma}}\right\|\right)^2 + \beta\left(\left\|\frac{x}{\sqrt{\tau}}\right\|^2 + \left\|\frac{y^*}{\sqrt{\sigma}}\right\|^2\right) \\
&\geq \beta(\tau^{-1}\|x\|^2 + \sigma^{-1}\|y^*\|^2) \\
&\geq \beta \min\{\tau^{-1}, \sigma^{-1}\}\|(x, y^*)\|_{\mathbf{X}}^2,
\end{aligned}
\tag{5.62}
$$

which confirms that $\boldsymbol{U}$ is strongly monotone. It remains to show that (5.60) is a realization of (5.54) with the above operators $\boldsymbol{\mathcal{K}}$ and $\boldsymbol{U}$. Define $(\forall n \in \mathbb{N})$ $\boldsymbol{x}_n = (x_n, y_n^*)$, $\boldsymbol{p}_n = (p_n, q_n^*)$ and $\boldsymbol{u}_n = \boldsymbol{U}\boldsymbol{x}_n$. Then we derive from (5.60) and (2.18) that

$$
(\forall n \in \mathbb{N}) \quad
\begin{cases}
x_n - p_n - \tau L^* y_n^* \in \tau A p_n, \\
y_n^* - q_n^* + \sigma L(2p_n - x_n) \in \sigma B^{-1} q_n^*.
\end{cases}
\tag{5.63}
$$

This yields $(\forall n \in \mathbb{N})\ \boldsymbol{u}_n - \boldsymbol{U}\boldsymbol{p}_n \in \boldsymbol{\mathcal{K}}\boldsymbol{p}_n$, i.e. $\boldsymbol{p}_n = (\boldsymbol{U}+\boldsymbol{\mathcal{K}})^{-1}\boldsymbol{u}_n$. Altogether, (5.60) corresponds to the iteration

$$
\begin{array}{l}
\text{for } n = 0, 1, \ldots \\
\left\lfloor
\begin{array}{l}
\boldsymbol{u}_n = \boldsymbol{U}\boldsymbol{x}_n \\
\boldsymbol{p}_n = (\boldsymbol{U} + \boldsymbol{\mathcal{K}})^{-1}\boldsymbol{u}_n \\
\boldsymbol{x}_{n+1} = \boldsymbol{x}_n + \lambda_n(\boldsymbol{p}_n - \boldsymbol{x}_n),
\end{array}
\right.
\end{array}
\tag{5.64}
$$

which is precisely (5.54) with $(\forall n \in \mathbb{N})\ \gamma_n = 1$. □

Remark 5.21. Here are a few observations regarding Example 5.20.

(i) We have derived weak convergence from Proposition 5.16(i). Using items (ii) or (iii) in Proposition 5.16 leads to alternative forms of (5.60) involving proximal parameters $(\gamma_n)_{n\in\mathbb{N}}$.

(ii) It is straightforward to derive a strongly convergent best approximation variant of (5.60) from Proposition 5.18 by following the same pattern as in the proof of Example 5.20, i.e. applying (5.55) to the operators $\boldsymbol{\mathcal{K}}$ and $\boldsymbol{U}$ of (5.61).

(iii) Algorithm (5.60) can be adapted to Problem 3.11 by applying it to the setting of (5.45) and using Example 2.37.

(iv) Let $f \in \Gamma_0(\mathcal{H})$ and $g \in \Gamma_0(\mathcal{G})$, and set $A = \partial f$ and $B = \partial g$ in Example 5.20, which corresponds to the primal–dual minimization setting of Problem 3.9. The specialization of Example 5.20 to this minimization problem appears in Condat (2013, Theorem 3.2), where (5.60) is called the *Chambolle–Pock algorithm* because it collapses to the algorithm proposed in Chambolle and Pock (2011, Algorithm I) in Euclidean spaces when $(\forall n \in \mathbb{N})\ \lambda_n = 1$; see Condat *et al.* (2023) for variations on this algorithm. The fact that the Chambolle–Pock algorithm is a renormed proximal point algorithm was first observed in He and Yuan (2012).

6. Douglas–Rachford splitting

6.1. *Preview*

The Douglas–Rachford splitting algorithm is an implicit alternating direction method designed in Douglas and Rachford (1956) to solve the matrix equation $Ax + Bx = f$, where A and B are positive-definite matrices arising from the discretization of partial differentiation operators. It is described by the iteration process

$$
\begin{array}{l}
\text{for } n = 0, 1, \ldots \\
\left\lfloor
\begin{array}{l}
x_{n+1/2} - x_n + Ax_{n+1/2} + Bx_n = f \\
x_{n+1} - x_n + Ax_{n+1/2} + Bx_{n+1} = f.
\end{array}
\right.
\end{array}
\tag{6.1}
$$

Lieutaud (1969*a*) (see also Lieutaud 1969*b*) proposed an infinite-dimensional nonlinear generalization of the method by showing that (6.1) can be extended to single-valued hemicontinuous monotone operators with $\operatorname{dom} A = \operatorname{dom} B = \mathcal{H}$. In particular, he established in Lieutaud (1969*a*) that, with the additional assumption that A or B is strongly monotone, $(x_n)_{n\in\mathbb{N}}$ converges strongly to some $x \in \mathcal{H}$ which satisfies $Ax+Bx = f$. The investigation of the method for general set-valued maximally monotone operators was initiated in Lions and Mercier (1979), with subsequent improvements in Bauschke and Combettes (2017), Bauschke and Moursi (2017), Combettes (2009), Eckstein and Bertsekas (1992) and Svaiter (2011). See also Xue (2023*a*) for further analysis.

To chart the path from the original Douglas–Rachford algorithm to its modern version for monotone set-valued operators, let us go back to the matrix setting. Upon eliminating the intermediate variables $(x_{n+1/2})_{n\in\mathbb{N}}$ in (6.1) and noting that $AJ_A = \mathrm{Id} - J_A$, we obtain

$$\begin{aligned}(\forall n \in \mathbb{N})\quad x_{n+1} &= J_B(x_n - AJ_A(x_n - Bx_n + f) + f)\\ &= J_B(Bx_n + J_A(x_n - Bx_n + f)).\end{aligned}\tag{6.2}$$

Now set $(\forall n \in \mathbb{N})\ x_n = J_B y_n$. Then we derive from (6.2) that

$$\begin{aligned}(\forall n \in \mathbb{N})\quad y_{n+1} &= BJ_B y_n + J_A(J_B y_n - BJ_B y_n + f)\\ &= y_n - J_B y_n + J_A(2J_B y_n - y_n + f),\end{aligned}\tag{6.3}$$

which leads to the recursion

$$\begin{array}{l}\text{for } n = 0, 1, \ldots\\ \left\lfloor\begin{array}{l} x_n = J_B y_n\\ z_n = J_A(2x_n - y_n + f)\\ y_{n+1} = y_n + z_n - x_n.\end{array}\right.\end{array}\tag{6.4}$$

As noted in Lions and Mercier (1979), unlike (6.1), this algorithm is well defined for arbitrary maximally monotone set-valued operators and is now referred to as the Douglas–Rachford splitting algorithm in this context.

Remark 6.1. In particular, upon setting $B = 0$ and $f = 0$ in (6.4) and assuming that $A\colon \mathcal{H} \to \mathcal{H}$ is hemicontinuous and strongly monotone, it follows from Lieutaud's result (Lieutaud 1969*a*) that the sequence $(x_n)_{n\in\mathbb{N}}$ generated by the recursion

$$(\forall n \in \mathbb{N})\quad x_{n+1} = J_A x_n \tag{6.5}$$

converges strongly to a zero of A. This is actually the first instance of convergence of the proximal point algorithm, which has been attributed to later work in the literature. The case when A and B are gradients of convex functions was also considered in Lieutaud (1969*a*) in connection with the minimization of the sum of two differentiable convex functions.

6.2. Weak convergence

We present results for a form of the Douglas–Rachford algorithm (6.4) which includes relaxation parameters and a dual inclusion problem.

Theorem 6.2. Let $A\colon \mathcal{H} \to 2^{\mathcal{H}}$ and $B\colon \mathcal{H} \to 2^{\mathcal{H}}$ be maximally monotone, let $(\lambda_n)_{n\in\mathbb{N}}$ be a sequence in $]0,2[$ such that $\sum_{n\in\mathbb{N}} \lambda_n(2-\lambda_n) = +\infty$, and let $\gamma \in\,]0,+\infty[$. Suppose that the set Z of solutions to the inclusion

$$\text{find } x \in \mathcal{H} \text{ such that } 0 \in Ax + Bx \tag{6.6}$$

is not empty and let Z^* be the set of solutions to the dual problem

$$\text{find } x^* \in \mathcal{H} \text{ such that } 0 \in -A^{-1}(-x^*) + B^{-1}x^*. \tag{6.7}$$

Let $y_0 \in \mathcal{H}$ and iterate

$$\begin{array}{l} \text{for } n = 0,1,\ldots \\ \left\lfloor \begin{array}{l} x_n = J_{\gamma B} y_n \\ x_n^* = \gamma^{-1}(y_n - x_n) \\ z_n = J_{\gamma A}(2x_n - y_n) \\ y_{n+1} = y_n + \lambda_n(z_n - x_n). \end{array} \right. \end{array} \tag{6.8}$$

Then there exists $y \in \mathcal{H}$ such that $y_n \rightharpoonup y$. Now set $x = J_{\gamma B} y$ and $x^* = {}^{\gamma}B y$. Then the following hold:

(i) $x_n \rightharpoonup x \in Z$.

(ii) $x_n^* \rightharpoonup x^* \in Z^*$.

Proof. We rely on the embedding of Example 3.17. Set

$$R_{\gamma A} = 2J_{\gamma A} - \mathrm{Id}, \quad R_{\gamma B} = 2J_{\gamma B} - \mathrm{Id} \quad \text{and} \quad \boldsymbol{\mathcal{M}} = \left(\frac{R_{\gamma A} \circ R_{\gamma B} + \mathrm{Id}}{2}\right)^{-1} - \mathrm{Id}. \tag{6.9}$$

Then it follows from (2.33) and Lemma 2.34(iii) that $(R_{\gamma A} \circ R_{\gamma B} + \mathrm{Id})/2$ is firmly nonexpansive and that $\boldsymbol{\mathcal{M}}$ is maximally monotone. In addition, Proposition 26.1(iii)(b) of Bauschke and Combettes (2017) asserts that

$$\varnothing \neq Z = J_{\gamma B}(\operatorname{zer} \boldsymbol{\mathcal{M}}), \tag{6.10}$$

while Proposition 26.1(iii)(c) of Bauschke and Combettes (2017) asserts that

$$\varnothing \neq Z^* = {}^{\gamma}B(\operatorname{zer} \boldsymbol{\mathcal{M}}). \tag{6.11}$$

Furthermore, we derive from (6.8) and (6.9) that

$$(\forall n \in \mathbb{N}) \quad y_{n+1} = y_n + \frac{\lambda_n}{2}(R_{\gamma A}(R_{\gamma B} y_n) - y_n) = y_n + \lambda_n(J_{\boldsymbol{\mathcal{M}}} y_n - y_n), \tag{6.12}$$

that is, $(y_n)_{n\in\mathbb{N}}$ is constructed by the proximal point algorithm (5.3) for $\boldsymbol{\mathcal{M}}$. Since (6.10) implies that $\operatorname{zer} \boldsymbol{\mathcal{M}} \neq \varnothing$, Theorem 5.1(i) asserts that

$$J_{\boldsymbol{\mathcal{M}}} y_n - y_n \to 0 \quad \text{and} \quad (\exists\, y \in \operatorname{zer} \boldsymbol{\mathcal{M}}) \quad y_n \rightharpoonup y. \tag{6.13}$$

In turn, (6.10) yields $x = J_{\gamma B} y \in Z$, while (6.8) yields

$$z_n - x_n = J_{\gamma A}(2x_n - y_n) - x_n = J_{\boldsymbol{\mathcal{M}}} y_n - y_n \to 0. \tag{6.14}$$

(i) Let us set

$$(\forall n \in \mathbb{N}) \quad z_n^* = \gamma^{-1}(2x_n - y_n - z_n). \tag{6.15}$$

Then (6.8) and (2.18) yield

$$(\forall n \in \mathbb{N}) \quad \begin{cases} (z_n, z_n^*) \in \operatorname{gra} A, \\ (x_n, x_n^*) \in \operatorname{gra} B, \\ x_n - z_n = \gamma(x_n^* + z_n^*). \end{cases} \tag{6.16}$$

Since Lemma 2.34(iii) asserts that $J_{\gamma B}$ is nonexpansive,

$$(\forall n \in \mathbb{N}) \quad \|x_n - x_0\| = \|J_{\gamma B} y_n - J_{\gamma B} y_0\| \leq \|y_n - y_0\|. \tag{6.17}$$

Hence, since $(y_n)_{n\in\mathbb{N}}$ is bounded, so is $(x_n)_{n\in\mathbb{N}}$. Now take $z \in \mathfrak{W}(x_n)_{n\in\mathbb{N}}$, say $x_{k_n} \rightharpoonup z$. Then it follows from (6.14), (6.13), (6.15) and (6.16) that

$$z_{k_n} \rightharpoonup z, \quad z_{k_n}^* \rightharpoonup \gamma^{-1}(z - y), \quad z_n - x_n \to 0 \quad \text{and} \quad z_n^* + x_n^* = \gamma^{-1}(x_n - z_n) \to 0. \tag{6.18}$$

In turn, Lemma 2.50 yields $z \in \operatorname{zer}(A + B) = Z$,

$$(z, \gamma^{-1}(z - y)) \in \operatorname{gra} A \quad \text{and} \quad (z, \gamma^{-1}(y - z)) \in \operatorname{gra} B. \tag{6.19}$$

Hence, (2.18) implies that

$$z = J_{\gamma B} y. \tag{6.20}$$

Thus, $x = J_{\gamma B} y$ is the unique weak sequential cluster point of the bounded sequence $(x_n)_{n\in\mathbb{N}}$ and therefore, by Lemma 4.1(ii), $x_n \rightharpoonup x$.

(ii) We have $y_n \rightharpoonup y \in \operatorname{zer} \boldsymbol{\mathcal{M}}$ and, by (i), $x_n \rightharpoonup x$. Hence, $x_n^* = \gamma^{-1}(y_n - x_n) \rightharpoonup \gamma^{-1}(y - x) = {}^{\gamma}B y = x^*$. In view of (6.11), the proof is complete. □

Remark 6.3. The convergence result of Lions and Mercier (1979) is that, for the unrelaxed scheme (6.4), $(y_n)_{n\in\mathbb{N}}$ converges weakly to a point $y \in \mathcal{H}$ such that $J_{\gamma B} y \in Z$; see Combettes (2004) and Eckstein and Bertsekas (1992) for the relaxed case. In the special case when $J_{\gamma B}$ is weakly sequentially continuous, as is the case when $\mathcal{H}$ is finite-dimensional, $x_n = J_{\gamma B} y_n \rightharpoonup J_{\gamma B} y \in Z$. The key fact that $(x_n)_{n\in\mathbb{N}}$ converges weakly to a point in $\operatorname{zer}(A + B)$ without any further assumption was first proved in Svaiter (2011) in the unrelaxed case. Theorem 6.2 was established in Bauschke and Combettes (2017, Theorem 26.11). The component of the proof given above up to (6.13) exploits an idea from Eckstein and Bertsekas (1992), that identifies the core iteration of (6.8) as an instantiation of the proximal point algorithm.

Remark 6.4. Connections between the Douglas–Rachford algorithms and the method of partial inverses of Section 5.4.4 are discussed in Lawrence and Spingarn (1987, Section 1); see also Eckstein and Bertsekas (1992, Section 5) and Mahey, Oualibouch and Dinh Tao (1995). Let us show that we can actually derive Theorem 5.13(ii) from Theorem 6.2. Let $(x_n)_{n\in\mathbb{N}}$, $(x_n^*)_{n\in\mathbb{N}}$, $(p_n)_{n\in\mathbb{N}}$ and $(p_n^*)_{n\in\mathbb{N}}$ be the sequence generated by (5.29) and set $(\forall n \in \mathbb{N})\ y_n = x_n + x_n^*$ and $z_n = \operatorname{proj}_V(2p_n - y_n)$. Then (5.29) yields

$$(\forall n \in \mathbb{N})\ \operatorname{proj}_V p_n^* + \operatorname{proj}_{V^\perp} p_n = \operatorname{proj}_V(y_n - p_n) + p_n - \operatorname{proj}_V p_n = p_n - z_n. \tag{6.21}$$

Altogether,

$$(\forall n \in \mathbb{N})\quad p_n = J_A y_n, \quad z_n = \operatorname{proj}_V(2p_n - y_n) \quad \text{and} \quad y_{n+1} = y_n + \lambda_n(z_n - p_n). \tag{6.22}$$

In view of Example 2.36, this recursion is precisely that of (6.8) for the operators (N_V, A) with $\gamma = 1$. We therefore derive the following from Theorem 6.2: $(y_n)_{n\in\mathbb{N}}$ converges weakly to a point $y \in \mathcal{H}$ and, if we set $x = J_A y$ and $x^* = y - J_A y$, then $p_n \rightharpoonup x \in \operatorname{zer}(N_V + A)$ and, by Example 2.15, $p_n^* \rightharpoonup x^* \in \operatorname{zer}(N_{V^\perp} + A^{-1})$. Furthermore, (6.19)–(6.20) implies that $(x, -x^*) = (x, x - y) \in \operatorname{gra} N_V$ and $(x, x^*) = (x, y - x) \in \operatorname{gra} A$. Thus, Example 2.15 yields $(x, x^*) \in \operatorname{gra} N_V \cap \operatorname{gra} A$ and (x, x^*) therefore solves (5.28). Finally, since equation (11) of Combettes (2009) asserts that $J_A y = \operatorname{proj}_V y$ and since proj_V is weakly continuous, we have $x_n = \operatorname{proj}_V(x_n + x_n^*) = \operatorname{proj}_V y_n \rightharpoonup \operatorname{proj}_V y = x$ and $x_n^* = \operatorname{proj}_{V^\perp} y_n \rightharpoonup \operatorname{proj}_{V^\perp} y = y - \operatorname{proj}_V y = x^*$. Let us add that, in this setting, the operator $\boldsymbol{\mathcal{M}}$ of (6.9) is just the partial inverse A_V.

Remark 6.5. The many application areas of the Douglas–Rachford algorithm (in its original two-operator form or transposed in product spaces) include road design (Bauschke, Koch and Phan 2016), equilibrium problems (Briceño-Arias 2012), biostatistics (Combettes and Müller 2021), signal recovery (Combettes and Pesquet 2007), traffic theory (Fukushima 1996), noise removal (Steidl and Teuber 2010) and compressive sensing (Yu, Peng, Han and Cui 2017); see also Lindstrom and Sims (2021) for additional references.

6.3. *Strong convergence*

As shown in Bùi and Combettes (2020*a*, Counterexample 2), the convergence of $(x_n)_{n\in\mathbb{N}}$ in Theorem 6.2(i) is only weak. The following version based on Theorem 5.3 furnishes strong convergence.

Theorem 6.6. Let $A\colon \mathcal{H} \to 2^{\mathcal{H}}$ and $B\colon \mathcal{H} \to 2^{\mathcal{H}}$ be maximally monotone, suppose that $\operatorname{zer}(A + B) \neq \varnothing$, let $y_0 \in \mathcal{H}$, let $(\lambda_n)_{n\in\mathbb{N}}$ be a sequence in $]0, 1]$ such that $\inf_{n\in\mathbb{N}} \lambda_n > 0$, and let $\gamma \in\ ]0, +\infty[$. Iterate

$$\begin{array}{l} \text{for } n = 0, 1, \ldots \\ \left\lfloor \begin{array}{l} x_n = J_{\gamma B} y_n \\ x_n^* = \gamma^{-1}(y_n - x_n) \\ z_n = J_{\gamma A}(2x_n - y_n) \\ y_{n+1} = \mathsf{Q}(y_0, y_n, y_n + \lambda_n(z_n - x_n)), \end{array} \right. \end{array} \tag{6.23}$$

where Q is defined in Lemma 4.6. Let Z and Z^* be the sets of solutions to (6.6) and (6.7), respectively. Then the following hold:

(i) $(x_n)_{n\in\mathbb{N}}$ converges strongly to a point in Z.

(ii) $(x_n^*)_{n\in\mathbb{N}}$ converges strongly to a point in Z^*.

Proof. Define $\boldsymbol{\mathcal{M}}$ as in (6.9) and set $y = \operatorname{proj}_{\operatorname{zer}\boldsymbol{\mathcal{M}}} y_0$, $x = J_{\gamma B} y$ and $x^* = \gamma^{-1}(y - x)$. Then it follows from (6.10) that $x \in Z$ and from (6.11) that $x^* \in Z^*$. Additionally, we derive from (6.23) that

$$(\forall n \in \mathbb{N}) \quad y_{n+1} = \mathsf{Q}(y_0, y_n, y_n + \lambda_n(J_{\boldsymbol{\mathcal{M}}} y_n - y_n)). \tag{6.24}$$

Hence, Theorem 5.3 yields $y_n \to y$ and, by continuity of $J_{\gamma B}$, $x_n = J_{\gamma B} y_n \to J_{\gamma B} y = x$. Finally, $x_n^* = \gamma^{-1}(y_n - x_n) \to \gamma^{-1}(y - x) = x^*$. ☐

Remark 6.7. The method of partial inverses of Theorem 5.13 may converge only weakly (Bùi and Combettes 2020*a*, Counterexample 4). A strongly convergent version can be designed using Remark 6.4 and Theorem 6.6.

6.4. *Special cases and variants*

6.4.1. *Minimization setting*

We illustrate an application of the Douglas–Rachford algorithm to primal–dual minimization.

Example 6.8. Let $f \in \Gamma_0(\mathcal{H})$ and $g \in \Gamma_0(\mathcal{H})$ be such that $Z = \operatorname{Argmin}(f+g) \neq \varnothing$ and $0 \in \operatorname{sri}(\operatorname{dom} f - \operatorname{dom} g)$. Set $Z^* = \operatorname{Argmin}(f^* \circ (-\mathrm{Id}) + g^*)$, let $(\lambda_n)_{n\in\mathbb{N}}$ be a sequence in $]0, 2[$ such that $\sum_{n\in\mathbb{N}} \lambda_n(2 - \lambda_n) = +\infty$, let $\gamma \in \,]0, +\infty[$, let $y_0 \in \mathcal{H}$, and iterate

$$\begin{array}{l} \text{for } n = 0, 1, \ldots \\ \left\lfloor \begin{array}{l} x_n = \operatorname{prox}_{\gamma g} y_n \\ x_n^* = \gamma^{-1}(y_n - x_n) \\ z_n = \operatorname{prox}_{\gamma f}(2x_n - y_n) \\ y_{n+1} = y_n + \lambda_n(z_n - x_n). \end{array} \right. \end{array} \tag{6.25}$$

Then it follows from Problem 3.9, Example 2.35 and Theorem 6.2 that there exists $(x, x^*) \in Z \times Z^*$ such that $x_n \rightharpoonup x$ and $x_n^* \rightharpoonup x^*$.

Remark 6.9. Relations between the Douglas–Rachford algorithm (6.25) and other methods have been noted in the literature.

(i) It is observed in Condat (2013, Section 3.1.1) that the Douglas–Rachford algorithm (6.25) can be viewed as a limiting case of the Chambolle–Pock algorithm (see Remark 5.21(iv)) by implementing it in the case when $\mathcal{G} = \mathcal{H}$, $L = \mathrm{Id}$ and $\sigma = 1/\tau = \gamma$. Note, however, that this setting violates the condition $\tau\sigma\|L\|^2 < 1$ used to prove weak convergence of (5.60) in Example 5.20.

(ii) Consider the setting of Problem 3.9 and note that the primal minimization problem (3.12) is equivalent to

$$\underset{(x,y)\in\operatorname{gra}L}{\text{minimize}}\; f(x)+g(y). \tag{6.26}$$

The (unscaled) *augmented Lagrangian* associated with (6.26) is the saddle function (see Example 2.21) on $(\mathcal{H}\oplus\mathcal{G})\oplus\mathcal{G}$ defined as

$$\begin{aligned} F\colon \mathcal{H}\oplus\mathcal{G}\oplus\mathcal{G} &\to \left]-\infty,+\infty\right] \\ (x,y,v^*) &\mapsto f(x)+g(y)+\langle Lx-y \mid v^*\rangle+\frac{1}{2}\|Lx-y\|^2. \end{aligned} \tag{6.27}$$

Iteration n of the alternating-direction method of multipliers (ADMM) consists in minimizing F over x for y_n and v_n^* fixed to get x_n, then over y for x_n and v_n^* fixed to get y_{n+1}, and then applying a proximal maximization step with respect to the Lagrange multiplier v^* for x_n and y_{n+1} fixed to get v_{n+1}^*. It was originally proposed in Glowinski and Marrocco (1974), refined in Gabay and Mercier (1976), and further developed in Boyd *et al.* (2010), Eckstein and Bertsekas (1992), Gabay (1983) and Glowinski and Le Tallec (1989). Given $y_0\in\mathcal{G}$ and $v_0^*\in\mathcal{G}$, ADMM iterates

$$\begin{array}{l} \text{for } n=0,1,\ldots \\ \left\lfloor \begin{array}{l} x_n \in \underset{x\in\mathcal{H}}{\operatorname{Argmin}} \left(f(x)+\langle Lx \mid v_n^*\rangle+\frac{1}{2}\|Lx-y_n\|^2\right) \\ d_n = Lx_n \\ y_{n+1} = \underset{y\in\mathcal{G}}{\operatorname{argmin}} \left(g(y)-\langle y \mid v_n^*\rangle+\frac{1}{2}\|d_n-y\|^2\right) \\ v_{n+1}^* = v_n^*+d_n-y_{n+1}. \end{array}\right. \end{array} \tag{6.28}$$

It should be emphasized that ADMM is not a splitting algorithm in our sense since the computation of x_n involves a minimization step which does not separate f and L, and can therefore be hard to execute. This step is also set-valued in general. Nonetheless, (6.28) can be interpreted as an application of the Douglas–Rachford algorithm (6.25) to the functions $f^*\circ(-L^*)$ (here again, note that f and L are not separated and that the typically non-explicit operator $\operatorname{prox}_{f^*\circ(-L^*)}$ intervenes) and g^* present in the dual problem (3.13) (Gabay 1983); see also Eckstein and Bertsekas (1992). This is merely an algorithmic identification and not a claim that ADMM converges. Convergence requires more restrictions on the problem, for instance finite-dimensionality of $\mathcal{H}$ and $\mathcal{G}$ and invertibility of $L^*\circ L$ in Eckstein and Bertsekas (1992, Section 5). For further analysis, see Banert, Boţ and Csetnek (2021), Boţ and Csetnek (2019) and Ryu, Liu and Yin (2019).

6.4.2. *Peaceman–Rachford splitting*

The first implicit alternating direction method (Birkhoff and Varga 1959) to solve the positive-definite matrix equation $Ax+Bx=f$ is the Peaceman–Rachford algorithm (Peaceman and Rachford 1955; see also Douglas 1955). It is described by the iterative process

$$
\begin{array}{l}
\text{for } n = 0, 1, \ldots \\
\left\lfloor
\begin{array}{l}
x_{n+1/2} - x_n + Ax_{n+1/2} + Bx_n = f \\
x_{n+1} - x_{n+1/2} + Ax_{n+1/2} + Bx_{n+1} = f.
\end{array}
\right.
\end{array}
\tag{6.29}
$$

Using the same arguments used to transition from (6.1) to (6.4), we rewrite (6.29) as

$$
\begin{array}{l}
\text{for } n = 0, 1, \ldots \\
\left\lfloor
\begin{array}{l}
x_n = J_B y_n \\
z_n = J_A(2x_n - y_n + f) \\
y_{n+1} = y_n + 2(z_n - x_n).
\end{array}
\right.
\end{array}
\tag{6.30}
$$

The strong convergence of $(x_n)_{n\in\mathbb{N}}$ to a solution to the equation $Ax + Bx = f$, where A and B are single-valued hemicontinuous monotone operators such that $\operatorname{dom} A = \operatorname{dom} B = \mathcal{H}$ and B is strongly monotone, was established in Lieutaud (1969*a*) and, with the additional assumption that $\mathcal{H}$ is finite-dimensional and the operators are continuous, in Kellogg (1969).

Algorithm (6.30) was first considered for general maximally monotone set-valued operators A and B in Lions and Mercier (1979). In the presence of a scaling parameter $\gamma \in \,]0, +\infty[$ and taking $f = 0$ without loss of generality, the Peaceman–Rachford algorithm becomes

$$
\begin{array}{l}
\text{for } n = 0, 1, \ldots \\
\left\lfloor
\begin{array}{l}
x_n = J_{\gamma B} y_n \\
z_n = J_{\gamma A}(2x_n - y_n) \\
y_{n+1} = y_n + 2(z_n - x_n).
\end{array}
\right.
\end{array}
\tag{6.31}
$$

Upon defining $\boldsymbol{\mathcal{M}}$ as in (6.9), we derive from (6.31) that

$$
(\forall n \in \mathbb{N}) \quad y_{n+1} = (2J_{\boldsymbol{\mathcal{M}}} - \mathrm{Id})y_n. \tag{6.32}
$$

We can view (6.31) as a limiting case of the Douglas–Rachford algorithm (6.8) in which the relaxation parameters $(\lambda_n)_{n\in\mathbb{N}}$ are allowed to be 2. This, of course, means that (6.31) operates outside of the setting of Theorem 5.1 and hence of the geometric framework of Theorem 4.2. As a result, the weak convergence of $(y_n)_{n\in\mathbb{N}}$ cannot be guaranteed without additional assumptions since (6.32) amounts to iterating a merely nonexpansive operator; see Lions and Mercier (1979, Remark 6) for a counterexample. Strong convergence of $(x_n)_{n\in\mathbb{N}}$ to a point in $\operatorname{zer}(A + B)$ takes place when B is strongly monotone (Lions and Mercier 1979, Remark 2). More generally, strong convergence occurs when B is uniformly monotone on bounded sets or when $\operatorname{int}\operatorname{Fix}(2J_{\gamma A} - \mathrm{Id})(2J_{\gamma B} - \mathrm{Id}) \neq \varnothing$ (Combettes 2009, Remark 2.2(iv)).

6.4.3. A three-operator splitting algorithm

An extension of the Douglas–Rachford algorithm (6.8) was proposed in Davis and Yin (2017) by adding a cocoercive operator to the inclusion (6.6).

Proposition 6.10. Let $\tau \in]0, +\infty[$, let $A\colon \mathcal{H} \to 2^{\mathcal{H}}$ and $B\colon \mathcal{H} \to 2^{\mathcal{H}}$ be maximally monotone, and let $C\colon \mathcal{H} \to \mathcal{H}$ be τ-cocoercive. Suppose that the set Z of solutions to the inclusion

$$\text{find } x \in \mathcal{H} \text{ such that } 0 \in Ax + Bx + Cx \tag{6.33}$$

is not empty and let Z^* be the set of solutions to the dual problem

$$\text{find } x^* \in \mathcal{H} \text{ such that } 0 \in -(A + C)^{-1}(-x^*) + B^{-1}x^*. \tag{6.34}$$

Let $\gamma \in]0, 2\tau[$, set $\delta = 2 - \gamma/(2\tau)$, let $(\lambda_n)_{n\in\mathbb{N}}$ be a sequence in $]0, \delta[$ such that $\sum_{n\in\mathbb{N}} \lambda_n(\delta - \lambda_n) = +\infty$, and let $y_0 \in \mathcal{H}$. Iterate

$$\begin{array}{l} \text{for } n = 0, 1, \ldots \\ \left\lfloor \begin{array}{l} x_n = J_{\gamma B}\, y_n \\ x_n^* = \gamma^{-1}(y_n - x_n) \\ r_n = y_n + \gamma C x_n \\ z_n = J_{\gamma A}(2x_n - r_n) \\ y_{n+1} = y_n + \lambda_n(z_n - x_n). \end{array} \right. \end{array} \tag{6.35}$$

Then there exists $y \in \mathcal{H}$ such that $y_n \rightharpoonup y$. Now set $x = J_{\gamma B} y$ and $x^* = {}^{\gamma}B y$. Then the following hold:

(i) $x_n \rightharpoonup x \in Z$.

(ii) $x_n^* \rightharpoonup x^* \in Z^*$.

Proof. Remarkably, we can closely follow the proof of Theorem 6.2. The key additional facts established in Davis and Yin (2017, Proposition 2.1 and Lemma 2.2) are that, for $\alpha = 1/\delta$,

$$T = J_{\gamma A} \circ (2J_{\gamma B} - \mathrm{Id} - \gamma C \circ J_{\gamma B}) + \mathrm{Id} - J_{\gamma B} \text{ is } \alpha\text{-averaged and } Z = J_{\gamma B}(\operatorname{Fix} T). \tag{6.36}$$

We write the maximally monotone operator $\boldsymbol{\mathcal{M}}$ of (3.26) as

$$\boldsymbol{\mathcal{M}} = \left(\mathrm{Id} + \frac{1}{2\alpha}(J_{\gamma A} \circ (2J_{\gamma B} - \mathrm{Id} - \gamma C \circ J_{\gamma B}) - J_{\gamma B})\right)^{-1} - \mathrm{Id} \tag{6.37}$$

and, in view of Example 3.16 and (6.36), work with the embedding $(\mathcal{H}, \boldsymbol{\mathcal{M}}, J_{\gamma B})$ of (6.33). Then $\varnothing \neq Z = J_{\gamma B}(\operatorname{zer} \boldsymbol{\mathcal{M}})$ and $(y_n)_{n\in\mathbb{N}}$ is produced by the proximal point algorithm $(\forall n \in \mathbb{N})\ y_{n+1} = y_n + \mu_n(J_{\boldsymbol{\mathcal{M}}} y_n - y_n)$, where $\mu_n = 2\alpha\lambda_n \in]0, 2[$. Using Theorem 5.1(i), we infer that $(y_n)_{n\in\mathbb{N}}$ converges weakly to a point $y \in \operatorname{zer} \boldsymbol{\mathcal{M}}$ and that $J_{\boldsymbol{\mathcal{M}}} y_n - y_n \to 0$. Hence, we derive from (6.36), (6.35) and (6.37) that

$$x = J_{\gamma B} y \in Z \quad \text{and} \quad z_n - x_n = 2\alpha(J_{\boldsymbol{\mathcal{M}}} y_n - y_n) \to 0, \tag{6.38}$$

and hence that

$$\|Cz_n - Cx_n\| \leq \alpha^{-1}\|z_n - x_n\| \to 0. \tag{6.39}$$

(i) Set $(\forall n \in \mathbb{N})\ z_n^* = \gamma^{-1}(2x_n - z_n - r_n) + Cz_n$. In view of (6.35) and (2.18),

$$(\forall n \in \mathbb{N}) \quad \begin{cases} (z_n, z_n^*) \in \operatorname{gra}(A + C), \\ (x_n, x_n^*) \in \operatorname{gra} B, \\ z_n^* + x_n^* = \gamma^{-1}(x_n - z_n) + Cz_n - Cx_n. \end{cases} \tag{6.40}$$

Next, fix $z \in \mathfrak{W}(x_n)_{n\in\mathbb{N}}$, say $x_{k_n} \rightharpoonup z$. Since $y_{k_n} \rightharpoonup y$, it follows from (6.38), (6.39), (6.40) and (6.35) that

$$z_{k_n} \rightharpoonup z, \quad z_{k_n}^* \rightharpoonup \gamma^{-1}(z - y), \quad z_n - x_n \to 0 \quad \text{and} \quad z_n^* + x_n^* \to 0. \tag{6.41}$$

By applying Lemma 2.50 to the maximally monotone operators $A + C$ (see Example 2.5 and Lemma 2.27(ii)) and B, we deduce from (6.40) and (6.41) that $z \in \operatorname{zer}(A + C + B) = Z$,

$$(z, \gamma^{-1}(z - y)) \in \operatorname{gra}(A + C) \quad \text{and} \quad (z, \gamma^{-1}(y - z)) \in \operatorname{gra} B. \tag{6.42}$$

In turn, (2.18) asserts that $z = J_{\gamma B}y$, making $x = J_{\gamma B}y$ the unique weak sequential cluster point of $(x_n)_{n\in\mathbb{N}}$ which is bounded since $(y_n)_{n\in\mathbb{N}}$ is. By Lemma 4.1(ii), $x_n \rightharpoonup x$.

(ii) Since $y_n \rightharpoonup y$ and $x_n \rightharpoonup x$, we have $x_n^* = \gamma^{-1}(y_n - x_n) \rightharpoonup \gamma^{-1}(y - x) = {}^{\gamma}By = x^* \in Z^*$ by (6.11). ☐

Remark 6.11. Here are a few comments on Proposition 6.10.

(i) The conclusion of Proposition 6.10(i) was first established in Davis and Yin (2017, Theorem 2.1.1(b)) with a different proof. See also Raguet (2019) for a discussion and connections with Raguet *et al.* (2013).

(ii) The duality result of Proposition 6.10(ii) is new.

(iii) A strongly convergent version of Proposition 6.10 can be obtained by adapting the proof of Theorem 6.6 to the presence of C, as was done above.

(iv) When $C = 0$, Proposition 6.10 produces the Douglas–Rachford setting of Theorem 6.2. When $B = 0$, (6.35) yields a special case of the forward–backward method of Combettes and Yamada (2015, Proposition 4.4(iii)) in which the proximal parameters are all equal to γ.

7. Tseng's forward–backward–forward splitting

7.1. Preview

In Section 5.4.1, we have discussed a Euler method for finding a zero of a single-valued operator $B\colon \mathcal{H} \to \mathcal{H}$ under a cocoercivity condition. Under the more general assumption that B is monotone and β-Lipschitzian, the Euler method is no longer appropriate, and we can use a scheme proposed by Antipin (1976) and

Korpelevič (1976) that involves a double activation of the operator B. Specifically, in this method, $\gamma \in \left]0, 1/\beta\right[$ and $x_0 \in \mathcal{H}$ are fixed and we iterate

$$\begin{array}{l}\text{for } n = 0, 1, \ldots \\ \left\lfloor \begin{array}{l} b_n^* = \gamma B x_n \\ m_n = x_n - b_n^* \\ m_n^* = B m_n \\ x_{n+1} = x_n - \gamma m_n^*. \end{array}\right.\end{array} \tag{7.1}$$

Clearly, the sequence $(m_n, m_n^*)_{n\in\mathbb{N}}$ lies gra B and it is straightforward to see that, by choosing $(\lambda_n)_{n\in\mathbb{N}}$ suitably in (4.32), we obtain (7.1). The convergence properties of the Antipin–Korpelevič method can therefore be deduced from the results of Section 4.4 applied to B.

Tseng's algorithm can be viewed as a generalization of (7.1) for the problem of finding a zero of $A + B$, where $A\colon \mathcal{H} \to 2^{\mathcal{H}}$ is maximally monotone and B is as above. It is called the forward–backward–forward algorithm because it performs a forward step on B, then a backward step on A, and finally another forward step on B. We are going to derive the convergence of Tseng's forward–backward–forward splitting algorithm from the principles of Section 4.4 and, more precisely, from the warped resolvent algorithm of Section 4.5.

7.2. *Fejérian algorithm*

We cast the forward–backward–forward algorithm as an instance of (4.34) and then prove its weak convergence via Theorem 4.12. This result was originally established in Tseng (2000, Theorem 3.4(b)), where different arguments were used.

Theorem 7.1. Let $\beta \in \left]0, +\infty\right[$, let $A\colon \mathcal{H} \to 2^{\mathcal{H}}$ be maximally monotone, let $B\colon \mathcal{H} \to \mathcal{H}$ be monotone and β-Lipschitzian, and suppose that $Z = \operatorname{zer}(A+B) \neq \varnothing$. Let $x_0 \in \mathcal{H}$, let $\varepsilon \in \left]0, 1/(\beta+1)\right[$, and let $(\gamma_n)_{n\in\mathbb{N}}$ be a sequence in $[\varepsilon, (1-\varepsilon)/\beta]$. Iterate

$$\begin{array}{l}\text{for } n = 0, 1, \ldots \\ \left\lfloor \begin{array}{l} b_n^* = \gamma_n B x_n \\ m_n = J_{\gamma_n A}(x_n - b_n^*) \\ x_{n+1} = m_n - \gamma_n B m_n + b_n^*. \end{array}\right.\end{array} \tag{7.2}$$

Then $(x_n)_{n\in\mathbb{N}}$ converges weakly to a point in Z.

Proof. Our objective is to apply Theorem 4.12 with

$$W = A + B, \quad C = 0, \quad \text{and} \quad (\forall n \in \mathbb{N}) \quad U_n = \gamma_n^{-1}\mathrm{Id} - B \quad \text{and} \quad q_n = w_n. \tag{7.3}$$

Since $C = 0$, let us rename $(w_n)_{n\in\mathbb{N}}$ as $(m_n)_{n\in\mathbb{N}}$. Example 2.3 and Lemma 2.27(ii) entail that W is maximally monotone. Moreover, a consequence of Lemma 2.48(i)–(ii) is that

$$(\forall n \in \mathbb{N}) \quad \gamma_n U_n \text{ is } \varepsilon\text{-strongly monotone and } 1/(2-\varepsilon)\text{-cocoercive.} \tag{7.4}$$

Additionally, we derive from Bùi and Combettes (2020*b*, Proposition 3.9) that

$$(\forall n \in \mathbb{N}) \quad \operatorname{ran} U_n \subset \operatorname{ran}(U_n + W + C) \quad \text{and} \quad U_n + W + C \text{ is injective.} \tag{7.5}$$

We also observe that

$$(\forall n \in \mathbb{N}) \quad J^{U_n}_{W+C} = J^{U_n}_{A+B} = \big(\gamma_n^{-1}\mathrm{Id} + A\big) \circ \big(\gamma_n^{-1}\mathrm{Id} - B\big) = J_{\gamma_n A} \circ (\mathrm{Id} - \gamma_n B). \tag{7.6}$$

Hence, the variables of (4.34) in this setting become

$$(\forall n \in \mathbb{N}) \quad \begin{cases} m_n = J_{\gamma_n A}(x_n - \gamma_n B x_n), \\ t_n^* = U_n x_n - U_n m_n, \\ \delta_n = \langle m_n - x_n \mid U_n m_n - U_n x_n \rangle. \end{cases} \tag{7.7}$$

Now set

$$(\forall n \in \mathbb{N}) \quad \lambda_n = \begin{cases} \dfrac{\gamma_n \|t_n^*\|^2}{\delta_n}, & \text{if } \delta_n > 0, \\ \varepsilon, & \text{otherwise.} \end{cases} \tag{7.8}$$

We derive from (7.4) that

$$(\forall n \in \mathbb{N}) \quad \delta_n = \langle m_n - x_n \mid U_n m_n - U_n x_n \rangle \geq \beta\varepsilon \|m_n - x_n\|^2, \tag{7.9}$$

which implies that

$$(\forall n \in \mathbb{N}) \quad \delta_n \leq 0 \Leftrightarrow m_n = x_n \Leftrightarrow t_n^* = 0. \tag{7.10}$$

A consequence of (7.4) is that, if $\delta_n > 0$,

$$\frac{\varepsilon}{\gamma_n} \leq \frac{\|U_n m_n - U_n x_n\|}{\|m_n - x_n\|} \leq \frac{\|U_n m_n - U_n x_n\|^2}{\langle m_n - x_n \mid U_n m_n - U_n x_n \rangle} \leq \frac{2-\varepsilon}{\gamma_n}, \tag{7.11}$$

and we therefore obtain from (7.8) that

$$\lambda_n = \frac{\gamma_n \|U_n m_n - U_n x_n\|^2}{\langle m_n - x_n \mid U_n m_n - U_n x_n \rangle} \in [\varepsilon, 2-\varepsilon]. \tag{7.12}$$

Hence, (4.34) and (7.10) yield

$$(\forall n \in \mathbb{N}) \quad d_n = \frac{\gamma_n}{\lambda_n} t_n^*. \tag{7.13}$$

Consequently, the sequence $(x_n)_{n\in\mathbb{N}}$ produced by (7.2) coincides with that of (4.34). We therefore appeal to Theorem 4.12(ii) to conclude since its condition (iib) holds thanks to (7.12), whereas its condition (iid) holds thanks to (7.4) and the fact that $(\gamma_n)_{n\in\mathbb{N}}$ lies in $[\varepsilon, (1-\varepsilon)/\beta]$. □

7.3. Haugazeau-like algorithm

We present a strongly convergent best approximation version of the forward–backward–forward method based on Theorem 4.14.

Theorem 7.2. Let $\beta \in \,]0,+\infty[$, let $A\colon \mathcal{H} \to 2^{\mathcal{H}}$ be maximally monotone, let $B\colon \mathcal{H} \to \mathcal{H}$ be monotone and β-Lipschitzian, and suppose that $Z = \operatorname{zer}(A+B) \neq \varnothing$. Let $x_0 \in \mathcal{H}$, let $\varepsilon \in \,]0, 1/(\beta+1)[$, and let $(\gamma_n)_{n\in\mathbb{N}}$ be a sequence in $[\varepsilon, (1-\varepsilon)/\beta]$. Iterate

$$
\begin{array}{l}
\text{for } n = 0,1,\ldots\\
\left\lfloor
\begin{array}{l}
b_n^* = \gamma_n B x_n\\
m_n = J_{\gamma_n A}(x_n - b_n^*)\\
r_n = \dfrac{1}{2}(x_n + m_n - \gamma_n B m_n + b_n^*)\\
x_{n+1} = \mathsf{Q}(x_0, x_n, r_n),
\end{array}
\right.
\end{array}
\tag{7.14}
$$

where Q is defined in Lemma 4.6. Then $(x_n)_{n\in\mathbb{N}}$ converges strongly to $\operatorname{proj}_Z x_0$.

Proof. We prove the claim as an application of Theorem 4.14 in the setting of (7.3). Let us use the same variables as in (7.7) and

$$
(\forall n \in \mathbb{N}) \quad \lambda_n = \begin{cases} \dfrac{\gamma_n \|t_n^*\|^2}{2\delta_n}, & \text{if } \delta_n > 0,\\ \varepsilon/2, & \text{otherwise.} \end{cases} \tag{7.15}
$$

Then, using the same arguments as in the proof of Theorem 4.12, we see that $(\lambda_n)_{n\in\mathbb{N}}$ lies in $[\varepsilon/2, 1]$ and that the sequence $(x_n)_{n\in\mathbb{N}}$ produced by (7.14) coincides with that of (4.44). Since conditions (iib) and (iid) in Theorem 4.14(ii) are fulfilled, we obtain the claim. ☐

7.4. *Special cases and variants*

7.4.1. *The monotone+skew algorithm*

The approach presented here was proposed in Briceño-Arias and Combettes (2011) to solve the monotone inclusion (3.7) and it was the first algorithm to fully split the operators A, B and L. Its methodology conforms to the programme of Framework 1.2: we use the embedding of Example 3.20 to transfer the initial 3-operator problem (3.7) in the primal space $\mathcal{H}$ to one involving the Kuhn–Tucker operator $\mathcal{K} = \boldsymbol{M} + \boldsymbol{S}$ of (3.10) in the larger primal–dual space $\mathbf{X} = \mathcal{H} \oplus \mathcal{G}$. The algorithmic strategy *per se* is then straightforward: since $\boldsymbol{M}$ is maximally monotone and $\boldsymbol{S}$ is monotone and Lipschitzian, we can apply Tseng's forward–backward–forward algorithm (Theorem 7.1) in $\mathbf{X}$ to find a Kuhn–Tucker point and hence a primal–dual solution.

We derive from Theorem 7.1 the weak convergence of the monotone+skew algorithm of Briceño-Arias and Combettes (2011, Theorem 3.1(ii)) (we can derive a strongly convergent version from Theorem 7.2 using the same arguments).

Proposition 7.3. Let $A\colon \mathcal{H} \to 2^{\mathcal{H}}$ and $B\colon \mathcal{G} \to 2^{\mathcal{G}}$ be maximally monotone, and assume that $0 \neq L \in \mathcal{B}(\mathcal{H},\mathcal{G})$. Suppose that the set Z of solutions to the primal inclusion

$$
\text{find } x \in \mathcal{H} \text{ such that } 0 \in Ax + L^*(B(Lx)) \tag{7.16}
$$

is not empty and let Z^* be the set of solutions to the dual inclusion

$$\text{find } y^* \in \mathcal{G} \text{ such that } 0 \in -L(A^{-1}(-L^*y^*)) + B^{-1}y^*. \tag{7.17}$$

Let $x_0 \in \mathcal{H}$, let $y_0^* \in \mathcal{G}$, let $\varepsilon \in \left]0, 1/(\|L\|+1)\right[$, let $(\gamma_n)_{n\in\mathbb{N}}$ be a sequence in $[\varepsilon, (1-\varepsilon)/\|L\|\,]$, and set

$$\begin{array}{l}\text{for } n = 0, 1, \ldots \\ \left\lfloor \begin{array}{l} y_{1,n} = x_n - \gamma_n L^* y_n^* \\ y_{2,n}^* = y_n^* + \gamma_n L x_n \\ m_{1,n} = J_{\gamma_n A} y_{1,n} \\ m_{2,n}^* = J_{\gamma_n B^{-1}} y_{2,n}^* \\ q_{1,n} = m_{1,n} - \gamma_n L^* m_{2,n}^* \\ q_{2,n}^* = m_{2,n}^* + \gamma_n L m_{1,n} \\ x_{n+1} = x_n - y_{1,n} + q_{1,n} \\ y_{n+1}^* = y_n^* - y_{2,n}^* + q_{2,n}^*. \end{array} \right.\end{array} \tag{7.18}$$

Then there exist $x \in Z$ and $y^* \in Z^*$ such that $-L^*y^* \in Ax$, $y^* \in B(Lx)$, $x_n \rightharpoonup x$ and $y_n^* \rightharpoonup y^*$.

Proof. Set $\mathbf{X} = \mathcal{H} \oplus \mathcal{G}$, define $\boldsymbol{M}$ and $\boldsymbol{S}$ as in (3.9), and set $(\forall n \in \mathbb{N})$ $\boldsymbol{x}_n = (x_n, y_n^*)$, $\boldsymbol{y}_n = (y_{1,n}, y_{2,n}^*)$, $\boldsymbol{m}_n = (m_{1,n}, m_{2,n}^*)$ and $\boldsymbol{q}_n = (q_{1,n}, q_{2,n}^*)$. Then, in view of Example 2.37, (7.18) becomes

$$\begin{array}{l}\text{for } n = 0, 1, \ldots \\ \left\lfloor \begin{array}{l} \boldsymbol{y}_n = \boldsymbol{x}_n - \gamma_n \boldsymbol{S}\boldsymbol{x}_n \\ \boldsymbol{m}_n = J_{\gamma_n \boldsymbol{M}} \boldsymbol{y}_n \\ \boldsymbol{q}_n = \boldsymbol{m}_n - \gamma_n \boldsymbol{S}\boldsymbol{m}_n \\ \boldsymbol{x}_{n+1} = \boldsymbol{x}_n - \boldsymbol{y}_n + \boldsymbol{q}_n, \end{array} \right.\end{array} \tag{7.19}$$

which we rewrite as an instance of (7.2), namely,

$$\begin{array}{l}\text{for } n = 0, 1, \ldots \\ \left\lfloor \begin{array}{l} \boldsymbol{b}_n^* = \gamma_n \boldsymbol{S}\boldsymbol{x}_n \\ \boldsymbol{m}_n = J_{\gamma_n \boldsymbol{M}}(\boldsymbol{x}_n - \boldsymbol{b}_n^*) \\ \boldsymbol{x}_{n+1} = \boldsymbol{m}_n - \gamma_n \boldsymbol{S}\boldsymbol{m}_n + \boldsymbol{b}_n^*. \end{array} \right.\end{array} \tag{7.20}$$

It then follows from Theorem 7.1 and Lemma 3.8 that $(\boldsymbol{x}_n)_{n\in\mathbb{N}}$ converges weakly to a point in $\operatorname{zer}(\boldsymbol{M} + \boldsymbol{S}) \subset Z \times Z^*$, as claimed. ☐

Remark 7.4. The methodology of Theorem 7.1 is to find a Kuhn–Tucker point, i.e. a zero of $\boldsymbol{M}+\boldsymbol{S}$. As noted in Briceño-Arias and Combettes (2011, Remark 2.9), this can also be achieved by using the Douglas–Rachford algorithm (6.8) which, upon setting $U = (\mathrm{Id} + \gamma^2 L^* \circ L)^{-1}$ and $V = (\mathrm{Id} + \gamma^2 L \circ L^*)^{-1}$, and taking $\gamma \in \left]0, +\infty\right[$ and a sequence $(\lambda_n)_{n\in\mathbb{N}}$ in $\left]0, 2\right[$ such that $\sum_{n\in\mathbb{N}} \lambda_n(2-\lambda_n) = +\infty$, assumes the form

$$
\begin{array}{l}
\text{for } n = 0, 1, \ldots \\
\left\lfloor
\begin{array}{l}
x_n = U\big(y_{1,n} - \gamma L^* y_{2,n}^*\big) \\
y_n^* = V\big(y_{2,n}^* + \gamma L y_{1,n}\big) \\
y_{1,n+1} = y_{1,n} + \lambda_n(J_{\gamma A}(2x_n - y_{1,n}) - x_n) \\
y_{2,n+1}^* = y_{2,n}^* + \lambda_n\big(J_{\gamma B^{-1}}\big(2y_n^* - y_{2,n}^*\big) - y_n^*\big).
\end{array}
\right.
\end{array}
\tag{7.21}
$$

Weak convergence of $(x_n, y_n^*)_{n\in\mathbb{N}}$ to a point in $Z \times Z^*$ follows from Theorem 6.2(i). The numerical effectiveness of (7.21) depends on the ease of implementation of the operators U and V. This approach was rediscovered in O'Connor and Vandenberghe (2014) in an image restoration application.

7.4.2. A Lagrangian approach to composite minimization

We revisit the setting of Problem 3.9, which was identified as an instance of Problem 3.7 and can therefore be solved using (7.18) with $A = \partial f$ and $B = \partial g$. Following Combettes (2018, Section 4.5), we explore a different route which amounts to employing the embedding $(\mathbf{X}, \boldsymbol{\mathcal{S}}_{\boldsymbol{F}}, \boldsymbol{\mathcal{T}})$, where $\mathbf{X} = \mathcal{H} \oplus \mathcal{G} \oplus \mathcal{G}$,

$$
\boldsymbol{\mathcal{S}}_{\boldsymbol{F}} \colon \quad
\begin{array}{rcl}
\mathbf{X} & \to & 2^{\mathbf{X}} \\
(x, y, v^*) & \mapsto & (\partial f(x) + L^* v^*) \times (\partial g(y) - v^*) \times \{-Lx + y\}
\end{array}
\tag{7.22}
$$

is the saddle operator of (3.24), and $\boldsymbol{\mathcal{T}} \colon \mathbf{X} \to \mathcal{H} \colon (x, y, v^*) \mapsto x$. Let us write $\boldsymbol{\mathcal{S}}_{\boldsymbol{F}} = \boldsymbol{M} + \boldsymbol{S}$, where

$$
\begin{cases}
\boldsymbol{M} \colon (x, y, v^*) \mapsto \partial f(x) \times \partial g(y) \times \{0\}, \\
\boldsymbol{S} \colon (x, y, v^*) \mapsto (L^* v^*, -v^*, -Lx + y).
\end{cases}
\tag{7.23}
$$

Then $\|\boldsymbol{S}\| = \sqrt{1 + \|L\|^2}$ and $(\forall n \in \mathbb{N})\ J_{\gamma_n \boldsymbol{M}} = \operatorname{prox}_{\gamma_n f} \times \operatorname{prox}_{\gamma_n g} \times \mathrm{Id}$. Hence, applying Theorem 7.1 to this decomposition in $\mathbf{X}$, we obtain the following realization of Framework 1.2.

Proposition 7.5. Let $f \in \Gamma_0(\mathcal{H})$, $g \in \Gamma_0(\mathcal{G})$ and $L \in \mathcal{B}(\mathcal{H}, \mathcal{G})$ be such that $0 \in \operatorname{sri}(L(\operatorname{dom} f) - \operatorname{dom} g)$. Suppose that the primal problem

$$
\underset{x \in \mathcal{H}}{\text{minimize}} \;\; f(x) + g(Lx)
\tag{7.24}
$$

admits solutions and consider the dual problem

$$
\underset{v^* \in \mathcal{G}}{\text{minimize}} \;\; f^*(-L^* v^*) + g^*(v^*).
\tag{7.25}
$$

Let $(x_0, y_0, v_0^*) \in \mathcal{H} \oplus \mathcal{G} \oplus \mathcal{G}$, let $\varepsilon \in\,]0, 1/(1 + \sqrt{1 + \|L\|^2})[$, and let $(\gamma_n)_{n\in\mathbb{N}}$ be a sequence in $[\varepsilon, (1-\varepsilon)/\sqrt{1+\|L\|^2}]$. Iterate

$$
\begin{array}{l}
\text{for } n = 0, 1, \ldots \\
\left\lfloor
\begin{array}{l}
r_n = \gamma_n(Lx_n - y_n) \\
m_{1,n} = \operatorname{prox}_{\gamma_n f}(x_n - \gamma_n L^* v_n^*) \\
m_{2,n} = \operatorname{prox}_{\gamma_n g}(y_n + \gamma_n v_n^*) \\
x_{n+1} = m_{1,n} - \gamma_n L^* r_n \\
y_{n+1} = m_{2,n} + \gamma_n r_n \\
v_{n+1}^* = v_n^* + \gamma_n(Lm_{1,n} - m_{2,n}).
\end{array}
\right.
\end{array}
\tag{7.26}
$$

Then $(x_n)_{n\in\mathbb{N}}$ and $(v_n^*)_{n\in\mathbb{N}}$ converge weakly to solutions to (7.24) and (7.25), respectively.

Remark 7.6. Let $(\mu_n)_{n\in\mathbb{N}}$ be a sequence in $[\varepsilon, (1-\varepsilon)\min\{1, 1/\|L\|\}/2]$. Algorithm (7.26) bears a certain resemblance with the iterative scheme

$$
\begin{array}{l}
\text{for } n = 0, 1, \ldots \\
\left\lfloor
\begin{array}{l}
p_n = v_n^* + \mu_n(Lx_n - y_n) \\
x_{n+1} = \operatorname{prox}_{\mu_n f}(x_n - \mu_n L^* p_n) \\
y_{n+1} = \operatorname{prox}_{\mu_n g}(y_n + \mu_n p_n) \\
v_{n+1}^* = v_n^* + \mu_n(Lx_{n+1} - y_{n+1})
\end{array}
\right.
\end{array}
\tag{7.27}
$$

proposed in Chen and Teboulle (1994) to solve (7.24)–(7.25) in a finite-dimensional setting.

Remark 7.7. In the finite-dimensional context of Eckstein (1994), the saddle operator (7.22) was split as $\boldsymbol{\mathcal{S}}_{\boldsymbol{F}} = \boldsymbol{M}_1 + \boldsymbol{M}_2$, where

$$
\begin{cases}
\boldsymbol{M}_1\colon (x, y, v^*) \mapsto (\partial f(x) + L^*v^*) \times \{0\} \times \{-Lx\}, \\
\boldsymbol{M}_2\colon (x, y, v^*) \mapsto \{0\} \times (\partial g(y) - v^*) \times \{y\}.
\end{cases}
\tag{7.28}
$$

Given $\gamma \in\,]0, +\infty[$, $\mu_1 \in \mathbb{R}$, $\mu_2 \in \mathbb{R}$ and $(x_0, y_0, v_0^*) \in \mathcal{H} \oplus \mathcal{G} \oplus \mathcal{G}$, applying the Douglas–Rachford algorithm (6.8) to find a zero of $\boldsymbol{M}_1 + \boldsymbol{M}_2$ leads to the algorithm (Eckstein 1994)

$$
\begin{array}{l}
\text{for } n = 0, 1, \ldots \\
\left\lfloor
\begin{array}{l}
x_{n+1} \in \underset{x\in\mathcal{H}}{\operatorname{Argmin}} \left(f(x) + \langle Lx \mid v_n^* \rangle + \dfrac{1}{2\gamma}\|Lx - y_n\|^2 + \dfrac{\gamma\mu_1^2}{2}\|x - x_n\|^2 \right) \\
y_{n+1} = \underset{y\in\mathcal{G}}{\operatorname{argmin}} \left(g(y) - \langle y \mid v_n^* \rangle + \dfrac{1}{2\gamma}\|Lx_{n+1} - y\|^2 + \dfrac{\gamma\mu_2^2}{2}\|y - y_n\|^2 \right) \\
v_{n+1}^* = v_n^* + \gamma^{-1}(Lx_{n+1} - y_{n+1}).
\end{array}
\right.
\end{array}
\tag{7.29}
$$

When $\mu_1 = \mu_2 = 0$, we recover the alternating direction method of multipliers (ADMM) discussed in Remark 6.9(ii). Just like ADMM, (7.29) necessitates a potentially complex minimization involving f and L jointly to construct x_{n+1}. By contrast, (7.26) achieves full splitting of f, g and L.

Remark 7.8. In view of Example 3.23, the above saddle operator formalism can be extended to the more general primal–dual inclusion pair of Problem 3.7. As in Proposition 7.5, a zero (x, y, v^*) of the saddle operator $\boldsymbol{\mathcal{S}}$ of (3.25) can be constructed by executing (7.26), where $\operatorname{prox}_{\gamma_n f}$ is replaced with $J_{\gamma_n A}$ and $\operatorname{prox}_{\gamma_n g}$ with $J_{\gamma_n B}$. In this setting, the weak limits x and v^* solve, respectively, the primal inclusion (3.7) and the dual inclusion (3.8).

7.4.3. Mixtures of composite, Lipschitzian and parallel-sum operators

The Kuhn–Tucker operator of Lemma 3.8 employed in Section 7.4.1 can be expressed in block format as

$$\boldsymbol{\mathcal{K}} = \boldsymbol{M} + \boldsymbol{S} = \underbrace{\begin{bmatrix} A & 0 \\ 0 & B^{-1} \end{bmatrix}}_{\text{monotone}} + \underbrace{\begin{bmatrix} 0 & L^* \\ -L & 0 \end{bmatrix}}_{\text{skew}}. \tag{7.30}$$

A Kuhn–Tucker point was obtained in Proposition 7.3 by applying the forward–backward–forward algorithm (7.2) to $\boldsymbol{M}$ and $\boldsymbol{S}$. In doing so, we did not exploit the linearity and skewness of $\boldsymbol{S}$, but just the fact that it is monotone and Lipschitzian. Let us observe that, if we fill the diagonal of $\boldsymbol{S}$ with monotone Lipschitzian operators $Q \colon \mathcal{H} \to \mathcal{H}$ and $D^{-1} \colon \mathcal{G} \to \mathcal{G}$, we obtain a new monotone and Lipschitzian operator $\boldsymbol{Q} \colon \mathbf{X} \to \mathbf{X}$. In lieu of (7.30), we then consider the decomposition

$$\boldsymbol{\mathcal{K}} = \boldsymbol{M} + \boldsymbol{Q} = \underbrace{\begin{bmatrix} A & 0 \\ 0 & B^{-1} \end{bmatrix}}_{\text{monotone}} + \underbrace{\begin{bmatrix} Q & L^* \\ -L & D^{-1} \end{bmatrix}}_{\text{monotone and Lipschitzian}}. \tag{7.31}$$

Using (2.62), we write

$$\boldsymbol{\mathcal{K}} = \begin{bmatrix} A + Q & L^* \\ -L & (B \,\square\, D)^{-1} \end{bmatrix} \tag{7.32}$$

and interpret it as a variant of the Kuhn–Tucker operator (3.10) associated with Problem 3.7 in which A is replaced with $A + Q$ and B with $B \,\square\, D$. In other words, the primal inclusion is to

$$\text{find } x \in \mathcal{H} \text{ such that } 0 \in Ax + L^*((B \,\square\, D)(Lx)) + Qx \tag{7.33}$$

and the dual inclusion is to

$$\text{find } y^* \in \mathcal{G} \text{ such that } 0 \in -L((A + Q)^{-1}(-L^*y^*)) + B^{-1}y^* + D^{-1}y^* \tag{7.34}$$

or, equivalently,

$$\text{find } y^* \in \mathcal{G} \text{ such that } (\exists x \in \mathcal{H}) \begin{cases} -L^*y^* \in Ax + Qx, \\ Lx \in B^{-1}y^* + D^{-1}y^*. \end{cases} \tag{7.35}$$

As in Lemma 3.8, for every $(x, y^*) \in \mathbf{X}$,

$$(x, y^*) \in \operatorname{zer} \mathcal{K} \quad \Rightarrow \quad \begin{cases} x \text{ solves (7.33)}, \\ y^* \text{ solves (7.35)}, \end{cases} \tag{7.36}$$

and we therefore recover the embedding principle of Framework 1.2.

Example 7.9. In the above setting, set $\mathbf{X} = \mathcal{H} \oplus \mathcal{G}$, let $\mathcal{K}$ be the Kuhn–Tucker operator of (7.32), and let $\mathcal{T} : \mathbf{X} \to \mathcal{H} : (x, y^*) \mapsto x$. Then $(\mathbf{X}, \mathcal{K}, \mathcal{T})$ is an embedding of (7.33).

The primal–dual inclusion problem (7.33)–(7.34) was first studied in Combettes and Pesquet (2012), where it was solved via Tseng's forward–backward–forward algorithm. Here is Theorem 3.1(ii)(c)–(d) of Combettes and Pesquet (2012), which describes this approach when the operators L, B and D above are deployed in a product space $\mathcal{G} = \mathcal{G}_1 \oplus \cdots \oplus \mathcal{G}_p$ in the spirit of Problem 3.11; further analysis of the asymptotic behaviour of the method in special cases can be found in Boţ and Hendrich (2014).

Proposition 7.10. Let $0 < p \in \mathbb{N}$, let $\mu \in \left]0, +\infty\right[$, let $A : \mathcal{H} \to 2^{\mathcal{H}}$ be maximally monotone, let $Q : \mathcal{H} \to \mathcal{H}$ be monotone and μ-Lipschitzian. For every $k \in \{1, \dots, p\}$, let $\nu_k \in \left]0, +\infty\right[$, let $\mathcal{G}_k$ be a real Hilbert space, let $B_k : \mathcal{G}_k \to 2^{\mathcal{G}_k}$ be maximally monotone, let $D_k : \mathcal{G}_k \to 2^{\mathcal{G}_k}$ be maximally monotone and such that $D_k^{-1} : \mathcal{G}_k \to \mathcal{G}_k$ is ν_k-Lipschitzian, and assume that $0 \neq L_k \in \mathcal{B}(\mathcal{H}, \mathcal{G}_k)$. Suppose that the set Z of solutions to the primal inclusion

$$\text{find } x \in \mathcal{H} \text{ such that } 0 \in Ax + \sum_{k=1}^{p} L_k^*((B_k \,\Box\, D_k)(L_k x)) + Qx \tag{7.37}$$

is not empty and let Z^* be the set of solutions to the dual inclusion

$$\begin{aligned} &\text{find } y_1^* \in \mathcal{G}_1, \dots, y_p^* \in \mathcal{G}_p \text{ such that} \\ &(\exists x \in \mathcal{H}) \begin{cases} -\sum_{k=1}^{p} L_k^* y_k^* \in Ax + Qx, \\ (\forall k \in \{1, \dots, p\}) \; L_k x \in B_k^{-1} y_k^* + D_k^{-1} y_k^*. \end{cases} \end{aligned} \tag{7.38}$$

Set

$$\beta = \max\{\mu, \nu_1, \dots, \nu_p\} + \sqrt{\sum_{k=1}^{p} \|L_k\|^2}, \tag{7.39}$$

let $x_0 \in \mathcal{H}$, let $(y_{1,0}^*, \dots, y_{p,0}^*) \in \mathcal{G}_1 \oplus \cdots \oplus \mathcal{G}_p$, let $\varepsilon \in \left]0, 1/(\beta + 1)\right[$, and let

$(\gamma_n)_{n\in\mathbb{N}}$ be a sequence in $[\varepsilon,(1-\varepsilon)/\beta]$. Iterate

$$
\begin{array}{l}
\text{for } n = 0,1,\ldots\\
\left\lfloor
\begin{array}{l}
y_{1,n} = x_n - \gamma_n\big(Qx_n + \sum_{k=1}^p L_k^* y_{k,n}^*\big)\\
m_{1,n} = J_{\gamma_n A}\, y_{1,n}\\
\text{for } k = 1,\ldots,p\\
\left\lfloor
\begin{array}{l}
y_{2,k,n}^* = y_{k,n}^* + \gamma_n\big(L_k x_n - D_k^{-1} y_{k,n}^*\big)\\
m_{2,k,n}^* = J_{\gamma_n B_k^{-1}} y_{2,k,n}^*\\
q_{2,k,n}^* = m_{2,k,n}^* + \gamma_n\big(L_k m_{1,n} - D_k^{-1} m_{2,k,n}^*\big)\\
y_{k,n+1}^* = y_{k,n}^* - y_{2,k,n}^* + q_{2,k,n}^*
\end{array}
\right.\\
q_{1,n} = m_{1,n} - \gamma_n\big(Qm_{1,n} + \sum_{k=1}^p L_k^* m_{2,k,n}^*\big)\\
x_{n+1} = x_n - y_{1,n} + q_{1,n}.
\end{array}
\right.
\end{array}
\tag{7.40}
$$

Then there exist $x \in Z$ and $(y_1^*,\ldots,y_p^*) \in Z^*$ such that $x_n \rightharpoonup x$, and, for every $k \in \{1,\ldots,p\}$, $y_{k,n}^* \rightharpoonup y_k^*$.

Proof. The duality between (7.37) and (7.38) follows as in Problem 3.11, by replacing A with $A+Q$ and $(B_k^{-1})_{1\le k\le p}$ with $(B_k^{-1}+D_k^{-1})_{1\le k\le p}$. Now set

$$
\begin{cases}
\mathcal{G} = \mathcal{G}_1 \oplus \cdots \oplus \mathcal{G}_p,\\
B\colon \mathcal{G} \to 2^{\mathcal{G}}\colon (y_1,\ldots,y_p) \mapsto B_1y_1 \times \cdots \times B_py_p,\\
D\colon \mathcal{G} \to 2^{\mathcal{G}}\colon (y_1,\ldots,y_p) \mapsto D_1y_1 \times \cdots \times D_py_p,\\
L\colon \mathcal{H} \to \mathcal{G}\colon x \mapsto (L_1x,\ldots,L_px),
\end{cases}
\tag{7.41}
$$

define $\boldsymbol{M}$ and $\boldsymbol{Q}$ as in (7.31), and set

$$
(\forall n \in \mathbb{N})\quad
\begin{cases}
\boldsymbol{x}_n = \big(x_n, y_{1,n}^*,\ldots,y_{p,n}^*\big),\\
\boldsymbol{m}_n = \big(m_{1,n}, m_{2,1,n}^*,\ldots,m_{2,p,n}^*\big).
\end{cases}
\tag{7.42}
$$

Then $\boldsymbol{M}$ is maximally monotone and $\boldsymbol{Q}$ is monotone and β-Lipschitzian (Combettes and Pesquet 2012, equation (3.11)) and, following the same steps as in the proof of Proposition 7.3, we rewrite (7.40) as

$$
\begin{array}{l}
\text{for } n = 0,1,\ldots\\
\left\lfloor
\begin{array}{l}
\boldsymbol{b}_n^* = \gamma_n \boldsymbol{Q}\boldsymbol{x}_n\\
\boldsymbol{m}_n = J_{\gamma_n \boldsymbol{M}}(\boldsymbol{x}_n - \boldsymbol{b}_n^*)\\
\boldsymbol{x}_{n+1} = \boldsymbol{m}_n - \gamma_n \boldsymbol{Q}\boldsymbol{m}_n + \boldsymbol{b}_n^*,
\end{array}
\right.
\end{array}
\tag{7.43}
$$

and conclude by invoking Theorem 7.1 and (7.36). □

Remark 7.11. In (7.37), suppose that $p = 1$, $\mathcal{G}_1 = \mathcal{H}$, $L_1 = \mathrm{Id}$, $B_1 = B$, $D_1 = \{0\}^{-1}$ and $\operatorname{zer}(A+B+Q) \neq \varnothing$. Let $x_0 \in \mathcal{H}$, let $y_0^* \in \mathcal{H}$, let $\varepsilon \in\,]0, 1/(\mu+2)[$, and let $(\gamma_n)_{n\in\mathbb{N}}$ be a sequence in $[\varepsilon,(1-\varepsilon)/(\mu+1)]$. Then we deduce from

Proposition 7.10 that the sequence $(x_n)_{n\in\mathbb{N}}$ generated by the iterations

$$
\begin{array}{l}
\text{for } n = 0, 1, \ldots \\
\left\lfloor
\begin{array}{l}
y_n = x_n - \gamma_n(Qx_n + y_n^*) \\
p_n = J_{\gamma_n A}\, y_n \\
q_n^* = J_{\gamma_n B^{-1}}(y_n^* + \gamma_n x_n) \\
x_{n+1} = x_n - y_n + p_n - \gamma_n(Qp_n + q_n^*) \\
y_{n+1}^* = q_n^* + \gamma_n(p_n - x_n).
\end{array}
\right.
\end{array}
\tag{7.44}
$$

converges weakly to a zero of $A + B + Q$. An alternative method to solve this inclusion is proposed in Ryu and Vũ (2020), with constant proximal parameters $(\gamma_n)_{n\in\mathbb{N}}$ and the feature that it coincides with the unrelaxed version of the Douglas–Rachford algorithm when $Q = 0$ (in the spirit of the method of Section 6.4.3 where Q is cocoercive).

Example 7.12. In Proposition 7.10, make the additional assumptions that $Q = 0$ and, for every $k \in \{1, \ldots, p\}$, $\mathcal{G}_k = \mathcal{H}$, $L_k = \mathrm{Id}$, and D_k^{-1} is strictly monotone. Then (7.37) collapses to

$$
\text{find } x \in \mathcal{H} \text{ such that } 0 \in Ax + \sum_{k=1}^{p} (B_k \,\square\, D_k)(x). \tag{7.45}
$$

It is shown in Combettes (2013*b*, Proposition 4.2) that (7.45) is an exact relaxation of the (possibly inconsistent) instance of the problem

$$
\text{find } x \in \mathcal{H} \text{ such that } 0 \in Ax \text{ and } (\forall k \in \{1, \ldots, p\})\ 0 \in B_k x \tag{7.46}
$$

in the sense that the solutions to (7.45) are the same as the solutions to (7.46) when the latter happen to exist.

The specialization of Proposition 7.10 to minimization is as follows. It features the ability to split infimal convolutions (see (2.7)) together with linearly composed functions.

Example 7.13 (Combettes and Pesquet 2012). Let $0 < p \in \mathbb{N}$, let $\mu \in \left]0, +\infty\right[$, let $f \in \Gamma_0(\mathcal{H})$, and let $h\colon \mathcal{H} \to \mathbb{R}$ be convex, differentiable, and such that ∇h is μ-Lipschitzian. For every $k \in \{1, \ldots, p\}$, let $\nu_k \in \left]0, +\infty\right[$, let $\mathcal{G}_k$ be a real Hilbert space, let $g_k \in \Gamma_0(\mathcal{G}_k)$, let $\ell_k \in \Gamma_0(\mathcal{G}_k)$ be $1/\nu_k$-strongly convex, and suppose that $0 \neq L_k \in \mathcal{B}(\mathcal{H}, \mathcal{G}_k)$. Let Z be the set of solutions to the primal problem

$$
\underset{x\in\mathcal{H}}{\text{minimize}}\ \ f(x) + \sum_{k=1}^{p} (g_k \,\square\, \ell_k)(L_k x) + h(x), \tag{7.47}
$$

let Z^* be the set of solutions to the dual problem

$$
\underset{y_1^*\in\mathcal{G}_1,\ldots,y_p^*\in\mathcal{G}_p}{\text{minimize}}\ \ (f^* \,\square\, h^*)\biggl(-\sum_{k=1}^{p} L_k^* y_k^*\biggr) + \sum_{k=1}^{p} \bigl(g_k^*(y_k^*) + \ell_k^*(y_k^*)\bigr), \tag{7.48}
$$

and suppose that

$$\operatorname{zer}\left(\partial f+\sum_{k=1}^{p} L_k^*\circ(\partial g_k \,\square\, \partial\ell_k)\circ L_k+\nabla h\right)\neq\varnothing. \tag{7.49}$$

Set

$$\beta=\max\{\mu,\nu_1,\dots,\nu_p\}+\sqrt{\sum_{k=1}^{p}\|L_k\|^2}, \tag{7.50}$$

let $x_0\in\mathcal{H}$, let $(y_{1,0}^*,\dots,y_{p,0}^*)\in\mathcal{G}_1\oplus\cdots\oplus\mathcal{G}_p$, let $\varepsilon\in\,]0,1/(\beta+1)[$, and let $(\gamma_n)_{n\in\mathbb{N}}$ be a sequence in $[\varepsilon,(1-\varepsilon)/\beta]$. Iterate

$$\begin{array}{l}
\text{for } n=0,1,\dots\\
\left\lfloor\begin{array}{l}
y_{1,n}=x_n-\gamma_n\big(\nabla h(x_n)+\sum_{k=1}^{p}L_k^*y_{k,n}^*\big)\\
m_{1,n}=\operatorname{prox}_{\gamma_n f}\,y_{1,n}\\
\text{for } k=1,\dots,p\\
\left\lfloor\begin{array}{l}
y_{2,k,n}^*=y_{k,n}^*+\gamma_n\big(L_kx_n-\nabla\ell_k^*(y_{k,n}^*)\big)\\
m_{2,k,n}^*=\operatorname{prox}_{\gamma_n g_k^*}\,y_{2,k,n}^*\\
q_{2,k,n}^*=m_{2,k,n}^*+\gamma_n\big(L_km_{1,n}-\nabla\ell_k^*(m_{2,k,n}^*)\big)\\
y_{k,n+1}^*=y_{k,n}^*-y_{2,k,n}^*+q_{2,k,n}^*
\end{array}\right.\\
q_{1,n}=m_{1,n}-\gamma_n\big(\nabla h(m_{1,n})+\sum_{k=1}^{p}L_k^*m_{2,k,n}^*\big)\\
x_{n+1}=x_n-y_{1,n}+q_{1,n}.
\end{array}\right.
\end{array} \tag{7.51}$$

Then there exist $x\in Z$ and $(y_1^*,\dots,y_p^*)\in Z^*$ such that $x_n\rightharpoonup x$, and, for every $k\in\{1,\dots,p\}$, $y_{k,n}^*\rightharpoonup y_k^*$.

Remark 7.14. Conditions under which (7.49) holds are provided in Combettes and Pesquet (2012, Proposition 4.3).

8. Forward–backward splitting

8.1. Preview

The forward–backward splitting method is a basic algorithm for solving Problem 3.1 when B is cocoercive. At iteration n, given a step size $\gamma_n\in\,]0,+\infty[$, a discrete dynamics associated with the Cauchy problem (5.1) with $M=A+B$ is

$$\frac{x_n-x_{n+1}}{\gamma_n}\in Ax_{n+1}+Bx_n. \tag{8.1}$$

It amounts to performing a forward Euler step relative to the operator B and a backward Euler step relative to the operator A. In view of (2.18), this means that $x_{n+1}=J_{\gamma_nA}(x_n-\gamma_nBx_n)$. This iteration scheme goes back to the gradient-projection method (Goldstein 1964, Levitin and Polyak 1966) for the constrained minimization of a smooth function (see Example 8.7 below) and its extension to variational inequalities (Bakušinskiĭ and Polyak 1974, Mercier 1979).

8.2. Fejérian algorithm

We establish a new, geometric proof of the convergence of a relaxed primal–dual version of the forward–backward algorithm found in Combettes and Yamada (2015, Proposition 4.4(iii)) for the primal result and in Bauschke and Combettes (2017, Theorem 26.14(ii)) for the dual result, where the proximal parameters $(\gamma_n)_{n\in\mathbb{N}}$ are constant. Related primal results and special cases can be found in Gabay (1983), Lemaire (1996, 1997), Mercier (1980) and Tseng (1991). The importance of cocoercivity in establishing weak convergence was first identified by Mercier (1979) in the context of variational inequalities and, more generally, in Mercier (1980).

Theorem 8.1. Let $\alpha \in\,]0,+\infty[$, let $A\colon \mathcal{H} \to 2^{\mathcal{H}}$ be maximally monotone, and let $B\colon \mathcal{H} \to \mathcal{H}$ be α-cocoercive. Let $\varepsilon \in\,]0, \alpha/(\alpha+1)[$, let $(\gamma_n)_{n\in\mathbb{N}}$ be a sequence in $[\varepsilon, (2-\varepsilon)\alpha]$, and let

$$(\forall n \in \mathbb{N}) \quad \varepsilon \leq \mu_n \leq (1-\varepsilon)\frac{4\alpha - \gamma_n}{2\alpha}. \tag{8.2}$$

Suppose that the set Z of solutions to the problem

$$\text{find } x \in \mathcal{H} \text{ such that } 0 \in Ax + Bx \tag{8.3}$$

is not empty and let Z^* be the set of solutions to the dual problem

$$\text{find } x^* \in \mathcal{H} \text{ such that } 0 \in -A^{-1}(-x^*) + B^{-1}x^*. \tag{8.4}$$

Let $x_0 \in \mathcal{H}$ and iterate

$$\begin{array}{l} \text{for } n = 0, 1, \ldots \\ \left\lfloor \begin{array}{l} b_n^* = \gamma_n B x_n \\ w_n = J_{\gamma_n A}(x_n - b_n^*) \\ x_{n+1} = x_n + \mu_n(w_n - x_n). \end{array} \right. \end{array} \tag{8.5}$$

Then the following hold:

(i) $(x_n)_{n\in\mathbb{N}}$ converges weakly to a point in Z.

(ii) Z^* contains a single point $\overline{x}^*$ and $(\forall z \in Z)\ Bz = \overline{x}^*$.

(iii) $(Bx_n)_{n\in\mathbb{N}}$ converges strongly to $\overline{x}^*$.

Proof. The proof hinges on an application of Theorem 4.12 with

$$W = A, \quad C = B \quad \text{and} \quad (\forall n \in \mathbb{N}) \quad U_n = \gamma_n^{-1}\mathrm{Id} - B \quad \text{and} \quad q_n = x_n. \tag{8.6}$$

In this setting

$$(\forall n \in \mathbb{N}) \quad J_{W+C}^{U_n} = J_{A+B}^{U_n} = (\gamma_n^{-1}\mathrm{Id} + A) \circ (\gamma_n^{-1}\mathrm{Id} - B) = J_{\gamma_n A} \circ (\mathrm{Id} - \gamma_n B) \tag{8.7}$$

and the variables of (4.34) become

$$(\forall n \in \mathbb{N}) \quad \begin{cases} w_n = J_{\gamma_n A}(x_n - \gamma_n B x_n), \\ t_n^* = \dfrac{x_n - w_n}{\gamma_n}, \\ \delta_n = \left(\dfrac{1}{\gamma_n} - \dfrac{1}{4\alpha}\right) \|w_n - x_n\|^2. \end{cases} \tag{8.8}$$

Furthermore, we derive from Bùi and Combettes (2020*b*, Proposition 3.9) that (7.5) holds. Now set

$$(\forall n \in \mathbb{N}) \quad \lambda_n = \frac{4\alpha\mu_n}{4\alpha - \gamma_n}. \tag{8.9}$$

Then (8.2) yields

$$(\forall n \in \mathbb{N}) \quad \varepsilon \leq \frac{4\alpha\varepsilon}{4\alpha - \varepsilon} \leq \lambda_n \leq \frac{4\alpha(1-\varepsilon)(4\alpha - \gamma_n)}{(4\alpha - \gamma_n)2\alpha} \leq 2 - \varepsilon. \tag{8.10}$$

We also deduce from (8.8) that

$$(\forall n \in \mathbb{N}) \quad \delta_n \leq 0 \;\Leftrightarrow\; w_n = x_n \;\Leftrightarrow\; t_n^* = 0. \tag{8.11}$$

Hence, (4.34) yields

$$(\forall n \in \mathbb{N}) \quad d_n = \frac{\mu_n}{\lambda_n}(x_n - w_n). \tag{8.12}$$

Altogether, we arrive at the conclusion that the sequence $(x_n)_{n\in\mathbb{N}}$ produced by (8.5) coincides with that of (4.34). Hence, by Theorem 4.12(i) and (8.10),

$$\sum_{n\in\mathbb{N}} \|d_n\|^2 < +\infty. \tag{8.13}$$

In turn, upon invoking (8.12), we obtain

$$w_n - x_n \to 0. \tag{8.14}$$

(i) In view of (8.10), condition (iib) in Theorem 4.12(ii) is fulfilled. On the other hand, since Lemma 2.48(iii) asserts that the operators $(\gamma_n U_n)_{n\in\mathbb{N}}$ are nonexpansive, (8.14) implies that $\|U_n w_n - U_n x_n\| \leq \|w_n - x_n\|/\varepsilon \to 0$, so that condition (iic) is also fulfilled. Thus, the assertion follows from Theorem 4.12(ii).

(ii) The strong monotonicity of B^{-1} implies that of $-A^{-1} \circ (-\mathrm{Id}) + B^{-1}$. Hence, Corollary 23.37(ii) of Bauschke and Combettes (2017) asserts that (8.4) admits a unique solution $\overline{x}^*$. Now let $z \in Z$. Then $-Bz \in Az$ and therefore $-z \in -A^{-1}(-Bz)$. Thus, $0 = -z + z \in -A^{-1}(-Bz) + B^{-1}(Bz)$, i.e. $Bz \in Z^* = \{\overline{x}^*\}$.

(iii) It follows from (i) and (8.14) that $(x_n)_{n\in\mathbb{N}}$ and $(w_n)_{n\in\mathbb{N}}$ are bounded. Now let $z \in Z$. We retrieve from (4.27) that

$$(\forall n \in \mathbb{N}) \quad \langle z - w_n \mid t_n^* \rangle \leq \langle x_n - w_n \mid Bx_n - Bz \rangle - \alpha\|Bx_n - Bz\|^2. \tag{8.15}$$

Hence, the Cauchy–Schwarz inequality, (2.32), (8.8) and (8.14) imply that

$$\begin{aligned}\alpha\|Bx_n - Bz\|^2 &\leq \|w_n - x_n\|\,\|Bx_n - Bz\| + \|w_n - z\|\,\|t_n^*\| \\ &\leq \frac{1}{\alpha}\|w_n - x_n\|\,\|x_n - z\| + \frac{1}{\gamma_n}\|w_n - z\|\,\|w_n - x_n\| \\ &\to 0. \end{aligned} \tag{8.16}$$

In view of (ii), $Bx_n \to Bz = \overline{x}^*$. □

The following examples address Example 3.2 and Example 3.3, respectively.

Example 8.2. Let $\alpha \in \,]0,+\infty[$, let $f \in \Gamma_0(\mathcal{H})$, let $B\colon \mathcal{H} \to \mathcal{H}$ be α-cocoercive, suppose that the set Z of solutions to the variational inequality

$$\text{find } x \in \mathcal{H} \text{ such that } (\forall y \in \mathcal{H})\ \langle x - y \mid Bx\rangle + f(x) \leq f(y) \tag{8.17}$$

is not empty, and let Z^* be the set of solutions to the dual problem

$$\text{find } x^* \in \mathcal{H} \text{ such that } 0 \in -\partial f^*(-x^*) + B^{-1}x^*. \tag{8.18}$$

Let $x_0 \in \mathcal{H}$, let $\varepsilon \in \,]0, \alpha/(\alpha+1)[$, let $(\gamma_n)_{n\in\mathbb{N}}$ be a sequence in $[\varepsilon, (2-\varepsilon)\alpha]$, and suppose that $(\mu_n)_{n\in\mathbb{N}}$ satisfies (8.2). Iterate

$$\begin{array}{l}\text{for } n = 0, 1, \ldots \\ \left\lfloor \begin{array}{l} b_n^* = \gamma_n Bx_n \\ w_n = \operatorname{prox}_{\gamma_n f}(x_n - b_n^*) \\ x_{n+1} = x_n + \mu_n(w_n - x_n). \end{array}\right.\end{array} \tag{8.19}$$

Then $(x_n)_{n\in\mathbb{N}}$ converges weakly to a point in Z and $(Bx_n)_{n\in\mathbb{N}}$ converges strongly to the unique point in Z^*.

Proof. Use Example 2.12 and Example 2.35 and set $A = \partial f$ in Theorem 8.1. □

Example 8.3. Let $\alpha \in \,]0,+\infty[$, let C be a nonempty closed convex subset of $\mathcal{H}$, let $B\colon \mathcal{H} \to \mathcal{H}$ be α-cocoercive, suppose that the set Z of solutions to the variational inequality

$$\text{find } x \in C \text{ such that } (\forall y \in C)\ \langle x - y \mid Bx\rangle \leq 0 \tag{8.20}$$

is not empty, and let Z^* be the set of solutions to the dual problem

$$\text{find } x^* \in \mathcal{H} \text{ such that } 0 \in -\partial\sigma_C(-x^*) + B^{-1}x^*. \tag{8.21}$$

Let $x_0 \in \mathcal{H}$, let $\varepsilon \in \,]0, \alpha/(\alpha+1)[$, let $(\gamma_n)_{n\in\mathbb{N}}$ be a sequence in $[\varepsilon, (2-\varepsilon)\alpha]$, and suppose that $(\mu_n)_{n\in\mathbb{N}}$ satisfies (8.2). Iterate

$$\begin{array}{l}\text{for } n = 0, 1, \ldots \\ \left\lfloor \begin{array}{l} b_n^* = \gamma_n Bx_n \\ w_n = \operatorname{proj}_C(x_n - b_n^*) \\ x_{n+1} = x_n + \mu_n(w_n - x_n). \end{array}\right.\end{array} \tag{8.22}$$

Then $(x_n)_{n\in\mathbb{N}}$ converges weakly to a point in Z and $(Bx_n)_{n\in\mathbb{N}}$ converges strongly to the unique point in Z^*.

Proof. Use Example 2.36 and (2.2), and set $f = \iota_C$ in Example 8.2. □

The following example focuses on the minimization in the setting of Problem 3.5(ii). This framework has found a multitude of applications, especially in the areas of signal processing and machine learning (Argyriou *et al.* 2012, Beck and Teboulle 2010, Chan, Setzer and Steidl 2008, Combettes and Wajs 2005, Combettes *et al.* 2018, Dexter, Tran and Webster 2022, Jenatton *et al.* 2011, Vaiter *et al.* 2018).

Example 8.4. Let $\beta \in \left]0, +\infty\right[$, let $f \in \Gamma_0(\mathcal{H})$ and let $g\colon \mathcal{H} \to \mathbb{R}$ be convex and differentiable. Suppose that ∇g is β-Lipschitzian and that the set Z of solutions to the problem

$$\underset{x\in\mathcal{H}}{\text{minimize}}\;\; f(x) + g(x) \tag{8.23}$$

is not empty, and let Z^* be the set of solutions to the dual problem

$$\underset{x^*\in\mathcal{H}}{\text{minimize}}\;\; f^*(-x^*) + g^*(x^*). \tag{8.24}$$

Let $x_0 \in \mathcal{H}$, let $\varepsilon \in \left]0, 1/(\beta+1)\right[$, let $(\gamma_n)_{n\in\mathbb{N}}$ be a sequence in $[\varepsilon, (2-\varepsilon)/\beta]$, and suppose that

$$(\forall n \in \mathbb{N}) \quad \varepsilon \leq \mu_n \leq (1-\varepsilon)\frac{4 - \beta\gamma_n}{2}. \tag{8.25}$$

Iterate

$$\begin{array}{l}
\text{for } n = 0, 1, \ldots \\
\left\lfloor \begin{array}{l}
b_n^* = \gamma_n \nabla g(x_n) \\
w_n = \operatorname{prox}_{\gamma_n f}(x_n - b_n^*) \\
x_{n+1} = x_n + \mu_n(w_n - x_n).
\end{array}\right.
\end{array} \tag{8.26}$$

Then $(x_n)_{n\in\mathbb{N}}$ converges weakly to a point in Z and $(\nabla g(x_n))_{n\in\mathbb{N}}$ converges strongly to the unique point in Z^*.

Proof. The claim is established by applying Theorem 8.1(i) with $A = \partial f$ (see Example 2.12) and $B = \nabla g$ (see Lemma 2.2). □

Remark 8.5. In some applications, it may be of interest to quantify the asymptotic behaviour of the function values $(f(x_n) + g(x_n))_{n\in\mathbb{N}}$ produced by (8.26). This topic has been the focus of a lot of interest since the publication of the influential papers by Beck and Teboulle (2009*b*,*a*) and Chambolle and Dossal (2015); see Garrigos, Rosasco and Villa (2023) and its bibliography for recent results on the unrelaxed implementation of (8.26) with constant proximal parameters.

The following example, taken from Combettes and Wajs (2005), models linear inverse problems in which the prior knowledge is modelled by penalizing the coefficients of the decomposition of the ideal solution in an orthonormal basis; see Combettes, Salzo and Villa (2018), Daubechies, Defrise and De Mol (2004) and Figueiredo and Nowak (2003) for special cases.

Example 8.6. Suppose that $\mathcal{H}$ is separable, let $(e_k)_{k\in\mathbb{K}\subset\mathbb{N}}$ be an orthonormal basis of $\mathcal{H}$, let $y \in \mathcal{G}$, suppose that $0 \neq L \in \mathcal{B}(\mathcal{H},\mathcal{G})$, and let $(\phi_k)_{k\in\mathbb{K}}$ be functions in $\Gamma_0(\mathbb{R})$ such that $(\forall k \in \mathbb{K})\ \phi_k \geq 0 = \phi_k(0)$. Suppose that the set Z of solutions to the problem

$$\underset{x\in\mathcal{H}}{\text{minimize}} \quad \sum_{k\in\mathbb{K}} \phi_k(\langle x \mid e_k\rangle) + \frac{1}{2}\|Lx - y\|^2 \tag{8.27}$$

is not empty. Let $x_0 \in \mathcal{H}$, let $\varepsilon \in \left]0, 1/(\|L\|^2+1)\right[$, let $(\gamma_n)_{n\in\mathbb{N}}$ be a sequence in $[\varepsilon, (2-\varepsilon)/\|L\|^2]$, and suppose that

$$(\forall n \in \mathbb{N}) \quad \varepsilon \leq \mu_n \leq (1-\varepsilon)\frac{4 - \|L\|^2\gamma_n}{2}. \tag{8.28}$$

Iterate

$$\begin{array}{l}\text{for } n = 0, 1, \ldots\\ \left\lfloor \begin{array}{l} b_n^* = \gamma_n L^*(Lx_n - y)\\ w_n = \sum_{k\in\mathbb{K}}(\operatorname{prox}_{\gamma_n\phi_k}\langle x_n - b_n^* \mid e_k\rangle)e_k\\ x_{n+1} = x_n + \mu_n(w_n - x_n).\end{array}\right.\end{array} \tag{8.29}$$

Then $(x_n)_{n\in\mathbb{N}}$ converges weakly to a point in Z.

Proof. Set $f\colon x \mapsto \sum_{k\in\mathbb{K}}\phi_k(\langle x \mid e_k\rangle)$ and $g\colon x \mapsto \|Lx - y\|^2/2$. Then, as shown in Combettes and Wajs (2005, Example 2.19), $f \in \Gamma_0(\mathcal{H})$ and $\operatorname{prox}_{\gamma f}\colon x \mapsto \sum_{k\in\mathbb{K}}(\operatorname{prox}_{\gamma_n\phi_k}\langle x \mid e_k\rangle)e_k$. On the other hand, g is convex and differentiable and $\nabla g\colon x \mapsto L^*(Lx - y)$ is $\|L\|^2$-Lipschitzian. Altogether, the conclusion follows from Example 8.4. □

Next, we specialize Example 8.4 to the gradient-projection method, which minimizes a smooth function over a convex set (see Example 3.6) and goes back to Goldstein (1964) and Levitin and Polyak (1966).

Example 8.7. Let $\beta \in \left]0, +\infty\right[$, let C be a nonempty closed convex subset of $\mathcal{H}$, and let $g\colon \mathcal{H} \to \mathbb{R}$ be convex and differentiable. Suppose that ∇g is β-Lipschitzian and that the set Z of solutions to the problem

$$\underset{x\in C}{\text{minimize}} \quad g(x) \tag{8.30}$$

is not empty, and let Z^* be the set of solutions to the dual problem

$$\underset{x^*\in\mathcal{H}}{\text{minimize}} \quad \sigma(-x^*) + g^*(x^*). \tag{8.31}$$

Let $x_0 \in \mathcal{H}$, let $\varepsilon \in \left]0, 1/(\beta+1)\right[$, let $(\gamma_n)_{n\in\mathbb{N}}$ be a sequence in $[\varepsilon, (2-\varepsilon)/\beta]$, and suppose that $(\mu_n)_{n\in\mathbb{N}}$ satisfies (8.25). Iterate

$$\begin{array}{l}\text{for } n = 0, 1, \ldots\\ \left\lfloor \begin{array}{l} b_n^* = \gamma_n\nabla g(x_n)\\ w_n = \operatorname{proj}_C(x_n - b_n^*)\\ x_{n+1} = x_n + \mu_n(w_n - x_n).\end{array}\right.\end{array} \tag{8.32}$$

Then $(x_n)_{n\in\mathbb{N}}$ converges weakly to a point in Z and $(\nabla g(x_n))_{n\in\mathbb{N}}$ converges strongly to the unique point in Z^*.

Proof. Set $f = \iota_C$ in Example 8.4. Alternatively, set $B = \nabla g$ in Example 8.3. □

Remark 8.8. Attouch, Peypouquet and Redont (2018) studied the backward–forward iterations

$$\begin{array}{l}\text{for } n = 0, 1, \dots \\ \left\lfloor \begin{array}{l} p_n = J_{\gamma A} x_n \\ q_n = p_n - \gamma B p_n \\ x_{n+1} = x_n + \mu_n (q_n - x_n), \end{array} \right.\end{array} \tag{8.33}$$

and showed them to be related to the forward–backward iterations applied to Yosida envelopes of B and A.

8.3. *Haugazeau-like algorithm*

As seen in Combettes and Wajs (2005, Remark 5.12), the strong convergence of $(x_n)_{n\in\mathbb{N}}$ in Theorem 8.1(i) may fail. Item (i) below on the strong convergence of a best approximation forward–backward algorithm extends Theorem 5.6(i) and Remark 5.5 of Combettes and Hirstoaga (2005), where $(\forall n \in \mathbb{N})\ \gamma_n = \gamma \in \left]0, 2\alpha\right[$ and $\mu_n \leq 1$.

Theorem 8.9. Let $\alpha \in \left]0, +\infty\right[$, let $A\colon \mathcal{H} \to 2^{\mathcal{H}}$ be maximally monotone, and let $B\colon \mathcal{H} \to \mathcal{H}$ be α-cocoercive. Let $\varepsilon \in \left]0, \min\{1/2, 2\alpha\}\right[$, let $(\gamma_n)_{n\in\mathbb{N}}$ be a sequence in $[\varepsilon, 2\alpha]$, and let

$$(\forall n \in \mathbb{N}) \quad \varepsilon \leq \mu_n \leq \frac{4\alpha - \gamma_n}{4\alpha}. \tag{8.34}$$

Suppose that the set Z of solutions to the problem

$$\text{find } x \in \mathcal{H} \text{ such that } 0 \in Ax + Bx \tag{8.35}$$

is not empty and let Z^* be the set of solutions to the dual

$$\text{find } x^* \in \mathcal{H} \text{ such that } 0 \in -A^{-1}(-x^*) + B^{-1}x^*. \tag{8.36}$$

Let $x_0 \in \mathcal{H}$ and iterate

$$\begin{array}{l}\text{for } n = 0, 1, \dots \\ \left\lfloor \begin{array}{l} b_n^* = \gamma_n B x_n \\ w_n = J_{\gamma_n A}(x_n - b_n^*) \\ x_{n+1} = \mathsf{Q}(x_0, x_n, x_n + \mu_n (w_n - x_n)), \end{array} \right.\end{array} \tag{8.37}$$

where Q is defined in Lemma 4.6. Then the following hold:

(i) $(x_n)_{n\in\mathbb{N}}$ converges strongly to $\mathrm{proj}_Z\, x_0$.

(ii) Z^* contains a single point $\overline{x}^*$ and $(Bx_n)_{n\in\mathbb{N}}$ converges strongly to $\overline{x}^*$.

Proof. We apply Theorem 4.14 in the setting of (8.6), using the same variables as in (8.8) and $(\lambda_n)_{n\in\mathbb{N}}$ defined as in (8.9). Then (8.11) holds and

$$(\forall n \in \mathbb{N}) \quad \varepsilon \leq \frac{4\alpha\varepsilon}{4\alpha - \varepsilon} \leq \lambda_n \leq 1. \tag{8.38}$$

Therefore the sequence $(x_n)_{n\in\mathbb{N}}$ produced by (8.37) coincides with that of (4.44). Hence, by Theorem 4.14(i),

$$w_n - x_n \to 0. \tag{8.39}$$

(i) This follows from Theorem 4.14(ii) since, as in the proof of Theorem 8.1(i), its conditions (iib) and (iid) are fulfilled.

(ii) Since B is continuous, (i) and Theorem 8.1(ii) imply that $Bx_n \to B(\mathrm{proj}_Z x_0) \in Z^*$, where Z^* is a singleton. □

8.4. *Special cases and variants*

8.4.1. *Projected Landweber method*

In inverse problems, constrained least-squares estimation has a long history (Benning and Burger 2018, Bertero *et al.* 1997, Eicke 1992, Neubauer 1988, Phillips 1962). We address the numerical solution of this problem from the viewpoint of the forward–backward algorithm to obtain a relaxed version of the projected Landweber method with iteration-dependent parameters.

Proposition 8.10. Let $\mathcal{G}$ be a real Hilbert space, suppose that $0 \neq L \in \mathcal{B}(\mathcal{H}, \mathcal{G})$, let $y \in \mathcal{G}$, and let C be a closed convex subset of $\mathcal{H}$ such that the set Z of solutions to the problem

$$\underset{x\in C}{\text{minimize}} \ \frac{1}{2}\|Lx - y\|^2 \tag{8.40}$$

is not empty. Let $x_0 \in \mathcal{H}$, let $\varepsilon \in \left]0, 1/(\|L\|^2 + 1)\right[$, let $(\gamma_n)_{n\in\mathbb{N}}$ be a sequence in $[\varepsilon, (2-\varepsilon)/\|L\|^2]$, and suppose that $(\mu_n)_{n\in\mathbb{N}}$ satisfies (8.28). Iterate

$$\begin{array}{l} \text{for } n = 0, 1, \ldots \\ \left\lfloor \begin{array}{l} b_n^* = \gamma_n L^*(Lx_n - y) \\ w_n = \mathrm{proj}_C(x_n - b_n^*) \\ x_{n+1} = x_n + \mu_n(w_n - x_n). \end{array} \right. \end{array} \tag{8.41}$$

Then $(x_n)_{n\in\mathbb{N}}$ converges weakly to a point in Z.

Proof. Apply Example 8.7 with $g\colon x \mapsto \|Lx - y\|^2/2$. □

Proposition 8.10 was established in Eicke (1992, Section 3.1) with $(\forall n \in \mathbb{N})\ \lambda_n = 1$ and $\gamma_n = \gamma \in \left]0, 2/\|L\|^2\right[$. There, it was also conjectured that the convergence was strong, which was disproved in Combettes and Wajs (2005, Remark 5.12). This motivates the following result.

Proposition 8.11. Let $\mathcal{G}$ be a real Hilbert space, suppose that $0 \neq L \in \mathcal{B}(\mathcal{H}, \mathcal{G})$, let $y \in \mathcal{G}$, let C be a closed convex subset of $\mathcal{H}$, and suppose that the set Z of solutions to (8.40) is not empty. Let $x_0 \in \mathcal{H}$, let $\varepsilon \in]0, \min\{1/2, 2/\|L\|^2\}[$, let $(\gamma_n)_{n\in\mathbb{N}}$ be a sequence in $[\varepsilon, 2/\|L\|^2]$, and suppose that $(\forall n \in \mathbb{N})\ \varepsilon \leq \mu_n \leq 1 - \|L\|^2\gamma_n/4$. Iterate

$$\begin{array}{l} \text{for } n = 0, 1, \ldots \\ \left\lfloor \begin{array}{l} b_n^* = \gamma_n L^*(Lx_n - y) \\ w_n = \operatorname{proj}_C(x_n - b_n^*) \\ x_{n+1} = \mathsf{Q}(x_0, x_n, x_n + \mu_n(w_n - x_n)), \end{array} \right. \end{array} \tag{8.42}$$

where Q is defined in Lemma 4.6(ii). Then $(x_n)_{n\in\mathbb{N}}$ converges strongly to $\operatorname{proj}_Z x_0$.

Proof. Follow the pattern of the proof of Proposition 8.10 and use Example 2.36 to apply Theorem 8.9(i) with $A = N_C$ and $B\colon x \mapsto L^*(Lx - y)$. □

Here is an application of Proposition 8.10 to the problem of finding the best approximation to a point from a linearly transformed convex set.

Example 8.12. Consider the setting of Proposition 8.10 with the assumption that $L(C)$ is closed, which guarantees that (8.40) admits solutions. Then $x_n \rightharpoonup x$, where x solves (8.40). Furthermore, if we set $p = Lx$, then $p = \operatorname{proj}_{L(C)} y$. Hence, upon rewriting (8.41) as

$$\begin{array}{l} \text{for } n = 0, 1, \ldots \\ \left\lfloor \begin{array}{l} q_n = Lx_n \\ b_n^* = \gamma_n L^*(q_n - y) \\ w_n = \operatorname{proj}_C(x_n - b_n^*) \\ x_{n+1} = x_n + \mu_n(w_n - x_n), \end{array} \right. \end{array} \tag{8.43}$$

and invoking the weak continuity of L, we conclude that $q_n \rightharpoonup \operatorname{proj}_{L(C)} y$.

Example 8.13. Let $\mathcal{G}$ be a real Hilbert space, and suppose that $0 \neq L \in \mathcal{B}(\mathcal{H}, \mathcal{G})$ and that $\operatorname{ran} L$ is closed. Additionally, let $x_0 \in \mathcal{H}$, let $\varepsilon \in]0, 1/(\|L\|^2 + 1)[$, and let $(\nu_n)_{n\in\mathbb{N}}$ be a sequence in $[\varepsilon, (2 - \varepsilon)/\|L\|^2]$. Iterate

$$\begin{array}{l} \text{for } n = 0, 1, \ldots \\ \left\lfloor \begin{array}{l} q_n = Lx_n \\ x_{n+1} = x_n - \nu_n L^* q_n, \end{array} \right. \end{array} \tag{8.44}$$

and let q be the minimal-norm element of $\operatorname{ran} L$. Then $q_n \rightharpoonup q$.

Proof. Apply Example 8.12 with $C = \mathcal{H}$ and $y = 0$. □

The next example is about a composite best approximation problem.

Example 8.14. Let $\mathcal{G}$ be a real Hilbert space, let $y \in \mathcal{G}$, and let $0 < p \in \mathbb{N}$. For every $k \in \{1, \ldots, p\}$, let $\mathcal{H}_k$ be a real Hilbert space, let C_k be a nonempty closed convex subset of $\mathcal{H}_k$, let $0 \neq L_k \in \mathcal{B}(\mathcal{H}_k, \mathcal{G})$, and let $x_{k,0} \in \mathcal{H}_k$. Suppose that $\sum_{k=1}^p L_k(C_k)$ is closed and set $\beta = \sum_{k=1}^p \|L_k\|^2$. Furthermore, let

$\varepsilon \in]0, 1/(\beta+1)[$, let $(\gamma_n)_{n\in\mathbb{N}}$ be a sequence in $[\varepsilon, (2-\varepsilon)/\beta]$, and suppose that $(\mu_n)_{n\in\mathbb{N}}$ satisfies (8.25). Iterate

$$\begin{array}{l}\text{for } n = 0, 1, \ldots \\ \left\lfloor \begin{array}{l} q_n = \sum_{k=1}^{p} L_k x_{k,n} \\ \text{for } k = 1, \ldots, p \\ \left\lfloor \begin{array}{l} b^*_{k,n} = \gamma_n L_k^*(q_n - y) \\ w_{k,n} = \operatorname{proj}_{C_k}(x_{k,n} - b^*_{k,n}) \\ x_{k,n+1} = x_{k,n} + \mu_n(w_{k,n} - x_{k,n}). \end{array}\right. \end{array}\right. \end{array} \tag{8.45}$$

Then $q_n \rightharpoonup \operatorname{proj}_{\sum_{k=1}^{p} L_k(C_k)} y$.

Proof. Set $\boldsymbol{\mathcal{H}} = \mathcal{H}_1 \oplus \cdots \oplus \mathcal{H}_p$, $\boldsymbol{C} = C_1 \times \cdots \times C_p$ and

$$\boldsymbol{L} \colon \boldsymbol{\mathcal{H}} \to \mathcal{G} \colon (x_k)_{1\le k\le p} \mapsto \sum_{k=1}^{p} L_k x_k. \tag{8.46}$$

Then $\operatorname{proj}_{\boldsymbol{C}} \colon (x_k)_{1\le k\le p} \mapsto (\operatorname{proj}_{C_k} x_k)_{1\le k\le p}$ (see Examples 2.36 and 2.37), $\|\boldsymbol{L}\|^2 = \beta$ and $\boldsymbol{L}^* \colon \mathcal{G} \to \boldsymbol{\mathcal{H}} \colon y^* \mapsto (L_1^* y^*, \ldots, L_p^* y^*)$. Altogether, the result is an application of Example 8.12 to $\boldsymbol{C}$ and $\boldsymbol{L}$ in $\boldsymbol{\mathcal{H}}$. □

As an application of Example 8.14, we address the problem of computing the best approximation from the Minkowski sum of closed convex sets; see Bauschke, Bùi and Wang (2019), Eaves (1984), Martínez-Legaz and Seeger (1994), Qin and An (2019), Seeger (1998), Wang, Zhang and Zhang (2020) and Won, Xu and Lange (2019) for instances of decompositions with respect to such sums.

Example 8.15. Let $z \in \mathcal{H}$ and $0 < p \in \mathbb{N}$. For every $k \in \{1, \ldots, p\}$, let C_k be a nonempty closed convex subset of $\mathcal{H}$ and let $x_{k,0} \in \mathcal{H}$. Suppose that $\sum_{k=1}^{p} C_k$ is closed, let $\varepsilon \in]0, 1/(p+1)[$, let $(\gamma_n)_{n\in\mathbb{N}}$ be a sequence in $[\varepsilon, (2-\varepsilon)/p]$, and suppose that $(\forall n \in \mathbb{N})\ \varepsilon \le \mu_n \le (1-\varepsilon)(2 - p\gamma_n/2)$. Iterate

$$\begin{array}{l}\text{for } n = 0, 1, \ldots \\ \left\lfloor \begin{array}{l} q_n = \sum_{k=1}^{p} x_{k,n} \\ b^*_n = \gamma_n(q_n - z) \\ \text{for } k = 1, \ldots, p \\ \left\lfloor \begin{array}{l} w_{k,n} = \operatorname{proj}_{C_k}(x_{k,n} - b^*_n) \\ x_{k,n+1} = x_{k,n} + \mu_n(w_{k,n} - x_{k,n}). \end{array}\right. \end{array}\right. \end{array} \tag{8.47}$$

Then $q_n \rightharpoonup \operatorname{proj}_{\sum_{k=1}^{p} C_k} z$.

Proof. Apply Example 8.14 with $\mathcal{G} = \mathcal{H}$, $y = z$ and $(\forall k \in \{1, \ldots, p\})\ \mathcal{H}_k = \mathcal{H}$ and $L_k = \mathrm{Id}$. □

8.4.2. Partial Yosida approximation to inconsistent common zero problems

We extend a framework proposed in Combettes (2004, Section 6.3), where no linear transformations were present. We start with the following composite common zero problem; see Byrne, Censor, Gibali and Reich (2012) for a special case.

Problem 8.16. Let $A\colon \mathcal{H} \to 2^{\mathcal{H}}$ be maximally monotone and let $0 < p \in \mathbb{N}$. For every $k \in \{1,\ldots,p\}$, let $\mathcal{G}_k$ be a real Hilbert space, let $B_k\colon \mathcal{G}_k \to 2^{\mathcal{G}_k}$ be maximally monotone, and suppose that $0 \neq L_k \in \mathcal{B}(\mathcal{H},\mathcal{G}_k)$. The objective is to

$$\text{find } x \in \operatorname{zer} A \text{ such that } (\forall k \in \{1,\ldots,p\})\ L_k x \in \operatorname{zer} B_k. \tag{8.48}$$

Example 8.17. Suppose that, in Problem 8.16, $A = N_C$, where C is a nonempty closed convex subset of $\mathcal{H}$, and, for every $k \in \{1,\ldots,p\}$, $B_k = N_{D_k}$, where D_k is a nonempty closed convex subset of $\mathcal{G}_k$. Then (8.48) is the *split feasibility problem* (Reich, Truong and Mai 2020)

$$\text{find } x \in C \text{ such that } (\forall k \in \{1,\ldots,p\})\ L_k x \in D_k. \tag{8.49}$$

Example 8.18. Suppose that, in Problem 8.16, $A = \partial f$, where $f \in \Gamma_0(\mathcal{H})$, and, for every $k \in \{1,\ldots,p\}$, $\mathcal{G}_k = \mathcal{H}$, $L_k = \mathrm{Id}$ and $B_k = \partial f_k$, where $f_k \in \Gamma_0(\mathcal{H})$. Then (8.48) becomes

$$\text{find } x \in (\operatorname{Argmin} f) \cap \bigcap_{k=1}^{p} \operatorname{Argmin} f_k. \tag{8.50}$$

Example 8.19. Suppose that, in Problem 8.16, $A = N_C$, where C is a nonempty closed convex subset of $\mathcal{H}$, and, for every $k \in \{1,\ldots,p\}$, $B_k = (\mathrm{Id} - F_k + r_k)^{-1} - \mathrm{Id}$, where $F_k\colon \mathcal{G}_k \to \mathcal{G}_k$ is firmly nonexpansive and $r_k \in \mathcal{G}_k$. Then (8.48) becomes

$$\text{find } x \in C \text{ such that } (\forall k \in \{1,\ldots,p\})\ F_k(L_k x) = r_k. \tag{8.51}$$

Note that the operators $(\mathrm{Id} - F_k + r_k)_{1\leq k\leq p}$ are firmly nonexpansive as well, which makes the operators $(B_k)_{1\leq k\leq p}$ maximally monotone by Lemma 2.34(iii). This formulation was investigated in Combettes and Woodstock (2022) in the context of recovering a signal in C from p nonlinear observations modelled as outputs of Wiener systems (see also Example 5.12).

Our focus here is on situations in which (8.48) is not guaranteed to have solutions; see Censor and Zaknoon (2018), Combettes and Bondon (1999), Combettes and Glaudin (2019) and Goldburg and Marks II (1985) for concrete illustrations. In such environments, it is natural to approximate it by a more general problem, which exhibits better regularity properties and admits solutions. We propose the following relaxation of Problem 8.16, in which $\operatorname{dom} A$ serves as a hard constraint.

Problem 8.20. Consider the setting of Problem 8.16 and let $(\rho_k)_{1\leq k\leq p}$ and $(\omega_k)_{1\leq k\leq p}$ be in $]0,+\infty[$. The objective is to solve the *partial Yosida approximation*

$$\text{find } x \in \mathcal{H} \text{ such that } 0 \in Ax + \sum_{k=1}^{p} \omega_k L_k^* ({}^{\rho_k} B_k (L_k x)) \tag{8.52}$$

to Problem 8.16.

The fact that Problem 8.20 is an appropriate relaxation of Problem 8.16 is supported by the following argument.

Proposition 8.21. Suppose that the set of solutions to Problem 8.16 is not empty. Then it coincides with the set of solutions to Problem 8.20.

Proof. Let $\overline{x}$ be a solution to Problem 8.16. Then (2.22) yields

$$0 = -\sum_{k=1}^{p} \omega_k L_k^*({}^{\rho_k}B_k(L_k\overline{x})) \in A\overline{x}, \tag{8.53}$$

which shows that $\overline{x}$ solves Problem 8.20. Now let x be a solution to Problem 8.20. Then

$$-\sum_{k=1}^{p} \omega_k L_k^*({}^{\rho_k}B_k(L_k x)) \in Ax. \tag{8.54}$$

It follows from (8.53), (8.54), the monotonicity of A and the cocoercivity of the operators $({}^{\rho_k}B_k)_{1\leq k\leq p}$ (see Example 2.7) that

$$\begin{aligned} 0 &\geq \left\langle x-\overline{x} \,\middle|\, \sum_{k=1}^{p} \omega_k L_k^*({}^{\rho_k}B_k(L_k x)) - \sum_{k=1}^{p} \omega_k L_k^*({}^{\rho_k}B_k(L_k\overline{x}))\right\rangle \\ &= \sum_{k=1}^{p} \omega_k \left\langle L_k x - L_k\overline{x} \,\middle|\, {}^{\rho_k}B_k(L_k x) - {}^{\rho_k}B_k(L_k\overline{x})\right\rangle \\ &\geq \sum_{k=1}^{p} \omega_k \rho_k \|{}^{\rho_k}B_k(L_k x) - {}^{\rho_k}B_k(L_k\overline{x})\|^2 \\ &= \sum_{k=1}^{p} \omega_k \rho_k \|{}^{\rho_k}B_k(L_k x)\|^2. \end{aligned} \tag{8.55}$$

Hence, we deduce from (2.22) that $(\forall k \in \{1,\ldots,p\})\ L_k x \in \operatorname{zer}{}^{\rho_k}B_k = \operatorname{zer} B_k$. In view of (8.54), we conclude that x solves Problem 8.16. □

Remark 8.22. It should be emphasized that Problem 8.20 is a relaxation of Problem 8.16, and not of the inclusion

$$\text{find } x \in \mathcal{H} \text{ such that } 0 \in Ax + \sum_{k=1}^{p} \omega_k L_k^*(B_k(L_k x)). \tag{8.56}$$

In particular, $\operatorname{zer}(A+{}^{\rho}B) \neq \operatorname{zer}(A+B)$ when $\operatorname{zer}(A+B) \neq \varnothing$. However, the problem of finding a zero of $A + {}^{\rho}B$ can be regarded as a regularization of that of finding a zero of $A + B$ in the sense that solutions to the former approaches a particular solution of the latter as $\rho \to 0$ (Mahey and Pham 1993, Mercier 1980, Moudafi 2000).

Example 8.23. Consider the setting of Example 8.17 and let $(\forall k \in \{1,\ldots,p\})$ $\rho_k = 1$. Then (8.52) relaxes the possibly inconsistent problem (8.49) to the problem

$$\underset{x\in C}{\text{minimize}} \sum_{k=1}^{p} \omega_k d_{D_k}^2(L_k x). \tag{8.57}$$

(i) Assume that, for every $k \in \{1,\ldots,p\}$, $\mathcal{G}_k = \mathcal{H}$ and $L_k = \mathrm{Id}$. Then (8.57) is the relaxed formulation of Combettes and Bondon (1999).

(ii) Assume that $\mathcal{H} = \mathbb{R}^N$, $C = \mathbb{R}^N$, and, for every $k \in \{1,\ldots,p\}$, $\mathcal{G}_k = \mathbb{R}$, $L_k \colon x \mapsto u_k^\top x$ with $u_k \in \mathbb{R}^N$, and $D_k = \{\eta_k\}$ with $\eta_k \in \mathbb{R}$. Let $U \in \mathbb{R}^{p\times N}$ be the matrix with rows $u_1^\top,\ldots,u_p^\top$ and set $y = (\eta_k)_{1\leq k\leq p}$. Then (8.49) amounts to solving the linear system $Ux = y$ and (8.57) to minimizing $x \mapsto \|Ux - y\|^2$. This least-squares relaxation was proposed by Legendre (1805) and rediscovered by Gauss (1809).

Example 8.24. Consider the setting of Example 8.18 and recall that $(\forall k \in \{1,\ldots,p\})$ ${}^{\rho_k}(\partial f_k) = \{\nabla({}^{\rho_k} f_k)\}$ (Bauschke and Combettes 2017, Example 23.3). Thus, (8.52) relaxes the possibly inconsistent problem (8.50) to the problem

$$\underset{x\in\mathcal{H}}{\text{minimize}}\ f(x) + \sum_{k=1}^{p} \omega_k({}^{\rho_k} f_k)(x). \tag{8.58}$$

This formulation arises in particular in federated learning (Pathak and Wainwright 2020).

Example 8.25. Consider the setting of Example 8.19 and let $(\forall k \in \{1,\ldots,p\})$ $\rho_k = 1$. Then it follows from Example 2.14 and (2.21) that (8.52) relaxes the possibly inconsistent problem (8.51) to the variational inequality problem (see Problem 3.3)

$$\text{find } x \in C \text{ such that } (\forall y \in C)\ \sum_{k=1}^{p} \omega_k \langle L_k(y - x) \mid F_k(L_k x) - r_k\rangle \geq 0, \tag{8.59}$$

which is precisely the relaxation of (8.51) studied in Combettes and Woodstock (2022).

Let us now solve Problem 8.20 with the forward–backward algorithm.

Proposition 8.26. Consider the setting of Problem 8.20, suppose that its set Z of solutions is not empty, and set

$$\alpha = \frac{1}{\displaystyle\sum_{k=1}^{p} \frac{\omega_k \|L_k\|^2}{\rho_k}}. \tag{8.60}$$

Let $x_0 \in \mathcal{H}$, let $\varepsilon \in \left]0, \alpha/(\alpha+1)\right[$, let $(\gamma_n)_{n\in\mathbb{N}}$ be a sequence in $[\varepsilon, (2-\varepsilon)\alpha]$, and suppose that $(\mu_n)_{n\in\mathbb{N}}$ satisfies (8.2). Iterate

$$\begin{array}{l} \text{for } n = 0, 1, \ldots \\ \left\lfloor \begin{array}{l} \text{for } k = 1, \ldots, p \\ \left\lfloor \begin{array}{l} y_{k,n} = L_k x_n \\ p_{k,n} = \rho_k^{-1}(y_{k,n} - J_{\rho_k B_k} y_{k,n}) \end{array} \right. \\ b_n^* = \gamma_n \sum_{k=1}^{p} \omega_k L_k^* p_{k,n} \\ w_n = J_{\gamma_n A}(x_n - b_n^*) \\ x_{n+1} = x_n + \mu_n(w_n - x_n). \end{array} \right. \end{array} \tag{8.61}$$

Then $(x_n)_{n\in\mathbb{N}}$ converges weakly to a point in Z.

Proof. Define

$$B = \sum_{k=1}^{p} \omega_k L_k^* \circ ({}^{\rho_k}B_k) \circ L_k. \tag{8.62}$$

Then it follows from Bauschke and Combettes (2017, Proposition 4.12) and Example 2.7 that B is α-cocoercive. Since (8.61) is a specialization of (8.5), Theorem 8.1(i) furnishes the desired conclusion. ☐

8.4.3. Backward–backward splitting

We focus on the following special case of Problem 8.20.

Problem 8.27. Let $A\colon \mathcal{H} \to 2^{\mathcal{H}}$ and $B\colon \mathcal{H} \to 2^{\mathcal{H}}$ be maximally monotone, and let $\rho \in \,]0, +\infty[$. The objective is to

$$\text{find } x \in \mathcal{H} \text{ such that } 0 \in Ax + {}^{\rho}Bx. \tag{8.63}$$

Proposition 8.28. Consider the setting of Problem 8.27 under the assumption that $Z = \operatorname{zer}(A + {}^{\rho}B) \neq \varnothing$. Let $x_0 \in \mathcal{H}$, let $\varepsilon \in \,]0, \rho/(\rho+1)[$, let $(\gamma_n)_{n\in\mathbb{N}}$ be a sequence in $[\varepsilon, (2-\varepsilon)\rho]$, and suppose that $(\mu_n)_{n\in\mathbb{N}}$ satisfies (8.2) with $\alpha = \rho$. Iterate

$$\begin{array}{l} \text{for } n = 0, 1, \ldots \\ \left\lfloor \begin{array}{l} p_n = \rho^{-1}(x_n - J_{\rho B} x_n) \\ w_n = J_{\gamma_n A}(x_n - \gamma_n p_n) \\ x_{n+1} = x_n + \mu_n(w_n - x_n). \end{array} \right. \end{array} \tag{8.64}$$

Then $(x_n)_{n\in\mathbb{N}}$ converges weakly to a point in Z.

Proof. Apply Proposition 8.26 with $p = 1$, $\mathcal{G}_1 = \mathcal{H}$, $L_1 = \mathrm{Id}$, $B_1 = B$, $\omega_1 = 1$ and $\rho_1 = \rho$. ☐

Example 8.29. In particular, if we execute (8.64) with, for every $n \in \mathbb{N}$, $\gamma_n = \rho$ and $\mu_n = 1$, then

$$(\forall n \in \mathbb{N}) \quad x_{n+1} = J_{\rho A}(J_{\rho B} x_n). \tag{8.65}$$

This recursion is known as the *backward–backward algorithm*, as it alternates two backward Euler steps. As derived above, it is a special case of (8.61) and therefore of the forward–backward algorithm (8.5). Its asymptotic behaviour has been studied by Bauschke *et al.* (2005) and Mercier (1980); see also Lions (1978) and Passty (1979) for ergodic convergence.

Example 8.30. Let f and g be functions in $\Gamma_0(\mathcal{H})$. In Problem 8.27, suppose that $A = \partial f$ and $B = \partial g$. Then, as in Example 8.24, (8.65) becomes

$$\underset{x \in \mathcal{H}}{\text{minimize}} \;\; f(x) + {}^{\rho}g(x) \tag{8.66}$$

and (8.65) reduces to the *alternating proximal point algorithm*

$$(\forall n \in \mathbb{N}) \quad x_{n+1} = \text{prox}_{\rho f}(\text{prox}_{\rho g} x_n). \tag{8.67}$$

This method was first investigated in Acker and Prestel (1980), with further developments in Bauschke *et al.* (2005).

Example 8.31. Let C and D be nonempty closed convex subsets of $\mathcal{H}$. In Example 8.30, suppose that $f = \iota_C$ and $g = \iota_D$. Then (8.67) is the problem of finding a point in C at minimal distance from D and (8.67) yields the *alternating projection method*

$$(\forall n \in \mathbb{N}) \quad x_{n+1} = \text{proj}_C(\text{proj}_D x_n), \tag{8.68}$$

which was first investigated in Cheney and Goldstein (1959*b*). Its weak convergence was established in Gubin *et al.* (1967, Theorem 2)

Example 8.32. Let $f \in \Gamma_0(\mathcal{H})$, $h \in \Gamma_0(\mathcal{H})$, $z \in \mathcal{H}$ and $\rho \in \,]0, +\infty[$. The problem is to

$$\underset{x \in \mathcal{H}, w \in \mathcal{H}}{\text{minimize}} \;\; f(x) + h(w) + \frac{1}{2\rho}\|x + w - z\|^2. \tag{8.69}$$

Following Combettes and Wajs (2005, Section 4.4), set $g\colon y \mapsto h(z - y)$. Then, with the change of variable $y = z - w$, the objective of (8.69) is to

$$\underset{x \in \mathcal{H}, y \in \mathcal{H}}{\text{minimize}} \;\; f(x) + g(y) + \frac{1}{2\rho}\|x - y\|^2, \tag{8.70}$$

which is precisely (8.66) in terms of the variable x. Now let $x_0 \in \mathcal{H}$, let $\varepsilon \in \,]0, \rho/(\rho + 1)[$, let $(\gamma_n)_{n\in\mathbb{N}}$ be a sequence in $[\varepsilon, (2 - \varepsilon)\rho]$, and let $(\mu_n)_{n\in\mathbb{N}}$ be a sequence in $[\varepsilon, 1]$. Applying algorithm (8.64) to $A = \partial f$ and $B = \partial g$, and noting that $J_{\rho B} = \text{prox}_{\rho g}\colon x \mapsto z - \text{prox}_{\rho h}(z - x)$ yields

$$\begin{array}{l} \text{for } n = 0, 1, \ldots \\ \left\lfloor \begin{array}{l} p_n = \rho^{-1}(x_n - z + \text{prox}_{\rho h}(z - x_n)) \\ w_n = \text{prox}_{\gamma_n f}(x_n - \gamma_n p_n) \\ x_{n+1} = x_n + \mu_n(w_n - x_n). \end{array} \right. \end{array} \tag{8.71}$$

It follows from Proposition 8.28 that $(x_n)_{n\in\mathbb{N}}$ converges weakly to a point x such that $(x, \text{prox}_{\rho h}(z - x))$ solves (8.69).

Next, we revisit the problem of projecting onto the Minkowski sum of two convex sets (see Example 8.15).

Example 8.33. Let C and D be nonempty closed convex subsets of $\mathcal{H}$ such that $C + D$ is closed, and let $z \in \mathcal{H}$. Upon setting $f = \iota_C$, $h = \iota_D$ and $\rho = 1$ in Example 8.32, (8.69) specializes to the problem of finding the projection of z onto $C + D$. Now let $x_0 \in C$, let $\varepsilon \in]0, 1/2[$, let $(\gamma_n)_{n\in\mathbb{N}}$ be a sequence in $[\varepsilon, 2 - \varepsilon]$, and let $(\mu_n)_{n\in\mathbb{N}}$ be a sequence in $[\varepsilon, 1]$. Then (8.71) assumes the form

$$\begin{array}{l} \text{for } n = 0, 1, \ldots \\ \left\lfloor \begin{array}{l} p_n = x_n - z + \operatorname{proj}_D(z - x_n) \\ w_n = \operatorname{proj}_C(x_n - \gamma_n p_n) \\ x_{n+1} = x_n + \mu_n(w_n - x_n), \end{array} \right. \end{array} \tag{8.72}$$

and it follows from Proposition 8.28 that $(x_n)_{n\in\mathbb{N}}$ converges weakly to a point x such that $\operatorname{proj}_{C+D} z = x + \operatorname{proj}_D(z - x)$. This best approximation algorithm was first obtained in Seeger (1998, Theorem 2.1) in the case when $(\forall n \in \mathbb{N})\ \gamma_n = \mu_n = 1$, that is,

$$(\forall n \in \mathbb{N}) \quad x_{n+1} = \operatorname{proj}_C(z - \operatorname{proj}_D(z - x_n)). \tag{8.73}$$

8.4.4. Dual implementation

We present a framework for solving strongly monotone composite inclusion problems by applying the forward–backward algorithm to the dual problem. The embedding underlying this approach is that of Example 3.22.

Problem 8.34. Let $\rho \in]0, +\infty[$, let $0 < p \in \mathbb{N}$, let $z \in \mathcal{H}$, and let $A : \mathcal{H} \to 2^{\mathcal{H}}$ be maximally monotone. For every $k \in \{1, \ldots, p\}$, let $B_k : \mathcal{G}_k \to 2^{\mathcal{G}_k}$ be maximally monotone, let $\nu_k \in]0, +\infty[$, let $D_k : \mathcal{G}_k \to 2^{\mathcal{G}_k}$ be maximally monotone and ν_k-strongly monotone, and suppose that $0 \neq L_k \in \mathcal{B}(\mathcal{H}, \mathcal{G}_k)$. Further, suppose that

$$z \in \operatorname{ran}\left(A + \sum_{k=1}^{p} L_k^* \circ (B_k \,\Box\, D_k) \circ L_k + \rho \mathrm{Id} \right). \tag{8.74}$$

The problem is to solve the primal inclusion

$$\text{find } x \in \mathcal{H} \text{ such that } z \in Ax + \sum_{k=1}^{p} L_k^*((B_k \,\Box\, D_k)(L_k x)) + \rho x, \tag{8.75}$$

together with the dual inclusion

$$\begin{array}{l} \text{find } y_1^* \in \mathcal{G}_1, \ldots, y_p^* \in \mathcal{G}_p \text{ such that } (\forall k \in \{1, \ldots, p\}) \\ 0 \in -L_k\left(J_{A/\rho}\left(\dfrac{1}{\rho}\left(z - \displaystyle\sum_{j=1}^{p} L_j^* y_j^* \right)\right)\right) + B_k^{-1} y_k^* + D_k^{-1} y_k^*. \end{array} \tag{8.76}$$

We refer to Combettes and Vũ (2014, Proposition 5.2(iv)) for sufficient conditions that guarantee (8.74). The mechanism to solve (8.75) dually hinges on the following properties.

Proposition 8.35 (Combettes and Vũ 2014). Consider the setting of Problem 8.34, and set

$$M = A + \sum_{k=1}^{p} L_k^* \circ (B_k \,\square\, D_k) \circ L_k \quad \text{and} \quad \overline{x} = J_{M/\rho}(z/\rho). \tag{8.77}$$

Then the following hold:

(i) $\overline{x}$ is the unique solution to the primal problem (8.75).

(ii) The dual problem (8.76) admits solutions and, if $(\overline{y}_k^*)_{1\leq k\leq p}$ solves (8.76), then

$$\overline{x} = J_{A/\rho}\left(\rho^{-1}\left(z - \sum_{k=1}^{p} L_k^* \overline{y}_k^*\right)\right). \tag{8.78}$$

We now apply the forward–backward algorithm of Theorem 8.1 to the dual inclusion (8.76) to construct a sequence $(x_n)_{n\in\mathbb{N}}$ which converges strongly to the solution to primal inclusion (8.75). The following result is an adaptation of Corollary 5.4 of Combettes and Vũ (2014).

Proposition 8.36. Consider the setting of Problem 8.34, and set

$$\nu = \min_{1\leq k\leq p} \nu_k \quad \text{and} \quad \alpha = \frac{1}{1/\nu + (1/\rho)\sum_{1\leq k\leq p}\|L_k\|^2}. \tag{8.79}$$

Let $\varepsilon \in \left]0, \alpha/(\alpha+1)\right[$, let $(\gamma_n)_{n\in\mathbb{N}}$ be a sequence in $[\varepsilon, (2-\varepsilon)\alpha]$, suppose that $(\mu_n)_{n\in\mathbb{N}}$ satisfies (8.2), and, for every $k \in \{1,\ldots,p\}$, let $y_{k,0}^* \in \mathcal{G}_k$. Iterate

$$\begin{array}{l}
\text{for } n = 0, 1, \ldots \\
\left\lfloor \begin{array}{l}
q_n = z - \sum_{k=1}^{p} L_k^* y_{k,n}^* \\
x_n = J_{A/\rho}(q_n/\rho) \\
\text{for } k = 1, \ldots, p \\
\left\lfloor \begin{array}{l}
w_{k,n} = y_{k,n}^* + \gamma_n\big(L_k x_n - D_k^{-1} y_{k,n}^*\big) \\
y_{k,n+1}^* = y_{k,n}^* + \mu_n\big(J_{\gamma_n B_k^{-1}} w_{k,n} - y_{k,n}^*\big).
\end{array}\right.
\end{array}\right.
\end{array} \tag{8.80}$$

Then the following hold for the solution $\overline{x}$ to (8.75) and for some solution $\overline{\boldsymbol{y}}^* = (\overline{y}_1^*, \ldots, \overline{y}_p^*)$ to (8.76):

(i) $(\forall k \in \{1,\ldots,p\})\ y_{k,n}^* \rightharpoonup \overline{y}_k^*$.

(ii) $x_n \to \overline{x}$.

Proof. We deduce from Bauschke and Combettes (2017, Proposition 22.11(ii)) that, for every $k \in \{1,\ldots,p\}$, D_k^{-1} is ν_k-cocoercive with $\operatorname{dom} D_k^{-1} = \mathcal{G}_k$. Let us set $\boldsymbol{\mathcal{G}} = \mathcal{G}_1 \oplus \cdots \oplus \mathcal{G}_p$ and

$$\begin{cases} T\colon \mathcal{H} \to \mathcal{H}\colon x \mapsto J_{\rho^{-1}A}(\rho^{-1}(z-x)),\\ \boldsymbol{A}\colon \boldsymbol{\mathcal{G}} \to 2^{\boldsymbol{\mathcal{G}}}\colon \boldsymbol{y}^* \mapsto \bigtimes_{1\leq k\leq p} B_k^{-1}y_k^*,\\ \boldsymbol{D}\colon \boldsymbol{\mathcal{G}} \to \boldsymbol{\mathcal{G}}\colon \boldsymbol{y}^* \mapsto \big(D_k^{-1}y_k^*\big)_{1\leq k\leq p},\\ \boldsymbol{L}\colon \mathcal{H} \to \boldsymbol{\mathcal{G}}\colon x \mapsto (L_k x)_{1\leq k\leq p},\\ \boldsymbol{B} = \boldsymbol{D} - \boldsymbol{L}\circ T\circ \boldsymbol{L}^*. \end{cases} \tag{8.81}$$

It follows from Lemmas 2.23 and 2.24 that $\boldsymbol{A}$ is maximally monotone, from (8.79) that $\boldsymbol{D}$ is ν-cocoercive, from Lemma 2.34(iii) that $-T$ is ρ-cocoercive, and hence from Bauschke and Combettes (2017, Proposition 4.12) that

$$\boldsymbol{B} = \boldsymbol{D} + \boldsymbol{L}\circ(-T)\circ\boldsymbol{L}^* \text{ is } 1/(1/\nu + \|\boldsymbol{L}\|^2/\rho)\text{-cocoercive.} \tag{8.82}$$

Since $\|\boldsymbol{L}\|^2 \leq \sum_{k=1}^p \|L_k\|^2$, (8.79) implies that $\boldsymbol{B}$ is α-cocoercive. Next, let us define $(\forall n \in \mathbb{N})$ $\boldsymbol{y}_n^* = (y_{k,n}^*)_{1\leq k\leq p}$ and $\boldsymbol{w}_n = (w_{k,n})_{1\leq k\leq p}$. Then, upon combining (8.81) and Example 2.37, (8.80) can be rewritten as

$$\begin{array}{l} \text{for } n = 0, 1, \ldots \\ \left\lfloor \begin{array}{l} \boldsymbol{w}_n = \boldsymbol{y}_n^* - \gamma_n \boldsymbol{B}\boldsymbol{y}_n^* \\ \boldsymbol{y}_{n+1}^* = \boldsymbol{y}_n^* + \mu_n\big(J_{\gamma_n\boldsymbol{A}}\,\boldsymbol{w}_n - \boldsymbol{y}_n^*\big), \end{array} \right. \end{array} \tag{8.83}$$

and the dual problem (8.76) as

$$\text{find } \boldsymbol{y}^* \in \boldsymbol{\mathcal{G}} \text{ such that } \boldsymbol{0} \in \boldsymbol{A}\boldsymbol{y}^* + \boldsymbol{B}\boldsymbol{y}^*. \tag{8.84}$$

(i) In view of the above, the claim follows from Theorem 8.1(i).

(ii) We derive from Proposition 8.35, (8.80) and (8.81) that

$$\overline{x} = T(\boldsymbol{L}^*\overline{\boldsymbol{y}}^*) \quad\text{and}\quad (\forall n \in \mathbb{N})\quad x_n = T(\boldsymbol{L}^*\boldsymbol{y}_n^*). \tag{8.85}$$

In turn, we deduce from the ρ-cocoercivity of $-T$, (i), the monotonicity of $\boldsymbol{D}$, and the Cauchy–Schwarz inequality that

$$\begin{aligned} (\forall n \in \mathbb{N})\quad \rho\|x_n - \overline{x}\|^2 &= \rho\|T(\boldsymbol{L}^*\boldsymbol{y}_n^*) - T(\boldsymbol{L}^*\overline{\boldsymbol{y}}^*)\|^2 \\ &\leq \langle \boldsymbol{L}^*(\boldsymbol{y}_n^* - \overline{\boldsymbol{y}}^*) \mid T(\boldsymbol{L}^*\overline{\boldsymbol{y}}^*) - T(\boldsymbol{L}^*\boldsymbol{y}_n^*)\rangle \\ &= \langle \boldsymbol{y}_n^* - \overline{\boldsymbol{y}}^* \mid (\boldsymbol{L}\circ T\circ\boldsymbol{L}^*)\overline{\boldsymbol{y}}^* - (\boldsymbol{L}\circ T\circ\boldsymbol{L}^*)\boldsymbol{y}_n^*\rangle \\ &\leq \langle \boldsymbol{y}_n^* - \overline{\boldsymbol{y}}^* \mid \boldsymbol{D}\boldsymbol{y}_n^* - \boldsymbol{D}\overline{\boldsymbol{y}}^*\rangle \\ &\quad - \langle \boldsymbol{y}_n^* - \overline{\boldsymbol{y}}^* \mid (\boldsymbol{L}\circ T\circ\boldsymbol{L}^*)\boldsymbol{y}_n^* - (\boldsymbol{L}\circ T\circ\boldsymbol{L}^*)\overline{\boldsymbol{y}}^*\rangle \\ &= \langle \boldsymbol{y}_n^* - \overline{\boldsymbol{y}}^* \mid \boldsymbol{B}\boldsymbol{y}_n^* - \boldsymbol{B}\overline{\boldsymbol{y}}^*\rangle \\ &\leq \delta\|\boldsymbol{B}\boldsymbol{y}_n^* - \boldsymbol{B}\overline{\boldsymbol{y}}^*\|, \end{aligned} \tag{8.86}$$

where, by (i),

$$\delta = \sup_{n\in\mathbb{N}} \|\boldsymbol{y}_n^* - \overline{\boldsymbol{y}}^*\| < +\infty. \tag{8.87}$$

Therefore, using (8.83) and Theorem 8.1(ii)–(iii), we conclude that $\|x_n - \overline{x}\| \to 0$. □

Here is an application to strongly convex minimization problems that arise in particular in mechanics (Ekeland and Temam 1974, Mercier 1980) and in signal processing (Combettes, Đinh Dũng and Vũ 2010, 2011, Potter and Arun 1993).

Example 8.37. Let $0 < p \in \mathbb{N}$, let $z \in \mathcal{H}$, let $f \in \Gamma_0(\mathcal{H})$, and let ${}^1(f^*)$ be the Moreau envelope of f^* (see (2.11)). For every $k \in \{1, \dots, p\}$, let $g_k \in \Gamma_0(\mathcal{G}_k)$, let $\nu_k \in \,]0, +\infty[$, let $h_k \in \Gamma_0(\mathcal{G}_k)$ be ν_k-strongly convex, and suppose that $0 \neq L_k \in \mathcal{B}(\mathcal{H}, \mathcal{G}_k)$. Define α as in (8.79) and suppose that

$$z \in \operatorname{ran}\left(\partial f + \sum_{k=1}^{p} L_k^* \circ (\partial g_k \,\Box\, \partial h_k) \circ L_k + \mathrm{Id}\right). \tag{8.88}$$

Then the primal problem

$$\underset{x \in \mathcal{H}}{\text{minimize}} \;\; f(x) + \sum_{k=1}^{p} (g_k \,\Box\, h_k)(L_k x) + \frac{1}{2}\|x - z\|^2 \tag{8.89}$$

admits a unique solution $\overline{x}$, namely

$$\overline{x} = \operatorname{prox}_{f + \sum_{k=1}^{p} (g_k \,\Box\, h_k) \circ L_k} z, \tag{8.90}$$

and the dual problem is

$$\underset{y_1^* \in \mathcal{G}_1, \dots, y_p^* \in \mathcal{G}_p}{\text{minimize}} \;\; {}^1(f^*)\left(z - \sum_{k=1}^{p} L_k^* y_k^*\right) + \sum_{k=1}^{p} (g_k^*(y_k^*) + h_k^*(y_k^*)). \tag{8.91}$$

Now let $\varepsilon \in \,]0, \alpha/(\alpha+1)[$, let $(\gamma_n)_{n \in \mathbb{N}}$ be a sequence in $[\varepsilon, (2-\varepsilon)\alpha]$, suppose that $(\mu_n)_{n \in \mathbb{N}}$ satisfies (8.2), and, for every $k \in \{1, \dots, p\}$, let $y_{k,0}^* \in \mathcal{G}_k$. Iterate

$$\begin{array}{l} \text{for } n = 0, 1, \dots \\ \left\lfloor \begin{array}{l} q_n = z - \sum_{k=1}^{p} L_k^* y_{k,n}^* \\ x_n = \operatorname{prox}_f q_n \\ \text{for } k = 1, \dots, p \\ \left\lfloor \begin{array}{l} w_{k,n} = y_{k,n}^* + \gamma_n\big(L_k x_n - \nabla h_k^*\big(y_{k,n}^*\big)\big) \\ y_{k,n+1}^* = y_{k,n}^* + \mu_n\big(\operatorname{prox}_{\gamma_n g_k^*} w_{k,n} - y_{k,n}^*\big). \end{array} \right. \end{array} \right. \end{array} \tag{8.92}$$

Then the following hold:

(i) There exists a solution $(\overline{y}_1^*, \dots, \overline{y}_p^*)$ to (8.91) such that $(\forall k \in \{1, \dots, p\})$ $y_{k,n}^* \rightharpoonup \overline{y}_k^*$.

(ii) $x_n \to \overline{x}$.

Proof. Apply Proposition 8.36 with $\rho = 1$, $A = \partial f$, and $(\forall k \in \{1, \dots, p\})$ $B_k = \partial g_k$ and $D_k = \partial h_k$; see Combettes and Vũ (2014, Example 5.6) for details. □

Remark 8.38. In Example 8.37, suppose that $\mathcal{H} = H_0^1(\Omega)$, where Ω is a bounded open domain in $\mathbb{R}^2$, $p = 1$, $\mathcal{G}_1 = L^2(\Omega) \oplus L^2(\Omega)$, $L_1 = \nabla$, $g_1 = \mu\|\cdot\|_{2,1}$ with $\mu \in]0, +\infty[$, and $h_1 = \iota_{\{0\}}$. Then (8.89) reduces to

$$\underset{x \in H_0^1(\Omega)}{\text{minimize}} \;\; f(x) + \mu \int_\Omega |\nabla x(\omega)|_2 \, d\omega + \frac{1}{2}\|x - z\|^2. \tag{8.93}$$

In mechanics, (8.93) has been studied for certain potentials f (Ekeland and Temam 1974). For instance, $f = 0$ yields Mossolov's problem and its dual analysis is carried out in Ekeland and Temam (1974, Section IV.3.1). In image processing, Mossolov's problem corresponds to the total variation denoising problem. Mercier (1980) proposed a dual projection algorithm to solve Mossolov's problem. In image processing, this approach was rediscovered in a discrete setting in Chambolle (2004, 2005).

8.4.5. Barycentric Dykstra-like algorithm

Using Proposition 8.36 and, thereby, the forward–backward algorithm, we obtain a method for computing the resolvent of a sum of maximally monotone operators. This result, which generalizes the barycentric Dykstra algorithm of Gaffke and Mathar (1989) for projecting onto an intersection of closed convex sets, was originally derived in Combettes (2009, Theorem 3.3) with different techniques.

Proposition 8.39. Let $0 < p \in \mathbb{N}$, let $z \in \mathcal{H}$, and, for every $k \in \{1, \dots, p\}$, let $A_k : \mathcal{H} \to 2^{\mathcal{H}}$ be maximally monotone. Suppose that

$$z \in \operatorname{ran}\left(\sum_{k=1}^{p} A_k + \mathrm{Id}\right) \tag{8.94}$$

and consider the inclusion problem

$$\text{find } x \in \mathcal{H} \text{ such that } z \in \sum_{k=1}^{p} A_k x + x. \tag{8.95}$$

Set $x_0 = z$ and $(\forall k \in \{1, \dots, p\})\ z_{k,0} = z$. Iterate

$$\begin{array}{l} \text{for } n = 0, 1, \dots \\ \left\lfloor \begin{array}{l} \text{for } k = 1, \dots, p \\ \left\lfloor\ r_{k,n} = J_{pA_k} z_{k,n} \right. \\ x_{n+1} = (1/p) \sum_{k=1}^{p} r_{k,n} \\ \text{for } k = 1, \dots, p \\ \left\lfloor\ z_{k,n+1} = z_{k,n} - r_{k,n} + x_{n+1}. \right. \end{array} \right. \end{array} \tag{8.96}$$

Then $x_n \to J_{\sum_{k=1}^{p} A_k} z$.

Proof. First, we observe that (8.94)–(8.95) is the special case of (8.74)–(8.75) in which $A = 0$ and, for every $k \in \{1, \ldots, p\}$, $\mathcal{G}_k = \mathcal{H}$, $B_k = A_k$, $L_k = \mathrm{Id}$ and $D_k = \{0\}^{-1}$. Moreover, the cocoercivity constant in (8.79) is $\alpha = 1/p$. With this scenario, implementing (8.80) with, for every $n \in \mathbb{N}$, $\mu_n = 1$ and $\gamma_n = 1/p$, and, for every $k \in \{1, \ldots, p\}$, $y^*_{k,0} = 0$ leads to the recursion

$$
\begin{array}{l}
\text{for } n = 0, 1, \ldots \\
\left\lfloor
\begin{array}{l}
x_n = z - \sum_{k=1}^{p} y^*_{k,n} \\
\text{for } k = 1, \ldots, p \\
\left\lfloor\; y^*_{k,n+1} = J_{A_k^{-1}/p}\big(y^*_{k,n} + x_n/p\big), \right.
\end{array}
\right.
\end{array}
\tag{8.97}
$$

and Proposition 8.36(ii) guarantees that $x_n \to J_{\sum_{k=1}^{p} A_k} z$. Alternatively, with the initialization $x_0 = z$, we rewrite (8.97) as

$$
\begin{array}{l}
\text{for } n = 0, 1, \ldots \\
\left\lfloor
\begin{array}{l}
\text{for } k = 1, \ldots, p \\
\left\lfloor\; y^*_{k,n+1} = J_{A_k^{-1}/p}\big(y^*_{k,n} + x_n/p\big) \right. \\
x_{n+1} = z - \sum_{k=1}^{p} y^*_{k,n+1}.
\end{array}
\right.
\end{array}
\tag{8.98}
$$

Let us introduce the variables $(\forall n \in \mathbb{N})(\forall k \in \{1, \ldots, p\})$ $z_{k,n} = p y^*_{k,n} + x_n$, where $z_{k,0} = x_0 = z$. Then (8.98) corresponds to the iterations

$$
\begin{array}{l}
\text{for } n = 0, 1, \ldots \\
\left\lfloor
\begin{array}{l}
x_{n+1} = z - \sum_{k=1}^{p} J_{A_k^{-1}/p}(z_{k,n}/p) \\
\text{for } k = 1, \ldots, p \\
\left\lfloor\; z_{k,n+1} = p\, J_{A_k^{-1}/p}(z_{k,n}/p) + x_{n+1}. \right.
\end{array}
\right.
\end{array}
\tag{8.99}
$$

By construction,

$$
(\forall n \in \mathbb{N}) \quad \sum_{k=1}^{p} z_{k,n} = pz. \tag{8.100}
$$

Hence, appealing to (2.21), (8.99) becomes

$$
\begin{array}{l}
\text{for } n = 0, 1, \ldots \\
\left\lfloor
\begin{array}{l}
x_{n+1} = (1/p) \sum_{k=1}^{p} J_{pA_k} z_{k,n} \\
\text{for } k = 1, \ldots, p \\
\left\lfloor\; z_{k,n+1} = z_{k,n} - J_{pA_k} z_{k,n} + x_{n+1}, \right.
\end{array}
\right.
\end{array}
\tag{8.101}
$$

which is precisely (8.96). □

Example 8.40. Consider the instantiation of Proposition 8.39 in which, for every $k \in \{1, \ldots, p\}$, $A_k = \partial f_k$, with $f_k \in \Gamma_0(\mathcal{H})$, and execute (8.96), which becomes

$$
\begin{array}{l}
\text{for } n = 0, 1, \ldots \\
\left\lfloor
\begin{array}{l}
\text{for } k = 1, \ldots, p \\
\left\lfloor\ r_{k,n} = \operatorname{prox}_{p f_k} z_{k,n} \right. \\
x_{n+1} = (1/p) \sum_{k=1}^{p} r_{k,n} \\
\text{for } k = 1, \ldots, p \\
\left\lfloor\ z_{k,n+1} = z_{k,n} - r_{k,n} + x_{n+1}. \right.
\end{array}
\right.
\end{array}
\tag{8.102}
$$

Then $x_n \to \operatorname{prox}_{\sum_{k=1}^{p} f_k} z$.

Our last example addresses the barycentric Dykstra algorithm *per se*. The original Dykstra algorithm was devised in Dykstra (1983) to project onto the intersection of closed convex cones (see also Han 1988 for general closed convex sets whose intersection has a nonempty interior) in Euclidean spaces using periodic applications of the projectors onto the individual sets. Convergence of this periodic scheme in the general case of arbitrary closed and convex sets in Hilbert spaces was established in Boyle and Dykstra (1986); see Bauschke and Combettes (2008) for an extension to monotone operators. The barycentric version described below, in which all the projectors are used at each iteration, was devised in Gaffke and Mathar (1989, Section 6). Its connection with the forward–backward algorithm is discussed in Combettes *et al.* (2010, Remark 3.8) and Combettes *et al.* (2011, Remark 2.3), and its asymptotic behaviour in the inconsistent case in Bauschke and Borwein (1994, Theorem 6.1).

Example 8.41. In Example 8.40, suppose that, for every $k \in \{1, \ldots, p\}$, $f_k = \iota_{C_k}$, where C_k is a nonempty closed convex subset of $\mathcal{H}$. Then algorithm (8.102) becomes

$$
\begin{array}{l}
\text{for } n = 0, 1, \ldots \\
\left\lfloor
\begin{array}{l}
\text{for } k = 1, \ldots, p \\
\left\lfloor\ r_{k,n} = \operatorname{proj}_{C_k} z_{k,n} \right. \\
x_{n+1} = (1/p) \sum_{k=1}^{p} r_{k,n} \\
\text{for } k = 1, \ldots, p \\
\left\lfloor\ z_{k,n+1} = z_{k,n} - r_{k,n} + x_{n+1}, \right.
\end{array}
\right.
\end{array}
\tag{8.103}
$$

and $x_n \to \operatorname{proj}_{\bigcap_{k=1}^{p} C_k} z$.

8.4.6. *Renorming*

We preface our discussion with a renormed version of Theorem 8.1.

Proposition 8.42. Let $\alpha \in \left]0, +\infty\right[$, let $\beta \in \left]0, +\infty\right[$, let $A \colon \mathcal{H} \to 2^{\mathcal{H}}$ be maximally monotone, let $B \colon \mathcal{H} \to \mathcal{H}$ be α-cocoercive, let $U \in \mathcal{B}(\mathcal{H})$ be self-adjoint and β-strongly monotone, and let $\mathcal{X}$ be the real Hilbert space obtained by endowing $\mathcal{H}$

with the scalar product $(x, y) \mapsto \langle Ux \mid y\rangle$. Let $\varepsilon \in \,]0, \alpha\beta/(\alpha\beta+1)[$, let $(\gamma_n)_{n\in\mathbb{N}}$ be a sequence in $[\varepsilon, (2-\varepsilon)\alpha\beta]$, and let $(\lambda_n)_{n\in\mathbb{N}}$ be a sequence in $[\varepsilon, 1]$. Suppose that the set Z of solutions to the problem

$$\text{find } x \in \mathcal{H} \text{ such that } 0 \in Ax + Bx \tag{8.104}$$

is not empty and let Z^* be the set of solutions to the dual problem

$$\text{find } x^* \in \mathcal{H} \text{ such that } 0 \in -A^{-1}(-x^*) + B^{-1}x^*. \tag{8.105}$$

Let $x_0 \in \mathcal{H}$ and iterate

$$\begin{array}{l} \text{for } n = 0, 1, \ldots \\ \left\lfloor \begin{array}{l} u_n^* = \gamma_n^{-1}Ux_n - Bx_n \\ w_n = \left(\gamma_n^{-1}U + A\right)^{-1}u_n^* \\ x_{n+1} = x_n + \lambda_n(w_n - x_n). \end{array} \right. \end{array} \tag{8.106}$$

Then the following hold:

(i) $(x_n)_{n\in\mathbb{N}}$ converges weakly to a point in Z.

(ii) Z^* contains a single point $\overline{x}^*$ and $(\forall z \in Z)\ Bz = \overline{x}^*$.

(iii) $(Bx_n)_{n\in\mathbb{N}}$ converges strongly to $\overline{x}^*$.

Proof. We derive from Lemma 2.25 and Example 2.39 that

$$(\forall n \in \mathbb{N}) \quad x_{n+1} = x_n + \lambda_n\big(J_{\gamma_n U^{-1}\circ A}\big(x_n - \gamma_n U^{-1}(Bx_n)\big) - x_n\big), \tag{8.107}$$

where $U^{-1} \circ A \colon \mathcal{X} \to 2^{\mathcal{X}}$ is maximally monotone, $U^{-1} \circ B \colon \mathcal{X} \to \mathcal{X}$ is $\alpha\beta$-cocoercive, and $\operatorname{zer}(A+B) = \operatorname{zer}(U^{-1} \circ (A+B))$. Hence the assertions follow from Theorem 8.1 applied to $U^{-1} \circ A$ and $U^{-1} \circ B$ in $\mathcal{X}$. □

Remark 8.43. In terms of the warped resolvents of Section 2.4.3, (8.106) can be condensed into

$$(\forall n \in \mathbb{N}) \quad x_{n+1} = x_n + \lambda_n\big(J^{U_n}_{\gamma_n(A+B)}x_n - x_n\big), \quad \text{where } U_n = U - \gamma_n B. \tag{8.108}$$

We present an approach proposed in Vũ (2013), which revisited the primal–dual setting of Combettes and Pesquet (2012) discussed in Proposition 7.10 by replacing the monotone Lipschitz property of the operators C and $(D_k^{-1})_{1\leq k\leq p}$ with the stronger cocoercivity property.

Proposition 8.44 (Vũ 2013). Let $0 < p \in \mathbb{N}$, let $\alpha \in \,]0, +\infty[$, let $A \colon \mathcal{H} \to 2^{\mathcal{H}}$ be maximally monotone, and let $C \colon \mathcal{H} \to \mathcal{H}$ be α-cocoercive. For every $k \in \{1, \ldots, p\}$, let $\beta_k \in \,]0, +\infty[$, let $\mathcal{G}_k$ be a real Hilbert space, let $B_k \colon \mathcal{G}_k \to 2^{\mathcal{G}_k}$ be maximally monotone, let $D_k \colon \mathcal{G}_k \to 2^{\mathcal{G}_k}$ be maximally monotone and β_k-strongly monotone, and suppose that $0 \neq L_k \in \mathcal{B}(\mathcal{H}, \mathcal{G}_k)$. Additionally, suppose that the set Z of solutions to the primal inclusion

$$\text{find } x \in \mathcal{H} \text{ such that } 0 \in Ax + \sum_{k=1}^{p} L_k^*((B_k \,\square\, D_k)(L_k x)) + Cx \tag{8.109}$$

is not empty, and let Z^* be the set of solutions to the dual inclusion

$$\text{find } y_1^* \in \mathcal{G}_1, \ldots, y_p^* \in \mathcal{G}_p \text{ such that}$$
$$(\exists x \in \mathcal{H}) \begin{cases} x \in (A+C)^{-1}\Big(-\sum_{k=1}^{p} L_k^* y_k^*\Big), \\ (\forall k \in \{1,\ldots,p\})\ L_k x \in B_k^{-1} y_k^* + D_k^{-1} y_k^*. \end{cases} \tag{8.110}$$

Let $\varepsilon \in \,]0,1[$, let $(\lambda_n)_{n\in\mathbb{N}}$ be a sequence in $[\varepsilon,1]$, let $x_0 \in \mathcal{H}$, let $(y_{1,0}^*,\ldots,y_{p,0}^*) \in \mathcal{G}_1 \oplus \cdots \oplus \mathcal{G}_p$, let $\tau \in \,]0,+\infty[$, and let $(\sigma_1,\ldots,\sigma_p) \in \,]0,+\infty[^p$. Set

$$\aleph = \min\{\alpha, \beta_1, \ldots, \beta_p\} \quad \text{and} \quad \beta = \frac{1 - \sqrt{\tau \sum_{k=1}^{p} \sigma_k \|L_k\|^2}}{\max\{\tau, \sigma_1, \ldots, \sigma_p\}} \tag{8.111}$$

and assume that

$$\aleph\beta > \frac{1}{2}. \tag{8.112}$$

Iterate

$$\begin{array}{l} \text{for } n = 0, 1, \ldots \\ \left\lfloor \begin{array}{l} x_n^* = \tau\big(\sum_{k=1}^{p} L_k^* y_{k,n}^* + Cx_n\big) \\ p_n = J_{\tau A}(x_n - x_n^*) \\ x_{n+1} = x_n + \lambda_n(p_n - x_n) \\ \text{for } k = 1, \ldots, p \\ \left\lfloor \begin{array}{l} y_{k,n} = \sigma_k\big(L_k(2p_n - x_n) - D_k^{-1} y_{k,n}^*\big) \\ q_{k,n}^* = J_{\sigma_k B_k^{-1}}\big(y_{k,n}^* + y_{k,n}\big) \\ y_{k,n+1}^* = y_{k,n}^* + \lambda_n\big(q_{k,n}^* - y_{k,n}^*\big). \end{array} \right. \end{array} \right. \end{array} \tag{8.113}$$

Then there exist $x \in Z$ and $(y_1^*,\ldots,y_p^*) \in Z^*$ such that $x_n \rightharpoonup x$, and, for every $k \in \{1,\ldots,p\}$, $y_{k,n}^* \rightharpoonup y_k^*$.

Proof. Set $\mathbf{X} = \mathcal{H} \oplus \mathcal{G}_1 \oplus \cdots \oplus \mathcal{G}_p$ and

$$\begin{cases} \boldsymbol{M}\colon \mathbf{X} \to 2^{\mathbf{X}}\colon (x, y_1^*, \ldots, y_p^*) \mapsto \\ \qquad \big(Ax + \sum_{k=1}^{p} L_k^* y_k^*\big) \times \big(-L_1 x + B_1^{-1} y_1^*\big) \times \cdots \times \big(-L_p x + B_p^{-1} y_p^*\big), \\ \boldsymbol{C}\colon \mathbf{X} \to \mathbf{X}\colon (x, y_1^*, \ldots, y_p^*) \mapsto \big(Cx, D_1^{-1} y_1^*, \ldots, D_p^{-1} y_p^*\big), \\ \boldsymbol{U}\colon \mathbf{X} \to \mathbf{X}\colon (x, y_1^*, \ldots, y_p^*) \mapsto \\ \qquad \big(\tau^{-1} x - \sum_{k=1}^{p} L_k^* y_k^*, -L_1 x + \sigma_1^{-1} y_1^*, \ldots, -L_p x + \sigma_p^{-1} y_p^*\big). \end{cases} \tag{8.114}$$

As in (5.61), $\boldsymbol{M}$ is maximally monotone, while $\boldsymbol{C}$ is $\aleph$-cocoercive. Furthermore, $\boldsymbol{U} \in \mathcal{B}(\mathcal{H})$ is self-adjoint and, as shown in Vũ (2013, equation (3.20)), (8.112) implies that it is β-strongly monotone. Now set $(\forall n \in \mathbb{N})$ $\boldsymbol{x}_n = (x_n, y_{1,n}^*, \ldots, y_{p,n}^*)$ and $\boldsymbol{w}_n = (p_n, q_{1,n}^*, \ldots, q_{p,n}^*)$. Then, adopting the same pattern as in the proof of

Example 5.20, we rewrite (8.113) as

$$\begin{array}{l}\text{for } n = 0, 1, \ldots \\ \left\lfloor \begin{array}{l} \boldsymbol{u}_n^* = \boldsymbol{U}\boldsymbol{x}_n - \boldsymbol{C}\boldsymbol{x}_n \\ \boldsymbol{w}_n = (\boldsymbol{U} + \boldsymbol{M})^{-1}\boldsymbol{u}_n^* \\ \boldsymbol{x}_{n+1} = \boldsymbol{x}_n + \lambda_n(\boldsymbol{w}_n - \boldsymbol{x}_n), \end{array} \right. \end{array} \tag{8.115}$$

and thus recover (8.106) with $(\forall n \in \mathbb{N})\ \gamma_n = 1 < 2\aleph\beta$. We therefore appeal to Proposition 8.42(i) to obtain the weak convergence of $(\boldsymbol{x}_n)_{n\in\mathbb{N}}$ to a point $(x, y_1^*, \ldots, y_p^*) \in \operatorname{zer}(\boldsymbol{M} + \boldsymbol{C})$. However, replacing A with $A + C$ and $(B_k^{-1})_{1\leq k\leq p}$ with $(B_k^{-1} + D_k^{-1})_{1\leq k\leq p}$ in Lemma 3.12(ii) yields $\operatorname{zer}(\boldsymbol{M} + \boldsymbol{C}) \subset Z \times Z^*$. □

Remark 8.45. In terms of Framework 1.2, the embedding underlying Proposition 8.44 employs $\mathsf{X} = \mathcal{H} \oplus \mathcal{G}_1 \oplus \cdots \oplus \mathcal{G}_p$, $\boldsymbol{\mathcal{M}} = \boldsymbol{M} + \boldsymbol{C}$ and $\boldsymbol{\mathcal{T}}\colon \mathsf{X} \to \mathcal{H}\colon (x, y_1^*, \ldots, y_p^*) \mapsto x$.

The following application to minimization revisits the setting of Example 7.13 and Remark 7.14.

Example 8.46. Let $0 < p \in \mathbb{N}$, let $\alpha \in \left]0, +\infty\right[$, let $f \in \Gamma_0(\mathcal{H})$, and let $h\colon \mathcal{H} \to \mathbb{R}$ be convex, differentiable, and such that ∇h is $1/\alpha$-Lipschitzian. For every $k \in \{1, \ldots, p\}$, let $\beta_k \in \left]0, +\infty\right[$, let $\mathcal{G}_k$ be a real Hilbert space, let $g_k \in \Gamma_0(\mathcal{G}_k)$, let $\ell_k \in \Gamma_0(\mathcal{G}_k)$ be β_k-strongly convex, and suppose that $0 \neq L_k \in \mathcal{B}(\mathcal{H}, \mathcal{G}_k)$. Let Z be the set of solutions to the primal problem

$$\underset{x\in\mathcal{H}}{\text{minimize}}\ \ f(x) + \sum_{k=1}^{p} (g_k \,\square\, \ell_k)(L_k x) + h(x), \tag{8.116}$$

let Z^* be the set of solutions to the dual problem

$$\underset{y_1^*\in\mathcal{G}_1,\ldots,y_p^*\in\mathcal{G}_p}{\text{minimize}}\ \ (f^* \,\square\, h^*)\left(-\sum_{k=1}^{p} L_k^* y_k^*\right) + \sum_{k=1}^{p} \big(g_k^*(y_k^*) + \ell_k^*(y_k^*)\big), \tag{8.117}$$

and suppose that

$$\operatorname{zer}\left(\partial f + \sum_{k=1}^{p} L_k^* \circ (\partial g_k \,\square\, \partial \ell_k) \circ L_k + \nabla h\right) \neq \varnothing. \tag{8.118}$$

Let $\varepsilon \in \left]0, 1\right[$, let $(\lambda_n)_{n\in\mathbb{N}}$ be a sequence in $[\varepsilon, 1]$, let $x_0 \in \mathcal{H}$, let $(y_{1,0}^*, \ldots, y_{p,0}^*) \in \mathcal{G}_1 \oplus \cdots \oplus \mathcal{G}_p$, let $\tau \in \left]0, +\infty\right[$, and let $(\sigma_1, \ldots, \sigma_p) \in \left]0, +\infty\right[^p$ be such that

(8.111)–(8.112) hold. Iterate

$$
\begin{array}{l}
\text{for } n = 0, 1, \ldots \\
\left\lfloor
\begin{array}{l}
x_n^* = \tau\big(\sum_{k=1}^{p} L_k^* y_{k,n}^* + \nabla h(x_n)\big) \\
p_n = \operatorname{prox}_{\tau f}(x_n - x_n^*) \\
x_{n+1} = x_n + \lambda_n(p_n - x_n) \\
\text{for } k = 1, \ldots, p \\
\left\lfloor
\begin{array}{l}
y_{k,n} = \sigma_k\big(L_k(2p_n - x_n) - \nabla \ell_k^*(y_{k,n}^*)\big) \\
q_{k,n}^* = \operatorname{prox}_{\sigma_k g_k^*}\big(y_{k,n}^* + y_{k,n}\big) \\
y_{k,n+1}^* = y_{k,n}^* + \lambda_n\big(q_{k,n}^* - y_{k,n}^*\big).
\end{array}
\right.
\end{array}
\right.
\end{array}
\tag{8.119}
$$

Then there exist $x \in Z$ and $(y_1^*, \ldots, y_p^*) \in Z^*$ such that $x_n \rightharpoonup x$, and, for every $k \in \{1, \ldots, p\}$, $y_{k,n}^* \rightharpoonup y_k^*$.

Proof. It follows from the arguments presented in Combettes and Pesquet (2012, Section 4) that this is an application of Proposition 8.44 with $A = \partial f$, $C = \nabla h$, and $(\forall k \in \{1, \ldots, p\})$ $B_k = \partial g_k$ and $D_k = \partial \ell_k$. □

Remark 8.47. If we make the additional assumptions that, for every $k \in \{1, \ldots, p\}$, $\ell_k = \iota_{\{0\}}$ and $\sigma_k = \sigma_1$, Example 8.46 was independently obtained in Condat (2013, Section 5). For this reason, (8.119) in this particular setting is called the *Condat–Vũ* algorithm.

8.5. Forward–backward–half-forward splitting

Let $A \colon \mathcal{H} \to 2^{\mathcal{H}}$ be maximally monotone, let $C \colon \mathcal{H} \to \mathcal{H}$ be cocoercive, and let $Q \colon \mathcal{H} \to \mathcal{H}$ be monotone and Lipschitzian. Then a zero of $M = A + C + Q$ can be constructed through the forward–backward–forward algorithms of Theorem 7.1 or Theorem 7.2, applied to A and the monotone and Lipschitzian operator $B = C + Q$. These algorithms require two applications of B, i.e. two applications of C and Q, at each iteration. However, the algorithms discussed so far require two applications of a monotone Lipschitzian operator per iteration, as in the Antipin–Korpelevič method of Section 7.1 and the forward–backward–forward methods of Sections 7.2 and 7.3, but only one application of a cocoercive operator, as in the Euler method of Section 5.4.1 and the forward–backward methods of Sections 8.2 and 8.3. It is therefore natural to ask whether one can find a zero of $A + C + Q$ using only one application of C per iteration. A positive answer to this question was given in Briceño-Arias and Davis (2018) with the following forward–backward–half-forward splitting algorithm. We provide a simple proof of its convergence using our geometric framework.

Proposition 8.48 (Briceño-Arias and Davis 2018). Let $\alpha \in \,]0, +\infty[$, let $\beta \in \,]0, +\infty[$, let $A \colon \mathcal{H} \to 2^{\mathcal{H}}$ be maximally monotone, let $C \colon \mathcal{H} \to \mathcal{H}$ be α-cocoercive, let $Q \colon \mathcal{H} \to \mathcal{H}$ be monotone and β-Lipschitzian, and suppose that the set of solutions Z to the inclusion

$$\text{find } x \in \mathcal{H} \text{ such that } 0 \in Ax + Cx + Qx \tag{8.120}$$

is not empty. Let $x_0 \in \mathcal{H}$, set $\chi = 4\alpha/(1 + \sqrt{1 + 16\alpha^2\beta^2})$, let $\varepsilon \in \left]0, \chi/(\chi + 1)\right[$, and let $(\gamma_n)_{n\in\mathbb{N}}$ be a sequence in $[\varepsilon, (1 - \varepsilon)\chi]$. Iterate

$$\begin{array}{l} \text{for } n = 0, 1, \ldots \\ \left\lfloor \begin{array}{l} c_n^* = \gamma_n C x_n \\ q_n^* = \gamma_n Q x_n \\ w_n = J_{\gamma_n A}(x_n - c_n^* - q_n^*) \\ x_{n+1} = w_n - \gamma_n Q w_n + q_n^*. \end{array} \right. \end{array} \tag{8.121}$$

Then $(x_n)_{n\in\mathbb{N}}$ converges weakly to a point in Z.

Proof. The claims will be established as an application of Theorem 4.12 with

$$W = A + Q, \quad \text{and} \quad (\forall n \in \mathbb{N}) \quad U_n = \gamma_n^{-1}\mathrm{Id} - C - Q \quad \text{and} \quad q_n = x_n. \tag{8.122}$$

In this setting, Proposition 3.9 of Bùi and Combettes (2020*b*) implies that (7.5) is satisfied, we have

$$\begin{aligned} (\forall n \in \mathbb{N}) \quad J_{W+C}^{U_n} &= \left(\gamma_n^{-1}\mathrm{Id} + A\right) \circ \left(\gamma_n^{-1}\mathrm{Id} - C - Q\right) \\ &= J_{\gamma_n A} \circ (\mathrm{Id} - \gamma_n(C + Q)), \end{aligned} \tag{8.123}$$

and the variables of (4.34) become

$$(\forall n \in \mathbb{N}) \quad \begin{cases} w_n = J_{\gamma_n A}(x_n - \gamma_n(Cx_n + Qx_n)), \\ t_n^* = \left(\gamma_n^{-1}\mathrm{Id} - Q\right)x_n - \left(\gamma_n^{-1}\mathrm{Id} - Q\right)w_n, \\ \delta_n = \left(\dfrac{1}{\gamma_n} - \dfrac{1}{4\alpha}\right)\|w_n - x_n\|^2 - \langle w_n - x_n \mid Qw_n - Qx_n\rangle. \end{cases} \tag{8.124}$$

Now set

$$(\forall n \in \mathbb{N}) \quad \lambda_n = \begin{cases} \dfrac{\gamma_n\|t_n^*\|^2}{\delta_n}, & \text{if } \delta_n > 0, \\ \varepsilon, & \text{otherwise,} \end{cases} \tag{8.125}$$

and note that the assumptions yield

$$\inf_{n\in\mathbb{N}} \lambda_n > 0 \quad \text{and} \quad \sup_{n\in\mathbb{N}} \lambda_n < 2. \tag{8.126}$$

As a consequence of (8.124) and the properties of Q, we have

$$\begin{aligned} (\forall n \in \mathbb{N}) \quad \delta_n \leq 0 &\Rightarrow \left(\frac{1}{\gamma_n} - \frac{1}{4\alpha} - \beta\right)\|w_n - x_n\|^2 \leq 0 \\ &\Leftrightarrow w_n = x_n \\ &\Leftrightarrow t_n^* = 0. \end{aligned} \tag{8.127}$$

Hence, (4.34) yields

$$(\forall n \in \mathbb{N}) \quad d_n = \frac{\gamma_n}{\lambda_n} t_n^* = \frac{1}{\lambda_n}(x_n - w_n + \gamma_n(Qw_n - Qx_n)). \tag{8.128}$$

As a result, the sequence $(x_n)_{n\in\mathbb{N}}$ produced by (8.121) coincides with that of (4.34). Hence, by Theorem 4.12(i) and (8.126), $\sum_{n\in\mathbb{N}} \|d_n\|^2 < +\infty$ which, in view of (8.128), yields

$$(\mathrm{Id} - \gamma_n Q)w_n - (\mathrm{Id} - \gamma_n Q)x_n \to 0. \tag{8.129}$$

However, since $\chi \leq 1/\beta$, $(\gamma_n)_{n\in\mathbb{N}}$ lies in $[\varepsilon, (1-\varepsilon)/\beta]$ and Lemma 2.48(i) implies that the operators $(\mathrm{Id} - \gamma_n Q)_{n\in\mathbb{N}}$ are ε-strongly monotone. Hence,

$$(\forall n \in \mathbb{N}) \quad \varepsilon\|w_n - x_n\|^2 \leq \langle w_n - x_n \mid (\mathrm{Id} - \gamma_n Q)w_n - (\mathrm{Id} - \gamma_n Q)x_n\rangle \tag{8.130}$$

and, by the Cauchy–Schwarz inequality and (8.129),

$$\|w_n - x_n\| \leq \varepsilon^{-1}\|(\mathrm{Id} - \gamma_n Q)w_n - (\mathrm{Id} - \gamma_n Q)x_n\| \to 0. \tag{8.131}$$

In turn, since C is $1/\alpha$-Lipschitzian, these facts confirm that

$$\begin{aligned} \|U_n w_n - U_n x_n\| &\leq \gamma_n^{-1}\|(\mathrm{Id} - \gamma_n Q)w_n - (\mathrm{Id} - \gamma_n Q)x_n\| + \|Cw_n - Cx_n\| \\ &\leq \varepsilon^{-1}\|(\mathrm{Id} - \gamma_n Q)w_n - (\mathrm{Id} - \gamma_n Q)x_n\| + \alpha^{-1}\|w_n - x_n\| \\ &\to 0. \end{aligned} \tag{8.132}$$

Thus, the assertion follows from Theorem 4.12(ii) since its conditions (iib) and (iic) are fulfilled. □

Remark 8.49. We complement Proposition 8.48 with a few commentaries.

(i) Suppose that $C = 0$. Then, since α can be arbitrarily large, $\chi = 1/\beta$ and (8.121) reverts to the forward–backward–forward algorithm (7.2).

(ii) Suppose that $Q = 0$. Then, since $\beta = 0$, $\chi = 2\alpha$ and (8.121) becomes an unrelaxed version of forward–backward algorithm (8.5).

(iii) Using the geometric pattern of the proof given above, a strongly convergent version of the forward–backward–half-forward algorithm can be derived from Theorem 4.14.

As an illustration, we extend the Lagrangian approach of Proposition 7.5.

Example 8.50. Let $f \in \Gamma_0(\mathcal{H})$, $g \in \Gamma_0(\mathcal{G})$ and $L \in \mathcal{B}(\mathcal{H}, \mathcal{G})$ be such that $0 \in \operatorname{sri}(L(\operatorname{dom} f) - \operatorname{dom} g)$. Let $\alpha \in \left]0, +\infty\right[$ and let $h\colon \mathcal{H} \to \mathbb{R}$ be convex and differentiable and such that ∇h is $1/\alpha$-Lipschitzian. Suppose that the primal problem

$$\underset{x\in\mathcal{H}}{\text{minimize}} \ \ f(x) + g(Lx) + h(x) \tag{8.133}$$

admits solutions and consider the dual problem

$$\underset{v^*\in\mathcal{G}}{\text{minimize}} \ \ (f^* \,\square\, h^*)(-L^*v^*) + g^*(v^*). \tag{8.134}$$

Let $(x_0, y_0, v_0^*) \in \mathcal{H} \oplus \mathcal{G} \oplus \mathcal{G}$, set $\chi = 4\alpha/(1 + \sqrt{1 + 16\alpha^2(1 + \|L\|^2)})$, let $\varepsilon \in]0, \chi/(\chi + 1)[$, and let $(\gamma_n)_{n\in\mathbb{N}}$ be a sequence in $[\varepsilon, (1 - \varepsilon)\chi]$. Iterate

$$
\begin{array}{l}
\text{for } n = 0, 1, \ldots \\
\left\lfloor
\begin{array}{l}
c_n^* = \gamma_n \nabla h(x_n) \\
q_{1,n}^* = \gamma_n L^* v_n^* \\
q_{2,n}^* = -\gamma_n v_n^* \\
q_{3,n}^* = \gamma_n (y_n - L x_n) \\
a_{1,n} = \operatorname{prox}_{\gamma_n f}\left(x_n - c_n^* - q_{1,n}^*\right) \\
a_{2,n} = \operatorname{prox}_{\gamma_n g}\left(y_n - q_{2,n}^*\right) \\
x_{n+1} = a_{1,n} + \gamma_n L^* q_{3,n}^* \\
y_{n+1} = a_{2,n} - \gamma_n q_{3,n}^* \\
v_{n+1}^* = v_n^* + \gamma_n (L a_{1,n} - a_{2,n}).
\end{array}
\right.
\end{array}
\tag{8.135}
$$

Then $(x_n)_{n\in\mathbb{N}}$ and $(v_n^*)_{n\in\mathbb{N}}$ converge weakly to solutions to (8.133) and (8.134), respectively.

Proof. We adapt the approach of Section 7.4.2. The saddle operator of (7.22)–(7.23) becomes $\boldsymbol{\mathcal{S}} = \boldsymbol{A} + \boldsymbol{C} + \boldsymbol{Q}$, where

$$
\begin{cases}
\boldsymbol{A} \colon (x, y, v^*) \mapsto \partial f(x) \times \partial g(y) \times \{0\}, \\
\boldsymbol{C} \colon (x, y, v^*) \mapsto (\nabla h(x), 0, 0), \\
\boldsymbol{Q} \colon (x, y, v^*) \mapsto (L^* v^*, -v^*, -Lx + y).
\end{cases}
\tag{8.136}
$$

As in Section 7.4.2, $\boldsymbol{A}$ is maximally monotone and $\boldsymbol{Q}$ is monotone and $\sqrt{1 + \|L\|^2}$-Lipschitzian. Further, by virtue of Lemma 2.2, $\boldsymbol{C}$ is α-cocoercive. Now set $(\forall n \in \mathbb{N})$ $\boldsymbol{x}_n = (x_n, y_n, v_n^*)$, $\boldsymbol{c}_n^* = (c_n^*, 0, 0)$, $\boldsymbol{q}_n^* = (q_{1,n}^*, q_{2,n}^*, q_{3,n}^*)$ and $\boldsymbol{w}_n = (a_{1,n}, a_{2,n}, v_n^* - q_{3,n}^*)$. Then (8.135) assumes the form

$$
\begin{array}{l}
\text{for } n = 0, 1, \ldots \\
\left\lfloor
\begin{array}{l}
\boldsymbol{c}_n^* = \gamma_n \boldsymbol{C} \boldsymbol{x}_n \\
\boldsymbol{q}_n^* = \gamma_n \boldsymbol{Q} \boldsymbol{x}_n \\
\boldsymbol{w}_n = J_{\gamma_n \boldsymbol{A}}(\boldsymbol{x}_n - \boldsymbol{c}_n^* - \boldsymbol{q}_n^*) \\
\boldsymbol{x}_{n+1} = \boldsymbol{w}_n - \gamma_n \boldsymbol{Q} \boldsymbol{w}_n + \boldsymbol{q}_n^*,
\end{array}
\right.
\end{array}
\tag{8.137}
$$

which is (8.121). Hence, by Proposition 8.48, $(x_n, y_n, v_n^*)_{n\in\mathbb{N}}$ converges weakly to a point $(x, y, v^*) \in \operatorname{zer} \boldsymbol{\mathcal{S}}$. □

Remark 8.51. Let $\alpha \in]0, +\infty[$, let $A \colon \mathcal{H} \to 2^{\mathcal{H}}$ and $B \colon \mathcal{G} \to 2^{\mathcal{G}}$ be maximally monotone, let $C \colon \mathcal{H} \to \mathcal{H}$ be α-cocoercive, and let $0 \neq L \in \mathcal{B}(\mathcal{H}, \mathcal{G})$. As in Remark 7.8, the saddle approach of Example 8.50 has a natural extension to the problem of finding a zero of $A + L^* \circ B \circ L + C$ and the dual problem of finding a zero of $-L \circ (A + C)^{-1} \circ (-L^*) + B^{-1}$. In this setting, the saddle operator is

$$
\begin{array}{rccl}
\boldsymbol{\mathcal{S}} \colon & \mathcal{H} \oplus \mathcal{G} \oplus \mathcal{G} & \to & 2^{\mathcal{H} \oplus \mathcal{G} \oplus \mathcal{G}} \\
& (x, y, v^*) & \mapsto & (Ax + Cx + L^* v^*) \times (By - v^*) \times \{-Lx + y\}.
\end{array}
\tag{8.138}
$$

Accordingly, it suffices to replace ∇h with C, $\text{prox}_{\gamma_n f}$ with $J_{\gamma_n A}$ and $\text{prox}_{\gamma_n g}$ with $J_{\gamma_n B}$ in (8.135) to find primal–dual solutions.

9. Block-iterative Kuhn–Tucker projective splitting

9.1. Preview

Unlike the methods described so far, those described in this section were explicitly designed by employing the geometric principle of Theorem 4.2. The terminology *projective splitting* was coined in Eckstein and Svaiter (2008) in the context of an algorithm to solve Problem 3.1 by choosing points in the graph of A and B to construct half-spaces containing an 'extended solution set'. In the language of Lemma 3.8, this set is actually the set of zeros of the Kuhn–Tucker operator (3.10), which collapses to

$$\text{zer}\,\boldsymbol{\mathcal{K}} = \{(x, x^*) \in \mathcal{H} \oplus \mathcal{H} \mid -x^* \in Ax \text{ and } x \in B^{-1}x^*\}. \tag{9.1}$$

Eckstein and Svaiter (2008) initiated a fruitful line of work towards more complex monotone inclusions (Alotaibi, Combettes and Shahzad 2014, 2015, Bednarczuk *et al.* 2018, Bùi 2022*b*, Combettes and Eckstein 2018, Eckstein 2017, Eckstein and Svaiter 2009, Johnstone and Eckstein 2019, 2020, 2021, 2022, Machado 2018, Machado and Sicre 2023, Sicre 2020). We use the term *Kuhn–Tucker projective splitting* to describe a method that operates through the principles of Framework 1.2, where $\boldsymbol{\mathcal{M}}$ is a Kuhn–Tucker operator. As we shall see, projective splitting algorithms have features quite different from those of the traditional methods of Sections 5–8 and they display an unprecedented level of flexibility in terms of implementation.

9.2. Primal–dual composite inclusions

Let us go back to the composite Problem 3.7. The sets of primal and dual solutions are, respectively,

$$Z = \text{zer}(A + L^* \circ B \circ L) \quad \text{and} \quad Z^* = \text{zer}(-L \circ A^{-1} \circ (-L^*) + B^{-1}). \tag{9.2}$$

Moreover, as pointed out in Example 3.20, an embedding of (3.7) is $(\mathbf{X}, \boldsymbol{\mathcal{K}}, \boldsymbol{\mathcal{T}})$, where $\mathbf{X} = \mathcal{H} \oplus \mathcal{G}$, $\boldsymbol{\mathcal{K}}$ is the Kuhn–Tucker operator of (3.10), that is,

$$\boldsymbol{\mathcal{K}}\colon \mathbf{X} \to 2^{\mathbf{X}}\colon (x, y^*) \mapsto (Ax + L^*y^*) \times (B^{-1}y^* - Lx), \tag{9.3}$$

and $\boldsymbol{\mathcal{T}}\colon \mathbf{X} \to \mathcal{H}\colon (x, y^*) \mapsto x$. The task is therefore to find a zero of $\boldsymbol{\mathcal{K}}$. This is the path followed in the monotone+skew approach of Section 7.4.1. However, this method requires knowledge of $\|L\|$ (or of a tight upper bound for it), which may be difficult to obtain in certain problems. The renormed algorithms of Example 5.20 and Boţ and Hendrich (2013), the saddle algorithm of Remark 8.51 and the minimal lifting algorithm of Aragón-Artacho *et al.* (2023) share the same potential limitation. On the other hand, the method of Proposition 5.15, which was derived from the method of partial inverses, requires the inversion of linear operators, a task that may also face implementation issues.

A strategy which circumvents the above shortcomings was proposed in Alotaibi *et al.* (2014), where the approach of Eckstein and Svaiter (2008) for solving Problem 3.1 was extended to Problem 3.7. More precisely, it employs the geometric principle of Proposition 4.10 as follows. Let us assume that, at iteration n, points $(a_n, a_n^*) \in \operatorname{gra} A$ and $(b_n, b_n^*) \in \operatorname{gra} B$ are available and set

$$\boldsymbol{m}_n = (a_n, b_n^*) \quad \text{and} \quad \boldsymbol{m}_n^* = (a_n^* + L^* b_n^*, b_n - La_n). \tag{9.4}$$

Then it is clear from (9.3) that $(\boldsymbol{m}_n, \boldsymbol{m}_n^*) \in \operatorname{gra} \boldsymbol{\mathcal{K}}$. Hence, given $\lambda_n \in \left]0,2\right[$, iteration n of algorithm (4.32) updates $(x_n, y_n^*) \in \mathbf{X}$ via the routine

$$\left\lfloor \begin{array}{l} (t_n, t_n^*) = (b_n - La_n, a_n^* + L^* b_n^*) \\ \tau_n = \|t_n\|^2 + \|t_n^*\|^2 \\ \text{if } \tau_n > 0 \\ \left\lfloor \begin{array}{l} \theta_n = \dfrac{\lambda_n}{\tau_n} \max\{0, \langle x_n \mid t_n^* \rangle + \langle t_n \mid y_n^* \rangle - \langle a_n \mid a_n^* \rangle - \langle b_n \mid b_n^* \rangle\} \end{array} \right. \\ \text{else } \theta_n = 0 \\ (x_{n+1}, y_{n+1}^*) = (x_n - \theta_n t_n^*, y_n^* - \theta_n t_n). \end{array} \right. \tag{9.5}$$

In view of Proposition 4.10(ii), the task is now to specify $(a_n, a_n^*) \in \operatorname{gra} A$ and $(b_n, b_n^*) \in \operatorname{gra} B$ so as to guarantee that $\boldsymbol{m}_n - (x_n, y_n^*) \rightharpoonup 0$ and $\boldsymbol{m}_n^* \to 0$, that is,

$$a_n - x_n \rightharpoonup 0, \quad b_n^* - y_n^* \rightharpoonup 0, \quad b_n - La_n \to 0 \quad \text{and} \quad a_n^* + L^* b_n^* \to 0. \tag{9.6}$$

Given γ_n and σ_n in $\left]0,+\infty\right[$, choosing

$$(a_n, a_n^*) = (J_{\gamma_n A}(x_n - \gamma_n L^* y_n^*), \gamma_n^{-1}(x_n - J_{\gamma_n A}(x_n - \gamma_n L^* y_n^*)) - L^* y_n^*) \tag{9.7}$$

and

$$(b_n, b_n^*) = (J_{\sigma_n B}(Lx_n + \sigma_n y_n^*), \sigma_n^{-1}(Lx_n - J_{\sigma_n B}(Lx_n + \sigma_n y_n^*)) + y_n^*) \tag{9.8}$$

satisfies this requirement, which leads to the following result.

Proposition 9.1 (Alotaibi *et al.* 2014). Let $A\colon \mathcal{H} \to 2^{\mathcal{H}}$ and $B\colon \mathcal{G} \to 2^{\mathcal{G}}$ be maximally monotone, and let $L \in \mathcal{B}(\mathcal{H}, \mathcal{G})$. Suppose that the set Z of solutions to the primal inclusion

$$\text{find } x \in \mathcal{H} \text{ such that } 0 \in Ax + L^*(B(Lx)) \tag{9.9}$$

is not empty and let Z^* be the set of solutions to the dual inclusion

$$\text{find } y^* \in \mathcal{G} \text{ such that } 0 \in -L(A^{-1}(-L^* y^*)) + B^{-1} y^*. \tag{9.10}$$

Let $\varepsilon \in \left]0,1\right[$, let $(\gamma_n)_{n\in\mathbb{N}}$ and $(\sigma_n)_{n\in\mathbb{N}}$ be sequences in $[\varepsilon, 1/\varepsilon]$, let $(\lambda_n)_{n\in\mathbb{N}}$ be a sequence in $[\varepsilon, 2-\varepsilon]$, let $x_0 \in \mathcal{H}$, and let $y_0^* \in \mathcal{G}$. Iterate

$$
\begin{array}{l}
\text{for } n = 0, 1, \ldots \\
\left\lfloor
\begin{array}{l}
a_n = J_{\gamma_n A}(x_n - \gamma_n L^* y_n^*) \\
l_n = L x_n \\
b_n = J_{\sigma_n B}(l_n + \sigma_n y_n^*) \\
t_n = b_n - L a_n \\
t_n^* = \gamma_n^{-1}(x_n - a_n) + \sigma_n^{-1} L^*(l_n - b_n) \\
\tau_n = \|t_n\|^2 + \|t_n^*\|^2 \\
\text{if } \tau_n > 0 \\
\left\lfloor\ \theta_n = \lambda_n(\gamma_n^{-1}\|x_n - a_n\|^2 + \sigma_n^{-1}\|l_n - b_n\|^2)/\tau_n \right. \\
\text{else } \theta_n = 0 \\
x_{n+1} = x_n - \theta_n t_n^* \\
y_{n+1}^* = y_n^* - \theta_n t_n.
\end{array}
\right.
\end{array}
\tag{9.11}
$$

Then $(x_n)_{n\in\mathbb{N}}$ converges weakly to a point $x \in Z$ and $(y_n^*)_{n\in\mathbb{N}}$ converges weakly to a point $y^* \in Z^*$.

Remark 9.2. Here are notable instantiations of Proposition 9.1.

(i) The first instance of (9.11) in the literature seems to be that of Dong (2005), where $\mathcal{H}$ and $\mathcal{G}$ are Euclidean spaces, $A = 0$, and $(\forall n \in \mathbb{N})\ \gamma_n = \sigma_n = 1$ and $\lambda_n = \lambda \in \,]0, 2[$. Convergence of the primal sequence $(x_n)_{n\in\mathbb{N}}$ was established by different means.

(ii) In the setting of Problem 3.1 (i.e. $\mathcal{G} = \mathcal{H}$ and $L = \mathrm{Id}$), (9.11) was studied in Eckstein and Svaiter (2008). Under the additional assumptions that $A + B$ is maximally monotone or that $\mathcal{H}$ is finite-dimensional, weak convergence was established in Eckstein and Svaiter (2008, Proposition 3) for a version of (9.11) which allows for an additional relaxation parameter in the definition of a_n.

Remark 9.3. So far, we have presented several methods to solve Problem 3.7; see Proposition 5.15, Example 5.20, Proposition 7.3 and Remark 8.51. Some features that distinguish the splitting algorithm (9.11) from them are as follows.

(i) At each iteration of (9.11), different proximal parameters γ_n and σ_n can be used for the operators A and B and, since ε is chosen by the user, their values can be arbitrarily large.

(ii) The execution of (9.11) does not require that $\|L\|$ or an approximation thereof be known, or the inversion of linear operators.

(iii) A variant of (9.11) exploiting the cocoercivity of some of the operators and activating them via Euler steps is discussed in Johnstone and Eckstein (2021).

(iv) The complexity of certain special cases and variants of (9.11) is investigated in Johnstone and Eckstein (2019) and Machado and Sicre (2023).

The following strongly convergent projective splitting algorithm results from Proposition 4.11.

Proposition 9.4 (Alotaibi *et al.* 2015). Let $A\colon \mathcal{H} \to 2^{\mathcal{H}}$ and $B\colon \mathcal{G} \to 2^{\mathcal{G}}$ be maximally monotone, and let $L \in \mathcal{B}(\mathcal{H}, \mathcal{G})$. Suppose that the set Z of solutions to the primal inclusion

$$\text{find } x \in \mathcal{H} \text{ such that } 0 \in Ax + L^*(B(Lx)) \tag{9.12}$$

is not empty and let Z^* be the set of solutions to the dual inclusion

$$\text{find } y^* \in \mathcal{G} \text{ such that } 0 \in -L(A^{-1}(-L^*y^*)) + B^{-1}y^*. \tag{9.13}$$

Let $\varepsilon \in\,]0,1[$, let $(\gamma_n)_{n\in\mathbb{N}}$ and $(\sigma_n)_{n\in\mathbb{N}}$ be sequences in $[\varepsilon, 1/\varepsilon]$, let $(\lambda_n)_{n\in\mathbb{N}}$ be a sequence in $[\varepsilon, 1]$, let $x_0 \in \mathcal{H}$, and let $y_0^* \in \mathcal{G}$. Iterate

$$
\begin{array}{l}
\text{for } n = 0, 1, \ldots \\
\left\lfloor
\begin{array}{l}
a_n = J_{\gamma_n A}(x_n - \gamma_n L^* y_n^*) \\
l_n = Lx_n \\
b_n = J_{\sigma_n B}(l_n + \sigma_n y_n^*) \\
t_n = b_n - La_n \\
t_n^* = \gamma_n^{-1}(x_n - a_n) + \sigma_n^{-1} L^*(l_n - b_n) \\
\tau_n = \|t_n\|^2 + \|t_n^*\|^2 \\
\text{if } \tau_n > 0 \\
\left\lfloor\ \theta_n = \lambda_n\big(\gamma_n^{-1}\|x_n - a_n\|^2 + \sigma_n^{-1}\|l_n - b_n\|^2\big)/\tau_n \right. \\
\text{else } \theta_n = 0 \\
r_n = x_n - \theta_n t_n^* \\
r_n^* = y_n^* - \theta_n t_n \\
\chi_n = \theta_n(\langle x_0 - x_n \mid t_n^*\rangle + \langle t_n \mid y_0^* - y_n^*\rangle) \\
\mu_n = \|x_0 - x_n\|^2 + \|y_0^* - y_n^*\|^2 \\
\nu_n = \theta_n^2 \tau_n \\
\rho_n = \mu_n \nu_n - \chi_n^2 \\
\text{if } \rho_n = 0 \text{ and } \chi_n \geq 0 \\
\left\lfloor
\begin{array}{l}
x_{n+1} = r_n \\
y_{n+1}^* = r_n^*
\end{array}
\right. \\
\text{if } \rho_n > 0 \text{ and } \chi_n \nu_n \geq \rho_n \\
\left\lfloor
\begin{array}{l}
x_{n+1} = x_0 - \theta_n(1 + \chi_n/\nu_n) t_n^* \\
y_{n+1}^* = y_0^* - \theta_n(1 + \chi_n/\nu_n) t_n
\end{array}
\right. \\
\text{if } \rho_n > 0 \text{ and } \chi_n \nu_n < \rho_n \\
\left\lfloor
\begin{array}{l}
x_{n+1} = x_n + (\nu_n/\rho_n)(\chi_n(x_0 - x_n) - \mu_n \theta_n t_n^*) \\
y_{n+1}^* = y_n^* + (\nu_n/\rho_n)(\chi_n(y_0^* - y_n^*) - \mu_n \theta_n t_n).
\end{array}
\right.
\end{array}
\right.
\end{array}
\tag{9.14}
$$

Then $(x_n)_{n\in\mathbb{N}}$ converges strongly to a point $x \in Z$ and $(y_n^*)_{n\in\mathbb{N}}$ converges strongly to a point $y^* \in Z^*$.

9.3. Block-iterative asynchronous method

We consider a refinement of Problem 3.11 in which the primal variable is specified in terms of finitely many coordinates, say $\boldsymbol{x} = (x_1, \ldots, x_m)$, where each x_i lies in a Hilbert space $\mathcal{H}_i$. Such coupled systems of inclusions arise in particular in multivariate optimization (Acker and Prestel 1980, Attouch, Bolte, Redont and Soubeyran 2008, Attouch *et al.* 2010, Combettes 2013*b*), domain decomposition methods (Alduncin 2023, Attouch *et al.* 2016, 2011), image processing (Aujol and Chambolle 2005, Briceño-Arias, Combettes, Pesquet and Pustelnik 2011, Chaux *et al.* 2013, Vese and Osher 2004), game theory (Belgioioso *et al.* 2021, Börgens and Kanzow 2021, Briceño-Arias and Combettes 2013, Bùi and Combettes 2022*a*), network flow problems (Bertsekas 1998, Bùi 2022*a*, Rockafellar 1984, 1995), machine learning (Briceño-Arias, Chierchia, Chouzenoux and Pesquet 2019, Jenatton *et al.* 2011, Micchelli, Morales and Pontil 2013, Villa, Rosasco, Mosci and Verri 2014), signal processing (Briceño-Arias and Combettes 2009), mean field games (Briceño-Arias *et al.* 2018), statistics (Combettes and Müller 2020, Yan and Bien 2021), tensor completion (Gandy, Recht and Yamada 2011, Mizoguchi and Yamada 2019) and semi-definite programming (Hu, Sotirov and Wolkowicz 2023, Oliveira, Wolkowicz and Xu 2018).

Problem 9.5. Let $I = \{1, \ldots, m\}$ and $K = \{1, \ldots, p\}$ be nonempty finite sets. For every $i \in I$ and every $k \in K$, let $\mathcal{H}_i$ and $\mathcal{G}_k$ be real Hilbert spaces, let $A_i : \mathcal{H}_i \to 2^{\mathcal{H}_i}$ and $B_k : \mathcal{G}_k \to 2^{\mathcal{G}_k}$ be maximally monotone, and let $L_{ki} \in \mathcal{B}(\mathcal{H}_i, \mathcal{G}_k)$. Set

$$\boldsymbol{\mathcal{H}} = \bigoplus_{i \in I} \mathcal{H}_i \quad \text{and} \quad \boldsymbol{\mathcal{G}} = \bigoplus_{k \in K} \mathcal{G}_k. \tag{9.15}$$

The objective is to solve the primal inclusion

$$\text{find } \boldsymbol{x} \in \boldsymbol{\mathcal{H}} \text{ such that } (\forall i \in I)\; 0 \in A_i x_i + \sum_{k \in K} L_{ki}^* \bigg(B_k \bigg(\sum_{j \in I} L_{kj} x_j \bigg) \bigg) \tag{9.16}$$

together with the dual inclusion

$$\text{find } \boldsymbol{y}^* \in \boldsymbol{\mathcal{G}} \text{ such that } (\exists\, \boldsymbol{x} \in \boldsymbol{\mathcal{H}}) \begin{cases} (\forall i \in I)\; x_i \in A_i^{-1}\bigg(-\sum_{k \in K} L_{ki}^* y_k^* \bigg), \\ (\forall k \in K)\; \sum_{i \in I} L_{ki} x_i \in B_k^{-1} y_k^*. \end{cases} \tag{9.17}$$

Remark 9.6. There is an oversight in the dual problem given in Combettes and Eckstein (2018, Problem 1): the correct formulation of the dual inclusion is (9.17).

The counterpart of Lemma 3.12 for Problem 9.5 is as follows.

Lemma 9.7. In the setting of Problem 9.5, set $\mathbf{X} = \boldsymbol{\mathcal{H}} \oplus \boldsymbol{\mathcal{G}}$, and let $\boldsymbol{Z}$ and $\boldsymbol{Z}^*$ be the sets of solutions to (9.16) and (9.17), respectively. Define the Kuhn–Tucker operator of Problem 9.5 as

$$\begin{aligned}\boldsymbol{\mathcal{K}}\colon \mathbf{X} \to 2^{\mathbf{X}}\colon (\boldsymbol{x}, \boldsymbol{y}^*) \mapsto\\ \left(A_1 x_1 + \sum_{k\in K} L_{k1}^* y_k^*\right) \times \cdots \times \left(A_m x_m + \sum_{k\in K} L_{km}^* y_k^*\right)\\ \times \left(-\sum_{i\in I} L_{1i} x_i + B_1^{-1} y_1^*\right) \times \cdots \times \left(-\sum_{i\in I} L_{pi} x_i + B_p^{-1} y_p^*\right)\end{aligned} \tag{9.18}$$

and the set of Kuhn–Tucker points as $\operatorname{zer}\boldsymbol{\mathcal{K}}$. Then the following hold:

(i) $\boldsymbol{\mathcal{K}}$ is maximally monotone.

(ii) $\operatorname{zer}\boldsymbol{\mathcal{K}}$ is a closed convex subset of $\boldsymbol{Z} \times \boldsymbol{Z}^*$.

(iii) $\boldsymbol{Z}^* \neq \varnothing \Leftrightarrow \operatorname{zer}\boldsymbol{\mathcal{K}} \neq \varnothing \Rightarrow \boldsymbol{Z} \neq \varnothing$.

Example 9.8. In the setting of Problem 9.5, set $\mathbf{X} = \boldsymbol{\mathcal{H}} \oplus \boldsymbol{\mathcal{G}}$, let $\boldsymbol{\mathcal{K}}$ be the Kuhn–Tucker operator of (9.18), and let $\boldsymbol{\mathcal{T}}\colon \mathbf{X} \to \boldsymbol{\mathcal{H}}\colon (\boldsymbol{x}, \boldsymbol{y}^*) \mapsto \boldsymbol{x}$. Then it follows from Lemma 9.7(ii) that $(\mathbf{X}, \boldsymbol{\mathcal{K}}, \boldsymbol{\mathcal{T}})$ is an embedding of (9.16).

When the monotone operators $(A_i)_{1\leq i\leq m}$ and $(B_k)_{1\leq k\leq p}$ are taken to be subdifferentials, Problem 9.5 specializes to a multivariate minimization problem under a suitable qualification condition.

Example 9.9. Define $\boldsymbol{\mathcal{H}}$ and $\boldsymbol{\mathcal{G}}$ as in Problem 9.5. For every $i \in I$ and every $k \in K$, let $f_i \in \Gamma_0(\mathcal{H}_i)$, let $g_k \in \Gamma_0(\mathcal{G}_k)$, and let $L_{ki} \in \mathcal{B}(\mathcal{H}_i, \mathcal{G}_k)$. Suppose that (existence of a Kuhn–Tucker point)

$$(\exists\, \boldsymbol{x} \in \boldsymbol{\mathcal{H}})(\exists\, \boldsymbol{y}^* \in \boldsymbol{\mathcal{G}}) \begin{cases} (\forall i \in I) & -\displaystyle\sum_{k\in K} L_{ki}^* y_k^* \in \partial f_i(x_i),\\ (\forall k \in K) & \displaystyle\sum_{i\in I} L_{ki} x_i \in \partial g_k^*(y_k^*). \end{cases} \tag{9.19}$$

The objective is to solve the primal minimization problem

$$\underset{\boldsymbol{x}\in\boldsymbol{\mathcal{H}}}{\text{minimize}} \;\; \sum_{i\in I} f_i(x_i) + \sum_{k\in K} g_k\left(\sum_{i\in I} L_{ki} x_i\right) \tag{9.20}$$

together with its dual problem

$$\underset{\boldsymbol{y}^*\in\boldsymbol{\mathcal{G}}}{\text{minimize}} \;\; \sum_{i\in I} f_i^*\left(-\sum_{k\in K} L_{ki}^* y_k^*\right) + \sum_{k\in K} g_k^*(y_k^*). \tag{9.21}$$

In an attempt to recast Problem 9.5 as a realization of Problem 3.7, let us define

$$\begin{cases} A\colon \mathcal{H} \to 2^{\mathcal{H}}\colon \boldsymbol{x} \mapsto A_1x_1 \times \cdots \times A_mx_m, \\ B\colon \mathcal{G} \to 2^{\mathcal{G}}\colon \boldsymbol{y} \mapsto B_1y_1 \times \cdots \times B_py_p, \\ L\colon \mathcal{H} \to \mathcal{G}\colon \boldsymbol{x} \mapsto \big(\sum_{i\in I} L_{1i}x_i, \ldots, \sum_{i\in I} L_{pi}x_i\big). \end{cases} \tag{9.22}$$

Upon injecting these operators into (9.11) and invoking Example 2.37, we obtain an algorithm that requires that $m+p$ resolvents be evaluated at each iteration. In large-scale problems, m and p can be huge and this requirement poses implementation issues as the only information flow within an iteration is from the m operators $(A_i)_{i\in I}$ calculations to the p operators $(B_k)_{k\in K}$ calculations. This results in an algorithm in which large blocks of calculations must be performed before any information is exchanged between subsystems. Thus, if some small subset of the subsystems represented by the operators $(A_i)_{i\in I}$ or $(B_k)_{k\in K}$ are more computation-intensive than others, load balancing can become problematic: most processors may have to sit idle while the remaining few complete their tasks. More generally, none of the methods discussed so far can handle block-processing or asynchronicity.

The algorithm we present now was conceived in Combettes and Eckstein (2018) around combined objectives which were beyond the reach of the existing splitting algorithms.

- *Block iterations.* At iteration n, it necessitates calculation of new points in the graphs of only some of the operators, say $(A_i)_{i\in I_n}$ and $(B_k)_{k\in K_n}$ with $I_n \subset I$ and $K_n \subset K$. The deterministic control sequences $(I_n)_{n\in\mathbb{N}}$ and $(K_n)_{n\in\mathbb{N}}$ dictate how frequently the various operators are used.
- *Asynchronicity.* A new point $(a_{i,n}, a^*_{i,n}) \in \operatorname{gra} A_i$ being incorporated into the calculations at iteration n may be based on data $x_{i,\pi_i(n)}$ and $(y^*_{k,\pi_i(n)})_{k\in K}$ available at some possibly earlier iteration $\pi_i(n) \leq n$. Therefore, the calculation of $(a_{i,n}, a^*_{i,n})$ could have been initiated at iteration $\pi_i(n)$, with its results becoming available only at iteration n. Likewise, for every $k \in K_n$, the computation of $(b_{k,n}, b^*_{k,n}) \in \operatorname{gra} B_k$ can be initiated at some iteration $\omega_k(n) \leq n$, based on $(x_{i,\omega_k(n)})_{i\in I}$ and $y^*_{k,\omega_k(n)}$.
- *Convergence.* It guarantees (weak or strong) convergence of the iterates to primal and dual solutions.

Remark 9.10. Regarding block iterations for Problem 9.5, a product space version of the Douglas–Rachford algorithm was introduced in Combettes and Pesquet (2015), which features random activation of the blocks. A random block-iterative version of the forward–backward algorithm was also proposed in Combettes and Pesquet (2015), which in Pesquet and Repetti (2015) led to algorithms for Problem 9.5 via the renorming techniques presented in Section 8.4.6; for specialized block-iterative forward–backward algorithms tailored for instances of Example 9.9, see Briceño-Arias *et al.* (2019), Liu and Wright (2015), Salzo and Villa (2022) and Traoré, Salzo and Villa (2023). These methods differ from the deterministic ones presented below in that they operate under stochastic assumptions on the underlying processes, have a less predictable computational load over the iterations, have less freedom in the choice of the proximal parameters, and offer only almost sure convergence guarantees; see also Bùi, Combettes and Woodstock (2022) for numerical comparisons.

Going back to (9.5) in the setting of (9.22) and Lemma 9.7, what is actually needed at iteration n to create the half-space containing zer $\boldsymbol{\mathcal{K}}$ are points

$$\begin{cases}(a_{i,n}, a^*_{i,n}) \in \operatorname{gra} A_i, & \text{for } i \in I,\\ (b_{k,n}, b^*_{k,n}) \in \operatorname{gra} B_k, & \text{for } k \in K.\end{cases} \tag{9.23}$$

The key observation is that not all of these points have to be new in order to obtain a new half-space. In other words, we can update only some of them while keeping old ones and still create a new half-space onto which the current primal–dual iterate $(\boldsymbol{x}_n, \boldsymbol{y}^*_n) = (x_{1,n}, \ldots, x_{m,n}, y^*_{1,n}, \ldots, y^*_{p,n})$ will be projected. How often the points in the individual graphs should be updated, and in which fashion, will be regulated by the following rules.

Assumption 9.11. Given $0 < R \in \mathbb{N}$, $(I_n)_{n\in\mathbb{N}}$ is a sequence of nonempty subsets of I, and $(K_n)_{n\in\mathbb{N}}$ is a sequence of nonempty subsets of K such that

$$I_0 = I, \quad K_0 = K \quad \text{and} \quad (\forall n \in \mathbb{N}) \begin{cases}\displaystyle\bigcup_{j=n}^{n+R-1} I_j = I,\\ \displaystyle\bigcup_{j=n}^{n+R-1} K_j = K.\end{cases} \tag{9.24}$$

Assumption 9.12. $T \in \mathbb{N}$ and, for every $i \in I$ and every $k \in K$, $(\pi_i(n))_{n\in\mathbb{N}}$ and $(\omega_k(n))_{n\in\mathbb{N}}$ are sequences in $\mathbb{N}$ such that $(\forall n \in \mathbb{N})$ $n - T \leq \pi_i(n) \leq n$ and $n - T \leq \omega_k(n) \leq n$.

With these considerations and by making selections for the updated points $(a_{i,n}, a^*_{i,n})_{i\in I_n}$ and $(b^*_{k,n}, b^*_{k,n})_{k\in K_n}$ akin to those of (9.7) and (9.8), we arrive at the following realization of (9.5).

Algorithm 9.13. Consider the setting of Problem 9.5, suppose that Assumptions 9.11 and 9.12 are in force, let $\varepsilon \in \,]0,1[$, and let $(\lambda_n)_{n\in\mathbb{N}}$ be a sequence in $[\varepsilon, 2-\varepsilon]$. For every $i \in I$, let $(\gamma_{i,n})_{n\in\mathbb{N}}$ be a sequence in $[\varepsilon, 1/\varepsilon]$ and let $x_{i,0} \in \mathcal{H}_i$.

For every $k \in K$, let $(\sigma_{k,n})_{n\in\mathbb{N}}$ be a sequence in $[\varepsilon, 1/\varepsilon]$ and let $y^*_{k,0} \in \mathcal{G}_k$. Iterate

$$
\begin{array}{l}
\text{for } n = 0, 1, \ldots \\
\left\lfloor
\begin{array}{l}
\text{for every } i \in I_n \\
\left\lfloor
\begin{array}{l}
l^*_{i,n} = \sum_{k\in K} L^*_{ki} y^*_{k,\pi_i(n)} \\
a_{i,n} = J_{\gamma_{i,\pi_i(n)} A_i}\big(x_{i,\pi_i(n)} - \gamma_{i,\pi_i(n)} l^*_{i,n}\big) \\
a^*_{i,n} = \gamma^{-1}_{i,\pi_i(n)}(x_{i,\pi_i(n)} - a_{i,n}) - l^*_{i,n}
\end{array}
\right. \\
\text{for every } i \in I \smallsetminus I_n \\
\left\lfloor \big(a_{i,n}, a^*_{i,n}\big) = \big(a_{i,n-1}, a^*_{i,n-1}\big) \right. \\
\text{for every } k \in K_n \\
\left\lfloor
\begin{array}{l}
l_{k,n} = \sum_{i\in I} L_{ki} x_{i,\omega_k(n)} \\
b_{k,n} = J_{\sigma_{k,\omega_k(n)} B_k}\big(l_{k,n} + \sigma_{k,\omega_k(n)} y^*_{k,\omega_k(n)}\big) \\
b^*_{k,n} = y^*_{k,\omega_k(n)} + \sigma^{-1}_{k,\omega_k(n)}(l_{k,n} - b_{k,n})
\end{array}
\right. \\
\text{for every } k \in K \smallsetminus K_n \\
\left\lfloor \big(b_{k,n}, b^*_{k,n}\big) = \big(b_{k,n-1}, b^*_{k,n-1}\big) \right. \\
\text{for every } i \in I \\
\left\lfloor t^*_{i,n} = a^*_{i,n} + \sum_{k\in K} L^*_{ki} b^*_{k,n} \right. \\
\text{for every } k \in K \\
\left\lfloor t_{k,n} = b_{k,n} - \sum_{i\in I} L_{ki} a_{i,n} \right. \\
\tau_n = \sum_{i\in I} \|t^*_{i,n}\|^2 + \sum_{k\in K} \|t_{k,n}\|^2 \\
\text{if } \tau_n > 0 \\
\left\lfloor
\begin{array}{l}
\theta_n = \dfrac{\lambda_n}{\tau_n} \max\Big\{0, \sum_{i\in I}\big(\langle x_{i,n} \mid t^*_{i,n}\rangle - \langle a_{i,n} \mid a^*_{i,n}\rangle\big) \\
\qquad\qquad + \sum_{k\in K}\big(\langle t_{k,n} \mid y^*_{k,n}\rangle - \langle b_{k,n} \mid b^*_{k,n}\rangle\big)\Big\}
\end{array}
\right. \\
\text{else } \theta_n = 0 \\
\text{for every } i \in I \\
\left\lfloor x_{i,n+1} = x_{i,n} - \theta_n t^*_{i,n} \right. \\
\text{for every } k \in K \\
\left\lfloor y^*_{k,n+1} = y^*_{k,n} - \theta_n t_{k,n}. \right.
\end{array}
\right.
\end{array}
\tag{9.25}
$$

Weak convergence is obtained by applying the principles of Proposition 4.10(ii).

Theorem 9.14 (Combettes and Eckstein 2018). Consider the setting of Problem 9.5 and Algorithm 9.13, and suppose that the Kuhn–Tucker operator $\mathcal{K}$ of (9.18) has zeros. Then, for every $i \in I$, $(x_{i,n})_{n\in\mathbb{N}}$ converges weakly to a point $x_i \in \mathcal{H}_i$ and, for every $k \in K$, $(y^*_{k,n})_{n\in\mathbb{N}}$ converges weakly to a point $y^*_k \in \mathcal{G}_k$. In addition, $(x_i)_{i\in I}$ solves the primal problem (9.16) and $(y^*_k)_{k\in K}$ solves the dual problem (9.17).

Remark 9.15. Here are a few comments on algorithm (9.13).

(i) The synchronous implementation is obtained by taking, for every $n \in \mathbb{N}$, every $i \in I_n$ and every $k \in K_n$, $\pi_i(n) = \omega_k(n) = n$.

(ii) We recover Theorem 4.3 of Alotaibi *et al.* (2014) (and in particular Proposition 9.4 when $m = p = 1$) in the special case when the implementation is synchronous, and at every iteration n, every operator is used (i.e. $I_n = I$ and $K_n = K$), with $\gamma_{i,n} = \gamma_n$ for every $i \in I$ and $\sigma_{k,n} = \sigma_n$ for every $k \in K$.

(iii) The specialization of Theorem 9.14 to the minimization setting of Example 9.9 is obtained by replacing each $J_{\gamma_{i,\pi_i(n)}A_i}$ with $\operatorname{prox}_{\gamma_{i,\pi_i(n)}f_i}$ and each $J_{\sigma_{k,\omega_k(n)}B_k}$ with $\operatorname{prox}_{\sigma_{k,\omega_k(n)}g_k}$. Numerical experiments are presented in Bùi *et al.* (2022) in the context of signal recovery and machine learning, and in Eckstein, Watson and Woodruff (2023) in the context of stochastic programming.

(iv) For the strongly convergent variant of Theorem 9.14 based on Proposition 4.11, see Combettes and Eckstein (2018, Theorem 15).

(v) When $m = 1$ and $A = 0$, a variant that takes into account the fact that some of the operators $(B_k)_{k\in K}$ may be monotone and Lipschitzian, and which activate them via Euler steps is presented in Johnstone and Eckstein (2022); see also Johnstone and Eckstein (2020).

10. Block-iterative saddle projective splitting

10.1. Preview

In all the algorithms discussed so far, each monotone operator has one of three properties: it is set-valued, single-valued and cocoercive, or single-valued and Lipschitzian. In addition, at each iteration, a set-valued operator is used once via its resolvents, a cocoercive operator once via a Euler step and a Lipschitzian operator twice via Euler steps. This is particularly the case in the forward–backward–half-forward algorithm of Section 8.5, the objective of which is to find a zero of

$$M = A + C + Q, \quad \text{where} \begin{cases} A\colon \mathcal{H} \to 2^{\mathcal{H}} & \text{is maximally monotone,} \\ C\colon \mathcal{H} \to \mathcal{H} & \text{is cocoercive,} \\ Q\colon \mathcal{H} \to \mathcal{H} & \text{is monotone and Lipschitzian.} \end{cases} \tag{10.1}$$

On the other hand, the Kuhn–Tucker projective splitting techniques of Section 9 activate all the operators via their resolvents (exceptions were noted in Remarks 9.3(iii) and 9.15(v), but they concern special cases of Problem 9.5). Furthermore, they are not designed to handle problems such as (7.37) or (8.109), which incorporate parallel sums.

In this section, following Bùi and Combettes (2022*b*), we unify all the problem formulations encountered in Sections 5–9 by including parallel sums in the system of monotone inclusions of Problem 9.5, and decomposing each operator in the resulting problem as in (10.1). In addition, nonlinear coupling operators $(R_i)_{i\in I}$ are incorporated.

Problem 10.1. Let $(\mathcal{H}_i)_{i\in I}$ and $(\mathcal{G}_k)_{k\in K}$ be finite families of real Hilbert spaces, and set

$$\boldsymbol{\mathcal{H}} = \bigoplus_{i\in I} \mathcal{H}_i \quad \text{and} \quad \boldsymbol{\mathcal{G}} = \bigoplus_{k\in K} \mathcal{G}_k. \tag{10.2}$$

For every $i \in I$ and every $k \in K$, suppose that the following are satisfied:

[a] $A_i\colon \mathcal{H}_i \to 2^{\mathcal{H}_i}$ is maximally monotone, $C_i\colon \mathcal{H}_i \to \mathcal{H}_i$ is cocoercive with constant $\alpha_i^c \in \left]0,+\infty\right[$, $Q_i\colon \mathcal{H}_i \to \mathcal{H}_i$ is monotone and Lipschitzian with constant $\alpha_i^\ell \in \left[0,+\infty\right[$, and $R_i\colon \boldsymbol{\mathcal{H}} \to \mathcal{H}_i$.

[b] $B_k^m\colon \mathcal{G}_k \to 2^{\mathcal{G}_k}$ is maximally monotone, $B_k^c\colon \mathcal{G}_k \to \mathcal{G}_k$ is cocoercive with constant $\beta_k^c \in \left]0,+\infty\right[$, and $B_k^\ell\colon \mathcal{G}_k \to \mathcal{G}_k$ is monotone and Lipschitzian with constant $\beta_k^\ell \in \left[0,+\infty\right[$.

[c] $D_k^m\colon \mathcal{G}_k \to 2^{\mathcal{G}_k}$ is maximally monotone, $D_k^c\colon \mathcal{G}_k \to \mathcal{G}_k$ is cocoercive with constant $\delta_k^c \in \left]0,+\infty\right[$, and $D_k^\ell\colon \mathcal{G}_k \to \mathcal{G}_k$ is monotone and Lipschitzian with constant $\delta_k^\ell \in \left[0,+\infty\right[$.

[d] $L_{ki} \in \mathcal{B}(\mathcal{H}_i, \mathcal{G}_k)$.

In addition,

[e] $\boldsymbol{R}\colon \boldsymbol{\mathcal{H}} \to \boldsymbol{\mathcal{H}}\colon \boldsymbol{x} \mapsto (R_i\boldsymbol{x})_{i\in I}$ is monotone and Lipschitzian with constant $\chi \in \left[0,+\infty\right[$.

The objective is to solve the primal problem

$$\begin{aligned}&\text{find } \boldsymbol{x} = (x_i)_{i\in I} \in \boldsymbol{\mathcal{H}} \text{ such that } (\forall i \in I)\ 0 \in A_i x_i + C_i x_i + Q_i x_i + R_i \boldsymbol{x}\\ &+ \sum_{k\in K} L_{ki}^*\Big(\big((B_k^m + B_k^c + B_k^\ell) \,\square\, (D_k^m + D_k^c + D_k^\ell)\big)\Big(\sum_{j\in I} L_{kj}x_j\Big)\Big)\end{aligned} \tag{10.3}$$

and the associated dual problem

$$\begin{aligned}&\text{find } \boldsymbol{y}^* = (y_k^*)_{k\in K} \in \boldsymbol{\mathcal{G}} \text{ such that } (\exists\, \boldsymbol{x} \in \boldsymbol{\mathcal{H}})\\ &\begin{cases}(\forall i \in I)\ -\displaystyle\sum_{k\in K} L_{ki}^* y_k^* \in A_i x_i + C_i x_i + Q_i x_i + R_i \boldsymbol{x},\\ (\forall k \in K)\ y_k^* \in \big((B_k^m + B_k^c + B_k^\ell) \,\square\, (D_k^m + D_k^c + D_k^\ell)\big)\Big(\displaystyle\sum_{i\in I} L_{ki}x_i\Big).\end{cases}\end{aligned} \tag{10.4}$$

Here is an instance of Problem 10.1 which is not captured by previous monotone inclusion models.

Example 10.2. We consider a game-theoretic minimax problem. Let I be a finite set and suppose that $\varnothing \neq J \subset I$. For every $i \in I$, the strategy x_i of player i belongs to a real Hilbert space $\mathcal{H}_i$. A strategy profile is a point

$$\boldsymbol{x} = (x_i)_{i \in I} \in \bigoplus_{i \in I} \mathcal{H}_i, \tag{10.5}$$

and the associated profile of the players other than $i \in I$ is $\boldsymbol{x}_{\smallsetminus i} = (x_j)_{j \in I \smallsetminus \{i\}}$. For every $i \in I$ and every

$$(x_i, \boldsymbol{y}) \in \mathcal{H}_i \oplus \bigoplus_{j \in I} \mathcal{H}_j, \tag{10.6}$$

we set $(x_i; \boldsymbol{y}_{\smallsetminus i}) = (y_1, \ldots, y_{i-1}, x_i, y_{i+1}, \ldots, y_p)$. Now set

$$\boldsymbol{\mathcal{U}} = \bigoplus_{i \in I \smallsetminus J} \mathcal{H}_i, \quad \boldsymbol{\mathcal{V}} = \bigoplus_{j \in J} \mathcal{H}_j \quad \text{and} \quad \boldsymbol{\mathcal{H}} = \boldsymbol{\mathcal{U}} \oplus \boldsymbol{\mathcal{V}}, \tag{10.7}$$

and, for every $i \in I$, let $f_i \in \Gamma_0(\mathcal{H}_i)$. Further, let $\boldsymbol{F} \colon \boldsymbol{\mathcal{H}} \to \mathbb{R}$ be differentiable with a Lipschitzian gradient and such that, for every $\boldsymbol{u} \in \boldsymbol{\mathcal{U}}$ and every $\boldsymbol{v} \in \boldsymbol{\mathcal{V}}$, the functions $-\boldsymbol{F}(\boldsymbol{u}, \cdot)$ and $\boldsymbol{F}(\cdot, \boldsymbol{v})$ are convex. We consider the multivariate minimax problem

$$\underset{\boldsymbol{u} \in \boldsymbol{\mathcal{U}}}{\text{minimize}} \ \underset{\boldsymbol{v} \in \boldsymbol{\mathcal{V}}}{\text{maximize}} \ \sum_{i \in I \smallsetminus J} f_i(u_i) + \boldsymbol{F}(\boldsymbol{u}, \boldsymbol{v}) - \sum_{j \in J} f_j(v_j). \tag{10.8}$$

Now define

$$(\forall i \in I) \quad \boldsymbol{h}_i \colon \boldsymbol{\mathcal{H}} \to \mathbb{R} \colon (\boldsymbol{u}, \boldsymbol{v}) \mapsto \begin{cases} \boldsymbol{F}(\boldsymbol{u}, \boldsymbol{v}), & \text{if } i \in I \smallsetminus J, \\ -\boldsymbol{F}(\boldsymbol{u}, \boldsymbol{v}), & \text{if } i \in J. \end{cases} \tag{10.9}$$

Then (10.8) can be put in the form

$$\text{find } \boldsymbol{x} \in \boldsymbol{\mathcal{H}} \text{ such that } (\forall i \in I)\ x_i \in \operatorname{Argmin} f_i + \boldsymbol{h}_i(\cdot; \boldsymbol{x}_{\smallsetminus i}). \tag{10.10}$$

Since

$$(\forall i \in I)(\forall \boldsymbol{x} \in \boldsymbol{\mathcal{H}}) \quad \nabla_i \boldsymbol{h}_i(\boldsymbol{x}) = \begin{cases} \nabla_i \boldsymbol{F}(\boldsymbol{x}), & \text{if } i \in I \smallsetminus J, \\ -\nabla_i \boldsymbol{F}(\boldsymbol{x}), & \text{if } i \in J, \end{cases} \tag{10.11}$$

the operator

$$\boldsymbol{R} \colon \boldsymbol{\mathcal{H}} \to \boldsymbol{\mathcal{H}} \colon \boldsymbol{x} \mapsto (\nabla_i \boldsymbol{h}_i(\boldsymbol{x}))_{i \in I} = ((\nabla_i \boldsymbol{F}(\boldsymbol{x}))_{i \in I \smallsetminus J}, (-\nabla_j \boldsymbol{F}(\boldsymbol{x}))_{j \in J}) \tag{10.12}$$

is monotone (Rockafellar 1970*b*, 1971) and Lipschitzian. Now, for every $i \in I$, set $A_i = \partial f_i$. Then, by Fermat's rule, (10.10) is equivalent to

$$\text{find } \boldsymbol{x} \in \boldsymbol{\mathcal{H}} \text{ such that } (\forall i \in I)\ 0 \in A_i x_i + R_i \boldsymbol{x}, \tag{10.13}$$

which shows that (10.8) is an instantiation of (10.3). Special cases of (10.8) under the above assumptions arise in Attouch *et al.* (2010), Combettes and Pesquet (2021), He and Monteiro (2015), Nemirovski (2004), Rockafellar (1995), Thekumparampil, Jain, Netrapalli and Oh (2019) and Yoon and Ryu (2021).

Our objective is to solve Problem 10.1 with the same level of flexibility and the same primal–dual convergence guarantees as in Theorem 9.14, that is, to achieve full splitting of all the operators using an asynchronous block-iterative algorithm without knowledge of the norms of the linear operators or inversion of linear operators. In addition, all the single-valued operators should be activated via Euler steps.

10.2. Saddle operator formulation

The approach adopted in Section 9 to break Problem 9.5 into manageable pieces hinged on the Kuhn–Tucker operator of Lemma 9.7 to obtain the embedding of Framework 1.2. This strategy does not appear to lead to a full splitting of Problem 10.1, as it contains a larger number of operators. We therefore require an embedding in a space $\mathbf{X}$ which is bigger than the primal–dual space $\mathcal{H}_1 \oplus \cdots \mathcal{H}_m \oplus \mathcal{G}_1 \oplus \cdots \oplus \mathcal{G}_p$ of Theorem 9.14. As discussed in Remark 8.51, saddle operators are defined on a bigger space than Kuhn–Tucker operators (e.g. $\mathcal{H} \oplus \mathcal{G} \oplus \mathcal{G}$ versus $\mathcal{H} \oplus \mathcal{G}$ in (8.138)) and their zeros still provide primal–dual solutions. Following Framework 1.2, as we did in Example 3.23, the methodology of *saddle projective splitting* is to introduce a saddle operator for Problem 10.1. We shall then devise asynchronous block-iterative splitting algorithms based on the geometric principles of Theorems 4.8 and 4.9 to find a zero of it, from which solutions to Problem 10.1 will be extracted. This is outlined in the following lemma.

Lemma 10.3 (Bùi and Combettes 2022*b*). Define $\boldsymbol{\mathcal{H}}$ and $\boldsymbol{\mathcal{G}}$ as in (10.2), set $\mathbf{X} = \boldsymbol{\mathcal{H}} \oplus \boldsymbol{\mathcal{G}} \oplus \boldsymbol{\mathcal{G}} \oplus \boldsymbol{\mathcal{G}}$ and define the saddle operator of Problem 10.1 as

$$\begin{aligned}
\boldsymbol{\mathcal{S}}\colon \mathbf{X} \to 2^{\mathbf{X}}\colon (\boldsymbol{x}, \boldsymbol{y}, \boldsymbol{z}, \boldsymbol{v}^*) \mapsto \\
&\Bigg(\bigtimes_{i\in I} \Big(A_i x_i + C_i x_i + Q_i x_i + R_i \boldsymbol{x} + \sum_{k\in K} L_{ki}^* v_k^* \Big), \\
&\bigtimes_{k\in K} \big(B_k^m y_k + B_k^c y_k + B_k^\ell y_k - v_k^* \big), \\
&\bigtimes_{k\in K} \big(D_k^m z_k + D_k^c z_k + D_k^\ell z_k - v_k^* \big), \\
&\bigtimes_{k\in K} \Big\{ y_k + z_k - \sum_{i\in I} L_{ki} x_i \Big\} \Bigg),
\end{aligned} \tag{10.14}$$

let $\boldsymbol{Z}$ be the set of solutions to (10.3) and let $\boldsymbol{Z}^*$ be the set of solutions to (10.4). Then the following hold:

(i) $\boldsymbol{\mathcal{S}}$ is maximally monotone.

(ii) $\operatorname{zer} \boldsymbol{\mathcal{S}}$ is closed and convex.

(iii) Suppose that $(\boldsymbol{x}, \boldsymbol{y}, \boldsymbol{z}, \boldsymbol{v}^*) \in \operatorname{zer} \boldsymbol{\mathcal{S}}$. Then $(\boldsymbol{x}, \boldsymbol{v}^*) \in \boldsymbol{Z} \times \boldsymbol{Z}^*$.

(iv) $\boldsymbol{Z}^* \neq \varnothing \Leftrightarrow \operatorname{zer} \boldsymbol{\mathcal{S}} \neq \varnothing \Rightarrow \boldsymbol{Z} \neq \varnothing$.

We thus obtain the following generalization of Example 3.23.

Example 10.4. In the setting of Problem 10.1, set

$$\mathsf{X} = \boldsymbol{\mathcal{H}} \oplus \boldsymbol{\mathcal{G}} \oplus \boldsymbol{\mathcal{G}} \oplus \boldsymbol{\mathcal{G}}, \tag{10.15}$$

let $\boldsymbol{\mathcal{S}}$ be the saddle operator of (10.14), and let

$$\boldsymbol{\mathcal{T}} \colon \mathsf{X} \to \boldsymbol{\mathcal{H}} \colon (\boldsymbol{x}, \boldsymbol{y}, \boldsymbol{z}, \boldsymbol{v}^*) \mapsto \boldsymbol{x}. \tag{10.16}$$

Then it follows from Lemma 10.3(iii) that $(\mathsf{X}, \boldsymbol{\mathcal{S}}, \boldsymbol{\mathcal{T}})$ is an embedding of (10.3).

Thus, to solve Problem 10.1 via Theorem 4.8, we need a decomposition of the saddle operator (10.14) as $\boldsymbol{\mathcal{S}} = \mathsf{W} + \mathsf{C}$, where $\mathsf{W} \colon \mathsf{X} \to 2^{\mathsf{X}}$ is maximally monotone and $\mathsf{C} \colon \mathsf{X} \to \mathsf{X}$ is α-cocoercive. This will be achieved with

$$\mathsf{C} \colon \mathsf{X} \to \mathsf{X} \colon (\boldsymbol{x}, \boldsymbol{y}, \boldsymbol{z}, \boldsymbol{v}^*) \mapsto \Big((C_i x_i)_{i \in I}, \big(B_k^{c} y_k\big)_{k \in K}, \big(D_k^{c} z_k\big)_{k \in K}, \boldsymbol{0} \Big) \tag{10.17}$$

and $\alpha = \min\{\alpha_i^{c}, \beta_k^{c}, \delta_k^{c}\}_{i \in I, k \in K}$. These considerations lead to the following implementation of (4.23).

Algorithm 10.5. In the setting of Problem 10.1, set

$$\alpha = \min\big\{\alpha_i^{c}, \beta_k^{c}, \delta_k^{c}\big\}_{\substack{i \in I \\ k \in K}}, \tag{10.18}$$

let $\sigma \in \left]1/(4\alpha), +\infty\right[$ and $\varepsilon \in \left]0, 1\right[$ be such that

$$\frac{1}{\varepsilon} > \sigma + \max\big\{\alpha_i^{\ell} + \chi, \beta_k^{\ell}, \delta_k^{\ell}\big\}_{\substack{i \in I \\ k \in K}}, \tag{10.19}$$

and let $(\lambda_n)_{n \in \mathbb{N}}$ be a sequence in $[\varepsilon, 2 - \varepsilon]$. For every $i \in I$, let $(\gamma_{i,n})_{n \in \mathbb{N}}$ be a sequence in $\big[\varepsilon, 1/\big(\alpha_i^{\ell} + \chi + \sigma\big)\big]$ and let $x_{i,0} \in \mathcal{H}_i$. For every $k \in K$, let $(\mu_{k,n})_{n \in \mathbb{N}}$ be a sequence in $\big[\varepsilon, 1/\big(\beta_k^{\ell} + \sigma\big)\big]$, let $(\rho_{k,n})_{n \in \mathbb{N}}$ be a sequence in $\big[\varepsilon, 1/\big(\delta_k^{\ell} + \sigma\big)\big]$, let $(\sigma_{k,n})_{n \in \mathbb{N}}$ be a sequence in $[\varepsilon, 1/\varepsilon]$, and let $\{y_{k,0}, z_{k,0}, v_{k,0}^*\} \subset \mathcal{G}_k$. Suppose that Assumptions 9.11 and 9.12 are in force and iterate

$$
\begin{array}{l}
\text{for } n = 0, 1, \ldots \\
\quad \text{for every } i \in I_n \\
\quad\quad l^*_{i,n} = Q_i x_{i,\pi_i(n)} + R_i \boldsymbol{x}_{\pi_i(n)} + \sum_{k\in K} L^*_{ki} v^*_{k,\pi_i(n)}; \\
\quad\quad a_{i,n} = J_{\gamma_{i,\pi_i(n)} A_i}\big(x_{i,\pi_i(n)} - \gamma_{i,\pi_i(n)}\big(l^*_{i,n} + C_i x_{i,\pi_i(n)}\big)\big); \\
\quad\quad a^*_{i,n} = \gamma^{-1}_{i,\pi_i(n)}(x_{i,\pi_i(n)} - a_{i,n}) - l^*_{i,n} + Q_i a_{i,n}; \\
\quad\quad \xi_{i,n} = \|a_{i,n} - x_{i,\pi_i(n)}\|^2; \\
\quad \text{for every } i \in I \smallsetminus I_n \\
\quad\quad a_{i,n} = a_{i,n-1};\ a^*_{i,n} = a^*_{i,n-1};\ \xi_{i,n} = \xi_{i,n-1}; \\
\quad \text{for every } k \in K_n \\
\quad\quad u^*_{k,n} = v^*_{k,\omega_k(n)} - B^\ell_k y_{k,\omega_k(n)};\ w^*_{k,n} = v^*_{k,\omega_k(n)} - D^\ell_k z_{k,\omega_k(n)}; \\
\quad\quad b_{k,n} = J_{\mu_{k,\omega_k(n)} B^m_k}\big(y_{k,\omega_k(n)} + \mu_{k,\omega_k(n)}\big(u^*_{k,n} - B^c_k y_{k,\omega_k(n)}\big)\big); \\
\quad\quad d_{k,n} = J_{\rho_{k,\omega_k(n)} D^m_k}\big(z_{k,\omega_k(n)} + \rho_{k,\omega_k(n)}\big(w^*_{k,n} - D^c_k z_{k,\omega_k(n)}\big)\big); \\
\quad\quad e^*_{k,n} = \sigma_{k,\omega_k(n)}\big(\sum_{i\in I} L_{ki} x_{i,\omega_k(n)} - y_{k,\omega_k(n)} - z_{k,\omega_k(n)}\big) \\
\quad\quad\quad\quad + v^*_{k,\omega_k(n)}; \\
\quad\quad q^*_{k,n} = \mu^{-1}_{k,\omega_k(n)}(y_{k,\omega_k(n)} - b_{k,n}) + u^*_{k,n} + B^\ell_k b_{k,n} - e^*_{k,n}; \\
\quad\quad t^*_{k,n} = \rho^{-1}_{k,\omega_k(n)}(z_{k,\omega_k(n)} - d_{k,n}) + w^*_{k,n} + D^\ell_k d_{k,n} - e^*_{k,n}; \\
\quad\quad \eta_{k,n} = \|b_{k,n} - y_{k,\omega_k(n)}\|^2 + \|d_{k,n} - z_{k,\omega_k(n)}\|^2; \\
\quad\quad e_{k,n} = b_{k,n} + d_{k,n} - \sum_{i\in I} L_{ki} a_{i,n}; \\
\quad \text{for every } k \in K \smallsetminus K_n \\
\quad\quad b_{k,n} = b_{k,n-1};\ d_{k,n} = d_{k,n-1};\ e^*_{k,n} = e^*_{k,n-1}; \\
\quad\quad q^*_{k,n} = q^*_{k,n-1};\ t^*_{k,n} = t^*_{k,n-1};\ \eta_{k,n} = \eta_{k,n-1}; \\
\quad\quad e_{k,n} = b_{k,n} + d_{k,n} - \sum_{i\in I} L_{ki} a_{i,n}; \\
\quad \text{for every } i \in I \\
\quad\quad p^*_{i,n} = a^*_{i,n} + R_i \boldsymbol{a}_n + \sum_{k\in K} L^*_{ki} e^*_{k,n}; \\
\quad \Delta_n = -(4\alpha)^{-1}\big(\sum_{i\in I} \xi_{i,n} + \sum_{k\in K} \eta_{k,n}\big) + \sum_{i\in I} \langle x_{i,n} - a_{i,n} \mid p^*_{i,n}\rangle \\
\quad\quad\quad + \sum_{k\in K}\big(\langle y_{k,n} - b_{k,n} \mid q^*_{k,n}\rangle + \langle z_{k,n} - d_{k,n} \mid t^*_{k,n}\rangle \\
\quad\quad\quad\quad + \langle e_{k,n} \mid v^*_{k,n} - e^*_{k,n}\rangle\big); \\
\quad \text{if } \Delta_n > 0 \\
\quad\quad \theta_n = \lambda_n \Delta_n / \big(\sum_{i\in I} \|p^*_{i,n}\|^2 + \sum_{k\in K}\big(\|q^*_{k,n}\|^2 + \|t^*_{k,n}\|^2 + \|e_{k,n}\|^2\big)\big); \\
\quad\quad \text{for every } i \in I \\
\quad\quad\quad x_{i,n+1} = x_{i,n} - \theta_n p^*_{i,n}; \\
\quad\quad \text{for every } k \in K \\
\quad\quad\quad y_{k,n+1} = y_{k,n} - \theta_n q^*_{k,n};\ z_{k,n+1} = z_{k,n} - \theta_n t^*_{k,n}; \\
\quad\quad\quad v^*_{k,n+1} = v^*_{k,n} - \theta_n e_{k,n}; \\
\quad \text{else} \\
\quad\quad \text{for every } i \in I \\
\quad\quad\quad x_{i,n+1} = x_{i,n}; \\
\quad\quad \text{for every } k \in K \\
\quad\quad\quad y_{k,n+1} = y_{k,n};\ z_{k,n+1} = z_{k,n};\ v^*_{k,n+1} = v^*_{k,n}.
\end{array}
\tag{10.20}
$$

10.3. Convergence

The convergence properties of Algorithm 10.5 are laid out in the following theorem.

Theorem 10.6 (Bùi and Combettes 2022*b*). Consider the setting of Problem 10.1 and Algorithm 10.5, and suppose that the saddle operator $\boldsymbol{\mathcal{S}}$ of (10.14) has zeros. Then, for every $i \in I$, $(x_{i,n})_{n\in\mathbb{N}}$ converges weakly to a point $x_i \in \mathcal{H}_i$ and, for every $k \in K$, $(v^*_{k,n})_{n\in\mathbb{N}}$ converges weakly to a point $v^*_k \in \mathcal{G}_k$. In addition, $(x_i)_{i\in I}$ solves the primal problem (10.3) and $(v^*_k)_{k\in K}$ solves the dual problem (10.4).

Remark 10.7. The strongly convergent variant of Theorem 10.6 based on Theorem 4.9 is proposed in Bùi and Combettes (2022*b*, Theorem 2(iv)).

Remark 10.8. A fact that has not be appreciated previously is that Theorem 10.6 contains as special cases various weak convergence results of Sections 7–8. Thus, suppose that

$$I = K = \{1\}, \quad R_1 = 0 \quad \text{and} \quad L_{11} = 0. \tag{10.21}$$

Then Problem 10.1 reduces to finding a zero of $A_1 + C_1 + Q_1$ (see (8.120)), (10.20) reduces to the forward–backward–half-forward algorithm (8.121), and Theorem 10.6 reduces to Proposition 8.48. This covers both the forward–backward–forward algorithm (7.2) for $C_1 = 0$ (Theorem 7.1) and the unrelaxed forward–backward algorithm (8.5) for $Q_1 = 0$ (Theorem 8.1). In a similar fashion, we can recover the multivariate forward–backward–forward algorithm of Combettes (2013*b*) by choosing

$$(\forall i \in I)(\forall k \in K) \quad C_i = R_i = 0 \quad \text{and} \quad B^c_k = B^\ell_k = D^c_k = D^\ell_k = 0. \tag{10.22}$$

Going back to the simple inclusion problem (8.120), Theorem 10.6 offers several other possibilities, for instance by implementing it with

$$\begin{aligned} &I = K = \{1\},\ A_1 = A,\ R_1 = C_1 = Q_1 = 0,\ L_{11} = \mathrm{Id},\\ &B^m_1 = 0,\ B^c_1 = C,\ B^\ell_1 = Q,\ \text{and}\ D^m_1 = D^c_1 = D^\ell_1 = \{0\}^{-1}. \end{aligned} \tag{10.23}$$

As mentioned earlier, Problem 10.1 encompasses all the problems discussed earlier. Theorem 10.6 can therefore be used to provide alternative algorithms to solve them in an asynchronous and block-iterative manner, and with operator-dependent proximal parameters (these features are absent from the algorithms of Sections 5–8). Here is an example.

Example 10.9. In Problem 10.1, suppose that

$$\begin{aligned} &I = \{1\},\ K = \{1,\dots,p\},\ A_1 = A,\ C_1 = R_1 = 0,\ Q_1 = Q \text{ and } (\forall k \in K)\\ &L_{k1} = L_k,\ B^m_k = B_k,\ B^c_k = B^\ell_k = 0,\ D^m_k = D_k \text{ and } D^c_k = D^\ell_k = 0. \end{aligned} \tag{10.24}$$

Then we obtain the primal–dual inclusions (7.37)–(7.38) of Proposition 7.10, and Theorem 10.6 furnishes a flexible alternative to Proposition 7.10 which, in addition, places no restriction on the operators $(D_k)_{k\in K}$, with the algorithm

$$
\begin{array}{l}
\text{for } n = 0, 1, \ldots \\
\left\lfloor
\begin{array}{l}
l_n^* = Qx_{\pi(n)} + \sum_{k\in K} L_k^* v_{k,\pi(n)}^*; \\
a_n = J_{\gamma_{\pi(n)}A}(x_{\pi(n)} - \gamma_{\pi(n)} l_n^*); \\
a_n^* = \gamma_{\pi(n)}^{-1}(x_{\pi(n)} - a_n) - l_n^* + Qa_n; \\
\text{for every } k \in K_n \\
\left\lfloor
\begin{array}{l}
b_{k,n} = J_{\mu_{k,\omega_k(n)}B_k}\big(y_{k,\omega_k(n)} + \mu_{k,\omega_k(n)} v_{k,\omega_k(n)}^*\big); \\
d_{k,n} = J_{\rho_{k,\omega_k(n)}D_k}\big(z_{k,\omega_k(n)} + \rho_{k,\omega_k(n)} v_{k,\omega_k(n)}^*\big); \\
e_{k,n}^* = \sigma_{k,\omega_k(n)}(L_k x_{\omega_k(n)} - y_{k,\omega_k(n)} - z_{k,\omega_k(n)}) + v_{k,\omega_k(n)}^*; \\
q_{k,n}^* = \mu_{k,\omega_k(n)}^{-1}(y_{k,\omega_k(n)} - b_{k,n}) + v_{k,\omega_k(n)}^* - e_{k,n}^*; \\
t_{k,n}^* = \rho_{k,\omega_k(n)}^{-1}(z_{k,\omega_k(n)} - d_{k,n}) + v_{k,\omega_k(n)}^* - e_{k,n}^*; \\
\eta_{k,n} = \|b_{k,n} - y_{k,\omega_k(n)}\|^2 + \|d_{k,n} - z_{k,\omega_k(n)}\|^2; \\
e_{k,n} = b_{k,n} + d_{k,n} - L_k a_n;
\end{array}
\right. \\
\text{for every } k \in K \smallsetminus K_n \\
\left\lfloor
\begin{array}{l}
b_{k,n} = b_{k,n-1};\ d_{k,n} = d_{k,n-1};\ e_{k,n}^* = e_{k,n-1}^*; \\
q_{k,n}^* = q_{k,n-1}^*;\ t_{k,n}^* = t_{k,n-1}^*;\ \eta_{k,n} = \eta_{k,n-1}; \\
e_{k,n} = b_{k,n} + d_{k,n} - L_k a_n;
\end{array}
\right. \\
p_n^* = a_n^* + \sum_{k\in K} L_k^* e_{k,n}^*; \\
\Delta_n = -(4\alpha)^{-1}\big(\|a_n - x_{\pi(n)}\|^2 + \sum_{k\in K}\eta_{k,n}\big) + \langle x_n - a_n \mid p_n^*\rangle \\
\qquad + \sum_{k\in K}\big(\langle y_{k,n} - b_{k,n} \mid q_{k,n}^*\rangle + \langle z_{k,n} - d_{k,n} \mid t_{k,n}^*\rangle \\
\qquad\qquad + \langle e_{k,n} \mid v_{k,n}^* - e_{k,n}^*\rangle\big); \\
\text{if } \Delta_n > 0 \\
\left\lfloor
\begin{array}{l}
\theta_n = \lambda_n\Delta_n/\big(\|p_n^*\|^2 + \sum_{k\in K}\big(\|q_{k,n}^*\|^2 + \|t_{k,n}^*\|^2 + \|e_{k,n}\|^2\big)\big); \\
x_{n+1} = x_n - \theta_n p_n^*; \\
\text{for every } k \in K \\
\left\lfloor
\begin{array}{l}
y_{k,n+1} = y_{k,n} - \theta_n q_{k,n}^*;\ z_{k,n+1} = z_{k,n} - \theta_n t_{k,n}^*; \\
v_{k,n+1}^* = v_{k,n}^* - \theta_n e_{k,n};
\end{array}
\right.
\end{array}
\right. \\
\text{else} \\
\left\lfloor
\begin{array}{l}
x_{n+1} = x_n; \\
\text{for every } k \in K \\
\lfloor\ y_{k,n+1} = y_{k,n};\ z_{k,n+1} = z_{k,n};\ v_{k,n+1}^* = v_{k,n}^*.
\end{array}
\right.
\end{array}
\right.
\end{array}
\tag{10.25}
$$

Remark 10.10. In the same vein as Example 10.9, we can solve the primal–dual inclusions (8.109)–(8.110) of Proposition 8.44 via Theorem 10.6 by making the modifications $C_1 = C$ and $Q_1 = 0$ in (10.24).

11. Extensions and variants

The flowchart in Figure 11.1 summarizes the articulation of the main splitting methods presented in the previous sections (a similar flowchart can be drawn for the chain of strong convergence results starting with the Haugazeau principle of Theorem 4.7, then Theorem 4.9, etc.). This flowchart suggests that any extension or variant of the main theorems of Section 4 (Theorems 4.2, 4.8 and 4.12) will lead to further splitting methods or, at least, different implementations of them. We discuss some of the possible variations on the basic geometric principles we have employed.

The basic operating principle of Theorem 4.2 is Fejér-monotonicity, i.e. its property (i). There are extensions of this notion which preserve the main weak convergence conclusions. For instance, the notion of quasi-Fejér monotonicity, introduced in Ermol'ev and Tuniev (1968) and studied in detail in Combettes (2001*b*), requires that there exist a summable sequence $(\varepsilon_n)_{n\in\mathbb{N}}$ in $[0,+\infty[$ such that

$$(\forall z \in Z)(\forall n \in \mathbb{N}) \quad \|x_{n+1} - z\|^2 \leq \|x_n - z\|^2 + \varepsilon_n. \tag{11.1}$$

It follows from Combettes (2001*b*, Section 3) that Theorem 4.2 remains valid if, for some sequence $(e_n)_{n\in\mathbb{N}}$ in $\mathcal{H}$ such that $\sum_{n\in\mathbb{N}} \lambda_n\|e_n\| < +\infty$, we use an approximate projection $p_n = \operatorname{proj}_{H_n} x_n + e_n$ in (4.1); see also Combettes and Pesquet (2015) for a stochastic version of this result that allows for random iteration modelling. This summable error framework can be propagated in Figure 11.1 to recover approximate implementation results from Boţ and Hendrich (2013), Combettes (2004, 2013*b*), Combettes and Pesquet (2012), Condat (2013), Rockafellar (1976*b*) and Vũ (2013).

- Cutting plane Fejér principle (Theorem 4.2)

  ⇓

  - Graph-based cuts (Theorem 4.8)
    - Section 9 (Block-iterative Kuhn–Tucker projective splitting)
    - Section 10 (Block-iterative saddle projective splitting)
    - Warped resolvent splitting (Theorem 4.12)

      ⇓

      - Section 5 (Proximal point algorithm)
      - Section 6 (Douglas–Rachford splitting)
      - Section 7 (Forward–backward–forward splitting)
      - Section 8 (Forward–backward splitting).

Figure 11.1. Articulation of the splitting methods.

Variable metric quasi-Fejér-monotonicity is an extension of (11.1) described by

$$(\forall z \in Z)(\forall n \in \mathbb{N}) \quad \|x_{n+1} - z\|^2_{U_{n+1}} \leq \|x_n - z\|^2_{U_n} + \varepsilon_n, \tag{11.2}$$

where $(U_n)_{n\in\mathbb{N}}$ is a sequence of strongly monotone operators in $\mathcal{B}(\mathcal{H})$ satisfying certain properties (Combettes and Vũ 2013). It follows from Combettes and Vũ (2013, Theorem 3.3) that the conclusions of Theorem 4.2 remain valid in this setting, which amounts to changing the metric of $\mathcal{H}$ at each iteration. See Chen and Rockafellar (1997) and Combettes and Vũ (2014) for applications to forward–backward splitting, Rockafellar (2024) for applications to multiplier methods and Raguet and Landrieu (2015) for considerations on the choice of the variable metrics. All the results derived from Theorem 4.2 can be revisited in this variable-metric context. Another extension of (11.1) of interest is the multi-step quasi-Fejér-monotonicity notion

$$(\forall z \in Z)(\forall n \in \mathbb{N}) \quad \|x_{n+1} - x\|^2 \leq \sum_{j=0}^{n} \mu_{n,j}\|x_j - x\|^2 + \varepsilon_n \tag{11.3}$$

of Combettes and Glaudin (2021, Lemma 2.2), where $(\mu_{n,j})_{n\in\mathbb{N},0\leq j\leq n}$ is an array in $[0,+\infty[$ satisfying certain properties. This setting led to deterministic block-iterative implementations of the forward–backward algorithm (Combettes and Glaudin 2021, Proposition 4.9) in the spirit of methods found in Mishchenko, Iutzeler and Malick (2020) and Mokhtari, Gürbüzbalaban and Ribeiro (2018) in the minimization case.

The hybrid proximal-extragradient/projection methods of Solodov (2004) and Solodov and Svaiter (1999*a*,*b*, 2001) revolve around a variant of Proposition 4.10 in which, at iteration n, (m_n, m_n^*) is merely required to be in the graph of a perturbed version of M, which permits us to recover certain iterative methods beyond the proximal point algorithm. See also Svaiter (2014) for more recent work along these lines, where approximate resolvents are used to recover an instance of the forward–backward algorithm.

As is apparent from Figure 11.1, many convergence results we have discussed follow from Theorem 4.12. We now present a perturbed extension of it in which, at iteration n, the warped resolvent is applied at a point $\widetilde{x}_n$ and not necessarily at the current iterate x_n. The special case when $C = 0$, $(q_n)_{n\in\mathbb{N}} = (w_n)_{n\in\mathbb{N}}$, and conditions (iib) and (iic) of Theorem 4.12 are fulfilled appears in Bùi and Combettes (2020*b*, Theorem 4.2).

Theorem 11.1. Let $\alpha \in]0,+\infty[$, let $W\colon \mathcal{H} \to 2^{\mathcal{H}}$ be maximally monotone, let $C\colon \mathcal{H} \to \mathcal{H}$ be α-cocoercive and such that $Z = \operatorname{zer}(W + C) \neq \varnothing$, let $x_0 \in \mathcal{H}$, and let $(\lambda_n)_{n\in\mathbb{N}}$ be a sequence in $]0,2[$. Further, for every $n \in \mathbb{N}$, let $\widetilde{x}_n \in \mathcal{H}$ and let $U_n\colon \mathcal{H} \to \mathcal{H}$ be an operator such that $\operatorname{ran} U_n \subset \operatorname{ran}(U_n + W + C)$ and $U_n + W + C$

is injective. Iterate

$$
\begin{array}{l}
\text{for } n = 0, 1, \ldots \\
\left\lfloor
\begin{array}{l}
w_n = J_{W+C}^{U_n} \widetilde{x}_n \\
w_n^* = U_n \widetilde{x}_n - U_n w_n - C w_n \\
q_n \in \mathcal{H} \\
t_n^* = w_n^* + C q_n \\
\delta_n = \langle x_n - w_n \mid t_n^* \rangle - \| w_n - q_n \|^2 / (4\alpha) \\
d_n = \begin{cases} \dfrac{\delta_n}{\|t_n^*\|^2} t_n^*, & \text{if } \delta_n > 0, \\ 0, & \text{otherwise} \end{cases} \\
x_{n+1} = x_n - \lambda_n d_n.
\end{array}
\right.
\end{array}
\tag{11.4}
$$

Suppose that $\widetilde{x}_n - x_n \to 0$. Then the conclusions of Theorem 4.12 remain valid if the condition $U_n w_n - U_n x_n \to 0$ in (iic) is replaced by $U_n w_n - U_n \widetilde{x}_n \to 0$.

Proof. Adapt the pattern of the proof of Theorem 4.12. □

Remark 11.2. The auxiliary sequence $(\widetilde{x}_n)_{n\in\mathbb{N}}$ in Theorem 11.1 adds considerable breadth to the scope of the algorithm, compared to that of Theorem 4.12. Here are some illustrations of the condition $\widetilde{x}_n - x_n \to 0$, where we assume that $\inf_{n\in\mathbb{N}} \lambda_n > 0$ and $\sup_{n\in\mathbb{N}} \lambda_n < 2$.

(i) At iteration n, $\widetilde{x}_n$ can model an additive perturbation of x_n, say $\widetilde{x}_n = x_n + e_n$. Here, the error sequence $(e_n)_{n\in\mathbb{N}}$ need only satisfy $\|e_n\| \to 0$ and not the usual summability condition $\sum_{n\in\mathbb{N}} \|e_n\| < +\infty$ required in the quasi-Fejérian splitting methods of Boţ and Hendrich (2013), Combettes (2001*b*, 2004, 2013*b*), Combettes and Pesquet (2012) and Vũ (2013).

(ii) In the spirit of inertial methods (Attouch and Cabot 2020, Beck and Teboulle 2009*b*, Chambolle and Dossal 2015, Combettes and Glaudin 2017, Polyak 1964), let $(\alpha_n)_{n\in\mathbb{N}}$ be a sequence in $\mathbb{R}$ and set $(\forall n \in \mathbb{N} \smallsetminus \{0\})\ \widetilde{x}_n = x_n + \alpha_n(x_n - x_{n-1})$. In these methods, $\alpha_n(x_n - x_{n-1}) \to 0$, which guarantees that $\|\widetilde{x}_n - x_n\| \to 0$, as required.

(iii) More generally, weak convergence results can be derived from Theorem 11.1 for iterations with memory, that is,

$$
(\forall n \in \mathbb{N}) \quad \widetilde{x}_n = \sum_{j=0}^{n} \mu_{n,j} x_j,
$$

$$
\text{where } (\mu_{n,j})_{0\le j\le n} \in \mathbb{R}^{n+1} \quad \text{and} \quad \sum_{j=0}^{n} \mu_{n,j} = 1. \tag{11.5}
$$

Here we have $\widetilde{x}_n - x_n \to 0$ if $(1 - \mu_{n,n})x_n - \sum_{j=0}^{n-1} \mu_{n,j} x_j \to 0$. In the case of standard inertial methods, weak convergence requires more stringent conditions on the weights $(\mu_{n,j})_{n\in\mathbb{N}, 0\le j\le n}$ (Combettes and Glaudin 2017).

(iv) As indicated in Figure 11.1, Theorem 9.14 on the Kuhn–Tucker projective splitting algorithm was derived from Proposition 4.10, hence from Theorem 4.8, and it does not appear possible to derive it from Theorem 4.12. However, as shown in Bùi (2022*b*, Corollary 4), Theorem 9.14 follows from Theorem 11.1 (implemented with $C = 0$ and $q_n = w_n$) through a suitable choice of the auxiliary sequence $(\widetilde{x}_n)_{n\in\mathbb{N}}$. This last example provides further confirmation of the effectiveness of warped resolvents.

Acknowledgement
The author thanks Minh N. Bùi for his careful proofreading of the paper and his suggestions.

References

F. Acker and M. A. Prestel (1980), Convergence d'un schéma de minimisation alternée, *Ann. Fac. Sci. Toulouse V. Sér. Math.* **2**, 1–9.

S. Adly, A. Hantoute and B. K. Le (2017), Maximal monotonicity and cyclic monotonicity arising in nonsmooth Lur'e dynamical systems, *J. Math. Anal. Appl.* **448**, 691–706.

S. Agmon (1954), The relaxation method for linear inequalities, *Canad. J. Math.* **6**, 382–392.

Y. Alber and I. Ryazantseva (2006), *Nonlinear Ill-Posed Problems of Monotone Type*, Springer.

G. Alduncin (2005), Composition duality principles for mixed variational inequalities, *Math. Comput. Modelling* **41**, 639–654.

G. Alduncin (2023), Multidomain optimal control of variational subpotential mixed evolution inclusions, *Appl. Math. Optim.* **88**, art. 35.

M. A. Alghamdi, A. Alotaibi, P. L. Combettes and N. Shahzad (2014), A primal–dual method of partial inverses for composite inclusions, *Optim. Lett.* **8**, 2271–2284.

M. Alimohammady, M. Ramazannejad and M. Roohi (2014), Notes on the difference of two monotone operators, *Optim. Lett.* **8**, 81–84.

A. Alotaibi, P. L. Combettes and N. Shahzad (2014), Solving coupled composite monotone inclusions by successive Fejér approximations of their Kuhn–Tucker set, *SIAM J. Optim.* **24**, 2076–2095.

A. Alotaibi, P. L. Combettes and N. Shahzad (2015), Best approximation from the Kuhn–Tucker set of composite monotone inclusions, *Numer. Funct. Anal. Optim.* **36**, 1513–1532.

W. N. Anderson Jr and G. E. Trapp (1976), A class of monotone operator functions related to electrical network theory, *Linear Algebra Appl.* **15**, 53–67.

A. S. Antipin (1976), On a method for convex programs using a symmetrical modification of the Lagrange function, *Èkonom. i Mat. Metody* **12**, 1164–1173.

K. Aoyama, Y. Kimura and W. Takahashi (2008), Maximal monotone operators and maximal monotone functions for equilibrium problems, *J. Convex Anal.* **15**, 395–409.

F. J. Aragón-Artacho, R. I. Boţ and D. Torregrosa-Belén (2023), A primal–dual splitting algorithm for composite monotone inclusions with minimal lifting, *Numer. Algorithms* **93**, 103–130.

A. Argyriou, R. Foygel and N. Srebro (2012), Sparse prediction with the k-support norm, in *Advances in Neural Information Processing Systems 25* (F. Pereira *et al.*, eds), Curran Associates, pp. 1457–1465.

H. Attouch and A. Cabot (2020), Convergence of a relaxed inertial proximal algorithm for maximally monotone operators, *Math. Program.* **A184**, 243–287.

H. Attouch and M. Théra (1996), A general duality principle for the sum of two operators, *J. Convex Anal.* **3**, 1–24.

H. Attouch, J. Bolte, P. Redont and A. Soubeyran (2008), Alternating proximal algorithms for weakly coupled convex minimization problems: Applications to dynamical games and PDE's, *J. Convex Anal.* **15**, 485–506.

H. Attouch, L. M. Briceño-Arias and P. L. Combettes (2010), A parallel splitting method for coupled monotone inclusions, *SIAM J. Control Optim.* **48**, 3246–3270.

H. Attouch, L. M. Briceño-Arias and P. L. Combettes (2016), A strongly convergent primal–dual method for nonoverlapping domain decomposition, *Numer. Math.* **133**, 433–470.

H. Attouch, G. Buttazzo and G. Michaille (2014), *Variational Analysis in Sobolev and BV Spaces*, second edition, SIAM.

H. Attouch, A. Cabot, P. Frankel and J. Peypouquet (2011), Alternating proximal algorithms for linearly constrained variational inequalities: Application to domain decomposition for PDE's, *Nonlinear Anal.* **74**, 7455–7473.

H. Attouch, J. Peypouquet and P. Redont (2018), Backward–forward algorithms for structured monotone inclusions in Hilbert spaces, *J. Math. Anal. Appl.* **457**, 1095–1117.

J.-P. Aubin and A. Cellina (1984), *Differential Inclusions: Set-Valued Maps and Viability Theory*, Springer.

J.-F. Aujol and A. Chambolle (2005), Dual norms and image decomposition models, *Int. J. Comput. Vision* **63**, 85–104.

J.-B. Baillon and G. Haddad (1977), Quelques propriétés des opérateurs angle-bornés et n-cycliquement monotones, *Israel J. Math.* **26**, 137–150.

A. B. Bakušinskiĭ and B. T. Polyak (1974), The solution of variational inequalities, *Soviet Math. Dokl.* **15**, 1705–1710.

S. Banert, R. I. Boţ and E. R. Csetnek (2021), Fixing and extending some recent results on the ADMM algorithm, *Numer. Algorithms* **86**, 1303–1325.

V. Barbu (2010), *Nonlinear Differential Equations of Monotone Types in Banach Spaces*, Springer.

S. Bartz, H. H. Bauschke, S. M. Moffat and X. Wang (2016), The resolvent average of monotone operators: Dominant and recessive properties, *SIAM J. Optim.* **26**, 602–634.

H. H. Bauschke and J. M. Borwein (1994), Dykstra's alternating projection algorithm for two sets, *J. Approx. Theory* **79**, 418–443.

H. H. Bauschke and P. L. Combettes (2001), A weak-to-strong convergence principle for Fejér-monotone methods in Hilbert spaces, *Math. Oper. Res.* **26**, 248–264.

H. H. Bauschke and P. L. Combettes (2008), A Dykstra-like algorithm for two monotone operators, *Pacific J. Optim.* **4**, 383–391.

H. H. Bauschke and P. L. Combettes (2017), *Convex Analysis and Monotone Operator Theory in Hilbert Spaces*, second edition, Springer.

H. H. Bauschke and W. M. Moursi (2017), On the Douglas–Rachford algorithm, *Math. Program.* **A164**, 263–284.

H. H. Bauschke, J. Bolte and M. Teboulle (2017), A descent lemma beyond Lipschitz gradient continuity: First-order methods revisited and applications, *Math. Oper. Res.* **42**, 330–348.

H. H. Bauschke, J. M. Borwein and P. L. Combettes (2003), Bregman monotone optimization algorithms, *SIAM J. Control Optim.* **42**, 596–636.

H. H. Bauschke, M. N. Bùi and X. Wang (2019), On sums and convex combinations of projectors onto convex sets, *J. Approx. Theory* **242**, 31–57.

H. H. Bauschke, P. L. Combettes and S. Reich (2005), The asymptotic behavior of the composition of two resolvents, *Nonlinear Anal.* **60**, 283–301.

H. H. Bauschke, F. Deutsch and H. Hundal (2009), Characterizing arbitrarily slow convergence in the method of alternating projections, *Int. Trans. Oper. Res.* **16**, 413–425.

H. H. Bauschke, V. R. Koch and H. M. Phan (2016), Stadium norm and Douglas–Rachford splitting: A new approach to road design optimization, *Oper. Res.* **64**, 201–218.

H. H. Bauschke, E. Matoušková and S. Reich (2004), Projection and proximal point methods: Convergence results and counterexamples, *Nonlinear Anal.* **56**, 715–738.

A. Beck and M. Teboulle (2009*a*), Fast gradient-based algorithms for constrained total variation image denoising and deblurring problems, *IEEE Trans. Image Process.* **18**, 2419–2434.

A. Beck and M. Teboulle (2009*b*), A fast iterative shrinkage-thresholding algorithm for linear inverse problems, *SIAM J. Imaging Sci.* **2**, 183–202.

A. Beck and M. Teboulle (2010), Gradient-based algorithms with applications to signal recovery problems, in *Convex Optimization in Signal Processing and Communications* (D. P. Palomar and Y. C. Eldar, eds), Cambridge University Press, pp. 42–88.

S. R. Becker and P. L. Combettes (2014), An algorithm for splitting parallel sums of linearly composed monotone operators, with applications to signal recovery, *J. Nonlinear Convex Anal.* **15**, 137–159.

E. M. Bednarczuk, A. Jezierska and K. E. Rutkowski (2018), Proximal primal–dual best approximation algorithm with memory, *Comput. Optim. Appl.* **71**, 767–794.

G. Belgioioso, A. Nedich and S. Grammatico (2021), Distributed generalized Nash equilibrium seeking in aggregative games on time-varying networks, *IEEE Trans. Automat. Control* **66**, 2061–2075.

R. Bellman, R. E. Kalaba and J. A. Lockett (1966), *Numerical Inversion of the Laplace Transform: Applications to Biology, Economics Engineering, and Physics*, Elsevier.

E. Beltrami (1972), A note regarding abstract operators and passive networks, *Quart. Appl. Math.* **30**, 369–370.

M. Benning and M. Burger (2018), Modern regularization methods for inverse problems, *Acta Numer.* **27**, 1–111.

C. Berge and A. Ghouila-Houri (1962), *Programmes, Jeux, et Réseaux de Transport*, Dunod. English translation: *Programming, Games and Transportation Networks*, Wiley (1965).

M. Bertero, D. Bindi, P. Boccacci, M. Cattaneo, C. Eva and V. Lanza (1997), Application of the projected Landweber method to the estimation of the source time function in seismology, *Inverse Problems* **13**, 465–486.

D. P. Bertsekas (1998), *Network Optimization: Continuous and Discrete Models*, Athena Scientific.

G. Birkhoff and R. S. Varga (1959), Implicit alternating direction methods, *Trans. Amer. Math. Soc.* **92**, 13–24.

E. Blum and W. Oettli (1994), From optimization and variational inequalities to equilibrium problems, *Math. Student* **63**, 123–145.

E. Börgens and C. Kanzow (2021), ADMM-type methods for generalized Nash equilibrium problems in Hilbert spaces, *SIAM J. Optim.* **31**, 377–403.

J. M. Borwein (2010), Fifty years of maximal monotonicity, *Optim. Lett.* **4**, 473–490.

R. I. Boţ (2010), *Conjugate Duality in Convex Optimization*, Springer.

R. I. Boţ and E. R. Csetnek (2019), ADMM for monotone operators: Convergence analysis and rates, *Adv. Comput. Math.* **45**, 327–359.

R. I. Boţ and C. Hendrich (2013), A Douglas–Rachford type primal–dual method for solving inclusions with mixtures of composite and parallel-sum type monotone operators, *SIAM J. Optim.* **23**, 2541–2565.

R. I. Boţ and C. Hendrich (2014), Convergence analysis for a primal–dual monotone+skew splitting algorithm with applications to total variation minimization, *J. Math. Imaging Vision* **49**, 551–568.

S. Boyd, N. Parikh, E. Chu, B. Peleato and J. Eckstein (2010), Distributed optimization and statistical learning via the alternating direction method of multipliers, *Found. Trends Machine Learn.* **3**, 1–122.

J. P. Boyle and R. L. Dykstra (1986), A method for finding projections onto the intersection of convex sets in Hilbert spaces, in *Advances in Order Restricted Statistical Inference*, Vol. 37 of Lecture Notes in Statistics, Springer, pp. 28–47.

K. Bredies, E. Chenchene, D. A. Lorenz and E. Naldi (2022), Degenerate preconditioned proximal point algorithms, *SIAM J. Optim.* **32**, 2376–2401.

L. M. Brègman (1965), The method of successive projection for finding a common point of convex sets, *Soviet Math. Dokl.* **6**, 688–692.

L. M. Brègman (1967), The relaxation method of finding the common point of convex sets and its application to the solution of problems in convex programming, *USSR Comput. Math. Math. Phys.* **7**, 200–217.

H. Brézis (1966), Les opérateurs monotones, *Séminaire Choquet: Initiation à l'Analyse* **5**, 1–33.

H. Brézis (1971), Monotonicity methods in Hilbert spaces and some applications to nonlinear partial differential equations, in *Contributions to Nonlinear Functional Analysis* (E. H. Zarantonello, ed.), Academic, pp. 101–156.

H. Brézis (1973), *Opérateurs Maximaux Monotones et Semi-Groupes de Contractions dans les Espaces de Hilbert*, North-Holland/Elsevier.

H. Brézis and F. Browder (1998), Partial differential equations in the 20th century, *Adv. Math.* **135**, 76–144.

H. Brézis and P. L. Lions (1978), Produits infinis de résolvantes, *Israel J. Math.* **29**, 329–345.

L. M. Briceño-Arias (2012), A Douglas–Rachford splitting method for solving equilibrium problems, *Nonlinear Anal.* **75**, 6053–6059.

L. M. Briceño-Arias and P. L. Combettes (2009), Convex variational formulation with smooth coupling for multicomponent signal decomposition and recovery, *Numer. Math. Theory Methods Appl.* **2**, 485–508.

L. M. Briceño-Arias and P. L. Combettes (2011), A monotone+skew splitting model for composite monotone inclusions in duality, *SIAM J. Optim.* **21**, 1230–1250.

L. M. Briceño-Arias and P. L. Combettes (2013), Monotone operator methods for Nash equilibria in non-potential games, in *Computational and Analytical Mathematics* (D. Bailey *et al.*, eds), Springer, pp. 143–159.

L. M. Briceño-Arias and D. Davis (2018), Forward–backward-half forward algorithm for solving monotone inclusions, *SIAM J. Optim.* **28**, 2839–2871.

L. M. Briceño-Arias and F. Roldán (2023), Primal–dual splittings as fixed point iterations in the range of linear operators, *J. Global Optim.* **85**, 847–866.

L. M. Briceño-Arias, G. Chierchia, E. Chouzenoux and J.-C. Pesquet (2019), A random block-coordinate Douglas–Rachford splitting method with low computational complexity for binary logistic regression, *Comput. Optim. Appl.* **72**, 707–726.

L. M. Briceño-Arias, P. L. Combettes, J.-C. Pesquet and N. Pustelnik (2011), Proximal algorithms for multicomponent image recovery problems, *J. Math. Imaging Vision* **41**, 3–22.

L. M. Briceño-Arias, J. Deride, S. López-Rivera and F. J. Silva (2023), A primal–dual partial inverse algorithm for constrained monotone inclusions: Applications to stochastic programming and mean field games, *Appl. Math. Optim.* **87**, art. 21.

L. M. Briceño-Arias, D. Kalise and F. J. Silva (2018), Proximal methods for stationary mean field games with local couplings, *SIAM J. Control Optim.* **56**, 801–836.

B. Brogliato and A. Tanwani (2020), Dynamical systems coupled with monotone set-valued operators: Formalisms, applications, well-posedness, and stability, *SIAM Rev.* **62**, 3–129.

B. Brogliato, R. Lozano, B. Maschke and O. Egeland (2007), *Dissipative Systems Analysis and Control: Theory and Applications*, second edition, Springer.

F. E. Browder (1963), The solvability of non-linear functional equations, *Duke Math. J.* **30**, 557–566.

F. E. Browder (1965), Multi-valued monotone nonlinear mappings and duality mappings in Banach spaces, *Trans. Amer. Math. Soc.* **118**, 338–351.

F. E. Browder (1968/1976), Nonlinear operators and nonlinear equations of evolution in Banach spaces, *Proc. Sympos. Pure Math.* **18**, 1–308.

R. E. Bruck (1973), The iterative solution of the equation $y \in x + Tx$ for a monotone operator T in Hilbert space, *Bull. Amer. Math. Soc.* **79**, 1258–1261.

R. E. Bruck (1974), A strongly convergent iterative solution of $0 \in U(x)$ for a maximal monotone operator U in Hilbert space, *J. Math. Anal. Appl.* **48**, 114–126.

R. E. Bruck (1975), An iterative solution of a variational inequality for certain monotone operators in Hilbert space, *Bull. Amer. Math. Soc.* **81**, 890–892. Corrigendum: **82**, 353 (1976).

R. C. Buck (1956), *Advanced Calculus*, first edition, McGraw-Hill.

M. N. Bùi (2022*a*), A decomposition method for solving multicommodity network equilibria, *Oper. Res. Lett.* **50**, 40–44.

M. N. Bùi (2022*b*), Projective splitting as a warped proximal algorithm, *Appl. Math. Optim.* **85**, art. 4.

M. N. Bùi and P. L. Combettes (2020*a*), The Douglas–Rachford algorithm converges only weakly, *SIAM J. Control Optim.* **58**, 1118–1120.

M. N. Bùi and P. L. Combettes (2020*b*), Warped proximal iterations for monotone inclusions, *J. Math. Anal. Appl.* **491**, art. 124315.

M. N. Bùi and P. L. Combettes (2021), Bregman forward–backward operator splitting, *Set-Valued Var. Anal.* **29**, 583–603.

M. N. Bùi and P. L. Combettes (2022*a*), Analysis and numerical solution of a modular convex Nash equilibrium problem, *J. Convex Anal.* **29**, 1007–1021.

M. N. Bùi and P. L. Combettes (2022*b*), Multivariate monotone inclusions in saddle form, *Math. Oper. Res.* **47**, 1082–1109.

M. N. Bùi, P. L. Combettes and Z. C. Woodstock (2022), Block-activated algorithms for multicomponent fully nonsmooth minimization, in *IEEE International Conference on Acoustics, Speech and Signal Processing (ICASSP 2022)*, pp. 5428–5432.

C. Byrne, Y. Censor, A. Gibali and S. Reich (2012), The split common null point problem, *J. Nonlinear Convex Anal.* **13**, 759–775.

M. K. Camlibel and J. M. Schumacher (2016), Linear passive systems and maximal monotone mappings, *Math. Program.* **B157**, 397–420.

A. Cauchy (1847), Méthode générale pour la résolution des systèmes d'équations simultanées, *C. R. Acad. Sci. Paris* **25**, 536–538.

I. Cederbaum (1962), On optimal operation of communication nets, *J. Franklin Inst.* **274**, 130–141.

A. Cegielski (2012), *Iterative Methods for Fixed Point Problems in Hilbert Spaces*, Vol. 2057 of Lecture Notes in Mathematics, Springer.

Y. Censor and M. Zaknoon (2018), Algorithms and convergence results of projection methods for inconsistent feasibility problems: A review, *Pure Appl. Funct. Anal.* **3**, 565–586.

Y. Censor and S. A. Zenios (1992), Proximal minimization algorithm with D-functions, *J. Optim. Theory Appl.* **73**, 451–464.

T. Chaffey and R. Sepulchre (2024), Monotone one-port circuits, *IEEE Trans. Automat. Control* **69**, 783–796.

T. Chaffey, S. Banert, P. Giselsson and R. Pates (2023*a*), Circuit analysis using monotone+skew splitting, *Eur. J. Control* **74**, art. 100854.

T. Chaffey, F. Forni and R. Sepulchre (2023*b*), Graphical nonlinear system analysis, *IEEE Trans. Automat. Control* **68**, 6067–6081.

A. Chambolle (2004), An algorithm for total variation minimization and applications, *J. Math. Imaging Vision* **20**, 89–97.

A. Chambolle (2005), Total variation minimization and a class of binary MRF models, in *Energy Minimization Methods in Computer Vision and Pattern Recognition (EMMCVPR 2005)*, Vol. 3757 of Lecture Notes in Computer Science, Springer, pp. 136–152.

A. Chambolle and C. Dossal (2015), On the convergence of the iterates of the 'fast iterative shrinkage/thresholding algorithm', *J. Optim. Theory Appl.* **166**, 968–982.

A. Chambolle and T. Pock (2011), A first-order primal–dual algorithm for convex problems with applications to imaging, *J. Math. Imaging Vision* **40**, 120–145.

A. Chambolle and T. Pock (2016), An introduction to continuous optimization for imaging, *Acta Numer.* **25**, 161–319.

R. H. Chan, S. Setzer and G. Steidl (2008), Inpainting by flexible Haar-wavelet shrinkage, *SIAM J. Imaging Sci.* **1**, 273–293.

S. H. Chan, X. Wang and O. A. Elgendy (2017), Plug-and-play ADMM for image restoration: Fixed-point convergence and applications, *IEEE Trans. Comput. Imaging* **3**, 84–98.

C. Chaux, M. El-Gheche, J. Farah, J.-C. Pesquet and B. Pesquet-Popescu (2013), A parallel proximal splitting method for disparity estimation from multicomponent images under illumination variation, *J. Math. Imaging Vision* **47**, 167–178.

G. Chen and M. Teboulle (1994), A proximal-based decomposition method for convex minimization problems, *Math. Program.* **64**, 81–101.

G. H.-G. Chen and R. T. Rockafellar (1997), Convergence rates in forward–backward splitting, *SIAM J. Optim.* **7**, 421–444.

E. W. Cheney and A. A. Goldstein (1959*a*), Newton's method for convex programming and Tchebycheff approximation, *Numer. Math.* **1**, 253–268.

W. Cheney and A. A. Goldstein (1959*b*), Proximity maps for convex sets, *Proc. Amer. Math. Soc.* **10**, 448–450.

G. Chierchia, E. Chouzenoux, P. L. Combettes and J.-C. Pesquet (2016), The proximity operator repository. Available at http://proximity-operator.net/.

C. Clason and T. Valkonen (2017), Primal–dual extragradient methods for nonlinear nonsmooth PDE-constrained optimization, *SIAM J. Optim.* **27**, 1314–1339.

G. Cohen (1987), Nash equilibria: Gradient and decomposition algorithms, *Large Scale Syst.* **12**, 173–184.

P. L. Combettes (2001*a*), Fejér-monotonicity in convex optimization, in *Encyclopedia of Optimization* (C. A. Floudas and P. M. Pardalos, eds), Vol. 2, Springer, pp. 106–114. Also available in second edition (2009), pp. 1016–1024.

P. L. Combettes (2001*b*), Quasi-Fejérian analysis of some optimization algorithms, in *Inherently Parallel Algorithms for Feasibility and Optimization* (D. Butnariu, Y. Censor and S. Reich, eds), Elsevier, pp. 115–152.

P. L. Combettes (2004), Solving monotone inclusions via compositions of nonexpansive averaged operators, *Optimization* **53**, 475–504.

P. L. Combettes (2009), Iterative construction of the resolvent of a sum of maximal monotone operators, *J. Convex Anal.* **16**, 727–748.

P. L. Combettes (2013*a*), Can one genuinely split $m > 2$ monotone operators?, in *Workshop on Algorithms and Dynamics for Games and Optimization (ADGO 2013).* Available at https://pcombet.math.ncsu.edu/2013open-pbs1.pdf.

P. L. Combettes (2013*b*), Systems of structured monotone inclusions: Duality, algorithms, and applications, *SIAM J. Optim.* **23**, 2420–2447.

P. L. Combettes (2018), Monotone operator theory in convex optimization, *Math. Program.* **B170**, 177–206.

P. L. Combettes (2023), Resolvent and proximal compositions, *Set-Valued Var. Anal.* **31**, art. 22.

P. L. Combettes and P. Bondon (1999), Hard-constrained inconsistent signal feasibility problems, *IEEE Trans. Signal Process.* **47**, 2460–2468.

P. L. Combettes and J. Eckstein (2018), Asynchronous block-iterative primal–dual decomposition methods for monotone inclusions, *Math. Program.* **B168**, 645–672.

P. L. Combettes and L. E. Glaudin (2017), Quasinonexpansive iterations on the affine hull of orbits: From Mann's mean value algorithm to inertial methods, *SIAM J. Optim.* **27**, 2356–2380.

P. L. Combettes and L. E. Glaudin (2019), Proximal activation of smooth functions in splitting algorithms for convex image recovery, *SIAM J. Imaging Sci.* **12**, 1905–1935.

P. L. Combettes and L. E. Glaudin (2021), Solving composite fixed point problems with block updates, *Adv. Nonlinear Anal.* **10**, 1154–1177.

P. L. Combettes and S. A. Hirstoaga (2005), Equilibrium programming in Hilbert spaces, *J. Nonlinear Convex Anal.* **6**, 117–136.

P. L. Combettes and C. L. Müller (2020), Perspective maximum likelihood-type estimation via proximal decomposition, *Electron. J. Statist.* **14**, 207–238.

P. L. Combettes and C. L. Müller (2021), Regression models for compositional data: General log-contrast formulations, proximal optimization, and microbiome data applications, *Statist. Biosci.* **13**, 217–242.

P. L. Combettes and J.-C. Pesquet (2007), A Douglas–Rachford splitting approach to nonsmooth convex variational signal recovery, *IEEE J. Select. Topics Signal Process.* **1**, 564–574.

P. L. Combettes and J.-C. Pesquet (2011), Proximal splitting methods in signal processing, in *Fixed-Point Algorithms for Inverse Problems in Science and Engineering*, Vol. 49 of Springer Optimization and its Applications, Springer, pp. 185–212.

P. L. Combettes and J.-C. Pesquet (2012), Primal–dual splitting algorithm for solving inclusions with mixtures of composite, Lipschitzian, and parallel-sum type monotone operators, *Set-Valued Var. Anal.* **20**, 307–330.

P. L. Combettes and J.-C. Pesquet (2015), Stochastic quasi-Fejér block-coordinate fixed point iterations with random sweeping, *SIAM J. Optim.* **25**, 1221–1248.

P. L. Combettes and J.-C. Pesquet (2020), Deep neural network structures solving variational inequalities, *Set-Valued Var. Anal.* **28**, 491–518.

P. L. Combettes and J.-C. Pesquet (2021), Fixed point strategies in data science, *IEEE Trans. Signal Process.* **69**, 3878–3905.

P. L. Combettes and B. C. Vũ (2013), Variable metric quasi-Fejér monotonicity, *Nonlinear Anal.* **78**, 17–31.

P. L. Combettes and B. C. Vũ (2014), Variable metric forward–backward splitting with applications to monotone inclusions in duality, *Optimization* **63**, 1289–1318.

P. L. Combettes and V. R. Wajs (2005), Signal recovery by proximal forward–backward splitting, *Multiscale Model. Simul.* **4**, 1168–1200.

P. L. Combettes and Z. C. Woodstock (2022), A variational inequality model for the construction of signals from inconsistent nonlinear equations, *SIAM J. Imaging Sci.* **15**, 84–109.

P. L. Combettes and I. Yamada (2015), Compositions and convex combinations of averaged nonexpansive operators, *J. Math. Anal. Appl.* **425**, 55–70.

P. L. Combettes, Đinh Dũng and B. C. Vũ (2010), Dualization of signal recovery problems, *Set-Valued Var. Anal.* **18**, 373–404.

P. L. Combettes, Đinh Dũng and B. C. Vũ (2011), Proximity for sums of composite functions, *J. Math. Anal. Appl.* **380**, 680–688.

P. L. Combettes, S. Salzo and S. Villa (2018), Consistent learning by composite proximal thresholding, *Math. Program.* **B167**, 99–127.

L. Condat (2013), A primal–dual splitting method for convex optimization involving Lipschitzian, proximable and linear composite terms, *J. Optim. Theory Appl.* **158**, 460–479.

L. Condat, D. Kitahara, A. Contreras and A. Hirabayashi (2023), Proximal splitting algorithms for convex optimization: A tour of recent advances, with new twists, *SIAM Rev.* **65**, 375–435.

H. B. Curry (1944), The method of steepest descent for non-linear minimization problems, *Quart. Appl. Math.* **2**, 258–261.

S. Dafermos (1980), Traffic equilibrium and variational inequalities, *Transport. Sci.* **14**, 42–54.

G. Darboux (1875), Mémoire sur les fonctions discontinues, *Ann. Sci. École Normale Sup.* **4**, 57–112.

I. Daubechies, M. Defrise and C. De Mol (2004), An iterative thresholding algorithm for linear inverse problems with a sparsity constraint, *Comm. Pure Appl. Math.* **57**, 1413–1457.

D. Davis and W. Yin (2017), A three-operator splitting scheme and its optimization applications, *Set-Valued Var. Anal.* **25**, 829–858.

C. A. Desoer and M. Vidyasagar (1975), *Feedback Systems: Input–Output Properties*, Academic.

C. A. Desoer and F. F. Wu (1974), Nonlinear monotone networks, *SIAM J. Appl. Math.* **26**, 315–333.

N. Dexter, H. Tran and C. G. Webster (2022), On the strong convergence of forward–backward splitting in reconstructing jointly sparse signals, *Set-Valued Var. Anal.* **30**, 543–557.

V. Doležal (1979*a*), Feedback systems described by monotone operators, *SIAM J. Control Optim.* **17**, 339–364.

V. Doležal (1979*b*), *Monotone Operators and Applications in Control and Network Theory*, Elsevier.

Q.-L. Dong, Y. J. Cho, S. He, P. M. Pardalos and T. M. Rassias (2022), *The Krasnosel'skiĭ–Mann Iterative Method: Recent Progress and Applications*, Springer.

Y. Dong (2005), An LS-free splitting method for composite mappings, *Appl. Math. Lett.* **18**, 843–848.

J. Douglas (1955), On the numerical integration of $\partial^2 u/\partial x^2 + \partial^2 u/\partial y^2 = \partial u/\partial t$ by implicit methods, *J. Soc. Indust. Appl. Math.* **3**, 42–65.

J. Douglas and H. H. Rachford (1956), On the numerical solution of heat conduction problems in two or three space variables, *Trans. Amer. Math. Soc.* **82**, 421–439.

R. J. Duffin (1946), Nonlinear networks I, *Bull. Amer. Math. Soc.* **52**, 833–838.

R. J. Duffin (1947), Nonlinear networks IIa, *Bull. Amer. Math. Soc.* **53**, 963–971.

R. J. Duffin (1948), Nonlinear networks IIb, *Bull. Amer. Math. Soc.* **54**, 119–127.

R. L. Dykstra (1983), An algorithm for restricted least squares regression, *J. Amer. Statist. Assoc.* **78**, 837–842.

B. C. Eaves (1984), Subdivisions from primal and dual cones and polytopes, *Linear Algebra Appl.* **62**, 277–285.

J. Eckstein (1993), Nonlinear proximal point algorithms using Bregman functions, with applications to convex programming, *Math. Oper. Res.* **18**, 202–226.

J. Eckstein (1994), Some saddle-function splitting methods for convex programming, *Optim. Methods Softw.* **4**, 75–83.

J. Eckstein (2017), A simplified form of block-iterative operator splitting and an asynchronous algorithm resembling the multi-block alternating direction method of multipliers, *J. Optim. Theory Appl.* **173**, 155–182.

J. Eckstein and D. P. Bertsekas (1992), On the Douglas–Rachford splitting method and the proximal point algorithm for maximal monotone operators, *Math. Program.* **55**, 293–318.

J. Eckstein and M. C. Ferris (1999), Smooth methods of multipliers for complementarity problems, *Math. Program.* **A86**, 65–90.

J. Eckstein and B. F. Svaiter (2008), A family of projective splitting methods for the sum of two maximal monotone operators, *Math. Program.* **111**, 173–199.

J. Eckstein and B. F. Svaiter (2009), General projective splitting methods for sums of maximal monotone operators, *SIAM J. Control Optim.* **48**, 787–811.

J. Eckstein, J.-P. Watson and D. L. Woodruff (2023), Projective hedging algorithms for multistage stochastic programming, supporting distributed and asynchronous implementation, *Oper. Res.* Available at doi:10.1287/opre.2022.0228.

B. Eicke (1992), Iteration methods for convexly constrained ill-posed problems in Hilbert space, *Numer. Funct. Anal. Optim.* **13**, 413–429.

I. Ekeland and R. Temam (1974), *Analyse Convexe et Problèmes Variationnels*, Dunod. English translation: *Convex Analysis and Variational Problems*, SIAM (1999).

I. I. Eremin (1968*a*), Methods of Fejér approximations in convex programming, *Mat. Zametki* **3**, 217–234.

I. I. Eremin (1968*b*), On the speed of convergence in the method of Fejér approximations, *Mat. Zametki* **4**, 53–62.

Y. M. Ermol'ev and A. D. Tuniev (1968), Random Fejér and quasi-Fejér sequences, *Theory of Optimal Solutions – Akad. Nauk Ukrainskoĭ SSR Kiev* **2**, 76–83.

F. Facchinei and J.-S. Pang (2003), *Finite-Dimensional Variational Inequalities and Complementarity Problems*, Springer.

F. Facchinei, A. Fischer and V. Piccialli (2007), On generalized Nash games and variational inequalities, *Oper. Res. Lett.* **35**, 159–164.

L. Fejér (1922), Über die Lage der Nullstellen von Polynomen, die aus Minimumforderungen gewisser Art entspringen, *Math. Ann.* **85**, 41–48.

G. Fichera (1963), Sul problema elastostatico di signorini con ambigue condizioni al contorno, *Atti Accad. Naz. Lincei Rend. Ser. VIII* **34**, 138–142.

M. A. T. Figueiredo and R. D. Nowak (2003), An EM algorithm for wavelet-based image restoration, *IEEE Trans. Image Process.* **12**, 906–916.

M. Fortin and R. Glowinski, eds (1983), *Augmented Lagrangian Methods: Applications to the Numerical Solution of Boundary-Value Problems*, North-Holland.

A. Froda (1929), *Sur la Distribution des Propriétés de Voisinage des Fonctions de Variables Réelles*, Hermann.

M. Fukushima (1996), The primal Douglas–Rachford splitting algorithm for a class of monotone mappings with application to the traffic equilibrium problem, *Math. Program.* **72**, 1–15.

D. Gabay (1983), Applications of the method of multipliers to variational inequalities, in *Augmented Lagrangian Methods: Applications to the Numerical Solution of Boundary-Value Problems* (M. Fortin and R. Glowinski, eds), North-Holland, pp. 299–331.

D. Gabay and B. Mercier (1976), A dual algorithm for the solution of nonlinear variational problems via finite elements approximations, *Comput. Math. Appl.* **2**, 17–40.

N. Gaffke and R. Mathar (1989), A cyclic projection algorithm via duality, *Metrika* **36**, 29–54.

S. Gandy, B. Recht and I. Yamada (2011), Tensor completion and low-n-rank tensor recovery via convex optimization, *Inverse Problems* **27**, art. 025010.

G. Garrigos, L. Rosasco and S. Villa (2023), Convergence of the forward–backward algorithm: Beyond the worst-case with the help of geometry, *Math. Program.* **A198**, 937–996.

C. F. Gauss (1809), *Theoria Motus Corporum Coelestium*, Perthes and Besser.

P. Gautam, D. R. Sahu, A. Dixit and T. Som (2021), Forward–backward-half forward dynamical systems for monotone inclusion problems with application to v-GNE, *J. Optim. Theory Appl.* **190**, 491–523.

A. Genel and J. Lindenstrauss (1975), An example concerning fixed points, *Israel J. Math.* **22**, 81–86.

A. Ghizzetti, ed. (1969), *Theory and Applications of Monotone Operators*, Edizioni Oderisi.

N. Ghoussoub (2009), *Self-Dual Partial Differential Systems and Their Variational Principles*, Springer.

P. Giselsson (2021), Nonlinear forward–backward splitting with projection correction, *SIAM J. Optim.* **31**, 2199–2226.

R. Glowinski and P. Le Tallec, eds (1989), *Augmented Lagrangian and Operator-Splitting Methods in Nonlinear Mechanics*, SIAM.

R. Glowinski and A. Marrocco (1974), Sur l'approximation, par éléments finis d'ordre un, et la résolution, par pénalisation-dualité, d'une classe de problèmes de Dirichlet non linéaires, *C. R. Acad. Sci. Paris* **A278**, 1649–1652. See also *RAIRO Anal. Numer.* **9**, 41–76 (1975).

R. Glowinski, S. J. Osher and W. Yin, eds (2016), *Splitting Methods in Communication, Imaging, Science, and Engineering*, Springer.

D. Goeleven (2017), *Complementarity and Variational Inequalities in Electronics*, Academic.

M. Goldburg and R. J. Marks II (1985), Signal synthesis in the presence of an inconsistent set of constraints, *IEEE Trans. Circuits Syst.* **32**, 647–663.

A. A. Goldstein (1964), Convex programming in Hilbert space, *Bull. Amer. Math. Soc.* **70**, 709–710.

M. Golomb (1935), Zur Theorie der nichtlinearen Integralgleichungen, Integralgleichungssysteme und allgemeinen Funktionalgleichungen, *Math. Z.* **39**, 45–75.

M. Golomb (1936), Über Systeme von nichtlinearen Integralgleichungen, *Publ. Math. Univ. Belgrade* **5**, 52–83.

E. G. Gol'shtein (1987), A general approach to decomposition of optimization systems, *Sov. J. Comput. Syst. Sci.* **25**, 105–114.

E. G. Gol'shtein and N. V. Tret'yakov (1979), Modified Lagrangians in convex programming and their generalizations, *Math. Program. Studies* **10**, 86–97.

E. G. Gol'shtein and N. V. Tret'yakov (1996), *Modified Lagrangians and Monotone Maps in Optimization*, Wiley.

C. W. Groetsch (1972), A note on segmenting Mann iterates, *J. Math. Anal. Appl.* **40**, 369–372.

L. G. Gubin, B. T. Polyak and E. V. Raik (1967), The method of projections for finding the common point of convex sets, *Comput. Math. Math. Phys.* **7**, 1–24.

O. Güler (1991), On the convergence of the proximal point algorithm for convex minimization, *SIAM J. Control Optim.* **29**, 403–419.

H. Hahn (1921), *Theorie der reellen Funktionen*, Springer.

S. P. Han (1988), A successive projection method, *Math. Program.* **40**, 1–14.

Y. Haugazeau (1967), Sur la minimisation de formes quadratiques avec contraintes, *C. R. Acad. Sci. Paris* **A264**, 322–324.

Y. Haugazeau (1968), Sur les inéquations variationnelles et la minimisation de fonctionnelles convexes. Thesis, Université de Paris.

B. He and X. Yuan (2012), Convergence analysis of primal–dual algorithms for a saddle-point problem: From contraction perspective, *SIAM J. Imaging Sci.* **5**, 119–149.

Y. He and R. D. C. Monteiro (2015), Accelerating block-decomposition first-order methods for solving composite saddle-point and two-player Nash equilibrium problems, *SIAM J. Optim.* **25**, 2182–2211.

M. R. Hestenes (1969), Multiplier and gradient methods, *J. Optim. Theory Appl.* **4**, 303–320.

H. Hu, R. Sotirov and H. Wolkowicz (2023), Facial reduction for symmetry reduced semidefinite and doubly nonnegative programs, *Math. Program.* **A200**, 475–529.

H. S. Hundal (2004), An alternating projection that does not converge in norm, *Nonlinear Anal.* **57**, 35–61.

H. Idrissi, O. Lefebvre and C. Michelot (1989), Applications and numerical convergence of the partial inverse method, in *Optimization*, Vol. 1405 of Lecture Notes in Mathematics, Springer, pp. 39–54.

R. Jenatton, J. Mairal, G. Obozinski and F. Bach (2011), Proximal methods for hierarchical sparse coding, *J. Mach. Learn. Res.* **12**, 2297–2334.

J. L. W. V. Jensen (1906), Sur les fonctions convexes et les inégalités entre les valeurs moyennes, *Acta Math.* **30**, 175–193.

P. R. Johnstone and J. Eckstein (2019), Convergence rates for projective splitting, *SIAM J. Optim.* **29**, 1931–1957.

P. R. Johnstone and J. Eckstein (2020), Projective splitting with forward steps only requires continuity, *Optim. Lett.* **14**, 229–247.

P. R. Johnstone and J. Eckstein (2021), Single-forward-step projective splitting: Exploiting cocoercivity, *Comput. Optim. Appl.* **78**, 125–166.

P. R. Johnstone and J. Eckstein (2022), Projective splitting with forward steps, *Math. Program.* **A191**, 631–670.

J. L. Joly and P. J. Laurent (1971), Stability and duality in convex minimization problems, *Rev. Fr. Inform. Rech. Opér.* **5**, 3–42.

R. I. Kačurovskiĭ (1960), Monotone operators and convex functionals, *Uspekhi Mat. Nauk* **15**, 213–215.

R. I. Kačurovskiĭ (1968), Non-linear monotone operators in Banach spaces, *Russian Math. Surveys* **23**, 117–165.

T. Kato (1980), *Perturbation Theory for Linear Operators*, second edition, Springer.

J. E. Kelley (1960), The cutting-plane method for solving convex programs, *J. Soc Indust. Appl. Math.* **8**, 703–712.

R. B. Kellogg (1969), A nonlinear alternating direction method, *Math. Comp.* **23**, 23–27.

D. Kinderlehrer and G. Stampacchia (1980), *An Introduction to Variational Inequalities and Their Applications*, Academic.

Y. Kōmura (1967), Nonlinear semi-groups in Hilbert space, *J. Math. Soc. Japan* **19**, 493–507.

G. M. Korpelevič (1976), The extragradient method for finding saddle points and other problems, *Èkonom. i Mat. Metody* **12**, 747–756.

M. A. Krasnosel'skiĭ (1955), Two remarks on the method of successive approximations, *Uspekhi Mat. Nauk* **10**, 123–127.

A. V. Kryanev (1973), The solution of incorrectly posed problems by methods of successive approximations, *Soviet Math. Dokl.* **14**, 673–676.

P. Latafat and P. Patrinos (2017), Asymmetric forward–backward-adjoint splitting for solving monotone inclusions involving three operators, *Comput. Optim. Appl.* **68**, 57–93.

P. J. Laurent and B. Martinet (1970), Méthodes duales pour le calcul du minimum d'une fonction convexe sur une intersection de convexes, in *Symposium on Optimization*, Vol. 132 of Lecture Notes in Mathematics, Springer, pp. 159–180.

J. Lawrence and J. E. Spingarn (1987), On fixed points of non-expansive piecewise isometric mappings, *Proc. London Math. Soc.* **55**, 605–624.

A. M. Legendre (1805), *Nouvelles Méthodes pour la Détermination des Orbites des Comètes*, Firmin Didot.

B. Lemaire (1989), The proximal algorithm, in *New Methods in Optimization and Their Industrial Uses* (J. P. Penot, ed.), Vol. 87 of International Series of Numerical Mathematics, Birkhäuser, pp. 73–87.

B. Lemaire (1996), Stability of the iteration method for nonexpansive mappings, *Serdica Math. J.* **22**, 331–340.

B. Lemaire (1997), Which fixed point does the iteration method select?, in *Recent Advances in Optimization*, Vol. 452 of Lecture Notes in Economics and Mathematical Systems, Springer, pp. 154–167.

A. Lenoir and P. Mahey (2017), A survey on operator splitting and decomposition of convex programs, *RAIRO-Oper. Res.* **51**, 17–41.

J. Leray and J.-L. Lions (1965), Quelques résultats de Višik sur les problèmes elliptiques nonlinéaires par les méthodes de Minty–Browder, *Bull. Soc. Math. France* **93**, 97–107.

E. S. Levitin and B. T. Polyak (1966), Constrained minimization methods, *USSR Comput. Math. Math. Phys.* **6**, 1–50.

J. Lieutaud (1969*a*), Approximations d'opérateurs monotones par des méthodes de splitting, in *Theory and Applications of Monotone Operators* (A. Ghizzetti, ed.), Edizioni Oderisi, pp. 259–264.

J. Lieutaud (1969*b*), Approximation d'opérateurs par des méthodes de décomposition. Thesis, Université de Paris.

S. Lindstrom and B. Sims (2021), Survey: Sixty years of Douglas–Rachford, *J. Aust. Math. Soc.* **110**, 333–370.

J.-L. Lions (1969), *Quelques Méthodes de Résolution des Problèmes aux Limites Non Linéaires*, Dunod.

J.-L. Lions, ed. (2010), *Numerical Analysis of Partial Differential Equations*, Springer.

P.-L. Lions (1978), Une méthode itérative de résolution d'une inéquation variationnelle, *Israel J. Math.* **31**, 204–208.

P.-L. Lions and B. Mercier (1979), Splitting algorithms for the sum of two nonlinear operators, *SIAM J. Numer. Anal.* **16**, 964–979.

J. Liu and S. J. Wright (2015), Asynchronous stochastic coordinate descent: Parallelism and convergence properties, *SIAM J. Optim.* **25**, 351–376.

H. Lu, R. M. Freund and Y. Nesterov (2018), Relatively smooth convex optimization by first-order methods, and applications, *SIAM J. Optim.* **28**, 333–354.

M. P. Machado (2018), On the complexity of the projective splitting and Spingarn's methods for the sum of two maximal monotone operators, *J. Optim. Theory Appl.* **178**, 153–190.

P. Machado and M. R. Sicre (2023), A projective splitting method for monotone inclusions: Iteration-complexity and application to composite optimization, *J. Optim. Theory Appl.* **198**, 552–587.

P. Mahey and D. T. Pham (1993), Partial regularization of the sum of two maximal monotone operators, *RAIRO Modélisation Math. Analyse Numér.* **27**, 375–392.

P. Mahey, S. Oualibouch and P. Dinh Tao (1995), Proximal decomposition on the graph of a maximal monotone operator, *SIAM J. Optim.* **5**, 454–466.

Y. Malitsky and M. K. Tam (2023), Resolvent splitting for sums of monotone operators with minimal lifting, *Math. Program.* **A201**, 231–262.

W. R. Mann (1953), Mean value methods in iteration, *Proc. Amer. Math. Soc.* **4**, 506–510.

B. Martinet (1970), Régularisation d'inéquations variationnelles par approximations successives, *Rev. Fr. Inform. Rech. Opér.* **4**, 154–158.

B. Martinet (1972), Détermination approchée d'un point fixe d'une application pseudo-contractante: Cas de l'application prox, *C. R. Acad. Sci. Paris* **A274**, 163–165.

J. E. Martínez-Legaz and A. Seeger (1994), A general cone decomposition theory based on efficiency, *Math. Program.* **65**, 1–20.

B. Mercier (1979), *Topics in Finite Element Solution of Elliptic Problems*, Vol. 63 of Lectures on Mathematics, Tata Institute of Fundamental Research, Bombay.

B. Mercier (1980), *Inéquations Variationnelles de la Mécanique*, Vol. 80.01 of Publications Mathématiques d'Orsay, Université de Paris-XI, Orsay, France.

C. A. Micchelli, J. M. Morales and M. Pontil (2013), Regularizers for structured sparsity, *Adv. Comput. Math.* **38**, 455–489.

W. Millar (1951), Some general theorems for non-linear systems possessing resistance, *London Edin. Dublin Philos. Mag.* **42**, 1150–1160.

G. J. Minty (1960), Monotone networks, *Proc. R. Soc. Lond. A* **57**, 194–212.

G. J. Minty (1961), Solving steady-state nonlinear networks of 'monotone' elements, *IRE Trans. Circuit Theory* **8**, 99–104.

G. J. Minty (1962), Monotone (nonlinear) operators in Hilbert space, *Duke Math. J.* **29**, 341–346.

G. J. Minty (1963), On a 'monotonicity' method for the solution of nonlinear equations in Banach spaces, *Proc. Nat. Acad. Sci. USA* **50**, 1038–1041.

G. J. Minty (1964), On the monotonicity of the gradient of a convex function, *Pacific J. Math.* **14**, 243–247.

G. J. Minty (1969), On some aspects of the theory of monotone operators, in *Theory and Applications of Monotone Operators* (A. Ghizzetti, ed.), Edizioni Oderisi, pp. 67–82.

K. Mishchenko, F. Iutzeler and J. Malick (2020), A distributed flexible delay-tolerant proximal gradient algorithm, *SIAM J. Optim.* **30**, 933–959.

T. Mizoguchi and I. Yamada (2019), Hypercomplex tensor completion via convex optimization, *IEEE Trans. Signal Process.* **67**, 4078–4092.

A. Mokhtari, M. Gürbüzbalaban and A. Ribeiro (2018), Surpassing gradient descent provably: A cyclic incremental method with linear convergence rate, *SIAM J. Optim.* **28**, 1420–1447.

J. J. Moreau (1962), Fonctions convexes duales et points proximaux dans un espace hilbertien, *C. R. Acad. Sci. Paris Sér. A Math.* **255**, 2897–2899.

J. J. Moreau (1965), Proximité et dualité dans un espace hilbertien, *Bull. Soc. Math. France* **93**, 273–299.

J. J. Moreau (1966–1967), *Fonctionnelles Convexes*, Vol. 2 of Séminaire Jean Leray sur les Équations aux Dérivées Partielles, Collège de France, Paris.

U. Mosco (1972), Dual variational inequalities, *J. Math. Anal. Appl.* **40**, 202–206.

T. S. Motzkin and I. J. Schoenberg (1954), The relaxation method for linear inequalities, *Canad. J. Math.* **6**, 393–404.

A. Moudafi (2000), On the regularization of the sum of two maximal monotone operators, *Nonlinear Anal.* **42**, 1203–1208.

A. Moudafi and M. Théra (1999), Proximal and dynamical approaches to equilibrium problems, in *Ill-Posed Variational Problems and Regularization Techniques*, Vol. 477 of Lecture Notes in Economics and Mathematical Systems, Springer, pp. 187–201.

A. Nemirovski (2004), Prox-method with rate of convergence $O(1/t)$ for variational inequalities with Lipschitz continuous monotone operators and smooth convex-concave saddle point problems, *SIAM J. Optim.* **15**, 229–251.

A. Neubauer (1988), Tikhonov-regularization of ill-posed linear operator equations on closed convex sets, *J. Approx. Theory* **53**, 304–320.

Q. V. Nguyen (2017), Forward–backward splitting with Bregman distances, *Vietnam J. Math.* **45**, 519–539.

D. O'Connor and L. Vandenberghe (2014), Primal–dual decomposition by operator splitting and applications to image deblurring, *SIAM J. Imaging Sci.* **7**, 1724–1754.

D. E. Oliveira, H. Wolkowicz and Y. Xu (2018), ADMM for the SDP relaxation of the QAP, *Math. Program. Comput.* **10**, 631–658.

N. Papadakis, G. Peyré and E. Oudet (2014), Optimal transport with proximal splitting, *SIAM J. Imaging Sci.* **7**, 212–238.

J. M. Papakonstantinou and R. A. Tapia (2013), Origin and evolution of the secant method in one dimension, *Amer. Math. Monthly* **120**, 500–518.

D. Pascali and S. Sburlan (1978), *Nonlinear Mappings of Monotone Type*, Editura Academiei, Bucharest.

G. B. Passty (1979), Ergodic convergence to a zero of the sum of monotone operators in Hilbert space, *J. Math. Anal. Appl.* **72**, 383–390.

R. Pathak and M. J. Wainwright (2020), FedSplit: An algorithmic framework for fast federated optimization, in *Advances in Neural Information Processing Systems 33* (H. Larochelle *et al.*, eds), Curran Associates, pp. 7057–7066.

D. W. Peaceman and H. H. Rachford (1955), The numerical solution of parabolic and elliptic differential equations, *J. Soc. Indust. Appl. Math.* **3**, 28–41.

T. Pennanen (2000), Dualization of generalized equations of maximal monotone type, *SIAM J. Optim.* **10**, 809–835.

T. Pennanen (2002), A splitting method for composite mappings, *Numer. Funct. Anal. Optim.* **23**, 875–890.

J.-C. Pesquet and A. Repetti (2015), A class of randomized primal–dual algorithms for distributed optimization, *J. Nonlinear Convex Anal.* **16**, 2453–2490.

J.-C. Pesquet, A. Repetti, M. Terris and Y. Wiaux (2021), Learning maximally monotone operators for image recovery, *SIAM J. Imaging Sci.* **14**, 1206–1237.

W. V. Petryshyn (1966), On the extension and solution of nonlinear operator equations, *Illinois J. Math.* **10**, 255–274.

D. L. Phillips (1962), A technique for the numerical solution of certain integral equations of the first kind, *J. Assoc. Comput. Mach.* **9**, 84–97.

R. S. Phillips (1959), Dissipative operators and hyperbolic systems of partial differential equations, *Trans. Amer. Math. Soc.* **90**, 193–254.

G. Pierra (1976), Éclatement de contraintes en parallèle pour la minimisation d'une forme quadratique, in *Optimization Techniques: Modeling and Optimization in the Service of Man 2*, Vol. 41 of Lecture Notes in Computer Science, Springer, pp. 200–218.

G. Pierra (1984), Decomposition through formalization in a product space, *Math. Program.* **28**, 96–115.

B. T. Polyak (1964), Some methods of speeding up the convergence of iteration methods, *USSR Comput. Math. Math. Phys.* **4**, 1–17.

L. C. Potter and K. S. Arun (1993), A dual approach to linear inverse problems with convex constraints, *SIAM J. Control Optim.* **31**, 1080–1092.

M. J. D. Powell (1969), A method for nonlinear constraints in minimization problems, in *Optimization* (R. Fletcher, ed.), Academic, pp. 283–298.

X. Qin and N. T. An (2019), Smoothing algorithms for computing the projection onto a Minkowski sum of convex sets, *Comput. Optim. Appl.* **74**, 821–850.

H. Raguet (2019), A note on the forward–Douglas–Rachford splitting for monotone inclusion and convex optimization, *Optim. Lett.* **13**, 717–740.

H. Raguet and L. Landrieu (2015), Preconditioning of a generalized forward–backward splitting and application to optimization on graphs, *SIAM J. Imaging Sci.* **8**, 2706–2739.

H. Raguet, J. Fadili and G. Peyré (2013), A generalized forward–backward splitting, *SIAM J. Imaging Sci.* **6**, 1199–1226.

E. Raik (1967), Fejér type methods in Hilbert space, *Eesti NSV Tead. Akad. Toimetised Füüs.-Mat.* **16**, 286–293.

E. Raik (1969), A class of iterative methods with Fejér-monotone sequences, *Eesti NSV Tead. Akad. Toimetised Füüs.-Mat.* **18**, 22–26.

H. J. Reich (1961), *Functional Circuits and Oscillators*, Van Nostrand.

S. Reich, M. T. Truong and T. N. H. Mai (2020), The split feasibility problem with multiple output sets in Hilbert spaces, *Optim. Lett.* **14**, 2335–2353.

A. Renaud and G. Cohen (1997), An extension of the auxiliary problem principle to nonsymmetric auxiliary operators, *ESAIM Control Optim. Calc. Var.* **2**, 281–306.

S. M. Robinson (1998), A reduction method for variational inequalities, *Math. Program.* **80**, 161–169.

S. M. Robinson (1999), Composition duality and maximal monotonicity, *Math. Program.* **85**, 1–13.

S. M. Robinson (2001), Generalized duality in variational analysis, in *Advances in Convex Analysis and Global Optimization* (N. Hadjisavvas and P. M. Pardalos, eds), Kluwer, pp. 205–219.

R. T. Rockafellar (1967), Duality and stability in extremum problems involving convex functions, *Pacific J. Math.* **21**, 167–187.

R. T. Rockafellar (1969), Convex functions and duality in optimization problems and dynamics, in *Mathematical Systems Theory and Economics I* (H. W. Kuhn and G. P. Szegö, eds), Springer, pp. 117–141.

R. T. Rockafellar (1970*a*), *Convex Analysis*, Princeton University Press.

R. T. Rockafellar (1970*b*), Monotone operators associated with saddle-functions and minimax problems, in *Nonlinear Functional Analysis, Part 1* (F. E. Browder, ed.), AMS, pp. 241–250.

R. T. Rockafellar (1971), Saddle-points and convex analysis, in *Differential Games and Related Topics* (H. W. Kuhn and G. P. Szegö, eds), North-Holland, pp. 109–127.

R. T. Rockafellar (1973), The multiplier method of Hestenes and Powell applied to convex programming, *J. Optim. Theory Appl.* **12**, 555–562.

R. T. Rockafellar (1974), *Conjugate Duality and Optimization*, SIAM.

R. T. Rockafellar (1976*a*), Augmented Lagrangians and applications of the proximal point algorithm in convex programming, *Math. Oper. Res.* **1**, 97–116.

R. T. Rockafellar (1976*b*), Monotone operators and the proximal point algorithm, *SIAM J. Control Optim.* **14**, 877–898.

R. T. Rockafellar (1984), *Network Flows and Monotropic Optimization*, Wiley.

R. T. Rockafellar (1995), Monotone relations and network equilibrium, in *Variational Inequalities and Network Equilibrium Problems* (F. Giannessi and A. Maugeri, eds), Plenum, pp. 271–288.

R. T. Rockafellar (2024), Generalizations of the proximal method of multipliers in convex optimization, *Comput. Optim. Appl.* **87**, 219–247.

R. T. Rockafellar and J. Sun (2019), Solving monotone stochastic variational inequalities and complementarity problems by progressive hedging, *Math. Program.* **A174**, 453–471.

R. T. Rockafellar and R. J. B. Wets (1991), Scenarios and policy aggregation in optimization under uncertainty, *Math. Oper. Res.* **16**, 1–29.

E. K. Ryu (2020), Uniqueness of DRS as the 2 operator resolvent-splitting and impossibility of 3 operator resolvent-splitting, *Math. Program.* **A182**, 233–273.

E. K. Ryu and B. C. Vũ (2020), Finding the forward–Douglas–Rachford-forward method, *J. Optim. Theory Appl.* **184**, 858–876.

E. K. Ryu, Y. Liu and W. Yin (2019), Douglas–Rachford splitting and ADMM for pathological convex optimization, *Comput. Optim. Appl.* **74**, 747–778.

E. K. Ryu, A. B. Taylor, C. Bergeling and P. Giselsson (2020), Operator splitting performance estimation: Tight contraction factors and optimal parameter selection, *SIAM J. Optim.* **30**, 2251–2271.

S. Salzo and S. Villa (2022), Parallel random block-coordinate forward–backward algorithm: A unified convergence analysis, *Math. Program.* **A193**, 225–269.

A. Seeger (1998), Alternating projection and decomposition with respect to two convex sets, *Math. Japon.* **47**, 273–280.

R. Shefi and M. Teboulle (2014), Rate of convergence analysis of decomposition methods based on the proximal method of multipliers for convex minimization, *SIAM J. Optim.* **24**, 269–297.

R. E. Showalter (1997), *Monotone Operators in Banach Space and Nonlinear Partial Differential Equations*, American Mathematical Society.

M. Sibony (1970), Méthodes itératives pour les équations et inéquations aux dérivées partielles non linéaires de type monotone, *Calcolo* **7**, 65–183.

M. R. Sicre (2020), On the complexity of a hybrid proximal extragradient projective method for solving monotone inclusion problems, *Comput. Optim. Appl.* **76**, 991–1019.

S. Singh, G. Weiss and M. Tucsnak (2022), A class of incrementally scattering-passive nonlinear systems, *Automatica* **142**, art. 110369.

M. V. Solodov (2004), A class of decomposition methods for convex optimization and monotone variational inclusions via the hybrid inexact proximal point framework, *Optim. Methods Softw.* **19**, 557–575.

M. V. Solodov and B. F. Svaiter (1999*a*), A hybrid approximate extragradient-proximal point algorithm using the enlargement of a maximal monotone operator, *Set-Valued Var. Anal.* **7**, 323–345.

M. V. Solodov and B. F. Svaiter (1999*b*), A hybrid projection-proximal point algorithm, *J. Convex Anal.* **6**, 59–70.

M. V. Solodov and B. F. Svaiter (2000), Forcing strong convergence of proximal point iterations in a Hilbert space, *Math. Program.* **A87**, 189–202.

M. V. Solodov and B. F. Svaiter (2001), A unified framework for some inexact proximal point algorithms, *Numer. Funct. Anal. Optim.* **22**, 1013–1035.

J. E. Spingarn (1983), Partial inverse of a monotone operator, *Appl. Math. Optim.* **10**, 247–265.

J. E. Spingarn (1985), Applications of the method of partial inverses to convex programming: Decomposition, *Math. Program.* **32**, 199–223.

J. E. Spingarn (1987), A projection method for least-squares solutions to overdetermined systems of linear inequalities, *Linear Algebra Appl.* **86**, 211–236.

G. Steidl and T. Teuber (2010), Removing multiplicative noise by Douglas–Rachford splitting methods, *J. Math. Imaging Vision* **36**, 168–184.

B. F. Svaiter (2011), On weak convergence of the Douglas–Rachford method, *SIAM J. Control Optim.* **49**, 280–287.

B. F. Svaiter (2014), A class of Fejér convergent algorithms, approximate resolvents and the hybrid proximal-extragradient method, *J. Optim. Theory Appl.* **162**, 133–153.

M. Teboulle (1992), Entropic proximal mappings with applications to nonlinear programming, *Math. Oper. Res.* **17**, 670–690.

M. Teboulle (2018), A simplified view of first order methods for optimization, *Math. Program.* **B170**, 67–96.

B. D. H. Tellegen (1948), The gyrator, a new electric network element, *Philips Res. Rept.* **3**, 81–101.

K. K. Thekumparampil, P. Jain, P. Netrapalli and S. Oh (2019), Efficient algorithms for smooth minimax optimization, in *Advances in Neural Information Processing Systems 32* (H. Wallach *et al.*, eds), Curran Associates.

C. Traoré, S. Salzo and S. Villa (2023), Convergence of an asynchronous block-coordinate forward–backward algorithm for convex composite optimization, *Comput. Optim. Appl.* **86**, 303–344.

P. Tseng (1990), Further applications of a splitting algorithm to decomposition in variational inequalities and convex programming, *Math. Program.* **48**, 249–263.

P. Tseng (1991), Applications of a splitting algorithm to decomposition in convex programming and variational inequalities, *SIAM J. Control Optim.* **29**, 119–138.

P. Tseng (2000), A modified forward–backward splitting method for maximal monotone mappings, *SIAM J. Control Optim.* **38**, 431–446.

M. M. Vaĭnberg (1956), *Variatsionnye Metody Issledovaniya Nelineinykh Operatorov*, Gosudarstv. Izdat. Tehn.-Teor. Lit., Moscow. English translation: *Variational Methods for the Study of Non-Linear Operators*, Holden-Day (1964).

M. M. Vaĭnberg (1959), New theorems for non-linear operators and equations, *Dokl. Akad. Nauk SSSR* **129**, 1199–1202.

M. M. Vaĭnberg (1960), On the convergence of the method of steepest descent for nonlinear equations, *Dokl. Akad. Nauk SSSR* **130**, 9–12.

M. M. Vaĭnberg (1961), On the convergence of the process of steepest descent for nonlinear equations, *Sibirsk. Mat. Zh.* **2**, 201–220.

M. M. Vaĭnberg (1972), *Variational Method and Method of Monotone Operators in the Theory of Nonlinear Equations*, Nauka. English translation: Wiley (1973).

M. M. Vaĭnberg and R. I. Kačurovskiĭ (1959), On the variational theory of nonlinear operators and equations, *Dokl. Akad. Nauk SSSR* **129**, 1199–1202.

S. Vaiter, G. Peyré and J. Fadili (2018), Model consistency of partly smooth regularizers, *IEEE Trans. Inform. Theory* **64**, 1725–1737.

A. F. Veinott (1967), The supporting hyperplane method for unimodal programming, *Oper. Res.* **15**, 147–152.

L. A. Vese and S. J. Osher (2004), Image denoising and decomposition with total variation minimization and oscillatory functions, *J. Math. Imaging Vision* **20**, 7–18.

S. Villa, L. Rosasco, S. Mosci and A. Verri (2014), Proximal methods for the latent group lasso penalty, *Comput. Optim. Appl.* **58**, 381–407.

M. I. Vishik (1961), Boundary-value problems for quasilinear strongly elliptic systems of equations having divergence form, *Soviet Math. Dokl.* **2**, 643–647.

B. C. Vũ (2013), A splitting algorithm for dual monotone inclusions involving cocoercive operators, *Adv. Comput. Math.* **38**, 667–681.

X. Wang, J. Zhang and W. Zhang (2020), The distance between convex sets with Minkowski sum structure: Application to collision detection, *Comput. Optim. Appl.* **77**, 465–490.

E. Winston and J. Z. Kolter (2020), Monotone operator equilibrium networks, in *Advances in Neural Information Processing Systems 33* (H. Larochelle *et al.*, eds), Curran Associates, pp. 10718–10728.

J.-H. Won, J. Xu and K. Lange (2019), Projection onto Minkowski sums with application to constrained learning, in *Proceedings of the 36th International Conference on Machine Learning* (K. Chaudhuri and R. Salakhutdinov, eds), Vol. 97 of Proceedings of Machine Learning Research, PMLR, pp. 3642–3651.

S. J. Wright and B. Recht (2022), *Optimization for Data Analysis*, Cambridge University Press.

F. Xue (2023*a*), Equivalent resolvents of Douglas–Rachford splitting and other operator splitting algorithms: A unified degenerate proximal point analysis, *Optimization*. Available at doi:10.1080/02331934.2023.2231005.

F. Xue (2023*b*), A generalized forward–backward splitting operator: Degenerate analysis and applications, *Comput. Appl. Math.* **42**, art. 9.

X. Yan and J. Bien (2021), Rare feature selection in high dimensions, *J. Amer. Statist. Assoc.* **116**, 887–900.

P. Yi and S. Ching (2020), Synthesis of recurrent neural dynamics for monotone inclusion with application to Bayesian inference, *Neural Networks* **131**, 231–241.

T. Yoon and E. K. Ryu (2021), Accelerated algorithms for smooth convex-concave minimax problems with $\mathcal{O}(1/k^2)$ rate on squared gradient norm, in *Proceedings of the 38th International Conference on Machine Learning* (M. Meila and T. Zhang, eds), Vol. 139 of Proceedings of Machine Learning Research, PMLR, pp. 12098–12109.

D. C. Youla (1987), Mathematical theory of image restoration by the method of convex projections, in *Image Recovery: Theory and Application* (H. Stark, ed.), Academic Press, pp. 29–77.

Y. Yu, J. Peng, X. Han and A. Cui (2017), A primal Douglas–Rachford splitting method for the constrained minimization problem in compressive sensing, *Circuits Syst. Signal Process.* **36**, 4022–4049.

G. Zames (1966*a*), On the input–output stability of time-varying nonlinear feedback systems part I: Conditions derived using concepts of loop gain, conicity, and positivity, *IEEE Trans. Automat. Control* **11**, 228–238.

G. Zames (1966*b*), On the input–output stability of time-varying nonlinear feedback systems part II: Conditions involving circles in the frequency plane and sector nonlinearities, *IEEE Trans. Automat. Control* **11**, 465–476.

G. Zames and P. L. Falb (1968), Stability conditions for systems with monotone and slope-restricted nonlinearities, *SIAM J. Control* **6**, 89–108.

W. I. Zangwill (1969), *Nonlinear Programming: A Unified Approach*, Prentice Hall.

E. H. Zarantonello (1960), Solving functional equations by contractive averaging. Mathematical Research Center technical summary report no. 160, University of Wisconsin, Madison.

E. H. Zarantonello (1964), The closure of the numerical range contains the spectrum, *Bull. Amer. Math. Soc.* **70**, 781–787.

E. H. Zarantonello, ed. (1971), *Contributions to Nonlinear Functional Analysis*, Academic.

E. Zeidler (1990), *Nonlinear Functional Analysis and its Applications II/B: Nonlinear Monotone Operators*, Springer.

Acta Numerica (2024), pp. 633–713 Printed in the United Kingdom
doi:10.1017/S0962492923000089

Numerical analysis of physics-informed neural networks and related models in physics-informed machine learning

Tim De Ryck
Seminar for Applied Mathematics, ETH Zürich,
Rämistrasse 101, 8092 Zürich, Switzerland
E-mail: tim.deryck@math.ethz.ch

Siddhartha Mishra
Seminar for Applied Mathematics & ETH AI Center, ETH Zürich,
Rämistrasse 101, 8092 Zürich, Switzerland
E-mail: siddhartha.mishra@math.ethz.ch

Physics-informed neural networks (PINNs) and their variants have been very popular in recent years as algorithms for the numerical simulation of both forward and inverse problems for partial differential equations. This article aims to provide a comprehensive review of currently available results on the numerical analysis of PINNs and related models that constitute the backbone of physics-informed machine learning. We provide a unified framework in which analysis of the various components of the error incurred by PINNs in approximating PDEs can be effectively carried out. We present a detailed review of available results on approximation, generalization and training errors and their behaviour with respect to the type of the PDE and the dimension of the underlying domain. In particular, we elucidate the role of the regularity of the solutions and their stability to perturbations in the error analysis. Numerical results are also presented to illustrate the theory. We identify training errors as a key bottleneck which can adversely affect the overall performance of various models in physics-informed machine learning.

2020 Mathematics Subject Classification: Primary 65M15
Secondary 68T07, 35A35

CONTENTS

1	Introduction	634
2	Physics-informed machine learning	635
3	Analysis	656
4	Approximation error	659
5	Stability	674
6	Generalization	684
7	Training	691
8	Conclusion	706
Appendix: Sobolev spaces		707
References		707

1. Introduction

Machine learning, particularly deep learning (Goodfellow, Bengio and Courville 2016), has permeated into every corner of modern science, technology and society, whether it is computer vision, natural language understanding, robotics and automation, image and text generation or protein folding and prediction.

The application of deep learning to computational science and engineering, in particular the numerical solution of (partial) differential equations, has gained enormous momentum in recent years. One notable avenue highlighting this application falls within the framework of *supervised learning*, i.e. approximating the solutions of PDEs by deep learning models from (large amounts of) data about the underlying PDE solutions. Examples include high-dimensional parabolic PDEs (Han, Jentzen and E 2018 and references therein), parametric elliptic (Schwab and Zech 2019, Kutyniok, Petersen, Raslan and Schneider 2022) or hyperbolic PDEs (De Ryck and Mishra 2023, Lye, Mishra and Ray 2020) and learning the solution operator directly from data (Chen and Chen 1995, Lu *et al.* 2021*b*, Li *et al.* 2021, Raonić *et al.* 2023 and references therein). However, generating and accessing large amounts of data on solutions of PDEs requires either numerical methods or experimental data, both of which can be (prohibitively) expensive. Consequently, there is a need for so-called *unsupervised* or *semi-supervised* learning methods which require only small amounts of data.

Given this context, it would be useful to solve PDEs using machine learning models and methods, directly from the underlying physics (governing equations) and without having to access data about the underlying solutions. Such methods can be loosely termed as constituting *physics-informed machine learning*.

The most prominent contemporary models in physics-informed machine learning are *physics-informed neural networks* or PINNs. The idea behind PINNs is very simple: as neural networks are universal approximators of a large variety of function classes (even measurable functions), we consider the strong form of the PDE residual within the *ansatz space* of neural networks, and minimize this residual via (stochastic) gradient descent to obtain a neural network that approximates the solution of the underlying PDE. This framework was already considered in the 1990s, in Dissanayake and Phan-Thien (1994), Lagaris, Likas and Fotiadis (1998), Lagaris, Likas and Papageorgiou (2000) and references therein, and these papers should be considered as the progenitors of PINNs. However, the modern version of PINNs was introduced, and named more recently in Raissi and Karniadakis (2018) and Raissi, Perdikaris and Karniadakis (2019). Since then, there has been an explosive growth in the literature on PINNs and they have been applied in numerous settings, in the solution of both forward and inverse problems for PDEs and related equations (SDEs, SPDEs); see Karniadakis *et al.* (2021) and Cuomo *et al.* (2022) for extensive reviews of the available literature on PINNs.

However, PINNs are not the only models within the framework of physics-informed machine learning. One can consider other forms of the PDE residual such as the variational form resulting in VPINNs (Kharazmi, Zhang and Karniadakis 2019), the weak form resulting in wPINNs (De Ryck, Mishra and Molinaro 2024*c*), or minimizing the underlying energy resulting in the DeepRitz method (E and Yu 2018). Similarly, one can use Gaussian processes (Rasmussen 2003), DeepONets (Lu *et al.* 2021*b*) or neural operators (Kovachki *et al.* 2023) as ansatz spaces to realize alternative models for physics-informed machine learning.

Given the exponentially growing literature on PINNs and related models in physics-informed machine learning, it is essential to ask whether these methods possess rigorous mathematical guarantees on their performance, and whether one can analyse these methods in a manner that is analogous to the huge literature on the analysis of traditional numerical methods such as finite differences, finite elements, finite volumes and spectral methods. Although the overwhelming focus of research has been on the widespread applications of PINNs and their variants in different domains in science and engineering, a significant number of papers rigorously analysing PINNs have emerged in recent years. The aim of this article is to review the available literature on the *numerical analysis of PINNs and related models that constitute physics-informed machine learning.* Our goal is to critically analyse PINNs and its variants with a view to ascertaining when they can be applied and what are the limits to their applicability. To this end, we start in Section 2 by presenting the formulation for physics-informed machine-learning in terms of the underlying PDEs, their different forms of residuals and the approximating ansatz spaces. In Section 3 we outline the main components of the underlying errors with physics-informed machine learning; the resulting approximation, stability, generalization and training errors are analysed in Sections 4, 5, 6 and 7, respectively.

2. Physics-informed machine learning

In this section we set the stage for the rest of the article. In Section 2.1 we introduce an abstract PDE setting and consider a number of important examples of PDEs that will be used in the numerical analysis later on. Next, in Section 2.2, different model classes are introduced. In Section 2.3 we introduce variations of physics-informed learning based on different formulations of PDEs, of which the resulting residuals will constitute the physics-informed loss function by being discretized (Section 2.4) and optimized (Section 2.5). Finally, a summary of the whole method is given in Section 2.6.

2.1. PDE setting

Throughout this article, we consider the following setting. Let X, Y, W be separable Banach spaces with norms $\|\cdot\|_X$, $\|\cdot\|_Y$ and $\|\cdot\|_W$, respectively, and analogously let $X^* \subset X$, $Y^* \subset Y$ and $W^* \subset W$ be closed subspaces with norms $\|\cdot\|_{X^*}$, $\|\cdot\|_{Y^*}$ and $\|\cdot\|_{W^*}$, respectively. Typical examples of such spaces include L^p and Sobolev spaces. We consider PDEs of the abstract form

$$\mathcal{L}[u] = f, \quad \mathcal{B}[u] = g, \tag{2.1}$$

where $\mathcal{L}\colon X^* \to Y^*$ is a *differential operator* and $f \in Y^*$ is an *input* or *source* function. The boundary conditions (including the initial condition for time-dependent PDEs) are prescribed by the *boundary operator* $\mathcal{B}\colon X^* \to W^*$ and the boundary data $g \in W^*$. We assume that for all $f \in Y^*$ there exists a unique $u \in X^*$ such that (2.1) holds. Finally, we assume that for all $u \in X^*$ and $f \in Y^*$, we have

$$\|\mathcal{L}[u]\|_{Y^*} < +\infty, \quad \|f\|_{Y^*} < +\infty, \quad \|\mathcal{B}[u]\|_{W^*} < +\infty. \tag{2.2}$$

We will also denote the domain of u as Ω, where either $\Omega = D \subset \mathbb{R}^d$ for time-independent PDEs or $\Omega = D \times [0,T] \subset \mathbb{R}^{d+1}$ for time-dependent PDEs.

2.1.1. Forward problems

The *forward problem* for the abstract PDE (2.1) amounts to the following: given a source f and boundary conditions g (and potentially another parameter function a such that $\mathcal{L} := \mathcal{L}_a$), find the solution u of the PDE. This can be summarized as finding an *operator* that maps an input function $v \subset \{f, g, a\} \subset \mathcal{X}$ to the corresponding solution $u \in X^*$ of the PDE

$$\mathcal{G}\colon \mathcal{X} \to X^*\colon v \mapsto \mathcal{G}[v] = u. \tag{2.3}$$

Note that it can also be interesting to determine the whole operator $\mathcal{G}$ rather than just the function u. Finding a good approximation of $\mathcal{G}$ is referred to as *operator learning*. A first concrete example of an operator would be the mapping between the input or source function f and the corresponding solution u, and hence $\mathcal{X} = Y^*$ and $v = f$:

$$\mathcal{G}\colon Y^* \to X^*\colon f \mapsto \mathcal{G}[f] = u. \tag{2.4}$$

Another common example for a time-dependent PDE, where $\mathcal{B}[u] = g$ entails the prescription of the initial condition $u(\cdot, 0) = u_0$, is the *solution operator* that maps the initial condition u_0 to the solution of the PDE

$$\mathcal{G}\colon W^* \to X^*\colon u_0 \mapsto \mathcal{G}[u_0] = u, \tag{2.5}$$

where $\mathcal{X} = W^*$ and $v = u_0$. Alternatively, one can also consider the mapping $u_0 \mapsto u(\cdot, T)$.

A final example of operator learning can be found in *parametric PDEs*, where the differential operator, source function, initial condition or boundary condition depends on scalar parameters or a parameter function. Although solving parametric PDEs is usually not called operator learning, it is evident that it can be identified as such, the only difference being that the input space is not necessarily infinite-dimensional.

In what follows we give examples of PDEs (2.1) for which we will consider the forward problem later on.

Example 2.1 (semilinear heat equation). We first consider the set-up of the semilinear heat equation. Let $D \subset \mathbb{R}^d$ be an open connected bounded set with a continuously differentiable boundary ∂D. The semilinear parabolic equation is then given by

$$\begin{cases} u_t = \Delta u + f(u) & \text{for all } x \in D,\ t \in (0,T), \\ u(x,0) = u_0(x) & \text{for all } x \in D, \\ u(x,t) = 0 & \text{for all } x \in \partial D,\ t \in (0,T). \end{cases} \tag{2.6}$$

Here, $u_0 \in C^k(D)$, $k \geq 2$, is the initial data, $u \in C^k([0,T] \times D)$ is the classical solution and $f\colon \mathbb{R} \times \mathbb{R}$ is the nonlinear source (reaction) term. We assume that the nonlinearity is globally Lipschitz, that is, there exists a constant $C_f > 0$ such that

$$|f(v) - f(w)| \leq C_f\,|v - w| \quad \text{for all } v, w \in \mathbb{R}. \tag{2.7}$$

In particular, the homogeneous linear heat equation with $f(u) \equiv 0$ and the linear source term $f(u) = \alpha u$ are examples of (2.6). Semilinear heat equations with globally Lipschitz nonlinearities arise in several models in biology and finance (Beck *et al.* 2021). The existence, uniqueness and regularity of the semilinear parabolic equations with Lipschitz nonlinearities such as (2.6) can be found in classical textbooks such as that of Friedman (1964).

Example 2.2 (Navier–Stokes equations). We consider the well-known incompressible Navier–Stokes equations (Temam 2001 and references therein)

$$\begin{cases} u_t + u \cdot \nabla u + \nabla p = \nu \Delta u & \text{in } D \times [0,T], \\ \operatorname{div}(u) = 0 & \text{in } D \times [0,T], \\ u(t=0) = u_0 & \text{in } D. \end{cases} \tag{2.8}$$

Here, $u\colon D \times [0,T] \to \mathbb{R}^d$ is the fluid velocity, $p\colon D \times [0,T] \to \mathbb{R}$ is the pressure and $u_0\colon D \to \mathbb{R}^d$ is the initial fluid velocity. The viscosity is denoted by $\nu \geq 0$. For the rest of the paper, we consider the Navier–Stokes equations (2.8) on the d-dimensional torus $D = \mathbb{T}^d = [0,1)^d$ with periodic boundary conditions.

Example 2.3 (viscous and inviscid scalar conservation laws). We consider the following one-dimensional version of *viscous scalar conservation laws* as a model problem for quasilinear, convection-dominated diffusion equations:

$$\begin{cases} u_t + f(u)_x = \nu u_{xx} & \text{for all } x \in (0,1),\ t \in [0,T], \\ u(x,0) = u_0(x) & \text{for all } x \in (0,1), \\ u(0,t) = u(1,t) \equiv 0 & \text{for all } t \in [0,T]. \end{cases} \tag{2.9}$$

Here, $u_0 \in C^k([0,1])$, for some $k \geq 1$, is the initial data and we consider zero Dirichlet boundary conditions. Note that $0 < \nu \ll 1$ is the viscosity coefficient. The flux function is denoted by $f \in C^k(\mathbb{R};\mathbb{R})$. One can follow standard textbooks such as that of Godlewski and Raviart (1991) to conclude that as long as $\nu > 0$, there exists a classical solution $u \in C^k([0,T)\times[0,1])$ of the viscous scalar conservation law (2.9).

It is often interesting to set $\nu = 0$ and consider the *inviscid scalar conservation law*

$$u_t + f(u)_x = 0. \tag{2.10}$$

In this case, the solution might no longer be continuous, as it can develop shocks, even for smooth initial data (Holden and Risebro 2015).

Example 2.4 (Poisson's equation). Finally, we consider a prototypical elliptic PDE. Let $\Omega \subset \mathbb{R}^d$ be a bounded open set that satisfies the exterior sphere condition, let $f \in C^1(\Omega) \cap L^\infty(\Omega)$ and let $g \in C(\partial\Omega)$. Then there exists $u \in C^2(\Omega) \cap C(\overline{\Omega})$ that satisfies Poisson's equation:

$$\begin{cases} -\Delta u = f & \text{in } \Omega, \\ u = g & \text{on } \partial\Omega. \end{cases} \tag{2.11}$$

Moreover, if f is smooth then u is smooth in Ω.

2.1.2. *Inverse problems*

Secondly, we consider settings where we do not have complete information on the inputs to the PDE (2.1) (e.g. initial data, boundary conditions and parameter coefficients). As a result, the forward problem cannot be solved uniquely. However, when we have access to (noisy) data for (observables of) the underlying solution, we can attempt to determine the unknown inputs of the PDEs and consequently its solution. Problems with the aim of recovering the PDE input based on data are referred to as *inverse problems*.

We will mainly consider one kind of inverse problem, namely the *unique continuation problem* (for time-dependent problems also known as the *data assimilation problem*). Recall the abstract PDE $\mathcal{L}[u] = f$ in Ω (2.1). Moreover, we adopt the assumptions (2.2) of the introduction to this subsection. The difference from the forward problem is that we now have incomplete information on the boundary conditions. As a result, a unique solution to the PDE does not exist. Instead, we assume that we have access to (possibly noisy) measurements of a certain observable

$$\Psi(u) = g \quad \text{in } \Omega', \tag{2.12}$$

where $\Omega' \subset \Omega$ is the observation domain, g is the measured data, and $\Psi\colon X^* \to Z^*$, where Z^* is some Banach space. We make the assumption that

$$\begin{aligned} &\|\Psi(u)\|_{Z^*} < +\infty \quad \text{for all } u \in X^*, \text{ with } \|u\|_{X^*} < +\infty, \\ &\|g\|_{Z^*} < +\infty. \end{aligned} \tag{2.13}$$

If Ω' includes the boundary of Ω, meaning that we have measurements on the boundary, then it is again feasible to just use the standard framework for forward problems. However, this is often not possible. In heat transfer problems, for example, the maximum temperature will be reached at the boundary. As a result, it might be too hot near the boundary to place a sensor there and we will only have measurements that are far enough away from the boundary. Fortunately, we can adapt the physics-informed framework to retrieve an approximation to u based on f and g.

In analogy with operator learning for forward problems, we can also attempt to entirely learn the *inverse operator*

$$\mathcal{G}^{-1}\colon Z^* \to X^*\colon \Psi[u] = g \mapsto \mathcal{G}^{-1}[g] = v, \tag{2.14}$$

where v could for instance be the boundary conditions, the parameter function a or the input/source function f.

Example 2.5 (heat equation). We revisit the heat equation $u_t - \Delta u = f$ with zero Dirichlet boundary conditions and $f \in L^2(\Omega)$, where $\Omega = D \times (0, T)$. We assume that $u \in H^1(((0,T); H^{-1}(D)) \cap L^2((0,T); H^1(D))$ will solve the heat equation in a weak sense. The heat equation would have been well-posed if the initial conditions, i.e. $u_0 = u(x, 0)$, had been specified. Recall that the aim of the data assimilation problem is to infer the initial conditions and the entire solution field at later times from some measurements of the solution in time. To model this, we consider the following *observables*:

$$\Psi[u] := u = g \quad \text{for all } (x,t) \in D' \times (0,T) =: \Omega', \tag{2.15}$$

for some open, simply connected observation domain $D' \subset D$ and for $g \in L^2(\Omega')$. Thus, solving the data assimilation problem for the heat equation amounts to finding the solution u of the heat equation in the whole space–time domain $\Omega = D \times (0, T)$, given data on the observation subdomain $\Omega' = D' \times (0, T)$.

Example 2.6 (Poisson equation). We revisit the Poisson equation (Example 2.4), but now with data on a subdomain of Ω:

$$-\Delta u = f \quad \text{in } \Omega \subset \mathbb{R}^d, \quad u|_{\Omega'} = g \quad \text{in } \Omega' \subset \Omega. \tag{2.16}$$

In this case the observable Ψ introduced in (2.12) is the identity $\Psi[u] = u$.

Example 2.7 (Stokes equation). Finally, we consider a simplified version of the Navier–Stokes equations, namely the stationary Stokes equation. Let $D \subset \mathbb{R}^d$ be an open, bounded, simply connected set with smooth boundary. We consider the Stokes equations as a model of stationary, highly viscous fluid:

$$\begin{cases} \Delta u + \nabla p = f & \text{for all } x \in D, \\ \operatorname{div}(u) = f_d & \text{for all } x \in D. \end{cases} \tag{2.17}$$

Here, $u\colon D \to \mathbb{R}^d$ is the velocity field, $p\colon D \to \mathbb{R}$ is the pressure and $f\colon D \to \mathbb{R}^d$, $f_d\colon D \to \mathbb{R}$ are source terms.

Note that the Stokes equation (2.17) is not well-posed as we are not providing any boundary conditions. In the corresponding data assimilation problem (Burman and Hansbo 2018 and references therein), we provide the following *data*:

$$\Psi[u] := u = g \quad \text{for all } x \in D', \tag{2.18}$$

for some open, simply connected set $D' \subset D$. Thus the data assimilation inverse problem for the Stokes equation amounts to inferring the velocity field u (and the pressure p), given f, f_d and g.

2.2. *Ansatz spaces*

We give an overview of different classes of parametric functions $\{u_\theta\}_{\theta \in \Theta}$ that are often used to approximate the true solution u of the PDE (2.1). We let n be the number of (real) parameters, such that the parameter space Θ is a subset of $\mathbb{R}^n$.

2.2.1. *Linear models*

We first consider models that linearly depend on the parameter θ, i.e. models that are a linear combination of a fixed set of functions $\{\phi_i\colon \Omega \to \mathbb{R}\}_{1 \le i \le n}$. For any $\theta \in \mathbb{R}^n$ the parametrized linear model u_θ is then defined as

$$u_\theta(x) = \sum_{i=1}^{n} \theta_i \phi_i(x). \tag{2.19}$$

This general model class constitutes the basis of many existing numerical methods for approximation solutions to PDEs, of which we give an overview below.

First, *spectral methods* use smooth functions that are globally defined, i.e. supported on almost all of Ω, and often form an orthogonal set; see e.g. Hesthaven, Gottlieb and Gottlieb (2007). The particular choice of functions generally depends on the geometry and boundary conditions of the considered problem: whereas trigonometric polynomials (Fourier basis) are a default choice on periodic domains, Chebyshev and Legendre polynomials might be more suitable choices on non-periodic domains. The optimal parameter vector θ^* is then determined by solving a linear system of equations. Assuming that $\mathcal{L}$ is a linear operator, this system is given by

$$\sum_{i=1}^{n} \theta_i \int_\Omega \psi_k \mathcal{L}(\phi_i) = \int_\Omega \psi_k f \quad \text{for } 1 \le k \le K, \tag{2.20}$$

where either $\psi_k = \phi_k$, resulting in the *Galerkin method*, or $\psi_k = \delta_{x_k}$ with $x_k \in \Omega$, resulting in the *collocation method*, or the more general case where the ψ_k and ϕ_k are distinct (and also not Dirac deltas), which is referred to as the *Petrov–Galerkin method*. In the collocation method, the choice of so-called collocation points $\{x_k\}_k$ depends on the functions ϕ_i. Boundary conditions are implemented through the choice of the functions ϕ_i or by adding more equations to the above linear system.

In marked contrast to the global basis functions of the spectral method stands the *finite element method* (FEM), where we consider locally defined polynomials. More precisely, we first divide Ω into (generally overlapping) subdomains Ω_i (subintervals in one dimension) of diameter at most h and then consider on each Ω_i a piecewise defined polynomial ϕ_i of degree at most p that is zero outside Ω_i. Much like with the spectral Galerkin method, we can then define a linear system of equations such as (2.20) that needs to be solved to find the optimal parameter vector θ^*. As ϕ_i is now non-smooth on Ω, we need to rewrite $\int_\Omega \phi_k \mathcal{L}(\phi_i)$ using integration by parts. Another key difference from the spectral method is that, due to the local support of the functions ϕ_i, the linear system we have to solve will be sparse and have structural properties related to the choice of grid and polynomials. The accuracy of the approximation can be improved by reducing h (h-FEM), increasing p (p-FEM), or both (hp-FEM). Alternatively, the finite element method can also be used in combination with a squared loss, resulting in the *least-squares finite element method* (Bochev and Gunzburger 2009). The finite element method is ubiquitous in scientific computing and many variants can be found.

Another choice for the functions ϕ_i is that of *radial basis functions* (RBFs); see e.g. Buhmann (2000). The key ingredient of RBFs is a radial function $\varphi\colon [0,\infty) \to \mathbb{R}$, for which it must hold for any choice of nodes $\{x_i\}_{1\le i\le n} \subset \Omega$ that both (i) the functions $x \mapsto \phi_i(x) := \varphi(\|x - x_i\|)$ are linearly independent and (ii) the matrix A with entries $A_{ij} = \varphi(\|x_i - x_j\|)$ is non-singular. Radial functions can be globally supported (e.g. Gaussians, polyharmonic splines) or compactly supported (e.g. bump functions). The optimal parameter θ^* can then be determined in a Galerkin, collocation or least-squares fashion. RBFs are particularly advantageous as they are mesh-free (as opposed to FEM), which is useful for complex geometries, and straightforward to use in high-dimensional settings, as we can use randomly generated nodes, rather than nodes on a grid.

Finally, RBFs with randomly generated nodes can be seen as an example of a more general *random feature model* (RFM), where the functions ϕ_i are randomly drawn from a chosen distribution over a function space (e.g. Rahimi and Recht 2008). In practice we often choose the function space to be parametric itself, such that generating ϕ_i reduces to randomly drawing finite vectors α_i and setting $\phi_i(x) := \phi(x;\alpha_i)$. Random feature models where ϕ_i is a neural network with randomly initialized weights and biases are closely connected to neural networks of which only the outer layer is trained. When the network is large then the latter model is often referred to as an *extreme learning machine* (ELM) (Huang, Zhu and Siew 2006).

2.2.2. *Nonlinear models*

Neural networks. A first, but very central, example of parametric models that depend nonlinearly on their parameters is that of *neural networks*. The simplest form of a neural network is a *multilayer perceptron* (MLP). MLPs get their name from their definition as a concatenation of affine maps and activation functions. As a consequence of this structure, the computational graph of an MLP is organized in layers. At the kth layer, the activation function is applied component-wise to obtain the value

$$z^k = \sigma(C_{k-1}(z^{k-1})) = \sigma(W_{k-1}z^{k-1} + b_{k-1}), \tag{2.21}$$

where z^0 is the input of the network, $W_k \in \mathbb{R}^{d_{k+1}\times d_k}$, $b_k \in \mathbb{R}^{d_{k+1}}$ and $C_k : \mathbb{R}^{d_k} \to \mathbb{R}^{d_{k+1}} : x \mapsto W_k x + b_k$ for $d_k \in \mathbb{N}$ is an affine map. A multilayer perceptron (MLP) with L layers, activation function σ, output function σ_o, input dimension d_{in} and output dimension d_{out} is then a function $f_\theta : \mathbb{R}^{d_{\text{in}}} \to \mathbb{R}^{d_{\text{out}}}$ of the form

$$f_\theta(x) = (\sigma_o \circ C_L \circ \sigma \circ \cdots \circ \sigma \circ C_0)(x) \quad \text{for all } x \in \mathbb{R}^{d_{\text{in}}}, \tag{2.22}$$

where σ is applied element-wise and $\theta = ((W_0, b_0), \ldots, (W_L, b_L))$. The dimension d_k of each layer is also called the *width* of that layer. The width of the MLP is defined as $\max_k d_k$. Furthermore, the matrices W_k are called the *weights* of the MLP and the vectors b_k are called the *biases* of the MLP. Together, they constitute the (trainable) parameters of the MLP, denoted by $\theta = ((W_0, b_0), \ldots, (W_L, b_L))$. Finally, an output function σ_o can be employed to increase the interpretability of the output, but is often equal to the identity in the setting of function approximation.

Remark 2.8. In some texts, the number of layers L corresponds to the number of affine maps in the definition of f_θ. Of these L layers, the first $L-1$ layers are called hidden layers. An L-layer network in this alternative definition hence corresponds to an $(L-1)$-layer network in our definition.

Many of the properties of a neural network depend on the choice of activation function. We discuss the most common activation functions below.

- The *Heaviside* function is the activation that was used in the McCulloch–Pitts neuron and is defined as

$$H(x) = \begin{cases} 0, & x < 0, \\ 1, & x \geq 0. \end{cases} \tag{2.23}$$

  Because the gradient is zero wherever it exists, it is not possible to train the network using a gradient-based approach. For this reason, the Heaviside function is no longer used.

- The *logistic* function is defined as

$$\rho(x) = \frac{1}{1 + \mathrm{e}^{-x}} \tag{2.24}$$

  and can be thought of as a smooth approximation of the Heaviside function as $\rho(\lambda x) \to H(x)$ for $\lambda \to \infty$ for almost every x. It is a so-called sigmoidal function, meaning that the function is smooth, monotonic and has horizontal asymptotes for $x \to \pm\infty$. Many early approximation-theoretical results were stated for neural networks with a sigmoidal activation function.

- The hyperbolic tangent or *tanh* function, defined by

$$\sigma(x) = \frac{\mathrm{e}^{x} - \mathrm{e}^{-x}}{\mathrm{e}^{x} + \mathrm{e}^{-x}}, \tag{2.25}$$

  is another smooth, sigmoidal activation function that is symmetric, unlike the logistic function. One problem with the tanh (and all other sigmoidal functions) is that the gradient vanishes away from zero, which might be problematic when using a gradient-based optimization approach.

- The rectified linear unit, rectifier function or *ReLU* function, defined as

$$\mathrm{ReLU}(x) = \max\{x, 0\}, \tag{2.26}$$

  is a very popular activation function. It is easy to compute, scale-invariant and reduces the vanishing gradient problem that sigmoidal functions have. One common issue with ReLU networks is that of 'dying neurons', caused by the fact that the gradient of ReLU is zero for $x < 0$. This issue can however be overcome by using the so-called *leaky ReLU* function

$$\mathrm{ReLU}_{\nu}(x) = \max\{x, -\nu x\}, \tag{2.27}$$

  where $\nu > 0$ is small. Other (smooth) adaptations are the Gaussian error linear unit, the sigmoid linear unit, exponential linear unit and softplus function.

Neural operators. In order to approximate operators, we need to allow the input and output of the learning architecture to be infinite-dimensional. A possible approach is to use *deep operator networks* (DeepONets), as proposed in Chen and Chen (1995) and Lu, Jin, Pang, Zhang and Karniadakis (2021*a*). Given m fixed sensor locations $\{x_j\}_{j=1}^{m} \subset \Omega$ and the corresponding *sensor values* $\{v(x_j)\}_{j=1}^{m}$ as input, a DeepONet can be formulated in terms of two (deep) neural networks: a *branch net* $\boldsymbol{\beta}\colon \mathbb{R}^m \to \mathbb{R}^p$ and a *trunk net* $\boldsymbol{\tau}\colon \Omega \to \mathbb{R}^{p+1}$. The branch and trunk nets are then combined to approximate the underlying nonlinear operator as the following *DeepONet*:

$$\mathcal{G}_\theta\colon \mathcal{X} \to L^2(\Omega), \quad \text{with} \quad \mathcal{G}_\theta[v](y) = \tau_0(y) + \sum_{k=1}^{p} \beta_k(v)\tau_k(y). \tag{2.28}$$

One alternative to DeepONets that also allows us to learn maps between infinite-dimensional spaces is that of *neural operators* (Li *et al.* 2020). Neural operators are inspired by the fact that under some general conditions the solution u to the PDE (2.1) can be represented in the form

$$u(x) = \int_D G(x, y) f(y)\,\mathrm{d}y, \tag{2.29}$$

where $G\colon \Omega \times \Omega \to \mathbb{R}$ is the Green's function. This kernel representation led Li *et al.* (2020) to propose replacing the formula for a hidden layer of a standard neural network with the following operator:

$$\mathcal{L}_\ell(v)(x) = \sigma\left(W_\ell v(x) + \int_D \kappa_\phi(x, y, a(x), a(y)) v(y)\,\mathrm{d}\mu_x(\mathrm{d}y) \right), \tag{2.30}$$

where W_ℓ and ϕ are to be learned from the data, κ_ϕ is a kernel and μ_x is a Borel measure for every $x \in D$. Li *et al.* (2020) modelled the kernel κ_ϕ as a neural network. More generally, neural operators are mappings of the form (Li *et al.* 2020, Kovachki, Lanthaler and Mishra 2021)

$$\mathcal{N}\colon \mathcal{X} \to \mathcal{Y}\colon u \mapsto \mathcal{Q} \circ \mathcal{L}_L \circ \mathcal{L}_{L-1} \circ \cdots \circ \mathcal{L}_1 \circ \mathcal{R}(u), \tag{2.31}$$

for a given depth $L \in \mathbb{N}$, where $\mathcal{R}\colon \mathcal{X}(D; \mathbb{R}^{d_\mathcal{X}}) \to \mathcal{U}(D; \mathbb{R}^{d_v})$, $d_v \geq d_u$, is a *lifting* operator (acting locally), of the form

$$\mathcal{R}(a)(x) = Ra(x), \quad R \in \mathbb{R}^{d_v \times d_\mathcal{X}}, \tag{2.32}$$

and $\mathcal{Q}\colon \mathcal{U}(D; \mathbb{R}^{d_v}) \to \mathcal{Y}(D; \mathbb{R}^{d_\mathcal{Y}})$ is a local *projection* operator, of the form

$$\mathcal{Q}(v)(x) = Qv(x), \quad Q \in \mathbb{R}^{d_\mathcal{Y} \times d_v}. \tag{2.33}$$

In analogy with canonical finite-dimensional neural networks, the layers $\mathcal{L}_1, \ldots, \mathcal{L}_L$ are nonlinear operator layers, $\mathcal{L}_\ell\colon \mathcal{U}(D; \mathbb{R}^{d_v}) \to \mathcal{U}(D; \mathbb{R}^{d_v})$, $v \mapsto \mathcal{L}_\ell(v)$, which we assume to be of the form

$$\mathcal{L}_\ell(v)(x) = \sigma(W_\ell v(x) + b_\ell(x) + (\mathcal{K}(a; \theta_\ell)v)(x)) \quad \text{for all } x \in D. \tag{2.34}$$

Here, the weight matrix $W_\ell \in \mathbb{R}^{d_v \times d_v}$ and bias $b_\ell(x) \in \mathcal{U}(D; \mathbb{R}^{d_v})$ define an affine pointwise mapping $W_\ell v(x) + b_\ell(x)$. We conclude the definition of a neural operator by stating that $\sigma\colon \mathbb{R} \to \mathbb{R}$ is a nonlinear activation function that is applied component-wise, and $\mathcal{K}$ is a *non-local* linear operator,

$$\mathcal{K}\colon \mathcal{X} \times \Theta \to L(\mathcal{U}(D; \mathbb{R}^{d_v}), \mathcal{U}(D; \mathbb{R}^{d_v})), \tag{2.35}$$

that maps the input field a and a parameter $\theta \in \Theta$ in the parameter set Θ to a bounded linear operator $\mathcal{K}(a, \theta)\colon \mathcal{U}(D; \mathbb{R}^{d_v}) \to \mathcal{U}(D; \mathbb{R}^{d_v})$. When we define the linear operators $\mathcal{K}(a, \theta)$ through an integral kernel, then (2.34) reduces again to (2.30). *Fourier neural operators* (FNOs) (Li *et al.* 2021) are special cases of such general neural operators in which this integral kernel corresponds to a convolution kernel, which on the periodic domain $\mathbb{T}^d$ leads to nonlinear layers $\mathcal{L}_\ell$ of the form

$$\mathcal{L}_\ell(v)(x) = \sigma(W_\ell v(x) + b_\ell(x) + \mathcal{F}^{-1}(P_\ell(k) \cdot \mathcal{F}(v)(k))(x)). \tag{2.36}$$

In practice, however, this definition cannot be used, as evaluating the Fourier transform requires the exact computation of an integral. The practical implementation (i.e. discretization) of an FNO maps from and to the space of trigonometric polynomials of degree at most $N \in \mathbb{N}$, denoted by L^2_N, and can be identified with a finite-dimensional mapping that is a composition of affine maps and nonlinear layers of the form

$$\mathfrak{L}_\ell(z)_j = \sigma\big(W_\ell v_j + b_{\ell,j}\mathcal{F}_N^{-1}(P_\ell(k) \cdot \mathcal{F}_N(z)(k)_j)\big), \tag{2.37}$$

where the $P_\ell(k)$ are coefficients that define a non-local convolution operator via the discrete Fourier transform $\mathcal{F}_N$; see Kovachki *et al.* (2021).

It has been observed that the proposed operator learning models above may not behave as operators when implemented on a computer, questioning the very essence of what operator learning should be. Bartolucci *et al.* (2023) contend that in addition to defining the operator at the continuous level, some form of continuous–discrete equivalence is necessary for an architecture to genuinely learn the underlying operator rather than just discretizations of it. To this end, they introduce the unifying mathematical framework of *representation-equivalent neural operators* (ReNOs) to ensure that operations at the continuous and discrete level are equivalent. Lack of this equivalence is quantified in terms of aliasing errors.

Gaussian processes. Deep neural networks and their operator versions are only one possible nonlinear surrogate model for the true solution u. Another popular class of nonlinear models is *Gaussian process regression* (Rasmussen 2003), which belongs to a larger class of so-called Bayesian models. Gaussian process regression (GPR) relies on the assumption that u is drawn from a Gaussian measure on a suitable function space, parametrized by

$$u_\theta(x) \sim \mathrm{GP}(m_\theta(x), k_\theta(x, x')), \tag{2.38}$$

where $m_\theta(x) = \mathbb{E}[u_\theta(x)]$ is the mean, and the underlying covariance function is given by $k_\theta(x, x') = \mathbb{E}[(u_\theta(x) - m_\theta(x))(u_\theta(x') - m_\theta(x'))]$.

Popular choices for the covariance function in (2.38) are the squared exponential kernel (which is a radial function as for RBFs) and Matérn covariance functions:

$$k_{\mathrm{SE}}(x, x') = \exp\left(-\frac{\|x - x'\|^2}{2\ell^2}\right), \tag{2.39}$$

$$k_{\mathrm{Matern}}(x, x') = \frac{2^{1-\nu}}{\Gamma(\nu)}\left(\sqrt{2\nu}\frac{\|x - x'\|}{\ell}\right)^\nu K_\nu\left(\sqrt{2\nu}\frac{\|x - x'\|}{\ell}\right). \tag{2.40}$$

Here $\|\cdot\|$ denotes the standard Euclidean norm, K_ν is the modified Bessel function of the second kind and ℓ the *characteristic length*, describing the length scale of the correlations between the points x and x'. The parameter vector then consists of the hyperparameters of m_θ and k_θ, such as ν and ℓ.

2.3. *Physics-informed residuals and loss functions*

With a chosen model class in place, the key to physics-informed learning is to find a *physics-informed loss functional* $\mathcal{E}_{\text{PIL}}[\cdot]$, independent of u, such that when it is minimized over the class of models $\{u_\theta : \theta \in \Theta\}$, the minimizer is a good approximation of the PDE solution u. In other words, we must find a functional $\mathcal{E}_{\text{PIL}}[\cdot]$ such that for $\theta^* = \operatorname{argmin}_\theta \mathcal{E}_{\text{PIL}}[u_\theta]$, the minimizer u_{θ^*} is a good approximation of u. We discuss three alternative formulations for a PDE, i.e. definitions of what it means to solve a PDE, all leading to different functionals and their corresponding *loss functions*:

- the classical formulation of the PDE, as in (2.1), in which a pointwise condition on a function and its derivatives is stated;
- the weak formulation of the PDE, which is a more general (weaker) condition than the classical formulation, often obtained by multiplying the classical formulation by a smooth test function and performing integration by parts;
- the variational formulation of the PDE, where the PDE solution can be defined as the minimizer of a functional.

Note that not all formulations are available for all PDEs.

2.3.1. *Classical formulation*

Physics-informed learning is based on the elementary observation that for the true solution of the PDE u it holds that $\mathcal{L}(u) = f$ and $\mathcal{B}(u) = g$, and the subsequent hope that if a model u_θ satisfies that $\mathcal{L}(u_\theta) - f \approx 0$ and $\mathcal{B}(u_\theta) - g \approx 0$, that then also $u \approx u_\theta$. To this end, a number of *residuals* are introduced. First and foremost, the *PDE residual* (or *interior residual*) measures how well the model u_θ satisfies the PDE

$$\mathcal{R}_{\text{PDE}}[u_\theta](z) = \mathcal{L}[u_\theta](z) - f(z) \quad \text{for all } z \in \Omega, \tag{2.41}$$

where we recall that $\Omega = D$ and $z = x$ for a time-independent PDE and $\Omega = D \times [0, T]$ and $z = (x, t)$ for a time-dependent PDE. In the first case, the boundary operator $\mathcal{B}$ in (2.1) uniquely refers to the spatial boundary conditions, such that $\mathcal{B}(u) = g$ corresponds to $D_x^k u = g$. The corresponding *spatial boundary residual* is then given by

$$\mathcal{R}_s[u_\theta](x) = D_x^k u_\theta(x) - g(x) \quad \text{for all } x \in \partial D. \tag{2.42}$$

For time-dependent PDEs the boundary operator $\mathcal{B}$ generally also prescribes the initial condition, i.e. $u(x, 0) = u_0(x)$ for all $x \in D$ for a given function u_0. This condition gives rise to the *temporal boundary residual*

$$\mathcal{R}_t[u_\theta](x) = u_\theta(x, 0) - u_0(x) \quad \text{for all } x \in D. \tag{2.43}$$

The physics-informed learning framework hence dictates that we should simultaneously minimize the residuals $\mathcal{R}_{\text{PDE}}[u_\theta]$ and $\mathcal{R}_s[u_\theta]$, and additionally $\mathcal{R}_t[u_\theta]$ for a time-dependent PDE. This is usually translated as the minimization problem

$$\min_{\theta \in \Theta} \mathcal{E}_{\text{PIL}}[u_\theta]^2, \tag{2.44}$$

where for a time-independent PDE the *physics-informed loss* $\mathcal{E}_{\text{PIL}}$ is given by

$$\mathcal{E}_{\text{PIL}}[u_\theta]^2 = \int_{D\times[0,T]} \mathcal{R}_{\text{PDE}}[u_\theta]^2 + \lambda_s \int_{\partial D} \mathcal{R}_s[u_\theta]^2, \tag{2.45}$$

where $\lambda_s > 0$ is a weighting hyperparameter, and for a time-dependent PDE the quantity $\mathcal{E}_{\text{PIL}}$ is given by

$$\mathcal{E}_{\text{PIL}}[u_\theta]^2 = \int_{D\times[0,T]} \mathcal{R}_{\text{PDE}}[u_\theta]^2 + \lambda_s \int_{\partial D\times[0,T]} \mathcal{R}_s[u_\theta]^2 + \lambda_t \int_0^T \mathcal{R}_t[u_\theta]^2, \tag{2.46}$$

where $\lambda_s, \lambda_t > 0$ are weighting hyperparameters. Since the boundary conditions (and initial condition) are enforced through an additional term in the loss function, we allow the possibility that the final model will only approximately satisfy the boundary conditions. This is generally referred to as having *soft boundary conditions*. On the other hand, if we choose the model class in such a way that all u_θ exactly satisfy the boundary conditions ($\mathcal{B}[u_\theta] = g$), we have implemented *hard boundary conditions*.

Remark 2.9. The residuals and loss functions defined above can readily be extended to systems of PDEs and more complicated boundary conditions. If the system of PDEs is given by $\mathcal{L}^{(i)}[u] = f^{(i)}$, then one can readily define multiple PDE residuals $\mathcal{R}_{\text{PDE}}^{(i)}$ and consider the sum of integrals of the square PDE residuals $\sum_i \int (\mathcal{R}_{\text{PDE}}^{(i)})^2$ in the physics-informed loss. The generalization to more complicated boundary conditions, including piecewise defined boundary conditions, proceeds in a completely analogous manner.

The minimization problem (2.44), solely based on the physics-informed loss functions defined in (2.45) and (2.46), is the most basic version of physics-informed learning and can be extended by adding additional terms to the objective in (2.44). For example, in some settings one might have access to measurements $\Psi(u) = g$ on a subdomain Ω', where Ψ is a so-called *observable*. This case is particularly relevant for inverse problems (Section 2.1.2). In this case it makes sense to add a *data term* to the optimization objective, defined in terms of a *data residual* $\mathcal{R}_d$:

$$\mathcal{R}_d[u_\theta](x) = \Psi[u](x) - g(x) \quad \text{for all } x \in \Omega'. \tag{2.47}$$

Note that the boundary residual can be seen as a special case of the data residual with $\Psi = \mathcal{B}$ and $\Omega' = \partial\Omega$. Additionally, it might happen in practice that we add a *parameter regularization term*, often the L^q-norm of the parameters θ, to the loss function to either aid the minimization process or improve the (generalization) properties of the final model. A generalization of the physics-informed minimization problem (2.44) could therefore be

$$\min_{\theta\in\Theta}\left(\mathcal{E}_{\mathrm{PIL}}[u_\theta]^2+\lambda_d\int_{\Omega'}\mathcal{R}_d[u_\theta]^2+\lambda_r\|\theta\|_q^q\right). \tag{2.48}$$

2.3.2. Weak formulation

In the previous subsection we considered PDE residuals and physics-informed loss functions that are based on the strong formulation of the PDE. However, for various classes of PDEs there might not be a smooth classical solution u for which it holds in every $x \in \Omega$ that $\mathcal{L}[u](x) = f(x)$. Fortunately, one can sometimes generalize the definition of 'a solution to a PDE' in a precise manner such that there are functions that satisfy this generalized definition after all. Such functions are called *weak solutions*. They are said to satisfy the *weak formulation* of the PDE, which can be obtained by multiplying the strong formulation of the PDE (2.1) with a smooth and compactly supported test function ψ and then performing integration by parts:

$$\int_\Omega u\mathcal{L}^*[\psi]=\int_\Omega f\psi, \tag{2.49}$$

where $\mathcal{L}^*$ is the (formal) adjoint of $\mathcal{L}$ (if it exists). Consequently, a function u is said to be a weak solution if it satisfies (2.49) for all (compactly supported) test functions ψ in some predefined set of functions. Note that u no longer needs to be differentiable but merely integrable.

Weak formulations such as (2.49) form the basis of many classical numerical methods, in particular the finite element method (FEM: Section 2.2.1). Often, however, there is no need to perform integration by parts until u is completely free of derivatives. In FEM, the weak formulation for second-order elliptic equations is generally obtained by performing integration by parts only once. For the Laplacian $\mathcal{L} = \Delta$ we then have

$$\int_\Omega \Delta u\psi=-\int_\Omega \nabla u\cdot\nabla\psi$$

(given suitable boundary conditions), which is a formulation with many useful properties in practice.

Other than using spectral methods (as in Section 2.2.1), there are now different avenues one can follow to obtain a *weak physics-informed loss function* based on (2.49). The first method draws inspiration from the Petrov–Galerkin method (Section 2.2.1) and considers a finite set of K test functions ψ_k for which (2.49) must hold. The corresponding loss function is then defined as

$$\mathcal{E}_{\mathrm{VPIL}}[u_\theta]^2=\sum_{k=1}^K\left(\int_\Omega u_\theta\mathcal{L}^*[\psi_k]-\int_\Omega f\psi_k\right)^2, \tag{2.50}$$

where additional terms can be added to take care of any boundary conditions. This method was first introduced for neural networks under the name *variational*

PINNs (VPINNs) (Kharazmi *et al.* 2019) and used sine and cosine functions or polynomials as test functions. In analogy with hp-FEM, the method was later adapted to use localized test functions, leading to hp-VPINNs (Kharazmi, Zhang and Karniadakis 2021). In this case the model u_θ is of limited regularity and might not be sufficiently many times differentiable for $\mathcal{L}[u_\theta]$ to be well-defined.

A second variant of a weak physics-informed loss function can be obtained by replacing ψ with a parametric model ψ_ϑ. In particular, we choose the ψ_ϑ to be the function that maximizes the squared weak residual, leading to the loss function

$$\mathcal{E}_{\mathrm{wPIL}}[u_\theta]^2 = \max_\vartheta \left(\int_\Omega u_\theta \mathcal{L}^*[\psi_\vartheta] - \int_\Omega f \psi_\vartheta \right)^2, \tag{2.51}$$

where again additional terms can be added to take care of any boundary conditions. Using $\mathcal{E}_{\mathrm{wPIL}}$ has the advantage over $\mathcal{E}_{\mathrm{VPIL}}$ that we do not need a basis of test functions $\{\psi_k\}_k$, which might be inconvenient in high-dimensional settings, and settings with complex geometries. On the other hand, minimizing $\mathcal{E}_{\mathrm{wPIL}}$ corresponds to solving a min-max optimization problem where the maximum and minimum are taken with respect to neural network training parameters, which is considerably more challenging than a minimization problem.

The weak physics-informed loss $\mathcal{E}_{\mathrm{wPIL}}$ was originally proposed to approximate scalar conservation laws with neural networks under the name *weak PINNs* (wPINNs) (De Ryck *et al.* 2024c). As there might be infinitely many weak solutions for a given scalar conservation law, we need to impose further requirements other than (2.49) to guarantee that a scalar conservation law has a unique solution. This additional challenge is discussed in the next example.

Example 2.10 (wPINNs for scalar conservation laws). We revisit Example 2.3 and recall that weak solutions of scalar conservation laws need not be unique (Holden and Risebro 2015). To recover uniqueness, we need to impose additional admissibility criteria in the form of *entropy conditions*. To this end, we consider the so-called *Kruzkhov entropy functions*, given by $|u-c|$ for any $c \in \mathbb{R}$, and the resulting entropy flux functions

$$Q\colon \mathbb{R}^2 \to \mathbb{R}\colon (u,c) \mapsto \operatorname{sgn}(u-c)(f(u)-f(c)). \tag{2.52}$$

We then say that a function $u \in L^\infty(\mathbb{R}\times\mathbb{R}_+)$ is an *entropy solution* of (2.10) with initial data $u_0 \in L^\infty(\mathbb{R})$ if u is a weak solution of (2.10) and if u satisfies

$$\partial_t|u-c| + \partial_x Q[u;c] \le 0, \tag{2.53}$$

in a distributional sense, or more precisely

$$\begin{aligned} &\int_0^T \int_{\mathbb{R}} (|u-c|\varphi_t + Q[u;c]\varphi_x)\,\mathrm{d}x\,\mathrm{d}t \\ &\quad - \int_{\mathbb{R}} (|u(x,T)-c|\phi(x,T) - |u_0(x)-c|\phi(x,0))\,\mathrm{d}x \ge 0 \end{aligned} \tag{2.54}$$

for all $\phi \in C_c^1(\mathbb{R} \times \mathbb{R}_+)$, $c \in \mathbb{R}$ and $T > 0$. It holds that entropy solutions are unique and continuous in time. For more details of classical, weak and entropy solutions for hyperbolic PDEs, we refer the reader to LeVeque (2002) and Holden and Risebro (2015).

This definition inspired De Ryck *et al.* (2024*c*) to define the *Kruzkhov entropy residual*

$$\mathcal{R}(v, \phi, c) := -\int_D \int_{[0,T]} (|v(x,t) - c|\partial_t \phi(x,t) + Q[v(x,t); c]\partial_x \phi(x,t))\,\mathrm{d}x\,\mathrm{d}t, \tag{2.55}$$

based on which the following weak physics-informed loss is defined:

$$\mathcal{E}_{\mathrm{wPIL}}[u_\theta] = \max_{\vartheta, c} \mathcal{R}(u_\theta, \phi_\vartheta, c), \tag{2.56}$$

where additional terms need to be added to take care of initial and boundary conditions (De Ryck *et al.* 2024*c*, Section 2.4).

2.3.3. *Variational formulation*

Finally, we discuss a third physics-informed loss function that can be formulated for *variational problems*. Variational problems are PDEs (2.1) where the differential operator can (at least formally) be written as the derivative of an *energy functional* I, in the sense that

$$\mathcal{L}[u] - f = I'[u] = 0. \tag{2.57}$$

The *calculus of variations* thus allows us to replace the problem of solving the PDE (2.1) to finding the minimum of the functional I, which might be much easier (Evans 2022). A first instance of this observation is known as *Dirichlet's principle*.

Example 2.11 (Dirichlet's principle). Let $\Omega \subset \mathbb{R}^d$ be open and bounded, and define $\mathcal{A} = \{w \in C^2(\Omega) : w = g \text{ on } \partial\Omega\}$. If we consider the functional $I(w) = \frac{1}{2}\int_\Omega |\nabla w|^2$, then for $u \in \mathcal{A}$ we have

$$I[u] = \min_{w \in \mathcal{A}} I[w] \iff \begin{cases} \Delta u = 0, & x \in \Omega, \\ u = g, & x \in \partial\Omega. \end{cases} \tag{2.58}$$

In other words, u minimizes the functional I over $\mathcal{A}$ if and only if u is a solution to Laplace's equation on Ω with Dirichlet boundary conditions g.

Dirichlet's principle can be generalized to functionals I of the form

$$I[w] = \int_\Omega L(\nabla w(x), w(x), x)\,\mathrm{d}x, \tag{2.59}$$

that are the integrals of a so-called *Lagrangian* L,

$$L : \mathbb{R}^d \times \mathbb{R} \times \Omega \to \mathbb{R} : (p, z, x) \mapsto L(p, z, x), \tag{2.60}$$

where $w : \overline{\Omega} \to \mathbb{R}$ is a smooth function that satisfies some prescribed boundary conditions. One can then verify that any smooth minimizer u of $I[\cdot]$ must satisfy the nonlinear PDE

$$-\sum_{i=1}^{d} \partial_{x_i}(\partial_{p_i} L(\nabla u, u, x)) + \partial_z L(\nabla u, u, x) = 0, \quad x \in \Omega. \tag{2.61}$$

This equation is the *Euler–Lagrange equation* associated with the energy functional I from (2.59). Revisiting Example 2.11, one can verify that Laplace's equation is the Euler–Lagrange equation associated with the energy functional based on the Lagrangian $L(p, z, x) = \frac{1}{2}|p|^2$. Dirichlet's principle can also be generalized further.

Example 2.12 (generalized Dirichlet's principle). Let $a^{ij} : \Omega \to \mathbb{R}$ be functions with $a^{ij} = a^{ji}$ and consider the elliptic PDE

$$-\sum_{i,j=1}^{d} \partial_{x_i}(a^{ij}\partial_{x_j} u) = f, \quad x \in \Omega. \tag{2.62}$$

The above PDE is the Euler–Lagrange equation associated with the functional of the form (2.59) and the Lagrangian

$$L(p, z, x) = \frac{1}{2}\sum_{i,j=1}^{d} a^{ij}(x) p_i p_j - z f(x). \tag{2.63}$$

More examples and conditions under which $I[\cdot]$ has a (unique) minimizer can be found in Evans (2022), for example.

Solving a PDE within the class of variational problems by minimizing its associated energy functional is known in the literature as the *Ritz (–Galerkin) method*, or as the *Rayleigh–Ritz method* for eigenvalue problems. More recently, E and Yu (2018) have proposed the *deep Ritz method*, which aims to minimize the energy functional over the set of deep neural networks:

$$\min_{\theta} \mathcal{E}_{\mathrm{DRM}}[u_\theta]^2, \quad \mathcal{E}_{\mathrm{DRM}}[u_\theta]^2 := I[u_\theta] + \lambda \int_{\partial\Omega} (\mathcal{B}[u_\theta] - g)^2, \tag{2.64}$$

where the second term was added because the neural networks used do not automatically satisfy the boundary conditions of the PDE (2.1).

2.4. *Quadrature rules*

All the loss functions discussed in the previous subsections are formulated as integrals in terms of the model u_θ. Very often it will not be feasible or computationally desirable to evaluate these integrals in an exact way. Instead, one can discretize the integral and approximate it using a sum. We recall some basic results on numerical quadrature rules for integral approximation.

For a mapping $g\colon \Omega \to \mathbb{R}^m$, $\Omega \subset \mathbb{R}^{\overline{d}}$, such that $g \in L^1(\Omega)$, we are interested in approximating the integral

$$\overline{g} := \int_\Omega g(y)\,\mathrm{d}y,$$

with $\mathrm{d}y$ denoting the Lebesgue measure on Ω. In order to approximate the above integral by a quadrature rule, we need the quadrature points $y_i \in \Omega$ for $1 \le i \le N$, for some $N \in \mathbb{N}$ as well as weights w_i, with $w_i \in \mathbb{R}_+$. Then a quadrature is defined by

$$\mathcal{Q}_N^\Omega[g] := \sum_{i=1}^N w_i g(y_i), \tag{2.65}$$

for weights w_i and quadrature points y_i. Depending on the choice of weights and quadrature points, as well as the regularity of g, the quadrature error is bounded by

$$|\overline{g} - \mathcal{Q}_N^\Omega[g]| \le C_{\mathrm{quad}} N^{-\alpha}, \tag{2.66}$$

for some $\alpha > 0$ and where C_{quad} depends on g and its properties.

As an elementary example, we mention the *midpoint rule*. For $M \in \mathbb{N}$, we partition Ω into $N \sim M^d$ cubes of edge length $1/M$ and we let $\{y_i\}_{i=1}^N$ denote the midpoints of these cubes. The formula and accuracy of the midpoint rule $\mathcal{Q}_N^\Omega$ are then given by

$$\mathcal{Q}_N^\Omega[g] := \frac{1}{N}\sum_{i=1}^N g(y_i), \quad |\overline{g} - \mathcal{Q}_M^\Omega[g]| \le C_g N^{-2/d}, \tag{2.67}$$

where $C_g \lesssim \|g\|_{C^2}$.

As long as the domain Ω is in reasonably low dimensions, i.e. $\overline{d} \le 4$, we can use the midpoint rule and more general standard (composite) Gauss quadrature rules on an underlying grid. In this case, the quadrature points and weights depend on the order of the quadrature rule (Stoer and Bulirsch 2002) and the rate α depends on the regularity of the underlying integrand. On the other hand, these grid-based quadrature rules are not suitable for domains in high dimensions. For moderately high dimensions, i.e. $4 \le \overline{d} \le 20$, we can use low-discrepancy sequences, such as the Sobol and Halton sequences, as quadrature points. As long as the integrand g is of bounded Hardy–Krause variation, the error in (2.66) converges at a rate $(\log(N))^{\overline{d}} N^{-1}$ (Longo, Mishra, Rusch and Schwab 2021). For problems in very high dimensions, $d \gg 20$, Monte Carlo quadrature is the numerical integration method of choice. In this case, the quadrature points are randomly chosen, independent and identically distributed (with respect to a scaled Lebesgue measure).

As an example, we discuss how the physics-informed loss based on the classical formulation for a time-dependent PDE (2.46) can be discretized. As the loss function (2.46) contains integrals over three different domains ($D \times [0,T]$, $\partial D \times [0,T]$ and D), we must consider three different quadratures. We consider quadratures

(2.65) for which the weights are the inverse of the number of grid points, such as for the midpoint or Monte Carlo quadrature, and we denote the quadrature points by $\mathcal{S}_i = \{(z_i)\}_{1\le i\le N_i} \subset D\times[0,T]$, $\mathcal{S}_s = \{y_i\}_{1\le i\le N_s} \subset \partial D\times[0,T]$ and $\mathcal{S}_t = \{x_i\}_{1\le i\le N_t} \subset D$. Following machine learning terminology, we refer to the set of all quadrature points $\mathcal{S} = (\mathcal{S}_i, \mathcal{S}_s, \mathcal{S}_t)$ as the *training set*. The resulting loss function $\mathcal{J}$ now depends not only on u_θ but also on $\mathcal{S}$, and is given by

$$\begin{aligned}\mathcal{J}(\theta,\mathcal{S}) &= \mathcal{Q}_{N_i}^{D\times[0,T]}[\mathcal{R}_{\mathrm{PDE}}^2] + \lambda_s \mathcal{Q}_{N_s}^{\partial D\times[0,T]}[\mathcal{R}_s^2] + \lambda_t \mathcal{Q}_{N_t}^{D}[\mathcal{R}_t^2] \\ &= \frac{1}{N_i}\sum_{i=1}^{N_i}(\mathcal{L}[u_\theta](z_i) - f(z_i))^2 + \frac{\lambda_s}{N_s}\sum_{i=1}^{N_s}(u_\theta(y_i) - u(y_i))^2 \\ &\quad + \frac{\lambda_t}{N_t}\sum_{i=1}^{N_t}(u_\theta(x_i,0) - u(x_i,0))^2. \end{aligned} \tag{2.68}$$

For any of the other loss functions defined in Section 2.3, a similar discretization can be defined. In practice, physics-informed learning thus comes down to solving the minimization problem given by

$$\theta^*(\mathcal{S}) := \operatorname*{argmin}_{\theta\in\Theta} \mathcal{J}(\theta,\mathcal{S}). \tag{2.69}$$

The final optimized or *trained* model is then given by $u^* := u_{\theta^*(\mathcal{S})}$.

2.5. *Optimization*

Since Θ is high-dimensional and the map $\theta \mapsto \mathcal{J}(\theta,\mathcal{S})$ is generally non-convex, solving the minimization problem (2.69) can be very challenging. Fortunately, the loss function $\mathcal{J}$ is almost everywhere differentiable, so gradient-based iterative optimization methods can be used.

The simplest example of such an algorithm is *gradient descent*. Starting from an initial guess θ_0, the idea is to take a small step in the parameter space Θ in the direction of the steepest descent of the loss function to obtain a new guess θ_1. Note that this comes down to taking a small step in the opposite direction to the gradient evaluated in θ_0. Repeating this procedure yields the iterative formula

$$\theta_{\ell+1} = \theta_\ell - \eta_\ell \nabla_\theta \mathcal{J}(\theta_\ell,\mathcal{S}) \quad \text{for all } \ell\in\mathbb{N}, \tag{2.70}$$

where the learning rate η_ℓ controls the size of the step and is generally quite small. The gradient descent formula yields a sequence of parameters $\{\theta_\ell\}_{\ell\in\mathbb{N}}$ that converge to a local minimum of the loss function under very general conditions. Because of the non-convexity of the loss function, convergence to a global minimum cannot be ensured. Another issue lies in the computation of the gradient $\nabla_\theta\mathcal{J}(\theta_\ell,\mathcal{S})$. A simple rewriting of the definition of this gradient (up to regularization term),

$$\nabla_\theta\mathcal{J}(\theta_\ell,\mathcal{S}) = \frac{1}{N}\sum_{i=1}^{N}\nabla_\theta\mathcal{J}_i(\theta_\ell), \quad \text{where } \mathcal{J}_i(\theta_\ell) = \mathcal{J}(\theta_\ell, \{(x_i, f(x_i))\}), \tag{2.71}$$

reveals that we actually need to evaluate $N = |\mathcal{S}|$ gradients. As a consequence, the computation of the full gradient will be very slow and memory intensive for large training data sets.

Stochastic gradient descent (SGD) overcomes this problem by only calculating the gradient in one data point instead of the full training data set. More precisely, for every ℓ a random index i_ℓ, with $1 \le i_\ell \le N$, is chosen, resulting in the following update formula:

$$\theta_{\ell+1} = \theta_\ell - \eta_\ell \nabla_\theta \mathcal{J}_{i_\ell}(\theta_\ell) \quad \text{for all } \ell \in \mathbb{N}. \tag{2.72}$$

Similarly to gradient descent, stochastic gradient descent provably converges to a local minimum of the loss function under rather general conditions, although the convergence will be slower. In particular, the convergence will be noisier and there is no guarantee that $\mathcal{J}(\theta_{\ell+1}) \le \mathcal{J}(\theta_\ell)$ for every ℓ. Because the cost of each iteration is much lower than that of gradient descent, it might still make sense to use stochastic gradient descent.

Mini-batch gradient descent tries to combine the advantages of gradient descent and SGD by evaluating the gradient of the loss function in a subset of $\mathcal{S}$ in each iteration (as opposed to the full training set for gradient descent and one training sample for SGD). More precisely, for every $\ell \in \mathbb{N}$ a subset $\mathcal{S}_\ell \subset \mathcal{S}$ of size m is chosen. These subsets are referred to as *mini-batches* and their size m as the mini-batch size. In contrast to SGD, these subsets are not entirely randomly selected in practice. Instead, we randomly partition the training set into $\lceil N/m \rceil$ mini-batches, which are then all used as a mini-batch $\mathcal{S}_\ell$ for consecutive ℓ. Such a cycle of $\lceil N/m \rceil$ iterations is called an *epoch*. After every epoch, the mini-batches are reshuffled, meaning that a new partition of the training set is drawn. In this way, every training sample is used once during every epoch. The corresponding update formula reads as

$$\theta_{\ell+1} = \theta_\ell - \eta_\ell \nabla_\theta \mathcal{J}(\theta_\ell, \mathcal{S}_\ell) \quad \text{for all } \ell \in \mathbb{N}. \tag{2.73}$$

By setting $m = N$ (resp. $m = 1$), we retrieve gradient descent (resp. SGD). For this reason, gradient descent is also often referred to as *full batch* gradient descent.

The performance of mini-batch gradient descent can be improved in many ways. One particularly popular example of such an improvement is the *ADAM* optimizer (Kingma and Ba 2015). ADAM, whose name is derived from *adaptive moment estimation*, calculates at each ℓ (exponential) moving averages of the first and second moment of the mini-batch gradient g_ℓ,

$$m_\ell = \beta_1 m_{\ell-1} + (1-\beta_1) g_\ell, \quad v_\ell = \beta_2 v_{\ell-1} + (1-\beta_2) g_\ell^2, \quad g_\ell = \nabla_\theta \mathcal{J}(\theta_\ell, \mathcal{S}_\ell), \tag{2.74}$$

where β_1 and β_2 are close to 1. However, one can calculate that these moving averages are biased estimators. This bias can be removed using the following correction:

$$\widehat{m}_\ell = \frac{m_\ell}{1-\beta_1^\ell}, \quad \widehat{v}_\ell = \frac{v_\ell}{1-\beta_2^\ell} \quad \text{for all } \ell \in \mathbb{N}, \tag{2.75}$$

where $\widehat{m}_\ell, \widehat{v}_\ell$ are unbiased estimators. One can obtain ADAM from the basic mini-batch gradient descent update formula (2.73) by replacing $\nabla_\theta \mathcal{J}(\theta_\ell, \mathcal{S}_\ell)$ with the moving average m_ℓ and by setting $\eta_\ell = \alpha/\sqrt{\widehat{v}_\ell + \epsilon}$, where $\alpha > 0$ is small and $\epsilon > 0$ is close to machine precision. This leads to the update formula

$$\theta_{\ell+1} = \theta_\ell - \alpha \frac{\widehat{m}_\ell}{\sqrt{\widehat{v}_\ell + \epsilon}} \quad \text{for all } \ell \in \mathbb{N}. \tag{2.76}$$

Compared to mini-batch gradient descent, the convergence of ADAM will be less noisy for two reasons. First, a fraction of the noise will be removed since a moving average of the gradient is used. Second, the step size of ADAM decreases if the gradient g_ℓ varies a lot, i.e. when $\widehat{v}_\ell$ is large. These design choices mean that ADAM performs well on a large number of tasks, making it one of the most popular optimizers in deep learning.

Finally, the interest in *second-order* optimization algorithms has been steadily growing over the past years. Most second-order methods used in practice are adaptations of the well-known Newton method,

$$\theta_{\ell+1} = \theta_\ell - \eta_\ell (\nabla^2_\theta \mathcal{J}(\theta, \mathcal{S}))^{-1} \nabla_\theta \mathcal{J}(\theta, \mathcal{S}) \quad \text{for all } \ell \in \mathbb{N}, \tag{2.77}$$

where $\nabla^2_\theta \mathcal{J}(\theta, \mathcal{S})$ denotes the Hessian of the loss function. The generalization to a mini-batch version is immediate. As the computation of the Hessian and its inverse is quite costly, we often use a *quasi*-Newton method that computes an approximate inverse by solving the linear system $\nabla^2_\theta \mathcal{J}(\theta, \mathcal{S}) h_\ell = \nabla_\theta \mathcal{J}(\theta, \mathcal{S})$. An example of a popular quasi-Newton method in deep learning for scientific computing is *L-BFGS* (Liu and Nocedal 1989). In many other application areas, however, first-order optimizers remain the dominant choice, for a myriad of reasons.

2.5.1. *Parameter initialization*

All iterative optimization algorithms above require an initial guess θ_0. Making this initial guess is referred to as parameter initialization or weights initialization (for neural networks). Although it is often overlooked, a bad initialization can cause a lot of problems, as will be discussed in Section 7.

Many initialization schemes have already been considered, in particular for neural networks. When the initial weights are too large, the corresponding gradients calculated by the optimization algorithm might be very large, leading to the *exploding gradient problem*. Similarly, when the initial weights are too small, the gradients might also be very close to zero, leading to the *vanishing gradient problem*. Fortunately, it is possible to choose the variance of the initial weights in such a way that the output of the network has the same order of magnitude as the input of the network. One such possible choice is *Xavier initialization* (Glorot and

Bengio 2010), meaning that the weights are initialized as

$$(W_k)_{ij} \sim N\left(0, \frac{2g^2}{d_{k-1}+d_k}\right) \quad \text{or} \quad (W_k)_{ij} \sim U\left(-g\sqrt{\frac{6}{d_{k-1}+d_k}}, g\sqrt{\frac{6}{d_{k-1}+d_k}}\right), \tag{2.78}$$

where N denotes the normal distribution, U is the uniform distribution, d_k is the output dimension of the kth layer, and the value g depends on the activation function. For the ReLU activation we set $g = \sqrt{2}$, and for the tanh activation function $g = 1$ is a valid choice. Note that this initialization has primarily been proposed with a supervised learning setting in mind, e.g. when using a square loss such as $\int_\Omega (u_\theta - u)$. For the different loss functions of Section 2.3, different initialization schemes might be better suited. This will be discussed in Section 7.

Finally, it is customary to retrain neural network with different starting values θ_0 (drawn from the same distribution) as the gradient-based optimizer might converge to a different local minimum for each θ_0. One can then choose the 'best' neural network based on some suitable criterion, or combine the different neural networks if the setting allows. Also note that this task can be performed in parallel.

2.6. *Summary of algorithm*

We summarize the physics-informed learning algorithm as follows.

(1) Choose a model class $\{u_\theta : \theta \in \Theta\}$ (Section 2.2).

(2) Choose a loss function based on the classical, weak or variational formulation of the PDE (Section 2.3).

(3) Generate a training set $\mathcal{S}$ based on a suitable quadrature rule to obtain a feasibly computable loss function $\mathcal{J}(\theta, \mathcal{S})$ (Section 2.4).

(4) Initialize the model parameters and run an optimization algorithm from Section 2.5 until an approximate local minimum $\theta^*(\mathcal{S})$ is reached. The resulting function $u^* := u_{\theta^*(\mathcal{S})}$ is the desired model for approximating the solution u of the PDE (2.1).

3. Analysis

As intuitive as the physics-informed learning framework of Section 2.6 might seem, there is *a priori* little theoretical guarantee that it will actually work well. The aim of the next sections is therefore to theoretically analyse physics-informed machine learning and to give an overview of the available mathematical guarantees.

One central element will be the development of error estimates for physics-informed machine learning. The relevant concept of error is the error emanating from approximating the solution u of (2.1) by the model $u^* = u_{\theta^*(\mathcal{S})}$:

$$\mathcal{E}(\theta) := \|u - u_\theta\|_X, \quad \mathcal{E}^* := \mathcal{E}(\theta^*(\mathcal{S})). \tag{3.1}$$

In general, we will choose $\|\cdot\|_X = \|\cdot\|_{L^2(\Omega)}$. Note that it is not possible to compute $\mathcal{E}$ during the training process. On the other hand, we monitor the so-called *training error* given by

$$\mathcal{E}_T(\theta, \mathcal{S})^2 := \mathcal{J}(\theta, \mathcal{S}), \quad \mathcal{E}_T^* := \mathcal{E}_T(\theta^*(\mathcal{S}), \mathcal{S}), \tag{3.2}$$

with $\mathcal{J}$ being the loss function defined in Section 2.4 as the discretized version of the (integral-form) loss functions in Section 2.3. Hence, the training error $\mathcal{E}_T^*$ can be readily computed, after training has been completed, from the loss function. We will also refer to the integral form of the physics-informed loss function (see Section 2.3) as the *generalization error* $\mathcal{E}_G$,

$$\mathcal{E}_G(\theta) := \mathcal{E}_{\mathrm{PIL}}[u_\theta], \quad \mathcal{E}_G^* := \mathcal{E}_G(\theta^*(\mathcal{S}), \mathcal{S}), \tag{3.3}$$

as it measures how well the performance of a model transfers or *generalizes* from the training set $\mathcal{S}$ to the full domain Ω.

Given these definitions, some fundamental theoretical questions arise immediately.

Q1. *Existence.* Does there exist a model $\widehat{u}$ in the hypothesis class for which the generalization error $\mathcal{E}_G(\widehat{\theta})$ is small? More precisely, given a chosen error tolerance $\epsilon > 0$, does there exist a model in the hypothesis class $\widehat{u} = u_{\widehat{\theta}}$ for some $\widehat{\theta} \in \Theta$ such that for the corresponding generalization error $\mathcal{E}_G(\widehat{\theta})$ it holds that $\mathcal{E}_G(\widehat{\theta}) < \epsilon$? As minimizing the generalization error (i.e. the physics-informed loss) is the key pillar of physics-informed learning, obtaining a positive answer to this question is of the utmost importance. Moreover, it would be desirable to relate the size of the parameter space Θ (or hypothesis class) to the accuracy ϵ. For example, for linear models (Section 2.2.1) we would want to know how many functions ϕ_i are needed to ensure the existence of $\widehat{\theta} \in \Theta$ such that $\mathcal{E}_G(\widehat{\theta}) < \epsilon$. Similarly, for neural networks, we wish to find estimates of how large a neural network (Section 2.2.2) should be (in terms of depth, width and modulus of the weights) in order to ensure this. We will answer this question by considering the *approximation error* of a model class in Section 4.

Q2. *Stability.* Given that a model u_θ has a small generalization error $\mathcal{E}_G(\theta)$, will the corresponding total error $\mathcal{E}(\theta)$ be small as well? In other words, is there a function $\delta\colon \mathbb{R}_+ \to \mathbb{R}_+$ with $\lim_{\epsilon\to 0} \delta(\epsilon) = 0$ such that $\mathcal{E}_G(\theta) < \epsilon$ implies that $\mathcal{E}(\theta) < \delta(\epsilon)$ for any $\epsilon > 0$? In practice, we will be even more ambitious, as we hope to find constants $C, \alpha > 0$ (independent of θ) such that

$$\mathcal{E}(\theta) \leq C\mathcal{E}_G(\theta)^\alpha. \tag{3.4}$$

This is again an essential ingredient, as obtaining a small generalization error $\mathcal{E}_G^*$ is *a priori* meaningless, given that we only care about the total error $\mathcal{E}^*$. The answer to this question first and foremost depends on the properties of the underlying PDE, and only to a lesser extent on the model class. We will discuss this question in Section 5.

Q3. *Generalization.* Given a small training error $\mathcal{E}_T^*$ and a sufficiently large training set $\mathcal{S}$, will the corresponding generalization error $\mathcal{E}_G^*$ also be small? This question is not uniquely tied to physics-informed machine learning and is closely tied to the accuracy of the chosen quadrature rule (Section 2.4). We will use standard arguments to answer this question in Section 6.

Q4. *Optimization.* Can $\mathcal{E}_T^*$ be made sufficiently close to $\min_\theta \mathcal{J}(\theta, \mathcal{S})$? This final question acknowledges the fact that it might be very hard to solve the minimization problem (2.69), even approximately. Especially for physics-informed neural networks there is a growing literature on the settings in which the optimization problem is hard to solve, or even infeasible, and how to potentially overcome these training issues. We theoretically analyse these phenomena in Section 7.

The above questions are of fundamental importance, as affirmative answers certify that the model that minimizes the optimization problem (2.69), denoted by u^*, is a good approximation of u in the sense that the error $\mathcal{E}^*$ is small. We show this by first proving an error decomposition for the generalization error $\mathcal{E}_G^*$, for which we need the following auxiliary result.

Lemma 3.1. For any $\theta_1, \theta_2 \in \Theta$ and training set $\mathcal{S}$, we have

$$\mathcal{E}_G(\theta_1) \le \mathcal{E}_G(\theta_2) + 2 \sup_{\theta \in \Theta} |\mathcal{E}_T(\theta, \mathcal{S}) - \mathcal{E}_G(\theta)| + \mathcal{E}_T(\theta_1, \mathcal{S}) - \mathcal{E}_T(\theta_2, \mathcal{S}). \tag{3.5}$$

Proof. Fix $\theta_1, \theta_2 \in \Theta$ and generate a random training set $\mathcal{S}$. The proof consists of the repeated adding, subtracting and removing of terms. We have

$$\begin{aligned}\mathcal{E}_G(\theta_1) &= \mathcal{E}_G(\theta_2) + \mathcal{E}_G(\theta_1) - \mathcal{E}_G(\theta_2)\\ &= \mathcal{E}_G(\theta_2) - (\mathcal{E}_T(\theta_1, \mathcal{S}) - \mathcal{E}_G(\theta_1)) + (\mathcal{E}_T(\theta_2, \mathcal{S}) - \mathcal{E}_G(\theta_2))\\ &\quad + \mathcal{E}_T(\theta_1, \mathcal{S}) - \mathcal{E}_T(\theta_2, \mathcal{S})\\ &\le \mathcal{E}_G(\theta_2) + 2 \max_{\theta \in \{\theta_2, \theta_1\}} |\mathcal{E}_T(\theta, \mathcal{S}) - \mathcal{E}_G(\theta)|\\ &\quad + \mathcal{E}_T(\theta_1, \mathcal{S}) - \mathcal{E}_T(\theta_2, \mathcal{S}).\end{aligned} \tag{3.6}$$

□

Assuming that we can answer Q1 in the affirmative, we can now take $\theta_1 = \theta^*(\mathcal{S})$ and $\theta_2 = \widehat{\theta}$ (as in Q1). Moreover, if we also assume that we can positively answer Q2, or more strongly we can find constants $C, \alpha > 0$ such that (3.4) holds, then we find the following error decomposition:

$$\mathcal{E}^* \le C\Big(\underbrace{\mathcal{E}_G(\widehat{\theta})}_{\text{approximation error}} \underbrace{+2 \sup_{\theta \in \Theta} |\mathcal{E}_T(\theta, \mathcal{S}) - \mathcal{E}_G(\theta)| +}_{\text{generalization gap}} \underbrace{\mathcal{E}_T^* - \mathcal{E}_T(\widehat{\theta})}_{\text{optimization error}} \Big)^\alpha. \tag{3.7}$$

The total error $\mathcal{E}^*$ is now decomposed into three error sources. The *approximation error* $\mathcal{E}(\widehat{\theta})$ should be provably small, by answering Q1. The second term of the right-hand side, called the *generalization gap*, quantifies how well the training error approximates the generalization error, corresponding to Q3. Finally, an *optimization error* is incurred due to the inability of the optimization procedure to find $\widehat{\theta}$ based on a finite training data set. Indeed one can see that if $\theta^*(\mathcal{S}) = \widehat{\theta}$, the optimization error vanishes. This source of error is addressed by Q4. Hence, an affirmative answer to Q1–Q4 leads to a small error $\mathcal{E}^*$.

We will present general results to answer questions Q1–Q4 and then apply them to the following prototypical examples to further highlight interesting phenomena.

- The Navier–Stokes equation (Example 2.2) as an example of a challenging but low-dimensional PDE.
- The heat equation (Example 2.1) as a prototypical example of a potentially very high-dimensional PDE. We will investigate whether one can overcome the *curse of dimensionality* through physics-informed learning for this equation and related PDEs.
- Inviscid scalar conservation laws (Example 2.3 with $\nu = 0$) will serve as an example of a PDE where the solutions might not be smooth and even discontinuous. As a result, the weak residuals from Section 2.3.2 will be needed.
- Poisson's equation (Example 2.4) as an example for which the variational formulation of Section 2.3.3 can be used.

4. Approximation error

In this section we answer the first question, Q1.

Question 1 (Q1). Does there exist a model $\widehat{u} = u_{\widehat{\theta}}$ in the hypothesis class for which the generalization error $\mathcal{E}_G(\widehat{\theta})$ is small?

The difficulty in answering this question lies in the fact that the generalization error $\mathcal{E}_G(\theta)$ in physics-informed learning is given by the physics-informed loss of the model $\mathcal{E}_{\mathrm{PIL}}(u_\theta)$ rather than just being $\|u - u_\theta\|_{L^2}^2$ as would be the case for regression. For neural networks, for example, the universal approximation theorems (e.g. Cybenko 1989) guarantee that any measurable function can be approximated by a neural network in supremum-norm. However, they do not imply an affirmative answer to Question 1: a neural network that approximates a function well in supremum-norm might be highly oscillatory, such that the derivatives of the network and that of the function are very different, giving rise to a large PDE residual. Hence, we require results on the existence of models that approximate functions in a *stronger norm* than the supremum-norm. In particular, the norm should quantify how well the derivatives of the model approximate those of the function.

For solutions of PDEs, a very natural norm that satisfies this criterion is the *Sobolev norm.* For $s \in \mathbb{N}$ and $q \in [1, \infty]$, the Sobolev space $W^{s,q}(\Omega; \mathbb{R}^m)$ is the space of all functions $f\colon \Omega \to \mathbb{R}^m$ for which f, as well as the (weak) derivatives of f up to order s, belong to $L^q(\Omega; \mathbb{R}^m)$. When more smoothness is available, one could replace the space $W^{s,\infty}(\Omega; \mathbb{R}^m)$ with the space of s times continuously differentiable functions $C^s(\Omega; \mathbb{R}^m)$; for more information see the Appendix.

Given an approximation result in some Sobolev (semi-) norm, we can derive the physics-informed loss of the model $\mathcal{E}_{\mathrm{PIL}}(u_\theta)$ if the following assumption holds true.

When the physics-informed loss is based on the strong (classical) formulation of the PDE (see Section 2.3.1), we assume that it can be bounded in terms of the errors related to all relevant partial derivatives, denoted by $D^{(k,\alpha)} := D_t^k D_x^\alpha := \partial_t^k \partial_{x_1}^{\alpha_1} \ldots \partial_{x_d}^{\alpha_d}$, for $(k, \alpha) \in \mathbb{N}_0^{d+1}$ for time-dependent PDEs. This assumption is valid for many classical solutions of PDEs.

Assumption 4.1. Let $k, \ell \in \mathbb{N}$, $q \in [1, +\infty]$, $C > 0$ be independent from d and let $\mathcal{X} \subset \{u_\theta\colon \theta \in \Theta\}$ and $\Omega = [0, T] \times D$. It holds for every $u_\theta \in \mathcal{X}$ that

$$\|\mathcal{L}[u_\theta]\|_{L^q(\Omega)} \leq C \cdot \mathrm{poly}(d) \cdot \sum_{\substack{(k',\alpha)\in\mathbb{N}_0^{d+1} \\ k'\leq k, \|\alpha\|_1 \leq \ell}} \|D^{(k',\alpha)}(u - u_\theta)\|_{L^q(\Omega)}. \tag{4.1}$$

Note that the corresponding assumption for operator learning can be obtained by replacing u with the operator $\mathcal{G}$ and u_θ with $\mathcal{G}_\theta$. We give a brief example to illustrate the assumption.

Example 4.2 (heat equation). For the heat equation we have

$$\begin{aligned} \|\mathcal{L}[u_\theta]\|_{L^q(\Omega)} &= \|\mathcal{L}[u_\theta] - \mathcal{L}[u]\|_{L^q(\Omega)} \\ &\leq \|\Delta(u_\theta - u)\|_{L^q(\Omega)} + \|\partial_t(u_\theta - u)\|_{L^q(\Omega)}. \end{aligned} \tag{4.2}$$

4.1. *First general result*

Note in particular that Assumption 4.1 is satisfied when

$$\|\mathcal{L}[u_\theta]\|_{L^q(\Omega)} \leq C \cdot \mathrm{poly}(d) \cdot \|u - u_\theta\|_{W^{k,q}(\Omega)}. \tag{4.3}$$

Hence, it suffices to prove that there exists $\widehat{\theta} \in \Theta$ for which $\|u - u_{\widehat{\theta}}\|_{W^{k,q}(\Omega)}$ can be made small.

For many linear models, such approximation results are widely available. For example, when we choose the (default enumeration of the) d-dimensional Fourier basis as functions ϕ_i (see Section 2.2.1), then we can use the following result (Hesthaven *et al.* 2007).

Theorem 4.3. Let $s, k \in \mathbb{N}_0$ with $s > d/2$ and $s \geq k$, and $u \in C^s(\mathbb{T}^d)$. Then there exists $\widehat{\theta} \in \mathbb{R}^N$ such that

$$\left\| u - \sum_{i=1}^{N^d} \widehat{\theta}_i \phi_i \right\|_{H^k(\mathbb{T}^d)} \leq C(s,d) N^{-(s-k)/d} \|u\|_{H^s(\mathbb{T}^d)}, \tag{4.4}$$

for a constant $C(s,d) > 0$ that only depends on s and d.

Analogous results are also available for neural networks. We state a result that is tailored to tanh neural networks (De Ryck, Lanthaler and Mishra 2021, Theorem 5.1), but more general results are also available (Gühring and Raslan 2021).

Theorem 4.4. Let $\Omega = [0,1]^d$. For every $N \in \mathbb{N}$ and every $f \in W^{s,\infty}(\Omega)$, there exists a tanh neural network $\widehat{f}$ with two hidden layers of width N such that for every $0 \leq k < s$ we have

$$\|f - \widehat{f}\|_{W^{k,\infty}(\Omega)} \leq C(\ln(cN))^k N^{-(s-k)/d}, \tag{4.5}$$

where $c, C > 0$ are independent of N and explicitly known.

Proof. The main ingredients are a piecewise polynomial approximation, the existence of which is guaranteed by the Bramble–Hilbert lemma (Verfürth 1999), and the ability of tanh neural networks to efficiently approximate polynomials, the multiplication operator and an approximate partition of unity. The full proof can be found in De Ryck *et al.* (2021, Theorem 5.1). □

Remark 4.5. In this section we have mainly focused on bounding the (strong) PDE residual, as it is generally the most difficult residual to bound, containing the highest derivatives. Indeed, when using the classical formulation, the bounds on the spatial (and temporal) boundary residuals generally follow from a (Sobolev) trace inequality. Similarly, loss functions resulting from the weak or variational formulation contain less high derivatives of the neural network and are therefore easier to bound.

As an example, we apply the above result to the Navier–Stokes equations.

Example 4.6 (Navier–Stokes). One can use Theorem 4.4 to prove the existence of neural networks with a small generalization error for the Navier–Stokes equations (Example 2.2). The existence and regularity of the solution to (2.8) depends on the regularity of u_0, as is stated by the following well-known theorem (Majda and Bertozzi 2002, Theorem 3.4). Other regularity results with different boundary conditions can be found in Temam (2001), for example.

Theorem 4.7. If $u_0 \in H^r(\mathbb{T}^d)$ with $r > d/2 + 2$ and $\operatorname{div} u_0 = 0$, then there exist $T > 0$ and a classical solution u to the Navier–Stokes equation such that $u(t=0) = u_0$ and $u \in C([0,T]; H^r(\mathbb{T}^d)) \cap C^1([0,T]; H^{r-2}(\mathbb{T}^d))$.

Based on this result one can prove that u is Sobolev-regular, that is, $u \in H^k(D \times [0,T])$ for some $k \in \mathbb{N}$, provided that r is large enough (De Ryck, Jagtap and Mishra 2024*a*).

Corollary 4.8. If $k \in \mathbb{N}$ and $u_0 \in H^r(\mathbb{T}^d)$ with $r > d/2 + 2k$ and $\operatorname{div} u_0 = 0$, then there exist $T > 0$ and a classical solution u to the Navier–Stokes equation such that $u \in H^k(\mathbb{T}^d \times [0,T])$, $\nabla p \in H^{k-1}(\mathbb{T}^d \times [0,T])$ and $u(t = 0) = u_0$.

Combining this regularity result with Theorem 4.4 then gives rise to the following approximation result for the Navier–Stokes equations (De Ryck *et al.* 2024*a*, Theorem 3.1).

Theorem 4.9. Let $n \geq 2$, $d, r, k \in \mathbb{N}$, with $k \geq 3$, and let $u_0 \in H^r(\mathbb{T}^d)$ with $r > d/2 + 2k$ and $\operatorname{div} u_0 = 0$. Let $T > 0$ be the time from Corollary 4.8 such that the classical solution of u exists on $\mathbb{T}^d \times [0,T]$. Then, for every $N > 5$, there exist tanh neural networks $\widehat{u}_j$, $1 \leq j \leq d$, and $\widehat{p}$, each with two hidden layers, of widths

$$3\left\lceil \frac{k+n-2}{2} \right\rceil \binom{d+k-1}{d} + \lceil TN \rceil + dN \quad \text{and} \quad 3\left\lceil \frac{d+n}{2} \right\rceil \binom{2d+1}{d} \lceil TN \rceil N^d,$$

such that

$$\|(\widehat{u}_j)_t + \widehat{u} \cdot \nabla \widehat{u}_j + (\nabla \widehat{p})_j - \nu \Delta \widehat{u}_j\|_{L^2(\Omega)} \leq C_1 \ln^2(\beta N) N^{-k+2}, \tag{4.6}$$

$$\|\operatorname{div} \widehat{u}\|_{L^2(\Omega)} \leq C_2 \ln(\beta N) N^{-k+1}, \tag{4.7}$$

$$\|(u_0)_j - \widehat{u}_j(t=0)\|_{L^2(\mathbb{T}^d)} \leq C_3 \ln(\beta N) N^{-k+1}, \tag{4.8}$$

for every $1 \leq j \leq d$, and where the constants β, C_1, C_2, C_3 are explicitly defined in the proof and can depend on k, d, T, u and p but not on N. The weights of the networks can be bounded by $O(N^\gamma \ln(N))$, where $\gamma = \max\{1, d(2 + k^2 + d)/n\}$.

4.2. *Second general result*

Combining results such as Theorem 4.3 or Theorem 4.4 with Assumption 4.1 thus allows us to prove that the generalization error $\mathcal{E}_G$ can be made arbitrarily small, thereby answering Question 1. There is, however, still room for improvement.

- For many model classes the main results available only state that the error $\|u - u_\theta\|_{L^q}$ can be made small. Is there a way to infer that if $\|u - u_\theta\|_{L^q}$ is small then $\|u - u_\theta\|_{W^{k,q}}$ is also small (under some assumptions)?
- Both Theorem 4.3 and Theorem 4.4 suffer from the *curse of dimensionality* (CoD), meaning that the number of parameters needed to reach an accuracy of ϵ scales exponentially with the input dimension d, namely as $\epsilon^{-(s-k)/d}$. Unless $s \approx k + d$, this will mean that an infeasibly large number of parameters is needed to guarantee a sufficiently small generalization error. Experimentally, however, we have observed for many PDEs that the CoD can be overcome. This raises the following question: Can we improve upon the approximation results stated in this section?

- Both Theorem 4.3 and Theorem 4.4 require the true solution u of the PDE to be sufficiently (Sobolev) regular. Is there a way to still prove that the approximation error is small if u is less regular, such as for inviscid scalar conservation laws?

As it turns out, the first two questions above can be answered in the same manner, which is the topic below.

We first show that, under some assumptions, we can indeed answer Question 1 even if we only have access to an approximation result for $\|u - u_\theta\|_{L^q}$ and not for $\|u - u_\theta\|_{W^{k,q}}$. We do this by using finite difference (FD) approximations. Depending on whether forward, backward or central differences are used, an FD operator might not be defined on the whole domain D; for example, for $f \in C([0, 1])$ the (forward) operator $\Delta_h^+[f] := f(x + h) - f(x)$ is not well-defined for $x \in (1 - h, 1]$. This can be solved by resorting to piecewise defined FD operators, e.g. a forward operator on $[0, 0.5]$ and a backward operator on $(0.5, 1]$. In a general domain Ω one can find a well-defined piecewise FD operator if Ω satisfies the following assumption, which is satisfied by many domains (e.g. rectangular or smooth domains).

Assumption 4.10. There exists a finite partition $\mathcal{P}$ of Ω such that for all $P \in \mathcal{P}$ there exists $\epsilon_P > 0$, and $v_P \in B_\infty^1 = \{x \in \mathbb{R}^{\dim(\Omega)} : \|x\|_\infty \leq 1\}$ such that for all $x \in P$ we have $x + \epsilon_P(v_P + B_\infty^1) \subset \Omega$.

Under this assumption we can prove an upper bound on the (strong) PDE residual of the model u_θ.

Theorem 4.11. Let $q \in [1, \infty]$, $r, \ell \in \mathbb{N}$ with $\ell \leq r$ and $u, u_\theta \in C^r(\Omega)$. If Assumptions 4.1 and 4.10 hold, then there exists a constant $C(r) > 0$ such that for any $\alpha \in \mathbb{N}_0^d$ with $\ell := \|\alpha\|_1$, it holds for all $h > 0$ that

$$\|\mathcal{L}(u_\theta)\|_{L^q} \leq C\big(\|u - u_\theta\|_{L^q} h^{-\ell} + (|u|_{C^r} + |u_\theta|_{C^r}) h^{r-\ell}\big). \tag{4.9}$$

Proof. The proof follows from approximating any $D^\alpha(u_\theta - u)$ using an rth order accurate finite-difference formula (which is possible thanks to Assumption 4.10), and combining this with Assumption 4.1. See also De Ryck and Mishra (2022*b*, Lemma B.1). □

Theorem 4.11 shows that $\|\mathcal{L}(u_\theta)\|_{L^q}$ is small if $\|u - u_\theta\|_{L^q}$ is small and $|u_\theta|_{C^r}$ does not increase too much in terms of the model size. It must be stated that these assumptions are not necessarily trivially satisfied. Indeed, assume that $\Theta \subset \mathbb{R}^n$, that $\|u - u_\theta\|_{L^q} \sim n^{-\alpha}$ and $|u_\theta|_{C^r} \sim n^\beta$, and finally that $n \sim h^{-\gamma}$ for $\alpha, \beta, \gamma \geq 0$. In this case, the upper bound of Theorem 4.11 is equivalent to

$$\|\mathcal{L}(u_\theta)\|_{L^q} \lesssim h^{\alpha\gamma} h^{-\ell} + h^{-\beta\gamma} h^{r-\ell}. \tag{4.10}$$

To make sure that the right-hand side is small for $h \to 0$, the inequality $\beta\ell < \alpha(r-\ell)$ should hold. Hence, either $\|u - u_\theta\|_{L^q}$ should converge fast (large α), $|u_\theta|_{C^r}$

should diverge slowly (small β) or not at all ($\beta = 0$), or the true solution u should be very smooth (large r). By way of example, we investigate what kind of bound is produced by Theorem 4.11 if the Fourier basis is used.

Example 4.12 (Fourier basis). Theorem 4.3 tells us that $\alpha = r/d$ and that $\beta = 0$. Hence the condition $\beta\ell < \alpha(r-\ell)$ is already satisfied. We now choose $\gamma = d$ such that the optimal rate is obtained, for which $h^{\alpha\gamma}h^{-\ell} = h^{r-\ell}$. The final convergence rate is hence $N^{-(r-\ell)/d}$, in agreement with that of Theorem 4.3.

Remark 4.13. Unlike in the previous example, Theorem 4.11 is not expected to produce optimal convergence rates, particularly when loose upper bounds for $|u_\theta|_{C^r}$ are used. However, this is not a problem when trying to prove that a model class can overcome the curse of dimensionality in the approximation error, as is the topic of the next section.

Finally, as noted at the beginning of this section, all results in the section so far all assume that the solution u of the PDE is sufficiently (Sobolev) regular. In the next example we show that one can also prove approximation results, thereby answering Question 1, when u has discontinuities.

Example 4.14 (scalar conservation laws). In Example 2.10 a physics-informed loss based on the (weak) Kruzkhov entropy residual was introduced for scalar conservation laws (Example 2.3). It consists of the term $\max_{\vartheta,c} \mathcal{R}(u_\theta, \phi_\vartheta, c)$ together with the integral of the squared boundary residual $\mathcal{R}_s[u_\theta]$ and the integral of the squared initial condition residual $\mathcal{R}_t[u_\theta]$. Moreover, as solutions to scalar conservation laws might not be Sobolev-regular, an approximation result cannot be directly proved based on Theorem 4.4. Instead, De Ryck *et al.* (2024*c*) proved that the relevant counterpart to Assumption 4.1 is the following lemma.

Lemma 4.15. Let $p, q > 1$ be such that $1/p + 1/q = 1$, or let $p = \infty$ and $q = 1$. Let u be the entropy solution of (2.10) and let $v \in L^q(D \times [0,T])$. Assume that f has Lipschitz constant L_f. Then it holds for any $\varphi \in W_0^{1,\infty}(D \times [0,T])$ that

$$\mathcal{R}(v, \varphi, c) \leq (1 + 3L_f)|\varphi|_{W^{1,p}(D\times[0,T])}\|u - v\|_{L^q(D\times[0,T])}. \tag{4.11}$$

Hence we now need an approximation result for $\|u - u_\theta\|_{L^q(D\times[0,T])}$ rather than for a higher-order Sobolev norm of $u - u_\theta$. The following holds (De Ryck *et al.* 2024*c*, Lemma 3.3).

Lemma 4.16. For every $u \in BV([0,1] \times [0,T])$ and $\epsilon > 0$ there is a tanh neural network $\widehat{u}$ with two hidden layers and at most $O(\epsilon^{-2})$ neurons such that

$$\|u - \widehat{u}\|_{L^1([0,1]\times[0,T])} \leq \epsilon. \tag{4.12}$$

Proof. Write $\Omega = [0,1] \times [0,T]$. For every $u \in BV(\Omega)$ and $\epsilon > 0$ there exists a function $\widetilde{u} \in C^\infty(\Omega) \cap BV(\Omega)$ such that $\|u - \widetilde{u}\|_{L^1(\Omega)} \lesssim \epsilon$ and $\|\nabla\widetilde{u}\|_{L^1(\Omega)} \lesssim \|u\|_{BV(\Omega)} + \epsilon$ (Bartels 2012). Then use the approximation techniques of De Ryck *et al.* (2021, 2024*a*) and the fact that $\|\widetilde{u}\|_{W^{1,1}(\Omega)}$ can be uniformly bounded in ϵ to find the existence of a tanh neural network $\widehat{u}$ with two hidden layers and at most $O(\epsilon^{-2})$ neurons that satisfies the mentioned error estimate. □

If we additionally know that u is piecewise smooth, for instance as in the solutions of convex scalar conservation laws (Holden and Risebro 2015), then one can use the results of Petersen and Voigtlaender (2018) to obtain the following result (De Ryck *et al.* 2024c, Lemma 3.4).

Lemma 4.17. Let $m, n \in \mathbb{N}$, $1 \leq q < \infty$ and let $u\colon [0,1] \times [0,T] \to \mathbb{R}$ be a function that is piecewise C^m-smooth and with a C^n-smooth discontinuity surface. Then there is a tanh neural network $\widehat{u}$ with at most three hidden layers and $O(\epsilon^{-2/m} + \epsilon^{-q/n})$ neurons such that

$$\|u - \widehat{u}\|_{L^q([0,1]\times[0,T])} \leq \epsilon. \tag{4.13}$$

Finally, De Ryck *et al.* (2024c) found that we have to consider test functions φ that might grow as $|\varphi|_{W^{1,p}} \sim \epsilon^{-3(1+2(p-1)/p)}$. Consequently, we will need to use Lemma 4.16 with $\epsilon \leftarrow \epsilon^{4+6(p-1)/p}$, leading to the following corollary.

Corollary 4.18. Assume the setting of Lemma 4.17, assume that $mq > 2n$, and let $p \in [1, \infty]$ be such that $1/p + 1/q = 1$. There is a fixed-depth tanh neural network $\widehat{u}$ with size $O(\epsilon^{-(4q+6)n})$ such that

$$\max_{c\in\mathcal{C}} \sup_{\varphi\in\overline{\Phi}_\epsilon} \mathcal{R}(\widehat{u},\varphi,c) + \lambda_s\|\mathcal{R}_s[\widehat{u}]\|^2_{L^2(\partial D\times[0,T])} + \lambda_t\|\mathcal{R}_t[\widehat{u}]\|^2_{L^2(D)} \leq \epsilon, \tag{4.14}$$

where

$$\overline{\Phi}_\epsilon = \left\{\varphi\colon |\varphi|_{W^{1,p}} = O\left(\epsilon^{-3(1+2(p-1)/p)}\right)\right\}.$$

Hence we have answered Question 1 in the affirmative for scalar conservation laws.

Proof. We find that $\widehat{u}$ will need to have a size of $O(\epsilon^{-(4q+6)/\beta})$ for it to hold that $\max_{c\in\mathcal{C}} \sup_{\varphi\in\Phi_\epsilon} \mathcal{R}(\widehat{u},\varphi,c) \leq \epsilon/3$, where we used that $q = p/(p-1)$. Since in the proof of Lemma 4.17 the network $\widehat{u}$ is constructed as an approximation of piecewise Taylor polynomials, the spatial and temporal boundary residuals ($\mathcal{R}_s$ and $\mathcal{R}_t$) are automatically minimized as well, given that Taylor polynomials provide approximations in the C^0-norm. □

4.3. *Neural networks overcome curse of dimensionality in approximation error*

We investigate whether (physics-informed) neural networks can overcome the curse of dimensionality (CoD) for the approximation error. Concretely, we want to prove the existence of a neural network with parameter $\widehat{\theta}$ for which $\mathcal{E}_G(\widehat{\theta}) < \epsilon$, without the size of the network growing exponentially in the input dimension d. We discuss two frameworks in which this can be done, both of which exploit the properties of the PDE.

- The first framework is tailored to time-dependent PDEs and considers initial conditions and PDE solutions that are Sobolev-regular (or continuously differentiable). The crucial assumption is that one must be able to approximate the initial condition u_0 with a neural network, without incurring the CoD, i.e. approximate it to accuracy ϵ with a network of size $O(d^{\alpha}\epsilon^{-\beta})$ with $\alpha, \beta > 0$ independent of d. This framework can also be used to prove results for (physics-informed) operator learning.
- The second framework circumvents this assumption by considering PDE solutions that belong to a smaller space, namely the so-called *Barron space*. It can be proved that functions in this space can be approximated without the curse of dimensionality. Hence, in this case the challenge lies in proving that the PDE solution is indeed a *Barron function*.

4.3.1. *Results based on Sobolev- and C^k-regularity*

We initially direct our focus onto the first framework. To start with, we give particular attention to the case where it is known that a neural network can efficiently approximate the solution to a time-dependent PDE at a fixed time. Such neural networks are usually obtained by emulating a classical numerical method. Examples include finite difference schemes, finite volume schemes, finite element methods, iterative methods and Monte Carlo methods (e.g. Jentzen, Salimova and Welti 2021, Opschoor, Petersen and Schwab 2020, Chen, Lu and Lu 2021, Marwah, Lipton and Risteski 2021). In these cases Theorem 4.11 cannot be directly applied, as there is no upper bound for $\|u - u_\theta\|_{L^q(D\times[0,T])}$. To allow for a further generalization to operator learning, we will write down the following assumption in terms of operators (see Section 2.1.1).

More precisely, for $\epsilon > 0$, we assume that we have access to an operator $\mathcal{U}^\epsilon : \mathcal{X} \times [0,T] \to \mathcal{H}$ that, for any $t \in [0,T]$, maps any initial condition/parameter function $v \in \mathcal{X}$ to a neural network $\mathcal{U}^\epsilon(v,t)$ that approximates the PDE solution $\mathcal{G}(v)(\cdot,t) = u(\cdot,t) \in L^q(D), q \in \{2,\infty\}$, at time t, as specified below. Moreover, we will assume that we know how its size depends on the accuracy ϵ.

Assumption 4.19. Let $q \in \{2,\infty\}$. For any $B, \epsilon > 0$, $\ell \in \mathbb{N}$, $t \in [0,T]$ and any $v \in \mathcal{X}$ with $\|v\|_{C^\ell} \le B$, there exist a neural network $\mathcal{U}^\epsilon(v,t)\colon D \to \mathbb{R}$ and a constant $C^B_{\epsilon,\ell} > 0$ such that

$$\|\mathcal{U}^\epsilon(v,t) - \mathcal{G}(v)(\cdot,t)\|_{L^q(D)} \le \epsilon \quad \text{and} \quad \max_{t\in[0,T]} \|\mathcal{U}^\epsilon(v,t)\|_{W^{\ell,q}(D)} \le C^B_{\epsilon,\ell}. \tag{4.15}$$

For vanilla neural networks and PINNs, however, we can always set $\mathcal{X} := \{v\}$, with $v := u_0$ or $v := a$ and $\mathcal{G}(v) := u$ in Assumption 4.19 above.

Under Assumption 4.19, the existence of space–time neural networks that minimize the generalization error $\mathcal{E}_G$ (i.e. the physics-informed loss) can be proved (De Ryck and Mishra 2022*b*).

Theorem 4.20. Let $s, r \in \mathbb{N}$, let $u \in C^{(s,r)}([0,T] \times D)$ be the solution of the PDE (2.1) and let Assumption 4.19 be satisfied. There exists a constant $C(s,r) > 0$ such that, for every $M \in \mathbb{N}$ and $\epsilon, h > 0$, there exists a tanh neural network $u_\theta \colon [0,T] \times D \to \mathbb{R}$ for which

$$\|u_\theta - u\|_{L^q([0,T]\times D)} \leq C(\|u\|_{C^{(s,0)}} M^{-s} + \epsilon), \tag{4.16}$$

and if additionally Assumptions 4.1 and 4.10 hold, then

$$\begin{aligned} &\|\mathcal{L}(u_\theta)\|_{L^2([0,T]\times D)} + \|u_\theta - u\|_{L^2(\partial([0,T]\times D))} \\ &\quad \leq C \cdot \mathrm{poly}(d) \cdot \ln^k(M) \big(\|u\|_{C^{(s,\ell)}} M^{k-s} + M^{2k} \big(\epsilon h^{-\ell} + C^B_{\epsilon,\ell} h^{r-\ell}\big)\big). \end{aligned} \tag{4.17}$$

Moreover, the depth is given by $\mathrm{depth}(u_\theta) \leq C \cdot \sup_{t\in[0,T]} \mathrm{depth}(\mathcal{U}^\epsilon(u(t)))$ and the width by $\mathrm{width}(u_\theta) \leq CM \cdot \sup_{t\in[0,T]} \mathrm{width}(\mathcal{U}^\epsilon(u(t)))$.

Proof. We only provide a sketch of the full proof (De Ryck and Mishra 2022*b*, Theorem 3.5). The main idea is to divide $[0,T]$ into M uniform subintervals and construct a neural network that approximates a Taylor approximation in time of u in each subinterval. In the formula obtained, we approximate the monomials and multiplications by neural networks and approximate the derivatives of u by finite differences and use the accuracy of finite difference formulas to find an error estimate in the $C^k([0,T], L^q(D))$-norm. We again use finite difference operators to prove that spatial derivatives of u are accurately approximated as well. The neural network will also approximately satisfy the initial/boundary conditions as

$$\|u_\theta - u\|_{L^2(\partial([0,T]\times D))} \lesssim C\mathrm{poly}(d)\|u_\theta - u\|_{H^1([0,T]\times D)},$$

which follows from a Sobolev trace inequality. □

As a next step, we use Assumption 4.19 to prove estimates for deep operator learning. Given the connection between DeepONets and FNOs (Kovachki *et al.* 2021, Theorem 36), we focus on DeepONets in the following. In order to prove this error estimate, we need to assume that the operator $\mathcal{U}^\epsilon$ from Assumption 4.19 is stable with respect to its input function, as specified in Assumption 4.21 below. Moreover, we will take the d-dimensional torus as domain $D = \mathbb{T}^d = [0, 2\pi)^d$ and assume periodic boundary conditions for simplicity in what follows. This is not a restriction, as for every Lipschitz subset of $\mathbb{T}^d$ there exists a (linear and continuous) $\mathbb{T}^d$-periodic extension operator for which the derivatives are also $\mathbb{T}^d$-periodic (Kovachki *et al.* 2021, Lemma 41).

Assumption 4.21. Assumption 4.19 is satisfied and let $p \in \{2, \infty\}$. For every $\epsilon > 0$ there exists a constant $C^\epsilon_{\mathrm{stab}} > 0$ such that, for all $v, v' \in \mathcal{X}$,

$$\|\mathcal{U}^\epsilon(v,T) - \mathcal{U}^\epsilon(v',T)\|_{L^2} \leq C^\epsilon_{\mathrm{stab}} \|v - v'\|_{L^p}. \tag{4.18}$$

In this setting, we prove a generic approximation result for DeepONets (De Ryck and Mishra 2022*b*, Theorem 3.10).

Theorem 4.22. Let $s, r \in \mathbb{N}, T > 0, \mathcal{A} \subset C^r(\mathbb{T}^d)$ and let $\mathcal{G}: \mathcal{A} \to C^{(s,r)}([0,T] \times \mathbb{T}^d)$ be an operator that maps a function u_0 to the solution u of the PDE (2.1) with initial condition u_0, let Assumptions 4.1, 4.10 and 4.21 be satisfied and let $p^* \in \{2, \infty\} \setminus \{p\}$. There exists a constant $C > 0$ such that for every $Z, N, M \in \mathbb{N}$, $\epsilon, \rho > 0$ there is a DeepONet $\mathcal{G}_\theta: \mathcal{A} \to L^2([0,T] \times \mathbb{T}^d)$ with Z^d sensors with accuracy

$$\begin{aligned}\|\mathcal{L}(\mathcal{G}_\theta(v))\|_{L^2([0,T]\times\mathbb{T}^d)} \leq CM^{k+\rho}\big[&\|u\|_{C^{(s,\ell)}}M^{-s} \\ &+ M^{s-1}N^\ell\big(\epsilon + C^\epsilon_{\text{stab}}Z^{-r+d/p^*} + C^{CB}_{\epsilon,r}N^{-r}\big)\big],\end{aligned} \tag{4.19}$$

for all v. Moreover,

$$\begin{aligned}&\text{width}(\boldsymbol{\beta}) = O(M(Z^d + N^d\text{width}(\mathcal{U}^\epsilon))), &&\text{depth}(\boldsymbol{\beta}) = \text{depth}(\mathcal{U}^\epsilon), \\ &\text{width}(\boldsymbol{\tau}) = O(MN^d(N + \ln(N))), &&\text{depth}(\boldsymbol{\tau}) = 3,\end{aligned} \tag{4.20}$$

where $\text{width}(\mathcal{U}^\epsilon) = \sup_{u_0\in\mathcal{A}} \text{width}(\mathcal{U}^\epsilon)(u_0))$ and similarly for $\text{depth}(\mathcal{U}^\epsilon)$.

Finally, we show how the results of this section can be applied to high-dimensional PDEs, for which it is not possible to obtain efficient approximation results using standard neural network approximation theory (see Theorem 4.4) as they will lead to convergence rates that suffer from the *curse of dimensionality* (CoD), meaning that the neural network size scales exponentially in the input dimension. In the literature it has been shown for some PDEs that their solution at a fixed time can be approximated to accuracy $\epsilon > 0$ with a network that has size $O(\text{poly}(d)\epsilon^{-\beta})$, with $\beta > 0$ independent of d, and therefore *overcomes the CoD*.

Example 4.23 (linear Kolmogorov PDEs). Linear Kolmogorov PDEs are a class of linear time-dependent PDEs, including the heat equation and the Black–Scholes equation, of the following form. Let $s, r \in \mathbb{N}$, $u_0 \in C^2_0(\mathbb{R}^d)$ and let $u \in C^{(s,r)}([0,T] \times \mathbb{R}^d)$ be the solution of

$$\mathcal{L}[u] = \partial_t u - \frac{1}{2}\text{Tr}(\sigma(x)\sigma(x)^\top\Delta_x[u]) - \mu^\top\nabla_x[u] = 0, \quad u(0,x) = u_0(x) \tag{4.21}$$

for all $(x,t) \in D \times [0,T]$, where $\sigma: \mathbb{R}^d \to \mathbb{R}^{d\times d}$ and $\mu: \mathbb{R}^d \to \mathbb{R}^d$ are affine functions. We make the assumption that $\|u\|_{C^{(s,2)}}$ grows at most polynomially in d, and that for every $\epsilon > 0$ there is a neural network $\widehat{u}_0$ of width $O(\text{poly}(d)\epsilon^{-\beta})$ such that $\|u_0 - \widehat{u}_0\|_{L^\infty(\mathbb{R}^d)} < \epsilon$.

In this setting, Grohs, Hornung, Jentzen and von Wurstemberger (2018), Berner, Grohs and Jentzen (2020) and Jentzen *et al.* (2021) construct a neural network that approximates $u(T)$ and overcomes the CoD by emulating Monte Carlo methods based on the Feynman–Kac formula. De Ryck and Mishra (2022*a*) have proved that PINNs overcome the CoD as well, in the sense that the network size grows as $O(\text{poly}(d\rho_d)\epsilon^{-\beta})$, where ρ_d is a PDE-dependent constant that for a subclass of Kolmogorov PDEs scales as $\rho_d = \text{poly}(d)$, such that the CoD is fully overcome. We demonstrate that the generic bounds of this section (Theorem 4.20) can be used to provide a (much shorter) proof of this result (De Ryck and Mishra 2022*b*, Theorem 4.2).

Theorem 4.24. For every $\sigma, \epsilon > 0$ and $d \in \mathbb{N}$, there is a tanh neural network u_θ of depth $O(\text{depth}(\widehat{u}_0))$ and width

$$O\left(\text{poly}(\text{d}\rho_d)\epsilon^{-(2+\beta)\frac{r+\sigma}{r-2}\frac{s+1}{s-1}-\frac{1+\sigma}{s-1}}\right)$$

such that

$$\|\mathcal{L}(u_\theta)\|_{L^2([0,T]\times[0,1]^d)} + \|u_\theta - u\|_{L^2(\partial([0,T]\times[0,1]^d))} \leq \epsilon. \tag{4.22}$$

Example 4.25 (nonlinear parabolic PDEs). Next, we consider nonlinear parabolic PDEs as in (4.23), which typically arise in the context of nonlinear diffusion–reaction equations that describe the change in space and time of some quantities, such as in the well-known *Allen–Cahn equation* (Allen and Cahn 1979). Let $s, r \in \mathbb{N}$, and for $u_0 \in \mathcal{X} \subset C^r(\mathbb{T}^d)$ let $u \in C^{(s,r)}([0,T] \times \mathbb{T}^d)$ be the solution of

$$\mathcal{L}(u)(x,t) = \partial_t u(t,x) - \Delta_x u(t,x) - F(u(t,x)) = 0, \quad u(0,x) = u_0(x), \tag{4.23}$$

for all $(t,x) \in [0,T] \times D$, with period boundary conditions, where $F\colon \mathbb{R} \to \mathbb{R}$ is a polynomial. As in Example 4.23, we assume that $\|u\|_{C^{(s,2)}}$ grows at most polynomially in d, and that for every $\epsilon > 0$ there is a neural network $\widehat{u}_0$ of width $O(\text{poly}(d)\epsilon^{-\beta})$ such that $\|u_0 - \widehat{u}_0\|_{L^\infty(\mathbb{T}^d)} < \epsilon$.

Hutzenthaler, Jentzen, Kruse and Nguyen (2020) have proved that ReLU neural networks overcome the CoD in the approximation of $u(T)$. Using Theorem 4.20 we can now prove that PINNs overcome the CoD for nonlinear parabolic PDEs (De Ryck and Mishra 2022*b*, Theorem 4.4).

Theorem 4.26. For every $\sigma, \epsilon > 0$ and $d \in \mathbb{N}$, there is a tanh neural network u_θ of depth $O(\text{depth}(\widehat{u}_0) + \text{poly}(d)\ln(1/\epsilon))$ and width

$$O\left(\text{poly}(d)\epsilon^{-(2+\beta)\frac{r+\sigma}{r-2}\frac{s+1}{s-1}-\frac{1+\sigma}{s-1}}\right)$$

such that

$$\|\mathcal{L}(u_\theta)\|_{L^2([0,T]\times\mathbb{T}^d)} + \|u - u_\theta\|_{L^2(\partial([0,T]\times\mathbb{T}^d))} \leq \epsilon. \tag{4.24}$$

Similarly, one can use the results of this section to obtain estimates for (physics-informed) DeepONets for nonlinear parabolic PDEs (4.23) such as the Allen–Cahn equation. In particular, a dimension-independent convergence rate can be obtained if the solution is smooth enough, improving upon the result of Lanthaler, Mishra and Karniadakis (2022), which incurred the CoD. For simplicity, we present results for $C^{(2,r)}$ functions, rather than $C^{(s,r)}$ functions, as we found that assuming more regularity did not necessarily improve the convergence rate further (De Ryck and Mishra 2022*b*, Theorem 4.5).

Theorem 4.27. Let $\mathcal{G}\colon \mathcal{X} \to C^r(\mathbb{T}^d)\colon u_0 \mapsto u(T)$ and $\mathcal{G}^*\colon \mathcal{X} \to C^{(2,r)}([0,T] \times \mathbb{T}^d)\colon u_0 \mapsto u$. For every $\sigma, \epsilon > 0$, there exists a DeepONet $\mathcal{G}^*_\theta$ such that

$$\|\mathcal{L}(\mathcal{G}^*_\theta)\|_{L^2([0,T]\times\mathbb{T}^d\times\mathcal{X})} \leq \epsilon. \tag{4.25}$$

Moreover, for $\mathcal{G}^*_\theta$ we have $O\big(\epsilon^{-\frac{(3+\sigma)d}{r-2}}\big)$ sensors and

$$\begin{aligned} \text{width}(\beta) &= O\big(\epsilon^{-1-\frac{(3+\sigma)(d+r(2+\beta))}{r-2}}\big), && \text{depth}(\beta) = O(\ln(1/\epsilon)), \\ \text{width}(\tau) &= O\big(\epsilon^{-1-\frac{(3+\sigma)(d+1)}{r-2}}\big), && \text{depth}(\tau) = 3. \end{aligned} \tag{4.26}$$

4.3.2. Results based on Barron regularity

A second framework that can be used to prove that neural networks can overcome the curse of dimensionality is that of Barron spaces. These spaces are named after the seminal work of Barron (1993), who showed that a function f with Fourier transform $\widehat{f}$ can be approximated to accuracy $1/\sqrt{m}$ by a shallow neural network with m neurons, as long as

$$\int_{\mathbb{R}^d} |\widehat{f}(\xi)| \cdot |\xi| \, \mathrm{d}\xi < \infty. \tag{4.27}$$

Later, the space of functions satisfying condition (4.27) was named a *Barron space*, and its definition was generalized in many different ways. One notable generalization is that of *spectral Barron spaces.*

Definition 4.28. The spectral Barron space with index s, denoted $\mathcal{B}^s(\mathbb{R}^d)$, is defined as the collection of functions $f\colon \mathbb{R}^d \to \mathbb{R}$ for which the spectral Barron norm is finite:

$$\|f\|_{\mathcal{B}^s(\mathbb{R}^d)} := \int_{\mathbb{R}^d} |\widehat{f}(\xi)| \cdot (1+|\xi|^2)^{s/2} \, \mathrm{d}\xi < \infty. \tag{4.28}$$

First, notice how the initial condition (4.27) of Barron (1993) corresponds to the spectral Barron space $\mathcal{B}^1(\mathbb{R}^d)$. Second, from this definition the difference between Barron spaces and Sobolev spaces becomes apparent: whereas the spectral Barron norm is the L^1-norm of $|\widehat{f}(\xi)| \cdot (1+|\xi|^2)^{s/2}$, the Sobolev $H^s(\mathbb{R}^d)$-norm is defined as the L^2-norm of that same quantity. Finally, it follows that $\mathcal{B}^r(\mathbb{R}^d) \subset \mathcal{B}^s(\mathbb{R}^d)$ for $r \geq s$ and that $\mathcal{B}^0(\mathbb{R}^d) \subset L^\infty(\mathbb{R}^d)$; see e.g. Chen, Lu, Lu and Zhou (2023).

An example of an approximation result for functions in spectral Barron spaces can be found in Lu, Lu and Wang (2021*c*) for functions with the softplus activation function.

Theorem 4.29. Let $\Omega = [0,1]^d$. For every $u \in \mathcal{B}^2(\Omega)$, there exists a shallow neural network $\widehat{u}$ with softplus activation function ($\sigma(x) = \ln(1+\exp(x))$), with width at most m, such that

$$\|u - \widehat{u}\|_{H^1(\Omega)} \leq \frac{\|u\|_{\mathcal{B}^2(\Omega)}(6\log(m)+30)}{\sqrt{m}}. \tag{4.29}$$

Moreover, an exact bound on the weights is given in Lu *et al.* (2021*c*, Theorem 2.2).

There are two key questions that must be answered before this theorem can be used to argue that neural networks can overcome the CoD in the approximation of PDE solutions.

- Under which conditions does it hold that $u \in \mathcal{B}^2(\Omega)$?
- How does $\|u\|_{\mathcal{B}^2(\Omega)}$ depend on d?

Answering these questions requires us to build a regularity theory for Barron spaces, which can be challenging as they are not Hilbert spaces such as the Sobolev spaces H^s. An important contribution for high-dimensional elliptic equations has been made in Lu *et al.* (2021*c*).

Example 4.30 (Poisson's equation). Let $s \geq 0$, let $f \in \mathcal{B}^{s+2}(\Omega)$ satisfy

$$\int_\Omega f(x)\,\mathrm{d}x = 0$$

and let u be the unique solution to Poisson's equation $-\Delta u = f$ with zero Neumann boundary conditions. Then we have $u \in \mathcal{B}^s(\Omega)$ and $\|u\|_{\mathcal{B}^s(\Omega)} \leq d\|f\|_{\mathcal{B}^{s+2}(\Omega)}$ (Lu *et al.* 2021*c*, Theorem 2.6).

Example 4.31 (static Schrödinger equation). Let $s \geq 0$, let $V \in \mathcal{B}^{s+2}(\Omega)$ satisfy $V(x) \geq V_{\min} > 0$ for all $x \in \Omega$ and let u be the unique solution to the static Schrödinger's equation $-\Delta u + Vu = f$ with zero Neumann boundary conditions. Then we have $u \in \mathcal{B}^s(\Omega)$ and $\|u\|_{\mathcal{B}^s(\Omega)} \leq C\|f\|_{\mathcal{B}^{s+2}(\Omega)}$ (Chen *et al.* 2023, Theorem 2.3).

Similar results for a slightly different definition of Barron spaces (E, Ma and Wu 2022) can be found in E and Wojtowytsch (2022*b*), for example. E *et al.* also showed that the Barron space is the *right* space for two-layer neural network models in the sense that optimal direct and inverse approximation theorems hold. Moreover, E and Wojtowytsch (2020, 2022*a*) explored the connection between Barron and Sobolev spaces and provided examples of functions that are and functions that (sometimes surprisingly) are not Barron functions.

Another slightly different definition can be found in Chen *et al.* (2021), where a Barron function g is defined as a function that can be written as an infinite-width neural network,

$$g = \int a\sigma(w^\top x + b)\rho(\mathrm{d}a, \mathrm{d}w, \mathrm{d}b), \tag{4.30}$$

where ρ is a probability distribution on the (now infinite) parameter space Θ. In particular, the neural network

$$\frac{1}{k}\sum_{i=1}^{k} a_i\sigma\left(w_i^\top x + b_i\right)$$

is a Barron function.

Definition 4.32. Fix $\Omega \subset \mathbb{R}^d$ and $R \in [0, +\infty]$. For a function g as in (4.30), we define its Barron norm on Ω with index $p \in [1, +\infty]$ and support radius R by

$$\|g\|_{\mathcal{B}_R^p(\Omega)} = \inf_{\rho \in \mathcal{A}_g} \left\{ \left(\int |a|^p \rho(\mathrm{d}a, \mathrm{d}w, \mathrm{d}b) \right)^{1/p} \right\}, \tag{4.31}$$

where $\mathcal{A}_g$ is the set of probability measures ρ supported on $\mathbb{R} \times \overline{B}_R^d \times \mathbb{R}$, where

$$\overline{B}_R^d = \{x \in \mathbb{R}^d : \|x\| \le R\},$$

such that

$$g = \int a\sigma(w^\top x + b)\rho(\mathrm{d}a, \mathrm{d}w, \mathrm{d}b).$$

The corresponding Barron space is then defined as

$$\mathcal{B}_R^p(\Omega) = \left\{ g : \|g\|_{\mathcal{B}_R^p(\Omega)} < \infty \right\}. \tag{4.32}$$

A number of interesting facts have been proved about this space, including the following.

- The Barron norms and spaces introduced in Definition 4.32 are independent of p, that is,

  $$\|g\|_{\mathcal{B}_R^p(\Omega)} = \|g\|_{\mathcal{B}_R^q(\Omega)},$$

  and hence

  $$\mathcal{B}_R^p(\Omega) = \mathcal{B}_R^q(\Omega) \quad \text{for any } p, q \in [1, +\infty].$$

  See Chen *et al.* (2021, Proposition 2.4) and also E *et al.* (2022, Proposition 1).
- Under some conditions on the activation function σ, the Barron space $\mathcal{B}_R^p(\Omega)$ is an algebra (Chen *et al.* 2021, Lemma 3.3).
- Neural networks can approximate Barron functions in Sobolev norm without incurring the curse of dimensionality.

We demonstrate the final point by slightly adapting Theorem 2.5 of Chen *et al.* (2021).

Theorem 4.33. Let σ be a smooth activation function, $\ell \in \mathbb{N}$, $R \ge 1$, let $\Omega \subset \mathbb{R}^d$ be open and let $f \in B_R^1(\Omega)$. Let μ be a probability measure on Ω and set $C_1 := \max_{m \le \ell} \|\sigma^{(m)}\|_\infty < \infty$. For any $k \in N$ there exist $\{(a_i, w_i, b_i)\}_{i=1}^k$ such that

$$\left\| \frac{1}{k} \sum_{i=1}^k a_i \sigma\left(w_i^T x + b_i\right) - f(x) \right\|_{H_\mu^\ell(\Omega)} \le \frac{2C_1(R\sqrt{ed})^\ell \|f\|_{B_R^1(\Omega)}}{\sqrt{k}}. \tag{4.33}$$

Proof. Since $f \in B^1_R(\Omega)$, there must exist a probability measure ρ supported on $\mathbb{R} \times \overline{B}^d_R \times \mathbb{R}$ such that

$$f(x) = \int a\sigma(w^T x + b)\rho(\mathrm{d}a, \mathrm{d}w, \mathrm{d}b), \qquad \int |a|^2 \,\mathrm{d}\rho \le \frac{4}{\sqrt{\pi}} \|f\|_{B^2(\Omega)}, \tag{4.34}$$

where $4/\sqrt{\pi}$ is a constant that is strictly larger than 1, which will be convenient later on. Define the error

$$\mathcal{E}_k = \frac{1}{k}\sum_{i=1}^{k} a_i \sigma\big(w_i^T x + b_i\big) - f(x). \tag{4.35}$$

We calculate that, for $\ell = |\alpha|_1$,

$$\begin{aligned}
&\mathbb{E}\big[\|D^\alpha \mathcal{E}_k\|^2_{L^2(\Omega_0)}\big] \\
&= \iint_{\Omega_0} \left(\frac{1}{k}\sum_{i=1}^{k} a_i \sigma^l(w_i^T x + b_i)\prod_{j=1}^{d} w_{ij}^{\alpha_j} - D^\alpha f(x)\right)^2 \mathrm{d}\mu\, \mathrm{d}\rho^k \\
&= \frac{1}{k}\iint_{\Omega_0} \left(a\sigma^l(w^T x + b)\prod_{j=1}^{d} w_j^{\alpha_j} - D^\alpha f(x)\right)^2 \mathrm{d}\mu\, \mathrm{d}\rho \\
&= \frac{1}{k}\int_{\Omega_0} \mathrm{Var}\left(a\sigma^l(w^T x + b)\prod_{j=1}^{d} w_j^{\alpha_j}\right) \mathrm{d}\mu \\
&\le \frac{1}{k}\int_{\Omega_0} \mathbb{E}\left[\left(a\sigma^l(w^T x + b)\prod_{j=1}^{d} w_j^{\alpha_j}\right)^2\right] \mathrm{d}\mu \\
&\le \frac{(C_1 R^l)^2}{k}\mathbb{E}[|a|^2] \le \frac{4(C_1 R^l)^2 \|f\|^2_{B^2(\Omega)}}{k\sqrt{\pi}}.
\end{aligned} \tag{4.36}$$

Then, using Lemma 2.1 of De Ryck *et al.* (2021) and the previous inequality, we find that

$$\mathbb{E}\big[\|\mathcal{E}_k\|^2_{H^\ell(\Omega)}\big] \le \sqrt{\pi}(ed)^\ell \max_{\substack{\alpha\in\mathbb{N}_0^d,\\ |\alpha|_1\le\ell}} \mathbb{E}[\|D^\alpha \mathcal{E}_k\|^2] \le \frac{4C_1^2(R^2 ed)^\ell \|f\|^2_{B^2(\Omega)}}{k}. \tag{4.37}$$

We can then conclude by using the fact that for a random variable Y that satisfies $\mathbb{E}[|Y|] \le \epsilon$ for some $\epsilon > 0$, it must hold that $\mathbb{P}(|Y| \le \epsilon) > 0$ (Grohs *et al.* 2018, Proposition 3.3). □

If we know how $\|f\|_{B^2(\Omega)}$ scales with d, then we can prove the existence of neural networks that overcome the CoD for physics-informed loss functions, under Assumption 4.1 and by following the approach of Section 4.1. More related works can be found in Bach (2017), Hong, Siegel and Xu (2021) and references therein.

5. Stability

Next, we investigate whether a small physics-informed loss implies a small total error, often the L^2-error, as formulated in the following question.

Question 2 (Q2). Given that a model u_θ has a small generalization error $\mathcal{E}_G(\theta)$, will the corresponding total error $\mathcal{E}(\theta)$ be small as well?

In the general results we will focus on physics-informed loss functions based on the strong (classical) formulation of the PDE, but we also give some examples when the loss is based on the weak solution (Example 5.8) or the variational solution (Example 5.11). We first look at stability results for forward problems (Section 5.1) and afterwards for inverse problems (Section 5.2).

5.1. Stability for forward problems

We investigate whether a small PDE residual implies that the total error (3.1) will be small as well (Question 2). Such a stability bound can be formulated as the requirement that for any $u, v \in X^*$, the differential operator $\mathcal{L}$ satisfies

$$\|u - v\|_X \le C_{\mathrm{PDE}} \|\mathcal{L}(u) - \mathcal{L}(v)\|_Y, \tag{5.1}$$

where the constant $C_{\mathrm{PDE}} > 0$ is allowed to depend on $\|u\|_{X^*}$ and $\|v\|_{X^*}$.

As a first example of a PDE with solutions satisfying (5.1), we consider a linear differential operator $\mathcal{L}\colon X \to Y$, i.e. $\mathcal{L}(\alpha u + \beta v) = \alpha\mathcal{L}(u) + \beta\mathcal{L}(v)$, for any $\alpha, \beta \in \mathbb{R}$. For simplicity, let $X^* = X$ and $Y^* = Y$. By the assumptions on the existence and uniqueness of the underlying linear PDE (2.1), there exists an *inverse* operator $\mathcal{L}^{-1}\colon Y \to X$. Note that the assumption (5.1) is satisfied if the inverse is bounded i.e. $\|\mathcal{L}^{-1}\|_{\mathrm{op}} \le C < +\infty$, with respect to the natural norm on linear operators from Y to X. Thus the assumption (5.1) on stability boils down to the boundedness of the inverse operator for linear PDEs. Many well-known linear PDEs possess such bounded inverses (Dautray and Lions 1992).

As a second example, we will consider a nonlinear PDE (2.1), but with a well-defined linearization, that is, there exists an operator $\overline{\mathcal{L}}\colon X^* \mapsto Y^*$, such that

$$\mathcal{L}(u) - \mathcal{L}(v) = \overline{\mathcal{L}}_{(u,v)}(u - v) \quad \text{for all } u, v \in X^*. \tag{5.2}$$

Again for simplicity, we will assume that $X^* = X$ and $Y^* = Y$. We further assume that the inverse of $\overline{\mathcal{L}}$ exists and is bounded in the following manner:

$$\|\overline{\mathcal{L}}_{(u,v)}^{-1}\|_{\mathrm{op}} \le C(\|u\|_X, \|v\|_X) < +\infty \quad \text{for all } u, v \in X, \tag{5.3}$$

with the norm of $\overline{\mathcal{L}}^{-1}$ being an operator norm, induced by linear operators from Y to X. Then a straightforward calculation shows that (5.3) suffices to establish the stability bound (5.1).

We summarize our findings with the following informal theorem.

Theorem 5.1. Let $\mathcal{L}$ be a linear operator or a nonlinear operator with linearization as in (5.3). If $\mathcal{L}$ has a bounded inverse operator, then equation (5.1) will hold, that is, a small physics-informed loss will imply a small total error.

We give examples of the above theorem, with explicit stability constants, for the semilinear heat equation (strong formulation), Navier–Stokes equation (strong formulation), scalar conservation laws (strong and weak formulation) and the Poisson equation (variational formulation). Further examples can be found in Mishra and Molinaro (2023) and Lu *et al.* (2021*c*).

Example 5.2 (semilinear heat equation). We address the stability (Question 2) of the semilinear heat equation (see Example 2.1). The following theorem (Mishra and Molinaro 2023, Theorem 3.1) ensures that one can indeed bound the total error $\mathcal{E}(\theta)$ in terms of the generalization error $\mathcal{E}_G(\theta)$. A generalization to linear Kolmogorov equations (including the Black–Scholes model) can be found in De Ryck and Mishra (2022*a*, Theorem 3.7).

Theorem 5.3. Let $u \in C^k(\overline{D} \times [0,T])$ be the unique classical solution of the semilinear parabolic equation (2.6) with the source f satisfying (2.7), and let $v \in C^2(\overline{D} \times [0,T])$. Then the total error (3.1) can be estimated as

$$\begin{aligned} \|u - v\|^2_{L^2(D\times[0,T])} \leq C_1 \big[&\|\mathcal{R}_{\mathrm{PDE}}[v]\|^2_{L^2(D\times[0,T])} + \|\mathcal{R}_t[v]\|^2_{L^2(D)} \\ &+ C_2 \|\mathcal{R}_s[v]\|_{L^2(\partial D\times[0,T])} \big], \end{aligned} \tag{5.4}$$

with constants given by

$$\begin{aligned} C_1 &= \sqrt{T + (1+2C_f)T^2 \mathrm{e}^{(1+2C_f)T}}, \\ C_2 &= \sqrt{T^{1/2}|\partial D|^{1/2} \big(\|u\|_{C^1([0,T]\times\partial D)} + \|v\|_{C^1([0,T]\times\partial D)} \big)}. \end{aligned} \tag{5.5}$$

Proof. First, note that for $w = v - u$ we have

$$\partial_t w = \Delta w + f(v) - f(u) + \mathcal{R}_{\mathrm{PDE}}[v], \quad w(x,0) = \mathcal{R}_t[v](x), \quad v = \mathcal{R}_s[v]. \tag{5.6}$$

Multiplying this PDE by w, integrating over D and performing integration by parts then gives rise to the inequality

$$\begin{aligned} \frac{1}{2}\frac{\mathrm{d}}{\mathrm{d}t} \int_D |w|^2 &\leq \int_{\partial D} \mathcal{R}_s[v] \nabla w \cdot \widehat{n}_D + \int_D w(f(v) - f(u) + \mathcal{R}_{\mathrm{PDE}}[v]) \\ &\lesssim \left(\int_{\partial D} |\mathcal{R}_s[v]|^2 \right)^{1/2} + \int_D |w|^2 + \int_D |\mathcal{R}_{\mathrm{PDE}}[v]|^2. \end{aligned} \tag{5.7}$$

Integrating the above inequality over $[0,\tau] \subset [0,T]$, applying Grönwall's inequality and then integrating once again over time yields the claimed result. The full version can be found in Mishra and Molinaro (2023, Theorem 3.1). □

Example 5.4 (radiative transfer equation). The study of radiative transfer is of vital importance in many fields of science and engineering including astrophysics, climate dynamics, meteorology, nuclear engineering and medical imaging (Modest 2003). The fundamental equation describing radiative transfer is a *linear partial integro-differential equation*, referred to as the *radiative transfer equation*. Under the assumption of a static underlying medium, it has the form (Modest 2003)

$$\frac{1}{c}u_t + \omega \cdot \nabla_x u + ku + \sigma\left(u - \frac{1}{s_d}\int_\Lambda \int_S \Phi(\omega, \omega', \nu, \nu')u(t, x, \omega', \nu')\,\mathrm{d}\omega'\,\mathrm{d}\nu'\right) = f, \tag{5.8}$$

with time variable $t \in [0, T]$, space variable $x \in D \subset \mathbb{R}^d$ (and $D_T = [0, T] \times D$), *angle* $\omega \in S = \mathbb{S}^{d-1}$, i.e. the d-dimensional sphere, and *frequency* (or group energy) $\nu \in \Lambda \subset \mathbb{R}$. The constants in (5.8) are the speed of light c and the surface area s_d of the d-dimensional unit sphere. The unknown of interest in (5.8) is the so-called *radiative intensity* $u\colon D_T \times S \times \Lambda \mapsto \mathbb{R}$, while $k = k(x, \nu)\colon D \times \Lambda \mapsto \mathbb{R}_+$ is the *absorption coefficient* and $\sigma = \sigma(x, \nu)\colon D \times \Lambda \mapsto \mathbb{R}_+$ is the *scattering coefficient*. The integral term in (5.8) involves the so-called *scattering kernel* $\Phi\colon S \times S \times \Lambda \times \Lambda \mapsto \mathbb{R}$, which is normalized as $\int_{S\times\Lambda} \Phi(\cdot, \omega', \cdot, \nu')\,\mathrm{d}\omega'\,\mathrm{d}\nu' = 1$, in order to account for the conservation of photons during scattering. The dynamics of radiative transfer are driven by a source (emission) term $f = f(x, \nu)\colon D \times \Lambda \mapsto \mathbb{R}$.

Although the radiative transfer equation (5.8) is linear, explicit solution formulas are only available in very special cases (Modest 2003). Hence, numerical methods are essential for the simulation of the radiative intensity. Fortunately, Mishra and Molinaro (2021) have provided an affirmative answer to Question 2 for the radiative transfer equations, so that one can use physics-informed techniques (based on the strong PDE residual) to retrieve an approximation of the solution of (5.8).

Theorem 5.5. Let $u \in L^2(\Omega)$ be the unique weak solution of the radiative transfer equation (5.8), with absorption coefficient $0 \le k \in L^\infty(D \times \Lambda)$, scattering coefficient $0 \le \sigma \in L^\infty(D\times\Lambda)$ and a symmetric scattering kernel $\Phi \in C^\ell(S\times\Lambda\times S\times\Lambda)$, for some $\ell \ge 1$, such that the function Ψ, given by

$$\Psi(\omega, \nu) = \int_{S\times\Lambda} \Phi(\omega, \omega', \nu, \nu')\,\mathrm{d}\omega'\,\mathrm{d}\nu', \tag{5.9}$$

is in $L^\infty(S \times \Lambda)$. For any sufficiently smooth model u_θ we have

$$\|u - u_\theta\|_{L^2}^2 \le C\big(\|\mathcal{R}_{\mathrm{PDE}}[u_\theta]\|_{L^2}^2 + \|\mathcal{R}_s[u_\theta]\|_{L^2}^2 + \|\mathcal{R}_t\|_{L^2}^2\big), \tag{5.10}$$

where $C > 0$ is a constant that only depends (in a monotonously increasing way) on T and the quantity $s_d^{-1}(\|\sigma\|_{L^\infty} + \|\Psi\|_{L^\infty})$, where s_d is the surface area of the d-dimensional unit sphere.

Example 5.6 (Navier–Stokes equations). Next we revisit Example 2.2 to show that neural networks with small PINN residuals will provide a good L^2-approximation of the true solution $u\colon \Omega = D \times [0, T] \to \mathbb{R}^d$, $p\colon \Omega \to \mathbb{R}$ of the Navier–Stokes equation (2.8) on the torus $D = \mathbb{T}^d = [0, 1)^d$ with periodic boundary conditions. The analysis can be readily extended to other boundary conditions, such as no-slip boundary conditions, i.e. $u(x,t) = 0$ for all $(x,t) \in \partial D \times [0,T]$, and no-penetration boundary conditions, i.e. $u(x,t) \cdot \widehat{n}_D = 0$ for all $(x,t) \in \partial D \times [0,T]$.

For neural networks (u_θ, p_θ), we define the following PINN-related residuals:

$$\begin{gathered} \mathcal{R}_{\mathrm{PDE}} = \partial_t u_\theta + (u_\theta \cdot \nabla) u_\theta + \nabla p_\theta - \nu \Delta u_\theta, \quad \mathcal{R}_{\mathrm{div}} = \operatorname{div} u_\theta, \\ \mathcal{R}_{s,u}(x) = u_\theta(x) - u_\theta(x+1), \quad \mathcal{R}_{s,p}(x) = p_\theta(x) - p_\theta(x+1), \\ \mathcal{R}_{s,\nabla u}(x) = \nabla u_\theta(x) - \nabla u_\theta(x+1), \quad \mathcal{R}_s = (\mathcal{R}_{s,u}, \mathcal{R}_{s,p}, \mathcal{R}_{s,\nabla u}), \\ \mathcal{R}_t = u_\theta(t=0) - u(t=0), \end{gathered} \tag{5.11}$$

where we drop the θ-dependence in the definition of the residuals for notational convenience.

The following theorem (De Ryck *et al.* 2024*a*) then bounds the L^2-error of the PINN in terms of the residuals defined above. We write $|\partial D|$ for the $(d-1)$-dimensional Lebesgue measure of ∂D and $|D|$ for the d-dimensional Lebesgue measure of D.

Theorem 5.7. Let $d \in \mathbb{N}$, $D = \mathbb{T}^d$ and $u \in C^1(D \times [0,T])$ be the classical solution of the Navier–Stokes equation (2.8). Let (u_θ, p_θ) be a PINN with parameters θ. Then the resulting L^2-error is bounded as follows:

$$\int_\Omega \|u(x,t) - u_\theta(x,t)\|_2^2 \,\mathrm{d}x\,\mathrm{d}t \le \mathcal{C} T \exp(T(2d^2 \|\nabla u\|_{L^\infty(\Omega)} + 1)), \tag{5.12}$$

where the constant $\mathcal{C}$ is defined as

$$\begin{aligned} \mathcal{C} = \|\mathcal{R}_t\|^2_{L^2(D)} + \|\mathcal{R}_{\mathrm{PDE}}\|^2_{L^2(\Omega)} &+ C_1 \sqrt{T} \big[\sqrt{|D|} \|\mathcal{R}_{\mathrm{div}}\|_{L^2(\Omega)} \\ &+ (1+\nu) \sqrt{|\partial D|} \|\mathcal{R}_s\|_{L^2(\partial D \times [0,T])} \big], \end{aligned} \tag{5.13}$$

and

$$C_1 = C_1 \big(\|u\|_{C^1}, \|u_\theta\|_{C^1}, \|p\|_{C^0}, \|p_\theta\|_{C^0} \big) < \infty.$$

Proof. The proof is in a similar spirit to that of Theorem 5.3 and can be found in De Ryck *et al.* (2024*a*). □

Example 5.8 (scalar conservation law). In this example we consider viscous scalar conservation laws, as introduced in Example 2.3. We first assume that the solution to it is sufficiently smooth, so that we can use the strong PDE residual as in Section 2.3.1. We can then prove the following stability bound (Mishra and Molinaro 2023, Theorem 4.1).

Theorem 5.9. Let $\Omega = (0,T) \times (0,1)$, $\nu > 0$ and let $u \in C^k(\Omega)$ be the unique classical solution of the viscous scalar conservation law (2.9). Let $u^* = u_{\theta^*}$ be the PINN generated according to Section 2.6. Then the total error is bounded by

$$\begin{aligned} \|u - u_\theta\|^2_{L^2(\Omega)} \le (T + C_1 T^2 \mathrm{e}^{C_1 T}) \big[&\|\mathcal{R}_{\mathrm{PDE}}[u_\theta]\|^2_{L^2(\Omega)} \\ &+ \|\mathcal{R}_t[u_\theta]\|^2_{L^2(D)} + 2C_2 \|\mathcal{R}_s[u_\theta]\|^2_{L^2(\partial D \times [0,T])} \\ &+ 2\nu \sqrt{T} C_3 \|\mathcal{R}_s[u_\theta]\|_{L^2(\partial D \times [0,T])} \big]. \end{aligned} \tag{5.14}$$

Here the constants are defined as

$$C_1 = 1 + 2|f''(\max\{\|u\|_{L^\infty}, \|u_\theta\|_{L^\infty}\})|\|u_x\|_{L^\infty}, \tag{5.15}$$

$$C_2 = \|\partial_x u\|_{C(\overline{\Omega})} + \|\partial_x u_\theta\|_{C(\overline{\Omega})}, \tag{5.16}$$

$$C_3 = C_3(\|f'\|_\infty, \|u_\theta\|_{C^0(\Omega)}). \tag{5.17}$$

The proof and further details can be found in Mishra and Molinaro (2023). A close inspection of the estimate (5.14) reveals that at the very least, the classical solution u of the PDE (2.9) needs to be in $L^\infty((0,T); W^{1,\infty}((0,1)))$ for the right-hand side of (5.14) to be bounded. This indeed holds as long as $\nu > 0$. However, it is well known (see Godlewski and Raviart (1991) and references therein) that if u^ν is the solution of (2.9) for viscosity ν, then, for some initial data,

$$\|u^\nu\|_{L^\infty((0,T);W^{1,\infty}((0,1)))} \sim \frac{1}{\sqrt{\nu}}. \tag{5.18}$$

Thus, in the limit $\nu \longrightarrow 0$, the constant C_1 can blow up (exponentially in time) and the bound (5.14) no longer controls the generalization error. This is not unexpected as the whole strategy of this paper relies on pointwise realization of residuals. However, the zero-viscosity limit of (2.9) leads to a scalar conservation law with discontinuous solutions (shocks), and the residuals are measures that do not make sense pointwise. Thus the estimate (5.14) also points out the limitations of a PINN for approximating discontinuous solutions.

The above discussion is also the perfect motivation to consider the weak residual, rather than the strong residual, for scalar conservation laws. De Ryck *et al.* (2024*c*) therefore introduce a weak PINN (wPINN) formulation for scalar conservation laws. The wPINN loss function also reflects the fact that *physically admissible* weak solutions should also satisfy an entropy condition, giving rise to an entropy residual. The details of this weak residual and the corresponding loss function were already discussed in Example 2.10. Using the famous doubling of variables argument of Kruzkhov, one can prove the following stability bound on the L^1-error with *wPINNs* (De Ryck *et al.* 2024*c*, Theorem 3.7).

Theorem 5.10. Assume that u is the piecewise smooth entropy solution of (2.9) with $\nu = 0$, essential range $\mathcal{C}$ and $u(0,t) = u(1,t)$ for all $t \in [0,T]$. There is a constant $C > 0$ such that, for every $\epsilon > 0$ and $u_\theta \in C^1(D \times [0,T])$, we have

$$\begin{aligned}\int_0^1 &|u_\theta(x,T) - u(x,T)|\,\mathrm{d}x \\ &\le C\biggl(\int_0^1 |u_\theta(x,0) - u(x,0)|\,\mathrm{d}x + \max_{c\in\mathcal{C},\varphi\in\Phi_\epsilon} \mathcal{R}(u_\theta,\varphi,c) \\ &\quad + (1 + \|u_\theta\|_{C^1})\ln(1/\epsilon)^3\epsilon + \int_0^T |u_\theta(1,t) - u_\theta(0,t)|\,\mathrm{d}t\biggr).\end{aligned} \tag{5.19}$$

Whereas the bound of Theorem 5.9 becomes unusable in the low viscosity regime ($\nu \ll 1$), the bound of Theorem 5.10 is still valid if the solution contains shocks.

Hence we can give an affirmative answer to Question 2 for scalar conservation laws using both the PDE residual (classical formulation, only for $\nu > 0$) and the weak residual (for $\nu = 0$).

Example 5.11 (Poisson's equation). We revisit Poisson's equation (Example 2.4) for $\Omega = [0,1]^d$, but now with zero Neumann conditions, i.e. $\partial u/\partial \nu = 0$ on $\partial\Omega$, and with $f \in L^2(\Omega)$ with $\int_\Omega f \, \mathrm{d}x = 0$. Following Section 2.3.3, the solution of Poisson's equation is a minimizer of the following loss function, which is equal to the energy functional or variational formulation of the PDE

$$I[w] := \frac{1}{2}\int_\Omega |\nabla w|^2 \, \mathrm{d}x + \frac{1}{2}\left(\int_\Omega w \, \mathrm{d}x\right)^2 - \int_\Omega f w \, \mathrm{d}x. \tag{5.20}$$

Now define $\widetilde{u} = \operatorname{argmin}_{w \in H^1(\Omega)} I[w]$ and define

$$\mathcal{E}_G(\theta) := I[u_\theta] - I[\widetilde{u}], \tag{5.21}$$

where we have slightly adapted the definition of the generalization error to reflect the fact that $I[\cdot]$ merely needs to be minimized and does not need to vanish to produce a good approximation. In this setting, one can prove the following stability result (Lu *et al.* 2021*c*).

Proposition 5.12. For any $w \in H^1(\Omega)$,

$$2(I[w] - I[\widetilde{u}]) \le \|w - \widetilde{u}\|^2_{H^1(\Omega)} \le 2\max\{2C_P + 1, 2\}(I[w] - I[\widetilde{u}]), \tag{5.22}$$

where C_P is the Poincaré constant on the domain Ω, that is, for any $v \in H^1(\Omega)$,

$$\left\| v - \int_\Omega v \, \mathrm{d}x \right\|^2_{L^2(\Omega)} \le C_P \|\nabla v\|^2_{L^2(\Omega)}. \tag{5.23}$$

In particular, by setting $w = u_\theta$ this implies that

$$\mathcal{E}(\theta)^2 := \|u_\theta - \widetilde{u}\|^2_{H^1(\Omega)} \le 2\max\{2C_P + 1, 2\}\mathcal{E}_G(\theta). \tag{5.24}$$

The examples and theorems above might give the impression that stability holds for all PDEs and loss functions. It is important to highlight that this is not the case. The next example discusses a negative answer to Question 2 for the Hamilton–Jacobi–Bellman (HJB) equation when using an L^2-based loss function (Wang, Li, He and Wang 2022*a*).

Example 5.13 (Hamilton–Jacobi–Bellman equation). Wang *et al.* (2022*a*) consider the class of HJB equations in which the cost rate function is formulated as $r(x,m) = a_1|m_1|^{\alpha_1} + \cdots + a_n|m_n|^{\alpha_n} - f(x,t)$, giving rise to HJB equations of the form

$$\begin{cases} \mathcal{L}[u] := \partial_t u(x,t) + \dfrac{1}{2}\sigma^2 \Delta u(x,t) - \displaystyle\sum_{i=1}^{n} A_i |\partial_{x_i} u|^{c_i} = f(x,t), \\ \mathcal{B}[u] := u(x,T) = g(x), \end{cases} \tag{5.25}$$

for $x \in \mathbb{R}^d$ and $t \in [0,T]$ and where $A_i = (a_i\alpha_i)^{-1/(\alpha_i-1)} - a_i(a_i\alpha_i)^{-\alpha_i/(\alpha_i-1)} \in (0,+\infty)$ and $c_i = \alpha_i/(\alpha_i - 1) \in (1,\infty)$. Their chosen form of the cost function is relevant in optimal control, e.g. in optimal execution problems in finance. One of the main results in Wang *et al.* (2022*a*) is that stability can only be achieved when the physics-informed loss is based on the L^p-norm of the PDE residual with $p \sim d$. They show that this linear dependence on d cannot be relaxed in the following theorem (Wang *et al.* 2022*a*, Theorem 4.3).

Theorem 5.14. There exists an instance of the HJB equation (5.25), with exact solution u, such that for any $\varepsilon > 0$, $A > 0$, $r \geq 1$, $m \in \mathbb{N} \cup \{0\}$ and $p \in [1, d/4]$, there exists a function $w \in C^\infty(\mathbb{R}^n \times (0,T])$ such that $\operatorname{supp}(w-u)$ is compact and

$$\|\mathcal{L}[w] - f\|_{L^p(\mathbb{R}^d\times[0,T])} < \varepsilon, \quad \mathcal{B}[w] = \mathcal{B}[u], \tag{5.26}$$

and yet simultaneously

$$\|w-u\|_{W^{m,r}(\mathbb{R}^d\times[0,T])} > A. \tag{5.27}$$

This shows that for high-dimensional HJB equations, the learned solution may be arbitrarily distant from the true solution u if an L^2-based physics-inspired loss based on the classical residual is used.

Another example showing that stability of physics-informed learning is not guaranteed can be found in the following two variants on physics-informed neural networks.

Example 5.15 (XPINNs and cPINNs). Jagtap and Karniadakis (2020) and Jagtap, Kharazmi and Karniadakis (2020) proposed two variants on PINNs, inspired by domain decomposition techniques, in which separate neural networks are trained on different subdomains. In order to ensure continuity of the different subnetworks over the boundaries of the subdomains, both methods add a first term to the loss function to enforce that the function values of the subnetworks are (approximately) equal on the boundaries. *Extended PINNs* (XPINNs) (Jagtap and Karniadakis 2020) then propose adding another additional term to make sure the strong PDE residual is (approximately) continuous over the boundaries of the subdomains. In contrast, *conservative PINNs* (cPINNs) (Jagtap *et al.* 2020) focus on PDEs of the form $u_t + \nabla_x f(u) = 0$ (i.e. conservation laws) and add an additional term to make sure that the fluxes based on f are (approximately) equal over the boundaries. De Ryck, Jagtap and Mishra (2024*b*) have given an affirmative answer to Question 2 for cPINNs for the Navier–Stokes equations, advection–diffusion equations and scalar conservation laws. On the other hand, XPINNs were not found to be stable (in the sense of Question 2) for prototypical PDEs such as Poisson's equation and the heat equation.

Finally, we highlight some other works that have proved stability results for physics-informed machine learning. Shin (2020) proved a consistency result for PINNs, for linear elliptic and parabolic PDEs, where they show that if $\mathcal{E}_G(\theta_m) \to 0$ for a sequence of neural networks $\{u_{\theta_m}\}_{m\in\mathbb{N}}$, then $\|u_{\theta_m} - u\|_{L^\infty} \to 0$, under the assumption that we add a specific $C^{k,\alpha}$-regularization term to the loss function, thus partially addressing Question 2 for these PDEs. However, this result does not provide quantitative estimates on the underlying errors. A similar result, with more quantitative estimates for advection equations is provided in Shin, Zhang and Karniadakis (2023).

5.2. *Stability for inverse problems*

Next, we extend our analysis to inverse problems, as introduced in Section 2.1.2. A crucial ingredient to answer Question 2 for inverse problems will be the assumption that solutions to the inverse problem, defined by (2.1) and (2.12), satisfy the following *conditional stability estimate*. Let $\widehat{X} \subset X^* \subset X = L^p(\Omega)$ be a Banach space. For any $u, v \in \widehat{X}$, the differential operator $\mathcal{L}$ and restriction operator Ψ satisfy

$$\|u - v\|_{L^p(E)} \le C_{\mathrm{pd}}\left(\|\mathcal{L}(u) - \mathcal{L}(v)\|_Y^{\tau_p} + \|\Psi(u) - \Psi(v)\|_Z^{\tau_d}\right) \tag{5.28}$$

for some $0 < \tau_p, \tau_d \le 1$, for any subset $\Omega' \subset E \subset \Omega$ and where C_{pd} can depend on $\|u\|_{\widehat{X}}$ and $\|v\|_{\widehat{X}}$. This bound (5.28) is termed a *conditional stability estimate* as it presupposes that the underlying solutions have sufficiently regularity as $\widehat{X} \subset X^* \subset X$.

Remark 5.16. We can extend the hypothesis for the inverse problem as follows.

- Allow the measurement set Ω' to intersect the boundary, i.e. $\partial\Omega' \cap \Omega \neq \emptyset$.
- Replace the bound (5.28) with the weaker bound

$$\|u - v\|_{L^p(E)} \le C_{\mathrm{pd}}\omega(\|\mathcal{L}(u) - \mathcal{L}(v)\|_Y + \|\Psi(u) - \Psi(v)\|_Z), \tag{5.29}$$

  with $\omega\colon \mathbb{R} \mapsto \mathbb{R}_+$ being a modulus of continuity.

We will prove a general estimate on the error due to a model u_θ in approximating the solution u of the inverse problem for PDE (2.1) with data (2.12). Following Mishra and Molinaro (2023) we set $\Omega' \subset E \subset \Omega$ and define the total error (3.1) as

$$\mathcal{E}(E;\theta) := \|u - u_\theta\|_{L^p(E)}. \tag{5.30}$$

We will bound the error in terms of the PDE residual (2.41) and the data residual (2.47) (Mishra and Molinaro 2022, Theorem 2.4).

Theorem 5.17. Let $u \in \widehat{X}$ be the solution of the inverse problem associated with PDE (2.1) and data (2.12). Assume that the stability hypothesis (5.28) holds for any $\Omega' \subset E \subset \Omega$. Let u_θ be any sufficiently smooth function. Assume that the residuals $\mathcal{R}_{\text{PDE}}$ and $\mathcal{R}_d$ are square-integrable. Then the following estimate on the generalization error (5.30) holds:

$$\|u - u_\theta\|_{L^2(E)} \leq C_{\text{pd}}\left(\|\mathcal{R}_{\text{PDE}}[u_\theta]\|_{L^2(\Omega)}^{\tau_p} + \|\mathcal{R}_d[u_\theta]\|_{L^2(\Omega')}^{\tau_d}\right), \tag{5.31}$$

with constants $C_{\text{pd}} = C_{\text{pd}}\left(\|u\|_{\widehat{X}}, \|u^*\|_{\widehat{X}}\right)$ as from (5.28).

Many concrete examples of PDEs where PINNs are used to find a unique continuation can be found in Mishra and Molinaro (2022). The following examples summarize their results for the Poisson equation and heat equation.

Example 5.18 (Poisson equation). We consider the inverse problem for the Poisson equation, as introduced in Example 2.6. In this case, the conditional stability (see (5.28)) is guaranteed by the three balls inequality (Alessandrini, Rondi, Rosset and Vessella 2009). Therefore a result like Theorem 5.17 holds. Note that this theorem only calculates the generalization error on $E \subset \Omega$. However, it can be guaranteed that the generalization error is small on the whole domain Ω and even in Sobolev norm, as follows from the next lemma (Mishra and Molinaro 2022, Lemma 3.3).

Lemma 5.19. For $f \in C^{k-2}(\Omega)$ and $g \in C^k(\Omega')$, with continuous extensions of the functions and derivatives up to the boundaries of the underlying sets and with $k \geq 2$, let $u \in H^1(\Omega)$ be the weak solution of the inverse problem corresponding to the Poisson's equation (2.16) and let u_θ be any sufficiently smooth model. Then the total error is bounded by

$$\|u - u_\theta\|_{H^1(D)} \leq C\left|\log\left(\|\mathcal{R}_{\text{PDE}}[u_\theta]\|_{L^2(\Omega)} + \|\mathcal{R}_d[u_\theta]\|_{L^2(\Omega')}\right)\right|^{-\tau} \tag{5.32}$$

for some $\tau \in (0, 1)$ and a constant $C > 0$ depending on u, u_θ and τ.

Example 5.20 (heat equation). We consider the data assimilation problem for the heat equation, as introduced in Example 2.5, which amounts to finding the solution u of the heat equation in the whole space–time domain $\Omega = D \times (0, T)$, given data on the observation subdomain $\Omega' = D' \times (0, T)$. For any $0 \leq \tau < T$, we define the error of interest for the model u_θ as

$$\mathcal{E}_\tau(\theta) = \|u - u_\theta\|_{C([\tau,T];L^2(D))} + \|u - u_\theta\|_{L^2((0,T);H^1(D))}. \tag{5.33}$$

The theory for this data assimilation inverse problem for the heat equation is classical, and several well-posedness and stability results are available. Our subsequent error estimates for physics-informed models rely on a classical result of Imanuvilov (1995), based on the well-known Carleman estimates. Using these results, one can state the following theorem (Mishra and Molinaro 2022, Lemma 4.3).

Theorem 5.21. For $f \in C^{k-2}(\Omega)$ and $g \in C^k(\Omega')$, with continuous extensions of the functions and derivatives up to the boundaries of the underlying sets and with $k \geq 2$, let $u \in H^1((0, T); H^{-1}(D)) \cap L^2((0, T); H^1(D))$ be the solution of the inverse problem corresponding to the heat equation and that satisfies (2.15). Then, for any $0 \leq \tau < T$, the error (5.33) corresponding to the sufficiently smooth model u_θ is bounded by

$$\mathcal{E}_\tau(\theta) \leq C\big(\|\mathcal{R}_{\mathrm{PDE}}[u_\theta]\|_{L^2(\Omega)} + \|\mathcal{R}_d[u_\theta]\|_{L^2(\Omega')} + \|\mathcal{R}_s[u_\theta]\|_{L^2(\partial D\times(0,T))}\big) \quad (5.34)$$

for some constant C depending on τ, u and u_θ.

Example 5.22 (Stokes equation). The effectiveness of PINNs in approximating inverse problems was recently showcased by Raissi, Yazdani and Karniadakis (2018), who proposed PINNs for the data assimilation problem with the Navier–Stokes equations (Example 2.2). As a first step towards rigorously analysing this, we follow Mishra and Molinaro (2022) and focus on the stationary Stokes equation as introduced in Example 2.7.

Recall that the data assimilation inverse problem for the Stokes equation amounts to inferring the velocity field u (and the pressure p)), given f, f_d and g. In particular, we wish to find solutions $u \in H^1(D;\mathbb{R}^d)$ and $p \in L^2_0(D)$ (i.e. square-integrable functions with zero mean), such that the following holds:

$$\begin{aligned} \int_D \nabla u \cdot \nabla v \,\mathrm{d}x + \int_D p\,\mathrm{div}(v)\,\mathrm{d}x &= \int_D f v\,\mathrm{d}x, \\ \int_D \mathrm{div}(u) w\,\mathrm{d}x &= \int_D f_d w\,\mathrm{d}x, \end{aligned} \quad (5.35)$$

for all test functions $v \in H^1_0(D;\mathbb{R}^d)$ and $w \in L^2(D)$.

Let $B_{R_1}(x_0)$ be the largest ball inside the observation domain $D' \subset D$. We will consider balls $B_R(x_0) \in D$ such that $R > R_1$ and estimate the following error for the model u_θ:

$$\mathcal{E}_R(\theta) := \|u - u_\theta\|_{L^2(B_R(x_0))}. \quad (5.36)$$

The well-posedness and conditional stability estimates for the data assimilation problem for the Stokes equation (2.17)–(2.18) have been extensively investigated in Lin, Uhlmann and Wang (2010) and references therein. Using these results, we can state the following estimate on $\mathcal{E}_R(\theta)$ (Mishra and Molinaro 2022, Lemma 6.2). The physics-informed residuals $\mathcal{R}_{\mathrm{PDE}}$ and $\mathcal{R}_{\mathrm{div}}$ are defined analogously to those in Example 5.6.

Theorem 5.23. Let $f \in C^{k-2}(D;\mathbb{R}^d)$, $f_d \in C^{k-1}(D)$ and $g \in C^k(D')$, with $k \geq 2$. Further, let $u \in H^1(D;\mathbb{R}^d)$ and $p \in H^1(D)$ be the solution of the inverse problem corresponding to the Stokes equations (2.17), that is, they satisfy (5.35) for all test functions $v \in H^1_0(D;\mathbb{R}^d)$, $w \in L^2(D)$ and satisfy the data (2.18). Let u_θ be a sufficiently smooth model and let $B_{R_1}(x_0)$ be the largest ball inside $D' \subset D$. Then there exists $\tau \in (0,1)$ such that the generalization error (5.36) for balls $B_R(x_0) \subset D$

with $R > R_1$ is bounded by

$$\mathcal{E}_R(\theta)^2 \leq C \cdot \max_{\gamma \in \{1,\tau\}} \left(\|\mathcal{R}_{\mathrm{PDE}}[u_\theta]\|_{L^2(D)}^2 + \|\mathcal{R}_{\mathrm{div}}[u_\theta]\|_{L^2(D)}^2\right)^\gamma \left(1 + \|\mathcal{R}_d[u_\theta]\|_{L^2(D)}^{2\tau}\right), \tag{5.37}$$

where C depends on u and u_θ.

Further examples can be found in Mishra and Molinaro (2022).

6. Generalization

Now we turn our attention to Question 3.

Question 3 (Q3). Given a small training error $\mathcal{E}_T^*$ and a sufficiently large training set $\mathcal{S}$, will the corresponding generalization error $\mathcal{E}_G^*$ also be small?

We will answer this question by proving that for any model u_θ,

$$\mathcal{E}_G(\theta) \leq \mathcal{E}_T(\theta) + \epsilon(\theta), \tag{6.1}$$

for some value $\epsilon(\theta) > 0$ that depends on the model class and the size of the training set. Using the terminology of the error decomposition (3.7), this will then imply that the *generalization gap* is small:

$$\sup_{\theta \in \Theta} |\mathcal{E}_T(\theta, \mathcal{S}) - \mathcal{E}_G(\theta)| \leq \sup_{\theta \in \Theta} \epsilon(\theta). \tag{6.2}$$

As $\epsilon(\theta)$ can depend on θ, we must choose the model (parameter) space Θ in a suitable way to avoid $\sup_{\theta \in \Theta} \epsilon(\theta)$ diverging.

In view of Section 2.4, one can see that the training error is nothing more than applying a suitable quadrature rule to each term in the generalization error. It is clear that the error bound for a quadrature rule immediately proves that the generalization error can be bounded in terms of the training error and the training set size, thereby answering Question 3, at least for deterministic quadrature rules. We can now combine this generalization result with the stability bounds from Section 5 to prove an upper bound on the total error of the optimized model $\mathcal{E}^*$ (3.1) (Mishra and Molinaro 2023, Theorem 2.6).

Theorem 6.1. Let $u \in X^*$ be the unique solution of the PDE (2.1) and assume that the stability hypothesis (5.1) holds. Let $u_\theta \in X^*$ be a model and let $\mathcal{S}$ be a training set of quadrature points corresponding to the quadrature rule (2.65) with order α, as in (2.66). Then, for any L^2-based generalization error $\mathcal{E}_G$, such as (2.45) or (2.46),

$$\mathcal{E}_G(\theta)^2 \leq \mathcal{E}_T(\theta)^2 + C_{\mathrm{quad}} N^{-\alpha}, \tag{6.3}$$

with constant C_{quad} stemming from (2.66), and which might depend on u_θ.

The approach described above only makes sense for low to moderately high dimensions ($d < 20$), since the convergence rate of numerical quadrature rules with deterministic quadrature decreases with increasing d. For problems in very high dimensions, $d \gg 20$, Monte Carlo quadrature is the numerical integration method of choice. In this case, the quadrature points are randomly chosen, independent and identically distributed (i.i.d.) (with respect to a scaled Lebesgue measure). Since in this case the optimized model is correlated with all training points, we must be careful when calculating the convergence rate. We can, however, prove error bounds that hold with a certain probability or, alternatively, that hold averaged over all possible training sets of the same size.

In this setting, we prove a general *a posteriori* upper bound on the generalization error. Consider $\mathcal{F}\colon \mathcal{D} \to \mathbb{R}$ (an operator or function) and a corresponding model $\mathcal{F}_\theta\colon \mathcal{D} \to \mathbb{R}$, $\theta \in \Theta$. Given a training set $\mathcal{S} = \{X_1, \dots X_N\}$, where $\{X_i\}_{i=1}^N$ are i.i.d. random variables on $\mathcal{D}$ (according to a measure μ), the training error $\mathcal{E}_T$ and generalization error $\mathcal{E}$ are

$$\mathcal{E}_T(\theta, \mathcal{S})^2 = \frac{1}{N}\sum_{i=1}^{N} |\mathcal{F}(z_i) - \mathcal{F}_\theta(z_i)|^2, \quad \mathcal{E}_G(\theta)^2 = \int_{\mathcal{D}} |\mathcal{F}_\theta(z) - \mathcal{F}(z)|^2 \, \mathrm{d}\mu(z), \tag{6.4}$$

where μ is a probability measure on $\mathcal{D}$. This setting allows us to bound all possible terms and residuals that were mentioned in Section 2.3.

- For the term resulting from the PDE residual (2.41), we can set $\mathcal{D} = \Omega$, $\mathcal{F} = 0$ and $\mathcal{F}_\theta = \mathcal{L}[u_\theta] - f$.
- For the data term (2.47), we can set $\mathcal{D} = \Omega'$, $\mathcal{F} = u$ and $\mathcal{F}_\theta = u_\theta$. Similarly, for the term arising from the spatial boundary conditions, we set $\mathcal{D} = \partial D$ (or $\mathcal{D} = \partial D \times [0, T]$), and for the term arising from the initial condition, we set $\mathcal{D} = D$.
- For operator learning with an input function space $\mathcal{X}$, we can set $\mathcal{D} = \Omega \times \mathcal{X}$, $\mathcal{F} = \mathcal{G}$ and $\mathcal{F}_\theta = \mathcal{G}_\theta$.
- Finally, for physics-informed operator learning, we can set $\mathcal{D} = \Omega \times \mathcal{X}$, $\mathcal{F} = 0$ and $\mathcal{F}_\theta = \mathcal{L}(\mathcal{G}_\theta)$.

With the above definitions in place, we can state the following theorem (De Ryck and Mishra 2022*b*, Theorem 3.11), which provides a computable *a posteriori* error bound on the expectation of the generalization error for a general class of approximators. We refer to Beck, Jentzen and Kuckuck (2022) and De Ryck and Mishra (2022*a*), for example, for bounds on n, c and $\mathfrak{L}$.

Theorem 6.2. For $R > 0$ and $N, n \in \mathbb{N}$, let $\Theta = [-R, R]^n$ be the parameter space, and for every training set $\mathcal{S}$, let $\theta^*(\mathcal{S}) \in \Theta$ be an (approximate) minimizer of $\theta \mapsto \mathcal{E}_T(\theta, \mathcal{S})^2$, assume that $\theta \mapsto \mathcal{E}(\theta, \mathcal{S})^2$ and $\theta \mapsto \mathcal{E}_T(\theta, \mathcal{S})^2$ are bounded by $c > 0$ and Lipschitz-continuous with Lipschitz constant $\mathfrak{L} > 0$. If $N \geq 2c^2 \mathrm{e}^8/(2R\mathfrak{L})^{n/2}$, then

$$\mathbb{E}[\mathcal{E}_G(\theta^*(\mathcal{S}))^2] \leq \mathbb{E}[\mathcal{E}_T(\theta^*(\mathcal{S}), \mathcal{S})^2] + \sqrt{\frac{2c^2(n+1)}{N} \ln(R\mathfrak{L}\sqrt{N})}. \tag{6.5}$$

Proof. The proof combines standard techniques, based on covering numbers and Hoeffding's inequality, with an error composition from De Ryck and Mishra (2022*a*). □

Due to its generality, Theorem 6.2 provides a satisfactory answer to Question 3 when a randomly generated training set is used. However, the two central assumptions, the existence of the constants $c > 0$ and $\mathfrak{L} > 0$, should not be swept under the rug. In particular, for neural networks it might initially be unclear how large these constants are.

For any type of neural network architecture of depth L, width W and weights bounded by R, we find that $N \sim LW(W+d)$. For tanh neural networks and operator learning architectures, we have $\ln(\mathfrak{L}) \sim L\ln(dRW)$, whereas for physics-informed neural networks and DeepONets we find that $\ln(\mathfrak{L}) \sim (k+\ell)L\ln(dRW)$, with k and ℓ as in Assumption 4.1 (Lanthaler *et al.* 2022, De Ryck and Mishra 2022*a*). Taking this into account, we also find that the imposed lower bound on n is not very restrictive. Moreover, the right-hand side of (6.5) depends at most polynomially on L, W, R, d, k, ℓ and c. For physics-informed architectures, however, upper bounds on c often depend exponentially on L (De Ryck and Mishra 2022*a*, De Ryck *et al.* 2024*a*).

Remark 6.3. As Theorem 6.2 is an *a posteriori* error estimate, one can use the network sizes of the trained networks for L, W and R. The sizes stemming from the approximation error estimates of the previous sections can be disregarded for this result. Moreover, instead of considering the expected values of $\mathcal{E}$ and $\mathcal{E}_T$ in (6.5), one can also prove that such an inequality holds with a certain probability (De Ryck and Mishra 2022*b*).

Next, we show how Theorems 6.1 and 6.2 can be used in practice as *a posteriori* error estimates.

Example 6.4 (Navier–Stokes). We first state the *a posteriori* estimate for the Navier–Stokes equation (Example 2.2). By using the midpoint rule (2.67) on the sets $D \times [0,T]$ with N_{int} quadrature points, D with N_t quadrature points and $\partial D \times [0,T]$ with N_s quadrature points, one can prove the following estimate (De Ryck *et al.* 2024*a*, Theorem 3.10).

Theorem 6.5. Let $T > 0$, $d \in \mathbb{N}$, let $(u,p) \in C^4(\mathbb{T}^d \times [0,T])$ be the classical solution of the Navier–Stokes equation (2.8) and let $(u_\theta, p_\theta) \in C^4(\mathbb{T}^d \times [0,T])$ be a model. Then the following error bound holds:

$$\begin{aligned}\mathcal{E}(\theta)^2 &= \int_\Omega \|u(x,t)-u_\theta(x,t)\|_2^2 \,\mathrm{d}x\,\mathrm{d}t \\ &\le O\big(\mathcal{E}_T(\theta,\mathcal{S}) + N_t^{-2/d} + N_{\mathrm{int}}^{-1/(d+1)} + N_s^{-1/d}\big). \end{aligned} \tag{6.6}$$

The exact constant implied in the O-notation depends on (u_θ, p_θ) and (u,p) and their derivatives (De Ryck *et al.* 2024*a*, Theorem 3.10). There are two interesting observations to be made from equation (6.6).

- When the training set is large enough, the total error scales with the *square root* of the training error, $\mathcal{E}(\theta) \sim \sqrt{\mathcal{E}_T(\theta,\mathcal{S})}$. This sublinear relation is also observed in practice; see Figure 6.1.
- The bound (6.6) reveals different convergence rates in terms of the training set sizes N_{int}, N_t and N_s. In particular, we need relatively many training points in the interior of the domain compared to its boundary. We will see that these ratios will be different for other PDEs.

De Ryck *et al.* (2024*a*) verified the above result by training a PINN to approximate the Taylor–Green vortex in two space dimensions, the exact solution of which is given by

$$u(t,x,y) = -\cos(\pi x)\sin(\pi y)\exp(-2\pi^2 \nu t), \tag{6.7}$$

$$v(t,x,y) = \sin(\pi x)\cos(\pi y)\exp(-2\pi^2 \nu t), \tag{6.8}$$

$$p(t,x,y) = -\frac{\rho}{4}[\cos(2\pi x)+\cos(2\pi y)]\exp(-4\pi^2 \nu t). \tag{6.9}$$

The spatio-temporal domain is $x,y \in [0.5, 4.5]^2$ and $t \in [0,1]$. The results are reported in Figure 6.1. As one can see, all error types decrease with increasing number of quadrature points. In particular, the relationship $\mathcal{E}(\theta) \sim \sqrt{\mathcal{E}_T(\theta,\mathcal{S})}$ is (approximately) observed as well.

In the next example we show that Theorem 6.2 can be used to show that the curse of dimensionality can also be overcome in the training set size.

Example 6.6 (high-dimensional heat equation). We consider the high-dimensional (linear) heat equation (Example 2.1), which is an example of a linear Kolmogorov equation, for which we have proved that the CoD can be overcome in the approximation error in Example 4.23. In this example, we show that the training set size only needs to grow at most polynomially in the input dimension d to achieve a certain accuracy. Applying Theorem 6.2 to every term of the right-hand side of the stability result for the heat equation (Theorem 5.3) then gives us the following result.

Theorem 6.7. Under the assumptions of Theorem 6.2 and given a training set of size N_{int} on $D \times [0,T]$, size N_t on D and size N_s on $\partial D \times [0,T]$, we find that there

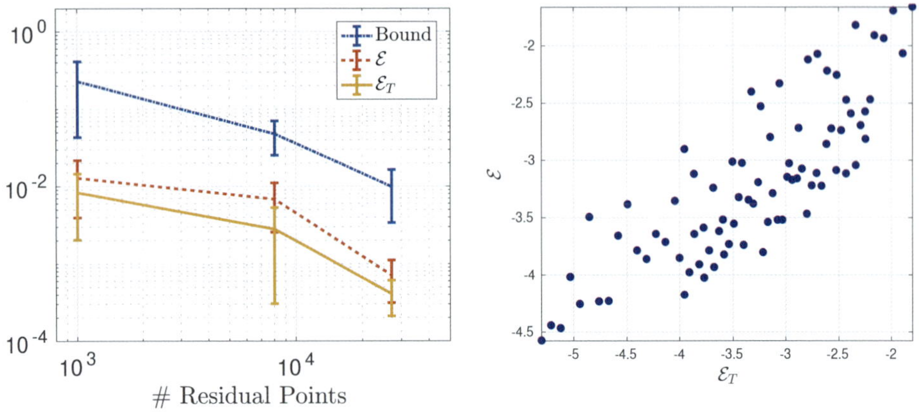

(a) (b)

Figure 6.1. Experimental results for the Navier–Stokes equation with $\nu = 0.01$. The total error $\mathcal{E}$ and the training error $\mathcal{E}_T$ are shown in terms of the number of residual points N_{int} ((a), and compared with the bound from Theorem 6.5) and also in terms of each other (b). Reproduced from De Ryck *et al.* (2024*a*), with permission of the IMA.

exist constants $C, \alpha > 0$ that are independent of d such that

$$\mathbb{E}[\mathcal{E}(\theta^*(\mathcal{S}))^2] \leq Cd^\alpha \left(\mathbb{E}[\mathcal{E}_T(\theta^*(\mathcal{S}), \mathcal{S})] + \frac{\ln(N_{\text{int}})}{N_{\text{int}}^2} + \frac{\ln(N_t)}{N_t^2} + \frac{\ln(N_s)}{N_s^4} \right). \quad (6.10)$$

A more general result for linear Kolmogorov equations, including a more detailed expression for C and α, can be found in De Ryck and Mishra (2022*a*). A comparison with the corresponding result for the Navier–Stokes equations (Theorem 6.5) reveals the following.

- Just as for the Navier–Stokes equations, the total error scales with the *square root* of the training error, $\mathcal{E}(\theta) \sim \sqrt{\mathcal{E}_T(\theta, \mathcal{S})}$ when the training set is large enough.
- For the Navier–Stokes equations (with $d = 2$ or $d = 3$) we found that we need to choose $N_{\text{int}} \gg N_s \gg N_t$. However, in this case we find that far fewer training points in the interior are needed, as $N_s \gg N_t \approx N_{\text{int}}$ is sufficient.

As a first example, Mishra and Molinaro (2022) consider the one-dimensional heat equation on the domain $[-1, 1]$ with initial condition $\sin(\pi x)$ and final time $T = 1$. The results can be seen in Figure 6.2. As one can see, both the generalization and training error decrease with increasing number of quadrature points. In particular, the relationship $\mathcal{E}(\theta) \sim \sqrt{\mathcal{E}_T(\theta, \mathcal{S})}$ is (approximately) observed as well.

As a second example, we consider the one-dimensional heat equation with parametric initial condition, and where the parameter space is very high-dimensional.

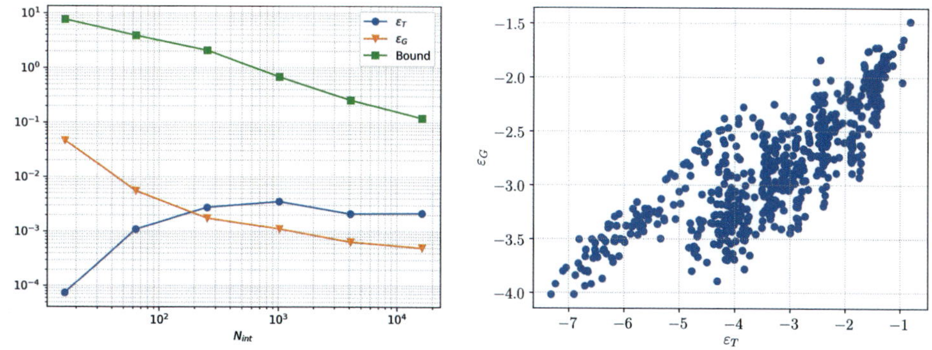

(a) (b)

Figure 6.2. Experimental results for the heat equation. The generalization error $\mathcal{E}_G$ and the training error $\mathcal{E}_T$ are shown in terms of the number of residual points N_{int} ((a), and compared with the bound from Theorem 6.7) and also in terms of each other (b). Figure from Molinaro (2023).

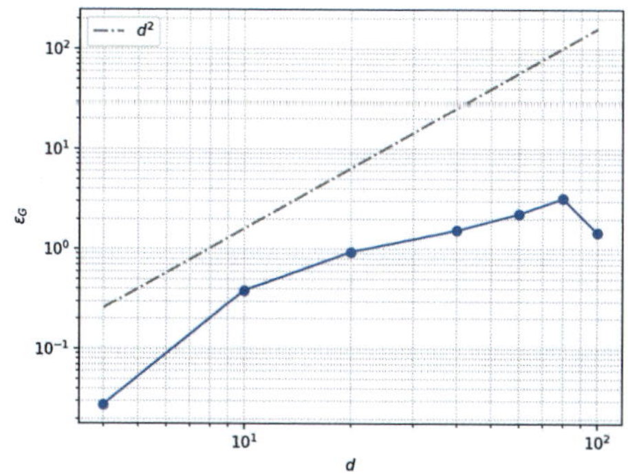

Figure 6.3. Relative generalization percentage error $\mathcal{E}_G$ in terms of the parameter dimension d for the heat equation with a parametrized initial condition. Figure from Molinaro (2023).

In more detail, the initial condition considered is parametrized by μ and given by

$$u_0(x;\mu) = \sum_{i=1}^{d} \frac{-1}{dm^2} \sin(m\pi(x-\mu_i)). \tag{6.11}$$

Figure 6.3 demonstrates that the generalization error $\mathcal{E}_G$ does not grow exponentially (but rather sub-quadratically) in the parameter dimension d, and that therefore the curse of dimensionality is mitigated. Even for very large d the relative generalization error is less than 2%.

Example 6.8 (scalar conservation laws). We again consider weak PINNs for the scalar conservation laws. Recall from the stability result (Theorem 5.8) that the generalization error is given by

$$\begin{aligned}
&\mathcal{E}(\theta, \varphi, c) \\
&\quad = -\int_0^1 \int_0^T (|u_{\theta^*(\mathcal{S})}(x,t) - c|\partial_t \varphi(x,t) + Q[u_{\theta^*(\mathcal{S})}(x,t); c]\partial_x \varphi(x,t))\,\mathrm{d}x\,\mathrm{d}t \\
&\qquad + \int_0^1 |u_\theta(x,0) - u(x,0)|\,\mathrm{d}x + \int_0^T |u_\theta(0,t) - u_\theta(1,t)|\,\mathrm{d}t. \qquad (6.12)
\end{aligned}$$

To this end, we consider the simplest case of random (Monte Carlo) quadrature and generate a set of collocation points,

$$\mathcal{S} = \{(x_i, t_i)\}_{i=1}^N \subset D \times [0,T],$$

where all (x_i, t_i) are i.i.d. drawn from the uniform distribution on $D \times [0,T]$. For a fixed $\theta \in \Theta$, $\varphi \in \Phi_\epsilon$, $c \in \mathcal{C}$ and for this data set $\mathcal{S}$, we can then define the *training error*

$$\begin{aligned}
&\mathcal{E}_T(\theta, \mathcal{S}, \varphi, c) \\
&\quad = -\frac{T}{N}\sum_{i=1}^N (|u_\theta(x_i,t_i) - c|\partial_t \varphi(x_i,t_i) + Q[u_\theta(x_i,t_i); c]\partial_x \varphi(x_i,t_i)) \\
&\qquad + \frac{T}{N}\sum_{i=1}^N |u_\theta(x_i,0) - u(x_i,0)| + \frac{T}{N}\sum_{i=1}^N |u_\theta(0,t_i) - u_\theta(1,t_i)|. \qquad (6.13)
\end{aligned}$$

During training, we then aim to obtain neural network parameters $\theta^*_{\mathcal{S}}$, a test function $\varphi^*_{\mathcal{S}}$ and a scalar $c^*_{\mathcal{S}}$ such that

$$\mathcal{E}_T\left(\theta^*_{\mathcal{S}}, \mathcal{S}, \varphi^*_{\mathcal{S}}, c^*_{\mathcal{S}}\right) \approx \min_{\theta \in \Theta} \max_{\varphi \in \Phi_\epsilon} \max_{c \in \mathcal{C}} \mathcal{E}_T(\theta, \mathcal{S}, \varphi, c) \qquad (6.14)$$

for some $\epsilon > 0$. We call the resulting neural network $u^* := u_{\theta^*_{\mathcal{S}}}$ a *weak PINN* (wPINN). If the network has width W, depth L and its weights are bounded by R, and the parameter $\epsilon > 0$ is as in Theorem 5.10, then we obtain the following theorem (De Ryck *et al.* 2024c, Theorem 3.9 and Corollary 3.10).

Theorem 6.9. Let $\mathcal{C}$ be the essential range of u and let $N, L, W \in \mathbb{N}$, $R \geq \max\{1, T, |\mathcal{C}|\}$ with $L \geq 2$ and $N \geq 3$. Moreover, let $u_\theta \colon D \times [0,T] \to \mathbb{R}$, $\theta \in \Theta$, be tanh neural networks with at most $L-1$ hidden layers, width at most W and weights and biases bounded by R. Assume that $\mathcal{E}_G$ and $\mathcal{E}_T$ are bounded by $B \geq 1$. It holds with a probability of at least $1-\delta$ that

$$\mathcal{E}_G\left(\theta^*_{\mathcal{S}}, \varphi^*_{\mathcal{S}}, c^*_{\mathcal{S}}\right) \leq \mathcal{E}_T\left(\theta^*_{\mathcal{S}}, \mathcal{S}, \varphi^*_{\mathcal{S}}, c^*_{\mathcal{S}}\right) + \frac{3BLW}{\sqrt{N}}\sqrt{\ln\left(\frac{C\ln(1/\epsilon)WRN}{\epsilon^3 \delta B}\right)}. \qquad (6.15)$$

7. Training

Despite the generally affirmative answers to our first three questions (Q1–Q3) in the previous sections, significant problems have been identified with physics-informed machine learning. Arguably, the foremost problem lies in *training* these frameworks with (variants of) gradient descent methods. It has been increasingly observed that PINNs and their variants are *slow, even infeasible*, to train even on certain model problems, with the training process either not converging, or converging to unacceptably large loss values (Krishnapriyan *et al.* 2021, Moseley, Markham and Nissen-Meyer 2023, Wang, Teng and Perdikaris 2021*a*, Wang, Yu and Perdikaris 2022*b*). These observations highlight the relevance of our fourth question.

Question 4 (Q4). Can the training error $\mathcal{E}_T^*$ be made sufficiently close to the true minimum of the loss function $\min_\theta \mathcal{J}(\theta, \mathcal{S})$?

To answer this question, we must find out the reason behind the issues observed with training physics-informed machine learning algorithms. Empirical studies such as that of Krishnapriyan *et al.* (2021) attribute failure modes to the non-convex loss landscape, which is much more complex when compared to the loss landscape of supervised learning. Others such as Moseley *et al.* (2023) and Dolean, Heinlein, Mishra and Moseley (2023) have implicated the well-known spectral bias (Rahaman *et al.* 2019) of neural networks as being a cause of poor training, whereas Wang *et al.* (2021*a*) and Wang, Wang and Perdikaris (2021*b*) used infinite-width NTK theory to propose that the subtle balance between the PDE residual and supervised components of the loss function could explain and possibly ameliorate training issues. Nevertheless, it is fair to say that there is still a paucity of principled analysis of the training process for gradient descent algorithms in the context of physics-informed machine learning.

In what follows, we first demonstrate that when an affirmative answer to Question 4 is available (or assumed), it becomes possible to provide *a priori error estimates* on the error of models learned by minimizing a physics-informed loss function (Section 7.1). Next, we derive precise conditions under which gradient descent for a physics-informed loss function can be approximated by a *simplified gradient descent* algorithm, which amounts to the gradient descent update for a *linearized* form of the training dynamics (Section 7.2). It will then turn out that the speed of convergence of the gradient descent is related to the *condition number* of an operator, which in turn is composed of the *Hermitian square* ($\mathcal{L}^*\mathcal{L}$) of the differential operator ($\mathcal{L}$) of the underlying PDE and a *kernel integral operator*, associated to the tangent kernel for the underlying model (Section 7.3). This analysis automatically suggests that *preconditioning* the resulting operator is necessary to alleviate training issues for physics-informed machine learning (Section 7.4). Finally, in Section 7.5, we discuss how different preconditioning strategies can overcome training bottlenecks, and also how existing techniques, proposed in the literature for improving training, can be viewed from this operator preconditioning perspective.

7.1. *Global minimum of the loss is a good approximation to the PDE solution*

We demonstrate that the results of the previous sections can be used to prove *a priori error estimates* for physics-informed learning, provided that one can find a global minimum of the physics-informed loss function (2.69).

We revisit the error decomposition (3.7) proposed in Section 3. There it was shown that if one can answer Question 2 in the affirmative, then, for any $\theta^*, \widehat{\theta} \in \Theta$,

$$\begin{aligned}\mathcal{E}(\theta^*) &\leq C\mathcal{E}_G(\theta^*)^\alpha \\ &\leq C\left(\mathcal{E}_G(\widehat{\theta}) + 2\sup_{\theta\in\Theta} |\mathcal{E}_T(\theta,\mathcal{S}) - \mathcal{E}_G(\theta)| + \mathcal{E}_T(\theta^*,\mathcal{S}) - \mathcal{E}_T(\widehat{\theta},\mathcal{S})\right)^\alpha. \quad (7.1)\end{aligned}$$

For any $\epsilon > 0$, we can now let $\widehat{\theta}$ be the parameter corresponding to the model from Question 1, and hence we find that $\mathcal{E}_G(\widehat{\theta}) < \epsilon$ if the model class is expressive enough. Next we can use the results of Section 6 to deduce a lower limit on the size of the training set $\mathcal{S}$ (in terms of ϵ) such that also $\sup_{\theta\in\Theta} |\mathcal{E}_T(\theta,\mathcal{S}) - \mathcal{E}_G(\theta)| < \epsilon$. Finally, if we assume that $\theta^* = \theta^*(\mathcal{S})$ is the global minimizer of the loss function $\theta \mapsto \mathcal{J}(\theta,\mathcal{S}) = \mathcal{E}_T(\theta,\mathcal{S})$ (2.69), then it must hold that $\mathcal{E}_T(\theta^*,\mathcal{S}) \leq \mathcal{E}_T(\widehat{\theta},\mathcal{S})$ and hence $\mathcal{E}_T(\theta^*,\mathcal{S}) - \mathcal{E}_T(\widehat{\theta},\mathcal{S}) \leq 0$. As a result, we can infer from (7.1) that

$$\mathcal{E}^* \leq C(3\epsilon)^\alpha. \quad (7.2)$$

Alternatively, we note that many of the *a posteriori* error bounds in Section 6 are of the form

$$\mathcal{E}(\theta^*) \leq C(\mathcal{E}_T(\theta^*,\mathcal{S}) + \epsilon)^\alpha \quad (7.3)$$

if the training set is large enough (depending on ϵ). As before,

$$\mathcal{E}_T(\theta^*,\mathcal{S}) \leq \mathcal{E}_T(\widehat{\theta},\mathcal{S}) \leq \mathcal{E}_G(\widehat{\theta}) + \sup_{\theta\in\Theta} |\mathcal{E}_T(\theta,\mathcal{S}) - \mathcal{E}_G(\theta)| \leq 2\epsilon, \quad (7.4)$$

such that in total we again find that $\mathcal{E}^* \leq C(3\epsilon)^\alpha$, in agreement with (7.2).

We first demonstrate this argument for the Navier–Stokes equations, and then show a similar result for the Poisson equation in which the curse of dimensionality is overcome.

Example 7.1 (Navier–Stokes equations). In Sections 4, 5 and 6 we found affirmative answers to questions Q1–Q3, and hence we should be able to apply the above argument under the assumptions that an exact global minimizer to the loss function can be found.

An *a posteriori* estimate as in (7.3) is provided by Theorem 6.5, but it is initially unclear whether this can also be used for an *a priori* estimate, given the dependence of the bound on the derivatives of the model. It turns out that this is possible for PINNs if the true solution is sufficiently smooth and the neural network and training set sufficiently large (De Ryck *et al.* 2024*a*, Corollary 3.9).

Corollary 7.2. Let $\epsilon > 0$, $T > 0$, $d \in \mathbb{N}$, $k > 6(3d+8) =: \gamma$, let $(u, p) \in H^k(\mathbb{T}^d \times [0,T])$ be the classical solution of the Navier–Stokes equation (2.8), let the hypothesis space consist of PINNs for which the upper bound on the weights is at least $R \geq \epsilon^{-1/(k-\gamma)} \ln(1/\epsilon)$, the width is at least $W \geq \epsilon^{-(d+1)/(k-\gamma)}$ and the depth is at least $L \geq 3$, let $(u_{\theta^*(\mathcal{S})}, p_{\theta^*(\mathcal{S})})$ be the PINN that solves (2.69) and let the training set $\mathcal{S}$ satisfy $N_t \geq \epsilon^{-d(1+\gamma/(k-\gamma))}$, $N_{\text{int}} \geq \epsilon^{-2(d+1)(1+\gamma/(k-\gamma))}$ and $N_s \geq \epsilon^{-2d(1+\gamma/(k-\gamma))}$. It holds that

$$\|u - u_{\theta^*(\mathcal{S})}\|_{L^2(D\times[0,T])} = O(\epsilon). \tag{7.5}$$

Proof. The results follow from combining results on the approximation error (Theorem 4.9), the stability (Theorem 5.7) and the generalization error (Theorem 6.5). □

Example 7.3 (Poisson's equation). We revisit Poisson's equation with the variation residual, as previously considered in Example 4.30 (Question 1) and Example 5.11 (Question 2). Recall that if $f \in \mathcal{B}^0(\Omega)$ (see Definition 4.28) satisfies $\int_\Omega f(x)\,\mathrm{d}x = 0$, then the corresponding unique solution u to Poisson's equation $-\Delta u = f$ with zero Neumann boundary conditions is $u \in \mathcal{B}^2(\Omega)$.

Now, assume that the training set $\mathcal{S}$ consists of N i.i.d. randomly generated points in Ω, and we optimize over a subset of shallow softplus neural networks of width m, corresponding to parameter space $\Theta^* \subset \Theta$; a more precise definition can be found in Lu *et al.* (2021*c*, eq. 2.13). As usual, we then define $u_{\theta^*(\mathcal{S})}$ as the minimizer of the discretized energy residual $\mathcal{J}(\theta, \mathcal{S})$. In this setting, one can prove the following *a priori* generalization result (Lu *et al.* 2021*c*, Remark 2.1).

Theorem 7.4. Assume the setting defined above. If we set $m = N^{1/3}$, then there exists a constant $C > 0$ (independent of m and N) such that

$$\mathbb{E}\big[\|u - u_{\theta^*(\mathcal{S})}\|^2_{H^1(\Omega)}\big] \leq C\frac{(\log(N))^2}{N^{1/3}}. \tag{7.6}$$

In the above theorem, we can see that the convergence rate is independent of d, such that the curse of dimensionality is mitigated. The constant C might depend on the Barron norm of u, and its dependence on d might therefore not be exactly clear.

Example 7.5 (scalar conservation law). Using Theorem 5.10 and the above bound on the generalization gap, we can prove the following rigorous upper bound (De Ryck *et al.* 2024*c*, Corollary 3.10) on the total L^1-error of the weak PINN, which we denote as

$$\mathcal{E}(\theta) = \int_D |u_\theta(x,T) - u(x,T)|\,\mathrm{d}x. \tag{7.7}$$

Corollary 7.6. Assume the setting of Theorem 6.9. It holds with a probability of at least $1 - \delta$ that

$$\mathcal{E}^* \leq C\left[\mathcal{E}_T^* + \frac{3BdLW}{\sqrt{N}}\sqrt{\ln\left(\frac{C\ln(1/\epsilon)WRN}{\epsilon^3\delta B}\right)} + (1 + \|u^*\|_{C^1})\ln(1/\epsilon)^3\epsilon\right], \quad (7.8)$$

where $\mathcal{E}^* := \mathcal{E}(\theta_{\mathcal{S}}^*)$ and $\mathcal{E}_T^* := \mathcal{E}_T\left(\theta_{\mathcal{S}}^*, \mathcal{S}, \varphi_{\mathcal{S}}^*, c_{\mathcal{S}}^*\right)$.

Unfortunately, this result in its current form is not strong enough to lead to an *a priori* generalization error. Using (7.4), one can easily ensure that $\mathcal{E}_T^*$ is small, and verify that B, R, N and L depend (at most) polynomially on ϵ^{-1} such that the second term in (7.8) can be made small for large N. The problem lies in the third term, as most available upper bounds on $\|u^*\|_{C^1}$ will be overestimates that grow with ϵ^{-1}.

7.2. *Characterization of gradient descent*

As mentioned at the beginning of Section 7, for some types of PDEs significant problems during the training of the model have been observed. To investigate this issue, we first consider a linearized version of the gradient descent (GD) update formula.

First, recall that physics-informed machine learning boils down to minimizing the physics-informed loss, as constructed in Section 2.4, that is, to find

$$\theta^*(\mathcal{S}) = \operatorname*{argmin}_{\theta\in\Theta} L(\theta), \quad (7.9)$$

where in this section we focus on loss functions of the form

$$L(\theta) = \underbrace{\frac{1}{2}\int_\Omega |\mathcal{L}u(x) - f(x)|^2\,\mathrm{d}x}_{R(\theta)} + \underbrace{\frac{\lambda}{2}\int_{\partial\Omega} |u(x) - g(x)|^2\,\mathrm{d}\sigma(x)}_{B(\theta)}. \quad (7.10)$$

As mentioned in Section 2.5, it is customary in machine learning (Goodfellow *et al.* 2016) that the non-convex optimization problem (7.9) is solved with (variants of) a gradient descent algorithm which takes the generic form

$$\theta_{k+1} = \theta_k - \eta\nabla_\theta L(\theta_k), \quad (7.11)$$

with descent steps $k > 0$, learning rate $\eta > 0$, loss L and the initialization θ_0 chosen randomly, following Section 2.5.1.

We want to investigate the *rate of convergence* to ascertain the computational cost of training. As the loss L (7.11) is non-convex, it is hard to rigorously analyse the training process in complete generality. We need to make certain assumptions on (7.11) to make the problem tractable. To this end, we fix the step k in (7.11) and start with the following Taylor expansion:

$$u(x;\theta_k) = u(x;\theta_0) + \nabla_\theta u(x;\theta_0)^\top(\theta_k - \theta_0) + \frac{1}{2}(\theta_k - \theta_0)^\top H_k(x)(\theta_k - \theta_0). \quad (7.12)$$

Here, $H_k(x) := \mathrm{Hess}_\theta(u(x;\tau_k\theta_0 + (1-\tau_k)\theta_k)$ is the Hessian of $u(\cdot\,,\theta)$ evaluated at intermediate values, with $0 \le \tau_k \le 1$. Now introducing the notation $\phi_i(x) = \partial_{\theta_i} u(x;\theta_0)$, we define the matrix $\mathbb{A} \in \mathbb{R}^{n\times n}$ and the vector $\mathbb{C}, \in \mathbb{R}^n$ as

$$\begin{aligned} \mathbb{A}_{i,j} &= \langle \mathcal{L}\phi_i, \mathcal{L}\phi_j\rangle_{L^2(\Omega)} + \lambda\langle \phi_i, \phi_j\rangle_{L^2(\partial\Omega)}, \\ \mathbb{C}_i &= \langle \mathcal{L}u_{\theta_0} - f, \mathcal{L}\phi_i\rangle_{L^2(\Omega)} + \lambda\langle u_{\theta_0} - u, \phi_i\rangle_{L^2(\partial\Omega)}. \end{aligned} \tag{7.13}$$

Substituting the above formulas in the GD algorithm (7.11), we can rewrite it identically as

$$\theta_{k+1} = \theta_k - \eta\nabla_\theta L(\theta_k) = (I - \eta\mathbb{A})\theta_k + \eta(\mathbb{A}\theta_0 + \mathbb{C}) + \eta\epsilon_k, \tag{7.14}$$

where ϵ_k is an error term that collects all terms that depend on the Hessians H_k and $\mathcal{L}H_k$, and is explicitly given by

$$\begin{aligned} 2\epsilon_k &= \langle \mathcal{L}u_{\theta_k} - f, \mathcal{L}H_k(\theta_k - \theta_0)\rangle_{L^2(\Omega)} + \lambda\langle u_{\theta_k} - u, H_k(\theta_k - \theta_0)\rangle_{L^2(\partial\Omega)} \\ &\quad + \langle (\theta_k - \theta_0)^\top \mathcal{L}H_k(\theta_k - \theta_0), \mathcal{L}\nabla_\theta u_{\theta_0}\rangle_{L^2(\Omega)} \\ &\quad + \lambda\langle (\theta_k - \theta_0)^\top H_k(\theta_k - \theta_0), \nabla_\theta u_{\theta_0}\rangle_{L^2(\partial\Omega)}. \end{aligned} \tag{7.15}$$

From this characterization of gradient descent, we clearly see that (7.11) is related to a *simplified* version of gradient descent given by

$$\widetilde{\theta}_{k+1} = (I - \eta\mathbb{A})\widetilde{\theta}_k + \eta(\mathbb{A}\widetilde{\theta}_0 + \mathbb{C}), \quad \widetilde{\theta}_0 = \theta_0, \tag{7.16}$$

modulo the *error* term ϵ_k defined in (7.15).

The following lemma (De Ryck, Bonnet, Mishra and de Bézenac 2023) shows that this simplified GD dynamics (7.16) approximates the full GD dynamics (7.11) to desired accuracy as long as the error term ϵ_k is small.

Lemma 7.7. Let $\delta > 0$ be such that $\max_k \|\epsilon_k\|_2 \le \delta$. If $\mathbb{A}$ is invertible and $\eta = c/\lambda_{\max}(\mathbb{A})$ for some $0 < c < 1$, then it holds for any $k \in \mathbb{N}$ that

$$\|\theta_k - \widetilde{\theta}_k\|_2 \le \kappa(\mathbb{A})\delta/c, \tag{7.17}$$

with $\lambda_{\min} := \min_j |\lambda_j(\mathbb{A})|$, $\lambda_{\max} := \max_j |\lambda_j(\mathbb{A})|$ and the *condition number*

$$\kappa(\mathbb{A}) = \lambda_{\max}(\mathbb{A})/\lambda_{\min}(\mathbb{A}). \tag{7.18}$$

The key assumption in Lemma 7.7 is the smallness of the error term ϵ_k (7.14) for all k. This is trivially satisfied for linear models $u_\theta(x) = \sum_k \theta_k\phi_k$ as $\epsilon_k = 0$ for all k in this case. From the definition of ϵ_k (7.15), we see that a more general sufficient condition for ensuring this smallness is to ensure that the Hessians of u_θ and $\mathcal{L}u_\theta$ (resp. H_k and $\mathcal{L}H_k$ in (7.12)) are small during training. This amounts to requiring *approximate linearity* of the parametric function $u(\cdot\,;\theta)$ near the initial value θ_0 of the parameter θ. For any differentiable parametrized function f_θ, its linearity is equivalent to the constancy of the associated *tangent kernel* (Liu, Zhu and Belkin 2020)

$$\Theta[f_\theta](x,y) := \nabla_\theta f_\theta(x)^\top \nabla_\theta f_\theta(y). \tag{7.19}$$

Hence, it follows that if the tangent kernel associated to u_θ and $\mathcal{L}u_\theta$ is (approximately) constant along the optimization path, then the error term ϵ_k will be small.

For neural networks this entails that the *neural tangent kernels* (NTK) $\Theta[u_\theta]$ and $\Theta[\mathcal{L}u_\theta]$ stay approximately constant along the optimization path. The following informal lemma from De Ryck *et al.* (2023), based on Wang *et al.* (2022*b*), confirms that this is indeed the case for wide enough neural networks.

Lemma 7.8. For a neural network u_θ with one hidden layer of width m and a linear differential operator $\mathcal{L}$, we have $\lim_{m\to\infty}\Theta[u_{\theta_k}] = \lim_{m\to\infty}\Theta[u_{\theta_0}]$ and $\lim_{m\to\infty}\Theta[\mathcal{L}u_{\theta_k}] = \lim_{m\to\infty}\Theta[\mathcal{L}u_{\theta_0}]$ for all k. Consequently, the error term ϵ_k (7.14) is small for wide neural networks, $\lim_{m\to\infty}\max_k \|\epsilon_k\|_2 = 0$.

Now, given the much simpler structure of (7.16), when compared to (7.11), one can study the corresponding gradient descent dynamics explicitly and obtain the following convergence theorem (De Ryck *et al.* 2023).

Theorem 7.9. Let $\mathbb{A}$ in (7.16) be invertible with condition number $\kappa(\mathbb{A})$ (7.18) and let $0 < c < 1$. Set $\eta = c/\lambda_{\max}(\mathbb{A})$ and $\vartheta = \theta_0 + \mathbb{A}^{-1}\mathbb{C}$. It holds for any $k \in \mathbb{N}$ that

$$\|\widetilde{\theta}_k - \vartheta\|_2 \le (1 - c/\kappa(\mathbb{A}))^k \|\theta_0 - \vartheta\|_2. \tag{7.20}$$

An immediate consequence of the quantitative convergence rate (7.20) is as follows: to obtain an error of size ε, i.e. $\|\widetilde{\theta}_k - \vartheta\|_2 \le \varepsilon$, we can readily calculate the number of GD steps $N(\varepsilon)$ as

$$N(\varepsilon) = \ln(\varepsilon/\|\theta_0 - \vartheta\|_2)/\ln(1 - c/\kappa(\mathbb{A})) = O\left(\kappa(\mathbb{A})\ln\frac{1}{\epsilon}\right). \tag{7.21}$$

Hence, for a fixed value c, large values of the condition number $\kappa(\mathbb{A})$ will severely impede convergence of the simplified gradient descent (7.16) by requiring a much larger number of steps.

7.3. *Training and operator preconditioning*

So far, we have established that, under suitable assumptions, the rate of convergence of the gradient descent algorithm for physics-informed machine learning boils down to the *conditioning* of the matrix $\mathbb{A}$ (7.13). However, at first sight, this matrix is not very intuitive and we want to relate it to the differential operator $\mathcal{L}$ from the underlying PDE (2.1). To this end, we first introduce the so-called *Hermitian square* $\mathcal{A}$, given by

$$\mathcal{A} = \mathcal{L}^*\mathcal{L}, \tag{7.22}$$

in the sense of operators, where $\mathcal{L}^*$ is the *adjoint operator* for the differential operator $\mathcal{L}$. Note that this definition implicitly assumes that the adjoint $\mathcal{L}^*$ exists and the Hermitian square operator $\mathcal{A}$ is defined on an appropriate function space. As an example, consider as differential operator the Laplacian, i.e. $\mathcal{L}u = -\Delta u$, defined for instance on $u \in H^1(\Omega)$; then the corresponding Hermitian square is $\mathcal{A}u = \Delta^2 u$, identified as the *bi-Laplacian* that is well-defined on $u \in H^2(\Omega)$.

Next, for notational simplicity, we consider a loss function that only consists of the integral of the squared PDE residual (2.41) and omit boundary terms in the following. Let $\mathcal{H}$ be the span of the functions $\phi_k := \partial_{\theta_k} u(\cdot; \theta_0)$. Define the maps $T \colon \mathbb{R}^n \to \mathcal{H}, v \to \sum_{k=1}^n v_k \phi_k$ and $T^* \colon L^2(\Omega) \to \mathbb{R}^n; f \to \{\langle \phi_k, f \rangle\}_{k=1,\ldots,n}$. We define the following scalar product on $L^2(\Omega)$:

$$\langle f, g \rangle_{\mathcal{H}} := \langle f, TT^* g \rangle_{L^2(\Omega)} = \langle T^* f, T^* g \rangle_{\mathbb{R}^n}. \tag{7.23}$$

Note that the maps T, T^* provide a correspondence between the continuous space (L^2) and discrete space ($\mathcal{H}$) spanned by the functions ϕ_k. This continuous–discrete correspondence allows us to relate the conditioning of the matrix $\mathbb{A}$ in (7.13) to the conditioning of the Hermitian square operator $\mathcal{A} = \mathcal{L}^*\mathcal{L}$ via the following theorem (De Ryck *et al.* 2023).

Theorem 7.10. It holds for the operator $\mathcal{A} \circ TT^* \colon L^2(\Omega) \to L^2(\Omega)$ that $\kappa(\mathbb{A}) \geq \kappa(\mathcal{A} \circ TT^*)$. Moreover, if the Gram matrix $\langle \phi, \phi \rangle_{\mathcal{H}}$ is invertible then equality holds, i.e. $\kappa(\mathbb{A}) = \kappa(\mathcal{A} \circ TT^*)$.

Thus we show that the conditioning of the matrix $\mathbb{A}$ that determines the speed of convergence of the simplified gradient descent algorithm (7.16) for physics-informed machine learning is intimately tied to the conditioning of the operator $\mathcal{A} \circ TT^*$. This operator, in turn, composes the Hermitian square of the underlying differential operator of the PDE (2.1), with the so-called *kernel integral operator* TT^*, associated with the (neural) tangent kernel $\Theta[u_\theta]$. Theorem 7.10 implies in particular that if the operator $\mathcal{A} \circ TT^*$ is ill-conditioned, then the matrix $\mathbb{A}$ is ill-conditioned and the gradient descent algorithm (7.16) for physics-informed machine learning will converge very slowly.

Remark 7.11. One can readily generalize Theorem 7.10 to the setting with boundary conditions, i.e. with $\lambda > 0$ in the loss (7.10). In this case one can prove for the operator $\mathcal{A} = \mathbb{1}_{\mathring{\Omega}} \cdot \mathcal{L}^*\mathcal{L} + \lambda \mathbb{1}_{\partial\Omega} \cdot \mathrm{Id}$, and its corresponding matrix $\mathbb{A}$ (as in (7.13)) that $\kappa(\mathbb{A}) \geq \kappa(\mathcal{A} \circ TT^*)$, where equality holds if the relevant Gram matrix is invertible. More details can be found in De Ryck *et al.* (2023, Appendix A.6).

It is instructive to compare physics-informed machine learning with standard supervised learning through the prism of the analysis presented here. It is straightforward to see that for supervised learning, i.e. when the physics-informed loss is replaced by the supervised loss $\frac{1}{2}\|u - u_\theta\|^2_{L^2(\Omega}$ by simply setting $\mathcal{L} = \mathrm{Id}$, the corresponding operator in Theorem 7.10 is simply the kernel integral operator TT^*, associated with the tangent kernel as $\mathcal{A} = \mathrm{Id}$. Thus the complexity in training physics-informed machine learning models is entirely due to the spectral properties of the Hermitian square $\mathcal{A}$ of the underlying differential operator $\mathcal{L}$.

7.4. *Preconditioning to improve training in physics-informed machine learning*

Having established in the previous section that, under suitable assumptions, the speed of training physics-informed machine learning models depends on the condition number of the operator $\mathcal{A} \circ TT^*$, or equivalently the matrix $\mathbb{A}$ (7.13), we now investigate whether this operator is ill-conditioned and, if so, how we can better condition it by reducing the condition number. The fact that $\mathcal{A} \circ TT^*$ (equivalently $\mathbb{A}$) is very poorly conditioned for most PDEs of practical interest will be demonstrated both theoretically and empirically below. This makes *preconditioning*, i.e. strategies to improve (reduce) the conditioning of the underlying operator (matrix), a key component in improving training for physics-informed machine learning models.

7.4.1. *General framework for preconditioning*

Intuitively, reducing the condition number of the underlying operator $\mathcal{A} \circ TT^*$ amounts to finding new maps $\widetilde{T}, \widetilde{T}^*$ for which the kernel integral operator $\widetilde{T}\widetilde{T}^*$ is approximately equal to $\mathcal{A}^{-1}$, i.e. choosing the architecture and initialization of the parametrized model u_θ such that the associated kernel integral operator $\widetilde{T}\widetilde{T}^*$ is an (approximate) Green's function for the Hermitian square $\mathcal{A}$ of the differential operator $\mathcal{L}$. For an operator $\mathcal{A}$ with well-defined eigenvectors ψ_k and eigenvalues ω_k, the ideal case $\widetilde{T}\widetilde{T}^* = \mathcal{A}^{-1}$ is realized when

$$\widetilde{T}\widetilde{T}^* \phi_k = \frac{1}{\omega_k}\psi_k.$$

This can be achieved by transforming ϕ linearly with a (positive definite) matrix $\mathbb{P}$ such that

$$(\mathbb{P}^\top \phi)_k = \frac{1}{\omega_k}\psi_k,$$

which corresponds to the change of variables $\mathcal{P}u_\theta := u_{\mathbb{P}\theta}$. Assuming the invertibility of $\langle \phi, \phi \rangle_{\mathcal{H}}$, Theorem 7.10 then shows that $\kappa(\mathcal{A} \circ \widetilde{T}\widetilde{T}^*) = \kappa(\widetilde{\mathbb{A}})$ for a new matrix $\widetilde{\mathbb{A}}$, which can be computed as

$$\widetilde{\mathbb{A}} := \langle \mathcal{L}\nabla_\theta u_{\mathbb{P}\theta_0}, \mathcal{L}\nabla_\theta u_{\mathbb{P}\theta_0} \rangle_{L^2(\Omega)} = \langle \mathcal{L}\mathbb{P}^\top \nabla_\theta u_{\theta_0}, \mathcal{L}\mathbb{P}^\top \nabla_\theta u_{\theta_0} \rangle_{L^2(\Omega)} = \mathbb{P}^\top \mathbb{A}\mathbb{P}. \tag{7.24}$$

This implies a general approach for preconditioning, namely linearly transforming the parameters of the model, i.e. considering $\mathcal{P}u_\theta := u_{\mathbb{P}\theta}$ instead of u_θ, which corresponds to replacing the matrix $\mathbb{A}$ with its preconditioned variant $\widetilde{\mathbb{A}} = \mathbb{P}^\top \mathbb{A}\mathbb{P}$. The new simplified GD update rule is then

$$\theta_{k+1} = \theta_k - \eta\widetilde{\mathbb{A}}(\theta_k - \theta_0) + \mathbb{C}. \tag{7.25}$$

Hence, finding $\widetilde{T}\widetilde{T}^* \approx \mathcal{A}^{-1}$, which is the aim of preconditioning, reduces to constructing a matrix $\mathbb{P}$ such that $1 \approx \kappa(\widetilde{\mathbb{A}}) \ll \kappa(\mathbb{A})$. We emphasize that $\widetilde{T}\widetilde{T}^*$ need not serve as the exact inverse of $\mathcal{A}$: even an approximate inverse can lead to significant performance improvements. This is the foundational principle of preconditioning.

Moreover, performing gradient descent using the transformed parameters $\widehat{\theta}_k := \mathbb{P}\theta_k$ yields

$$\widehat{\theta}_{k+1} = \mathbb{P}\theta_{k+1} = \mathbb{P}\theta_k - \eta\mathbb{P}\mathbb{P}^\top \nabla_\theta L(\mathbb{P}\theta_k) = \widehat{\theta}_k - \eta\mathbb{P}\mathbb{P}^\top \nabla_\theta L(\widehat{\theta}_k). \tag{7.26}$$

Given that any positive definite matrix can be written as $\mathbb{P}\mathbb{P}^\top$, this shows that linearly transforming the parameters is equivalent to preconditioning the gradient of the loss by multiplying by a positive definite matrix. Hence, parameter transformations are all we need in this context.

7.4.2. *Preconditioning linear physics-informed machine learning models*

We now rigorously analyse the effect of preconditioning linear parametrized models of the form $u_\theta(x) = \sum_k \theta_k \phi_k(x)$, where $\phi_1, \ldots, \phi_n$ are any smooth functions, as introduced in Section 2.2.1. A corresponding preconditioned model, as explained above, would have the form $\widetilde{u}_\theta(x) = \sum_k (\mathbb{P}\theta)_k \phi_k(x)$, where $\mathbb{P} \in \mathbb{R}^{n\times n}$ is the preconditioner. We motivate the choice of this preconditioner with a simple, yet widely used example.

Our differential operator is the one-dimensional Laplacian $\mathcal{L} = \mathrm{d}^2/\mathrm{d}x^2$, defined on the domain $(-\pi, \pi)$, for simplicity with periodic zero boundary conditions. Consequently, the corresponding PDE is the Poisson equation (Example 2.4). As the machine learning model, we choose $u_\theta(x) = \sum_{k=-K}^{K} \theta_k \phi_k(x)$ with

$$\phi_0(x) = \frac{1}{\sqrt{2\pi}}, \quad \phi_{-k}(x) = \frac{1}{\sqrt{\pi}}\cos(kx), \quad \phi_k(x) = \frac{1}{\sqrt{\pi}}\sin(kx) \tag{7.27}$$

for $1 \le k \le K$. This model corresponds to the widely used learnable *Fourier features* in the machine learning literature (Tancik *et al.* 2020) or *spectral methods* in numerical analysis (Hesthaven *et al.* 2007). We can readily verify that the resulting matrix $\mathbb{A}$ (7.13) is given by

$$\mathbb{A} = \mathbb{D} + \lambda v v^\top, \tag{7.28}$$

where $\mathbb{D}$ is a diagonal matrix with $\mathbb{D}_{kk} = k^4$ and $v := \phi(\pi)$, that is, v is a vector with $v_{-k} = (-1)^k/\sqrt{\pi}$, $v_0 = 1/\sqrt{2\pi}$, $v_k = 0$ for $1 \le k \le K$. Preconditioning solely based on $\mathcal{L}^*\mathcal{L}$ would correspond to finding a matrix $\mathbb{P}$ such that $\mathbb{P}\mathbb{D}\mathbb{P}^\top = \mathrm{Id}$. However, given that $\mathbb{D}_{00} = 0$, this is not possible. We therefore set $\mathbb{P}_{kk} = 1/k^2$ for $k \neq 0$ and $\mathbb{P}_{00} = \gamma \in \mathbb{R}$. The preconditioned matrix is therefore

$$\widetilde{\mathbb{A}}(\lambda, \gamma) = \mathbb{P}\mathbb{D}\mathbb{P}^\top + \lambda \mathbb{P}v(\mathbb{P}v)^\top. \tag{7.29}$$

The conditioning of the unpreconditioned and preconditioned matrices considered above are summarized in the following theorem (De Ryck *et al.* 2023).

Theorem 7.12. The following statements hold for all $K \in \mathbb{N}$.

(1) The condition number of the unpreconditioned matrix above satisfies

$$\kappa(\mathbb{A}(\lambda)) \ge K^4. \tag{7.30}$$

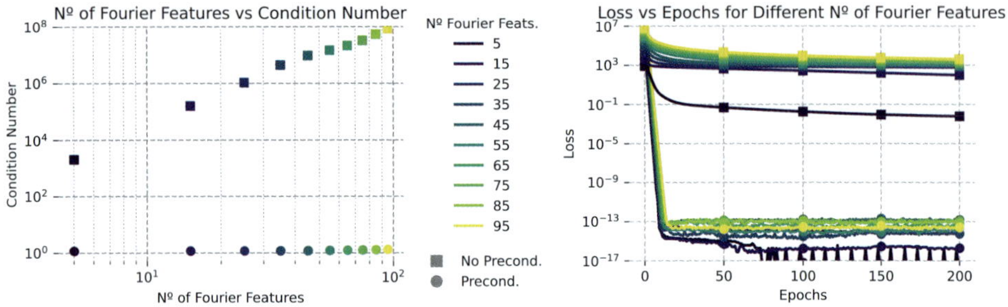

(a) (b)

Figure 7.1. Poisson equation with Fourier features. (a) Optimal condition number vs. number of Fourier features. (b) Training for the unpreconditioned and preconditioned Fourier features. Figure from De Ryck *et al.* (2023).

(2) There exists a constant $C(\lambda, \gamma) > 0$ that is independent of K such that

$$\kappa(\widetilde{\mathbb{A}}(\lambda, \gamma) \leq C. \tag{7.31}$$

(3) It holds that $\kappa(\widetilde{\mathbb{A}}(2\pi/\gamma^2, \gamma)) = 1 + O(1/\gamma)$ and hence

$$\lim_{\gamma \to +\infty} \kappa(\widetilde{\mathbb{A}}(2\pi/\gamma^2, \gamma)) = 1. \tag{7.32}$$

We observe from Theorem 7.12 that (i) the matrix $\mathbb{A}$, which governs gradient descent dynamics for approximating the Poisson equation with learnable Fourier features, is very poorly conditioned, and (ii) we can (optimally) precondition it by *rescaling* the Fourier features based on the eigenvalues of the underlying differential operator (or its Hermitian square).

These conclusions are also observed empirically. In Figure 7.1(a) we plot the condition number of the matrix $\mathbb{A}$, minimized over λ, as a function of maximum frequency K and verify that this condition number increases as K^4, as predicted by Theorem 7.12. Consequently, as shown in Figure 7.1(b), where we plot the loss function in terms of increasing training epochs, the model is very hard to train with large losses (particularly for higher values of K), showing a very slow decay of the loss function as the number of frequencies is increased. On the other hand, in Figure 7.1(a) we also show that the condition number (minimized over λ) of the preconditioned matrix (7.29) remains constant with increasing frequency and is very close to the optimal value of 1, verifying Theorem 7.12. As a result, we observe from Figure 7.1(b) that the loss in the preconditioned case decays exponentially fast as the number of epochs is increased. This decay is independent of the maximum frequency of the model. The results demonstrate that the preconditioned version of the Fourier features model can learn the solution of the Poisson equation efficiently, in contrast to the failure of the unpreconditioned model to do so. Entirely analogous results are obtained with the Helmholtz equation (De Ryck *et al.* 2023).

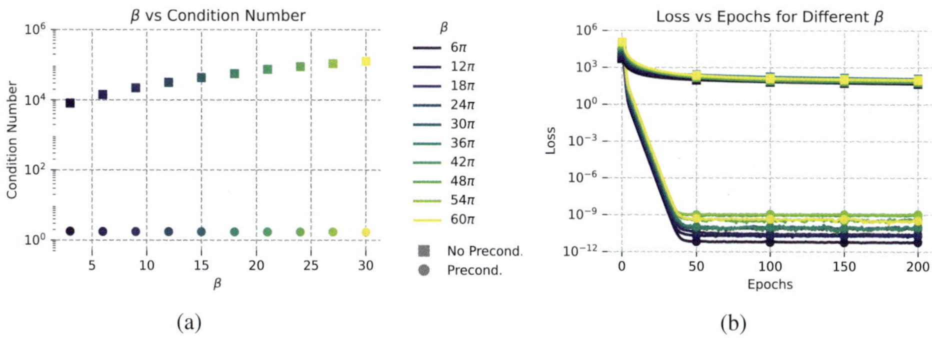

Figure 7.2. Linear advection equation with Fourier features. (a) Optimal condition number vs. β. (b) Training for the unpreconditioned and preconditioned Fourier features. Figure from De Ryck *et al.* (2023).

As a different example, we consider the linear advection equation $u_t + \beta u_x = 0$ on the one-dimensional spatial domain $x \in [0, 2\pi]$ and with 2π-periodic solutions in time with $t \in [0, 1]$. As in Krishnapriyan *et al.* (2021), our focus in this case is to study how physics-informed machine learning models train when the advection speed $\beta > 0$ is increased. To empirically evaluate this example, we again choose learnable time-dependent Fourier features as the model and precondition the resulting matrix $\mathbb{A}$ (7.13). In Figure 7.2(a) we see that the condition number of $\mathbb{A}(\beta) \sim \beta^2$ grows quadratically with advection speed. On the other hand, the condition number of the preconditioned model remains constant. Consequently, as shown in Figure 7.2(b), the unpreconditioned model trains very slowly (particularly for increasing values of the advection speed β), with losses remaining high despite being trained for a large number of epochs. In complete contrast, the preconditioned model trains very fast, irrespective of the values of the advection speed β.

7.4.3. Preconditioning nonlinear physics-informed machine learning models

Next, De Ryck *et al.* (2023) have investigated the conditioning of nonlinear models (Section 2.2.2) by considering the Poisson equation on $(-\pi, \pi)$ and learning its solution with neural networks of the form $u_\theta(x) = \Phi_\theta(x)$. It was observed that most of the eigenvalues of the resulting matrix $\mathbb{A}$ (7.13) are clustered near zero. Consequently, it is difficult to analyse the condition number *per se*. However, they also observed that there are only a couple of large non-zero eigenvalues of $\mathbb{A}$, indicating a very uneven spread of the spectrum. It is well known in classical numerical analysis (Trefethen and Embree 2005) that such spread-out spectra are very poorly conditioned and that this will impede training with gradient descent. This is corroborated in Figure 7.3(b), where the physics-informed MLP trains very slowly. Moreover, it turns out that preconditioning also localizes the spectrum (Trefethen and Embree 2005). This is attested in Figure 7.3(a), where we see

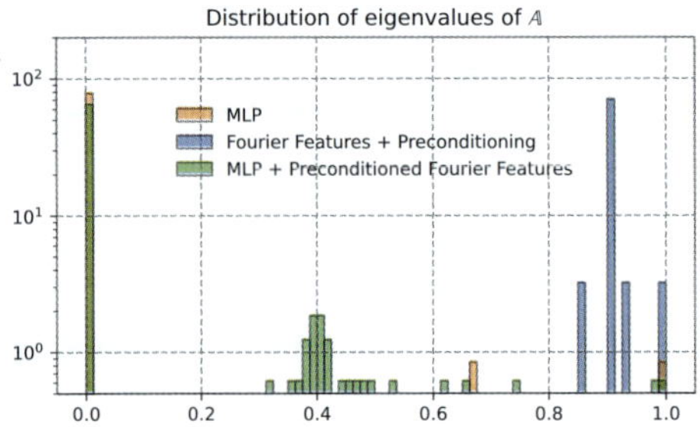

(a)

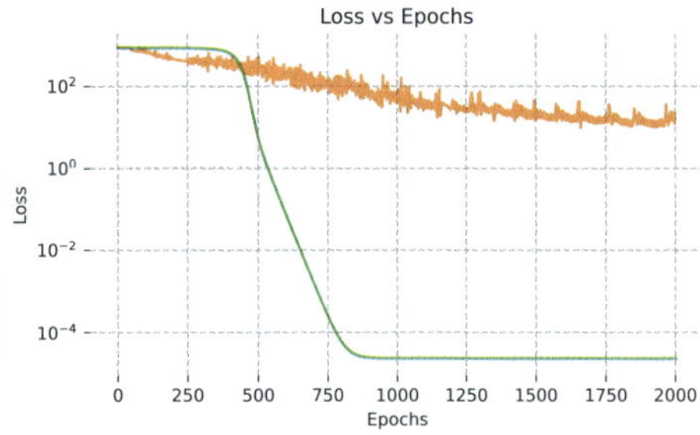

(b)

Figure 7.3. Poisson equation with MLPs. (a) Histogram of normalized spectrum (eigenvalues multiplied with learning rate). (b) Loss vs. number of epochs. Figure from De Ryck *et al.* (2023).

the localized spectrum of the preconditioned Fourier features model considered previously, which is also correlated with its fast training (Figure 7.1(b)).

However, preconditioning $\mathbb{A}$ for nonlinear models such as neural networks can, in general, be hard. Here we consider a simple strategy by coupling the MLP with Fourier features ϕ_k defined above, i.e. setting

$$u_\theta = \Phi_\theta\left(\sum_k \alpha_k \phi_k\right).$$

Intuitively, we must choose α_k carefully to control $(\mathrm{d}^{2n}/\mathrm{d}x^{2n})u_\theta(x)$, as it will include terms such as

$$\left(\sum_{k\neq 0} \alpha_k(-k)^{2n}\phi_k\right)\Phi_\theta^{(2n)}\left(\sum_k \alpha_k\phi_k\right).$$

Hence, such rescaling could better condition the Hermitian square of the differential operator. To test this for Poisson's equation, we choose $\alpha_k = 1/k^2$ (for $k \neq 0$) in this FF-MLP model and present the eigenvalues in Figure 7.3(a), which shows that although there are still quite a few (near-) zero eigenvalues, the number of non-zero eigenvalues is significantly increased, leading to a much more even spread in the spectrum when compared to the unpreconditioned MLP case. This possibly accounts for the fact that the resulting loss function decays much more rapidly, as shown in Figure 7.3(b), when compared to unpreconditioned MLP.

7.5. Strategies for improving training in physics-informed machine learning

Given the difficulties encountered in training physics-informed machine learning models, several *ad hoc* strategies have been proposed in the recent literature to improve training. It turns out that many of these strategies can also be interpreted in terms of preconditioning.

7.5.1. Choice of λ

The parameter λ in a physics-informed loss such as (2.46) plays a crucial role, as it balances the relative contributions of the physics-informed loss R and the supervised loss at the boundary B. Given the analysis of the previous sections, it is natural to suggest that this parameter should be chosen as

$$\lambda^* := \min_{\lambda} \kappa(\mathbb{A}(\lambda)) \tag{7.33}$$

in order to obtain the smallest condition number of $\mathbb{A}$ and accelerate convergence. Another strategy is suggested by Wang *et al.* (2021*a*), who suggest a learning rate annealing algorithm to iteratively update λ throughout training as follows:

$$\lambda_a^* := \frac{\max_k |(\nabla_\theta R(\theta))_k|}{(2K+1)^{-1}\sum_k |(\nabla_\theta B(\theta))_k|}. \tag{7.34}$$

Similarly, Wang *et al.* (2022*b*) propose setting

$$\lambda_b^* := \frac{\mathrm{Tr}(K_{rr}(n))}{\mathrm{Tr}(K_{uu}(n))} \approx \frac{\int_\Omega \nabla_\theta \mathcal{L}u_\theta^\top \nabla_\theta \mathcal{L}u_\theta}{\int_{\partial\Omega} \nabla_\theta u_\theta^\top \nabla_\theta u_\theta}; \tag{7.35}$$

we refer to Wang *et al.* (2022*b*) for the exact definition of K_{rr} and K_{uu}.

In the below example, we compare λ^* (7.33), λ_a^* (7.34) and λ_b^* (7.35) for the Poisson equation.

Example 7.13 (Poisson's equation). We compare the three proposed strategies above for the one-dimensional Poisson equation and a linear model using Fourier features up to frequency K. First of all, De Ryck *et al.* (2023) computed that $\lambda^* \sim K^2$. Next, to calculate λ_a^* we observe that $(\nabla_\theta R(\theta))_k = k^4\theta_k$ for $k \neq \ell$ and $(\nabla_\theta R(\theta))_\ell = \ell^4(\theta_k - 1)$. We find that $\max_k |(\nabla_\theta R(\theta))_k| \sim K^4$ at initialization. Next we calculate that (given that at initialization $\theta_m \sim \mathcal{N}(0,1)$ i.i.d.)

$$\mathbb{E}\sum_k |(\nabla_\theta B(\theta))_k| = \sum_{k=0}^K \mathbb{E}\left|\sum_{m=0}^K \theta_m(-1)^{k+m}\right| = \sum_{k=0}^K \mathbb{E}\left|\sum_{m=0}^K \theta_m\right| = (K+1)^{3/2} \tag{7.36}$$

This brings us to the rough estimate that

$$(2K+1)^{-1}\sum_k |(\nabla_\theta B(\theta))_k| \sim \sqrt{K}.$$

So in total we find that $\lambda_a^* \sim K^{3.5}$. Finally, for λ_b^* one can make the estimate that

$$\lambda_b^* \approx \frac{2\pi\sum_{k=1}^K k^4}{K+1} \sim K^4, \tag{7.37}$$

Hence it turns out that applying these different strategies leads to different scalings of λ with respect to increasing K for the Fourier features model. Despite these differences, it was observed that the resulting condition numbers $\kappa(\mathbb{A}(\lambda^*))$, $\kappa(\mathbb{A}(\lambda_a^*))$ and $\kappa(\mathbb{A}(\lambda_b^*))$ are actually very similar.

7.5.2. *Hard boundary conditions*

From the very advent of PINNs (Lagaris *et al.* 1998, 2000), several authors have advocated modifying machine learning models such that the boundary conditions in PDE (2.1) can be imposed exactly and the boundary loss in (2.46) is zero. Such *hard* imposition of boundary conditions (BCs) has been empirically shown to aid training; see e.g. Dong and Ni (2021), Moseley *et al.* (2023), Dolean *et al.* (2023) and references therein. This can be explained by saying that implementing hard boundary conditions leads to a different matrix $\mathbb{A}$, by changing the operator TT^*.

We first consider a toy model to demonstrate these phenomena in a straightforward way. We again consider the Poisson equation $-\Delta u = -\sin$ with zero Dirichlet boundary conditions (see Example 2.4). We choose as our model

$$u_\theta(x) = \theta_{-1}\cos(x) + \theta_0 + \theta_1\sin(x), \tag{7.38}$$

and report the comparison of the condition number between various variants of soft and hard boundary conditions from De Ryck *et al.* (2023).

- *Soft boundary conditions.* If we set the optimal λ using (7.33), we find that the condition number for the optimal λ^* is given by $3 + 2\sqrt{2} \approx 5.83$.
- *Hard boundary conditions: variant 1.* A first common method to implement hard boundary conditions is to multiply the model by a function $\eta(x)$ so that the product exactly satisfies the boundary conditions, regardless of u_θ. In our setting, we could consider $\eta(x)u_\theta(x)$ with $\eta(\pm\pi) = 0$. For $\eta = \sin$ the total model is given by

  $$\eta(x)u_\theta(x) = -\frac{\theta_{-1}}{2}\cos(2x) + \frac{\theta_{-1}}{2} + \theta_0\sin(x) + \frac{\theta_1}{2}\sin(2x), \tag{7.39}$$

  and gives rise to a condition number of 4. Different choices of η will inevitably lead to different condition numbers.
- *Hard boundary conditions: variant 2.* Another option would be to subtract $u_\theta(\pi)$ from the model so that the boundary conditions are exactly satisfied. This corresponds to the model

  $$u_\theta(x) - u_\theta(\pi) = \theta_{-1}(\cos(x) + 1) + \theta_1\sin(x). \tag{7.40}$$

  Note that this implies that one can discard θ_0 as parameter, leaving only two trainable parameters. The corresponding condition number is 1.

Hence, in this example the condition number for hard boundary conditions is strictly smaller than for soft boundary conditions, that is,

$$\kappa(\mathbb{A}_{\text{hard BC}}) < \min_\lambda \kappa(\mathbb{A}_{\text{soft BC}}(\lambda)). \tag{7.41}$$

This phenomenon can also be observed for other PDEs such as the linear advection equation (De Ryck *et al.* 2023).

7.5.3. Second-order optimizers

There are many empirical studies which demonstrate that first-order optimizers such as (stochastic) gradient descent or ADAM are not suitable for physics-informed machine learning, and we need to use second-order (quasi-) Newton-type optimizers such as L-BGFS in order to make training of physics-informed machine learning models feasible. As it turns out, one can prove that for linear physics-informed models the Hessian of the loss is identical to the matrix $\mathbb{A}$ (7.13). Given any loss $\mathcal{J}(\theta)$, the gradient flow of Newton's method is given by

$$\frac{\mathrm{d}\theta(t)}{\mathrm{d}t} = -\gamma H[\mathcal{J}(\theta(t))]^{-1}\nabla_\theta \mathcal{J}(\theta(t)), \tag{7.42}$$

where H is the Hessian. In our case, we consider the physics-informed loss $L(\theta) = R(\theta) + \lambda B(\theta)$ and consider the model $u_\theta(x) = \sum_\ell \theta_\ell \phi_\ell(x)$, which corresponds to a linear model or a neural network in the NTK regime (e.g. Lemma 7.8). Using $\partial_{\theta_i}\partial_{\theta_j} u_\theta = 0$, we calculate

$$\begin{aligned}
\partial_{\theta_i}\partial_{\theta_j} L(\theta) &= \int_\Omega (\mathcal{L}\partial_{\theta_i} u_\theta(x)) \cdot \mathcal{L}\partial_{\theta_j} u_\theta(x)\,\mathrm{d}x + \lambda \int_{\partial\Omega} \partial_{\theta_i} u_{\theta(t)}(x) \cdot \partial_{\theta_j} u_\theta(x)\,\mathrm{d}x \\
&= \int_\Omega \mathcal{L}\phi_i(x) \cdot \mathcal{L}\phi_j(x)\,\mathrm{d}x + \lambda \int_{\partial\Omega} \phi_i(x) \cdot \phi_j(x)\,\mathrm{d}x \\
&= \mathbb{A}_{ij}.
\end{aligned} \tag{7.43}$$

Therefore, in this case (quasi-) Newton methods automatically compute an (approximate) inverse of the Hessian and hence precondition the matrix $\mathbb{A}$, relating the use of (quasi-) Newton-type optimizers to preconditioning operators in this context. See Müller and Zeinhofer (2023) for further analysis of this connection.

7.5.4. Domain decomposition

Domain decomposition (DD) is a widely used technique in numerical analysis to precondition linear systems that arise out of classical methods such as finite elements (Dolean, Jolivet and Nataf 2015). Recently there have been attempts to use DD-inspired methods within physics-informed machine learning (see Moseley *et al.* 2023, Dolean *et al.* 2023 and references therein), although no explicit link with preconditioning the models was established. This link was established in De Ryck *et al.* (2023) for the linear advection equation, where they computed that the condition number of $\mathbb{A}$ depends quadratically on the advection speed β. They argued that considering N identical submodels on subintervals of $[0, T]$ essentially comes down to rescaling β. Since the condition number scales as β^2, splitting the time domain in N parts will then reduce the condition number of each individual submodel by a factor of N^2. This is also intrinsically connected to what happens in *causal learning* based training of PINNs (Wang, Sankaran and Perdikaris 2024), which therefore can also be viewed as a strategy for improving the condition number.

8. Conclusion

In this article we have provided a detailed review of available theoretical results on the numerical analysis of physics-informed machine learning models, such as PINNs and its variants, for the solution of forward and inverse problems for large classes of PDEs. We formulated physics-informed machine learning in terms of minimizing suitable forms (strong, weak, variational) of the *residual* of the underlying PDE within suitable subspaces of Ansatz spaces of *parametric functions*. PINNs are a special case of this general formulation corresponding to the strong form of the PDE residual and neural networks as Ansatz spaces. The residual is minimized with variants of the (stochastic) gradient descent algorithm.

Our analysis revealed that as long as the solutions to the PDE are *stable* in a suitable sense, the overall (total) error, which is the mismatch between the exact PDE solution and the approximation produced (after training) of the physics-informed machine learning model, can be estimated in terms of the following constituents: (i) the approximation error, which measures the smallness of the PDE residual within the Ansatz space (or hypothesis class), (ii) the generalization gap, which measures the *quadrature* error, and (iii) the optimization error, which quantifies how well the optimization problem is solved. Detailed analysis of each component of the error is provided, both in general terms as well as exemplified for many prototypical PDEs. Although a lot of the analysis is PDE-specific, we can discern many features that are shared across PDEs and models, which we summarize below.

- The (weak) stability of the underlying PDE plays a key role in the analysis, and this can be PDE-specific. In general, error estimates can depend on how the solution of the underlying PDEs changes with respect to perturbations. This is also consistent with the analysis of traditional numerical methods.
- Sufficient regularity of the underlying PDE solution is often required in the analysis, although this regularity can be lessened by changing the form of the PDE.
- The key bottleneck in physics-informed machine learning is due to the inability of the models to train properly. Generically, training PINNs and their variants can be difficult, as it depends on the spectral properties of the Hermitian square of the underlying differential operator, although suitable preconditioning might ameliorate training issues.

Finally, theory does provide some guidance to practitioners about how and when PINNs and their variants can be used. Succinctly, regularity dictates the form (strong vs. weak), and physics-informed machine learning models are expected to outperform traditional methods on high-dimensional problems (see Mishra and Molinaro 2021 as an example) or problems with complex geometries, where grid generation is prohibitively expensive, or inverse problems, where data supplements the physics. Physics-informed machine learning will be particularly attractive when coupled with operator learning models such as neural operators, as the underlying space is infinite-dimensional. More research on physics-informed operator learning models is highly desirable.

Appendix: Sobolev spaces

Let $d \in \mathbb{N}$, $k \in \mathbb{N}_0$, $1 \leq p \leq \infty$ and let $\Omega \subseteq \mathbb{R}^d$ be open. For a function $f\colon \Omega \to \mathbb{R}$ and a (multi-) index $\alpha \in \mathbb{N}_0^d$, we let

$$D^\alpha f = \frac{\partial^{|\alpha|} f}{\partial x_1^{\alpha_1} \cdots \partial x_d^{\alpha_d}} \tag{A.1}$$

denote the classical or distributional (i.e. weak) derivative of f. We let $L^p(\Omega)$ denote the usual Lebesgue space, and we define the Sobolev space $W^{k,p}(\Omega)$ as

$$W^{k,p}(\Omega) = \{f \in L^p(\Omega)\colon D^\alpha f \in L^p(\Omega) \text{ for all } \alpha \in \mathbb{N}_0^d \text{ with } |\alpha| \leq k\}. \tag{A.2}$$

We define the following seminorms on $W^{k,p}(\Omega)$: for $p < \infty$,

$$|f|_{W^{m,p}(\Omega)} = \Bigg(\sum_{|\alpha|=m} \|D^\alpha f\|_{L^p(\Omega)}^p \Bigg)^{1/p} \quad \text{for } m = 0, \ldots, k, \tag{A.3}$$

and for $p = \infty$,

$$|f|_{W^{m,\infty}(\Omega)} = \max_{|\alpha|=m} \|D^\alpha f\|_{L^\infty(\Omega)} \quad \text{for } m = 0, \ldots, k. \tag{A.4}$$

Based on these seminorms, we can define the following norms: for $p < \infty$,

$$\|f\|_{W^{k,p}(\Omega)} = \Bigg(\sum_{m=0}^{k} |f|_{W^{m,p}(\Omega)}^p \Bigg)^{1/p}, \tag{A.5}$$

and for $p = \infty$,

$$\|f\|_{W^{k,\infty}(\Omega)} = \max_{0 \leq m \leq k} |f|_{W^{m,\infty}(\Omega)}. \tag{A.6}$$

The space $W^{k,p}(\Omega)$ equipped with the norm $\|\cdot\|_{W^{k,p}(\Omega)}$ is a Banach space.

We let $C^k(\Omega)$ denote the space of functions that are k times continuously differentiable, and equip this space with the norm $\|f\|_{C^k(\Omega)} = \|f\|_{W^{k,\infty}(\Omega)}$.

References

G. Alessandrini, L. Rondi, E. Rosset and S. Vessella (2009), The stability of the Cauchy problem for elliptic equations, *Inverse Problems* **25**, art. 123004.

S. M. Allen and J. W. Cahn (1979), A microscopic theory for antiphase boundary motion and its application to antiphase domain coarsening, *Acta Metallurg.* **27**, 1085–1095.

F. Bach (2017), Breaking the curse of dimensionality with convex neural networks, *J. Mach. Learn. Res.* **18**, 629–681.

A. R. Barron (1993), Universal approximation bounds for superpositions of a sigmoidal function, *IEEE Trans. Inform. Theory* **39**, 930–945.

S. Bartels (2012), Total variation minimization with finite elements: Convergence and iterative solution, *SIAM J. Numer. Anal.* **50**, 1162–1180.

F. Bartolucci, E. de Bézenac, B. Raonić, R. Molinaro, S. Mishra and R. Alaifari (2023), Representation equivalent neural operators: A framework for alias-free operator learning, in *Advances in Neural Information Processing Systems 36* (A. Oh *et al.*, eds), Curran Associates, pp. 69661–69672.

C. Beck, S. Becker, P. Grohs, N. Jaafari and A. Jentzen (2021), Solving the Kolmogorov PDE by means of deep learning, *J. Sci. Comput.* **88**, 1–28.

C. Beck, A. Jentzen and B. Kuckuck (2022), Full error analysis for the training of deep neural networks, *Infin. Dimens. Anal. Quantum Probab. Relat. Top.* **25**, art. 2150020.

J. Berner, P. Grohs and A. Jentzen (2020), Analysis of the generalization error: Empirical risk minimization over deep artificial neural networks overcomes the curse of dimensionality in the numerical approximation of Black–Scholes partial differential equations, *SIAM J. Math. Data Sci.* **2**, 631–657.

P. B. Bochev and M. D. Gunzburger (2009), *Least-Squares Finite Element Methods*, Vol. 166 of Applied Mathematical Sciences, Springer.

M. D. Buhmann (2000), Radial basis functions, *Acta Numer.* **9**, 1–38.

E. Burman and P. Hansbo (2018), Stabilized nonconfirming finite element methods for data assimilation in incompressible flows, *Math. Comp.* **87**, 1029–1050.

T. Chen and H. Chen (1995), Universal approximation to nonlinear operators by neural networks with arbitrary activation functions and its application to dynamical systems, *IEEE Trans. Neural Netw.* **6**, 911–917.

Z. Chen, J. Lu and Y. Lu (2021), On the representation of solutions to elliptic PDEs in Barron spaces, in *Advances in Neural Information Processing Systems 34* (M. Ranzato *et al.*, eds), Curran Associates, pp. 6454–6465.

Z. Chen, J. Lu, Y. Lu and S. Zhou (2023), A regularity theory for static Schrödinger equations on $\mathbb{R}^d$ in spectral Barron spaces, *SIAM J. Math. Anal.* **55**, 557–570.

S. Cuomo, V. S. di Cola, F. Giampaolo, G. Rozza, M. Raissi and F. Piccialli (2022), Scientific machine learning through physics-informed neural networks: Where we are and what's next, *J. Sci. Comput.* **92**, art. 88.

G. Cybenko (1989), Approximation by superpositions of a sigmoidal function, *Math. Control Signals Systems* **2**, 303–314.

R. Dautray and J. L. Lions (1992), *Mathematical Analysis and Numerical Methods for Science and Technology*, Vol. 5, *Evolution Equations I*, Springer.

T. De Ryck and S. Mishra (2022*a*), Error analysis for physics-informed neural networks (PINNs) approximating Kolmogorov PDEs, *Adv. Comput. Math.* **48**, art. 79.

T. De Ryck and S. Mishra (2022*b*), Generic bounds on the approximation error for physics-informed (and) operator learning, in *Advances in Neural Information Processing Systems 35* (S. Koyejo *et al.*, eds), Curran Associates, pp. 10945–10958.

T. De Ryck and S. Mishra (2023), Error analysis for deep neural network approximations of parametric hyperbolic conservation laws, *Math. Comp.* Available at https://doi.org/10.1090/mcom/3934.

T. De Ryck, F. Bonnet, S. Mishra and E. de Bézenac (2023), An operator preconditioning perspective on training in physics-informed machine learning. Available at arXiv:2310.05801.

T. De Ryck, A. D. Jagtap and S. Mishra (2024*a*), Error estimates for physics-informed neural networks approximating the Navier–Stokes equations, *IMA J. Numer. Anal.* **44**, 83–119.

T. De Ryck, A. D. Jagtap and S. Mishra (2024*b*), On the stability of XPINNs and cPINNs. In preparation.

T. De Ryck, S. Lanthaler and S. Mishra (2021), On the approximation of functions by tanh neural networks, *Neural Networks* **143**, 732–750.

T. De Ryck, S. Mishra and R. Molinaro (2024*c*), wPINNs: Weak physics informed neural networks for approximating entropy solutions of hyperbolic conservation laws, *SIAM J. Numer. Anal.* **62**, 811–841.

M. W. M. G. Dissanayake and N. Phan-Thien (1994), Neural-network-based approximations for solving partial differential equations, *Commun. Numer. Methods Engrg* **10**, 195–201.

V. Dolean, A. Heinlein, S. Mishra and B. Moseley (2023), Multilevel domain decomposition-based architectures for physics-informed neural networks. Available at arXiv:2306.05486.

V. Dolean, P. Jolivet and F. Nataf (2015), *An Introduction to Domain Decomposition Methods: Algorithms, Theory, and Parallel Implementation*, SIAM.

S. Dong and N. Ni (2021), A method for representing periodic functions and enforcing exactly periodic boundary conditions with deep neural networks, *J. Comput. Phys.* **435**, art. 110242.

W. E and S. Wojtowytsch (2020), On the Banach spaces associated with multi-layer ReLU networks: Function representation, approximation theory and gradient descent dynamics. Available at arXiv:2007.15623.

W. E and S. Wojtowytsch (2022*a*), Representation formulas and pointwise properties for Barron functions, *Calc. Var. Partial Differential Equations* **61**, 1–37.

W. E and S. Wojtowytsch (2022*b*), Some observations on high-dimensional partial differential equations with Barron data, in *Proceedings of the Second Conference on Mathematical and Scientific Machine Learning*, Vol. 145 of Proceedings of Machine Learning Research, PMLR, pp. 253–269.

W. E and B. Yu (2018), The deep Ritz method: A deep learning-based numerical algorithm for solving variational problems, *Commun. Math. Statist.* **6**, 1–12.

W. E, C. Ma and L. Wu (2022), The Barron space and the flow-induced function spaces for neural network models, *Constr. Approx.* **55**, 369–406.

L. C. Evans (2022), *Partial Differential Equations*, Vol. 19 of Graduate Studies in Mathematics, American Mathematical Society.

A. Friedman (1964), *Partial Differential Equations of the Parabolic Type*, Prentice Hall.

X. Glorot and Y. Bengio (2010), Understanding the difficulty of training deep feedforward neural networks, in *Proceedings of the Thirteenth International Conference on Artificial Intelligence and Statistics*, Vol. 9 of Proceedings of Machine Learning Research, PMLR, pp. 249–256.

E. Godlewski and P. A. Raviart (1991), *Hyperbolic Systems of Conservation Laws*, Ellipsis.

I. Goodfellow, Y. Bengio and A. Courville (2016), *Deep Learning*, MIT Press.

P. Grohs, F. Hornung, A. Jentzen and P. von Wurstemberger (2018), A proof that artificial neural networks overcome the curse of dimensionality in the numerical approximation of Black–Scholes partial differential equations. Available at arXiv:1809.02362.

I. Gühring and M. Raslan (2021), Approximation rates for neural networks with encodable weights in smoothness spaces, *Neural Networks* **134**, 107–130.

J. Han, A. Jentzen and W. E (2018), Solving high-dimensional partial differential equations using deep learning, *Proc. Nat. Acad. Sci.* **115**, 8505–8510.

J. S. Hesthaven, S. Gottlieb and D. Gottlieb (2007), *Spectral Methods for Time-Dependent Problems*, Vol. 21 of Cambridge Monographs on Applied and Computational Mathematics, Cambridge University Press.

H. Holden and N. H. Risebro (2015), *Front Tracking for Hyperbolic Conservation Laws*, Vol. 152 of Applied Mathematical Sciences, Springer.

Q. Hong, J. W. Siegel and J. Xu (2021), *A priori* analysis of stable neural network solutions to numerical PDEs. Available at arXiv:2104.02903.

G.-B. Huang, Q.-Y. Zhu and C.-K. Siew (2006), Extreme learning machine: Theory and applications, *Neurocomput.* **70**, 489–501.

M. Hutzenthaler, A. Jentzen, T. Kruse and T. A. Nguyen (2020), A proof that rectified deep neural networks overcome the curse of dimensionality in the numerical approximation of semilinear heat equations, *SN Partial Differ. Equ. Appl.* **1**, 1–34.

O. Y. Imanuvilov (1995), Controllability of parabolic equations, *Sb. Math.* **186**, 109–132.

A. D. Jagtap and G. E. Karniadakis (2020), Extended physics-informed neural networks (XPINNs): A generalized space-time domain decomposition based deep learning framework for nonlinear partial differential equations, *Commun. Comput. Phys.* **28**, 2002–2041.

A. D. Jagtap, E. Kharazmi and G. E. Karniadakis (2020), Conservative physics-informed neural networks on discrete domains for conservation laws: Applications to forward and inverse problems, *Comput. Methods Appl. Mech. Engrg* **365**, art. 113028.

A. Jentzen, D. Salimova and T. Welti (2021), A proof that deep artificial neural networks overcome the curse of dimensionality in the numerical approximation of Kolmogorov partial differential equations with constant diffusion and nonlinear drift coefficients, *Commun. Math. Sci.* **19**, 1167–1205.

G. E. Karniadakis, I. G. Kevrekidis, L. Lu, P. Perdikaris, S. Wang and L. Yang (2021), Physics informed machine learning, *Nat. Rev. Phys.* **3**, 422–440.

E. Kharazmi, Z. Zhang and G. E. Karniadakis (2019), Variational physics-informed neural networks for solving partial differential equations. Available at arXiv:1912.00873.

E. Kharazmi, Z. Zhang and G. E. Karniadakis (2021), hp-VPINNs: Variational physics-informed neural networks with domain decomposition, *Comput. Methods Appl. Mech. Engrg* **374**, art. 113547.

D. P. Kingma and J. Ba (2015), Adam: A method for stochastic optimization, in *Proceedings of the 3rd International Conference on Learning Representations (ICLR 2015)*.

N. B. Kovachki, Z. Li, B. Liu, K. Azizzadenesheli, K. Bhattacharya, A. M. Stuart and A. Anandkumar (2023), Neural operator: Learning maps between function spaces with applications to PDEs, *J. Mach. Learn. Res.* **24**, 1–97.

N. Kovachki, S. Lanthaler and S. Mishra (2021), On universal approximation and error bounds for Fourier neural operators, *J. Mach. Learn. Res.* **22**, 13237–13312.

A. S. Krishnapriyan, A. Gholami, S. Zhe, R. M. Kirby and M. W. Mahoney (2021), Characterizing possible failure modes in physics-informed neural networks, in *Advances in Neural Information Processing Systems 34* (M. Ranzato *et al.*, eds), Curran Associates, pp. 26548–26560.

G. Kutyniok, P. Petersen, M. Raslan and R. Schneider (2022), A theoretical analysis of deep neural networks and parametric PDEs, *Constr. Approx.* **55**, 73–125.

I. E. Lagaris, A. Likas and D. I. Fotiadis (1998), Artificial neural networks for solving ordinary and partial differential equations, *IEEE Trans. Neural Netw.* **9**, 987–1000.

I. E. Lagaris, A. Likas and G. D. Papageorgiou (2000), Neural-network methods for boundary value problems with irregular boundaries, *IEEE Trans. Neural Netw.* **11**, 1041–1049.

S. Lanthaler, S. Mishra and G. E. Karniadakis (2022), Error estimates for DeepONets: A deep learning framework in infinite dimensions, *Trans. Math. Appl.* **6**, tnac001.

R. J. LeVeque (2002), *Finite Volume Methods for Hyperbolic Problems*, Vol. 31 of Cambridge Texts in Applied Mathematics, Cambridge University Press.

Z. Li, N. Kovachki, K. Azizzadenesheli, B. Liu, K. Bhattacharya, A. Stuart and A. Anandkumar (2020), Neural operator: Graph kernel network for partial differential equations. Available at arXiv:2003.03485.

Z. Li, N. B. Kovachki, K. Azizzadenesheli, B. Liu, K. Bhattacharya, A. Stuart and A. Anandkumar (2021), Fourier neural operator for parametric partial differential equations, in *International Conference on Learning Representations (ICLR 2021)*.

C.-L. Lin, G. Uhlmann and J.-N. Wang (2010), Optimal three-ball inequalities and quantitative uniqueness for the Stokes system, *Discrete Contin. Dyn. Syst.* **28**, 1273–1290.

C. Liu, L. Zhu and M. Belkin (2020), On the linearity of large non-linear models: When and why the tangent kernel is constant, in *Advances in Neural Information Processing Systems 33* (H. Larochelle *et al.*, eds), Curran Associates, pp. 15954–15964.

D. C. Liu and J. Nocedal (1989), On the limited memory BFGS method for large scale optimization, *Math. Program.* **45**, 503–528.

M. Longo, S. Mishra, T. K. Rusch and C. Schwab (2021), Higher-order quasi-Monte Carlo training of deep neural networks, *SIAM J. Sci. Comput.* **43**, A3938–A3966.

L. Lu, P. Jin, G. Pang, Z. Zhang and G. E. Karniadakis (2021*a*), Learning nonlinear operators via DeepONet based on the universal approximation theorem of operators, *Nat. Mach. Intell.* **3**, 218–229.

Y. Lu, H. Chen, J. Lu, L. Ying and J. Blanchet (2021*b*), Machine learning for elliptic PDEs: Fast rate generalization bound, neural scaling law and minimax optimality, in *International Conference on Learning Representations (ICLR 2022)*.

Y. Lu, J. Lu and M. Wang (2021*c*), *A priori* generalization analysis of the deep Ritz method for solving high dimensional elliptic partial differential equations, in *Proceedings of Thirty Fourth Conference on Learning Theory*, Vol. 134 of Proceedings of Machine Learning Research, PMLR, pp. 3196–3241.

K. O. Lye, S. Mishra and D. Ray (2020), Deep learning observables in computational fluid dynamics, *J. Comput. Phys.* **410**, art. 109339.

A. J. Majda and A. L. Bertozzi (2002), *Vorticity and Incompressible Flow*, Vol. 55 of Cambridge Texts in Applied Mathematics, Cambridge University Press.

T. Marwah, Z. Lipton and A. Risteski (2021), Parametric complexity bounds for approximating PDEs with neural networks, in *Advances in Neural Information Processing Systems 34* (M. Ranzato *et al.*, eds), Curran Associates, pp. 15044–15055.

S. Mishra and R. Molinaro (2021), Physics informed neural networks for simulating radiative transfer, *J. Quant. Spectrosc. Radiative Transfer* **270**, art. 107705.

S. Mishra and R. Molinaro (2022), Estimates on the generalization error of physics-informed neural networks for approximating a class of inverse problems for PDEs, *IMA J. Numer. Anal.* **42**, 981–1022.

S. Mishra and R. Molinaro (2023), Estimates on the generalization error of physics-informed neural networks for approximating PDEs, *IMA J. Numer. Anal.* **43**, 1–43.

M. F. Modest (2003), *Radiative Heat Transfer*, Elsevier.

R. Molinaro (2023), Applications of deep learning to scientific computing. PhD thesis, ETH Zürich.

B. Moseley, A. Markham and T. Nissen-Meyer (2023), Finite basis physics-informed neural networks (FBPINNs): A scalable domain decomposition approach for solving differential equations, *Adv. Comput. Math.* **49**, art. 62.

J. Müller and M. Zeinhofer (2023), Achieving high accuracy with PINNs via energy natural gradient descent, in *Proceedings of the 40th International Conference on Machine Learning* (A. Krause *et al.*, eds), Vol. 202 of Proceedings of Machine Learning Research, PMLR, pp. 25471–25485.

J. A. A. Opschoor, P. C. Petersen and C. Schwab (2020), Deep ReLU networks and high-order finite element methods, *Anal. Appl.* **18**, 715–770.

P. Petersen and F. Voigtlaender (2018), Optimal approximation of piecewise smooth functions using deep ReLU neural networks, *Neural Networks* **108**, 296–330.

N. Rahaman, A. Baratin, D. Arpit, F. Draxler, M. Lin, F. A. Hamprecht, Y. Bengio and A. Courville (2019), On the spectral bias of neural networks, in *Proceedings of the 36th International Conference on Machine Learning (ICML 2019)*, Vol. 97 of Proceedings of Machine Learning Research, PMLR, pp. 5301–5310.

A. Rahimi and B. Recht (2008), Uniform approximation of functions with random bases, in *2008 46th Annual Allerton Conference on Communication, Control, and Computing*, IEEE, pp. 555–561.

M. Raissi and G. E. Karniadakis (2018), Hidden physics models: Machine learning of nonlinear partial differential equations, *J. Comput. Phys.* **357**, 125–141.

M. Raissi, P. Perdikaris and G. E. Karniadakis (2019), Physics-informed neural networks: A deep learning framework for solving forward and inverse problems involving nonlinear partial differential equations, *J. Comput. Phys.* **378**, 686–707.

M. Raissi, A. Yazdani and G. E. Karniadakis (2018), Hidden fluid mechanics: A Navier–Stokes informed deep learning framework for assimilating flow visualization data. Available at arXiv:1808.04327.

B. Raonić, R. Molinaro, T. D. Ryck, T. Rohner, F. Bartolucci, R. Alaifari, S. Mishra and E. de Bézenac (2023), Convolutional neural operators for robust and accurate learning of PDEs, in *Advances in Neural Information Processing Systems 36* (A. Oh *et al.*, eds), Curran Associates, pp. 77187–77200.

C. E. Rasmussen (2003), Gaussian processes in machine learning, in *Summer School on Machine Learning*, Springer, pp. 63–71.

C. Schwab and J. Zech (2019), Deep learning in high dimension: Neural network expression rates for generalized polynomial chaos expansions in uq, *Anal. Appl.* **17**, 19–55.

Y. Shin (2020), On the convergence of physics informed neural networks for linear second-order elliptic and parabolic type PDEs, *Commun. Comput. Phys.* **28**, 2042–2074.

Y. Shin, Z. Zhang and G. E. Karniadakis (2023), Error estimates of residual minimization using neural networks for linear PDEs, *J. Mach. Learn. Model. Comput.* **4**, 73–101.

J. Stoer and R. Bulirsch (2002), *Introduction to Numerical Analysis*, Springer.

M. Tancik, P. Srinivasan, B. Mildenhall, S. Fridovich-Keil, N. Raghavan, U. Singhal, R. Ramamoorthi, J. Barron and R. Ng (2020), Fourier features let networks learn high frequency functions in low dimensional domains, in *Advances in Neural Information Processing Systems 33* (H. Larochelle *et al.*, eds), Curran Associates, pp. 7537–7547.

R. Temam (2001), *Navier–Stokes Equations: Theory and Numerical Analysis*, American Mathematical Society.

N. Trefethen and M. Embree (2005), *Spectra and Pseudo-Spectra*, Princeton University Press.

R. Verfürth (1999), A note on polynomial approximation in Sobolev spaces, *ESAIM Math. Model. Numer. Anal.* **33**, 715–719.

C. Wang, S. Li, D. He and L. Wang (2022*a*), Is L^2 physics informed loss always suitable for training physics informed neural network?, in *Advances in Neural Information Processing Systems 35* (S. Koyejo *et al.*, eds), Curran Associates, pp. 8278–8290.

S. Wang, S. Sankaran and P. Perdikaris (2024), Respecting causality for training physics-informed neural networks, *Comput. Methods Appl. Mech. Engrg* **421**, art. 116813.

S. Wang, Y. Teng and P. Perdikaris (2021*a*), Understanding and mitigating gradient flow pathologies in physics-informed neural networks, *SIAM J. Sci. Comput.* **43**, A3055–A3081.

S. Wang, H. Wang and P. Perdikaris (2021*b*), Learning the solution operator of parametric partial differential equations with physics-informed DeepONets, *Sci. Adv.* **7**, eabi8605.

S. Wang, X. Yu and P. Perdikaris (2022*b*), When and why PINNs fail to train: A neural tangent kernel perspective, *J. Comput. Phys.* **449**, art. 110768.

Acta Numerica (2024), pp. 715–840 Printed in the United Kingdom
doi:10.1017/S0962492924000023

Optimal experimental design: Formulations and computations

Xun Huan
University of Michigan, 1231 Beal Ave, Ann Arbor, MI 48109, USA
Email: xhuan@umich.edu

Jayanth Jagalur
Center for Applied Scientific Computing, Lawrence Livermore National Laboratory, 7000 East Ave, Livermore, CA 94550, USA
Email: jagalur1@llnl.gov

Youssef Marzouk
Massachusetts Institute of Technology, 77 Massachusetts Ave, Cambridge, MA 02139, USA
Email: ymarz@mit.edu

Questions of 'how best to acquire data' are essential to modelling and prediction in the natural and social sciences, engineering applications, and beyond. Optimal experimental design (OED) formalizes these questions and creates computational methods to answer them. This article presents a systematic survey of modern OED, from its foundations in classical design theory to current research involving OED for complex models. We begin by reviewing criteria used to formulate an OED problem and thus to encode the goal of performing an experiment. We emphasize the flexibility of the Bayesian and decision-theoretic approach, which encompasses information-based criteria that are well-suited to nonlinear and non-Gaussian statistical models. We then discuss methods for estimating or bounding the values of these design criteria; this endeavour can be quite challenging due to strong nonlinearities, high parameter dimension, large per-sample costs, or settings where the model is implicit. A complementary set of computational issues involves optimization methods used to find a design; we discuss such methods in the discrete (combinatorial) setting of observation selection and in settings where an exact design can be continuously parametrized. Finally we present emerging methods for sequential OED that build non-myopic design policies, rather than explicit designs; these methods naturally adapt to the outcomes of past experiments in proposing new experiments, while seeking coordination among all experiments to be performed. Throughout, we highlight important open questions and challenges.

2020 Mathematics Subject Classification: Primary 62-02, 62-08, 62B15, 62K05
Secondary 62L05, 94A17, 65M32

CONTENTS

1	Introduction	716
2	Optimal design criteria	720
3	Numerical approximation of design criteria	748
4	Design optimization methods	773
5	Sequential optimal experimental design	789
6	Outlook	812
References		819

1. Introduction

Acquiring data to inform models and guide decisions is an essential part of scientific enquiry, engineering design and even policy making. Seldom can we construct a useful model in isolation from data. Rather, data must be used to infer parameters of models, to assess whether models can provide useful predictions, and to spur a wide variety of model improvements. In this setting, it is natural to consider how to acquire data *efficiently*. Experimental data and field measurements can be costly or time-consuming to acquire; other information sources, similarly, may be expensive to query. Yet we face a multitude of choices in designing such queries. Where to place a sensor? What experimental conditions to impose? What quantity to observe? Do measurements need to be very precise, or would a noisier measurement suffice? When should measurements be made? And how much data should be collected? More broadly, what *combination* of observations would be most informative or useful, and how should we precisely define notions of 'informative' or 'useful' in the first place? Crucially, these notions must allow experimental choices to be made *before* data are acquired.

Optimal experimental design (OED) aims to answer these questions, through mathematical formulations that formalize and tailor them to the ultimate goals of acquiring data. A model developer may have many possible goals, and hence there are many possible criteria for what comprises a good experimental design. Another essential aspect of OED involves numerical algorithms, e.g. for evaluating suitable design criteria, for optimizing over possible experimental configurations, and possibly doing so in a 'closed-loop' sequential fashion. Collectively, these endeavours lie at the intersection of many fields: statistical inference and decision theory, information theory, Monte Carlo methods, continuous and combinatorial optimization, dynamic programming and stochastic control, and even reinforcement learning.

Every OED problem has two essential ingredients: an *experiment*, which is the source of data, and a mathematical *model*. The role of the model is to simulate what might happen in candidate experiments, and to assess how the results of such experiments might improve the model and its predictions. Formally, the model is a statistical model; in some applications, evaluating this statistical model also involves significant numerical simulation. An underlying presumption of OED is that it is *worthwhile* to perform many calculations involving the model, in order to plan experiments that are more efficient and effective. These calculations might be far less expensive (by some metric) than performing experiments, or there might be other impediments to experimentation that make finding an optimal design in advance, or building an online optimal design policy for online settings, worth the effort.

The notion of an 'experiment' should be construed quite broadly, and certainly not limited to laboratory experiments in the traditional sense. A high-fidelity simulation of a complex model, used to produce data to calibrate the parameters of a simpler model, constitutes an experiment. Arranging a network of sensors, or planning the path of an airborne vehicle carrying instruments, also constitutes designing an experiment. Application domains in which OED is used are similarly vast. OED has long been an essential part of statistical modelling, from the design of clinical trials (Berry, Carlin, Lee and Müller 2010) to the design of computer experiments (Sacks, Welch, Mitchell and Wynn 1989); the latter is closely related to the classical problem of design for regression (Elfving 1952, Kiefer and Wolfowitz 1959, Kiefer 1961*a*). But OED can also be applied to problems involving parameter inference in ordinary or partial differential equations (Huan and Marzouk 2013), to a wide range of inverse problems (Haber, Horesh and Tenorio 2008, Alexanderian, Petra, Stadler and Ghattas 2016*b*, Ruthotto, Chung and Chung 2018, Alexanderian 2021, Helin, Hyvönen and Puska 2022), and to myriad other 'complex' models of data-generating processes – in astronomy (Loredo 2011), systems biology (Liepe, Filippi, Komorowski and Stumpf 2013), aeroelasticity (Riley *et al.* 2019), reliability testing (Weaver and Meeker 2021), and beyond.

The evolution of experimental design has a rich and fascinating history. Early twentieth-century approaches to experimental design were motivated by agricultural experiments and similar applications. Statistical methods in this setting often involved hypothesis testing, and a good design was one that maximized the sensitivity of the test. Notable works by Fisher and his collaborators during this period introduced concepts such as balance, orthogonality, blocking and aliasing (Fisher 1936, Craig and Fisher 1936, Yates 1933, 1937, 1940, Bose 1939, Bose and Nair 1939). Wald recognized that these ideas were relevant to many other fields, and in a seminal paper (Wald 1943) introduced formal notions of the efficiency of a design. With this framework, he was able to explain the success of designs based on Latin squares, commonly used in agricultural experimentation. Following Wald's work, many core ideas of modern OED emerged towards the middle of the twentieth century (Elfving 1952, Lindley 1956, Kiefer 1958, Stone 1959, Kiefer and Wolfowitz 1959, Kiefer 1959, 1961*a*), some of them influenced by the emerging discipline of decision theory. Kiefer distinguished various design criteria by giving them meaningful names, and thus originated current 'alphabetic optimality' terminology (Kiefer 1958). Much later, Kiefer also showed inter-relationships among various design criteria by introducing more general notions of 'universal' optimality (Kiefer 1974). Lindley's work in the same period (Lindley 1956) aligns more closely with Bayesian statistical thinking, and deserves special mention in light of the current popularity of information-theoretic design criteria. These approaches remained largely intractable half a century ago, but with advances in computing power and algorithms, these more general and arguably more rigorous design criteria have become feasible to implement. For a more detailed history of OED, we refer readers to Wynn (1984) and to the texts by Fedorov (1972), Shah and Sinha (1989) and Pukelsheim (2006).

In recent years, interest in OED has expanded from the statistics literature into the uncertainty quantification and applied mathematics communities, where, as noted earlier, an animating goal has been to advance OED methodologies for 'complex' models. Here many forms of complexity are relevant: high-dimensional parameters, computationally intensive likelihood functions that involve the evaluation of ordinary or partial differential equations, strong nonlinearity in the dependence of observables on parameters, and the associated non-Gaussianity of posterior distributions in the Bayesian setting. Additional complexities can arise due to data-generating processes that evolve in time, which present the opportunity to design and implement experiments adaptively, i.e. where the results of previous experiments influence the next; this is the setting of *sequential* optimal experimental design. Another thread relevant to modern OED has come from the computer science literature, where much attention has been paid to the optimization of *set functions*; methods here underpin a combinatorial approach to OED, where the problem is cast as choosing a subset of a given 'ground set' of feasible candidate experiments. And in recent years, tools from machine learning and in particular deep learning have become quite useful for OED, for instance by offering new ways of evaluating complex design criteria in high dimensional, non-Gaussian settings. Of particular interest are expressive machine learning models and learning algorithms for the underlying density (or density ratio) estimation tasks. We will discuss all of these threads, and many more, in the ensuing sections.

We mention here several other excellent surveys that may be of interest to the reader. Steinberg and Hunter (1984) and Pukelsheim (2006) (updated from the original 1993 version) provide comprehensive reviews of non-Bayesian OED for linear models. Ford, Titterington and Kitsos (1989) discuss design for nonlinear models, while DasGupta (1995) discusses Bayesian designs, mostly for linear models. Atkinson, Donev and Tobias (2007) provide extensive coverage of linear optimal design theory, with some extension to nonlinear and Bayesian methods. The paper of Chaloner and Verdinelli (1995) is an influential review from a statistical perspective, highlighting features of the Bayesian and decision-theoretic approach to OED. We view the Bayesian approach as very natural in the setting of design, and will largely adopt such a perspective here. Clyde (2001) presents an overview of Bayesian OED formulations with various choices of utility.

More recent reviews have placed a greater emphasis on computation. Ryan, Drovandi, McGree and Pettitt (2016) discuss Bayesian formulations of OED and then survey computational algorithms for realizing Bayesian designs, emphasizing Monte Carlo methods for estimating design criteria and for searching through design space. Alexanderian (2021) reviews OED for Bayesian inverse problems, emphasizing formulations and algorithms for the infinite-dimensional (function space) setting. Rainforth, Foster, Ivanova and Smith (2023) provide a survey highlighting recent machine learning and reinforcement learning tools in OED, including sequential design. Strutz and Curtis (2024) present a review of variational OED methodologies and their application to geophysical and in particular seismological problems. There may be other recent reviews of which we are unaware, and we apologize for such omissions.

1.1. Scope and organization of this article

This article aims to provide a broad, comprehensive survey of methodologies for optimal experimental design. Our perspective covers both formulations, i.e. the many ways of *posing* an OED problem, and computations, i.e. numerical algorithms for *finding* optimal designs, or tractable approximations of optimal designs, for a range of problem settings. The second topic in particular is multi-faceted – we will draw upon Monte Carlo methods, algorithms for inference and density estimation, and a variety of optimization approaches – but naturally enjoys a close interplay with the first. Our goal is to guide readers who are new to OED from the basic ideas to the research frontier, and to illuminate open issues and challenges at that frontier. Indeed, we believe that the present moment is ripe for a survey of the field: the problems that remain are quite challenging, and a synthesis of ideas and approaches from many different intellectual communities (some of which have been rather disconnected) is needed to make progress.

The article is organized as follows. We begin in Section 2 by describing possible design criteria for OED, each encoding a different notion of what constitutes a ‘good’ experiment. The applicability of these criteria ranges from the rather specific (e.g. linear-Gaussian problems) to the very general. Many are rooted in information-theoretic and/or decision-theoretic formulations. We also mention alternative design heuristics that have been used in practice, and clarify distinctions between the OED problem and several related but different problems, such as Bayesian optimization.

In Section 3 we turn to the first step of computation: numerical algorithms for estimating or bounding the values of these design criteria. We survey Monte Carlo schemes, as well as variational approximations and density estimation methods, for this purpose. Some of these algorithms are applicable to so-called ‘implicit’ Bayesian models, where evaluations of the likelihood function or prior density may be unavailable. We also discuss challenges associated with high-dimensional parameters and data, and dimension reduction schemes that mitigate these challenges. In Section 4 we discuss strategies and algorithms for efficiently optimizing design criteria in various settings. On one hand, we address problems where an exact design of interest is represented by continuous (real-valued) variables. On the other hand, we discuss settings where the design optimization problem is discrete and combinatorial, e.g. a form of subset selection, and continuous relaxations thereof.

Sections 2–4 focus on the all-at-once ('batch') design of experiments, that is, 'fixed' or static designs that cannot be adjusted as the outcomes of experiments are realized. In contrast, Section 5 turns to the problem of sequential OED, where design decisions are naturally adapted according to the outcomes of previous experiments, while taking into consideration the information to be gained from future experiments. We present sequential OED in a fully Bayesian setting, leveraging the formalism of Markov decision processes. We then highlight recent computational methods for solving this challenging problem, which make use of dynamic programming, various reinforcement learning techniques and information bounds.

We close in Section 6 with a discussion of open questions and opportunities for future work.

2. Optimal design criteria

We begin by addressing the central question of any OED formulation: In what sense should one deem a candidate design to be 'good'? More specifically, by what considerations should one candidate design be deemed better than another? Answering these questions is essential to the notion of *optimal* design. The answers are formalized by choosing a quantitative design criterion – a function that can then be maximized in order to identify the corresponding optimal design. In this section, we will discuss a wide variety of design criteria and the goals they encode.

To formulate these criteria, we must first specify a *statistical model* for the observations Y obtained under a candidate design. We need such a model in order to predict the outcomes of candidate experiments, and to relate these outcomes to the ensuing estimation or prediction tasks that motivated the acquisition of data in the first place. We will initially consider parametric statistical models. Any such model is a family of probability distributions for Y, indexed by (unknown) model parameters $\theta \in \mathbf{\Theta} \subseteq \mathbb{R}^p$ and by the choice of design $\xi \in \Xi$:

$$\mathcal{M} = \{F_{Y;\theta,\xi} : \theta \in \mathbf{\Theta},\ \xi \in \Xi\}. \tag{2.1}$$

This model encodes, for any given value of θ and ξ, a complete probabilistic description of the resulting observations, via the cumulative distribution function $F_{Y;\theta,\xi}$. Further, if every element of $\mathcal{M}$ is absolutely continuous with respect to Lebesgue measure, we can also write the statistical model as a family of conditional probability density functions, i.e. $\mathcal{M} = \{p_{Y|\theta,\xi}(y|\theta,\xi) : \theta \in \mathbf{\Theta},\ \xi \in \Xi\}$. We will make this simplifying assumption from here on, with the understanding that the conditional density of Y could be replaced by a conditional probability measure $\mu(\mathrm{d}y|\theta,\xi)$ as needed.

The parameters θ in the model are unknown and hence the object of estimation or inference. On the other hand, ξ can be controlled by the experimenter. For instance, if candidate Y corresponded to observations of a spatiotemporal process, ξ might represent the spatial and temporal coordinates of a chosen set of observations. In a regression model, ξ encodes the values of the covariates (i.e. independent variables) at which observations are obtained. Different ways of representing ξ will give rise to different optimization problems, which we will discuss in Section 4.

The statistical model immediately yields a likelihood function $\theta \mapsto p_{Y|\theta,\xi}(y|\theta,\xi)$, and the corresponding symmetric, positive semi-definite $p \times p$ Fisher information matrix

$$F(\theta,\xi) := \mathbb{E}_{Y|\theta,\xi}[\nabla_\theta \log p_{Y|\theta,\xi}(Y|\theta,\xi) \otimes \nabla_\theta \log p_{Y|\theta,\xi}(Y|\theta,\xi)], \tag{2.2}$$

which is a central object in estimation theory (Lehmann and Casella 1998). Below we will discuss various design criteria based on the Fisher information matrix.

Alternatively, one can take a Bayesian approach and endow the unknown parameters, now denoted as Θ, with a prior distribution. Let this distribution have (Lebesgue) density p_Θ on $\mathbb{R}^p$. We will always assume that the prior density is functionally independent of the design ξ. The posterior density of Θ is then given by Bayes' rule:

$$p_{\Theta|Y,\xi}(\theta|y,\xi) = \frac{p_{Y|\Theta,\xi}(y|\theta,\xi)\, p_\Theta(\theta)}{p_{Y|\xi}(y|\xi)}. \tag{2.3}$$

In the Bayesian paradigm, the prior and posterior distributions represent, respectively, our states of knowledge before and after having observed $Y = y$. The marginal density of the observations,

$$p_{Y|\xi}(y|\xi) = \int p_{Y|\Theta,\xi}(y|\theta,\xi)\, p_\Theta(\theta)\,\mathrm{d}\theta,$$

appearing in the denominator of (2.3), is called the evidence or marginal likelihood. We will also call this distribution the *prior predictive*, as it reflects our probabilistic prediction of future values of Y given only the prior on Θ, the statistical model and a chosen design. Having observed a particular value of the data, say y^*, at some chosen design ξ, the *posterior predictive* density of the data Y for a new design ξ^+ is

$$p_{Y|\xi^+,y^*,\xi}(y|\xi^+,y^*,\xi) = \int p_{Y|\Theta,\xi}(y|\theta,\xi^+)\, p_{\Theta|Y,\xi}(\theta|y^*,\xi)\,\mathrm{d}\theta.$$

Many design criteria discussed below will explicitly take advantage of this Bayesian formulation of the inference problem. Indeed, we will see that it is useful to have the ability to incorporate prior information on Θ – in general, but especially in nonlinear design settings – and that the Bayesian approach to OED is natural for decision-theoretic reasons as well.

2.1. *Design criteria for the linear-Gaussian setting*

A rich variety of design criteria, both Bayesian and non-Bayesian, have been developed for linear-Gaussian models. This class of statistical models has numerous practical applications: certainly linear regression, but also *linear inverse problems*, where observations depend *indirectly* on the parameters through the action of some linear operator. Canonical linear inverse problems include deconvolution, computerized tomography and source inversion, among many others (Kaipio and Somersalo 2006). Design criteria in this setting are often quite explicit and tractable, and also serve as a building block for certain nonlinear design approaches.

We specify a general linear-Gaussian model as

$$Y = G\theta + \mathcal{E}, \tag{2.4}$$

where $G \in \mathbb{R}^{n\times p}$ represents the linear 'forward' operator, mapping parameters to data in $\mathbb{R}^n$, and $\mathcal{E}$ is a Gaussian random variable with full-rank covariance matrix $\Gamma_{Y|\Theta} \in \mathbb{R}^{n\times n}$ that does not depend on θ. We let $\mathcal{E}$ have mean zero; choosing otherwise would not affect the developments below. In general, both G and $\Gamma_{Y|\Theta}$ can depend on the design ξ; that is, we have $G(\xi)$ and $\Gamma_{Y|\Theta}(\xi)$. The linear-Gaussian model can be summarized as $Y|\theta,\xi \sim \mathcal{N}(G(\xi)\theta, \Gamma_{Y|\Theta}(\xi))$.

2.1.1. *Classical alphabetic optimality*

The Fisher information matrix associated with (2.4) is

$$F(\theta,\xi) = F(\xi) = G(\xi)^{\top}\Gamma_{Y|\Theta}(\xi)^{-1}G(\xi). \tag{2.5}$$

It is thus independent of the value of the parameters θ; this property of linear models greatly simplifies the construction of design criteria. Note also that the inverse of the Fisher information matrix, F^{-1}, when it exists, is the covariance of the least-squares estimate and hence the maximum likelihood estimate, $\hat{\theta}(y)$, of θ. Many classical design criteria are therefore chosen to be scalar-valued functionals of the matrix F. Indeed, we must somehow 'scalarize' F to produce a useful optimization objective, and various scalarizations encode different goals. We recall several of these so-called 'alphabetic optimality' criteria as follows.

A-optimal design seeks

$$\xi^* \in \arg\max_{\xi\in\Xi} \operatorname{tr}(F(\xi)).$$

When F is invertible, which is invariably the situation of interest in classical design and what we shall assume in the remainder of this subsection, the optimization problem above is equivalent to $\min_{\xi\in\Xi} \operatorname{tr}(F^{-1}(\xi))$, which can be interpreted as minimizing the average variance of the p components of $\hat{\theta}$. *D-optimal* design, similarly, seeks

$$\xi^* \in \arg\min_{\xi\in\Xi} \log\det(F^{-1}(\xi)),$$

which minimizes the volume (in $\mathbb{R}^p$) of the smallest $100(1-\alpha)\%$ confidence ellipsoid for θ, for any confidence level $1-\alpha$. A useful distinguishing feature of D-optimal designs is that they are invariant under linear reparametrization (and hence rescaling) of θ; that is, if $\theta' = M\theta$ for some invertible matrix M, then a design that is D-optimal for θ is also D-optimal for θ'. This is not true, in general, for other optimality criteria.

While the A- and D-optimality criteria explicitly involve all the eigenvalues of F, E-optimal designs minimize the maximum eigenvalue of $F^{-1}(\xi)$, $\lambda_{\max}(F^{-1}(\xi))$ (equivalently, maximize $\lambda_{\min}(F(\xi)) = 1/\lambda_{\max}(F^{-1}(\xi))$). Such designs thus minimize the variance of $q^\top\hat{\theta}$ among all $q \in \mathbb{R}^p$ satisfying a norm constraint, that is, they minimize $\max_{\|q\|=w} \mathrm{Var}(q^\top\hat{\theta}) = \max_{\|q\|=w} q^\top F(\xi)^{-1} q$ for any $w > 0$.

A-, D- and E-optimality are perhaps the essential building-block design criteria for linear models, but there are numerous others. Some focus on the estimation of a linear combination of the parameters θ, or a subset of the elements of θ. Suppose, for example, that we are primarily interested in $A^\top\theta$, where $A \in \mathbb{R}^{p\times s}$, $s < p$, and A has rank s. Then D_A-optimality generalizes D-optimality by seeking

$$\xi^* \in \arg\min_{\xi\in\Xi} \log\det(A^\top F^{-1}(\xi)A). \tag{2.6}$$

This objective is justified by noting that $A^\top F^{-1}(\xi)A$ is the covariance matrix of $A^\top\hat{\theta}$. If we put $A = [I_s \;\; 0]^\top$, then the design criterion (2.6) focuses on the first s elements of θ; this is called D_S-optimality.

An analogous generalization of A-optimality, for some matrix $L \in \mathbb{R}^{p\times p}$, is called L-optimality (Atkinson *et al.* 2007):

$$\xi^* \in \arg\min_{\xi\in\Xi} \mathrm{tr}(F^{-1}(\xi)L).$$

If L is symmetric and has rank $s \leq p$, then it can be expressed as $L = AA^\top$ for $A \in \mathbb{R}^{p\times s}$. Then, using the cyclic property of the trace, the L-optimality objective can be rewritten as $\mathrm{tr}(A^\top F^{-1}(\xi)A)$. If $A^\top$ is a row vector in this setting (i.e. $s = 1$), then the criterion seeks to minimize the variance of a single linear combination of the parameters, and it is called c-optimality.

Other criteria instead seek to control the variance of predictions of the linear model. Consider, specifically, a regression model on some compact domain $\mathcal{X}$, where each row of G is given by the evaluation of a feature vector $f\colon \mathcal{X} \to \mathbb{R}^p$; that is, the ith row of G is $G(i,:) = f^\top(x_i)$ for some $x_i \in \mathcal{X}$ that is in the support of the design ξ. The G-optimality criterion considers the variance of the predicted response at any point $x \in \mathcal{X}$, $f^\top(x)F^{-1}(\xi)f(x)$, and seeks a design ξ that will minimize the maximum value of this variance, that is,

$$\xi^* \in \operatorname*{argmin}_{\xi\in\Xi} \max_{x\in\mathcal{X}} f^\top(x)F^{-1}(\xi)f(x). \tag{2.7}$$

Variants of this criterion that target the *average* variance of the predicted response over a region, rather than its maximum, are called I-optimality or V-optimality. For a much more comprehensive discussion of classical alphabetic optimality criteria and their properties, we refer to Hedayat (1981) and Shah and Sinha (1989) as well as Atkinson *et al.* (2007, Chapter 10).

We should note here that the assumption of normality on Y is generally not needed for these criteria to apply, and for the statistical interpretations given above to hold. Rather, we need only that $\mathbb{E}[Y] = G\theta$ and $\mathrm{Cov}(Y) = \Gamma_{Y|\Theta}$. The best linear unbiased estimator still follows from the least-squares solution in this setting, and its performance is bounded by the Fisher information matrix (cf. Cramér–Rao bounds). In many situations (including the usual setting for G-optimality described above), it is further assumed that $\Gamma_{Y|\Theta} = \sigma^2 I$, that is, the observational errors are uncorrelated and have constant variance.

2.1.2. *Bayesian alphabetic optimality*

In the Bayesian setting, the parameters θ of the linear model (2.4) are endowed with a prior distribution. Since these parameters are now modelled as random variables, we write them as uppercase Θ and require that $\mathcal{E}$ and Θ are independent. Choosing a conjugate Gaussian prior, $\Theta \sim \mathcal{N}(\mu_\Theta, \Gamma_\Theta)$, we obtain a posterior distribution that is again Gaussian; it can be written in closed form as $\Theta|y,\xi \sim \mathcal{N}(\mu_{\Theta|Y}(y,\xi), \Gamma_{\Theta|Y}(\xi))$, where

$$\Gamma_{\Theta|Y}(\xi) := \left(G(\xi)^\top \Gamma_{Y|\Theta}(\xi)^{-1} G(\xi) + \Gamma_\Theta^{-1}\right)^{-1}, \tag{2.8}$$

$$\mu_{\Theta|Y}(y,\xi) := \Gamma_{\Theta|Y,\xi}\left(G(\xi)^\top \Gamma_{Y|\Theta}(\xi)^{-1} y + \Gamma_\Theta^{-1}\mu_\Theta\right). \tag{2.9}$$

The posterior mean therefore depends on the realized value of the data y, but the posterior covariance matrix does not.

The Bayesian analogue to classical alphabetic optimality uses design criteria that are functions of the *posterior covariance matrix*. Prior knowledge – and more generally the 'balance' of information between the prior and the likelihood, where the latter may be affected by the number of observations – will therefore affect the optimal design, since the design criteria depend on the dispersion (shape and scale) of the posterior.

For instance, Bayesian A-optimality seeks designs that minimize the trace of the posterior covariance matrix,

$$\xi^* \in \arg\min_{\xi\in\Xi} \mathrm{tr}(\Gamma_{\Theta|Y}(\xi)) = \mathrm{tr}((F(\xi) + \Gamma_\Theta^{-1})^{-1}).$$

Note that, in contrast with classical A-optimality, this objective no longer requires F to be invertible, as long as the prior covariance Γ_Θ is chosen to have full rank. The same is true of all other Bayesian alphabetic optimality criteria, making these criteria well-suited to designs with fewer than p support points, and to ill-posed inverse problems generally (Haber *et al.* 2008). Bayesian D-optimality seeks to minimize the log-determinant of the posterior covariance matrix,

$$\xi^* \in \arg\min_{\xi\in\Xi} \log\det(\Gamma_{\Theta|Y}(\xi)). \tag{2.10}$$

Similarly, Bayesian D_A-optimality seeks to minimize the log-determinant of the covariance of the posterior predictive distribution of $A^\top\theta$, for some matrix $A \in \mathbb{R}^{p\times s}$ with rank $s < p$; hence the goal is to minimize $\log\det(A^\top\Gamma_{\Theta|Y}(\xi)A)$. See Attia, Alexanderian and Saibaba (2018) for an application of this criterion, and of a Bayesian analogue of classical L-optimality. Bayesian E-optimal design minimizes the maximum eigenvalue of $\Gamma_{\Theta|Y}(\xi)$, and so on. For an extensive discussion of Bayesian alphabetic optimality criteria and their interpretations, we refer the reader to Chaloner and Verdinelli (1995) and DasGupta (1995). We will also revisit several of these criteria from the more general viewpoint of decision theory in Section 2.2; the decision-theoretic formulation lets us derive many Bayesian alphabetic optimality criteria from specific utility functions and, crucially, enables generalization to nonlinear models.

In the Bayesian setting, it is also natural to consider linear models with unknown variance parameters, e.g. $\Gamma_{Y|\theta} = \sigma^2 I$ with unknown σ^2, and to endow these variance parameters with suitable priors. In general, this extension modifies the optimality criteria discussed above – with some exceptions, such as Bayesian A-optimality using a conjugate inverse Gamma prior for σ^2; for more information, see Chaloner and Verdinelli (1995).

2.1.3. Designs as probability measures

So far, we have deliberately remained somewhat non-specific in describing how the design ξ enters the statistical models (2.1) or (2.4), other than to think of ξ as representing all the chosen 'coordinates' or locations of the observations on some continuous domain $\mathcal{X}$, or the indices of observations selected from some countable set of candidates. One rather elegant way of formalizing these examples is to cast the design as a *probability measure* on some domain $\mathcal{X}$. This viewpoint, originating in Kiefer and Wolfowitz (1959), is widely adopted in the classical literature on optimal design.

To explain this perspective, let us first consider the discrete case, where $\mathcal{X}$ is a countable and perhaps finite set of observation indices, $\mathcal{X} = \{1, 2, 3, \ldots\}$. Suppose that we wish to choose n observations in total. If n_i observations are taken at each point $i = 1, 2, 3, \ldots$ and $\sum_i n_i = n$, then we can write $\xi_i = n_i/n$ and consider $(\xi_i)_i$ to be a probability measure (Chaloner and Verdinelli 1995). This class of 'quantized' measures, with integer n_i and hence weights that are multiples of $1/n$, is called an *exact design*. If each point can only be observed once, i.e. if the selection is without replacement, then we further require $n_i \in \{0, 1\}$. It is often useful to relax the integer constraint, such that ξ is any probability measure over the set of candidate indices; in this case, ξ is called a *continuous* or approximate design. Then Ξ is the set of all probability mass functions over $\mathcal{X}$.

Now consider the setting of continuous observation indices, on a compact set $\mathcal{X} \subseteq \mathbb{R}^d$, for $d \geq 1$. This setting allows candidate observations to be indexed by some continuous coordinates, e.g. angles for a tomography problem, real-valued spatial coordinates for a sensor placement problem, or any other covariates in a generic regression problem. The notion of a continuous *design* extends naturally: ξ is simply a probability measure on $\mathcal{X}$, and Ξ is the set of all such probability measures. An exact design here would be a finite mixture of Dirac measures with quantized weights, that is, a measure supported on a finite collection of points in $\mathcal{X}$ with mixture weights that are multiples of $1/n$, i.e. $\xi = \frac{1}{n}\sum_{i=1}^{n}\delta_{x_i}$ for $x_i \in \mathcal{X}$.

A nice consequence of this general viewpoint is that many quantities relevant to the preceding design criteria can be written as integrals with respect to ξ. Consider a linear regression model with features $f\colon \mathcal{X} \to \mathbb{R}^p$ and uncorrelated observational errors. If the design ξ is supported on n equally weighted points $x_i \in \mathcal{X}$, $i = 1, \dots, n$, then the Fisher information matrix of the model is

$$F(\xi) = \frac{1}{\sigma^2}\sum_{i=1}^{n} f(x_i)f(x_i)^\top. \tag{2.11}$$

This expression follows from (2.5) by setting

$$G = [f(x_1)^\top; f(x_2)^\top; \dots; f(x_n)^\top] \in \mathbb{R}^{n\times p}$$

and $\Gamma_{Y|\theta} = \sigma^2 I_n$. If the design has continuous support, then we simply have instead

$$F(\xi) = \frac{n}{\sigma^2}\int_{\mathcal{X}} f(x)f(x)^\top \xi(\mathrm{d}x),$$

where the n above ensures that the scaling of the integral is consistent with (2.11).

It is natural to wonder how to reconcile this continuous viewpoint with the existence of classical optimal designs supported on a *finite* set of points in $\mathcal{X}$ (Atwood 1969). In other words, when optimizing some design criterion over the set of all probability measures on the infinite set $\mathcal{X}$, why should a minimizer be supported only on a finite number of locations? In fact, as explained in Atwood (1969) and Kiefer (1961*b*), under conditions satisfied by any of the design criteria in Section 2.1.1, there exists an optimal design supported on at most $p(p+1)/2$ points. Intuition for the result follows from Carathéodory's theorem, compactness of $\mathcal{X}$, and the fact that F is a $p \times p$ symmetric matrix: the optimal information matrix F can always be expressed as a convex combination of at most $p(p+1)/2$ rank-one matrices, each produced by a single-point design. The need for F to be of full rank, on the other hand, imposes a lower bound of p on the number of points in an optimal design. In the case of Bayesian alphabetic optimality criteria for linear models, similar upper bounds for the number of support points in an optimal design have also been derived (Chaloner 1984). In the nonlinear setting, however, the Fisher information matrix depends on the parameter θ (see Section 2.1.4). A common approach, discussed below, is then to average a local design criterion over a distribution on θ. Because now (infinitely) many information matrices $F(\theta, \xi)$ are involved, upper bounds on the number of support points do not in general hold (Atkinson *et al.* 2007, Chaloner and Larntz 1989).

An important theme in classical design theory has been the construction of so-called 'equivalence theorems' for continuous designs. The first such result was due to Kiefer and Wolfowitz (1960), who showed that any continuous D-optimal design is also G-optimal, and vice versa. This result was substantially generalized by Kiefer (1974) and Whittle (1973). The resulting 'general equivalence theorem' relies on the fact that the design criteria to be minimized are convex functionals ϕ of the probability measure ξ. Under this condition, with some further assumptions on the regularity of ϕ, the equivalence theorem provides multiple equivalent conditions for the optimality of a design ξ^*, some of which correspond to verifying that the directional derivatives of ϕ (with respect to feasible designs ξ) are zero at ξ^*. The latter are useful to check optimality of a given design measure (in the continuous case), and have been employed in algorithms (Wynn 1972). Variants of the general equivalence theorem have been established for Bayesian alphabetic optimality in linear models (Chaloner 1984, Pilz 1991) and for certain local optimality criteria averaged over the prior in nonlinear models (Chaloner and Larntz 1989). There is a vast array of results along this theme, which we will not attempt to survey here. Instead we refer the reader to the comprehensive framework in Pukelsheim (2006), which tackles design optimality for linear models using tools of convex analysis, and the historical perspective in Wynn (1984).

Since our interest is largely in nonlinear problems (as well as infinite-dimensional linear problems), we will generally resort to numerical methods for optimizing over ξ, which must in any case employ *tractable* finite-dimensional parametrizations of candidate designs. Moreover, *only an exact design can be realized in an experiment*, and hence some notion of 'rounding' is needed if a problem is initially solved from a continuous design perspective. We will discuss these considerations further in Section 4.

2.1.4. Challenges of nonlinear design

In problems where dependence on the model parameters θ is nonlinear, the Fisher information matrix F will generally vary with θ. Since θ is unknown, it is then unclear where to evaluate $F(\theta, \xi)$ and any of the associated design criteria.

A relatively crude approach is simply to choose some reference or 'best-guess' parameter value θ_0 and proceed; this is known as 'local' design, but it clearly ignores parameter uncertainty and the impact of nonlinearity, and may have sharp dependence on the choice of θ_0. One could iterate this approach, by estimating θ after collecting data from a first local design, and then using the estimated parameter value to produce a new local design, and so on (Korkel, Bauer, Bock and Schloder 1999). A more principled alternative is a minimax formulation (Fedorov and Hackl 1997, King and Wong 2000, Berger and Wong 2009). If $\phi(\theta, \xi)$ is any 'local' design criterion (containing the parameter-dependent Fisher information matrix $F(\theta, \xi)$, for instance, $\phi(\theta, \xi) = \log\det F^{-1}(\theta, \xi)$), then we seek

$$\xi^* \in \operatorname*{argmin}_{\xi\in\Xi} \max_{\theta\in\boldsymbol{\Theta}} \phi(\theta, \xi).$$

An interpretation of this objective is that it seeks the design with best performance for the worst-case parameter value, i.e. the value of θ that is most challenging to estimate (in the sense of ϕ). This idea has been explored by Sun and Yeh (2007) and Siade, Hall and Karelse (2017), among others, but generally leads to a rather difficult optimization problem. Typically ϕ is chosen to be a classical alphabetic optimality criterion (as in the example of D-optimality above), and thus the only 'prior' information on θ in these formulations is via the set $\boldsymbol{\Theta}$.

Another natural alternative is to introduce a prior distribution p_Θ for θ and to average any local design criterion $\phi(\theta, \xi)$ over this prior. This approach is widely adopted, due in no small part to its computational tractability; see Pronzato and Walter (1985) and Fedorov and Hackl (1997). If ϕ is solely based on the Fisher information, however, then such a formulation is only 'pseudo-Bayesian' (Atkinson *et al.* 2007, Ryan *et al.* 2016). Indeed, in some such works, the prior is used as a means of handling the parameter dependence arising from nonlinearity but then discarded for subsequent analysis. Ryan *et al.* (2016) argue that any 'fully Bayesian' design criterion must be a functional of the posterior distribution. Interestingly, however, Walker (2016) shows that prior expectation of the trace of the Fisher information matrix,

$$\int \operatorname{tr}(F(\theta, \xi)) p_\Theta(\theta) \, \mathrm{d}\theta,$$

has an information-theoretic interpretation. Overstall (2022) and Prangle, Harbisher and Gillespie (2023) both explore using this quantity as a design criterion in nonlinear problems, and show that it has a decision-theoretic justification as well.

As a step in the 'fully Bayesian' direction, given a nonlinear model and a Gaussian prior p_Θ with full rank covariance matrix Γ_Θ, one could seek, as proposed in Chaloner and Verdinelli (1995),

$$\xi^* \in \arg\max_{\xi \in \Xi} \int \log\det\bigl(F(\theta, \xi) + \Gamma_\Theta^{-1}\bigr) p_\Theta(\theta) \, \mathrm{d}\theta. \tag{2.12}$$

This can be loosely interpreted as minimizing the average, over the prior predictive distribution of Y, of the log-determinant of the covariance of a Gaussian approximation of each resulting posterior. Yet this interpretation is rather imprecise, as for a general nonlinear model, $F(\theta', \xi) + \Gamma_\Theta^{-1}$ can be very far from the precision matrix of the posterior distribution that results from a realization of the data $y \sim p_{Y|\theta,\xi}(\cdot|\theta', \xi)$. Criteria such as this are best viewed as approximations of an expected utility function arising from a more principled decision-theoretic formulation, which we discuss next.

2.2. *Decision- and information-theoretic formulations*

The decision-theoretic approach to OED was formalized by Lindley (1956) (see also Stone 1959, Raiffa and Schlaifer 1961) and has two primary ingredients: a utility function u, chosen to reflect the purpose of the experiment, and the Bayesian principle of averaging over what is uncertain (Berger 1985). In this framework, any design criterion takes the form of an *expected utility*, to be maximized with respect to ξ:[1]

$$U(\xi) = \mathbb{E}_{Y,\Theta|\xi}[u(\xi, Y, \Theta)] = \iint p(y, \theta|\xi)\, u(\xi, y, \theta)\, \mathrm{d}\theta\, \mathrm{d}y. \tag{2.13}$$

Here, the utility function $u(\xi, y, \theta)$ quantifies the value of an experiment performed with a design ξ that yields observations y, if the true parameter value is θ. Since the outcome of the experiment is not known when selecting ξ, and since the parameter value is also uncertain, we average over the joint prior distribution of Y and Θ. This process yields the expected utility $U(\xi)$. Many choices of utility function u have been proposed and explored in the literature. We review some of the possibilities below.

2.2.1. *Expected information gain in parameters*

The influential paper of Lindley (1956) proposed using the expected gain in Shannon information, from prior to posterior, as an optimal experimental design criterion. It is evocative to think of this quantity in at least two ways, namely

$$U_{\mathrm{KL}}(\xi) = \mathbb{E}_{Y|\xi}[D_{\mathrm{KL}}(p_{\Theta|Y,\xi}||p_\Theta)] \tag{2.14}$$

$$= H(\Theta) - H(\Theta|Y, \xi), \tag{2.15}$$

where D_{KL} denotes the Kullback–Leibler (KL) divergence, or relative entropy, from prior to posterior,

$$D_{\mathrm{KL}}(p_{\Theta|y,\xi}||p_\Theta) = \int p(\theta|y, \xi) \log \frac{p(\theta|y, \xi)}{p(\theta)}\, \mathrm{d}\theta, \tag{2.16}$$

and the two terms in (2.15) are the entropy and conditional entropy, respectively:

$$H(\Theta) = -\int p(\theta) \log p(\theta)\, \mathrm{d}\theta, \tag{2.17}$$

$$H(\Theta|Y, \xi) = -\iint p(y, \theta|\xi) \log p(\theta|y, \xi)\, \mathrm{d}\theta\, \mathrm{d}y. \tag{2.18}$$

Equality of the expressions (2.14) and (2.15) is easily verified. Note that U_{KL} is always non-negative: conditioning reduces entropy *on average* (not necessarily for each realization $Y = y$, but when averaging over values of y), with zero entropy reduction $H(\Theta) = H(\Theta|Y, \xi)$ if and only if Θ and $Y|\xi$ are independent. Similarly the KL divergence, whose expectation yields (2.14), is always non-negative and reaches zero if and only if the two distributions being compared are identical (Cover and Thomas 2006). Some intuition for maximizing this criterion is that the design ξ yielding data Y that *most* reduce Shannon entropy is, in this information-theoretic

[1] Beginning in this section, we will drop subscripts from probability density functions when the arguments are explicit, letting these arguments make the choice of density clear.

sense, the most informative. Another intuition is that an optimal experiment should maximize the 'change' (here, quantified by the KL divergence) from the prior to the posterior, averaged over the prior predictive distribution of the data Y.

Lindley's original rationale for the design criterion U_{KL} was not rooted in decision theory, but the criterion was later given a decision-theoretic justification by Bernardo (1979); see also the discussion in Prangle *et al.* (2023). Bernardo frames the task of inference as a decision problem, where making a decision amounts to returning a probability density function for the parameters of interest Θ. He argues that the utility function for this decision should be a *proper, local scoring rule* (Gneiting and Raftery 2007), and that these desiderata in turn dictate that u must specifically be a *logarithmic* scoring rule. In the language of (2.13) above, this means that one should choose

$$u^{\mathrm{score}}(\xi, y, \theta) = \log p(\theta|y, \xi) - \log p(\theta). \tag{2.19}$$

Substituting this utility into (2.13) immediately yields (2.14) and (2.15).

Note also that choosing the utility to be the KL divergence from prior to posterior, which depends explicitly on y and ξ but not on θ,

$$u^{\mathrm{div}}(\xi, y, \theta) = D_{\mathrm{KL}}(p_{\Theta|y,\xi}||p_{\Theta}) = u^{\mathrm{div}}(\xi, y), \tag{2.20}$$

and substituting this utility into (2.13), yields the same expected utility U_{KL} (2.14). We also call U_{KL} the *expected information gain* (EIG) in Θ, from prior to posterior. It is useful for subsequent computations (see Section 3) to write the EIG more explicitly as follows:

$$\begin{aligned}
U_{\mathrm{KL}}(\xi) &= \iint p(y, \theta|\xi) \log \frac{p(\theta|y, \xi)}{p(\theta)} \, \mathrm{d}\theta \, \mathrm{d}y && (2.21)\\
&= \iint p(y, \theta|\xi) \log \frac{p(y|\theta, \xi)}{p(y|\xi)} \, \mathrm{d}\theta \, \mathrm{d}y && (2.22)\\
&= \iint p(y, \theta|\xi) \log \frac{p(y, \theta|\xi)}{p(y|\xi)p(\theta)} \, \mathrm{d}\theta \, \mathrm{d}y && (2.23)\\
&=: \mathcal{I}(Y; \Theta|\xi).
\end{aligned}$$

In all of these expressions, the joint prior predictive density of Y and Θ can be factored as $p(y, \theta|\xi) = p(y|\theta, \xi)p(\theta)$, i.e. as a product of likelihood and prior. Moving from (2.21) to (2.22) is an application of Bayes' rule (2.3). Both (2.21) and (2.22) make clear that some kind of posterior calculation is necessary: the former involves the *normalized* posterior density $p(\theta|y, \xi)$, while the latter involves the posterior normalizing constant $p(y|\xi)$. Moreover, these expressions must be evaluated for a range of y values – i.e. for 'all possible' posterior distributions – via the outer expectation. The last expression above, (2.23), shows that EIG is equivalent to the *mutual information* (MI) between the parameters and observations given the design, $\mathcal{I}(Y; \Theta|\xi)$. Henceforth we will use the terms EIG and MI interchangeably.

Expanding (2.22) into two terms also shows that the EIG is a difference of entropies of the data, parallelling (2.15),

$$U_{\mathrm{KL}}(\xi) = H(Y|\xi) - H(Y|\Theta, \xi). \tag{2.24}$$

For some statistical models, $H(Y|\Theta, \xi)$ is a constant function of ξ. One example is the case of a nonlinear model with additive noise, $Y = G(\Theta, \xi) + \mathcal{E}$, where $\mathcal{E}$ is independent of Θ and its distribution does not depend on ξ; then the entropy $H(Y|\Theta = \theta, \xi) = H(\mathcal{E})$ and hence does not depend on θ or ξ. Maximizing EIG then *specializes* to maximizing the entropy of the prior marginal distribution of Y. This design strategy is called 'maximum entropy sampling'; see Shewry and Wynn (1987) and Sebastiani and Wynn (2000).

The EIG objective has additional desirable properties. For one, it is invariant under bijective transformations of θ; this property follows from the invariance of the KL divergence to such transformations, and thus includes rescalings of the parameters as well as more complex reparametrizations. We also emphasize that there are *no assumptions of linearity or Gaussianity* in the motivation for the EIG objective, and in any of its expressions above.

In the linear-Gaussian case, however, maximizing EIG is equivalent to seeking a Bayesian D-optimal design (2.10). To see this, note that when Θ and $Y|\xi$ are jointly Gaussian (as in Section 2.1.2), the EIG or MI can be written in closed form as

$$\mathcal{I}(Y; \Theta|\xi) = \frac{1}{2}(\log\det\Gamma_\Theta - \log\det\Gamma_{\Theta|Y}(\xi)) \tag{2.25}$$

$$= \frac{1}{2}(\log\det\Gamma_Y(\xi) - \log\det\Gamma_{Y|\Theta}(\xi)), \tag{2.26}$$

where Γ_Y is the marginal (prior predictive) covariance of Y,

$$\Gamma_Y(\xi) = G(\xi)\Gamma_\Theta G(\xi)^\top + \Gamma_{Y|\Theta}(\xi).$$

From (2.25), it is apparent that maximizing EIG with respect to ξ is equivalent to minimizing the log-determinant of the posterior covariance. If, additionally, the observational error covariance is independent of the design, then $\Gamma_{Y|\Theta}(\xi) = \Gamma_{Y|\Theta}$, and an equivalent goal, via (2.26), is to maximize the log-determinant of the marginal covariance Γ_Y; this is a specific (linear-Gaussian) case of the maximum entropy sampling described above.

It is interesting to note that the Bayesian D-optimality criterion for linear-Gaussian models can be derived from other utility functions, besides (2.19) and (2.20); see Chaloner and Verdinelli (1995, Section 2.2) for details.

2.2.2. *Other utility functions*

Having defined an information- and decision-theoretic design criterion for inference of the model parameters Θ, it is natural to extend this construction to other goals.

Suppose we are interested in only a subset of the model parameters. Partitioning Θ as $\Theta = (\Theta_1, \Theta_2)$, information gain in Θ_1 is captured by the KL divergence from its prior *marginal* distribution to its posterior *marginal* distribution:

$$u(\xi, y) = D_{\mathrm{KL}}(p_{\Theta_1|y,\xi} || p_{\Theta_1}), \tag{2.27}$$

where

$$p(\theta_1) = \int p(\theta_1, \theta_2)\, \mathrm{d}\theta_2 \quad \text{and} \quad p(\theta_1|y, \xi) = \int p(\theta_1, \theta_2|y, \xi)\, \mathrm{d}\theta_2.$$

The outer expectation over Y in (2.13), yielding the *expected* information gain in Θ_1 via (2.13), must still account for uncertainty in both blocks of Θ. The optimal design criterion in this case becomes

$$U(\xi) = \iiint p(y|\theta_1, \theta_2, \xi) p(\theta_1, \theta_2) \log \frac{p(\theta_1|y, \xi)}{p(\theta_1)}\, \mathrm{d}\theta_1\, \mathrm{d}\theta_2\, \mathrm{d}y \tag{2.28}$$

$$= \iiint p(y|\theta_1, \theta_2, \xi) p(\theta_1, \theta_2) \log \frac{p(y|\theta_1, \xi)}{p(y|\xi)}\, \mathrm{d}\theta_1\, \mathrm{d}\theta_2\, \mathrm{d}y. \tag{2.29}$$

The second expression is analogous to (2.22) in that it uses densities for the data Y, but now even the numerator of the density ratio involves marginalization, as

$$p(y|\theta_1, \xi) = \int p(y|\theta_1, \theta_2, \xi) p(\theta_2|\theta_1)\, \mathrm{d}\theta_2. \tag{2.30}$$

Compared to the EIG in Θ, evaluating this objective therefore requires an additional integration over Θ_2. Note also that (2.28) and (2.29) are equivalent to the MI, $\mathcal{I}(Y; \Theta_1|\xi)$. In this formulation, Θ_2 could represent a variety of possible 'nuisance' parameters in the statistical model, i.e. any parameters that are uncertain but simply not the modeller's immediate object of interest (Feng and Marzouk 2019, Alexanderian, Petra, Stadler and Sunseri 2021). Special examples include the parameters of a discrepancy model (Kennedy and O'Hagan 2001) designed to capture model error, or the background medium in an inverse scattering problem (Borges and Biros 2018).

A generalization of the preceding formulation is to consider the EIG in some (generally nonlinear) *function* of the parameters Θ, $Z = \Psi(\Theta)$, where $\Psi\colon \boldsymbol{\Theta} \to \mathbb{R}^q$ for some $q \leq p$. This can be thought of as a 'goal-oriented' objective (just like Bayesian $\mathrm{D_A}$-optimality in the linear-Gaussian case) where the function Ψ encodes the true quantity of interest. Now we seek to maximize the expected KL divergence from the prior predictive distribution of Z to its posterior predictive distribution:

$$U(\xi) = \mathbb{E}_{Y|\xi}[D_{\mathrm{KL}}(p_{Z|Y,\xi} || p_Z)]$$

$$= \iint p(y, z|\xi) \log \frac{p(z|y, \xi)}{p(z)}\, \mathrm{d}z\, \mathrm{d}y \tag{2.31}$$

$$= \iint p(y, z|\xi) \log \frac{p(y|z, \xi)}{p(y|\xi)} \,\mathrm{d}z \,\mathrm{d}y \tag{2.32}$$

$$= \mathcal{I}(Y; Z|\xi).$$

Notably, this objective is also the original object of interest in Bernardo (1979). While it is straightforward to write down, it raises significant computational challenges, *beyond* those associated with calculating EIG in the parameters alone. For generic Ψ, $p(\theta)$ and $p(y|\theta, \xi)$, we do not have simple expressions for the prior density $p(z)$ or the posterior density $p(z|y, \xi)$ of Z (even up to a normalizing constant) appearing in (2.31). Nor do we have easy access to the marginal likelihood $p(y|z, \xi)$ to instead evaluate (2.32). Numerical approximations are needed, involving density estimation or approximate Bayesian computation; see Section 3. Of course, in specific cases (such as linear Ψ with Gaussian priors), some aspects of the expressions above become more tractable. We note also that if Ψ is bijective, then EIG in Z is identical to EIG in Θ, since (as noted earlier) the information gain objective is invariant under bijective transformations; otherwise the EIG in Z is smaller (Bernardo 1979, Theorem 1).

The optimal design obtained by maximizing the EIG in some Z can differ drastically from the design maximizing EIG in Θ. Figure 2.1 illustrates these contrasts for sensor placement in a time-dependent advection–diffusion problem (Zhong, Shen, Catanach and Huan 2024). The parameter Θ is the unknown source location, endowed with a uniform prior on $[0, 1]^2$. The source emits a scalar quantity that diffuses and is advected towards the top-right of the $[0, 1]^2$ domain, with the advection velocity increasing linearly in time, from a value of zero at $t = 0$. The design entails placing a single sensor, with coordinates $\xi = (\xi_1, \xi_2) \in [0, 1]^2$, that measures the concentration of the scalar at time $t_1 > 0$. Figure 2.1(a) shows a map of EIG in Θ as a function of sensor location (estimated numerically; see Section 3). We see that the optimal measurement location is not unique, but lies roughly 0.2 units of distance from the centre of the domain; the slight asymmetry is due to the direction of advection. Figure 2.1(b), in contrast, shows maps of EIG for four different choices of Z; each Z is the predicted concentration of the passive scalar at a future time $t_2 > t_1$ and at the specific location marked by a red star in each panel. We see that the optimal sensor locations, maximizing these goal-oriented EIG criteria, are markedly different from those in Figure 2.1(a).

If the quantity of interest Z is random given θ, for example if it is described by a conditional density $p(z|\theta)$, then the formulation above still applies. Computations may actually be easier: the problem of estimating $p(z)$ or $p(z|y, \xi)$ is *smoothed* by the kernel $p(z|\theta)$, and the restriction that $q \leq p$ is lifted as long as the conditional density $p(z|\theta)$ on $\mathbb{R}^q$ exists. A special case is when $Z = Y|\xi^+$, that is, we wish to maximize information gain in the model prediction for some given design ξ^+. We call this future prediction Y^+, to distinguish it from the potential outcomes Y of the experiment being currently designed. In this case, the utility u can be set to

$$u(\xi, y) = D_{\mathrm{KL}}(p_{Y^+|y,\xi,\xi^+} || p_{Y^+|\xi^+}), \tag{2.33}$$

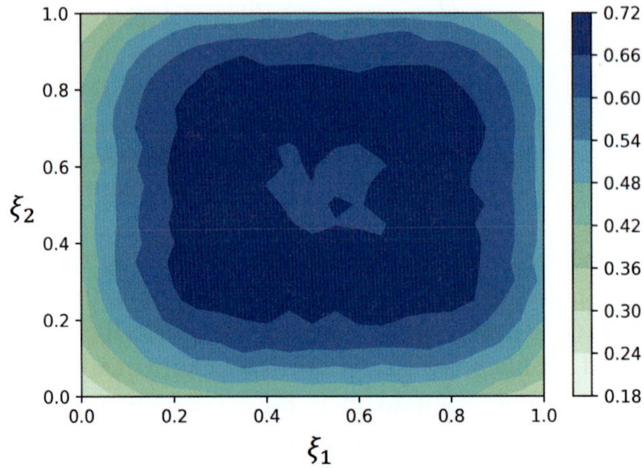

(a) EIG in Θ

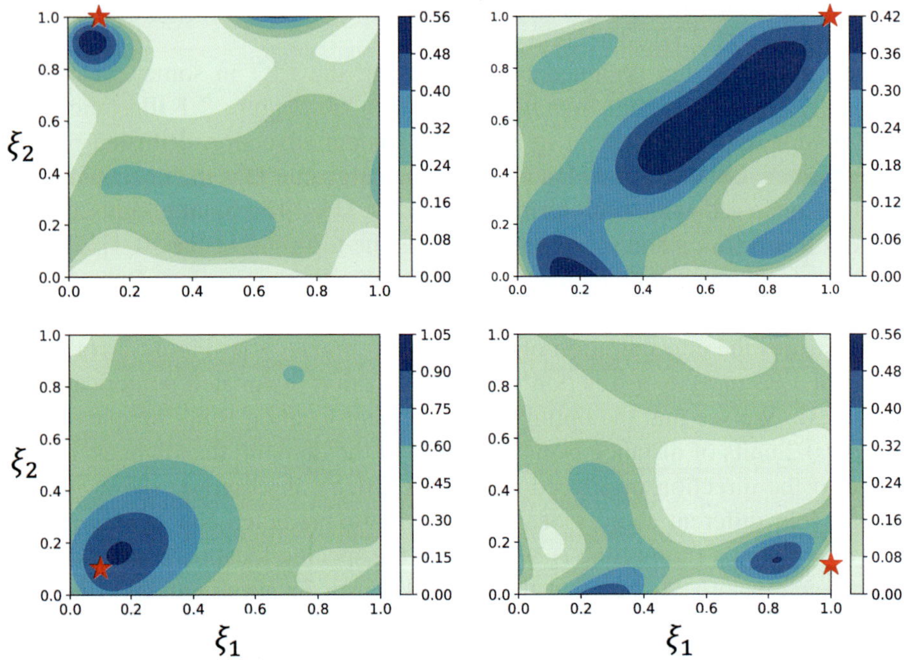

(b) EIG in various Z

Figure 2.1. Optimal sensor placement in a time-dependent advection–diffusion problem. Each figure shows a map of expected information gain (EIG) in a chosen quantity, as a function of the sensor location $\xi \in [0, 1]^2$. Measurements are made at time $t_1 > 0$, and advection is towards the top right. (a) EIG in the unknown source location Θ. (b) EIG for different quantities of interest Z, where each Z is the predicted concentration at some future time $t_2 > t_1$, at the location marked by the red star. The optimal designs, maximizing EIG in each case, differ significantly.

which yields, as an expected utility, the expected information gain in Y^+:

$$U(\xi) = \iint p(y, y^+|\xi, \xi^+) \log \frac{p(y^+|y, \xi, \xi^+)}{p(y^+|\xi^+)} \, \mathrm{d}y^+ \, \mathrm{d}y \tag{2.34}$$

$$= \iiint p(y|\theta, \xi) p(y^+|\theta, \xi^+) p(\theta) \log \frac{p(y^+|y, \xi, \xi^+)}{p(y^+|\xi^+)} \, \mathrm{d}y^+ \, \mathrm{d}y \, \mathrm{d}\theta \tag{2.35}$$

$$= \mathcal{I}(Y; Y^+|\xi, \xi^+). \tag{2.36}$$

The penultimate line above reflects the fact that Y and Y^+ are conditionally independent given θ, and hence their joint prior predictive can be expanded and factored by introducing the parameter θ explicitly.

A final information-theoretic utility that we will consider arises from problems of *model discrimination* in the Bayesian setting (Myung and Pitt 2009). Suppose we have a countable set of models $\mathcal{M}_m$, $m = 1, 2, \ldots$, each with its own parameters, $\theta_m \in \mathbf{\Theta}_m \subseteq \mathbb{R}^{p_m}$, and its own prior on parameters, $p(\theta_m)$. Suppose there is also a (discrete) prior distribution over the model indicators m. As a utility, we choose the relative entropy from this prior to the posterior distribution over model indicators,

$$u(\xi, y) = \sum_m P(m|y, \xi) \log \frac{P(m|y, \xi)}{P(m)}. \tag{2.37}$$

Following (2.13), we must take an expectation over the prior predictive distribution of Y to obtain an expected utility. Now, however, because there are multiple possible models, the prior predictive distribution is itself a mixture of the prior predictives of each model:

$$p(y|\xi) = \sum_m P(m) p(y|m, \xi) = \sum_m P(m) \int_{\mathbf{\Theta}_m} p(y|\theta_m, \xi) p(\theta_m) \, \mathrm{d}\theta_m, \tag{2.38}$$

where conditioning on θ_m also implies conditioning on m at the same time. The EIG in the model indicator m follows by combining (2.37) and (2.38): $U(\xi) = \int u(\xi, y) p(y|\xi) \, \mathrm{d}y$. As noted in Ryan *et al.* (2016), this design approach applies in the $\mathcal{M}$-closed framework for model selection (see Bernardo and Smith 2000, Chapter 6); that is, the true (data-generating) model is assumed to be within the set of models considered, and one must assign a prior weight $P(m)$ to the event that each model is true. Examples of Bayesian OED for model discrimination can be found in Myung and Pitt (2009), Cavagnaro, Myung, Pitt and Kujala (2010), McGree, Drovandi and Pettitt (2012), Drovandi, McGree and Pettitt (2014), Aggarwal, Demkowicz and Marzouk (2016) and Hainy, Price, Restif and Drovandi (2022).

While most of the preceding discussion has focused on information-theoretic utilities, the decision-theoretic framework described at the start of Section 2.2 is certainly not limited to utility functions of this kind. As described in Ryan *et al.* (2016) and Chaloner and Verdinelli (1995), another natural utility is a quadratic loss, motivated by the desire to extract a point estimate of θ from the posterior. For instance, let $\bar{\theta}(y,\xi) := \mathbb{E}[\Theta|y,\xi]$ denote the posterior mean. Then we can write

$$u(\xi, y, \theta) = -(\theta - \bar{\theta}(y,\xi))^\top B(\theta - \bar{\theta}(y,\xi)) \tag{2.39}$$

for some symmetric positive semi-definite matrix $B \in \mathbb{R}^{p\times p}$. The expected utility is then

$$U(\xi) = -\iint (\theta - \bar{\theta}(y,\xi))^\top B(\theta - \bar{\theta}(y,\xi))p(y,\theta|\xi)\,\mathrm{d}\theta\,\mathrm{d}y, \tag{2.40}$$

which is the negative Bayes risk of the posterior mean under a *weighted squared error* loss. Maximizing the expected utility over ξ thus minimizes this risk. As noted in Chaloner and Verdinelli (1995), in the case of a linear-Gaussian model, this formulation reverts to Bayesian A-optimal design with weight matrix B, i.e. $\min_\xi \operatorname{tr}(B\Gamma_{\Theta|Y}(\xi))$.

Another family of non-information-theoretic utilities involves scalar functionals of the posterior covariance matrix; these are not strictly motivated by point estimation, but rather can be seen as a more computationally tractable alternative to information-theoretic utilities that require calculation of posterior normalizing constants. Ryan *et al.* (2016) specifically suggest using as a utility the determinant of the posterior precision matrix,

$$u(\xi, y) = \frac{1}{\det(\mathrm{Cov}(\Theta|y,\xi))}, \tag{2.41}$$

and then, as usual, averaging this quantity over the prior predictive of Y to obtain an expected utility:

$$U(\xi) = \int (\det(\mathrm{Cov}(\Theta|y,\xi)))^{-1} p(y|\xi)\,\mathrm{d}y \tag{2.42}$$

$$= \iint (\det(\mathrm{Cov}(\Theta|y,\xi)))^{-1} p(y|\theta,\xi)p(\theta)\,\mathrm{d}\theta\,\mathrm{d}y. \tag{2.43}$$

We emphasize that this criterion is intended for nonlinear/non-Gaussian problems, and thus calculation of the posterior covariance for different realizations of Y is not a computationally trivial undertaking. It is instructive to compare this criterion to the similar but cruder heuristic (2.12), which is motivated by a series of Gaussian approximations as described in Chaloner and Verdinelli (1995).

To close this section, we point the reader to a more general formalism for what comprises a 'valid' notion of information gain from a statistical experiment, due to Ginebra (2007). In this formalism, an information measure must satisfy a *minimal* set of requirements: (i) it is real-valued, (ii) it returns zero for a 'totally non-informative experiment', where Y is independent of Θ, and (iii) it satisfies sufficiency ordering (Blackwell 1951, 1953, Le Cam 1964). The last requirement can be understood as follows. Let $Y|\theta,\xi_1$ and $Y|\theta,\xi_2$ be the outcomes of two different experiments, for the same parameter value θ, and let η be an independent random variable with fixed and known distribution, introducing auxiliary randomness. If there exists a function W such that $W(Y,\eta)|\theta,\xi_1$ has the same distribution as $Y|\theta,\xi_2$

for all θ, then the experiment with design ξ_1 is said to be 'sufficient for' or 'always at least as informative as' the experiment with design ξ_2. That is to say, the data from ξ_1 can generate data from ξ_2 with an additional randomization mechanism and without knowing θ. In such a situation, ξ_1 is preferred. This generalized notion of an information measure broadly encompasses several commonly used objectives in OED, including mutual information. We refer readers to Ginebra (2007) for an extended discussion, including connections to likelihood ratio and posterior-to-prior ratio statistics.

2.3. Design criteria for infinite-dimensional problems

Infinite-dimensional statistical models arise in the Bayesian approach to inverse problems (Stuart 2010, Dashti and Stuart 2017, Knapik, van der Vaart and van Zanten 2011) and, more broadly, in non-parametric estimation and non-parametric Bayesian procedures (Giné and Nickl 2021). These problems can be understood as estimation or inference of *functions*; in other words, the parameter θ of the statistical model now belongs to a function space, rather than to $\mathbb{R}^p$ for finite p. Application domains of such models are vast, and we will not attempt to review them here. Instead we will focus on two classes of problems where the integration of OED with infinite-dimensional models has proved to be particularly fruitful.

2.3.1. Inverse problems in the Bayesian setting

The infinite-dimensional setting is natural for inverse problems involving partial differential equations, where the parameter to be learned is typically an initial condition, a source term, a boundary condition, or a heterogeneous coefficient – and thus a function of space or time. An important research theme, at the intersection of applied mathematics and statistics, has been to create statistical formulations of inverse problems that are well-defined in the infinite-dimensional setting (Stuart 2010). This is necessary, for instance, to create consistent Bayesian models for inverse problems – Bayesian models that have a well-defined limit as the discretization of the underlying functions is refined (Bui-Thanh, Ghattas, Martin and Stadler 2013). Another important practical result of these efforts is algorithms with discretization-invariant (and hence dimension-independent) performance, for example Markov chain Monte Carlo methods whose sampling efficiency does not deteriorate with grid refinement (Hairer, Stuart and Voss 2011, Cotter, Roberts, Stuart and White 2013, Cui, Law and Marzouk 2016, Rudolf and Sprungk 2018, Villa, Petra and Ghattas 2021).

OED for infinite-dimensional Bayesian inverse problems is well explored in the setting of Gaussian priors, particularly for linear inverse problems (Alexanderian, Petra, Stadler and Ghattas 2014), and summarized in the recent review by Alexanderian (2021). To explain these developments, we first briefly sketch the setting and refer the reader to Stuart (2010) and Dashti and Stuart (2017) for full details and precise results. The parameter Θ is modelled as a random variable taking values in an infinite-dimensional separable Hilbert space $(\mathcal{H}, \langle\cdot,\cdot\rangle_{\mathcal{H}})$. Θ is further assumed to be Gaussian, which means that the real scalar-valued random variable $\langle v, \Theta\rangle_{\mathcal{H}}$ is Gaussian for any $v \in \mathcal{H}$. The mean of Θ can be defined as an element $\bar{\theta}$ of $\mathcal{H}$ satisfying $\langle\bar{\theta}, v\rangle_{\mathcal{H}} = \mathbb{E}[\langle\Theta, v\rangle_{\mathcal{H}}]$ for all $v \in \mathcal{H}$. The covariance operator of Θ is the positive, self-adjoint and compact linear operator $C\colon \mathcal{H} \to \mathcal{H}$ defined through $\langle v, Cw\rangle_{\mathcal{H}} = \mathbb{E}[\langle\Theta - \bar{\theta}, v\rangle_{\mathcal{H}}\langle\Theta - \bar{\theta}, w\rangle_{\mathcal{H}}]$ for all $v, w \in \mathcal{H}$. Because Θ takes values in $\mathcal{H}$, the trace of C is finite; it is then said that C is of *trace class*. We can write this Gaussian (prior) measure of Θ as $\mu = \mathcal{N}(\bar{\theta}, C)$, and will assume that C is strictly positive.

A common assumption on the statistical model for the observations $Y \in \mathbb{R}^n$ is that they result from the action of a (possibly nonlinear) 'forward' operator $G_\xi\colon \mathcal{H} \to \mathbb{R}^n$ perturbed with additive Gaussian noise: $Y = G_\xi(\Theta) + \mathcal{E}$, where $\mathcal{E} \sim \mathcal{N}(0, \Gamma_{Y|\Theta})$ is independent of Θ. For any fixed $Y = y$, this in turn defines a likelihood function that is proportional to

$$\mathcal{L}_\xi^y\colon \theta \mapsto -\frac{1}{2}\exp((y - G_\xi(\theta))^\top \Gamma_{Y|\Theta}^{-1}(y - G_\xi(\theta))).$$

Under appropriate conditions on G_ξ and the prior μ, detailed in Stuart (2010), the posterior distribution of Θ, i.e. the distribution of Θ given $Y = y$, denoted by μ_ξ^y, is well-defined and dominated by the prior measure, $\mu_\xi^y \ll \mu$. One can then write μ_ξ^y in terms of its Radon–Nikodym derivative with respect to μ:

$$\frac{\mathrm{d}\mu_\xi^y}{\mathrm{d}\mu}(\theta) \propto \mathcal{L}_\xi^y(\theta). \tag{2.44}$$

The KL divergence from prior to posterior, $D_{\mathrm{KL}}(\mu_\xi^y \| \mu)$, can also be defined under these conditions.

With this background in hand, we can summarize several design criteria that have been proposed for infinite-dimensional Bayesian inverse problems. When the forward operator G is *linear*, the posterior measure μ_ξ^y is again Gaussian, with a covariance operator $C_{\mathrm{pos}}(\xi)$ that is independent of y. In this setting, Alexanderian *et al.* (2014) propose minimizing the trace of the posterior covariance operator with respect to ξ: $\min_{\xi\in\Xi} \operatorname{tr}(C_{\mathrm{pos}}(\xi))$. This is the infinite-dimensional version of Bayesian A-optimality; the objective is well-defined because the posterior covariance $C_{\mathrm{pos}}(\xi)$ is also of trace class, under the conditions noted above. A marginalized version of infinite-dimensional Bayesian A-optimality, focused on the covariance of a subset of variables of interest, was used for design in Alexanderian *et al.* (2021). This can be compared to L-optimality in the finite-dimensional setting.

The analogue of Bayesian D-optimality is somewhat less straightforward, as the eigenvalues of the posterior covariance operator $C_{\mathrm{pos}}(\xi)$ accumulate at zero, and hence minimizing the log-determinant of this operator is not meaningful (Alexanderian 2021). Instead, Alexanderian, Gloor and Ghattas (2016*a*) use the correspondence between D-optimality and maximizing EIG in linear problems (see the discussion at the end of Section 2.2.1) to derive an alternative objective for Bayesian D-optimality in the linear setting. Specifically, they start with the EIG, taking advantage of the fact that the KL divergence from prior to posterior is well-defined, under conditions summarized above. Specializing this quantity and its expectation over Y to the linear-Gaussian setting, the objective thus obtained is $\frac{1}{2}\log\det(\mathrm{Id}+\widetilde{H})$, where $\widetilde{H}$ is the prior-preconditioned Hessian operator of the negative log-likelihood (Alexanderian *et al.* 2016*a*, Theorem 1). This expression coincides with the log-determinant of the posterior covariance in the finite-dimensional case. Efficient ways to estimate this objective, leveraging low rank structure, are discussed in Alexanderian and Saibaba (2018). We note also that $\widetilde{H}$ is a central quantity in dimension reduction for Bayesian inverse problems, and will appear again in Section 3.

For nonlinear forward operators, a full treatment of EIG in the infinite-dimensional setting has not (to our knowledge) been used as a design criterion. This may be largely due to computational tractability, though some theoretical gaps (e.g. checking that the mutual information between Y and the infinite-dimensional Θ is well-defined; see Duncan 1970) may remain. Instead, researchers have focused on simpler design criteria and their further approximations. For instance, Alexanderian (2021), motivated by the finite-dimensional approach in Haber, Horesh and Tenorio (2009), discusses design that minimizes the Bayes risk of the posterior mode $\hat{\theta}(y)$ under the squared error loss defined by the inner product on $\mathcal{H}$, i.e. $\|\cdot\|^2=\langle\,,\,\rangle_{\mathcal{H}}$. This is analogous to (2.39) but using the posterior mode (also called the *maximum a posteriori* (MAP) estimate (Dashti, Law, Stuart and Voss 2013)) $\hat{\theta}(y)$ rather than the posterior mean, as the former is typically more computationally tractable. Another closely related heuristic suggested in Alexanderian (2021) is to minimize the trace of the posterior covariance operator, in expectation over the data Y:

$$\min_{\xi\in\Xi}\int_{\mathcal{H}}\int_{\mathbb{R}^n}\mathrm{tr}(C_{\mathrm{pos}}(y,\xi))\mathcal{L}_{\xi}^{y}(\theta)\,\mathrm{d}y\,\mu(\mathrm{d}\theta). \tag{2.45}$$

This is essentially the 'Bayesian A-posterior precision' expected utility described in Ryan *et al.* (2016, Section 3.1.2). As the posterior covariance operator is difficult to approximate in nonlinear inverse problems (not to mention *many* posterior covariances, one for each realization of the data used to evaluate the integral in (2.45)), Alexanderian *et al.* (2016*b*) instead propose replacing $\mathrm{tr}(C_{\mathrm{pos}}(y,\xi))$ above with the trace of the inverse Hessian of a Laplace approximation of the posterior at the MAP point $\hat{\theta}(y)$. This approximation is reasonable when the posterior is 'close' to Gaussian (Schillings, Sprungk and Wacker 2020, Helin and Kretschmann 2022, Spokoiny 2023).

Computing any of these design criteria in the setting of infinite-dimensional Bayesian inverse problems is a computationally challenging undertaking, due to the high discretization dimension used to represent the parameter θ in practice, as well as the cost of forward operator evaluations. Considerable computational ingenuity is required; an essential step towards mitigating the impact of high discretization dimension is to take advantage of low-rank structure in the prior-preconditioned Hessian, and to exploit randomized numerical linear algebra methods for computing eigendecompositions, estimating the trace, and so on.

2.3.2. *Gaussian process regression*

Gaussian process (GP) regression, also known as kriging, is a ubiquitous tool in spatial statistics, time series modelling, machine learning, engineering design, surrogate modelling and countless other applications. Rasmussen and Williams (2006) and Gramacy (2020) provide excellent expositions of both applications and some theoretical foundations. Experimental design for GP regression has thus received considerable attention.

From the perspective of the previous section, GP regression can be viewed as a linear Bayesian 'inverse problem' on function space, with a trivial forward operator: a selection operator. The underlying true function, $\theta^*\colon \mathcal{X} \to \mathbb{R}$, for $\mathcal{X} \subseteq \mathbb{R}^d$, is observed directly through evaluation at a finite collection of points $x_i \in \mathcal{X}$, $i = 1, \ldots, n$, perhaps with additive Gaussian noise, e.g. $Y_i = \theta^*(x_i) + \mathcal{E}_i$ with $\mathcal{E}_i \sim \mathcal{N}(0, \sigma^2)$. Here we let the prior model for the true function be a Gaussian process Θ on $\mathcal{X}$, with mean function $m(x) := \mathbb{E}[\Theta(x)]$ and positive semi-definite covariance function $c(x, x') := \mathbb{E}[(\Theta(x) - m(x))(\Theta(x') - m(x'))]$, which defines the prior covariance operator C via

$$(Cv)(x)\colon x \mapsto \int_{\mathcal{X}} c(x, x')v(x')\,\mathrm{d}x',$$

for functions $v \in L^2(\mathcal{X})$. Given a collection of observations $y_s := (y_i)_{i=1}^n \in \mathbb{R}^n$, taken at corresponding covariate values $x_s := (x_i)_{i=1}^n$, performing GP regression entails conditioning Θ on these data. The posterior distribution, describing this conditioned process, remains Gaussian,

$$\Theta|(x_s, y_s) \sim \mathcal{N}(m_{\mathrm{pos}}, C_{\mathrm{pos}}), \tag{2.46}$$

with the posterior mean m_{pos} and the posterior covariance function c_{pos} (yielding C_{pos}) expressible in closed form:

$$m_{\mathrm{pos}}(x) = m(x) + \alpha^\top c(x_s, x) \quad \text{and} \quad c_{\mathrm{pos}}(x, x'; x_s) = c(x, x') - c(x_s, x)^\top R c(x_s, x'),$$

where the matrix $R^{-1} \in \mathbb{R}^{n \times n}$ has entries $[R^{-1}]_{ij} = c(x_i, x_j) + \delta_{ij}\sigma^2$, the coefficient vector $\alpha \in \mathbb{R}^n$ has entries $\alpha_i = R[i,:](y_s - m(x_s))$, and $c(x_s, x) := (c(x_1, x), \ldots, c(x_n, x))\colon \mathcal{X} \to \mathbb{R}^n$ (Rasmussen and Williams 2006, Gramacy 2020).

Since the purpose of GP regression is generally to make predictions about θ^* at unseen values of the covariates x, most design criteria involve the (posterior) *predictive* variance, i.e. the variance of $\Theta|(x_s, y_s)$. In this setting, however, the 'parameter' is the process Θ, and hence the boundary between parameters and predictions is rather blurred.

Another perspective on GP regression follows intuitively from finite-dimensional distributions of the process Θ, which are always multivariate Gaussian (both before and after conditioning on the data). For a finite number of sites $x_{\mathcal{S}} := (x_i)_{i=1}^m \in \mathcal{X}$, $\Theta(x_{\mathcal{S}}) := (\Theta(x_i))_{i \in \mathcal{S}}$ is simply a multivariate normal random vector. We can observe some components of this vector, and we wish to use these observations to predict other components.

Maximum entropy sampling (Shewry and Wynn 1987) originates with this discretized perspective. Let $x_s \subset x_{\mathcal{S}}$ denote the $n < m$ distinct locations selected for a candidate design and let $x_{s^c} = x_{\mathcal{S}} \setminus x_s$ denote its complement. Then the chain rule for entropy yields

$$H(\Theta(x_{\mathcal{S}})) = H(\Theta(x_s)) + H(\Theta(x_{s^c})|\Theta(x_s)). \tag{2.47}$$

Since the left-hand side of (2.47) is fixed, minimizing entropy in the predictions at unobserved sites x_{s^c} given the observations (the second term on the right) can be accomplished by maximizing the first term on the right. Thus finding an optimal design is cast as maximizing the entropy of the model predictions (the *joint* entropy of these predictions) at the observed locations, $H(\Theta(x_s))$. (Recall that we also discussed maximum entropy sampling for general parametric statistical models in Section 2.2.1.) More explicitly, the optimization problem is typically posed with some cardinality constraint on the number of observations, e.g. $|x_s| \leq n$:

$$\underset{x_s \subset x_{\mathcal{S}},\ |x_s| \leq n}{\operatorname{argmax}} \quad H(\Theta(x_s)). \tag{2.48}$$

The objective above can be understood as a *set function*, i.e. a function of all subsets of $x_{\mathcal{S}}$. In the present case, since $\Theta(x_s)$ is a Gaussian vector, closed-form expressions for the entropy are immediately available. The problem is equivalent to finding the principal submatrix of $\operatorname{Cov}(\Theta(x_{\mathcal{S}}))$ that has largest determinant. This problem is NP-hard, but many practical algorithms have been developed to tackle it (Ko, Lee and Queyranne 1995).

A crude approximation to maximum entropy sampling is to choose the elements of x_s one at a time, in a greedy fashion: beginning with $x_s = \emptyset$ and $x_{s^c} = x_{\mathcal{S}}$, at each iteration select from x_{s^c} the point with maximum predictive variance,[2] by ranking the diagonal elements of $\operatorname{Cov}(\Theta(x_{s^c})|\Theta(x_s))$. Then add this point to x_s and repeat. Seo, Wallat, Graepel and Obermayer (2000) and subsequent papers call this approach 'active learning MacKay', after MacKay (1992). Its performance can be far from optimal, however, as the entropy objective is not submodular (see Section 4).

An alternative design approach, advocated by Krause, Singh and Guestrin (2008) (see also Caselton and Zidek 1984) is to maximize MI, rather than entropy.

[2] Recall that the entropy of a univariate Gaussian random variable is an increasing function of its variance.

Specifically, the problem is posed as

$$\underset{x_s \subset x_{\mathcal{S}},\ |x_s| \leq n}{\operatorname{argmax}} \quad \mathcal{I}(\Theta(x_s); \Theta(x_{s^c})). \tag{2.49}$$

Greedy approaches are typically applied to this problem, for reasons of computational tractability. Section 4 will discuss these algorithmic considerations in much more detail. Here, however, we will note that greedy approaches tend to work far better for MI maximization than for entropy maximization, as MI is *submodular* (Krause *et al.* 2008, Nemhauser, Wolsey and Fisher 1978, Fisher, Nemhauser and Wolsey 1978). Krause *et al.* (2008, Section 4.1) provide some useful intuition contrasting greedy selection via MI and greedy selection via predictive entropy. A key consideration in the set-up above is to ensure also that the objective is *monotone* increasing for $|x_s| \leq n$, which is required for optimization guarantees to hold. Krause *et al.* (2008) shows that $\mathcal{I}(\Theta(x_s); \Theta(x_{s^c}))$ is approximately monotone in this regime as long as the discretization of the underlying domain $\mathcal{X}$, via $x_{\mathcal{S}}$, is sufficiently fine. Beck and Guillas (2016) present improvements to greedy MI maximization tailored to computer model emulation.

A rather different class of design approaches is more rooted in the continuous view of GPs, seeking the observation locations that minimize the resulting posterior predictive variance, *integrated* over the domain of the process, $\mathcal{X}$. Letting $x_s \subset \mathcal{X}$ denote a finite collection of observation locations (not necessarily chosen from a finite candidate set), the objective to be minimized, over feasible x_s, can be written as

$$\int_{\mathcal{X}} c_{\text{pos}}(x, x; x_s)\, \mathrm{d}x. \tag{2.50}$$

This objective has appeared in many papers (Sacks *et al.* 1989, Seo *et al.* 2000, Santner, Williams and Notz 2018, Gorodetsky and Marzouk 2016) and has been variously called the integrated mean-squared error (IMSE) criterion, the integrated mean-squared prediction error (IMSPE) criterion, or the integrated variance (IVAR) criterion. With discrete candidate sets and a greedy one-point-at-a-time approach to constructing x_s (see below), it is also called 'active learning Cohn' (ALC), after Cohn, Ghahramani and Jordan (1996). In the language of classical design, minimizing (2.50) can be understood as a kind of (Bayesian) V-optimality, in that one minimizes the predictive variance integrated over a region. We should also note that (2.50) is precisely $\operatorname{tr}(C_{\text{pos}})$, i.e. the trace of the posterior covariance operator, which is the infinite-dimensional notion of A-optimality discussed in Section 2.3.1. In practice, the integral (2.50) is approximated by a large set of points chosen uniformly over $\mathcal{X}$, or perhaps non-uniformly to reflect some desired weight. Both discrete selection methods (see Seo *et al.* 2000) and methods that optimize over the continuous coordinates of n points $x_s = (x_1, \ldots, x_n)$ have been explored in the literature. For the latter, see Sacks *et al.* (1989) and Gorodetsky and Marzouk (2016). Here, as with maximum entropy sampling, one can also optimize for all n elements of x_s simultaneously (a 'full batch' design procedure), or proceed in a greedier but sub-optimal fashion: select a subset of design points to minimize (2.50), 'freeze' these points and update c_{pos} accordingly, and repeat for the next subset. Batch approaches are more computationally demanding, but generally yield better performance; see demonstrations in Gorodetsky and Marzouk (2016). We will discuss related optimization issues further in Section 4.

Finally, we mention several approaches that rely on spectral decompositions of the GP, specifically the Karhunen–Loève representation of $\Theta \sim \mathcal{N}(m, C)$ (Karhunen 1947, Loève 1948). The idea is to find the leading eigenvalues and eigenfunctions (λ_i, ϕ_i) of the prior covariance operator C, and to write the GP as

$$\Theta(x) = m(x) + \sum_{i=1}^{\infty} \sqrt{\lambda_i}\phi_i(x)\zeta_i, \tag{2.51}$$

where the scalar-valued random variables ζ_i are standard Gaussian and mutually independent. If the eigensystem is truncated to $r < \infty$ eigenpairs $(\lambda_i, \phi_i)_{i=1}^r$, then GP regression is reduced to parametric regression with coefficients $(\zeta_1, \ldots, \zeta_r)$. Then any standard Bayesian alphabetic optimality criterion can be applied; see Fedorov and Flanagan (1997), Fedorov and Müller (2007) and Harari and Steinberg (2014). See also Spöck (2012) for a related approach based on the polar spectral representation of Θ, assuming that the process is stationary and isotropic. The truncated eigendecomposition can also be used to approximate the integrated posterior variance objective (2.50); see Fedorov (1996).

One outstanding issue in model-based design for GP regression is that hyperparameters of the prior covariance function (controlling, for example, the scale of the prior variance, correlation lengths and smoothness) must often be learned from data as well. The methods discussed above all take the prior covariance as fixed, and thus ignore the 'outer' hyperparameter learning process – as well as the impact that uncertainty in the hyperparameters has on the predictive distribution. In some applications, moreover, the main interest is not in prediction of an unknown function, but rather in learning the parameters γ of the covariance function itself (Pardo-Igúzquiza 1998); it is natural to expect that optimal designs for this purpose should differ from optimal design for prediction. A common way of learning γ is an 'empirical Bayes' approach, which provides a point estimate of the covariance parameters by maximizing the log-marginal likelihood, $\hat{\gamma} \in \operatorname{argmax}_\gamma \log p(y_s|x_s, \gamma)$, where

$$\log p(y_s|x_s, \gamma) = -\frac{1}{2} y_s^\top R_\gamma y_s - \frac{1}{2}\log\det R_\gamma^{-1} - \frac{n}{2}\log 2\pi,$$

and the notation R_γ emphasizes that γ controls the matrix $R \in \mathbb{R}^{n\times n}$ defined earlier via $[R^{-1}]_{ij} = c(x_i, x_j) + \delta_{ij}\sigma^2$. Here, γ may include parameters of the covariance function c as well as the noise variance σ^2. Solving this optimization problem is easier, computationally, than treating γ in a fully Bayesian way. That said, there are many papers and software packages (Gramacy 2022) that do the latter, endowing γ with a prior distribution and inferring it jointly with Θ; usually, this task requires Markov chain Monte Carlo (MCMC) or sequential Monte Carlo methods, as the joint distribution of Θ and γ is non-Gaussian. These fully Bayesian approaches are therefore computationally demanding – even more so in the setting of design. In principle, one could use the joint posterior distribution of (Θ, γ) to find designs that maximize EIG in Θ, γ or both, following the criteria developed in Sections 2.2.1–2.2.2. We are not aware of methods that completely realize this approach. Instead, a variety of more practical computational schemes have been devised for experimental design in GP regression when the prior covariance parameters are uncertain.

Zhu and Stein (2005) focus on design for parameters of the covariance function only, using the Fisher information matrix derived from the marginal likelihood. Since dependence on these parameters γ is generally nonlinear, the authors propose using either 'local' D-optimal design, a minimax approach or a prior-averaged D-optimality criterion (cf. Section 2.1.4). It is interesting to note that the resulting designs involve points that are non-uniformly spaced over $\mathcal{X}$; intuitively, such point sets are useful for learning correlation lengths in the prior covariance function (Gramacy 2020). Zhu and Stein (2006) then suggest a two-stage design process, where some fraction of the design points are chosen according to a criterion focused on estimation of the covariance parameters γ, while the rest are chosen to improve prediction of Θ; uncertainty in the covariance parameters, given some asymptotic approximations, is accounted for in the design criterion for the latter. Spöck and Pilz (2010) cast design for prediction, using a spectral decomposition of the GP, within a minimax formulation over a compact set of parametrized covariance functions. On the other hand, the local, sequential design scheme of Gramacy and Apley (2015) interleaves design for prediction with local Fisher-information matrix-based design for the length-scale parameter in γ. And simpler sequential schemes, where batches of standard (e.g. IMSE) design for Θ are interleaved with maximum likelihood estimation of covariance parameters, are pursued in Harari and Steinberg (2014) and Gorodetsky and Marzouk (2016). Further perspective on such sequential schemes is given in Gramacy (2020, Section 6.2). Indeed, it is very natural to interleave updates of the covariance function parameters with actual observations, in a sequential design fashion. In the purely Gaussian case (i.e. with a fixed covariance function), the realized values of y_s would have *no* impact on subsequent designs, since the posterior covariance and entropy depend only on x_s, not on y_s. When parameters of the covariance function must *also* be inferred, however, we are in the nonlinear design setting and hence there is value to feedback: y_s informs the covariance parameters, and in turn these parameters reshape the predictive uncertainty of the Gaussian process for the next stage.

We will discuss closed-loop sequential design much more systematically in Section 5. Here we will mention just a few more instantiations in the setting of GP regression. Riis *et al.* (2022) perform myopic sequential design using criteria based on the marginal posterior of Θ, where marginalization over the kernel hyperparameters is performed with MCMC samples. Hoang, Low, Jaillet and Kankanhalli (2014) develop a non-myopic sequential design policy that can be understood as approximating a sequential variant of maximum entropy sampling; the authors argue that this policy naturally balances effort between informing covariance hyperparameters and directly reducing uncertainty in the prediction of Θ itself.

2.4. *Related problems and their distinctions*

To help orient the reader, here we discuss several classes of problems in the broader literature that have some conceptual overlap with optimal experimental design, but also some essential differences.

Space-filling and other non-model based designs. Space-filling designs, as the name suggests, spread design points throughout the domain $\mathcal{X}$ so that one can reasonably assess variations of a generic response or the parameters of an associated statistical model. In their simplest form, these methods do not attempt to exploit the structure of a statistical model for the response, or make any assumptions on such a model; they are hence *not* model-based designs, in contrast to the focus of this article. Spread in the design space is achieved by formulating an optimization problem, cases of which are primarily distinguished by whether the focus is exclusively on the distance among the design points in ξ (*maximin distance* design) or the distance to all points in the ground set $\mathcal{X}$ (*minimax distance* design) (Johnson, Moore and Ylvisaker 1990). These notions have clear analogies to the well-known problem of sphere packing (Zong 1999). Pure space-filling designs tend to have poor projection properties, failing to retain their optimality properties when viewed in subspaces. This is undesirable in circumstances when the response is insensitive to one or more of the design variables. In such cases, Latin hypercube designs (McKay, Beckman and Conover 1979), which ensure that any of their projections along a single coordinate axis yields a *maximin distance* design, are a suitable alternative. Space-filling properties in larger subspaces can be induced through orthogonal array extensions of Latin hypercube design (Owen 1992, Tang 1993). Further, 'maximum projection' (MaxPro) designs have been developed to achieve space-filling properties on *all* possible subsets of factors (Joseph, Gul and Ba 2015, 2020).

Other approaches using entropy maximization have also been suggested for space-filling design (Jourdan and Franco 2010). Note that this is not akin to the maximum entropy sampling methods we previously discussed in Section 2.3.2. Here, 'entropy' is that of the empirical distribution of design points on $\mathcal{X}$, and is used as a design criterion to be maximized. The core idea is to relate the space-filling quality of the design to the uniform distribution on $\mathcal{X}$, motivated by the fact that the uniform distribution has maximum entropy among all distributions with prescribed finite support. A similar idea is pursued through designs that seek to minimize the *discrepancy* (Niederreiter 1992) of a set of design points. This notion relates to the much broader topic of low-discrepancy sequences and quasi-Monte Carlo methods for integration (Caflisch 1998, Dick, Kuo and Sloan 2013). We refer the reader also to Pronzato and Müller (2012) and Santner *et al.* (2018) for a more comprehensive discussion of space-filling designs targeting computer experiments.

Other standard non-model-based design strategies include factorial designs, with blocking and fractional variants, as well as composite designs; see Atkinson *et al.* (2007, Chapter 7) for more.

Active learning. Active learning is a term originating in the computer science and machine learning communities, referring to a diverse array of algorithms for choosing which data to 'label', usually in a supervised learning setting (Dasgupta 2011). In statistical terms, we can understand this setting as regression with real or discrete-valued outcomes y_i (where the latter case is classification). In so-called 'pool-based active learning', there is a large pool of candidate covariates or feature values $\{x_i\}_i$, referred to as the unlabelled data (Schein and Ungar 2007). To label a chosen data point is to obtain its associated outcome, thus creating the pair (x_{i^*}, y_{i^*}) for some chosen index i^*. We can thus think of this problem as OED with a countable and even finite design space $\mathcal{X}$, corresponding to which indices should be chosen from the unlabelled pool. Other learning scenarios might select covariates x_i from an infinite set (e.g. with $\mathcal{X}$ now a region of $\mathbb{R}^d$); alternatively, one might be presented with a stream of successive x_i and be required to choose, on-the-fly, whether to label the current value (Settles 2009). In any of these cases, active learning usually refers to a *sequential* version of the OED problem where data to be labelled are selected one at a time or in a batch, then labelled, then used to update a model, and then the process repeats. Most often, these iterations take a *greedy* approach (see Section 5).

An important point of differentiation among active learning methods is *by what mechanism* this selection occurs. One class of selection methods, known as uncertainty sampling, selects the unlabelled data point(s) for which the model's current predictions are most uncertain. The notions of uncertainty used here vary widely, and can include many heuristics that do not have a statistical justification. This is an important distinction from OED. If the notion of uncertainty is tied to the posterior predictive distribution of a Bayesian model, however, then we can recover maximum entropy sampling, discussed in Sections 2.2.1 and 2.3.2 above. Indeed, predictive entropy is commonly used to rank candidate points in active learning (Lewis 1995). Another widely used class of selection methods is 'query-by-committee' (Freund, Seung, Shamir and Tishby 1997), where disagreement among an ensemble of models is used to rank candidate points, such that points with greater disagreement are chosen. Yet another class of criteria ranks candidate points by how much their label would reduce uncertainty in the predictions of the model being trained. An example of the latter, where uncertainty is captured by integrated variance, is the ALC approach discussed in Section 2.3.2. Other selection methods focus on more tailored prediction goals; for instance, Blanchard and Sapsis (2021) describe selection criteria favouring regions of the input space that yield unusual output values, to help build regression models capable of predicting extreme events.

We should also emphasize that many active learning methods are not based on explicit design criteria, but rather on other heuristics (Settles 2009). Moreover, even criterion-based active learning methods usually focus on reducing uncertainty in predictions, rather than on improving estimation or inference of the *parameters of a statistical model*. Again, this is an important distinction from many OED problems.

Bayesian optimization. Bayesian optimization (BO) (Močkus 1975, Jones, Schonlau and Welch 1998, Wang, Jin, Schmitt and Olhofer 2023) is widely used in applications from engineering design to machine learning, to name just a few. It is essentially a derivative-free optimization method, used to maximize 'black-box' (and often computationally expensive) objective functions, i.e. functions that can be evaluated pointwise but whose derivatives cannot be directly evaluated. Gaussian process regression is a key ingredient of modern BO. GP regression is used to build an approximation of the objective function (via the mean of the GP) and an estimate of uncertainty in the predicted value of this objective (via the variance of the GP), and both are refined over the course of the optimization iterations. A design-type question then arises in choosing where (i.e. at what points in the input domain $\mathcal{X}$) to evaluate the objective. In BO, this question is typically resolved by defining a real-valued, easy-to-evaluate 'acquisition function' over $\mathcal{X}$ and finding its maxima. A point at which the acquisition function is maximized is then taken to be the next evaluation point for the objective. The realized value of the objective at each iteration affects the acquisition function at the next stage, and hence the design process is sequential and adaptive. The acquisition process is generally formulated in a myopic way, though there are a few exceptions (Lam and Willcox 2017, Wu and Frazier 2019).

Many acquisition functions have been proposed in the literature, beginning with Jones *et al.* (1998) and in the decades since; these functions generally balance a notion of 'exploration' (learning the objective in unseen places, where the GP model has large predictive variance) with 'exploitation' (evaluating the objective at points where it is expected to be larger than the current best value). A prominent choice is the 'expected improvement' function and its many variants (Močkus 1975, Zhan and Xing 2020). Yet many other choices are possible, with batch/parallel (Chevalier and Ginsbourger 2013, Wang, Clark, Liu and Frazier 2020) and even multi-fidelity (Song, Chen and Yue 2019) schemes. BO is a large and active field which we cannot hope to survey here; instead we point the reader to a few recent reviews and tutorials (Shahriari *et al.* 2016, Frazier 2018, Wang *et al.* 2023). To set it in context relative to OED, however, we emphasize that BO and OED are essentially different: the goal of BO is to *maximize* a function, not to predict the value of the function over all of $\mathcal{X}$ or even some *a priori* chosen subset of $\mathcal{X}$. The resulting sets of evaluation points thus differ, both in their configuration and in their purpose, from the GP regression designs discussed in Section 2.3.2.

Data summarization. Reducing or somehow 'summarizing' large data sets is a problem of frequent practical interest, motivated by considerations of both storage and computation. The cost of most Bayesian inference algorithms, for example, scales at least linearly with the size of the data. Random subsampling of data is of course an option, but a more effective approach is to identify a small weighted subset of the data that is somehow representative of the full dataset. This is the notion of a *coreset*, which originated in computational geometry and computer science (Agarwal, Har-Peled and Varadarajan 2005, Feldman and Langberg 2011). It has since been formulated in a statistical setting, and notably the Bayesian setting (Huggins, Campbell and Broderick 2016, Campbell and Broderick 2018, 2019, Campbell and Beronov 2019), where the idea is to find a weighted data subset of given size that least changes the likelihood or the posterior distribution from its original full-data version. Ostensibly this problem seems similar to OED, but a crucial difference is that coresets are generally identified *after* the data Y are realized, and depend on the realized values of data. In optimal design, on the other hand, a design must be chosen *before* $Y|\xi$ are observed.

3. Numerical approximation of design criteria

Now we turn to one of the central computational questions of optimal design: how to approximate, numerically, the value of a chosen design criterion at any candidate design? This issue is not particularly vexing for standard alphabetic optimality criteria and linear models, where closed-form expressions are generally available. Evaluating these expressions can become costly when parameters Θ are high- or infinite-dimensional, however, and we discuss dimension reduction methods relevant to this setting in Section 3.4. The information-theoretic design criteria introduced in Section 2.2, on the other hand – which have become a mainstay of modern OED due to their flexibility and their applicability to complex nonlinear models – can be very challenging to evaluate, even when the parameter dimension is low. This section will discuss a variety of computational approaches for approximating such objectives, resting on nested Monte Carlo estimation (Section 3.1), approximation of the relevant densities within tractable families (Section 3.2) and more general methods for constructing variational bounds (Section 3.3). Some of these methods differ with regard to which aspects of the underlying Bayesian model are assumed to be computationally accessible: for example, is the problem in the 'standard' setting where the likelihood function can be evaluated, or is it in the 'implicit model' setting where likelihood evaluations and possibly prior density evaluations are unavailable? Throughout this section, we will comment on the applicability of the methods being discussed to either setting.

3.1. Nested Monte Carlo estimators

For a generic expected utility design criterion, as formulated in (2.13), a standard Monte Carlo estimator of the expectation employs pairs of samples $(y^{(i)}, \theta^{(i)})$ drawn from the joint prior of parameters and observations given the design ξ, $p(y, \theta|\xi)$:

$$U(\xi) = \mathbb{E}_{Y,\Theta|\xi}[u(\xi, Y, \Theta)] \approx \frac{1}{N}\sum_{i=1}^{N} u(\xi, y^{(i)}, \theta^{(i)}). \tag{3.1}$$

These samples are typically obtained by first drawing $\theta^{(i)} \sim p(\theta)$ from the prior and then drawing $y^{(i)} \sim p(y|\theta^{(i)}, \xi)$ from the conditional density of the observations. The utility function u must then be evaluated at the samples generated: $u(\xi, y^{(i)}, \theta^{(i)})$. This may not be easy to do. For example, evaluating either of the utility functions u^{score} (2.19) or u^{div} (2.20), which render U equal to the expected information gain (EIG) in Θ, requires evaluating a *normalized* posterior density, where the normalizing constant may change for each realization of y and each value of ξ. One way forward is to estimate these normalizing constants by *another* Monte Carlo simulation, which gives rise to *nested Monte Carlo* (NMC) estimators.

As a canonical/representative case, we describe NMC approaches to estimating the EIG in parameters Θ, from prior to posterior, i.e. the U_{KL} defined in (2.14) and the expressions thereafter. A widely used estimator, proposed in Ryan (2003), employs the form of U_{KL} given in (2.22): we simply replace the posterior normalizing constant $p(y|\xi)$ with a Monte Carlo estimate, that is,

$$\begin{aligned} U_{\text{KL}}(\xi) &= \iint p(y, \theta|\xi) \log \frac{p(y|\theta, \xi)}{p(y|\xi)} \, \mathrm{d}\theta \, \mathrm{d}y \\ &= \iint p(y, \theta|\xi) \log \frac{p(y|\theta, \xi)}{\int p(y|\tilde{\theta}, \xi) p(\tilde{\theta}) \, \mathrm{d}\tilde{\theta}} \, \mathrm{d}\theta \, \mathrm{d}y \\ &\approx \frac{1}{N}\sum_{i=1}^{N}\left(\log p(y^{(i)}|\theta^{(i)}, \xi) - \log\left[\frac{1}{M}\sum_{j=1}^{M} p(y^{(i)}|\tilde{\theta}^{(i,j)}, \xi)\right]\right) =: \widehat{U}_{\text{KL}}^{N,M}(\xi). \end{aligned} \tag{3.2}$$

Here the 'outer loop' sample pairs $\{(y^{(i)}, \theta^{(i)})\}_{i=1}^{N}$ are drawn from $p(y, \theta|\xi)$ as before, but we also require an *independent* collection of samples from the prior,

$$\{\tilde{\theta}^{(i,j)}\}_{i=1,j=1}^{i=N,j=M} \sim p(\theta),$$

amounting to M independent samples for each outer loop iteration. Overall, evaluating $\widehat{U}_{\text{KL}}^{N,M}$ requires (i) the ability to sample from the prior, (ii) the ability to sample from the statistical model for $Y|\xi$, and (iii) the ability to evaluate the likelihood function.

Properties of the estimator $\widehat{U}_{\text{KL}}^{N,M}$ have been analysed in Ryan (2003), Beck *et al.* (2018) and Rainforth *et al.* (2018). It is biased at finite M, but asymptotically unbiased and consistent. More precisely, the leading order terms of its bias and variance are given by

$$\mathbb{E}\big[\widehat{U}_{\mathrm{KL}}^{N,M}(\xi)\big] - U_{\mathrm{KL}}(\xi) = \frac{C_1(\xi)}{M} + O\left(\frac{1}{M^2}\right), \tag{3.3}$$

$$\mathrm{Var}\big[\widehat{U}_{\mathrm{KL}}^{N,M}(\xi)\big] = \frac{C_2(\xi)}{N} + \frac{C_3(\xi)}{NM} + O\left(\frac{1}{NM^2}\right), \tag{3.4}$$

where C_1, C_2, C_3 are design-dependent constants. The constant C_1 is always positive (see Beck *et al.* 2018, Proposition 1), and thus the NMC estimator $\widehat{U}_{\mathrm{KL}}^{N,M}$ is, to leading order, *positively* biased. Intuitively, bias arises because the Monte Carlo estimator of $p(y|\xi)$ (which is unbiased) is transformed by a nonlinear function.

It is useful then to ask how to optimally allocate the sample sizes M and N to minimize the mean-square error

$$\mathrm{MSE} = \mathbb{E}\big[\big(\widehat{U}_{\mathrm{KL}}^{N,M} - U_{\mathrm{KL}}\big)^2\big]$$

for any given budget. The total number of samples drawn is $W = (M + 1)N$; similarly, if cost lies in evaluating the statistical model $p(y|\theta, \xi)$ for any new value of θ, then computing $\widehat{U}_{\mathrm{KL}}^{N,M}$ incurs $W = (M + 1)N$ evaluations. Letting $\alpha^2 = M/N$ denote the ratio of inner-to-outer loop sample sizes, one can show that the optimal value of this ratio for any given W scales as $\alpha_*^2 = O(W^{-1/3})$ (Beck *et al.* 2018, Feng and Marzouk 2019, Rainforth *et al.* 2018). In other words, the ratio should decrease slowly as the computational budget increases. This scaling translates to setting $M = O(\sqrt{N})$, and an optimal convergence rate of MSE $= O(W^{-2/3})$. Significantly, this is *slower* than the standard Monte Carlo rate! Put another way, the computational effort required to achieve an MSE of ϵ^2 is $O(\epsilon^{-3})$, rather than $O(\epsilon^{-2})$.

Many improvements upon the 'vanilla' NMC estimator (3.2) have been proposed. A straightforward idea is to use importance sampling to estimate the evidence $p(y|\xi)$ in the inner loop, i.e. to sample from some possibly y-dependent biasing distribution rather than the prior p_Θ. To make this explicit, we write only the outer loop of (3.2) and replace the log-evidence term with a plugin estimate,

$$U_{\mathrm{KL}}(\xi) \approx \widehat{U}_{\mathrm{KL}}^{N,\mathrm{is}}(\xi) := \frac{1}{N}\sum_{i=1}^{N}\big(\log p(y^{(i)}|\theta^{(i)}, \xi) - \log \widehat{p}(y^{(i)}|\xi)\big), \tag{3.5}$$

where

$$\widehat{p}(y^{(i)}|\xi) = \frac{1}{M}\sum_{j=1}^{M} p(y^{(i)}|\tilde{\theta}^{(i,j)}, \xi)w^{(i,j)}, \quad \tilde{\theta}^{(i,j)} \overset{\text{i.i.d.}}{\sim} q^{i,\xi}, \quad w^{(i,j)} = \frac{p(\tilde{\theta}^{(i,j)})}{q^{i,\xi}(\tilde{\theta}^{(i,j)})}. \tag{3.6}$$

Here the superscripts on the density $q^{i,\xi}$ of the biasing distribution emphasize that it can depend on the outer-loop index i, and thus on the point $y^{(i)}$ where the evidence is being evaluated. The biasing distribution generally depends on the design ξ as well. The question then becomes to how to choose this biasing distribution, for each summand of the outer loop.

One natural choice, proposed in Beck *et al.* (2018), is to use a Laplace approximation of the posterior distribution $p(\theta|y^{(i)}, \xi)$ associated with each outer-loop sample. The Laplace approximation seeks the point of highest posterior density, $\theta^i_{\text{map}} \equiv \theta_{\text{map}}(y^{(i)}, \xi) \in \operatorname{argmax}_\theta \log p(\theta|y^{(i)}, \xi)$, and builds a Gaussian approximation centred at this point, with covariance equal to the negative Hessian of $\log p(\theta|y^{(i)}, \xi)$ evaluated at θ^i_{map}. In general, we define

$$\Sigma_{\text{map}}(y, \xi) = \left(\nabla^2_\theta \log p(\theta|y, \xi)|_{\theta=\theta_{\text{map}}(y,\xi)}\right)^{-1}.$$

The biasing distribution suggested by Beck *et al.* (2018) is then

$$q^{i,\xi} = \mathcal{N}\left(\theta^i_{\text{map}}, \Sigma_{\text{map}}(y^{(i)}, \xi)\right). \tag{3.7}$$

This choice aims to focus samples of the biasing distribution onto regions of higher posterior density, thus reducing the variance of the evidence estimate $\widehat{p}(y^{(i)}|\xi)$ and hence both the bias and the variance of estimates of U_{KL}, as demonstrated in Beck *et al.* (2018). As is typically the case with importance sampling, choosing a better biasing distribution reduces the magnitudes of the constants C_1, C_2, C_3 in (3.3)–(3.4), but does not change the rates of convergence of the estimator with M, N.

Computing a Laplace approximation efficiently, which must be done here N times (once for each outer-loop sample), generally requires gradients of the log-posterior density, $\nabla_\theta \log p(\theta|y^{(i)}, \xi)$, and a good approximation of the Hessian of this log-density (Bui-Thanh *et al.* 2013, Schillings and Schwab 2016). We also note that, in general, $\theta^i_{\text{map}} \neq \theta^{(i)}$; that is, the posterior mode does not coincide with the data-generating value of the parameter. Englezou, Waite and Woods (2022) propose a simplification of Laplace-based importance sampling that instead centres the biasing distribution at the data-generating value of the parameter, drawn from the outer loop. Specifically, both the mean of the Gaussian and the position of the Hessian evaluation in (3.7) are set to $\theta^{(i)}$, rather than θ^i_{map}. Doing so avoids the cost of numerical optimization to find θ^i_{map}, and may yield only a modest increase in the MSE of the estimator at finite sample sizes (Englezou *et al.* 2022). An (approximate) Hessian of the log-posterior, however, is still required. We should also emphasize that choosing a Gaussian approximation as a biasing distribution can become unstable when the posterior is strongly non-Gaussian, and particularly if the posterior has heavy tails. Here other choices of biasing – for instance, using the mean and covariance matrix of the Laplace approximation to parametrize a heavier-tailed multivariate-t distribution – could be more robust (Owen 2013).

Another approach to importance sampling, developed in Feng and Marzouk (2019), is derivative-free. It is a multiple importance sampling scheme that proceeds iteratively over the outer-loop index i: at any given i, past inner-loop samples $\left\{(\tilde{\theta}^{(k,j)})_{j=1}^M\right\}_{k<i}$ and their associated likelihood evaluations are used to create a *mixture* biasing distribution $q^{i,\xi}$ tailored to estimating the current evidence $p(y^{(i)}|\xi)$.

To make this process efficient and maximize re-use of information, the outer-loop iterations are ordered from largest to smallest prior density of $\theta^{(i)}$, $p(\theta^{(i)})$. Again, this approach can substantially decrease the bias and variance of EIG estimates relative to a vanilla NMC scheme.

A different way of accelerating nested Monte Carlo involves multilevel formulations (Giles 2015). Goda, Hironaka and Iwamoto (2020) introduce a multilevel Monte Carlo estimator of U_{KL}, where the level controls the number of inner-loop samples. They also develop an antithetic coupling for the inner-loop estimates that reduces variance, and show that the overall construction reduces the computational complexity required to achieve an MSE of ϵ^2 to $O(\epsilon^{-2})$, improving on the $O(\epsilon^{-3})$ complexity of the standard NMC estimator. In other words, we recover the optimal Monte Carlo rate. Goda *et al.* (2020) also show how importance sampling, e.g. the Laplace approximation-based importance scheme discussed above, can be incorporated within their multilevel Monte Carlo estimator to reduce constants of the error terms. Beck, Dia, Espath and Tempone (2020) also introduce a multilevel scheme for EIG, varying not only the number of inner-loop samples but also the discretization/approximation of some underlying partial differential equation (PDE) model used to define the likelihood.

All of the estimators discussed so far are consistent, that is, they converge in probability to the true EIG as the relevant sample sizes (e.g. M and N, or their analogues in a multilevel scheme) are sent to infinity. Other schemes proposed in the literature, with the goal of improving computational efficiency, are not consistent. For instance, Long, Scavino, Tempone and Wang (2013) propose replacing the utility $u^{\mathrm{div}}(\xi, y)$ (2.20), whose expectation yields the EIG, with an approximation of the Kullback–Leibler (KL) divergence from the prior to a Laplace approximation $\mathcal{N}(\theta_{\mathrm{map}}(y,\xi), \Sigma_{\mathrm{map}}(y,\xi))$ of the posterior $p(\theta|y,\xi)$. Their construction introduces a series of additional approximations – replacing the true MAP estimate $\theta_{\mathrm{map}}(y,\xi)$ with the data-generating value of the parameter θ, using the Gauss–Newton approximation of the Hessian at this point rather than at the MAP (the inverse of which yields a 'covariance' matrix $\widetilde{\Sigma}(\theta,\xi)$ that does not depend on y), and taking further asymptotic approximations of the relevant integrals – to obtain an approximation of EIG that involves only integration over the prior,

$$\mathrm{EIG}(\xi) \approx \int \left(-\frac{1}{2} \log \det \widetilde{\Sigma}(\theta,\xi) - \frac{p}{2}(\log 2\pi + 1) - \log p(\theta) \right) p(\theta)\, \mathrm{d}\theta. \tag{3.8}$$

As a design criterion, we note that this EIG approximation is similar in structure to (2.12). The error of this approximation can be related to the Gaussianity of the posterior *and* the number of independent repeated trials N_{tr} of the experiments specified by ξ, and is bounded by $O(1/N_{\mathrm{tr}})$ in probability (Long *et al.* 2013). Intuitively, repeated trials cause the posterior to concentrate, and this concentration controls the error of integral approximations leading to (3.8). In practice, for small N_{tr} and a non-Gaussian problem, this error can be large. An alternative and perhaps more straightforward way of using the Laplace approximation is proposed by Overstall, McGree and Drovandi (2018); here the approach is simply to replace the estimate of the log-evidence term $\log p(y^{(i)}|\xi)$ in (3.5) with the log-evidence of the standard Laplace approximation of the posterior $p(\theta|y^{(i)})$. The explicit likelihood term and the outer Monte Carlo sum in (3.5) are unchanged. The bias of this EIG approximation increases from zero as the posterior departs from Gaussianity, but compared to the scheme in Long *et al.* (2013), it does not rely *directly* on posterior concentration.

A different family of approximations follows by replacing computationally expensive aspects of statistical model $p(y|\theta, \xi)$ with a computationally cheaper 'surrogate'. For instance, suppose that the data arise from a nonlinear forward operator $G_\xi : \mathbb{R}^p \to \mathbb{R}^n$ perturbed with additive noise $\mathcal{E}$, i.e. $Y = G_\xi(\theta) + \mathcal{E}$. This set-up corresponds to a discretization of the Bayesian inverse problems discussed in Section 2.3.1. Here, evaluating the function G_ξ often involves solving a set of PDEs or integral equations, and it is natural to replace this solution with a cheaper approximation $\widetilde{G}_\xi \approx G_\xi$, which in turn induces an approximation $\widetilde{p}(y|\theta, \xi)$ of the likelihood and an approximation $\widetilde{p}(\theta|y, \xi)$ of the posterior distribution via (2.3). Inserting any of these approximations into (2.15) and equivalent expressions thus yields an approximate EIG. By the same token, using evaluations of $\widetilde{p}(y|\theta, \xi)$ and samples $\tilde{y}^{(i)}$ drawn from this approximate model in any of the NMC estimators discussed above will yield consistent estimates of this *approximate* EIG.

Huan and Marzouk (2013) proposed such a procedure, using polynomial approximations $\widetilde{G}_\xi$ of G_ξ built via sparse quadrature. But a wide variety of other approximation schemes and formats are possible: direct function approximation, whether via polynomials, Gaussian processes or neural networks (Herrmann, Schwab and Zech 2020), but also reduced-order models (Benner, Gugercin and Willcox 2015), or even coarser numerical discretizations of the model giving rise to the parameter-to-observable map G_ξ. The impact of such approximations on the posterior distribution is by now reasonably well understood, especially when one can build a family of approximations $\widetilde{G}^\ell_\xi$, indexed by ℓ, that converge (in some appropriate sense) to G_ξ as $\ell \to \infty$ (Marzouk and Xiu 2009, Stuart 2010, Stuart and Teckentrup 2018, Sprungk 2020). The impact of such approximations on the EIG has only recently been analysed, however. Duong, Helin and Rojo-Garcia (2023) show that the difference between the EIG and its approximation is controlled by the prior expectation of likelihood perturbations under the KL divergence. As a consequence, in the setting of Gaussian likelihoods that we sketched here, Duong *et al.* (2023, Theorem 4.4) show that closeness of $\widetilde{G}^\ell_\xi(\theta)$ to $G_\xi(\theta)$ in a prior-weighted L^2 sense, uniformly over designs $\xi \in \Xi$, guarantees uniform control over the error in the approximate EIG $U^\ell_{\mathrm{KL}}(\xi)$ and convergence of the maximizers of U^ℓ_{KL} as $\ell \to \infty$.

We close this section with a caution. Even the consistent NMC estimators of $U_{\mathrm{KL}}(\xi)$ discussed so far are biased at finite inner-loop sample sizes. This bias can be significant but, more importantly, will vary with ξ in general. Figure 3.1, adapted from Feng and Marzouk (2019), illustrates this phenomenon for a four-dimensional linear-Gaussian problem (thus allowing comparison with exact solutions). EIG in

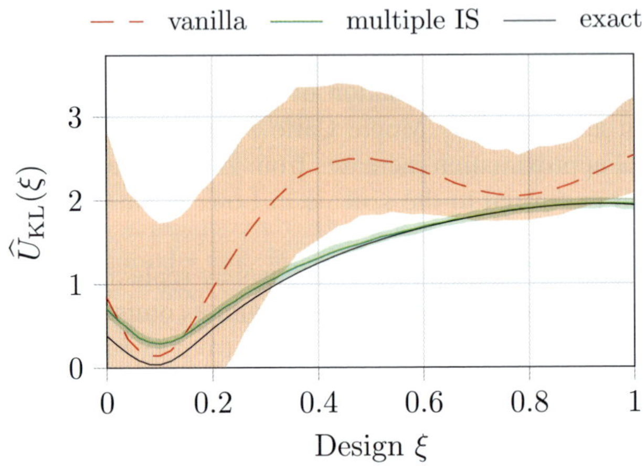

Figure 3.1. Estimated EIG as a function of a scalar design parameter ξ for a linear-Gaussian model, using vanilla NMC (red) or an improved multiple importance sampling scheme (green), compared to the true EIG (black). Shaded areas represent the interval containing 95% of 2000 independent estimates of EIG at each ξ; red dashed and solid green lines are the means of these estimates. Figure adapted from Feng and Marzouk (2019).

a subset of the model parameters (2.28)–(2.29) is estimated using an adaptation of the vanilla NMC estimator $\widehat{U}_{\mathrm{KL}}^{N,M}$ (3.2) (in red), and with the multiple importance sampling scheme developed in Feng and Marzouk (2019) (in green). Sample sizes are *fixed* for all ξ, and the true EIG is shown in black. We see that the bias of the vanilla estimator actually obscures the location of the true maximum. The bias is largest where posterior concentration is maximized, as this is where the prior-weighted estimates of the evidence have greatest variance. The multiple importance scheme fares better, but there is generally no guarantee regarding stability of the maxima for any finite sample size. Care is thus needed to adjust the approximation of EIG in conjunction with the optimization procedure. We will revisit this issue in Section 4.

3.2. *Mutual information bounds via density approximations*

As anticipated in the discussion following (2.21)–(2.22), one of the core computational tasks of the NMC estimators discussed in Section 3.1 is to estimate the posterior normalizing constant $p(y|\xi)$ across many different values of y. The vanilla NMC approach does this entirely independently for each value of y, as does the Laplace-based importance sampling method. (The adaptive importance sampling scheme of Feng and Marzouk (2019), on the other hand, could be said to 'borrow' information from other values of y to create a local biasing distribution.) A rather different way of approaching this problem is to approximate the marginal density $p(y|\xi)$, or similarly, the *normalized* posterior density $p(\theta|y,\xi)$, directly, e.g. in some parametric family of densities.

Let us recall (2.21)–(2.23) in a more concise form:

$$\mathcal{I}(Y;\Theta|\xi) = \mathbb{E}_{Y,\Theta|\xi}\left[\log\frac{p(y|\theta,\xi)}{p(y|\xi)}\right] \tag{3.9}$$

$$= \mathbb{E}_{Y,\Theta|\xi}\left[\log\frac{p(\theta|y,\xi)}{p(\theta)}\right]. \tag{3.10}$$

In either (3.9) or (3.10), one could in principle seek to approximate the density in the numerator, the density in the denominator, or both. Suppose that we replace $p(y|\xi)$ in (3.9) with some approximating probability density function $q_{\mathrm{mar}}(y|\xi)$ (where the subscript stands for 'marginal'). Then, as noted in Barber and Agakov (2003), Poole *et al.* (2019) and Foster *et al.* (2019),

$$\begin{aligned}\mathcal{I}(Y;\Theta|\xi) &= \mathbb{E}_{Y,\Theta|\xi}\left[\log\frac{p(y|\theta,\xi)\,q_{\mathrm{mar}}(y|\xi)}{q_{\mathrm{mar}}(y|\xi)\,p(y|\xi)}\right]\\ &= \mathbb{E}_{Y,\Theta|\xi}\left[\log\frac{p(y|\theta,\xi)}{q_{\mathrm{mar}}(y|\xi)}\right] - D_{\mathrm{KL}}(p_{Y|\xi}\|q_{Y|\xi})\\ &\leq \mathbb{E}_{Y,\Theta|\xi}\left[\log\frac{p(y|\theta,\xi)}{q_{\mathrm{mar}}(y|\xi)}\right] = \mathbb{E}_{\Theta}[D_{\mathrm{KL}}(p_{Y|\Theta,\xi}\|q_{Y|\xi})],\end{aligned} \tag{3.11}$$

where the inequality follows from the non-negativity of the KL divergence. Hence, for *any* approximation $q_{\mathrm{mar}}(y|\xi)$ of the marginal density of $Y|\xi$, (3.11) is an upper bound on the mutual information (EIG).

Similarly, if we replace $p(\theta|y,\xi)$ in (3.10) with some approximating probability density function $q_{\mathrm{pos}}(\theta|y,\xi)$ (where the subscript denotes 'posterior'), we obtain

$$\begin{aligned}\mathcal{I}(Y;\Theta|\xi) &= \mathbb{E}_{Y,\Theta|\xi}\left[\log\frac{q_{\mathrm{pos}}(\theta|y,\xi)\,p(\theta|y,\xi)}{p(\theta)\,q_{\mathrm{pos}}(\theta|y,\xi)}\right]\\ &= \mathbb{E}_{Y,\Theta|\xi}\left[\log\frac{q_{\mathrm{pos}}(\theta|y,\xi)}{p(\theta)}\right] + \mathbb{E}_{Y|\xi}[D_{\mathrm{KL}}(p_{\Theta|Y,\xi}\|q_{\Theta|Y,\xi})]\\ &\geq \mathbb{E}_{Y,\Theta|\xi}\left[\log\frac{q_{\mathrm{pos}}(\theta|y,\xi)}{p(\theta)}\right] = \mathbb{E}_{Y,\Theta|\xi}[\log q_{\mathrm{pos}}(\Theta|Y,\xi)] + H(\Theta).\end{aligned} \tag{3.12}$$

Hence, for any approximation $q_{\mathrm{pos}}(\theta|y,\xi)$ of the posterior density, (3.12) is a *lower bound* on the mutual information (EIG). This bound is sometime called the Barber–Agakov bound, after Barber and Agakov (2003). Note that all expectations above, and specifically in (3.11) and (3.12), are with respect to the *true* distribution $p_{Y,\Theta|\xi}$.

Evaluating the upper bound (3.11) requires the ability to evaluate the likelihood function, and thus it does not apply to the implicit model setting where the likelihood is intractable. Evaluating the lower bound, on the other hand, only requires access to the prior density (or the differential entropy of the prior) and the ability to find a tractable approximation $q_{\text{pos}}(\theta|y,\xi)$.

Foster *et al.* (2019) were the first to suggest using these mutual information bounds in OED, and in practice selected the approximations q from simple parametric families of densities (e.g. Gaussian, uniform) that were tailored to the design problem at hand. Once such a family $\mathcal{Q}$ is specified, the best member of the family, i.e. the density yielding the closest approximation of the EIG, can be found by tightening the bound. Specifically, for the marginal approximation, we seek (for any given design ξ)

$$q^*_{\text{mar}} \in \operatorname*{argmax}_{q\in\mathcal{Q}} \mathbb{E}_{Y|\xi}[\log q(Y|\xi)], \tag{3.13}$$

which minimizes the upper bound (3.11), while for the posterior approximation we seek

$$q^*_{\text{pos}} \in \operatorname*{argmax}_{q\in\mathcal{Q}} \mathbb{E}_{Y,\Theta|\xi}[\log q(\Theta|Y,\xi)], \tag{3.14}$$

which maximizes the lower bound (3.12). In practice, the expectations in (3.13) or (3.14) are approximated using samples from the model; for example, (3.13) becomes

$$\hat{q}_{\text{mar}} \in \operatorname*{argmax}_{q\in\mathcal{Q}} \sum_{i=1}^{M} \log q(y^{(i)}|\xi), \quad y^{(i)} \sim p(y|\xi), \tag{3.15}$$

and analogously for (3.14). Inspecting (3.15), it is apparent that identifying a member of the variational family in this way is none other than *maximum likelihood estimation* of either the marginal density $p(y|\xi)$ or conditional density $p(\theta|y,\xi)$.[3]

With this link in mind, we can immediately generalize the machinery used to construct good density approximations. One powerful approach for estimating both joint and conditional densities rests on transportation of measure (Villani 2009, Marzouk, Moselhy, Parno and Spantini 2016, Spantini, Bigoni and Marzouk 2018). Given some generic target distribution π on $\mathbb{R}^d$, the idea behind transport methods is to find an invertible transformation $S : \mathbb{R}^d \to \mathbb{R}^d$ that pushes forward π to a simple reference distribution ρ on $\mathbb{R}^d$ whose density can be easily evaluated. Abusing notation (by not distinguishing measures from densities), we can write this as $S_\sharp\pi = \rho$, which means that $\rho(A) = \pi(S^{-1}(A))$ for any ρ-measurable set A. Crucially, for any diffeomorphism $\tilde{S}$ on $\mathbb{R}^d$, the distribution $\tilde{S}^\sharp\rho := \tilde{S}^{-1}_\sharp\rho$, called the *pullback* of ρ under $\tilde{S}$, has a closed-form expression for its density:

$$\tilde{S}^\sharp\rho = (\rho\circ\tilde{S}) \det\nabla\tilde{S}, \tag{3.16}$$

[3] 'Maximum likelihood' here refers to the density estimation problem immediately at hand, i.e. estimating the best $q \in \mathcal{Q}$ given samples, and should not be confused with the idea of estimating θ by maximizing $\theta \mapsto p(y|\theta,\xi)$.

which is guaranteed to be positive and to integrate to one. Density estimation can thus be recast as the problem of finding a map $\tilde{S}$ in some suitable class such that $\tilde{S}^\sharp\rho$ is 'close' to π (Wang and Marzouk 2022).

In practice, these models can be quite expressive. Any measure absolutely continuous with respect to the Lebesgue measure on $\mathbb{R}^d$ can be represented as the pullback, under some map, of a Gaussian ρ on $\mathbb{R}^d$; in fact, for any pair of equivalent continuous measures (ρ, π), there exist infinitely many transport maps S that achieve $S^\sharp\rho = \pi$. Normalizing flows (Kobyzev, Prince and Brubaker 2020, Papamakarios *et al.* 2021) are a special case of this formulation, i.e. a particular class of transport map parametrizations that guarantee invertibility, differentiability, and easy evaluation of the Jacobian determinant $\det \nabla S$. But many other representations are useful. Monotone triangular maps (Bogachev, Kolesnikov and Medvedev 2005), for instance, can represent arbitrary absolutely continuous distributions and be parametrized in a way that endows the maximum likelihood estimation problems (3.13)–(3.14) with optimization guarantees; see Baptista, Marzouk and Zahm (2023*b*) for details. Continuous optimal transport maps can also be estimated by first solving a discrete optimal transport problem (i.e. between empirical measures) and smoothing the result (Manole, Balakrishnan, Niles-Weed and Wasserman 2021, Pooladian and Niles-Weed 2021) or by parametrizing a differentiable convex potential (Huang, Chen, Tsirigotis and Courville 2020).

In the context of OED, it is useful to employ *block-triangular* maps (Baptista, Hosseini, Kovachki and Marzouk 2023*a*) (a class which includes strictly triangular maps but infinitely many other choices), as they naturally capture conditional densities. Consider the following block arrangement for a map $S\colon \mathbb{R}^{n+p} \to \mathbb{R}^{n+p}$:

$$S(y,\theta) = \begin{bmatrix} S^Y(y) \\ S^\Theta(y,\theta) \end{bmatrix}, \tag{3.17}$$

where $S^Y : \mathbb{R}^n \to \mathbb{R}^n$ and $S^\Theta\colon \mathbb{R}^{n+p} \to \mathbb{R}^p$. Let the reference distribution ρ on $\mathbb{R}^{n+p}$ factor as $\rho = \rho_n \otimes \rho_p$, where the subscripts denote the dimension of the factors. (The standard Gaussian distribution on $\mathbb{R}^{n+p}$ naturally factorizes in this way, but any distribution with appropriate block-independence would suffice.) Then, as shown in Baptista *et al.* (2023*a*) and Marzouk *et al.* (2016), if $S_\sharp p_{Y,\Theta} = \rho$, then $S^Y_\sharp p_Y = \rho_n$ and $(\theta \mapsto S^\Theta(y,\theta))_\sharp p_{\Theta|Y=y} = \rho_p$ for any $y \in \operatorname{supp} p_Y$. Thus the two component functions of the map capture the evidence (Y-marginal) and the posterior, respectively. Note also that (3.17) can easily be extended to depend on the design, that is,

$$S(y,\theta;\xi) = \begin{bmatrix} S^Y(\xi,y) \\ S^\Theta(\xi,y,\theta) \end{bmatrix}. \tag{3.18}$$

Suppose now that we obtain a maximum likelihood estimate $\widehat{S}$ of a map of the form (3.18), given samples from $p(y,\theta|\xi)$ and some tractable class of candidate block-triangular maps $\mathcal{S}$, for instance as described in Baptista *et al.* (2023*b*):

$$\widehat{S}(\cdot\,;\xi) \in \arg\max_{S\in\mathcal{S}} \sum_{i=1}^{M} \log S^{\sharp}\rho(y^{(i)},\theta^{(i)}). \tag{3.19}$$

The map $\widehat{S}$ yields a plug-in estimate of the desired densities, via its two component functions:

$$\hat{q}_{\mathrm{mar}}(y|\xi) = \rho_n(\widehat{S}^Y(\xi,y)) \det \nabla_y \widehat{S}^Y(\xi,y),$$
$$\hat{q}_{\mathrm{pos}}(\theta|y,\xi) = \rho_p(\widehat{S}^\Theta(\xi,y,\theta)) \det \nabla_\theta \widehat{S}^\Theta(\xi,y,\theta).$$

(For a statistical convergence analysis of transport-based density estimators, in a general non-parametric setting, see Wang and Marzouk 2022.) In fact, since the reference ρ is a product distribution, (3.19) splits into two separate optimization problems such that the component functions S^Y and S^Θ can be estimated separately, and hence only the component needed for the desired variational bound needs to be learned. The resulting density estimates $\hat{q}_{\mathrm{mar}}$ or $\hat{q}_{\mathrm{pos}}$ can then be substituted into (3.11) or (3.12), respectively. The transport approach effectively defines the approximating class of densities $\mathcal{Q}$ in (3.13) or (3.14) as the set of all densities that can be expressed as $S^\sharp\rho$ for $S \in \mathcal{S}$.

Several recent instantiations of this transport approach in OED have parametrized the maps as normalizing flows; see Kennamer, Walton and Ihler (2023), Orozco, Herrmann and Chen (2024) and Dong *et al.* (2024). In particular, the lower component function S^Θ can be represented as a *conditional* normalizing flow, which is essentially a structured invertible function of θ that is parametrized by y. Another canonical choice of S^Θ is a conditional Brenier map (Carlier, Chernozhukov and Galichon 2016), which can be understood as a family of L^2-optimal transport maps from $p_{\Theta|Y=y}$ to ρ_p, parametrized by y. The essential requirements are to respect the overall block structure of (3.17) in an invertible, differentiable map that pushes forward $p_{Y,\Theta}$ to a block-independent reference distribution.

Figure 3.2, adapted from Li, Baptista and Marzouk (2024*a*), shows an application of these transport-based density estimators to the nonlinear Mössbauer spectroscopy example described in Feng and Marzouk (2019, Section 4.2), for a fixed design ξ. The orange violin plots and circles show repeated independent estimates of the upper bound (3.11), while the blue violin plots and circles illustrate repeated independent estimates of the lower bound (3.12). These estimates are produced by learning the appropriate transport map component, S^Y or S^Θ, via the adaptive semi-parametric procedure described in Baptista *et al.* (2023*b*), which naturally enlarges the family of maps being considered as the sample size available for estimation increases. Here the maps S are strictly triangular, and hence approximations of the Knothe–Rosenblatt rearrangement; see Rosenblatt (1952), Knothe (1957) and Santambrogio (2015, Section 2.3). The horizontal axis shows the total number of independent samples drawn from the model, comprising two batches: one to estimate the map and a separate batch to estimate the outer expectations $\mathbb{E}_{Y,\Theta|\xi}$ in

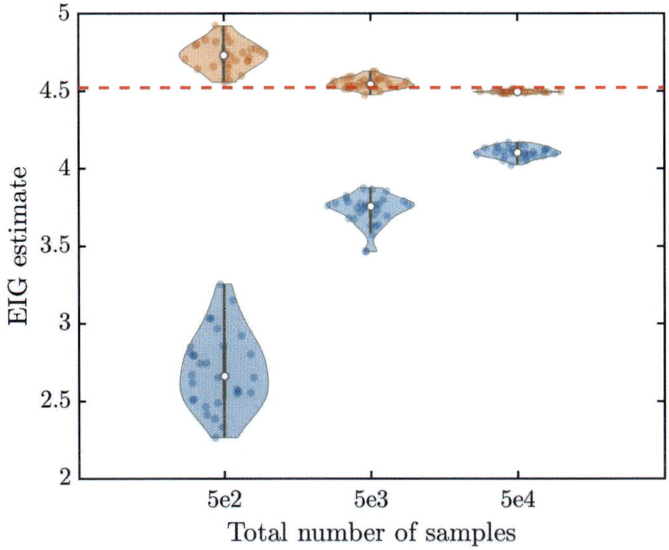

Figure 3.2. Variational upper (orange) and lower (blue) bounds on the EIG in a nonlinear design problem, compared to a biased estimate obtained via NMC (dashed red line). See the discussion in Section 3.2. Figure adapted from Li *et al.* (2024*u*).

(3.11) and (3.12). (Asymptotically optimal allocations of this sample budget are discussed in Li *et al.* 2024*a*.) The upper and lower bounds 'sandwich' the true EIG as the sample size increases.

It is useful to understand the source and nature of the randomness in these results. Each estimate of a transport map yields a density estimate $\hat{q}$, and plugging *any* such estimate into (3.11) and (3.12) yields a guaranteed upper or lower bound, as appropriate. Approximating the expectations in (3.11) and (3.12) with samples, however, yields an unbiased *estimate* of the bound. There is no guarantee that each realization of the estimator will be above or below the true EIG, though the figure suggests that no realizations cross this threshold at the sample sizes considered here. Overall, however, fluctuation in the estimates reflects randomness in both $\hat{q}$ and in the Monte Carlo approximation of the outer expectation. The red dashed line in Figure 3.2 is an NMC estimate of the EIG obtained with 1.58×10^6 samples, which is two to four orders of magnitude greater than the number of samples used in the variational estimators. The NMC estimate appears to be positively biased, even at this large sample size.

As noted earlier, estimating the posterior density to construct a lower bound for EIG is well suited to implicit models, since (3.12) does not require the ability to evaluate $p(y|\theta, \xi)$. But the flexibility of using such density estimates also applies to other information-based design criteria discussed in Section 2.2.2. For instance, the expected marginal information gain objective in (2.28) can be bounded from below by estimating $p(\theta_1|y,\xi)$, i.e. by replacing this marginal posterior density with some $q(\theta_1|y,\xi)$. Baptista *et al.* (2024) uses this approach to assess the information value of different observables in a phase-field model (a nonlinear PDE), where high-dimensional random initial conditions take the role of the 'nuisance parameters' Θ_2. If the prior marginal density $p(\theta_1)$ is not available in closed form, it can also be estimated from samples, with the caveat that if $p(\theta_1|y,\xi)$ and $p(\theta_1)$ are *both* replaced with approximations, the resulting EIG approximation will be neither an upper nor a lower bound for the true marginal EIG in general (Foster *et al.* 2019).

In the case of goal-oriented OED (2.31)–(2.32), the need for learning both the numerator and denominator in the associated density ratios is more acute: for a generic Ψ, none of the densities in (2.31) or (2.32) may be readily available. But as long as we can simulate from $p(z,y|\xi)$ – which can be done by drawing $\theta^{(i)} \sim p(\theta)$, drawing $y^{(i)} \sim p(y|\theta^{(i)},\xi)$ and evaluating $z^{(i)} = \Psi(\theta^{(i)})$ – the density estimation techniques discussed above will apply. If we choose to use transport in this setting, then two block-triangular maps will be relevant:

$$S_1(y,z;\xi) = \begin{bmatrix} S_1^Y(\xi,y) \\ S_1^Z(\xi,y,z) \end{bmatrix} \quad \text{and} \quad S_2(z,y;\xi) = \begin{bmatrix} S_2^Z(\xi,z) \\ S_2^Y(\xi,z,y) \end{bmatrix}. \tag{3.20}$$

Approximating the densities in (2.31) would use the lower component of S_1 and the upper component of S_2; conversely, approximating the densities in (2.32) would use the upper component of S_1 and the lower component of S_2.

In closing, we note that transport approximations of the densities relevant to an EIG calculation can be built by means other than maximum likelihood estimation. For instance, Koval, Herzog and Scheichl (2024) first prescribe some density over the continuous real-valued design variables ξ, and then build triangular transport maps directly from unnormalized evaluations of the joint density $p(y,\theta|\xi)p(\xi)$, using a functional tensor-train format (Cui, Dolgov and Zahm 2023). This construction method does not correspond to tightening a variational bound, in contrast to maximizing the likelihood, but the error in EIG due to the map approximation can nonetheless be analysed.

Finally we note that another use for the variational approximations discussed in this subsection – specifically the approximation $q_{\text{pos}}(\theta|y,\xi)$ – is as a biasing distribution in NMC estimation of EIG. This idea, proposed in Foster *et al.* (2019), fits directly into (3.6) by setting $q^{i,\xi}(\theta) = q_{\text{pos}}(\theta|y^{(i)},\xi)$, and recovers the consistency guarantees of NMC.

3.3. More general variational bounds for mutual information

So far we have discussed bounds on mutual information that follow from parametrizing a density, or a transport map from which a density estimate can be derived. These bounds are variational in the sense that they can be tightened by solving an optimization problem over some class of densities or transport maps. For the bounds discussed so far, if the class of densities or transports is sufficiently rich, in principle there is a member of the class that corresponds to the true density and the bound will attain the true mutual information.

But there are many other variational bounds for mutual information, based on different approximation formats and more general classes of functions (i.e. not just densities or invertible transport maps). Such bounds are in fact an important topic in information theory, with many applications in machine learning. And several have found recent use in OED. Poole *et al.* (2019) provide a unifying framework for understanding many variational bounds, which inspires some of the discussion below.

To provide some context, we first recall a classical non-parametric estimator of mutual information (which is not variational). The k-nearest-neighbour estimator of Kraskov, Stögbauer and Grassberger (2004) (KSG) takes joint sample pairs $(y^{(i)}, \theta^{(i)}) \sim p_{Y,\Theta}$ as input, and uses the statistics of nearest-neighbour distances in both the joint space and in the marginal directions to construct an estimator. Its construction is related to k-nearest-neighbour estimators of the Shannon entropy (Kozachenko and Leonenko 1987). The KSG estimator has been used for OED (Terejanu, Upadhyay and Miki 2012), but is known to scale poorly with dimension. Rigorous statistical analyses of this estimator, e.g. proofs of consistency for fixed k and bounds on the rate of convergence of the MSE with sample size N, have appeared relatively recently (Gao, Oh and Viswanath 2018). For instance, Gao *et al.* (2018, Corollary 2) establish an upper bound on the MSE that is, up to polylogarithmic factors, $O(N^{-2/(n+p)})$, where n is the dimension of Y and p is the dimension of X. They suggest that this convergence rate cannot be refined even if the densities of interest are assumed to be Hölder-smooth.

Non-parametric estimators of the entropy (rather than the mutual information) have been used to construct lower bounds for mutual information in Ao and Li (2020). Here the idea is to write $\mathcal{I}(Y;\Theta|\xi) = H(Y|\xi) - H(Y|\Theta,\xi)$ and then to seek an upper bound for the conditional entropy $H(Y|\Theta,\xi)$. Ao and Li develop an upper bound that requires only (unconditional) entropy estimation, performed via a k-nearest-neighbours approach (Kozachenko and Leonenko 1987).

3.3.1. Variational lower bounds and approximate density ratios

Now we turn our discussion back to variational bounds, parametrized by *learnable functions*. Dropping our earlier requirement that the function be a properly normalized density or a smooth invertible transport map, Poole *et al.* (2019) develop the following bound, which holds for any function $f\colon \mathbb{R}^p \times \mathbb{R}^n \to \mathbb{R}$ and any positive function $a\colon \mathbb{R}^n \to \mathbb{R}_{>0}$:

$$\begin{aligned}\mathcal{I}(Y;\Theta) &\geq \mathbb{E}_{\Theta,Y}[f(\Theta,Y)] - \mathbb{E}_Y\left[\frac{\mathbb{E}_\Theta[\mathrm{e}^{f(\Theta,Y)}]}{a(Y)} + \log a(Y) - 1\right] \\ &=: \mathcal{L}^{\mathrm{TUBA}}(f,a). \end{aligned} \tag{3.21}$$

Here 'TUBA' denotes 'tractable unnormalized Barber–Agakov (BA)', for reasons we will explain shortly. The function f is known as the 'critic'. This bound is tightened by maximizing simultaneously over f and a. Since the mutual information will generally depend on the design ξ, the critic and the function a can both depend on the design as well, but here we have temporarily dropped dependence on ξ to lighten notation.[4] We can link this bound to the BA bound in (3.12), which uses variational approximation over a class of normalized densities, by writing

$$q(\theta|y) = \frac{p(\theta)}{Z(y)} e^{f(\theta,y)} \tag{3.22}$$

and

$$Z(y) = \mathbb{E}_{\Theta}[\exp f(\Theta, y)]. \tag{3.23}$$

Hence f parametrizes an approximation $q(\theta|y)/p(\theta)$ of the true posterior-to-prior density ratio $p(\theta|y)/p(\theta)$, and Z is the y-dependent normalizing constant. The bound $\mathcal{L}^{\mathrm{TUBA}}(f, a)$ is then tight when

$$f(\theta, y) = \log p(y|\theta) + c(y) \quad \text{and} \quad a(y) = Z(y),$$

where c is any function of y. One specific case is when $c = 0$, and hence $f(\theta, y) = \log p(y|\theta)$ and $a(y) = p(y)$.

Poole *et al.* (2019) point out that the lower bound for mutual information introduced earlier by Nguyen, Wainwright and Jordan (2010), often called the NWJ bound, is a special case of (3.21) that follows from setting $a = e$:

$$\mathcal{I}(Y; \Theta) \geq \mathbb{E}_{\Theta,Y}[f(\Theta, Y)] - \frac{1}{e}\mathbb{E}_{Y}[Z(Y)] =: \mathcal{L}^{\mathrm{NWJ}}(f). \tag{3.24}$$

In this case, the unique optimal critic is

$$f^*(\theta, y) = 1 + \log \frac{p(\theta|y)}{p(\theta)}.$$

Unbiased estimates of these lower bounds, for any f and a, can be constructed by simple Monte Carlo estimation of the expectations therein. Similarly, one can produce unbiased estimates of the gradient of each bound with respect to f and a (or some parametrization thereof). The modern approach is to parametrize f and a with deep neural networks, and to maximize over the network parameters, with the hope that the resulting class of functions is rich enough to yield a good approximation of the optimal critic. To our knowledge, the statistical properties of these nonlinear M-estimators are not well understood, save perhaps for the NWJ estimator in the more classical setting where the critic is constrained to lie in a reproducing kernel Hilbert space and parametrized with kernels (Nguyen *et al.* 2010). A more general observation is that while these bounds can in principle

[4] For the remainder of this section, we will keep notation lighter by not explicitly writing dependence of the mutual information, the critic, the likelihood and other quantities on ξ, with the understanding that everything is appropriately conditioned on ξ, since we are considering the mutual information at a particular design.

become tight, their estimators exhibit high variance, especially when the true value of the mutual information is large; see McAllester and Stratos (2020) for a discussion.

For completeness, we should note that the classical Donsker–Varadhan bound (Donsker and Varadhan 1983), a variational lower bound on KL divergence, can be adapted to mutual information to yield

$$\mathcal{I}(Y;\Theta) \geq \mathbb{E}_{\Theta,Y}[f(\Theta,Y)] - \log \mathbb{E}_Y[Z(Y)] =: \mathcal{L}^{\mathrm{DV}}(f). \tag{3.25}$$

When the critic f is parametrized by a neural network, this is the 'mutual information neural estimation' (MINE) approach of Belghazi *et al.* (2018). The challenge in using this bound is that the second term on the right yields a *nested* expectation. Hence, in practice, when estimating the bound with Monte Carlo as in Belghazi *et al.* (2018), we revert to the NMC setting of Section 3.1, where estimates are biased at finite inner-loop sample sizes. Then, as emphasized in Poole *et al.* (2019), we will obtain (to leading order) a positively biased estimator of a lower bound. The mean of this estimator is neither an upper nor a lower bound on the desired $\mathcal{I}(Y;\Theta)$.

3.3.2. *Multi-sample lower bounds*

A different class of lower bounds are called 'multi-sample' in that they involve not only an expectation over the joint distribution of $\Theta, Y \sim p_{\Theta,Y}$ but simultaneously another expectation over $\Theta_{2:M} \sim \prod_{j=2}^{M} r_\Theta$, where r_Θ could be different from the marginal p_Θ.

The simplest among these is the so-called 'prior contrastive estimator' (PCE) proposed in Foster *et al.* (2020):

$$\mathcal{I}(Y;\Theta) \geq \mathbb{E}_{Y_1|\Theta_1}\mathbb{E}_{\Theta_1}\mathbb{E}_{\Theta_{2:M}}\left[\log \frac{p(Y_1|\Theta_1)}{\frac{1}{M}\sum_{j=1}^{M} p(Y_1|\Theta_j)}\right] =: \mathcal{L}^{\mathrm{PCE}}(M), \tag{3.26}$$

where the expectation is over $\Theta_1, Y_1 \sim p_{\Theta,Y}$ and $\Theta_{2:M} \overset{\text{i.i.d.}}{\sim} p_\Theta$. Keep in mind that to *estimate* the right-hand side of (3.26), one must draw many independent and identically distributed (i.i.d.) realizations of $(Y_1, \Theta_1, \Theta_{2:M})$. Suppose that we draw N such (outer-loop) realizations, with sample index i. The result is very much like the NMC estimator (3.2), with one crucial modification: each outer loop sample of the parameters, $\Theta_1^{(i)}$, is used once more, within the inner-loop estimate of the evidence for the corresponding $Y_1^{(i)}$. Doing so guarantees that $\mathcal{L}^{\mathrm{PCE}}(M)$ is a lower bound of the mutual information for any $M > 1$; moreover, like NMC, the bound becomes tight as $M \to \infty$ (Foster *et al.* 2020, Theorem 1). Intuitively, recycling the outer-loop parameter sample makes the denominator of (3.26) over-estimate the evidence and hence under-estimate the mutual information. The samples $\Theta_{2:M}^{(i)}$ are called contrastive samples, in that, unlike the original sample $\Theta_1^{(i)}$, they are not responsible for $Y_1^{(i)}$.

Just as the variance and bias of the NMC estimators in Section 3.1 were improved by importance sampling, a y-dependent, properly normalized, biasing distribution $q(\theta|y)$ can be used within a contrastive estimator as well. (The biasing distribution will generally depend on ξ too, but recall that we are not explicitly writing dependence on ξ in this section, to keep notation simple.) This leads to the idea of an 'adaptive contrastive estimator' (ACE) proposed by Foster *et al.* (2020),

$$\mathcal{I}(Y;\Theta) \geq \mathbb{E}_{Y_1,\Theta_1}\mathbb{E}_{\Theta_{2:M}}\left[\log \frac{p(Y_1|\Theta_1)}{\frac{1}{M}\sum_{j=1}^{M}\frac{p(Y_1|\Theta_j)p(\Theta_j)}{q(\Theta|Y_1)}}\right] =: \mathcal{L}^{\mathrm{ACE}}(M,q), \tag{3.27}$$

where now the expectation is over $\Theta_1, Y_1 \sim p_{\Theta,Y}$ and $\Theta_{2:M} \overset{\text{i.i.d.}}{\sim} q_{\Theta|Y_1}$. The optimal biasing distribution in this case is $q(\theta|y) = p(\theta|y)$, which leads to zero-variance estimates of the evidence $p(Y_1)$. Seeking this q is exactly the idea described at the end of Section 3.2 – combining a variational approximation of the posterior density with nested Monte Carlo – with only the addition of a contrastive recycling of the outer-loop Θ_1. With the optimal choice of q, this bound is tight for any finite M; otherwise, for sub-optimal q, it becomes tight as $M \to \infty$.

Both (3.26) and (3.27) require the ability to evaluate the likelihood, and thus in contrast with the variational bounds (3.12), (3.21) and (3.24), they are not suited to implicit models. A further modification can make (3.26) likelihood-free. It introduces a critic function f as follows:

$$\mathcal{I}(Y;\Theta) \geq \mathbb{E}_{Y_1,\Theta_1}\mathbb{E}_{\Theta_{2:M}}\left[\log \frac{\exp f(\Theta_1, Y_1)}{\frac{1}{M}\sum_{j=1}^{M}\exp f(\Theta_j, Y_1)}\right] =: \mathcal{L}^{\mathrm{NCE}}(M,f), \tag{3.28}$$

where the expectation is again over $\Theta_1, Y_1 \sim p_{\Theta,Y}$ and $\Theta_{2:M} \overset{\text{i.i.d.}}{\sim} p_\Theta$. This is essentially the 'information noise-contrastive estimation' (InfoNCE) approach proposed in van den Oord, Li and Vinyals (2018). $\mathcal{L}^{\mathrm{NCE}}(M,f)$ is a lower bound of the mutual information for all f and M (Poole *et al.* 2019). An amalgam of (3.27) and (3.28), incorporating both a biasing distribution q and a critic f, is called 'likelihood-free ACE' in Foster *et al.* (2020). We note that both $\mathcal{L}^{\mathrm{PCE}}(M)$ and $\mathcal{L}^{\mathrm{NCE}}(M,f)$ are bounded above by $\log(M)$ – no matter how the critic is chosen in the latter. Hence if $\mathcal{I}(Y;\Theta) > \log(M)$, these contrastive bounds cannot become tight; caution is therefore needed when the mutual information is high or the contrastive sample size M is small. Note also that the $\log(M)$ limit does not apply to (3.27) and to other variants that use importance sampling, because in principle there exists a biasing distribution that yields a zero-variance estimate of the evidence, and in such a case even $M = 1$ is sufficient.

An alternative way of estimating the PCE and InfoNCE bounds given only a single set of i.i.d. sample pairs $(Y_i,\Theta_i)_{i=1}^M \sim p_{Y,\Theta}$ is to rotate the role of the 'data-generating sample' and the 'contrastive samples' through the set. We can take the ith pair as the 'data-generating sample' and use the remaining $M-1$ samples as 'contrastive samples' to obtain

$$\log \frac{\exp f(\Theta_i, Y_i)}{\frac{1}{M}\sum_{j=1}^{M} \exp f(\Theta_j, Y_i)}.$$

Repeating this for $i = 1, \ldots, M$ so that each pair has the opportunity to be the 'data-generating sample', and then taking the average, we obtain the estimator,

$$\frac{1}{M}\sum_{i=1}^{M} \log \frac{\exp f(\Theta_i, Y_i)}{\frac{1}{M}\sum_{j=1}^{M} \exp f(\Theta_j, Y_i)}. \tag{3.29}$$

As shown in Poole *et al.* (2019), the expectation of this estimator is a lower bound of the true mutual information, for any critic f:

$$\mathcal{I}(Y;\Theta) \geq \mathbb{E}_{(Y_i,\Theta_i)_{i=1}^{M}}\left[\frac{1}{M}\sum_{i=1}^{M} \log \frac{\exp f(\Theta_i, Y_i)}{\frac{1}{M}\sum_{j=1}^{M} \exp f(\Theta_j, Y_i)}\right]. \tag{3.30}$$

The expectation is also bounded above by $\log M$, in the same way as (3.26) and (3.28) above. Inequality (3.30) is in fact the form of InfoNCE bound originally proposed in van den Oord *et al.* (2018). It is in essence the same as (3.28) but using correlated, rather than independent, terms to estimate the expectation of

$$\log\left(\exp f(\Theta_1, Y_1) \Big/ \frac{1}{M}\sum_{j=1}^{M} \exp f(\Theta_j, Y_1)\right).$$

If the conditional density $p(y|\theta)$ is tractable, then we can replace f with an optimal critic, $f(\theta, y) = \log p(y|\theta)$, such that (3.30) becomes

$$\mathcal{I}(Y;\Theta) \geq \mathbb{E}_{(Y_i,\Theta_i)_{i=1}^{M}}\left[\frac{1}{M}\sum_{i=1}^{M} \log \frac{p(Y_i|\Theta_i)}{\frac{1}{M}\sum_{j=1}^{M} p(Y_i|\Theta_j)}\right], \tag{3.31}$$

which can be understood as a 'sample re-use' analogue of the PCE bound (3.26) above. Interestingly, the corresponding estimator (i.e. (3.29) with the log-likelihood substituting for f) was used in Huan and Marzouk (2013), before it was formally understood to be a lower bound in expectation. There, the rationale was to lower the cost of nested Monte Carlo by reducing the number of expensive model evaluations from $O(NM)$ to $O(M)$, in a setting where evaluations of the likelihood $p(y|\theta)$ are costly for each new value of θ, but cheap for new values of y given a fixed θ.

3.3.3. Deploying variational bounds

The relative merits of the variational bounds discussed above are very much a current subject of research, involving both empirical comparisons and some theory (Czyż *et al.* 2023, Song and Ermon 2020, Letizia, Novello and Tonello 2023). In general, the TUBA, NWJ and DV bounds can become tight with increasingly good choices of critic, but their Monte Carlo estimates generally have much higher variance than estimates of multi-sample bounds such as PCE and InfoNCE. The latter bounds, on the other hand, are hampered by the fact they cannot become tight for arbitrarily large mutual information when using finite contrastive sample sizes. We also emphasize that new variants of these variational bounds continue to be proposed; we have attempted to cover the key constructions relevant to OED, but do not claim to be comprehensive. We refer to Song and Ermon (2020) and Czyż *et al.* (2023) for empirical evaluations of the mutual information estimators (and estimators of mutual information bounds) discussed here, along with many others. Important settings for these evaluations include low versus high values of mutual information, heavy versus light tails, various forms of non-Gaussianity, high dimensionality, sparse interactions, and so forth. As emphasized by Czyż *et al.* (2023), evaluations limited to the Gaussian setting, while convenient in that exact analytical results are available for comparison, are not fully representative.

Deploying the bounds discussed in Sections 3.2 and 3.3 requires optimization. Specifically, it is useful to have an *unbiased estimate* of the gradient of each bound with respect to parameters representing the critic f (and possibly a) or the density approximation q, along with an unbiased estimate of the gradient of the bound with respect to the design ξ, for any given critic or density approximation. Explicit forms of these gradient estimates, for several bounds above, are detailed in Kleinegesse and Gutmann (2021). Given these gradients, one can *simultaneously* maximize the bound with respect to both the variational parameters and the design, e.g. via stochastic gradient ascent. We will return to this idea in Section 4.3.

3.4. *Low-dimensional structure*

So far, we have discussed a variety of methods that, essentially, approximate the densities or density ratios appearing in expressions for the mutual information (expected information gain). A different and complementary class of approximation methods involve *dimension reduction*. Here we will discuss methods for Bayesian OED that exploit possible low dimensionality in the *update* from prior to posterior; such methods rest on the idea that the posterior is well approximated by a distribution that departs from the prior only on a subspace of dimension $r \ll p$. At the same time, we will consider dimension reduction for the observation vector $y \in \mathbb{R}^n$; the central idea is that conditioning on y could be replaced by conditioning on the *projection* of y onto a subspace of dimension $s \ll n$.

Both of these notions of low dimensionality appear quite often in Bayesian inverse problems, where the data are related to the parameters θ by the action of a forward operator that is somehow smoothing, e.g. in that it suppresses information about finer scales (Cui *et al.* 2014). Low dimensionality more generally results from limited informativeness of the data in certain directions of $\boldsymbol{\Theta}$ and/or redundancy among the components of y. There is additional potential for dimension reduction in *goal-oriented* problems, as in Bayesian $\mathrm{D_A}$-optimality and (2.31)–(2.32), where the experimenter wishes to learn about a specific quantity of interest $Z = \Psi(\Theta)$.

Identifying and exploiting this low dimensionality can lead to much more computationally efficient approximations of the mutual information, as we shall discuss below.

3.4.1. Linear models

We first consider dimension reduction, of both parameters and data, in the context of finite-dimensional Bayesian linear-Gaussian models. Recall the problem set-up and notation from the start of Section 2.1.2. The mutual information in this setting can already be written in closed form as a ratio of log-determinants, as in (2.25) and (2.26). It is nonetheless useful to express this quantity in a way that reveals the structure and *intrinsic dimensionality* of the linear-Gaussian model.

The central objects in such a construction are generalized eigenvalue problems involving combinations of the prior covariance matrix Γ_Θ, the noise covariance $\Gamma_{Y|\Theta}$, the posterior covariance $\Gamma_{\Theta|Y}$ (2.8), the negative Hessian of the log-likelihood, i.e. $H := G^\top \Gamma_{Y|\Theta}^{-1} G$, and the marginal covariance of the data, $\Gamma_Y = G\Gamma_\Theta G^\top + \Gamma_{Y|\Theta}$. Several different generalized eigenvalue problems involving these matrices can be written, all closely related via simple transformations. Perhaps the most direct is the symmetric definite generalized eigenproblem

$$Hw_j = \sigma_j \Gamma_\Theta^{-1} w_j, \tag{3.32}$$

where, since $H \succeq 0$ and $\Gamma_\Theta^{-1} \succ 0$, we have $\sigma_j \geq 0$. This eigenproblem can be understood as balancing informativeness of the likelihood with informativeness of the prior, via the Rayleigh ratio $w^\top Hw / w^\top \Gamma_\Theta^{-1} w$: an eigendirection $w_j \in \mathbb{R}^p$ associated with $\sigma_j > 1$ is more constrained by the likelihood than by the prior, and thus more important to the update from prior to posterior. See Spantini *et al.* (2015) for thorough interpretations and demonstrations of this eigenstructure. Closely related to (3.32) are two other $p \times p$ parameter-space matrix pencils: $(\Gamma_\Theta - \Gamma_{\Theta|Y}, \Gamma_{\Theta|Y})$ and $(\Gamma_\Theta, \Gamma_{\Theta|Y})$. The first has eigenvalue–eigenvector pairs (σ_j, v_j), where $v_j = \Gamma_\Theta^{-1} w_j$, and the second has eigenvalue–eigenvector pairs $(1 + \sigma_j, v_j)$. See Jagalur-Mohan and Marzouk (2021, Proposition 10) for a proof of this equivalence. The leading eigendirections v_j (ordered from largest to smallest σ_j) are the directions along which the ratio of posterior to prior variance is minimized; see Spantini *et al.* (2015, Section 3.1).

In the linear-Gaussian setting, there is a duality between spectral structure in the parameter space and in the data space. Consider the $n \times n$ data-space matrix pencils $(\Gamma_Y - \Gamma_{Y|\Theta}, \Gamma_{Y|\Theta})$ and $(\Gamma_Y, \Gamma_{Y|\Theta})$. Jagalur-Mohan and Marzouk (2021, Proposition 10) shows that $(\Gamma_Y - \Gamma_{Y|\Theta}, \Gamma_{Y|\Theta})$ and $(\Gamma_\Theta - \Gamma_{\Theta|Y}, \Gamma_{\Theta|Y})$ have identical non-trivial generalized eigenvalues, $\sigma_j > 0$, and that $(\Gamma_Y, \Gamma_{Y|\Theta})$ and $(\Gamma_\Theta, \Gamma_{\Theta|Y})$ have identical eigenvalues that are strictly greater than one, i.e. $1 + \sigma_j > 1$. The eigenvectors u_j of the two data-space pencils are identical, with $w_j = \frac{1}{\alpha_j} \Gamma_\Theta G^\top u_j$ for all j with $\sigma_j > 0$, where α_j is a scaling parameter that can be set to $\alpha_j = \sqrt{\sigma_j}$ to obtain $\langle U_i, U_j \rangle_{\Gamma_{Y|\Theta}} = \delta_{ij}$, given $\langle W_i, W_j \rangle_{\Gamma_\Theta^{-1}} = \delta_{ij}$.

Spantini *et al.* (2015, Theorem 2.3) considers optimal approximations of the posterior covariance $\Gamma_{\Theta|Y}$ that take the form of *low-rank updates* of the prior covariance. Optimality is cast in terms of minimizing a class of loss functions $\ell(M)$ over the manifold of all symmetric positive definite (SPD) matrices M; this class includes the geodesic distance from M to $\Gamma_{\Theta|Y}$ on SPD manifold, and the KL divergence or Hellinger distance from $\mathcal{N}(\mu, M)$ to $\mathcal{N}(\mu, \Gamma_{\Theta|Y})$, where $\mu \in \mathbb{R}^p$ is an arbitrary (common) mean vector. The approximations sought are of the form $\widetilde{\Gamma}^r_{\Theta|Y} \in \{\Gamma_\Theta - KK^T > 0, \text{ rank}(K) \le r\}$, and the optimal approximation for any $r \le n$, simultaneously minimizing all loss functions in the class, is given by

$$KK^\top = \sum_{i=1}^{r} \frac{\sigma_i}{1+\sigma_i} w_i w_i^\top;$$

hence it follows from the leading eigenpairs of the parameter-space pencil (H, Γ_Θ^{-1}). The decay of the eigenvalue sequence $(\sigma_j)_{j \ge 1}$, common to all of the matrix pencils above, captures the intrinsic dimensionality of the Bayesian model.

These optimal approximation results are related to the approximation of a more central object of interest in OED: the mutual information. Using the fact that the generalized eigenvectors v_j simultaneously diagonalize Γ_Θ and $\Gamma_{\Theta|Y}$, or that the generalized eigenvectors u_j simultaneously diagonalize Γ_Y and $\Gamma_{Y|\Theta}$, the mutual information between parameters Θ and data Y (2.25)–(2.26) can be written as

$$\mathcal{I}(Y; \Theta) = \frac{1}{2} \sum_j \log(1 + \sigma_j). \tag{3.33}$$

This expression for mutual information has appeared in many places in recent literature, e.g. Alexanderian *et al.* (2016*a*) and Giraldi, Le Maître, Hoteit and Knio (2018). We note also that $\mathcal{I}(Y; \Theta)$ can be written using the squared canonical correlation scores (Bach and Jordan 2002) between Y and Θ; these scores follow from generalized eigenvalue problems that are slightly different from those discussed above, but closely related. Truncating the sum in (3.33) after $r < \min(n, p)$ terms yields a low-rank approximation of the mutual information. From a computational perspective, since the MI is dominated by the largest generalized eigenvalues (though the log function slows this decay), problems with quickly decaying spectra and hence low intrinsic dimension are easier to approximate: one must compute only the leading eigenvalues of *any* of the matrix pencils described above.

Truncating (3.33) in this way in fact corresponds to the mutual information obtained from an *optimal* r-dimensional projection of the data Y, of the parameters Θ, or of both simultaneously. Specifically, Giraldi *et al.* (2018) show that the leading eigenvectors $U_{1:r} = [u_1 \cdots u_r]$ of the data-space matrix pencils define low-dimensional projections of the data that are optimal at any given dimension $r \le n$, in the sense of maximizing $\mathcal{I}(A_r^\top Y; \Theta)$ over matrices $A \in \mathbb{R}^{n \times r}$. The resulting mutual information is

$$\mathcal{I}\big(U_{1:r}^\top Y;\Theta\big) = \frac{1}{2}\sum_{j=1}^{r}\log(1+\sigma_j).$$

Similarly, replacing Θ with the r-dimensional projection $V_{1:r}^\top\Theta$ yields

$$\mathcal{I}\big(Y;V_{1:r}^\top\Theta\big) = \frac{1}{2}\sum_{j=1}^{r}\log(1+\sigma_j).$$

Going further, one can show that the same, optimal value of mutual information is also achieved by

$$\mathcal{I}\big(U_{1:r}^\top Y;V_{1:r}^\top\Theta\big) = \frac{1}{2}\sum_{j=1}^{r}\log(1+\sigma_j).$$

We also note that this truncated mutual information is equivalent to the mutual information that would be obtained if the forward operator G in (2.4) were replaced by certain 'projected' versions. Define

$$\mathbb{P}_{\text{obs}} := \Gamma_{Y|\Theta}U_{1:r}U_{1:r}^\top \quad \text{and} \quad \mathbb{P}_{\text{param}} := \Gamma_\Theta V_{1:r}V_{1:r}^\top = W_{1:r}W_{1:r}^\top\Gamma_\Theta^{-1}.$$

Then one can show that

$$\mathbb{P}_{\text{obs}}G = G\mathbb{P}_{\text{param}} = \mathbb{P}_{\text{obs}}G\mathbb{P}_{\text{param}}, \tag{3.34}$$

and that these projected forward operators achieve the mutual information

$$\frac{1}{2}\sum_{j=1}^{r}\log(1+\sigma_j).$$

We emphasize, however, that the equivalence of these three projected models, and more broadly the strict duality between parameter and observation reduction discussed above, is specific to the linear-Gaussian setting.

The expression (3.33) naturally appears in infinite-dimensional formulations of linear Bayesian inverse problems, as discussed in Section 2.3.1. Indeed, finding the leading eigenpairs of (3.32) is equivalent to constructing a low-rank approximation of the prior-preconditioned Hessian $\widetilde{H}$ as discussed in Alexanderian *et al.* (2016*a*). To see this intuitively in finite dimensions, note that $\widetilde{H} = \Gamma_\Theta^{1/2}H\Gamma_\Theta^{1/2}$, and that its (simple, not generalized) eigenvalue–eigenvector pairs are $\big(\sigma_j, \Gamma_\Theta^{-1/2}w_j\big)$, where (σ_j, w_j) are the eigenpairs of (H, Γ_Θ^{-1}). The algorithms proposed in Alexanderian and Saibaba (2018) (see also Ghattas and Willcox 2021) compute these low-rank approximations efficiently in a discretization-invariant manner, when the forward operator G is described by partial differential equations.

Now we turn to the *goal-oriented* linear setting. Our interest here lies not in learning the parameters Θ *per se*, but rather in informing a quantity of interest

$Z = A^\top \Theta$, where $A \in \mathbb{R}^{p \times s}$ has full column rank $s < p$. Maximizing the mutual information $\mathcal{I}(Y; Z)$ is equivalent to maximizing the Bayesian $\mathrm{D_A}$-optimality criterion mentioned in Section 2.1.2, and is a linear special case of (2.31)–(2.32). Now, however, there is further potential for dimension reduction, stemming from the interaction of A with other elements of the problem. Intuitively, the update from the prior predictive distribution to posterior predictive distribution of Z (and hence the mutual information) should be well captured by directions in the parameter space $\mathbb{R}^p$ that are simultaneously informed by the data (relative to the prior) *and* influential on Z. We can expect the latter consideration to help 'screen out' parameter directions that are dampened by $A^\top$. (Consider, as a limiting example, elements of the kernel of $A^\top$.)

This idea is formalized by Spantini *et al.* (2017). Beginning with the linear-Gaussian model $Y = G\Theta + \mathcal{E}$ (2.4), a Gaussian prior distribution $\Theta \sim \mathcal{N}(0, \Gamma_\Theta)$ (zero mean for simplicity), and the specification of $Z = A^\top \Theta$, Spantini *et al.* (2017, Lemma 2.2) introduce an *equivalent* linear-Gaussian model that directly relates Y and Z,

$$Y = GA_\dagger Z + \Delta, \tag{3.35}$$

where $A_\dagger := \Gamma_\Theta A \Gamma_Z^{-1}$, $\Gamma_Z := A^\top \Gamma_\Theta A$, $Z \sim \mathcal{N}(0, \Gamma_Z)$ is the prior induced on Z, and

$$\Delta \sim \mathcal{N}(0, \Gamma_\Delta) \quad \text{with} \quad \Gamma_\Delta := \Gamma_{Y|\Theta} + G(\Gamma_\Theta - \Gamma_\Theta A \Gamma_Z^{-1} A^\top \Gamma_\Theta) G^\top.$$

Crucially, these choices render the 'effective noise' Δ *independent* of Z. Now the low-rank approximation methods discussed earlier for the Y–Θ system can be applied directly to the Y–Z system above.

For example, as a goal-oriented counterpart to (3.32), one can compute the leading eigenpairs of the matrix pencil $(A_\dagger^\top G^\top \Gamma_\Delta^{-1} G A_\dagger, \Gamma_Z^{-1})$. Letting $(\tilde{\sigma}_j)_{j \geq 1}$ denote the generalized eigenvalues of this pencil, arranged in descending order, we can then write the mutual information as $\mathcal{I}(Y; Z) = \frac{1}{2} \sum_j \log(1 + \tilde{\sigma}_j)$ and truncate this sum as desired to obtain a low-rank approximation; see Wu, Chen and Ghattas (2023*a*) for more details and a first application of this approach in OED. As demonstrated in Spantini *et al.* (2017), $(\tilde{\sigma}_j)$ can decay much more quickly than the spectrum (σ_j) of (3.32), yielding a more efficient approximation than would be obtained by first approximating the posterior covariance matrix of Θ and then applying A. Moreover, analogously to the discussion above, one can obtain optimal *goal-oriented* projections of the data using the leading eigenvectors of $(\Gamma_Y, \Gamma_\Delta)$. Each of these eigenproblems optimally balances all the ingredients of the goal-oriented inference problem: the structure of the forward operator, the scale and structure of the observational noise covariance and prior covariance, and the ultimate prediction goals.

In closing, we should point out that (as in Section 3.3) we have not explicitly noted dependence of the quantities above on the design ξ. Recall, however, that our generic linear-Gaussian model (2.4) can be written more explicitly as $Y|\Theta, \xi \sim$

$\mathcal{N}(G(\xi)\Theta, \Gamma_{Y|\Theta}(\xi))$. Hence all of the eigenvalues, eigenspaces and projectors described above may depend on ξ via G and $\Gamma_{Y|\Theta}$. The details of this dependence are problem-dependent, but it is of interest to understand how to exploit any smoothness or other regularity that may be present, e.g. by differentiating the eigenvalues σ_j with respect to ξ. To our knowledge, algorithmic approaches in this vein are in their infancy.

3.4.2. Nonlinear models

Extending the low-dimensional structure discussed in Section 3.4.1 to nonlinear models is considerably more complex, and a subject of ongoing research. The tools we will highlight here originate in dimension reduction for Bayesian inverse problems and Bayesian statistical models more generally: likelihood-informed subspaces (Cui *et al.* 2014, Cui and Tong 2022), certified dimension reduction (Zahm *et al.* 2022, Li, Marzouk and Zahm 2024*b*), and related efforts. Our perspective is similar to that taken in the linear case: find the subspace of the parameter space where the posterior differs 'most' from the prior, which is equivalent to finding the subspace that is best informed by the data; and find the subspace of the data that is most informative of the parameters.

We focus here on the approximation of mutual information, highlighted in Baptista, Marzouk and Zahm (2022), which makes the intuitive notions just stated more precise. In the nonlinear case, it is generally difficult to find *optimal* low-dimensional projections. Instead, one can develop gradient-based upper bounds on the error induced by projecting Y and Θ onto lower-dimension subspaces, i.e. find upper bounds on the difference

$$\mathcal{I}(Y;\Theta) - \mathcal{I}\big(U_{1:s}^\top Y; V_{1:r}^\top \Theta\big)$$

for some matrices $U_{1:s} \in \mathbb{R}^{n\times s}$ and $V_{1:r} \in \mathbb{R}^{p\times r}$, and then minimize these bounds over U and V. (Note that the matrices U and V discussed in this section are generally different from the U and V found in Section 3.4.1.)

The key assumption underlying this analysis is that the joint distribution $p_{Y,\Theta}$ must satisfy a *subspace logarithmic Sobolev inequality*; see Baptista *et al.* (2022, Definition 2) or Zahm *et al.* (2022, Theorem 2.10). Letting $Z = (Y, \Theta)$ be a random variable taking values in $\mathbb{R}^{n+p}$, and letting $W \in \mathbb{R}^{(n+p)\times(n+p)}$ be a unitary matrix, the essence of this assumption is that any conditional distribution $p(Z_t|Z_\perp = z_\perp)$ defined by the decomposition $W = [W_t \ W_\perp]$, with $W_t \in \mathbb{R}^{(n+p)\times t}$, $Z_t = W_t^\top Z$ and $Z_\perp = W_\perp^\top Z$, satisfies a logarithmic Sobolev inequality with constant $C(p(Z_t|Z_\perp = z_\perp))$ bounded above by some $\overline{C} < \infty$, for all W, t and $z_\perp$. As shown in Zahm *et al.* (2022), a *sufficient* condition for a distribution to satisfy the subspace logarithmic Sobolev inequality is that it has convex support and that its log-density is a bounded perturbation of a strongly convex function; see Zahm *et al.* (2022, Examples 2.5–2.9) and Baptista *et al.* (2022, Examples 1–2). This allows, for example, Gaussian mixtures (and hence certain multi-modal distributions with strictly positive densities) but not distributions with exponential tails.

Now define the diagnostic matrices

$$H_\Theta = \iint (\nabla_\theta \nabla_y \log p_{Y|\Theta}(y|\theta))^\top (\nabla_\theta \nabla_y \log p_{Y|\Theta}(y|\theta)) p_{Y,\Theta}(y,\theta)\, \mathrm{d}\theta\, \mathrm{d}y, \quad (3.36)$$

$$H_Y = \iint (\nabla_\theta \nabla_y \log p_{Y|\Theta}(y|\theta))(\nabla_\theta \nabla_y \log p_{Y|\Theta}(y|\theta))^\top p_{Y,\Theta}(y,\theta)\, \mathrm{d}\theta\, \mathrm{d}y, \quad (3.37)$$

where $\nabla_\theta \nabla_y \log p_{Y|\Theta} \in \mathbb{R}^{n\times p}$. We then have the following theorem, adapted from Baptista *et al.* (2022).

Theorem 3.1. Let $p_{Y,\Theta}$ satisfy a subspace logarithmic Sobolev inequality with constant $\overline{C} < \infty$. Then, for any unitary matrices $V = [V_{1:r}\ V_{r+1:p}] \in \mathbb{R}^{p\times p}$ and $U = [U_{1:s}\ U_{s+1:n}] \in \mathbb{R}^{n\times n}$, we have

$$\mathcal{I}(Y;\Theta) - \mathcal{I}\big(U_{1:s}^\top Y; V_{1:r}^\top \Theta\big) \le \overline{C}^2 \big(\mathrm{tr}\big(V_{r+1:p}^\top H_\Theta V_{r+1:p}\big) + \mathrm{tr}\big(U_{s+1:n}^\top H_Y U_{s+1:n}\big)\big). \quad (3.38)$$

Crucially, the right-hand side of (3.38) can be minimized for any dimensions $r \le p$ and $s \le n$ by choosing $V_{1:r}$ to be the leading eigenvectors of H_Θ, and $U_{1:s}$ to be the leading eigenvectors of H_Y. This choice yields the bound

$$\mathcal{I}(Y;\Theta) - \mathcal{I}\big(U_{1:s}^\top Y; V_{1:r}^\top \Theta\big) \le \overline{C}^2 \left(\sum_{i=r+1}^{p} \lambda_i(H_\theta) + \sum_{i=s+1}^{n} \lambda_i(H_Y) \right). \quad (3.39)$$

Fast decay of the eigenvalues λ_i of the two diagnostic matrices thus provides more opportunity for dimension reduction, with controlled error.

Computationally, this dimension reduction approach can be viewed as a first step towards approximating a mutual information design objective: first, find low-dimensional projections of the data and parameters, $Y_s = U_{1:s}^\top Y$ and $\Theta_r = V_{1:r}^\top \Theta$, for some r and s, by finding the dominant eigenspaces of H_Y and H_Θ. (Note that, unlike in the linear-Gaussian case, the spectra of these two matrices are generally different.) Then, use any of the techniques discussed in Sections 3.1–3.3 to estimate or bound $\mathcal{I}(Y_s;\Theta_r)$. By the data-processing inequality (or the positivity of the right-hand side of (3.38)) we always have $\mathcal{I}(Y_s;\Theta_r) \le \mathcal{I}(Y;\Theta)$. For example, Li *et al.* (2024*a*) use the transport-based density estimators of Section 3.2 to bound the projected mutual information $\mathcal{I}(Y_s;\Theta_r)$. The density estimation problems to be solved are thus of reduced dimension, and involve estimating transport maps that act on lower-dimensional spaces. Alternatively, Wu, O'Leary-Roseberry, Chen and Ghattas (2023*b*) apply $V_{1:r}$ for parameter dimension reduction as proposed above, but use a simpler dimension reduction scheme for the data (principal component analysis based on the marginal covariance $\mathrm{Cov}(Y)$); both reductions are a prelude to constructing a reduced-dimensional surrogate $\widetilde{G}$ for the forward operator in a Bayesian inverse problem (see Section 3.1), and then using this surrogate within a standard nested Monte Carlo estimator of the mutual information.

In Theorem 3.1, for simplicity, we restricted the matrices U and V to be unitary. This constraint can effectively be relaxed by preconditioning the problem. Note that with any change of variables $\overline{\Theta} = A\Theta$ and $\overline{Y} = BY$, the left-hand side of (3.38) does not change, but the right-hand side does, via the diagnostic matrices and the log-Sobolev constant. It is unclear how to choose an optimal change of variables, as the log-Sobolev constant is in general very hard to compute, but one heuristic suggested in Baptista *et al.* (2022, Section 4), for the case of a Gaussian prior and Gaussian additive noise, is to 'whiten' the problem so that $\Gamma_{\overline{\Theta}}$ and $\Gamma_{\overline{Y}|\overline{\Theta}}$ become identity matrices. In the case of a linear-Gaussian model, this change of variables leads the proposed method to recover the optimal subspaces of Section 3.4.1. We also note that the change of variables need not be linear; any bijective transformation will leave the mutual information terms on the left-hand side of (3.38) unchanged. Thus, in principle one could compose a nonlinear change of variables with linear dimension reduction to effectively achieve nonlinear dimension reduction!

Another remark is that the bounds in Theorem 3.1 are generally not tight. One can intuitively see this in the linear-Gaussian setting where, after the linear preconditioning described above, we have $\sigma_i = \lambda_i(H_\Theta) = \lambda_i(H_Y)$. Then we can compare (3.33), where the exact truncation error involves *logarithms* of trailing eigenvalues $\log(1 + \sigma_i)$, with (3.39), which sums the trailing eigenvalues directly. These quantities are not the same, and the difference is analysed in Baptista *et al.* (2022, Section 4.3); see also Zahm *et al.* (2022, Section 2.3). There is reason to believe that recent dimensional improvements to log-Sobolev inequalities (Eskenazis and Shenfeld 2024) could help tighten the bounds used here.

4. Design optimization methods

Now we will discuss methods for optimizing (generally, maximizing) OED criteria over possible designs $\xi \in \Xi$. Different representations of candidate designs and choices of design criterion can yield very different classes of optimization problems. This section will therefore review a broad array of optimization approaches, ranging from algorithms for combinatorial optimization (including continuous relaxations thereof) to gradient-based algorithms for intrinsically continuous problems. Section 4.1 focuses on exact designs for linear models, as linearity lends the problem special structure; here, we discuss various discrete algorithms, as well as convex continuous relaxations that are followed by rounding. Section 4.2 discusses exact designs for more general nonlinear problems, with design criteria treated as set functions. Section 4.3 turns to designs that are parametrized by real variables (e.g. sensor positions or times, dosages), and discusses how a variety of derivative-free or gradient-based optimization approaches interact with estimators or bounds for decision-theoretic design criteria.

4.1. Optimization methods for linear design

We state a prototypical linear design problem, by recalling the linear regression model with features $f \colon \mathcal{X} \to \mathbb{R}^p$ and uncorrelated observational errors. If the design ξ is supported on n equally weighted points $x_i \in \mathcal{X} \in \mathbb{R}^d$, $i = 1, \ldots, n$, then the Fisher information matrix of the model is given by (2.11); as pointed out earlier, this expression follows from (2.5) by setting

$$G = \left[f(x_1)^\top; f(x_2)^\top; \ldots; f(x_n)^\top \right] \in \mathbb{R}^{n \times p}$$

and $\Gamma_{Y|\theta} = \sigma^2 I_n$. The classical design objectives discussed in Section 2.1 can be viewed as functions of the Fisher information matrix.

We will use the index set $\mathcal{V} := \{1, \ldots, m\}$ to represent the support of the design space. It is instructive to think of $\mathcal{V}$ as a specific instance of $\mathcal{X}$ in the setting of a discrete design domain. In the simplest setting, the task of finding an optimal *exact design* amounts to selecting a set $\mathcal{S} \subseteq \mathcal{V}$ of specified cardinality n that maximizes the chosen objective, i.e. any of the design criteria for linear models given in Section 2.1. More complex cases can involve constraints on the overall budget and variable costs associated with each individual design instance.

Strictly speaking, the set $\mathcal{S}$ is better characterized as a multiset, where elements can be duplicated. In a conventional set, elements are typically assumed to be different from one another. The notion of a multiset is more general, and accommodates the 'with repetition' scenario where a given design point can be chosen multiple times. (Repeated experiments at the same design point are not necessarily redundant, since they may yield different outcomes due to statistically independent noise.) The contrasting scenario is 'without repetition', where each point is selected at most once; this is applicable, for instance, in the setting of deterministic computer experiments.

4.1.1. Local search and exchange algorithms

Algorithms guided by local search and exchange heuristics are non-sequential in nature. (We discuss sequential algorithms later in Section 4.1.3.) In the non-sequential setting, we start with an initial design of the required size, and at each iteration attempt to improve the quality of the design by deleting, adding or exchanging points as guided by certain rules. The non-sequential nature of these algorithms implies that the resulting solution sets are non-nested, meaning that optimal design set of cardinality n may look quite different from the optimal set of cardinality $n-1$ or $n+1$.

In the context of exchange algorithms, the approach of Fedorov (1972) has been quite influential. Fedorov's exchange method starts with an arbitrary initial set $\mathcal{S}_0$ of n candidate points, and at each iteration t attempts to improve the objective by exchanging one of the points,

$$\mathcal{S}_t \leftarrow \{j\} \cup \mathcal{S}_{t-1} \setminus \{i\},$$

where $i \in \mathcal{S}_{t-1}$ and $j \in \mathcal{V} \setminus \mathcal{S}_{t-1}$. The search at each iteration is exhaustive: the improvement resulting from every possible exchange is calculated, and the best exchange is selected. The process is continued as long as an exchange improves the design objective.

Fedorov's original work focused exclusively on D-optimal design, although its applicability to other design criteria has also been demonstrated. While the exchange algorithm is relatively simple, it is accompanied by a sizeable computational overhead due to the exhaustive search at each step. Cook and Nachtsheim (1980) proposed a modified Fedorov exchange procedure, where they consider each design point in turn, perhaps in random order, carrying out any beneficial exchange as soon as it is discovered. Johnson and Nachtsheim (1983) suggested further improvements by focusing the search on fewer points from the current design set, specifically those that have the lowest predictive variance.

The underlying ideas of the exchange algorithm can be extended to the setting where design points are selected from a continuous compact space, without an underlying candidate set (Meyer and Nachtsheim 1995). A preliminary step here is to specify the expected utility of changing one factor or coordinate at a time, while holding the others constant. This step can be made more efficient by using Gaussian process emulators, as demonstrated by Overstall and Woods (2017). These ideas in principle are also applicable to nonlinear design problems.

Although the original Fedorov exchange algorithm was proposed several decades ago, rigorous analyses establishing *approximation guarantees* have only recently emerged. Here, an approximation guarantee is a provable bound on the ratio between the value of the objective function evaluated at the solution returned by the algorithm, and the optimal solution. It is thus a bound on the worst-case performance of the algorithm (in a relative sense). This is a typical way to assess the performance of an algorithm that produces approximate solutions to optimization problems, in particular NP-hard problems (Hochba 1997, Vazirani 2001).

Madan, Singh, Tantipongpipat and Xie (2019) establish approximation guarantees for local search algorithms for D-optimal design and A-optimal design; their results for the case of A-optimality are restricted to the less general 'with repetition' setting. They show that the algorithms are asymptotically optimal when n/p is large, and that in the case of A-optimal design there could be arbitrarily bad local optima. They also propose approximate local search algorithms, where exchanges are made only when the objective improves by a factor of $1+\delta$, leading to improved running times with a slight degradation in the approximation guarantees. Lau and Zhou (2022) improve upon these results using novel analysis tools, and provide approximation guarantees for D/A/E-optimal designs. In particular they show that Fedorov's exchange method for A-optimal design works well as long as there exists an near-optimal solution with a well-conditioned design matrix.

4.1.2. Continuous convex relaxation approaches

The combinatorial linear design problem is strictly speaking an integer program, whose exact solution is often intractable. One elegant solution approach is to perform a continuous relaxation of the design variables into a convex program that can be solved using established techniques (Boyd and Vandenberghe 2004,

Ben-Tal and Nemirovski 2001). A solution to the original combinatorial problem is then obtained by appropriately rounding the convex solution to an integer solution. Various polynomial-time (and possibly randomized) rounding algorithms, which convert the convex solution to an integer solution for the combinatorial problem, have been proposed.

Consider the linear design problem, where ϕ denotes an operator that acts on the Fisher information matrix to form the design criterion (e.g. $\phi = -\mathrm{tr}$ for A-optimality, $\phi = -\log\det$ for D-optimality). The relaxed continuous optimization problem takes the form

$$\underset{\xi=(\xi_1,\dots,\xi_n)}{\operatorname{argmin}} \ \phi\left(\sum_{i=1}^{m} \xi_i f(x_i) f(x_i)^\top\right) \quad \text{subject to} \quad \|\xi\|_1 \le n. \tag{4.1}$$

Written in this way, the traditional optimality criteria discussed in Section 2.1.1 are all convex functions of the Fisher information matrix. The convex constraint $\|\xi\|_1 \le n$ ensures that only n points are selected. If we further specify $0 \le \xi_i \le 1$, then we constrain each point to be selected at most once, corresponding to the without-replacement scenario. More complex constraints, e.g. encoding varied costs associated with selecting each point and a ceiling on the cumulative cost (i.e. a knapsack constraint), can easily be added to (4.1). The fractional solution weights from this convex program have natural analogues to the notion of designs as probability measures, specifically to *continuous designs*, as discussed in Section 2.1.3. However, the notion of a continuous design arose organically from the work of early experimental design researchers, independently of these modern continuous relaxation formulations.

There are a number of convex programming relaxation approaches for the D/A/E-optimal design problems, differing from each other in how the rounding algorithm provides integer solutions. Even though these design criteria share attributes such as convexity, they behave differently in terms of approximability. Allen-Zhu, Li, Singh and Wang (2017) connect experimental design to matrix sparsification (Spielman and Srivastava 2008, Batson, Spielman and Srivastava 2009) and use regret minimization methods (Allen-Zhu, Liao and Orecchia 2015) to obtain approximate designs for popular optimality criteria. Singh and Xie (2020) devise an approximation algorithm for D-optimal design where the rounding strategy uses approximate positively correlated distributions. Nikolov, Singh and Tantipongpipat (2022) develop an approximation algorithm for D/A-optimal design using proportional volume sampling; interestingly, they also show that the same approach will not work for E-optimal design. Lau and Zhou (2022) modify the iterative randomized rounding algorithm based on the regret minimization framework of Lau and Zhou (2020) for D/A/E-optimal design problems with knapsack constraints. Approximation algorithms using more involved relaxation techniques have enabled D-optimal design under partition (Nikolov and Singh 2016) and matroid (Madan, Nikolov, Singh and Tantipongpipat 2020) constraints.

4.1.3. *Sequential algorithms*

Sequential design methods arrive at a solution set either by the gradual addition of candidate points to a smaller design set, or by the sequential deletion of candidate points from a larger design set. These are called 'forward' and 'backward' procedures respectively (Atkinson *et al.* 2007). Typically the greedy heuristic guides the selection of candidate points at each step. In some cases, designs obtained using sequential procedures can be further improved using the non-sequential techniques outlined in Section 4.1.1. For the case of D/A-optimal designs, Madan *et al.* (2019) analyse the sequential forward procedure, and provide approximation guarantees that retain a specificity to the initialized set. The dependence on the initial set is rather undesirable, which can be ameliorated by leveraging distributed computing resources and running the algorithms with different initializations. Sequential design algorithms and greedy heuristics have been more generally studied using the language of set functions for combinatorial problems; we will discuss these methods more deeply in Section 4.2.

4.2. *Combinatorial approaches for discrete design variables*

In this section we formulate experimental design problems as optimization of set functions. The latter topic has been studied extensively (Wolsey and Nemhauser 1999, Lovász 2007, Schrijver 2003, Papadimitriou and Steiglitz 1998), and has a rich mathematical structure. We discuss some fundamental properties of set functions and their relevance to popular design criteria, including criteria for nonlinear design. We also discuss algorithms for their optimization.

4.2.1. *Background*

Previously we defined the candidate index set as $\mathcal{V} := \{1, \dots, m\}$, $m \in \mathbb{Z}_{>0}$. Let its power set (i.e. the set of all subsets) be denoted by $2^{\mathcal{V}}$. An important property of a certain class of set functions is submodularity: any real-valued set function $\psi\colon 2^{\mathcal{V}} \to \mathbb{R}$ such that $\psi(\emptyset) = 0$ is submodular (Fujishige 2005, Bach 2013) if and only if, for all subsets $\mathcal{A}, \mathcal{B} \subseteq \mathcal{V}$, we have

$$\psi(\mathcal{A}) + \psi(\mathcal{B}) \geq \psi(\mathcal{A} \cup \mathcal{B}) + \psi(\mathcal{A} \cap \mathcal{B}). \tag{4.2}$$

A function is *supermodular* if its negation is submodular, and it is *modular* if it is both supermodular and submodular. Many objective functions that arise in experimental design are naturally submodular. Consider, for instance, sensor placement where each sensor has a certain coverage area, and our interest is in optimizing the collective locations of a fixed number of sensors. There is a natural *diminishing returns* property that accompanies such objectives, which is characteristic of submodular functions. An alternative but equivalent (Fujishige 2005, Bach 2013) definition of submodularity using first-order differences highlights this *diminishing returns* property and is often easier to demonstrate in practice: the set function ψ is submodular if and only if, for all $\mathcal{A}, \mathcal{B} \subseteq \mathcal{V}$ and $v \in \mathcal{V}$ such that $\mathcal{A} \subseteq \mathcal{B}$ and $v \notin \mathcal{B}$, we have

$$\psi(\mathcal{A} \cup \{v\}) - \psi(\mathcal{A}) \geq \psi(\mathcal{B} \cup \{v\}) - \psi(\mathcal{B}).$$

The term $\rho_v(\mathcal{A}) := \psi(\mathcal{A} \cup \{v\}) - \psi(\mathcal{A})$ is the incremental gain associated with the element v when added to the set $\mathcal{A}$. It refers to the amount of change in the objective when the individual item v is added to an existing pool of items in set $\mathcal{A}$. The definition is applicable by extension when we add a set $\mathcal{B}$ to set $\mathcal{A}$.

In certain cases when the function is not strictly submodular, it is helpful to understand how much it deviates from submodularity, or in other circumstances to quantify how close a set function is to being modular. Such characterizations also prove useful in analysing the performance of many optimization algorithms. The notion of curvature introduced by Conforti and Cornuéjols (1984) in the context of non-negative set-functions is a bound on the intrinsic value of an item against its value in conjunction with all items in the candidate set. Formally, curvature is defined as the scalar c:

$$c := \max_{v \in \mathcal{V}} \frac{\rho_v(\emptyset) - \rho_v(\mathcal{V} \setminus \{v\})}{\rho_v(\emptyset)} = 1 - \min_{v \in \mathcal{V}} \frac{\rho_v(\mathcal{V} \setminus \{v\})}{\rho_v(\emptyset)}.$$

The classical notion of curvature (Conforti and Cornuéjols 1984) measures how close a submodular set function is to being modular, while the notion of generalized curvature (Bian, Buhmann, Krause and Tschiatschek 2017) measures how close a set function – not necessarily submodular – is to being supermodular. The submodularity ratio introduced by Das and Kempe (2011) for a general non-negative set function is a lower bound on the ratio of the sum of incremental gains associated with elements in a set, to the incremental gain associated with the set itself. Intuitively it captures 'how close' to submodularity is the function in question. Formally it is defined as the scalar γ:

$$\gamma := \min_{\mathcal{B} \subseteq \mathcal{V}, \mathcal{A} \cap \mathcal{B} = \emptyset} \frac{\sum_{v \in \mathcal{A}} \rho_v(\mathcal{B})}{\rho_{\mathcal{A}}(\mathcal{B})}.$$

In Bian *et al.* (2017) the submodularity ratio and the generalized curvature together quantify how close a set function is to being modular.

The counterpart of the submodularity ratio termed the supermodularity ratio (Tzoumas, Carlone, Pappas and Jadbabaie 2021, Bogunovic, Zhao and Cevher 2018, Karaca and Kamgarpour 2018, Jagalur-Mohan and Marzouk 2021) has also proved useful in analysing many algorithms. It is defined as the scalar η:

$$\eta := \min_{\mathcal{B} \subseteq \mathcal{V}, \mathcal{A} \cap \mathcal{B} = \emptyset} \frac{\rho_{\mathcal{A}}(\mathcal{B})}{\sum_{v \in \mathcal{A}} \rho_v(\mathcal{B})}.$$

In Jagalur-Mohan and Marzouk (2021) the product of supermodularity and submodularity ratios measures deviation from modularity. We will revisit these notions and their implications for performance guarantees in Section 4.2.3. We now discuss set function attributes of widely used experimental design criteria.

4.2.2. Set function attributes of design criteria

In Section 2.1.1 we considered several classical alphabetic optimality design criteria. The D-optimality criterion defined using the log determinant is submodular. Using strictly linear algebraic techniques, it can be shown that the log determinant of a principal submatrix is a submodular function with respect to the indices defining the submatrices (Gantmacher and Kreĭn 1960, Kotelyanskiĭ 1950, Fan 1967, 1968, Kelmans and Kimelfeld 1983, Johnson and Barrett 1985). Since the log determinant evaluations may be non-positive, however, most existing algorithms for submodular maximization are not directly applicable. The A-optimality criterion in general is not submodular, as shown in Krause *et al.* (2008). We can, however, qualify its non-submodular nature by bounding the submodularity ratio γ and curvature c for the Bayesian A-optimal design objective (Bian *et al.* 2017).

In Section 2.2 we introduced a variety of information-theoretic design criteria. Specifically, we defined the expected information gain (EIG) in parameters and showed that it is equivalent to the mutual information between the parameters and the observations given the design, $\mathcal{I}(Y;\Theta|\xi)$. Suppose that selecting a design corresponds to choosing individual components of the $\mathbb{R}^m$-valued random variable Y. Then we can recast the optimization problem as finding a selection operator $\mathcal{P} \in \mathbb{R}^{m\times n}$, $n < m$, $\mathcal{P} := [e_{i_1}, \dots, e_{i_n}]$, where e_{i_j} are distinct canonical unit vectors in $\mathbb{R}^m$:

$$\mathcal{P}_{\text{opt}} = \operatorname*{argmax}_{\mathcal{P}\in\mathbb{R}^{m\times n}} \mathcal{I}(\mathcal{P}^\top Y;\Theta). \tag{4.3}$$

Given a desired number of observations $n < m$, we seek a selection operator $\mathcal{P}_{\text{opt}}$ such that the mutual information between the inference parameter Θ and the selected observations $Y_{\mathcal{P}_{\text{opt}}} := \mathcal{P}_{\text{opt}}^\top Y$ is maximized. Jagalur-Mohan and Marzouk (2021) show that the mutual information between parameters Θ and subsets of data $Y_{\mathcal{P}}$ is a submodular function if the observations are conditionally independent. Interestingly, this property holds even when the underlying joint distribution is non-Gaussian. In the setting of Bayesian inverse problems with additive noise, correlated noise renders the mutual information design objective non-submodular. When the inverse problem is linear, it is possible to quantify the non-submodularity of the information-theoretic objective by bounding the submodularity ratio γ and supermodularity ratio η. It was shown in Jagalur-Mohan and Marzouk (2021) that those parameters can be both lower-bounded by $\log \zeta_{\min}/\log \zeta_{\max}$, where ζ is any generalized eigenvalue of the definite pair $(\Gamma_Y, \Gamma_{Y|\Theta})$.

4.2.3. Greedy algorithms for cardinality-constrained designs

The case of cardinality-constrained optimization commonly arises in OED, with the prototypical example being sensor placement. When the design objective is monotone and non-negative, the greedy heuristic of successively picking the candidate corresponding to the highest incremental gain performs well despite its simplicity. Nemhauser *et al.* (1978) were the first to analyse the greedy heuristic for the class of submodular functions, and showed that the algorithm has a constant factor $1 - 1/e$ approximation guarantee. Nemhauser and Wolsey (1978) further showed that the approximation guarantee cannot be improved in general by any other polynomial-time algorithm.

In the above paragraph, the term 'constant factor' refers to the approximation guarantee not depending on the particular instance of the function and only requiring the function to be submodular. By incorporating parameters that capture more specific attributes of the function, however, more expressive results can be obtained and offer useful insights into performance of the greedy heuristic in different scenarios. For instance, Conforti and Cornuéjols (1984) proved the more refined guarantee $\frac{1}{c}(1 - e^{-c})$ for submodular functions, where $c \in [0, 1]$ is the curvature. When the function is known to have a small curvature, the improved performance of the greedy heuristic is easily explained. Now suppose the function is not submodular but has a submodularity ratio $\gamma \in [0, 1]$; Das and Kempe (2011) proved that the greedy heuristic has a $1 - e^{-\gamma}$ approximation guarantee. Bian *et al.* (2017) improved upon that result by incorporating the notion of a generalized curvature $\alpha \in [0, 1]$ to obtain the factor $\frac{1}{\alpha}(1 - e^{-\alpha\gamma})$. As we alluded to in Section 4.2.2, these parameters can be concretely bounded for many experimental design objectives.

The greedy algorithm has a $O(mn)$ complexity, where m is the size of the candidate set and n is the desired cardinality. Closely related variants of the greedy heuristic can be better choices depending on the context and needs. Robertazzi and Schwartz (1989) explored an accelerated version wherein the computed incremental gains are stored and exploited in the successive step, possibly reducing the overall number of function evaluations. This is much like the modified Fedorov algorithm we discussed previously in Section 4.1.1. To reduce the complexity further and make it independent of the cardinality constraint, Mirzasoleiman *et al.* (2015) analyse a randomized version of the greedy heuristic, termed stochastic greedy. This algorithm achieves, in expectation, a $(1 - 1/e - \epsilon)$ approximation guarantee relative to the optimum solution. The number of function evaluations does not depend on the cardinality constraint, but linearly on the size of the candidate set, thus reducing the complexity substantially.

To benefit from modern HPC platforms, Mirzasoleiman, Karbasi, Sarkar and Krause (2013) proposed a two-stage parallellized version which reduces the number of function evaluations per parallel process. The approximation guarantee for the algorithm, however, in general depends on the size of the candidate set and cardinality constraint, which can only be overcome in special cases. Jagalur-Mohan and Marzouk (2021) analysed a batch version of stochastic and distributed greedy algorithms with applicability to non-submodular objectives. The heuristic involved choosing multiple candidates in each step but relying solely on the incremental gains associated with individual candidates. This reduces the computational overhead but with reduced approximation guarantees.

4.2.4. Beyond greedy: algorithms for more general design problems

As mentioned earlier in passing, many experimental design problems involve constraints more complex than simple cardinality bounds. Consider, for instance, sensor placement with non-uniform costs $c(v)$ associated with placing the sensors. Here the goal could be to maximize the design objective $\psi(\cdot)$ while ensuring that the cumulative cost is within a budget $\sum_{v\in\mathcal{S}} c(v) \leq b$:

$$\max_{\mathcal{S}\subset\mathcal{V}} \psi(\mathcal{S}) \quad \text{subject to} \quad \sum_{v\in\mathcal{S}} c(v) \leq b.$$

This is the well-known knapsack problem, where the goal is to maximize the set function ψ subject to a non-negative modular constraint. A natural modification of the greedy algorithm in this setting is to pick the element maximizing the benefit-to-cost ratio, at each step. Surprisingly, for the case when ψ is submodular, it can be shown that either the output of the standard greedy algorithm or the index set returned by the cost-benefit greedy algorithm will be within $(1 - 1/e)/2$ of the optimal solution (Leskovec *et al.* 2007). A stronger result $(1 - 1/e)$ is possible via a more computationally involved algorithm that exhaustively enumerates all subsets of size 3, and augments them using the cost-benefit greedy algorithm (Sviridenko 2004).

Although we exclusively focused on non-negative monotone functions in Section 4.2.3, several design criteria can be non-monotone. In Section 2.3.2 we discussed an information-theoretic design objective arising in sensor placement that is symmetric submodular but strictly speaking non-monotone: $\mathcal{I}(\Theta(x_s); \Theta(x_{s^c}))$. For the maximization of such submodular objectives with cardinality constraints, Buchbinder, Feldman, Naor and Schwartz (2014) have analysed discrete random greedy and continuous double greedy algorithms. We refer the interested readers to this work for more details on such methods.

An important tool in the field of submodular optimization is the use of *extensions*, particularly the multilinear extension (Vondrák 2008) in the context of maximization. The multilinear extension of the set function ψ is the function $\Psi\colon [0, 1]^m \to \mathbb{R}$ defined as

$$\Psi(\xi) = \sum_{\mathcal{S}\subseteq\mathcal{V}} \psi(\mathcal{S}) \prod_{i\in\mathcal{S}} \xi_i \prod_{i\in\mathcal{V}\setminus\mathcal{S}} (1 - \xi_i).$$

An intuitive interpretation of this extension is to think of the original set function as defined over the corners of the hypercube $\{0, 1\}^m$, while the extension is valid over the entire unit cube $[0, 1]^m$. $\Psi(\xi)$ is the expected value of ψ over sets, where for any set $\mathcal{S}$, each element i is included independently with probability ξ_i, and not included with probability $1 - \xi_i$. If we write $\mathcal{S} \sim \xi$ to indicate that $\mathcal{S}$ is the random subset sampled according to ξ, then the multilinear extension is simply $\Psi(\xi) = \mathbb{E}_{\mathcal{S}\sim\xi}[\psi(\mathcal{S})]$. The multilinear extension (Vondrák 2008) is quite different from the Lovász extension (Lovász 1983); the latter maps any discrete submodular function to its continuous convex counterpart while the former maps it to its continuous concave counterpart. Note that these continuous extensions are quite different from the convex relaxation formulations we described in Section 4.1.2! Many of the more general approaches to maximizing submodular functions under a wide class of constraints rely on the multilinear extension. The approach here involves first approximately solving the problem

$$\max \Psi(\xi) \quad \text{subject to} \quad \xi \in \mathcal{X} \subseteq [0,1]^m$$

and then rounding the continuous solution to obtain a near-optimal set (Vondrák 2010, Chekuri, Vondrák and Zenklusen 2014, Vondrák, Chekuri and Zenklusen 2011, Calinescu, Chekuri, Pál and Vondrák 2011, Sviridenko, Vondrák and Ward 2017). We refer the interested reader to the cited works for details of the algorithms. For a broader survey on submodular maximization, see Krause and Golovin (2014).

4.3. *Optimizing real-valued design variables*

In this section we consider OED problems where the design is naturally parametrized by real-valued coordinates. These coordinates could be, for example, the spatial locations of a finite collection of sensors, times at which measurements should be taken, the values of certain experimental controls (pressure, temperature), and so on. In Section 2.1.3 we formalized this continuous-parameter case by letting these coordinates be elements of a set $\mathcal{X} \subseteq \mathbb{R}^d$ and considering the design ξ to be a probability measure on $\mathcal{X}$. An exact n-point design can then be understood as a mixture of Dirac measures, $\xi = \frac{1}{n}\sum_{i=1}^{n} \delta_{x_i}$, where each $x_i \in \mathcal{X}$. Here we will focus on continuous optimization methods for such exact designs, and assume the number of support points n to be prescribed. The design measure ξ is thus entirely parametrized by $(x_1, \dots, x_n) \in \bigtimes_{i=1}^{n} \mathcal{X} \subseteq \mathbb{R}^D$, where $D = nd$. To keep notation intuitive in this setting, we will *elide* the notions of the design measure and its parametrization and simply write $\xi = (x_1, \dots, x_n)$. In other words, throughout this section we will consider ξ to be a vector in $\mathcal{X}^n = \bigtimes_{i=1}^{n} \mathcal{X} = \Xi$. This way, the design remains exact, and gradients with respect to ξ are straightforward to define.

Of course, one could discretize $\mathcal{X}^n$ and employ approaches from Section 4.2 to find a design within this discretized space. This approach may be convenient when dealing with complicated constraints on the set of feasible designs or when the number of desired experiments n is not fixed (as in the case of knapsack constraints). For any naïve discretization of $\mathcal{X}^n$, however, the number of candidate designs would grow exponentially with nd. Instead, we will focus here on optimization algorithms that make use of a continuous (Euclidean) parametrization of ξ.

When the design criterion can be evaluated exactly (i.e. in closed form) for any given ξ, the problem of optimizing over a real-valued ξ in some sense becomes standard – though of course it inherits the natural difficulties associated with how the parametrized design enters the objective, or with the geometry of the feasible domain $\mathcal{X}$ (e.g. nonlinear constraints, non-convexity). These issues are generally quite problem-dependent, however, and are not specific to OED.

The decision-theoretic OED objectives formulated in Section 2.2, on the other hand, do raise certain cross-cutting challenges for optimization. The expected utility U, which is the objective function to be maximized, typically must be estimated using Monte Carlo techniques. The optimal design problem can thus be written as

$$\xi^* \in \arg\max_{\xi \in \Xi} \mathbb{E}_{W \mid \xi}[\widehat{U}(\xi, W)], \tag{4.4}$$

where $\widehat{U}$ is an estimator of U, and the random variable W is the source of stochasticity in this estimator. For example, the Monte Carlo estimator of a general expected utility U given in (3.1),

$$\widehat{U}(\xi, W) = \frac{1}{N}\sum_{i=1}^{N} u(\xi, Y^{(i)}, \Theta^{(i)}), \tag{4.5}$$

with $\Theta^{(i)} \sim p(\theta)$ and $Y^{(i)} \sim p(y|\theta^{(i)}, \xi)$, has $W = (Y^{(i)}, \Theta^{(i)})_{i=1}^N$. If, at any given design ξ, all we can compute are estimates $\widehat{U}(\xi, W)$ for different realizations of $W \sim p(W|\xi)$, the optimization problem (4.4) is essentially a *stochastic approximation* (SA) problem.

Implicit in writing (4.4) is the assumption that we are content to maximize $\mathbb{E}[\widehat{U}]$: that is, either $\widehat{U}(\xi, W)$ is an unbiased estimator of the desired utility $U(\xi)$ at any ξ or the bias of $\mathbb{E}[\widehat{U}] - U$ is small and acceptable to the experimenter. In Sections 4.3.3 and 4.3.4 we will discuss additional techniques that can be deployed when this is not the case, as in the case of mutual information maximization with biased nested Monte Carlo estimators or variational bounds. Starting below, however, we will use the notation $\bar{U}(\xi) := \mathbb{E}_{W \mid \xi}[\widehat{U}(\xi, W)]$ for the objective in (4.4) to make the possible presence of bias clear.

Many nonlinear optimization methods can be applied to (4.4), and we do not attempt to survey this vast literature here. Instead, we will briefly recall a number of derivative-free and gradient-based approaches that have been used in the context of OED with continuous design variables. We call a method 'derivative-free' if it only requires evaluations of $\widehat{U}$ (even if these evaluations are used to estimate derivatives) and we call a method 'gradient-based' if it requires additional derivative information as an input, e.g. unbiased estimates of the gradient of $\bar{U}$ with respect to ξ. In both cases, we assume that the objective $\bar{U}$ is differentiable, but we make no assumptions on convexity or other structure. We also note that estimating $\nabla_\xi \bar{U}$ may, in many cases, require evaluating derivatives of the log-likelihood or an underlying simulation model with respect to ξ, and that this task may not be straightforward for PDE-based forward operators or intractable likelihoods.

4.3.1. *Derivative-free methods*

Larson, Menickelly and Wild (2019) provide a comprehensive survey of derivative-free optimization methods, focusing on local optimization, with methods for solving stochastic problems of the form (4.4) specifically discussed in Larson *et al.* (2019, Section 6). A key desideratum for algorithms for this setting is that they perform well with *noisy* objective evaluations. Below we briefly point out several classes of derivative-free optimization methods that do exactly this, and that are therefore useful for OED. We make no attempt to be comprehensive, and instead refer the reader to the preceding survey for more information.

Bayesian optimization. We introduced Bayesian optimization (BO) in Section 2.4 to elucidate its differences from OED for Gaussian processes. But BO can be quite useful as a tool *within* OED – specifically, as a means of solving (4.4) when ξ is of moderate dimension. BO (Močkus 1975, Jones *et al.* 1998, Wang *et al.* 2023) is essentially a derivative-free algorithm for global optimization, well suited to 'black-box' objective functions that are expensive to evaluate and potentially noisy. BO uses Gaussian process regression to construct and refine a model for the objective (in the case of OED, the function $\bar{U}$); this regression naturally handles noisy pointwise evaluations and smooths the underlying function estimate. Both the level of noise and the smoothness of the reproducing kernel Hilbert space (RKHS) containing this estimate (the posterior mean of the Gaussian process) can be adjusted by learning hyperparameters of the covariance function of the Gaussian process. Examples of the application of BO to OED include, among others, works by Weaver, Williams, Anderson-Cook and Higdon (2016), Overstall and Woods (2017), Xu and Liao (2020) and Zhong *et al.* (2024). For more information on BO, we refer to the surveys by Shahriari *et al.* (2016) and Frazier (2018).

Stochastic approximation methods based on finite differences. A prototypical deterministic optimization method is gradient ascent:

$$\xi_{k+1} = \xi_k + \alpha_k g(\xi_k), \tag{4.6}$$

where $g(\xi_k) := \nabla_\xi \bar{U}(\xi_k)$ is the gradient of the objective $\bar{U}(\xi) = \mathbb{E}_{W|\xi}[\widehat{U}(\xi, W)]$ evaluated at ξ_k, and $\{\alpha_k\}$ is a sequence of scalar step sizes with $\alpha_k > 0$. The Kiefer–Wolfowitz algorithm (Kiefer and Wolfowitz 1952) estimates $g(\xi_k)$ by applying centred differences to unbiased estimates of $\bar{U}$. Specifically, for the ith component of g, where $i = 1 \ldots D$, we have

$$g_i(\xi_k) \approx \frac{\widehat{U}(\xi_k + e_i\Delta_k, w_i^+) - \widehat{U}(\xi_k - e_i\Delta_k, w_i^-)}{2\Delta_k}, \tag{4.7}$$

where w_i^+ and w_i^- are independent draws from $p_{W|\xi_k}$, e_i is the unit vector in coordinate i, and the scalar $\Delta_k > 0$ is a difference parameter. Hence we need to evaluate the estimator $\widehat{U}$ $2D$ times to compute each gradient estimate. Under appropriate conditions on the step size sequence $\{\alpha_k\}$ and difference sequence $\{\Delta_k\}$, the objective function $\bar{U}$ and the noise W, Kiefer–Wolfowitz iterations converge to a first-order critical point of $\bar{U}$ almost surely; see e.g. Blum (1954) and Bhatnagar, Prasad and Prashanth (2013). Simultaneous perturbation stochastic approximation (SPSA) (Spall 1998*a*,*b*) is similar in form to Kiefer–Wolfowitz, except that it always uses a single centred difference at each iteration, the direction of which is chosen from a particular probability distribution. SPSA thus uses only *two* independent realizations of $\widehat{U}$ at each step k, independent of the dimension of ξ. An intuitive justification for SPSA is that error in the gradient produced by restricting the direction of the finite differences 'averages out' over a large number of iterations (Spall 1998*b*).

Model-based methods. Model-based methods for derivative-free optimization use pointwise evaluations of the objective function (and constraints, if applicable) to form local approximations (called models) that can be analysed to produce the next optimization step. There is an extensive literature on these methods, with many effective algorithms; for thorough reviews, see Larson *et al.* (2019) and Conn, Scheinberg and Vicente (2009). Local models can take the form of low-order polynomials or radial basis function approximations, constructed via interpolation or regression on carefully chosen point sets. Updating of these models is often set within a trust region framework that carefully manages the point sets, the trust region radius, and the acceptance of each optimization step. Rigorous convergence guarantees have been developed for most prevalent algorithms (Larson *et al.* 2019).

Model-based derivative-free optimization has been applied in settings where only noisy estimates $\widehat{U}$ of the objective are available. The models fit by these methods (e.g. via local regression) are stochastic, and hence certain modifications to the algorithms and certainly to their convergence analyses are required, as explained in Larson *et al.* (2019, Section 6.3). On the algorithmic side, one modification that has been proposed is to introduce adaptive schemes for Monte Carlo sampling that balance noise in $\widehat{U}$ with other errors (Shashaani, Hashemi and Pasupathy 2018).

Direct search methods. Direct search methods (Torczon 1997, Larson *et al.* 2019) are also good choices for (4.4). In general, these methods do not explicitly build local approximations of the objective or attempt to approximate its gradient, but rather compute optimization steps based on the *relative ordering* of the evaluated function values. A classical pattern search approach is the Nelder–Mead nonlinear simplex (Nelder and Mead 1965), which has been explored for OED by Huan and Marzouk (2013). Modifications to Nelder–Mead for strongly noisy objectives include re-sampling schemes to account for the impact of noise on rankings (Chang 2012). Other direct search methods, such as generalized pattern search methods (Audet and Dennis 2002, Audet 2004), have variants for stochastic objectives as well (Sriver, Chrissis and Abramson 2009).

4.3.2. *Gradient-based methods*

Robbins–Monro. The Robbins–Monro (RM) algorithm (Robbins and Monro 1951), the progenitor of modern stochastic gradient descent algorithms, follows (4.6) but requires access to an unbiased estimator $\hat{g}$ of the gradient $\nabla_\xi \bar{U}$. That is, we need a random $\hat{g}$ satisfying

$$\mathbb{E}[\hat{g}(\xi)] = g(\xi) := \nabla_\xi \bar{U}(\xi).$$

One way of constructing such an estimator is to reparametrize $\widehat{U}$ such that the distribution over which we take the expectation is functionally independent of the design variable ξ. Such a reparametrization is always feasible (Mohamed, Rosca, Figurnov and Mnih 2020). For example, in (4.5) where $W = (Y^{(i)}, \Theta^{(i)})_{i=1}^N$, we initially have $\mathbb{E}_{W|\xi} = \mathbb{E}_{Y^{(1:N)}|\theta^{(1:N)},\xi}\,\mathbb{E}_{\Theta^{(1:N)}}$, but it is easy to replace the design-dependent observations Y in this expectation with random variables that do not depend on ξ. If the observations are described as $Y = G_\xi(\theta) + \mathcal{E}$, for instance, with an observation noise $\mathcal{E}$ whose distribution is independent of the design, this goal is immediately achieved: we can replace the original expectation with $\mathbb{E}_{\mathcal{E}^{(1:N)}}\,\mathbb{E}_{\Theta^{(1:N)}}$ and reparametrize the expected utility estimator accordingly. More fundamentally, however, the source of randomness in any statistical model is always a transformation of some *independent* random input. Call this random input $\check{W} \equiv (\check{W}^{(i)})_{i=1}^N \sim p_{\check{W}}$. (In the example we just raised, $\check{W}^{(i)} = (\mathcal{E}^{(i)}, \Theta^{(i)})$.) After reparametrization, the estimator (4.5) of the expected utility is rewritten as

$$\check{U}(\xi, \check{W}) = \frac{1}{N}\sum_{i=1}^{N} \check{u}(\xi, \check{W}^{(i)}), \tag{4.8}$$

and satisfies $\mathbb{E}_{\check{W}}[\check{U}(\xi, \check{W})] = \mathbb{E}_{W|\xi}[\widehat{U}(\xi, W)] = \bar{U}(\xi)$. Since $\check{W}$ is independent of ξ, we can then write

$$\nabla_\xi \bar{U}(\xi) = \nabla_\xi \mathbb{E}_{\check{W}}[\check{U}(\xi, \check{W})] = \mathbb{E}_{\check{W}}[\nabla_\xi \check{U}(\xi, \check{W})] = \mathbb{E}_{\check{W}}[\hat{g}(\xi, \check{W})],$$

where

$$\hat{g}(\xi, \check{W}) := \nabla_\xi \check{U}(\xi, \check{W}) = \frac{1}{N}\sum_{i=1}^{N} \nabla_\xi \check{u}(\xi, \check{W}^{(i)}) \tag{4.9}$$

is hence an unbiased estimator of the gradient.

Under appropriate conditions on the objective, the noise and the step size sequence $\{\alpha_k\}$, it can be shown that RM converges almost surely to the global optimum of $\bar{U}$ in convex problems (Kushner and Yin 2003) or to the first-order critical set in certain non-convex problems (Mertikopoulos, Hallak, Kavis and Cevher 2020). Many improvements on the original RM algorithm have been devised and widely deployed, including Polyak–Ruppert averaging (Ruppert 1988, Polyak and Juditsky 1992), which can improve robustness and convergence rates by averaging iterates along the optimization path, and Nesterov acceleration (Nesterov 1983). RM-type algorithms have seen extensive use in OED, for example by Huan and Marzouk (2014), Foster *et al.* (2019), Carlon *et al.* (2020), Foster *et al.* (2020), Foster, Ivanova, Malik and Rainforth (2021), Kleinegesse and Gutmann (2020), Ivanova *et al.* (2021), Shen and Huan (2021), Zhang, Bi and Zhang (2021) and Shen and Huan (2023).

Sample-average approximation. Sample-average approximation (SAA) (also referred to as the retrospective method or the sample-path method) (Shapiro 1991,

Healy and Schruben 1991, Gürkan, Özge and Robinson 1994, Kleywegt, Shapiro and Homem-de Mello 2002) takes a different strategy: it seeks to reduce the stochastic objective to a deterministic one by fixing the randomness throughout the entire optimization process. This again requires reparametrizing the distribution over which we take the expectation (i.e. $\mathbb{E}_{W|\xi}$) to be independent of the design variable ξ, as in the Monte Carlo estimator $\check{U}$ of (4.8). The SAA problem then becomes

$$\xi^*_{\text{SAA}} \in \arg\max_{\xi \in \Xi} \check{U}(\xi, \check{W} = w_s), \tag{4.10}$$

where w_s is a realization of $\check{W}$ that is *fixed* across Ξ. SAA can also be viewed as an application of common random numbers, where holding the sample of $\check{W}$ fixed essentially correlates the estimates of $\bar{U}$ across the design space and yields a smoother objective surface. (For further discussion of the relationships between common random numbers, stochastic finite difference approximations to the gradient as in (4.7) and stochastic gradient estimates as in (4.9), see Asmussen and Glynn (2007, Chapter VII.2).) The optimization problem (4.10) is *deterministic*, and hence any standard deterministic optimization algorithm can be adopted. For instance, Huan and Marzouk (2014) applied SAA to a nonlinear OED problem with an EIG objective, and used the Broyden–Fletcher–Goldfarb–Shanno (BFGS) quasi-Newton scheme (Nocedal and Wright 2006, Chapter 6) to find optimal designs. Stochastic bounds on the optimality gap $\bar{U}(\xi^*) - \bar{U}(\xi^*_{\text{SAA}})$ can also be obtained (Norkin, Pflug and Ruszczynski 1998, Mak, Morton and Wood 1999).

4.3.3. *Unbiased EIG gradient estimators*

As discussed near the start of Section 4.3, the algorithms described thus far focus on the objective $\bar{U}$; in other words, they assume that we have an unbiased estimator of the desired objective at any ξ (or, practically speaking, that the bias is 'small enough' to be ignored). Section 4.3.2 then describes how to obtain an unbiased estimator of the gradient of $\bar{U}$. But we have not yet addressed what to do when $\bar{U}$ departs from the true design objective U.

For the mutual information (EIG) objectives discussed throughout Section 3, we do not have unbiased estimators; this is due to the presence of nested expectations and hence nested Monte Carlo estimators, or the use of density or density-ratio estimators within a nonlinear function (i.e. the logarithm). Yet directly applying any of the optimization methods that we have just discussed to a biased estimator of EIG can lead to arbitrarily large departures from the true maximizer (recall the example of Figure 3.1), as the objectives at hand may have multiple local maxima over Ξ and the bias itself may vary significantly as a function of ξ. Similarly, stochastic gradient-based methods will converge to critical points specified by the biased gradient, i.e. points where $\nabla_\xi \bar{U}(\xi) = 0$, which again differ from points where $\nabla_\xi U(\xi) = 0$ if we cannot enforce $\nabla_\xi \bar{U}(\xi) = \nabla_\xi U(\xi)$ for all ξ.

To remedy this situation for gradient-based optimization methods, Goda, Hironaka, Kitade and Foster (2022) propose an unbiased estimator of the gradient of the EIG (2.14) with respect to ξ, i.e. of $\nabla_\xi U_{\mathrm{KL}}(\xi)$. Their approach builds on the multilevel nested Monte Carlo estimator of Goda *et al.* (2020) by combining it with the random truncation approach of Rhee and Glynn (2015), which generally aims to 'de-bias' a consistent sequence of estimators. Goda *et al.* (2022) use the Rhee and Glynn scheme to randomly truncate the infinite telescoping sum in the inner loop of nested Monte Carlo, in a way that guarantees unbiasedness of the overall gradient estimator, allowing it to satisfy the requirements of the RM algorithm.

A different estimator of $\nabla_\xi U_{\mathrm{KL}}(\xi)$ is proposed by Ao and Li (2024). Here, the authors first use the reparametrization trick to express the observations Y as a function of the parameters Θ and additional design-independent random variables $\mathcal{E}$, i.e. as $Y = h(\Theta, \mathcal{E}, \xi)$, and then apply the following identity to rewrite the gradient of the log-evidence term in (3.9),

$$\nabla_\xi \log p_{Y|\xi}(y^{(i)}|\xi) = -\mathbb{E}_{\Theta|y^{(i)},\xi}\left[\nabla_\xi \log p_{Y|\Theta,\xi}\left(h(\theta^{(i)}, \mathcal{E}^{(i)}, \xi) \mid \Theta, \xi\right)\right],$$

where $y^{(i)} = h(\theta^{(i)}, \mathcal{E}^{(i)}, \xi)$ is any realization of Y. MCMC sampling from the posterior $p(\theta|y^{(i)}, \xi)$ yields an estimate of $\nabla_\xi \log p(y^{(i)}|\xi)$, which is then embedded in an outer loop of standard Monte Carlo sampling over the joint distribution of Θ and $\mathcal{E}$. (One must also estimate the gradient of the expected log-likelihood term in (3.9), but this is straightforward to do in an unbiased way using the reparametrization trick.) In so far as the inner posterior samples are exact, the resulting nested estimator of $\nabla_\xi U_{\mathrm{KL}}(\xi)$ is unbiased. Of course, finite-length MCMC chains produce biased estimates of the associated expectations, but the authors demonstrate empirically that this bias can be made negligible. Moreover, other posterior sampling schemes could be used instead.

4.3.4. *Simultaneous bound tightening and design optimization*

Focusing again on design objectives that involve EIG in parameters, predictions or some other quantity of interest, an important theme of Sections 3.2 and 3.3 was the construction of variational lower bounds for the EIG, i.e. functions $\mathcal{L}(f, \xi) \leq \mathrm{EIG}(\xi)$. Most of these lower bounds (with some exceptions, e.g. $\mathcal{L}^{\mathrm{PCE}}$ (3.26)) are parametrized by learnable functions f, which could take the form of probability densities, transport maps or more generic 'critic' functions. Moreover, most of these learnable bounds can become tight for an appropriate choice of f (again with some exceptions, e.g. certain contrastive multi-sample bounds in Section 3.3.2). In these cases, it is then natural to maximimize *simultaneously* over f and ξ, as proposed in Foster *et al.* (2020) and Kleinegesse and Gutmann (2020, 2021):

$$\xi^* \in \operatorname*{argmax}_{\xi \in \Xi} \max_f \mathcal{L}(f, \xi). \tag{4.11}$$

This approach simultaneously seeks to tighten the lower bound and to find a good design. The maximizer should in principle be a maximizer of the true EIG.

Recall that all of the variational bounds $\mathcal{L}$ were expressed as expectations. To apply a stochastic gradient technique to (4.11), unbiased estimates of gradients $\nabla_{\xi}\mathcal{L}(f,\xi)$ and $\nabla_{f}\mathcal{L}(f,\xi)$ are required. As noted in Section 3.3.3, these estimates are generally available through simple Monte Carlo estimation, without the need for nested Monte Carlo or other complexities. Simultaneous maximization (4.11) via stochastic gradient techniques has been used in the setting of batch OED by Foster *et al.* (2020), Kleinegesse and Gutmann (2020, 2021) and Zhang *et al.* (2021), and in the setting of sequential experimental design by Ivanova *et al.* (2021) and Shen, Dong and Huan (2023), which we will discuss in the next section.

A related approach to leveraging bounds on EIG is proposed by Zheng, Hayden, Pacheco and Fisher (2020). Here, the authors take advantage of the ability to obtain *both* lower and upper bounds (in expectation) for the EIG – e.g. the prior-contrastive estimator (3.26) as the lower bound and the standard NMC estimator (3.2) as the upper bound. Both of these bounds can be refined and made tight by increasing the inner-loop sample size M. Zheng *et al.* (2020) thus develop an adaptive refinement strategy, in which the bounds are tightened over the course of optimization, that achieves regret-style guarantees.

5. Sequential optimal experimental design

Sequential experimental design is concerned with the planning of multiple experiments that are conducted in a sequence, where the *results* of previous experiments in the sequence can inform the design of subsequent experiments. This is the crucial difference between sequential design and the ‘batch’ (also called ‘static’) design approaches that were the focus of previous sections. Here we will focus on a Bayesian approach to sequential experimental design.

One straightforward sequential design procedure is to apply the batch optimal experimental design (OED) framework and methods from Sections 2–4 to one experiment, or subset of experiments, at a time: optimally choose the next design, perform that experiment, update the prior to the posterior based on the outcome of the experiment, and repeat the process for the next. Such a procedure is called *greedy* or *myopic*, because it does not take into account future experiments when finding the (immediate) next experiment; it involves no ‘lookahead’. Yet greedy design is conceptually simple to implement, especially if computational tools for batch design already exist, and it is flexible for situations where the total number of desired experiments is unknown. As a result, a large body of sequential experimental design research has been based on some form of greedy design (Box 1992, Dror and Steinberg 2008, Cavagnaro *et al.* 2010, Solonen, Haario and Laine 2012, Drovandi, McGree and Pettitt 2013, Drovandi *et al.* 2014, Kim *et al.* 2014, Hainy, Drovandi and McGree 2016, Kleinegesse, Drovandi and Gutmann 2021).

Of course, to improve coordination among the experiments, one could design *all* the experiments simultaneously and thus revert to an overall batch design. But doing so would forgo the opportunity to adapt to new observations; in other words, it would allow for coordination among the experiments, but *no feedback*. A hallmark of sequential experimental design is precisely the idea of feedback.

We note that the greedy design approach described in this section differs from the greedy combinatorial algorithms for solving batch design problems discussed in Section 4. In particular, the latter are not for sequential experimental design, as they do not involve observing the results of experiments between design decisions, and thus do not incorporate any feedback. Rather, they still seek a static design for a single batch of experiments, but break the search into stages to control the dimension of the design space and avoid combinatorial scaling of computational complexity.

In the remainder of this section, we will present a sequential optimal experimental design formulation that includes both lookahead and feedback; see e.g. Müller *et al.* (2007), von Toussaint (2011, VII.G) and Huan (2015, Chapter 3). We refer to this formulation as *sequential OED* (sOED). sOED generalizes both greedy and batch design (Shen and Huan 2023, Section 2.3). The key ideas of sOED are that (i) each design should be selected while taking into consideration *all remaining experiments* to be performed, and (ii) designs should be given in the form of *functions*, called policies, that adaptively specify the next experiment in the sequence given the current state of information. We will formalize these ideas using the framework of Markov decision processes, to be presented shortly.

As we shall see, the numerical solution of the sOED problem is rather challenging, and there have been relatively few attempts to solve it in great generality. For example, Carlin, Kadane and Gelfand (1998), Gautier and Pronzato (2000), Pronzato and Thierry (2002), Brockwell and Kadane (2003), Christen and Nakamura (2003), Murphy (2003), Wathen and Christen (2006), Müller, Duan and Garcia Tec (2022) and Tec, Duan and Müller (2023) have all made advances largely limited to discrete settings, or did not employ a Bayesian framework with information-theoretic design criteria. In this section we will survey recent progress towards realizing fully Bayesian sOED, including methods that make use of dynamic programming (Huan 2015, Huan and Marzouk 2016), reinforcement learning (Blau, Bonilla, Chades and Dezfouli 2022, Shen and Huan 2023) and information bounds (Foster *et al.* 2021, Ivanova *et al.* 2021, Shen *et al.* 2023).

5.1. *Background*

We focus on the design of a finite number of experiments, indexed by $k = 0, 1, \dots, N-1$. We assume N is known and fixed. The Bayesian update for the kth experiment then becomes

$$p(\theta|y_k, \xi_k, I_k) = \frac{p(y_k|\theta, \xi_k, I_k)\, p(\theta|I_k)}{p(y_k|\xi_k, I_k)}, \tag{5.1}$$

where ξ_k and y_k, respectively, are the design and the value of the observation realized in the kth experiment, and $I_k := [\xi_0, y_0, \dots, \xi_{k-1}, y_{k-1}]$ is the 'background information' sequence composed of the history of designs and observations from all experiments preceding the current one, and $I_0 = \emptyset$. The prior density $p(\theta|I_k)$ represents the state of knowledge about the uncertain parameters Θ before the kth experiment, and the posterior density $p(\theta|y_k, \xi_k, I_k)$ represents the updated state of knowledge after having observed the outcome of the kth experiment. Equation (5.1) uses the simplification $p(\theta|\xi_k, I_k) = p(\theta|I_k)$, since the prior density does not depend on the pending choice of design. The posterior after the kth experiment $p(\theta|y_k, \xi_k, I_k) = p(\theta|I_{k+1})$ then becomes the prior for the $(k+1)$th experiment, and (5.1) can be applied recursively. In (5.1) we present the general setting where the density of the current data $p(y_k|\theta, \xi_k, I_k)$ may depend on the design and observations from previous experiments, i.e. on I_k. In many situations, however, Y_k is conditionally independent of the other observations in the sequence given θ and ξ_k, in which case its density simplifies to $p(y_k|\theta, \xi_k, I_k) = p(y_k|\theta, \xi_k)$.

We note that the experiments in the sequence do not need be of the same type, as long as they share the same parameter Θ. The spaces of possible designs $\Xi_k \ni \xi_k$ and observations $\mathcal{Y}_k \ni y_k$ can differ from one experiment to the next, and even change dimension. Similarly, the conditional density of the data in (5.1) is, in more explicit notation, $p_{Y_k|\Theta,\xi_k,I_k}(y_k|\theta, \xi_k, I_k)$ and thus has a k dependence as well. For example, in learning the properties of a fluid, a first experiment might entail selecting the shape of an obstacle to place in the flow and observing the velocity in its wake, while the next experiment might involve choosing where on the surface of this obstacle to place a pressure sensor.

5.2. *Formulation as a Markov decision process*

Sequential experimental design can be modelled through a Markov decision process (MDP) defined by a tuple $(\mathcal{S}, \{\mathcal{A}_k\}_k, s_0, \{r_k(\cdot)\}_k, \{T_k(\cdot)\}_k)$ consisting of a state space $\mathcal{S}$ with states $s_k \in \mathcal{S}$, action spaces $\mathcal{A}_k$ comprising possible actions (which here are designs) $\xi_k \in \mathcal{A}_k$,[5] an initial state s_0, scalar-valued reward functions $r_k(s_k, \xi_k, y_k)$ that evaluate the instantaneous reward when taking action ξ_k and observing y_k at state s_k, and state transition kernels $T_k(\mathcal{S}_{k+1}|s_k, \xi_k)$ that evaluate the probability of transitioning to any set of states $\mathcal{S}_{k+1} \subseteq \mathcal{S}$ at stage $k+1$ having taken action ξ_k at state s_k. In the context of experimental design, the action being taken is the selection of a design; thus we use the terms 'action' and 'design' interchangeably.

State. The state of the system before the kth experiment is described by $s_k = \{s_k^b, s_k^p\}$, a quantity that summarizes all information deemed relevant to future design decisions. We split s_k into a 'belief state' s_k^b, representing the state of knowledge/uncertainty in Θ, and a 'physical state' s_k^p comprising any other deterministic

[5] $\mathcal{A}_k \equiv \Xi_k$, i.e. spaces of candidate designs indexed by k, but we use $\mathcal{A}_k$ in this section to be consistent with the usual notation in the MDP literature.

variables that may be relevant to the design process. Since Θ is not directly observed and can only be inferred from observations Y_k, this set-up can be viewed as a partially observed MDP (POMDP) on Θ or a belief-MDP on s_k (Kaelbling, Littman and Cassandra 1998).

The belief state is simply the posterior distribution of Θ, given all past experimental designs and realized observations, I_k. For Θ taking values on $\mathbb{R}^p$, the belief state s_k^b is thus a probability distribution on $\mathbb{R}^p$. Numerically, it can be represented by, for example, a density function approximation, or a set of weighted particles, or by tracking I_k directly. Tracking I_k is easiest to implement since it does not require additional calculations translating $(\xi_i)_{i<k}$ and $(y_i)_{i<k}$ to another representation, but the dimension of I_k grows with k, though it is bounded for finite N. In general, the set of possible posterior distributions that can be realized is uncountably infinite,[6] and hence this setting differs from a discrete or finite-state system.

Maintaining only a belief state, in the form of the posterior, does not suffice to preserve the Markov property of the system if the likelihood depends on the history of past experiments I_k, as presented in (5.1). This can be fixed by introducing a physical state. With regard to the I_k-dependence in $p(y_k|\theta, \xi_k, I_k)$, the physical state essentially extracts and tracks relevant features from I_k that allow the likelihood to be evaluated or the observations Y_k simulated. An example of the physical state would be the known position of a mobile sensor platform, which might evolve from one stage of the design sequence to the next. Note that if I_k is adopted as the belief state, then information about the physical state is already contained in I_k, even if only implicitly.

Action (design) and policy. Sequential experimental design is adaptive in nature. Whereas a batch design problem seeks a single design ξ (see Section 2), sequential design now looks for a strategy, called a *policy*, describing *how to choose* the design depending on the current state. The policy is a collection of functions $\pi = \{\mu_k : \mathcal{S} \to \mathcal{A}_k, k = 0, \ldots, N-1\}$, where the policy function μ_k returns the design for the kth experiment given the current state, $\xi_k = \mu_k(s_k)$. In general, μ_k differs from experiment to experiment for finite N.[7] Some intuition for this fact follows by considering that even when starting from the same belief and physical state, a 'good' design can be quite different depending on how many experiments remain in the overall sequence. We focus on deterministic policies in this formulation, although stochastic policies that evaluate the probability of choosing different candidate designs can also be adopted.

[6] Unless both ξ_k and Y_k are discrete.

[7] In contrast, in the infinite-horizon setting, the policy must be stationary, i.e. independent of k. Thus only a single μ needs to be found, making infinite-horizon problems generally easier to tackle.

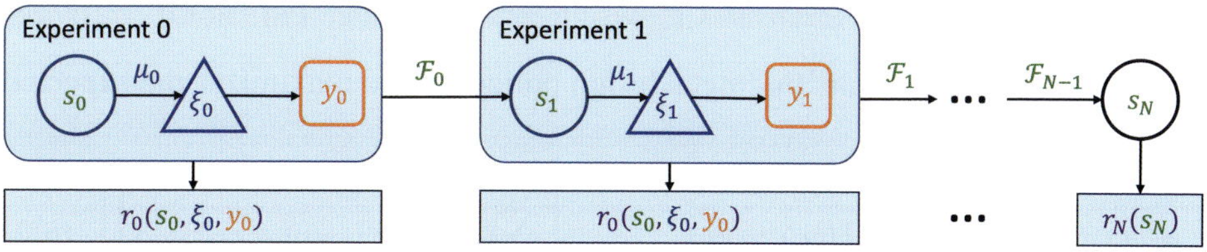

Figure 5.1. In the MDP progression of sequential experimental design, we start with an initial state s_0 (i.e. initial prior and physical state), evaluate the policy function at the state $\mu_0(s_0)$ to obtain the design ξ_0 for experiment 0, conduct the experiment to obtain its outcome y_0 and immediate reward $r_0(s_0, \xi_0, y_0)$, update the state to the new state via the transition dynamics $s_1 = \mathcal{F}_0(s_0, \xi_0, y_0)$ (i.e. updated posterior and physical state) and repeat for the next experiments. Once the last experiment $N - 1$ is completed, the terminal state $s_N = \mathcal{F}_{N-1}(s_{N-1}, \xi_{N-1}, y_{N-1})$ can be computed along with the corresponding terminal reward $r_N(s_N)$. Figure adapted from Shen and Huan (2023).

State transition dynamics. When an experiment is performed, the state changes according to a transition kernel $T_k(\mathcal{S}_{k+1}|s_k, \xi_k)$ describing the probability of transitioning from the current state s_k, having chosen design ξ_k and observed the outcome of resulting experiment, to any set of states at stage $k + 1$, $\mathcal{S}_{k+1} \subseteq \mathcal{S}$. This kernel is generally intractable to evaluate, but it can instead be *simulated*, by sampling from the prior predictive distribution of y_k given the design ξ_k and then applying Bayes' rule (5.1). Of course, both of these steps make use of the statistical model for y_k. We denote the latter transition dynamics by $s_{k+1} = \mathcal{F}_k(s_k, \xi_k, y_k)$; the function $\mathcal{F}_k$ encapsulates the transition from prior to posterior, given values of ξ_k and the realized data y_k, following (5.1).

The physical state, if present, evolves according to a model for the relevant physical process.

Reward (utility). Here $r_k(s_k, \xi_k, y_k)$ denotes the real-valued reward immediately obtained after performing the kth experiment, which may depend on the state, design and observation values. For example, the reward can reflect the cost of carrying out the kth experiment and/or the information gained in Θ as a result. On the other hand, $r_N(s_N)$ denotes the terminal reward, which reflects any rewards that can only be quantified after all experiments are completed. The choice of reward functions $(r_k)_{k=0}^{N}$ is quite flexible and can be based on the wide range of optimal design utilities discussed in Section 2.2. In the next section we will highlight the reward functions corresponding to u^{div} (2.20), which reflects information gain in the model parameters Θ.

The overall progression of an MDP for the sequential design of N experiments is depicted in Figure 5.1, adapted from Shen and Huan (2023).

5.3. Problem statement

The sOED problem entails finding a design policy π^* that maximizes the expected utility $U(\pi)$:

$$\pi^* \in \underset{\pi=\{\mu_0,\ldots,\mu_{N-1}\}}{\operatorname{argmax}} \left\{ U(\pi) := \mathbb{E}_{Y_{0:N-1}|\pi,s_0}\left[\sum_{k=0}^{N-1} r_k(s_k,\xi_k,y_k) + r_N(s_N) \right] \right\} \tag{5.2}$$
$$\text{subject to} \quad \xi_k = \mu_k(s_k) \in \mathcal{A}_k,$$
$$s_{k+1} = \mathcal{F}_k(s_k,\xi_k,y_k), \quad \text{for } k = 0,\ldots,N-1.$$

Here $Y_{0:N-1} = Y_0, Y_1, \ldots, Y_{N-1}$. The initial state s_0 is assumed known; if it is not known, then another expectation can be taken over s_0. Optionally, a discount factor $\gamma \in (0, 1]$ can multiply the reward at each successive stage (hence yielding terms $\gamma^k r_k$) to artificially reduce the value of rewards obtained further in the future. The expected utility here is also known as the *expected return* (or expected total reward) in MDP terminology; we will use these terms interchangeably.

In the field of reinforcement learning (Kaelbling, Littman and Moore 1996, Sutton and Barto 2018), problem (5.2) corresponds to a model-based planning problem, described by a finite-horizon belief-state MDP with continuous action and observation spaces. Embedded in each transition $\mathcal{F}_k$ is a step of Bayesian inference, which can be quite expensive to perform, especially in nonlinear and non-Gaussian settings with computationally intensive likelihoods.

5.4. Solution approaches

In general, problem (5.2) cannot be solved analytically. Different numerical strategies must be adopted to find an approximate solution. Before we highlight representative approaches from recent literature, we introduce two types of value function that are central to the MDP formulation, as many solution strategies are built on approximating these functions.

The *action-value functions* (or *Q-functions*) corresponding to a policy π are

$$Q_k^\pi(s_k,\xi_k) = \mathbb{E}_{Y_{k:(N-1)}|\pi,s_k,\xi_k}\left[r_k(s_k,\xi_k,y_k) + \sum_{t=k+1}^{N-1} r_t(s_t,\mu_t(s_t),y_t) + r_N(s_N) \right] \tag{5.3}$$

$$= \mathbb{E}_{Y_k|s_k,\xi_k}\left[r_k(s_k,\xi_k,y_k) + Q_{k+1}^\pi(s_{k+1},\mu_{k+1}(s_{k+1})) \right], \tag{5.4}$$

$$Q_N^\pi(s_N) = r_N(s_N), \tag{5.5}$$

for $k = 0,\ldots,N-1$ and subject to $s_{t+1} = \mathcal{F}_t(s_t,\xi_t,y_t)$. The value of the Q-function $Q_k^\pi(s_k,\xi_k)$ is the expected remaining cumulative reward (i.e. the expected sum of all remaining rewards) for performing the kth experiment at design ξ_k from state s_k and thereafter following policy π.

The *state-value functions* (or *V-functions*) corresponding to a policy π are

$$V_k^\pi(s_k) = \mathbb{E}_{Y_{k:(N-1)}|\pi,s_k}\left[\sum_{t=k}^{N-1} r_t(s_t,\mu_t(s_t),y_t) + r_N(s_N)\right] \tag{5.6}$$

$$= \mathbb{E}_{Y_k|\pi,s_k}\left[r_k(s_k,\mu_k(s_k),y_k) + V_{k+1}^\pi(s_{k+1})\right], \tag{5.7}$$

$$V_N^\pi(s_N) = r_N(s_N), \tag{5.8}$$

for $k = 0,\dots,N-1$ and subject to $s_{t+1} = \mathcal{F}_t(s_t,\xi_t,y_t)$. The value of the V-function $V_k^\pi(s_k)$ is the expected remaining cumulative reward starting from a given state s_k and following policy π for all remaining experiments. The expected utility in (5.2) can be succinctly written as $U(\pi) = V_0^\pi(s_0)$.

The V-function and Q-function are related to each other via

$$V_k^\pi(s_k) = Q_k^\pi(s_k,\mu_k(s_k)), \tag{5.9}$$

and thus often only one of the two value functions needs to be solved for. We note that both value functions can also be expressed in the recursive forms (5.4) and (5.7). When the policy is the optimal policy π^* from (5.2), the recursive relations become the well-known *Bellman optimality equations*:

$$V_k^*(s_k) = \max_{\xi_k\in\mathcal{A}_k} \mathbb{E}_{Y_k|\xi_k,s_k}\left[r_k(s_k,\xi_k,y_k) + V_{k+1}^*(s_{k+1})\right], \tag{5.10}$$

$$V_N^*(s_N) = r_N(s_N), \tag{5.11}$$

for $k = 0,\dots,N-1$.

Furthermore, if the reward terms are chosen to create the sequential analogue of u^{div} in (2.20), i.e. to capture the total expected information gain (EIG) in the model parameter Θ from *all* experiments, two natural formulations arise. A *terminal formulation* places in the terminal reward a single Kullback–Leibler (KL) divergence from the initial prior to the final posterior:

$$r_k(s_k,\xi_k,y_k) = 0, \quad k = 0,\dots,N-1, \tag{5.12}$$

$$r_N(s_N) = D_{\text{KL}}(p_{\Theta|I_N}||p_\Theta). \tag{5.13}$$

An *incremental formulation* instead captures all incremental KL divergence terms from each intermediate experiment's prior to its corresponding posterior:

$$r_k(s_k,\xi_k,y_k) = D_{\text{KL}}(p_{\Theta|I_{k+1}}||p_{\Theta|I_k}), \quad k = 0,\dots,N-1, \tag{5.14}$$

$$r_N(s_N) = 0. \tag{5.15}$$

As pointed out in Theorem 1 of Shen and Huan (2021, 2023) and Theorem 1 of Foster *et al.* (2021), the expected utility $U_T(\cdot)$ produced by substituting (5.12)–(5.13) into (5.2), and the expected utility $U_I(\cdot)$ produced by substituting (5.14)–(5.15) into (5.2), are *equal*: $U_T(\pi) = U_I(\pi)$, for any given policy π. We can also show this equality using the chain rule of mutual information. Starting from the incremental formulation, each immediate reward's contribution to (5.2) is

$$
\begin{aligned}
&\mathbb{E}_{Y_{0:N-1}|\pi,s_0}\left[D_{\mathrm{KL}}\left(p_{\Theta|I_{k+1}}||p_{\Theta|I_k}\right)\right] \\
&= \mathbb{E}_{Y_{0:k-1}|\pi,s_0}\mathbb{E}_{Y_k|Y_{0:k-1},\pi,s_0}\left[\mathbb{E}_{\Theta|I_{k+1}}\left[\log\frac{p(\theta|I_{k+1})}{p(\theta|I_k)}\right]\right] \\
&= \mathbb{E}_{Y_{0:k-1}|\pi,s_0}\mathbb{E}_{\Theta,Y_k|\xi_k,I_k}\left[\log\frac{p(\theta,y_k|\xi_k,I_k)}{p(\theta|I_k)p(y_k|\xi_k,I_k)}\right] \\
&= \mathbb{E}_{Y_{0:k-1}|\xi_{0:k-1}}\left[D_{\mathrm{KL}}\left(p_{\Theta,Y_k|\xi_k,I_k}||p_{\Theta|I_k}\otimes p_{Y_k|\xi_k,I_k}\right)\right] \\
&= \mathcal{I}(\Theta;Y_k|\xi_k,I_k),
\end{aligned} \tag{5.16}
$$

where all ξ_k follow from the given policy π. In the first equality above, the expectation over $Y_{k+1:N-1}$ collapses since the immediate reward does not depend on these observations. In the second equality, we use the fact that

$$\mathbb{E}_{Y_k|Y_{0:k-1},\pi,s_0} = \mathbb{E}_{Y_k|Y_{0:k-1},\xi_{0:k}} = \mathbb{E}_{Y_k|\xi_k,I_k}.$$

Summing the conditional mutual information terms (5.16) according to (5.2) yields

$$U_I(\pi) = \sum_{k=0}^{N-1}\mathcal{I}(\Theta;Y_k|\xi_k,I_k) = \mathcal{I}(\Theta;Y_{0:N-1}|\xi_{0:N-1}) = U_T(\pi), \tag{5.17}$$

via the chain rule of mutual information.

We note that the incremental reward functions can be augmented with additional terms, e.g. rewards reflecting the costs of candidate designs, without affecting the correspondence of these two ways of expressing total EIG.

Computationally, the terminal formulation requires only a single KL divergence estimate per trajectory, at its terminal point. (A 'trajectory' is a realization of the sequence of designs and observations, also known as an 'episode' in MDP terminology.) In contrast, the incremental formulation needs many more intermediate KL divergence calculations, which can be quite costly. On the other hand, the terminal formulation is a case of *delayed reward*, where the feedback from the reward occurs only at the completion of all experiments, which can make learning of the intermediate value functions more difficult. Lastly, we note that a greedy design strategy requires calculating *all* intermediate posteriors and KL divergence terms, which is computationally much more expensive than the sOED terminal formulation.

5.4.1. Approximate dynamic programming (ADP-sOED)

While approximate dynamic programming (ADP) can refer to a range of computational techniques for finding an optimal policy (Powell 2011), we refer here to the approach introduced in Huan (2015) and Huan and Marzouk (2016), which centres on the idea of numerically approximating the optimal V-functions $V_k^*(s_k)$ that satisfy the Bellman optimality equations (5.10) and (5.11), using some parametrized $\tilde{V}_k^*(s_k)$. We call this method ADP-sOED.

In their work, the belief state s_k^b is represented by discretizing the posterior density function on an adaptive tensor-product grid, which is expanded/shrunk and refined/coarsened based on local density values. Alternatively, Huan (2015) also explores representing all possible posterior distributions simultaneously, by constructing a triangular transport map (El Moselhy and Marzouk 2012, Marzouk *et al.* 2016) jointly on (ξ_k, Y_k, Θ), where ξ_k is sampled from a probability distribution with full support on $\mathcal{A}_k$ and (Y_k, Θ) are drawn from the corresponding prior predictives at stage k. See more discussion of such constructions in Section 3.2. In either case, a linear architecture $\tilde{V}_k^* = \sum_{i=1}^m \beta_{k,i}\psi_{k,i}(s_k)$ is used to approximate the optimal V-functions, where features $\psi_{k,i}(s_k)$ are selected to be polynomials of the physical state and of the posterior moments.

Once the representations of s_k^b and $\tilde{V}_k^*$ are chosen, a procedure known as approximate value iteration (also referred to as backward induction, especially for finite-horizon settings) is used to build the approximate V-functions. The main steps of the overall algorithm are summarized as follows.

1 *Trajectory simulation.* Generate trajectories induced by the current approximate V-functions $\tilde{V}_k^*$, by choosing the design $\tilde{\xi}_k^*$ at each stage via

$$\tilde{\xi}_k^* \in \arg\max_{\xi_k \in \mathcal{A}_k} \mathbb{E}_{Y_k|\xi_k,s_k}\big[r_k(s_k,\xi_k,y_k) + \tilde{V}_{k+1}^*(\mathcal{F}_k(s_k,\xi_k,y_k))\big]. \tag{5.18}$$

To realize the corresponding state trajectories $(s_k^{(i)})_{k=0}^N$ sequentially, where i indexes trajectories, an observation $y_k^{(i)}$ is simulated at each stage using the statistical model $p\big(y_k|\theta^{(i)}, \tilde{\xi}_k^{*,(i)}, I_k^{(i)}\big)$, the current design, and a realization $\theta^{(i)}$ of Θ generated from the prior at $k = 0$ and then fixed for that entire trajectory; in other words, $\theta^{(i)}$ serves as the true value of Θ for that trajectory. Belief states are updated using the grid or transport representations noted above. An exploration policy, which simply generates random designs, can be used if $\tilde{V}_k^*$ is not yet available, and otherwise may provide supplementary trajectory samples.

2 *Approximate value iteration (backward induction).* Start from the final stage $k = N - 1$ and evaluate (5.10) at the sample states $s_{N-1}^{(i)}$ generated in step 1,

$$\tilde{V}_{\text{tr}}^{(i)} = \max_{\xi_k \in \mathcal{A}_k} \mathbb{E}_{Y_k|\xi_k,s_k}\big[r_k\big(s_k^{(i)},\xi_k,y_k\big) + \tilde{V}_{k+1}^*\big(\mathcal{F}_k\big(s_k^{(i)},\xi_k,y_k\big)\big)\big], \tag{5.19}$$

and then use these evaluations as training points to update the approximate V-functions $\tilde{V}_k^*$ via linear regression:

$$\big\{s_k^{(i)}, \tilde{V}_{\text{tr}}^{(i)}\big\} \to \tilde{V}_k^*(s_k). \tag{5.20}$$

Once $\tilde{V}_k^*$ is updated, repeat the same process stepping backwards from $k = N-2, N-3, \ldots$ to $k = 0$, thus completing the update for all of the V-function approximations.

3 *Refinement.* Optionally repeat steps 1–2 to improve the pool of trajectory samples and hence (state, value) pairs for training, using the newly updated V-function approximations. Once iterations are terminated, a final set of functions $\tilde{V}_k^*$ is returned, which can be used to evaluate approximate optimal design actions through (5.18).

The approximation structure in (5.18) is known as *one-step lookahead* due to its invocation of an approximation function after one step of dynamic programming[8] (Bertsekas 2005). This is not to be confused with a greedy or myopic design strategy that is based on truncating the problem *horizon*; doing so would not incorporate value from any future experiments beyond that truncated horizon.

It is important to note that both (5.18) and (5.20) require solving a stochastic approximation problem (e.g. using the Robbins–Monro algorithm (Robbins and Monro 1951)), since the expectation therein is typically estimated using Monte Carlo. In fact, since the policy is only implicitly represented by the V-functions, using the final policy still involves solving (5.18). As a result, ADP-sOED is quite computationally expensive.

5.4.2. *Actor–critic policy gradient (PG-sOED)*

In response to the heavy computations required by ADP-sOED, rooted in approximating the optimal V-functions in lieu of the policy, faster methods have been developed by explicitly representing the policy and extracting its gradient, generally known as policy gradient (PG) approaches (Sutton, McAllester, Singh and Mansour 1999). One such method is the PG-based sOED (PG-sOED) introduced in Shen and Huan (2023) (and its earlier version, Shen and Huan 2021), which makes use of actor–critic techniques (Konda and Tsitsiklis 1999, Peters and Schaal 2008) – specifically, actor–critic techniques for deterministic policies with deep neural network parametrizations (Silver *et al.* 2014, Lillicrap *et al.* 2016, Mnih *et al.* 2015).

Suppose that each policy function μ_k is given a parametric representation μ_{k,w_k} with parameters w_k. Collecting these functions in

$$\pi_w := \{\mu_{0,w_0}, \mu_{1,w_1}, \ldots, \mu_{N-1,w_{N-1}}\} \quad \text{with} \quad w := \{w_0, w_1, \ldots, w_{N-1}\},$$

the sOED problem searching within the new parametrized policy space now entails solving

$$\max_w U(\pi_w).$$

[8] For example, two-step lookahead (Bertsekas 2005, p. 304) would take the form of

$$\tilde{\xi}_k^* \in \arg\max_{\xi_k \in \mathcal{A}_k} \mathbb{E}_{Y_k|\xi_k,s_k}\Big[r_k(s_k,\xi_k,y_k) + \max_{\xi_{k+1}\in\mathcal{A}_{k+1}} \mathbb{E}_{Y_{k+1}|\xi_{k+1},s_{k+1}}[r_{k+1}(s_{k+1},\xi_{k+1},y_{k+1}) + \hat{V}_{k+2}^*(\mathcal{F}_{k+1}(s_{k+1},\xi_{k+1},y_{k+1}))]\Big]$$

for some approximate V-function $\hat{V}_{k+2}^*$ two steps ahead.

An expression for the policy gradient can be derived (Shen and Huan 2023, Theorem 2):

$$\nabla_w U(\pi_w) = \sum_{k=0}^{N-1} \mathbb{E}_{Y_k|\pi_w,s_0}\left[\nabla_w \mu_{k,w_k}(s_k)\nabla_{\xi_k} Q_k^{\pi_w}(s_k,\xi_k)\right] \tag{5.21}$$

with $\xi_k = \mu_{k,w_k}(s_k)$. The appearance of $\nabla_{\xi_k} Q_k^{\pi_w}$ in (5.21) further motivates a parametrization of the Q-functions, similarly denoted by

$$Q_\eta^{\pi_w} := \left\{Q_{0,\eta_0}^{\pi_w}, Q_{1,\eta_1}^{\pi_w}, \ldots, Q_{N,\eta_N}^{\pi_w}\right\} \quad \text{with} \quad \eta := \{\eta_0, \eta_1, \ldots, \eta_N\}.$$

Simultaneously learning the parametrized policy (the *actor* π_w) and the Q-functions (the *critic* $Q_\eta^{\pi_w}$) makes this an *actor–critic* method. Overall, access to the policy gradient opens the door to a wide range of gradient-based optimization methods to iteratively improve the policy π_w *en route* to maximizing $U(\pi_w)$.

To this end, Shen and Huan (2023) use the policy gradient to develop several numerical methods for solving (5.2). In what follows, the background information sequence I_k is always used to represent the state, as discussed in Section 5.2. First, a Monte Carlo estimator of (5.21) is formed via

$$\nabla_w U(\pi_w) \approx \frac{1}{M}\sum_{i=1}^{M}\sum_{k=0}^{N-1} \nabla_w \mu_{k,w_k}\left(s_k^{(i)}\right)\nabla_{\xi_k} Q_k^{\pi_w}\left(s_k^{(i)}, \xi_k^{(i)}\right) \tag{5.22}$$

with $\xi_k^{(i)} = \mu_{k,w_k}\left(s_k^{(i)}\right)$ computed from the current policy. Here for the ith trajectory, a 'true' data-generating $\theta^{(i)}$ is drawn from the prior s_0^b and used to generate all subsequent $y_k^{(i)}$ via the models $p\left(y_k|\theta^{(i)}, \xi_k^{(i)}, I_k^{(i)}\right)$ for that entire trajectory.

Second, deep neural networks (DNNs) are used to parametrize both the policy and the Q-functions, following the ideas of deep Q-networks (DQN) (Mnih *et al.* 2015) and deep deterministic policy gradient (DDPG) (Lillicrap *et al.* 2016). The policy π_w, and hence all μ_{k,w_k} for $k = 0, \ldots, N-1$, are represented by a single DNN called the policy network. The consolidated policy network has an input layer that takes in the state, specifically in the form of the information sequence I_k and the current experimental stage k, as depicted in Figure 5.2. The stage k can be represented either directly as an integer or via its one-hot encoding (i.e. the N-dimensional unit vector). Designs ξ_t and observations y_t for future experiments that have not yet taken place ($t \geq k$) are padded with zeros. The output layer returns ξ_k. The gradient of such a DNN-based policy, which is needed to evaluate (5.22), can be computed efficiently via back-propagation. Note that the policy network is therefore not trained in a supervised learning manner, but rather by improving w *en route* to maximizing $U(\pi_w)$ in solving (5.2).

Similarly, the consolidated Q-function $Q_\eta^{\pi_w}$ is represented by a separate, single DNN called the Q-network. The Q-network is trained in a supervised learning manner from the Monte Carlo trajectory samples by minimizing a quadratic loss

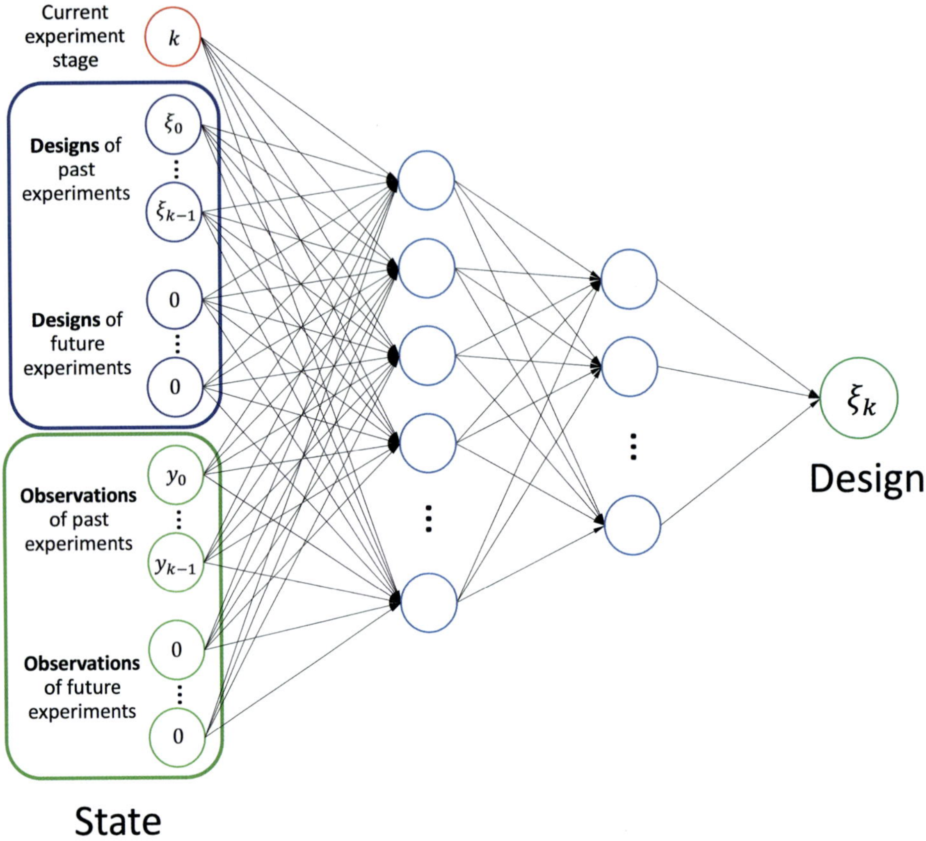

Figure 5.2. A policy is a mapping from state to design. In this DNN representation of the policy, its input entails the current experiment stage, and designs of past experiments and their resulting observations. The designs and observations of future experiments that have not yet taken place are padded with zeros.

derived from (5.4):

$$\ell(\eta) = \frac{1}{M}\sum_{i=1}^{M}\sum_{k=0}^{N-1}\left[Q^{\pi_w}_{k,\eta_k}\left(s_k^{(i)},\xi_k^{(i)}\right) - \left(r_k\left(s_k^{(i)},\xi_k^{(i)},y_k^{(i)}\right) + Q^{\pi_w}_{k+1}\left(s_{k+1}^{(i)},\xi_{k+1}^{(i)}\right)\right)\right]^2 \tag{5.23}$$

with $\xi_k^{(i)} = \mu_{k,w_k}\left(s_k^{(i)}\right)$ computed from the current policy. Note that the last term in (5.23), $Q^{\pi_w}_{k+1}\left(s_{k+1}^{(i)},\xi_{k+1}^{(i)}\right)$, is the *true* Q-function as defined in (5.3)–(5.5) and thus does not depend on η. The true Q-function is usually not available, however, and therefore this term is typically replaced by the Q-function approximation at the current iteration, $Q^{\pi_w}_{k+1,\eta_{k+1}}\left(s_{k+1}^{(i)},\xi_{k+1}^{(i)}\right)$, but with its contribution to the η-gradient of the loss in (5.23) ignored. The trained Q-network $Q^{\pi_w}_{\eta}$ from (5.23) is then used to replace the true Q-function in evaluations of (5.22). Notably, this PG computation does not require evaluating gradients of the likelihood (or the underlying simulation model) with respect to ξ_k, since the gradient operator now only needs to act on the Q-network.

Third, the information-based immediate and/or terminal rewards r_k and r_N, which require evaluation of the KL divergence based on the belief state, are calculated by directly approximating integrals involving the relevant unnormalized densities, via numerical quadrature. This technique, however, is only practical for low-dimensional θ (e.g. $p \leq 4$).

Assembling these numerical methods, an optimal policy is sought using gradient-based optimization such as stochastic gradient ascent. The main steps of the overall algorithm are summarized as follows.

1 *Trajectory simulation.* Generate trajectories. For each trajectory, first draw θ from the prior, then for each of the $k = 0, \dots, N-1$ experiments in the trajectory, sequentially compute ξ_k from the current policy and draw y_k from its statistical model. Calculate the associated trajectory of states; for example, if using I_k as the state, simply store I_N, from which all $I_{k<N}$ can be easily extracted. Compute the corresponding immediate and terminal rewards r_k and r_N.

2 *Value function update.* Update the Q-network by finding an η that minimizes the loss in (5.23).

3 *Policy update.* Estimate $\nabla_w U(w)$ through (5.22) but using the Q-network, and update the policy network through, for example, gradient ascent $w = w + \alpha \nabla_w U(w)$, where α is the learning rate.

4 *Refinement.* Repeat steps 1–3 to improve the trajectory sample pool with the newly updated policy network and Q-network. Once terminated, a final policy network is returned.

Because of the many complex approximations in these algorithms, it is useful to validate them in simple settings where an optimal policy can be derived analytically. To this end, we show how ADP-sOED and PG-sOED perform on a two-experiment linear-Gaussian benchmark problem (Huan and Marzouk 2016, Shen and Huan 2023). The model is a simplified version of (2.4):

$$Y_k = \xi_k \Theta + \mathcal{E}_k \tag{5.24}$$

with $\mathcal{E}_k \sim \mathcal{N}(0, 1^2)$ and no physical state. The benchmark entails $N = 2$ experiments, with prior $\Theta \sim \mathcal{N}(0, 3^2)$ and designs constrained to lie in $d_k \in [0.1, 3]$. The conjugate prior ensures that all subsequent posteriors are Gaussian, following (2.8) and (2.9). The rewards are set to

$$r_k(s_k, \xi_k, y_k) = 0, \quad k = 0, 1, \tag{5.25}$$

$$r_N(s_N) = D_{\mathrm{KL}}(p_{\Theta|I_N} \| p_\Theta) - 2\big(\log \sigma_N^2 - \log 2\big)^2, \tag{5.26}$$

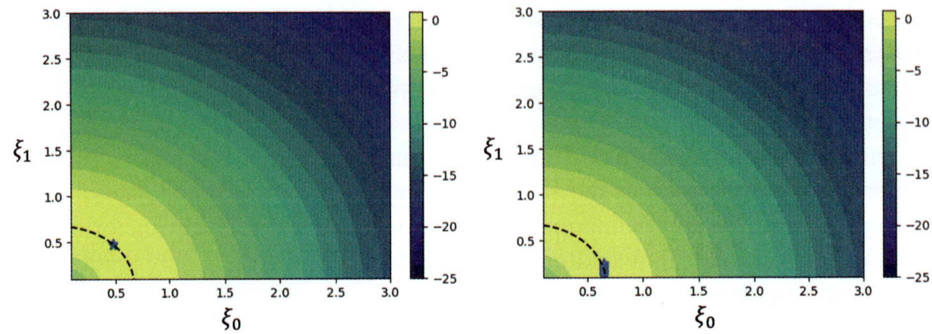

(a) ADP-sOED (b) PG-sOED

Figure 5.3. Expected utility contours for the linear-Gaussian benchmark with the dashed curve showing the set of optimal designs. The cluster of points in each plot indicates the designs selected by (a) ADP-sOED and (b) PG-sOED policies under repeated trials. Both algorithms arrived at the optimal set and achieved expected utility values consistent with the analytic optimal policy: $U(\pi^*_{\text{ADP}}) \approx 0.775$, $U(\pi^*_{\text{PG}}) \approx 0.775$, $U(\pi^*) \approx 0.783$.

where σ_N^2 is the variance of the final posterior, and the additive penalty in r_N is purposefully inserted to make the problem more challenging. A derivation of the resulting optimal policies can be found in Huan (2015, Appendix B), with $U(\pi^*) \approx 0.783$.

In this particular linear-Gaussian problem, sOED is equivalent to batch OED. Upon substituting the terminal reward (5.26) into the sOED objective (5.2) and taking an expectation over $Y_0, Y_1|\pi$, the objective depends explicitly *only* on the final posterior variance. The portion of the objective following from the KL divergence term in (5.26) only involves the posterior variance, as shown by (2.25)–(2.26), and the portion resulting from the penalty term in (5.26) only involves the posterior variance by construction. While this variance depends on the designs chosen by the policy, in the linear-Gaussian setting it is independent of the realized values of $Y_{0:N-1}$, $y_{0:N-1}$. Consequently, feedback or adaptation in response to $y_{0:N-1}$ would not affect the expected utility, and sOED and batch OED therefore coincide. See Huan (2015, Appendix B) for a detailed derivation of this equivalence.

Figure 5.3 illustrates the expected utility for different design choices via coloured contours, with the dashed curve illustrating the set of optimal designs. The cluster of points in each plot indicates the designs selected by the ADP-sOED and PG-sOED policies under repeated trials. Both algorithms arrive at the optimal set and achieve expected utility values consistent with the analytic optimal policy: $U(\pi^*_{\text{ADP}}) \approx 0.775$, $U(\pi^*_{\text{PG}}) \approx 0.775$, $U(\pi^*) \approx 0.783$. The computational times reported in Table 5.1, adapted from Shen and Huan (2023), are obtained using a single 2.6 GHz CPU on a MacBook Pro laptop. The timing values reflect 30 gradient ascent updates for PG-sOED in the training stage, and one policy update (the minimum

Table 5.1. Comparison of computational costs between ADP-sOED and PG-sOED for the linear-Gaussian benchmark. Data adapted from Shen and Huan (2023).

	Training time (s)	Forward model evaluations	Testing time (s)
ADP-sOED	837	5.3×10^8	24 396
PG-sOED	24	3.1×10^6	4

needed) for ADP-sOED. PG-sOED is orders of magnitude faster than ADP-sOED, especially in testing times, making it suitable for applications that have real-time requirements. This drastic difference is due to ADP-sOED being a value-based (critic-only) approach wherein the policy (actor) is not explicitly represented, such that evaluating the policy requires solving a (stochastic) optimization problem. In contrast, applying PG-sOED requires only a single forward pass through the policy network, without any additional optimization runs or forward model evaluations.

Figures 5.4–5.5, both adapted from Shen and Huan (2023), illustrate the application of PG-sOED to a problem of mobile sensor guidance in a two-dimensional advection–diffusion partial differential equation. The location of the centre of a source term (e.g. emitting some contaminant) in the advection–diffusion problem is uncertain, along with the strength and radius of the source. Figure 5.4 shows an example of the contaminant plume concentration evolving in time, with the advection velocity pointing up and to the right. The design variables are the displacements of the mobile sensor from one stage to the next. The inference goal is to learn the unknown source location, source strength and radius of the source, from noisy measurements of contaminant concentration at successive times. All four of these parameters are endowed with uniform priors, and the velocity field advecting the contaminant field is assumed known. The rewards for the sOED problem include the joint information gain (KL divergence from the prior to the posterior after four experiments) in all four parameters, instituted in the terminal formulation manner, along with a negative stagewise reward corresponding to a quadratic penalty on sensor movement from one stage to the next. Figure 5.5 illustrates an application of the resulting PG-sOED policy for one particular realization of the experimental sequence. The top row shows marginal posterior densities of the source location (θ_x, θ_y), while the bottom row shows marginals of the source radius θ_h and source strength θ_s. The movements of the sensor, i.e. the chosen designs, are visualized in the top row (red dots and lines) along with the true source location (fixed purple star).

Recall that the objective (5.2) being maximized involves an expectation over Θ and over all values of $Y_{0:N-1}$ realized under the policy. While Figure 5.5 shows a single trajectory of experiments, a policy is designed to work well, on average, over all possible trajectories. Thus the most comprehensive way of assessing the

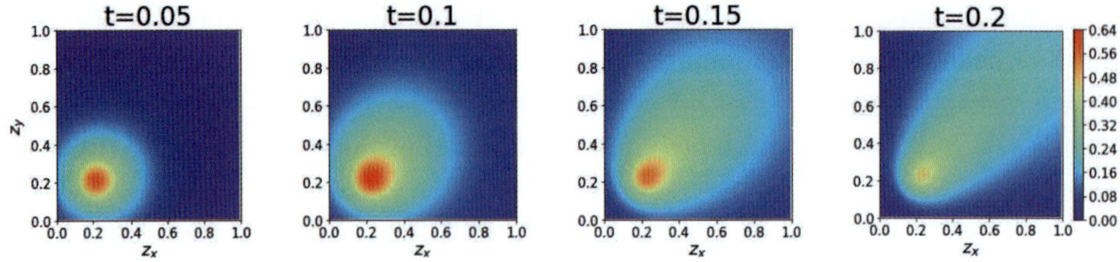

Figure 5.4. An example time-evolution of a convection–diffusion field. The contours show the concentration of the plume. Figure adapted from Shen and Huan (2023).

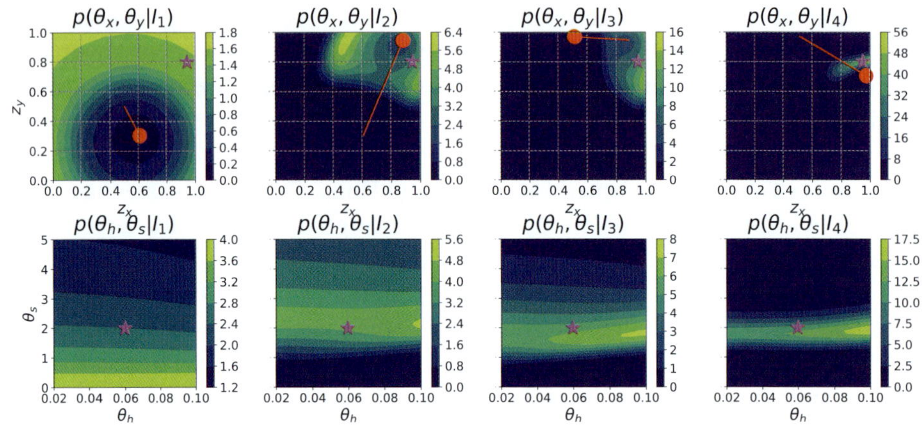

Figure 5.5. Sequence of marginal posterior densities (unknown source locations θ_x, θ_y on the first row, unknown source width and strength θ_h, θ_s on the second row) from an example trajectory instance using PG-sOED. The purple star represents the true data-generating θ value, the red dot represents the physical state (vehicle location) and the red line segment tracks the vehicle displacement (design) from the preceding location. Figure adapted from Shen and Huan (2023).

effectiveness of a policy is to study the reward it produces over many realized trajectories. Figure 5.6, adapted from Shen and Huan (2023), presents histograms of the total rewards obtained from 10^4 trajectories using the batch, greedy and PG-sOED policies. Their expected values, indicated by the vertical black lines, are respectively $U(\pi^*_{\text{batch}}) \approx 2.856$, $U(\pi^*_{\text{greedy}}) \approx 3.057$ and $U(\pi_{\text{PG}}) \approx 3.435$, with PG-sOED achieving the highest expected reward. In this example, greedy design tends to 'chase after' the most recent estimate of the source location. Batch design can plan ahead and take advantage of knowing where the plume will advect in future experiments, but is unable to adapt to new measurements. PG-sOED can do both. Further details of the comparison can be found in Shen and Huan (2023).

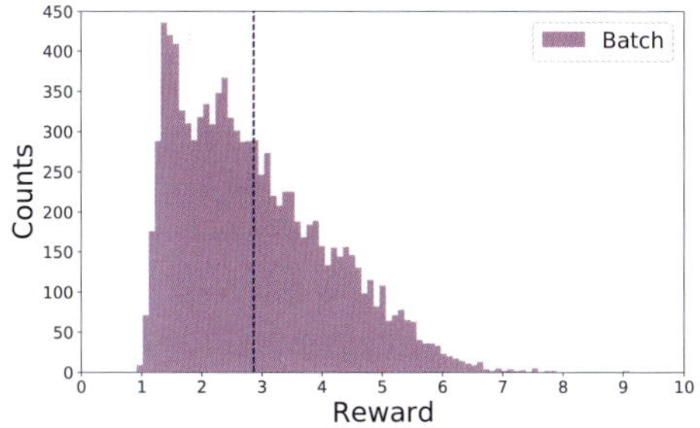

(a) Batch

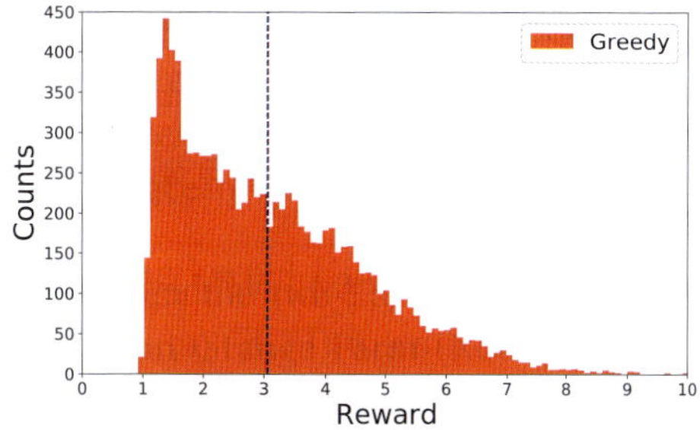

(b) Greedy

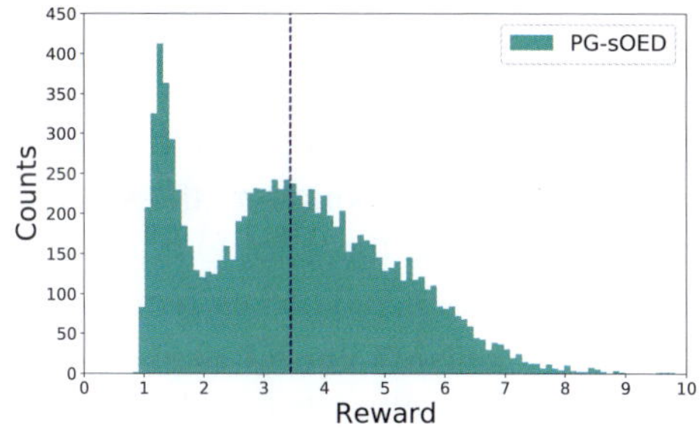

(c) PG-sOED

Figure 5.6. Histograms of total rewards from 10^4 test trajectories generated using batch, greedy and PG-sOED policies, with respective expected total reward values (indicated by vertical lines) $U(\pi^*_{\text{batch}}) \approx 2.856$, $U(\pi^*_{\text{greedy}}) \approx 3.057$ and $U(\pi_{\text{PG}}) \approx 3.435$. Figure adapted from Shen and Huan (2023).

5.4.3. Methods that leverage information bounds

Another prominent series of sequential experimental design algorithms, developed around the same time as those discussed in Section 5.4.2, leverage various information bounds. Here we briefly summarize these methods.

DAD and related methods. Foster *et al.* (2021) introduced a technique they named deep adaptive design (DAD). Like PG-sOED, DAD performs the bulk of its computation offline, constructing a DNN-based policy that can quickly produce the next design online as data are realized. Unlike PG-sOED, however, DAD does not use an actor–critic approach. For a discussion of the relative merits of actor–critic algorithms and more direct PG approaches, see Konda and Tsitsiklis (1999) and Sutton and Barto (2018, Chapter 13). The expected utility $U(\pi)$ in DAD is the total EIG in the parameters Θ from prior to posterior after N experiments, which we call $U_{\text{KL}}(\pi)$ and is equivalent to the mutual information $\mathcal{I}(Y_{0:N-1};\Theta|\pi)$. DAD *directly* targets maximization of the EIG, by first expressing it in terms of log-density ratio of the observations $Y_{0:N-1}$, that is,

$$U_{\mathrm{KL}}(\pi) = \mathcal{I}(Y_{0:N-1};\Theta|\pi) = \mathbb{E}_{Y_{0:N-1}|\pi,s_0}[D_{\mathrm{KL}}(p_{\Theta|I_N}||p_\Theta)]$$
$$= \mathbb{E}_{Y_{0:N-1}|\Theta,\pi,s_0}\mathbb{E}_\Theta\left[\log\frac{p(Y_{0:N-1}|\Theta,\xi_{0:N-1})}{p(Y_{0:N-1}|\xi_{0:N-1})}\right], \quad (5.27)$$

where $\xi_k = \mu_k(s_k)$ follows the given policy π. A sequential version of the prior contrastive estimator (PCE) (cf. (3.26)), called the sequential PCE (sPCE), is then introduced, taking the form

$$\mathcal{I}(Y_{0:N-1};\Theta|\pi) \geq \mathbb{E}_{Y_{0:N-1}|\pi,\Theta_1}\mathbb{E}_{\Theta_1}\mathbb{E}_{\Theta_{2:M}}\left[\log\frac{p(Y_{0:N-1}|\Theta_1,\xi_{0:N-1})}{\frac{1}{M}\sum_{j=1}^{M}p(Y_{0:N-1}|\Theta_j,\xi_{0:N-1})}\right]$$
$$=: \mathcal{L}^{\mathrm{sPCE}}(\pi;M), \quad (5.28)$$

where Θ_1 is the data-generating parameter for $\xi_{0:N-1}, y_{0:N-1}$ in the equation above with $\xi_k = \mu_k(s_k)$ following the given policy π; the subscripts for Y and ξ refer to the experiment (stage) index, while those for Θ refer to the multi-sample index of sPCE as introduced in Section 3.3.2. The key here is that the parameter value of the *data-generating* Θ_1 is included in the Monte Carlo estimate of the evidence in the denominator; including this value ensures that (5.28) is a lower bound to $\mathcal{I}(Y_{0:N-1};\Theta|\pi)$ for any π, which becomes tight as $M \to \infty$ (Foster *et al.* 2021, Theorem 2). (When Θ_1 is excluded from the evidence estimate, (5.28) reverts to the standard nested Monte Carlo estimator of EIG.) Approximating the expectations in sPCE with Monte Carlo sampling from the prior and from the statistical models for successive $(Y_k)_{k=0}^{N-1}$ yields in the end a *negatively biased Monte Carlo estimator* of $\mathcal{I}(Y_{0:N-1};\Theta|\pi)$.

When the policy π is parametrized as π_w, DAD seeks a policy that satisfies

$$\max_w \mathcal{L}^{\mathrm{sPCE}}(\pi_w;M).$$

The gradient of the sPCE bound can be obtained by moving the gradient operator ∇_w inside the expectations. This may require first reparametrizing the stochasticity in the observations to be independent of the policy. (For instance, with a statistical model $Y_k = G_k(\Theta,\xi_k)+\mathcal{E}_k$, an expectation over Y_k can be rewritten as an expectation over $\mathcal{E}_k$; if the distribution of $\mathcal{E}_k$ is functionally independent of the design ξ_k, then the goal is already achieved. More generally, if the distribution of $\mathcal{E}_k$ does depend on ξ_k, then the expectation can always be rewritten in terms of some other random variable whose distribution is independent of ξ_k and hence of π.) The sPCE gradient is then

$$\nabla_w\mathcal{L}^{\mathrm{sPCE}}(\pi_w;M) = \mathbb{E}_{\mathcal{E}_{1:N}}\mathbb{E}_{\Theta_{1:M}}\left[\nabla_w\left(\log\frac{p(Y_{0:N-1}|\Theta_1,\xi_{0:N-1})}{\frac{1}{M}\sum_{j=1}^{M}p(Y_{0:N-1}|\Theta_j,\xi_{0:N-1})}\right)\right] \quad (5.29)$$

with $\xi_k = \mu_{k,w_k}(s_k)$ following the given policy π_w. With this gradient in hand, one can use any gradient-based optimization scheme, such as stochastic gradient ascent, to find a parametrized policy that maximizes the estimated lower bound. With regard to how to choose the parametrization π_w, Foster *et al.* (2021) recommend a 'pooling-emitter' DNN architecture, motivated by the permutation invariance of the EIG under certain conditions (e.g. if all the observations in the sequence are conditionally independent given the parameters, and described by the same parametric statistical model).

Blau *et al.* (2022) build on the sPCE bound from DAD, but seek a stochastic policy via the randomized ensembled double Q-learning approach of Chen, Wang, Zhou and Ross (2021), which is an actor–critic method. They demonstrate applicability of this approach to discrete design spaces.

iDAD. All of the sequential design methods discussed so far require the ability to explicitly evaluate the likelihood (i.e. to evaluate the probability densities $p(y_k|\theta, \xi_k, I_k)$). In certain problems, however, the likelihood may be only implicitly defined and otherwise intractable to compute; yet sampling y_k from the model may still be possible. To accommodate such models in sOED, a variation of DAD, called implicit deep adaptive design (iDAD), has been developed by Ivanova *et al.* (2021).

iDAD introduces several likelihood-free lower bounds. The main idea behind these bounds can be understood as approximating the posterior-to-prior or likelihood-to-evidence density ratio through a parametrized 'critic' function, denoted here by $f\colon \boldsymbol{\Theta} \times \Xi_{0:N-1} \times \mathcal{Y}_{0:N-1} \to \mathbb{R}$. Note that this notion of a critic *differs* from the 'critic' in the actor–critic methods of Section 5.4.2 (where the role of the critic was played by the Q-function). In these lower bound expressions, any choice of f yields a tractable lower bound for the EIG; thus f can act as the 'knob' to maximize the lower bound. One example is the sequential design version of the NWJ bound (*cf.* (3.24))

$$\begin{aligned}\mathcal{I}(Y_{0:N-1}; \Theta|\pi) &\geq \mathbb{E}_{\Theta}\mathbb{E}_{Y_{0:N-1}|\Theta,\pi}[f(\Theta, I_N)] - \frac{1}{e}\mathbb{E}_{Y_{0:N-1}|\pi}\mathbb{E}_{\Theta}[\exp f(\Theta, I_N)] \\ &=: \mathcal{L}^{\text{sNWJ}}(\pi; f), \end{aligned} \tag{5.30}$$

while another is the InfoNCE bound (cf. (3.28))

$$\begin{aligned}\mathcal{I}(Y_{0:N-1}; \Theta|\pi) &\geq \mathbb{E}_{\Theta_1}\mathbb{E}_{Y_{0:N-1}|\Theta_1,\pi}\mathbb{E}_{\Theta_{2:M}}\left[\log \frac{\exp f(\Theta_0, I_N)}{\frac{1}{M}\sum_{j=1}^{M}\exp f(\Theta_j, I_N)}\right] \\ &=: \mathcal{L}^{\text{sNCE}}(\pi; M, f). \end{aligned} \tag{5.31}$$

Both bounds can be tight for an optimal selection of the critic function f and as $M \to \infty$ (Ivanova *et al.* 2021). Importantly, these bounds no longer involve the densities $p(y_k|\theta, \xi_k, I_k)$. Upon parametrizing the policy π as π_w and critic f as f_ϕ, a policy can be found by maximizing (tightening) the lower bound simultaneously over w and ϕ. For example, in the NWJ case:

$$\max_{w,\phi} \mathcal{L}^{\text{sNWJ}}(\pi_w; f_\phi).$$

Similar to DAD, the gradient of the lower bound with respect to both the policy and critic parameters can be obtained by bringing the gradient inside the expectation upon reparametrizing the observations, allowing for stochastic gradient ascent updates to tighten the bound. We note that while iDAD does not require explicit likelihoods and relies only on a simulator of Y_k, it does require access to derivatives of the output of this simulator with respect to the design ξ_k, and with respect to all previous experiments' designs and outcomes, I_k.

vsOED. The variational sequential OED (vsOED) method in Shen *et al.* (2023) derives a lower bound of the EIG by employing variational approximations of the relevant posterior distributions and then substituting these approximations in 'one-point' approximations of the KL divergence in the reward terms. The framework generalizes the objective functions used in previous approaches in that it simultaneously accommodates multiple models, nuisance parameters, predictive quantities of interest and implicit likelihoods.

Let $\mathcal{M}_m$ be a countable set of models indexed by $m = 1, 2, \ldots$, where each model has its own parameters $\theta_m \in \boldsymbol{\Theta}_m \subseteq \mathbb{R}^{p_m}$ and predictive quantity of interest (QoI) $z_m = \Psi_m(\theta_m)$, with $\Psi_m \colon \boldsymbol{\Theta}_m \to \mathbb{R}^{q_m}$ for some $q_m \leq p_m$. Suppose also that we have a prior distribution $P_M(m)$ over the model indicator M, a prior $p_{\Theta_m}(\theta_m|m)$ for the parameters Θ_m of each model, and a prior $p_{Z_m}(z_m)$ for each model's QoI Z_m induced by p_{Θ_m} and Ψ_m. A combined reward function that includes a weighted combination of information gains in these random variables can then be formed. For example, the incremental formulation in (5.14) and (5.15) becomes

$$\begin{aligned} r_k(s_k, \xi_k, y_k) = \alpha_M D_{\mathrm{KL}}(P_{M|I_{k+1}} || P_{M|I_k}) \\ + \mathbb{E}_{M|I_{k+1}} \big[\alpha_\Theta D_{\mathrm{KL}}(p_{\Theta_m|I_{k+1}} || p_{\Theta_m|I_k}) \\ + \alpha_Z D_{\mathrm{KL}}(p_{Z_m|I_{k+1}} || p_{Z_m|I_k}) \big], \quad k = 0, \ldots, N-1, \quad (5.32) \\ r_N(s_N) = 0, \quad (5.33) \end{aligned}$$

where $\alpha_M \in [0, 1]$ (for the model indicator), $\alpha_\Theta \in [0, 1]$ (for the model parameters) and $\alpha_Z \in [0, 1]$ (for the predictive QoIs) are the weights of KL contributions from these variables. The terminal formulation can be constructed in a similar manner.

To compute the sOED objective in (5.2) with the rewards (5.32) and (5.33), an expectation needs to be taken over $Y_{0:N-1}|\pi, s_0$. This requires sampling trajectories. For each trajectory, a model indicator and corresponding parameter value are drawn from the priors,

$$m_0^{(i)} \sim P_M, \quad \theta_{m,0}^{(i)} \sim p\big(\theta_{m,0}|m_0^{(i)}\big),$$

with a corresponding QoI sample

$$z_{m,0}^{(i)} = \Psi_{m_0^{(i)}}\big(\theta_{m,0}^{(i)}\big).$$

Then, $m_0^{(i)}$ and $\theta_{m,0}^{(i)}$ generate a trajectory $I_N^{(i)}$. For any such trajectory, we substitute the 'oracle' values of the model indicator and model parameters that generated

I_N into the integrands appearing in the incremental reward (5.32), to produce a 'one-point' approximation $\tilde{r}_k$ of each r_k,

$$\tilde{r}_k(s_k, \xi_k, y_k) = \alpha_M \log \frac{P(m_0|I_{k+1})}{P(m_0|I_k)} + \alpha_\Theta \log \frac{p(\theta_{m,0}|I_{k+1})}{p(\theta_{m,0}|I_k)} + \alpha_Z \log \frac{p(z_{m,0}|I_{k+1})}{p(z_{m,0}|I_k)}, \quad k = 0, \ldots, N-1, \tag{5.34}$$

$$\tilde{r}_N(s_N) = 0, \tag{5.35}$$

where we recall that $I_{k+1} = (I_k, \xi_k, y_k)$. Replacing r_k with $\tilde{r}_k$ in (5.2) yields exactly the same expected utility. Specifically, letting U_{KL} denote the expected utility formed when using the full KL rewards in (5.32) and (5.33), and letting $\tilde{U}_{\text{KL}}$ denote the expected utility formed when using the one-point approximations in (5.34) and (5.35), Shen *et al.* (2023, Theorem 2) show that $\tilde{U}_{\text{KL}}(\pi) = U_{\text{KL}}(\pi)$ for any policy π. In other words, using the one-point approximation does not alter the sOED problem. The approximations $\tilde{r}_k$ can also be viewed as one-point estimators of the expectation of r_k.

Evaluating the probability terms in (5.34), however, remains highly challenging. To make the computation tractable, each of the probability terms is approximated within a parametrized family of distributions – taking a variational inference approach – leading to the following *variational* one-point approximations:

$$\tilde{r}_k(s_k, \xi, y_k; \phi) = \alpha_M \log \frac{q(m_0|I_{k+1}; \phi_M)}{q(m_0|I_k; \phi_M)} + \alpha_\Theta \log \frac{q(\theta_{m,0}|I_{k+1}; \phi_{\Theta_m})}{q(\theta_{m,0}|I_k; \phi_{\Theta_m})} + \alpha_Z \log \frac{q(z_{m,0}|I_{k+1}; \phi_{Z_m})}{q(z_{m,0}|I_k; \phi_{Z_m})}, \quad k = 0, \ldots, N-1, \tag{5.36}$$

$$\tilde{r}_N(s_N) = 0, \tag{5.37}$$

with $\phi = \{\phi_M, \phi_{\Theta_m}, \phi_{Z_m}\}$ being the variational parameters describing the approximations q above. We use the notation q_ϕ to represent succinctly the complete set of approximating distributions. Letting $\mathcal{L}^{\text{vsOED}}(\pi; q_\phi)$ denote the expected utility function obtained when using (5.36) and (5.37), Shen *et al.* (2023, Theorem 3) show that $\mathcal{L}^{\text{vsOED}}(\pi; q_\phi) \leq \tilde{U}_{\text{KL}}(\pi) = U_{\text{KL}}(\pi)$ for any policy π. That is, adopting the variational approximation forms a lower bound to the expected utility.

To solve the problem, Shen *et al.* (2023) use an actor–critic approach as in PG-sOED, parametrizing the policy π as π_w and the corresponding Q-function (critic) Q^{π_w} as $Q_\eta^{\pi_w}$. A policy can then be found by maximizing (tightening) the lower bound simultaneously over w and ϕ:

$$\max_{w,\phi} \mathcal{L}^{\text{vsOED}}(\pi_w; q_\phi).$$

(Recall that η is not directly optimized, but updated via (5.23).) This can be achieved by gradient ascent updates, where gradients with respect to w and ϕ are found following similar steps to those used to obtain (5.21). Notably, with the introduction of the approximating distributions q_ϕ, evaluation of the reward no longer requires explicit likelihoods and thus the method can be used for implicit

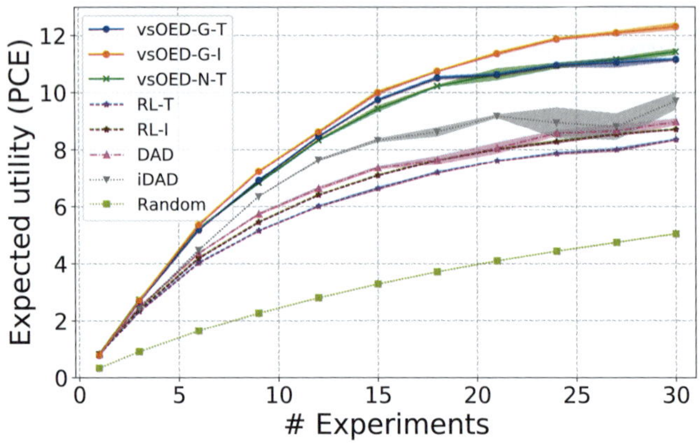

Figure 5.7. Expected utility lower bounds achieved by various sequential experimental design policies for a source location problem. vsOED-G and vsOED-N denote the vsOED method using Gaussian mixture models and normalizing flows, respectively, as variational distributions q. Suffixes -T and -I indicate whether the algorithm uses the terminal or incremental formulation. Figure adapted from Shen *et al.* (2023).

models. Unlike DAD and iDAD, vsOED does not require access to gradients of the log-likelihood or of an underlying simulator, due to the use of a Q-network in the same way as PG-sOED.

Test problems in Shen *et al.* (2023) illustrate how vsOED can handle multiple models, predictive QoIs and implicit likelihoods. For example, Figure 5.7, adapted from that paper, compares the EIG policies obtained via several sequential experimental design approaches on a source location problem introduced in Foster *et al.* (2021), for a range of design horizons N. Results are obtained by fixing the total number of trajectory samples available to each algorithm. The policies produced by each algorithm are then re-evaluated to assess their achieved EIG using a common estimator, namely the PCE lower bound estimator.

5.5. *Open questions in sequential optimal design*

While the approaches discussed in this section represent promising recent advances in sequential OED, there is a need for further development in many important directions. We briefly mention a few below.

One rather immediate need is for more efficient state representations – in particular, representations of the belief state s_k^b (the evolving posterior distribution) that achieve the natural 'compression' induced by Bayesian logic. Many of the approaches described in this section simply track I_k, which has the advantages of being trivial to update (by appending the newly arrived ξ_k and y_k) and free of approximation error. Moreover, I_k contains all the information necessary to compute any intermediate and final physical states (though they may not be accessible until evaluated via the transition dynamics for the physical state). A key disadvantage of using I_k, however, is that its dimension grows linearly with k, i.e. the number of experiments completed. A resulting numerical difficulty can already be seen in the policy network architectures used for PG-sOED (Figure 5.2) and related methods, where the state input layer needs to accommodate a variable-length I_k. This is crudely handled by creating a layer of maximum possible size and zero-padding the inapplicable entries. The efficiency of such a padding technique degrades as the design horizon N increases (i.e. there are many more stages where a large fraction of the inputs are set to zero). Additionally, an I_k of unbounded size introduces conceptual and technical difficulties in the infinite-horizon setting.

In essence, tracking I_k – the trivial sufficient statistics of the posterior – delays engaging with the actual process of Bayesian inference and the computation of posteriors. I_k is therefore not as 'compressed' as the posterior, and the correspondence of I_k values to posterior distributions is generally non-injective (multiple I_k may lead to the same posterior). In other words, I_k carries additional, unnecessary information beyond that which is needed to perform the Bayesian update (Jaynes and Bretthorst 2003, Cox 1946). Using state representations that directly capture the posterior – and that simultaneously allow numerical accuracy, scalability to high-dimensional parameter spaces and easy online updating – would therefore be quite desirable. Inspiration should arise from other research areas that must maintain efficient online posterior representations, such as sequential Bayesian inference, data assimilation and streaming variational inference.

More broadly, more efficient and scalable computational methods are needed to realize sOED with computationally intensive models, high-dimensional parameters and data, and complex design objectives. Indeed, the development of sOED is rather new compared to batch OED, and its demonstrations have thus far been confined to rather simple formulations and models. An important goal is to expand sOED's capabilities so that it can tackle problems of the complexity now attained in batch design studies. To this end, accurate calculations of the posterior and the associated information measures – which occur *repeatedly* in sOED – will need to be pushed to higher dimensions. More expressive function representations and advanced reinforcement learning techniques will be needed to improve the *sample efficiency* of constructing a policy.

Along these lines, there is an enormous need for numerical analysis and approximation theory in these settings. From an approximation perspective, little is known about properties of *optimal* policies and value functions for sOED, outside of perhaps the simplest (e.g. linear-Gaussian) cases. How smooth are these functions? Do they enjoy anisotropic dependence on their inputs? Are they well approximated by low-rank tensors, or ridge functions, or in some other format that is particularly tractable? What neural network architectures are hence most suitable? Understanding these questions is necessary to characterize the optimality gaps emerging from current *ad hoc* choices of representation and network architecture. This understanding will also help to create more tailored and effective policy and Q-function approximations, ideally endowed with error bounds and performance guarantees. At the same time, convergence guarantees and rates for the reinforcement learning techniques used here, namely policy gradient and actor–critic methods, should be strengthened for the case of sequential experimental design in a Bayesian setting.

There is also a need to explore more sophisticated sequential design formulations: optimal stopping of sequential experiments, interleaving of batch and sequential designs, and policies that are robust to horizon changes.

Finally, as sOED is generally more computationally demanding than simpler batch and greedy designs, it would be extremely valuable to develop ways of choosing the appropriate design strategy *a priori*. Here we would advocate for inexpensive ways of estimating or bounding the *benefit of feedback* and the *benefit of lookahead*, before actually solving the sOED problem, and then balancing these potential upsides with their computational costs.

6. Outlook

The preceding sections have presented a broad overview of the current state of the art in optimal experimental design (OED). Research continues in all of the threads we have discussed: (i) devising design criteria that are sufficiently expressive to capture diverse experimental goals, with complex models; (ii) estimating these criteria in efficient and structure-exploiting ways, in high dimensions and/or in the presence of strong non-Gaussianity; (iii) maximizing a chosen design criterion given different parametrizations of feasible designs, whether continuous or discrete, and developing guarantees for the associated optimization algorithms; (iv) advancing formulations and algorithms for optimal sequential design. In this final section we will highlight some broader questions and issues that have not yet been discussed. Many of these issues remain rather open-ended, and thus we believe they comprise fertile ground for ongoing research.

6.1. Model misspecification

OED is intrinsically model-based: all of the methods we have discussed use a statistical model, perhaps augmented with prior information, to *predict* the outcomes of experiments at candidate designs and to *assess* how these outcomes might improve knowledge of model parameters, reduce uncertainty in model-based predictions, and so on. In the general decision-theoretic terms of Section 2.2, models are needed to define the utility function u and to specify the distribution $p_{Y,\Theta}$ over which we take expectations to yield the expected utility U.

This reliance on models gives OED great power, but raises the question of what happens when models are, inevitably, misspecified – by which we mean that our statistical model $\mathcal{M}$ (2.1) of the data-generating process, for any given design, does not adequately capture how the data are generated in the actual experiment. In other words, $\mathcal{M}$ might not contain the distribution that generated the observed data.

A natural way to address misspecification is simply to augment the model, so that it can more closely capture the data-generating distribution. One line of work, developed in the setting of Bayesian inverse problems, is the Bayesian approximation error approach of Kaipio and Kolehmainen (2013), Kaipio and Somersalo (2007) and Alexanderian, Nicholson and Petra (2022), which augments the noise model – for example, in the case of additive Gaussian noise, enlarging the covariance matrix of this noise – to account for error in the forward operator. Another approach is to explicitly introduce new 'nuisance' parameters: rather than considering only $p(y|\theta,\xi)$, one could introduce a richer model $p(y|\theta,\eta,\xi)$, where $\eta \in \boldsymbol{H}$ are additional parameters that capture previously un-modelled phenomena, such that there exists some $(\theta^*,\eta^*) \in \boldsymbol{\Theta} \times \boldsymbol{H}$ for which $p(y|\theta^*,\eta^*,\xi)$ matches the data-generating distribution for *all* designs $\xi \in \Xi$. This viewpoint underlies several recent efforts, e.g. Alexanderian *et al.* (2022) and Sargsyan, Najm and Ghanem (2015). Here η could represent variability or imprecision in the placement or timing of observations, background stochasticity or uncertain initial conditions, the impact of unresolved scales, etc. In a fully Bayesian setting, the design formulation discussed following (2.27) is relevant: the utility function can be chosen to depend on the posterior (and prior) marginal distributions of θ, with the impact of the additional uncertain parameters η handled via integration over the prior $p(\eta)$. There are two equivalent ways of writing the posterior marginal of θ:

$$p(\theta|y,\xi) = \int p(\theta,\eta|y,\xi)\,\mathrm{d}\eta, \tag{6.1}$$

$$\text{where} \quad p(\theta,\eta|y,\xi) \propto p(y|\theta,\eta,\xi)p(\theta|\eta)p(\eta),$$

and

$$p(\theta|y,\xi) \propto p(y|\theta,\xi)p(\theta), \tag{6.2}$$

$$\text{where} \quad p(y|\theta,\xi) = \int p(y|\theta,\eta,\xi)p(\eta|\theta)\,\mathrm{d}\eta.$$

In the first, we perform inference for both θ and η, then focus attention on the posterior marginal of θ. In the second, we first create a marginal likelihood for θ and then perform inference with this integrated quantity.

Although these viewpoints of *inference* are equivalent, there are two distinct ways of modelling subsequent *posterior predictions*. From a standard hierarchical Bayesian perspective, the distribution of predictions Y^+ at a design ξ^+ (which may or may not be equal to ξ) follows from the *joint* posterior of θ and η:

$$p_1(y^+|\xi^+,y,\xi) = \iint p(y^+|\theta,\eta,\xi^+)\,p(\theta,\eta|y,\xi)\,\mathrm{d}\theta\,\mathrm{d}\eta. \tag{6.3}$$

In this setting, both θ and η were uncertain before observing y, but since they both influenced the observed value of y, conditioning on this observation updates one's knowledge of both parameters. Crucially, the unknown value of the nuisance parameter η that influences subsequent predictions is assumed to be the same as the one that affected y (just as θ is common to both stages). A different model for the predictions $Y^+|\xi^+$, however, is that they arise from an independent, *newly realized* η (here called η^+, for emphasis) that has *not* been conditioned on previous observations:

$$p_2(y^+|\xi^+, y, \xi) = \iint p(y^+|\theta, \eta^+, \xi^+)\, p(\theta|y, \xi)\, p(\eta^+)\, d\theta\, d\eta^+. \tag{6.4}$$

Embedded inadequacy models (Sargsyan *et al.* 2015, Sargsyan, Huan and Najm 2019, Morrison, Oliver and Moser 2018), which view uncertainty in η^+ as somehow irreducible, can be understood according to this prediction model (6.4). This model is also somewhat related to the viewpoint in Koval, Alexanderian and Stadler (2020) and Alexanderian *et al.* (2022). We emphasize that, in this setting, the posterior *marginal* $p(\theta|y, \xi)$ is the meaningful representation of uncertainty in the parameters θ resulting from an experiment, even if η is assumed irreducible; therefore it appears in the integrand of (6.4). It is not meaningful to simply average the conditional posterior $p(\theta|\eta, y, \xi)$, or some functional thereof, over $p(\eta)$.

In any case, whether the goal is parameter inference, or prediction following (6.3) or (6.4), the ability to perform OED with implicit models as discussed throughout Section 3 is quite useful. In particular, the posterior marginal $p(\theta|y, \xi)$ or the marginal likelihood of θ, $p(y|\theta, \xi)$ are likely part of the utility function u; these densities are generally intractable and thus need to be estimated from samples. If the utility involves comparing prior and posterior predictions of Y^+, then again the densities (6.3) and (6.4) are likely intractable.

The approaches just discussed, however, are at best a partial solution to the problem of model misspecification, as they essentially rely on the modeller being able to create a 'better' statistical model for the data. In many situations, doing so may not be feasible. Approaches to this more challenging (and general) situation are very much a subject of ongoing research. They are, at least in spirit, related to earlier work on *robust Bayesian analysis*, which, in the words of Berger (1994), 'studies ... the sensitivity of Bayesian answers to uncertain inputs'. Research on robust Bayesian methods first flourished several decades ago, but much less has been done to address the sensitivity of Bayesian OED. Some relevant analysis is found in Duong *et al.* (2023), which considers perturbations to the likelihood $p(y|\theta, \xi)$ (in the sense of the Kullback–Leibler divergence) and elucidates the rate at which the associated mutual information and its maximizers converge as the likelihood perturbations grow smaller. In a rather different (and parametric) approach, Chowdhary, Tong, Stadler and Alexanderian (2023) consider Bayesian linear inverse problems in infinite dimensions and show how to compute derivatives of the mutual information/expected information gain with respect to a finite set of 'auxiliary' model parameters, i.e. parameters that are held fixed during the inference procedure. Attia, Leyffer and Munson (2023) propose a robust Bayesian A-optimal design formulation, again for linear (or linearized) models, where parametric families of prior and noise covariance matrices are specified and the design criterion is maximized for the worst-case element of these families.

Go and Isaac (2022) instead take a non-parametric approach to misspecification, and propose a robust OED formulation rooted in distributionally robust optimization (Rahimian and Mehrotra 2019, Kuhn, Esfahani, Nguyen and Shafieezadeh-Abadeh 2019). The robustness considered therein is with respect to the choice of prior p_Θ only: the authors introduce an ambiguity set specified by the Kullback–Leibler divergence, $\mathcal{Q}_\epsilon = \{q\colon D_{\mathrm{KL}}(q||p_\Theta) \leq \epsilon\}$, and seek the design that maximizes the *worst-case* mutual information, over priors drawn from $\mathcal{Q}_\epsilon$:

$$\xi^* = \arg\max_{\xi\in\Xi} \inf_{q_\Theta\in\mathcal{Q}_\epsilon} \mathbb{E}_{Y|\xi}[D_{\mathrm{KL}}(p_{\Theta|Y,\xi}||q_\Theta)], \tag{6.5}$$

where the expectation is over $Y|\xi \sim \int p(y|\theta,\xi)\, q_\Theta(\theta)\,\mathrm{d}\theta$. This problem is not directly tractable, so Go and Isaac (2022) propose an approximation that is well-behaved for sufficiently small ϵ. We suggest that there is ample opportunity for further work at this intersection of distributional robustness and OED. For instance, it is important to consider robustness to other aspects of the joint distribution $p_{Y,\Theta|\xi}$, especially the likelihood (Zhang *et al.* 2022) and the forward operator therein. It would also be natural to consider ambiguity sets based on other divergences or distances, e.g. Wasserstein distances. And it remains to understand how to set the radius ϵ of any such ambiguity set, and to relate this value to other information one might have about the nature of the model misspecification.

We also note that the goal of 'robustifying' OED is very much related to recent efforts to formulate robust notions of *inference* under model misspecification (see Kleijn and van der Vaart 2012, Bochkina 2019), and can benefit from advances in this direction. These two threads are inextricably linked: a 'robustified' OED might do a better job of producing data, but this data must then be interpreted through a model. Reverting solely to the misspecified model for this second phase makes little sense. A multitude of interesting approaches to inference under model misspecification have been proposed in recent years, e.g. coarsened inference (Miller and Dunson 2019), power posteriors (Grünwald and van Ommen 2017), reweighing data to reduce the effect of outliers, data contamination and other forms of misspecification (Dewaskar, Tosh, Knoblauch and Dunson 2023), averaging posteriors over bootstrapped datasets (Huggins and Miller 2023, Pompe and Jacob 2021), modularized inference (Carmona and Nicholls 2020, Jacob, Murray, Holmes and Robert 2017), and many others. It would be of interest to understand how these methods could both mitigate the impact of model misspecification on OED procedures and help to better process the resulting data.

It may also be natural to depart from reliance solely on optimality criteria and to re-introduce some flavour of *randomness* or *space filling* to designs, as a means of hedging against misspecification of the models that produced the design criterion. This approach may be of particular interest in the sequential OED setting, where model error could be assessed and to some extent quantified at each stage of experimentation, in a way that informs subsequent rounds of design.

6.2. *Risk-aware design criteria*

The design formulations described in Section 2.2 and addressed throughout this paper focused primarily on the *expectation* of a utility function $u(\xi, Y, \Theta)$, where randomness in the utility is induced by the randomness in Y and Θ. Yet the notion of *risk* (see Royset 2022 for a recent review), aimed at quantifying 'hazard' or undesirable outcomes, is also relevant to OED. For instance, an experimenter might be interested in characterizing – and controlling – the probability and/or severity of an experimental outcome with very low utility (e.g. low information gain). Doing so requires moving beyond the expected value.

Popular risk measures include mean-plus-deviation and mean-plus-variance (Markowitz 1952), quantiles and superquantiles (also called the value-at-risk and the conditional value-at-risk (CVaR), respectively), worst-case risk, entropic risk, and many more; see Rockafellar and Uryasev (2013) and Shapiro, Dentcheva and Ruszczynski (2021, Chapter 6). Desiderata for candidate risk measures include the notion of coherence (Artzner, Delbaen, Eber and Heath 1999), with consideration for robustness, elicitability and backtesting (He, Kou and Peng 2022). A 'risk quadrangle' system, characterizing the relationships among measures of risk, regret, deviation and error, has also been proposed (Rockafellar and Uryasev 2013, Rockafellar and Royset 2015).

To our knowledge, risk measures have not been widely applied in the general setting of (nonlinear) design with generic y- and θ-dependent utility functions. Some initial work in this direction appears in Shen (2023, Chapters 4, 5), which explores using a variance-penalized expected utility (i.e. a mean-plus-variance risk) for nonlinear OED. In the batch setting, the objective becomes

$$U(\xi) = \mathbb{E}_{Y,\Theta|\xi}[u(\xi, Y, \Theta)] - \lambda \operatorname{Var}_{Y,\Theta|\xi}[u(\xi, Y, \Theta)], \tag{6.6}$$

where $\lambda > 0$ is a scaling parameter reflecting the experimenter's degree of risk-aversion (larger λ) or tolerance. Estimating this objective involves estimating the second moment of the utility, $\mathbb{E}_{Y,\Theta|\xi}[u^2(\xi, Y, \Theta)]$. When u is chosen as in (2.19) or (2.20), i.e. to reflect information gain in the parameters Θ, Shen (2023) constructs a consistent nested Monte Carlo estimator of this term and hence of the overall objective (6.6), and characterizes its bias and variance to leading order. Shen (2023) also demonstrates this variance-penalized utility for sequential OED.

A different form of risk-aware OED has been proposed by Kouri, Jakeman and Gabriel Huerta (2022), who address the setting of classical (non-Bayesian) linear design for regression, where risk is now associated with the distribution of the variance of the predicted response, $v_\xi(x) := f^\top(x)F^{-1}(\xi)f(x)$, over *inputs/covariates* $x \in \mathcal{X}$, given a probability measure μ on $\mathcal{X}$ and Fisher information matrix F. (Recall our notation from Section 2.1.1.) In other words, the predictive variance $v_\xi(X)$ becomes random once the choice of input is treated as random, $X \sim \mu$. This formulation can be understood as a nonlinear interpolation between G-optimality (2.7) (choosing ξ to minimize the maximum over $x \in \mathcal{X}$, and hence the worst case, of $v_\xi(x)$) and I-optimality (minimizing the average over $X \sim \mu$ of the predictive variance, $\mathbb{E}_\mu[v_\xi(X)]$). Specifically, Kouri *et al.* (2022) propose an *R-optimality* criterion that involves minimizing the CVaR (also called the average value-at-risk) of $v_\xi(X)$ at confidence level β:

$$\xi^* \in \operatorname*{argmin}_{\xi \in \Xi} \mathrm{CVaR}_\beta(v_\xi(X)),$$

where, for any random variable Z and $\beta \in [0,1]$,

$$\mathrm{CVaR}_\beta(Z) := \frac{1}{1-\beta}\int_\beta^1 q_\alpha(Z)\,\mathrm{d}\alpha,$$

and $q_\alpha(Z) := \inf\{t \in \mathbb{R} \mid \mathbb{P}[X \le t] \ge \alpha\}$ is the upper α-quantile of Z (Rockafellar and Uryasev 2002). Thus the R-optimality criterion seeks to mitigate the risk of large prediction variances by minimizing the average of the *upper tail* of predictive variances, arising over the domain $\mathcal{X}$. Kouri *et al.* (2022) show that the R-optimality criterion satisfies standard properties of classical design criteria, e.g. that it is an 'information function' in the sense of Pukelsheim (2006, Chapter 5.8), on the set of positive semi-definite matrices F, and also show how to compute its gradient (or subgradient) to facilitate optimization over designs ξ. In a nonlinear extension, the authors handle the θ-dependence of the information matrix F by averaging the R-optimality criterion over samples of θ.

6.3. *OED in practice*

This article has focused on mathematical/statistical formulations of the optimal design problem, and on computational methods for producing designs according to these formulations. We have not emphasized specific applications, but nonetheless it remains crucial to appreciate the interplay between the reality of practical applications, ways of formulating an OED problem, and the many *modelling* choices therein. These modelling choices certainly may have computational implications.

For instance, what if the experiments we perform have a non-zero chance of failing: a computational simulation could abort or fail to converge, or a laboratory experiment could be interrupted or cancelled, for a variety of reasons. If the probability of failure is known, or can be parametrized and learned, it can become part of the statistical model of the experiment and hence the utility function, with failed experiments returning zero or negative utility. Similarly, if the designs realized in practice might not precisely match the intended design, this mismatch too can become part of the model; one could put $\xi' = \xi + \eta$, for some random variable η, or one could convolve the utility with a kernel $q(\xi'|\xi)$.

Sequential OED, as always, raises more complex possibilities. The non-myopic formulations discussed in Section 5 involve assessing information gain from future experiments, but what if the environment changes before these experiments can be executed? The experimental horizon might be cut short, the space of feasible designs Ξ_k at some future stage might change due to supply or personnel constraints, or the timing of planned experiments might have to be altered. Robust policies able to hedge against these possibilities would be highly valuable.

Another important question for OED, whether batch or sequential, is how to 'validate' the designs that are produced. We can interpret this question in many ways, but an initial version is to ask how to verify that the designs produced by an OED procedure are in fact optimal. One possible approach, following the definition of expected utility in (2.13), is to generate prior samples of Θ and Y at different values of ξ and to scrutinize the results of estimation or inference. For example, when maximizing the expected information gain (2.14), we can check if on average (over the Y samples), the posteriors at the optimal design have lower entropy than those at other designs. Similarly, we could evaluate the Bayes risk of point estimators at different designs. Yet such checks are ultimately *internal* – i.e. based only on simulation – and only reveal potential inconsistencies or errors in calculating and maximizing the intended expected utility.

Notions of *external* validation, involving conducting the actual experiments being designed, are conceptually much more challenging. One difficulty arises from the fact that any experiment necessarily produces Y values, and hence realized information gains, based on nature's 'true' data-generating distribution. In the well-specified case, this distribution corresponds to *one* value of Θ; otherwise it corresponds to *no* value of Θ. While we could perform repeated experiments for a given design, or for a variety of feasible designs, it is generally not possible or meaningful to perform real experiments for 'other' values of Θ or other data-generating distributions. In other words, we cannot sample from the prior. The real-world setting thus differs fundamentally from the OED methodology itself. In a Bayesian OED procedure, we introduce a prior distribution over Θ in order to reflect our belief or lack of knowledge about the parameters, *not* some intrinsic variability in the data-generating process. And the way in which we define optimality of the design requires the specification of this prior. At the same time, this situation is perhaps as it should be: when a decision (here the choice of design) is made under uncertainty, the correct decision is one conditioned only on the information available at the time of decision-making. Correctness should not be judged by how things actually turned out to be.

A broader practical challenge here is to encode the many complexities of real-world experimentation into the elements of an OED formulation that we have discussed throughout this article. For instance, in human subject experiments, the space of feasible designs Ξ must adhere to institutional review board requirements, and to other regulatory and ethical standards. Laboratory protocols for biological experiments, intended to ensure safety and minimize the potential for contamination, should similarly be reflected in Ξ. In Bayesian design, the choice of prior is crucial, and in some settings creating a suitable prior might involve techniques of elicitation (O'Hagan *et al.* 2006). Even the design criterion could be refined to reflect specialized, expert knowledge. For instance, the information-based objectives we have discussed could be shaped or augmented using past examples of how experiments have been selected; here, an interesting approach could involve *inverse* reinforcement learning (Ng and Russell 2000), which seeks to estimate unknown rewards from data describing how an agent selected actions (experiments) under varying conditions. The ability to tailor and constrain elements of the OED problem in these ways are not only of interest mathematically: they might help improve the overall trust that practitioners place in OED methodologies, and stimulate the adoption of OED in new applications.

Acknowledgements
We are grateful to all of our students, postdocs and other collaborators – too many to list by name – who have shaped our thinking on optimal experimental design and related subjects, and who contributed to the work described here. XH acknowledges support from the US Department of Energy (DOE), Office of Science, Office of Advanced Scientific Computing Research (ASCR), under award numbers DE-SC0021397 and DE-SC0021398. JJ and YMM acknowledge support from DOE ASCR under award number DE-SC0023188. JJ further acknowledges that this work was performed under the auspices of the DOE by Lawrence Livermore National Laboratory under contract DE-AC52-07NA27344, Release number LLNL-JRNL-861483. YMM further acknowledges support from DOE ASCR awards DE-SC0023187 and DE-SC0021226, from the Air Force Office of Scientific Research under award FA9550-20-1-0397, and from the Office of Naval Research under award N00014-20-1-2595. We also thank our families for their patience and support throughout the writing of this article.

References

P. K. Agarwal, S. Har-Peled and K. R. Varadarajan (2005), Geometric approximation via coresets, *Combin. Comput. Geom.* **52**, 1–30.

R. Aggarwal, M. J. Demkowicz and Y. M. Marzouk (2016), Information-driven experimental design in materials science, in *Information Science for Materials Discovery and Design* (T. Lookman, F. Alexander and K. Rajan, eds), Springer, pp. 13–44.

A. Alexanderian (2021), Optimal experimental design for infinite-dimensional Bayesian inverse problems governed by PDEs: A review, *Inverse Problems* **37**, art. 043001.

A. Alexanderian and A. K. Saibaba (2018), Efficient D-optimal design of experiments for infinite-dimensional Bayesian linear inverse problems, *SIAM J. Sci. Comput.* **40**, A2956–A2985.

A. Alexanderian, P. J. Gloor and O. Ghattas (2016*a*), On Bayesian A- and D-optimal experimental designs in infinite dimensions, *Bayesian Anal.* **11**, 671–695.

A. Alexanderian, R. Nicholson and N. Petra (2022), Optimal design of large-scale nonlinear Bayesian inverse problems under model uncertainty. Available at arXiv:2211.03952.

A. Alexanderian, N. Petra, G. Stadler and O. Ghattas (2014), A-optimal design of experiments for infinite-dimensional Bayesian linear inverse problems with regularized ℓ_0-sparsification, *SIAM J. Sci. Comput.* **36**, A2122–A2148.

A. Alexanderian, N. Petra, G. Stadler and O. Ghattas (2016*b*), A fast and scalable method for A-optimal design of experiments for infinite-dimensional Bayesian nonlinear inverse problems, *SIAM J. Sci. Comput.* **38**, A243–A272.

A. Alexanderian, N. Petra, G. Stadler and I. Sunseri (2021), Optimal design of large-scale Bayesian linear inverse problems under reducible model uncertainty: Good to know what you don't know, *SIAM/ASA J. Uncertain. Quantif.* **9**, 163–184.

Z. Allen-Zhu, Y. Li, A. Singh and Y. Wang (2017), Near-optimal design of experiments via regret minimization, in *Proceedings of the 34th International Conference on Machine Learning (ICML 2017)*, Vol. 70 of Proceedings of Machine Learning Research, PMLR, pp. 126–135.

Z. Allen-Zhu, Z. Liao and L. Orecchia (2015), Spectral sparsification and regret minimization beyond matrix multiplicative updates, in *Proceedings of the 47th Annual ACM Symposium on Theory of Computing (STOC 2015)*, ACM, pp. 237–245.

Z. Ao and J. Li (2020), An approximate KLD based experimental design for models with intractable likelihoods, in *Proceedings of the 23rd International Conference on Artificial Intelligence and Statistics*, Vol. 108 of Proceedings of Machine Learning Research, PMLR, pp. 3241–3251.

Z. Ao and J. Li (2024), On estimating the gradient of the expected information gain in Bayesian experimental design, in *Proceedings of the 38th AAAI Conference on Artificial Intelligence* (M. Wooldridge, J. Dy and S. Natarajan, eds), AAAI Press, pp. 20311–20319.

P. Artzner, F. Delbaen, J. Eber and D. Heath (1999), Coherent measures of risk, *Math. Finance* **9**, 203–228.

S. Asmussen and P. W. Glynn (2007), *Stochastic Simulation: Algorithms and Analysis*, Springer.

A. C. Atkinson, A. N. Donev and R. D. Tobias (2007), *Optimum Experimental Designs, with SAS*, Oxford University Press.

A. Attia, A. Alexanderian and A. K. Saibaba (2018), Goal-oriented optimal design of experiments for large-scale Bayesian linear inverse problems, *Inverse Problems* **34**, art. 095009.

A. Attia, S. Leyffer and T. Munson (2023), Robust A-optimal experimental design for Bayesian inverse problems. Available at arXiv:2305.03855.

C. L. Atwood (1969), Optimal and efficient designs of experiments, *Ann. Math. Statist.* **40**, 1570–1602.

C. Audet (2004), Convergence results for generalized pattern search algorithms are tight, *Optim. Engrg* **5**, 101–122.

C. Audet and J. E. Dennis (2002), Analysis of generalized pattern searches, *SIAM J. Optim.* **13**, 889–903.

F. Bach (2013), Learning with submodular functions: A convex optimization perspective, *Found. Trends Mach. Learn.* **6**, 145–373.

F. R. Bach and M. I. Jordan (2002), Kernel independent component analysis, *J. Mach. Learn. Res.* **3**, 1–48.

R. Baptista, L. Cao, J. Chen, O. Ghattas, F. Li, Y. M. Marzouk and J. T. Oden (2024), Bayesian model calibration for block copolymer self-assembly: Likelihood-free inference and expected information gain computation via measure transport, *J. Comput. Phys.* **503**, art. 112844.

R. Baptista, B. Hosseini, N. B. Kovachki and Y. Marzouk (2023*a*), Conditional sampling with monotone GANs: From generative models to likelihood-free inference. Available at arXiv:2006.06755.

R. Baptista, Y. Marzouk and O. Zahm (2022), Gradient-based data and parameter dimension reduction for Bayesian models: An information theoretic perspective. Available at arXiv:2207.08670.

R. Baptista, Y. Marzouk and O. Zahm (2023*b*), On the representation and learning of monotone triangular transport maps, *Found. Comput. Math.* Available at doi:10.1007/s10208-023-09630-x.

D. Barber and F. Agakov (2003), The IM algorithm: A variational approach to information maximization, in *Advances in Neural Information Processing Systems 16*, MIT Press, pp. 201–208.

J. D. Batson, D. A. Spielman and N. Srivastava (2009), Twice-Ramanujan sparsifiers, in *Proceedings of the 41st Annual ACM Symposium on Theory of Computing (STOC 2009)*, ACM, pp. 255–262.

J. Beck and S. Guillas (2016), Sequential design with mutual information for computer experiments (MICE): Emulation of a tsunami model, *SIAM/ASA J. Uncertain. Quantif.* **4**, 739–766.

J. Beck, B. M. Dia, L. Espath and R. Tempone (2020), Multilevel double loop Monte Carlo and stochastic collocation methods with importance sampling for Bayesian optimal experimental design, *Int. J. Numer. Methods Engrg* **121**, 3482–3503.

J. Beck, B. M. Dia, L. F. Espath, Q. Long and R. Tempone (2018), Fast Bayesian experimental design: Laplace-based importance sampling for the expected information gain, *Comput. Methods Appl. Mech. Engrg* **334**, 523–553.

M. I. Belghazi, A. Baratin, S. Rajeswar, S. Ozair, Y. Bengio, A. Courville and R. D. Hjelm (2018), Mutual information neural estimation, in *Proceedings of the 35th International Conference on Machine Learning (ICML 2018)*, Vol. 80 of Proceedings of Machine Learning Research, PMLR, pp. 531–540.

A. Ben-Tal and A. Nemirovski (2001), *Lectures on Modern Convex Optimization*, SIAM.

P. Benner, S. Gugercin and K. Willcox (2015), A survey of projection-based model reduction methods for parametric dynamical systems, *SIAM Rev.* **57**, 483–531.

J. O. Berger (1985), *Statistical Decision Theory and Bayesian Analysis*, Springer Series in Statistics, Springer.

J. O. Berger (1994), An overview of robust Bayesian analysis (with discussion), *Test* **3**, 5–124.

M. P. F. Berger and W. K. Wong (2009), *An Introduction to Optimal Designs for Social and Biomedical Research*, Wiley.

J. M. Bernardo (1979), Expected information as expected utility, *Ann. Statist.* **7**, 686–690.

J. M. Bernardo and A. F. M. Smith (2000), *Bayesian Theory*, Wiley.

S. M. Berry, B. P. Carlin, J. J. Lee and P. Müller (2010), *Bayesian Adaptive Methods for Clinical Trials*, Chapman & Hall/CRC.

D. P. Bertsekas (2005), *Dynamic Programming and Optimal Control*, Vol. 1, Athena Scientific.

S. Bhatnagar, H. L. Prasad and L. A. Prashanth (2013), *Stochastic Recursive Algorithms for Optimization*, Springer.

A. A. Bian, J. M. Buhmann, A. Krause and S. Tschiatschek (2017), Guarantees for greedy maximization of non-submodular functions with applications, in *Proceedings of the 34th International Conference on Machine Learning (ICML 2017)* (D. Precup and Y. W. Teh, eds), Vol. 70 of Proceedings of Machine Learning Research, PMLR, pp. 498–507.

D. Blackwell (1951), Comparison of experiments, in *Proceedings of the 2nd Berkeley Symposium on Mathematical Statistics and Probability*, University of California Press, pp. 93–102.

D. Blackwell (1953), Equivalent comparisons of experiments, *Ann. Math. Statist.* **24**, 265–272.

A. Blanchard and T. Sapsis (2021), Output-weighted optimal sampling for Bayesian experimental design and uncertainty quantification, *SIAM/ASA J. Uncertain. Quantif.* **9**, 564–592.

T. Blau, E. V. Bonilla, I. Chades and A. Dezfouli (2022), Optimizing sequential experimental design with deep reinforcement learning, in *Proceedings of the 39th International Conference on Machine Learning (ICML 2022)* (K. Chaudhuri *et al.*, eds), Vol. 162 of Proceedings of Machine Learning Research, PMLR, pp. 2107–2128.

J. R. Blum (1954), Multidimensional stochastic approximation methods, *Ann. Math. Statist.* **25**, 737–744.

N. Bochkina (2019), Bernstein–von Mises theorem and misspecified models: A review, in *Foundations of Modern Statistics* (D. Belomestny *et al.*, eds), Vol. 425 of Springer Proceedings in Mathematics & Statistics, Springer, pp. 355–380.

V. I. Bogachev, A. V. Kolesnikov and K. V. Medvedev (2005), Triangular transformations of measures, *Sbornik Math.* **196**, art. 309.

I. Bogunovic, J. Zhao and V. Cevher (2018), Robust maximization of non-submodular objectives, in *Proceedings of the Twenty-First International Conference on Artificial Intelligence and Statistics* (A. Storkey and F. Perez-Cruz, eds), Vol. 84 of Proceedings of Machine Learning Research, PMLR, pp. 890–899.

C. Borges and G. Biros (2018), Reconstruction of a compactly supported sound profile in the presence of a random background medium, *Inverse Problems* **34**, art. 115007.

R. C. Bose (1939), On the construction of balanced incomplete block designs, *Ann. Eugen.* **9**, 353–399.

R. C. Bose and K. R. Nair (1939), Partially balanced incomplete block designs, *Sankhyā* **4**, 337–372.

G. E. P. Box (1992), Sequential experimentation and sequential assembly of designs, *Qual. Engrg* **5**, 321–330.

S. P. Boyd and L. Vandenberghe (2004), *Convex Optimization*, Cambridge University Press.

A. E. Brockwell and J. B. Kadane (2003), A gridding method for Bayesian sequential decision problems, *J. Comput. Graph. Statist.* **12**, 566–584.

N. Buchbinder, M. Feldman, J. Naor and R. Schwartz (2014), Submodular maximization with cardinality constraints, in *Proceedings of the Twenty-Fifth Annual ACM-SIAM Symposium on Discrete Algorithms*, SIAM, pp. 1433–1452.

T. Bui-Thanh, O. Ghattas, J. Martin and G. Stadler (2013), A computational framework for infinite-dimensional Bayesian inverse problems I: The linearized case, with application to global seismic inversion, *SIAM J. Sci. Comput.* **35**, A2494–A2523.

R. E. Caflisch (1998), Monte Carlo and quasi-Monte Carlo methods, *Acta Numer.* **7**, 1–49.

G. Calinescu, C. Chekuri, M. Pál and J. Vondrák (2011), Maximizing a monotone submodular function subject to a matroid constraint, *SIAM J. Comput.* **40**, 1740–1766.

T. Campbell and B. Beronov (2019), Sparse variational inference: Bayesian coresets from scratch, in *Advances in Neural Information Processing Systems 32* (H. Wallach *et al.*, eds), Curran Associates, pp. 11461–11472.

T. Campbell and T. Broderick (2018), Bayesian coreset construction via greedy iterative geodesic ascent, in *Proceedings of the 35th International Conference on Machine Learning (ICML 2018)*, Vol. 80 of Proceedings of Machine Learning Research, PMLR, pp. 698–706.

T. Campbell and T. Broderick (2019), Automated scalable Bayesian inference via Hilbert coresets, *J. Mach. Learn. Res.* **20**, 551–588.

G. Carlier, V. Chernozhukov and A. Galichon (2016), Vector quantile regression: An optimal transport approach, *Ann. Statist.* **44**, 1165–1192.

B. P. Carlin, J. B. Kadane and A. E. Gelfand (1998), Approaches for optimal sequential decision analysis in clinical trials, *Biometrics* **54**, 964–975.

A. G. Carlon, B. M. Dia, L. Espath, R. H. Lopez and R. Tempone (2020), Nesterov-aided stochastic gradient methods using Laplace approximation for Bayesian design optimization, *Comput. Methods Appl. Mech. Engrg* **363**, art. 112909.

C. U. Carmona and G. K. Nicholls (2020), Semi-modular inference: Enhanced learning in multi-modular models by tempering the influence of components, in *Proceedings of the 23rd International Conference on Artificial Intelligence and Statistics*, Vol. 108 of Proceedings of Machine Learning Research, PMLR, pp. 4226–4235.

W. F. Caselton and J. V. Zidek (1984), Optimal monitoring network designs, *Statist. Probab. Lett.* **2**, 223–227.

D. R. Cavagnaro, J. I. Myung, M. A. Pitt and J. V. Kujala (2010), Adaptive design optimization: A mutual information-based approach to model discrimination in cognitive science, *Neural Comput.* **22**, 887–905.

K. Chaloner (1984), Optimal Bayesian experimental design for linear models, *Ann. Statist.* **12**, 283–300.

K. Chaloner and K. Larntz (1989), Optimal Bayesian design applied to logistic regression experiments, *J. Statist. Plann. Infer.* **21**, 191–208.

K. Chaloner and I. Verdinelli (1995), Bayesian experimental design: A review, *Statist. Sci.* **10**, 273–304.

K.-H. Chang (2012), Stochastic Nelder–Mead simplex method: A new globally convergent direct search method for simulation optimization, *European J. Oper. Res.* **220**, 684–694.

C. Chekuri, J. Vondrák and R. Zenklusen (2014), Submodular function maximization via the multilinear relaxation and contention resolution schemes, *SIAM J. Comput.* **43**, 1831–1879.

X. Chen, C. Wang, Z. Zhou and K. Ross (2021), Randomized ensembled double Q-learning: Learning fast without a model, in *9th International Conference on Learning Representations (ICLR 2021)*. Available at https://openreview.net/forum?id=AY8zfZm0tDd.

C. Chevalier and D. Ginsbourger (2013), Fast computation of the multi-points expected improvement with applications in batch selection, in *Learning and Intelligent Optimization*, Vol. 7997 of Lecture Notes in Computer Science, Springer, pp. 59–69.

A. Chowdhary, S. Tong, G. Stadler and A. Alexanderian (2023), Sensitivity analysis of the information gain in infinite-dimensional Bayesian linear inverse problems. Available at arXiv:2310.16906.

J. A. Christen and M. Nakamura (2003), Sequential stopping rules for species accumulation, *J. Agric. Biol. Environ. Statist.* **8**, 184–195.

M. A. Clyde (2001), Experimental design: Bayesian designs, in *International Encyclopedia of the Social & Behavioral Sciences* (N. J. Smelser and P. B. Baltes, eds), Science Direct, pp. 5075–5081.

D. A. Cohn, Z. Ghahramani and M. I. Jordan (1996), Active learning with statistical models, *J. Artificial Intelligence Res.* **4**, 129–145.

M. Conforti and G. Cornuéjols (1984), Submodular set functions, matroids and the greedy algorithm: Tight worst-case bounds and some generalizations of the Rado–Edmonds theorem, *Discrete Appl. Math.* **7**, 251–274.

A. R. Conn, K. Scheinberg and L. N. Vicente (2009), *Introduction to Derivative-Free Optimization*, SIAM.

R. D. Cook and C. J. Nachtsheim (1980), A comparison of algorithms for constructing exact D-optimal designs, *Technometrics* **22**, 315–324.

S. L. Cotter, G. O. Roberts, A. M. Stuart and D. White (2013), MCMC methods for functions: Modifying old algorithms to make them faster, *Statist. Sci.* **28**, 424–446.

T. A. Cover and J. A. Thomas (2006), *Elements of Information Theory*, second edition, Wiley.

R. T. Cox (1946), Probability, frequency and reasonable expectation, *Amer. J. Phys.* **14**, 1–13.

C. C. Craig and R. A. Fisher (1936), The design of experiments, *Amer. Math. Monthly* **43**, 180.

T. Cui and X. T. Tong (2022), A unified performance analysis of likelihood-informed subspace methods, *Bernoulli* **28**, 2788–2815.

T. Cui, S. Dolgov and O. Zahm (2023), Scalable conditional deep inverse Rosenblatt transports using tensor trains and gradient-based dimension reduction, *J. Comput. Phys.* **485**, art. 112103.

T. Cui, K. J. H. Law and Y. M. Marzouk (2016), Dimension-independent likelihood-informed MCMC, *J. Comput. Phys.* **304**, 109–137.

T. Cui, J. Martin, Y. M. Marzouk, A. Solonen and A. Spantini (2014), Likelihood-informed dimension reduction for nonlinear inverse problems, *Inverse Problems* **30**, art. 114015.

P. Czyż, F. Grabowski, J. Vogt, N. Beerenwinkel and A. Marx (2023), Beyond normal: On the evaluation of mutual information estimators, in *Advances in Neural Information Processing Systems 36* (A. Oh *et al.*, eds), Curran Associates, pp. 16957–16990.

A. Das and D. Kempe (2011), Submodular meets spectral: Greedy algorithms for subset selection, sparse approximation and dictionary selection, in *Proceedings of the 28th International Conference on Machine Learning (ICML 2011)* (L. Getoor and T. Scheffer, eds), ACM, pp. 1057–1064.

A. DasGupta (1995), Review of optimal Bayes designs. Technical report, Purdue University, West Lafayette, IN.

S. Dasgupta (2011), Two faces of active learning, *Theoret. Comput. Sci.* **412**, 1767–1781.

M. Dashti and A. M. Stuart (2017), The Bayesian approach to inverse problems, in *Handbook of Uncertainty Quantification* (R. Ghanem, D. Higdon and H. Owhadi, eds), Springer, pp. 311–428.

M. Dashti, K. J. Law, A. M. Stuart and J. Voss (2013), MAP estimators and their consistency in Bayesian nonparametric inverse problems, *Inverse Problems* **29**, art. 095017.

M. Dewaskar, C. Tosh, J. Knoblauch and D. B. Dunson (2023), Robustifying likelihoods by optimistically re-weighting data. Available at arXiv:2303.10525.

J. Dick, F. Y. Kuo and I. H. Sloan (2013), High-dimensional integration: The quasi-Monte Carlo way, *Acta Numer.* **22**, 133–288.

J. Dong, C. Jacobsen, M. Khalloufi, M. Akram, W. Liu, K. Duraisamy and X. Huan (2024), Variational Bayesian optimal experimental design with normalizing flows. Available at arXiv:2404.13056.

M. D. Donsker and S. R. S. Varadhan (1983), Asymptotic evaluation of certain Markov process expectations for large time IV, *Commun. Pure Appl. Math.* **36**, 183–212.

H. A. Dror and D. M. Steinberg (2008), Sequential experimental designs for generalized linear models, *J. Amer. Statist. Assoc.* **103**, 288–298.

C. C. Drovandi, J. M. McGree and A. N. Pettitt (2013), Sequential Monte Carlo for Bayesian sequentially designed experiments for discrete data, *Comput. Statist. Data Anal.* **57**, 320–335.

C. C. Drovandi, J. M. McGree and A. N. Pettitt (2014), A sequential Monte Carlo algorithm to incorporate model uncertainty in Bayesian sequential design, *J. Comput. Graph. Statist.* **23**, 3–24.

T. E. Duncan (1970), On the calculation of mutual information, *SIAM J. Appl. Math.* **19**, 215–220.

D.-L. Duong, T. Helin and J. R. Rojo-Garcia (2023), Stability estimates for the expected utility in Bayesian optimal experimental design, *Inverse Problems* **39**, art. 125008.

T. A. El Moselhy and Y. M. Marzouk (2012), Bayesian inference with optimal maps, *J. Comput. Phys.* **231**, 7815–7850.

G. Elfving (1952), Optimum allocation in linear regression theory, *Ann. Math. Statist.* **23**, 255–262.

Y. Englezou, T. W. Waite and D. C. Woods (2022), Approximate Laplace importance sampling for the estimation of expected Shannon information gain in high-dimensional Bayesian design for nonlinear models, *Statist. Comput.* **32**, art. 82.

A. Eskenazis and Y. Shenfeld (2024), Intrinsic dimensional functional inequalities on model spaces, *J. Funct. Anal.* **286**, art. 110338.

K. Fan (1967), Subadditive functions on a distributive lattice and an extension of Szász's inequality, *J. Math. Anal. Appl.* **18**, 262–268.

K. Fan (1968), An inequality for subadditive functions on a distributive lattice, with application to determinantal inequalities, *Linear Algebra Appl.* **1**, 33–38.

V. V. Fedorov (1972), *Theory of Optimal Experiments*, Academic Press.

V. V. Fedorov (1996), Design of spatial experiments: Model fitting and prediction. Technical report, Oak Ridge National Laboratory, Oak Ridge, TN.

V. V. Fedorov and D. Flanagan (1997), Optimal monitoring network design based on Mercer's expansion of covariance kernel, *J. Combin. Inform. System Sci.* **23**, 237–250.

V. V. Fedorov and P. Hackl (1997), *Model-Oriented Design of Experiments*, Vol. 125 of Lecture Notes in Statistics, Springer.

V. V. Fedorov and W. G. Müller (2007), Optimum design for correlated fields via covariance kernel expansions, in *mODa 8: Advances in Model-Oriented Design and Analysis* (J. López-Fidalgo, J. M. Rodríguez-Díaz and B. Torsney, eds), Contributions to Statistics, Physica, Springer, pp. 57–66.

D. Feldman and M. Langberg (2011), A unified framework for approximating and clustering data, in *Proceedings of the 43rd Annual ACM Symposium on Theory of Computing (STOC 2011)*, ACM, pp. 569–578.

C. Feng and Y. M. Marzouk (2019), A layered multiple importance sampling scheme for focused optimal Bayesian experimental design. Available at arXiv:1903.11187.

M. L. Fisher, G. L. Nemhauser and L. A. Wolsey (1978), An analysis of approximations for maximizing submodular set functions II, *Math. Program.* **8**, 73–87.

R. A. Fisher (1936), Design of experiments, *Brit. Med. J.* **1**(3923), 554.

I. Ford, D. M. Titterington and C. P. Kitsos (1989), Recent advances in nonlinear experimental design, *Technometrics* **31**, 49–60.

A. Foster, D. R. Ivanova, I. Malik and T. Rainforth (2021), Deep adaptive design: Amortizing sequential Bayesian experimental design, in *Proceedings of the 38th International Conference on Machine Learning (ICML 2021)* (M. Meila and T. Zhang, eds), Vol. 139 of Proceedings of Machine Learning Research, PMLR, pp. 3384–3395.

A. Foster, M. Jankowiak, E. Bingham, P. Horsfall, Y. W. Teh, T. Rainforth and N. Goodman (2019), Variational Bayesian optimal experimental design, in *Advances in Neural Information Processing Systems 32* (H. Wallach *et al.*, eds), Curran Associates, pp. 14036–14047.

A. Foster, M. Jankowiak, M. O'Meara, Y. W. Teh and T. Rainforth (2020), A unified stochastic gradient approach to designing Bayesian-optimal experiments, in *Proceedings of the 23rd International Conference on Artificial Intelligence and Statistics*, Vol. 108 of Proceedings of Machine Learning Research, PMLR, pp. 2959–2969.

P. I. Frazier (2018), Bayesian optimization, *INFORMS TutORials in Operations Research* **2018**, 255–278.

Y. Freund, H. S. Seung, E. Shamir and N. Tishby (1997), Selective sampling using the query by committee algorithm, *Mach. Learn.* **28**, 133–168.

S. Fujishige (2005), *Submodular Functions and Optimization*, Vol. 58 of Annals of Discrete Mathematics, second edition, Elsevier.

F. R. Gantmacher and M. G. Kreĭn (1960), *Oszillationsmatrizen, Oszillationskerne und kleine Schwingungen mechanischer Systeme*, Vol. 5, Akademie.

W. Gao, S. Oh and P. Viswanath (2018), Demystifying fixed k-nearest neighbor information estimators, *IEEE Trans. Inform. Theory* **64**, 5629–5661.

R. Gautier and L. Pronzato (2000), Adaptive control for sequential design, *Discuss. Math. Probab. Statist.* **20**, 97–113.

O. Ghattas and K. Willcox (2021), Learning physics-based models from data: Perspectives from inverse problems and model reduction, *Acta Numer.* **30**, 445–554.

M. B. Giles (2015), Multilevel Monte Carlo methods, *Acta Numer.* **24**, 259–328.

E. Giné and R. Nickl (2021), *Mathematical Foundations of Infinite-Dimensional Statistical Models*, Cambridge University Press.

J. Ginebra (2007), On the measure of the information in a statistical experiment, *Bayesian Anal.* **2**, 167–212.

L. Giraldi, O. P. Le Maître, I. Hoteit and O. M. Knio (2018), Optimal projection of observations in a Bayesian setting, *Comput. Statist. Data Anal.* **124**, 252–276.

T. Gneiting and A. E. Raftery (2007), Strictly proper scoring rules, prediction, and estimation, *J. Amer. Statist. Assoc.* **102**, 359–378.

J. Go and T. Isaac (2022), Robust expected information gain for optimal Bayesian experimental design using ambiguity sets, in *Proceedings of the 38th Conference on Uncertainty in Artificial Intelligence*, Vol. 180 of Proceedings of Machine Learning Research, PMLR, pp. 718–727.

T. Goda, T. Hironaka, W. Kitade and A. Foster (2022), Unbiased MLMC stochastic gradient-based optimization of Bayesian experimental designs, *SIAM J. Sci. Comput.* **44**, A286–A311.

T. Goda, T. Hironaka and T. Iwamoto (2020), Multilevel Monte Carlo estimation of expected information gains, *Stoch. Anal. Appl.* **38**, 581–600.

A. Gorodetsky and Y. Marzouk (2016), Mercer kernels and integrated variance experimental design: Connections between Gaussian process regression and polynomial approximation, *SIAM/ASA J. Uncertain. Quantif.* **4**, 796–828.

R. B. Gramacy (2020), *Surrogates: Gaussian Process Modeling, Design, and Optimization for the Applied Sciences*, CRC Press.

R. B. Gramacy (2022), plgp: Particle learning of Gaussian processes. Available at https://cran.r-project.org/package=plgp.

R. B. Gramacy and D. W. Apley (2015), Local Gaussian process approximation for large computer experiments, *J. Comput. Graph. Statist.* **24**, 561–578.

P. Grünwald and T. van Ommen (2017), Inconsistency of Bayesian inference for misspecified linear models, and a proposal for repairing it, *Bayesian Anal.* **12**, 1069–1103.

G. Gürkan, A. Y. Özge and S. M. Robinson (1994), Sample-path optimization in simulation, in *Proceedings of the 1994 Winter Simulation Conference (WSC '94)* (D. A. Sadowski *et al.*, eds), ACM, pp. 247–254.

E. Haber, L. Horesh and L. Tenorio (2008), Numerical methods for experimental design of large-scale linear ill-posed inverse problems, *Inverse Problems* **24**, art. 055012.

E. Haber, L. Horesh and L. Tenorio (2009), Numerical methods for the design of large-scale nonlinear discrete ill-posed inverse problems, *Inverse Problems* **26**, art. 025002.

M. Hainy, C. C. Drovandi and J. M. McGree (2016), Likelihood-free extensions for Bayesian sequentially designed experiments, in *mODa 11: Advances in Model-Oriented Design and Analysis* (J. Kunert, C. Müller and A. Atkinson, eds), Contributions to Statistics, Springer, pp. 153–161.

M. Hainy, D. J. Price, O. Restif and C. Drovandi (2022), Optimal Bayesian design for model discrimination via classification, *Statist. Comput.* **32**, art. 25.

M. Hairer, A. M. Stuart and J. Voss (2011), Signal processing problems on function space: Bayesian formulation, stochastic PDEs and effective MCMC methods, in *The Oxford Handbook of Nonlinear Filtering* (D. Crisan and B. Rozovskii, eds), Oxford University Press, pp. 833–873.

O. Harari and D. M. Steinberg (2014), Optimal designs for Gaussian process models via spectral decomposition, *J. Statist. Plann. Infer.* **154**, 87–101.

X. D. He, S. Kou and X. Peng (2022), Risk measures: Robustness, elicitability, and backtesting, *Annu. Rev. Statist. Appl.* **9**, 141–166.

K. Healy and L. W. Schruben (1991), Retrospective simulation response optimization, in *Proceedings of the 1991 Winter Simulation Conference (WSC '91)* (B. L. Nelson *et al.*, eds), IEEE Computer Society, pp. 901–906.

A. Hedayat (1981), Study of optimality criteria in design of experiments, in *Statistics and Related Topics: International Symposium Proceedings* (M. Csorgo, ed.), Elsevier Science, pp. 39–56.

T. Helin and R. Kretschmann (2022), Non-asymptotic error estimates for the Laplace approximation in Bayesian inverse problems, *Numer. Math.* **150**, 521–549.

T. Helin, N. Hyvönen and J.-P. Puska (2022), Edge-promoting adaptive Bayesian experimental design for X-ray imaging, *SIAM J. Sci. Comput.* **44**, B506–B530.

L. Herrmann, C. Schwab and J. Zech (2020), Deep neural network expression of posterior expectations in Bayesian PDE inversion, *Inverse Problems* **36**, art. 125011.

T. N. Hoang, B. K. H. Low, P. Jaillet and M. Kankanhalli (2014), Nonmyopic ε-Bayes-optimal active learning of Gaussian processes, in *Proceedings of the 31st International Conference on Machine Learning (ICML 2014)*, Vol. 32 of Proceedings of Machine Learning Research, PMLR, pp. 739–747.

D. S. Hochba (1997), Approximation algorithms for NP-hard problems, *ACM SIGACT News* **28**, 40–52.

X. Huan (2015), Numerical approaches for sequential Bayesian optimal experimental design. PhD thesis, Massachusetts Institute of Technology.

X. Huan and Y. M. Marzouk (2013), Simulation-based optimal Bayesian experimental design for nonlinear systems, *J. Comput. Phys.* **232**, 288–317.

X. Huan and Y. M. Marzouk (2014), Gradient-based stochastic optimization methods in Bayesian experimental design, *Int. J. Uncertain. Quantif.* **4**, 479–510.

X. Huan and Y. M. Marzouk (2016), Sequential Bayesian optimal experimental design via approximate dynamic programming. Available at arXiv:1604.08320.

C.-W. Huang, R. T. Q. Chen, C. Tsirigotis and A. Courville (2020), Convex potential flows: Universal probability distributions with optimal transport and convex optimization. Available at arXiv:2012.05942.

J. Huggins, T. Campbell and T. Broderick (2016), Coresets for scalable Bayesian logistic regression, in *Advances in Neural Information Processing Systems 29* (D. Lee *et al.*, eds), Curran Associates, pp. 4080–4088.

J. H. Huggins and J. W. Miller (2023), Reproducible model selection using bagged posteriors, *Bayesian Anal.* **18**, 79–104.

D. R. Ivanova, A. Foster, S. Kleinegesse, M. U. Gutmann and T. Rainforth (2021), Implicit deep adaptive design: Policy-based experimental design without likelihoods, in *Advances in Neural Information Processing Systems 34* (M. Ranzato *et al.*, eds), Curran Associates, pp. 25785–25798.

P. E. Jacob, L. M. Murray, C. C. Holmes and C. P. Robert (2017), Better together? Statistical learning in models made of modules. Available at arXiv:1708.08719.

J. Jagalur-Mohan and Y. Marzouk (2021), Batch greedy maximization of non-submodular functions: Guarantees and applications to experimental design, *J. Mach. Learn. Res.* **22**, 11397–11458.

E. T. Jaynes and G. L. Bretthorst (2003), *Probability Theory: The Logic of Science*, Cambridge University Press.

C. R. Johnson and W. W. Barrett (1985), Spanning-tree extensions of the Hadamard–Fischer inequalities, *Linear Algebra Appl.* **66**, 177–193.

M. E. Johnson and C. J. Nachtsheim (1983), Some guidelines for constructing exact D-optimal designs on convex design spaces, *Technometrics* **25**, 271–277.

M. E. Johnson, L. M. Moore and D. Ylvisaker (1990), Minimax and maximin distance designs, *J. Statist. Plann. Infer.* **26**, 131–148.

D. R. Jones, M. Schonlau and W. J. Welch (1998), Efficient global optimization of expensive black-box functions, *J. Global Optim.* **13**, 455–492.

V. R. Joseph, E. Gul and S. Ba (2015), Maximum projection designs for computer experiments, *Biometrika* **102**, 371–380.

V. R. Joseph, E. Gul and S. Ba (2020), Designing computer experiments with multiple types of factors: The MaxPro approach, *J. Qual. Technol.* **52**, 343–354.

A. Jourdan and J. Franco (2010), Optimal Latin hypercube designs for the Kullback–Leibler criterion, *AStA Adv. Statist. Anal.* **94**, 341–351.

L. P. Kaelbling, M. L. Littman and A. R. Cassandra (1998), Planning and acting in partially observable stochastic domains, *Artif. Intell.* **101**, 99–134.

L. P. Kaelbling, M. L. Littman and A. W. Moore (1996), Reinforcement learning: A survey, *J. Artif. Intell. Res.* **4**, 237–285.

J. Kaipio and V. Kolehmainen (2013), Approximate marginalization over modelling errors and uncertainties in inverse problems, in *Bayesian Theory and Applications* (P. Damien *et al.*, eds), Oxford University Press, pp. 644–672.

J. Kaipio and E. Somersalo (2006), *Statistical and Computational Inverse Problems*, Vol. 160 of Applied Mathematical Sciences, Springer.

J. Kaipio and E. Somersalo (2007), Statistical inverse problems: Discretization, model reduction and inverse crimes, *J. Comput. Appl. Math.* **198**, 493–504.

O. Karaca and M. Kamgarpour (2018), Exploiting weak supermodularity for coalition-proof mechanisms, in *2018 IEEE Conference on Decision and Control (CDC)*, IEEE, pp. 1118–1123.

K. Karhunen (1947), Über lineare Methoden in der Wahrscheinlichkeitsrechnung, *Am. Acad. Sci. Fennicade, Ser. A, I* **37**, 3–79.

A. K. Kelmans and B. N. Kimelfeld (1983), Multiplicative submodularity of a matrix's principal minor as a function of the set of its rows and some combinatorial applications, *Discrete Math.* **44**, 113–116.

N. Kennamer, S. Walton and A. Ihler (2023), Design amortization for Bayesian optimal experimental design, in *Proceedings of the 37th AAAI Conference on Artificial Intelligence* (B. Williams, Y. Chen and J. Neville, eds), AAAI Press, pp. 8220–8227.

M. C. Kennedy and A. O'Hagan (2001), Bayesian calibration of computer models, *J. R. Statist. Soc. Ser. B. Statist. Methodol.* **63**, 425–464.

J. Kiefer (1958), On the nonrandomized optimality and randomized nonoptimality of symmetrical designs, *Ann. Math. Statist.* **29**, 675–699.

J. Kiefer (1959), Optimum experimental designs, *J. R. Statist. Soc. Ser. B. Statist. Methodol.* **21**, 272–304.

J. Kiefer (1961*a*), Optimum designs in regression problems II, *Ann. Math. Statist.* **32**, 298–325.

J. Kiefer (1961*b*), Optimum experimental designs V, with applications to systematic and rotatable designs, in *Proceedings of the Fourth Berkeley Symposium on Mathematical Statistics and Probability*, Vol. 1, University of California Press, pp. 381–405.

J. Kiefer (1974), General equivalence theory for optimum designs (approximate theory), *Ann. Statist.* **2**, 849–879.

J. Kiefer and J. Wolfowitz (1952), Stochastic estimation of the maximum of a regression function, *Ann. Math. Statist.* **23**, 462–466.

J. Kiefer and J. Wolfowitz (1959), Optimum designs in regression problems, *Ann. Math. Statist.* **30**, 271–294.

J. Kiefer and J. Wolfowitz (1960), The equivalence of two extremum problems, *Canad. J. Statist.* **12**, 363–366.

W. Kim, M. A. Pitt, Z.-L. Lu, M. Steyvers and J. I. Myung (2014), A hierarchical adaptive approach to optimal experimental design, *Neural Comput.* **26**, 2565–2492.

J. King and W.-K. Wong (2000), Minimax D-optimal designs for the logistic model, *Biometrics* **56**, 1263–1267.

B. J. K. Kleijn and A. W. van der Vaart (2012), The Bernstein–von-Mises theorem under misspecification, *Electron. J. Statist.* **6**, 354–381.

S. Kleinegesse and M. U. Gutmann (2020), Bayesian experimental design for implicit models by mutual information neural estimation, in *Proceedings of the 37th International Conference on Machine Learning (ICML 2020)* (H. Daumé and A. Singh, eds), Vol. 119 of Proceedings of Machine Learning Research, PMLR, pp. 5316–5326.

S. Kleinegesse and M. U. Gutmann (2021), Gradient-based Bayesian experimental design for implicit models using mutual information lower bounds. Available at arXiv:2105.04379.

S. Kleinegesse, C. Drovandi and M. U. Gutmann (2021), Sequential Bayesian experimental design for implicit models via mutual information, *Bayesian Anal.* **16**, 773–802.

A. J. Kleywegt, A. Shapiro and T. Homem-de Mello (2002), The sample average approximation method for stochastic discrete optimization, *SIAM J. Optim.* **12**, 479–502.

B. T. Knapik, A. W. van der Vaart and J. H. van Zanten (2011), Bayesian inverse problems with Gaussian priors, *Ann. Statist.* **39**, 2626–2657.

H. Knothe (1957), Contributions to the theory of convex bodies, *Michigan Math. J.* **4**, 39–52.

C.-W. Ko, J. Lee and M. Queyranne (1995), An exact algorithm for maximum entropy sampling, *Oper. Res.* **43**, 684–691.

I. Kobyzev, S. J. Prince and M. A. Brubaker (2020), Normalizing flows: An introduction and review of current methods, *IEEE Trans. Pattern Anal. Mach. Intell.* **43**, 3964–3979.

V. R. Konda and J. N. Tsitsiklis (1999), Actor–critic algorithms, in *Advances in Neural Information Processing Systems 12* (S. Solla *et al.*, eds), MIT Press, pp. 1008–1014.

S. Korkel, I. Bauer, H. G. Bock and J. P. Schloder (1999), A sequential approach for nonlinear optimum experimental design in DAE systems, in *Scientific Computing in Chemical Engineering II* (F. Keil *et al.*, eds), Springer, pp. 338–345.

D. M. Kotelyanskiĭ (1950), On the theory of nonnegative and oscillating matrices, *Ukrains'kyi Matematychnyi Zhurnal* **2**, 94–101.

D. P. Kouri, J. D. Jakeman and J. Gabriel Huerta (2022), Risk-adapted optimal experimental design, *SIAM/ASA J. Uncertain. Quantif.* **10**, 687–716.

K. Koval, A. Alexanderian and G. Stadler (2020), Optimal experimental design under irreducible uncertainty for linear inverse problems governed by PDEs, *Inverse Problems* **36**, art. 075007.

K. Koval, R. Herzog and R. Scheichl (2024), Tractable optimal experimental design using transport maps. Available at arXiv:2401.07971.

L. F. Kozachenko and N. N. Leonenko (1987), A statistical estimate for the entropy of a random vector, *Probl. Inf. Transm.* **23**, 9–16.

A. Kraskov, H. Stögbauer and P. Grassberger (2004), Estimating mutual information, *Phys. Rev. E* **69**, art. 066138.

A. Krause and D. Golovin (2014), Submodular function maximization, in *Tractability, Practical Approaches to Hard Problems* (L. Bordeaux, Y. Hamadi and P. Kohli, eds), Cambridge University Press, pp. 71–104.

A. Krause, A. Singh and C. Guestrin (2008), Near-optimal sensor placements in Gaussian processes: Theory, efficient algorithms and empirical studies, *J. Mach. Learn. Res.* **9**, 235–284.

D. Kuhn, P. M. Esfahani, V. A. Nguyen and S. Shafieezadeh-Abadeh (2019), Wasserstein distributionally robust optimization: Theory and applications in machine learning, in *Operations Research & Management Science in the Age of Analytics*, INFORMS, pp. 130–166.

H. J. Kushner and G. G. Yin (2003), *Stochastic Approximation and Recursive Algorithms and Applications*, second edition, Springer.

R. Lam and K. Willcox (2017), Lookahead Bayesian optimization with inequality constraints, in *Advances in Neural Information Processing Systems 30* (I. Guyon *et al.*, eds), Curran Associates, pp. 1890–1900.

J. Larson, M. Menickelly and S. M. Wild (2019), Derivative-free optimization methods, *Acta Numer.* **28**, 287–404.

L. C. Lau and H. Zhou (2020), A spectral approach to network design, in *Proceedings of the 52nd Annual ACM SIGACT Symposium on Theory of Computing (STOC 2020)*, ACM, pp. 826–839.

L. C. Lau and H. Zhou (2022), A local search framework for experimental design, *SIAM J. Comput.* **51**, 900–951.

L. Le Cam (1964), Sufficiency and approximate sufficiency, *Ann. Math. Statist.* **35**, 1419–1455.

E. L. Lehmann and G. Casella (1998), *Theory of Point Estimation*, Springer Texts in Statistics, Springer.

J. Leskovec, A. Krause, C. Guestrin, C. Faloutsos, J. VanBriesen and N. Glance (2007), Cost-effective outbreak detection in networks, in *Proceedings of the 13th ACM SIGKDD International Conference on Knowledge Discovery and Data Mining*, ACM, pp. 420–429.

N. A. Letizia, N. Novello and A. M. Tonello (2023), Variational f-divergence and derangements for discriminative mutual information estimation. Available at arXiv:2305.20025.

D. D. Lewis (1995), A sequential algorithm for training text classifiers: Corrigendum and additional data, *SIGIR Forum* **29**, 13–19.

F. Li, R. Baptista and Y. Marzouk (2024*a*), Expected information gain estimation via density approximations: Sample allocation and dimension reduction. Forthcoming.

M. T. C. Li, Y. Marzouk and O. Zahm (2024*b*), Principal feature detection via ϕ-Sobolev inequalities. To appear in *Bernoulli*. Available at https://bernoullisociety.org/publications/bernoulli-journal/bernoulli-journal-papers.

J. Liepe, S. Filippi, M. Komorowski and M. P. H. Stumpf (2013), Maximizing the information content of experiments in systems biology, *PLoS Comput. Biol.* **9**, e1002888.

T. P. Lillicrap, J. J. Hunt, A. Pritzel, N. Heess, T. Erez, Y. Tassa, D. Silver and D. Wierstra (2016), Continuous control with deep reinforcement learning, in *Proceedings of the 4th International Conference on Learning Representations (ICLR 2016)* (Y. Bengio and Y. LeCun, eds).

D. V. Lindley (1956), On a measure of the information provided by an experiment, *Ann. Math. Statist.* **27**, 986–1005.

M. Loève (1948), Fonctions aléatoires du second ordre, in *Processus Stochastique et Mouvement Brownien* (P. Lévy, ed.), Gauthier-Villars.

Q. Long, M. Scavino, R. Tempone and S. Wang (2013), Fast estimation of expected information gains for Bayesian experimental designs based on Laplace approximations, *Comput. Methods Appl. Mech. Engrg* **259**, 24–39.

T. J. Loredo (2011), Rotating stars and revolving planets: Bayesian exploration of the pulsating sky, in *Bayesian Statistics 9: Proceedings of the Ninth Valencia International Meeting* (J. M. Bernardo, M. J. Bayarri and J. O. Berger, eds), Oxford University Press, pp. 361–392.

L. Lovász (1983), Submodular functions and convexity, in *Mathematical Programming The State of the Art: Bonn 1982* (A. Bachem, B. Korte and M. Grötschel, eds), Springer, pp. 235–257.

L. Lovász (2007), *Combinatorial Problems and Exercises*, second edition, American Mathematical Society.

D. J. C. MacKay (1992), Information-based objective functions for active data selection, *Neural Comput.* **4**, 590–604.

V. Madan, A. Nikolov, M. Singh and U. Tantipongpipat (2020), Maximizing determinants under matroid constraints, in *2020 IEEE 61st Annual Symposium on Foundations of Computer Science (FOCS)*, IEEE, pp. 565–576.

V. Madan, M. Singh, U. Tantipongpipat and W. Xie (2019), Combinatorial algorithms for optimal design, in *Proceedings of the 32nd Conference on Learning Theory*, Vol. 99 of Proceedings of Machine Learning Research, PMLR, pp. 2210–2258.

W.-K. Mak, D. P. Morton and R. K. Wood (1999), Monte Carlo bounding techniques for determining solution quality in stochastic programs, *Oper. Res. Lett.* **24**, 47–56.

T. Manole, S. Balakrishnan, J. Niles-Weed and L. Wasserman (2021), Plugin estimation of smooth optimal transport maps. Available at arXiv:2107.12364.

H. Markowitz (1952), Portfolio selection, *J. Finance* **7**, 77–91.

Y. Marzouk and D. Xiu (2009), A stochastic collocation approach to Bayesian inference in inverse problems, *Commun. Comput. Phys.* **6**, 826–847.

Y. Marzouk, T. Moselhy, M. Parno and A. Spantini (2016), Sampling via measure transport: An introduction, in *Handbook of Uncertainty Quantification* (R. Ghanem, D. Higdon and H. Owhadi, eds), Springer, pp. 1–41.

D. McAllester and K. Stratos (2020), Formal limitations on the measurement of mutual information, in *Proceedings of the 23rd International Conference on Artificial Intelligence and Statistics*, Vol. 108 of Proceedings of Machine Learning Research, PMLR, pp. 875–884.

J. McGree, C. Drovandi and A. Pettitt (2012), A sequential Monte Carlo approach to the sequential design for discriminating between rival continuous data models. Technical report, Queensland University of Technology, Brisbane, Australia.

M. D. McKay, R. J. Beckman and W. J. Conover (1979), A comparison of three methods for selecting values of input variables in the analysis of output from a computer code, *Technometrics* **21**, 239–245.

P. Mertikopoulos, N. Hallak, A. Kavis and V. Cevher (2020), On the almost sure convergence of stochastic gradient descent in non-convex problems, in *Advances in Neural Information Processing Systems 33* (H. Larochelle *et al.*, eds), Curran Associates, pp. 1117–1128.

R. K. Meyer and C. J. Nachtsheim (1995), The coordinate-exchange algorithm for constructing exact optimal experimental designs, *Technometrics* **37**, 60–69.

J. W. Miller and D. B. Dunson (2019), Robust Bayesian inference via coarsening, *J. Amer. Statist. Assoc.* **114**, 1113–1125.

B. Mirzasoleiman, A. Badanidiyuru, A. Karbasi, J. Vondrák and A. Krause (2015), Lazier than lazy greedy, in *Proceedings of the 29th AAAI Conference on Artificial Intelligence (AAAI 2015)*, pp. 1812–1818.

B. Mirzasoleiman, A. Karbasi, R. Sarkar and A. Krause (2013), Distributed submodular maximization: Identifying representative elements in massive data, in *Advances in Neural Information Processing Systems 26* (C. J. Burges *et al.*, eds), Curran Associates, pp. 2049–2057.

V. Mnih, K. Kavukcuoglu, D. Silver, A. A. Rusu, J. Veness, M. G. Bellemare, A. Graves, M. Riedmiller, A. K. Fidjeland, G. Ostrovski, S. Petersen, C. Beattie, A. Sadik, I. Antonoglou, H. King, D. Kumaran, D. Wierstra, S. Legg and D. Hassabis (2015), Human-level control through deep reinforcement learning, *Nature* **518**, 529–533.

J. Močkus (1975), On Bayesian methods for seeking the extremum, in *Optimization Techniques IFIP Technical Conference* (G. I. Marchuk, ed.), Springer, pp. 400–404.

S. Mohamed, M. Rosca, M. Figurnov and A. Mnih (2020), Monte Carlo gradient estimation in machine learning, *J. Mach. Learn. Res.* **21**, 5183–5244.

R. E. Morrison, T. A. Oliver and R. D. Moser (2018), Representing model inadequacy: A stochastic operator approach, *SIAM/ASA J. Uncertain. Quantif.* **6**, 457–496.

P. Müller, D. A. Berry, A. P. Grieve, M. Smith and M. Krams (2007), Simulation-based sequential Bayesian design, *J. Statist. Plann. Infer.* **137**, 3140–3150.

P. Müller, Y. Duan and M. Garcia Tec (2022), Simulation-based sequential design, *Pharma. Statist.* **21**, 729–739.

S. A. Murphy (2003), Optimal dynamic treatment regimes, *J. R. Statist. Soc. Ser. B. Statist. Methodol.* **65**, 331–366.

J. I. Myung and M. A. Pitt (2009), Optimal experimental design for model discrimination, *Psychol. Rev.* **116**, 499–518.

J. A. Nelder and R. Mead (1965), A simplex method for function minimization, *Comput. J.* **7**, 308–313.

G. L. Nemhauser and L. A. Wolsey (1978), Best algorithms for approximating the maximum of a submodular set function, *Math. Oper. Res.* **3**, 177–188.

G. L. Nemhauser, L. A. Wolsey and M. L. Fisher (1978), An analysis of approximations for maximizing submodular set functions I, *Math. Program.* **14**, 265–294.

Y. Nesterov (1983), A method of solving a convex programming problem with convergence rate $\mathcal{O}(1/k^2)$, *Soviet Math. Doklady* **27**, 372–376.

A. Ng and S. Russell (2000), Algorithms for inverse reinforcement learning, in *Proceedings of the 17th International Conference on Machine Learning (ICML 2000)*, Morgan Kaufmann, pp. 663–670.

X. Nguyen, M. J. Wainwright and M. I. Jordan (2010), Estimating divergence functionals and the likelihood ratio by convex risk minimization, *IEEE Trans. Inform. Theory* **56**, 5847–5861.

H. Niederreiter (1992), *Random Number Generation and Quasi-Monte Carlo Methods*, SIAM.

A. Nikolov and M. Singh (2016), Maximizing determinants under partition constraints, in *Proceedings of the 48th Annual ACM Symposium on Theory of Computing (STOC 2016)*, ACM, pp. 192–201.

A. Nikolov, M. Singh and U. Tantipongpipat (2022), Proportional volume sampling and approximation algorithms for A-optimal design, *Math. Oper. Res.* **47**, 847–877.

J. Nocedal and S. J. Wright (2006), *Numerical Optimization*, Springer.

V. Norkin, G. Pflug and A. Ruszczynski (1998), A branch and bound method for stochastic global optimization, *Math. Program.* **83**, 425–450.

A. O'Hagan, C. E. Buck, A. Daneshkhah, J. R. Eiser, P. H. Garthwaite, D. J. Jenkinson, J. E. Oakley and T. Rakow (2006), *Uncertain Judgements: Eliciting Experts' Probabilities*, Wiley.

R. Orozco, F. J. Herrmann and P. Chen (2024), Probabilistic Bayesian optimal experimental design using conditional normalizing flows. Available at arXiv:2402.18337.

A. M. Overstall (2022), Properties of Fisher information gain for Bayesian design of experiments, *J. Statist. Plann. Infer.* **218**, 138–146.

A. M. Overstall and D. C. Woods (2017), Bayesian design of experiments using approximate coordinate exchange, *Technometrics* **59**, 458–470.

A. M. Overstall, J. M. McGree and C. C. Drovandi (2018), An approach for finding fully Bayesian optimal designs using normal-based approximations to loss functions, *Statist. Comput.* **28**, 343–358.

A. B. Owen (1992), Orthogonal arrays for computer experiments, integration and visualization, *Statist. Sinica* **2**, 439–452.

A. B. Owen (2013), Monte Carlo theory, methods and examples. Available at https://artowen.su.domains/mc/.

C. H. Papadimitriou and K. Steiglitz (1998), *Combinatorial Optimization: Algorithms and Complexity*, Courier Corporation.

G. Papamakarios, E. Nalisnick, D. J. Rezende, S. Mohamed and B. Lakshminarayanan (2021), Normalizing flows for probabilistic modeling and inference, *J. Mach. Learn. Res.* **22**, 2617–2680.

E. Pardo-Igúzquiza (1998), Maximum likelihood estimation of spatial covariance parameters, *Math. Geol.* **30**, 95–108.

J. Peters and S. Schaal (2008), Natural actor–critic, *Neurocomput.* **71**, 1180–1190.

J. Pilz (1991), *Bayesian Estimation and Experimental Design in Linear Regression Models*, Wiley.

B. T. Polyak and A. B. Juditsky (1992), Acceleration of stochastic approximation by averaging, *SIAM J. Control Optim.* **30**, 838–855.

E. Pompe and P. E. Jacob (2021), Asymptotics of cut distributions and robust modular inference using posterior bootstrap. Available at arXiv:2110.11149.

A.-A. Pooladian and J. Niles-Weed (2021), Entropic estimation of optimal transport maps. Available at arXiv:2109.12004.

B. Poole, S. Ozair, A. Van Den Oord, A. Alemi and G. Tucker (2019), On variational bounds of mutual information, in *Proceedings of the 36th International Conference on Machine Learning (ICML 2019)*, Vol. 97 of Proceedings of Machine Learning Research, PMLR, pp. 5171–5180.

W. B. Powell (2011), *Approximate Dynamic Programming: Solving the Curses of Dimensionality*, second edition, Wiley.

D. Prangle, S. Harbisher and C. S. Gillespie (2023), Bayesian experimental design without posterior calculations: An adversarial approach, *Bayesian Anal.* **18**, 133–163.

L. Pronzato and W. G. Müller (2012), Design of computer experiments: Space filling and beyond, *Statist. Comput.* **22**, 681–701.

L. Pronzato and É. Thierry (2002), Sequential experimental design and response optimisation, *Statist. Methods Appl.* **11**, 277–292.

L. Pronzato and E. Walter (1985), Robust experiment design via stochastic approximation, *Math. Biosci.* **75**, 103–120.

F. Pukelsheim (2006), *Optimal Design of Experiments*, SIAM.

H. Rahimian and S. Mehrotra (2019), Distributionally robust optimization: A review. Available at arXiv:1908.05659.

H. Raiffa and R. Schlaifer (1961), *Applied Statistical Decision Theory*, Wiley.

T. Rainforth, R. Cornish, H. Yang, A. Warrington and F. Wood (2018), On nesting Monte Carlo estimators, in *Proceedings of the 35th International Conference on Machine Learning (ICML 2018)*, Vol. 80 of Proceedings of Machine Learning Research, PMLR, pp. 4267–4276.

T. Rainforth, A. Foster, D. R. Ivanova and F. B. Smith (2023), Modern Bayesian experimental design, *Statist. Sci.* **39**, 100–114.

C. E. Rasmussen and C. K. I. Williams (2006), *Gaussian Processes for Machine Learning*, MIT Press.

C. H. Rhee and P. W. Glynn (2015), Unbiased estimation with square root convergence for SDE models, *Oper. Res.* **63**, 1026–1043.

C. Riis, F. Antunes, F. Hüttel, C. Lima Azevedo and F. Pereira (2022), Bayesian active learning with fully Bayesian Gaussian processes, in *Advances in Neural Information Processing Systems 35* (S. Koyejo *et al.*, eds), Curran Associates, pp. 12141–12153.

Z. B. Riley, R. A. Perez, G. W. Bartram, S. M. Spottswood, B. P. Smarslok and T. J. Beberniss (2019), Aerothermoelastic experimental design for the AEDC/VKF Tunnel C: Challenges associated with measuring the response of flexible panels in high-temperature, high-speed wind tunnels, *J. Sound Vib.* **441**, 96–105.

H. Robbins and S. Monro (1951), A stochastic approximation method, *Ann. Math. Statist.* **22**, 400–407.

T. Robertazzi and S. Schwartz (1989), An accelerated sequential algorithm for producing D-optimal designs, *SIAM J. Sci. Statist. Comput.* **10**, 341–358.

R. T. Rockafellar and J. O. Royset (2015), Measures of residual risk with connections to regression, risk tracking, surrogate models, and ambiguity, *SIAM J. Optim.* **25**, 1179–1208.

R. T. Rockafellar and S. Uryasev (2002), Conditional value-at-risk for general loss distributions, *J. Bank. Finance* **26**, 1443–1471.

R. T. Rockafellar and S. Uryasev (2013), The fundamental risk quadrangle in risk management, optimization and statistical estimation, *Surv. Oper. Res. Manag. Sci.* **18**, 33–53.

M. Rosenblatt (1952), Remarks on a multivariate transformation, *Ann. Math. Statist.* **23**, 470–472.

J. O. Royset (2022), Risk-adaptive approaches to learning and decision making: A survey. Available at arXiv:2212.00856.

D. Rudolf and B. Sprungk (2018), On a generalization of the preconditioned Crank–Nicolson Metropolis algorithm, *Found. Comput. Math.* **18**, 309–343.

D. Ruppert (1988), Efficient estimations from a slowly convergent Robbins–Monro process. Technical report, Cornell University. Available at http://ecommons.cornell.edu/bitstream/handle/1813/8664/TR000781.pdf?sequence=1.

L. Ruthotto, J. Chung and M. Chung (2018), Optimal experimental design for inverse problems with state constraints, *SIAM J. Sci. Comput.* **40**, B1080–B1100.

E. G. Ryan, C. C. Drovandi, J. M. McGree and A. N. Pettitt (2016), A review of modern computational algorithms for Bayesian optimal design, *Int. Statist. Rev.* **84**, 128–154.

K. J. Ryan (2003), Estimating expected information gains for experimental designs with application to the random fatigue-limit model, *J. Comput. Graph. Statist.* **12**, 585–603.

J. Sacks, W. J. Welch, T. J. Mitchell and H. P. Wynn (1989), Design and analysis of computer experiments, *Statist. Sci.* **4**, 118–128.

F. Santambrogio (2015), *Optimal Transport for Applied Mathematicians: Calculus of Variations, PDEs, and Modeling*, Vol. 87 of Progress in Nonlinear Differential Equations and their Applications, Springer.

T. J. Santner, B. J. Williams and W. I. Notz (2018), *The Design and Analysis of Computer Experiments*, second edition, Springer.

K. Sargsyan, X. Huan and H. N. Najm (2019), Embedded model error representation for Bayesian model calibration, *Int. J. Uncertain. Quantif.* **9**, 365–394.

K. Sargsyan, H. N. Najm and R. G. Ghanem (2015), On the statistical calibration of physical models, *Int. J. Chem. Kinet.* **47**, 246–276.

A. I. Schein and L. H. Ungar (2007), Active learning for logistic regression: An evaluation, *Mach. Learn.* **68**, 235–265.

C. Schillings and C. Schwab (2016), Scaling limits in computational Bayesian inversion, *ESAIM Math. Model. Numer. Anal.* **50**, 1825–1856.

C. Schillings, B. Sprungk and P. Wacker (2020), On the convergence of the Laplace approximation and noise-level-robustness of Laplace-based Monte Carlo methods for Bayesian inverse problems, *Numer. Math.* **145**, 915–971.

A. Schrijver (2003), *Combinatorial Optimization: Polyhedra and Efficiency*, Springer.

P. Sebastiani and H. P. Wynn (2000), Maximum entropy sampling and optimal Bayesian experimental design, *J. R. Statist. Soc. Ser. B. Statist. Methodol.* **62**, 145–157.

S. Seo, M. Wallat, T. Graepel and K. Obermayer (2000), Gaussian process regression: Active data selection and test point rejection, in *Proceedings of the IEEE-INNS-ENNS International Joint Conference on Neural Networks (IJCNN 2000)*, Springer, pp. 27–34.

B. Settles (2009), Active learning literature survey. Computer Sciences Technical Report 1648, University of Wisconsin–Madison.

K. R. Shah and B. K. Sinha (1989), *Theory of Optimal Designs*, Vol. 54 of Lecture Notes in Statistics, Springer.

B. Shahriari, K. Swersky, Z. Wang, R. P. Adams and N. de Freitas (2016), Taking the human out of the loop: A review of Bayesian optimization, *Proc. IEEE* **104**, 148–175.

A. Shapiro (1991), Asymptotic analysis of stochastic programs, *Ann. Oper. Res.* **30**, 169–186.

A. Shapiro, D. Dentcheva and A. Ruszczynski (2021), *Lectures on Stochastic Programming: Modeling and Theory*, third edition, SIAM.

S. Shashaani, F. S. Hashemi and R. Pasupathy (2018), ASTRO-DF: A class of adaptive sampling trust-region algorithms for derivative-free stochastic optimization, *SIAM J. Optim.* **28**, 3145–3176.

W. Shen (2023), Reinforcement learning based sequential and robust Bayesian optimal experimental design. PhD thesis, University of Michigan.

W. Shen and X. Huan (2021), Bayesian sequential optimal experimental design for nonlinear models using policy gradient reinforcement learning. Available at arXiv:2110.15335.

W. Shen and X. Huan (2023), Bayesian sequential optimal experimental design for nonlinear models using policy gradient reinforcement learning, *Comput. Methods Appl. Mech. Engrg* **416**, art. 116304.

W. Shen, J. Dong and X. Huan (2023), Variational sequential optimal experimental design using reinforcement learning. Available at arXiv:2306.10430.

M. C. Shewry and H. P. Wynn (1987), Maximum entropy sampling, *J. Appl. Statist.* **14**, 165–170.

A. J. Siade, J. Hall and R. N. Karelse (2017), A practical, robust methodology for acquiring new observation data using computationally expensive groundwater models, *Water Resour. Res.* **53**, 9860–9882.

D. Silver, G. Lever, N. Heess, T. Degris, D. Wierstra and M. Riedmiller (2014), Deterministic policy gradient algorithms, in *Proceedings of the 31st International Conference on Machine Learning (ICML 2014)*, Vol. 32 of Proceedings of Machine Learning Research, PMLR, pp. 387–395.

M. Singh and W. Xie (2020), Approximation algorithms for D-optimal design, *Math. Oper. Res.* **45**, 1512–1534.

A. Solonen, H. Haario and M. Laine (2012), Simulation-based optimal design using a response variance criterion, *J. Comput. Graph. Statist.* **21**, 234–252.

J. Song and S. Ermon (2020), Understanding the limitations of variational mutual information estimators, in *Proceedings of the 8th International Conference on Learning Representations (ICLR 2020)*. Available at https://openreview.net/forum?id=B1x62TNtDS.

J. Song, Y. Chen and Y. Yue (2019), A general framework for multi-fidelity Bayesian optimization with Gaussian processes, in *Proceedings of the 22nd International Conference on Artificial Intelligence and Statistics*, Vol. 89 of Proceedings of Machine Learning Research, PMLR, pp. 3158–3167.

J. C. Spall (1998*a*), Implementation of the simultaneous perturbation algorithm for stochastic optimization, *IEEE Trans. Aerosp. Electron. Systems* **34**, 817–823.

J. C. Spall (1998*b*), An overview of the simultaneous perturbation method for efficient optimization, *Johns Hopkins APL Tech. Digest* **19**, 482–492.

A. Spantini, D. Bigoni and Y. Marzouk (2018), Inference via low-dimensional couplings, *J. Mach. Learn. Res.* **19**, 2639–2709.

A. Spantini, T. Cui, K. Willcox, L. Tenorio and Y. Marzouk (2017), Goal-oriented optimal approximations of Bayesian linear inverse problems, *SIAM J. Sci. Comput.* **39**, S167–S196.

A. Spantini, A. Solonen, T. Cui, J. Martin, L. Tenorio and Y. Marzouk (2015), Optimal low-rank approximations of Bayesian linear inverse problems, *SIAM J. Sci. Comput.* **37**, A2451–A2487.

D. A. Spielman and N. Srivastava (2008), Graph sparsification by effective resistances, in *Proceedings of the 40th Annual ACM Symposium on Theory of Computing (STOC 2008)*, ACM, pp. 563–568.

G. Spöck (2012), Spatial sampling design based on spectral approximations to the random field, *Environ. Model. Softw.* **33**, 48–60.

G. Spöck and J. Pilz (2010), Spatial sampling design and covariance-robust minimax prediction based on convex design ideas, *Stoch. Env. Res. Risk Assessment* **24**, 463–482.

V. Spokoiny (2023), Dimension free nonasymptotic bounds on the accuracy of high-dimensional Laplace approximation, *SIAM/ASA J. Uncertain. Quantif.* **11**, 1044–1068.

B. Sprungk (2020), On the local Lipschitz stability of Bayesian inverse problems, *Inverse Problems* **36**, art. 055015.

T. A. Sriver, J. W. Chrissis and M. A. Abramson (2009), Pattern search ranking and selection algorithms for mixed variable simulation-based optimization, *European J. Oper. Res.* **198**, 878–890.

D. M. Steinberg and W. G. Hunter (1984), Experimental design: Review and comment, *Technometrics* **26**, 71–97.

M. Stone (1959), Application of a measure of information to the design and comparison of regression experiments, *Ann. Math. Statist.* **30**, 55–70.

D. Strutz and A. Curtis (2024), Variational Bayesian experimental design for geophysical applications: Seismic source location, amplitude versus offset inversion, and estimating CO_2 saturations in a subsurface reservoir, *Geophys. J. Int.* **236**, 1309–1331.

A. Stuart and A. Teckentrup (2018), Posterior consistency for Gaussian process approximations of Bayesian posterior distributions, *Math. Comp.* **87**, 721–753.

A. M. Stuart (2010), Inverse problems: A Bayesian perspective, *Acta Numer.* **19**, 451–559.

N.-Z. Sun and W. W. Yeh (2007), Development of objective-oriented groundwater models, 2: Robust experimental design, *Water Resour. Res.* **43**, 1–14.

R. S. Sutton and A. G. Barto (2018), *Reinforcement Leaning*, second edition, MIT Press.

R. S. Sutton, D. McAllester, S. P. Singh and Y. Mansour (1999), Policy gradient methods for reinforcement learning with function approximation, in *Advances in Neural Information Processing Systems 12* (S. Solla *et al.*, eds), MIT Press, pp. 1057–1063.

M. Sviridenko (2004), A note on maximizing a submodular set function subject to a knapsack constraint, *Oper. Res. Lett.* **32**, 41–43.

M. Sviridenko, J. Vondrák and J. Ward (2017), Optimal approximation for submodular and supermodular optimization with bounded curvature, *Math. Oper. Res.* **42**, 1197–1218.

B. Tang (1993), Orthogonal array-based Latin hypercubes, *J. Amer. Statist. Assoc.* **88**, 1392–1397.

M. Tec, Y. Duan and P. Müller (2023), A comparative tutorial of Bayesian sequential design and reinforcement learning, *Amer. Statist.* **77**, 223–233.

G. Terejanu, R. R. Upadhyay and K. Miki (2012), Bayesian experimental design for the active nitridation of graphite by atomic nitrogen, *Experimental Therm. Fluid Sci.* **36**, 178–193.

V. Torczon (1997), On the convergence of pattern search algorithms, *SIAM J. Optim.* **7**, 1–25.

V. Tzoumas, L. Carlone, G. J. Pappas and A. Jadbabaie (2021), LQG control and sensing co-design, *IEEE Trans. Automat. Control* **66**, 1468–1483.

A. van den Oord, Y. Li and O. Vinyals (2018), Representation learning with contrastive predictive coding. Available at arXiv:1807.03748.

V. V. Vazirani (2001), *Approximation Algorithms*, Springer.

U. Villa, N. Petra and O. Ghattas (2021), HIPPYlib: An extensible software framework for large-scale inverse problems governed by PDEs I: Deterministic inversion and linearized Bayesian inference, *ACM Trans. Math. Software* **47**, 1–34.

C. Villani (2009), *Optimal Transport: Old and New*, Springer.

U. von Toussaint (2011), Bayesian inference in physics, *Rev. Modern Phys.* **83**, 943–999.

J. Vondrák (2008), Optimal approximation for the submodular welfare problem in the value oracle model, in *Proceedings of the Fortieth Annual ACM Symposium on Theory of Computing (STOC 2008)*, ACM, pp. 67–74.

J. Vondrák (2010), Submodularity and curvature: The optimal algorithm, *RIMS Kokyuroku Bessatsu* **B23**, 253–266.

J. Vondrák, C. Chekuri and R. Zenklusen (2011), Submodular function maximization via the multilinear relaxation and contention resolution schemes, in *Proceedings of the Forty-Third Annual ACM Symposium on Theory of Computing (STOC 2011)*, ACM, pp. 783–792.

A. Wald (1943), On the efficient design of statistical investigations, *Ann. Math. Statist.* **14**, 134–140.

S. G. Walker (2016), Bayesian information in an experiment and the Fisher information distance, *Statist. Probab. Lett.* **112**, 5–9.

J. Wang, S. C. Clark, E. Liu and P. I. Frazier (2020), Parallel Bayesian global optimization of expensive functions, *Oper. Res.* **68**, 1850–1865.

S. Wang and Y. Marzouk (2022), On minimax density estimation via measure transport. Available at arXiv:2207.10231.

X. Wang, Y. Jin, S. Schmitt and M. Olhofer (2023), Recent advances in Bayesian optimization, *ACM Comput. Surv.* **55**, 1–36.

J. K. Wathen and J. A. Christen (2006), Implementation of backward induction for sequentially adaptive clinical trials, *J. Comput. Graph. Statist.* **15**, 398–413.

B. P. Weaver and W. Q. Meeker (2021), Bayesian methods for planning accelerated repeated measures degradation tests, *Technometrics* **63**, 90–99.

B. P. Weaver, B. J. Williams, C. M. Anderson-Cook and D. M. Higdon (2016), Computational enhancements to Bayesian design of experiments using Gaussian processes, *Bayesian Anal.* **11**, 191–213.

P. Whittle (1973), Some general points in the theory of optimal experimental design, *J. R. Statist. Soc. Ser. B. Statist. Methodol.* **35**, 123–130.

L. A. Wolsey and G. L. Nemhauser (1999), *Integer and Combinatorial Optimization*, Wiley.

J. Wu and P. Frazier (2019), Practical two-step lookahead Bayesian optimization, in *Advances in Neural Information Processing Systems 32* (H. Wallach *et al.*, eds), Curran Associates, pp. 9813–9823.

K. Wu, P. Chen and O. Ghattas (2023*a*), An offline–online decomposition method for efficient linear Bayesian goal-oriented optimal experimental design: Application to optimal sensor placement, *SIAM J. Sci. Comput.* **45**, B57–B77.

K. Wu, T. O'Leary-Roseberry, P. Chen and O. Ghattas (2023*b*), Large-scale Bayesian optimal experimental design with derivative-informed projected neural network, *J. Sci. Comput.* **95**, art. 30.

H. P. Wynn (1972), Results in the theory and construction of D-optimum experimental designs, *J. R. Statist. Soc. Ser. B. Statist. Methodol.* **34**, 133–147.

H. P. Wynn (1984), Jack Kiefer's contributions to experimental design, *Ann. Statist.* **12**, 416–423.

Z. Xu and Q. Liao (2020), Gaussian process based expected information gain computation for Bayesian optimal design, *Entropy* **22**, art. 258.

F. Yates (1933), The principles of orthogonality and confounding in replicated experiments, *J. Agric. Sci.* **23**, 108–145.

F. Yates (1937), The design and analysis of factorial experiments. Technical Communication no. 35, Imperial Bureau of Soil Science.

F. Yates (1940), Lattice squares, *J. Agric. Sci.* **30**, 672–687.

O. Zahm, T. Cui, K. Law, A. Spantini and Y. Marzouk (2022), Certified dimension reduction in nonlinear Bayesian inverse problems, *Math. Comp.* **91**, 1789–1835.

D. Zhan and H. Xing (2020), Expected improvement for expensive optimization: A review, *J. Global Optim.* **78**, 507–544.

J. Zhang, S. Bi and G. Zhang (2021), A scalable gradient-free method for Bayesian experimental design with implicit models, in *Proceedings of the 24th International Conference on Artificial Intelligence and Statistics*, Vol. 130 of Proceedings of Machine Learning Research, PMLR, pp. 3745–3753.

X. Zhang, J. Blanchet, Y. Marzouk, V. A. Nguyen and S. Wang (2022), Distributionally robust Gaussian process regression and Bayesian inverse problems. Available at arXiv:2205.13111.

S. Zheng, D. Hayden, J. Pacheco and J. W. Fisher (2020), Sequential Bayesian experimental design with variable cost structure, in *Advances in Neural Information Processing Systems 33* (H. Larochelle *et al.*, eds), Curran Associates, pp. 4127–4137.

S. Zhong, W. Shen, T. Catanach and X. Huan (2024), Goal-oriented Bayesian optimal experimental design for nonlinear models using Markov chain Monte Carlo. Available at arXiv:2403.18072.

Z. Zhu and M. L. Stein (2005), Spatial sampling design for parameter estimation of the covariance function, *J. Statist. Plann. Infer.* **134**, 583–603.

Z. Zhu and M. L. Stein (2006), Spatial sampling design for prediction with estimated parameters, *J. Agric. Biol. Environ. Statist.* **11**, 24–44.

C. Zong (1999), *Sphere Packings*, Springer.

Acta Numerica (2024), pp. 841–908 Printed in the United Kingdom
doi:10.1017/S0962492923000053

The Moment-SOS hierarchy: Applications and related topics

Jean B. Lasserre
LAAS-CNRS and Toulouse School of Economics (TSE), University of Toulouse, 7 Avenue du Colonel Roche, BP54200, 31031 Toulouse cédex 4, France
E-mail: lasserre@laas.fr

The Moment-SOS hierarchy, first introduced in optimization in 2000, is based on the theory of the S-moment problem and its dual counterpart: polynomials that are positive on S. It turns out that this methodology can also be used to solve problems with positivity constraints '$f(\mathbf{x}) \geq 0$ for all $\mathbf{x} \in S$' or linear constraints on Borel measures. Such problems can be viewed as specific instances of the *generalized moment problem* (GMP), whose list of important applications in various domains of science and engineering is almost endless. We describe this methodology in optimization and also in two other applications for illustration. Finally we also introduce the Christoffel function and reveal its links with the Moment-SOS hierarchy and positive polynomials.

2020 Mathematics Subject Classification: Primary 14Q30, 44A60, 90C22, 90C23
Secondary 42C05, 47A57, 49L25

CONTENTS

1	Introduction	842
2	Notation, definitions and some preliminaries	845
3	The Moment-SOS hierarchy in optimization	853
4	The Moment-LP hierarchy	868
5	A Moment-SOS hierarchy of upper bounds	872
6	Some applications of the Moment-SOS hierarchy	875
7	Positive polynomials and the Christoffel function	887
8	Conclusion	900
References		901

1. Introduction

The Moment-SOS hierarchy was initially designed to help solve *polynomial* optimization problems (POPs), that is, optimization problems of the form

$$\mathbf{P}\colon\ f^* = \inf_{\mathbf{x}}\{f(\mathbf{x})\colon \mathbf{x} \in S\}, \tag{1.1}$$

where f is a polynomial and $S \subset \mathbb{R}^d$ is a basic semi-algebraic set, that is,

$$S := \{\mathbf{x} \in \mathbb{R}^d\colon\ g_j(\mathbf{x}) \geq 0,\ j = 1, \dots, m\}, \tag{1.2}$$

for some polynomials g_j, $j = 1, \dots, m$. Crucially, the description of $\mathbf{P}$ is entirely *algebraic* via its polynomial data f, g_j, $j = 1, \dots, m$. (However, semi-algebraic functions can also be tolerated to a certain extent.)

As $\mathbf{P}$ is a particular case of *non-linear programming* (NLP), what is so specific about $\mathbf{P}$ in (1.1)? The answer depends on the meaning of f^* in (1.1). If we are only interested in a *local* minimum, then the whole arsenal of efficient methods of NLP can be used to solve $\mathbf{P}$ and its algebraic features are not really exploited.

On the other hand, if f^* in (1.1) is understood as the *global* minimum of $\mathbf{P}$, then the situation is totally different. Why? First, to eliminate any ambiguity in the meaning of f^*, rewrite (1.1) as

$$\mathbf{P}\colon\ f^* = \sup\{\lambda\colon\ f(\mathbf{x}) - \lambda \geq 0,\ \forall\, \mathbf{x} \in S\}, \tag{1.3}$$

because then indeed f^* is necessarily the *global minimum* of $\mathbf{P}$. In full generality $\mathbf{P}$ is very difficult to solve as it is NP-hard. The reason is:

Given $\lambda \in \mathbb{R}$, checking whether '$f(\mathbf{x}) - \lambda \geq 0$ for all $\mathbf{x} \in S$' is difficult.

Indeed, by its very nature this positivity constraint is *global* and therefore cannot be handled by standard NLP optimization algorithms, which use only local information around a current iterate $\mathbf{x} \in S$. Therefore, to compute f^* in (1.3) it is necessary to handle the positivity constraint '$f(\mathbf{x}) - \lambda \geq 0$ for all $\mathbf{x} \in S$' in some efficient manner. Fortunately, if the data are algebraic, then the following hold.

(1) Powerful *positivity certificates* from real algebraic geometry (*Positivstellensätze* in German) are available.

(2) Some of these positivity certificates have an efficient practical implementation via *linear programming* (LP) or *semidefinite programming* (SDP). In particular, and crucially, testing whether a given polynomial is a sum of squares (SOS) simply reduces to solving a single semidefinite program (which can be done in time polynomial in the input size of the polynomial, up to arbitrary fixed precision[1]).

After the pioneering works of Shor (1987) and Nesterov (2000), Lasserre (2000, 2001) and Parrilo (2000, 2003) were the first to provide a systematic use of these

[1] In fact see O'Donnell (2017) for an update and more details.

two key ingredients in optimization and control, with convergence guarantees. It is also worth mentioning another closely related pioneering work, namely the celebrated SDP relaxation of Goemans and Williamson (1995), which provides a 0.878 approximation guarantee for Max-Cut, a famous problem in non-convex combinatorial optimization (and probably the simplest one). In fact it is perhaps the first famous example of such a successful application of the powerful SDP convex optimization technique to provide *guaranteed* good approximations to a notoriously difficult non-convex optimization problem. It turns out that this SDP relaxation is precisely the first semidefinite relaxation in the Moment-SOS hierarchy (a.k.a. the Lasserre hierarchy) when applied to the Max-Cut problem. The spectacular success story of SDP relaxations has been at the origin of a flourishing research activity in combinatorial optimization and computational complexity. In particular, the study of LP and SDP relaxation techniques in hardness of approximation is at the core of a central topic in combinatorial optimization and computational complexity, namely proving/disproving Khot's famous Unique Games Conjecture[2] (UGC) in theoretical computer science (TCS).

Another (and equivalent) 'definition' of the global optimum f^* of $\mathbf{P}$ reads

$$f^* = \inf_{\phi \in \mathscr{M}(S)_+} \left\{ \int_S f \, \mathrm{d}\phi : \phi(S) = 1 \right\}, \tag{1.4}$$

where the 'inf' is over the set $\mathscr{M}(S)_+$ of (positive) Borel measures ϕ on S. Indeed, as $f \geq f^*$ on S and ϕ is a probability measure on S,

$$\int_S f \, \mathrm{d}\phi \geq \int_S f^* \, \mathrm{d}\phi = f^*,$$

so the infimum in (1.4) is not smaller than f^*. On the other hand, for an arbitrary $\mathbf{x} \in S$, its value $f(\mathbf{x})$ is also obtained via $\int_S f \, \mathrm{d}\phi$, where ϕ is the Dirac probability measure $\delta_{\{\mathbf{x}\}}$ at $\mathbf{x} \in S$, and hence the infimum in (1.4) is not larger than f^*. In particular, if $\mathbf{x}^* \in S$ is a global minimizer of $\mathbf{P}$, then the Dirac measure $\phi^* := \delta_{\{\mathbf{x}^*\}}$ at $\mathbf{x}^*$ is an optimal solution of (1.4).

In fact (1.3) is the LP dual of (1.4), where λ is the dual variable associated with the constraint $\phi(S) = 1$ in (1.4). In other words, standard LP duality between the two conic programs (1.3) and (1.4) nicely captures a convex duality between the 'S-moment problem' in real and functional analysis, and 'polynomials positive on S' in real algebraic geometry (more details are given later).

Moreover, problem (1.4) is a very particular instance (and even the simplest instance) of the more general *generalized moment problem* (GMP) defined by

$$\inf_{\phi_j \in \mathscr{M}(S_j)_+} \left\{ \sum_{j=1}^{p} \int_{S_j} f_j \, \mathrm{d}\phi_j : \sum_{j=1}^{p} f_{ij} \, \mathrm{d}\phi_j \geq b_i, \ i = 1, \ldots, s \right\}, \tag{1.5}$$

[2] For this conjecture and its theoretical and practical implications, Subhash Khot was awarded the prestigious Nevanlinna prize at ICM 2014 in Seoul (Khot 2010, 2014).

for some given functions $f_j, f_{ij}\colon \mathbb{R}^{d_j} \to \mathbb{R}$, $i = 1,\ldots,s$, and sets $S_j \subset \mathbb{R}^{d_j}$, $j = 1,\ldots,p$. The GMP (1.5) is an infinite-dimensional LP with dual

$$\sup_{\lambda_1,\ldots,\lambda_s\geq 0}\left\{\sum_{i=1}^{s}\lambda_i\, b_i : f_j - \sum_{i=1}^{s}\lambda_i\, f_{ij} \geq 0 \text{ on } S_j,\ j = 1,\ldots,p\right\}. \tag{1.6}$$

Therefore it should be no surprise that the Moment-SOS hierarchy, initially developed for global optimization, also applies to solving the GMP. This is particularly interesting as moments and positive polynomials are at the intersection of several areas of mathematics (Landau 1987), and the list of important applications of the GMP is almost endless; see e.g. Landau (1987), Lasserre (2009*b*) and references therein, and see also Section 6, where for illustration we describe two particular applications.

Finally, since its birth in 2000 and in view of its many potential applications, the Moment-SOS hierarchy has gained attention from various research communities, with many different contributions, as follows.

(i) Its basic application in many (and diverse) areas after modelling the problem as an instance of the GMP. For illustration, two examples are described in Section 6; see also Henrion, Korda and Lasserre (2021), Korda, Henrion and Lasserre (2022) and references therein.

(ii) Its adaptation and extension to other domains, e.g. operations research (Parekh 2023), and entanglement, i.e. violation of Bell inequalities in quantum information, where its non-commutative version (the NPA hierarchy described in Navascués, Pironio and Acín (2012)) is also attracting a lot of attention; see also Burgdorf, Klep and Povh (2016), Parekh and Thompson (2021) and references therein.

(iii) Its detailed analysis by the TCS research community for hardness of approximation in combinatorial optimization (e.g. in relation to issues around the Unique Games Conjecture). See e.g. Barak and Steurer (2014), Raghavendra, Schramm and Steurer (2018) and Bafna *et al.* (2021).

(iv) The analysis of its rate of convergence with very interesting recent results on specific sets; see e.g. Slot and Laurent (2021), Slot (2022), Bach and Rudi (2023) and references therein.

(v) The development of algorithmic improvements to improve scalability of the standard Moment-SOS hierarchy. One direction is to take into account several types of sparsity or symmetries often present in large-scale optimization problems, as explained in Section 3.8. Another is to promote alternatives (e.g. first-order methods, second-order cone programming) to the costly interior point algorithm for semidefinite programming; see e.g. Ahmadi and Majumdar (2019), Yurtsever *et al.* (2021) and Ngoc Hoang Anh Mai *et al.* (2022).

Structure of the paper. For ease of exposition and clarity, we have not provided the proof of most results in the form of theorems and lemmas. However, at the end of each section we have included a subsection entitled 'Notes and sources', with pointers to articles for detailed proofs, and sometimes a discussion and comments on the results.

After introducing some notation and definitions, in Section 3 we describe the Moment-SOS hierarchy of *lower bounds* which converge to the global minimum in polynomial optimization. In Section 4 we provide a brief description of an alternative, the Moment-LP hierarchy. Section 5 describes the (less known) Moment-SOS hierarchy of *upper bounds* which also converge to the global minimum. Section 6 is devoted to other applications of the Moment-SOS hierarchy. For illustration, we describe how to apply the Moment-SOS hierarchy in two such applications and provide a (non-exhaustive) list of references to other applications in various fields. Finally, Section 7 is devoted to the Christoffel function (a classical tool from the theory of orthogonal polynomials and approximation) to reveal its connections to optimization and the Moment-SOS hierarchy.

2. Notation, definitions and some preliminaries

2.1. *Notation and definitions*

Let $\mathbb{R}[\mathbf{x}]$ denote the ring of polynomials in the variables $\mathbf{x} = (x_1, \ldots, x_d)$ and let $\mathbb{R}[\mathbf{x}]_n$ be the vector space of polynomials of degree at most n (whose dimension is $s(d) := \binom{d+n}{d}$). For every $n \in \mathbb{N}$, let $\mathbb{N}^d_n := \{\alpha \in \mathbb{N}^d : |\alpha| (= \sum_i \alpha_i) \leq n\}$, and let $\mathbf{v}_n(\mathbf{x}) = (\mathbf{x}^\alpha)$, $\alpha \in \mathbb{N}^d_n$, be the vector of monomials of the canonical basis $(\mathbf{x}^\alpha)$ of $\mathbb{R}[\mathbf{x}]_n$. Given a closed set $\mathcal{X} \subseteq \mathbb{R}^n$, let $\mathcal{P}(\mathcal{X}) \subset \mathbb{R}[\mathbf{x}]$ (resp. $\mathcal{P}_n(\mathcal{X}) \subset \mathbb{R}[\mathbf{x}]_n$) be the convex cone of polynomials (resp. polynomials of degree at most n) that are non-negative on $\mathcal{X}$. A polynomial $f \in \mathbb{R}[\mathbf{x}]_n$ is written

$$\mathbf{x} \mapsto f(\mathbf{x}) = \sum_{\alpha \in \mathbb{N}^d} f_\alpha \, \mathbf{x}^\alpha = \langle \mathbf{f}, \mathbf{v}_n(\mathbf{x}) \rangle,$$

with vector of coefficients $\mathbf{f} = (f_\alpha) \in \mathbb{R}^{s(n)}$ in the canonical basis of monomials $(\mathbf{x}^\alpha)_{\alpha \in \mathbb{N}^d}$. For real symmetric matrices, let $\langle \mathbf{B}, \mathbf{C} \rangle := \operatorname{trace}(\mathbf{B}\,\mathbf{C})$. We write $\mathbf{B} \succeq 0$ when $\mathbf{B}$ is positive semidefinite and $\mathbf{B} \succ 0$ when $\mathbf{B}$ is positive definite. Let $\mathcal{S}^n$ denote the space of real $n \times n$ symmetric matrices and let $\mathcal{S}^n_+$ denote its subset of positive semidefinite matrices.

For a closed set $S \subset \mathbb{R}^d$, let $\mathscr{M}(S)$ denote the space of finite signed Borel measures on S, and let $\mathscr{M}(S)_+ \subset \mathscr{M}(S)$ (resp. $\mathscr{P}(S)$) denote the convex cone of finite non-negative Borel measures (resp. probability measures) on S. The support $\operatorname{supp}(\mu)$ of a Borel measure μ on $\mathbb{R}^d$ is the smallest closed set $\Omega \subset \mathbb{R}^d$ such that $\mu(\mathbb{R}^d \setminus \Omega) = 0$.

Riesz linear functional. Given a sequence $\boldsymbol{\phi} = (\phi_\alpha)_{\alpha\in\mathbb{N}^d}$ (in bold), its associated Riesz linear functional is the linear mapping $\phi\colon \mathbb{R}[\mathbf{x}] \to \mathbb{R}$ (not in bold) defined by

$$f\left(=\sum_\alpha f_\alpha\,\mathbf{x}^\alpha\right) \mapsto \phi(f) = \sum_{\alpha\in\mathbb{N}^d} f_\alpha\,\phi_\alpha = \langle \mathbf{f}, \boldsymbol{\phi}\rangle. \tag{2.1}$$

A sequence $\boldsymbol{\phi}$ has a *representing* measure if its associated Riesz linear functional ϕ is a (positive) Borel measure on $\mathbb{R}^d$, in which case

$$\phi_\alpha = \int_{\mathbb{R}^d} \mathbf{x}^\alpha\,\mathrm{d}\phi \quad \text{for all } \boldsymbol{\alpha} \in \mathbb{N}^d.$$

Given a sequence $\boldsymbol{\phi} = (\phi_\alpha)_{\alpha\in\mathbb{N}^d}$ and a polynomial $g \in \mathbb{R}[\mathbf{x}]$, $\mathbf{x} \mapsto g(\mathbf{x}) = \sum_\gamma g_\gamma\,\mathbf{x}^\gamma$, define the new sequence $g \cdot \boldsymbol{\phi}$ by

$$(g\cdot\boldsymbol{\phi})_\alpha := \phi(\mathbf{x}^\alpha\,g) = \sum_{\gamma\in\mathbb{N}^d} g_\gamma\,\phi_{\alpha+\gamma} \quad \text{for all } \boldsymbol{\alpha} \in \mathbb{N}^d.$$

Therefore its associated Riesz linear functional, denoted by $g \cdot \phi$, satisfies

$$g\cdot\phi(f) = \phi(g\,f) \quad \text{for all } f \in \mathbb{R}[\mathbf{x}].$$

In particular, if $\boldsymbol{\phi}$ has a representing measure ϕ and g is non-negative, then the Riesz linear functional $g \cdot \phi$ is a representing measure, that is,

$$g\cdot\phi(f) = \phi(g\,f) = \int_{\mathbb{R}^d} f\,g\,\mathrm{d}\phi \quad \text{for all } f \in \mathbb{R}[\mathbf{x}].$$

Moment matrix. The (degree-n) *moment* matrix associated with a sequence $\boldsymbol{\phi} = (\phi_\alpha)_{\alpha\in\mathbb{N}^d}$ (or equivalently with the Riesz linear functional ϕ) is the real symmetric matrix $\mathbf{M}_n(\boldsymbol{\phi})$ (or $\mathbf{M}_n(\phi)$) with rows and columns indexed by $\mathbb{N}^d_n$, and whose entry $(\boldsymbol{\alpha},\boldsymbol{\beta})$ is just $\phi_{\alpha+\beta}$, for every $\boldsymbol{\alpha},\boldsymbol{\beta} \in \mathbb{N}^d_n$. Thus $\mathbf{M}_n(\boldsymbol{\phi})$ depends only on moments ϕ_α of degree at most $2n$. Alternatively, if we introduce the real symmetric matrices $(\mathbf{B}^1_\alpha) \subset \mathcal{S}^{s(n)}$ defined by

$$\mathbf{v}_n(\mathbf{x})\,\mathbf{v}_n(\mathbf{x})^\top = \sum_{\alpha\in\mathbb{N}^d_{2n}} \mathbf{B}^1_\alpha\,\mathbf{x}^\alpha \quad \text{for all } \mathbf{x} \in \mathbb{R}^d, \tag{2.2}$$

then $\mathbf{M}_n(\boldsymbol{\phi}) = \sum_{\alpha\in\mathbb{N}^d_{2n}} \phi_\alpha\,\mathbf{B}^1_\alpha$. Moreover, if $\boldsymbol{\phi}$ has a representing measure ϕ, then $\mathbf{M}_n(\boldsymbol{\phi}) \succeq 0$ because $\langle \mathbf{f}, \mathbf{M}_n(\boldsymbol{\phi})\,\mathbf{f}\rangle = \int f^2\mathrm{d}\phi \geq 0$ for all $f \in \mathbb{R}[\mathbf{x}]_n$.

A measure whose moments are all finite is *moment determinate* if there is no other measure with the same moments.

Localizing matrix. With $\boldsymbol{\phi}$ as above and $g \in \mathbb{R}[\mathbf{x}]$ (with $g(\mathbf{x}) = \sum_\gamma g_\gamma\mathbf{x}^\gamma$), the localizing matrix associated with $\boldsymbol{\phi}$ and g is the moment matrix $\mathbf{M}_n(g\cdot\boldsymbol{\phi})$ associated with the sequence $g \cdot \boldsymbol{\phi}$. That is, $\mathbf{M}_n(g\cdot\boldsymbol{\phi})$ is the real symmetric matrix with rows and columns indexed by $\mathbb{N}^d_n$, and whose entry $(\boldsymbol{\alpha},\boldsymbol{\beta})$ is just $(g\cdot\phi)_{\alpha+\beta}$, that is, $\mathbf{M}_n(g\cdot\boldsymbol{\phi})(\alpha,\boldsymbol{\beta}) = \sum_\gamma g_\gamma\phi_{\alpha+\beta+\gamma}$, for every $\boldsymbol{\alpha},\boldsymbol{\beta} \in \mathbb{N}^d_n$.

Alternatively, letting $d_g := \lceil \deg(g)/2 \rceil$, and introducing the real symmetric matrices $\mathbf{B}^g_\alpha \in \mathcal{S}^{s(n)}$, $\alpha \in \mathbb{N}^d$, defined by

$$g(\mathbf{x})\, \mathbf{v}_n(\mathbf{x})\, \mathbf{v}_n(\mathbf{x})^\top = \sum_{\alpha \in \mathbb{N}^d_{2(n+d_g)}} \mathbf{B}^g_\alpha\, \mathbf{x}^\alpha \quad \text{for all } \mathbf{x} \in \mathbb{R}^d, \tag{2.3}$$

we obtain

$$\mathbf{M}_n(g \cdot \boldsymbol{\phi}) = \sum_{\alpha \in \mathbb{N}^d_{2(n+d_g)}} \phi_\alpha\, \mathbf{B}^g_\alpha.$$

If $\boldsymbol{\phi}$ has a representing measure ϕ whose support is contained in the set $\{\mathbf{x} : g(\mathbf{x}) \geq 0\}$, then $\mathbf{M}_n(g \cdot \boldsymbol{\phi}) \succeq 0$ for all n, because

$$\langle \mathbf{f}, \mathbf{M}_n(g \cdot \boldsymbol{\phi})\, \mathbf{f} \rangle = g \cdot \phi(f^2) = \phi(f^2 g) = \int f^2\, g\, \mathrm{d}\phi \geq 0 \quad \text{for all } f \in \mathbb{R}[\mathbf{x}]_n.$$

2.2. *SOS polynomials and quadratic modules*

A polynomial $f \in \mathbb{R}[\mathbf{x}]$ is a sum of squares (SOS) if there exist $s \in \mathbb{N}$, and $f_1, \ldots, f_s \in \mathbb{R}[\mathbf{x}]$, such that $f(\mathbf{x}) = \sum_{k=1}^s f_k(\mathbf{x})^2$, for all $\mathbf{x} \in \mathbb{R}^d$. Let $\Sigma[\mathbf{x}]$ (resp. $\Sigma[\mathbf{x}]_n$) denote the set of SOS polynomials (resp. SOS polynomials of degree at most $2n$). Of course, every SOS polynomial is non negative. However, the converse is not true.

Membership in $\Sigma[\mathbf{x}]_n$. Checking whether a given polynomial f is non-negative on $\mathbb{R}^d$ is difficult, whereas – and this is crucial for the Moment-SOS hierarchy – checking whether f is SOS is much easier and can be done efficiently. Indeed, since the degree of f must be even for f to be SOS, we let $f \in \mathbb{R}[\mathbf{x}]_{2n}$ be defined by

$$\mathbf{x} \mapsto f(\mathbf{x}) = \sum_{\alpha \in \mathbb{N}^d_{2n}} f_\alpha\, \mathbf{x}^\alpha.$$

Then $f \in \mathbb{R}[\mathbf{x}]_{2n}$ is SOS if and only if there exists a real symmetric matrix $\mathbf{X}^\top = \mathbf{X}$ of size $s(n) = \binom{d+n}{d}$, such that

$$\mathbf{X} \succeq 0 \quad \text{and} \quad f_\alpha = \langle \mathbf{X}, \mathbf{B}^1_\alpha \rangle \quad \text{for all } \alpha \in \mathbb{N}^d_{2n}, \tag{2.4}$$

where the matrices $\mathbf{B}^1_\alpha$ were introduced in (2.2). It turns out that (2.4) defines the feasible set of what is called a *semidefinite program.*[3]

[3] A semidefinite program is a convex and conic optimization problem which can be solved (up to fixed arbitrary precision) in time polynomial in its input size; see e.g. Anjos and Lasserre (2012) and also O'Donnell (2017).

Quadratic module. Introduce the constant polynomial $\mathbf{x} \mapsto g_0(\mathbf{x}) := 1$ for all $\mathbf{x} \in \mathbb{R}^d$ (also denoted $g_0 = \mathbf{1}$). Given a family $(g_1, \ldots, g_m) \subset \mathbb{R}[\mathbf{x}]$, we associate the *quadratic module* $Q(g)\,(= Q(g_1, \ldots, g_m)) \subset \mathbb{R}[\mathbf{x}]$ defined by

$$Q(g) := \left\{ \sum_{j=0}^{m} \sigma_j\, g_j : \sigma_j \in \Sigma[\mathbf{x}],\ j = 0, \ldots, m \right\}, \tag{2.5}$$

and its degree-$2n$ *truncated* version

$$Q_n(g) := \left\{ \sum_{j=0}^{m} \sigma_j\, g_j : \sigma_j \in \Sigma[\mathbf{x}]_{n-d_j},\ j = 0, \ldots, m \right\}, \tag{2.6}$$

where $d_j := \lceil \deg(g_j)/2 \rceil$, $j = 0, \ldots, m$. Observe that $Q_n(g) \subset \mathbb{R}[\mathbf{x}]_{2n}$, because indeed in (2.6), $\deg(\sigma_j\, g_j) \leq 2n$, for all $j = 0, \ldots, m$. Obviously both $Q(g)$ and its truncated version $Q_n(g)$ are convex cones of $\mathbb{R}[\mathbf{x}]$.

Definition 2.1. The quadratic module $Q(g)$ is said to be *Archimedean* if there exists $M > 0$ such that the quadratic polynomial $\mathbf{x} \mapsto M - \|\mathbf{x}\|^2$ belongs to $Q(g)$ (i.e. belongs to $Q_n(g)$ for some n).

If $Q(g)$ is Archimedean, then necessarily the set

$$S := \{\mathbf{x} \in \mathbb{R}^d : g_j(\mathbf{x}) \geq 0,\ j = 1, \ldots, m\} \tag{2.7}$$

is compact but the reverse is not true. The Archimedean condition depends on the *representation* of S and can be seen as an *algebraic certificate* that S is compact.

Dual cone. The dual cone $Q_n^*(g)$ of $Q_n(g)$ is the convex cone of $\mathbb{R}^{s(2n)}$ defined by

$$Q_n^*(g) = \{\boldsymbol{\phi} \in \mathbb{R}^{s(2n)} : \mathbf{M}_{n-d_j}(g_j \cdot \boldsymbol{\phi}) \succeq 0,\ j = 0, \ldots, m\}, \tag{2.8}$$

where $\mathbf{M}_n(g_j \cdot \boldsymbol{\phi})$ is the localizing matrix associated with the polynomial g_j and the sequence $\boldsymbol{\phi}$, defined in Section 2.1.

For more details on the above notions of moment and localizing matrix and the quadratic module, as well as their use in potential applications, the interested reader is referred to Laurent (2008) and Lasserre (2009*b*). As we will see, both convex cones $Q_n(g)$ and $Q_n^*(g)$ play a crucial role in the Moment-SOS hierarchy of lower bounds.

2.3. *Certificates of positivity (Positivstellensätze)*

Below we describe particular certificates of positivity that are important because they provide a theoretical justification (or rationale) for convergence of the so-called SDP and LP relaxations for global optimization. In particular, the one below in (2.9) is at the core of the Moment-SOS hierarchy of lower bounds.

Theorem 2.2 (Putinar 1993). Let $S \subset \mathbb{R}^d$ be as in (2.7) and assume that $Q(g)$ is Archimedean.

(i) If a polynomial $f \in \mathbb{R}[\mathbf{x}]$ is (strictly) positive on S, then $f \in Q(g)$, that is,

$$f = \sum_{j=0}^{m} \sigma_j \, g_j, \tag{2.9}$$

for some SOS polynomials $\sigma_j \in \Sigma[\mathbf{x}]$, $j = 0, \ldots, m$ (and thus $f \in Q_n(g)$ for some $2n \geq \deg(f)$).

(ii) A sequence $\boldsymbol{\phi} = (\phi_\alpha)_{\alpha \in \mathbb{N}^d} \subset \mathbb{R}$ has a representing Borel measure on S if and only if $\phi(f^2 \, g_j) \geq 0$ for all $f \in \mathbb{R}[\mathbf{x}]$, and all $j = 0, \ldots, m$, or equivalently if and only if $\mathbf{M}_n(g_j \cdot \boldsymbol{\phi}) \succeq 0$ for all $j = 0, \ldots, m$, and all $n \in \mathbb{N}$.

In fact Theorem 2.2 is a refinement of a theorem by Schmüdgen from two years earlier.

Theorem 2.3 (Schmüdgen 1991). Let the basic semi-algebraic set $S \subset \mathbb{R}^d$ in (2.7) be compact.

(i) If a polynomial $f \in \mathbb{R}[\mathbf{x}]$ is (strictly) positive on S, then

$$f = \sum_{\alpha \in \{0,1\}^m} \sigma_\alpha \, g_1^{\alpha_1} \cdots g_m^{\alpha_m}, \tag{2.10}$$

for some SOS polynomials $\sigma_\alpha \in \Sigma[\mathbf{x}]$, $\boldsymbol{\alpha} \in \{0, 1\}^m$.

(ii) A sequence $\boldsymbol{\phi} = (\phi_\alpha)_{\alpha \in \mathbb{N}^d} \subset \mathbb{R}$ has a representing Borel measure on S if and only if $\phi(f^2 \, g_1^{\alpha_1} \cdots g_m^{\alpha_m}) \geq 0$ for all $f \in \mathbb{R}[\mathbf{x}]$, and all $\boldsymbol{\alpha} \in \{0, 1\}^m$, or equivalently if and only if $\mathbf{M}_n(g_1^{\alpha_1} \cdots g_m^{\alpha_m} \cdot \boldsymbol{\phi}) \succeq 0$ for all $\boldsymbol{\alpha} \in \{0, 1\}^m$, and all $n \in \mathbb{N}$.

Observe that (2.9) is of the same flavour as (2.10) but much simpler, as it involves only $m + 1$ SOS polynomials $\sigma_j \in \Sigma[\mathbf{x}]$ (as opposed to 2^m SOS σ_α in (2.10)). On the other hand, in Theorem 2.3 the only condition is that the set S is compact, whereas in Theorem 2.2 the quadratic module $Q(g)$ should also be Archimedean (an additional condition on the representation of S).

The reader may have noticed that Theorems 2.2 and 2.3 have two facets, (i) and (ii): the former is the algebraic facet (certificate of positivity), while the latter with a real analysis flavour is related to the S-moment problem. Both facets are a nice illustration of the duality between moments and positive polynomials.

We next provide a Nichtnegativstellensatz (a theorem of non-negativity) of the author (Lasserre 2011), which is instrumental in proving convergence of the hierarchy of upper bounds in Section 5.

Theorem 2.4. Let $S \subset \mathbb{R}^d$ be a compact set (not necessarily basic semi-algebraic), and let ϕ be a Borel measure with $\mathrm{supp}(\phi) = S$, and with moment sequence

$\boldsymbol{\phi} = (\phi_\alpha)_{\alpha\in\mathbb{N}^d}$. If $f \in \mathbb{R}[\mathbf{x}]$, then

$$f \geq 0 \text{ on } S \quad \Leftrightarrow \quad \mathbf{M}_n(f \cdot \boldsymbol{\phi}) \succeq 0 \quad \text{for all } n \in \mathbb{N}. \tag{2.11}$$

Theorem 2.4 states that to decide whether f is non-negative on S, one must check whether countably many linear matrix inequalities (LMIs) $\mathbf{M}_n(f \cdot \boldsymbol{\phi}) \succeq 0$, $n \in \mathbb{N}$, hold. Each constraint $\mathbf{M}_n(f \cdot \boldsymbol{\phi}) \succeq 0$ is indeed an LMI on the coefficients of the polynomial f because each entry of $\mathbf{M}_n(f \cdot \boldsymbol{\phi})$ is *linear* in the coefficients $\mathbf{f}$ of f. Therefore, identifying $f \in \mathbb{R}[\mathbf{x}]_k$ with its vector $\mathbf{f} \in \mathbb{R}^{s(k)}$ of coefficients, for each $n \in \mathbb{N}$, the convex cone $\Omega_n \subset \mathbb{R}[\mathbf{x}]_k$, defined by

$$\Omega_n := \{\mathbf{f} \in \mathbb{R}^{s(k)} : \mathbf{M}_n(f \cdot \boldsymbol{\phi}) \succeq 0\},$$

is a spectrahedron that contains the convex cone $\mathcal{P}_k(S)$ of polynomials of degree at most k that are non-negative on S. In addition, $\Omega_{n+1} \subset \Omega_n$ for all n, so that the sequence $(\Omega_n)_{n\in\mathbb{N}}$ of outer approximations of $\mathcal{P}_k(S)$ is monotone non-increasing. Moreover, it converges to $\mathcal{P}_k(S)$, i.e. $\bigcap_{n\in\mathbb{N}} \Omega_n = \mathcal{P}_k(S)$.

So in contrast to Theorems 2.2 and 2.3, Theorem 2.4 is valid for arbitrary compact sets $S \subset \mathbb{R}^d$ and non-negative (as opposed to positive) polynomials on S. On the other hand, its practical use in over-approximating the convex cone $\mathcal{P}_k(S)$ by Ω_n requires knowledge of moments $\boldsymbol{\phi} = (\phi_\alpha)_{\alpha\in\mathbb{N}^d}$ of a measure ϕ with $\mathrm{supp}(\phi) = S$. This is only possible for specific sets and measures. Examples of such special sets include the unit box, unit Euclidean ball, unit sphere, canonical simplex, discrete cube $\{-1, 1\}^d$, and their image by an affine transformation.

LP-based certificate. We next introduce another certificate of positivity which does not use SOS. Given $g_1, \ldots, g_m \in \mathbb{R}[\mathbf{x}]$, introduce the notation $\mathbf{g}^\alpha \in \mathbb{R}[\mathbf{x}]$, and $(1 - \mathbf{g}^\alpha) \in \mathbb{R}[\mathbf{x}]$, with

$$\mathbf{x} \mapsto \mathbf{g}^\alpha(\mathbf{x}) := g_1(\mathbf{x})^{\alpha_1} \cdots g_m(\mathbf{x})^{\alpha_m} \quad \text{for all } \mathbf{x} \in \mathbb{R}^d,$$
$$\mathbf{x} \mapsto (\mathbf{1} - \mathbf{g})^\alpha(\mathbf{x}) := (1 - g_1(\mathbf{x}))^{\alpha_1} \cdots (1 - g_m(\mathbf{x}))^{\alpha_m} \quad \text{for all } \mathbf{x} \in \mathbb{R}^d,$$

and the convex cone $\mathcal{L}_n(g) \subset \mathbb{R}[\mathbf{x}]$ defined by

$$\mathcal{L}_n(g) := \left\{ \sum_{(\alpha,\beta)\in\mathbb{N}^{2m}_n} c_{\alpha\beta}\, \mathbf{g}^\alpha\, (\mathbf{1} - \mathbf{g})^\beta : \mathbf{c} = (c_{\alpha\beta}) \geq 0 \right\}. \tag{2.12}$$

Theorem 2.5 (Krivine 1964*a*,*b*, Vasilescu 2003). Let $S \subset \mathbb{R}^d$ as in (2.7) be compact and such that (possibly after scaling) $0 \leq g_j(\mathbf{x}) \leq 1$ for all $\mathbf{x} \in S$, $j = 1, \ldots, m$. Assume also that $[\mathbf{1}, g_1, \ldots, g_m]$ generates $\mathbb{R}[\mathbf{x}]$.

(i) If a polynomial $f \in \mathbb{R}[\mathbf{x}]$ is (strictly) positive on S, then $f \in \mathcal{L}_n(g)$ for some n, that is,

$$f = \sum_{(\alpha,\beta)\in\mathbb{N}^{2m}} c_{\alpha\beta}\, \mathbf{g}^\alpha\, (\mathbf{1} - \mathbf{g})^\beta \tag{2.13}$$

for some non-negative vector $\mathbf{c} = (c_{\alpha\beta})_{(\alpha,\beta)\in\mathbb{N}^{2m}_n}$.

(ii) A sequence $\boldsymbol{\phi} = (\phi_\gamma)_{\gamma\in\mathbb{N}^d} \subset \mathbb{R}$ has a representing Borel measure on S if and only if $\phi(\mathbf{g}^\alpha\,(\mathbf{1}-\mathbf{g})^\beta) \geq 0$ for all $\boldsymbol{\alpha}, \boldsymbol{\beta} \in \mathbb{N}^m$.

Remark 2.6. Interestingly, as for Theorems 2.2 and 2.3, Theorem 2.5 also has two facets. The algebraic facet (i) is concerned with representation of polynomials that are positive on S, while facet (ii) is concerned with the S-moment problem in real analysis. Hence Theorem 2.5 is another illustration of the duality between *polynomials positive on S* and the *S-moment problem*.

2.4. *Practical implementation of Positivstellensätze via SDP or LP*

In addition to being interesting in their own right, Theorems 2.2(i) and 2.5(i) have another distinguishing feature. Both have a *practical implementation* that allows us to perform interesting computations. Indeed:

- testing membership in $Q_n(g)$ is just solving a single semidefinite program, whereas
- testing membership in $\mathcal{L}_n(g)$ is just solving a single linear program.

Testing membership in $Q_n(g)$. This is crucial for a practical and efficient implementation of the Moment-SOS hierarchy. Fortunately it can be done by solving a semidefinite program. Namely, let $f \in \mathbb{R}[\mathbf{x}]_k$ and recall that $d_j = \lceil \deg(g_j)/2 \rceil$, $j = 0, \ldots, m$. Then testing whether $f \in Q_n(g)$ (where necessarily $2n \geq k$) reduces to solving

$$f_\alpha = \sum_{j=0}^{m} \langle \mathbf{X}_j, \mathbf{B}_\alpha^{g_j} \rangle \quad \text{for all } \boldsymbol{\alpha} \in \mathbb{N}^d_{2n},\ \mathbf{X}_j \in \mathcal{S}^{n-d_j}, \tag{2.14}$$

$$\mathbf{X}_j \succeq 0, \quad j = 0, \ldots, m, \tag{2.15}$$

where the real symmetric matrices $\mathbf{B}_\alpha^{g_j} \in \mathcal{S}^{s(n-d_j)}$ are defined in (2.3) (here with $n - d_j$ instead of n). Each real symmetric matrix $\mathbf{X}_j$ is a Gram matrix of a polynomial σ_j, $j = 0, \ldots, m$. Next, (2.14) are *linear* equality constraints on the unknown entries of $\mathbf{X}_j$, while (2.15) is a positive semidefinite constraint on $(\mathbf{X}_j)_{j=0}^m$ to ensure that every σ_j is an SOS. (In (2.14), $f_\alpha = 0$ whenever $k < |\alpha| \leq 2n$ because $f \in \mathbb{R}[\mathbf{x}]_k$.) Observe that multiplying (2.14) by $\mathbf{x}^\alpha$ and summing up yields

$$\begin{aligned} f(\mathbf{x}) &= \sum_{\alpha\in\mathbb{N}^d_{2n}} f_\alpha\, \mathbf{x}^\alpha \\ &= \sum_{j=0}^{m} \left\langle \mathbf{X}_j, \sum_{\alpha\in\mathbf{B}^d_{2n}} \mathbf{B}_\alpha^{g_j}\, \mathbf{x}^\alpha \right\rangle \end{aligned}$$

$$
\begin{aligned}
&= \sum_{j=0}^{m} \langle \mathbf{X}_j, \mathbf{v}_{n-d_j}(\mathbf{x})\mathbf{v}_{n-d_j}(\mathbf{x})^\top \rangle \, g_j(\mathbf{x}) \quad \text{(see (2.3))} \\
&= \sum_{j=0}^{m} \underbrace{\mathbf{v}_{n-d_j}(\mathbf{x})^\top \mathbf{X}_j \mathbf{v}_{n-d_j}(\mathbf{x})}_{\sigma_j(\mathbf{x})} \, g_j(\mathbf{x}).
\end{aligned}
$$

Hence checking whether (2.14)–(2.15) has a solution reduces to solving a semidefinite program.

Testing membership in $\mathcal{L}_n(g)$. Obviously testing whether some polynomial $f \in \mathbb{R}[\mathbf{x}]_k$ is in $\mathcal{L}_n(g)$ reduces to solving an LP problem. Indeed, with n such that $s := (\max_j \deg(g_j))^n \geq k$, it amounts to finding a non-negative vector $\mathbf{c} = (c_{\alpha,\beta})$, $(\boldsymbol{\alpha}, \boldsymbol{\beta}) \in \mathbb{N}_n^{2m}$, such that

$$
f_\gamma = \left(\sum_{(\boldsymbol{\alpha},\boldsymbol{\beta}) \in \mathbb{N}_n^{2m}} c_{\alpha\beta} \prod_{j=1}^{m} g_j(\mathbf{x})^{\alpha_j} \, (1 - g_j(\mathbf{x}))^{\beta_j} \right)_\gamma \qquad \text{for all } \boldsymbol{\gamma} \in \mathbb{N}_s^d,
$$

and with $f_\gamma = 0$ whenever $|\boldsymbol{\gamma}| > k$. Clearly, the above constraints are *linear* in the unknown coefficients $c_{\alpha\beta} \geq 0$, so checking the existence of such a vector $\mathbf{c} \geq 0$ reduces to solving an LP problem.

2.5. *Notes and sources*

Most of the material is from Lasserre (2009*b*, 2015). Good references for exhaustive results on positive polynomials and moment problems are Laurent (2008), Marshall (2008), Prestel and Denzel (2001), Schmüdgen (2017), Nie (2023) and Kočvara, Mourrain and Riener (2023); see also Blekherman, Parrilo and Thomas (2012) for related material on convex algebraic geometry.

Section 2.3. Theorem 2.4 is from Lasserre (2011, 2013). Interestingly, it is also valid for some non-compact sets, such as the positive orthant $\mathbb{R}^d_+$ or even the whole space $\mathbb{R}^d$. For the former, the measure ϕ can be chosen to be the exponential measure $\mathrm{d}\phi = \exp(-\sum_i x_i)\,\mathrm{d}\mathbf{x}$, while for the latter we may choose the Gaussian measure $\mathrm{d}\phi = \exp(-\|\mathbf{x}\|^2/2)\,\mathrm{d}\mathbf{x}$. In both cases we obtain a monotone sequence $(\boldsymbol{\Omega}_n)_{n\in\mathbb{N}}$ of outer approximations, which converge to $\mathcal{P}(\mathbb{R}^d_+)$ and $\mathcal{P}(\mathbb{R}^d)$ respectively.

Section 2.4. It is worth mentioning that other certificates of positivity (via convex cones of positive polynomials) have also been defined to overcome (or at least mitigate) the computational burden associated with testing membership in $Q_n(g)$ in (2.14)–(2.15) (via semidefinite programming). For instance, membership in corresponding alternative convex cones can be checked by linear programming for DSOS and second-order cone programming for SDSOS (Ahmadi and Majumdar 2019); see also Majumdar, Hall and Ahmadi (2020). An alternative described in Section 3.8 is to consider a sparse version of Theorem 2.2 when $\mathbf{P}$ exhibits some sparsity pattern. As we will see, it yields a sparsity-adapted version of the

Moment-SOS hierarchy which can handle non-convex POPs with more than a thousand variables; see also Magron and Wang (2023).

3. The Moment-SOS hierarchy in optimization

Consider the polynomial optimization problem (POP) **P** in (1.1), and assume that its associated feasible set $S \subset \mathbb{R}^d$ is compact.

3.1. The Moment-SOS hierarchy

The underlying principle behind the Moment-SOS hierarchy is quite simple and proceeds in two steps.

Viewing **P** *in its equivalent formulation* (1.3) *(real algebraic glasses).*

Step 1. We replace the hard constraint '$f - \lambda \geq 0$ on S' with the equivalent constraint $f \in Q(g)$ (where $Q(g)$ is the quadratic module defined in (2.5)). Indeed,

$$\begin{aligned} f^* &= \sup_\lambda \{\lambda \colon f - \lambda \geq 0 \text{ on } S\} \\ &= \sup_\lambda \{\lambda \colon f - \lambda > 0 \text{ on } S\} \\ &= \sup_\lambda \{\lambda \colon f - \lambda \in Q(g)\}, \end{aligned} \tag{3.1}$$

where the second equality follows from Theorem 2.2(i) if the quadratic module $Q(g)$ is Archimedean. However, (3.1) is still an infinite-dimensional problem.

Step 2. Next, with $n_0 := \max[\lceil \deg(f)/2 \rceil, \max_j \lceil \deg(g_j)/2 \rceil]$, and $n \geq n_0$, we replace (3.1) with the more restrictive constraint

$$\tau_n^* = \sup_\lambda \{\lambda \colon f - \lambda \in Q_n(g)\} \quad (n \geq n_0) \tag{3.2}$$

$$= \sup_{\lambda, \sigma_j} \left\{ \lambda \colon f - \lambda = \sum_{j=0}^m \sigma_j \, g_j;\ \sigma_j \in \Sigma[\mathbf{x}]_{n-d_j},\ \forall j \right\}, \tag{3.3}$$

so that $f^* \geq \tau_n^*$ for all $n \geq n_0$. A crucial feature of (3.3) is that it is a finite-dimensional convex optimization problem, and more precisely a *semidefinite program*. Therefore (3.3) can be solved (up to arbitrary fixed precision) in time polynomial in its input size.

So, solving (3.2) for increasing values of $n \in \mathbb{N}$, we obtain a monotone non-decreasing sequence $(\tau_n^*)_{n \geq n_0}$ of lower bounds on the global minimum of f^* of **P**.

Viewing **P** *in its equivalent formulation* (1.4) *(real analysis glasses).* First observe that since $f \in \mathbb{R}[\mathbf{x}]$,

$$\int_S f \, d\phi = \sum_{\alpha \in \mathbb{N}^d} f_\alpha \int_S \mathbf{x}^\alpha \, d\phi = \sum_{\alpha \in \mathbb{N}^d} f_\alpha \, \phi_\alpha.$$

Therefore

$$\begin{aligned} f^* &= \inf_{\phi \in \mathcal{M}(S)_+} \left\{ \int f \, \mathrm{d}\phi \colon \phi(S) = 1 \right\} \\ &= \inf_{\boldsymbol{\phi}=(\phi_\alpha)_{\alpha\in\mathbb{N}^d}} \{ \langle \mathbf{f}, \boldsymbol{\phi} \rangle \colon \phi_0 = 1;\ \boldsymbol{\phi} \text{ has a representing measure supported on } S \}. \end{aligned}$$

Again we proceed in two steps.

Step 1. If $Q(g)$ is Archimedean, then by invoking Theorem 2.2(ii), we may replace the constraint '$\boldsymbol{\phi}$ has a representing measure supported on S' with the equivalent constraint '$\mathbf{M}_n(g_j \cdot \boldsymbol{\phi}) \succeq 0$ for all $j = 0, \ldots, m$, and all $n \in \mathbb{N}$'. However, the optimization problem

$$f^* = \inf_{\boldsymbol{\phi}=(\phi_\alpha)} \{ \langle \mathbf{f}, \boldsymbol{\phi} \rangle \colon \phi_0 = 1;\ \mathbf{M}_n(g_j \cdot \boldsymbol{\phi}) \succeq 0,\ j = 0, \ldots, m,\ n \in \mathbb{N} \} \tag{3.4}$$

is still infinite-dimensional.

Step 2. Next we replace (3.4) with its truncated versions

$$\tau_n = \inf_{\boldsymbol{\phi}=(\phi_\alpha)} \{ \langle \mathbf{f}, \boldsymbol{\phi} \rangle \colon \phi_0 = 1;\ \mathbf{M}_{n-d_j}(g_j \cdot \boldsymbol{\phi}) \succeq 0,\ j = 0, \ldots, m \}, \tag{3.5}$$

where $n \geq n_0$ with $n_0 := \max[\lceil \deg(f)/2 \rceil, \max_j \lceil \deg(g_j)/2 \rceil]$. For each fixed n, (3.5) is a finite-dimensional semidefinite program.

Clearly, $\tau_n \leq \tau_{n+1} \leq f^*$ for all $n \geq n_0$, and therefore, solving (3.5) for increasing values of $n \in \mathbb{N}$, we obtain a monotone non-decreasing sequence $(\tau_n)_{n \geq n_0}$ of lower bounds on the global minimum of f^* of $\mathbf{P}$.

In fact the semidefinite program (3.3) is the dual of the semidefinite program (3.5), and by *weak duality* in convex optimization,

$$\tau_n^* \leq \tau_n \leq f^* \quad \text{for all } n \geq n_0. \tag{3.6}$$

As is clear from its formulation, (3.5) is a semidefinite *relaxation* of $\mathbf{P}$ as its constraints are only necessary conditions for $\boldsymbol{\phi}$ to have a representing measure ϕ on S. We call (3.5) a *Moment-relaxation* of $\mathbf{P}$.

On the other hand, its dual (3.3) is a *reinforcement* (or *strengthening*) of $\mathbf{P}$ (viewed as the maximization problem (1.3)) as we have replaced '$f - \lambda \geq 0$ on S' with the sufficient condition '$f - \lambda \in Q_n(g)$'. We call (3.3) an *SOS-reinforcement* (or *SOS-strengthening*) of $\mathbf{P}$, whence the name 'Moment-SOS hierarchy' for (3.3)–(3.5). In both cases we obtain a lower bound τ_n (resp. τ_n^*) on f^*.

When $\tau_n = f^*$ (resp. $\tau_n^* = f^*$) for some n, we say that the degree-n Moment-relaxation (3.5) (resp. the degree-n SOS-reinforcement (3.3)) of $\mathbf{P}$ is *exact*. In addition, if $\tau_n = f^*$ and an optimal solution $\boldsymbol{\phi}^*$ of (3.5) satisfies $\operatorname{rank}(\mathbf{M}_n(\boldsymbol{\phi}^*)) = 1$, then $\boldsymbol{\phi}^*$ is simply the vector of moments up to degree $2n$ of the Dirac measure $\delta_{\{\mathbf{x}^*\}}$ at a global minimizer $\mathbf{x}^* \in S$. In particular, the subvector $\phi^*(x_i)_{i=1,\ldots,d}$ of first-order moments is just the vector $\mathbf{x}^*$.

In summary, the ultimate goal of the Moment-relaxation (3.5) is to obtain, at some step n, an optimal solution $\boldsymbol{\phi}^* \in \mathbb{R}^{s(2n)}$ which is the vector of moments (up to degree $2n$) of the Dirac measure $\delta_{\{\mathbf{x}^*\}}$ at a minimizer $\mathbf{x}^* \in S$.

When **P** has a unique global minimizer $\mathbf{x}^* \in S$, this happens generically; in the case of finitely many global minimizers it will happen generically that at some step n, $\boldsymbol{\phi}^*$ is the vector of moments of a convex combination of Dirac measures at such global minimizers; see Theorem 3.3 and Lemma 3.4 below.

Remark 3.1.

(a) As explained above, it is more appropriate to call (3.3) an SOS-reinforcement of **P** instead of an *SOS-relaxation* of **P**, as is sometimes written in the literature. Of course, it is also the dual of the Moment-relaxation (3.5) of **P**. (Relaxing in the primal is equivalent to reinforcing in the dual.)

(b) We call (1.3) (resp. (3.5)) a *primal* formulation (resp. a *primal* semidefinite relaxation) of **P** because in solving **P** we are mainly interested in an optimal solution $\mathbf{x}^* \in S$ (a global minimizer of **P**), so the primary variable of interest is $\mathbf{x} \in S \subset \mathbb{R}^d$. If $\mathbf{x}^* \in S$ is an optimal solution of **P** then $\boldsymbol{\phi}^* = ((\mathbf{x}^*)^\alpha)_{\alpha \in \mathbb{N}^d}$ is an optimal solution of (3.4).

Then (3.2) is the *dual* of the semidefinite relaxation (3.5), and we will see that indeed, when $\tau_n^* = f^*$ for some n and (3.2) has an optimal solution $\sigma_j^* \in \Sigma[\mathbf{x}]_{n-d_j}$, $j = 0, \ldots, m$, then $\sigma_j^*(\mathbf{x}^*) = \lambda_j^* \geq 0$ for all $j = 1, \ldots, m$, where the λ_j^* are optimal Karush–Kahn–Tucker (KKT) Lagrange multipliers associated with a global minimizer $\mathbf{x}^* \in S$.

Computational considerations. The Moment-relaxation (3.5) is a semidefinite program with

- $\binom{d+2n}{d}$ variables ϕ_α,
- $m + 1$ moment-localizing matrices of size $\binom{d+n-d_j}{d}$, $j = 0, \ldots, m$,

while the SOS-strengthening (3.3) is a semidefinite program with

- $1 + \sum_{j=0}^m \binom{d+2n-2d_j}{d}$ variables $(\lambda, (\sigma_0)_\alpha, \ldots, (\sigma_m)_\alpha)$,
- $m + 1$ semidefinite constraints for matrices of size $\binom{d+n-d_j}{d}$, $j = 0, \ldots, m$,
- $\binom{d+2n}{d}$ equality constraints.

For fixed dimension d, the size of matrices and the number of variables in the primal and dual semidefinite programs are polynomial in n. Therefore in principle they can be solved efficiently (up to arbitrary fixed precision) in time polynomial in their input size.[4] However, in view of their non-modest size and the current status of semidefinite solvers, such semidefinite programs can be solved only for POPs

[4] More details can be found in O'Donnell (2017), for example.

of modest dimension d and small degree-n relaxation. So in its canonical form (3.3)–(3.5), the Moment-SOS hierarchy is limited to POPs of modest dimension. However, fortunately we find the following.

- Practice reveals that finite convergence often occurs at low degree n.
- As is often the case for many POPs of large dimension d, some *sparsity pattern* or *symmetries* are present. It turns out that they can be exploited to define appropriate Moment-relaxations (resp. SOS-strengthenings) of **P** whose size is still compatible with current SDP solvers; see Section 3.8 for more details.
- Another possibility is to neglect the costly interior-point methods of SDP solvers and solve the semidefinite programs (3.3) and (3.5) by first-order methods; see e.g. Yurtsever *et al.* (2021) and Ngoc Hoang Anh Mai, Lasserre and Magron (2023).

More details are provided in Section 3.8.

3.2. Convergence of the Moment-SOS hierarchy

Observe that if $S \subset \mathbb{R}^d$ is compact then S is contained in the Euclidean ball of radius M for some $M > 0$, and in many applications M is known. Therefore the quadratic constraint $M^2 - \|\mathbf{x}\|^2 \geq 0$ is redundant when $\mathbf{x} \in S$.

For a practical implementation of the Moment-SOS hierarchy, it is indeed always recommended to add the redundant constraint $g_1(\mathbf{x}) := M^2 - \|\mathbf{x}\|^2 \geq 0$ in the definition (2.7) of S. Moreover, to avoid possible numerical ill-conditioning if M is large, it is even recommended to scale problem **P** in such a manner that $S \subset \mathbf{B}_1 := \{\mathbf{x} \colon \|\mathbf{x}\| \leq 1\}$ so that $g_1(\mathbf{x}) = 1 - \|\mathbf{x}\|^2$. Hence we state this formally.

Assumption 3.2. $S \subset \mathbb{R}^d$ defined in (2.7) is compact with $\mathbf{x} \mapsto g_1(\mathbf{x}) = 1 - \|\mathbf{x}\|^2$ (so that $S \subset \mathbf{B}_1$).

The reason for doing this is because under Assumption 3.2, the quadratic module $Q(g)$ in (2.5) is guaranteed to be Archimedean, a crucial property for convergence of the Moment-SOS hierarchy. (In general, proving that $Q(g)$ is Archimedean may not be trivial.)

Theorem 3.3. With $S \subset \mathbb{R}^d$ as in (2.7), let Assumption 3.2 hold and let τ_n (resp. τ_n^*) be as in (3.5) (resp. (3.3)) for all $n \geq n_0$. Then the following hold.

(a) We have $\tau_n^* = \tau_n$ for all $n \geq n_0$, and for every $n \geq n_0$, the semidefinite relaxation (3.5) has an optimal solution $\boldsymbol{\phi}^n = (\phi_\alpha^n)_{\alpha \in \mathbb{N}^d_{2n}}$. Moreover, both sequences $(\tau_n^*)_{n \geq n_0}$ and $(\tau_n)_{n \geq n_0}$ are monotone non-decreasing, and

$$\lim_{n\to\infty} \tau_n^* = \lim_{n\to\infty} \tau_n = f^*. \tag{3.7}$$

(b) If $\boldsymbol{\phi}^n$ is an optimal solution of (3.5), let $v := \max_{j=1,\ldots,m} \lceil \deg(g_j)/2 \rceil$. If

$$\operatorname{rank} \mathbf{M}_t(\boldsymbol{\phi}^n) = \operatorname{rank} \mathbf{M}_{t-v}(\boldsymbol{\phi}^n) \ (=: s), \tag{3.8}$$

for some $v \le t \le n$, then $\tau_n^* = \tau_n = f^*$ and from the vector $\boldsymbol{\phi}^n$ one may extract $\mathbf{x}^*(\ell) \in S$, $\ell = 1, \ldots, s$, where each $\mathbf{x}^*(\ell) \in S$ is a global minimizer of **P**, that is, $f(\mathbf{x}^*(\ell)) = f^*$, $\ell = 1, \ldots, s$.

(c) If $\mathrm{int}(S) \neq \emptyset$, then for every $n \geq n_0$, the SOS-strengthening (3.3) of **P** has an optimal solution $(\tau_n^*, \sigma_0^*, \ldots, \sigma_m^*)$.

Convergence of minimizers. Theorem 3.3 states that the sequence $(\tau_n)_{n \geq n_0}$ of optimal values converges to the global minimum f^* of **P** as the degree n increases, and moreover extraction of minimizers is also obtained if the degree-n Moment-relaxation is exact ($\tau_n = f^*$) and the flat extension condition (3.8) holds. But what about the sequence of minimizers $(\boldsymbol{\phi}^n)_{n \geq n_0}$ when the convergence is only asymptotic (as opposed to finite)?

Lemma 3.4. Let the sequence $(\boldsymbol{\phi}^n)_{n \geq n_0}$ with $\boldsymbol{\phi}^n = (\phi_\alpha^n)_{\alpha \in \mathbb{N}_{2n}^d}$ be as in Theorem 3.3(a). If $\mathbf{x}^* \in S$ is the unique global minimizer of **P**, then

$$\lim_{n \to \infty} \phi_\alpha^n \left(= \lim_{n \to \infty} \phi^n(\mathbf{x}^\alpha) \right) = (\mathbf{x}^*)^\alpha \quad \text{for all } \alpha \in \mathbb{N}^d. \tag{3.9}$$

In particular, $\lim_{n \to \infty} \phi^n(x_i) = x_i^*$ for every $i = 1, \ldots, d$.

So when **P** has a unique global minimizer and convergence of the Moment-relaxation (3.5) is only asymptotic (as opposed to finite), Lemma 3.4 states that the vector of degree-1 moments $(\phi^n(x_i))_{i=1,\ldots,d}$ converges to the unique global minimizer $\mathbf{x}^* \in S$ as n increases.

Notice that (3.9) is also interesting even if finite convergence takes place at some n, because we may already obtain a good approximation of $\mathbf{x}^* \in S$ from the degree-1 moments of $\boldsymbol{\phi}^t$ for $t < n$.

Equality constraints. Of course, in (2.7) we may tolerate equality constraints $g_j(\mathbf{x}) \geq 0$ and $g_{j+1}(\mathbf{x}) \geq 0$ with $g_{j+1} = -g_j$, $j \in J$, for some subset $J \subset \{1, \ldots, m\}$, in which case we simply write $g_j(\mathbf{x}) = 0$, $j \in J$ (and remove the constraint $g_{j+1} \geq 0$). The resulting modifications are as follows.

- In (3.3) the unknown SOS weight $\sigma_j \in \Sigma[\mathbf{x}]_{n-d_j}$ is now a polynomial in $\mathbb{R}[\mathbf{x}]_{2(n-d_j)}$ and no longer an SOS.
- In (3.5) the positive semidefinite constraint $\mathbf{M}_{n-d_j}(g_j \cdot \boldsymbol{\phi}) \succeq 0$ now reads as the equality constraints $\mathbf{M}_{n-d_j}(g_j \cdot \boldsymbol{\phi}) = 0$ on the variables (ϕ_α).
- Theorem 3.3(a,b) remains valid, whereas Theorem 3.3(c) needs some adjustment since $\mathrm{int}(S) = \emptyset$. For instance, if the ideal $\langle g_j \rangle_{j \in J} \subset \mathbb{R}[\mathbf{x}]$ generated by the polynomials g_j associated with the equality constraints is real radical, then for n sufficiently large, the SOS-strengthening (3.3) of **P** has an optimal solution $(\tau_n^*, \sigma_0^*, \ldots, \sigma_m^*)$.

Pseudo-Boolean case. An important case is when $S \subset \{-1,1\}^d$ (or equivalently $\{0,1\}^d$ after a simple linear transformation), that is, $J = \{1,\ldots,d\}$ and

$$S = \left\{\mathbf{x} \in \mathbb{R}^d : x_j^2 - 1 = 0, j \in J;\ g_j(\mathbf{x}) \geq 0,\ j \notin J\right\}, \tag{3.10}$$

of which the celebrated Max-Cut problem is a particular case (no inequality constraint). Then the ideal

$$\left\langle x_j^2 - 1\right\rangle_{j\in J} = \left\langle x_1^2 - 1, \ldots, x_d^2 - 1\right\rangle$$

is indeed real radical and Theorem 3.3 applies. Of course, in this case it follows that $\tau_n = f^*$ whenever $n \geq d + \max_j d_j$, and therefore the semidefinite relaxation (3.5) is not interesting as it contains 2^d variables ϕ_α. But the interest of Theorem 3.3 is that (3.8) may take place for $n \ll d$. For instance, in most random instances of Max-Cut problems with $d = 11$, we observe $f^* = \tau_2$ (and even $f^* = \tau_1$ in several cases).

3.3. *A global optimality condition for polynomial optimization*

Theorem 3.3(a) guarantees that asymptotically as n increases, we recover the global optimum f^*, and moreover, by Theorem 3.3(b), *finite* convergence takes place whenever the so-called *flatness condition* (3.8) holds at some degree n. In the latter case we can say more. Indeed, and remarkably, one can provide a global optimality condition for non-convex POPs, of the same flavour as the celebrated KKT optimality conditions for convex optimization, and under the same second-order sufficiency condition.

We first recall the well-known standard first-order necessary and second-order sufficient KKT optimality conditions in non-linear programming (NLP).

First-order necessary KKT optimality conditions. With S as in (2.7), let $\mathbf{x}^* \in S$ be a local minimizer of $\mathbf{P}$, and let $I(\mathbf{x}^*) := \{j \in \{1,\ldots,m\} : g_j(\mathbf{x}^*) = 0\}$ be the set of *active* constraints at $\mathbf{x}^* \in S$. With $\mathbb{S}^{d-1} := \{\mathbf{x} \in \mathbb{R}^d : \|\mathbf{x}\| = 1\}$, let

$$(\mathbf{x}^*)^\perp := \{\mathbf{u} \in \mathbb{S}^{d-1} : \langle \mathbf{u}, \nabla g_j(\mathbf{x}^*)\rangle = 0,\ \forall\, j \in I(\mathbf{x}^*)\}.$$

If the gradients $\nabla g_j(\mathbf{x}^*)$, $j \in I(\mathbf{x}^*)$, are linearly independent, there exists $\boldsymbol{\lambda}^* \in \mathbb{R}_+^m$ such that

$$\nabla f(\mathbf{x}^*) - \sum_{j=1}^m \lambda_j^* \nabla g_j(\mathbf{x}^*) = 0, \quad \lambda_j^* g_j(\mathbf{x}^*) = 0,\ j = 1,\ldots,m. \tag{3.11}$$

Moreover, *strict complementarity* holds if $\lambda_j^* > 0$ whenever $g_j(\mathbf{x}^*) = 0$. Next, observe that if in addition f and $-g_j$ are convex, then the Lagrangian

$$\mathbf{x} \mapsto L(\mathbf{x}) := f(\mathbf{x}) - f^* - \sum_{j=1}^m \lambda_j^* g_j(\mathbf{x}) \quad \text{for all } \mathbf{x} \in \mathbb{R}^d$$

is convex, non-negative, and satisfies $\nabla L(\mathbf{x}^*) = 0$. Hence $\mathbf{x}^*$ is also a global minimizer of L on the whole space $\mathbb{R}^d$, and $L(\mathbf{x}) \geq L(\mathbf{x}^*) = 0$ for all $\mathbf{x} \in \mathbb{R}^d$. This is a very strong property of the convex case.

Second-order sufficient KKT optimality condition holds at a local minimizer $\mathbf{x}^* \in S$ *of* $\mathbf{P}$. If (i) (3.11) and strict complementarity hold at $(\mathbf{x}^*, \boldsymbol{\lambda}^*)$, and (ii) in addition,

$$\left\langle \mathbf{u}, \left(\nabla^2 f(\mathbf{x}^*) - \sum_{j=1}^{m} \lambda_j^* \nabla^2 g_j(\mathbf{x}^*) \right) \mathbf{u} \right\rangle > 0 \quad \text{for all } \mathbf{u} \in (\mathbf{x}^*)^{\perp}. \tag{3.12}$$

If (3.11), strict complementarity and (3.12) hold at a global minimizer $\mathbf{x}^* \in S$ of $\mathbf{P}$, then we obtain a remarkable certificate of global optimality.

Theorem 3.5 (certificate of global optimality). With S as in (2.7), let $\mathbf{x}^* \in S$ be a global minimizer of $\mathbf{P}$, and assume the following.

(i) The gradients $\nabla g_j(\mathbf{x}^*)$, $j \in I(\mathbf{x}^*)$, are linearly independent (so that (3.11) holds for some $\boldsymbol{\lambda}^* \in \mathbb{R}_+^m$) and strict complementarity holds at $(\mathbf{x}^*, \boldsymbol{\lambda}^*)$.

(ii) Second-order sufficient condition (3.12) holds at $(\mathbf{x}^*, \boldsymbol{\lambda}^*)$.

Then there exists $n \in \mathbb{N}$ such that the SOS-strengthening (3.3) is exact, that is,

$$f(\mathbf{x}) - f^* = \sum_{j=0}^{m} \sigma_j^*(\mathbf{x})\, g_j(\mathbf{x}) \quad \text{for all } \mathbf{x} \in \mathbb{R}^d, \tag{3.13}$$

$$\sigma_j^*(\mathbf{x}^*)\, g_j(\mathbf{x}^*) = 0, \quad j = 0, \ldots, m, \tag{3.14}$$

for some SOS polynomials $\sigma_j^* \in \Sigma[\mathbf{x}]_{n-d_j}$. Moreover, let $\hat{\boldsymbol{\lambda}} = (\hat{\lambda}_j) \in \mathbb{R}_+^m$ with $\hat{\lambda}_j := \sigma_j^*(\mathbf{x}^*)$, $j = 1, \ldots, m$. Then the pair $(\mathbf{x}^*, \hat{\boldsymbol{\lambda}}) \in S \times \mathbb{R}_+^m$ satisfies (3.11) and (3.12).

As an immediate consequence of Theorem 3.5, finite convergence of the Moment-SOS hierarchy takes place at degree n, i.e. $\tau_n^* = \tau_n = f^*$.

Remark 3.6. We claim that Theorem 3.5, which provides an algebraic certificate of global optimality, is the perfect analogue for *non-convex* polynomial optimization of the KKT optimality conditions for *convex* optimization. Indeed, if f and $-g_j$ are all convex, then any local optimizer $\mathbf{x}^* \in S$ is a global minimizer and then (3.11) implies

$$\mathbf{x} \mapsto L(\mathbf{x}) = f(\mathbf{x}) - f^* - \sum_{j=1}^{m} \lambda_j^* g_j(\mathbf{x}) \geq 0 \quad \text{for all } \mathbf{x} \in \mathbb{R}^d, \tag{3.15}$$

$$L(\mathbf{x}^*) = 0. \tag{3.16}$$

But this of course is valid because f and $-g_j$ are convex, a very specific case. In general a global minimizer $\mathbf{x}^* \in S$ is *not* a global minimizer of the Lagrangian L on $\mathbb{R}^d$.

On the other hand, Theorem 3.5 states that if $\mathbf{x}^* \in S$ is a global minimizer, then under the standard second-order sufficient KKT optimality condition in NLP, $\mathbf{x}^*$ is also a global minimizer of the *extended Lagrangian*

$$\hat{L} := f - f^* - \sum_{j=1}^{m} \sigma_j^* g_j$$

over the whole space $\mathbb{R}^d$. Indeed,

$$\hat{L}(\mathbf{x}) = f(\mathbf{x}) - f^* - \sum_{j=1}^{m} \sigma_j^*(\mathbf{x})\, g_j(\mathbf{x}) = \sigma_0^*(\mathbf{x})\ (\geq 0) \quad \text{for all } \mathbf{x} \in \mathbb{R}^d, \tag{3.17}$$

$$\hat{L}(\mathbf{x}^*) = 0. \tag{3.18}$$

So in (3.17) the extended Lagrangian $\hat{L}$ looks like the standard Lagrangian L in (3.15) except that the scalar weight λ_j^* is now replaced by the SOS polynomial weight σ_j^*. Moreover, the scalar $\hat{\lambda}_j = \sigma_j^*(\mathbf{x}^*)$ is a standard Lagrange–KKT multiplier associated with the constraint $g_j \geq 0$ (like λ_j^* in (3.15)).

Interestingly, if the constraint $g_j \geq 0$ is not active at $\mathbf{x}^* \in S$ (i.e. $g_j(\mathbf{x}^*) > 0$), then $\hat{\lambda}_j = \sigma_j^*(\mathbf{x}^*) = 0$, but in general the SOS polynomial σ_j^* is *not* the trivial polynomial equal to zero. In fact, suppose that the constraint $g_j \geq 0$ is important even if it is not active at a global minimizer $\mathbf{x}^*$, meaning that if we remove that constraint in the definition (2.7) of S, then the new global minimum of the modified problem $\mathbf{P}$ is strictly smaller than f^*. Then, quite remarkably, $\sigma_j^* \neq 0$. In other words, a non-trivial SOS multiplier σ_j^* in Putinar's certificate of global optimality (3.13) identifies $g_j \geq 0$ as an important constraint, even if it is not active at global minimizers.

3.4. *Genericity*

In view of the remarkable form of Theorem 3.5, we may wonder how 'generic' the results of Theorem 3.5 might be. It turns out that Theorem 3.5 holds generically in a rigorous sense.

More precisely, let $m \in \mathbb{N}$ and $r_j \in \mathbb{N}$, $j = 0, \ldots, m$, be fixed, and consider the family of POPs whose (possibly empty) feasible set $S \subset \mathbb{R}^d$ is as in (2.7) for some polynomials $g_j \in \mathbb{R}[\mathbf{x}]_{r_j}$, and its criterion is some polynomial $f \in \mathbb{R}[\mathbf{x}]_{r_0}$.

Recall that $s(t) := \binom{d+t}{t}$. So a vector $\boldsymbol{\theta} \in \mathbb{R}^a$ with $a := \sum_{j=0}^{m} s(r_j)$ completely specifies an instance $\mathbf{P}(\boldsymbol{\theta})$ of such a problem $\mathbf{P}$. The next result is due to Nie (2014).

Theorem 3.7. There exists an integer L and finitely many real polynomials $\varphi_1, \ldots, \varphi_L \in \mathbb{R}[\boldsymbol{\theta}]$ in the coefficients $\boldsymbol{\theta}$ of the polynomials $f, g_1, \ldots, g_m$, such that if $\varphi_\ell(\boldsymbol{\theta}) \neq 0$ for all $\ell = 1, \ldots, L$, then (3.11), strict complementarity and second-order sufficient KKT optimality condition (3.12) hold at any global minimizer of problem $\mathbf{P}(\boldsymbol{\theta})$.

As a result, there exists $n \in \mathbb{N}$ such that (3.13) and (3.14) hold at every global minimizer $\mathbf{x}^* \in S$ of $\mathbf{P}(\boldsymbol{\theta})$, that is, finite convergence of the Moment-SOS hierarchy is generic.

3.5. *The convex case*

In this section we consider the particular case when $\mathbf{P}$ is a convex program,[5] that is, when the polynomials f and $-g_j$ are all convex, and thus the set S in (2.7) is convex. This class of problems is very important as they are considered 'easy' or at least 'easier' than non-convex problems. Indeed, as any local optimum of $\mathbf{P}$ is a global optimum, then $\mathbf{P}$ can be solved by several powerful local optimization algorithms.

So as the Moment-SOS hierarchy is able to solve difficult non-convex problems $\mathbf{P}$, a natural question is: How does the Moment-SOS hierarchy behave when $\mathbf{P}$ is a convex program?

The reason why such a question is relevant is because if the Moment-SOS hierarchy were not efficient in solving a convex problem, then one might raise reasonable doubts as to its efficiency in solving more difficult problems!

SOS-convex programs. Let us first consider the class of SOS-convex polynomials.

Definition 3.8. A polynomial $f \in \mathbb{R}[\mathbf{x}]$ is SOS-convex if its Hessian $\nabla^2 f$ is an SOS-matrix polynomial, that is, $\nabla^2 f = L\,L^\top$ for some real matrix polynomial $L \in \mathbb{R}[\mathbf{x}]^{d \times s}$ (for some integer s). In particular, every SOS-convex polynomial is convex and all quadratic convex polynomials are SOS-convex.

We have the following characterizations of SOS-convexity.

Theorem 3.9. Let $f \in \mathbb{R}[\mathbf{x}]$. The following four propositions are equivalent:

(i) f is SOS-convex,

(ii) $\nabla^2 f$ is SOS,

(iii) $(\mathbf{x}, \mathbf{y}) \mapsto f(\mathbf{x})/2 + f(\mathbf{y})/2 - f((\mathbf{x} + \mathbf{y})/2)$ is SOS,

(iv) $(\mathbf{x}, \mathbf{y}) \mapsto f(\mathbf{x}) - f(\mathbf{y}) - \langle \nabla f(\mathbf{y}), (\mathbf{x} - \mathbf{y}\rangle$ is SOS.

Notice that if f and $-g_j$ are convex, then necessarily their degree is either one or even. Importantly, SOS-convexity can be checked numerically by solving a semidefinite program (e.g. following Theorem 3.9(iii) and Section 2.2). The next result states that the Moment-SOS hierarchy somehow 'recognizes' easy SOS-convex problems.

[5] The set S in (2.7) may be convex even if the $-g_j$ are not convex (e.g. they can be quasi-convex). *Convex programming* usually refers to the case where f and the $-g_j$ are all convex.

Theorem 3.10. Let S be as in (2.7) and let Slater's condition hold (i.e. there exists $\mathbf{x}_0 \in S$ such that $g_j(\mathbf{x}_0) > 0$ for all j). If f and $-g_j$ are all SOS-convex, then with $n' := \max[\deg(f)/2, \max_j[\deg(g_j)/2]]$,

$$f - f^* = \sigma_0^* + \sum_{j=1}^{m} \lambda_j^* g_j \tag{3.19}$$

for some scalars $\lambda_j^* \geq 0$ and some $\sigma_0^* \in \Sigma[\mathbf{x}]_{n'}$. In addition,

$$f^* = \min_{\phi}\{\phi(f) \colon \phi(\mathbf{1}) = 1;\ \mathbf{M}_{n'}(\boldsymbol{\phi}) \succeq 0;\ \phi(g_j) \geq 0,\ j = 1, \ldots, m\}. \tag{3.20}$$

Moreover, $\phi^*(x_i) = x_i^*$, for all $i = 1, \ldots, d$, where $\boldsymbol{\phi}^*$ is an optimal solution of (3.20), and $\mathbf{x}^* \in S$ is a local (hence global) minimizer of $\mathbf{P}$.

It is also fairly straightforward to check that in (3.19), $\boldsymbol{\lambda}^* = (\lambda_j^*)_{1 \leq j \leq m}$ are Lagrange–KKT multipliers at an optimal solution $\mathbf{x}^* \in S$ of $\mathbf{P}$.

Next, observe that the semidefinite Moment-relaxation (3.20) is a particular case of (3.5) where the constraints $\mathbf{M}_{n-d_j}(g_j \cdot \boldsymbol{\phi}) \succeq 0$ are replaced by the simpler $\mathbf{M}_0(g_j \cdot \boldsymbol{\phi}) \succeq 0$ (i.e. the scalar linear inequality constraint $\phi(g_j) \geq 0$).

However, if we were unaware that f and $-g_j$ were SOS-convex and we solved (3.5) as a general POP, then we would still obtain $f^* = \tau_{n_0}$, that is, the first Moment-relaxation of the hierarchy would be exact. In other words, the Moment-SOS hierarchy has recognized that $\mathbf{P}$ was a convex (easy) problem.

The reason why the Moment-relaxation (3.5) can be replaced by the simpler (3.20) is because linear functionals $\phi \in \mathbb{R}[\mathbf{x}]_{2n}^*$ such that $\mathbf{M}_n(\boldsymbol{\phi}) \succeq 0$ have a nice property when acting on SOS-convex polynomials.

Lemma 3.11 (Jensen's inequality for linear functionals). Let $\phi \in \mathbb{R}[\mathbf{x}]_{2n}^*$ be such that $\mathbf{M}_n(\boldsymbol{\phi}) \succeq 0$, $\phi(\mathbf{1}) = 1$, and let $\mathbf{x}^* := (\phi(x_1), \ldots, \phi(x_d)) \in \mathbb{R}^d$. Then

$$\phi(f) \geq f(\mathbf{x}^*) \quad \text{for every SOS-convex polynomial } f \in \mathbb{R}[\mathbf{x}]_{2n}. \tag{3.21}$$

So let $\boldsymbol{\phi}^*$ be an optimal solution of

$$\tau_n' = \min_{\phi}\{\phi(f) \colon \phi(\mathbf{1}) = 1;\ \mathbf{M}_{n'}(\boldsymbol{\phi}) \succeq 0;\ \phi(g_j) \geq 0,\ j = 1, \ldots, m\}.$$

Of course, $\tau_n' \leq f^*$, as τ_n' is the optimal value of a relaxation of $\mathbf{P}$. As f and $-g_j$ are SOS-convex, and with $\mathbf{x}^* := (\phi^*(x_1), \ldots, \phi^*(x_d)) \in \mathbb{R}^d$,

$$\tau_n' = \phi^*(f) \geq f(\mathbf{x}^*), \quad 0 \leq \phi^*(g_j) \leq g_j(\mathbf{x}^*), \quad j = 1, \ldots, m,$$

which implies $\mathbf{x}^* \in S$ and $f(\mathbf{x}^*) \leq \tau_n' \leq f^*$, so that $\mathbf{x}^*$ is a global minimizer of $\mathbf{P}$.

General convex POPs. In the more general case of convex POPs, we also obtain finite convergence under some strict convexity assumption at every global minimizer $\mathbf{x}^* \in S$.

Theorem 3.12. With $S \subset \mathbb{R}^d$ as in (2.7), assume that $Q(g)$ is Archimedean, Slater's condition holds, and f and $-g_j$ are convex, $j = 1, \dots, m$. If $\nabla^2 f(\mathbf{x}^*) \succ 0$ at every global minimizer $\mathbf{x}^* \in S$ (assumed to be finitely many), then finite convergence takes place, that is, the Moment-relaxation (3.5) of **P** is exact at some degree n. Moreover, the SOS-strengthening (3.3) of **P** is also exact, and both (3.5) and (3.3) have an optimal solution $\boldsymbol{\phi}^n$ and $(\lambda^*, \sigma_0^*, \dots, \sigma_m^*)$, respectively.

So again without specifying that **P** is convex, the Moment-SOS hierarchy will converge in finitely many steps. However, in contrast to Theorem 3.9, in Theorem 3.12 we do not specify at which step n finite convergence takes place.

3.6. *General versus* ad hoc

We would like to emphasize that Theorem 3.5 is a fairly general global optimality condition that holds *generically* for POPs, hence with non-convex criterion and non-convex (and possibly disconnected) feasible sets S, and even with mixed-integer variables. The only requirement is to be able to translate all constraints of the problem into polynomial inequality and equality constraints.

Usually generality is at the price of reduced efficiency, and the usual algorithmic practice of optimization suggests using *ad hoc* algorithms, i.e. algorithms tailored to the type of problem to be solved. Indeed, for instance, if $x_i \in \{0, 1\}$, $i \in I$, for some I, it is not a good idea to model this constraint with the equality constraints $x_i^2 - x_i = 0$, $i \in I$, and then use standard first-order or second-order methods to obtain an (only local) optimum. We typically use Branch and Bound (or Branch and Cut) methods.

Remarkably, the Moment-SOS hierarchy does not suffer from its generality in just describing any POP by a set of polynomial inequality and equality constraints. (Of course, some descriptions may be more interesting than others.) Indeed, for instance, for SOS-convex programs and in particular convex quadratically constrained quadratic programs (convex QCQPs), Theorem 3.9 ensures that finite convergence takes place at the first step of the hierarchy, without the need to specify that the POP is SOS-convex. Similarly, if f and $-g_j$ are all convex, and $\nabla^2 f(\mathbf{x}) \succ 0$ at all global minimizers $\mathbf{x} \in S$, then finite convergence also takes place.

Of course, again, we do not claim that the Moment-SOS hierarchy is the most efficient algorithm for solving such convex problems, and indeed other efficient algorithms exist. But this remark simply emphasizes that, in some way, the Moment-SOS hierarchy recognizes easy problems (convex programs are usually considered easier to solve) as finite convergence takes place quickly. On the other hand, the Moment-SOS hierarchy has also been recognized by the theoretical computer science research community as a meta-algorithm which provides the best lower bounds for many combinatorial optimization problems, and in particular problems with $\{0, 1\}$ (or $\{-1, 1\}$) variables such as Max-Cut and its variants, which are notoriously difficult NP-hard problems. It is now considered an important tool for proving/disproving Khot's celebrated Unique Games Conjecture.

Table 3.1. Rates of convergence for the hierarchy of lower bounds (Slot 2022).

S	Error $f^* - \tau_n$	Certificate	Reference
Archimedean	$O(1/n^c)$	Putinar	Baldi and Mourrain (2022)
Compact	$O(1/n^c)$	Schmüdgen	Schweighofer (2005)
$\mathbb{S}^{d-1}$	$O(1/n^2)$	Putinar	Fang and Fawzi (2021)
$\mathbf{B}^d$	$O(1/n^2)$	Putinar	Slot (2022)
$[-1, 1]^d$	$O(1/n^2)$	Schmüdgen	Laurent and Slot (2023)
Δ^d	$O(1/n^2)$	Schmüdgen	Slot (2022)

3.7. Rates of convergence

In this section we provide rates of convergence for the Moment-SOS hierarchy of lower bounds on general compact basic semi-algebraic sets. Those rates have been refined for specific sets like the unit sphere $\mathbb{S}^{d-1}$, the unit ball $\mathbf{B}^d = \{\mathbf{x} \in \mathbb{R}^d : \|\mathbf{x}\|^2 \leq 1\}$, the box $[-1, 1]^d$, and the simplex $\Delta^d = \{\mathbf{x} \in \mathbb{R}^d_+ : 1 - \mathbf{e}^\top \mathbf{x} \geq 0\}$.

Table 3.1 is taken from Slot (2022). The first column is related to the set $S \subset \mathbb{R}^d$, and 'Archimedean' means that the quadratic module $Q(g)$ in (2.5) associated with S, is Archimedean (an algebraic certificate of its compactness used in Putinar's Positivstellensatz). The third column, 'Certificate', specifies whether in the Moment-SOS hierarchy we use Putinar's certificate (Theorem 2.2) or the more costly Schmüdgen's certificate (Theorem 2.3) for the semidefinite relaxation (3.3). In the notation $O(1/n^c)$, n is the order (or degree) of the semidefinite relaxations (3.3)–(3.5), and c is some positive constant. The rate $O(1/n^c)$ means that $f^* - \tau_n \leq M/n^c$ for some constant $M > 0$, where f^* is the global minimum and $\tau_n\, (\leq f^*)$ is as in (3.5). Finally, for optimization of trigonometric polynomials (and hence for POPs on the box $[0, 1]^d$ as well) and under some condition on global minimizers (isolated and with positive definite Hessian), an exponential rate of convergence has been provided in Bach and Rudi (2023). This shows that beyond general results such as those in Table 3.1, there is hope for even faster rate convergence, at the price of some additional conditions on the minimizers or the set S.

3.8. Handling sparsity

As already mentioned, in its canonical form (3.3)–(3.5), the Moment-SOS hierarchy is limited to problems **P** of modest dimension, even though for fixed dimension d the size parameters of each Moment-relaxation (3.5) are polynomial in the degree n. This is because, (3.5) being a semidefinite program, efficient algorithms based on interior-point methods are still very time-consuming. Fortunately, practice reveals that Moment-relaxations (3.5) of low degree n already provide tight lower bounds on f^*, and are sometimes *exact*. In addition, large-scale problems **P** usually exhibit some *sparsity* pattern or symmetries. For instance, as is typical for applications in large dimension d, (i) each constraint $g_j \geq 0$ in (2.7) sees only some small subset of variables $\{x_i : i \in I_k\}$ with $I_k \subset \{1, \ldots, d\}$, and (ii) the criterion f of **P** is very often a sum $\sum_k f_k$ of (low degree) polynomials f_k, where each polynomial f_k only sees variables $\{x_i : i \in I_k\}$. This type of sparsity is called *correlative sparsity*. Also, another type of sparsity called *term sparsity* occurs when all polynomials f and g_j in the description (2.7) of **P** contain a few monomials only. It turns out that correlative and term sparsity can be exploited so as to yield a new sparsity-adapted Moment-SOS hierarchy of semidefinite relaxations of **P**. These two types of sparsity can even be combined for further efficiency; see e.g. Wang, Magron, Lasserre and Ngoc Hoang Anh Mai (2022) and Magron and Wang (2023).

For the sake of completeness, we now briefly describe how correlative sparsity allows us to define an appropriate sparsity-adapted Moment-SOS hierarchy that can handle large-scale POPs.

Assumption 3.13. With **P** as in (1.1) with $S \subset \mathbb{R}^d$ as in (2.7):

- $I_0 := \{1, \ldots, d\} = \bigcup_{k=1}^p I_k$ (with possible overlaps),
- $\mathbb{R}[\mathbf{x}; I_k]$ is the ring of polynomials in the variables $\{x_i : i \in I_k\}$,
- $f = \sum_{k=1}^p f_k$ with $f_k \in \mathbb{R}[\mathbf{x}; I_k]$, $k = 1, \ldots, p$,
- for each $j = 1, \ldots, m$, $g_j \in \mathbb{R}[\mathbf{x}; I_k]$ for some $k \in \{1, \ldots, p\}$, so let

$$J_k := \{j : g_j \in \mathbb{R}[\mathbf{x}; I_k]\}, \quad k = 1, \ldots, p.$$

Of course, as we will see next, Assumption 3.13 is interesting when the cardinal $\#I_k$ of I_k is small for every $k = 1, \ldots, p$.

Observe that if $g \in \mathbb{R}[\mathbf{x}; I_k]$, then in the expansion $g(\mathbf{x}) = \sum_{\alpha \in \mathbb{N}^d} g_\alpha \mathbf{x}^\alpha$, $\alpha_i = 0$ if $i \notin I_k$, whenever $g_\alpha \neq 0$. So let

$$\mathbb{N}^{(k)} := \{\boldsymbol{\alpha} \in \mathbb{N}^d : \alpha_i = 0, \ \forall\, i \notin I_k\}, \quad k = 1, \ldots, p,$$
$$\mathbb{N}_n^{(k)} := \left\{\boldsymbol{\alpha} \in \mathbb{N}^{(k)} : \sum_i \alpha_i \leq n\right\}, \quad k = 1, \ldots, p.$$

Next, given $\boldsymbol{\phi} = (\phi_\alpha)_{\alpha \in \mathbb{N}^d}$, define $\mathbf{M}_n(\boldsymbol{\phi}; I_k)$ to be the submatrix of $\mathbf{M}_n(\boldsymbol{\phi})$ whose rows and columns are associated with monomials $(\mathbf{x}^\alpha)$, $\boldsymbol{\alpha} \in \mathbb{N}_n^{(k)}$. Similarly, when $j \in J_k$, the localizing matrix $\mathbf{M}_n(g_j \cdot \boldsymbol{\phi}; I_k)$ is the submatrix of $\mathbf{M}_n(g_j \cdot \boldsymbol{\phi})$ whose rows and columns are associated with monomials $(\mathbf{x}^\alpha)$, $\boldsymbol{\alpha} \in \mathbb{N}_{n-d_j}^{(k)}$.

When Assumption 3.13 holds, it is quite natural to define Moment-relaxations

$$\tau_n^{\text{sparse}} = \inf_{\phi=(\phi_\alpha)} \Big\{ \sum_{k=1}^{p} \phi(f_k) : \phi_0 = 1; \quad \mathbf{M}_n(\boldsymbol{\phi}; I_k) \succeq 0,\ k = 1, \ldots, p; \quad \mathbf{M}_{n-d_j}(g_j \cdot \boldsymbol{\phi}; I_k) \succeq 0,\ \forall j \in J_k;\ k = 1, \ldots, p \Big\}. \tag{3.22}$$

The reason why (3.22) is appealing is because the size of the matrix $\mathbf{M}_n(\boldsymbol{\phi}; I_k)$ is $s_k(n) := \binom{\#I_k+n}{n}$ while that of $\mathbf{M}_{n-d_j}(g_j \cdot \boldsymbol{\phi})$ is $s(n-d_j)$. Therefore, if $\#I_k \ll d$, then $s_k(n) \ll s(n)$ and $s_k(n-d_j) \ll s(n-d_j)$. So for a semidefinite solver it is always definitely better to have several (and perhaps even many) small-size matrices rather than a single large-size matrix constrained to be positive semidefinite.

It is quite straightforward to obtain that $\tau_n^{\text{sparse}} \le f^*$ for all $n \ge n_0$. Indeed, $\tau_n^{\text{sparse}} \le \tau_n \le f^*$ (where τ_n is the optimal value of the degree-n Moment-relaxation (3.5)). Moreover, the sequence $(\tau_n^{\text{sparse}})_{n \ge k_0}$ is monotone non-decreasing and, being bounded above, converges to some $\gamma \le f^*$.

Again assume (possibly after scaling) that the quadratic polynomial $1 - \|\mathbf{x}\|^2$ is in the quadratic module $Q_1(g)$, and therefore we can and do add the p redundant constraints

$$1 - \sum_{i \in I_k} x_i^2 \ge 0, \quad k = 1, \ldots, p, \tag{3.23}$$

to the definition of S.

Theorem 3.14 (Lasserre 2006). Let $S \subset \mathbb{R}^d$ be as in (2.7) with constraints (3.23) in its definition (2.7), and consider the hierarchy of semidefinite relaxations (3.22) with optimal value τ_n^{sparse}. Then $\tau_n^{\text{sparse}} \uparrow \gamma \le f^*$, as n increases. Moreover, if for every $k = 2, \ldots, p$,

$$I_k \cap \left(\bigcup_{j=1}^{k-1} I_j \right) \subseteq I_\ell, \tag{3.24}$$

for some $\ell \in \{1, \ldots, k-1\}$, then $\gamma = f^*$.

The condition (3.24) is called the *running intersection property*. It has the following important property: suppose that we are given p probability measures ϕ_j on $\mathbb{R}^{\#I_j}$, $j = 1, \ldots, p$, which are compatible, that is, such that for all pairs (i, j) with $I_i \cap I_j \neq \emptyset$,

$$\int \mathbf{x}^\alpha \, \mathrm{d}\phi_i(\mathbf{x} \in I_i \cap I_j) = \int \mathbf{x}^\alpha \, \mathrm{d}\phi_j(\mathbf{x} \in I_i \cap I_j) \quad \text{for all } \boldsymbol{\alpha} \in \mathbb{N}^{\#I_i \cap I_j}.$$

If (3.24) holds then there exists a probability measure ϕ on $\mathbb{R}^d$ such that

$$\int \mathbf{x}^\alpha \, \mathrm{d}\phi = \int \mathbf{x}^\alpha \, \mathrm{d}\phi_j \quad \text{for all } \boldsymbol{\alpha} \in \mathbb{N}^{\#I_j}, \quad j = 1, \ldots, p.$$

That is, given *local* measures ϕ_j on $\mathbb{R}^{\#I_j}$ that are compatible, we can reconstruct a *global* measure on $\mathbb{R}^d$ whose marginal on $\mathbb{R}^{\#I_j}$ is ϕ_j, for all $j = 1, \ldots, p$. Therefore the local information provided by the ϕ_j corresponds to partial (but consistent) knowledge of global information that we do not necessarily need.

On the dual side of positive polynomials we have the following sparse version of Putinar's Positivstellensatz.

Theorem 3.15 (sparse Positivstellensatz (Lasserre 2006)). Let $S \subset \mathbb{R}^d$ be as in (2.7) with constraints (3.23) in its definition (2.7), and let $f = \sum_{k=1}^{p} f_k$ with $f_k \in \mathbb{R}[\mathbf{x}; I_k]$, $k = 1, \ldots, p$. If (3.24) holds and $f > 0$ on S, then

$$f(\mathbf{x}) = \sum_{k=1}^{p} \left(\sigma_{0,k}(\mathbf{x}) + \sum_{j \in J_k} \sigma_{j,k}(\mathbf{x})\, g_j(\mathbf{x}) \right) \quad \text{for all } \mathbf{x} \in \mathbb{R}^d, \tag{3.25}$$

for some SOS polynomials $\sigma_{0,k}, \sigma_{j,k} \in \Sigma[\mathbf{x}, I_k]$, $j \in J_k$, $k = 1, \ldots, p$.

Theorem 3.15 provides a sparsity-adapted certificate of positivity *à la* Putinar where the SOS weight σ_j associated with a constraint $g_j \geq 0$, $j \in J_k$, 'sees' only the variables $\{x_i : i \in I_k\}$.

To cite a few examples, such sparsity-adapted semidefinite relaxations have been implemented for solving the optimal power flow (OPF) problem in the management of large-scale electricity networks (Molzhan and Hiskens 2015, Molzahn and Josz 2018), in geometric perception (Yang and Carlone 2023), robotics (Rosen, Carlone, Bandera and Leonard 2019) and sensor network localization (Nie 2009), as well as in Nie and Demmel (2008).

3.9. *Notes and sources*

Section 3 is mainly based on Lasserre (2009*b*, 2015), where the reader can find all proofs (or references to papers with proofs).

Section 3.1. The Moment-SOS hierarchy was first proposed in Lasserre (2000, 2001).

Sections 3.3–3.4. Theorems 3.5 and 3.7 are due to Nie (2014). Further, Nie (2013) shows that the flatness condition (3.8) at an optimal solution of the Moment-relaxation (3.5) also holds generically. A refinement of these results is provided in Baldi and Mourrain (2022). Those results are important as they guarantee that the Moment-SOS hierarchy has finite convergence, generically (in the sense of Theorem 3.7), and that one may extract global minimizers from an optimal solution of the semidefinite relaxation (3.5).

Sections 3.5–3.6. These are mainly taken from Lasserre (2015, Chapter 13). See also Lasserre (2009*a*) and de Klerk and Laurent (2011).

Section 3.7. This is essentially based on Slot (2022).

Section 3.8. Correlative sparsity was first proposed as a heuristic in Kojima, Kim and Maramatsu (2005) and Waki, Kim, Kojima and Maramatsu (2006), while its proof of convergence was provided in Lasserre (2006). Term sparsity was first described in Wang, Li and Xia (2019) and further exploited in the TSSOS hierarchy of Wang, Magron and Lasserre (2021). Finally, the combination of correlative and term sparsity which yields the CS-TSSOS hierarchy is described in Wang *et al.* (2022). The interested reader is referred to the book by Magron and Wang (2023), which, among other things, describes various forms of sparsity and associated Moment-relaxations based on appropriate cones of positive polynomials.

Finally let us mention that a lower bound $\tau_n \leq f^*$, obtained by the Moment-SOS hierarchy even at low degree n, can also be useful to 'gauge' how far from f^* is the value $f(\hat{\mathbf{x}})$ of a feasible solution $\hat{\mathbf{x}} \in S$ obtained by some numerical (local) optimization algorithm. Indeed, if $f(\hat{\mathbf{x}}) - \tau_n$ is not too large, then in some way τ_n certifies that the local optimization algorithm has produced a good feasible solution.

4. The Moment-LP hierarchy

As seen in Section 3, the Moment-SOS hierarchy is based on the use of Putinar's certificate of positivity (2.9) and its convergence relies on Theorem 2.2. We next provide a hierarchy of *LP relaxations* whose associated sequence of optimal values also converges to the global optimum from below. Similarly as for the Moment-SOS hierarchy, the Moment-LP hierarchy is also based on a positivity certificate, namely that in (2.13), and its convergence relies on Theorem 2.5.

Assumption 4.1. With $S \subset \mathbb{R}^d$ as in (2.7), assume that S is compact, $0 \leq g_j \leq 1$ on S, for every $j = 1, \ldots, m$, and the polynomials $\{\mathbf{1}, g_1, \ldots, g_m\}$ generate $\mathbb{R}[\mathbf{x}]$.

As S is compact, one can always rescale the g_j (and possibly add redundant constraints) to make the new definition of S satisfy Assumption 4.1. For more details the interested reader is referred to Lasserre (2009*b*).

Next, with the same notation $\mathbf{g}$ and $\mathbf{1} - \mathbf{g}$ as in (2.13), and $n \in \mathbb{N}$, introduce the following linear program (LP):

$$\rho_n = \min_{\boldsymbol{\phi}} \{\phi(f) \colon \phi(\mathbf{1}) = 1;\ \phi(\mathbf{g}^{\alpha}\,(\mathbf{1} - \mathbf{g})^{\beta}) \geq 0,\ (\boldsymbol{\alpha}, \boldsymbol{\beta}) \in \mathbb{N}^{2m}_n\}, \tag{4.1}$$

where $\boldsymbol{\phi} = (\phi_\gamma)_{\gamma \in \mathbb{N}^d_{s_n}}$ with $s_n := \max_{(\alpha,\beta) \in \mathbb{N}^{2m}_n} \deg(\mathbf{g}^{\alpha}(\mathbf{1} - \mathbf{g})^{\beta})$.

By its very nature (4.1) is a linear program, and it is a relaxation of $\mathbf{P}$ because the constraints in (4.1) are only *necessary* conditions on $\boldsymbol{\phi}$ to be moments of a probability measure supported on S; see Theorem 2.5. The dual of (4.1) is the linear program

$$\rho^*_n = \max_{c_{\alpha\beta} \geq 0, \lambda} \left\{ \lambda \colon f - \lambda = \sum_{(\alpha,\beta) \in \mathbb{N}^{2m}_n} c_{\alpha\beta}\, \mathbf{g}^{\alpha}\,(\mathbf{1} - \mathbf{g})^{\beta} \right\}. \tag{4.2}$$

In a similar manner, just as (3.3) was an SOS-strengthening of **P** in (1.3), the LP (4.2) is now an LP-strengthening of **P** in (1.3). Of course, from duality for linear programs, $\rho_n = \rho_n^*$ for all n.

Example 4.2. To better visualize the LP (4.1), consider the toy example where $S = [0, 1] = \{x : x \geq 0;\ (1-x) \geq 0\} \subset \mathbb{R}$. Then, for $n = 2$, $\boldsymbol{\phi} = (\phi_j)_{0\leq j\leq 2}$, and

$$\phi(1) = \phi_0, \quad \phi(x) = \phi_1, \quad \phi(1-x) = \phi_0 - \phi_1, \quad \phi(x^2) = \phi_2,$$
$$\phi(x(1-x)) = \phi_1 - \phi_2, \quad \phi((1-x)^2) = \phi_0 - 2\phi_1 + \phi_2,$$

so that with $f \in \mathbb{R}[x]_2$, $x \mapsto f(x) = \sum_{k=0}^{2} f_k\, x^k$,

$$\rho_2 = \min_{\boldsymbol{\phi}} \left\{ \sum_{k=0}^{2} \phi_k\, f_k : \phi_0 = 1;\ \phi_1 \geq 0;\ \phi_0 - \phi_1 \geq 0; \right.$$
$$\left. \phi_1 - \phi_2 \geq 0;\ \phi_0 - 2\phi_1 + \phi_2 \geq 0 \right\}.$$

Similarly,

$$\rho_2^* = \max_{\mathbf{c}\geq 0,\lambda} \{\lambda : f(x) - \lambda = c_{00} + c_{10}\, x + c_{01}(1-x)$$
$$+ c_{20}\, x^2 + c_{11} x(1-x) + c_{02}(1-x)^2,\ \forall x \in \mathbb{R}\},$$

or equivalently

$$\rho_2^* = \max_{\mathbf{c}\geq 0,\lambda} \{\lambda : f_0 - \lambda = c_{00} + c_{01};$$
$$f_1 = c_{10} - c_{01} + c_{11} + c_{02};\ f_2 = c_{20} - c_{11} + c_{02}\}.$$

Equality constraints are treated as for the Moment-SOS hierarchy. For instance, in (4.1), a Boolean constraint $x_i^2 = x_i$ of **P** translates into the moment equality constraints $\phi(x_i^k) = \phi(x_i)$ for all $k \leq n$.

Theorem 4.3 (Lasserre 2009*b*). With S as in (2.7), let Assumption 4.1 hold. Then, as n increases, the sequences $(\rho_n)_{n\in\mathbb{N}}$ and $(\rho_n^*)_{n\in\mathbb{N}}$ are monotone non-decreasing and converge to the global minimum f^* of **P**.

4.1. The case of a convex polytope

We now assume that $S \subset \mathbb{R}^d$ is a convex polytope (with non-empty interior), that is, for each $j = 1, \ldots, m$, $g_j \in \mathbb{R}[\mathbf{x}]_1$ (g_j is a polynomial of degree 1). In this case Theorem 2.5 takes the following specific form.

Theorem 4.4 (Handelman 1988). Let $S \subset \mathbb{R}^d$ be as in (2.7) with non-empty interior and with all g_j of degree 1, and assume that S is compact (hence S is a convex polytope). If $f \in \mathbb{R}[\mathbf{x}]$ is positive on S, then there exists $n \in \mathbb{N}$ and a non-negative vector $\mathbf{c} = (c_\alpha)_{\alpha\in\mathbb{N}_n^m}$, such that

$$f = \sum_{\alpha\in\mathbb{N}_n^m} c_\alpha\,\mathbf{g}^\alpha. \tag{4.3}$$

So the obvious analogue of (4.1) for a convex polytope now reads

$$\rho_n = \min_{\boldsymbol{\phi}}\,\{\phi(f)\colon\ \phi(\mathbf{1}) = 1;\ \phi(\mathbf{g}^\alpha) \geq 0,\ \alpha \in \mathbb{N}_n^m\} \tag{4.4}$$

where $\boldsymbol{\phi} = (\phi_\gamma)_{\gamma\in\mathbb{N}_{sn}^d}$ with $s_n := \max_{\alpha\in\mathbb{N}_n^m} \deg(\mathbf{g}^\alpha)$. The dual of (4.4) reads

$$\rho_n^* = \max_{c_\alpha\geq 0,\lambda}\left\{\lambda\colon\ f - \lambda = \sum_{\alpha\in\mathbb{N}_n^m} c_\alpha\,\mathbf{g}^\alpha\right\}. \tag{4.5}$$

So an analogue of Theorem 4.3 reads as follows.

Theorem 4.5. Let $S \subset \mathbb{R}^d$ be as in (2.7) with non-empty interior and with all g_j of degree 1, and assume that S is compact (hence S is a convex polytope). Let ρ_n (resp. ρ_n^*) be as in (4.4) (resp. (4.5)). Then, as n increases, both sequences $(\rho_n)_{n\in\mathbb{N}}$ and $(\rho_n^*)_{n\in\mathbb{N}}$ are monotone non-decreasing and converge to the global minimum f^* of $\mathbf{P}$.

On 0/1 *discrete problems and RLT.* Consider discrete optimization problems $\mathbf{P}$ $\min\{f(\mathbf{x})\colon \mathbf{x}\in S\}$ for which the set of feasible solutions is of the form

$$\begin{aligned} S &= \{\mathbf{x}\in\{0,1\}^d\colon\ g_j(\mathbf{x})\geq 0,\ j=1,\dots,m\}\\ &= \{\mathbf{x}\in\mathbb{R}^d\colon\ g_j(\mathbf{x})\geq 0,\ j=1,\dots,m;\ x_i^2 - x_i = 0,\ i=1,\dots,d\}. \end{aligned}$$

This formulation includes many combinatorial optimization problems, including the celebrated Max-Cut problem and its variants on $\{-1,1\}^d$ (after a simple linear transformation). The so-called reformation–linearization technique (RLT) (Sherali and Adams 1990, 1999) solves $\mathbf{P}$ when f and the g_j are all linear. In the RLT we 'lift' $\mathbf{P}$ to a space of higher dimension. Namely, with $t \leq d$ fixed, we proceed as follows.

- We define *order-t bound-factor* constraints

$$A_{J_1,J_2}(\mathbf{x}) := \prod_{i\in J_1} x_i \prod_{j\in J_2}(1-x_j) \geq 0 \quad \text{for all } (J_1,J_2),$$

  where $J_1, J_2 \subseteq \{1,\dots,d\}$, $J_1\cap J_2 = \emptyset$ and $|J_1\cup J_2| = t$.
- For every $j = 1,\dots,m$, we use the additional *constraint factor-based restrictions*

$$A_{J_1,J_2}\,g_j(\mathbf{x}) \geq 0 \quad \text{for all } (J_1,J_2).$$

- Then, in each such constraint, we replace every occurrence of the power x_i^k with x_i, and 'linearize' the resulting polynomial constraint, that is, every occurrence of the non-linear monomial $\prod_{i\in J} x_i$ is replaced by a variable y_J constrained to be non-negative.

We end up with a linear program in a space of higher dimension $s(t+1)$ which is an LP relaxation of **P**. In the RLT construction of this LP relaxation (which ignores constraints of the form $\mathbf{g}^\alpha \geq 0$), we recognize a particular case of the Moment-LP relaxation (4.1) of **P** in the presence of Boolean constraints $x_i^2 = x_i$, for all $i = 1, \ldots, d$. So it is fair to say that RLT, the first systematic construction of a 'hierarchy' of LP relaxations for 0/1 programs, was implicitly based on positivity certificates of the flavour (2.13), i.e. *à la* Krivine–Vasilescu.

4.2. Contrasting the Moment-SOS hierarchy with the Moment-LP hierarchy

At first glance we are tempted to favour the Moment-LP hierarchy because state-of-the-art LP solvers are very efficient and can solve potentially very large-scale (even huge) LPs, whereas in its canonical form (3.3)–(3.5) the Moment-SOS hierarchy is limited to POPs of modest dimension and small degree of relaxation n, unless some sparsity or symmetries can be exploited. Unfortunately the Moment-LP hierarchy has some serious drawbacks that also limit its application to problems of modest dimension. Indeed, except for discrete and linear POPs, finite convergence is impossible in general, even for convex problems!

However, for discrete problems with 0/1 variables, the Moment-LP hierarchy can be combined with *ad hoc* heuristics. For instance, we may try to solve such 0/1 problems with Branch and Bound methods where, at each node of the search tree, a lower bound associated with the node is computed by solving an appropriate Moment-LP relaxation of the discrete subproblem associated with the node in the Branch and Bound strategy.

Finite convergence is not possible in general. For clarity and simplicity of exposition, we illustrate this claim in the case where S is a convex polytope.

Proposition 4.6. Let $S \subset \mathbb{R}^d$ be a convex polytope and consider the LP-strengthening (4.5) of **P**. If **P** has finitely many global minimizers and (4.5) is exact for some degree n, then necessarily every global minimizer $\mathbf{x}^*$ is a vertex of S.

Proof. Let $0 \leq \mathbf{c}^* \neq 0$ be an optimal solution of the degree-n LP relaxation (4.5), and assume that (4.5) is exact, i.e. $\rho_n^* = f^*$. Then

$$f(\mathbf{x}) - f^* = \sum_{\alpha \in \mathbb{N}_n^m} c_\alpha^* \, g_1(\mathbf{x})^{\alpha_1} \cdots g_m(\mathbf{x})^{\alpha_m} \quad \text{for all } \mathbf{x} \in \mathbb{R}^d. \tag{4.6}$$

In particular, at every global minimizer $\mathbf{x}^* \in S$,

$$0 = f(\mathbf{x}^*) - f^* = \sum_{\alpha \in \mathbb{N}_n^m} c_\alpha^* \, g_1(\mathbf{x}^*)^{\alpha_1} \cdots g_m(\mathbf{x}^*)^{\alpha_m} \quad \text{for all } \mathbf{x} \in \mathbb{R}^d. \tag{4.7}$$

So assume that there exists a global minimizer $\mathbf{x}^* \in S$ which is not a vertex, and let $I(\mathbf{x}^*) := \{j \in \{1, \ldots, m\} \colon g_j(\mathbf{x}^*) = 0\}$ be the set of active constraints at $\mathbf{x}^* \in S$. Observe that $I(\mathbf{x}^*) \neq \emptyset$ because otherwise we would have $g_j(\mathbf{x}^*) > 0$ for all $j = 1, \ldots, m$, which in turn by (4.7) implies $\mathbf{c}^* = 0$, in contradiction to $\mathbf{c}^* \neq 0$. So

(4.7) already rules out the possibility of having a global minimizer in int(S). Next, from (4.7) we may infer

$$c_\alpha^* > 0 \Rightarrow \alpha_j > 0 \quad \text{for some } j \in I(\mathbf{x}^*). \tag{4.8}$$

As $\mathbf{x}^*$ is not a vertex and $\mathbf{P}$ has only finitely many global minimizers, the set $\{\mathbf{y} \in S\colon I(\mathbf{y}) = I(\mathbf{x}^*)\}$ contains a point $\hat{\mathbf{y}} \in S$ which is not a global minimizer, i.e. $f(\hat{\mathbf{y}}) > f^*$. But then (4.6) combined with (4.8) yields the contradiction

$$0 = \sum_{\alpha \in \mathbb{N}_n^m} c_\alpha^* \, g_1(\hat{\mathbf{y}})^{\alpha_1} \cdots g_m(\hat{\mathbf{y}})^{\alpha_m} = f(\mathbf{y}) - f^* > 0. \qquad \square$$

So Proposition 4.6 implies that for all problems $\mathbf{P}$ (on convex polytopes) with finitely many minimizers, a necessary condition for some degree-n LP-strengthening (4.5) to be exact is that every global minimizer is a vertex of S. In particular, this condition rules out most convex POPs on a polytope, with a non-linear criterion f as in general a local (hence global) minimizer is *not* a vertex of S. (However, if f is linear, i.e. if $\mathbf{P}$ is a linear program, then (4.5) is exact with $n = 1$.)

A similar conclusion is also valid for POPs on even more general basic semi-algebraic sets S and the LP-strengthening (4.2). Indeed, as in Proposition 4.6 and for the same reasons, if such a POP has finitely many minimizers, then a necessary condition for (4.2) to be exact at some degree n, is that for every global minimizer $\mathbf{x}^* \in S$, the set $\{\mathbf{y} \in S\colon I(\mathbf{y}) = I(\mathbf{x}^*)\}$ contains only global minimizers of $\mathbf{P}$. Such a condition is very restrictive and rules out most problems $\mathbf{P}$, in particular convex problems!

4.3. Notes and sources

Section 4 is mainly taken from Lasserre (2002*b*, 2015, Chapter 9). In Laurent (2003) the Moment-SOS hierarchy for 0/1 variables is described with specific notation proper to graph theory and is embedded in the family of *lift-and-project* hierarchies, which include the Lovász–Schrijver and Sherali–Adams hierarchies. In particular, Laurent (2003) shows that the Moment-SOS dominates the other lift-and-project hierarchies.

If LP-hierarchies are not efficient when used alone to solve optimization problems, they can still be useful when associated with other techniques of discrete optimization, for instance as in Aloise and Hansen (2011) when used in conjunction with Branch and Bound.

5. A Moment-SOS hierarchy of upper bounds

In this section we consider another (less known) Moment-SOS hierarchy which provides a monotone non-increasing sequence $(\kappa_n)_{n\in\mathbb{N}}$ of *upper bounds* on the global minimum f^* of $\mathbf{P}$ defined in (1.3). For each 'degree' n, the upper bound $\kappa_n \geq f^*$ is now computed by solving a very specific semidefinite program as it has only a single variable. In fact its dual reduces to computing the smallest generalized eigenvalue of a pair of moment matrices.

5.1. A first multivariate formulation

Consider problem **P** in (1.3) with feasible set $S \subset \mathbb{R}^d$ as in (2.7), and let μ be a Borel (reference) measure on S with $\mathrm{supp}(\mu) = S$.

Assumption 5.1.

(i) The set $S \subset \mathbb{R}^d$ is compact with non-empty interior.

(ii) The vector of moments $\boldsymbol{\mu} = (\mu_\alpha)_{\alpha\in\mathbb{N}^d}$ is available in closed form, or can be computed efficiently.

Of course, in view of Assumption 5.1(ii), the set S has to be rather specific, and indeed typical sets of this type are the unit box $[-1, 1]^d$, the Euclidean unit ball $\mathbf{B}(0, 1) = \{\mathbf{x}\colon \|\mathbf{x}\| \le 1\}$, the unit sphere $\mathbb{S}^{d-1} = \{\mathbf{x}\colon \|\mathbf{x}\| = 1\}$, the canonical simplex $\Delta^d = \{\mathbf{x} \in \mathbb{R}^d_+\colon \mathbf{e}^\top\mathbf{x} \le 1\}$, the discrete hypercube $\{-1, 1\}^d$, and their image by an affine transformation. Even though such sets S are rather specific, the associated problems **P** cover many interesting NP-hard optimization problems.

Theorem 5.2. Let Assumption 5.1 hold, and with $n \in \mathbb{N}$ fixed, consider the semidefinite problems

$$\kappa_n = \inf_{\sigma\in\Sigma[\mathbf{x}]_n} \left\{ \int f\,\sigma\,\mathrm{d}\mu\colon \int \sigma\,\mathrm{d}\mu = 1 \right\}, \quad n \in \mathbb{N}, \tag{5.1}$$

$$\kappa_n^* = \sup_\lambda \{\lambda\colon \lambda\,\mathbf{M}_n(\mu) \preceq \mathbf{M}_n(f\cdot\mu)\}, \quad n \in \mathbb{N}. \tag{5.2}$$

Then

$$\kappa_n = \kappa_n^*, \quad f^* \le \kappa_{n+1} \le \kappa_n \quad \text{for all } n \in \mathbb{N}, \quad \lim_{n\to\infty} \kappa_n = f^*. \tag{5.3}$$

Crucial to the proof of convergence, which can be found in Lasserre (2011, Theorem 4.2), is the Nichtnegativstellensatz Theorem 2.4. It turns out that (5.2) is just computing the smallest generalized eigenvalue associated with the pair of symmetric matrices $(\mathbf{M}_n(f\cdot\mu), \mathbf{M}_n(\mu))$, for which specialized software exists. If both matrices are expressed in a polynomial basis of $\mathbb{R}[\mathbf{x}]_n$ formed by polynomials that are orthonormal with respect to μ, then $\mathbf{M}_n(\mu)$ becomes the identity matrix **I**, and τ_n^* is just the smallest eigenvalue of $\mathbf{M}_n(f\cdot\mu)$.

Computational consideration. It is straightforward to fill up entries of both matrices $\mathbf{M}_n(\mu)$ and $\mathbf{M}_n(f\cdot\mu)$, so the main effort is in computing the generalized eigenvalue of the pair of (symmetric) matrices $(\mathbf{M}_n(f\cdot\mu), \mathbf{M}_n(\mu))$, which can be done via standard software for eigenvalue computation. However, and even if they are symmetric, computing κ_n is quite challenging because of the size $O(n^d)$ of the matrices as n increases.

Remark 5.3. A refinement of (5.1) is to consider polynomial densities σ nonnegative on S (instead of SOS). That is, we replace (5.1) with

$$\hat{\kappa}_n = \inf_{\sigma \in Q_n(g)} \left\{ \int f\,\sigma\,\mathrm{d}\mu : \int \sigma\,\mathrm{d}\mu = 1 \right\}, \quad n \in \mathbb{N}, \tag{5.4}$$

where $Q_n(g)$ is the degree-$2n$ truncated quadratic module associated with the generators g that define S; see (2.6). For instance, if S is the unit box,

$$Q_n(g) = \left\{ \sigma_0 + \sum_{j=1}^{d} \sigma_j\,(1 - x_j^2);\ \sigma_0 \in \Sigma[\mathbf{x}]_n;\ \sigma_j \in \Sigma[\mathbf{x}]_{n-1},\ j = 1, \ldots, m \right\}.$$

Of course, $f^* \leq \hat{\kappa}_n \leq \kappa_n$ for all n and therefore, in view of Theorem 5.2, $\hat{\kappa}_n \downarrow f^*$ as n increases.

5.2. *An alternative univariate formulation*

Let the (univariate) Borel measure $\#\mu$ be the pushforward of μ on $\mathbb{R}$ by the mapping f, that is,

$$\#\mu(B) = \mu(f^{-1}(B)) \quad \text{for all } B \in \mathcal{B}(\mathbb{R}), \tag{5.5}$$

where $\mathcal{B}(\mathbb{R})$ is the usual Borel σ-algebra generated by the open sets of $\mathbb{R}$. By construction of $\#\mu$, $\mathrm{supp}(\#\mu) = f(S)$, and therefore

$$f^* = \min_z \{z : z \in \mathrm{supp}(\#\mu)\}, \tag{5.6}$$

Moreover, the moments $(\#\mu_j)_{j\in\mathbb{N}}$ of $\#\mu$ satisfy

$$\#\mu_j = \int_{f(S)} z^j\,\mathrm{d}\#\mu(z) = \int_S f(\mathbf{x})^j\,\mathrm{d}\mu(\mathbf{x}) \quad \text{for all } j \in \mathbb{N}. \tag{5.7}$$

As all moments μ_α of μ are available, the $\#\mu_j$ can be obtained exactly, for instance, by expanding the polynomial f^j in the canonical basis $(\mathbf{x}^\alpha)$,

$$\mathbf{x} \mapsto f(\mathbf{x})^j = \sum_{\alpha\in\mathbb{N}^d} \theta_\alpha^{(j)}\,\mathbf{x}^\alpha, \quad \#\mu_j = \sum_\alpha \theta_\alpha^{(j)}\,\mu_\alpha.$$

However, notice that even though the above expansion is always possible, it can become very tedious if j is large, even for modest dimension d.

Theorem 5.4. Let $\#\mu$ be the measure on $\mathbb{R}$ in (5.5) (the pushforward of μ by f), and let

$$\rho_n := \inf_{\sigma\in\Sigma[z]_n} \left\{ \int z\,\sigma\,\mathrm{d}\#\mu : \int \sigma\,\mathrm{d}\#\mu = 1 \right\}, \quad n \in \mathbb{N}, \tag{5.8}$$

$$\rho_n^* := \sup_\lambda \{\lambda : \lambda\,\mathbf{M}_n(\#\mu) \preceq \mathbf{M}_n(z \cdot \#\mu)\}, \quad n \in \mathbb{N}. \tag{5.9}$$

Then $(\rho_n)_{n\in\mathbb{N}}$ is a monotone non-increasing sequence such that $\rho_n \downarrow f^*$ as n increases. In addition, and letting $d_f = \deg(f)$, we obtain $\rho_n \geq \kappa_{nd_f}$ for every

$n \in \mathbb{N}$ (where κ_n is defined in (5.1)), because if $\sigma \in \Sigma[z]_n$ is a feasible solution of (5.8) then $\sigma \circ f \in \Sigma[\mathbf{x}]_{nd_f}$ is a feasible solution of (5.1) with the same value.

A detailed proof can be found in Lasserre (2021). There is a striking difference between the hierarchies of upper bounds $(\kappa_n)_{n\in\mathbb{N}}$ in (5.1) and $(\rho_n)_{n\in\mathbb{N}}$ in (5.8). The latter involves *univariate* moment and localizing matrices. Both matrices are Hankel matrices of size $O(n)$ (in contrast to multivariate Hankel-type matrices of size $O(n^d)$ for computing κ_n). Therefore, computing the generalized eigenvalue ρ_n is much easier than computing κ_n.

On the other hand, filling up all entries of the Hankel moment matrix $\mathbf{M}_n(\#\mu)$ is in principle easy but tedious. Indeed, if f^j is expanded in the monomial basis then its integration (5.7) with respect to μ is straightforward. However, as already noted, such an expansion can be quite costly if j is not small (even for modest dimension d).

Rates of convergence. It is worth noting that, just as for the hierarchy of lower bounds, $O(1/n^2)$ rates of convergence have also been obtained for the hierarchy of upper bounds (5.1)–(5.2) on the sets $\mathbb{S}^{d-1}$, $\mathbf{B}(0, 1)$, $[-1, 1]^d$ and Δ^d; see Slot (2022, Table 2, p. 2615).

5.3. *Notes and sources*

Section 5 is essentially based on Lasserre (2011, 2013); the univariate formulation is from Lasserre (2021). In a series of papers, de Klerk, Laurent and collaborators (e.g. de Klerk and Laurent 2011, Slot and Laurent 2021, Laurent and Slot 2023) have obtained rates of convergence $\kappa_n \downarrow f^*$ (multivariate) and $\rho_n \downarrow f^*$ (univariate) as n grows, by playing with various reference measures μ on S and a clever choice of appropriate families of densities. The approach is also interesting in its own right as it is a mix of several sophisticated techniques, including polynomial kernels and asymptotics for roots of some distinguished orthogonal polynomials. Moreover, it turns out that such techniques have also been useful in obtaining rates of convergence for the Moment-SOS hierarchy of lower bounds on specific sets S; for more details the interested reader is referred to Slot (2022) and references therein.

6. Some applications of the Moment-SOS hierarchy

In this section we briefly describe how the Moment-SOS hierarchy can be used to help solve several problems in various fields of science and engineering. In brief, problems where the Moment-SOS hierarchy is a relevant tool are those which have an equivalent formulation as an instance of the so-called *generalized moment problem* (GMP) whose description is only through polynomials and semi-algebraic sets (i.e. GMPs with algebraic data). Indeed, the list of potential applications of the GMP is almost endless, with polynomial optimization being its simplest instance.

As is to be expected from what we have seen for optimization, the GMP can be formulated in a primal form (via moments) or a dual form (via polynomials). Its primal form is an infinite-dimensional and linear (hence convex) optimization problem on measure spaces, which reads

$$\text{GMP:}\quad \inf_{\phi_1,\ldots,\phi_s}\left\{\sum_{j=1}^{s}\int f_j\,\mathrm{d}\phi_j : \sum_{j=1}^{s}\int h_{kj}\,\mathrm{d}\phi_j \geq b_k,\ k\in\Gamma;\right.$$
$$\left.\operatorname{supp}(\phi_j)\subset S_j,\ j=1,\ldots,s\right\}, \tag{6.1}$$

where all functions f_j, h_{kj} are polynomials, and the ϕ_j are Borel measures whose supports $S_j \subset \mathbb{R}^{r_j}$, $j = 1, \ldots, s$, are all basic semi-algebraic sets.

The reader will note that all constraints are *linear* constraints linking moments of the involved measures ϕ_j, $j = 1, \ldots, s$ (whence the name of generalized moment problem). So the GMP is an infinite-dimensional LP on spaces of measures.

Of course, the GMP in (6.1) can be extended to more general functions and sets, but for a practical application of the Moment-SOS hierarchy, we need *algebraic data* (polynomials and basic semi-algebraic sets). Notice also that formulation (1.4) of a polynomial optimization problem is the simplest instance of the GMP in which there is only one unknown measure ϕ and only one (equality) moment constraint $\phi(S) = \int 1\mathrm{d}\phi = 1$.

The dual GMP* of (6.1) is also an infinite-dimensional LP, and when $p := \#\Gamma < \infty$, it reads

$$\text{GMP}^*:\quad \sup_{\lambda\in\mathbb{R}^p_+}\left\{\sum_{k=1}^{p}\lambda_k\,b_k :\right.$$
$$\left. f_j(\mathbf{x}) - \sum_{k=1}^{p}\lambda_k\,h_{kj}(\mathbf{x}) \geq 0,\ \forall\,\mathbf{x}\in S_j,\ j=1,\ldots,s\right\}. \tag{6.2}$$

Moment equality constraints are also tolerated in (6.1), in which case the associated dual variable λ_k in (6.2) is not constrained to be non-negative. As is the case in some important applications, the set Γ is also allowed to be (countably) infinite. Finally, the objective function of (6.1) can also be a convex function of finitely many moments of measures ϕ_j (e.g. $-\log\det(\mathbf{M}_n(\phi_j))$ of the moment matrix $\mathbf{M}_n(\phi_j)$), though we do not discuss this further.

As we see next in two examples, in some applications the problem to solve is already in the form of a GMP (or GMP*), whereas in other applications, some equivalent formulation of the problem is an instance of the GMP.

Strategy of the Moment-SOS hierarchy. Roughly speaking, to apply the Moment-SOS hierarchy to the GMP (6.1), we proceed as follows.

- We replace the measures $(\phi_j)_{j=1,\ldots,s}$ with degree-$2n$ truncated pseudo-moment vectors $\boldsymbol{\phi}_j = (\phi_{j,\alpha})$, $\alpha \in \mathbb{N}^{r_j}_{2n}$, $j = 1, \ldots, s$.

- We impose semidefinite constraints on the moment and localizing matrices associated with each $\boldsymbol{\phi}_j$ and each set S_j, which by Theorem 2.2(ii) are necessary conditions for $\boldsymbol{\phi}_j$ to be moments of a measure on S_j.

Then, as the moment constraints and the criterion are just linear on the pseudo-moment vectors $\boldsymbol{\phi}_j$, for each fixed n we end up with a finite-dimensional semidefinite relaxation that provides a lower bound on the optimal value of the GMP. Similarly, to apply the Moment-SOS hierarchy to GMP* in (6.2), we proceed as follows.

- For each $j = 1, \ldots, s$, we replace the positivity constraint

$$f_j - \sum_{k=1}^{p} \lambda_k \, h_{kj} \geq 0 \quad \text{on } S_j$$

with a Putinar certificate of positivity of degree $2n$; see Theorem 2.2(i). For instance, if $S_j = [-1, 1]^d$ then the above positivity constraint reads

$$f_j - \sum_{k=1}^{p} \lambda_k \, h_{kj} = \sum_{j=0}^{d} \sigma_j \, (1 - x_j^2), \quad \sigma_j \in \Sigma[\mathbf{x}]_{n-d_j}, \; j = 0, \ldots, d.$$

Then, as the criterion $\sum_k \lambda_k b_k$ is linear, we end up with a finite-dimensional semidefinite program, which is the dual of the program on pseudo-moment vectors. In fact, a primal–dual semidefinite solver will solve both of them at the same time.

For completeness, below are two illustrative applications of the above strategy. In the first example, in computational geometry and probability, the problem itself is described as a GMP, while in the second example, in optimal control, an alternative and so-called 'weak formulation' of the optimal control problem is an instance of the GMP.

6.1. *Illustration in probability and computational geometry*

Let $S \subset \mathbb{R}^d$ be a compact set and suppose that $S \subset \mathbf{B} := [-1, 1]^d$ (possibly after scaling). The goal is to approximate the Lebesgue volume vol(S) of S, as closely as desired. This is known to be a very hard problem. In fact, even if S is convex, approximating its volume is quite hard; see e.g. Dyer and Frieze (1988), the discussion in Henrion, Lasserre and Savorgnan (2009) and references therein.

Let λ be the Lebesgue measure on $[-1, 1]^d$, so that its (infinite) vector of moments $\boldsymbol{\lambda} = (\lambda_\alpha)_{\alpha \in \mathbb{N}^d}$ is available in closed form. Let $\mathbf{1} \in \mathbb{R}[\mathbf{x}]$ be the constant polynomial (equal to 1 for all $\mathbf{x}$), and for two measures μ, ν on $\mathbb{R}^d$, the notation $\nu \leq \mu$ stands for $\nu(B) \leq \mu(B)$ for all $B \in \mathcal{B}(\mathbb{R}^d)$.

Proposition 6.1. We have

$$\text{vol}(S) = \max_{\phi \in \mathscr{M}(S)_+, \nu \in \mathscr{M}(\mathbf{B})_+} \{\phi(\mathbf{1}) \colon \phi + \nu = \lambda\},$$

and $\phi^* := 1_S \, \lambda$ is the unique optimal solution.

Proof. As $\phi+\nu = \lambda$, $\phi \leq \lambda$ and since $\mathrm{supp}(\phi) \subseteq S$, we have $\phi(\mathbf{1}) = \phi(S) \leq \lambda(S) = \mathrm{vol}(S)$. Next, with $\phi^* := 1_S\,\lambda \in \mathscr{M}(S)_+$, and $\nu^* := \mu - \phi^* \in \mathscr{M}(\mathbf{B})_+$, we obtain $\phi^*(\mathbf{1}) = \lambda(S) = \mathrm{vol}(S)$. □

The above formulation of vol(S) as an optimization problem is not yet in the form of the GMP (6.1). But notice that since $\mathbf{B}$ is compact,

$$\phi + \nu = \lambda \;\Leftrightarrow\; \phi_\alpha + \nu_\alpha = \lambda_\alpha \quad \text{for all } \alpha \in \mathbb{N}^d,$$

and therefore

$$\mathrm{vol}(S) = \max_{\phi\in\mathscr{M}(S)_+,\nu\in\mathscr{M}(\mathbf{B})_+} \{\phi(\mathbf{1})\colon \phi_\alpha + \nu_\alpha = \lambda_\alpha,\ \forall\,\alpha \in \mathbb{N}^d\}, \tag{6.3}$$

which is an instance of the GMP (6.1) with $\Gamma = \mathbb{N}^d$ (a countable set). We next see how to implement the moment-SOS hierarchy. Let $g_0 = \mathbf{1}$,

$$S = \{\mathbf{x} \in \mathbb{R}^d\colon g_j(\mathbf{x}) \geq 0,\ j = 1,\dots,m\},$$

and recall that $d_j = \lceil \deg(g_j)/2 \rceil$ for all $j = 0,\dots,m$. For each $n \in \mathbb{N}$, consider the optimization problem

$$\begin{aligned}\tau_n = \max_{\phi,\nu}\{\phi_0\colon \phi_\alpha + \nu_\alpha = \lambda_\alpha,\ \forall\,\alpha \in \mathbb{N}^d_{2n};\\ \mathbf{M}_{n-d_j}(g_j\cdot\phi) \succeq 0,\ j = 0,\dots,m;\ \mathbf{M}_n(\nu) \succeq 0\}.\end{aligned} \tag{6.4}$$

For each fixed n, (6.4) is a semidefinite program and an obvious relaxation of (6.3) so that $\tau_n \geq \mathrm{vol}(S)$ for all n.

Theorem 6.2 (Henrion *et al.* 2009). The sequence of optimal values $(\tau_n)_{n\in\mathbb{N}}$ is monotone non-increasing, bounded below, and $\lim_{n\to\infty} \tau_n = \mathrm{vol}(S)$.

So (6.4), indexed by $n \in \mathbb{N}$, provides a hierarchy of semidefinite relaxations of (6.3) such that $(\tau_n)_{n\in\mathbb{N}}$ converges (from above) to the desired value vol(S) as n increases. However, in its basic form (6.4), its convergence is quite slow. To see why, consider the dual of (6.4), which is the semidefinite program

$$\begin{aligned}\tau_n^* = \min_{p\in\mathbb{R}[\mathbf{x}]_{2n}} \left\{\int_{\mathbf{B}} p\,\mathrm{d}\lambda\colon p - 1 = \sum_{j=0}^m \sigma_j\,g_j\right.\\ \left. p \in \Sigma[\mathbf{x}]_n;\ \sigma_j \in \Sigma[\mathbf{x}]_{n-d_j};\ j = 0,\dots,m\right\}.\end{aligned} \tag{6.5}$$

It turns out that if S has non-empty interior then $\tau_n = \tau_n^*$ for all n. Next observe that

$$p \in \Sigma[\mathbf{x}]_n \quad\text{and}\quad p - 1 = \sum_{j=0}^m \sigma_j\,g_j \;\Rightarrow\; p \geq 1_S \quad \text{for all } \mathbf{x} \in \mathbf{B},$$

and since $\int p\mathrm{d}\lambda \downarrow 1_S\mathrm{d}\lambda$ as n grows, in the dual (6.5) we search for a degree-$2n$ SOS polynomial p that tries to approximate from above the indicator function of

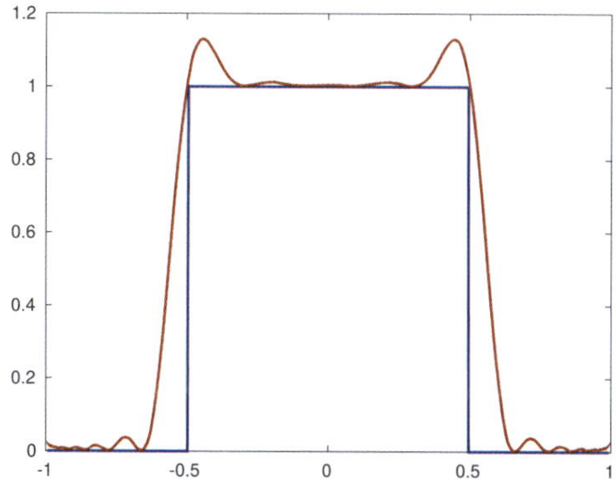

Figure 6.1. $S = [-0.5, 0.5] \subset [0, 1]$; polynomial approximation (in red) of 1_S on $[0, 1]$ with Gibbs phenomenon.

S for all $\mathbf{x} \in \mathbf{B}$. As is well known, for such polynomial approximations a typical Gibbs phenomenon (oscillations) occurs at points of discontinuities, which makes the convergence quite slow; see Figure 6.1.

Fortunately one can significantly attenuate (or even remedy) this problem. Indeed, as we know in advance the (unique) optimal solution $\phi^* = 1_S\lambda$ of (6.3), every available additional information on ϕ^* in terms of linear constraints on its moments can be added to (6.3) without changing its optimal value and solution. While such additional redundant constraints do not change (6.3), they have a dramatic impact on the relaxations (6.4) and yield a fairly significant acceleration of their convergence. This is indeed the case if we add additional moment constraints (satisfied by ϕ^*) coming from Stokes' theorem.

Stokes constraints. Let us see how it works for the case where $S = \{\mathbf{x} : g(\mathbf{x}) \geq 0\}$ for some polynomial $g \in \mathbb{R}[\mathbf{x}]$ with compact sublevel set S. As g vanishes on ∂S, by Stokes' theorem,

$$\int_S \mathrm{Div}(\mathbf{x}\, g(\mathbf{x})\, \mathbf{x}^\alpha)\, \mathrm{d}\mathbf{x} = \int_{\partial S} \langle \vec{n}_\mathbf{x}, \mathbf{x} \rangle g(\mathbf{x}) \mathbf{x}^\alpha \, \mathrm{d}\sigma(\mathbf{x}) = 0 \quad \text{for all } \alpha \in \mathbb{N}^d,$$

where $\vec{n}_\mathbf{x}$ is the outward pointing normal at $\mathbf{x} \in \partial S$. Hence each $\alpha \in \mathbb{N}^d$ provides us with the moment constraint

$$\phi^*(s_\alpha) := \phi^*(\mathrm{Div}(\mathbf{x}\, g(\mathbf{x}) \mathbf{x}^\alpha) = 0 \quad \text{on } \phi^*,$$

because $\mathbf{x} \mapsto s_\alpha(\mathbf{x}) := \mathrm{Div}(\mathbf{x}\, g(\mathbf{x})\, \mathbf{x}^\alpha)$ is a polynomial (of degree $\deg(g) + |\alpha| + 1$). Hence, for every $n \in \mathbb{N}$, the additional moment constraints

$$\phi(s_\alpha) = 0 \quad \text{for all } \alpha : |\alpha| \leq 2n - 1 - \deg(g) \tag{6.6}$$

can be included in the semidefinite relaxation (6.4). The effect on the dual (6.5) is to change the initial constraint $p - 1 = \sigma_0 + \sigma_1 g$ to

$$p + q - 1 = \sigma_0 + \sigma_1 g,$$

where $q := \sum_{|\alpha|\leq 2n-\deg(g)-1} \theta_\alpha\, s_\alpha$ for the dual variables (θ_α) associated with (6.6). Hence, p is *no longer* required to approximate 1_S from above! For more details on volume computation via the moment-SOS hierarchy, the interested reader is referred to Henrion *et al.* (2009), Tacchi, Weisser, Lasserre and Henrion (2021) and Tacchi, Lasserre and Henrion (2023). In particular, this technique has also been implemented in Tacchi *et al.* (2021) to approximate the volume of certain non-convex sets $S \subset \mathbb{R}^{100}$ where the description of S exhibits some structured sparsity, as described in Section 3.8.

Remark 6.3. One can consider (6.3) with a measure λ that is not Lebesgue measure on $\mathbf{B}$. For example, if λ is the Gaussian measure $\exp(-\|\mathbf{x}\|^2)\,\mathrm{d}\mathbf{x}$ on $\mathbb{R}^d$, or the exponential measure $\exp(-\sum_j x_j)\,\mathrm{d}\mathbf{x}$ on $\mathbb{R}^d_+$, one can approximate the value $\lambda(S)$ for non-compact semi-algebraic sets as closely as desired; see e.g. Lasserre (2017).

Application in probability. Suppose that X is a $\mathbb{R}^d$-valued random vector whose distribution is only partially known through a few of its moments $\boldsymbol{m} = (m_\alpha)_{\alpha\in\Gamma}$, where $\Gamma \subset \mathbb{N}^d$ is a finite set (typically the index set of moments up to order $3, 4$). Next, let $S \subset \mathbb{R}^d$ be a given compact basic semi-algebraic set with non-empty interior. The goal is to provide the best upper bound on $\mathrm{Prob}(X \in S)$, under the partial knowledge of $\boldsymbol{m} = (m_\alpha)_{\alpha\in\Gamma}$, that is, compute

$$\rho = \max_{\mu\in\mathscr{M}(\mathbb{R}^d)_+} \left\{\mu(S) \colon \int \mathbf{x}^\alpha\,\mathrm{d}\mu = m_\alpha,\ \forall\,\boldsymbol{\alpha} \in \Gamma\right\}. \tag{6.7}$$

Observe that (6.7) is an instance of the GMP (6.1) but with non-polynomial data because with $\mu \in \mathscr{M}(\mathbb{R}^d)_+$, $\mu(S) = \mu(1_S)$ and 1_S is not a polynomial. We need to consider that $\mu = \phi + \nu$ with $\phi \in \mathscr{M}(S)_+$ (in which case $\phi(S) = \phi(\mathbf{1}) = \phi_0$), and hence for every $n \in \mathbb{N}$, consider the optimization problem

$$\begin{aligned} \tau_n = \max_{\boldsymbol{\phi},\nu} \{\phi_0 \colon\ & \phi_\alpha + \nu_\alpha = m_\alpha,\ \forall\,\boldsymbol{\alpha} \in \Gamma; \\ & \mathbf{M}_{n-d_j}(g_j \cdot \boldsymbol{\phi}) \succeq 0,\ j = 0, \ldots, m;\ \mathbf{M}_n(\boldsymbol{\nu}) \succeq 0\}, \end{aligned} \tag{6.8}$$

which is a relaxation of (6.7) and a variant of (6.4) (in fact even easier because now Γ is a finite set instead of the countable set $\mathbb{N}^d$). For instance, with $\Gamma = \mathbb{N}^d_{2t}$ for some fixed t, the dual of (6.8) reads

$$\begin{aligned} \tau^*_n = \min_p \Bigg\{ \sum_{\alpha\in\Gamma} p_\alpha\, m_\alpha \colon\ & p - 1 = \sum_{j=0}^m \sigma_j\, g_j \\ & p \in \Sigma[\mathbf{x}]_t;\ \sigma_j \in \Sigma[\mathbf{x}]_{n-d_j};\ j = 0, \ldots, m \Bigg\}. \end{aligned} \tag{6.9}$$

The difference from (6.5) is that now, even when n changes, we still search for a degree-$2t$ polynomial $p \geq 1_S$ (where the degree $2t$ is fixed by the number of moments in Γ). In this case we do not have a Gibbs phenomenon because the degree of p is fixed.

Next, let

$$\Phi := \left\{ \mu \in \mathscr{M}(\mathbb{R}^d)_+ : \int \mathbf{x}^\alpha \, \mathrm{d}\mu = m_\alpha, \ \forall \alpha \in \Gamma \right\}.$$

If there is some $\mu \in \Phi$ with a strictly positive density with respect to Lebesgue measure, and with all moments finite, then each relaxation (6.9) is solvable and there is no duality gap, i.e. $\tau_n = \tau_n^*$, and there is an optimal solution $p^* \in \mathbb{R}[\mathbf{x}]_t$ for all n. Moreover, $\tau_n \downarrow \tau \geq \rho$ as $n \to \infty$. Finally, the bound τ may be sharp, i.e. $\tau = \rho$, if at some step n, the moment matrices of an optimal solution $(\boldsymbol{\phi}, \boldsymbol{\nu})$ of (6.8) satisfy some 'flatness' property; for more details see e.g. Lasserre (2002*a*) and Bertsimas and Popescu (2005).

6.2. *Illustration in optimal control of dynamical systems*

In this section we briefly describe how to apply the Moment-SOS hierarchy to help solve optimal control problems (OCPs) with algebraic data, i.e. those problems whose description is through polynomials and basic semi-algebraic sets. Consider the optimal control problem

$$\begin{aligned} \text{OCP:} \quad J(\mathbf{x}_0, 0) := \min_{\mathbf{u}} \bigg\{ \int_0^1 & h(\mathbf{x}(t), \mathbf{u}(t)) \, \mathrm{d}t + H(\mathbf{x}(1)) : \\ & \dot{\mathbf{x}}(t) = f(\mathbf{x}(t), \mathbf{u}(t)), \ \forall t \in (0,1), \\ & \mathbf{x}(t) \in X, \ \mathbf{u}(t) \in U, \ \forall t \in (0,1), \\ & \mathbf{x}(0) = \mathbf{x}_0 \bigg\}, \end{aligned} \tag{6.10}$$

where h, H, f are polynomials and $X \subset \mathbb{R}^d$ and $U \subset \mathbb{R}^m$ are basic semi-algebraic set, and $\mathbf{x}_0 \in X$ is the initial condition.

Equation (6.10) describes a dynamical system whose evolution in the time interval $[0, 1]$ of its state $\mathbf{x}(t) \in \mathbb{R}^d$, $t \in [0, 1]$, is governed by a controlled ODE with vector field $f : \mathbb{R}^d \times \mathbb{R}^m \to \mathbb{R}^d$, and control $\mathbf{u}(t) \in U$ for all $t \in (0, 1)$. The goal is to approximate an optimal (or close to optimal) control trajectory $t \mapsto \mathbf{u}^*(t) \in \mathbb{R}^m$, $t \in [0, 1]$, which minimizes the functional $\int_0^1 h(\mathbf{x}(t), \mathbf{u}(t)) \, \mathrm{d}t + H(\mathbf{x}(1))$. Here we will not discuss the appropriate function spaces in which we have to search the state and control trajectories $\mathbf{x}(t), \mathbf{u}(t)$. Instead we show:

(i) how to model (6.10) as a particular instance of the GMP (6.1), and

(ii) how to define an appropriate Moment-SOS hierarchy for solving (6.10).

Note that in contrast to POPs, where we search for a point $\mathbf{x}^* \in S \subset \mathbb{R}^d$, we now search for maps $(\mathbf{x}^*, \mathbf{u}^*)$: $[0, 1] \to \mathbb{R}^d \times \mathbb{R}^m$, a much more difficult problem which is already infinite-dimensional in its description.

Strategy. As for (static) optimization problems (1.1), where in the Moment-SOS hierarchy we search for a probability measure (the Dirac measure $\delta_{\{\mathbf{x}^*\}}$ at a global minimizer $\mathbf{x}^* \in S$), here we will also search for a measure μ, now supported on an optimal state-control trajectory $\{(\mathbf{x}^*(t), \mathbf{u}^*(t))\}$ from time $t = 0$ up to time $t = 1$.

By integrating polynomial test functions along feasible trajectories, the ordinary differential equation that governs the dynamical system will provide *linear* constraints on moments of μ and moments of a terminal measure ν on X at time $t = 1$ (because the vector field f in (6.10) is a polynomial). The state and control constraints in (6.10) translate into support constraints on μ and ν.

Modelling (6.10) *as a GMP via occupation measures.* The idea is to look at (6.10) via its impact on the evaluation of test functions all along feasible trajectories. Let $\{(\mathbf{x}(t), \mathbf{u}(t)) : t \in [0, 1]\}$ be an admissible trajectory, and let $(\mathbf{x}, t) \mapsto w(\mathbf{x}, t)$ be an arbitrary *test* function in $\mathscr{C}^1(X \times [0, 1])$. Then observe that

$$\begin{aligned} w(\mathbf{x}(1), 1) - w(\mathbf{x}(0), 0) &= \int_0^1 \mathrm{d}w(\mathbf{x}(t), t) \\ &= \int_0^1 \frac{\partial w(\mathbf{x}(t), t)}{\partial t} + \langle \nabla_{\mathbf{x}} w(\mathbf{x}(t), t), f(\mathbf{x}(t), \mathbf{u}(t)) \rangle \, \mathrm{d}t. \end{aligned} \tag{6.11}$$

Introduce the measures μ on $X \times U \times [0, 1]$, and ν_0, ν on $X \times [0, 1]$:

$$\mu(A \times B \times C) = \int_{C \cap [0,1]} 1_{A \cap X}(\mathbf{x}(t)) \, 1_{B \cap U}(\mathbf{u}(t)) \, \mathrm{d}t, \tag{6.12}$$

$$\nu(A \times C) = 1_{(A \cap X) \times (C \cap \{1\})}(\mathbf{x}, t), \tag{6.13}$$

$$\nu_0(A \times C) = 1_{(A \cap X) \times (C \cap \{0\})}(\mathbf{x}, t), \tag{6.14}$$

for all Borel sets $A \in \mathcal{B}(X)$, $B \in \mathcal{B}(U)$ and $C \in \mathcal{B}([0, 1])$. The measure μ is called the *occupation measure* up to time 1, while ν_0 (resp. ν) is called the occupation measure at time $t = 0$ (resp. at time $t = 1$), all associated with the trajectory $\{(\mathbf{x}(t), \mathbf{u}(t)) : t \in [0, 1]\}$. Another equivalent characterization of μ is via its *disintegration*

$$\mathrm{d}\mu(\mathbf{x}, \mathbf{u}, t) = \delta_{\{(\mathbf{x}(t), \mathbf{u}(t))\}}(d(\mathbf{x}, \mathbf{u})) \, 1_{[0,1]}(t) \, \mathrm{d}t, \tag{6.15}$$

into:

- its *marginal* $1_{[0,1]}(t)\, \mathrm{d}t$ on $[0, 1]$, and
- its *conditional probability* $\delta_{\{(\mathbf{x}(t), \mathbf{u}(t))\}}(d(\mathbf{x}, \mathbf{u}))$ on $X \times U$, given $t \in [0, 1]$ (which is the Dirac measure at the point $(\mathbf{x}(t), \mathbf{u}(t)) \in X \times U$).

It is important to observe that the support of μ is the *graph* $\{(t, \mathbf{x}(t), \mathbf{u}(t)) : t \in [0, 1]\}$ of state-control trajectories $(\mathbf{x}(t), \mathbf{u}(t))$. Ideally we search for the measure

$$\mathrm{d}\mu^*(\mathbf{x}, \mathbf{u}, t) = \delta_{\{(\mathbf{x}^*(t), \mathbf{u}^*(t))\}}(d(\mathbf{x}, \mathbf{u})) \, 1_{[0,1]}(t) \, \mathrm{d}t,$$

whose support is exactly the graph of optimal state-control trajectories $(\mathbf{x}^*(t), \mathbf{u}^*(t))$, $t \in [0, 1]$, when the latter exist.

One reason for introducing occupation measures is that the *time* integral (6.11) reads as the *spatial* integral

$$\int w\, \mathrm{d}\nu - \int w\, \mathrm{d}\nu_0 = \int \frac{\partial w}{\partial t}(\mathbf{x}, t) + \langle \nabla_{\mathbf{x}} w(\mathbf{x}, t), f(\mathbf{x}, \mathbf{u}) \rangle\, \mathrm{d}\mu(\mathbf{x}, \mathbf{u}, t), \qquad (6.16)$$

where the variables $(\mathbf{x}, \mathbf{u}, t)$ are now treated as *independent* variables. The respective dependence of $(\mathbf{x}, \mathbf{u})$ on t is implicit through the support of μ.

Next, introduce the operator $\mathcal{L}\colon \mathscr{C}^1(X \times [0, 1]) \to \mathscr{C}(X \times U \times [0, 1])$:

$$w \mapsto \mathcal{L}w := \frac{\partial w}{\partial t} + \langle \nabla_{\mathbf{x}} w, f \rangle,$$

and its adjoint $\mathcal{L}^*\colon \mathscr{C}(X \times U \times [0, 1])^* \to \mathscr{C}^1(X \times [0, 1])^*$ by

$$\mu \mapsto \mathcal{L}^*\mu := -\frac{\partial \mu}{\partial t} - \sum_{i=1}^{d} \frac{\partial (f_i \mu)}{\partial x_i} = -\frac{\partial \mu}{\partial t} - \operatorname{div}(f\mu),$$

where derivatives of measures are understood in a weak sense via their actions on smooth test functions (and the change of signs comes from integration by parts). Then (6.16) reads

$$\langle w, \nu \rangle - \langle w, \nu_0 \rangle = \langle \mathcal{L}w, \mu \rangle = \langle w, \mathcal{L}^*\mu \rangle,$$

and as it must be valid for all test functions w in a dense subset $\mathcal{D} \subset \mathscr{C}^1(X \times [0, 1])$, we obtain the equation $\mathcal{L}^*\mu = \nu - \nu_0$, that is,

$$\frac{\partial \mu}{\partial t} + \operatorname{div}(f\,\mu) + \nu = \nu_0. \qquad (6.17)$$

Equation (6.17) is a linear transport equation (transporting ν_0 to ν) which is classical in fluid mechanics, statistical physics and PDEs. It is known under several names, including the *equation of conservation of mass*, the *advection equation* and *Liouville's equation*.

This observation allows us to define the so-called *measure-valued weak formulation* of the OCP:

$$\begin{aligned}\rho = \inf_{\mu,\nu} \Bigg\{ &\int h\, \mathrm{d}\mu + \int H\, \mathrm{d}\nu : \\ &\int \frac{\partial w}{\partial t} + \langle \nabla_{\mathbf{x}} w, f \rangle\, \mathrm{d}\mu = \int w\, \mathrm{d}\nu - \int w\, \mathrm{d}\nu_0, \ \forall\, w \in \mathcal{D}; \\ &\mu \in \mathscr{M}(X \times U \times [0, 1])_+,\ \nu \in \mathscr{M}(X \times \{1\})_+ \Bigg\},\end{aligned} \qquad (6.18)$$

introduced by Vinter (1993).

Observe that if $\mathcal{D}$ is a countable set of polynomials, then (6.18) is an instance of the GMP in (6.1) and of course a relaxation of (6.10) so that $\rho \le J(\mathbf{x}_0, 0)$. The dual of (6.18) reads

$$\rho^* = \sup_{w\in\mathscr{C}^1(X\times[0,1])} \left\{ \int w\,\mathrm{d}\nu_0 \;(= w(\mathbf{x}_0, 0)) : \right.$$
$$h + \mathcal{L}w \geq 0,\ \forall\,(\mathbf{x},\mathbf{u},t) \in X \times U \times [0,1];$$
$$\left. w(\mathbf{x},1) \leq H(\mathbf{x}),\ \forall\,\mathbf{x} \in X \right\}. \tag{6.19}$$

It turns out that under some convexity assumptions, $\rho = J(\mathbf{x}_0, 0)$, that is, the measure-valued weak formulation (6.18) is equivalent to the strong formulation (6.10). In the dual (6.19) we approximate the optimal value function $J\colon X\times[0,1] \to \mathbb{R}$ all along an optimal trajectory $\{(\mathbf{x}^*(t), t)\colon t \in [0,1]; \mathbf{x}^*(0) = \mathbf{x}_0\}$ (but *not* for all $(\mathbf{x}, t) \in X \times [0,1]$). For more details see e.g. Lasserre, Henrion, Prieur and Trélat (2008), Korda *et al.* (2022) and references therein.

Next, to implement the moment SOS hierarchy we first select a countable set of test functions $\mathcal{D}$, namely the set of monomials

$$\mathcal{D} := \{(\mathbf{x}^\alpha\,\mathbf{u}^\beta\,t^k)\colon \alpha \in \mathbb{N}^d, \boldsymbol{\beta} \in \mathbb{N}^m, k \in \mathbb{N}\},$$

which is dense in $\mathscr{C}^1(X \times [0,1])$. Then, for every $n \in \mathbb{N}$, let $\mathcal{D}_n := \{(\mathbf{x}^\alpha\,\mathbf{u}^\beta\,t^k) \in \mathcal{D}\colon |\alpha + \boldsymbol{\beta}| + k \leq 2n\}$, and consider the optimization problem

$$\rho_n = \min_{\mu,\nu} \left\{ \mu(h) + \nu(H) : \right.$$
$$\mu\left(\frac{\partial w}{\partial t} + \langle \nabla_{\mathbf{x}} w, f\rangle\right) = \nu(w) - \nu_0(w),\ \forall\, w \in \mathcal{D}_n;$$
$$\mathbf{M}_n(\boldsymbol{\mu}),\ \mathbf{M}_n(\boldsymbol{\nu}) \succeq 0;\ \mathbf{M}_n((1-t)\boldsymbol{\nu}) = 0;$$
$$\mathbf{M}_{n-d_g}(g\cdot\boldsymbol{\mu}),\ \mathbf{M}_{n-d_g}(g\cdot\boldsymbol{\nu}) \succeq 0,\ \forall\, g \in G;$$
$$\left. \mathbf{M}_{n-1}(t(1-t)\cdot\boldsymbol{\mu}) \succeq 0,\ \mathbf{M}_{n-d_\theta}(\theta\cdot\boldsymbol{\mu}) \succeq 0,\ \forall\,\theta \in \Theta \right\}. \tag{6.20}$$

where $X = \{\mathbf{x}\colon g(\mathbf{x}) \geq 0,\ g \in G\}$, $U = \{\mathbf{u}\colon \theta(\mathbf{u}) \geq 0,\ \theta \in \Theta\}$, $d_g = \lceil \deg(g)/2 \rceil$, $g \in G$, and $d_\theta = \lceil \deg(\theta)/2 \rceil$, $\theta \in \Theta$.

So the sequence of optimal values $(\rho_n)_{n\in\mathbb{N}}$ is monotone non-decreasing, and under the convexity assumptions alluded to above, $\rho_n \uparrow J(\mathbf{x}_0, 0)$ as $n \to \infty$.

Reconstruction of optimal trajectories from moments. So far, by solving the semidefinite relaxations (6.20) we obtain a sequence $(\rho_n)_{n\in\mathbb{N}}$ of lower bounds on the optimal value $J(\mathbf{x}_0, 0)$ of the initial OCP (6.10). But from the vector of pseudo-moments $(\boldsymbol{\mu}^n, \boldsymbol{\nu}^n)$ optimal solution of (6.20) for some degree n, can we retrieve or approximate optimal trajectories $(\mathbf{x}^*(t), \mathbf{u}^*(t))$ when they exist, or provide ε-suboptimal trajectories otherwise?

Again, and ideally, when n is sufficiently large, we expect (μ^n, ν^n) to provide quite good approximations of the moments of the measures

$$\mathrm{d}\mu(\mathbf{x}, \mathbf{u}, t) = \delta_{\{\mathbf{x}^*(t),\mathbf{u}^*(t))\}}(d(\mathbf{x}, \mathbf{u}))\, 1_{[0,1]}(t)\, \mathrm{d}t$$

and ν, supported respectively on the trajectories $(\mathbf{x}^*(t), \mathbf{u}^*(t))$ and on the point $(\mathbf{x}^*(1), 1) \in X \times \{1\}$. In Section 7.3 we describe an efficient strategy via the Christoffel function, a tool from approximation theory and orthogonal polynomials, particularly well suited to identifying the support of a measure solely from knowledge of its moments.

6.3. *Other applications*

Here we provide the reader with some references to other applications of the Moment-SOS hierarchy. The purpose of this list, which is not exhaustive, is to convince the reader that the Moment-SOS hierarchy is indeed a versatile tool, widely applicable when the problem data are algebraic, and provided that some sparsity or symmetries can be exploited when the problem size demands.

- Control and stochastic control: Pauwels, Henrion and Lasserre (2017), Henrion and Garulli (2005), Parrilo (2000), Parrilo and Lall (2003), Aylward, Parrilo and Slotine (2008), Parrilo (2003), Korda, Henrion and Jones (2014), Fantuzzi, Goluskin, Huang and Chernyshenko (2016) and Goluskin and Fantuzzi (2019).
- For convex computation of the region of attraction for dynamical systems, see e.g. Henrion *et al.* (2021, Chapter 10), and for analysis and control of some types on non-linear PDEs, see e.g. Marx, Weisser, Henrion and Lasserre (2020), Henrion *et al.* (2021, Chapter 11) and Korda *et al.* (2022).
- Tensor computation: Nie (2023), Nie and Yang (2020), Fan, Nie and Zhou (2018) and Nie (2017).
- Algorithmic game theory: Stein, Ozdaglar and Parrilo (2008), Laraki and Lasserre (2012) and Nie and Tang (2023).
- Management of energy networks, and in particular for solving the optimal power flow problem for (large) electricity networks. The Moment-SOS hierarchy has been able to handle problems with thousands of variables by exploiting some inherent sparsity, in the spirit of Section 3.8; see e.g. Molzhan and Hiskens (2015), Tian, Wei and Tan (2015), Molzahn and Josz (2018), Cicconet and Almeida (2019) and Haussmann, Liers, Stingl and Vera (2018).
- Computer science, e.g. for coding and packing problems: Bachoc and Vallentin (2008), Bachoc, Passuello and Vallentin (2013), Dostert and Vallentin (2020) and de Laat *et al.* (2022).

- Computer vision, geometric perception and pattern recognition: Yang and Carlone (2020), Yang, Shi and Carlone (2020), Yang and Pavone (2023) and Probst, Paudel, Chhatkuli and Van Gool (2019).
- Mathematical finance, for portfolio optimization and option pricing: Lasserre, Priéto-Rumeau and Zervos (2006) and Gepp, Harris and Vanstone (2020). When the evolution in time is modelled by Itô's stochastic differential equations, a weak formulation of the problem via occupation measures is almost identical to that of OCPs, the only difference being that a second-order differential operator appears in the infinitesimal generator. It is also the case for computing exit-time distribution (of a given set) in stochastic models, as described in Lasserre and Priéto-Rumeau (2004), for example.
- The Internet of Things (IoT): Sedighi, Mishra, Shankar and Ottersten (2021).
- Computer graphics and geometry processing: Marschner, Palmer, Zhang and Solomon (2020, 2021).
- Signal processing: Marmin, Castella, Pesquet and Duval (2021) and de Castro, Gamboa, Henrion and Lasserre (2017).
- Optimal design in statistics: de Castro *et al.* (2019).
- Physics, for bounding ground-state energy of interacting particle systems: de Laat (2020).
- Chemistry, for deriving bounds on stochastic chemical kinetic systems: Dowdy and Barton (2018).
- Traffic networks, for bounding travel time: Xiangfeng Ji, Xuegang (Jeff) Ban, Jian Zhang and Bin Ran (2019).
- Engineering: Courtier *et al.* (2022).
- Machine learning, for certification of robustness for neural networks: Latorre, Rolland and Cevher (2020) and Chen Tong, Lasserre, Magron and Pauwels (2021).
- Quantum information (e.g. for several problems in entanglement theory): Eisert, Hyllus, Gühne and Curty (2004), Bacari, Gogolin, Wittek and Acín (2020), Parekh and Thompson (2021), Doherty, Parrilo and Spedalieri (2005) and Selby *et al.* (2023).
- Data analysis of citation networks: e.g. Wittek, Darányi and Nelhans (2017).
- Radar and wireless communications: Jie Pan and Fu Jiang (2020).
- Medical applications of cancer treatment: Moussa, Fiacchini and Alamir (2020).
- Truss topology design: Tyburec, Zeman, Kruzik and Henrion (2021).

6.4. *Notes and sources*

Section 6.1 is based on Lasserre (2002*a*) and Henrion *et al.* (2009), while Section 6.2 is based on Lasserre *et al.* (2008). Infinite-dimensional LP formulations of optimal control problems can be traced back to works of L. C. Young, A. F. Filippov, R. V. Gamkrelidze and J. Warga; see Fattorini (1999) for a historical survey. The novelty is the observation that such problems (with algebraic data) can be approximated numerically by semidefinite relaxations. In particular, it should be noted that while state constraints '$\mathbf{x}(t) \in X$ for all t' are usually considered a source of additional difficulties for classical numerical methods, they pose no problem for the Moment-SOS hierarchy as they simply appear in the support of the occupation measure.

The Moment-SOS hierarchy approach to analysis and control of some nonlinear PDEs in Korda *et al.* (2022) follows the same principles as for solving OCPs. Namely, we consider a measure-valued weak formulation of the problem, similar to that in (6.18) for OCPs, using test functions and occupation measures. Appropriate conditions are required for the weak formulation to be equivalent to the initial (strong) formulation. For instance, for the Burgers equation, additional entropy constraints (due to Kruzkhov) on the occupation measures are needed; see Marx *et al.* (2020) and Henrion *et al.* (2021, Chapter 11).

The reconstruction technique of state and control trajectories based on the Christoffel function is detailed in Marx *et al.* (2021). In particular, this technique has been used with success in Marx *et al.* (2020) to recover solutions to Burgers' PDE from moments of the measure supported on their graph. A (remarkably accurate) approximation of such moments has been obtained by solving semidefinite relaxations of the Moment-SOS hierarchy applied to the weak measure-valued formulation of the Burgers equation (and in a spirit similar to (6.20) for optimal control problems). The role of the Christoffel function is treated in more detail in Section 7, including its remarkable ability to recover a function solely from knowledge of the moments of the (degenerate) measure supported on its graph.

7. Positive polynomials and the Christoffel function

In this section we introduce the Christoffel–Darboux (CD) kernel and the Christoffel function (CF), which are classical tools from the fields of orthogonal polynomials and approximation theory. In addition to being interesting in their own right, they have proved to be useful in understanding and interpreting the Moment-SOS hierarchy of lower bounds. Further, the CF also appears in a certain distinguished representation of polynomials that are positive on a semi-algebraic set $S \subset \mathbb{R}^d$, as in (2.7), extensively used in the Moment-SOS hierarchy. In particular, every SOS polynomial p in the interior of the convex cone $\Sigma[\mathbf{x}]_n$ of degree-$2n$ SOS polynomials is the reciprocal of the CF of some linear functional ϕ in $\mathbb{R}[\mathbf{x}]_{2n}^*$. If $n = 2$, then ϕ has a clear interpretation in terms of a Gaussian measure, but in the general case, the link between p and ϕ is only partially understood and remains to be interpreted.

7.1. *Christoffel–Darboux kernel and Christoffel function*

The CF is usually defined for a measure μ with moments $\boldsymbol{\mu} = (\mu_\alpha)_{\alpha\in\mathbb{N}^d}$, whose support $S \subset \mathbb{R}^d$ is compact and such that its moment matrix $\mathbf{M}_n(\boldsymbol{\mu})$ (or equivalently, $\mathbf{M}_n(\mu)$) is positive definite for every degree $n \in \mathbb{N}$. However, it can also be defined for a Riesz linear functional $\phi \in \mathbb{R}[\mathbf{x}]^*$ (with $\boldsymbol{\phi} = (\phi_\alpha)_{\alpha\in\mathbb{N}^d}$) such that $\mathbf{M}_n(\boldsymbol{\phi}) \succ 0$ for every $n \in \mathbb{N}$, not necessarily coming from a measure μ.

So after fixing some ordering of monomials in $\mathbb{N}^d$, and since $\mathbf{M}_n(\boldsymbol{\phi}) \succ 0$ for every n, let $(P_\alpha)_{\alpha\in\mathbb{N}^d}$ be a family of polynomials that are *orthonormal* with respect to ϕ, that is, such that

$$\phi(P_\alpha\, P_\beta) = \delta_{\alpha=\beta} \quad \text{for all } \boldsymbol{\alpha}, \boldsymbol{\beta} \in \mathbb{N}^d, \tag{7.1}$$

where $\delta_{\cdot=\cdot}$ is the usual Kronecker symbol (with value 1 if $\boldsymbol{\alpha} = \boldsymbol{\beta}$ and 0 otherwise).

For every $n \in \mathbb{N}$, the Christoffel–Darboux (CD) kernel $K_n^\phi : \mathbb{R}^d \times \mathbb{R}^d \to \mathbb{R}$ is then defined by

$$(\mathbf{x}, \mathbf{y}) \mapsto K_n^\phi(\mathbf{x}, \mathbf{y}) := \sum_{\alpha\in\mathbb{N}_n^d} P_\alpha(\mathbf{x})\, P_\alpha(\mathbf{y}) \quad \text{for all } \mathbf{x}, \mathbf{y} \in \mathbb{R}^d,\ n \in \mathbb{N}, \tag{7.2}$$

and the Christoffel function (CF) $\Lambda_n^\phi : \mathbb{R}^d \to \mathbb{R}_+$ is defined by

$$\mathbf{x} \mapsto \Lambda_n^\phi(\mathbf{x}) := K_n^\phi(\mathbf{x}, \mathbf{x})^{-1} \quad \text{for all } \mathbf{x} \in \mathbb{R}^d,\ n \in \mathbb{N}, \tag{7.3}$$

that is, the CF is the reciprocal of the 'diagonal' of the CD kernel. Hence, by construction, $1/\Lambda_n^\phi$ is an SOS polynomial of degree $2n$.

A reproducing property. Let $p \in \mathbb{R}[\mathbf{x}]_n$, and since $(P_\alpha)_{\alpha\in\mathbb{N}_n^d}$ form a basis of $\mathbb{R}[\mathbf{x}]_n$, write

$$\mathbf{x} \mapsto p(\mathbf{x}) = \sum_{\alpha\in\mathbb{N}_n^d} p_\alpha\, P_\alpha(\mathbf{x}) \quad \text{for all } \mathbf{x} \in \mathbb{R}^d,$$

for some vector of coefficients $\mathbf{p} = (p_\alpha)_{\alpha\in\mathbb{N}_n^d}$ in $\mathbb{R}^{s(n)}$ (with $s(n) = \binom{n+d}{d}$).

With $\mathbf{x} \in \mathbb{R}^d$ fixed, $\mathbf{y} \mapsto K_n^\phi(\mathbf{x}, \mathbf{y}) \in \mathbb{R}[\mathbf{y}]_n$, and we have

$$\begin{aligned} \phi(K_n^\phi(\mathbf{x}, \cdot)\, p) &= \phi\Bigg(\Bigg(\sum_{\alpha\in\mathbb{N}_n^d} P_\alpha(\mathbf{x})\, P_\alpha(\mathbf{y})\Bigg) \cdot \Bigg(\sum_{\beta\in\mathbb{N}_n^d} p_\beta\, P_\beta(\mathbf{y})\Bigg)\Bigg) \\ &= \sum_{\alpha\in\mathbb{N}_n^d} p_\alpha\, P_\alpha(\mathbf{x}) = p(\mathbf{x}) \quad \text{for all } p \in \mathbb{R}[\mathbf{x}]_n, \end{aligned} \tag{7.4}$$

where we have used that

$$\phi(p_\beta P_\beta(\mathbf{y})\, P_\alpha(\mathbf{x}) P_\alpha(\mathbf{y})) = p_\beta\, P_\alpha(\mathbf{x})\, \phi(P_\beta\, P_\alpha) = p_\beta\, P_\alpha(\mathbf{x})\, \delta_{\beta=\alpha}.$$

For this reason, if μ is a measure on $S \subset \mathbb{R}^d$, and $L^2(\mu)$ is the Hilbert space of square-integrable functions with respect to μ, with scalar product

$$\langle f, g \rangle = \int_S f\, g\, \mathrm{d}\mu \quad \text{for all } f, g \in L^2(\mu),$$

then $(\mathbb{R}[\mathbf{x}]_n, \langle \cdot, \cdot \rangle) \subset L^2(\mu)$ is called a reproducing kernel Hilbert space (RKHS) with kernel K_n^μ, because

$$\int_S K_n^\mu(\mathbf{x}, \mathbf{y})\, p(\mathbf{y})\, \mathrm{d}\mu(\mathbf{y}) = p(\mathbf{x}) \quad \text{for all } p \in \mathbb{R}[\mathbf{x}]_n.$$

Alternative formulations of the CF. The CF can be defined by

$$\Lambda_n^\phi(\boldsymbol{\xi})^{-1} = \mathbf{v}_n(\boldsymbol{\xi})^\top \mathbf{M}_n(\boldsymbol{\phi})^{-1}\, \mathbf{v}_n(\boldsymbol{\xi}) \quad \text{for all } \boldsymbol{\xi} \in \mathbb{R}^d \tag{7.5}$$

(the ABC theorem in Simon (2008)), and it also has the variational formulation

$$\Lambda_n^\phi(\boldsymbol{\xi}) = \min_{p \in \mathbb{R}[\mathbf{x}]_n} \{\phi(p^2) \colon p(\boldsymbol{\xi}) = 1\} \quad \text{for all } \boldsymbol{\xi} \in \mathbb{R}^d. \tag{7.6}$$

In particular, observe that (7.6) can be rewritten as

$$\Lambda_n^\phi(\boldsymbol{\xi}) = \min_{\mathbf{p} \in \mathbb{R}^{s(n)}} \{\mathbf{p}^\top \mathbf{M}_n(\boldsymbol{\phi})\mathbf{p} \colon \langle \mathbf{p}, \mathbf{v}_n(\boldsymbol{\xi}) \rangle = 1\} \quad \text{for all } \boldsymbol{\xi} \in \mathbb{R}^d,$$

which is a convex quadratic optimization problem that can be solved efficiently even for large dimension d. After some algebra, the unique optimal solution $p^* \in \mathbb{R}[\mathbf{x}]_n$ of (7.6) reads

$$\mathbf{x} \mapsto p^*(\mathbf{x}) = \frac{K_n^\phi(\boldsymbol{\xi}, \mathbf{x})}{K_n^\phi(\boldsymbol{\xi}, \boldsymbol{\xi})} \quad \text{for all } \mathbf{x} \in \mathbb{R}^d.$$

7.2. *Some useful properties of the CF*

A crucial property of the CFs $(\Lambda_n^\mu)_{n \in \mathbb{N}}$ associated with a measure μ on a compact set $S \subset \mathbb{R}^d$ is to identify the support of μ. Indeed, its decay with the degree n exhibits the following interesting dichotomy:

- for all $\boldsymbol{\xi} \in \operatorname{supp}(\mu)$, $\Lambda_n^\mu(\boldsymbol{\xi})^{-1}$ grows at most as a polynomial in n,
- for all $\boldsymbol{\xi} \notin \operatorname{supp}(\mu)$, $\Lambda_n^\mu(\boldsymbol{\xi})^{-1}$ grows at least as an exponential in n.

This property has been exploited in data analysis to provide a simple and easy-to-use tool (with no tuning of parameters), e.g. to detect outliers, with performance similar to (and sometimes better than) state-of-the-art techniques; see Lasserre, Pauwels and Putinar (2022) and Lasserre and Pauwels (2016).

Next, let μ have a density f with respect to Lebesgue measure on S. Under some additional regularity properties of μ and its support S,

$$\lim_{n \to \infty} s(n)\, \Lambda_n^\mu(\boldsymbol{\xi}) = f(\boldsymbol{\xi}) / \mu_E(\boldsymbol{\xi}), \tag{7.7}$$

uniformly on compact subsets of $\operatorname{int}(S)$, where μ_E is the density of a so-called *equilibrium measure* of S.

Equilibrium measure. A Borel measure μ supported on a compact set $S \subset \mathbb{R}^d$ satisfies the Bernstein–Markov property if there exists a sequence of positive numbers $(M_n)_{n\in\mathbb{N}}$ such that for all n and $p \in \mathbb{R}[\mathbf{x}]_n$,

$$\sup_{\mathbf{x}\in S} |p(\mathbf{x})| \leq M_n \cdot \left(\int_S p^2 \, \mathrm{d}\mu \right)^{1/2} \quad \text{and} \quad \lim_{n\to\infty} \log(M_n)/n = 0 \tag{7.8}$$

(see e.g. Lasserre *et al.* (2022, Section 4.3.3)). The Bernstein–Markov property allows a qualitative description of asymptotics of the Christoffel function as n grows.

The notion of *equilibrium measure*, associated with a given set, originates from logarithmic potential theory (working in $\mathbb{C}$ in the univariate case) to minimize some energy functional. For instance, the (Chebyshev) measure $\mathrm{d}\mu := \mathrm{d}x/\pi\sqrt{1-x^2}$ is the equilibrium measure of the interval $[-1, 1]$. Some generalizations have been obtained in the multivariate case via pluripotential theory in $\mathbb{C}^d$. In particular, if $S \subset \mathbb{R}^d \subset \mathbb{C}^d$ is compact then its equilibrium measure (let us denote it by λ_S) is equivalent to Lebesgue measure on compact subsets of $\mathrm{int}(S)$. It has an even explicit expression if S is convex and symmetric about the origin; see e.g. Baran (1995) and Bedford and Taylor (1986, Theorems 1.1, 1.2). Moreover, if μ is a Borel measure on S and (S, μ) has the Bernstein–Markov property (7.8), then the sequence of measures

$$\nu_n = \frac{\mu}{s(n)\Lambda_n^{\mu}(\mathbf{x})},$$

$n \in \mathbb{N}$, converges to λ_S for the weak-$\star$ topology, and therefore in particular

$$\lim_{n\to\infty} \int_S \mathbf{x}^{\alpha} \, \mathrm{d}\nu_n = \lim_{n\to\infty} \int_S \frac{\mathbf{x}^{\alpha} \, \mathrm{d}\mu(\mathbf{x})}{s(n)\Lambda_n^{\mu}(\mathbf{x})} = \int_S \mathbf{x}^{\alpha} \, \mathrm{d}\lambda_S \quad \text{for all } \boldsymbol{\alpha} \in \mathbb{N}^d. \tag{7.9}$$

(See e.g. Lasserre *et al.* (2022, Theorem 4.4.4).) In addition, if the compact $S \subset \mathbb{R}^d$ is regular, then (S, λ_S) has the Bernstein–Markov property. For a brief account of equilibrium measures, see Baran (1995), Bedford and Taylor (1986) and the discussion in Lasserre *et al.* (2022, Sections 4–5, pp. 56–60), while for more detailed expositions see some of the references therein.

7.3. The CF for interpolation and approximation

In this section we briefly address the following issue, which is interesting in its own right and also central to the recovery of an optimal (or ε-optimal) trajectory $\{\mathbf{x}(t) \colon t \in [0, 1]\}$ in optimal control problems, solely from knowledge of the moments of the occupation measure supported on the graph $\{(t, \mathbf{x}(t)) \colon t \in [0, 1]\}$; see Section 6.2.

So with $X \subset \mathbb{R}$, let μ be a measure on $[0, 1] \times X$, defined by

$$\mathrm{d}\mu(t, x) = \delta_{\{f(t)\}}(\mathrm{d}x) \, 1_{[0,1]}(t) \, \mathrm{d}t,$$

for some unknown measurable function $f \colon [0, 1] \to X$, that is, μ is supported on the graph $\{(t, f(t)) : t \in [0,1]\}$ of f. The goal is to recover f solely from knowledge of the moments $\boldsymbol{\mu} = (\mu_{ij})_{(i,j)\in\mathbb{N}^2}$, where

$$\mu_{ij} = \int t^i x^j \, \mathrm{d}\mu(t,x) = \int_0^1 t^i f(t)^j \, \mathrm{d}t, \quad (i,j) \in \mathbb{N}^2.$$

We propose using the Christoffel function Λ_n^μ to recover f from $\boldsymbol{\mu}$ because, as seen earlier, Λ_n^μ is a good tool for identifying the support of μ, and in this case the support is precisely the graph of the unknown function f to recover. Here, observe that μ is a degenerate measure on $[0,1] \times X$, that is, its support has Lebesgue measure zero on $[0,1] \times X$. Therefore its moment matrix $\mathbf{M}_n(\mu)$ can be ill-conditioned and even singular if f is a polynomial (because then the vector of coefficients of $f \in \mathbb{R}[t]$ is in the kernel of $\mathbf{M}_n(\mu)$ when n is sufficiently large). So we first 'perturbate' (or regularize) $\mathbf{M}_n(\mu)$ to $\mathbf{M}_n(\mu) + \varepsilon\mathbf{I}$ with $\mathbf{I}$ the identity matrix and some small regularization parameter $\varepsilon > 0$, and we define a new perturbated Christoffel function $\hat{\Lambda}_n^\mu$ by

$$(t,x) \mapsto \hat{\Lambda}_n^\mu(t,x)^{-1} := \mathbf{v}_n(t,x)^\top (\mathbf{M}_n(\mu) + \varepsilon\mathbf{I})^{-1} \mathbf{v}_n(t,x) \quad \text{for all } (t,x) \in \mathbb{R}^2. \tag{7.10}$$

We then define the following n-approximant $f_n : [0,1] \to X$ of f by

$$t \mapsto f_n(t) := \arg\min_{x \subset X} \hat{\Lambda}_n^\mu(t,x)^{-1}, \quad t \in [0,1]. \tag{7.11}$$

(If there are several minimizers in (7.11) then just take the smallest one as a tie-breaker rule.) For every fixed $t \in [0,1]$, $f_n(t)$ can be computed efficiently as $x \mapsto \hat{\Lambda}_n^\mu(t,x)^{-1}$ is a *univariate* SOS polynomial in x.

Next, as n increases, pointwise convergence (except at points of discontinuity) and L^1-norm convergence to f are proved in Marx *et al.* (2021). Observe that the f_n approximant (7.11) is *not* a polynomial, and since it is semi-algebraic, it is able to approximate some discontinuous functions quite well with no Gibbs phenomenon.

For instance, in Figure 7.1(a) we may observe a typical Gibbs phenomenon (oscillations) when approximating the (discontinuous) step function $t \mapsto f(t) = 0$ if $t \in [0, 1/2]$ and $f(t) = 1$ if $t \in (1/2, 1]$ (in red) by a polynomial $p^* \in \mathbb{R}[t]_n$ (in black) that minimizes the integral of the mean squared error, that is,

$$p^* = \arg\min_{p\in\mathbb{R}[t]_n} \int_0^1 (p-f)^2 \, \mathrm{d}t$$

(even with degree $n = 12$). This L^2-norm approximation of f is a standard application of the CD kernel K_n^ν associated with the univariate measure $\nu = \mathrm{d}t$ on $[0,1]$. On the other hand, with $\varepsilon > 0$ very small and f_n as in (7.11), the step function is recovered almost exactly (in black) with no Gibbs phenomenon and with small degree $n = 4$. This is what we may call a non-standard application of the CD kernel, as we consider the degenerate bivariate measure μ on $[0,1] \times X$ instead of the univariate measure $f(t)\,\mathrm{d}t$ on $[0,1]$.

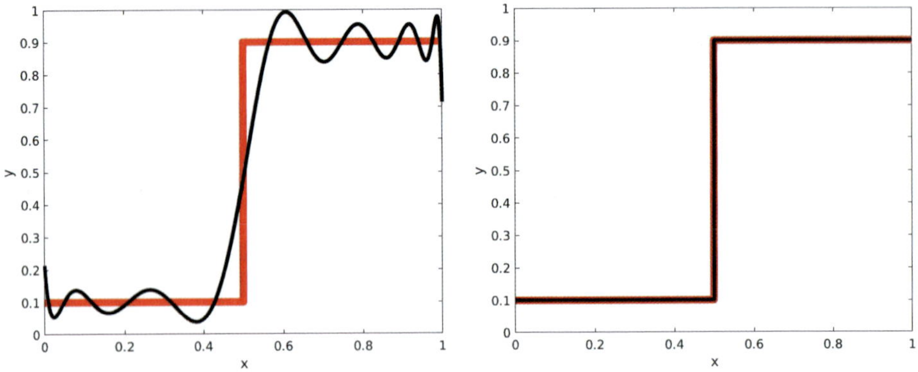

(a) (b)

Figure 7.1. (a) Degree-12 polynomial approximation of step function with Gibbs phenomenon. (b) Step function approximated by f_4 in (7.11). © Springer Nature, reproduced with permission from Marx *et al.* (2021).

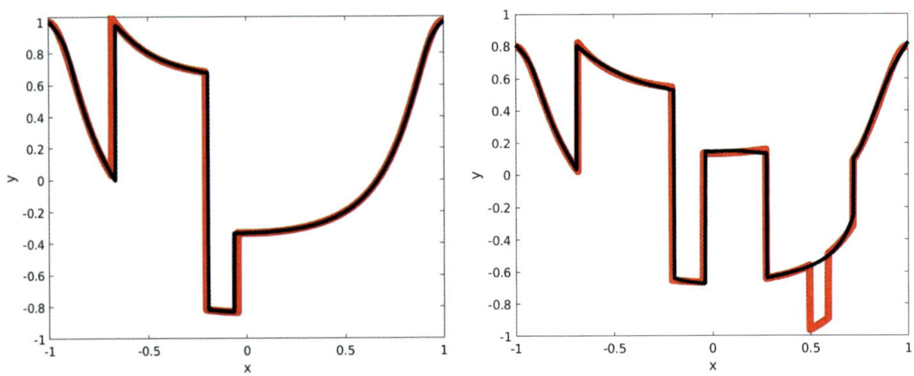

(a) (b)

Figure 7.2. Two Eckhoff functions (Eckhoff 1993) approximated by f_{10} in (7.11). © Springer Nature, reproduced with permission from Marx *et al.* (2021).

Similarly, in Figure 7.2, two discontinuous Eckhoff functions from Eckhoff (1993) (in red) are also recovered (in black) with very good precision via f_n in (7.11) with $n = 10$, and again with no Gibbs phenomenon; for more details the reader is referred to Marx *et al.* (2021).

Application to optimal control. As already mentioned in Section 6.2, such an approximation technique can be used to recover the graph of functions supported on trajectories $\{(\mathbf{x}^*(t), \mathbf{u}^*(t)) \colon t \in [0, 1]\}$, i.e. the optimal solutions of optimal control problems (6.10) described in Section 6.2. Indeed, when applying the Moment-SOS hierarchy to solve (6.10), at an optimal solution of (6.20), we obtain approximate moments up to degree $2n$, of the measure

$$d\mu^*(\mathbf{x}, \mathbf{u}, t) = \delta_{\{(\mathbf{x}^*(t), \mathbf{u}^*(t))\}}\, 1_{[0,1]}(t)\, dt,$$

supported on the graph of the map $(\mathbf{x}^*, \mathbf{u}^*)\colon [0, 1] \to \mathbb{R}^d \times \mathbb{R}^m$.

For instance, to recover the particular trajectory $\{x_i^*(t)\colon t \in [0, 1]\}$ for some coordinate $i \in \{1, \ldots, d\}$, we proceed as follows.

- We extract the sub-matrix $\mathbf{M}_n^{(x_i,t)}$ of $\mathbf{M}_n(\mu)$ obtained by restricting to rows and columns indexed by monomials $(x_i^k\, t^j)$, $(k, j) \in \mathbb{N}_n^2$ (i.e. $\mathbf{M}_n^{(x_i,t)}$ is the degree-n moment matrix of the marginal μ_i of μ on (x_i, t)).
- We compute the perturbed Christoffel function in (7.10) associated with μ_i, that is,

$$\hat{\Lambda}_n^{\mu_i}(x_i, t)^{-1} = \mathbf{v}_n(x_i, t)^\top (\mathbf{M}_n^{(x_i,t)} + \varepsilon\, \mathbf{I})^{-1} \mathbf{v}_n(x_i, t),$$

  and then the f_n approximant of the function $x_i(t)$ is obtained via (7.11).

The same procedure is repeated for all coordinates $x_i^*(t)$, $i \in \{1, \ldots, d\}$, of $\mathbf{x}^*(t)$, and all coordinates $u_j^*(t)$, $j \in \{1, \ldots, m\}$, of $\mathbf{u}^*(t)$, independently.

7.4. *Christoffel function and positive polynomials*

First notice that by construction, the reciprocal of a Christoffel function is an SOS polynomial. Next, with $S \subset \mathbb{R}^d$ as in (2.7), recall the convex cone

$$Q_n(g) := \left\{ \sum_{j=0}^m \sigma_j\, g_j \colon \sigma_j \in \Sigma[\mathbf{x}]_{n-d_j},\ j = 0, \ldots, m \right\}, \tag{7.12}$$

which is a degree-$2n$ *truncated* version of the quadratic module $Q(g)$ (with $d_j = \lceil \deg(g_j)/2 \rceil$ and $g_0 = \mathbf{1}$). For every polynomial $p = \sum_j \sigma_j g_j \in Q_n(g)$, the SOS weights σ_j provide p with an algebraic certificate of its positivity on S.

Recall that the dual of $Q_n(g)$ is the convex cone

$$Q_n^*(g) = \{\boldsymbol{\phi} \in \mathbb{R}^{s(2n)}\colon \mathbf{M}_{n-d_j}(g_j \cdot \boldsymbol{\phi}) \succeq 0,\ j = 0, \ldots, m\}, \tag{7.13}$$

where $\mathbf{M}_n(g_j \cdot \boldsymbol{\phi})$ is the localizing matrix associated with the polynomial g_j and the sequence $\boldsymbol{\phi}$ (or equivalently the moment matrix associated with the sequence $g_j \cdot \boldsymbol{\phi}$), defined in Section 2.1. We saw in Section 3 that $Q_n(g)$ and its dual $Q_n^*(g)$ are crucial to the construction of the Moment-SOS hierarchy of lower bounds. It turns out that there is a nice one-to-one correspondence between the respective interiors of $Q_n(g)$ and $Q_n^*(g)$, stated in terms of Christoffel functions.

Theorem 7.1. If $p \in \operatorname{int}(Q_n(g))$, then there exists $\boldsymbol{\phi} \in \operatorname{int}(Q_n^*(g))$ such that

$$p(\mathbf{x}) = \sum_{j=0}^m \Lambda_{n-d_j}^{g_j \cdot \boldsymbol{\phi}}(\mathbf{x})^{-1}\, g_j(\mathbf{x}) \quad \text{for all } \mathbf{x} \in \mathbb{R}^d, \tag{7.14}$$

or, equivalently,

$$\mathrm{int}(Q_n(g)) = \left\{ \sum_{j=0}^{m} \left(\Lambda_{n-d_j}^{g_j \cdot \boldsymbol{\phi}}\right)^{-1} g_j : \boldsymbol{\phi} \in \mathrm{int}(Q_n^*(g)) \right\}. \tag{7.15}$$

Theorem 7.1 is an interpretation from Lasserre (2022) of a duality result of Nesterov (2000). Remarkably, it states that every p in the interior of $Q_n(g)$ has a distinguished certificate of its positivity on S, with very specific SOS weights $\sigma_j = (\Lambda_{n-d_j}^{g_j \cdot \boldsymbol{\phi}})^{-1}$ in its Putinar's representation (2.9). Indeed, those weights all come from a *unique* element $\boldsymbol{\phi} \in \mathrm{int}(Q_n^*(g))$ and its Christoffel functions associated with the Riesz linear functionals $g_j \cdot \boldsymbol{\phi}$, $j = 0, \ldots, m$. It also turns out that those weights have an extremal property: consider the optimization problem

$$\begin{aligned} \rho_n = \inf_{\boldsymbol{\phi} \in \mathbb{R}^{s(2n)}} \Bigg\{ -\sum_{j=0}^{m} \log\det(\mathbf{M}_{n-d_j}(g_j \cdot \boldsymbol{\phi})) &: \\ \phi(p) = 1,\ \mathbf{M}_{n-d_j}(g_j \cdot \boldsymbol{\phi}) \succeq 0,\ \forall\, j = 0, \ldots, m & \Bigg\}. \end{aligned} \tag{7.16}$$

It is a convex optimization problem which has an explicit dual, namely

$$\begin{aligned} \rho_n^* = \sup_{\mathbf{Q}_j} \Bigg\{ \sum_{j=0}^{m} \log\det(\mathbf{Q}_j) &: \mathbf{Q}_j \succeq 0,\ \forall\, j = 0, \ldots, m \\ p(\mathbf{x}) \sum_{j=0}^{m} s(n-d_j) &= \sum_{j=0}^{m} g_j(\mathbf{x}) \cdot \mathbf{v}_{n-d_j}(\mathbf{x})^{\top} \mathbf{Q}_j \mathbf{v}_{n-d_j}(\mathbf{x}),\ \forall\, \mathbf{x} \in \mathbb{R}^d \Bigg\}, \end{aligned} \tag{7.17}$$

where the supremum is taken over real symmetric matrices $\mathbf{Q}_j$ of respective size $s(n-d_j)$, $j = 0, \ldots, m$. The criterion to maximize in (7.17) is minus the log-barrier of the convex cone $Q_n(g)$.

Theorem 7.2. With $n \in \mathbb{N}$ fixed, problems (7.16) and (7.17) have the same finite optimal value $\rho_n = \rho_n^*$ if and only if $p \in \mathrm{int}(Q_n(g))$. Moreover, both have a unique optimal solution $\boldsymbol{\phi}_{2n}^* \in \mathbb{R}^{s(2n)}$ and $(\mathbf{Q}_j^*)_{j=0,\ldots,m}$ respectively, which satisfy $\mathbf{Q}_j^* = \mathbf{M}_{n-d_j}(g_j \cdot \boldsymbol{\phi}_{2n}^*)^{-1}$ for all $j = 0, \ldots, m$. As a consequence,

$$\begin{aligned} p(\mathbf{x}) &= \frac{1}{\sum_{j=0}^{m} s(n-d_j)} \sum_{j=0}^{m} g_j(\mathbf{x})\, \mathbf{v}_{n-d_j}(\mathbf{x})^{\top} \mathbf{M}_{n-d_j}(g_j \cdot \boldsymbol{\phi}_{2n}^*)^{-1} \mathbf{v}_{n-d_j}(\mathbf{x}) \\ &= \frac{1}{\sum_{j=0}^{m} s(n-d_j)} \sum_{j=0}^{m} g_j(\mathbf{x})\, \Lambda_{n-d_j}^{g_j \cdot \boldsymbol{\phi}_{2n}^*}(\mathbf{x})^{-1} \quad \text{for all } \mathbf{x} \in \mathbb{R}^d. \end{aligned} \tag{7.18}$$

Notice that $\boldsymbol{\phi}$ in (7.14) is just $(\sum_{j=0}^{m} s(n-d_j))\, \boldsymbol{\phi}_{2n}^*$ with $\boldsymbol{\phi}_{2n}^*$ as in (7.18).

Of course, Theorem 7.1 immediately raises the following question: Given $p \in \operatorname{int}(Q_n(g))$, what is this linear functional $\phi \in \mathbb{R}[\mathbf{x}]_{2n}^*$ with associated moment sequence $\boldsymbol{\phi} \in \mathbb{R}^{s(2n)}$ in Theorem 7.1? It turns out that there is a simple and remarkable answer for special sets S and the constant polynomial $p = \mathbf{1}$.

Relating the constant polynomial and the equilibrium measure. Let $S \subset \mathbb{R}^d$ in (2.7) be a compact set with non-empty interior, generated by a finite set $\tilde{G} = \{g_1, \ldots, g_m\} \subset \mathbb{R}[\mathbf{x}]$ of polynomials. Let $G \subset \mathbb{R}[\mathbf{x}]$ be a certain finite set of polynomials formed with some products of polynomials in $\tilde{G}$. We provide three examples.

- If $S \subset \mathbb{R}^d$ is the Euclidean unit ball, then $\tilde{G} = \{g\}$, $G = \{\mathbf{1}, g\}$, with $\mathbf{x} \mapsto g(\mathbf{x}) = 1 - \|\mathbf{x}\|^2$. Then the equilibrium measure μ is proportional to
$$\mathrm{d}\mathbf{x}\left(1 - \|\mathbf{x}\|^2\right)^{-1/2}.$$
- If S is the unit box $[-1, 1]^d$, then $\tilde{G} = \{g_1, \ldots, g_d\}$ with $g_j(\mathbf{x}) = 1 - x_j^2$, $j = 1, \ldots, d$, and $G = \{g_{\boldsymbol{\varepsilon}} : \boldsymbol{\varepsilon} \in \{0, 1\}^d\}$, where
$$\mathbf{x} \mapsto g_{\boldsymbol{\varepsilon}}(\mathbf{x}) := \prod_{j=1}^{d} g_j(\mathbf{x})^{\varepsilon_j} \quad \text{for all } \mathbf{x} \in \mathbb{R}^d.$$
The equilibrium measure μ of S is proportional to
$$\mathrm{d}\mathbf{x} \prod_{j=1}^{d} \left(1 - x_j^2\right)^{-1/2}.$$
- If $S \subset \mathbb{R}^d$ is the canonical simplex, then $\tilde{G} = \{g_1, \ldots, g_{d+1}\}$ with $g_j(\mathbf{x}) = x_j$, $j = 1, \ldots, d$, $g_{d+1}(\mathbf{x}) = 1 - \sum_j x_j$ and $G = \{g_{\boldsymbol{\varepsilon}} : \boldsymbol{\varepsilon} \in \{0, 1\}^{d+1};\ |\boldsymbol{\varepsilon}| \in 2\mathbb{N}\}$, where
$$\mathbf{x} \mapsto g_{\boldsymbol{\varepsilon}}(\mathbf{x}) := \prod_{j=1}^{d+1} g_j(\mathbf{x})^{\varepsilon_j} \quad \text{for all } \mathbf{x} \in \mathbb{R}^d.$$
The equilibrium measure μ of S is proportional to
$$\mathrm{d}\mathbf{x}\left(\left(1 - \sum_j x_j\right) \prod_{j=1}^{d} x_j\right)^{-1/2}.$$

For every $g \in G$ let $t_g := \lceil \deg(g)/2 \rceil$. In addition, given $n \in \mathbb{N}$, let
$$G_n := \{g \in G : \deg(g) \leq 2n\}$$
so that $G_n = G$ if $n \geq \lceil d/2 \rceil$.

Theorem 7.3 (Lasserre 2023, Lasserre and Xu 2023). Let $S \subset \mathbb{R}^d$ be the Euclidean unit ball, the unit box or the simplex, and let μ be its equilibrium measure.

Then, for all integer n,

$$1 = \frac{1}{\sum_{g \in G_n} s(t - t_g)} \sum_{g \in G_n} g(\mathbf{x}) \, \Lambda_{n-t_g}^{g \cdot \mu}(\mathbf{x})^{-1} \quad \text{for all } \mathbf{x} \in \mathbb{R}^d. \tag{7.19}$$

So, remarkably, the constant polynomial $p = \mathbf{1} \in \text{int}(Q_n(g))$, for all n, is strongly related to the equilibrium measure μ of S. Its corresponding element $\boldsymbol{\phi} \in \text{int}(Q_n^*(g))$, in Theorem 7.1, is the moment vector $\boldsymbol{\mu} \in \mathbb{R}^{s(2n)}$ of μ.

In addition, for every n, the polynomials $(g/\Lambda_{n-t_g}^{g \cdot \mu})_{g \in G_n}$ (all non-negative on S) provide S with a polynomial *partition of unity*. We have called (7.19) a *generalized polynomial Pell's equation* solved by the Christoffel functions $(\Lambda_{n-t_g}^{g \cdot \mu})_{g \in G_n}$ (and the polynomials $g \in G_n$), because (7.19) is an exact multivariate generalization of the polynomial Pell's equation[6]

$$1 = T_n(x)^2 + (1 - x^2)\, U_{n-1}(x)^2 \quad \text{for all } x \in \mathbb{R}, \tag{7.20}$$

satisfied by the univariate Chebyshev polynomials of the first kind $(T_n)_{n \in \mathbb{N}}$ and Chebyshev polynomials of the second kind $(U_n)_{n \in \mathbb{N}}$, orthogonal with respect to the measures $\mathrm{d}x/\sqrt{1 - x^2}$ and $\sqrt{1 - x^2}\, \mathrm{d}x$, respectively. Indeed, after normalization to orthonormal polynomials, and summing up (7.20) over n, we obtain

$$1 = \frac{1}{s(n) + s(n-1)} \left(\Lambda_n^{\mu}(x)^{-1} + g(x)\, \Lambda_{n-1}^{g \cdot \mu}(x)^{-1} \right) \quad \text{for all } x \in \mathbb{R},\ n \in \mathbb{N}, \tag{7.21}$$

where $g(x) = 1 - x^2$, and $\mathrm{d}\mu(x) = \mathrm{d}x/\pi\sqrt{1 - x^2}$ is the equilibrium measure of the interval $S = [-1, 1]$. The term 'generalized' is justified because in (7.19) we have a sum of squares in $\mathbb{R}[\mathbf{x}]$ and several generators $g \in G_n$, instead of two single squares in $\mathbb{Z}[x]$ and a single generator g in (7.20). But formally, (7.19) is of exactly the same flavour as (7.21).

Remark 7.4. When $S = \mathbb{R}^d$, there is still a nice well-known and somewhat related fact. Let $p \in \mathbb{R}[\mathbf{x}]_2$ be a quadratic polynomial which is strictly positive on $\mathbb{R}^d$. With $\mathbf{v}_1(\mathbf{x}) = (1, x_1, \ldots, x_d)$, p is written as

$$\mathbf{x} \mapsto p(\mathbf{x}) := \mathbf{v}_1(\mathbf{x})^{\top} \mathbf{Q}\, \mathbf{v}_1(\mathbf{x}) \quad \text{for all } \mathbf{x} \in \mathbb{R}^d,$$

for some (unique) Gram matrix $\mathbf{Q} \succ 0$. It is well known that $\mathbf{Q}^{-1}$ is the moment matrix $\mathbf{M}_1(\mu)$ of a Gaussian measure μ on $\mathbb{R}^d$, and therefore

$$p(\mathbf{x}) = \Lambda_1^{\mu}(\mathbf{x})^{-1} \quad \text{for all } \mathbf{x} \in \mathbb{R}^d.$$

This is another particular case (but in a non-compact setting) where one can identify the linear functional $\boldsymbol{\phi}$ in Theorem 7.1 (now with $Q_1(g) = \Sigma[\mathbf{x}]_1$ and $\Sigma_1[\mathbf{x}]_1^* = \{\boldsymbol{\phi} \in \mathbb{R}^{2d+1} : \mathbf{M}_1(\boldsymbol{\phi}) \succeq 0\}$). For instance, the scaled Hermite polynomials of degree at most 1,

$$\widehat{H}_0(\mathbf{x}) = (2\pi)^{-d/4}, \quad \widehat{H}_j(\mathbf{x}) = (2\pi)^{-d/4} x_j, \quad j = 1, \ldots, d,$$

[6] A triple (F, g, H) of polynomials in $\mathbb{Z}[x]$ satisfies (polynomial) Pell's equation if $F^2 + g\, H^2 = 1$.

are orthonormal with respect to the Gaussian (product measure)

$$d\mu(\mathbf{x}) = \exp\Big(-\sum_j x_j^2/2\Big)\,d\mathbf{x},$$

and

$$\Lambda_1^\mu(\mathbf{x})^{-1} = \sum_{j=0}^{d} \widehat{H}_j(\mathbf{x})^2 = (2\pi)^{-d/2}(1+\|\mathbf{x}\|^2).$$

7.5. Comparing Moment-SOS hierarchies of upper and lower bounds

To compare the two Moment-SOS hierarchies of upper and lower bounds for solving the POP

$$\mathbf{P}\colon\ f^* = \min\{f(\mathbf{x})\colon \mathbf{x}\in S\},$$

we express them in the same language of *polynomial densities* with respect to a reference finite Borel probability measure μ whose support is exactly the set $S\subset\mathbb{R}^d$ (assumed to be compact with non-empty interior). Let $\mathscr{P}(S)$ be the space of probability measures on S, and let $(P_\alpha)_{\alpha\in\mathbb{N}^d}$ be a family of polynomials that are orthonormal with respect to μ.

Moment-SOS hierarchy of lower bounds. With $\phi\in\mathbb{R}[\mathbf{y}]_{2n}^*$ arbitrary, and from the reproducing property (7.4) of K_{2n}^μ, observe that

$$\begin{aligned}\phi(f) &= \phi\Big(\int_S \sum_{\alpha\in\mathbb{N}^d_{2n}} P_\alpha(\mathbf{y})P_\alpha(\mathbf{x})\,f(\mathbf{x})\,d\mu(\mathbf{x})\Big)\\ &= \sum_{\alpha\in\mathbb{N}^d_{2n}} \phi(P_\alpha)\int_S P_\alpha(\mathbf{x})\,f(\mathbf{x})\,d\mu(\mathbf{x})\\ &= \int_S f(\mathbf{x})\Big(\sum_{\alpha\in\mathbb{N}^d_{2n}} \phi(P_\alpha)\,P_\alpha(\mathbf{x})\Big)\,d\mu(\mathbf{x})\\ &= \int_S f(\mathbf{x})\,\sigma_\phi(\mathbf{x})\,d\mu(\mathbf{x}),\end{aligned}$$

where the degree-$2n$ polynomial

$$\mathbf{x}\mapsto\sigma_{\boldsymbol{\phi}}(\mathbf{x}) := \sum_{\alpha\in\mathbb{N}^d_{2n}} \phi(P_\alpha)\,P_\alpha(\mathbf{x}) \tag{7.22}$$

is a *signed density* with respect to μ.

Next, recall that in the semidefinite relaxation (3.5) of the Moment-SOS hierarchy of lower bounds on f^*, we search for a linear functional $\phi\in\mathbb{R}[\mathbf{x}]_{2n}^*$ that satisfies

$$\phi(\mathbf{1}) = 1,\quad \mathbf{M}_{n-d_j}(g_j\cdot\boldsymbol{\phi})\succeq 0\quad \text{for all } j=0,\dots,m,$$

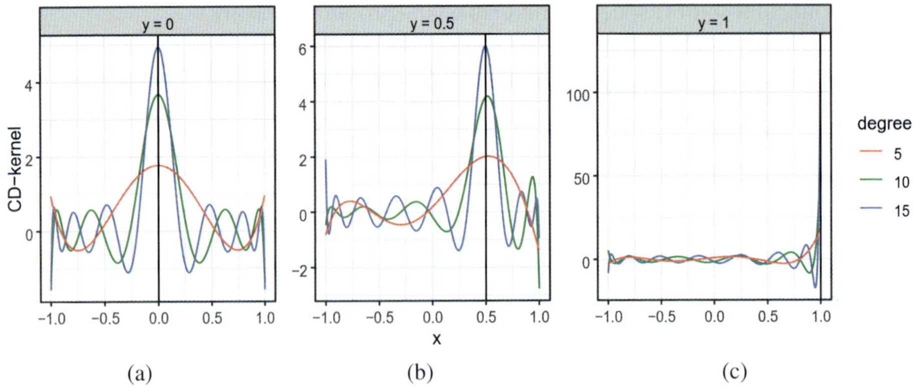

Figure 7.3. $S = [-1, 1]$, $\mu = \mathrm{d}x/2$; the signed density $x \mapsto \sigma_{\phi^*}(x) = K^{\mu}_{2n}(x, y)$ with (a) $y = 0$, (b) $y = 0.5$ and (c) $y = 1$. © Cambridge University Press, reproduced with permission from Lasserre *et al.* (2022).

and that minimizes $\langle \mathbf{f}, \boldsymbol{\phi} \rangle = \phi(f) = \int_S f\, \sigma_{\boldsymbol{\phi}}\, \mathrm{d}\mu$. Moreover, observe that

$$1 = \phi(\mathbf{1}) = \int_S \sigma_\phi\, \mathrm{d}\mu,$$

which means that σ_ϕ is a signed *probability* density. Therefore we have proved the following.

Theorem 7.5. Let μ be a finite Borel measure whose support is S in (2.7), and consider the Moment-SOS hierarchy of semidefinite relaxations (3.5) for solving $\mathbf{P}$. Then, with n fixed, (3.5) reads

$$\min_{\boldsymbol{\phi} \in \mathbb{R}^{s(2n)}} \left\{ \int_S f\, \sigma_\phi\, \mathrm{d}\mu \colon \phi(\mathbf{1}) = 1;\ \mathbf{M}_{n-d_j}(g_j \cdot \boldsymbol{\phi}) \succeq 0,\ j = 0, \dots, m \right\}, \tag{7.23}$$

where σ_ϕ is the signed probability density with respect to μ in (7.22).

So, again, solving the semidefinite relaxation (3.5) in the Moment-SOS hierarchy is searching for a polynomial signed probability density $\sigma_\phi \in \mathbb{R}[\mathbf{x}]_{2n}$ of the form (7.22), and as already mentioned, when the relaxation (3.5) is exact, $\phi^* = \delta_{\{\mathbf{y}\}}$, where $\mathbf{y} \in S$ is a global minimizer of f. Then the associated polynomial signed probability density $\sigma_{\phi^*} \in \mathbb{R}[\mathbf{x}]_{2n}$ reads

$$\mathbf{x} \mapsto \sigma_{\phi^*}(\mathbf{x}) = \sum_{\alpha \in \mathbb{N}^d_{2n}} P_\alpha(\mathbf{y})\, P_\alpha(\mathbf{x}) = K^{\mu}_{2n}(\mathbf{y}, \mathbf{x}).$$

It is interesting to see σ_{ϕ^*} in Figure 7.3 for the toy example where $S = [-1, 1]$ and $\mu = \mathrm{d}x/2$. Indeed, σ_{ϕ^*} has a peak at $x = y$, and thus mimics the Dirac measure at y (as far as moments up to degree $2n$ are concerned).

Table 7.1. Hierarchies of upper and lower bounds interpreted as searching for respective positive probability density $\sum_\alpha \sigma_\alpha P_\alpha$ and signed density $\sum_\alpha \phi(P_\alpha)P_\alpha$, with respect to μ.

Lower bounds	Upper bounds
Primal	Primal
$\tau_n = \inf_{\boldsymbol{\phi}} \int_S \big(\sum_{\alpha\in\mathbb{N}^d_{2n}} \phi(P_\alpha)\,P_\alpha\big)\, f\, d\mu$	$\hat{\kappa}_n = \inf_{\sigma,\psi_j} \int_S \big(\sum_{\alpha\in\mathbb{N}^d_{2n}} \sigma_\alpha\, P_\alpha\big)\, f\, d\mu$
s.t. $\phi_0\,(=\phi(\mathbf{1})) = 1;$	s.t. $\sigma_0 = 1;$
$\mathbf{M}_n(g_j\cdot\boldsymbol{\phi}) \succeq 0,\ 0 \le j \le m.$	$\sum_{\alpha\in\mathbb{N}^d_{2n}} \sigma_\alpha\, P_\alpha = \sum_{j=0}^m \psi_j\, g_j.$
	$\psi_j \in \Sigma[\mathbf{x}]_{n-d_j}, 0 \le j \le m$
Dual	Dual
$\tau_n^* = \sup_{\lambda,\psi_j}\ \lambda$	$\hat{\kappa}_n^* = \sup_{\lambda,\boldsymbol{\phi}}\ \lambda$
s.t. $f - \lambda = \sum_{j=0}^m \psi_j\, g_j$	s.t. $f - \lambda = \sum_{\alpha\in\mathbb{N}^d_{2n}} \phi(P_\alpha)\, P_\alpha$
$\psi_j \in \Sigma[\mathbf{x}]_{n-d_j},\ 0 \le j \le m$	$\mathbf{M}_n(g_j\cdot\boldsymbol{\phi}) \succeq 0,\ 0 \le j \le m$

Comparing with the Moment-SOS hierarchy of upper bounds. Let μ be the same (reference) measure on S as in Theorem 7.5. By construction, the (refined) hierarchy of upper bounds $(\hat{\kappa}_n)_{n\in\mathbb{N}}$ in (5.4) is searching for a *positive* probability density $\sigma \in Q_n(g)$. Hence $\mathbf{x} \mapsto \sigma(\mathbf{x}) = \sum_\alpha \sigma_\alpha\, P_\alpha(\mathbf{x})$, with

$$1 = \int_S \sigma\, d\mu = \sum_\alpha \sigma_\alpha \int_S P_\alpha\, d\mu = \sigma_0,$$

as $P_0 = \mathbf{1}$ (because μ is a probability measure), and

$$\sigma \in Q_n(g) \Rightarrow \sum_{\alpha\in\mathbb{N}^d_{2n}} \sigma_\alpha\, P_\alpha = \sum_{j=0}^m \psi_j\, g_j, \quad \psi_j \in \Sigma[\mathbf{x}]_{n-d_j},\ j = 0,\dots,m.$$

As we can see, Table 7.1 exhibits a complete symmetry between the primal and dual formulations of the respective Moment-SOS hierarchies of lower bounds and upper bounds, when the involved polynomials are expressed in the orthonormal basis $(P_\alpha)_{\alpha\in\mathbb{N}^d}$.

7.6. Notes and sources

Sections 7.1–7.2. For more details and historical background on the CD kernel and the Christoffel function, the interested reader is referred to Simon (2008), Nevai and Freud (1986), Lasserre *et al.* (2022) and references therein.

Section 7.3. This is based on Marx *et al.* (2021) and Henrion and Lasserre (2022).

Section 7.4. This is essentially from Lasserre (2022, 2023) and Lasserre and Xu (2023). Remarkably, the generalized Pell's equation establishes links between seemingly unrelated fields such as optimization, positivity certificates, conic duality on one hand, and orthogonal polynomials and equilibrium measures on the other hand. It is likely that the generalized Pell's equation is valid only for sets with specific geometries and with an appropriate set of generators. Indeed, from the proof in Lasserre and Xu (2023), a property of Gegenbauer polynomials (in particular a summation property) is crucial. However, for more general basic semi-algebraic sets S, there is an even weaker result that links the constant polynomial $\mathbf{1} \in Q_n(g)$ and moments $\boldsymbol{\phi} \in Q_n^*(g)$ of the equilibrium measure μ of S; see Lasserre and Xu (2023).

Section 7.5. Here we interpret both Moment-SOS hierarchies of upper and lower bounds in the common language of densities with respect to a reference measure μ whose support is S. In contrast to the hierarchy of upper bounds, finite convergence for the hierarchy of lower bounds is possible (and in fact takes place generically) because a *signed* density with respect to μ may have all its moments up to order $2n$ equal to those of the Dirac measure at a global minimizer, which is not possible for a *positive* density with respect to μ on S (with non-empty interior) as in the hierarchy of upper bounds.

8. Conclusion

We have described the Moment-SOS hierarchy methodology for polynomial optimization (hierarchies of lower and upper bounds). We have also used it to solve the generalized moment problem (GMP) with algebraic data, whose list of applications in many areas of science and engineering is almost endless. The basic principle behind the Moment-SOS hierarchy is quite simple, and for illustration we have described its application to two problems (viewed as instances of the GMP) in computational geometry and optimal control.

It is a powerful methodology but the computational cost of its basic formulation can be quite heavy, even for problems of modest dimension. Fortunately, large-scale problems often exhibit sparsity or symmetries in their formulation, and we have also described how such properties can be exploited to define a sparsity-adapted Moment-SOS hierarchy whose associated computational burden can be drastically reduced.

Much remains to be done in several research directions, some of which have been briefly mentioned. As it is at the intersection of several disciplines, it is very likely that we will see even more contributions in the coming years.

Acknowledgements
This work is supported by the ANR-NuSCAP 20-CE-48-0014 programme and also the AI Interdisciplinary Institute ANITI funding through the French programme 'Investing for the Future PI3A' under grant agreement ANR-19-PI3A-0004. This research is also part of the DesCartes programme, and is supported by the National Research Foundation, Prime Minister's Office, Singapore, under its Campus for Research Excellence and Technological Enterprise (CREATE) programme.

References

A. A. Ahmadi and A. Majumdar (2019), DSOS and SDSOS optimization: More tractable alternatives to sum of squares and semidefinite optimization, *SIAM J. Appl. Algebra Geometry* **3**, 193–230.

D. Aloise and P. Hansen (2011), Evaluating a branch-and-bound RLT-based algorithm for minimum sum-of-squares clustering, *J. Global Optim.* **49**, 449–465.

M. Anjos and J. B. Lasserre, eds (2012), *Handbook on Semidefinite, Conic and Polynomial Optimization*, Vol. 166 of International Series in Operations Research and Management Science, Springer.

E. M. Aylward, P. A. Parrilo and J.-J. E. Slotine (2008), Stability and robustness analysis of nonlinear systems via contraction metrics and SOS programming, *Automatica* **44**, 2163–2170.

F. Bacari, C. Gogolin, P. Wittek and A. Acín (2020), Verifying the output of quantum optimizers with ground-state energy lower bounds, *Phys. Rev. Res.* **2**, art. 043163.

F. Bach and A. Rudi (2023), Exponential convergence of Sum-of-Squares hierarchies for trigonometric polynomials, *SIAM J. Optim.* **33**, 2137–2159.

C. Bachoc and F. Vallentin (2008), New upper bounds for kissing numbers from semidefinite programming, *J. Amer. Math. Soc.* **21**, 909–924.

C. Bachoc, A. Passuello and F. Vallentin (2013), Bounds for projective codes from semidefinite programming, *Adv. Math. Commun.* **7**, 127–145.

M. Bafna, B. Barak, P. K. Kothari, T. Schramm and D. Steurer (2021), Playing unique games on certified small-set expanders, in *Proceedings of the 53rd Annual ACM SIGACT Symposium on Theory of Computing*, Association for Computing Machinery (ACM), pp. 1629–1642.

L. Baldi and B. Mourrain (2022), Exact moment representation in Polynomial Optimization. Available at arXiv:2012:14652.

B. Barak and D. Steurer (2014), Sum-of-squares proofs and the quest towards optimal algorithms, in *Proceedings of the International Congress of Mathematicians (ICM 2014)* (S. Y. Sang *et al.*, eds), Kyung Moon Sa, pp. 509–533.

M. Baran (1995), Complex equilibrium measure and Bernstein type theorems for compact sets in $\mathbb{R}^n$, *Proc. Amer. Math. Soc.* **123**, 485–494.

E. Bedford and B. A. Taylor (1986), The complex equilibrium measure of a symmetric convex set in $\mathbb{R}^n$, *Trans. Amer. Math. Soc.* **294**, 705–717.

D. Bertsimas and I. Popescu (2005), Optimal inequalities in probability theory: A convex optimization approach, *SIAM J. Optim.* **15**, 780–804.

G. Blekherman, P. Parrilo and R. Thomas, eds (2012), *Semidefinite Optimization and Convex Algebraic Geometry*, MOS-SIAM Series on Optimization, SIAM.

S. Burgdorf, I. Klep and J. Povh (2016), *Optimization of Polynomials in Non-Commuting Variables*, SpringerBriefs in Mathematics, Springer.

Chen Tong, J. B. Lasserre, V. Magron and E. Pauwels (2021), Semialgebraic representation of monotone deep equilibrium models and application to certification, in *Advances in Neural Information Processing Systems 34 (NeurIPS 2021)* (M. Ranzato *et al.*, eds), Curran Associates, pp. 27146–27159.

F. Cicconet and K. C. Almeida (2019), Moment-SOS relaxation of the medium term hydrothermal dispatch problem, *Int. J. Elec. Power Energy Systems* **104**, 124–133.

N. E. Courtier, R. Drummond, P. Ascensio, L. D. Couto and D. A. Howey (2022), Discretisation-free battery fast-charging optimisation using the measure-moment approach, in *2022 European Control Conference (ECC)*, IEEE, pp. 628–634.

Y. de Castro, F. Gamboa, D. Henrion and J. B. Lasserre (2017), Exact solutions to super resolution on semi-algebraic domains in higher dimensions, *IEEE Trans. Inform. Theory* **63**, 621–630.

Y. de Castro, F. Gamboa, D. Henrion, R. Hess and J. B. Lasserre (2019), Approximate optimal designs for multivariate polynomial regression, *Ann. Statist.* **47**, 125–157.

E. de Klerk and M. Laurent (2011), On the Lasserre hierarchy of semidefinite programming relaxations of convex polynomial optimization problems, *SIAM J. Optim.* **21**, 824–832.

D. de Laat (2020), Moment methods in energy minimization: New bounds for Riesz minimal energy problems, *Trans. Amer. Math. Soc.* **373**, 1407–1453.

D. de Laat, F. C. Machado, F. M. de Oliveira Filho and F. Vallentin (2022), k-point semidefinite programming bounds for equiangular lines, *Math. Program.* **194**, 533–567.

A. C. Doherty, P. A. Parrilo and F. Spedalieri (2005), Detecting multipartite entanglement, *Phys. Rev. A* **71**, art. 032333.

M. Dostert and F. Vallentin (2020), New dense superball packings in three dimensions, *Adv. Geom.* **20**, 473–482.

G. R. Dowdy and P. I. Barton (2018), Dynamics bounds on stochastic chemical kinetic systems using semidefinite programming, *J. Chem. Phys.* **149**, art. 74103.

M. E. Dyer and A. M. Frieze (1988), The complexity of computing the volume of a polyhedron, *SIAM J. Comput.* **17**, 967–974.

K. S. Eckhoff (1993), Accurate and efficient reconstruction of discontinuous functions from truncated series expansions, *Math. Comp.* **61**(204), 745–763.

J. Eisert, P. Hyllus, O. Gühne and M. Curty (2004), Complete hierarchies of efficient approximations to problems in entanglement theory, *Phys. Rev. A* **70**, art. 062317.

J. Fan, J. Nie and A. Zhou (2018), Tensor eigenvalue complementarity problems, *Math. Program. Ser. A* **170**, 507–539.

K. Fang and H. Fawzi (2021), The sum-of-squares hierarchy on the sphere and applications in quantum information theory, *Math. Program.* **190**, 331–360.

G. Fantuzzi, D. Goluskin, D. Huang and S. I. Chernyshenko (2016), Bounds for deterministic and stochastic dynamical systems using sum-of-squares optimization, *SIAM J. Appl. Dyn. Systems* **15**, 1962–1988.

H. O. Fattorini (1999), *Infinite Dimensional Optimization*, Cambridge University Press.

A. Gepp, G. Harris and B. Vanstone (2020), Financial applications of semidefinite programming: A review and call for interdisciplinary research, *Account. Finance* **60**, 3527–3555.

M. X. Goemans and D. P. Williamson (1995), Improved approximation algorithms for maximum cut and satisfiability problems using semidefinite programming, *J. Assoc. Comput. Mach.* **42**, 1115–1145.

D. Goluskin and G. Fantuzzi (2019), Bounds on mean energy in the Kuramoto–Sivashinsky equation computed using semidefinite programming, *Nonlinearity* **32**, 1705–1730.

D. Handelman (1988), Representing polynomials by positive linear functions on compact convex polyhedra, *Pacific J. Math.* **132**, 35–62.

D. Haussmann, F. Liers, M. Stingl and J. C. Vera (2018), Deciding robust feasibility and infeasibility using a set containment approach: An application to stationary passive gas network operations, *SIAM J. Optim.* **28**, 2489–2517.

D. Henrion and A. Garulli, eds (2005), *Positive Polynomials in Control*, Vol. 312 of Lecture Notes on Control and Information Sciences, Springer.

D. Henrion and J. B. Lasserre (2022), Graph recovery from incomplete moment information, *Constr. Approx.* **56**, 165–187.

D. Henrion, M. Korda and J. B. Lasserre (2021), *The Moment-SOS Hierarchy: Lectures in Probability, Statistics, Computational Geometry, Control and Nonlinear PDEs*, Vol. 4 of Optimization and Applications, World Scientific.

D. Henrion, J. B. Lasserre and C. Savorgnan (2009), Approximate volume and integration for basic semi algebraic sets, *SIAM Rev.* **51**, 722–743.

Jie Pan and Fu Jiang (2020), Low complexity beamspace super resolution for DOA estimation of linear array, *Sensors* **20**, art. 2222.

S. Khot (2010), Inapproximability of NP-complete problems, discrete Fourier analysis, and geometry, in *Proceedings of the 2010 International Congress of Mathematicians (ICM 2010)* (R. Bhatia *et al.*, eds), Vol. IV, Hindustan Book Agency, pp. 2676–2697.

S. Khot (2014), Hardness of approximation, in *Proceedings of the 2014 International Congress of Mathematicians (ICM 2014)* (S. Y. Sang *et al.*, eds), Kyung Moon Sa, pp. 711–728.

M. Kojima, S. Kim and M. Maramatsu (2005), Sparsity in sums of squares of squares of polynomials, *Math. Program.* **103**, 45–62.

M. Korda, D. Henrion and C. N. Jones (2014), Convex computation of the maximum controlled invariant set for polynomial control systems, *SIAM J. Control Optim.* **52**, 2944–2969.

M. Korda, D. Henrion and J. B. Lasserre (2022), Moments and convex optimization for analysis and control of nonlinear PDEs, in *Numerical Control: Part A* (E. Trélat and E. Zuazua, eds), Vol. 23 of Handbook of Numerical Analysis, Elsevier, pp. 339–366.

M. Kočvara, B. Mourrain and C. Riener, eds (2023), *Polynomial Optimization, Moments, and Applications*, Springer Optimization and its Applications, Springer.

J. L. Krivine (1964*a*), Anneaux préordonnés, *J. Anal. Math.* **12**, 307–326.

J. L. Krivine (1964*b*), Quelques propriétés des préordres dans les anneaux commutatifs unitaires, *C.R. Acad. Sci. Paris* **258**, 3417–3418.

H. Landau (1987), Classical background of the moment problem, in *Moments in Mathematics*, Vol. 37 of Proceedings of Symposia in Applied Mathematics, American Mathematical Society (AMS), pp. 1–15.

R. Laraki and J. B. Lasserre (2012), Semidefinite programming for min-max problems and games, *Math. Program. Ser. A* **131**, 305–332.

J. B. Lasserre (2000), Optimisation globale et théorie des moments, *C.R. Acad. Sci. Paris, Sér. I* **331**, 929–934.

J. B. Lasserre (2001), Global optimization with polynomials and the problem of moments, *SIAM J. Optim.* **11**, 796–817.

J. B. Lasserre (2002*a*), Bounds on measures satisfying moment conditions, *Ann. Appl. Prob.* **12**, 1114–1137.

J. B. Lasserre (2002*b*), Semidefinite programming vs. LP relaxations for polynomial programming, *Math. Oper. Res.* **27**, 347–360.

J. B. Lasserre (2006), Convergent SDP-relaxations in polynomial optimization with sparsity, *SIAM J. Optim.* **17**, 822–843.

J. B. Lasserre (2009*a*), Convexity in semi-algebraic geometry and polynomial optimization, *SIAM J. Optim.* **19**, 1995–2014.

J. B. Lasserre (2009*b*), *Moments, Positive Polynomials and their Applications*, Imperial College Press.

J. B. Lasserre (2011), A new look at nonnegativity on closed sets and polynomial optimization, *SIAM J. Optim.* **21**, 864–885.

J. B. Lasserre (2013), The K-moment problem for continuous functionals, *Trans. Amer. Math. Soc.* **365**, 2489–2504.

J. B. Lasserre (2015), *An Introduction to Polynomial and Semi-Algebraic Optimization*, Cambridge University Press.

J. B. Lasserre (2017), Computing Gaussian & exponential measures of semi-algebraic sets, *Adv. Appl. Math.* **91**, 137–163.

J. B. Lasserre (2021), Connecting optimization with spectral analysis of tri-diagonal matrices, *Math. Program. Ser. A* **190**, 795–809.

J. B. Lasserre (2022), A disintegration of the Christoffel function, *Comptes Rendus Math.* **360**, 1071–1079.

J. B. Lasserre (2023), Pell's equation, sum-of-squares and equilibrium measures on a compact set, *Comptes Rendus Math.* **361**, 935–952.

J. B. Lasserre and E. Pauwels (2016), Sorting out typicality via the inverse moment matrix SOS polynomial, in *Advances in Neural Information Processing Systems 29 (NIPS2016)* (D. Lee *et al.*, eds), Curran Associates, pp. 190–198.

J. B. Lasserre and T. Priéto-Rumeau (2004), SDP vs. LP relaxations for the Moment approach in some performance evaluation problems, *Stoch. Models* **20**, 439–456.

J. B. Lasserre and Y. Xu (2023), A generalized Pell's equation for a class of multivariate orthogonal polynomials. Technical report, LAAS-CNRS, Toulouse, France. Available at hal-04163153v1. To appear in *Trans. Amer. Math. Soc.*

J. B. Lasserre, D. Henrion, C. Prieur and E. Trélat (2008), Nonlinear optimal control via occupation measures and LMI-relaxations, *SIAM J. Control Optim.* **47**, 1649–1666.

J. B. Lasserre, E. Pauwels and M. Putinar (2022), *The Christoffel Function for Data Analysis*, Cambridge University Press.

J. B. Lasserre, T. Priéto-Rumeau and M. Zervos (2006), Pricing a class of exotic options via moments and SDP relaxations, *Math. Finance* **16**, 469–494.

F. Latorre, P. Rolland and V. Cevher (2020), Lipschitz constant estimation for neural networks via sparse polynomial optimization, in *8th International Conference on Learning Representations (ICLR 2020)*, OpenReview.

M. Laurent (2003), A comparison of the Sherali–Adams, Lovász–Schrijver and Lasserre relaxations for 0-1 programming, *Math. Oper. Res.* **28**, 470–496.

M. Laurent (2008), Sums of squares, moment matrices and optimization over polynomials, in *Emerging Applications of Algebraic Geometry* (M. Putinar and S. Sullivant, eds), Vol. 149 of The IMA Volumes in Mathematics and its Applications, Springer, pp. 157–270.

M. Laurent and L. Slot (2023), An effective version of Schmüdgen's Positivstellensatz for the hypercube, *Optim. Lett.* **17**, 515–530.

V. Magron and J. Wang (2023), *Sparse Polynomial Optimization: Theory and Practice*, Series on Optimization and its Applications, World Scientific.

A. Majumdar, G. Hall and A. A. Ahmadi (2020), A survey of recent scalability improvements for semidefinite programming with applications in machine learning, control, and robotics, *Annu. Rev. Control Robot Auton. Syst.* **3**, 331–60.

A. Marmin, M. Castella, J.-C. Pesquet and L. Duval (2021), Sparse signal reconstruction for nonlinear models via piecewise rational optimization, *Signal Process.* **179**, art. 107835.

Z. Marschner, D. Palmer, P. Zhang and J. Solomon (2020), Hexahedral mesh repair via sum-of-squares relaxations, *Comput. Graph. Forum* **39**, 133–147.

Z. Marschner, D. Palmer, P. Zhang and J. Solomon (2021), Sum-of-squares geometry processing, *ACM Trans. Graphics* **40**, 1–13.

M. Marshall (2008), *Positive Polynomials and Sums of Squares*, Vol. 146 of AMS Math. Surveys and Monographs, American Mathematical Society (AMS).

S. Marx, E. Pauwels, T. Weisser, D. Henrion and J. B. Lasserre (2021), Semi-algebraic approximation using Christoffel–Darboux kernel, *Constr. Approx.* **54**, 391–429.

S. Marx, T. Weisser, D. Henrion and J. B. Lasserre (2020), A moment approach for entropy solutions to nonlinear hyperbolic PDEs, *Math. Control Related Fields* **10**, 113–140.

D. Molzahn and C. Josz (2018), Lasserre hierarchy for large-scale polynomial optimization in real and complex variabales, *SIAM J. Optim.* **28**, 1017–1048.

D. Molzhan and I. Hiskens (2015), Sparsity-exploiting moment-based relaxations of the optimal power flow problem, *IEEE Trans. Power Systems* **30**, 3168–3180.

K. Moussa, M. Fiacchini and M. Alamir (2020), Robust optimal scheduling of combined chemo- and immunotherapy: Considerations on chemotherapy detrimental effects, in *Proceedings of the 2020 American Control Conference (ACC)*, IEEE, pp. 4252–4257.

M. Navascués, S. Pironio and A. Acín (2012), SDP relaxations for non-commutative polynomial optimization, in *Handbook on Semidefinite, Conic and Polynomial Optimization* (M. Anjos and J. B. Lasserre, eds), Vol. 166 of International Series in Operations Research and Management Science, Springer, pp. 601–634.

Y. Nesterov (2000), Squared functional systems and optimization problems, in *High Performance Optimization* (H. Frenk *et al.*, eds), Kluwer, pp. 405–440.

P. Nevai and G. Freud (1986), Orthogonal polynomials and christoffel functions, *J. Approx. Theory* **48**, 3–167.

Ngoc Hoang Anh Mai, J. B. Lasserre and V. Magron (2023), A hierarchy of spectral relaxations for polynomial optimization, *Math. Program. Comput* **15**, 651–701.

Ngoc Hoang Anh Mai, J. B. Lasserre, V. Magron and J. Wang (2022), Exploiting constant trace property in large scale polynomial optimization, *ACM Trans. Math. Software* **48**, art. 40.

J. Nie (2009), Sum of squares method for sensor network localization, *Comput. Optim. Appl.* **43**, 151–179.

J. Nie (2013), Certifying convergence of Lasserre's hierarchy via flat truncation, *Math. Program. Ser. A* **142**, 485–510.

J. Nie (2014), Optimality conditions and finite convergence of Lasserre's hierarchy, *Math. Program. Ser. A* **146**, 97–121.

J. Nie (2017), Low rank symmetric tensor approximations, *SIAM J. Matrix Anal. Appl.* **38**, 1517–1540.

J. Nie (2023), *Moment and Polynomial Optimization*, Vol. 31 of MOS/SIAM Series in Optimization, SIAM.

J. Nie and J. Demmel (2008), Sparse SOS relaxations for minimizing functions that are summations of small polynomials, *SIAM J. Optim.* **19**, 1534–1558.

J. Nie and X. Tang (2023), Convex generalized nash equilibrium problems and polynomial optimization, *Math. Program.* **198**, 1485–1518.

J. Nie and Z. Yang (2020), Hermitian tensor decompositions, *SIAM J. Matrix Anal. Appl.* **41**, 1115–1144.

R. O'Donnell (2017), SoS is not fully automatizable, even approximately, in *8th Innovations in Theoretical Computer Science Conference (ITCS 2017)* (C. H. Papadimitriou, ed.), Vol. 67 of Leibniz International Proceedings in Informatics (LIPIcs), Schloss Dagstuhl–Leibniz-Zentrum für Informatik, pp. 59:1–59:10.

O. Parekh (2023), Synergies between operations research and quantum information science, *INFORMS J. Comput.* **35**, 266–273.

O. Parekh and K. Thompson (2021), Application of the level-2 quantum Lasserre hierarchy in quantum approximation algorithms, in *48th International Colloquium on Automata, Languages, and Programming (ICALP 2021)* (N. Bansal *et al.*, eds), Vol. 198 of Leibniz International Proceedings in Informatics (LIPIcs), Schloss Dagstuhl–Leibniz-Zentrum für Informatik, pp. 102:1–102:20.

P. A. Parrilo (2000), Structured semidefinite programs and semialgebraic geometry methods in robustness and optimization. PhD thesis, California Institute of Technology, Pasadena, CA.

P. A. Parrilo (2003), Semidefinite programming relaxations for semialgebraic problems, *Math. Program.* **96**, 293–320.

P. A. Parrilo and S. Lall (2003), Semidefinite programming relaxations and algebraic optimization in control, *Eur. J. Control* **9**, 307–321.

E. Pauwels, D. Henrion and J. B. Lasserre (2017), Positivity certificates in optimal control, in *Geometric and Numerical Foundations of Movements* (J. P. Laumond *et al.*, eds), Vol. 117 of Springer Tracts in Advanced Robotics, Springer, pp. 113–132.

A. Prestel and C. N. Denzel (2001), *Positive Polynomials*, Springer Monographs in Mathematics, Springer.

T. Probst, D. D. Paudel, A. Chhatkuli and L. Van Gool (2019), Convex relaxations for consensus and non-minimal problems in 3D vision, in *Proceedings of the 2019 IEEE International Conference on Computer Vision (ICCV)*, IEEE, pp. 10233–10242.

M. Putinar (1993), Positive polynomials on compact semi-algebraic sets, *Indiana Univ. Math. J.* **42**, 969–984.

P. Raghavendra, T. Schramm and D. Steurer (2018), High dimensional estimation via sum-of-squares proofs, in *Proceedings of the International Congress of Mathematicians (ICM 2018)* (B. Sirakov *et al.*, eds), World Scientific, pp. 3389–3423.

D. M. Rosen, L. Carlone, A. S. Bandera and J. J. Leonard (2019), Se-Sync: A certifiably correct algorithm for synchronization over the special Euclidean group, *Int. J. Robotics Research* **38**, 95–125.

K. Schmüdgen (1991), The K-moment problem for compact semi-algebraic sets, *Math. Ann.* **289**, 203–206.

K. Schmüdgen (2017), *The Moment Problem*, Graduate Texts in Mathematics, Springer.

M. Schweighofer (2005), On the complexity of Schmüdgen's Positivstellensatz, *J. Complexity* **20**, 529–543.

S. Sedighi, K. V. Mishra, B. Shankar and M. R. B. Ottersten (2021), Localization with 1-Bit passive radars in Narrow Band IoT-applications using multivariate polynomial optimization, *IEEE Trans. Signal Process.* **69**, 2525–2540.

J. H. Selby, A. B. Sainz, V. Magron, L. Czekaj and M. Horodecki (2023), Correlations constrained by composite measurements, *Quantum* **7**, art. 1080.

H. D. Sherali and W. P. Adams (1990), A hierarchy of relaxations between the continuous and convex hull representations for zero-one programming problems, *SIAM J. Discrete Math.* **3**, 411–430.

H. D. Sherali and W. P. Adams (1999), *A Reformulation-Linearization Technique for Solving Discrete and Continuous Nonconvex Problems*, Kluwer.

N. Z. Shor (1987), Quadratic optimization problems, *Tekhnicheskaya Kibernetika* **1**, 128–139.

B. Simon (2008), The Christoffel–Darboux kernel, in *Perspectives in Partial Differential Equations, Harmonic Analysis and Applications*, Vol. 79 of Proceedings of Symposia in Pure Mathematics, American Mathematical Society (AMS), pp. 295–335.

L. Slot (2022), Sum-of-Ssquares hierarchies for polynomial optimization and the Christoffel–Darboux kernel, *SIAM J. Optim.* **32**, 2612–2635.

L. Slot and M. Laurent (2021), Near optimal analysis of Lasserre's univariate measure-based bounds for multivariate polynomial optimization, *Math. Program.* **188**, 443–460.

N. Stein, A. Ozdaglar and P. A. Parrilo (2008), Separable and low-rank continuous games, *Int. J. Game Theory* **37**, 457–474.

M. Tacchi, J. B. Lasserre and D. Henrion (2023), Stokes, Gibbs and volume computation of semi-algebraic sets, *Discrete Comput. Geom.* **69**, 260–283.

M. Tacchi, T. Weisser, J. B. Lasserre and D. Henrion (2021), Exploiting sparsity in semi-algebraic set volume computation, *Found. Comput. Math.* **22**, 161–209.

J. Tian, H. Wei and J. Tan (2015), Global optimization for power dispatch problems based on theory of moments, *Int. J. Electr. Power Energy Syst.* **71**, 184–194.

M. Tyburec, J. Zeman, M. Kruzik and D. Henrion (2021), Global optimality in minimum compliance topology optimization of frames and shells by moment-sum-of-squares hierarchy, *Struct. Multidiscipl. Optim.* **64**, 1963–1981.

F.-H. Vasilescu (2003), Spectral measures and moment problems, in *Spectral Theory and its Applications*, Vol. 2 of Theta Series in Advanced Mathematics, Theta, pp. 173–215.

R. Vinter (1993), Convex duality and nonlinear optimal control, *SIAM J. Control Optim.* **31**, 518–538.

S. Waki, S. Kim, M. Kojima and M. Maramatsu (2006), Sums of squares and semidefinite programming relaxations for polynomial optimization problems with structured sparsity, *SIAM J. Optim.* **17**, 218–242.

J. Wang, H. Li and B. Xia (2019), A new sparse SOS decomposition algorithm based on term sparsity, in *Proceedings of the 2019 International Symposium on Symbolic and Algebraic Computation (ISSAC '19)*, Association for Computing Machinery (ACM), pp. 347–354.

J. Wang, V. Magron and J. B. Lasserre (2021), TSSOS: A moment-SOS hierarchy that exploits term sparsity, *SIAM J. Optim.* **31**, 30–58.

J. Wang, V. Magron, J. B. Lasserre and Ngoc Hoang Anh Mai (2022), CS-TSSOS: Correlative and term sparsity for large-scale polynomial optimization, *ACM Trans. Math. Software* **48**, 1–26.

P. Wittek, S. Darányi and G. Nelhans (2017), Ruling out static latent homophily in citation networks, *Scientometrics* **110**, 765–777.

Xiangfeng Ji, Xuegang (Jeff) Ban, Jian Zhang and Bin Ran (2019), Moment-based travel time reliability assessment with Lasserre's relaxation, *Transp. B* **7**, 401–422.

H. Yang and L. Carlone (2020), In perfect shape: Certifiably optimal 3D shape reconstruction from 2D landmarks, in *Proceedings of the 2020 IEEE/CVF Conference on Computer Vision and Pattern Recognition (CVPR)*, IEEE, pp. 618–627.

H. Yang and L. Carlone (2023), Certifiably optimal outlier-robust geometric perception: Semidefinite relaxations and scalable global optimization, *IEEE Trans. Pattern Anal. Mach. Intell.* **45**, 2816–2834.

H. Yang and M. Pavone (2023), Object pose estimation with statistical guarantees: Conformal keypoint detection and geometric uncertainty propagation, in *Proceedings of the 2023 IEEE/CVF Conference on Computer Vision and Pattern Recognition (CVPR)*, IEEE, pp. 8947–8958.

H. Yang, J. Shi and L. Carlone (2020), TEASER: Fast and certifiable point cloud registration, *IEEE Trans. Robotics* **37**, 314–333.

A. Yurtsever, J. A. Tropp, O. Fercoq, M. Udell and V. Cevher (2021), Scalable semidefinite programming, *SIAM J. Math. Data Sci.* **3**, 171–200.